MEGAN KOCH

							0
							2 He 4.003
		IIIA	IVA	VA	VIA	VIIA	
		5 B 10.81	6 C 12.011	7 N 14.007	8 O 15.9994	9 F 18.998	10 Ne 20.18
IB	IIB	13 Al 26.98	14 Si 28.09	15 P 30.97	16 S 32.06	17 Cl 35.453	18 Ar 39.95
28 Ni 58.71	29 Cu 63.55	30 Zn 65.38	31 Ga 69.72	32 Ge 72.59	33 As 74.92	34 Se 78.96	35 Br 79.90	36 Kr 83.80
46 Pd 106.4	47 Ag 107.87	48 Cd 112.40	49 In 114.82	50 Sn 118.69	51 Sb 121.75	52 Te 127.60	53 I 126.90	54 Xe 131.30
78 Pt 195.09	79 Au 196.97	80 Hg 200.59	81 Tl 204.37	82 Pb 207.2	83 Bi 208.98	84 Po (210)	85 At (210)	86 Rn (222)

63 Eu 151.96	64 Gd 157.25	65 Tb 158.93	66 Dy 162.50	67 Ho 164.93	68 Er 167.26	69 Tm 168.93	70 Yb 173.04	71 Lu 174.97
95 Am (243)	96 Cm (247)	97 Bk (247)	98 Cf (251)	99 Es (254)	100 Fm (253)	101 Md (258)	102 No (255)	103 Lr (256)

General Chemistry
Principles and Modern Applications

General Chemistry
Principles and Modern Applications
Third Edition
Ralph H. Petrucci
California State College, San Bernardino

Macmillan Publishing Co., Inc.
New York

Collier Macmillan Publishers
London

The following publications have been prepared to accompany this text and may be used with it.

Student Solution Supplement
Student Study Guide by Robert K. Wismer, Millersville State College
Experiments in General Chemistry/A Laboratory Program by
  Gerald S. Weiss, Robert K. Wismer, and Thomas C. Greco,
  Millersville State College

Macmillan Publishing Co., Inc.
866 Third Avenue, New York, New York 10022

Collier Macmillan Canada, Inc.

Library of Congress Cataloging in Publication Data

Petrucci, Ralph H.
  General chemistry.

  Includes index.
  1. Chemistry.   I. Title.
QD31.2.P48   1982   540   81-6006
ISBN 0-02-395010-2 (Hardbound)   AACR2
ISBN 0-02-977640-6 (International Edition)

Printing: 345678   Year: 345678

ISBN   0-02-395010-2

Preface

As in earlier editions I have made certain assumptions about the students who will study general chemistry from this text. Typical students of general chemistry, I believe, are not preparing for careers in chemistry but are acquiring the chemical background required in other fields—biology, medicine, and engineering, to name but a few. Some may study chemistry simply to add to their general education in the sciences. My objective in preparing this third edition has been to present chemistry to these students in as understandable a manner as I can. To do so, I have attempted to strike the balances between principles and applications, between qualitative and quantitative discussions, and between rigor and simplification that seem most appropriate for these students. In all of this I have been guided by comments and suggestions from users of previous editions and reviewers of the current one. Changes from earlier editions have been made in content, organization, and special features.

Generally, applications are considered in a context where students will have acquired an understanding of the basic principles involved. This means that applications are emphasized in the latter third of the book. However, there are numerous applications that are considered in earlier chapters, such as an overview of analytical chemistry (Chapter 3); ideas concerning industrial chemistry (Chapter 4); the structure of the atmosphere (Chapter 5); synthetic fuels (Chapter 6); an introduction to descriptive chemistry (Chapter 8); colloids (Chapter 12); heterogeneous catalysis (Chapter 13); the Haber process (Chapter 14); heat engines and thermal pollution (Chapter 15); fractional precipitation in quantitative analysis (Chapter 16); and the electrometric determination of pH (Chapter 19). In addition, applications are noted throughout the text in the form of marginal notes and end-of-chapter exercises. The *Instructor's Manual* includes a chapter-by-chapter guide to these exercises, citing some of the specific extensions of theory and/or practical applications to be found there. Finally, if an instructor wishes to consider certain practical applications earlier in the course, this is possible by simultaneously assigning material from Chapter 26. Suggestions for doing this are also provided in the *Instructor's Manual*.

As with practical applications, traditional descriptive chemistry is featured in the latter third of the book (Chapters 20–26); but some topics are considered in earlier chapters, starting with an introduction to descriptive chemistry in Chapter 8. In Chapter 11, within the context of intermolecular forces, the physical properties of numerous elements and compounds are considered. Topics related to qualitative analysis of cations appear in several chapters (16, 18, 20, 21, and 22), each at a point appropriate to the principles involved.

Other new or expanded topics in this edition include the use of formal charges in writing Lewis structures (Chapter 9); the Born–Haber cycle for ionic compounds (Chapter 11); the dependence of rate constants on temperature (Chapter 13) and of equilibrium constants on temperature (Chapter 14); factors affecting acid/base strength (Chapter 17); and factors affecting nuclear stability (Chapter 23).

Reorganization of textual material includes discussing gases early (Chapter 5) and treating liquids, solids, and intermolecular forces in a single later chapter (Chapter 11); dividing the treatment of thermodynamics into an early chapter on thermochemistry (Chapter 6) and a later one on the second and third laws of thermodynamics (Chapter 15); distributing the discussion of equilibrium over several chapters: one on basic principles (Chapter 14) and three on solution equilibria (Chapters 16, 17, and 18); combining the chemistry of natural resources and industrial processes with environmental chemistry in a single concluding chapter (Chapter 26). Molar concentration and solution stoichiometry are considered with reaction stoichiometry (Chapter 4); solution properties are presented later (Chapter 12); and, for those who wish to approach solution stoichiometry from the standpoint of equivalent weight and normality, these topics are presented in the context of acid-base and oxidation-reduction chemistry (Chapters 18 and 19).

Even more so than with previous editions, I have attempted to provide opportunities for instructors to reorder certain topics according to their preferences. For example, those who wish to combine the material on thermodynamics into a single presentation may defer Chapter 6 until Chapter 15 is reached. If an instructor wishes to limit organic chemistry to a discussion of bonding, structure, and nomenclature and to take up these topics following chemical bonding, this can be done. A number of chapters conclude with a section that presents a specific application or an extension of concepts presented earlier in the chapter. These sections may be considered in full, deemphasized, or deleted at the instructor's discretion.

In this edition each chapter concludes with some *study aids*—a brief summary, a set of learning objectives, definitions of important new terms, two sets of exercises (Exercises and Additional Exercises), and a set of Self-Test Questions. Each end-of-chapter definition of a term is represented by a **boldface** page number in the index. The combination of these index listings and end-of-chapter definitions thus constitutes a glossary of the entire text. As in previous editions, exercises that either are more difficult or require an extension of concepts presented in the text are designated by a star ★. Answers to many of the Exercises and Self-Test Questions are provided at the end of the book; complete solutions are available in a separate *Student Solution Supplement*. Solutions to the Additional Exercises are available in a separate *Instructor's Manual*.

Also available to accompany this text are a *Student Study Guide* and a laboratory manual, *Experiments in General Chemistry*. The study guide was written by Professor Robert K. Wismer and the laboratory manual by Professors Gerald S. Weiss, Robert K. Wismer, and Thomas G. Greco, all of Millersville State College. The study guide is organized around the *Learning Objectives* in the textbook and features brief discussions of these objectives, drill problems, self-quizzes, and sample tests. The laboratory manual contains thirty-seven experiments that parallel the text, including a final group of six experiments on qualitative analysis.

I wish to thank all of the individuals who have helped with this third edition. They have reviewed what I have written with care—at times in great detail—and they have shared freely their own views of teaching chemistry. If I have been successful in improving this text over previous editions, it is in large part as a result of their contributions.

The following have provided critiques of the second edition and/or portions of the manuscript for the third edition: Luther K. Brice, Jr. (Virginia Polytechnic Institute and State University); K. R. Fountain (Northeast Missouri State University); Milton E. Fuller (California State University, Hayward); Henry Heikkinen (University of Maryland); William H. McMahan (Mississippi State University); Randall J. Remmel (University of Alabama in Birmingham); Donald E. Sands (University of Kentucky); W. P. Tappmeyer (University of Missouri, Rolla); and Milton J. Wieder (Metropolitan State College). Those who have read and commented on the entire manuscript for the third edition are Robert C. Brasted (University of Minnesota); Jimmie G. Edwards (University of Toledo); Lawrence Epstein (University of Pittsburgh); Joseph M. Kanamueller (Western Michigan University); Curtis T. Sears (Georgia State University); and Richard S. Treptow (Chicago State University). My colleague, Kenneth A. Mantei, was a steady source of helpful ideas for dealing with some of the stickier points of kinetics, thermodynamics, and equilibrium; and, as in previous editions, James D. Crum offered advice on matters dealing with organic and biochemistry. Special thanks are due to Robert K. Wismer who, in addition to his own efforts in producing the accompanying study guide and contributing to the laboratory manual, provided a detailed review of the second edition and extensive comments on the manuscript of the third edition.

I am happy to have had the opportunity to work with Elisabeth Belfer, Kate Moran, and Gregory Payne of Macmillan Publishing Company. The results of their efforts may not be so apparent to the reader, but I am keenly aware of them.

Finally, my greatest debt is again to my wife and family who, throughout this revision, have had to endure even more than the usual neglect of a preoccupied textbook writer. They have done so—at the same time that they have been asked to do photocopying, mailing, proofreading, and miscellaneous other tasks—with great patience and good humor.

San Bernardino, California   R. H. P.

Contents

7   Electrons in Atoms   144

8   Atomic Properties and the Periodic Table   172

9   Chemical Bonding I: Basic Concepts   196

14 Principles of Chemical Equilibrium 344

15 Thermodynamics, Spontaneous Change, and Equilibrium 371

16 Solubility Equilibria in Aqueous Solutions 393

21 The Chemistry of Transition Elements 529

22 Complex Ions and Coordination Compounds 548

23 Nuclear Chemistry 571

1

Matter—Its Properties and Measurement

Fire, known since ancient times, is a powerful agent for producing chemical changes. Fire was first used for cooking foods. Later it was used for baking pottery, making glass, and smelting ores to produce metals—first copper and later lead, tin, and iron. Other chemical processes known since ancient times include making butter, cheese, wine, beer, and soap; tanning hides; and dyeing fabrics.

Despite these many early applications, the principles of chemistry have been established only in recent times, primarily in the nineteenth and twentieth centuries. It is from this theoretical base that our ability to produce profound changes in our environment has developed. For example, from organic chemistry comes the knowledge of how to exploit the natural resource petroleum, both as a source of fuels and for the production of plastics, drugs, and pesticides.

Of more recent origin is our recognition that certain chemical processes (e.g., the production of smog) have a detrimental effect on the environment. An important challenge facing chemists today is to develop the processes and materials needed by modern society and simultaneously to minimize their environmental impact. Both of these objectives require a firm understanding of chemical principles, and this is one reason why principles are emphasized together with applications in this text.

1-1 Properties of Matter

It is easier to describe matter intuitively than to define it precisely. Let us say that **matter** is any object or material that occupies space, and that the quantity of matter is measured by a property called **mass** (described further in Section 1-5). Mass is only one of many properties or characteristics by which one sample of matter can be identified and distinguished from others. Properties of matter can be grouped into two categories: physical and chemical.

PHYSICAL PROPERTIES AND PHYSICAL CHANGE. Color, luster, and hardness are only a few of the many **physical properties** that may be used to describe the appearance of an object. A process in which an object changes its physical appearance but not its basic identity is called a **physical change.** A cube of copper metal can be flattened into a very thin foil; copper is malleable. Copper can also be drawn out into a fine wire; it is ductile. The melting of ice and the boiling of water are further examples of physical change.

It is usually possible to distinguish between physical and chemical changes in terms of the behavior of bulk matter, but when we focus on submicroscopic matter—atoms and molecules—the distinction sometimes becomes blurred.

FIGURE 1-1
A classification scheme for matter.

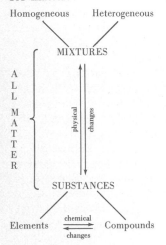

Every sample of matter is either a substance or a mixture. If a substance, it is either an element or a compound; if a mixture, either homogeneous or heterogeneous. Transformations between elements and compounds involve chemical changes; conversions between substances and mixtures, physical changes.

CHEMICAL PROPERTIES AND CHEMICAL CHANGE. Paper burns, iron rusts, and wood rots. In each case the object changes not only in physical appearance but also in its basic identity. In a **chemical change** a sample of matter is transformed into completely different materials. The abilities of materials to undergo chemical changes are called **chemical properties.**

1-2 Classification of Matter

Matter exists in countless different forms, and it is necessary to develop broad categories for its description. For example, the scheme in Figure 1-1 classifies matter into the categories *substance* and *mixture*. Particularly important among substances are those that cannot be made into simpler materials either by physical or chemical changes, that is, not by heating, cooling, crushing, exposing to acids, and so on. These substances are the **chemical elements.** At present, 106 different elements are known. They range from such common materials as iron, copper, silver, and gold to uncommon ones such as lutetium, promethium, and thulium. About 90 of the elements can be extracted from natural sources; the rest have been created through nuclear processes. A complete listing of the elements is presented on the inside back cover. A special tabular arrangement known as the periodic table (discussed in Chapter 8) is shown on the inside front cover.

Chemical compounds comprise a second class of substances. These are chemical combinations of two or more elements. Clearly, the potential number of different combinations of the 106 elements is enormous. The number of chemical compounds now known is in the millions, ranging in complexity from ordinary water to the protein hemoglobin. Chemical compounds retain their identities during physical changes but can be separated into their component elements by appropriate *chemical* changes.

The composition and properties of an element or a compound are uniform throughout a given sample *and from one sample to another*. Elements and compounds are said to be *pure* and are called **substances.** Some mixtures of substances also have compositions and properties that are uniform throughout a given sample *but variable from one sample to another*. These are **homogeneous mixtures** or **solutions.** A homogeneous mixture can be separated into its two or more components by appropriate *physical* changes. Ordinary air is a solution of several gases, principally the *elements* nitrogen and oxygen. Seawater is a solution of the *compounds* water, sodium chloride (salt), and a host of others.

At times scientific definitions may be at variance with practical definitions. For instance, if a sample of homogenized milk is viewed through a microscope, globules of fat can be seen dispersed in a watery medium. Homogenized milk is a heterogeneous mixture.

In some mixtures—sand and water, for example—the components separate into physically distinct regions. As a result the composition and physical properties vary from one part of the mixture to another. Such mixtures are said to be **heterogeneous.** Samples of matter ranging from a glass of iced tea to a slab of concrete to the leaf of a plant are heterogeneous.

1-3 The Scientific Method

What most distinguishes science from other intellectual activity is the manner in which scientific knowledge is acquired together with the way in which this knowledge can be used to *predict* future events. The time required for a rocket to travel from the earth to the moon can be predicted far more accurately than can the time required to drive an automobile from New York City to Washington. This is because the scientific basis of rocket propulsion is so well understood.

The Greek approach to acquiring knowledge was based on **deduction.** Starting from certain basic premises or assumptions, the Greeks developed means for establishing conclusions that must follow. This is the logical method that they employed with such success in the study of geometry. However, in the deductive method the validity of the basic assumptions always remains in doubt.

A significant development of the seventeenth century was a recognition of the importance of experimentation in the discovery of scientific facts. Rather than start with certain basic assumptions, the seventeenth-century scientist, exemplified by Galileo, Francis Bacon, Descartes, Boyle, Hooke, and Newton, made careful observations of phenomena and then formulated natural laws to summarize them. This process of formulating a general statement, a **natural law,** from a series of observations is called **induction.**

There is an inherent problem in the method of induction. When have enough observations been made to justify a generalization (a natural law): A city dweller has observed only a few dozen sheep in her lifetime and all have had a white coat. Is she justified in making the statement "All sheep are white"?

To test a natural law a scientist designs a controlled situation, an **experiment,** to see if the conclusions deduced from the natural law agree with actual experience. If a natural law stands the test of repeated experimentation, confidence in it grows. If agreement between predicted and observed behavior is imperfect, the natural law must be modified or limited in scope. The success of a natural law then is judged by how effective it is in summarizing observations and in predicting natural phenomena. However, no natural law can be accepted as an *absolute* truth because there is always the possibility that some experiment may be devised that would refute it.

A tentative explanation of a natural law is called a **hypothesis.** A hypothesis can be tested by experimentation, and if it survives this testing it is often referred to as a **theory.** The term *theory* can also be used in a broader sense; It is a conceptual framework or model (a way of looking at things) that can be used to explain and to make further predictions about natural phenomena. Sometimes, differing or conflicting theories are advanced to explain the same phenomenon. A choice among them is usually made as a result of experimentation designed to indicate which theory is most successful in its predictions. Also, the theory that involves the smallest number of assumptions, the simplest theory, is usually preferred. Over a period of time, as new evidence accumulates, most scientific theories undergo modification; some are discarded.

By way of illustration, consider the development of Dalton's atomic theory. Dalton's conception or model was that all matter is composed of minute, indivisible particles called atoms. To formulate his theory he had to rely on two natural laws of chemical combination that had been discovered previously. One of these was the law of conservation of mass: matter is neither created nor destroyed in a chemical reaction. The other was the law of definite composition: the proportions in which elements are combined in a compound are independent of the source of the compound. And Dalton's theory was useful in explaining laws of chemical combination discovered subsequently. These laws and Dalton's theory are discussed in Chapter 2.

The sum of all the activities described in the preceding paragraphs—observations, experimentation, and the formulation of laws, hypotheses, and theories—is called the **scientific method.** Single-minded devotion to the scientific method cannot alone guarantee success in scientific investigations, however. Occasionally, it is necessary for someone to break away from established patterns of thinking to discover the key to a scientific puzzle. These are the developments called scientific breakthroughs (see, for example, the discussion of the quantum theory in Chapter 7). And always, it is necessary to be alert to unexpected observations. A number of great scientific discoveries (e.g., x rays, radioactivity, penicillin) were made by accident in the course of other investigations. But accidental discoveries do not really happen by accident. As noted by Louis Pasteur (1822–1895), "Chance favors the prepared mind."

1-4 The Need for Measurement

In scientific work observations that can be assigned numerical values, **quantitative** observations, are generally preferred over verbal qualitative statements. The use of mathematics in describing the laws of nature enhances the scope of these laws and the precision with which they can be applied.

Even in nonscientific work, and from earliest times, people have found a need to make measurements and to express these measured quantities in convenient units. They needed to determine when to plant crops, how to exchange goods, or how to formulate recipes for primitive manufacturing. Interestingly enough, some of the quantities that ancient people most needed to measure—mass, length, and time—are also the basic properties of concern to scientists. Of course, the precision of these measurements and the units for expressing them have changed greatly over the years.

The yard, in England, was based on the arm length of the reigning monarch. Thus, it varied by as much as several inches from time to time. It was finally standardized in 1592 during the reign of Queen Elizabeth I.

Some early units of measurement were based on human features and some on common objects or materials. For example, the yard was introduced in the fifteenth century as the distance from a person's nose to the tip of the middle finger of the extended arm. An inch was defined as the length of "three barleycorns, round and dry, when laid together." Obviously, such quantities cannot be reproduced exactly, since the physical dimensions of human beings (and barleycorns) vary rather widely. Unless a unit of measurement can be defined precisely, it cannot be used for scientific work.

1-5 The English and Metric Systems of Measurement

The system of measurement used in certain English-speaking countries has as its basic unit of mass, the standard **pound (lb),** and as its basic unit of length, the standard **yard (yd).** The **English system** has been defined with sufficient precision to be used in modern manufacturing and commerce, but it is not particularly useful in scientific work. Primarily, this is because there is no regularity in the different units that may be used to express a measured quantity. Length may be expressed in inches, feet, yards, and miles; but the number of inches in 1 foot (12) is not the same as the number of feet in 1 yard (3) or the number of yards in 1 mile (1760). As a

Interest in the metric system in the United States originated in 1821 with John Quincy Adams, then Secretary of State, and continues to this day. Although the metric standards have been adopted as the *official* standards in the United States, the English system is still used widely for measurements in commerce and industry. Today, the United States is an "island in a metric world." Nevertheless, eventual adoption of the metric system in the United States seems assured.

TABLE 1-1
English and metric equivalents

Metric		English
	mass	
1 kg	=	**2.205 lb**
453.6 g	=	1 lb
	length	
1.609 km	=	1 mi
1 m	=	39.37 in.
2.54 cm	=	1 in.
	volume	
3.785 L	=	**1 gal**
1 L	=	1.057 qt

result, calculations requiring a conversion between units (e.g., inches to miles) are sometimes difficult to perform.

Prior to the French Revolution the system of measurement in France varied from one province to another, and commercial transactions between provinces were difficult to conduct. The need for a uniform system of measurement was clear; in about 1790 a commission of scientists was established to propose one. The commission chose as the standard unit of length, called a **meter (m),** a length equal to 1/10,000,000 of the distance at sea level from the North Pole to the equator along the meridian passing through Paris. The fact that the original measurement was in error is interesting but not significant. The standard that was adopted was the distance between two marks on a certain platinum–iridium metal bar kept in the International Bureau of Weights and Measures in Sèvres, near Paris. The system of measurement based on the meter is is called the **metric system.** It was adopted officially in France in 1840.

The metric system is a decimal system. The several different multiple and submultiple units by which a measured property may be expressed differ from one another by factors of ten. For example,

mega means *one million times* the base unit

kilo means *one thousand times* the base unit

centi means *one hundredth of* the base unit

milli means *one thousandth of* the base unit

micro means *one millionth of* the base unit

nano means *one billionth of* the base unit

Thus, 1 *kilo*meter (km) is 1000 m (about 0.6 mile); 1 *centi*meter (cm) is $\frac{1}{100}$ of 1 m (about 0.4 inch); and 1 *milli*meter is $\frac{1}{1000}$ of 1 m.

Some of the important relationships in the metric system and a comparison to the English system are presented in Figure 1-2. A few equivalent quantities in the two systems are listed in Table 1-1 and used in calculations in Section 1-9 and in the Exercises.

FIGURE 1-2
A comparison of the metric and English systems.

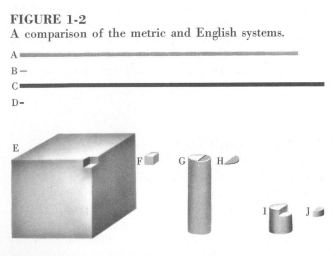

Line A:	A length of 1 yd
Line B:	A length of 1 in.—**36 in. = 1 yd**
Line C:	A length of 1 m—**1 m = 39.37 in.**
Line D:	A length of 1 cm—**100 cm = 1 m** and **2.54 cm = 1 in.**
Cube E:	A cube, 10 cm (1 decimeter, dm) on an edge, having a volume of 1000 cm^3 (1 dm^3), but with the upper right front corner (cube F) removed—**1 dm^3 = 1000 cm^3 = 1 liter (1 L)**
Cube F:	A cube, 1 cm on an edge, having a volume of 1 cm^3 (1 cubic centimeter)—**1 cm^3 = 1 ml (1 milliliter)**
Cylinder G:	A steel cylinder having a mass of 1 kg, but with a sector (H), representing $\frac{1}{1000}$ of the total mass, removed.
Sector H:	A mass of 1 g—**1000 g = 1 kg**
Cylinder I:	A steel cylinder having a mass of 1 lb, but with a sector (J), representing $\frac{1}{16}$ of the total mass, removed—**2.205 lb = 1 kg**
Sector J:	A mass of 1 oz—**16 oz = 1 lb**

The symbol ∝ means "proportional to." It can always be replaced by an equality sign and a proportionality constant.
The proportionality constant in equation (1.1), *g*, is called the acceleration due to gravity. Its significance is explored further in Appendix B.

FIGURE 1-3
Determination of mass by the principle of weighing.

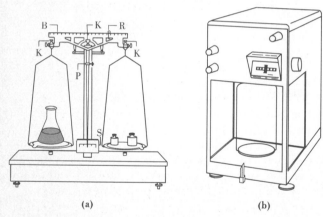

(a)   (b)

(a) **The principle of weighing.** A balance condition is achieved when the beam (B) is in a horizontal position and the pointer (P) rests at the center of the scale (S). When this occurs, the masses of the "unknown" and the standards are equal. The final adjustment of the balance point is made by moving a light metal wire called a rider (R) along the beam. The beam and pans pivot about fulcrums or pivot points called knife edges (K).
(b) **A modern single-pan balance** differs in several ways from the two-pan balance in (a). (1) Imagine that the balance in (a) is viewed "end on" so that only one pan is visible. (2) Now imagine that the pan out of view is replaced by a single large constant weight. (3) An appropriate set of weights is added *above* the single pan to just balance the constant weight. (4) When an unknown is placed on the pan, the balance condition is upset. This condition can be restored, however, by *removing* from the set above the pan a number of weights with mass equal to the unknown. The total mass removed is registered on the dials on the face of the balance.

MASS. Mass describes the quantity of matter in an object. The **kilogram (kg)** was originally defined as the mass of 1000 cubic centimeters (cm³) of water at 4°C and normal atmospheric pressure. It is now taken to be the mass of a cylindrical bar of platinum–iridium metal kept at the International Bureau of Weights and Measures. The kilogram is a fairly large unit for most applications in chemistry, so the unit **gram (g)** is more commonly employed (see Figure 1-2).

Weight, which describes the force of gravity on an object, is directly proportional to the mass of the object. This fact can be represented through a simple mathematical equation.

$$W \propto m \qquad \text{and} \qquad W = g \cdot m \qquad (1.1)$$

Although a given quantity of matter has a fixed mass (m), no matter where or how the measurement is made, its weight (W) may vary because g varies slightly from one point on earth to another. Thus, an object weighed first in Leningrad and then in Panama decreases in weight by about 0.4%, even though its mass remains constant. The terms *weight* and *mass* are often used interchangeably, but you must remember that mass is the basic measure of the quantity of matter. Figure 1-3 illustrates how mass is measured by the principle of weighing.

VOLUME. Volume is an important property, but it is not as fundamental as mass because volume varies with temperature and pressure, whereas mass does not. Volume has the unit (length)³. The basic unit of volume in the metric system is the **cubic meter (m³).** Another commonly used unit is the **cubic centimeter (cm³),** and still another is the **liter (L).** One liter is defined as a volume of 1000 cm³, which means that one **milliliter (1 ml)** is exactly equal to one cubic centimeter (1 cm³). The liter is also equal to one **cubic decimeter (1 dm³).** These volume units are illustrated in Figure 1-2.

1-6 SI Units

The metric standards described in Section 1-5 have been modified to overcome certain difficulties. First is the difficulty in comparing objects with a standard when the standard is one of a kind (such as the standard meter in Sèvres). Furthermore, such standards are subject to change. For example, the metal bar that has served as the standard meter changes in length as the temperature changes.

In the past physicists have used a version of the metric system called the mks (meter–kilogram–second) system, whereas chemists have preferred the cgs (centimeter–gram–second) system. This variability is eliminated in the SI system.

Such difficulties can be overcome by basing standards of measurement on *natural* universal constants. Then, any scientist, at any time, may set up standards and be assured that they will be the same as those obtained by others. This has been provided for in an international agreement adopted in 1960 and called the International System of Units or **SI units** (from the French, Le Système International d'Unités). In the SI system the unit of length corresponding to 1 meter is defined as a length equal to 1,650,763.73 wavelengths of a particular orange-red light emitted by a lamp containing krypton-86 gas. The unit of time, the second, is defined as the duration of 9,192,631,770 periods of a particular radiation emitted by cesium-133 atoms. The unit of mass, the kilogram, cannot be defined in terms of a basic physical constant, so it remains the mass of a cylindrical bar of metal maintained at Sèvres.

Another aspect of the SI system is that to facilitate communication among scientists in different disciplines and in different nations, certain base units and derived units are preferred over others. In time it is expected that the SI convention will be adopted and used consistently in all scientific work. However, as with any new development affecting the activities of large numbers of individuals, a transition period is necessary. In this text both familiar metric units and SI units are employed, and where they differ this fact is mentioned. Of the familiar units introduced to this point, the liter and milliliter are not SI units. Their SI counterparts are the cubic decimeter (dm^3) and the cubic centimeter (cm^3), respectively (recall Figure 1-2). A more complete description of the SI system is presented in Appendix C, to which you will need to refer from time to time. Some common SI prefixes are listed in Table 1-2.

TABLE 1-2
Some common SI prefixes

Multiple	Prefix
10^6	mega (M)
10^3	kilo (k)
10^{-1}	deci (d)
10^{-2}	centi (c)
10^{-3}	milli (m)
10^{-6}	micro (μ)[a]
10^{-9}	nano (n)

[a] The Greek letter μ (pronounced "mew").

1-7 Density

Density is obtained by dividing the mass of an object by its volume.

$$\text{density } (d) = \frac{\text{mass } (m)}{\text{volume } (V)} \tag{1.2}$$

A property whose magnitude depends on the quantity of material being observed is an **extensive** property. Both mass and volume are extensive properties. Any property that is independent of the quantity of material is an **intensive** property. Density, which is the ratio of mass to volume, is an intensive property. Intensive properties are generally preferred for scientific work because of their independence of the quantity of matter being studied.

The mass of 1000 cm^3 of water at 4°C and normal atmospheric pressure is almost exactly (but slightly less than) 1 kg. The density of water under these conditions is 1000 g/1000 cm^3 = 1.000 g/cm^3. Because volume varies with temperature while mass remains constant, density is a function of temperature. At 20°C the density of water is 0.998 g/cm^3. Some other densities at 20°C are

The SI units of density are kg/m^3 or g/cm^3, but you may occasionally encounter density expressed in g/ml or, for gases, in g/L.

ethyl alcohol, 0.789 g/cm^3; carbon tetrachloride, 1.59 g/cm^3; aluminum, 2.70 g/cm^3; iron, 7.86 g/cm^3; lead, 11.34 g/cm^3; gold, 19.3 g/cm^3

Numerical calculations involving the density concept are presented in Section 1-9.

There is an old riddle that goes: "What weighs more, a ton of bricks or a ton of feathers?" The correct answer is that they weigh the same, and anyone who answers in this way has demonstrated insight into the meaning of mass—a measure simply of the quantity of matter in an object. One who answers that the bricks weigh more than the feathers has confused the concepts of mass and density. Matter in a brick is more concentrated than in feathers, that is, confined to a smaller volume; bricks are more dense than feathers.

1-8 Temperature

Temperature is a property that is indeed difficult to define, even though we have an intuitive idea of what temperature is. To say that **temperature** is the degree of "hotness" of an object is not very precise, yet it does convey a certain meaning. If two objects of different temperature are brought into contact, the warmer object becomes colder and the colder one becomes warmer. Eventually, both objects come to the same degree of "hotness"—the same temperature.

Temperature can be measured in terms of the effect that its change produces on some other measurable property, for example, length. One common temperature-measuring device, a thermometer, is based on the length of a liquid column in a thin capillary bore in a glass tube. As the temperature of the thermometer changes, so does the length of the liquid column, increasing as the temperature increases.

To set up a scale of temperatures requires agreement upon certain fixed points of temperature and a degree of temperature change. Two commonly used fixed points are the temperature at which ice melts (the ice point) and the temperature at which liquid water boils (the steam point), both under normal atmospheric pressure.

On the Fahrenheit temperature scale the ice point is $32°F$, the steam point is $212°F$, and the interval between is divided into 180 equal parts, called degrees Fahrenheit. On the Celsius (centigrade) scale the ice point is $0°C$, the steam point is $100°C$, and the interval is divided into 100 equal parts, called degrees Celsius. Figure 1-4 compares the Fahrenheit and Celsius temperature scales. Also presented in Figure 1-4 are the equations used to convert temperatures between these two scales and examples of their use. The basic SI unit of temperature is the kelvin (K), which is introduced in Section 5-3.

FIGURE 1-4
A comparison of temperature scales.

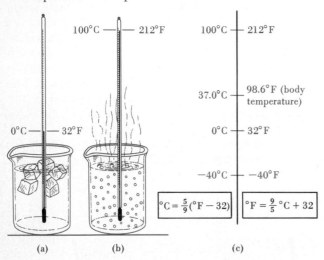

(a) Ice point. (b) Steam point. (c) Comparison of Fahrenheit and Celsius (formerly centigrade) temperature scales. Also shown are equations for converting between °C and °F. For example, to determine the Celsius equivalent of 68°F,

$$°C = \frac{5}{9}(°F - 32) = \frac{5}{9}(68 - 32) = \frac{5}{9}(36) = 20$$

The Fahrenheit equivalent of 30°C is

$$°F = \frac{9}{5}°C + 32 = \frac{9}{5}(30) + 32 = 86$$

1-9 Problem Solving—The Conversion Factor (Factor-Label) Method

Throughout the text we will emphasize properties that can be expressed through numbers—quantitative measurements. But a number by itself is usually meaningless. A measured quantity must be accompanied by a **unit.** The unit indicates the standard against which the measured quantity is to be compared. A metal rod 9 m in length is nine times as long as the standard meter.

A number of the calculations of beginning chemistry involve a conversion of measured quantities from one set of units to another, through the use of conversion factors. Consider this well-known fact, expressed as a simple mathematical equation.

1 yd = 36 in.

Divide each side of the equation by 1 yd.

$$\frac{1 \text{ yd}}{1 \text{ yd}} = \frac{36 \text{ in.}}{1 \text{ yd}}$$

The numerator and denominator on the left side are identical; they cancel.

$$1 = \frac{36 \text{ in.}}{1 \text{ yd}} \tag{1.3}$$

The numerator and denominator on the right side of equation (1.3) represent the *same length*. It is for this reason that the ratio of the numerator (36 in.) and the denominator (1 yd) is equal to 1. *A conversion factor must always have the numerator and denominator representing equivalent quantities.*

Consider the question: How many inches are there in 6 yd? The measured quantity is 6 yd and multiplying this quantity by 1 does not change its value.

6 yd × 1 = 6 yd

Now replace the 1 by its equivalent—the conversion factor (1.3). Cancel the unit, yd, and carry out the required multiplication.

$$6 \text{ yd} \times \underbrace{\frac{36 \text{ in.}}{1 \text{ yd}}}_{\substack{\text{this factor} \\ \text{converts} \\ \text{yd to in.}}} = 216 \text{ in.}$$

Next consider the question: How many yards are there in 540 in.? We cannot use exactly the same factor (1.3) as previously, for if we did the result would be nonsensical.

$$540 \text{ in.} \times \frac{36 \text{ in.}}{1 \text{ yd}} = 19,400 \text{ in.}^2/\text{yd}$$

Factor (1.3) must be rearranged to 1 yd/36 in.

$$540 \text{ in.} \times \underbrace{\frac{1 \text{ yd}}{36 \text{ in.}}}_{\substack{\text{this factor} \\ \text{converts} \\ \text{in. to yd}}} = 15 \text{ yd} \tag{1.4}$$

This second illustration emphasizes two important points.

1. There are two ways of writing a conversion factor—in one form or its reciprocal (inverse). Since a conversion factor is equivalent to 1, its value is not changed by inversion, but
2. A conversion factor must be used in such a way as to produce the desired cancellation of units.

A convenient way to think about calculations involving conversion factors is that

information sought = information given × conversion factor(s) \hfill (1.5)

Consider how expression (1.5) is used to answer this question: What is the length of 22 in., expressed in cm? The answer to this question must consist of *two* parts—a number and a unit. The required unit is suggested in the statement of the problem—cm. The numerical part of the answer must be determined by calculation, and we refer to this as "number" or "no." The *information sought* is *no. cm*. The *information given,* that is, the quantity that is to be multiplied by a conversion factor, is also determined by a close reading of the problem; it is *22 in*. Thus, expression (1.5) takes the form

no. cm = 22 in. × conversion factor

Sometimes the statement of the problem includes the necessary conversion factor(s), and sometimes you are expected to know or to be able to derive the factor(s) you need. *The key to problem solving by the conversion factor method lies in knowing where to find and how to use conversion factors.*

In the present case the necessary relationship is 1 in. = 2.54 cm; the unit "in." must cancel and the unit "cm" must remain.

$$\text{no. cm} = 22 \text{ in.} \times \underbrace{\frac{2.54 \text{ cm}}{1 \text{ in.}}}_{\substack{\text{this factor} \\ \text{converts} \\ \text{in. to cm}}} = 56 \text{ cm}$$

FIGURE 1-5
A comparison of one square foot and one square meter.

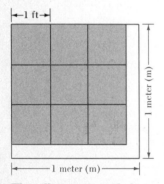

|←1 ft→|
|———— 1 meter (m) ————|
(vertical: 1 meter (m))

This illustration is helpful in visualizing the calculation of Example 1-2.

An alternative method is to convert the length 1.00 m to feet,

$$\text{no. ft} = 1.00 \text{ m}$$
$$\times \frac{39.37 \text{ in.}}{1.00 \text{ m}}$$
$$\times \frac{1 \text{ ft}}{12 \text{ in.}}$$
$$= 3.28 \text{ ft}$$

and square the result.

$$\text{no. ft}^2 = 3.28 \text{ ft} \times 3.28 \text{ ft}$$
$$= 10.8 \text{ ft}^2$$

Example 1-1   What is the distance 20 miles (mi), expressed in kilometers?

The starting point is a form of the general equation (1.5). Information is sought in the unit kilometers (km) and the given information is the distance, 20 mi.

no. km = 20 mi × conversion factors

From this point the problem can be solved in a number of ways, depending on the conversion factors that are known, for example, miles → yards → inches → meters → kilometers. An alternative set of conversion factors is miles → feet → inches → meters → kilometers.

$$\text{no. km} = 20 \text{ mi} \times \frac{5280 \text{ ft}}{1 \text{ mi}} \times \frac{12 \text{ in.}}{1 \text{ ft}} \times \frac{1 \text{ m}}{39.37 \text{ in.}} \times \frac{1 \text{ km}}{1000 \text{ m}} = 32 \text{ km}$$

$$\text{(mi} \longrightarrow \text{ft} \longrightarrow \text{in.} \longrightarrow \text{m} \longrightarrow \text{km)}$$

SIMILAR EXAMPLES: Exercises 19 through 24.

Example 1-2   How many square feet (ft²) correspond to an area of 1.00 square meter (m²)?

An area of 1.00 m² is represented in Figure 1-5; it can be thought of as a square with sides 1 m long. Also depicted are the length, 1 ft, and the area, 1.00 ft². There are somewhat more than 9 ft² in 1 m².

Equation (1.5) is written as follows:

$$\text{no. ft}^2 = 1.00 \text{ m}^2 \times \underbrace{\left(\frac{39.37 \text{ in.}}{1 \text{ m}}\right)\left(\frac{39.37 \text{ in.}}{1 \text{ m}}\right)}_{\substack{\text{to convert} \\ \text{m}^2 \text{ to in.}^2}} \times \underbrace{\left(\frac{1 \text{ ft}}{12 \text{ in.}}\right)\left(\frac{1 \text{ ft}}{12 \text{ in.}}\right)}_{\substack{\text{to convert} \\ \text{in.}^2 \text{ to ft}^2}}$$

This is the same as writing

$$\text{no. ft}^2 = 1.00 \text{ m}^2 \times \frac{(39.37)^2 \text{ in.}^2}{1 \text{ m}^2} \times \frac{1 \text{ ft}^2}{(12)^2 \text{ in.}^2} = 10.8 \text{ ft}^2$$

Illustrated here is the fact that it may be necessary at times to raise conversion factors to powers higher than the first.

SIMILAR EXAMPLES: Exercises 26, 27, 28.

Example 1-3   What is the velocity of 55 mi/h expressed in meters per second (m/s)?

The conversions required here are a bit more difficult than those considered previously, since the terms given and sought are both expressed as *ratios* of units. What is required is a conversion from miles to meters in the *numerator* and hours to seconds in the *denominator*. However, if the necessary conversion factors are set up with care—each factor must have a numerator and denominator that are equivalent, and each factor must produce the desired cancellation of units—the correct result is readily achieved.

$$\text{no. } \frac{m}{s} = \frac{55 \text{ mi}}{1 \text{ h}} \times \frac{5280 \text{ ft}}{1 \text{ mi}} \times \frac{12 \text{ in.}}{1 \text{ ft}} \times \frac{2.54 \text{ cm}}{1 \text{ in.}} \times \frac{1 \text{ m}}{100 \text{ cm}} \times \frac{1 \text{ h}}{60 \text{ min}} \times \frac{1 \text{ min}}{60 \text{ s}}$$

$$= 25 \frac{m}{s}$$

SIMILAR EXAMPLES: Exercises 25, 29.

Example 1-4   A block of wood having the dimensions 105 cm × 5.1 cm × 6.2 cm weighs 2.26 kg. What is the density of the wood, expressed in grams per cubic centimeter?

The volume of a rectangular block (a parallelepiped) is simply the product of the length (*l*), the width (*w*), and the height (*h*).

$$V = 105 \text{ cm} \times 5.1 \text{ cm} \times 6.2 \text{ cm} = 3300 \text{ cm}^3$$

The mass of the block must be expressed in grams.

$$m = 2.26 \text{ kg} \times \frac{1000 \text{ g}}{1 \text{ kg}} = 2260 \text{ g}$$

The density of the wood is

$$d = \frac{m}{V} = \frac{2260 \text{ g}}{3300 \text{ cm}^3} = 0.68 \text{ g/cm}^3$$

SIMILAR EXAMPLES: Exercises 34, 36.

Example 1-5   Several irregularly shaped pieces of zinc, weighing 30.0 g, are dropped into a graduated cylinder containing 20.0 cm³ of water. The water level rises to 24.2 cm³. What is the density of the zinc?

The volume of metal is simply the difference of the two water levels pictured in Figure 1-6.

$$\text{volume of zinc} = 24.2 \text{ cm}^3 - 20.0 \text{ cm}^3 = 4.2 \text{ cm}^3$$

$$\text{density} = \frac{\text{mass}}{\text{volume}} = \frac{30.0 \text{ g}}{4.2 \text{ cm}^3} = 7.1 \text{ g/cm}^3$$

SIMILAR EXAMPLE: Exercise 39.

Example 1-6   What is the volume, in liters, occupied by 50.0 kg of ethanol at 20°C? The density of ethanol at 20°C is 0.789 g/cm³.

This problem illustrates how density can be thought of as a conversion factor between mass and volume: 1.00 cm³ of ethanol = 0.789 g of ethanol.

The information given is *50.0 kg of ethanol* and what is sought is *number of liters of ethanol*. The series of conversions required is kg → g → cm³ → L.

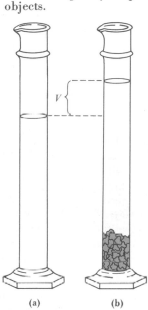

FIGURE 1-6
Measurement of the volume of irregularly shaped objects.

(a) A graduated cylinder is filled to a certain level with a liquid.
(b) Irregularly shaped objects are added to the liquid. They sink to the bottom of the cylinder and cause the liquid level to rise to an extent *V* that corresponds to their total volume.

An alternative method involves solving the density equation for volume, $V = m/d$, and substituting the appropriate information.

$$V = \frac{50,000 \text{ g}}{0.789 \text{ g/cm}^3}$$
$$= 63,400 \text{ cm}^3$$
$$= 63.4 \text{ L}$$

FIGURE 1-7
The concept of equivalence.

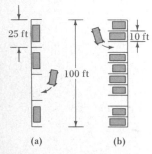

(a)   (b)

(a) For a parallel parking arrangement we can say that each automobile *is equivalent to* 25 ft. That is, 1 automobile $\backsim$ 25 ft.

no. automobiles
$$= 100 \text{ ft} \times \frac{1 \text{ automobile}}{25 \text{ ft}}$$
$$= 4 \text{ automobiles}$$

(b) In this parking arrangement (perpendicular) each automobile is equivalent to 10 ft of curb space. How many automobiles can be parked along the 100-ft section of curb?

no. L ethanol $= 50.0 \text{ kg ethanol} \times \dfrac{1000 \text{ g ethanol}}{1 \text{ kg ethanol}} \times \dfrac{1.00 \text{ cm}^3 \text{ ethanol}}{0.789 \text{ g ethanol}}$

$\times \dfrac{1 \text{ L ethanol}}{1000 \text{ cm}^3 \text{ ethanol}}$

$= 63.4 \text{ L ethanol}$

SIMILAR EXAMPLES: Exercises 35, 37, 38.

EQUIVALENCE AND EQUALITY. An additional insight into the use of conversion factors is suggested by Figure 1-7, which establishes an equivalence between an automobile and the curb space required to park it. In Figure 1-7a we need a conversion factor based on the relationship 1 automobile $\backsim$ 25 ft, where an equivalence sign ($\backsim$) has been used rather than an equality sign ($=$). An automobile is not literally the same thing as 25 ft of curb space. *For the purposes of this calculation* we consider them to be *equivalent:* For every 25 ft of curb space we are able to park one automobile (regardless of its actual length). A different equivalence between the automobile and curb space exists for the perpendicular parking arrangement in Figure 1-7b. We will use the sign $=$ when an equality exists and $\backsim$ for an equivalence, but conversion factors are set up and used in the same way in either case.

PERCENT AS A CONVERSION FACTOR. The term **percent** means, literally, parts per hundred, that is, parts of one constituent to 100 parts of the whole. For example, the statement that a seawater sample contains 3.5% sodium chloride, by mass, means that in a 100-g sample of the seawater there will be found 3.5 g of sodium chloride. Like a number of other relationships, this one should be stated as an equivalence:

3.5 g sodium chloride $\backsim$ 100 g seawater   (1.6)

A common difficulty encountered by students in using the notion of percent can be put as a question: When do you multiply and when do you divide by percent? Expressing percentages as conversion factors should help to clarify this matter. The factors are always set up to require multiplication, as illustrated in Examples 1-7 and 1-8.

Example 1-7 215 g of seawater containing 3.5% sodium chloride, by mass, is evaporated to dryness. How much sodium chloride is present in the solid residue?

The equivalence (1.6) is changed to a conversion factor with "g sodium chloride" in the numerator and "g seawater" in the denominator.

no. g sodium chloride $= 215 \text{ g seawater} \times \dfrac{3.5 \text{ g sodium chloride}}{100 \text{ g seawater}}$

$= 7.5 \text{ g sodium chloride}$

SIMILAR EXAMPLES: Exercises 41, 42.

Example 1-8 125 g of sodium chloride is to be produced by evaporating to dryness a quantity of seawater containing 3.5% sodium chloride, by mass. The seawater has a density of 1.03 g/cm^3. How many *liters* of seawater must be taken for this purpose?

In this example we need a conversion factor with "g seawater" in the numerator and "g sodium chloride" in the denominator. Also required are conversion factors for g seawater $\rightarrow$ cm^3 seawater $\rightarrow$ L seawater.

no. L seawater $= 125 \text{ g sodium chloride} \times \dfrac{100 \text{ g seawater}}{3.5 \text{ g sodium chloride}}$

$\times \dfrac{1 \text{ cm}^3 \text{ seawater}}{1.03 \text{ g seawater}} \times \dfrac{1 \text{ L seawater}}{1000 \text{ cm}^3 \text{ seawater}}$

$= 3.5 \text{ L seawater}$

SIMILAR EXAMPLES: Exercises 43, 44.

1-10 Significant Figures

Precision refers to the degree of reproducibility of a measured quantity, that is, the closeness of agreement among the values obtained when the same quantity is measured several times.

An important idea that was used implicitly in each of the preceding examples must now be stated explicitly: When a measured or calculated quantity is written down, some indication of the **precision** of the measurement must be given as well. For example, suppose that the same object is weighed on two different balances—one a relatively crude platform balance and the other a sophisticated analytical balance. Typical results might be

	Platform balance	Analytical balance
measured quantity	10.3 g	10.3107 g
uncertainty	±0.1 g	±0.0001 g
mass	10.3 ± 0.1 g	10.3107 ± 0.0001 g
precision	low or poor (one part in 103)	high (one part in 103,107)

On the platform balance the determination of mass is reproducible only to the nearest one-tenth gram (±0.1 g), whereas on the analytical balance the measurement is reproducible to the nearest one-tenth *milli*gram (±0.0001 g). The notations "10.3 ± 0.1 g" and "10.3107 ± 0.0001 g" indicate the precision of these measurements very clearly. This is a type of notation encountered frequently in laboratory work and reports in scientific journals. But this notation is a bit cumbersome to write and to use in numerical calculations.

An alternative approach is to assume that when a number is written down, all the digits preceding the last are known with certainty and that there is an *uncertainty of about one unit in the last digit shown*. Thus, the number 10.3 is "between 10.2 and 10.4," whereas the number 10.3107 is "between 10.3106 and 10.3108." The number 10.3 is said to consist of *three* significant figures, whereas 10.3107 consists of *six* significant figures. To designate the number of significant figures in a quantity is to give an indication of the confidence with which the number is known. The greater the number of significant figures used to express a quantity, the smaller the uncertainty (and the greater the precision) in its measurement.

Determining the numbers of significant figures in 4.006, 12.012, and 10.070 is not difficult: They are *4*, *5*, and *5*, respectively. The applicable rules are

1. All nonzero digits are *significant* (i.e., 4.006, 12.012, and 10.070).
2. Zeros placed between nonzero digits are *significant* (i.e., 4.006, 12.012, and 10.070).
3. Zeros at the end of a number to the *right* of the decimal point are *significant* (i.e., 10.070).

What are the numbers of significant figures in 0.00002 and 0.000020? The number 0.00002 has only *one* significant figure, since

4. Zeros to the left of the first nonzero digit are *not significant*. (They simply locate the decimal point.)

The number 0.000020 has *two* significant figures, based on rules 3 and 4.

Finally, what are the numbers of significant figures in 750 and 20,000? We cannot be certain whether the number 750 is meant to indicate 750 ± 10 (in which case there are *two* significant figures) or 750 ± 1 (in which case there are *three* significant figures). This ambiguity can be stated as follows:

5. Zeros appearing at the *end* of a number and to the *left* of the decimal point *may or may not be significant*.

A way of resolving this difficulty is to use exponential notation in expressing a number (see Appendix A). Thus, the number 20,000 is expressed differently depending on the precision with which it is known:

One significant figure	Two significant figures	Three significant figures
2×10^4	2.0×10^4	2.00×10^4

Precision can be neither gained nor lost during arithmetic operations. This requirement is generally met by this simple rule for multiplication and division of numbers: *The result may carry no more significant figures than the least precisely known quantity involved in the calculation.* In the chain multiplication that follows, the result should be rounded off to *three* significant figures.*

$$14.80 \quad \times \quad 12.10 \quad \times \quad 5.05 \quad = 904.354000 = 904 = 9.04 \times 10^2$$

(4 sig. fig.)   (4 sig. fig.)   (3 sig. fig.)   (3 sig. fig.)

In the addition or subtraction of numbers, the uncertainty in the sum or difference is the same as that of the least precisely known quantity. Consider the sum

$$\begin{array}{r} 115.016 \text{ g} \\ 12.0 \quad\text{ g} \\ 3.5182 \text{ g} \\ \hline 130.5342 \text{ g} = 130.5 \text{ g} \end{array}$$

Although the least precisely known quantity, "12.0," carries only three significant figures, the sum has four. The limitation here is not on significant figures but on the fact that the sum cannot be expressed any more precisely than ± 0.1, the absolute precision with which "12.0" is stated.

There are two situations when a number appearing in a calculation may be *exact*. This may occur by definition (3 ft = 1 yd; 2.54 cm = 1 in.) or as a result of counting rather than measurement (*two* hydrogen atoms in a water molecule). Exact numbers do not affect the rules on significant figures.

Example 1-9 The density of methanol at 20°C is determined by the series of measurements pictured in Figure 1-8. Express this density in the appropriate number of significant figures.

The mass of water required to fill the pycnometer at 20°C is

$$35.552 \text{ g} - 25.601 \text{ g} = 9.951 \text{ g}$$

The volume of water, and hence that of the pycnometer, is

$$V = \frac{m}{d} = \frac{9.951 \text{ g}}{0.99823 \text{ g/cm}^3} = 9.969 \text{ cm}^3$$

To prove that the result of the chain multiplication is known only to three significant figures, despite the large number of digits appearing in the product, proceed as follows. Multiply 14.80 by 12.10 and then multiply the product of these two numbers, first by 5.04, then by 5.05, and finally by 5.06 (because the number 5.05 is really 5.05 ± 0.01). The results obtained are 902.5632, 904.3540, and 906.1448. The first two digits are identical for all three products, "90...," but the third digit varies. Only the first three figures are significant, and the product should be expressed as $904 = 9.04 \times 10^2$.

*The rule to follow in "rounding off" is to increase the final digit by one unit if the digit dropped is greater than 5 and to leave the final digit unchanged if the digit dropped is less than 5. For example, to three significant figures, 15.56 rounds off to 15.6, and 15.54 rounds off to 15.5. If the digit dropped is 5, the final remaining digit is increased by one unit if necessary to make it *even;* otherwise, it is left unchanged. Thus, to three significant figures, 15.55 is rounded off to 15.6 and 15.45 is rounded off to 15.4.

FIGURE 1-8
Determination of density
with a pycnometer.

The calculations required
in this density determination are the subject of
Example 1-9.

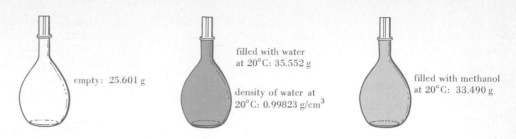

empty: 25.601 g

filled with water
at 20°C: 35.552 g

density of water at
20°C: 0.99823 g/cm³

filled with methanol
at 20°C: 33.490 g

The mass of methanol required to fill the pycnometer at 20°C is

33.490 g − 25.601 g = 7.889 g

The density of methanol at 20°C is

$$d = \frac{m}{V} = \frac{7.889 \text{ g}}{9.969 \text{ cm}^3} = 0.7914 \text{ g/cm}^3$$

SIMILAR EXAMPLES: Additional Exercises 4, 7.

Summary

Chemistry is a study of matter, and one of the first needs in this study is to classify matter into useful categories. The scheme introduced in this chapter considers matter as being either a substance—element or compound—or a mixture—homogeneous or heterogeneous. Among the many aspects of matter of interest to chemists are the distinctive characteristics or properties of matter and how these may be changed by physical and chemical means. Like other branches of modern science, chemistry makes use of the scientific method. This is a series of activities that starts with observations and culminates with theories to explain and predict natural phenomena. Many important discoveries are still made "by chance," however.

The need for precise measurement in chemistry is essential, and a uniform system for expressing the results of measurements is also important. The SI system provides this uniformity, but because SI units have not yet been universally adopted, other units of measurement are used in this text as well. Still another requirement is for methods of performing calculations involving measured quantities. The method introduced in the chapter requires recognizing relationships between quantities and using these relationships to establish conversion factors. The desired result is obtained by multiplying a given quantity by one or more conversion factors. In this method algebraic cancellation of units is used as a guide to ensure that conversion factors are formulated properly. Finally, measurements cannot be performed with certainty. It is necessary to express measurements in such a way as to indicate the uncertainties that exist in them. This requires the concept of significant figures and the use of certain simple rules in handling significant figures in arithmetic operations.

Learning Objectives

As a result of studying Chapter 1, you should be able to

1. Identify common materials using the terms *element, compound, homogeneous mixture,* and *heterogeneous mixture.*

2. Write the names and chemical symbols of the more common elements, including the first 20 (see the inside front and back covers).

3. Distinguish between physical and chemical properties and simple physical and chemical changes.

4. Describe the principal features of the scientific method and its limitations.

5. State the basic units of mass, length, and volume in the metric system and the common prefixes used in association with these.

6. Relate at least one unit each of mass, length, and volume in the English system to a corresponding unit in the metric system.

7. Describe the relationship of the SI to the older metric system, particularly the ways in which the systems resemble and differ from one another.

8. Convert Fahrenheit temperature to Celsius, and Celsius to Fahrenheit.

9. Write a conversion factor from a relationship between two quantities.

10. Express density and percent composition in the form of conversion factors.

11. Use conversion factors in a general problem-solving method.

12. State the number of significant figures in a numerical quantity.

13. Express the result of a calculation with the appropriate number of significant figures.

Some New Terms

A **chemical change** is a transformation of one or more substance(s) into one or more new substance(s).

A **chemical property** is a statement of a type of chemical change a substance is expected to undergo: for example, to combine with oxygen, dissolve in an acid.

A **compound** is a substance made up of two or more elements. It does not change its identity in physical changes but can be broken down into its constituent elements by chemical changes.

Density is a physical property obtained by dividing the mass of a substance or object by its volume (i.e., mass per unit volume).

An **element** is one of a group of fundamental substances that cannot be broken down into simpler substances.

An **extensive property** is one whose value depends on the quantity of matter observed. Examples of such properties are mass and volume.

A **heterogeneous mixture** is one in which the components separate into physically distinct regions of differing properties.

A **homogeneous mixture (solution)** is a mixture of elements and/or compounds that has a uniform composition within a given sample but a varying composition from one sample to another.

An **intensive property** is *independent* of the quantity of matter involved in the observation. Examples of such properties are density and temperature.

Mass is a measure of the amount of inertia possessed by an object. (Inertia is the tendency to remain at rest or in constant motion unless acted upon by an external force. The more mass in an object, the greater its inertia.)

Matter is anything that occupies space and has mass.

A **natural law** is a general statement that can be used to summarize observations of natural phenomena.

A **physical change** is one in which the physical appearance of a substance changes but its basic identity, that is, its chemical composition, remains unchanged.

A **physical property** is a characteristic that helps to identify a substance and which the substance can display without undergoing a fundamental change in its identity.

Scientific method refers to the general sequence of activities—observation, experimentation, and the formulation of laws and theories—that lead to the advancement of scientific knowledge.

Significant figures are those digits in an experimentally measured quantity that establish the precision with which the quantity is known.

A **substance** has constant composition and properties throughout a given sample and from one sample to another; all substances are either elements or compounds.

A **theory** is a conceptual framework with which one is able to *explain* one or a group of related natural laws.

Weight refers to the force exerted on an object when it is placed in a gravitational field (the "force of gravity"). The terms *weight* and *mass* are often used synonomously.

Exercises

Properties and classification of matter

1. Indicate whether the following properties are physical or chemical.
   (a) An iron nail is attracted to a magnet.
   (b) A silver spoon is tarnished in air.
   (c) Ice floats on liquid water.
   (d) Rubber objects disintegrate in a smog-filled environment.

2. Indicate whether each sample of matter listed is a substance or a mixture, and, if a mixture, whether homogeneous or heterogeneous.
   (a) a cube of sugar   (b) vegetable soup
   (c) premium gasoline   (d) mayonnaise
   (e) iodized salt   (f) tap water
   (g) ice

3. What type of change—physical or chemical—is necessary

to bring about each of the following separations? (*Hint:* Refer to the inside back cover for a listing of elements.)
   (a) hydrogen and oxygen gases from water
   (b) pure water from seawater
   (c) nitrogen and oxygen gases from air

4. Suggest physical changes by which the following mixtures can be separated
   (a) salt and sand
   (b) iron filings and wood chips
   (c) mineral oil and water

5. Indicate which of the following are extensive and which are intensive quantities.
   (a) the mass of air in a balloon
   (b) the temperature of melting ice
   (c) the length of time required to boil water
   (d) the color of light given off by a neon lamp

Scientific method

6. Is it possible to predict how many experiments are required to verify a natural law? Explain.

7. What are the principal reasons why one theory might be adopted over a conflicting one?

8. An important premise of science is that there exists an underlying order to nature. Einstein described this belief in the words "God is subtle but He is not malicious." Explain more fully what he probably meant by this remark.

Exponential arithmetic (see Appendix A)

9. Express the following numbers in exponential notation.
   (a) 7500   (b) 317,000
   (c) 8,152,000   (d) 100,000
   (e) 0.0062   (f) 0.0500
   (g) 0.00000038   (h) 0.100
   (i) 3

10. Express the following numbers in common decimal form.
   (a) 4.12×10^2 (b) 6.65×10^{-1}
   (c) 92×10^{-3} (d) 6.28×10^5
   (e) 4.0×10^0 (f) 2.98×10^{10}
   (g) 1.93×10^{-6} (h) 830×10^{-2}
   (i) 1.235×10^{-5}

11. A variety of measured or estimated quantities follow. Express each value in exponential form.
   (a) speed of light in vacuum: 186 thousand miles per second
   (b) mass of air in the atmosphere: 5 to 6 quadrillion tons
   (c) solar radiation received by the earth: 173 thousand trillion watts
   (d) diameter of a typical aerosol smog particle: one millionth of a meter
   (e) average diameter of a human cell: ten millionths of a meter

12. Perform the following calculations, expressing each number and the answer in exponential form.
   (a) $300 \times 6000 =$ (b) $42 \times 40 \times 4100 =$
   (c) $0.052 \times 0.0070 =$ (d) $0.0040 \times 550 =$
   (e) $\dfrac{4500}{0.0080} =$ (f) $\dfrac{120 \times 700 \times 0.10}{0.040 \times 2.5} =$

13. Express the result of each of the following calculations in exponential form.
   (a) $0.052 + (53 \times 6.0 \times 10^{-4}) =$
   (b) $\dfrac{20 + 380 + (1.60 \times 10^2)}{2.8 \times 10^{-1}} =$
   (c) $\dfrac{(1.2 \times 10^{-3})^2}{0.040 + (2.0 \times 10^{-2})} =$
   (d) $\dfrac{[(6.0 \times 10^3) + (4.4 \times 10^4)]^2}{(2.2 \times 10^3)^2 + 160,000} =$

Significant figures

14. Indicate whether each of the following is an exact number or a measured quantity subject to uncertainty.
   (a) the number of oranges in one dozen
   (b) the number of gallons of gasoline to fill an automobile gas tank
   (c) the distance between the earth and the sun
   (d) the number of days in the month of January
   (e) the area of a city lot

15. How many significant figures are shown in each of the following numbers? If indeterminate, give the range of significant figures possible.
   (a) 478   (b) 0.035
   (c) 750.06   (d) 1380.0
   (e) 44,000   (f) 0.0009
   (g) 8.0030   (h) 2

16. Rewrite each of the following numbers to consist of *four* significant figures.
   (a) 1418.2   (b) 303.51
   (c) 0.014045   (d) 156.251
   (e) 180,000   (f) 17.6050
   (g) 1.5×10^3

17. Perform the following calculations, retaining the appropriate number of significant figures in each result.
   (a) $512 \times 176 =$
   (b) $15.60 \times 10^3 \times 2.5 \times 10^5 =$
   (c) $\dfrac{3.58 \times 10^3}{1.8 \times 10^6} =$
   (d) $44.34 + 26.2 + 1.06 =$
   (e) $(1.561 \times 10^3) - (1.80 \times 10^2) + (2.02 \times 10^4) =$

Systems of measurement

18. Perform the following conversions within the metric system of measurement.
   (a) $2.76 \text{ kg} = \underline{\hspace{1cm}} \text{ g}$ (b) $8160 \text{ mm} = \underline{\hspace{1cm}} \text{ m}$
   (c) $368 \text{ mg} = \underline{\hspace{1cm}} \text{ kg}$ (d) $725 \text{ ml} = \underline{\hspace{1cm}} \text{ L}$

(e) 16.7 cm = ____ mm (f) 0.323 L = ____ cm³
(g) 2.67 g = ____ mg (h) 0.67 km = ____ m

19. Perform the following conversions within the English system of measurement.
   (a) 15.50 ft = ____ in. (b) 384 oz = ____ lb
   (c) 12.0 yd = ____ in. (d) 1.5 h = ____ s
   (e) 1401 ft = ____ yd (f) 2.30 mi = ____ ft

20. Perform the following conversions between the English and metric systems.
   (a) 12 in. = ____ cm (b) 16 ft = ____ m
   (c) 14 oz = ____ g (d) 55 kg = ____ lb
   (e) 22.5 m = ____ ft (f) 3500 mg = ____ oz

21. The English unit, the rod, is equal to 16.5 ft. What is this length expressed in meters?

22. A certain brand of coffee is offered for sale at $8.86 for a 3-lb can or $6.53 for a 1-kg can. Which is the better buy?

23. A sprinter runs the 100-yd dash in 9.3 s. What would be his time for a 100-m run if he ran at the same rate?

24. The unit of length, the furlong, is used in horseracing. The units of length, the chain and the link, are used in surveying. There are 8 furlongs in 1 mi, 10 chains in 1 furlong, and 100 links in 1 chain. To three significant figures, what is the length of 1 link, in inches?

25. An English unit of mass used in pharmaceutical work is the grain (gr). 15 gr = 1.0 g. A standard aspirin tablet contains 5.0 gr of aspirin. A 165-lb person takes two aspirin tablets.
   (a) What is the quantity of aspirin taken, expressed in milligrams?
   (b) What is the dosage rate of the aspirin, expressed in milligrams of aspirin per kilogram of body weight?

26. Determine the number of square meters (m²) in 1 square kilometer (km²).

27. What is the volume, in cubic meters, of a box that measures 28 in. by 40 in. by 16 in.?

*28. In the English system of measurement the *acre* is an important unit for measuring land area. The corresponding unit in the metric system is the *hectare*. There are 640 acres in 1 square mile (mi²), and 1 hectare is defined as 1 square hectometer (hm²); 1 hm = 100 m.
   (a) How many square feet are there in 1 acre?
   (b) How many hectares are there in 1 acre?

*29. An airplane flying at the speed of sound is said to be at Mach 1. Mach 1.5 is 1.5 times the speed of sound; and so on. If the speed of sound in air is given as 1130 ft/s, what is the speed of an airplane, in km/h, flying at Mach 1.27?

Temperature scales

30. Use the equations in Figure 1-4 to perform the following conversions between temperature scales.

   (a) 40°C = ____ °F (b) 77°F = ____ °C
   (c) 1232°F = ____ °C (d) −176°C = ____ °F

31. A table of climatic data lists the highest and lowest temperatures on record for San Bernardino, California, as 118°F and 17°F, respectively. What are these temperatures on the Celsius scale?

32. A class in home economics is given an assignment in candy making. The candy recipe calls for a sugar mixture to be brought to a "soft ball" stage (234 to 240°F). A student borrows a thermometer of range −10 to 110°C from the chemistry laboratory to do this assignment. Will this thermometer serve the purpose?

33. The absolute zero of temperature, introduced in Section 5-3, is −273.15°C. What is the absolute zero of temperature on the Fahrenheit scale?

Density

34. A 1.50-L sample of pure glycerol has a mass of 1892 g. What is the density of glycerol?

35. Ethylene glycol, an antifreeze, has a density of 1.11 g/cm³ at 20°C.
   (a) What is the mass, in grams, of 2.50 × 10² cm³ of the liquid?
   (b) What is the volume, in liters, occupied by 1.00 kg of the liquid?
   (c) What is the mass, in pounds, of 1.00 gal of the liquid?

36. To determine the density of a liquid, a 250.0-ml volumetric flask is weighed when empty (110.4 g) and again when filled to the mark with liquid (308.4 g). What is the density of the liquid?

37. It is desired to determine the volume of liquid that can be contained in an irregularly shaped glass vessel. The vessel is weighed when empty and found to have a mass of 80.3 g. Filled with liquid carbon tetrachloride (density = 1.59 g/cm³), the vessel weighs 245.8 g. What is the volume capacity of the vessel?

38. The following densities are given at 20°C: water, 0.998 g/cm³; iron, 7.86 g/cm³; aluminum, 2.70 g/cm³. Arrange the following items in terms of *increasing* mass.
   (a) A rectangular bar of iron, 175 cm × 1.0 cm × 0.50 cm.
   (b) A sheet of aluminum foil, 140.0 cm × 20.0 cm × 1.00 mm.
   (c) 1.00 L of water.

39. To determine the approximate mass of a small spherical shot of copper metal, the following experiment is performed. 100 pieces of the shot are counted out and added to 8.4 ml of water in a graduated cylinder; the total volume becomes 8.8 ml. The density of copper metal is 8.92 g/cm³. Determine the approximate mass of a single piece of shot, assuming that all the pieces are of nearly the same dimensions.

Percent composition

40. In a class of 50 students the results of a particular examination were 6 A, 11 B, 23 C, 7 D, and 3 F. What was the percent distribution of grades, that is, % A, % B, and so on?

41. A particular fertilizer is listed as containing 6.8% phosphorus, by mass. What mass, in grams, of phosphorus is contained in a 5.0-lb bag of the fertilizer?

42. A water solution that is 10.0% ethanol, by mass, has a density of 0.980 g/cm^3 at 25°C. What mass, in grams, of ethanol is contained in 3.50 L of this solution?

43. A sample of solid sodium chloride is said to be 98.0% pure. How much of this solid sample must be taken to contain 25.0 g of sodium chloride?

44. A solution containing 12.0% sodium hydroxide, by mass, has a density of 1.131 g/cm^3. What volume, in liters, of this solution must be used in an application requiring 0.350 kg of sodium hydroxide?

Additional Exercises

1. Human behavior cannot be studied quite as readily as the phenomena of natural science. Nevertheless, there are certain "laws" that are applicable to a variety of human activities. Explain what is meant by the "law of averages" and the "law of diminishing returns."

2. Use the concept of significant figures to criticize the manner in which the following information was stated in an official report: "In 1974, for example, there were approximately 1,573,006 students enrolled in California's four sectors of higher education."

3. Perform the following conversions between the English and metric systems of measurement.
   (a) 60 mi/h = _____ km/h   **(b)** 44 ft/s = _____ km/h
   (c) 125 in.^{3} = _____ cm^3 **(d)** 20 lb/in.^{2} = _____ g/cm^2
   (e) 62 lb/ft^3 = _____ g/cm^3

4. It is necessary to determine the density of a solution to *four* significant figures. The volume of solution can be measured to the nearest 0.1 ml.
   (a) What is the minimum volume of sample that can be used for the measurement?
   (b) Assuming the minimum volume sample determined in part (a), how accurately must the sample be weighed (i.e., to the nearest 0.1 g, 0.01 g, ...) if the density of the solution is greater than 1.00 g/cm^3? If the density is less than 1.00 g/cm^3?

*5. The temperature −40° on the Celsius scale is the same as −40°F. Can you verify this fact by calculation? Is there any other temperature for which the numerical values of the Celsius and Fahrenheit scales are the same?

*6. A Fahrenheit and a Celsius thermometer are immersed in the same medium, whose temperature is to be measured. At what Celsius temperature will the reading on the Fahrenheit thermometer be
   (a) twice that on the Celsius thermometer?
   (b) three times that on the Celsius thermometer?
   (c) one-fifth that on the Celsius thermometer?
   (d) 100° more than that on the Celsius thermometer?

*7. A pycnometer weighs 25.60 g empty and 35.55 g when filled with water at 20°C. The density of the water at this temperature is 0.998 g/cm^3. When 10.20 g of lead is placed in the pycnometer and the pycnometer again filled with water at 20°C, the total mass is 44.83 g. What is the density of lead?

*8. The standard kilogram mass pictured in Figure 1-2 was cut from a cylindrical bar of steel with a diameter of 4.00 cm. The density of the steel is 7.86 g/cm^3. How long was the section that was cut?

*9. The volume of seawater on earth is estimated to be 330,000,000 mi^3. Assuming that seawater is 3.5% sodium chloride, by mass, and that the density of seawater is 1.03 g/cm^3, what is the approximate mass of sodium chloride dissolved in the seawater on earth, expressed in tons?

*10. The diameter of metal wire is often referred to by its American wire gauge number. A 20-gauge wire has a diameter of 0.03196 in. What length of wire, in meters, is there in a 1.00-lb spool of 20-gauge copper wire? The density of copper is 8.92 g/cm^3.

*11. The principal source of magnesium metal is seawater, from which it is extracted by the Dow process (see Section 26-7). Magnesium occurs in seawater to the extent of 1.4 g of magnesium per kilogram of seawater. The annual production of magnesium in the United States is approximately 10^5 tons. What volume, in cubic miles, of seawater must be processed annually to yield this much magnesium? (Density of seawater = 1.03 g/cm^3.)

*12. The Antarctic, Greenland, and other ice caps contain approximately 7.2 million mi^3 of ice.
   (a) Given that the density of ice is 0.92 g/cm^3, together with other appropriate conversion factors, determine the mass of this ice, in tons.
   (b) When ice is melted, its volume decreases by about 10%. If all the polar ice were to melt completely, estimate the increase in sea level that would result from the additional liquid water entering the oceans. The oceans of the world cover approximately 1.4 × 10^8 mi^2.

*13. A typical rate of deposit of dust ("dustfall") from air that is not significantly polluted might be 10 tons per square mile per month. What is this dustfall, expressed in milligrams per square meter per hour?

*14. When water is used for irrigation purposes, its volume is often expressed in acre-feet. One acre-foot is a volume of water sufficient to cover 1 acre of land to a depth of 1 ft (640 acres = 1 mi²). The principal lake in the California State Water Project is Lake Oroville, whose water storage capacity is listed as 3.54×10^6 acre-feet. Express this volume in (a) ft³; (b) m³; (c) gal.

Self-Test Questions

For questions 1 through 6 select the single item that best completes each statement.

1. Of the following masses, that which is expressed to the nearest milligram is (a) 14.7 g; (b) 14.72 g; (c) 14.721 g; (d) 14.7213 g.

2. The greatest length of the following group is (a) 4.0 m; (b) 140 in.; (c) 12 ft; (d) 0.001 km.

3. The highest temperature of the following group is (a) −250°F; (b) 20°C; (c) 217°F; (d) 105°C.

4. Of the following substances, the greatest density is that of
   (a) 1000 g of water at 4°C   .
   (b) 100.0 cm³ of chloroform, which weighs 148.9 g
   (c) a 10.0-cm³ piece of wood weighing 9.50 g
   (d) an alcohol–water mixture of density 0.83 g/cm³

5. The largest volume of the following group is that of
   (a) 380 g of water at 4°C
   (b) 600 g of chloroform at 20°C (density = 1.5 g/cm³)
   (c) 0.50 L of milk
   (d) 100 cm³ of steel (density = 7.86 g/cm³)

6. Of the following numbers, the one with three significant figures is (a) 16.07; (b) 0.0140; (c) 1.070; (d) 0.016; (e) 200.

7. Describe briefly the distinction between the following pairs of terms.
   (a) element and compound
   (b) homogeneous and heterogeneous mixture
   (c) mass and density.

8. Use exponential notation and the appropriate number of significant figures to express the result of the following calculation.

$(19.541 + 1.05 - 3.6) \times 651 = ?$

9. A 55.0-gal drum weighs 75.0 lb when empty. What will be its total mass when filled with ethyl alcohol (density = 0.789 g/cm³; 1 gal = 3.78 L; 1 lb = 454 g)?

10. Describe several ways in which the SI system of units resembles and differs from the traditional metric system.

2 Development of the Atomic Theory

The word *atom*, not surprisingly, is of Greek origin [Gr. *átomos* (*á*, not + *tomos*, to cut)—uncut, undivided, or indivisible].

One of the oldest scientific concepts is that all matter can be broken down until finally the smallest possible particles are reached, particles that cannot be subdivided further. The Greek philosopher Democritus (ca. 460–370 B.C.) considered these particles to be in constant motion but able to fit together into stable combinations. Supposedly, the particular characteristics or properties of a material resulted from the different sizes, shapes, and arrangements of these particles. Today, we refer to these ultimate particles—these building blocks of all matter—as **atoms.** Many scientific discoveries have resulted from attempts to learn more about the basic nature of atoms, but we will emphasize only those matters important to an understanding of future topics.

2-1 Dalton's Atomic Theory

The development of the atomic theory involved contributions from many individuals, but there was one man whom we associate particularly with the origin of modern atomic theory. He was the English schoolteacher and chemist John Dalton (1776–1844). In what way, we may ask, was Dalton's contribution unique? Its uniqueness was of two parts: Dalton was the first to make use of *chemical* as well as physical evidence in formulating his ideas about atoms. Also, he based assumptions on *quantitative* data, not merely qualitative observations or speculation.

EXPERIMENTAL BASIS. Two types of experimental evidence—two natural laws—serve as the basis of Dalton's atomic theory.

The chemist Antoine Lavoisier (1743–1794) conducted a series of investigations designed to clarify the mechanism of combustion and related processes. He proved that in all these processes oxygen in the air combines with the materials undergoing change. Figure 2–1 depicts one of his experiments, in which liquid mercury was combined with oxygen to form red mercuric oxide (mercury calx). When the calx was recovered and reheated, it decomposed to produce liquid mercury and a quantity of oxygen gas equal in volume to that of the air consumed in the formation of the calx. In his scientific investigations Lavoisier used the idea that *the total mass of materials present after a chemical reaction is the same as before the reaction.* His careful *quantitative* experiments always produced results in accordance with this principle—the **law of conservation of mass.**

21

FIGURE 2-1

Lavoisier's experiment on heating mercury with air.

The original mercury level in the air container was at A, but after a number of days it rose to B and remained there. The difference in level between A and B represented the volume of air consumed by the mercury in forming the red powder (mercury calx). As proof of this fact Lavoisier collected the mercury calx and reheated it. The red powder decomposed into liquid mercury and a volume of gas (oxygen) equal to that of the air consumed in the original experiment.

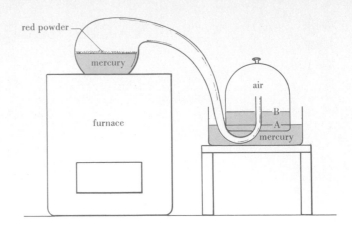

The validity of the law of definite composition was hotly debated by two contemporary French chemists, Proust and Claude Berthollet. Berthollet believed that compounds were merely mixtures in which the constituent elements could exist in varying proportions. Experimental evidence favored Proust's view, which became firmly established. Dalton's atomic theory was critically dependent on the law of definite composition.

The law of conservation of mass is the first law of chemical combination. A second law is the **law of definite composition** (also known as the **law of definite proportions**). A chemical compound, no matter what its origin or its method of preparation, always has the same composition, that is, the same proportions by mass of its constituent elements. Joseph Proust (1754–1826) performed a great number of analyses to show the constancy of composition of chemical compounds. For example, in 1799 he analyzed samples of natural copper carbonate from different places and also samples that had been synthesized in the laboratory. He found them all to have the same composition.

Example 2-1   The following data were obtained by heating strips of magnesium metal in oxygen gas to produce the white powder, magnesium oxide. Show that these data are consistent with the law of definite composition.

	Before heating, g magnesium	After heating, g magnesium oxide	Ratio: g magnesium/g magnesium oxide
strip 1	0.62	1.02	0.62/1.02 = 0.61*
strip 2	0.48	0.79	0.48/0.79 = 0.60*
strip 3	0.36	0.60	0.36/0.60 = 0.60*

According to the law of definite composition, the ratio of mass of magnesium to mass of magnesium oxide should have a constant value, regardless of the particular sample considered. The last column of values in the table demonstrates that it does. Within the accuracy of the measurements used (masses determined to ±0.01 g), the law of definite composition is verified.

SIMILAR EXAMPLES: Exercises 3, 4, 5.

DALTON'S ASSUMPTIONS.   Dalton's atomic theory was developed during the period 1803–1808 and was based on three principal assumptions.

1. Each chemical element is composed of minute indivisible, indestructible particles called atoms. Atoms can be neither created nor destroyed during a chemical change.

---

*These ratios, multiplied by 100, yield the percent magnesium in magnesium oxide, that is, 61%, 60%, and 60%, respectively.

FIGURE 2-2
The compound magnesium
oxide—as it might have
been conceived by Dalton.

If the mass of a magnesium atom is x and that of oxygen is y, the ratio of mass of magnesium to mass of oxygen, based on

one "atom" of compound $= (x/y)$

two "atoms" of compound $= (2x/2y) = 2/2 \cdot (x/y) = (x/y)$

three "atoms" of compound $= (3x/3y) = 3/3 \cdot (x/y) = (x/y)$

and so on.

Thus, regardless of the size of the sample taken, the ratio of mass of magnesium to mass of oxygen (and hence the percent composition of the compound) is fixed—the compound has a definite composition.

FIGURE 2-3
An experiment to determine the percent composition of water.

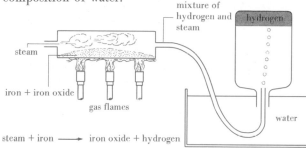

steam + iron ⟶ iron oxide + hydrogen

Steam is decomposed into its elements, hydrogen and oxygen. The hydrogen is collected over water and weighed. The oxygen combines with iron to form iron oxide. The increase in mass of the iron corresponds to the mass of oxygen present in the steam that decomposed. The sum of the masses of the hydrogen and the oxygen gives the total mass of steam decomposed.

$$\text{hydrogen, \%} = \frac{\text{mass hydrogen}}{\text{mass hydrogen} + \text{mass oxygen}} \times 100$$

$$\text{oxygen, \%} = \frac{\text{mass oxygen}}{\text{mass hydrogen} + \text{mass oxygen}} \times 100$$

Results of the analysis of water reported in Dalton's time were 87% oxygen and 13% hydrogen. This indicated a mass ratio of oxygen to hydrogen of 7:1, that is, an atomic weight of 7 for oxygen.

2. All atoms of a given element are alike in mass (weight) and other properties, but the atoms of one element are different from those of all other elements. **3.** In chemical compounds, atoms of different elements are united in simple numerical ratios: for example, one atom of A to one of B (AB), one atom of A to two of B (AB$_2$).

If the atoms of an element are indestructible (assumption 1), then the *very same* atoms must be present after a chemical reaction as were present before the reaction. The total mass of reactants and products must be the same. *Dalton's theory explains the law of conservation of mass.*

If all atoms of an element are alike in mass (assumption 2), and if atoms unite in *fixed* numerical ratios (assumption 3), the percentage composition of the compound must have a unique value, regardless of the size of the sample analyzed or its origin. *Dalton's theory also explains the law of definite composition.*

The combination of magnesium and oxygen to produce magnesium oxide, as it might have been conceived by Dalton, is illustrated in Figure 2-2.

THE "ATOMIC WEIGHT PROBLEM." For Dalton's theory to be of use in predicting new phenomena, it was necessary to assign characteristic masses to atoms. These masses became known as **atomic weights.** But because they are so inconceivably small, it is impossible to isolate and weigh individual atoms. Dalton was well aware of this fact and tried simply to establish *relative* atomic weights. If an atom of hydrogen, for example, is taken to have a mass of 1 unit, what must be the mass of an oxygen atom by comparison?

To answer this question requires

1. Selecting a compound that contains only hydrogen and oxygen (Dalton chose water, the only such compound known to him).
2. Determining the proportions by mass of oxygen and hydrogen in the compound.
3. Knowing the chemical *formula* of the compound, that is, the ratio of number of atoms of hydrogen to oxygen.

Figure 2-3 suggests an experiment for the analysis of water (steam). The composition of water is 88.81% oxygen and 11.19% hydrogen; that is, the mass of oxygen is about *eight* times that of hydrogen. This suggests a relative atomic weight of oxygen equal to 8. But there is an assumption involved here: that hydrogen and oxygen combine in the ratio 1:1—one atom of hydrogen for each oxygen atom.

FIGURE 2-4
The "atomic
weight problem."

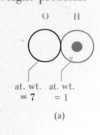

at. wt.   at. wt.
= 7   = 1

(a)

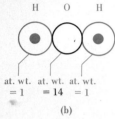

at. wt.   at. wt.   at. wt.
= 1   = 14   = 1

(b)

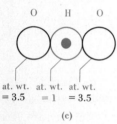

at. wt.   at. wt.   at. wt.
= 3.5   = 1   = 3.5

(c)

Analyses available to Dalton (87% O and 13% H) suggested that water contained seven times the mass of oxygen as hydrogen. Dalton assumed the simplest formula (a), and as a result assigned an atomic weight of 7 to oxygen. If he had assumed formula (b), his result would have been 14; if formula (c), 3.5.

	Compound (a)	Compound (b)	Compound (c)
$\dfrac{\text{Mass of oxygen}}{\text{Mass of hydrogen}}:$	$\dfrac{7}{1} = 7$	$\dfrac{14}{2 \times 1} = 7$	$\dfrac{2 \times 3.5}{1} = 7$

Modern data indicate (b) is correct and that the atomic weight of oxygen is 16, not 14.

Dalton developed a rather complicated set of pictorial symbols to denote the different kinds of atoms. Chemical compounds were represented by appropriate combinations of symbols. Dalton did not use any special term to describe these combinations of atoms, referring simply to atoms of a compound. Avogadro, in 1811, introduced the term *molecule* for a stable combination of small numbers of atoms. The term is used on occasion in this chapter, and its full significance is explored in the next and subsequent chapters.

Figure 2-4 shows how the relative atomic weight of oxygen differs according to the assumption made about the chemical formula of water. A chemical formula describes the relative *numbers* of the different atoms in a compound. If Dalton had known the relative weights of atoms, he could have deduced chemical formulas; but he needed chemical formulas in order to establish relative atomic weights. Dalton tried to resolve this dilemma with an assumption that proved *incorrect*—the rule of greatest simplicity. He thought that if two elements (A and B) form a single compound, they combine in a 1:1 ratio. If two compounds exist, one has the combining ratio of 1:1 (AB) and the other, either 1:2 (AB$_2$) or 2:1 (A$_2$B); if three compounds are formed they are AB, AB$_2$, A$_2$B; and so on. Since Dalton knew of only one hydrogen–oxygen compound—water—he assumed (incorrectly) the formula OH.

Example 2-2 The ratio of iron to oxygen in ordinary iron oxide is 2.33:1, by mass, and 2:3, by number of atoms. Given the atomic weight of 16 for oxygen (O), what is the atomic weight of iron (Fe)?

Each of the three ratios (1), (2), and (3) expresses the ratio of iron to oxygen on a mass basis.

$$\underset{(1)}{\frac{2 \times \text{at. wt. Fe}}{3 \times \text{at. wt. O}}} = \underset{(2)}{\frac{2 \times \text{at. wt. Fe}}{3 \times 16}} = \underset{(3)}{2.33}$$

$$\text{at. wt. Fe} = \frac{2.33 \times 3 \times 16}{2} = 56$$

SIMILAR EXAMPLES: Exercises 8, 9.

LAW OF MULTIPLE PROPORTIONS. Despite its other failures, the rule of greatest simplicity did suggest an important law of chemical combination. If two elements form more than a single compound, the masses of one element combined with a fixed mass of the second are in the ratio of small whole numbers. This law, called the **law of multiple proportions,** was first stated by Dalton in 1805. He observed that twice as much

FIGURE 2-5
The law of multiple proportions illustrated—
with Dalton's symbols and atomic weights.

ethylene

C   H

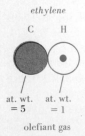

at. wt.   at. wt.
= 5   = 1

olefiant gas

methane

H   C   H

at. wt.   at. wt.   at. wt.
= 1   = 5   = 1

carburetted hydrogen

Per gram of hydrogen in olefiant gas there is 5 g of carbon, that is,

$$\frac{5 \text{ g carbon}}{1 \text{ g hydrogen}}$$

Per gram of hydrogen in carburetted hydrogen there is 2.5 g of carbon, that is,

$$\frac{5 \text{ g carbon}}{2 \text{ g hydrogen}} = \frac{2.5 \text{ g carbon}}{1 \text{ g hydrogen}}$$

$$\text{ratio} = \frac{5 \text{ g carbon}/1 \text{ g hydrogen}}{2.5 \text{ g carbon}/1 \text{ g hydrogen}} = \frac{2}{1}$$

hydrogen is combined with a given mass of carbon in methane gas (carburetted hydrogen) as in ethylene gas (olefiant gas). Dalton assigned the formula CH_2 to methane and CH to ethylene. (The correct formulas based on present knowledge are CH_4 and C_2H_4.) His reasoning is suggested by Figure 2-5.

2-2 Cathode Rays

The first significant investigations in the field of atomic physics involved the study of electric discharge through gases, reported by Faraday in 1838. The type of apparatus used by Faraday is diagrammed in Figure 2-6. Metal plates called electrodes are sealed into the ends of a glass tube having a side-arm opening. One electrode, the **cathode,** is connected to the negative terminal of a source of electric current at high voltage (several thousand volts); the other, the **anode,** is connected to the positive terminal. As long as the tube is filled with air, no electric current flows. Air is a very poor conductor of electricity.

The glass tube can be evacuated by connecting it to a vacuum pump. During the evacuation a purple glow develops within the tube, extending from the anode almost to the cathode. The cathode glows, and between the cathode and the column of purple light a dark region exists, the Faraday dark space. These phenomena occur when the pressure of the residual gas in the tube is reduced to about one thousandth of normal atmospheric pressure, the limit to which mechanical vacuum pumps could be operated in Faraday's time.

With the advent of better vacuum pumps it became possible to reduce the pressure within discharge tubes to perhaps one millionth of atmospheric pressure. Additional phenomena were observed at these greatly reduced pressures. In 1858, Plucker reported that the Faraday dark space enlarged as the air pressure was reduced, that the region of the cathode glow was extended, and that the glass tube itself emitted a phosphorescent glow. In 1869, Hittorf performed an experiment in which an object was mounted in a discharge tube and observed to cast a shadow. This suggested that the glowing of the tube was caused by rays emanating from the cathode and traveling in straight lines. Thus began a long and exciting series of experiments into the nature of cathode rays.

PROPERTIES OF CATHODE RAYS. Here are some of the more significant properties of cathode rays determined by Plucker, Hittorf, Crookes, and others.

1. Cathode rays are emitted from the cathode in an

FIGURE 2-6
Electric discharge in an evacuated chamber.

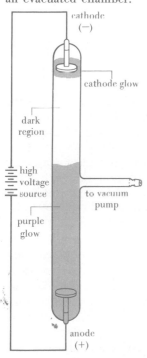

cathode (−)

cathode glow

dark region

high voltage source

purple glow

to vacuum pump

anode (+)

Before physicists could attack the problem of atomic structure they had to acquire a basic understanding of electricity and magnetism. A complete discussion of these topics is beyond the scope of this text, but a brief overview is provided in Appendix B.

William Crookes (1832–1919) became so closely associated with cathode ray experiments that electric discharge tubes are often called Crookes' tubes.

FIGURE 2-7
Deflection of cathode rays in a magnetic field.

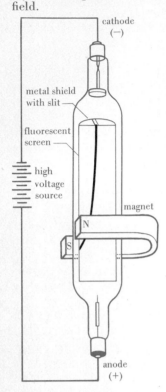

cathode
(−)

metal shield with slit

fluorescent screen

high voltage source

magnet

N

S

anode
(+)

Cathode rays are invisible. Only through their impact on a fluorescent material can they be detected. The beam of cathode rays is deflected as it enters the field of the horseshoe magnet. The deflection of the beam corresponds to that anticipated for negatively charged particles.

The coulomb (C) is the SI unit of electrical charge.

"We have in the cathode rays matter in a new state, a state in which the subdivision of matter is carried much further than in the ordinary gaseous state: a state in which all matter . . . is one and the same kind; this matter being the substance from which all the chemical elements are built up."

J. J. THOMSON

evacuated tube when electric current is passed. (Electric current is essential.)

2. The rays travel in straight lines.

3. The rays, upon striking glass or certain other materials, cause them to fluoresce (give off light). Only in this way can they be seen; the rays themselves are invisible.

4. Cathode rays are deflected by electric and magnetic fields in the manner expected for *negatively charged* particles (see Figure 2-7).

5. The properties of cathode rays are independent of the electrode material (i.e., whether iron, platinum, etc.).

INVESTIGATIONS OF J. J. THOMSON. During the period 1894–1897, J. J. Thomson (1856–1940) conducted a series of investigations that established the particle nature of cathode rays. His first studies determined with great accuracy the speeds of cathode rays. He found these to be only a small fraction of the speed of light, thus ruling out the possibility of their being electromagnetic radiation. Thomson's more significant experiments, however, were designed to determine the ratio of electric charge (e) to mass (m), that is, e/m. His experimental design and a brief analysis of it are presented in Figure 2-8.

The average value Thomson obtained for e/m for cathode rays was about 2×10^8 coulombs per gram. This value is about 2000 times greater than the e/m ratio calculated for hydrogen liberated in the electrolysis of water. For various reasons Thomson assumed cathode rays to have about the same electrical charge as that associated with hydrogen atoms in the electrolysis of water. This meant that they should possess only about $\frac{1}{2000}$ of the mass of a hydrogen atom. The assumed small size of cathode ray particles and the fact that the value of e/m was independent of the cathode material led Thomson to the following conclusion: Cathode ray particles are negatively charged *fundamental* particles of matter that must be found in *all* atoms. Cathode ray particles are the basic units of negative electric charge for which, in 1874, Stoney had proposed the term **electron.**

CHARGE ON THE ELECTRON. J. J. Thomson's investigations provided a precise *ratio* of charge to mass for an electron. Thomson speculated on probable values of e and m, but neither the charge nor the mass could be determined by his method alone. An independent evaluation of either e or m was clearly needed. From a precise measurement of one of these and from a knowledge of e/m, the value of the other could be determined. The electronic charge e was chosen as the

FIGURE 2-8
Thomson's apparatus for determining the charge-to-mass ratio, e/m, for cathode rays.

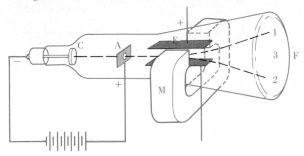

Code: C, cathode; A, anode (perforated to allow the passage of a narrow beam of cathode rays); E, electrically charged condenser plates; M, magnet; F, fluorescent screen.

Path 1:   In the presence of an electric field only, the cathode ray beam is deflected upward, striking the end screen at a point such as 1.

Path 2:   In the presence of a magnetic field only, the cathode ray beam is deflected downward, striking the screen at point 2.

Path 3:   The cathode ray beam can be made to strike the end screen at point 3, undeflected, if the forces on the particles exerted by the electric and magnetic fields are just counterbalanced.

Determination of e/m: From a knowledge of the strengths of the electric and magnetic fields producing path 3 and the radius of curvature of path 2, a value of e/m can be obtained. Precise measurements yield a result of

-1.759×10^{8} coulombs per gram

(Because cathode rays carry a negative charge, the sign of the charge-to-mass ratio is also negative.)

> "Here, then, is direct, unimpeachable proof that the electron is not a "statistical mean," but that rather the electrical charges found on ions all have either exactly the same value or else small exact multiples of that value."
>
> ROBERT MILLIKAN

FIGURE 2-9
Millikan's oil drop experiment.

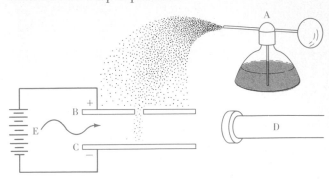

A spray of oil droplets is produced by the atomizer (A). These enter the apparatus through a tiny hole in the top plate of an electrical condenser. The motion of the droplets is observed with a telescope equipped with a micrometer eyepiece (D). Ions are produced by ionizing radiation, such as x rays, from a suitable source (E). Some of the oil droplets acquire an electric charge by adsorbing ions (attaching ions to their surface).

The fall of a droplet between the charged condenser plates (B and C) is either speeded up or slowed down to an extent that depends on the sign and magnitude of the charge on the droplet. By analyzing data from large numbers of droplets, Millikan concluded that the magnitude of the charge, q, on a droplet is always an integral multiple of the electronic charge, e. That is, $q = n \cdot e$ (where $n = 1, 2, 3$, and so on).

quantity most amenable to measurement, and the definitive measurements were made by Robert Millikan (1868–1953) at the University of Chicago during the period 1906–1914.

Millikan's famous "oil drop" experiment is suggested by Figure 2-9. Millikan found that the electric charge on all oil droplets that had acquired a charge could be expressed as $n \times e$, where n is a positive or negative integer and e represents the smallest observable unit of electric charge. The currently accepted value of the electronic charge e is -1.60219×10^{-19} C. Combining the results of Millikan and Thomson, one obtains the mass of an electron: 9.110×10^{-28} g.

2-3 Canal Rays (Positive Rays)

In addition to characterizing the fundamental unit of negative electric charge, the electron, cathode ray research also yielded evidence of a fundamental unit of *positive* charge. In 1886, Eugen Goldstein performed a series of experiments in which he discovered a new

FIGURE 2-10
The "plum pudding" model of atomic structure.

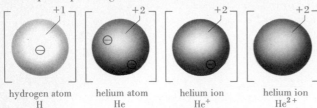

hydrogen atom
H

helium atom
He

helium ion
He⁺

helium ion
He²⁺

Thomson's conception of the atom was that of a "cloud" of positive electric charge with a sufficient number of electrons embedded within to neutralize the positive charge. Thus, a hydrogen atom was thought to consist of a positive cloud of charge $+1$, containing one electron (-1); helium, a positive cloud of charge $+2$, containing two electrons (-2); and so on. Normal atoms are electrically neutral. If an atomic species carries a *net* electric charge it is said to be an **ion.** The loss of one electron by a helium atom, through a collision with cathode rays, for example, results in the production of an ion with a net charge of $+1$, that is, He^+. The loss of both electrons results in the ion He^{2+} with a net charge of $+2$.

FIGURE 2-11
A canal ray tube.

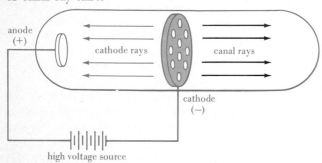

anode
(+)

cathode rays

canal rays

cathode
(−)

high voltage source

The distinctive feature of this tube is the perforated cathode. Cathode rays stream toward the anode. Their collisions with residual gas atoms dislodge electrons from the atoms, producing positively charged ions. These ions are attracted to the cathode (−), but some of the ions pass through the holes in the cathode and appear as a stream of particles (black arrows) on the other side. These beams of positive ions are called positive rays or canal rays.

Some early models of color television sets were found to produce a high level of x radiation by the impact of the electron beam on the phosphor materials coated on the picture tube. This situation was almost identical to the circumstances under which Roentgen discovered x rays.

type of particles called **canal rays** or positive rays. Some of the properties observed for these rays are the following:

1. The particles are deflected by electric and magnetic fields in a way that reveals their positive charge.
2. The ratio of charge to mass, e/m, for positive rays is considerably smaller than for electrons.
3. The e/m ratio of the positive rays depends on the nature of the gas in the tube. The highest e/m ratio is obtained with hydrogen gas. For other gases, e/m is an integral fraction (e.g., $\frac{1}{4}$, $\frac{1}{20}$) of the ratio for hydrogen.
4. The e/m ratio of positive rays produced when hydrogen gas is present in the tube is identical to the e/m ratio for hydrogen produced by the electrolysis of water.

These observations can be explained by an atomic model proposed by J. J. Thomson—the "plum pudding" model, illustrated in Figure 2-10. An explanation of the production of positive rays (ions) is provided by Figure 2-11. A conclusion consistent with the properties of canal rays is that all atoms carry fundamental units of positive charge, one in the hydrogen atom and larger numbers in other atoms. The fundamental unit of positive charge is now called the **proton.**

2-4 X Rays

A number of cathode ray experimenters, including apparently J. J. Thomson himself, occasionally observed objects *outside* the cathode ray tubes to glow when the tubes were run. It remained for Wilhelm Roentgen to show that the impact of cathode rays (electrons) on a surface produces a type of radiation which, in its turn, can cause certain substances to glow at a distance from the cathode ray tube. Because of the unknown nature of this radiation, the term *x ray* was coined, a name still in common use.

Within a few weeks of his initial discovery in 1895, Roentgen achieved a rather complete characterization of x rays. He found them to be undeflected by electric and magnetic fields and to have a very high penetrating power through matter. All properties of x rays suggested them to be electromagnetic radiation with wavelengths of approximately 1 angstrom unit $(1 \text{ Å} = 10^{-10} \text{ m})$.*

---

*A complete discussion of wavelength and other properties of electromagnetic radiation is presented in Chapter 7.

FIGURE 2-12
The production of x rays.

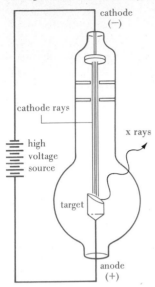

Fluorescence refers to the emission of visible light by certain materials when they are struck by a beam of energetic radiation (such as ultraviolet light or cathode rays).

Radiation emanating from radioactive materials can dislodge electrons when it strikes atoms of matter; ions are produced. Ionizing power refers to the number of ions produced by a given quantity of radiation.

Practical applications followed closely on the heels of the discovery of x rays. Because x rays have different penetrating powers for different types of matter, they can be used to photograph the interior of objects. Roentgen's original announcement of x rays was made on December 28, 1895. On January 20, 1896, in Dartmouth, New Hampshire, x rays were used to assist in setting a person's broken arm. This was something of a record time for turning a scientific discovery into a practical application.

Roentgen found that an especially effective source of x rays resulted from the impact of cathode rays on a dense metal anode called a *target*. Figure 2-12 is a diagram of an x ray tube based on this principle.

2-5 Radioactivity

The discovery of x rays, which resulted from research on cathode rays, in turn brought about the discovery of radioactivity. This discovery, by the French physicist Henri Becquerel, came within a few months of Roentgen's. In his original experiments Roentgen used the glass walls of a cathode ray tube as a source of x rays. The impact of cathode rays upon glass produces a fluorescence of the glass as well as the emission of x rays. Becquerel associated the emission of x rays with the property of fluorescence and asked the question: Will naturally fluorescent materials produce x rays? To answer this question he proceeded as follows.

A photographic plate was covered with thick black paper. A layer of a particular crystalline substance (a double sulfate of uranium and potassium) was placed on the outside of the paper, and the entire assembly was placed in sunlight. As expected, the photographic plate became exposed. Becquerel thought that sunlight caused the substance to fluoresce or glow and that some of this fluorescent radiation (x rays) penetrated the paper and exposed the photographic plate. On one occasion when he attempted to repeat this experiment, the sky became overcast, and Becquerel put the experimental assembly into a desk drawer, where it remained for several days. Before resuming the experiment he thought to replace the photographic plate, expecting it to be slightly exposed because of the length of time that had elapsed. Much to his surprise he found that the plate was *strongly* exposed, as much so as in his initial experiments. Becquerel hypothesized that the radiation responsible for exposing the photographic plate was not associated with the fluorescence at all, and he designed some experiments to test this hypothesis. He found that the radiation was emitted continuously by the crystalline material used in his experiments, in particular by the element uranium. Thus, radioactivity was discovered.

RADIOACTIVE ELEMENTS AND THEIR RADIATIONS.   Ernest Rutherford was able to show that two types of radiation existed. One type, which Rutherford called **alpha rays (α rays),** has a high ionizing power but a low penetration through matter. Alpha rays can be stopped by a sheet of ordinary paper. The other type is of a lower ionizing power but greater penetrating power. These rays can pass through aluminum foil up to 3 mm thick. Rutherford called this radiation **beta rays (β rays).**

Alpha rays are *particles* carrying two fundamental units of *positive* charge and having a mass equal to that of a helium atom; thus, an α particle is identical to the ion He^{2+} (see, again, Figure 2-10). Beta rays are *negatively* charged particles with the same charge-to-mass ratio, e/m, as electrons. They are indistinguishable from electrons. A third form of radiation has an extremely high penetrating power and is not deflected by electric and magnetic fields. This electromagnetic radiation is known as **gamma rays (γ rays).**

By the early 1900s several additional radioactive elements were discovered (e.g., thorium, radium, and polonium), principally through the work of Marie and Pierre Curie in France. And, in collaboration with Frederick Soddy, Rutherford made another profound discovery concerning radioactivity: The chemical properties of a radioactive element *change* as it undergoes radioactive decay. This observation could only be explained by assuming that radioactivity involves basic changes at the *subatomic* level—that in radioactive decay one element replaces another. Thus, with this discovery one of Dalton's basic assumptions—that atoms of an element are indivisible and unchangeable—toppled. Additional aspects of radioactivity and nuclear chemistry are considered in Chapter 23.

2-6 The Nuclear Atom

The phenomena of cathode ray production, light emission, and ionization all could be explained, after a fashion, by Thomson's plum pudding model (1898). However, this model could not be used to make quantitative predictions regarding these phenomena. Neither was the model consistent with other observations being made at about the same time.

SCATTERING OF ALPHA PARTICLES. In 1909, at Rutherford's suggestion, Hans Geiger and Ernest Marsden initiated a series of experiments in which very thin foils of gold and other metals (10^{-4} to 10^{-5} cm thick) were used as targets for α particles derived from a radioactive substance. The experimental apparatus used is suggested by Figure 2-13.

The radioactive substance was enclosed in a lead block in such a way that only a narrow beam of α particles could escape. The presence of α particles was detected by the scintillations or flashes of light they produced on a zinc sulfide screen mounted on the end of a telescope. Geiger and Marsden made the following observations.

1. The majority of the α particles penetrated the metal foil undeflected.
2. A few (about one in every 20,000) suffered rather serious deflections as they penetrated the foil.
3. A similar number did not pass through the foil at all but "bounced back" in the direction from which they had come.

Using Thomson's atomic model, Rutherford reasoned that the positive charge of the atom was so dif-

FIGURE 2-13
The scattering of alpha particles by metal foil.

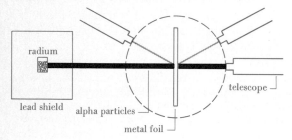

"It is about as incredible as if you had fired a 15-in. shell at a piece of tissue paper and it came back and hit you."
ERNEST RUTHERFORD

FIGURE 2-14

Rutherford's interpretation of the scattering of alpha particles by metal foil.

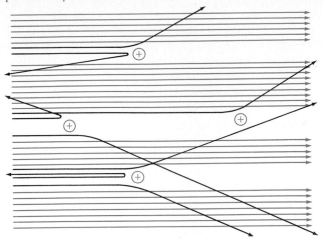

Each arrow represents an α particle. The symbol ⊕ represents an atomic nucleus.

Analogies to the Rutherford nuclear atom abound; here is one. Imagine the atom itself to be about the size of a football stadium. The nucleus would correspond roughly to an object the size of a marble in the center of the playing field, and electrons, to a sprinkling of marbles (perhaps a few dozen) on the playing field and in the stands.

fuse that the positively charged α particles should pass through this weak electric field largely undeflected. The severe deflections experienced by some of the particles, particularly those that bounced back from the foils, astounded Rutherford.

This type of behavior would be expected only if the positive charge and mass of an atom were highly concentrated in a small region. Rutherford called this the **nucleus.** The approach of an α particle to a nucleus of high positive charge and mass would lead to repulsive forces strong enough to reverse the direction of the particle while leaving the nucleus practically unmoved (see Figure 2-14). In this way the idea of a *nuclear* atom originated.

RUTHERFORD'S ATOMIC MODEL.   The main features of the atom postulated by Rutherford were these.

1. Most of the mass and all of the positive charge of an atom are centered in a very small region called the nucleus—the atom is mostly empty space.
2. The magnitude of the charge on the nucleus is different for different atoms and is approximately one half the numerical value of the atomic weight of the element.
3. There must be a number of electrons outside the nucleus of an atom that is equal to the number of units of nuclear charge (to account for the fact that the atom is electrically neutral.)

2-7 Isotopes

In 1912, J. J. Thomson designed some experiments to determine the charge-to-mass (e/m) ratios of positive ions produced in a canal ray tube. The results he obtained with neon gas were quite unexpected. He was able to explain them in only one way: In ordinary neon gas about 91% of the atoms have a "normal" mass and about 9% of the atoms are 22/20 heavier than "normal." Thomson's discovery was that atoms of the same element *may differ slightly in mass.* These differing atoms are called **isotopes.** Some years earlier Soddy proposed this term to designate atomic species which, although having different radioactive properties, have identical chemical properties. With the discovery of radioactivity, one of the assumptions of Dalton's atomic theory required modification—that atoms of an element are unchangeable. The discovery of isotopes required modification of a second assumption—that all atoms of an element are alike in mass.

2-8 Protons and Neutrons

Analysis of α particle scattering experiments suggested that positive charges on the nuclei of atoms exist as multiples of some fundamental unit of positive charge. Rutherford was actually able to determine characteristic nuclear charges for a number of elements. In 1913, H. G. J. Moseley, one of a group of brilliant scientists whose careers were launched under Rutherford's direction, reported the results of experiments in which different elements or their compounds were used as targets in an x ray tube. He found that the x ray wavelength varied depending on the target material. Moseley was able to correlate these wavelengths through a mathematical equation in which it was necessary to assign to each element a unique integral (whole) number. These integers, which he called atomic numbers, proved to be identical to the nuclear charges described by Rutherford.

Rutherford was the first to establish, in 1919, the independent existence of particles carrying a fundamental unit of positive charge, **protons.** This resulted from studies on the passage of α particles through air. Rutherford found that scintillations (flashes of light) could be detected on a zinc sulfide screen at distances from a radium source considerably greater than α particles were expected to travel before being absorbed. He concluded that upon striking the nuclei of nitrogen atoms in air α particles ejected protons and it was these that reached the zinc sulfide screen.

The concept of the constitution of the nucleus favored by Rutherford and a number of other physicists was this: The nucleus contains a number of protons equal to the atomic number and a sufficient number of *neutral* particles, called neutrons, to account for the observed mass of an atom.

In the early 1930s, in experiments performed by bombarding beryllium and boron with α particles, a very penetrating radiation was obtained. It seemed to have properties similar to γ rays, but was more energetic. In 1932, Chadwick showed that the properties of this radiation were more readily explained by assuming a beam of neutral particles having masses slightly greater than protons. In this way **neutrons,** first postulated 12 years earlier, were finally discovered. An atomic model involving protons, neutrons, and electrons is pictured in Figure 2-15.

FIGURE 2-15
The nuclear atom—illustrated by the helium atom.

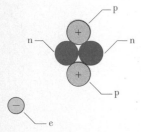
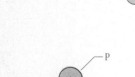

2-9 Summary of the Properties of Fundamental Particles

There are a number of fundamental particles now known in addition to the three described in this chapter, but the chemical behavior of an element is derived just from these three—the proton, the neutron, and the electron. Electric charges and masses of these three particles are presented in Table 2-1 in two sets of units. One is the metric system, and the other a special atomic unit with which the properties of fundamental particles can be related to one another.

TABLE 2-1
Properties of three fundamental particles

	Electric charge		Mass	
	Metric (C)	Atomic	Metric (g)	Atomic (amu)[a]
proton	$+1.602 \times 10^{-19}$	$+1$	1.673×10^{-24}	1.0073
neutron	0	0	1.675×10^{-24}	1.0087
electron	-1.602×10^{-19}	-1	9.110×10^{-28}	0.00055

[a]amu, atomic mass unit.

Consider the electric charge carried by an electron. This is the smallest unit of electric charge that can exist—we might call it an atomic unit of electricity. The proton also carries an atomic unit of electric charge, but positive in sign. It is convenient also to think of protons and neutrons as possessing an atomic unit of mass. As can be seen in Table 2-1, the **atomic mass unit (amu)** has been so defined that proton and neutron masses are just slightly greater than 1 amu and not quite equal to one another. By comparison the mass of an electron is seen to be extremely small.

Example 2-3 From the data given in Table 2-1 calculate the e/m ratio for the electron and compare this with the value listed in Figure 2-8.

charge: -1.602×10^{-19} C

mass: 9.110×10^{-28} g

charge-to-mass ratio: $e/m = \dfrac{-1.602 \times 10^{-19} \text{ C}}{9.110 \times 10^{-28} \text{ g}} = -1.759 \times 10^{8}$ C/g

This ratio has the same value as that listed in Figure 2-8.

SIMILAR EXAMPLES: Exercises 22, 23.

The number of protons in the nucleus of an atom is referred to as the **atomic number,** or proton number, Z. The number of electrons in an electrically neutral atom is also equal to the atomic number, Z. The total mass of an atom is determined very nearly by the total number of protons and neutrons in its nucleus. This total is called the **mass number,** A. The number of neutrons in an atom, the neutron number, is given by the quantity $A - Z$.

2-10 Chemical Elements

The term element refers to a substance with atoms all of a single kind. To the chemist the kind of atom is specified by its atomic number, since this is the property that determines its chemical behavior. At present all the atoms from $Z = 1$ to $Z = 106$ are known: There are 106 chemical elements. Each chemical element has been given a name and a distinctive symbol. For most elements the symbol is simply the abbreviated form of the English name consisting of one or two letters, for example,

oxygen = O   nitrogen = N   neon = Ne   magnesium = Mg

The symbol for tungsten (W) is based on its German name, wolfram.

Some elements that have been known for a long time have symbols based on their Latin names, for example,

iron = Fe (ferrum)   copper = Cu (cuprum)   lead = Pb (plumbum)

A few elements have symbols based on the Latin name of one of their compounds, the elements themselves having been discovered only in relatively recent times, for example,

sodium = Na (natrium = sodium carbonate)

potassium = K (kalium = potassium carbonate)

A complete listing of the elements may be found on the inside back cover.

In addition to identifying an element by its symbol, information on the composition of its atomic nuclei may be included along with the symbol. The symbol $_{Z}^{A}X$ signifies

$^{14}_{7}N$, $^{16}_{8}O$, $^{24}_{12}Mg$, $^{56}_{26}Fe$, $^{238}_{92}U$,   and so on.

What is represented by the symbol $^{A}_{Z}X$ is an atomic species called a **nuclide** of the element X, having an atomic number Z and a mass number A. One of the nuclides shown above, that of the element oxygen, is an atom with eight protons and eight neutrons in its nucleus and eight electrons outside the nucleus.

All atoms of a given element must have the same atomic number, but they may have different mass numbers. The different nuclides of an element are referred to collectively as isotopes of the element. In Section 2-7, it was pointed out that there is one type of neon atom with a mass $^{22}_{20}$ as great as the predominant atomic species. Actually three different nuclides exist; there are three isotopes of neon. By symbol, these are $^{20}_{10}Ne$, $^{21}_{10}Ne$, and $^{22}_{10}Ne$. The natural abundances of these nuclides are 90.9, 0.3, and 8.8%, respectively. Sometimes the mass numbers of isotopes are incorporated into the names of elements, such as neon-20, carbon-12, and oxygen-16. In a neutral atom the number of electrons must be equal to the number of protons, Z. But if an atom either loses or gains electrons, it acquires a net electric charge; it becomes an **ion.** The species $^{20}_{10}Ne^{+}$ and $^{20}_{10}Ne^{2+}$ are ions. The first one has ten protons, ten neutrons, and nine electrons; the second, ten protons, ten neutrons, and eight electrons.

It is possible to arrange four numbers as subscripts and superscripts about a chemical symbol, that is $^{a}_{b}X^{c}_{d}$. In this scheme a is the mass number, b is the atomic number, and c is the net electric charge, as in $^{20}_{10}Ne^{2+}$. The fourth number, the subscript to the right (d), represents the number of atoms in a molecule. Its use is introduced in Chapter 3.

Example 2-4 Indicate the numbers of protons, neutrons, and electrons in $^{35}_{17}Cl$ and $^{80}_{35}Br^{-}$.

If the species is a neutral atom, the number of electrons is equal to the number of protons, which in turn is equal to Z, the subscript numeral in the symbol $^{A}_{Z}X$. An atom becomes an ion only by losing or gaining *electrons:* An ion has the same number of protons as the atom from which it is formed, but a different number of electrons.

$^{35}_{17}Cl$: $Z = 17$, $A = 35$ A neutral atom:
number of protons $= 17$; number of electrons $= 17$; number of neutrons $=$ $A - Z = 35 - 17 = 18$.

$^{80}_{35}Br^{-}$: $Z = 35$, $A = 80$ An ion with a net charge of -1:
number of protons $= 35$; number of electrons $= 36$ (one electron must be *gained* to yield an ion having a charge of -1); number of neutrons $= A - Z = 80 - 35 = 45$.

SIMILAR EXAMPLES: Exercises 28, 29.

Example 2-5 Write appropriate symbols for **(a)** an atom of strontium-90; **(b)** the species containing 29 protons, 34 neutrons, and 27 electrons.
(a) Look up the element strontium on the inside back cover. It has the symbol Sr and the atomic number 38. The mass number is 90.

$^{90}_{38}Sr$

(b) The element with $Z = 29$ is copper (Cu). The mass number is equal to the number of protons plus the number of neutrons: $29 + 34 = 63$. Because the species has only 27 electrons, it must be an ion with a net charge of $+2$.

$^{63}_{29}Cu^{2+}$

SIMILAR EXAMPLES: Exercises 28, 29.

2-11 Atomic Weights

By international agreement a single atom of the nuclide $^{12}_{6}C$ (carbon-12) is arbitrarily assigned a mass of 12.00000 amu. For the reason discussed in Section 23-6, the masses of other nuclides cannot be obtained simply by totaling the masses of their fundamental particles. However, the ratio of the mass of any other nuclide to that of

FIGURE 2-16
The Bainbridge mass spectrograph.

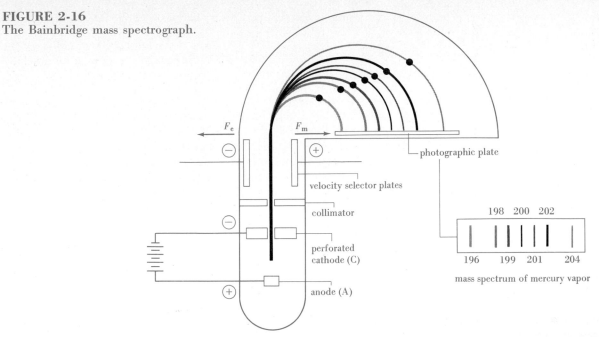

mass spectrum of mercury vapor

In this spectrograph, devised by Bainbridge in 1933, positive gaseous ions are produced and accelerated in the region between the anode (A) and cathode (C). Some of these positive ions pass through the opening in the cathode and enter the space between the velocity selector plates, where they experience two forces, electrostatic and magnetic. The electrical field exerts a force to the left, F_e, which depends only on the electric field strength and the magnitude of the charge on the positive ions (but not on the velocity of ions). The magnetic field is perpendicular to the plane of the page and exerts a force to the right, F_m. The poles of the magnet, not pictured here, would be above and below the page (similar to the arrangement in Figure 2-8). The force F_m is directly proportional to the magnetic field strength, the magnitude of the charge on the positive ions, *and* the velocity of the ions. Ions with low velocities are attracted into the negatively charged condenser plate and discharged. Those with high velocities are drawn into the positive plate (because of the effect of the magnetic field). For some particular velocity the ions pass through the velocity selector plates undeflected. Once these ions have gotten beyond the plates, they are affected only by the magnetic field.

All ions carrying the same charge experience equal forces in the magnetic field, but the lighter ones are bent into circular paths of smaller radii than are the heavier ones. As a result the detector (photographic plate) is struck at the extreme left by the lightest ions and at the extreme right by the heaviest. Furthermore, the more ions of a particular mass present in the sample under analysis, the more intense the corresponding spot on the photographic plate. The photographic record of the analysis of a gaseous sample consists of a number of exposed regions or "lines." This collection of lines is called the **mass spectrum** of the sample. The mass spectrum of mercury vapor is indicated in this figure.

Oxygen was formerly taken as the atomic weight standard. However, because physicists assigned an atomic weight of 16.0000 to the nuclide $^{16}_{8}O$, whereas chemists assigned this value to the naturally occurring *mixture* of oxygen isotopes, a number of troublesome discrepancies arose. The definition of atomic weights based on $^{12}_{6}C$ is unambiguous.

carbon-12 can be established using a **mass spectrometer.** This is a device in which a beam of gaseous ions is separated into components of differing mass by passage through electric and magnetic fields. The separated components are focused on a suitable measuring device, where their presence is detected and recorded. (If the record is photographic, the instrument is called a **mass spectrograph.**) The principle of mass spectrometry is illustrated in Figure 2-16.

Example 2-6 With mass spectral data it can be established that the ratio of the mass of $^{16}_{8}O$ to $^{12}_{6}C$ is 1.3329. What is the mass of the $^{16}_{8}O$ atom?

The ratio of the masses is $^{16}_{8}O/^{12}_{6}C = 1.3329$. The mass of the $^{16}_{8}O$ atom is 1.3329 times the mass of $^{12}_{6}C$.

mass of $^{16}_{8}O = 1.3329 \times 12.00000$ amu $= 15.9948$ amu

SIMILAR EXAMPLES: Exercises 33, 39.

In a table of international atomic weights, the atomic weight listed for carbon is 12.011; yet our atomic weight standard was taken as 12.00000. It is important to note that the atomic weight standard is based on the pure nuclide $^{12}_{6}C$, whereas naturally occurring carbon contains small amounts of $^{13}_{6}C$ (and traces of $^{14}_{6}C$). The existence of these heavier isotopes causes the atomic weight to be greater than 12. Atomic weights of the elements are always given for a *mixture* of isotopes in their naturally occurring proportions. These atomic weights are calculated from mass spectrometric data as illustrated in Example 2-7.

Example 2-7 The mass spectrum of carbon shows that 98.892% of carbon atoms are $^{12}_{6}C$ with a mass of 12.00000 amu and 1.108% are $^{13}_{6}C$ with a mass of 13.00335 amu. Calculate the atomic weight of naturally occurring carbon.

$$\begin{pmatrix}\text{contribution to} \\ \text{atomic weight} \\ \text{by } ^{12}_{6}C\end{pmatrix} = \begin{pmatrix}\text{fraction of} \\ \text{all C atoms} \\ \text{that are } ^{12}_{6}C\end{pmatrix} \times \begin{pmatrix}\text{mass of} \\ \text{a } ^{12}_{6}C \text{ atom}\end{pmatrix}$$
$$= 0.98892 \times 12.00000 = 11.867$$

$$\begin{pmatrix}\text{contribution to} \\ \text{atomic weight} \\ \text{by } ^{13}_{6}C\end{pmatrix} = \begin{pmatrix}\text{fraction of} \\ \text{all C atoms} \\ \text{that are } ^{13}_{6}C\end{pmatrix} \times \begin{pmatrix}\text{mass of} \\ \text{a } ^{13}_{6}C \text{ atom}\end{pmatrix}$$
$$= 0.01108 \times 13.00335 = 0.1441$$

$$\begin{pmatrix}\text{atomic weight} \\ \text{of naturally} \\ \text{occurring carbon}\end{pmatrix} = \begin{pmatrix}\text{contribution} \\ \text{by } ^{12}_{6}C\end{pmatrix} + \begin{pmatrix}\text{contribution} \\ \text{by } ^{13}_{6}C\end{pmatrix}$$
$$= 11.867 + 0.144 = 12.011$$

SIMILAR EXAMPLES: Exercises 35, 36, 37.

If an atomic weight scale is set up in the manner suggested by Example 2-7, the values obtained are all pure numbers, relative to 12.00000 for $^{12}_{6}C$. To attach a unit of mass to the atomic weight of an element, we may define the **gram atomic weight.** This is a quantity of an element whose mass, expressed in grams, is numerically equal to its atomic weight. The gram atomic weights of oxygen, carbon, and hydrogen, for example, are 15.9994, 12.011, and 1.0080 g, respectively.

2-12 Postscript: Do Atoms Exist?

In this chapter we have traced certain aspects of the development of the atomic theory, from the speculations of the ancient Greeks to some sophisticated experiments of modern times. We will have occasion throughout the text to pursue additional aspects of the subject of atomic structure, but we will also come to the realization that the final word on the atomic theory can never be written. So, we may ask, when will there be sufficient evidence to prove the existence of atoms? One recent set of experimental observations which seems to go a long way toward substantiating the atomic theory is suggested by Figure 2-17. Do atoms exist? At this point the evidence is most convincing.

FIGURE 2-17
Visibility of individual atoms.

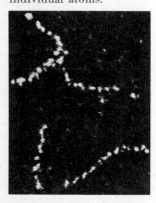

This photograph was obtained with a high resolution scanning electron microscope at the University of Chicago by A. V. Crewe, J. Wall, and J. Langmore [*Science,* 168:1340 (1970)], and is reproduced with Professor Crewe's permission. The bright spots result from individual thorium atoms, arranged in chains, in a complex substance containing thorium. The blurring of the spots is probably caused by the vibrational motion of the thorium atoms. A more recent photograph, showing uranium atoms, is featured on the cover of this book.

Summary

The most important of the early discoveries concerning the nature of chemical change were the law of conservation of mass and the law of definite composition. These laws served as the basis of Dalton's atomic theory. Dalton's theory did not meet with immediate success, however, because of some faulty assumptions made in the matter of assigning chemical formulas and atomic weights.

More fruitful were a series of investigations of the discharge of electricity through gases—cathode ray research. Cathode rays (electrons) were found to be fundamental particles of all matter and also to be fundamental units of negative electrical charge. Cathode ray research was directly responsible for the discovery of x rays, which in turn led to the discovery of radioactivity. Experiments with positive ions formed in cathode ray tubes yielded evidence of a fundamental unit of positive charge, led to the discovery of isotopes, and served as the basis of mass spectrometry. From studies of the scattering of α particles by thin metal foils, Rutherford and his coworkers made one of the most significant contributions in all of atomic physics—the concept of the nuclear atom. This concept, followed by later discoveries of the proton and neutron, made possible a description of atomic structure in terms of protons, neutrons, and electrons.

By assigning an atomic mass of 12.00000 amu to carbon-12, the masses of other atoms can now be determined with a mass spectrometer. Knowledge of the masses of the different isotopes of an element and their relative abundances leads directly to the atomic weight of an element. Equipped with reliable atomic weights, we can now deal with mass relationships in chemical compounds and reactions, the principal subject of Chapters 3 and 4.

Learning Objectives

As a result of studying Chapter 2, you should be able to

1. Describe and illustrate the laws of conservation of mass, definite composition, and multiple proportions.

2. State the basic assumptions of Dalton's atomic theory.

3. List some of the characteristic properties of cathode rays and of positive rays (canal rays).

4. Describe the production of x rays, the phenomenon of radioactivity, and the characteristics of α, β, and γ radiation.

5. Describe J. J. Thomson's determination of the charge-to-mass ratio of cathode rays, Millikan's determination of the charge on an electron, and Rutherford's studies of α particle scattering by thin metal foils.

6. Outline the general features of Rutherford's model of the nuclear atom.

7. Do simple calculations involving the masses and charges of protons, neutrons, and electrons.

8. List the numbers of protons, neutrons, and electrons present in atoms and ions, using the symbolism $^A_Z X$.

9. Explain the operation of a mass spectrometer and calculate atomic masses from experimentally determined mass ratios.

10. Calculate the atomic weight of an element from the known masses and relative abundances of its naturally occurring isotopes.

Some New Terms

An **atom** is the smallest particle of matter that characterizes an element.

An **atomic mass unit (amu)** is a basic unit for expressing the masses of individual atoms. One amu is $\frac{1}{12}$ of the mass of a $^{12}_6 C$ atom; the masses of the proton and of the neutron are just slightly greater than 1 amu.

The **atomic number, Z,** is the number of protons in the nucleus of an atom. It is also the number of electrons outside the nucleus of an electrically neutral atom.

The **atomic weight** of an element is the mass of the naturally occurring mixture of isotopes of the element, relative to an arbitrarily assigned mass of 12.00000 for carbon-12.

Cathode rays are negatively charged particles (electrons) emitted at the negative electrode (cathode) in the passage of electricity through gases at low pressures.

Chemical symbols are abbreviations of one or two letters assigned to the chemical elements (e.g., N = nitrogen and Ne = neon).

Electrons are particles carrying the fundamental unit of negative electrical charge and found outside the nuclei of all atoms.

An **ion** is an electrically charged species consisting of a single atom or a group of atoms. It is formed when a neutral atom or a group of atoms either gains or loses electrons.

Isotopes are atoms of an element which, because of differing numbers of neutrons in their nuclei, have different masses.

The **law of conservation of mass** states that matter can neither be created nor destroyed in ordinary physical or chemical processes.

The **law of definite composition (definite proportions)** states that a chemical compound has a unique composition in terms of its constituent elements, that is regardless of the source of the compound.

The **law of multiple proportions** deals with the proportions in which two elements combine when they are able to form more than a single compound: For each compound the ratio of the mass of one element to a fixed mass of the second is formulated. Then, for any pair of compounds, these ratios are themselves in the ratio of small whole numbers, such as $2.66:1/1.33:1 = 2:1$.

The **mass number**, A, is the total of the number of protons and neutrons in the nucleus of an atom.

A **mass spectrograph (mass spectrometer)** is a device used to separate and to measure precisely the quantities and masses of the different ions in a beam of positively charged gaseous ions.

Neutrons are electrically neutral fundamental particles of matter found in all atomic nuclei except that of the simple hydrogen atom, protium, $_1^1H$.

Nuclide is a term used to designate a specific atomic species, as represented by the symbolism $_Z^A X$.

Positive rays are beams of positively charged gaseous ions produced through collisions of cathode rays with residual gas atoms in cathode ray tubes.

Protons are fundamental particles carrying the basic unit of positive electric charge and found in the nuclei of all atoms.

Radioactivity is a phenomenon associated with unstable atomic nuclei in which small particles of matter (α or β particles) and/or electromagnetic radiation (γ rays) are emitted.

Exercises

Law of conservation of mass

1. When an iron object rusts, its mass increases. When a match is burned, its mass decreases. Do these observations violate the law of conservation of mass? Explain.

2. By calculation, show whether the law of conservation of mass is obeyed, within the limits of experimental error, in each of the following examples.
   (a) 10.00 g of zinc dust is mixed with 2.00 g of powdered sulfur and the mixture is heated carefully. The result is 6.08 g of white zinc sulfide and 5.92 g of unreacted zinc.
   (b) 10.00 g of calcium carbonate is dissolved in 100.0 cm³ of hydrochloric acid (density = 1.148 g/cm³). The products are 120.40 g of solution (a mixture of hydrochloric acid and calcium chloride) and 2.22 L of carbon dioxide gas, which has a density of 1.9769 g/L.

Law of definite composition

3. In a series of experiments samples of pure carbon weighing 1.00, 1.50, and 1.80 g, respectively, were burned completely in an excess of air. In each case the sole product, carbon dioxide gas, was captured and weighed. The masses obtained were 3.66, 5.50, and 6.60 g, respectively.
   (a) Do these data establish that carbon dioxide has a definite composition?
   (b) What is the composition of carbon dioxide, expressed in percent carbon and percent oxygen, by mass?

4. In one experiment 1.00 g of sodium metal was allowed to react with 10.00 g of chlorine gas. The sodium was completely consumed and 2.54 g of sodium chloride (salt) was produced. In a second experiment 1.00 g of chlorine was allowed to react with 10.00 g of sodium. The chlorine was consumed completely and 1.65 g of sodium chloride was produced. Show that these results are consistent with the law of definite composition.

5. In one experiment 1.50 g of hydrogen gas was allowed to react with an excess of oxygen gas, yielding 13.41 g of water. In a second experiment a sample of water was decomposed by electrolysis, resulting in 2.20 g of hydrogen and 17.46 g of oxygen. Show that these results are consistent with the law of definite composition.

Dalton's atomic theory

6. Suppose that the two elements A and B have atomic weights of 40 and 80, respectively.
   (a) What are the formulas that Dalton would have assigned if *three* compounds of the two elements were known?
   (b) What would be the percentage composition, by mass, of each of these compounds?

7. A compound of hydrogen and oxygen unknown in Dalton's time was hydrogen peroxide, which consists of 94% oxygen and 6% hydrogen, by mass. Had Dalton known of the existence of this compound, what formula do you think he would have assigned to it and to water (which is 89% O and 11% H)?

8. The atomic weight of oxygen is 16.0. Use information presented in Example 2-1 and Figure 2-2 to estimate the atomic weight of magnesium.

9. In an experiment of the type illustrated in Figure 2-1, Lavoisier obtained 45 grains of mercury calx (mercury

oxide). When this material was later decomposed, 41.5 grains of liquid mercury was recovered.

(a) If this were the only oxide of mercury known in Dalton's time, what formula would Dalton have assigned to this oxide?

(b) What would Dalton have deduced the atomic weight of mercury to be? (Recall that Dalton assumed the atomic weight of oxygen to be 7.)

(c) Use the correct formula of the oxide (HgO), the correct atomic weight of oxygen (16), and Lavoisier's data to deduce the atomic weight of mercury.

(d) Why would Dalton's value of the atomic weight of mercury [determined in part (b)] have been so much in error?

Law of multiple proportions

10. The formulas used by Dalton in establishing the law of multiple proportions were CH_2 for methane and CH for ethylene (see Figure 2-5). Show that the law is just as well established by using modern formulas for these gases (methane, CH_4, and ethylene, C_2H_4) and modern atomic weights for carbon and oxygen.

11. Use a table of modern atomic weights and the method of Figure 2-5 to show that the law of multiple proportions applies to each of the following pairs of compounds: (a) CO and CO_2; (b) Na_2O and Na_2O_2; (c) PCl_3 and PCl_5.

12. Show that the following data are in accord with the law of multiple proportions.

	Methane	Acetylene	Ethylene	Ethane
carbon, % by mass	75.0	92.3	85.7	80.0
hydrogen, % by mass	25.0	7.7	14.3	20.0

13. Dalton was familiar with three oxides of nitrogen. To one (nitrous gas) he assigned the formula NO. What formulas do you suppose he assigned to the other two? Show that the formulas for the three compounds are consistent with the law of multiple proportions. (Dalton assumed atomic weights of N = 5 and O = 7.)

14. Mercury combines with oxygen to form two different compounds. One of these contains 96.2% mercury, and the other, 92.6% mercury, by mass. Show that these data are consistent with the law of multiple proportions, and speculate on the formulas of these two oxides.

Fundamental particles

15. Cite the evidence that establishes most convincingly that electrons are fundamental particles of all matter.

16. List several significant differences between cathode rays and canal rays, stressing the manner of production, electric charge, mass, and so on.

17. The investigation of electrical discharge in gases led to the discovery of both negative and positive particles of matter. Explain why the negative particles proved to be fundamental particles of all matter but the positive particles did not.

18. Why couldn't the same methods that had been used to characterize electrons be used to isolate and detect neutrons?

Fundamental charges and charge-to-mass ratios

19. The following observations are made for a series of 10 oil drops in an experiment similar to Millikan's (see Figure 2-9). Drop 1 carries a charge of 1.28×10^{-18} C; drops 2 and 3 each carry $\frac{1}{2}$ the charge of drop 1; drop 4 carries $\frac{1}{4}$ the charge of drop 1; drop 5 carries a charge four times as great as drop 1; drops 6 and 7 have charges three times that of drop 1; drops 8 and 9 have charges twice that of drop 1; and drop 10 has the same charge as drop 1. Are these data consistent with the value of the electronic charge given in the text (1.602×10^{-19} C)? Could Millikan have inferred the charge on the electron from this particular series of data? Explain.

20. It is now known that static electric charges are caused by the transfer of electrons.

(a) How many excess electrons are present on an object with a charge of -5.5×10^{-15} C?

(b) How many electrons are deficient from an object with a net charge of $+6.4 \times 10^{-12}$ C?

21. What is the net charge, in coulombs, associated with 1.00×10^{12} (a) hydrogen atoms ($_1^1H$); (b) fluoride ions (F^-); (c) $_{10}^{22}Ne^{2+}$ ions?

22. Use data from Table 2-1 to verify the following statements made in this chapter.

(a) The mass of electrons is about $\frac{1}{2000}$ that of hydrogen atoms.

(b) The charge-to-mass ratio, e/m, for positive rays is considerably smaller than for electrons.

23. Arrange the following species in order of increasing absolute magnitude of charge to mass, e/m: proton, electron, neutron, α particle, the atom $_{18}^{40}Ar$, the ion $_{17}^{37}Cl^-$. (The absolute magnitude refers to the value of the e/m ratio without regard to its sign. Note also whether any two of the species listed have identical e/m ratios.)

Atomic models

24. Use the atomic model of J. J. Thomson (see Figure 2-10) to draw pictures of the following gaseous atoms and ions. (a) He; (b) O; (c) N^+; (d) F^-.

25. Represent the species given in Exercise 24 by the Rutherford model of the atom (see Figure 2-15).

Atomic number, mass number, nuclides, and isotopes

26. Describe the significance of each term in the symbol $^A_Z X$.

27. Although the symbol $^{35}_{17}Cl$ may be preferred to the simpler ^{35}Cl, explain why the two symbols actually convey the same information. Do the symbols $^{35}_{17}Cl$ and $_{17}Cl$ have the same meaning?

28. Complete the following table.

Name	Symbol	Number protons	Number electrons	Number neutrons	Mass number
sodium	$^{23}_{11}Na$	11	11	12	23
—	^{40}K	—	—	—	—
silicon	—	—	—	14	—
—	—	37	—	—	85
—	—	—	33	42	—
—	$^{20}Ne^{2+}$	—	—	—	—
—	—	—	—	—	80
—	—	—	—	126	—

What minimum amount of information is required to characterize completely an atomic species (nuclide)?

29. Arrange the following species in order of **(a)** increasing number of electrons; **(b)** increasing number of neutrons; **(c)** increasing mass.

$^{112}_{50}Sn, \quad ^{22}_{10}Ne, \quad ^{122}_{52}Te, \quad ^{59}_{29}Cu, \quad ^{120}_{48}Cd, \quad ^{58}_{27}Co$

30. The mass numbers of the isotopes of hydrogen, called protium, deuterium, and tritium, are 1, 2, and 3, respectively. What are the basic differences among these three types of hydrogen atoms? Which do you think occurs in greatest natural abundance? Explain.

Atomic mass units, atomic masses

31. For the nuclide $^{90}_{38}Sr$, express
  (a) the percentage, by number, of the fundamental particles in the atom that are neutrons
  (b) the approximate percentage of the mass of the atom contributed by protons

32. What is the mass, in grams, of each of the following?
  (a) 1.00×10^{12} atoms of chlorine-35 having individual masses of 34.96885 amu
  (b) 1.00×10^{12} atoms of chlorine-37 having individual masses of 36.96590 amu
  (c) 1.00×10^{12} atoms of a mixture of chlorine-35 and chlorine-37 in their naturally occurring abundances: 75.53% chlorine-35, 24.47% chlorine-37.

33. The following data on atomic masses are given in a handbook: **(a)** ^7_3Li, 7.01601 amu; **(b)** $^{19}_9F$, 18.99840 amu; **(c)** $^{84}_{36}Kr$, 83.9115 amu. What is the ratio of each of these masses to that of $^{12}_6C$?

Atomic weights

34. A table of atomic weights lists a value of 63.546 for copper. Which of the following is probably true concerning the masses of *individual* copper atoms: that *all, some,* or *none* have a mass of 63.546 amu? Explain your reasoning.

35. Gallium has two principal isotopes, with the following masses and relative abundances: $^{69}_{31}Ga$, 68.9257 amu, 60.4%; $^{71}_{31}Ga$, 70.9249 amu, 39.6%. Calculate the atomic weight of gallium and compare with the value listed on the inside back cover.

36. In naturally occurring uranium, 99.27% of the atoms are $^{238}_{92}U$ with mass 238.05 amu; 0.72% are $^{235}_{92}U$ with mass 235.04 amu; and 0.006% are $^{234}_{92}U$ with mass 234.04 amu. Calculate the atomic weight of naturally occurring uranium and compare with the value listed on the inside back cover.

37. The two principal isotopes of lithium have masses of 6.01513 and 7.01601 amu, respectively. The atomic weight of lithium is 6.941.
  (a) Which of these two isotopes is more abundant?
  (b) What is the approximate ratio of atoms of the more abundant to the less abundant isotope, that is, 2:1, 3:1, 4:1, . . . ?
  ★**(c)** Calculate the percent abundances of the two isotopes.

★**38.** The two naturally occurring isotopes of nitrogen have masses of 14.0031 and 15.0001 amu, respectively. Use the atomic weight listed for nitrogen to determine the percent $^{15}_7N$ in naturally occurring nitrogen.

Mass spectrometry

39. The following ratios of masses were obtained with a mass spectrometer: **(a)** $^{16}_8O : ^{12}_6C = 1.3329$; **(b)** $^{22}_{10}Ne : ^{16}_8O = 1.3749$; **(c)** $^{40}_{18}Ar : ^{22}_{10}Ne = 1.8172$. Determine the mass of an $^{40}_{18}Ar$ atom, in atomic mass units.

40. The three isotopes of hydrogen described in Exercise 30 are all capable of combining with chlorine atoms to form the simple diatomic molecules, HCl. The naturally occurring isotopes of chlorine are described in Exercise 32.
  (a) How many different kinds of HCl molecules are possible?
  (b) What are the mass numbers of these different molecules (i.e., the sum of the mass numbers of the two atoms in the molecules)?
  (c) Which is the most abundant of the possible HCl molecules? Which is the second most abundant?
  (d) Sketch the mass spectrum you would expect to obtain for HCl molecules on the assumption that all the positive ions obtained are $(HCl)^+$. (Refer to Figure 2-16.)

Additional Exercises

1. Given that one oxide of copper has 79.9% copper and 20.1% oxygen, by mass, suggest the percentage composition of a plausible copper–oxygen compound having a higher percentage of copper.

2. In 1815, the English chemist William Prout advanced the idea that all matter is comprised of a single fundamental substance—hydrogen. What evidence can you cite from this chapter in support of Prout's view?

3. All of the following radioactive nuclides have applications in medical science. Write their symbols in the form $_Z^A X$. **(a)** cobalt-60; **(b)** phosphorus-32; **(c)** iodine-131; **(d)** sulfur-35.

4. Given the following species: $^{24}Mg^{2+}$; ^{47}Cr; $^{59}Co^{2+}$; $^{35}Cl^-$; $^{124}Sn^{2+}$; ^{226}Th; ^{90}Sr.
   (a) Which contains the same number of neutrons as electrons?
   (b) In which do protons contribute more than 50% of the mass?
   (c) Which has a number of neutrons equal to the number of protons plus one-half the number of electrons?

5. Determine the approximate value of the charge-to-mass ratio, e/m, in coulombs per gram, for the ions $_{10}^{22}Ne^+$ and $_8^{18}O^{2+}$. Why are these values only approximate and not exact?

6. Estimate the density of matter in a proton, using Rutherford's estimate of the diameter of a nucleus as 10^{-13} cm and the mass of a proton given in Table 2-1. (Assume that an atomic nucleus has a spherical shape.)

7. The highest density of any actual substance is about 22 g/cm^3 (osmium metal). What does a comparison of this value and the density of the proton calculated in Additional Exercise 6 suggest about the amount of empty space in matter?

*8. The mass of a $_6^{12}C$ atom is taken to be exactly 12.00000 amu. Are there likely to be any other nuclides with an exact integral (whole number) mass, expressed in amu? Explain.

*9. The atomic weight of oxygen listed on the inside back cover is 15.9994. A chemistry textbook printed 30 years ago lists a value of 16.0000. How do you account for this discrepancy? Would you expect other atomic weights listed in the 30-year-old text to be the same, generally higher, or generally lower than in the current text? Explain. (*Hint:* See the marginal note on page 35.)

*10. Suppose it were decided to redefine the atomic weight scale by choosing as an arbitrary standard having an atomic weight of 35.00000 the naturally occurring *mixture* of chlorine isotopes.
   (a) What would be the atomic weights of helium, sodium, and iodine on this new atomic weight scale?
   (b) Why do you suppose these three elements have nearly integral (whole number) atomic weights based on $_6^{12}C$ but not based on naturally occurring chlorine?

*11. Based on the densities of the lines in the mass spectrum of a sample of krypton gas, the following conclusions were reached about the several isotopes observed.

(1) Somewhat more than 50% of the atoms were $_{36}^{84}Kr$.
(2) The number of $_{36}^{82}Kr$ and $_{36}^{83}Kr$ atoms were essentially equal.
(3) The number of $_{36}^{86}Kr$ atoms was 1.50 times greater than the number of $_{36}^{82}Kr$ atoms.
(4) The number of $_{36}^{80}Kr$ atoms was 19.6% of the number of $_{36}^{82}Kr$ atoms.
(5) The number of $_{36}^{78}Kr$ atoms was 3.0% of the number of $_{36}^{82}Kr$ atoms.

The atomic masses of the isotopes cited above are $_{36}^{78}Kr$, 77.9204 amu; $_{36}^{80}Kr$, 79.9164 amu; $_{36}^{82}Kr$, 81.9135 amu; $_{36}^{83}Kr$, 82.9141 amu; $_{36}^{84}Kr$, 83.9115 amu; $_{36}^{86}Kr$, 85.9106 amu. The atomic weight of krypton is 83.80. Use the data presented here to calculate the percent abundances of the six naturally occurring isotopes of krypton.

Self-Test Questions

For questions 1 through 6 select the single item that best completes each statement.

1. Of the following assumptions or results of Dalton's atomic theory, the only one that remains essentially correct in most cases is
   (a) All atoms of an element are identical in mass.
   (b) Atoms are indivisible and indestructible.
   (c) Oxygen has an atomic weight of 7.
   (d) Atoms of elements combine in the ratios of small whole numbers to form compounds.

2. Cathode rays **(a)** may be positively or negatively charged particles; **(b)** have properties identical to β particles; **(c)** are a form of electromagnetic radiation; **(d)** have masses that depend on the matter from which they are derived.

3. Rutherford's experiments on the scattering of α particles by thin metal foils established that **(a)** the mass and charge of an atom are concentrated in a nucleus; **(b)** electrons are fundamental particles of all matter; **(c)** all electrons have the same charge; **(d)** atoms are electrically neutral.

4. The species that has the same number of electrons as $^{32}_{16}S$ is **(a)** $^{35}_{17}Cl^-$; **(b)** $^{34}_{16}S^+$; **(c)** $^{40}_{18}Ar^{2+}$; **(d)** $^{35}_{16}S^{2-}$.

5. All of the following masses are possible for an *individual* carbon atom except one. That impossible one is **(a)** 12.00000 amu; **(b)** 12.01115 amu; **(c)** 13.00335 amu; **(d)** 14.00324 amu.

6. There are *two* principal isotopes of indium (atomic weight = 114.82). One of these, $^{113}_{49}In$, has an atomic mass of 112.9043 amu. The second isotope is most likely to be **(a)** $^{111}_{49}In$; **(b)** $^{112}_{49}In$; **(c)** $^{114}_{49}In$; **(d)** $^{115}_{49}In$.

7. When a strip of magnesium metal is burned in air, it gives off a brilliant glow and produces a white powder that weighs more than the original metal. When a strip of magnesium metal is ignited in a photoflash bulb, a brilliant glow is also produced, but in this case the bulb weighs the same before and after it is flashed. Explain the difference in these observations.

8. The mass of the nuclide $^{84}_{36}Kr$ = 83.9115 amu. If the atomic weight scale were *redefined* so that $^{84}_{36}Kr$ = 84.00000, what would be the mass of $^{12}_{6}C$?

9. What is the charge-to-mass ratio, e/m, in coulombs per gram, of a chloride ion, $^{37}_{17}Cl^-$? The mass of $^{37}_{17}Cl$ is 36.966 amu. The charge on an electron is -1.602×10^{-19} C, and the relationship between the units, amu and g, is 1.673×10^{-24} g = 1.0073 amu.

10. There are two principal isotopes of silver, $^{107}_{47}Ag$ and $^{109}_{47}Ag$. The atomic weight of naturally occurring silver is 107.87, with 51.82% of the atoms being $^{107}_{47}Ag$. The mass of an atom of $^{107}_{47}Ag$ is 106.9 amu. What is the mass, in amu, of an atom of $^{109}_{47}Ag$?

3

Stoichiometry I: Elements and Compounds

The word stoichiometry is derived from the Greek *stoicheion,* meaning element. Literally, stoichiometry means to measure the elements. The term is generally used more broadly, however, to include a wide variety of measurements and relationships involving substances and mixtures of chemical interest. Our discussion in this chapter is concerned with the stoichiometry of elements and compounds and with certain related subjects.

The ideas discussed in this chapter are basic to an understanding of many of the topics that will occur later in the text. However, these later topics, as they are encountered, may also provide additional insight into the concepts introduced here.

3-1 Avogadro's Number and the Concept of the Mole

Everyday examples of mass substituting for number of units of a desired quantity are numerous. With relatively large objects, such as eggs and oranges, if a certain number is desired, they may be counted directly. However, to plant a lawn, even though a certain number of grass seeds is required, we do not count them out—they are sold by the pound.

Avogadro's number is generally expressed with three significant figures when used in numerical calculations, that is, 6.02×10^{23}.

The most basic relationships among chemical quantities, as we will learn in this and following chapters, involve relative *numbers* of atoms, ions, or molecules, not masses or weights. Yet, although we recognize the importance of numbers of chemical units, there is no way by which we may count them literally. We must resort to some other measurement that is related to numbers of atoms, and one of the most convenient measurements for this purpose is that of mass. Thus, we need to establish a relationship between the *measured* mass of an element and some *known* but *uncountable* number of atoms contained in that mass.

Atomic weights can be established by comparing the masses of a large number of atoms of one kind and an *equal* number of atoms of the atomic weight standard, $^{12}_{6}C$. But what number of atoms should be taken for the purpose of establishing atomic weights? The number that proves most useful is the number of atoms present in exactly 12.00000 g of $^{12}_{6}C$. This number, which has the value of 6.0225×10^{23}, is called **Avogadro's number,** *N.* Another term that is nearly synonymous with Avogadro's number is the **mole** (abbreviated **mol**).

A mole of a substance is an amount of substance that contains the same number of elementary units as there are $^{12}_{6}C$ atoms in 12.00000 g $^{12}_{6}C$.

If a substance contains atoms all of a single nuclide, we may write

1 mol $^{12}_{6}C$ consists of 6.0225×10^{23} $^{12}_{6}C$ atoms and weighs 12.00000 g

1 mol $^{16}_{8}O$ consists of 6.0225×10^{23} $^{16}_{8}O$ atoms and weighs 15.9948 g

and so on.

The masses 12.011 g C, 15.9994 g O, and so on, were referred to in Section 2-11 as the gram atomic weight. Here we are redefining these quantities as one mole, a term that is used more generally. The mole has been adopted as the SI unit for an amount of substance.

Avogadro's number (6.0225×10^{23}) is an enormous number. Suppose that it were desired to represent an entire mole of atoms by piling up peas (each having a volume of about 0.1 cm^3). The required pile would cover the entire United States to a depth of about 6 km.

Most elements, however, are composed of a mixture of two or more isotopes. The atoms to be "counted out" to yield one mole are not all of the same mass. They must be taken in the proportions in which they occur naturally. Thus, in 1 mol of carbon, most of the atoms are carbon-12, but some are carbon-13 (and a very few are carbon-14).

1 mol of carbon consists of 6.0225×10^{23} C atoms and weighs 12.011 g

1 mol of oxygen consists of 6.0225×10^{23} O atoms and weighs 15.9994 g

and so on.

A new way of thinking about atomic weights emerges from the statements above. *The atomic weight of an element is equal numerically to the mass in grams of* 1 *mol* (6.0225×10^{23}) *of atoms of the element.* Figure 3-1 is a pictorial analogy of the meaning of 1 mol of atoms.

3-2 Some Illustrative Examples Using the Mole Concept

Example 3-1   How many atoms are present in 2.80 mol of iron metal?

This problem can be solved by a single-step application of the general problem-solving method (recall equation 1.5). The conversion factor is derived from the definition of a mole: 1 mol Fe $\backsimeq 6.02 \times 10^{23}$ Fe atoms.

$$\text{no. Fe atoms} = 2.80 \text{ mol Fe} \times \frac{6.02 \times 10^{23} \text{ Fe atoms}}{1 \text{ mol Fe}}$$

$$= 16.9 \times 10^{23} \text{ Fe atoms} = 1.69 \times 10^{24} \text{ Fe atoms}$$

SIMILAR EXAMPLE: Exercise 2.

Example 3-2   How many moles of magnesium are represented by a collection of 3.05×10^{20} Mg atoms?

Again the sole conversion factor required is based on the definition of a mole: 1 mol Mg $\backsimeq 6.02 \times 10^{23}$ Mg atoms. The factor is written with the unit mol Mg in the numerator and the unit Mg atoms in the denominator. This provides for the proper cancellation of units.

$$\text{no. mol Mg} = 3.05 \times 10^{20} \text{ Mg atoms} \times \frac{1 \text{ mol Mg}}{6.02 \times 10^{23} \text{ Mg atoms}}$$

$$= 0.507 \times 10^{-3} \text{ mol Mg} = 5.07 \times 10^{-4} \text{ mol Mg}$$

SIMILAR EXAMPLE: Exercise 3.

In the two examples just considered

From this point on in the text the cancellation of units will not be routinely shown in calculations. In every case, though, you should assure yourself that the proper cancellation will occur.

FIGURE 3-1
An attempt to picture a mole of atoms.

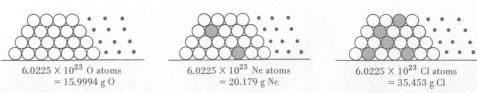

6.0225×10^{23} O atoms
$= 15.9994$ g O

6.0225×10^{23} Ne atoms
$= 20.179$ g Ne

6.0225×10^{23} Cl atoms
$= 35.453$ g Cl

If atoms could be piled up in the manner suggested here, it would take an enormously large pile to contain one mole of atoms (see marginal note at end of Section 3-1). In this representation, because the relative abundances of ^{17}O and ^{18}O are so small, the oxygen atoms are all shown to be alike. In the case of neon, about one of every ten atoms is of the heavier isotope, ^{22}Ne. In chlorine roughly three fourths of the atoms are ^{35}Cl and one fourth are ^{37}Cl.

1. A single-step conversion was possible.
2. The solutions actually were not dependent on the particular elements chosen (that is, the numerical answers would have been 1.69×10^{24} and 5.07×10^{-4} regardless of the elements of choice).
3. The situations described were only hypothetical—atoms cannot be counted directly. Some other property must have been measured.

The examples that follow indicate how other factors usually enter into calculations involving the mole concept.

Example 3-3   A sample of sodium metal has a mass of 12.5 g. **(a)** How many moles of sodium metal is this? **(b)** How many Na atoms are present in the sample?

In part (a) we must convert from grams to moles, and in part (b), from moles to number of atoms. Two conversion factors are required, which can be obtained from the relationships $23.0 \text{ g Na} \backsim 1 \text{ mol Na} \backsim 6.02 \times 10^{23}$ Na atoms.

(a) no. mol Na $= 12.5 \text{ g Na} \times \dfrac{1 \text{ mol Na}}{23.0 \text{ g Na}} = 0.543$ mol Na

(b) We can, of course, simply multiply the result of part (a) by the appropriate conversion factor (Avogadro's number). Instead, we have chosen below to begin at the same point as in part (a). This is done to illustrate that the calculation of intermediate results is often not necessary.

no. Na atoms $= 12.5 \text{ g Na} \times \dfrac{1 \text{ mol Na}}{23.0 \text{ g Na}} \times \dfrac{6.02 \times 10^{23} \text{ Na atoms}}{1 \text{ mol Na}}$

$$(\text{g Na} \quad \rightarrow \quad \text{mol Na} \quad \rightarrow \quad \text{Na atoms})$$

$$= 3.27 \times 10^{23} \text{ Na atoms}$$

SIMILAR EXAMPLE: Exercise 5.

Example 3-4   How many iron atoms are present in a stainless steel ball bearing having a radius of 0.100 in.? The stainless steel contains 85.6% Fe, by mass, and has a density of 7.75 g/cm^3.

Several conversion factors are applied in a stepwise fashion.

1. Determine the volume of the steel ball in cubic centimeters. (Use the formula for the volume of a sphere.)

$$V = \frac{4}{3}\pi r^3 = \frac{4}{3}\pi \left(0.100 \text{ in.} \times \frac{2.54 \text{ cm}}{1 \text{ in.}}\right)^3 = \frac{4}{3}(3.14)(0.254)^3 \text{ cm}^3$$

$$= 0.0686 \text{ cm}^3$$

2. Use the conversion factors indicated in parentheses to convert successively from
   (a) cubic centimeters to grams of stainless steel (density);
   (b) grams of stainless steel to grams of iron (percent composition);
   (c) grams of iron to moles of iron (atomic weight of iron);
   (d) moles of iron to number of iron atoms (Avogadro's number).

no. Fe atoms $= 0.0686 \text{ cm}^3 \text{ steel} \times \dfrac{7.75 \text{ g steel}}{1.00 \text{ cm}^3 \text{ steel}} \times \dfrac{85.6 \text{ g Fe}}{100 \text{ g steel}}$

$$(\text{cm}^3 \text{ steel} \quad \rightarrow \quad \text{g steel} \quad \rightarrow \quad \text{g Fe}$$

$$\times \dfrac{1 \text{ mol Fe}}{55.8 \text{ g Fe}} \times \dfrac{6.02 \times 10^{23} \text{ Fe atoms}}{1 \text{ mol Fe}}$$

$$\rightarrow \quad \text{mol Fe} \quad \rightarrow \quad \text{Fe atoms})$$

$$= 4.91 \times 10^{21} \text{ Fe atoms}$$

SIMILAR EXAMPLES: Exercises 8, 10, 11.

Calculations involving the mole concept, even those of some complexity as in Example 3-4, can be solved by the general problem-solving method. But, as always, care must be taken to introduce conversion factors in such a way as to achieve the required cancellation of units.

3-3 Chemical Compounds

FIGURE 3-2
Formula units of ionic and covalent compounds.

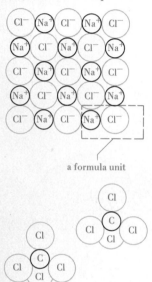

a formula unit

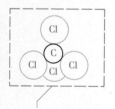

a formula unit (molecule)

NaCl illustrates the situation encountered with solid ionic compounds; CCl_4 is an example of a covalent compound. An ionic compound forms between a metal and a nonmetal; covalent compounds result from combinations of nonmetals. The distinction between metals and nonmetals is introduced in Section 3-4 and is explored more fully in Chapter 8.

Chemical compounds are substances composed of two or more elements. They are denoted by combinations of symbols called chemical formulas. A **chemical formula** is a symbolic representation of

1. The elements present in a compound.
2. The relative numbers of atoms of each element.

A **formula unit** is the *smallest* collection of atoms from which the formula can be derived.

In the following formula the elements present are denoted by their symbols and the relative numbers of atoms by *subscript* numerals (where no subscript is written, the number 1 is understood).

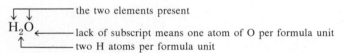

Here are three additional formula units.

NaCl	$MgCl_2$	CCl_4
sodium chloride	magnesium chloride	carbon tetrachloride

An important distinction between a formula unit of NaCl and one of CCl_4 is suggested by Figure 3-2. A formula unit of CCl_4 can exist as a separate unit or entity, whereas the formula unit NaCl is simply a pair of atoms (ions) selected from a much larger array of atoms (ions). The situation with $MgCl_2$ is similar to that of NaCl. A combination of atoms that can exist as an individual identifiable unit possessing a unique set of measurable properties is called a **molecule.** The formula unit and molecule of carbon tetrachloride are identical; both are represented as CCl_4. It is inappropriate to speak of a molecule of solid sodium chloride.

The situation with the compound hydrogen peroxide is again somewhat different. The smallest collection of atoms that can exist, a *molecule,* contains two atoms of hydrogen and two of oxygen: H_2O_2. But the smallest collection of atoms that discloses the combining ratio (relative numbers of atoms) of hydrogen and oxygen is HO. This collection, a *formula unit,* does not exist as a stable entity. The molecule contains *two* formula units.

A chemical formula based on the formula unit is called the **simplest** or **empirical formula.** The formula based on an actual molecule is called the **molecular formula.** And there are three possible relationships to consider.

1. The empirical and molecular formulas may be identical, as with CCl_4.
2. The molecular formula may be a *multiple* of the empirical formula. (The molecular formula, H_2O_2, is twice the empirical formula, HO.)
3. A compound in the solid state may have an empirical formula (such as NaCl, $MgCl_2$, or $NaNO_3$) and no molecular formula.

Formula Weight and Molecular Weight. Once the formula unit has been identified, it is a simple matter to establish the **formula weight** of a compound. *It is the mass of the formula unit relative to an assigned mass of* 12.00000 *amu for a* $^{12}_{6}C$ *atom.* Since atomic weights are also relative to $^{12}_{6}C$, formula weights can be determined by an appropriate summation of atomic weights. Thus, for sodium chloride, NaCl:

one formula unit of NaCl consists of one Na^+ and one Cl^-

formula weight NaCl = at. wt. Na + at. wt. Cl

formula weight NaCl = 22.99 + 35.45 = 58.44

And for magnesium chloride, $MgCl_2$:

formula weight $MgCl_2$ = at. wt. Mg + (2 × at. wt. Cl)

formula weight $MgCl_2$ = 24.30 + (2 × 35.45) = 95.20

> The atomic weights of an atom and its ion are practically identical. This is because the atom and ion differ only in the number of electrons they contain, and electrons contribute very little to the mass of an atom.

If a compound consists of discrete molecules, it is also appropriate to define a **molecular weight.** *It is the mass of a molecule relative to an assigned mass of* 12.00000 *amu for a* $^{12}_{6}C$ *atom.* To determine the molecular weight of carbon tetrachloride, CCl_4, for example, we note that

1 molecule of CCl_4 consists of one C atom and 4 Cl atoms

molecular weight CCl_4 = at. wt. C + (4 × at. wt. Cl)

molecular weight CCl_4 = 12.01 + (4 × 35.45) = 153.8

Although it is always acceptable to speak of the formula weight of a compound, the term *molecular weight* is valid only if discrete molecules of the compound actually exist. When the term is applied to compounds such as NaCl, $MgCl_2$, and $NaNO_3$ in the solid state, what is actually meant is formula weight. If the formula unit and a molecule of a compound are identical (as in CCl_4), the formula weight and the molecular weight are identical. Where molecules of a compound consist of two or more formula units, the molecular weight is a corresponding *multiple* of the formula weight.

Analogous to the situation described for atoms in Section 2-11, a unit of mass (gram) can be attached to formula and molecular weights, leading to the "gram formula weight" or the "gram molecular weight." However, these terms are neither as useful nor as basic as the concept "mole of a compound," which is introduced next.

Mole of a Compound. Based on the distinction between formula unit and molecule, a mole of a chemical compound can be described in one of two ways.

1. A mole is an amount of compound containing Avogadro's number (6.0225×10^{23}) of formula units.
2. For a compound that is composed of discrete molecules, a mole is Avogadro's number of molecules.

> Other frequently encountered terms are *mole weight, molar weight,* and *molar mass.* All mean simply the mass of 1 mol of a substance.

By analogy to the relationship between a mole of atoms and the atomic weight suggested by Figure 3-1, we note the following: *The mass in grams of* 1 mol (6.0225×10^{23}) *of formula units (molecules) of a compound is numerically equal to the formula weight (molecular weight) of the compound.* This statement leads to simple expressions of the type

1 mol $MgCl_2$ ≏ 95.20 g $MgCl_2$ ≏ 6.0225×10^{23} $MgCl_2$ formula units

and

1 mol CCl_4 ≏ 153.8 g CCl_4 ≏ 6.0225×10^{23} CCl_4 molecules

Example 3-5   How many Cl^- ions are present in 50.0 g $MgCl_2$?

$$\text{no. } Cl^- \text{ ions} = 50.0 \text{ g } MgCl_2 \times \frac{1 \text{ mol } MgCl_2}{95.2 \text{ g } MgCl_2} \times \frac{6.02 \times 10^{23} \text{ f.u. } MgCl_2}{1 \text{ mol } MgCl_2}$$

$$\times \frac{2 \, Cl^- \text{ ions}}{1 \text{ f.u. } MgCl_2}$$

$$= 6.32 \times 10^{23} \, Cl^- \text{ ions}$$

where we have abbreviated formula unit as f.u.

SIMILAR EXAMPLE: Exercise 9.

Example 3-6   How many liters of liquid CCl_4 ($d = 1.59$ g/cm^3) must be measured out to contain 1.00×10^{25} CCl_4 molecules?

The following series of conversions is required: molecules $\rightarrow$ mol $\rightarrow$ g $\rightarrow$ cm^3 $\rightarrow$ L.

$$\text{no. L } CCl_4 = 1.00 \times 10^{25} \, CCl_4 \text{ molecules} \times \frac{1 \text{ mol } CCl_4}{6.02 \times 10^{23} \, CCl_4 \text{ molecules}}$$

$$\text{(molecules } CCl_4 \qquad \rightarrow \qquad \text{mol } CCl_4$$

$$\times \frac{153.8 \text{ g } CCl_4}{1 \text{ mol } CCl_4} \times \frac{1 \text{ cm}^3 \, CCl_4}{1.59 \text{ g } CCl_4} \times \frac{1 \text{ L } CCl_4}{1000 \text{ cm}^3 \, CCl_4}$$

$$\rightarrow \qquad \text{g } CCl_4 \qquad \rightarrow \qquad \text{cm}^3 \, CCl_4 \qquad \rightarrow \qquad \text{L } CCl_4)$$

$$= 1.61 \text{ L } CCl_4$$

SIMILAR EXAMPLES: Exercises 11, 12.

MOLE OF AN ELEMENT—A SECOND LOOK.   In Section 3-1 we defined a mole of an element in terms of Avogadro's number of *atoms*. This is the appropriate definition and the only definition possible when describing certain elements, like iron, magnesium, sodium, and copper. In these elements enormous numbers of individual spherical atoms are clustered together, much like marbles in a can. But with some elements—hydrogen, oxygen, nitrogen, fluorine, chlorine, bromine, iodine, phosphorus, and sulfur, for example—atoms of the same kind are joined together to form molecules, and bulk samples of the elements are composed of collections of molecules. For these elements it is appropriate to speak of 1 mol of molecules and the molecular weight of the element. Molecular forms of the elements just cited are

$$H_2 \qquad O_2 \qquad N_2 \qquad F_2 \qquad Cl_2 \qquad Br_2 \qquad I_2 \qquad P_4 \qquad S_8$$

To say a mole of hydrogen is ambiguous; one should say either a mole of hydrogen atoms or a mole of hydrogen molecules. Better still is to write 1 mol H or 1 mol H_2. Furthermore, we may write 1.008 g H/mol H and 2.016 g H_2/mol H_2. The situation for the element sulfur is pictured in Figure 3-3.

3-4 Chemical Nomenclature

We need to consider some additional applications of the mole concept and will do so later in this chapter. Another need is to say something about writing chemical formulas and naming chemical compounds—a subject known as **chemical nomenclature.**

To name the 106 elements is not a difficult matter, nor is the assignment of 106 distinctive symbols. But to name and write formulas of chemical compounds is an altogether different question. It is essential that no two chemical substances have exactly the same name, yet at the same time there should be some similarities in the

FIGURE 3-3
What is a "mole of sulfur"?

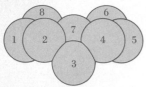

Experimental evidence indicates that in solid sulfur atoms are joined together into puckered rings with eight members. A molecule of sulfur is composed of eight atoms, represented as S_8. In a sample of solid sulfur, there exist eight times as many atoms as there are molecules. If by a "mole of sulfur" we mean 1 mol of sulfur atoms, we can measure out $\frac{1}{8}$ mol of the S_8 molecules and describe this amount as "1 mol S." On the other hand, "1 mol S_8" means that we measure out 1 mol of the S_8 molecules and obtain thereby 8 mol of individual sulfur atoms.

The distinction made here between 1 mol H and 1 mol H_2 is very much like the distinction between one dozen socks and one dozen pairs of socks (i.e., the H atom is analogous to a single sock and the H_2 molecule to a pair of socks).

This impossible situation would arise if all compounds were referred to by such trivial names as water (H_2O) and ammonia (NH_3).

names of similar compounds. Otherwise, the task of naming millions of different compounds would be next to impossible.

What is needed is a *systematic* method of assigning names—a system of nomenclature. But the matter is complicated by the fact that a single system is insufficient. Several systems are introduced in this text, each at an appropriate time. In the present discussion we restrict ourselves to some of the simpler aspects of nomenclature.

BINARY IONIC COMPOUNDS. **Binary compounds** are those formed between *two* elements, and these are the easiest compounds to name. First, one element is named, and this is followed by the name of the second element, whose ending is changed to "ide."

NaCl = sodium chloride

name unchanged / "ide" ending

joining of nonmetal & metal

$MgCl_2$ = magnesium chloride

A conscious decision was required in these examples: which element to name first.

Both in writing formulas and in naming compounds, the more metallic element is placed first, followed by the more nonmetallic one.

Of course, to apply this rule requires knowledge of which elements are metallic and which, nonmetallic. For the present let us simply say that the metallic elements are those whose atoms display a tendency to lose one or more electrons to form positive ions, called **cations.** Nonmetal atoms tend to gain one or more electrons to form negative ions, called **anions.**

The tabular arrangement of elements shown on the inside front cover and in Figure 3-4 is known as the periodic table. In this table the elements in the right-hand

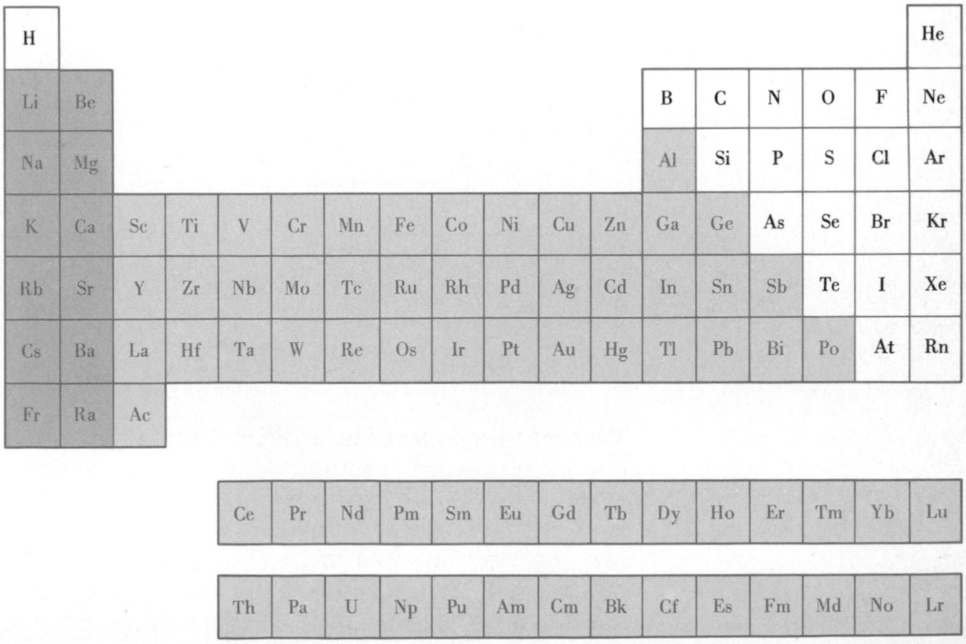

FIGURE 3-4
Metals, nonmetals, and the periodic table.

In this tabular arrangement the nonmetals are shown in black. They are found mostly in the right-hand portion of the table. The metallic elements are shown in color; the two groups of elements at the extreme left of the table are the most metallic of all. The special case of hydrogen (a nonmetal grouped with the metals) and further refinements of the classification scheme are discussed in Chapter 8. Refer to the periodic table on the inside front cover for the atomic numbers and atomic weights of the elements.

TABLE 3-1
Some simple ions[a]

Name	Symbol	Name	Symbol
Positive ions (cations)		copper(II) or cupric	Cu^{2+}
lithium	Li^+	zinc	Zn^{2+}
sodium	Na^+	silver	Ag^+
potassium	K^+	mercury(I) or mercurous[b]	Hg_2^{2+}
rubidium	Rb^+	mercury(II) or mercuric	Hg^{2+}
cesium	Cs^+	lead(II) or plumbous	Pb^{2+}
magnesium	Mg^{2+}		
calcium	Ca^{2+}	**Negative ions (anions)**	
strontium	Sr^{2+}	hydride[c]	H^-
barium	Ba^{2+}	nitride	N^{3-}
aluminum	Al^{3+}	oxide	O^{2-}
chromium(II) or chromous	Cr^{2+}	sulfide	S^{2-}
chromium(III) or chromic	Cr^{3+}	fluoride	F^-
iron(II) or ferrous	Fe^{2+}	chloride	Cl^-
iron(III) or ferric	Fe^{3+}	bromide	Br^-
copper(I) or cuprous	Cu^+	iodide	I^-

[a] The arrangement of ions in this table is similar to the arrangement of the elements in the periodic table (see Figure 3-4).

[b] Unlike other common ions, which are monatomic, the mercurous ion is *diatomic*. The ion is referred to as mercury(I) because two ions with *single* positive charges (Hg^+) are joined together to form the species Hg_2^{2+}. Formulas and names thus take the form: Hg_2Cl_2 = mercury(I) chloride and $HgCl_2$ = mercury(II) chloride.

[c] Hydrogen may also appear as a positive ion (H^+) in water solutions or other mixtures, but not in pure compounds.

portion are nonmetals. The rest of the elements are metals, with the elements of the first two columns on the far left being most metallic of all. The theoretical bases of the periodic table and the concepts of metals and nonmetals are discussed in Chapter 8. For the present, simply use Figure 3-4 as a guide in identifying metals and nonmetals.

Compounds of positive and negative ions—ionic compounds—must contain these ions in such proportions that the formula unit is electrically neutral—one Na^+ to one Cl^-, one Mg^{2+} to *two* Cl^-, one Mg^{2+} to one O^{2-}, and so on. That is, the total negative charge must equal the total positive charge in the ionic compound.

With the ideas introduced in the preceding paragraphs and the characteristic ionic charges presented in Table 3-1, it is a simple matter to write names and formulas for a host of binary ionic compounds. There is one minor difficulty, however: How do we write distinctive names and formulas in those cases where the metal can appear in more than one ionic form? Two methods suggested in Table 3-1 are made explicit below for the two chlorides of iron.

Formula	Systematic name	Common name
$FeCl_2$	iron(II) chloride	*ferrous* chloride
$FeCl_3$	iron(III) chloride	*ferric* chloride

The ion Fe^{2+} must be combined with *two* Cl^- ions in a neutral formula unit, and the ion Fe^{3+} with *three* Cl^- ions, leading to the formulas $FeCl_2$ and $FeCl_3$. The preferred method of naming these compounds is to refer to the Fe^{2+} ion as iron(II) and to the Fe^{3+} ion as iron(III). By an older system of nomenclature that is still in common use, Fe^{2+} is called ferrous ion, and Fe^{3+} ferric ion. This system uses the "ous" ending for the cation of lower charge and the "ic" ending for the cation of higher charge. Also, this system uses Latin rather than English names for some of the common metals (e.g., iron = ferrum).

BINARY COVALENT COMPOUNDS. When two nonmetallic elements combine, this occurs through a *sharing* rather than a transfer of electrons; no ions are formed. Nevertheless, it is generally possible to consider one of the two elements to be the more metallic. Both in writing the name and the formula of a binary covalent compound, the same order is followed: the more metallic element appears first, followed by the more nonmetallic one.

$$HCl \quad = \quad \text{hydrogen chloride}$$

— name unchanged —
"ide" ending —

$$H_2O \quad = \quad \text{hydrogen oxide (water)}$$

Only one binary compound of hydrogen and chlorine is possible—HCl. Some pairs of elements can form two or more compounds and each must have its own name. For example,

CO = carbon <u>mo</u>noxide

CO_2 = carbon <u>di</u>oxide

In this system prefixes are used to designate the relative numbers of atoms of each type in the molecule.

mono = 1; di = 2; tri = 3; tetra = 4; penta = 5; hexa = 6.

Although the prefix mon(o)- is used to name CO, general practice is to use *no* prefix when there is just one atom of a given type per molecule (e.g., NO = nitrogen oxide, not *mono*nitrogen *mono*xide). Several other binary covalent compounds are named by this system in Table 3-2. Finally, it should be noted that several substances have common names that are so well established that they are often not named according to this system. For example, H_2O = water; NH_3 = ammonia; N_2O = nitrous oxide; NO = nitric oxide.

Example 3-7 Name the following compounds: **(a)** NaBr, **(b)** MgF_2, **(c)** CS_2.
(a) In Table 3-1 we find the ions, Na^+ (sodium) and Br^- (bromide). NaBr is a binary ionic compound with the name: sodium bromide.
(b) Again we find the ions Mg^{2+} and F^- in Table 3-1. MgF_2 is magnesium fluoride. Note that it is both unnecessary and misleading to call this compound magnesium *di*fluoride. We do not need to signify the composition of the formula unit through the name of the compound; this is accomplished just by knowing the charges on the ions.
(c) Both C and S are nonmetals; CS_2 is a binary covalent compound called carbon disulfide.

SIMILAR EXAMPLES: Exercises 25a, b, 27.

Example 3-8 Write acceptable formulas for the following compounds: **(a)** potassium sulfide, **(b)** diboron tetrabromide.
(a) Potassium sulfide is an ionic compound consisting of the ions K^+ and S^{2-}. The neutral formula unit contains *two* K^+ ions and *one* S^{2-} ion. The correct formula is K_2S.

Again, we discover a need to assess the metallic/nonmetallic character of the elements, a subject explored in Chapter 8.

TABLE 3-2
Use of prefixes in nomenclature

Formula	Name[a]
BCl_3	boron trichloride
CCl_4	carbon tetrachloride
CO	carbon monoxide
CO_2	carbon dioxide
NO	nitrogen oxide
NO_2	nitrogen dioxide
N_2O	dinitrogen oxide
N_2O_3	dinitrogen trioxide
N_2O_4	dinitrogen tetroxide
N_2O_5	dinitrogen pentoxide
SF_6	sulfur hexafluoride

[a] Where the prefix ends in an "a" or "o" and the element name begins with an "a" or "o," the final vowel of the prefix is often *dropped for ease of pronunciation*. For example, carbon monoxide (*not* carbon monooxide) and dinitrogen tetroxide (*not* dinitrogen tetraoxide). However, PI_3 is phosphorus triiodide (not triodide) and SI_4 is sulfur tetraiodide (not tetriodide).

TABLE 3-3
Some common polyatomic ions

Name	Formula	Typical compound
Cation		
ammonium	$NH_4{}^+$	NH_4Cl
Anions		
acetate	$C_2H_3O_2{}^-$	$NaC_2H_3O_2$
carbonate	$CO_3{}^{2-}$	Na_2CO_3
hydrogen carbonate (or bicarbonate)	$HCO_3{}^-$	$NaHCO_3$
hypochlorite	ClO^-	$NaClO$
chlorite	$ClO_2{}^-$	$NaClO_2$
chlorate	$ClO_3{}^-$	$NaClO_3$
perchlorate	$ClO_4{}^-$	$NaClO_4$
chromate	$CrO_4{}^{2-}$	Na_2CrO_4
cyanide	CN^-	$NaCN$
hydroxide	OH^-	$NaOH$
nitrite	$NO_2{}^-$	$NaNO_2$
nitrate	$NO_3{}^-$	$NaNO_3$
permanganate	$MnO_4{}^-$	$NaMnO_4$
phosphate	$PO_4{}^{3-}$	Na_3PO_4
hydrogen phosphate	$HPO_4{}^{2-}$	Na_2HPO_4
dihydrogen phosphate	$H_2PO_4{}^-$	NaH_2PO_4
sulfite	$SO_3{}^{2-}$	Na_2SO_3
sulfate	$SO_4{}^{2-}$	Na_2SO_4
hydrogen sulfate (or bisulfate)	$HSO_4{}^-$	$NaHSO_4$
thiosulfate	$S_2O_3{}^{2-}$	$Na_2S_2O_3$

(b) From the name given we see that the formula must be based on a unit consisting of *two* boron atoms and *four* bromine atoms. The correct formula is B_2Br_4.

SIMILAR EXAMPLES: Exercises 28a, b, 29.

TERNARY COMPOUNDS.   A **ternary compound** is made up of *three* different elements. A particularly common type of ternary compound is one that involves a combination of a simple monatomic (one atom) ion with a polyatomic ion. A **polyatomic ion** contains two or more atoms. A number of common polyatomic ions and representative compounds containing these ions are listed in Table 3-3, from which we can infer that

1. Polyatomic anions are more commonly encountered than are polyatomic cations. A common polyatomic cation is $NH_4{}^+$.

2. An element common to many polyatomic anions is *oxygen*. The oxygen is combined with another nonmetal atom. (Such ions are commonly called oxyanions.)

3. Certain nonmetals (e.g., Cl, N, P, S) form a series of polyatomic anions containing different numbers of oxygen atoms. Their names are generally related to their oxygen content, according to the scheme

The relationship of names and formulas for such polyatomic anions is considered from another standpoint in Chapter 9.

hypo_____ite _____ite _____ate per_____ate

least O ◄─────────────────────────► most O

4. Some series of polyatomic anions contain varying numbers of hydrogen atoms and are named accordingly. For example, HPO_4^{2-} is the *hydrogen phosphate* ion and $H_2PO_4^-$ is the *dihydrogen phosphate* ion.

5. The prefix "thio" signifies that a sulfur atom has been substituted for an oxygen atom. (The sulfate ion has *one* S and *four* O atoms; thiosulfate ion has *two* S and *three* O atoms.)

Example 3-9 Use information from Tables 3-1 and 3-3 to supply names for the following compounds: **(a)** Cs_2SO_4; **(b)** $ZnCO_3$; **(c)** $Ca(H_2PO_4)_2$.
(a) The formula unit consists of *two* Cs^+ and *one* SO_4^{2-} ion. This compound is called cesium sulfate.
(b) *One* Zn^{2+} and *one* CO_3^{2-} constitute one formula unit of zinc carbonate.
(c) The polyatomic anion in this compound is $H_2PO_4^-$, the dihydrogen phosphate ion. Two of these ions are present for every Ca^{2+} ion in the compound called calcium dihydrogen phosphate.

SIMILAR EXAMPLE: Exercise 25.

Example 3-10 Use information from Tables 3-1 and 3-3 to supply correct formulas for the following compounds: **(a)** potassium cyanide, **(b)** ammonium chromate, **(c)** calcium hypochlorite.
(a) This compound consists of potassium (K^+) and cyanide (CN^-) ions in a 1:1 ratio. The formula of the compound is KCN.
(b) Two ammonium (NH_4^+) ions must be present for every chromate (CrO_4^{2-}) ion. The parentheses around the NH_4, followed by the subscript 2, signifies *two* NH_4^+ ions in a formula unit of ammonium chromate. The correct formula is $(NH_4)_2CrO_4$. (This formula is read as "N—H—4, taken twice, C—R—O—4.")
(c) Here there is *one* simple cation, Ca^{2+}, and *two* polyatomic anions, ClO^-, in the formula unit, leading to the formula $Ca(ClO)_2$.

SIMILAR EXAMPLE: Exercise 28.

COMPOUNDS OF GREATER COMPLEXITY. Let us note a few additional ideas represented through the following formulas, which are somewhat more complex than those we have been considering.

$CuSO_4 \cdot 5H_2O$. This compound is of a type known as a hydrate. A **hydrate** is a substance in which a formula unit has associated with it a certain number of water molecules. The formula shown here signifies *five* H_2O molecules per formula unit of $CuSO_4$. The compound is called copper(II) sulfate *penta*hydrate. Its formula weight is that of $CuSO_4$ (160) *plus* that associated with five H_2O (90) = 250.

$K_4[Fe(CN)_6]$. This compound, called potassium ferrocyanide or potassium hexacyanoferrate(II), belongs to a class known as coordination compounds. It consists of simple ions—K^+—and complex ions—$[Fe(CN)_6]^{4-}$. The nature of complex ions and coordination compounds is explored fully in Chapter 22. For the present all that we need to recognize is that

1. There are 4 K, 1 Fe, 6 C, and 6 N atoms, or a total of 17 atoms, in a formula unit.
2. The formula weight of the compound is $(4 \times 39.1) + 55.85 + (6 \times 12.0) + (6 \times 14.0) = 368.2$.

3-5 Composition of Chemical Compounds

There are several ways in which the composition of a chemical compound can be described, and these alternative forms are all derivable from the chemical formula. For example, the once commonly used insecticide DDT has the molecular formula

$C_{14}H_9Cl_5$

*Note how important the parentheses are in this formula. If they were omitted one would have $CaClO_2$—an *incorrect* formula for calcium *chlorite*. (What is the formula of calcium chlorite?)*

The atoms within the square brackets are those present in the complex ion. In $[Fe(CN)_6]^{4-}$ there are six CN^- ions bonded to one Fe^{2+} ion.

Per mole of DDT there are 14 moles of carbon atoms, 9 moles of hydrogen atoms, and 5 moles of chlorine atoms. That is,

1 mol $C_{14}H_9Cl_5$ ⇌ 14 mol C ⇌ 9 mol H ⇌ 5 mol Cl

and these equivalences can be used to write conversion factors

$$\frac{14 \text{ mol C}}{1 \text{ mol } C_{14}H_9Cl_5} ⇌ 1; \quad \frac{9 \text{ mol H}}{1 \text{ mol } C_{14}H_9Cl_5} ⇌ 1; \quad \frac{5 \text{ mol Cl}}{14 \text{ mol C}} ⇌ 1; \quad \text{and so on.}$$

Two of these conversion factors based on the chemical formula of DDT are applied in Example 3-11. Another is used in Example 3-12 to establish the percent composition, by mass, of DDT.

Example 3-11   **(a)** How many moles of H atoms are contained in 50.0 g DDT $(C_{14}H_9Cl_5)$? **(b)** How many moles of Cl atoms are present *for every gram* of C in DDT?

(a) no. mol H = 50.0 g $C_{14}H_9Cl_5$ × $\dfrac{1 \text{ mol } C_{14}H_9Cl_5}{354 \text{ g } C_{14}H_9Cl_5}$ × $\dfrac{9 \text{ mol H}}{1 \text{ mol } C_{14}H_9Cl_5}$

$\qquad\qquad = 1.27$ mol H

The percentage of C and H in DDT can be calculated similarly, leading to 47.5% C and 2.4% H. Alternatively, two of the percentages can be calculated and the remaining one obtained by difference. For example,

% H
$= 100 - \% \text{ C} - \% \text{ Cl}$
$= 100 - 47.5 - 50.1$
$= 2.4\%$

(b) no. mol Cl = 1.00 g C × $\dfrac{1 \text{ mol C}}{12.0 \text{ g C}}$ × $\dfrac{5 \text{ mol Cl}}{14 \text{ mol C}}$ = 0.0298 mol Cl

SIMILAR EXAMPLES: Exercises 13, 14, 15.

Example 3-12   What is the percent Cl, by mass, in DDT $(C_{14}H_9Cl_5)$?
    For every mole of $C_{14}H_9Cl_5$ (354 g) there are present 5 mol of Cl atoms (5 × 35.45 = 177 g). Percent is simply a mass ratio, multiplied by 100.

$$\% \text{ Cl, by mass} = \frac{(5 \times 35.45) \text{ g Cl}}{354 \text{ g } C_{14}H_9Cl_5} \times 100 = 50.1\%$$

SIMILAR EXAMPLES: Exercises 16, 18.

If the percent composition, by mass, of a compound is known, this information can be used to determine the *empirical formula*. If an independent measurement of the molecular weight exists, the *true molecular formula* can also be established. The basic principle of the method used in Example 3-13 is this: The ratios of numbers of atoms of each type in a compound are independent of whether a single formula unit, a mole of compound, or any arbitrary mass of compound is chosen for study. The sample size selected in Example 3-13 is 100 g, since the composition of the compound is given in percent (parts per hundred).

Example 3-13   The compound methyl benzoate, used in the manufacture of perfumes, is found to consist of 70.58% C, 5.93% H, and 23.49% O, by mass. What is the empirical formula of this compound? By other experiments the molecular weight of the compound is found to be 136. What is the molecular formula of the compound?
    In a 100-g sample of the compound the masses of the three elements are: 70.58 g C, 5.93 g H, and 23.49 g O. Each of these masses can be converted to a number of moles of atoms, as follows:

$$\text{no. mol C} = 70.58 \text{ g C} \times \frac{1 \text{ mol C}}{12.01 \text{ g C}} = 5.88 \text{ mol C}$$

$$\text{no. mol H} = 5.93 \text{ g H} \times \frac{1 \text{ mol H}}{1.01 \text{ g H}} = 5.87 \text{ mol H}$$

$$\text{no. mol O} = 23.49 \text{ g O} \times \frac{1 \text{ mol O}}{16.00 \text{ g O}} = 1.47 \text{ mol O}$$

The relative numbers of moles of atoms of each type in 100 g of the compound and the relative numbers of atoms in one formula unit are the same, that is,

$C_{5.88}H_{5.87}O_{1.47}$

But formula units contain integral (whole) numbers of atoms. The subscripts shown can be converted to integers by dividing each of them by the smallest (1.47).

$$C_{5.88/1.47}H_{5.87/1.47}O_{1.47/1.47} = C_{4.00}H_{3.99}O$$
$$= C_4H_4O \quad \text{(empirical formula)}$$

At times, the ratios obtained at this point may not be whole numbers. If they differ from whole numbers by only a slight amount, as is the case here, this is due to experimental error. These ratios can simply be rounded off to whole numbers. In some cases the ratios must be multiplied by a common factor to convert them to whole numbers. For example, a carbon–hydrogen compound containing 91.25% C and 8.75% H has the empirical formula

$C_{7.60}H_{8.68}$
$= C_{7.60/7.60}H_{8.68/7.60}$
$= C_{1.00}H_{1.14}$
$= C_{(1.00 \times 7)}H_{(1.14 \times 7)}$
$= C_7H_{7.98}$
$= C_7H_8$

The formula weight of the compound is $[(4 \times 12.01) + (4 \times 1.01) + 16.00] = 68.08$. Since the experimentally determined molecular weight (136) is almost exactly twice the formula weight, the molecular formula is $C_8H_8O_2$

SIMILAR EXAMPLES: Exercises 20, 21, 22.

3-6 Analytical Chemistry

AN OVERVIEW. To analyze means to separate something into its constituent parts in order to learn more of the nature of these constituents. The branch of chemistry that deals with the analysis of chemical compounds and mixtures is called **analytical chemistry.** At times all that we seek is a knowledge of what substances are present in a sample. Does this brand of paint contain lead compounds? Is there nitrate ion present in this sample of drinking water? What metals comprise this steel alloy? To answer such questions one uses the methods of **qualitative analysis.** If our concern is with the *actual* quantity of one or more of the constituents of a sample, a **quantitative analysis** is required. Typical examples would include determining the percentage of C, H, O, and N in a physiologically active compound extracted from a tropical plant, the percentage of iron in an iron ore sample, or the number of parts per million (ppm) of mercury in a fish.

The number of different methods employed by analytical chemists is almost without limit, but these methods can generally be categorized as involving chemical methods or instrumental methods. **Chemical methods** are based on subjecting the sample of interest to chemical reactions. In some way the outcome of these reactions is dependent on the composition of the sample. Quite often it is possible simply to measure certain physical properties of the sample, usually with sophisticated instruments. This type of analysis, known as **instrumental analysis,** is especially important. It can yield accurate results on small samples, often when the constituents of interest are present only in trace amounts, and sometimes without having to destroy the sample being analyzed.

Chemical methods of analysis can be further subdivided into two broad categories: volumetric and gravimetric analysis. In **volumetric analysis** chemical reactions are conducted in solutions, and the key measurements involve solution volumes. An introduction to this subject is provided in Section 4-4. In **gravimetric analysis** the principal instrument is the analytical balance and the key data are the masses or weights of substances. Although additional insight into gravimetric analysis will come with increased knowledge of chemical reactions, we have already acquired sufficient background to explore some aspects of this subject.

COMBUSTION ANALYSIS. This is an analytical procedure that can sometimes be used to provide the percentage composition data needed to establish the empirical formula of a compound (recall Example 3-13). It is especially applicable to organic compounds (compounds of carbon; see Chapter 24). As illustrated in Figure 3-5, a weighed sample of the compound is heated in a stream of oxygen gas in a furnace.

FIGURE 3-5
Apparatus for combustion analysis.

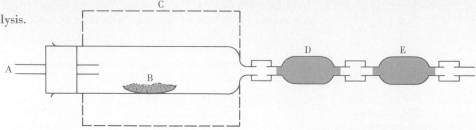

Oxygen gas (A) is passed into the combustion tube containing the sample to be analyzed (B). This portion of the apparatus is enclosed within a high temperature furnace (C). Products of the combustion are absorbed as they leave the furnace—water vapor by magnesium perchlorate (D) and carbon dioxide gas by sodium hydroxide (E).

FIGURE 3-6
Principle of combustion analysis—Example 3-14 visualized.

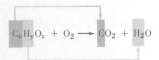

The basis of combustion analysis is in being able to trace what happens to all the atoms in the compound being analyzed (represented here as $C_xH_yO_z$). All of the C atoms in the compound appear as C atoms in CO_2 (and there is no other source of C atoms for the CO_2). All of the H atoms in the compound appear as H atoms in H_2O (and again, there is no other source of H atoms). Oxygen atoms in the CO_2 and H_2O come both from the compound being analyzed and from oxygen gas consumed in the combustion. The quantity of oxygen in the compound must be determined indirectly.

The hydrogen in the compound is converted to water vapor, which is absorbed by a substance such as magnesium perchlorate. The carbon contained in the compound is converted to carbon dioxide gas, which is absorbed in sodium hydroxide (to produce sodium carbonate). The increases in mass of these absorbers correspond to the masses of water and carbon dioxide produced. The basic principle of combustion analysis is suggested by Figure 3-6. Typical results are those presented in Example 3-14.

Example 3-14   A 0.2060-g sample of the carbon–hydrogen–oxygen compound described in Example 3-13 is subjected to combustion analysis. The quantities of CO_2 and H_2O produced are 0.5327 and 0.1091 g, respectively. Use these data to determine the percentage composition of the compound.

First we determine the number of grams of carbon in 0.5327 g CO_2 and the number of grams of hydrogen in 0.1091 g H_2O. In each case the conversions are: g compound → mol compound → mol element → g element.

$$\text{no. g C} = 0.5327 \text{ g } CO_2 \times \frac{1 \text{ mol } CO_2}{44.01 \text{ g } CO_2} \times \frac{1 \text{ mol C}}{1 \text{ mol } CO_2} \times \frac{12.01 \text{ g C}}{1 \text{ mol C}} = 0.1454 \text{ g C}$$

(g CO_2 →   mol CO_2 →   mol C   →   g C)

$$\text{no. g H} = 0.1091 \text{ g } H_2O \times \frac{1 \text{ mol } H_2O}{18.02 \text{ g } H_2O} \times \frac{2 \text{ mol H}}{1 \text{ mol } H_2O} \times \frac{1.008 \text{ g H}}{1 \text{ mol H}} = 0.01221 \text{ g H}$$

(g H_2O →   mol H_2O →   mol H   →   g H)

All the carbon in the CO_2 and the hydrogen in the H_2O come from the compound being analyzed. Thus, the percentages of carbon and hydrogen in the compound are:

$$\% \text{ C} = \frac{0.1454 \text{ g C}}{0.2060 \text{ g cpd.}} \times 100 = 70.58\%$$

$$\% \text{ H} = \frac{0.01221 \text{ g H}}{0.2060 \text{ g cpd.}} \times 100 = 5.93\%$$

$$\% \text{ O} = 100 - \% \text{ C} - \% \text{ H} = 100 - 70.58 - 5.93 = 23.49\%$$

SIMILAR EXAMPLES: Exercises 35, 36.

PRECIPITATION ANALYSIS.   In precipitation analysis a component of the sample being analyzed deposits from solution as an insoluble material (it precipitates from solution). This precipitate is then treated so as to yield a *pure* solid of *known composition*. From the measured masses of this solid and of the original sample, the percentage of the component in the sample may be determined. The determination of the

FIGURE 3-7
Determination of tin in brass—Example 3-15 visualized.

Brass is an alloy of copper and zinc with small amounts of tin, lead, and iron. When a weighed sample of brass is treated with nitric acid (a water solution of HNO_3), the copper, zinc, lead, and iron dissolve and appear in aqueous solution in their ionic forms. Tin is converted to an insoluble oxide with an unknown amount of water of hydration ($SnO_2 \cdot x\,H_2O$). This precipitate is filtered off from the solution, washed, dried, and then heated to drive off all the water of hydration. The result is *pure* SnO_2, which is weighed.

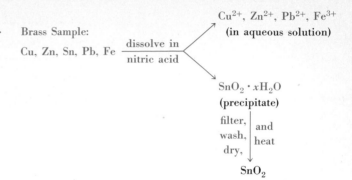

percent tin in a sample of brass is illustrated in Figure 3-7 and Example 3-15.

Example 3-15 A 2.568-g sample of brass, when treated in the manner outlined in Figure 3-7, yields 0.1330 g of pure SnO_2. What is the percent Sn in the brass sample?
First, determine the mass of tin in 0.1330 g SnO_2.

$$\text{no. g Sn} = 0.1330\ \text{g } SnO_2 \times \frac{1\ \text{mol } SnO_2}{150.7\ \text{g } SnO_2} \times \frac{1\ \text{mol Sn}}{1\ \text{mol } SnO_2} \times \frac{118.7\ \text{g Sn}}{1\ \text{mol Sn}}$$

$$= 0.1048\ \text{g Sn}$$

Since all the tin is derived from the brass sample,

$$\%\ Sn = \frac{0.1048\ \text{g Sn}}{2.568\ \text{g brass}} \times 100 = 4.08\%$$

SIMILAR EXAMPLES: Exercises 39, 40.

ATOMIC WEIGHT DETERMINATIONS. In Chapter 2 we indicated how mass spectra can be used to determine atomic weights. Mass spectrometric measurements are the most precise and accurate possible for this purpose, but prior to their development atomic weight determinations were based on the analysis of chemical compounds. Example 3-16 illustrates how precipitation analysis might be used to establish the atomic weight of an element.

The compound MCl_2 is an ionic compound. Its formula unit consists of one M^{2+} and two Cl^- ions. The reaction of Ag^+ and Cl^- to produce a precipitate of AgCl is discussed in Chapter 4.

Example 3-16 The chlorine present in an 0.5250-g sample of MCl_2 is precipitated as 0.5070 g of silver chloride, AgCl. What is the atomic weight of the element M?
These are the key facts to note.

1. All the chlorine in the compound MCl_2 appears in the compound AgCl.
2. The atomic weights of the elements Ag and Cl are known.

Let us proceed through the following steps.
1. Determine the number of grams of Cl in the AgCl.

$$\text{no. g Cl} = 0.5070\ \text{g AgCl} \times \frac{1\ \text{mol AgCl}}{143.3\ \text{g AgCl}} \times \frac{1\ \text{mol Cl}}{1\ \text{mol AgCl}} \times \frac{35.45\ \text{g Cl}}{1\ \text{mol Cl}} = 0.1254\ \text{g Cl}$$

2. The number of grams of Cl in the MCl_2 is also 0.1254.
3. Determine the number of grams of M in the MCl_2.

0.5250 g MCl_2 − 0.1254 g Cl = 0.3996 g M

4. Determine the number of moles of Cl in the MCl_2.

$$\text{no. mol Cl} = 0.1254\ \text{g Cl} \times \frac{1\ \text{mol Cl}}{35.45\ \text{g Cl}} = 3.537 \times 10^{-3}\ \text{mol Cl}$$

5. Determine the number of moles of **M** in the MCl_2.

$$\text{no. mol M} = 3.537 \times 10^{-3} \text{ mol Cl} \times \frac{1 \text{ mol M}}{2 \text{ mol Cl}} = 1.768 \times 10^{-3} \text{ mol M}$$

6. The atomic weight of **M** is the number of grams of **M** in the sample of MCl_2 divided by the number of moles of **M**.

$$\text{at. wt. M} = \frac{0.3996 \text{ g M}}{1.768 \times 10^{-3} \text{ mol M}} = 226.0 \text{ g M/mol M}$$

SIMILAR EXAMPLES: Exercises 43, 44.

Summary

A concept central to the study of chemistry introduced in this chapter is the mole. The mole describes an amount of substance in terms of the *number* of elementary units of the substance present. To use the concept of the mole, then, requires that the units being counted be clearly established—atoms, ions, formula units, molecules. Moreover, since amounts of substances must still be measured in terms of mass or volume, it is necessary to know how a mole of substance is related to these other quantities. The key relationships presented in this chapter are the masses of a mole of atoms (atomic weight), of a mole of formula units (formula weight), and of a mole of molecules (molecular weight).

The composition of a chemical compound may be expressed through its chemical formula. The percentage of each element present in a compound can be determined readily from the formula of the compound. Also, the formula can be derived from an experimental measurement of the percentage composition of a compound. Formulas obtained in this way are always empirical formulas—the simplest formulas that can be written. In some cases the empirical formula also represents the composition of a molecule of a substance; in others, the molecular formula is an integral multiple of the empirical formula.

Chemical compounds are designated by names as well as by formula. The relationship between names and formulas—nomenclature—is introduced in this chapter for some simple compounds. Also introduced in this chapter is the branch of chemistry that deals with the experimental determination of the compositions of compounds and mixtures—analytical chemistry.

Learning Objectives

As a result of studying Chapter 3, you should be able to

1. Write Avogadro's number and explain the meaning of a mole of an element and a mole of a compound.

2. Write formulas for the molecular forms of some common nonmetallic elements and distinguish between a "mole of atoms" and a "mole of molecules."

3. Distinguish between the empirical and molecular formulas of a compound.

4. Calculate the numbers of atoms, ions, formula units, or molecules in a substance from a given mass, or vice versa.

5. Write the names and symbols of common cations and anions and the names and formulas of binary ionic and binary covalent compounds.

6. Name some of the more common polyatomic ions, and write names and formulas of compounds containing these ions.

7. Use chemical formulas as a source of conversion factors for stoichiometric calculations.

8. Determine the percent composition of a compound from its formula.

9. Determine the empirical formula of a compound from its experimentally measured percent composition.

10. Establish the percent composition of a compound from the results of a combustion analysis.

11. Determine the quantity of a constituent in a compound or mixture from the results of a precipitation analysis.

12. Use appropriate analytical chemical data to establish the atomic weight of an element.

Some New Terms

Analytical chemistry deals with the analysis of compounds and mixtures. Samples are separated into their component parts to determine what is present (qualitative) and how much (quantitative analysis).

Avogadro's number, N, has a value of 6.0225×10^{23}, usually rounded off to 6.02×10^{23}. It is the number of elementary units in one mole.

A **chemical formula** is a symbolic representation of the relative numbers of atoms of each of the elements present in a compound.

Combustion analysis is an analytical procedure in which the percent composition of a compound is related to the amounts of CO_2 and H_2O produced when the compound is burned in oxygen.

An **empirical formula** is the simplest formula that can be written for a compound. It expresses the ratios of numbers of atoms of the elements present in the smallest whole numbers possible.

A **formula unit** is the smallest collection of atoms from which the formula of a compound can be established.

The **formula weight** of a compound is the mass of a formula unit of the compound compared to the nuclide $^{12}_{6}C$. Alternatively, the formula weight is the mass in grams of one mole of formula units of a compound.

A **hydrate** is a compound in which a certain number of water molecules are associated with each formula unit, for example, $CuSO_4 \cdot 5\ H_2O$.

A **mole** is an amount of substance that contains Avogadro's number of elementary units, that is, 6.02×10^{23} atoms, formula units, or molecules.

A **molecular formula** denotes the numbers of the different atoms present in a molecule. In some cases the molecular formula is the same as the empirical formula; in others it is an integral multiple of the empirical formula.

The **molecular weight** of a compound is the mass of a molecule of the compound compared to the nuclide $^{12}_{6}C$. Alternatively, the molecular weight is the mass in grams of one mole of molecules of a compound.

A **molecule** is a combination of atoms that can exist as an individual identifiable unit possessing a unique set of measurable properties.

Nomenclature refers to the writing of chemical names and formulas by some systematic method. Methods are introduced in this chapter for various binary (two-element) and ternary (three-element) compounds.

A **polyatomic ion** contains two or more atoms. The most commonly encountered are polyatomic anions (negative ions) that contain a nonmetal such as Cl, N, P, or S in combination with some number of oxygen atoms (oxyanions).

Stoichiometry refers to measurements and relationships involving substances and mixtures of chemical interest.

Exercises

Terminology

1. Use ideas introduced in this chapter to explain the distinction between each pair of related terms.
   (a) a mole of hydrogen atoms and a mole of hydrogen molecules
   (b) formula unit and molecule
   (c) empirical formula and true molecular formula
   (d) cation and anion
   (e) binary and ternary compound
   (f) qualitative and quantitative analysis

Avogadro's number and the mole

2. What is the number of atoms in each of the following samples of matter? (a) 2.50 mol Fe; (b) 0.015 mol Ar; (c) 4.0×10^{-10} mol Pu

3. How many *sulfur* atoms are present in each of the following quantities? (a) 4.5 mol S; (b) 2.8 mol S_8; (c) 5.0×10^{-5} mol H_2S; (d) 1.40 mol CS_2

4. In a collection of 2.48×10^{25} molecules of C_2H_5OH, what is the number of moles of (a) C atoms; (b) H atoms; (c) O atoms?

5. Calculate the quantities indicated.
   (a) the number of atoms in 125 g Mg
   (b) the mass, in grams, of 3.75 mol O_2 gas
   (c) the mass, in kg, of 3.25×10^{28} Fe atoms
   (d) the molar volume (volume occupied by 1 mol) of sodium metal, Na (density of Na = 0.971 g/cm³)
   (e) the number of $^{81}_{35}Br$ atoms in 75.0 cm³ of liquid Br_2 (density of liquid Br_2 = 3.12 g/cm³; natural abundance of $^{81}_{35}Br$ = 49.46%)

6. In 1.0 mol of the compound Cs_2S
   (a) What fraction of the total *number* of ions is S^{2-}?
   (b) What fraction of the total *mass* is contributed by S^{2-}?

7. A certain alloy of tin, lead, and bismuth contains these elements in the atomic proportions 2:5:3, respectively. What is the mass, in grams, of a sample of this alloy containing a total of 1.00×10^{25} atoms?

8. How many Ag atoms are present in a piece of sterling silver jewelry weighing 1.40 oz? Sterling silver contains 92.5% Ag, by mass.

9. A mixture consists of 35.0% $MgCl_2$ and 65.0% KCl, by mass. What is the total number of chloride ions (Cl^-) in 450 g of this mixture?

10. During a severe air pollution episode in Ghent, Belgium, in 1972 the concentration of lead (Pb) in suspended particulate matter rose to a maximum of 3.01 μg Pb/m^3. How many Pb atoms would be present in a 500-ml sample of this air (the approximate lung capacity of a human adult)?

11. In rhombic sulfur, sulfur atoms are joined together in groups of eight, producing the molecules S_8 (see Figure 3-3). If the density of rhombic sulfur is 2.07 g/cm^3, determine for a crystal having a volume of 4.15 mm^3: **(a)** the number of moles of S_8 present; **(b)** the total number of sulfur atoms.

12. A public water supply was found to contain 1 part per billion (ppb) by mass of chloroform, $CHCl_3$. (Consider this to be essentially 1.00 g $CHCl_3$ per 10^9 g of water.)

   (a) How many chloroform molecules would be present in a glassful of this water (250 g)?

   (b) If the chloroform present in this glassful of water could be isolated, would this amount be detectable on an ordinary analytical balance that measures mass to about ±0.0001 g?

Chemical formulas

13. Some of the following statements are correct and some are not when applied to the compound $C_6H_{12}O_6$. Determine which is the case for each and explain your reasoning.

   (a) The percents, by mass, of C and O in $C_6H_{12}O_6$ are the same as in CO.

   (b) The ratio of number of H atoms to number of O atoms is the same as in water.

   (c) The element present in highest percent by mass is oxygen.

   (d) The proportions of carbon and oxygen, by mass, are equal.

14. The amino acid methionine has the molecular formula $C_5H_{11}NO_2S$. Determine **(a)** the molecular weight of methionine; **(b)** the number of C atoms in 100.0 g of the amino acid.

15. The compound trinitrotoluene (TNT) has the formula $C_7H_5N_3O_6$. Determine **(a)** the total number of atoms in one formula unit; **(b)** the ratio of O atoms to N atoms; **(c)** the ratio, by mass, of hydrogen to carbon in the compound; **(d)** the element present in the greatest proportion, by mass.

Percent composition of compounds

16. Determine the percent by mass of each of the elements in the pain-killing drug codeine, $C_{18}H_{21}NO_3$.

17. Without performing detailed calculations, indicate which of the following compounds has the greatest percent by mass of sulfur: SO_2, SO_3, $MgSO_4$, Li_2S. Explain your reasoning.

18. All of the materials listed below are of value in fertilizers because they supply the element nitrogen. Which of these is potentially the richest source of nitrogen on a mass basis? Urea, $CO(NH_2)_2$; ammonium nitrate, NH_4NO_3; or guanidine, $HNC(NH_2)_2$.

19. Ammonium sulfate, $(NH_4)_2SO_4$, is a commonly used fertilizer. Instructions for fertilizing a mature avocado tree call for 1 lb of actual nitrogen per year. How many pounds of ammonium sulfate would be required?

Chemical formulas from percent composition

20. Analysis of an oxide of chlorine yields the results 38.8% Cl and 61.2% O. What is the empirical formula of this oxide?

21. A gaseous compound of boron and hydrogen is found to consist of 78.5% B and 21.5% H, by mass. By an independent experiment the molecular weight is determined to be 27.5 g/mol. What is the true molecular formula of this compound?

22. The stimulant caffeine has a molecular weight of 194.2 and the percent composition, by mass, 49.5% C, 5.2% H, 28.8% N, and 16.5% O. What is the molecular formula of caffeine?

23. Cobalt forms two oxides. One has 21.4% O, by mass, and the other, 28.9% O. What are the formulas of these oxides? Propose acceptable systematic names for these compounds.

24. By appropriate chemical reactions it is possible to substitute one or more Cl atoms for H atoms in the substance benzene, C_6H_6. A particular compound resulting from such a substitution is found to contain 48.2% Cl, by mass. What is the formula for this compound?

Nomenclature

25. Name the following compounds: **(a)** KI; **(b)** $CaCl_2$; **(c)** $Mg(NO_3)_2$; **(d)** K_2CrO_4; **(e)** Cs_2SO_4; **(f)** Cr_2O_3; **(g)** $FeSO_4$; **(h)** ZnS; **(i)** $Ca(HCO_3)_2$; **(j)** KCN; **(k)** K_2HPO_4; **(l)** NH_4I; **(m)** $Cu(OH)_2$.

26. Listed below are several pairs of compounds. From the information given about one of the compounds, supply the missing information about the other.

SnF_2, _____	$SnCl_4$, stannic chloride
PbO, lead(II) oxide	_____, lead(IV) acetate
$CoSO_4$, cobalt(II) sulfate	_____, cobalt(III) sulfate
KIO_3, potassium iodate	KIO_4, _____
AuCl, _____	$AuCl_3$, auric chloride

27. Assign plausible names to the following interhalogen compounds: **(a)** ICl; **(b)** ICl_3; **(c)** ClF_3; **(d)** BrF_5.

28. Write correct chemical formulas for the following: **(a)** calcium oxide; **(b)** strontium fluoride; **(c)** aluminum sulfate; **(d)** ammonium chromate; **(e)** magnesium hydroxide; **(f)** potassium carbonate; **(g)** zinc acetate; **(h)** mercury(II) nitrate; **(i)** iron(III) oxide; **(j)** chromium(II) chloride; **(k)** lithium sulfide; **(l)** calcium dihy-

drogen phosphate; **(m)** magnesium perchlorate; **(n)** potassium hydrogen sulfate.

29. Write a plausible formula for each of the following: **(a)** chlorine dioxide; **(b)** silicon tetrafluoride; **(c)** tricarbon disulfide; **(d)** diboron tetrabromide.

Hydrates

30. What is the percent, by mass, of water in the hydrate $ZnSO_4 \cdot 7\,H_2O$?

31. Without performing detailed calculations, indicate which of the following hydrates has the greatest percent, by mass, of water.

    (a) $CuSO_4 \cdot 5\,H_2O$;   **(b)** $Cr_2(SO_4)_3 \cdot 18\,H_2O$;
    (c) $MgCl_2 \cdot 6\,H_2O$;   **(d)** $LiC_2H_3O_2 \cdot 2\,H_2O$.

32. A certain hydrate of magnesium sulfate, $MgSO_4 \cdot x\,H_2O$, weighing 5.018 g is heated until all the water of hydration is driven off. The resulting anhydrous compound, $MgSO_4$, weighs 2.449 g. What is the formula of the hydrate?

33. Anhydrous sodium sulfate, Na_2SO_4, can absorb water vapor and be converted to the *deca*hydrate (10-hydrate), $Na_2SO_4 \cdot 10\,H_2O$. By how many grams would the mass of a 1.00-g sample of the thoroughly dried Na_2SO_4 increase if exposed to sufficient water vapor to be converted to the decahydrate?

34. A certain hydrate is found to have the following percent composition, by mass: 20.3% Cu; 8.95% Si; 36.3% F; 34.5% H_2O. What is the empirical formula of this hydrate?

Combustion analysis

35. An 0.1510-g sample of a solid hydrocarbon is subjected to combustion analysis. Produced as a result of the combustion are 0.5184 g CO_2 and 0.0849 g H_2O. By a separate experiment the molecular weight of the hydrocarbon is found to be 128.
    (a) What is the percent composition of the hydrocarbon?
    (b) What is the empirical formula of the compound?
    (c) What is the molecular formula of the compound?

36. An 0.4590-g sample of the carbon–oxygen–hydrogen compound, *n*-butanol, yields 1.0904 g CO_2 and 0.5580 g H_2O as a result of combustion analysis. What is the empirical formula of this compound?

Additional Exercises

1. If a sample of $MgBr_2$ is to contain 1.48×10^{23} Br^- ions:
    (a) How many Mg^{2+} ions will the sample contain?
    (b) How many formula units of $MgBr_2$ will be present?
    (c) What will be the mass of the sample?

2. The mineral spodumene has the chemical formula $Li_2O \cdot Al_2O_3 \cdot 4\,SiO_2$. Given the fact that the percentage of

***37.** The substance dimethylhydrazine is a carbon–hydrogen–nitrogen compound used in rocket fuels. When a 0.208-g sample is burned completely, 0.305 g CO_2 and 0.249 g H_2O are collected. A separate 0.350-g sample is treated in such a way as to convert its nitrogen content to 0.163 g N_2. What is the empirical formula of dimethylhydrazine?

***38.** A certain hydrocarbon, C_xH_y, is burned and found to produce 1.955 g CO_2 for every 1.000 g H_2O. What is the empirical formula of this hydrocarbon?

Precipitation analysis

39. A 0.1565-g sample of KI is dissolved in water, and all the iodide present is precipitated as AgI. How many grams of pure, dry AgI are obtained?

40. A particular type of brass contains the elements copper, tin, lead, and zinc. A sample of this brass weighing 1.502 g is treated in such a way as to convert the tin to 0.215 g SnO_2, the lead to 0.101 g $PbSO_4$, and the zinc to 0.216 g $Zn_2P_2O_7$. What is the percent, by mass, of each element in the brass sample?

***41.** A 1.150-g sample of $ZnSO_4 \cdot x\,H_2O$ is dissolved in water and the sulfate ion is precipitated by adding an excess of $BaCl_2$ solution. The mass of pure, dry $BaSO_4$ obtained is 0.9335 g. What is the formula of the zinc sulfate hydrate?

Atomic weight determinations

42. Two compounds of chlorine with the element X are found to have molecular weights and percents Cl, by mass, as follows: mol. wt. = 137, 77.5% Cl; mol. wt. = 208, 85.1% Cl. What is the element X, and what are the formulas of these compounds?

43. A sample of the compound MSO_4 weighing 0.1304 g reacts with barium chloride and yields 0.2528 g $BaSO_4$. What must be the atomic weight of the element M?

44. The metal M forms the sulfate $M_2(SO_4)_3$. A sample of this sulfate weighing 0.605 g is converted to 1.239 g $BaSO_4$. What is the atomic weight of M?

***45.** An 0.415-g sample of a metal oxide with the formula M_2O_3 is converted to the sulfide, MS, yielding 0.457 g. What is the atomic weight of the metal M?

^6_3Li atoms in naturally occurring lithium compounds is 7.40%, how many ^6_3Li atoms are present in a 68.0-g sample of spodumene?

3. What is the percent, by mass, of boron in the mineral axinite, $HCa_3Al_2BSi_4O_{16}$?

4. For the compound, $Ge[S(CH_2)_4CH_3]_4$, determine
   (a) the total number of atoms in one formula unit
   (b) the ratio, by number, of C atoms to H atoms
   (c) the ratio, by mass, of Ge to S
   (d) the number of grams of sulfur in 1 mol of the compound
   (e) the number of grams of compound required to contain 1.00 g Ge

5. The food flavor-enhancer monosodium glutamate (MSG) has the composition 13.6% Na, 35.5% C, 4.8% H, 8.3% N, 37.8% O, by mass. What is the empirical formula for MSG?

6. A swimming pool supply store sells three different brands of "liquid chlorine" for use in purifying water in home swimming pools. All cost $1 per gallon and all are water solutions of sodium hypochlorite. Brand A contains 10% hypochlorite (ClO) by mass; brand B, 7% available chlorine (Cl) by mass; and brand C, 14% sodium hypochlorite (NaClO) by mass. Which of the three brands would you buy?

7. The substance chlorophyll contains magnesium to the extent of 2.72% by mass. Assuming one Mg atom per chlorophyll molecule, what is the molecular weight of chlorophyll?

*8. When 2.750 g of the oxide of lead Pb_3O_4 is heated strongly, decomposition occurs, producing 0.0640 g of oxygen gas and 2.686 g of a second oxide of lead. What is the empirical formula of the second oxide?

*9. A gaseous hydrocarbon mixture consists of 60.0% by mass of C_3H_8 and 40.0% of a second hydrocarbon, C_xH_y. When 10.0 g of this mixture is burned, it yields 29.0 g CO_2 and 18.8 g H_2O as the only products. What is the formula of the unknown hydrocarbon?

*10. A 0.510-g mixture of methane, CH_4, and ethane, C_2H_6, is burned completely, yielding 1.468 g CO_2. What is the percent composition of this mixture (a) by mass; (b) on a mole basis?

*11. A thoroughly dried 1.00-g sample of Na_2SO_4 is exposed to the atmosphere for a period of time and found to gain 0.425 g in mass. What is the percent, by mass, of $Na_2SO_4 \cdot 10 H_2O$ in the resulting mixture of the hydrate and anhydrous compound?

*12. The atomic weight of bismuth is to be determined by converting the compound $Bi(C_6H_5)_3$ to Bi_2O_3. If 5.610 g $Bi(C_6H_5)_3$ yields 2.969 g Bi_2O_3, what is the atomic weight of Bi?

Self-Test Questions

For questions 1 through 5 select the single item that best completes each statement.

1. One *mole* of fluorine gas, F_2 (a) weighs 19.0 g; (b) contains 6.02×10^{23} F atoms; (c) contains 1.20×10^{24} F atoms; (d) weighs 6.02×10^{23} g.

2. Three of the following formulas might be either empirical (simplest) or molecular formulas, but one of the four must be a molecular formula. That one is (a) N_2O; (b) N_2O_4; (c) NH_3; (d) Mg_3N_2.

3. The compound $C_7H_7NO_2$ (a) contains 17 atoms per mole; (b) contains equal percentages of C and H, by mass; (c) contains twice the percent by mass of O as of N; (d) contains twice the percent by mass of N as of H.

4. The greatest number of N atoms is found in (a) 50.0 g N_2O; (b) 17 g NH_3; (c) 150 cm³ of liquid pyridine, C_6H_5N ($d = 0.983$ g/cm³); (d) 1 mol N_2.

5. The compound XF_3 is found to consist of 65% F, by mass. The *atomic weight* of X must be (a) 8; (b) 11; (c) 31; (d) 35.

6. The liquid $CHBr_3$ has a density of 2.89 g/cm³. What volume of this liquid should be measured out to contain a total of 3.40×10^{24} molecules of $CHBr_3$?

7. Supply the missing name or formula in each of the following.
   (a) CaI_2 = ____
   (b) ____   = iron(III) sulfate
   (c) ____   = sulfur trioxide
   (d) ____   = bromine pentafluoride
   (e) NH_4CN = ____
   (f) $Ca(ClO_2)_2$ = ____
   (g) ____   = lithium hydrogen carbonate

8. An important copper-containing mineral is the mixed hydroxide and carbonate, which is known as malachite, $CuCO_3 \cdot Cu(OH)_2$.
   (a) What is the percent Cu, by mass, in this mineral?
   (b) When malachite is heated strongly ("roasted"), gaseous carbon dioxide and water are driven off, yielding copper(II) oxide. What mass of copper(II) oxide, in grams, is produced *per kilogram* of malachite that is roasted?

9. A particular oxide of nitrogen has the percent composition, by mass, 25.9% N and 74.1% O. What is the empirical formula of this compound?

10. A certain hydrate of sodium sulfite contains almost exactly 50% H_2O, by mass. What must be the formula of this hydrate?

4

Stoichiometry II:
Chemical Reactions

Since chemical reactions are the essence of chemistry, a treatment of the principles of chemical reactions is central to a study of chemistry. It is important to realize, however, that a total understanding of chemical reactions is not quickly achieved. Instead, you should experience a growing familiarity with the nature of chemical reactions as we proceed through the text.

Our consideration of chemical reactions at this point concentrates on a few practical questions: What is a chemical reaction? How does one determine if a chemical reaction has occurred? How can a chemical reaction be described symbolically? What is the quantitative significance of a symbolic chemical equation? How does one use chemical reactions as a means of analyzing or synthesizing materials?

4-1 Experimental Evidence for Chemical Reactions

A chemical reaction is a process in which new chemical substances, called **products,** are produced from a set of original substances, called **reactants.** Sometimes the properties of the products are sufficiently different from those of the reactants that visual evidence is provided; we *see* something happen. But often the fact that new substances have been produced may be undetectable by human senses. Chemical analysis, sometimes employing sophisticated instruments, may be required to prove that a chemical reaction has indeed occurred.

Consider the addition of a water solution of silver nitrate to a water solution of sodium chloride. Both silver nitrate and sodium chloride are soluble in water, but the mixture yields a substance that is not. A voluminous white **precipitate** appears. The precipitate proves to be silver chloride. The **formation of a precipitate** provides visual evidence that a chemical reaction has occurred.

When a dilute solution of hydrochloric acid is added to solid calcium carbonate (marble chips), a vigorous effervescence occurs. Bubbles of carbon dioxide gas rise from the reaction mixture. The **evolution of a gas** signifies a chemical reaction.

A **color change** may also indicate that a chemical reaction has occurred. When the two colorless solutions, aqueous sodium sulfite and aqueous sulfuric acid, are added to a deep purple colored solution of potassium permanganate, the solution is decolorized (or at most retains a very pale pink color). Color in a substance originates from the nature of its chemical bonds. A change in color signifies an alteration of chemical bonds—a chemical reaction.

Many chemical reactions are accompanied by the **evolution or absorption of heat.** For example, the heat liberated when barium oxide reacts with a small quantity of water is sufficient to raise the temperature of the water to its boiling point. A reaction in which heat is liberated is called an **exothermic** reaction. Some chemical reactions proceed by *absorbing* heat from the surroundings; these are called **endothermic** reactions. More is said in later chapters about energy changes accompanying chemical reactions.

4-2 The Chemical Equation

A symbolic representation of a chemical reaction is called a **chemical equation.** In a chemical equation the formulas of the reactants are written on the left side and those of the products, on the right. The two sides are joined by an equal sign ($=$) or an arrow ($\rightarrow$). In writing a chemical equation it is usually necessary to proceed in three steps.

1. The names of the reactants and products are written down. The result is a word equation.
2. Chemical formulas are substituted for names. The result is called a skeleton equation.
3. The skeleton equation is *balanced*. This final step, the balancing of an equation, is a most important one if the equation is to be of maximum value.

These three steps are illustrated in the following example.

1. Word equation: nitrogen + oxygen $\longrightarrow$ nitrogen oxide
2. Skeleton equation: N_2 + O_2 $\longrightarrow$ NO
3. Balanced equation: N_2 + O_2 $\longrightarrow$ $2\,NO$

Molecules of nitrogen (N_2) and oxygen (O_2) each contain two atoms, and a molecule of nitrogen oxide contains one atom of nitrogen and one of oxygen. For every molecule of N_2 and O_2 entering into the reaction, not one, but *two* molecules of NO must be produced (otherwise, one atom each of N and O would disappear). In balancing an equation, **stoichiometric coefficients** (numbers) are placed in front of formulas in such a manner as to conform to the following idea.

In the process of balancing equations, only the coefficients may be changed, not the subscripts in the formulas.

The total number of atoms of each type remains unchanged in a chemical reaction; atoms can neither be created nor destroyed in a chemical reaction.

At this point you should be able to substitute the correct formulas, O_2, CO_2, and H_2O, for oxygen, carbon dioxide, and water, respectively. But you are not expected to know the formula for propane. The relationship between names and formulas for hydrocarbons is considered in Chapter 24.

Example 4-1 Propane gas, C_3H_8, is easily liquefied, stored, and transported for use as a fuel. When it burns completely, the sole products are carbon dioxide and water. Write a balanced equation to represent this combustion reaction.

word equation: propane + oxygen $\longrightarrow$ carbon dioxide + water

skeleton equation: C_3H_8 + O_2 $\longrightarrow$ CO_2 + H_2O

Since *three* C and *eight* H atoms are represented through the formula on the left side, the coefficients 3 and 4 are required on the right—three molecules of CO_2 contain *three* C atoms and four molecules of H_2O, *eight* H atoms.
Balance C:

$$C_3H_8 + O_2 \longrightarrow 3\,CO_2 + H_2O \quad \text{(equation not balanced)}$$

Balance H:

$$C_3H_8 + O_2 \longrightarrow 3\,CO_2 + 4\,H_2O \quad \text{(equation not balanced)}$$

The equation is still not balanced, however. The total number of O atoms on the right is 10—six in the three CO_2 molecules and four in the four H_2O molecules. The coefficient 5 is required on the left. (Five O_2 molecules contain *ten* O atoms.)

Balance O:

$$C_3H_8 + 5\,O_2 \longrightarrow 3\,CO_2 + 4\,H_2O \quad \text{(balanced)}$$

SIMILAR EXAMPLES: Exercises 4, 6.

Example 4-2 Methanol, CH_3OH, can be obtained from wood or coal and used as a fuel. Its complete combustion produces carbon dioxide and water. Write a balanced equation for this combustion reaction.

skeleton equation: $CH_3OH + O_2 \longrightarrow CO_2 + H_2O$ (equation not balanced)

Balance C:

The carbon atoms are balanced in the skeleton equation—one on each side.

Balance H:

$$CH_3OH + O_2 \longrightarrow CO_2 + 2\,H_2O \quad \text{(equation not balanced)}$$

Balance O:

Four O atoms are shown on the right side. To obtain 4 on the left side requires that one of the four be found in the molecule CH_3OH and the remaining three in oxygen molecules. ($\frac{3}{2}\,O_2$ means $1\frac{1}{2}$ O_2 molecules—3 O atoms.)

$$CH_3OH + \tfrac{3}{2}\,O_2 \longrightarrow CO_2 + 2\,H_2O \quad \text{(balanced)}$$

Final adjustment of coefficients

Although fractional coefficients are acceptable in some circumstances, general practice is to remove fractional coefficients by multiplying all coefficients in the balanced equation by the same whole number—in this case "2."

$$2\,CH_3OH + (2 \times \tfrac{3}{2})\,O_2 \longrightarrow 2\,CO_2 + (2 \times 2)\,H_2O$$

$$2\,CH_3OH + 3\,O_2 \longrightarrow 2\,CO_2 + 4\,H_2O \quad \text{(balanced)}$$

SIMILAR EXAMPLES: Exercises 4, 6.

The state of matter or physical form in which the reactants and products occur can also be represented in a chemical equation. This is done by using the following symbols.

(g) = gas   (l) = liquid   (s) = solid   (aq) = aqueous (water) solution

Thus, for the reaction of hydrogen and oxygen gases to form liquid water we may write

$$2\,H_2(g) + O_2(g) \longrightarrow 2\,H_2O(l) \tag{4.1}$$

NET IONIC EQUATIONS. The reaction of water solutions of silver nitrate and sodium chloride can be represented by the equation

$$AgNO_3(aq) + NaCl(aq) \longrightarrow AgCl(s) + NaNO_3(aq) \tag{4.2}$$

What are the actual species present in the aqueous solutions of $AgNO_3$, NaCl, and $NaNO_3$ described in equation (4.2)? Figure 3-2 shows that the pure compound NaCl is composed of ions—Na^+ and Cl^-. When this compound is dissolved in water, the ions become dissociated from one another. It is appropriate to think of Na^+ and Cl^- in water solution as if they were separate constituents. The compounds $AgNO_3$ and

---

Margin notes:

When one of the reactants or products is an element in the free (uncombined) form, it is generally helpful to balance that element last. In Example 4-1 this means balancing oxygen last.

This is the same procedure as multiplying through an algebraic equation by a constant.

The symbol (c) is sometimes used to represent the crystalline form of a substance, but the symbol (s) serves essentially the same purpose.

$NaNO_3$ are also ionic, and their ions become dissociated from one another in water solution. Thus, we may write an *ionic* equation.

$$Ag^+(aq) + \cancel{NO_3^-(aq)} + \cancel{Na^+(aq)} + Cl^-(aq) \longrightarrow$$
$$AgCl(s) + \cancel{Na^+(aq)} + \cancel{NO_3^-(aq)} \quad (4.3)$$

A still further refinement is to note that any species that appears on both sides of an equation is not directly involved in the reaction. The "spectator" ions in equation (4.3) can be eliminated (noted by cancellation signs), yielding a *net* ionic equation.

$$Ag^+(aq) + Cl^-(aq) \longrightarrow AgCl(s) \qquad (4.4)$$

The following equation shows that copper metal displaces silver metal from a solution containing silver ions.

$$Cu(s) + Ag^+(aq) \longrightarrow Cu^{2+}(aq) + Ag(s) \quad \text{(not balanced)} \qquad (4.5)$$

Although this ionic equation has the same number of atoms of each type on the two sides, *it is not balanced*. There must be a balance of electric charge as well. *Electric charge can neither be created nor destroyed in a chemical reaction.* In equation (4.5) one unit of positive charge is shown on the left and two on the right. This situation is corrected in equation (4.6).

$$Cu(s) + 2\,Ag^+(aq) \longrightarrow Cu^{2+}(aq) + 2\,Ag(s) \quad \text{(balanced)} \qquad (4.6)$$

Writing and balancing net ionic equations is an important skill for the student of chemistry to develop. This activity takes on more meaning, however, only as one learns how to distinguish ionic from covalent compounds, dissociated from undissociated molecules in aqueous solutions, and soluble from insoluble compounds. The concept of net ionic equations is introduced here simply to establish that electric charge as well as numbers of atoms must be balanced in a chemical equation. This principle is illustrated through Example 4-3.

Example 4-3 When hydrogen sulfide gas is passed into a water solution containing the ion Bi^{3+}, a dark brown precipitate of bismuth sulfide, Bi_2S_3, is formed, accompanied by an increase in the number of H^+ in solution. Balance the equation for this reaction.

Electric charges are shown in the following equation (it is written in ionic form). The final equation must show a balance both in numbers of atoms *and in electric charges*. The stepwise balancing of this equation is suggested below.

$$Bi^{3+}(aq) + H_2S(aq) \longrightarrow Bi_2S_3(s) + H^+(aq) \quad \text{(not balanced)}$$

Balance Bi:

$$2\,Bi^{3+}(aq) + H_2S(aq) \longrightarrow Bi_2S_3(s) + H^+(aq) \quad \text{(not balanced)}$$

Balance S:

$$2\,Bi^{3+}(aq) + 3\,H_2S(aq) \longrightarrow Bi_2S_3(s) + H^+(aq) \quad \text{(not balanced)}$$

Balance H:

$$2\,Bi^{3+}(aq) + 3\,H_2S(aq) \longrightarrow Bi_2S_3(s) + 6\,H^+(aq) \quad \text{(balanced)}$$

Proof of balance of electric charge

$$\underbrace{2 \times (+3)}_{\text{charge on } Bi^{3+}} = \underbrace{6 \times (+1)}_{\text{charge on } H^+}$$

left   right

SIMILAR EXAMPLE: Exercise 5.

Actually, an imbalance of electric charge is also a violation of the law of conservation of mass. Creation or destruction of charge corresponds to the creation or destruction of electrons, and electrons do have mass.

In this example electric charge was balanced automatically when balancing the atoms. This is not always the case. Ionic equations of greater complexity are considered in Chapter 19.

REACTION CONDITIONS To balance an equation, it is not necessary to know the conditions under which a reaction occurs. However, this information is necessary if one wishes to carry out the reaction in the laboratory or chemical plant. Reaction conditions are often written above or below the arrow. For example, the capital Greek letter delta, Δ, means that an elevated temperature is required; that is, the reaction mixture must be heated.

$$2\ Ag_2O(s) \xrightarrow{\Delta} 4\ Ag(s) + O_2(g)$$

Reaction conditions may be stated even more explicitly, as in the BASF (Badische Anilin- & Soda Fabrik) process for the synthesis of methanol from carbon monoxide and hydrogen. This reaction occurs at 350°C, under a total pressure of 340 atm, and on the surface of a mixture of ZnO and Cr_2O_3 (acting as a catalyst).

Gas pressure is discussed in Chapter 5; the function of a catalyst in a chemical reaction, in Chapter 13.

$$CO(g) + 2\ H_2(g) \xrightarrow[\substack{340\ atm \\ ZnO,\ Cr_2O_3}]{350°C} CH_3OH(g)$$

TYPES OF CHEMICAL REACTIONS. In discussing chemical reactions it is useful to establish categories of reactions. Three important types, all discussed fully later in the text, are precipitation reactions (Chapter 16), acid–base reactions (Chapters 17 and 18), and oxidation–reduction reactions (Chapter 19). Although these are classifications we wish ultimately to use, other descriptive terms are also available. The reactions described in Examples 4-1 and 4-2 are both combustion reactions. Carbon–hydrogen compounds (hydrocarbons) and carbon–hydrogen–oxygen compounds both produce carbon dioxide and water as their sole products when they burn completely in oxygen gas (recall the discussion of combustion analysis in Section 3-6).

Four other terms used to classify chemical reactions and examples of their use are

1. Combination (or synthesis) reactions in which a more complex substance is formed from two or more simpler substances (either elements or compounds).

$$2\ H_2(g) + O_2(g) \longrightarrow 2\ H_2O(l)$$

2. Decomposition reactions in which a substance is broken down into two or more simpler ones.

$$CaCO_3(s) \longrightarrow CaO(s) + CO_2(g)$$

3. Displacement (or single replacement) reactions in which one element replaces another in a compound. In the following reaction zinc displaces hydrogen.

$$Zn(s) + 2\ HCl(aq) \longrightarrow ZnCl_2(aq) + H_2(g)$$

4. Metathesis (or double replacement) reactions in which an exchange occurs between two reactants. In the following example, NO_3^- and Cl^- are exchanged between Ag^+ and Na^+. When combined, Ag^+ and Cl^- form insoluble AgCl.

$$AgNO_3(aq) + NaCl(aq) \longrightarrow AgCl(s) + NaNO_3(aq)$$

4-3 Quantitative Significance of the Chemical Equation

Because atoms can neither be created nor destroyed in a chemical reaction, *the total mass of all the reactants entering into a chemical reaction is equal to the total mass of all the products formed.* To illustrate this idea we proceed as follows.

The balanced equation

$$2\ H_2(g) + O_2(g) \longrightarrow 2\ H_2O(l)$$

means that

2 molecules H_2 + 1 molecule O_2 $\longrightarrow$ 2 molecules H_2O

or that

$2x$ molecules H_2 + x molecules O_2 $\longrightarrow$ $2x$ molecules H_2O

Suppose that we let $x = 6.02 \times 10^{23}$—Avogadro's number. Then x molecules is equal to *1 mol*. Thus, the balanced equation

$$2\,H_2(g) + O_2(g) \longrightarrow 2\,H_2O(l)$$

also means that

2 mol H_2 + 1 mol O_2 $\longrightarrow$ 2 mol H_2O

Mass of reactants	Mass of products
mass 2 mol H_2 + mass 1 mol O_2 $\longrightarrow$	mass 2 mol H_2O
$2 \times (2.02)$ g H_2 + 32.00 g O_2 $\longrightarrow$	$2 \times (18.02)$ g H_2O
36.04 g reactants $\longrightarrow$	36.04 g products

Can you see that this result—grams reactants = grams products—could not have been obtained if the original equation had not been balanced?

CONVERSION FACTORS FROM THE BALANCED EQUATION. A balanced equation contains a wealth of information that can be applied to calculations concerning chemical reactions. For example, for the reaction

$$2\,H_2(g) + O_2(g) \longrightarrow 2\,H_2O(l) \tag{4.1}$$

2 mol H_2O $\backsim$ 2 mol H_2

meaning that *two* moles H_2O are *produced* for every *two* moles H_2 *consumed;*

2 mol H_2O $\backsim$ 1 mol O_2

meaning that *two* moles H_2O are *produced* for every *one* mole O_2 *consumed;*

2 mol H_2 $\backsim$ 1 mol O_2

meaning that *two* moles H_2 are *consumed* for every *one* mole O_2 *consumed.*

Two examples illustrating the use of conversion factors from the chemical equation follow.

Example 4-4   How much H_2O, in moles, results from burning an excess of H_2 in 3.3 mol O_2? (Refer to equation 4.1.)

The statement "an excess of H_2" signifies that there is more than enough H_2 available to permit the complete conversion of 3.3 mol O_2 to H_2O. The necessary conversion factor is derived from the expression 2 mol H_2O $\backsim$ 1 mol O_2.

$$\text{no. mol } H_2O = 3.3 \text{ mol } O_2 \times \frac{2 \text{ mol } H_2O}{1 \text{ mol } O_2} = 6.6 \text{ mol } H_2O$$

SIMILAR EXAMPLES: Exercises 7, 8.

Example 4-5   What mass of H_2, in grams, must react with excess O_2 to produce 5.40 g H_2O?

Several new features are involved in this problem.

1. O_2 is in excess instead of H_2.
2. Information is sought about the required quantity of one of the *reactants* (H_2) instead of the product (H_2O).
3. Information is given and sought in the unit *gram* rather than mole.

A more accurate form for equation (4.7) would be the ionic form, since HCl(aq) is dissociated into H^+ and Cl^-, and $AlCl_3$(aq) into Al^{3+} and Cl^-. In this case Cl^- becomes a mere "spectator" ion and the balanced equation is

$$2\,Al(s) + 6\,H^+(aq) \longrightarrow \\ 2\,Al^{3+}(aq) + 3\,H_2(g)$$

Our principal reason for not using the ionic form is that we have not yet learned how to predict which species in aqueous solution are present as ions.

FIGURE 4-1
The reaction $2\,Al(s) + 6\,HCl(aq) \longrightarrow$
$2\,AlCl_3(aq) + 3\,H_2(g)$.

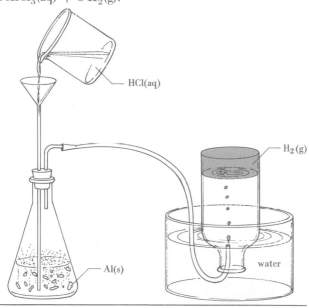

HCl(aq) is introduced to the flask on the left through a long funnel. The reaction of HCl(aq) and Al(s) occurs within the flask. The liberated H_2(g) is conducted to a gas collection apparatus where it displaces water. Hydrogen is only very slightly soluble in water.

Two factors relating grams and moles and one factor derived from the balanced equation are required in the conversions: $g\,H_2O \rightarrow mol\,H_2O \rightarrow mol\,H_2 \rightarrow g\,H_2$.

$$\text{no. g}\,H_2 = 5.40\ g\,H_2O \times \frac{1\ mol\,H_2O}{18.0\ g\,H_2O}$$

$$\times \frac{2\ mol\,H_2}{2\ mol\,H_2O} \times \frac{2.02\ g\,H_2}{1\ mol\,H_2}$$

$$= 0.606\ g\,H_2$$

SIMILAR EXAMPLES: Exercises 9, 10.

To continue our discussion of calculations based on the chemical equation, we shift our attention to the new reaction pictured in Figure 4-1.

$$2\,Al(s) + 6\,HCl(aq) \longrightarrow \\ 2\,AlCl_3(aq) + 3\,H_2(g) \quad (4.7)$$

Example 4-6 A small piece of pure aluminum metal having a volume of $0.650\ cm^3$ reacts with an excess of hydrochloric acid. What is the mass of hydrogen obtained? (The density of aluminum is $2.70\ g/cm^3$.)

No matter how we choose to solve this problem, our focus must be on equation (4.7), which provides the all-important conversion relationship $3\ mol\,H_2 \leftrightharpoons 2\ mol\,Al$. But there are several other factors required as well, resulting in the overall conversion: $cm^3\,Al \rightarrow g\,Al \rightarrow mol\,Al \rightarrow mol\,H_2 \rightarrow g\,H_2$.

$$\text{no. g}\,H_2 = 0.650\ cm^3\,Al \times \frac{2.70\ g\,Al}{cm^3\,Al} \times \frac{1\ mol\,Al}{27.0\ g\,Al}$$

$$(cm^3\,Al \quad \rightarrow \quad g\,Al \quad \rightarrow \quad mol\,Al$$

$$\times \frac{3\ mol\,H_2}{2\ mol\,Al} \times \frac{2.02\ g\,H_2}{1\ mol\,H_2}$$

$$\rightarrow \quad mol\,H_2 \quad \rightarrow \quad g\,H_2)$$

$$= 0.197\ g\,H_2$$

SIMILAR EXAMPLE: Exercise 14.

Example 4-7 An alloy consisting of 95.0% Al and 5.0% Cu, by mass, is to be used in reaction (4.7) to produce H_2(g). Assuming that all of the Al and none of the Cu dissolves, what mass of the aluminum alloy is required to produce $1.75\ g\,H_2$?

Here we are told the quantity of a product that is desired and we must calculate the quantity of a reactant required. However, this reactant is not pure. That is, the alloy that reacts with HCl(aq) is only 95.0% Al. The percent composition of the alloy supplies an essential conversion factor.

Because the alloy contains only 95.0% Al, the mass of alloy required for the reaction must be *greater* than the mass of Al consumed. The factor to convert from g Al to g alloy must have a value greater than 1. This fact, together with the requirement of the cancellation of units, guides us in writing: 100 g alloy/95.0 g Al (not 95.0 g Al/100 g alloy).

$$\text{no. g alloy} = 1.75\,\text{g H}_2 \times \frac{1\,\text{mol H}_2}{2.02\,\text{g H}_2} \times \frac{2\,\text{mol Al}}{3\,\text{mol H}_2} \times \frac{27.0\,\text{g Al}}{1\,\text{mol Al}} \times \frac{100\,\text{g alloy}}{95.0\,\text{g Al}}$$

$$(\text{g H}_2 \quad\rightarrow\quad \text{mol H}_2 \quad\rightarrow\quad \text{mol Al} \quad\rightarrow\quad \text{g Al} \quad\rightarrow\quad \text{g alloy})$$

$$= 16.4\,\text{g alloy}$$

SIMILAR EXAMPLES: Exercises 11, 12, 13.

Example 4-8 What is the minimum volume of a hydrochloric acid solution, 26.9% HCl by mass with a density of 1.14 g/cm^3, required to dissolve 1.00 g Al in reaction (4.7)?

Here we compare one reactant (Al) to another (HCl). The required conversions are

$$\text{g Al} \longrightarrow \text{mol Al} \longrightarrow \text{mol HCl} \longrightarrow \text{g HCl} \longrightarrow \text{g HCl soln} \longrightarrow \text{cm}^3\,\text{HCl soln}$$

$$\text{no. cm}^3\,\text{HCl soln} = 1.00\,\text{g Al} \times \frac{1\,\text{mol Al}}{27.0\,\text{g Al}} \times \frac{6\,\text{mol HCl}}{2\,\text{mol Al}} \times \frac{36.5\,\text{g HCl}}{1\,\text{mol HCl}}$$

$$(\text{g Al} \quad\rightarrow\quad \text{mol Al} \quad\rightarrow\quad \text{mol HCl} \quad\rightarrow\quad \text{g HCl}$$

$$\times \frac{100\,\text{g HCl soln}}{26.9\,\text{g HCl}} \times \frac{1\,\text{cm}^3\,\text{HCl soln}}{1.14\,\text{g HCl soln}}$$

$$\rightarrow \quad \text{g HCl soln} \quad\rightarrow\quad \text{cm}^3\,\text{HCl soln})$$

$$= 13.2\,\text{cm}^3\,\text{HCl soln}$$

SIMILAR EXAMPLE: Exercise 23b.

4-4 Chemical Reactions in Solutions

We need to explore more fully an important aspect of chemical reactions: *Some of the reactants and/or products may exist in solution.* From Section 1-2 we recall that a solution is a *homogeneous* mixture of two or more substances (elements or compounds). That is, a solution has uniform composition and properties throughout. One solution component, the one that determines whether the mixture will exist as a solid, liquid, or gas, is called the **solvent.** The other component(s) are called **solute(s).** The symbolism NaCl(aq), for example, describes a solution in which water is the solvent and sodium chloride, the solute. In seawater, water is again the solvent, but there are many solutes, of which NaCl is simply the most abundant.

The quantity of solute that can be dissolved in a solvent varies widely. As a result, it is necessary to specify the exact composition or concentration of a solution if calculations are to be made on chemical reactions in solution. In Example 4-8 information about the composition of a particular hydrochloric acid solution was given in terms of the solution density and its percent composition by mass. But this is generally not the most convenient way in which to describe the composition of a solution. More useful is a concentration unit based, as is the balanced chemical equation itself, on the concept of the mole.

In the SI system the term liter (L) is discouraged, and its equivalent, the cubic decimeter (dm^3), has been adopted. Thus, in SI units, molar concentration is expressed as mol dm^{-3}. Several other concentration units and their usage are introduced in Chapter 12.

MOLAR CONCENTRATION (MOLARITY). The composition or concentration of a solution expressed as the *number of moles of solute per liter of solution* is called the **molar concentration** or **molarity (M).**

$$\text{molar concentration }(M) = \frac{\text{number of moles solute}}{\text{number liters solution}} \tag{4.8}$$

For example, 0.500 mol of urea, CO(NH$_2$)$_2$, dissolved in 1.000 L of water solution corresponds to a molar concentration of

FIGURE 4-2
Preparation of
0.250 M Na$_2$SO$_4$—
Example 4-9 illustrated.

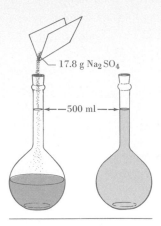

17.8 g Na$_2$SO$_4$

500 ml

The desired solution requires that 0.125 mol Na$_2$SO$_4$ (17.8 g) be dissolved in exactly 500 ml of water solution. One type of container used in the preparation of solutions is the volumetric flask. The flask pictured here contains 500.0 $\pm$ 0.2 ml when filled to the mark. The procedure is to dissolve the solute in less than 500 ml of water and, when dissolving is complete, to fill the container exactly to the mark with water.

$$\frac{0.500 \text{ mol CO(NH}_2)_2}{1.000 \text{ L soln}} = 0.500 \ M \text{ CO(NH}_2)_2$$

and 0.300 mol of ethanol, C$_2$H$_5$OH, dissolved in 100.0 cm^3 (0.1000 L) of aqueous solution corresponds to a molar concentration of

$$\frac{0.300 \text{ mol C}_2\text{H}_5\text{OH}}{0.1000 \text{ L soln}} = 3.00 \ M \text{ C}_2\text{H}_5\text{OH}$$

The preparation of solutions on the molar concentration scale is not quite as simple a matter as is implied by the two preceding examples, however. Molar quantities cannot be measured out directly; they must be related to other measurements, usually mass or volume.

Example 4-9 It is desired to prepare exactly 500.0 ml of an 0.250 M Na$_2$SO$_4$ solution in water. What is the mass of Na$_2$SO$_4$ required for this purpose? (See Figure 4-2.)

The information given is a solution volume and concentration; the information sought is the number of grams of solute. In the standard problem-solving method the required conversions are from ml soln $\rightarrow$ L soln $\rightarrow$ mol Na$_2$SO$_4$ $\rightarrow$ g Na$_2$SO$_4$. Molar concentration is a conversion factor between the volume of solution in liters and the number of moles of solute dissolved in solution. That is, molarity is a conversion factor derived from the chemical equivalence 1 L soln $\backsim$ 0.250 mol Na$_2$SO$_4$.

An alternative method is based on the definition of molar concentration. Substitute into equation (4.8) values for the molarity (M) and the solution volume (V). Solve for the number of moles of solute.

$$\text{no. g Na}_2\text{SO}_4 = 500.0 \text{ ml soln} \times \frac{1 \text{ L soln}}{1000 \text{ ml soln}} \times \frac{0.250 \text{ mol Na}_2\text{SO}_4}{\text{L soln}} \times \frac{142 \text{ g Na}_2\text{SO}_4}{1 \text{ mol Na}_2\text{SO}_4}$$

$$\text{(ml soln} \quad \rightarrow \quad \text{L soln} \quad \rightarrow \quad \text{mol Na}_2\text{SO}_4 \quad \rightarrow \quad \text{g Na}_2\text{SO}_4)$$

$$= 17.8 \text{ g Na}_2\text{SO}_4$$

SIMILAR EXAMPLE: Exercise 17.

Example 4-10 What volume of pure ethanol, C$_2$H$_5$OH ($d = 0.789$ g/cm^3), must be dissolved in water to produce exactly 250.0 ml of 0.150 M C$_2$H$_5$OH?

Our solution takes the form of the method used in Example 4-9, but we must add a factor to convert from g C$_2$H$_5$OH to cm^3 C$_2$H$_5$OH. This factor is derived from the density of C$_2$H$_5$OH.

$$\text{no. cm}^3 \text{ C}_2\text{H}_5\text{OH} = 250.0 \text{ ml soln} \times \frac{1 \text{ L soln}}{1000 \text{ ml soln}} \times \frac{0.150 \text{ mol C}_2\text{H}_5\text{OH}}{\text{L soln}}$$

$$\times \frac{46.1 \text{ g C}_2\text{H}_5\text{OH}}{1 \text{ mol C}_2\text{H}_5\text{OH}} \times \frac{1.00 \text{ cm}^3 \text{ C}_2\text{H}_5\text{OH}}{0.789 \text{ g C}_2\text{H}_5\text{OH}}$$

$$= 2.19 \text{ cm}^3 \text{ C}_2\text{H}_5\text{OH}$$

SIMILAR EXAMPLES: Exercises 17, 19.

FIGURE 4-3
Preparing a solution by dilution—Example 4-12 illustrated.

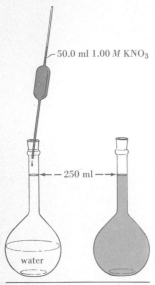

— 50.0 ml 1.00 M KNO$_3$

— 250 ml —

water

250 ml 0.200 M KNO$_3$

A pipet is used to dispense 50.0 ml of 1.00 M KNO$_3$ into a small quantity of water in a 250.0-ml volumetric flask. Following this, more water is added to bring the solution volume to the mark on the flask.

In the preparation of a solution by dilution, fairly simple volumetric glassware is used; no mass determinations are required (except in the preparation of the original stock solution).

Many practical applications dealing with solutions require that equation (4.8) be used two (or more) times. This is the case, for example, if two solutions are mixed and the final concentration is to be calculated, as in Example 4-11.

Example 4-11   24.5 ml of 1.50 M NaOH is added to 20.5 ml of 0.850 M NaOH. What is the molarity of the final solution? (Assume that the volumes are additive, that is, that the final solution volume is exactly 45.0 ml.)

We determine the number of moles of NaOH in each solution, add them together, and divide by the final volume (in liters).

$$\text{no. mol NaOH} = \left(0.0245 \text{ L soln} \times \frac{1.50 \text{ mol NaOH}}{\text{L soln}} \right)$$

$$+ \left(0.0205 \text{ L soln} \times \frac{0.850 \text{ mol NaOH}}{\text{L soln}} \right)$$

$$= (0.0368 + 0.0174) = 0.0542 \text{ mol NaOH}$$

$$\text{molar concentration of NaOH} = \frac{0.0542 \text{ mol NaOH}}{0.0450 \text{ L soln}} = 1.20 \text{ } M \text{ NaOH}$$

SIMILAR EXAMPLE: Exercise 18.

More common than the example just considered is the situation described in Example 4-12, in which a desired solution is prepared by adding water to a more concentrated solution. This procedure is often used in the chemical laboratory, where stock solutions of fairly high concentrations are stored and other solutions prepared by an appropriate dilution. The basic principle of this method, illustrated through Figure 4-3, is that

all the solute present in the original, more concentrated solution
appears in the final, diluted solution.   (4.9)

Statement (4.9) is all that is needed (together with the definition of molar concentration) in working dilution problems. However, some prefer a method based on an extension of equation (4.8). Since

$$\text{molar concentration } (M) = \frac{\text{number of moles of solute}}{\text{volume } (V) \text{ of soln in L}}$$

then

$$\text{no. mol solute} = \text{molarity } (M) \times \text{volume } (V)$$

When a solution is diluted, the number of moles of solute *remains constant* between the original solution and the final, diluted solution. Thus,

$$M_{\text{orig}} \times V_{\text{orig}} = \text{no. mol solute} = M_{\text{final}} \times V_{\text{final}}$$

and

$$M_{\text{orig}} \times V_{\text{orig}} = M_{\text{final}} \times V_{\text{final}} \tag{4.10}$$

Example 4-12. What volume of 1.000 M KNO_3 must be diluted with water to prepare 250.0 ml of 0.200 M KNO_3? (See Figure 4-3.)

Consider the two solutions separately. First, we calculate the amount of solute that must be present in the *final solution*.

$$\text{no. mol } KNO_3 = 250.0 \text{ ml soln} \times \frac{1 \text{ L soln}}{1000 \text{ ml soln}} \times \frac{0.200 \text{ mol } KNO_3}{\text{L soln}}$$

$$= 0.0500 \text{ mol } KNO_3$$

An alternative solution based on equation (4.10) requires substitutions for M_{orig}, M_{final}, and V_{final}.

$(1.000 \ M) \times V_{orig}$
$\quad = 0.200 \ M \times 0.250 \text{ L}$

V_{orig}
$\quad = \dfrac{0.200 \ M}{1.000 \ M} \times 0.250 \text{ L}$
$\quad = 0.0500 \text{ L} = 50.0 \text{ ml}$

Since all the solute in the final, diluted solution must come from the original, more concentrated solution, let us now ask this question: What volume of 1.000 M KNO_3 must be taken to contain 0.0500 mol KNO_3?

$$\text{no. ml soln} = 0.0500 \text{ mol } KNO_3 \times \frac{1 \text{ L soln}}{1.000 \text{ mol } KNO_3} \times \frac{1000 \text{ ml soln}}{1 \text{ L soln}}$$

$$= 50.0 \text{ ml soln}$$

SIMILAR EXAMPLES: Exercises 20, 21.

SOLUTION STOICHIOMETRY. Several of the ideas considered in this chapter can now be combined to answer the question posed in Example 4-13.

Example 4-13   What volume of 0.1060 M $AgNO_3$(aq) must react with 10.00 ml of 0.09720 M K_2CrO_4(aq) to precipitate all the chromate as Ag_2CrO_4?

$2 \ AgNO_3(aq) + K_2CrO_4(aq) \longrightarrow Ag_2CrO_4(s) + 2 \ KNO_3(aq)$

This three-step approach is perhaps simplest: (1) Determine the number of moles of K_2CrO_4 that react. (2) Use the balanced equation to determine the number of moles of $AgNO_3$ required to react with the K_2CrO_4. (3) Calculate the volume of 0.1060 M $AgNO_3$ containing the required amount of $AgNO_3$.

Step 1:

$$\text{no. mol } K_2CrO_4 = 0.01000 \text{ L soln} \times \frac{0.09720 \text{ mol } K_2CrO_4}{\text{L soln}}$$

$$= 9.720 \times 10^{-4} \text{ mol } K_2CrO_4$$

Step 2:

$$\text{no. mol } AgNO_3 = 9.720 \times 10^{-4} \text{ mol } K_2CrO_4 \times \frac{2 \text{ mol } AgNO_3}{1 \text{ mol } K_2CrO_4}$$

$$= 1.944 \times 10^{-3} \text{ mol } AgNO_3$$

Step 3:

$$\text{no. ml } AgNO_3(aq) = 1.944 \times 10^{-3} \text{ mol } AgNO_3 \times \frac{1 \text{ L soln}}{0.1060 \text{ mol } AgNO_3} \times \frac{1000 \text{ ml soln}}{1 \text{ L soln}}$$

$$= 18.34 \text{ ml } AgNO_3(aq)$$

SIMILAR EXAMPLES: Exercises 22, 23a.

A more common situation in the chemical laboratory is that described in Example 4-14. A chemical reaction is carried out between two solutions, one of known concentration and the other, unknown. The experimental data can be used to determine the concentration of the unknown.

Example 4-14   The electrolyte in a lead storage battery is an aqueous solution of H_2SO_4 (called sulfuric acid). A 5.00-ml sample of a battery acid requires 46.40 ml of 0.875 M

FIGURE 4-4
An acid-base titration—
Example 4-14 illustrated.

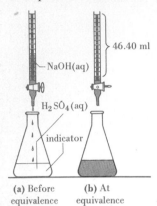

(a) Before
equivalence
point

(b) At
equivalence
point

The 5.00 ml H_2SO_4(aq) that is to participate in reaction (4.11) is added to a flask and diluted with water. A few drops of an appropriate acid-base indicator (phenolphthalein) are added. A solution of 0.875 M NaOH is contained in a long graduated tube from which the flow of liquid can be controlled with a stopcock. This device is called a buret. The buret is filled to the 0.00 ml mark.
(a) Solution from the buret is added, first rapidly and then dropwise.
(b) At the precise point where the reaction of the H_2SO_4 has been completed (the equivalence point) the acid-base indicator changes color. The buret reading at this point (46.40 ml) gives the volume of 0.875 M NaOH required for the titration.

NaOH for its complete reaction (neutralization). What is the molar concentration of H_2SO_4 in the acid? (See Figure 4-4.)

$$H_2SO_4(aq) + 2\,NaOH(aq) \longrightarrow Na_2SO_4(aq) + 2\,H_2O(l) \tag{4.11}$$

Let us again consider a three-step approach: (1) Determine the number of moles of NaOH in 46.40 ml of 0.875 M NaOH. (2) Determine the number of moles of H_2SO_4 that react with this NaOH. (3) Calculate the molarity of the H_2SO_4(aq).

Step 1:

$$\text{no. mol NaOH} = 46.40 \text{ ml} \times \frac{1 \text{ L}}{1000 \text{ ml}} \times \frac{0.875 \text{ mol NaOH}}{\text{L}} = 0.04060 \text{ mol NaOH}$$

Step 2:

$$\text{no. mol } H_2SO_4 = 0.04060 \text{ mol NaOH} \times \frac{1 \text{ mol } H_2SO_4}{2 \text{ mol NaOH}} = 0.02030 \text{ mol } H_2SO_4$$

Step 3:
Since the 0.02030 mol H_2SO_4 is derived from a 5.00-ml (0.00500-L) sample,

$$\text{molar concentration} = \frac{0.02030 \text{ mol } H_2SO_4}{0.00500 \text{ L soln}} = 4.06 \, M \, H_2SO_4$$

SIMILAR EXAMPLES: Exercises 24, 25.

The calculation involved in Example 4-14 is not difficult, but the experimental procedure necessary to obtain the data for the calculation—a procedure called **titration**— is rather exacting. In Example 4-14, how can we be assured that as NaOH(aq), a colorless solution, is slowly added to the 5.00 ml of H_2SO_4(aq), also a colorless solution, the reaction is completed when exactly 46.40 ml has been added—and not 46.35 or 46.45 or 46.50 . . . ? This can be done by having present in the H_2SO_4(aq) a trace of a substance called an **indicator,** which changes color at the precise point (called the equivalence point) where all the H_2SO_4(aq) has reacted. Thus, the key to titration reactions is knowing how an indicator works and being able to select an appropriate indicator. We will return to this subject in Sections 18-3 and 18-4.

4-5 Some Complexities in the Stoichiometry of Chemical Reactions

The majority of situations dealing with the stoichiometry of chemical reactions can be handled by the methods introduced up to this point. There are, however, some additional points that need to be considered.

DETERMINING THE LIMITING REAGENT. Typically, examples presented previously have included statements as to which reactant (also called a reagent) was in excess. This was done to indicate that the outcome of the reaction was determined by another reactant. That is, some of the reactant in excess remained after the reaction was completed. On the other hand, the reactant that determined the outcome—the limiting reagent—was consumed completely. Instances may occur in which the limiting reagent is not indicated explicitly. In such cases the limiting reagent must be determined by calculation. The principle involved is illustrated in Figure 4-5.

Example 4-15 What is the number of moles of $Fe(OH)_3$(s) that can be produced by allowing 1.0 mol Fe_2S_3, 2.0 mol H_2O, and 3.0 mol O_2 to react?

$$2\,Fe_2S_3(s) + 6\,H_2O(l) + 3\,O_2(g) \longrightarrow 4\,Fe(OH)_3(s) + 6\,S(s)$$

FIGURE 4-5
An analogy to determining the limiting reagent in a chemical reaction—assembling a hand out experiment.

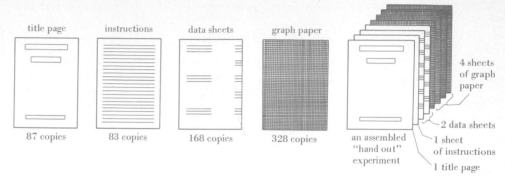

From the number of copies of each type of sheet (analogous to moles of reactants) calculate how many complete hand outs (analogous to moles of products) can be assembled. Do you get 82? Which is the "limiting reagent?"

Based on the title page we would say that no more than 87 complete copies of the hand out are possible, but based on the instruction page we would say no more than 83. Two data sheets are required per hand out. There are enough data sheets for $168/2 = 84$ hand outs, but no more than 83 hand outs are possible because of the limited number of instruction sheets. Finally, because four sheets of graph paper are required per hand out, we conclude that only $328/4 = 82$ hand outs are possible. The excess pages are the title page (5 copies), the instruction sheet (one copy), and the data page (4 copies).

The number of moles of Fe_2S_3 (1.0 mol) is less than of H_2O (2.0 mol) and O_2 (3.0 mol), but this *does not* automatically mean that Fe_2S_3 is the limiting reagent. *The amounts of the three reactants must be compared to the proportions in which they combine.*

$$2 \text{ mol } Fe_2S_3 \backsimeq 6 \text{ mol } H_2O \backsimeq 3 \text{ mol } O_2$$

For example, the reaction of 3.0 mol O_2 requires this much H_2O.

$$\text{no. mol } H_2O = 3 \text{ mol } O_2 \times \frac{6 \text{ mol } H_2O}{3 \text{ mol } O_2} = 6 \text{ mol } H_2O$$

Only 2.0 mol H_2O is available. Some of the O_2 must remain unreacted. *$O_2(g)$ is in excess.* Now that we have established that $O_2(g)$ is in excess, is there a sufficient amount of Fe_2S_3 available to react with 2.0 mol H_2O?

$$\text{no. mol } Fe_2S_3 = 2.0 \text{ mol } H_2O \times \frac{2 \text{ mol } Fe_2S_3}{6 \text{ mol } H_2O} = 0.67 \text{ mol } Fe_2S_3$$

There is 1.0 mol Fe_2S_3 available—more than is required to react with the available H_2O. *Fe_2S_3 is also in excess, and H_2O is the limiting reagent.* The amount of $Fe(OH)_3$ that will be obtained then is determined by the amount of H_2O available.

$$\text{no. mol } Fe(OH)_3 = 2.0 \text{ mol } H_2O \times \frac{4 \text{ mol } Fe(OH)_3}{6 \text{ mol } H_2O} = 1.3 \text{ mol } Fe(OH)_3$$

SIMILAR EXAMPLES: Exercises 28, 29.

Another approach to this question takes the exact form outlined in Figure 4-5. That is, if there were an excess of H_2O and O_2, 2.0 mol $Fe(OH)_3$ would be formed. With an excess of Fe_2S_3 and O_2, 1.3 mol $Fe(OH)_3$ would be formed; and with an excess of Fe_2S_3 and H_2O, 4.0 mol $Fe(OH)_3$. The correct answer is the smallest of the three—1.3 mol $Fe(OH)_3$.

A more common situation is that of Example 4-16. Here the quantities of the available reactants must be converted to a mole basis before a comparison can be made to identify the limiting reagent.

Example 4-16 What mass of PbI_2 will precipitate if 2.85 g $Pb(NO_3)_2$ is added to 225 ml of 0.0550 *M* KI(aq)?

$$Pb(NO_3)_2(aq) + 2\,KI(aq) \longrightarrow PbI_2(s) + 2\,KNO_3(aq)$$

$$\text{no. mol Pb(NO}_3)_2 \text{ available} = 2.85 \text{ g Pb(NO}_3)_2 \times \frac{1 \text{ mol Pb(NO}_3)_2}{331 \text{ g Pb(NO}_3)_2}$$

$$= 8.61 \times 10^{-3} \text{ mol Pb(NO}_3)_2$$

$$\text{no. mol KI available} = 0.225 \text{ L} \times \frac{0.0550 \text{ mol KI}}{\text{L}} = 1.24 \times 10^{-2} \text{ mol KI}$$

Now let us determine the number of moles of KI required to react with 8.61×10^{-3} mol $Pb(NO_3)_2$.

$$\text{no. mol KI required} = 8.61 \times 10^{-3} \text{ mol Pb(NO}_3)_2 \times \frac{2 \text{ mol KI}}{1 \text{ mol Pb(NO}_3)_2}$$

$$= 1.72 \times 10^{-2} \text{ mol KI}$$

There is only 1.24×10^{-2} mol KI available. *KI is the limiting reagent.* The final step is to calculate the mass of $PbI_2(s)$ produced by the reaction of 1.24×10^{-2} mol KI with an excess of $Pb(NO_3)_2$.

$$\text{no. g PbI}_2 = 1.24 \times 10^{-2} \text{ mol KI} \times \frac{1 \text{ mol PbI}_2}{2 \text{ mol KI}} \times \frac{461 \text{ g PbI}_2}{1 \text{ mol PbI}_2} = 2.86 \text{ g PbI}_2$$

SIMILAR EXAMPLES: Exercises 29, 31.

SIMULTANEOUS AND CONSECUTIVE REACTIONS.   Some stoichiometric calculations require that two or more chemical equations be used, each equation furnishing a conversion factor. In some cases the reactions occur at the same time (simultaneously) and in others they occur in succession (consecutively). Example 4-17 deals with a pair of simultaneous reactions, and Example 4-18, with three consecutive reactions.

Example 4-17   A 0.710-g sample of a magnalium alloy consisting of 70% Al and 30% Mg reacts with an excess of HCl(aq). What mass of $H_2(g)$ is produced?

$$2\,Al(s) + 6\,HCl(aq) \longrightarrow 2\,AlCl_3(aq) + 3\,H_2(g)$$

$$Mg(s) + 2\,HCl(aq) \longrightarrow MgCl_2(aq) + H_2(g)$$

These equations indicate how each of the metals reacts with hydrochloric acid. We need to determine the mass of each metal, the mass of hydrogen produced by each metal, and then the sum of the two masses of hydrogen.
Step 1:

$$\text{no. g Al} = 0.710 \text{ g alloy} \times \frac{70 \text{ g Al}}{100 \text{ g alloy}} = 0.50 \text{ g Al}$$

$$\text{no. g Mg} = 0.710 \text{ g alloy} \times \frac{30 \text{ g Mg}}{100 \text{ g alloy}} = 0.21 \text{ g Mg}$$

Step 2:

$$\text{no. g H}_2 = 0.50 \text{ g Al} \times \frac{1 \text{ mol Al}}{27.0 \text{ g Al}} \times \frac{3 \text{ mol H}_2}{2 \text{ mol Al}} \times \frac{2.02 \text{ g H}_2}{1 \text{ mol H}_2} = 0.056 \text{ g H}_2$$

$$\text{no. g H}_2 = 0.21 \text{ g Mg} \times \frac{1 \text{ mol Mg}}{24.3 \text{ g Mg}} \times \frac{1 \text{ mol H}_2}{1 \text{ mol Mg}} \times \frac{2.02 \text{ g H}_2}{1 \text{ mol H}_2} = 0.017 \text{ g H}_2$$

Step 3:

$$\text{total no. g H}_2 = 0.056 \text{ g H}_2 + 0.017 \text{ g H}_2 = 0.073 \text{ g H}_2$$

SIMILAR EXAMPLES: Exercises 32, 33.

Example 4-18   Sodium chlorate, $NaClO_3$, can be produced as follows:

$$2\,KMnO_4 + 16\,HCl \longrightarrow 2\,KCl + 2\,MnCl_2 + 8\,H_2O + 5\,Cl_2$$

$$6\,Cl_2 + 6\,Ca(OH)_2 \longrightarrow Ca(ClO_3)_2 + 5\,CaCl_2 + 6\,H_2O$$

$$Ca(ClO_3)_2 + Na_2SO_4 \longrightarrow CaSO_4 + 2\,NaClO_3$$

Assuming an excess of all other reactants, how many grams $NaClO_3$ can be prepared from 42.5 g HCl?

The key substances are shown in color in the equations. One approach would be to determine, as a first step, the quantity of Cl_2 produced in the first reaction from the given quantity of HCl (the limiting reagent). In the second step, one would determine the quantity of $Ca(ClO_3)_2$ obtained if all the Cl_2 produced in the first reaction entered into the second reaction. And as a final step, one would determine the quantity of $NaClO_3$ derived in the third reaction from the $Ca(ClO_3)_2$. Actually, it is not necessary to solve for any intermediate answers. The three steps are combined into the single setup shown. An important point to note in this calculation is that not all of the Cl associated with HCl in the first step appears in $NaClO_3$ in the third step. Some appears as $MnCl_2$ in the first step, and a good deal more as $CaCl_2$ in the second step.

In this setup the calculation through point (*a*) represents the number of moles of Cl_2 produced in the first reaction; through (*b*), the number of moles of $Ca(ClO_3)_2$ in the second reaction; and through (*c*), the number of moles of $NaClO_3$ in the third.

$$\text{no. g NaClO}_3 = 42.5\ \text{g HCl} \times \frac{1\ \text{mol HCl}}{36.5\ \text{g HCl}} \times \underbrace{\frac{5\ \text{mol Cl}_2}{16\ \text{mol HCl}}}_{(a)} \times \underbrace{\frac{1\ \text{mol Ca(ClO}_3)_2}{6\ \text{mol Cl}_2}}_{(b)}$$

$$\times \underbrace{\frac{2\ \text{mol NaClO}_3}{1\ \text{mol Ca(ClO}_3)_2}}_{(c)} \times \frac{106\ \text{g NaClO}_3}{1\ \text{mol NaClO}_3}$$

$$= 12.9\ \text{g NaClO}_3$$

SIMILAR EXAMPLES: Exercises 35, 36, 37.

4-6 Industrial Chemistry

Large-scale manufacturing processes for the production of chemicals employ the same principles as those considered in this chapter and elsewhere in the text. Yet many factors must be considered that go beyond the usual textbook examples. In this concluding section of Chapter 4 we attempt to capture some of the flavor of industrial chemistry.

HYDRAZINE AND ITS USES.   Hydrazine is a nitrogen–hydrogen compound having the formula N_2H_4. It is an oily, colorless liquid that freezes at 1.5°C and boils at 113.5°C. Liquid hydrazine has a density of 1.0083 g/cm³ at 20°C. The nature of chemical bonding in hydrazine is discussed in Section 9-4, and its physical and chemical properties are explored further in Section 20-7.

The principal use of hydrazine and certain compounds derived from it is as rocket fuels, but it is also used in fuel cells (see Section 19-8), in the treatment of water in boilers to remove dissolved oxygen gas, and in the plastics industry.

One widely used method for the manufacture of hydrazine (the Raschig process) is illustrated in Figure 4-6, which brings out several important aspects of industrial chemistry.

STEPWISE REACTIONS, INTERMEDIATES, THE NET CHEMICAL REACTION.   An industrial process is usually carried out in stages or steps. The Raschig process involves three reaction steps. Sodium hypochlorite (NaOCl), produced in the first reaction, and chloramine (NH_2Cl), produced in the second reaction, are **intermediates** in the

FIGURE 4-6
The commercial production of hydrazine (N_2H_4).

Production Scheme:

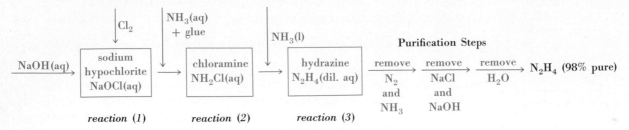

reaction (1)   *reaction (2)*   *reaction (3)*

The Net Chemical Reaction:

reaction (1) $2\,NaOH(aq) + Cl_2 \longrightarrow NaOCl(aq) + NaCl(aq) + H_2O$
reaction (2) $NaOCl(aq) + NH_3(aq) \longrightarrow NH_2Cl(aq) + NaOH(aq)$

reaction (3) $NH_2Cl(aq) + NH_3(l) + NaOH(aq) \xrightarrow{\Delta} N_2H_4(aq) + NaCl(aq) + H_2O$

(1) + (2) + (3) $2\,NaOH(aq) + Cl_2 + NaOCl(aq) + NH_3(aq) + NH_2Cl(aq) + NH_3(l) + NaOH(aq) \longrightarrow$
              $NaOCl(aq) + NaCl(aq) + H_2O + NH_2Cl(aq) + NaOH(aq) + N_2H_4(aq) + NaCl(aq) + H_2O$

cancellations: $2\,NaOH(aq) + Cl_2 + \cancel{NaOCl(aq)} + NH_3(aq) + \cancel{NH_2Cl(aq)} + NH_3(l) + \cancel{NaOH(aq)} \longrightarrow$
              $\cancel{NaOCl(aq)} + NaCl(aq) + H_2O + \cancel{NH_2Cl(aq)} + \cancel{NaOH(aq)} + N_2H_4(aq) + NaCl(aq) + H_2O$

Net equation: $2\,NaOH(aq) + Cl_2 + 2\,NH_3 \longrightarrow N_2H_4(aq) + 2\,NaCl(aq) + 2\,H_2O$

Side Reaction: $2\,NH_2Cl(aq) + N_2H_4(aq) \xrightarrow{Cu^{2+}} 2\,NH_4Cl(aq) + N_2(g)$

process. Their presence is crucial to the overall process, but they are consumed immediately after their formation, leading to the desired product, hydrazine (N_2H_4), in the third reaction. The **net chemical reaction** represents the overall change that results from the process. In the Raschig process the original starting materials are NaOH, Cl_2, and NH_3, and the end products are N_2H_4, NaCl, and H_2O. As illustrated in Figure 4-6, a net chemical equation can be written by adding together equations for individual, consecutive reactions in such a way as to "cancel out" intermediate species.

Some chemicals—about 90% of all hydrochloric acid [HCl(aq)], for example—are produced primarily as by-products of other manufacturing processes.

BY-PRODUCTS, SIDE REACTIONS.   The net equation for the production of N_2H_4 indicates that two other products are formed along with the N_2H_4—NaCl and H_2O. Substances formed along with the desired product are called **by-products** of a process. Often, the economic value of the by-products may determine the commercial feasibility of the total manufacturing process (although the by-products have no great value in the present case).

Almost as inevitable as the formation of by-products through the main chemical reactions is their formation through one or more competing reactions that reduce the amount of desired product; these reactions are called **side reactions.** Some of the hydrazine, as it is produced in the third reaction, combines with the intermediate chloramine to produce ammonium chloride (NH_4Cl) and nitrogen gas. This side reaction is speeded up or catalyzed by the presence of Cu^{2+} or other heavy metal ions in solution. Removal of these ions helps to minimize the importance of the side reaction.

PURIFICATION OF PRODUCT.   Rarely is the desired product of an industrial chemical process produced in a state of sufficient purity for its intended use—the product must be purified. The hydrazine produced in the Raschig process is in the form of a dilute water solution, in which dissolved NaCl, NaOH, and NH_3 are also found. The substances that are obtainable as solids—NaCl and NaOH—are crystallized from the hydrazine solution. The other substances—NH_3 and H_2O—are removed by distillation. The ultimate product after these several purification steps is hydrazine of better than 98% purity. Ammonia recovered in the purification process is reused in the main reactions, illustrating still another principle of industrial chemistry: *materials are recycled whenever possible.*

THEORETICAL YIELD, ACTUAL YIELD, PERCENT YIELD. The quantity of product resulting from given quantities of the initial reactants may be calculated from the balanced net equation with the methods developed in this chapter. This *calculated* quantity is called the **theoretical yield.** A measurement of the quantity of product that is *actually* produced in a chemical process is called the **actual yield.** The **percent yield** is defined as

$$\text{percent yield} = \frac{\text{actual yield}}{\text{theoretical yield}} \times 100 \qquad (4.12)$$

The actual yield of a chemical reaction is always *less than* the theoretical yield, and the percent yield, less than 100%. The reasons are fairly obvious: The desired reaction may not go to completion; side reactions may reduce the yield of product; material may be lost in the purification steps. Other factors (cost of materials, labor, equipment, and so forth) being equal, one would choose the process giving the highest percent yield if alternative processes yielding the same product are possible. Rarely, however, are all other factors equal.

Example 4-19 A chemical plant using the Raschig process obtains 0.299 kg of 98.0% N_2H_4 for every 1.00 kg Cl_2 that is reacted with excess NaOH and NH_3. What are the **(a)** theoretical, **(b)** actual, and **(c)** percent yields of *pure* N_2H_4?
(a) To determine the **theoretical yield** we use the *net* equation for the overall process (see Figure 4-6).

$$\text{no. kg } N_2H_4 = 1.00 \text{ kg } Cl_2 \times \frac{1000 \text{ g } Cl_2}{1 \text{ kg } Cl_2} \times \frac{1 \text{ mol } Cl_2}{70.9 \text{ g } Cl_2} \times \frac{1 \text{ mol } N_2H_4}{1 \text{ mol } Cl_2}$$

$$\times \frac{32.0 \text{ g } N_2H_4}{1 \text{ mol } N_2H_4} \times \frac{1 \text{ kg } N_2H_4}{1000 \text{ g } N_2H_4}$$

$$= 0.451 \text{ kg } N_2H_4$$

(b) The **actual yield** of 98.0% N_2H_4 is 0.299 kg. Based on 100% purity for the N_2H_4, we would write

$$\text{no. kg } N_2H_4 = 0.299 \text{ kg product} \times \frac{98.0 \text{ kg } N_2H_4}{100 \text{ kg product}} = 0.293 \text{ kg } N_2H_4$$

(c) Percent yield $= \dfrac{\text{actual yield}}{\text{theoretical yield}} \times 100 = \dfrac{0.293 \text{ kg } N_2H_4}{0.451 \text{ kg } N_2H_4} \times 100 = 65.0\%$

SIMILAR EXAMPLES: Exercises 39, 40, 41.

REACTION CONDITIONS. The precise conditions under which a chemical process is carried out are not always made clear by balanced equations alone, nor are the reasons for these conditions made clear. In the Raschig process the formation of chloramine (the second reaction) proceeds very rapidly, but the conversion of chloramine and ammonia to hydrazine (the third reaction) occurs more slowly. To speed

The effect of temperature on the speed or rate of a chemical reaction is discussed in Section 13-8.

up this conversion, elevated temperatures (about 130°C) are used for the third reaction.

At the same time that the main reaction is speeded up, the side reaction must be minimized. This is accomplished in two ways. First, a large excess of NH_3 is used—perhaps 20 or 30 mol NH_3 per mole of NH_2Cl. This means that NH_2Cl is more likely to react with NH_3 (the third reaction) than it is to react with N_2H_4 (the side reaction). Second, a protein-based material—gelatin, albumin, or glue—is added to the reaction mixture. Protein molecules bind with metal ions in solution (Cu^{2+}, for example) and interfere with the ability of these ions to catalyze the side reaction. The side reaction occurs much more slowly in the presence of an inhibitor.

The effects of catalysts and inhibitors on the rates of chemical reactions are discussed in Section 13-9.

Even the fact that NH_3 in the third reaction is introduced as the pure anhydrous (dry) liquid (rather than in aqueous solution as in the second reaction) has its reasons: (1) The heat given off when the NH_3 dissolves in the aqueous solution of chloramine is sufficient to raise the reaction mixture to the desired temperature of 130°C (thus avoiding the need to consume fuel to heat the mixture). (2) The less extraneous water that is added to the reaction mixture, the less that will have to be removed later in the purification steps.

Summary

When brought together, certain substances (reactants) may undergo transformation into a new set of substances (products). At the macroscopic (visible) level, evidence of a chemical reaction can take a variety of forms—evolution of a gas, formation of a precipitate, heat effect, color change. At times, though, subtle changes occur that can only be detected with sophisticated instruments.

Once the reactants and products of a chemical reaction have been identified, it is possible to represent the reaction through a chemical equation. In a chemical equation symbols and formulas are used to represent the reactants and products. To be of maximum use, a chemical equation must be balanced atomically, and, where necessary, for electrical charge as well. Additional types of information that can be conveyed through a chemical equation are the physical states or forms in which reactants and products appear and the reaction conditions employed. A wide variety of calculations (stoichiometric calculations) can be based on the balanced chemical equation. The principal feature shared by all such calculations is that the equation yields conversion factors. Molecular weights, densities, and percent composition are often required as well in these calculations.

Many chemical reactions are carried out in solution, and for this reason it proves convenient to describe the composition or concentration of solutions on a molar basis. The molarity or molar concentration scale indicates the number of moles of solute per liter of solution. From this definition it is possible to perform calculations relating molar concentration, solution volume, and amount of solute. This may be done for individual solutions or for situations in which solutions are mixed, diluted by adding more solvent, or concentrated by removing solvent. The use of molar concentration as a conversion factor in stoichiometric calculations is a straightforward matter. Of practical importance is the procedure for carrying out a reaction between two solutions—titration.

Additional features sometimes arise in stoichiometric calculations. It may be necessary to identify the single reactant, called the limiting reagent, that determines the amount of product formed (other reactants being in excess). Two or more reactions may occur at the same time, or a series of reactions may occur in succession. The conduct of reactions on an industrial scale bears some resemblance to situations described in textbooks, but there are significant differences as well. The failure of most chemical reactions to yield as much product as that calculated from the balanced equation is common in industrial processes (and often in the laboratory). That is, the actual yield of a reaction is generally less than the theoretical yield, and the percent yield is less than 100%.

Learning Objectives

As a result of studying Chapter 4, you should be able to

1. List the kinds of experimental evidence that indicate the occurrence of a chemical reaction.

2. Write word equations and symbolic equations to represent chemical reactions for which the reactants and products are stated.

3. Balance a symbolic chemical equation atomically, and, where necessary, for electrical charge.

4. Apply the terms *combination, decomposition, displacement,* and *metathesis* to describe types of reactions.

5. Write balanced equations for reactions involving the complete combustion of carbon–hydrogen compounds (hydrocarbons) and carbon–hydrogen–oxygen compounds.

6. Derive from balanced chemical equations conversion factors for use in stoichiometric calculations.

7. Solve a variety of problems based on the balanced chemical equation, with quantities given and/or sought in units of moles, mass, volume, density, percent by mass, and so on.

8. Define the molar concentration (*M*) scale and perform calculations involving the amount of solute, solution volume, and the molarity of a solution.

9. Do calculations relating to the mixing of solutions or to the addition or removal of solvent from a solution.

10. Use solution volumes and molar concentrations in stoichiometric calculations for reactions involving solutions.

11. Use titration data to determine the composition of a sample of matter, such as the concentration of a solution.

12. Determine the limiting reagent in a chemical reaction.

13. Calculate the amount of product when two or more reactions occur simultaneously or consecutively.

14. Combine a series of chemical equations to obtain a single net equation for a product formed through a stepwise process.

15. Define the terms *actual yield, theoretical yield,* and *percent yield* of a reaction, and calculate these quantities from appropriate data.

Some New Terms

Balancing an equation refers to placing numbers, **stoichiometric coefficients,** in front of the symbols and formulas in a chemical equation. In this way the numbers of atoms of each kind are made equal on the two sides of the equation. In **net ionic equations,** electrical charges must also be balanced.

By-products are those substances produced along with the principal reaction product in a commercial chemical process.

A **chemical equation** is a symbolic representation of a chemical reaction; that is, symbols and formulas are substituted for the names of reactants and products.

A **chemical reaction** is a process in which one set of substances (reactants) is transformed into another set of substances (products).

Dilution is the process of reducing the concentration of a solution by adding more solvent.

An **intermediate** is the product of one reaction that is consumed in a following reaction in a process that proceeds through several steps.

The **limiting reagent** in a chemical reaction is the reactant that is consumed completely. The amount of product(s) formed is dependent on the amount of the limiting reagent.

Molar concentration or molarity (*M*) refers to the composition or concentration of a solution expressed as number of moles of solute per liter of solution.

The **net chemical reaction** is the overall chemical change that occurs in a process that proceeds through two or more steps.

A **side reaction** is a chemical reaction that occurs at the same time as the principal reaction in a chemical process. Usually, the existence of side reactions reduces the yield of the primary reaction product.

Titration is a procedure for carrying out a chemical reaction between two solutions by the controlled addition of one solution (from a buret) to the other.

The **yield** of a chemical reaction refers to the quantity of a desired substance associated with a chemical process. The **theoretical yield** is the quantity *calculated* from the balanced chemical equation. The **actual yield** is the *measured* quantity of product for a given process. **Percent yield** is the percent of the theoretical yield that is actually obtained; that is, percent yield = (actual yield/theoretical yield) × 100.

Exercises

Evidence for chemical reactions

1. From the following observations, indicate whether chemical reactions are involved. If so, write appropriate balanced chemical equations.

   (a) Clear colorless aqueous solutions of sodium nitrate and potassium chloride are mixed. Analysis of the resulting clear colorless aqueous solution reveals the presence of Na^+, K^+, NO_3^-, and Cl^- ions only.

   (b) Water solutions of radium chloride and sodium sulfate are mixed. A white precipitate is formed which is found to be highly radioactive. A portion of the remaining solution is treated with silver nitrate and yields a white precipitate. [*Hint:* Recall that radium is a radioactive element (Section 2-5).]

   (c) Hydrogen gas is passed over Fe_2O_3 at 400°C. Water vapor is formed, together with a black residue—a compound consisting of 72.3% Fe and 27.7% O.

Writing and balancing chemical equations

2. Write balanced equations for each of the following.

   (a) magnesium metal + oxygen gas $\longrightarrow$ solid magnesium oxide

   (b) solid sulfur + oxygen gas $\longrightarrow$ gaseous sulfur dioxide

   (c) methane gas (CH_4) + oxygen gas $\longrightarrow$ carbon dioxide gas + liquid water

   (d) aqueous silver sulfate + aqueous barium iodide $\longrightarrow$ solid barium sulfate + solid silver iodide

3. From the following descriptions of chemical reactions, write appropriate balanced equations.

   (a) Steam (gaseous water) reacts with carbon at high temperatures to produce carbon monoxide and hydrogen gases.

   (b) Aluminum metal displaces copper(II) ion from aqueous solution, producing aluminum ion and copper metal.

   (c) Zinc sulfide (ZnS) dissolves in an aqueous solution containing hydrogen ion (H^+), producing hydrogen sulfide gas (H_2S) and a water solution containing zinc ion.

   (d) At high temperatures the gases chlorine and water react to produce hydrogen chloride and oxygen gases.

4. Balance the following equations by inspection.

   (a) $Na_2SO_4(s) + C(s) \longrightarrow Na_2S(s) + CO_2(g)$

   (b) $Cl_2(aq) + H_2O(l) \longrightarrow HCl(aq) + HOCl(aq)$

   (c) $PCl_3(l) + H_2O(l) \longrightarrow H_3PO_3(aq) + HCl(aq)$

   (d) $P_2H_4(l) \longrightarrow PH_3(g) + P_4(g)$

   (e) $PbO(s) + NH_3(g) \longrightarrow Pb(s) + N_2(g) + H_2O(l)$

   (f) $NO_2(g) + H_2O(l) \longrightarrow HNO_3(aq) + NO(g)$

   (g) $S_2Cl_2 + NH_3 \longrightarrow N_4S_4 + NH_4Cl + S_8$

   (h) $Mg_3N_2 + H_2O \longrightarrow Mg(OH)_2 + NH_3$

   (i) $SO_2Cl_2 + HI \longrightarrow H_2S + H_2O + HCl + I_2$

   (j) $S_8 + NaOH \longrightarrow Na_2S + Na_2S_2O_3 + H_2O$

5. Balance the following equations written in *ionic* form.

   (a) $Zn(s) + Ag^+(aq) \longrightarrow Zn^{2+}(aq) + Ag(s)$

   (b) $Mn^{2+}(aq) + H_2S(g) \longrightarrow MnS(s) + H^+(aq)$

   (c) $Al(s) + H^+(aq) \longrightarrow Al^{3+}(aq) + H_2(g)$

   (d) $S_2O_3{}^{2-}(aq) + H^+(aq) \longrightarrow$
   $$H_2O(l) + S(s) + SO_2(g)$$

   (e) $MnO_2(s) + H^+(aq) + Cl^-(aq) \longrightarrow$
   $$Mn^{2+}(aq) + H_2O(l) + Cl_2(g)$$

6. Write balanced equations to represent the complete combustion of the following compounds: (a) C_5H_{12}; (b) C_6H_6; (c) $C_6H_{12}O_6$; (d) $C_2H_6O_2$; (e) C_3H_7OH

Stoichiometry of chemical reactions

7. Iron metal reacts with chlorine gas according to the equation

$$2 Fe(s) + 3 Cl_2(g) \longrightarrow 2 FeCl_3(s)$$

How much $Cl_2(g)$, in moles, is required to convert 4.4 mol Fe to $FeCl_3$?

8. It is desired to produce 4.50 mol PCl_3 by the reaction

$$6 Cl_2 + P_4 \longrightarrow 4 PCl_3$$

How many moles of Cl_2 and of P_4 are required?

9. The reaction of calcium hydride with water can be used to prepare small quantities of hydrogen gas.

$$CaH_2(s) + 2 H_2O(l) \longrightarrow Ca(OH)_2(aq) + 2 H_2(g)$$

   (a) How much $H_2(g)$, in moles, results from the reaction of 125 g CaH_2 with an excess of water?

   (b) What mass of H_2O is consumed in the reaction of 125 g CaH_2?

   (c) What mass of CaH_2 must be allowed to react with an excess of water to produce 1.00×10^{24} molecules of $H_2(g)$?

10. A laboratory method for preparing oxygen gas involves the decomposition of potassium chlorate.

$$2 KClO_3(s) \xrightarrow{\Delta} 2 KCl(s) + 3 O_2(g)$$

Upon decomposing a 10.0-g sample of $KClO_3(s)$, (a) how many moles of $O_2(g)$, (b) how many molecules of $O_2(g)$, and (c) how many grams of KCl are produced?

11. A 12.0-g sample of a Na_2CO_3–Na_2SO_4 mixture containing 52.5% Na_2CO_3 reacts with an excess of hydrochloric acid. Only the Na_2CO_3 reacts. What mass of CO_2 is produced?

$$Na_2CO_3(s) + 2 HCl(aq) \longrightarrow$$
$$2 NaCl(aq) + H_2O(l) + CO_2(g)$$

12. An iron ore sample is impure Fe_2O_3. When Fe_2O_3 is heated with an excess of carbon, iron metal is produced. From a sample of ore weighing 752 kg, 453 kg of pure iron is obtained. What is the percent, by mass, of Fe_2O_3 in the ore sample? (*Hint:* Determine the quantity of Fe_2O_3 required to produce 453 kg of iron.)

$$Fe_2O_3(s) + 3\ C(s) \xrightarrow{\Delta} 2\ Fe(l) + 3\ CO(g)$$

13. Silver oxide decomposes completely at temperatures in excess of 300°C, yielding metallic silver and oxygen gas. A 1.60-g sample of *impure* silver oxide yields 0.104 g O_2. Assuming that Ag_2O is the only source of O_2, what is the percent, by mass, of Ag_2O in the sample?

14. A piece of aluminum foil measuring 4.0 in. × 3.0 in. × 0.025 in. is dissolved in an excess of hydrochloric acid. What mass of H_2, in grams, is produced? (Density of aluminum = 2.70 g/cm³.) (*Hint:* Use equation 4.7.)

15. How many milliliters of a water solution of $KMnO_4$ containing 15.8 g/L must be used to complete the conversion of 7.50 g KI to iodine (I_2) by the reaction

$$2\ KMnO_4 + 10\ KI + 8\ H_2SO_4 \longrightarrow$$
$$6\ K_2SO_4 + 2\ MnSO_4 + 5\ I_2 + 8\ H_2O$$

Molar concentration

16. What are the molar concentrations of the solutes listed below when dissolved in water?
   (a) 1.50 mol C_2H_5OH in 4.80 L of solution
   (b) 12.5 g CH_3OH in 50.0 ml of solution
   (c) 10.0 ml of pure glycerol, $C_3H_8O_3$ (density = 1.26 g/cm³), in 250.0 ml of solution
   (d) 125 g $CO(NH_2)_2$, 98.6% pure, in 500.0 ml of solution
   (e) 15.0 mg of ether, $(C_2H_5)_2O$, in 3.00 gal of water (1 gal = 3.78 L).

17. How much
   (a) KCl, in moles, is required to prepare 1.50×10^2 L of 0.275 M KCl(aq)?
   (b) Na_2SO_4, in grams, is required to produce 325 ml of 0.115 M Na_2SO_4(aq)?
   (c) methanol, CH_3OH (d = 0.792 g/cm³), in cm³, must be dissolved in water to produce 2.50 L of 0.150 M CH_3OH(aq)?
   (d) ethanol, C_2H_5OH (d = 0.789 g/cm³), in gal, must be dissolved in water to produce 55.0 gal of 2.10 M C_2H_5OH(aq)? (1 gal = 3.78 L.)

18. Two sucrose solutions are mixed. One has a volume of 385 ml and a concentration of 1.50 M $C_{12}H_{22}O_{11}$; the other, 615 ml and a concentration of 1.25 M $C_{12}H_{22}O_{11}$. What is the molar concentration of $C_{12}H_{22}O_{11}$ in the final solution? (Assume the solution volumes to be additive.)

19. What volume of a concentrated hydrochloric acid solution (36.0% HCl, by mass; d = 1.26 g/ml) is required to produce 15.0 L of 0.500 M HCl?

20. What volume of 1.000 M KOH must be diluted with water to prepare 2.00 L of 0.278 M KOH?

21. Water is evaporated from 70.0 ml of 0.485 M $MgSO_4$ solution until the solution volume becomes 45.0 ml. What is the molar concentration of $MgSO_4$ in the resulting solution?

Chemical reactions in solutions

22. What volume of 0.1060 M NaOH is required for the titration (i.e., the complete reaction) of 25.00 ml of 0.1252 M HNO_3?

$$HNO_3(aq) + NaOH(aq) \longrightarrow NaNO_3(aq) + H_2O(l)$$

23. The neutralization of hydrochloric acid by calcium hydroxide proceeds as follows:

$$Ca(OH)_2(s) + 2\ HCl(aq) \longrightarrow CaCl_2(aq) + 2\ H_2O$$

   (a) What mass of $Ca(OH)_2$, in grams, is required to neutralize 325 ml of 0.410 M HCl?
   (b) What mass of $Ca(OH)_2$, in kilograms, is required to neutralize 152 L of an HCl solution that is 30.12% HCl, by mass, and has a density of 1.15 g/cm³.

24. Household ammonia is an aqueous solution of NH_3. 31.20 ml of 1.000 M HCl is required to react completely with the NH_3 present in a 5.00-ml sample of a particular brand of household ammonia.

$$NH_3(aq) + HCl(aq) \longrightarrow NH_4Cl(aq)$$

   (a) What is the molar concentration of NH_3 in the sample?
   (b) Assuming a density of 0.96 g/ml for the ammonia solution, what is its percent NH_3, by mass?

25. For use in a variety of titration reactions in the laboratory, it is desired to prepare 20 L of a solution of about 0.25 M HCl. The solution is to have a concentration known to four significant figures. A two-step approach is required. First, a solution that has an approximate concentration of 0.25 M is prepared by dilution of concentrated hydrochloric acid. Then, the resulting HCl(aq) is titrated with an NaOH solution of known concentration, and the molar concentration of the HCl is calculated.
   (a) What volume of concentrated hydrochloric acid (d = 1.19 g/cm³; 38% HCl, by mass) must be diluted to 20 L with water to prepare an 0.25 M HCl solution?
   (b) A 25.00-ml sample of the approximately 0.25 M HCl prepared in part (a) requires exactly 30.10 ml of 0.2000 M NaOH for its titration. What is the exact molar concentration of the diluted HCl(aq)?

$$HCl(aq) + NaOH(aq) \longrightarrow NaCl(aq) + H_2O(l)$$

   (c) Why is a titration necessary? That is, why could not the final solution be prepared simply by an appropriate dilution of the concentrated hydrochloric acid?

26. An iron ore sample weighing 0.8515 g is dissolved in HCl(aq) and all the iron is converted to $FeCl_2$(aq). This solution is then titrated with exactly 36.10 ml of 0.0410 M $K_2Cr_2O_7$. What must be the percent Fe, by mass, in the ore sample?

$$6\,FeCl_2(aq) + K_2Cr_2O_7(aq) + 14\,HCl(aq) \longrightarrow$$
$$6\,FeCl_3(aq) + 2\,CrCl_3(aq) + 2\,KCl(aq) + 7\,H_2O(l)$$

27. A method of adjusting the concentration of an HCl(aq) solution is to allow the solution to react with a small quantity of magnesium metal.

$$Mg(s) + 2\,HCl(aq) \longrightarrow MgCl_2(aq) + H_2(g)$$

How much Mg, in milligrams, must be added to 500.0 ml of 1.012 M HCl to reduce the concentration to exactly 1.000 M HCl?

Determining the limiting reagent

28. The following is encountered as a side reaction in the manufacture of rayon from wood pulp.

$$3\,CS_2 + 6\,NaOH \longrightarrow 2\,Na_2CS_3 + Na_2CO_3 + 3\,H_2O$$

(a) How many moles each of the products—Na_2CS_3, Na_2CO_3, and H_2O—are produced by allowing 1.00 mol of each reactant—CS_2 and NaOH—to react?

(b) What mass of Na_2CS_3 is produced by allowing 100.0 cm^3 of liquid CS_2 ($d = 1.26$ g/cm^3) and 3.50 mol NaOH to react?

29. Ammonia gas can be generated by heating together the solids NH_4Cl and $Ca(OH)_2$. In addition to NH_3, $CaCl_2$ and H_2O are formed as products. If a mixture containing 15.0 g each of NH_4Cl and $Ca(OH)_2$ is heated, what mass of NH_3 is formed? (*Hint:* Start by writing a balanced equation for the reaction.)

30. A mixture of 4.800 g H_2 and 36.40 g O_2 is allowed to react.

(a) Write a balanced equation for this reaction.

(b) What substances are present after the reaction, and in what quantities?

(c) Show that the total mass of substances present before and after the reaction remains constant.

31. Chlorine gas can be produced in the laboratory by the reaction indicated. A 61.3-g sample that is 96% $K_2Cr_2O_7$ is allowed to react with 320 ml of a hydrochloric acid solution having a density of 1.15 g/cm^3 and containing 30% HCl, by mass. How much Cl_2 is produced, expressed in grams?

$$K_2Cr_2O_7 + 14\,HCl \longrightarrow$$
$$2\,KCl + 2\,CrCl_3 + 7\,H_2O + 3\,Cl_2(g)$$

Simultaneous reactions

32. A mixture is known to contain 26.0% $MgCO_3$ and 74.0% $Mg(OH)_2$, by mass. How much HCl, in grams, is required to dissolve a 140.0-g sample of this mixture?

$$MgCO_3(s) + 2\,HCl(aq) \longrightarrow$$
$$MgCl_2(aq) + H_2O(l) + CO_2(g)$$
$$Mg(OH)_2(s) + 2\,HCl(aq) \longrightarrow MgCl_2(aq) + 2\,H_2O(l)$$

33. A particular natural gas sample consists of 68.2% propane (C_3H_8) and 31.8% butane (C_4H_{10}), by mass. The combustion of this gaseous mixture yields CO_2(g) and H_2O(l) as the sole products. How many moles of CO_2(g) would be produced by the complete combustion of 225 g of this gaseous mixture?

34. A particular organic liquid is believed to be either pure methyl alcohol (CH_3OH), pure ethyl alcohol (C_2H_5OH), or a mixture of the two. A 0.220-g sample of the liquid is burned in an excess of oxygen and yields 0.352 g CO_2. Is the liquid a pure alcohol or a mixture of the two?

Consecutive reactions

35. Dichlorodifluoromethane, a widely used refrigerant, can be prepared by the following reactions:

$$CH_4 + 4\,Cl_2 \longrightarrow CCl_4 + 4\,HCl$$
$$CCl_4 + 2\,HF \longrightarrow CCl_2F_2 + 2\,HCl$$

How many moles of Cl_2 must be consumed to produce 15.0 mol CCl_2F_2?

36. A 5.00-ml sample of liquid benzene, C_6H_6 ($d = 0.879$ g/cm^3), is burned completely, yielding CO_2 and H_2O as the only products. The CO_2 formed is passed into a barium hydroxide solution and yields a precipitate of barium carbonate. What mass of $BaCO_3$ is obtained?

$$CO_2(g) + Ba(OH)_2(aq) \longrightarrow BaCO_3(s) + H_2O$$

37. The following process has been used to obtain iodine from oil-field brines in California.

$$NaI + AgNO_3 \longrightarrow AgI + NaNO_3$$
$$2\,AgI + Fe \longrightarrow FeI_2 + 2\,Ag$$
$$2\,FeI_2 + 3\,Cl_2 \longrightarrow 2\,FeCl_3 + 2\,I_2$$

How much $AgNO_3$, in grams, is required in the first step for every gram of I_2 produced in the third step?

38. Sodium bromide can be prepared commercially by the following series of reactions. How much Fe, in kilograms, is consumed to produce 1.00×10^3 kg NaBr?

$$Fe + Br_2 \longrightarrow FeBr_2$$
$$FeBr_2 + Br_2 \longrightarrow Fe_3Br_8 \qquad \text{(not balanced)}$$
$$Fe_3Br_8 + Na_2CO_3 \longrightarrow NaBr + CO_2 + Fe_3O_4$$
$$\text{(not balanced)}$$

Theoretical, actual, and percent yields

39. Azobenzene, an important intermediate in the manufacture of dyes, can be prepared from nitrobenzene by reaction with triethylene glycol in the presence of zinc and potassium hydroxide. In one particular reaction 0.10 L of nitrobenzene

($d = 1.20$ g/cm³) and 0.30 L of triethylene glycol ($d = 1.12$ g/cm³) yielded 55 g of azobenzene. What was the percent yield of this reaction?

$$2\,C_6H_5NO_2 + 4\,C_6H_{14}O_4 \xrightarrow[\text{KOH}]{\text{Zn}}$$

nitro- triethylene
benzene glycol

$$(C_6H_5N)_2 + 4\,C_6H_{12}O_4 + 4\,H_2O$$
azobenzene

40. A laboratory method of producing acetyl chloride (C_2H_3OCl) from acetic acid ($C_2H_4O_2$) involves reaction with PCl_3, with an expected yield of 70%. How many grams of commercial acetic acid, which contains 97% $C_2H_4O_2$ by mass, must be allowed to react with an excess of PCl_3 to produce 50.0 g of acetyl chloride?

$$C_2H_4O_2 + PCl_3 \longrightarrow C_2H_3OCl + H_3PO_3 \quad \text{(not balanced)}$$

41. A laboratory experiment is described in which copper metal is carried through a series of transformations, resulting finally in pure copper once again. The reactions are represented below. If the percent yield of the copper-containing substance in each step is 98%, how much copper would be recovered in the final step, starting with a 1.00-g sample in the first step?

$$Cu(s) + 4\,HNO_3(aq) \longrightarrow$$
$$Cu(NO_3)_2(aq) + 2\,H_2O + 2\,NO_2(g)$$

$$Cu(NO_3)_2(aq) + 2\,NaHCO_3(aq) \longrightarrow$$
$$CuCO_3(s) + 2\,NaNO_3(aq) + H_2O + CO_2(g)$$

$$CuCO_3(s) + H_2SO_4(aq) \longrightarrow$$
$$CuSO_4(aq) + H_2O + CO_2(g)$$

$$Zn(s) + CuSO_4(aq) \longrightarrow Cu(s) + ZnSO_4(aq)$$

Industrial chemistry

42. A method employed to produce sodium sulfate, a substance used extensively in the textile industry, involves heating a mixture of ordinary salt and sulfuric acid.

$$2\,NaCl + H_2SO_4 \xrightarrow{\Delta} Na_2SO_4 + 2\,HCl(g)$$

The sulfuric acid is in the form of a concentrated aqueous solution having a density of 1.73 g/cm³ and containing 80% H_2SO_4, by mass. How many liters of the sulfuric acid solution must be used to complete the reaction of 1.00×10^3 kg of salt?

43. In a particular plant using the Raschig process for N_2H_4 (refer to Figure 4-6), the mole ratio of NH_3 to NH_2Cl used in the third reaction is 30:1. What is the maximum quantity of NH_3, in kilograms, that is recoverable in the purification steps for every 1.00 kg Cl_2 that reacts?

44. Acrylonitrile, CH_2CHCN, is used in the production of synthetic fibers, plastics, and rubber goods. The Sohio process produces acrylonitrile by reacting propylene, air, and ammonia. The reaction is

$$CH_2CHCH_3 + NH_3 + O_2 \longrightarrow CH_2CHCN + H_2O$$
$$\text{(not balanced)}$$

The process yields 0.73 lb of acrylonitrile (CH_2CHCN) per lb of propylene (CH_2CHCH_3).
  (a) Balance the equation for the reaction.
  (b) What is the percent yield of the process?
  (c) At the percent yield calculated in part (b), what is the minimum mass of NH_3 required to produce 1.00 ton (2000 lb) of acrylonitrile by this process?

Additional Exercises

1. Balance the following equations by inspection.
  (a) $SiCl_4(l) + H_2O(l) \longrightarrow SiO_2(s) + HCl(g)$
  (b) $CaC_2(s) + H_2O(l) \longrightarrow Ca(OH)_2(s) + C_2H_2(g)$
  (c) $Na_2HPO_4(s) \longrightarrow Na_4P_2O_7(s) + H_2O(l)$
  (d) $NCl_3(g) + H_2O(l) \longrightarrow NH_3(g) + HOCl(aq)$
  (e) $CS_2(l) + NaOH(aq) \longrightarrow$
$$Na_2CS_3(aq) + Na_2CO_3(aq) + H_2O$$
  (f) $NO_2(g) + H_2(g) \longrightarrow NH_3(g) + H_2O(g)$
  (g) $Pb(NO_3)_2(s) \longrightarrow PbO(s) + NO_2(g) + O_2(g)$
  (h) $CaCO_3(s) + H^+(aq) \longrightarrow$
$$Ca^{2+}(aq) + H_2O(l) + CO_2(g)$$
  (i) $Al_2O_3(s) + H^+(aq) \longrightarrow Al^{3+}(aq) + H_2O(l)$
  (j) $Zn(s) + H^+(aq) + NO_3^-(aq) \longrightarrow$
$$Zn^{2+}(aq) + H_2O(l) + N_2O(g)$$

2. Synthesis gas, a mixture of CO(g) and H_2(g), derives its name from the usefulness of this mixture as a starting material for the synthesis of a variety of organic chemical compounds. Synthesis gas can be produced by reacting a hydrocarbon, C_xH_y, with steam—H_2O(g)—or with oxygen—O_2(g). Write a balanced chemical equation to represent the formation of synthesis gas from **(a)** C_9H_{20} using steam; **(b)** C_5H_{12} using oxygen.

3. Pure silver metal results when silver carbonate is decomposed by heating.

$$Ag_2CO_3(s) \xrightarrow{\Delta} Ag(s) + CO_2(g) + O_2(g) \quad \text{(not balanced)}$$

In one particular reaction 14.8 g Ag is obtained. How many grams of Ag_2CO_3 must have been decomposed?

4. Given the reaction

$$3\,Fe(s) + 4\,H_2O(g) \xrightarrow{\Delta} Fe_3O_4(s) + 4\,H_2(g)$$

  (a) How many moles of H_2(g) can be produced by reacting 125 g Fe with an excess of H_2O(g) [steam]?

(b) How many grams of H_2O would be consumed in the conversion of 115 g Fe to Fe_3O_4?

(c) In one particular experiment 3.20 mol $H_2(g)$ was produced as a result of this reaction. How many grams of Fe_3O_4 must also have been produced?

5. What volume of 0.500 *M* $CO(NH_2)_2$ solution must be diluted with water to produce 1.00 L of a solution with a concentration of 1.00 mg N per milliliter?

6. A seawater sample contains 2.8% NaCl, by mass, and has a density of 1.03 g/cm^3. A saturated solution of NaCl in water is about 5.45 *M* NaCl. How much water would have to be evaporated from 1.00×10^3 L of the seawater sample before sodium chloride would precipitate from the solution? (A saturated solution contains the maximum amount of dissolved solute possible. When water is evaporated from a solution, the molar concentration of the solution increases until the solution becomes saturated. Then, the molar concentration remains constant as solute precipitates from the solution.)

7. 99.8 ml of 12.0% KI solution having a density of 1.093 g/ml is added to 96.7 ml of 14.0% $Pb(NO_3)_2$ solution having a density of 1.134 g/ml. What mass of PbI_2 is formed? (Both solution compositions are in percent, *by mass*.)

$$Pb(NO_3)_2(aq) + 2\ KI(aq) \longrightarrow PbI_2(s) + 2\ KNO_3(aq)$$

8. Either of the following reactions can be used to produce small quantities of $Cl_2(g)$ in the laboratory.

(1) $K_2Cr_2O_7 + 14\ HCl \longrightarrow$
$$2\ CrCl_3 + 2\ KCl + 7\ H_2O + 3\ Cl_2(g)$$

(2) $MnO_2 + 4\ HCl \longrightarrow MnCl_2 + 2\ H_2O + Cl_2(g)$

The HCl is used in the form of concentrated hydrochloric acid that is 36.0% HCl, by mass. The cost of the acid is $0.40/lb; the cost of the MnO_2 is $1.10/lb; and that of $K_2Cr_2O_7$ is $2.40/lb. Which reaction, (1) or (2), would be the cheaper source of $Cl_2(g)$?

9. Assuming that acetic acid, $HC_2H_3O_2$, is the only substance present that reacts with NaOH, calculate the molar concentration of acetic acid in ordinary vinegar if a 10.00-ml sample of the vinegar requires 19.32 ml of 0.500 *M* NaOH for its complete titration.

$$HC_2H_3O_2(aq) + NaOH(aq) \longrightarrow NaC_2H_3O_2(aq) + H_2O(l)$$

*10. It is desired to determine the acetylsalicylic acid content of a series of aspirin tablets by titration with NaOH.

$$HC_9H_7O_4(aq) + NaOH(aq) \longrightarrow NaC_9H_7O_4(aq) + H_2O(l)$$

Each of the tablets is expected to contain about 0.32 g of acetylsalicylic acid, $HC_9H_7O_4$. What molar concentration of NaOH must be used if titration volumes of about 22 ml are desired. (This procedure ensures good precision in measurements and allows the titration of two samples with the contents of a 50-ml buret.)

*11. An 0.155-g sample of an aluminum–magnesium alloy is dissolved in an excess of hydrochloric acid, producing 0.0163 g H_2. What is the percent of magnesium in the alloy?

$$Mg + 2\ HCl \longrightarrow MgCl_2 + H_2$$

$$2\ Al + 6\ HCl \longrightarrow 2\ AlCl_3 + 3\ H_2$$

*12. The principal method of manufacturing ethyl alcohol, C_2H_5OH, also yields diethyl ether, $(C_2H_5)_2O$, as a by-product. The *complete combustion* of a 1.005-g sample of the product of this process yields 1.963 g CO_2. What must be the percent, by mass, of C_2H_5OH and of $(C_2H_5)_2O$ in this sample?

*13. Hydrogen produced by the decomposition of water has considerable potential as an alternative fuel to gasoline (see p. 136, "the hydrogen economy"). A critical problem is in developing a reaction cycle—a series of chemical reactions—that has as its single net reactant, H_2O, and as its ultimate products, H_2 and O_2. That is, the cycle must yield this net balanced equation: $2\ H_2O \rightarrow 2\ H_2 + O_2$. Demonstrate that this condition is met by the Fe/Cl cycle.

$$FeCl_2 + H_2O \xrightarrow{650°C} Fe_3O_4 + HCl + H_2$$

$$Fe_3O_4 + HCl + Cl_2 \xrightarrow{200°C} FeCl_3 + H_2O + O_2$$

$$FeCl_3 \xrightarrow{420°C} FeCl_2 + Cl_2$$

*14. An important objection to the use of coal as a fuel is that when the coal is burned, sulfur found in coal produces SO_2, an atmospheric pollutant. One method currently being investigated for the removal of pyritic sulfur (sulfur present as FeS_2) involves treatment of coal with an aqueous solution containing Fe^{3+}.

(1) $2\ Fe^{3+}(aq) + FeS_2 \longrightarrow 3\ Fe^{2+}(aq) + 2\ S$

(2) $14\ Fe^{3+}(aq) + 8\ H_2O + FeS_2 \longrightarrow$
$$15\ Fe^{2+}(aq) + 2\ SO_4^{2-}(aq) + 16\ H^+(aq)$$

The free sulfur (S) produced in the first reaction is extracted from the coal with a suitable solvent. Sulfate ion, SO_4^{2-}, produced in the second reaction simply remains in the aqueous solution. By experiment, it has been found that for every mole of FeS_2 consumed in reaction (1) about 1.4 mol FeS_2 is consumed in reaction (2). How many moles of $Fe^{3+}(aq)$ would be consumed in treating 1.00×10^6 g of a coal sample containing 1.54% S in the form of FeS_2?

*15. Under appropriate conditions, copper sulfate, potassium chromate, and water react to form a product containing Cu^{2+}, CrO_4^{2-}, and OH^- ions. Analysis of the compound yields 48.7% Cu, 35.6% CrO_4^{2-}, and 15.7% OH^-.

(a) Derive the empirical formula of the compound.

(b) Write a plausible equation for the reaction.

Self-Test Questions

For questions 1 through 6 select the single item that best completes each statement.

1. For the reaction $2\,H_2S + SO_2 \longrightarrow 3\,S + 2\,H_2O$

   (a) 3 mol S is produced for every mole of H_2S that reacts.
   (b) 1 mol SO_2 is consumed for every mole of H_2S that reacts.
   (c) 1 mol H_2O is produced for every mole of H_2S that reacts.
   (d) the number of moles of products is independent of how many moles of reactants are used.

2. 1.0 mol of calcium cyanamide ($CaCN_2$) and 1.0 mol of water are allowed to react.

   $$CaCN_2 + 3\,H_2O \longrightarrow CaCO_3 + 2\,NH_3$$

   The number of moles of NH_3 produced is   (a) 3.0; (b) 2.0;   (c) 1.0;   (d) less than 1.0.

3. If the reaction of 1.00 mol $NH_3(g)$ and 1.00 mol $O_2(g)$

   $$4\,NH_3(g) + 5\,O_2(g) \longrightarrow 4\,NO(g) + 6\,H_2O(l)$$

   is carried to completion,   (a) all the $O_2(g)$ is consumed;   (b) 4.0 mol $NO(g)$ is produced;   (c) 1.5 mol $H_2O(l)$ is produced;   (d) none of these.

4. To prepare a solution that is 0.50 M KCl starting with 100 ml of 0.40 M KCl,   (a) add 0.75 g KCl;   (b) add 20 ml of water;   (c) add 0.10 mol KCl;   (d) evaporate 10 ml of water.

5. To complete the titration of 10.00 ml 0.0500 M NaOH

   $$2\,NaOH(aq) + H_2SO_4(aq) \longrightarrow Na_2SO_4(aq) + 2\,H_2O$$

   requires   (a) 50.0 ml of 0.0100 M H_2SO_4;   (b) 25.0 ml of 0.0100 M H_2SO_4;   (c) 100.0 ml of 0.0100 M H_2SO_4;   (d) 10.0 ml of 0.100 M H_2SO_4.

6. In the reaction of 2.0 mol CCl_4 with an excess of HF, 1.70 mol CCl_2F_2 is obtained.

   $$CCl_4 + 2\,HF \longrightarrow CCl_2F_2 + 2\,HCl$$

   (a) The theoretical yield of the reaction is 1.70 mol CCl_2F_2.
   (b) The theoretical yield of the reaction is 1.0 mol CCl_2F_2.
   (c) The percent yield of the reaction is 85%.
   (d) The theoretical yield of the reaction depends on how large an excess of HF is used.

7. Write a balanced chemical equation to represent each of the following reactions.

   (a) The decomposition, by heating, of solid mercury(II) nitrate to produce pure liquid mercury, nitrogen dioxide gas, and oxygen gas.
   (b) The reaction of aqueous sodium carbonate solution with aqueous hydrochloric acid (hydrogen chloride) to produce water, carbon dioxide gas, and aqueous sodium chloride.
   (c) The complete combustion of benzoic acid, a compound consisting of 68.8% C, 4.95% H, and 26.2% O, by mass.

8. What volume of 0.0102 M $Ba(OH)_2(aq)$, in milliliters, is required to titrate 10.00 ml of 0.0526 M $HNO_3(aq)$?

   $$2\,HNO_3(aq) + Ba(OH)_2(aq) \longrightarrow Ba(NO_3)_2(aq) + 2\,H_2O$$

9. How many grams of Na must be reacted with 125 ml H_2O to produce an NaOH solution that is 0.250 M? (Assume that the final solution volume is 125 ml.)

   $$2\,Na(s) + 2\,H_2O(l) \longrightarrow 2\,NaOH(aq) + H_2(g)$$

10. A nearly 100% yield is essential for a chemical reaction that is to be used to *analyze* a chemical compound, but is almost never expected for a reaction that is to be used to synthesize a compound. Explain why this is so.

5 Gases

Familiar samples of bulk matter exist as solids, liquids, or gases. This observation hardly needs mention; we understand it intuitively. However, we do need to describe the properties of these physical forms, or states, of matter in some detail.

The simplest of the three states of matter to understand is the gaseous state. Most of the ideas presented in this chapter predate the twentieth century, and the development of modern chemistry paralleled closely the growth of knowledge of the gaseous state. The behavior of gases figured prominently in the discovery of the basic laws of chemical combination and in Dalton's atomic theory and its further testing. The stoichiometry of chemical reactions can be expanded in some interesting ways following the introduction of a few key ideas about gases. Finally, the study of gases provides a basis for one of the great theories of science, the kinetic molecular theory. This theory will provide us with new insights, particularly of the concept of temperature.

5-1 Properties of a Gas

Gases may be characterized in many ways. All gases expand to fill and assume the shapes of the containers in which they are placed. All gaseous substances diffuse into one another and mix in all proportions; that is, all gaseous mixtures are homogeneous solutions. Gases are invisible in the sense that there are no visible particles of a gas. Some gases are colored, such as gaseous chlorine, bromine, and iodine; some are combustible, such as hydrogen; and some are chemically inert, such as helium and neon.

Four basic properties determine the physical behavior of a gas: amount of gas, gas volume, temperature, and pressure. From numerical values of three of these it is possible to calculate a value for the fourth. This is done through a mathematical equation called an **equation of state.** In principle at least, many other properties of a gas can be calculated from an equation of state. We have already discussed to some extent the properties of mass, volume, and temperature. A brief discussion of gas pressure is presented next.

5-2 Gas Pressure

The observation that a rubber balloon expands when inflated with a gas is a familiar one, but what maintains the balloon in its distended shape? A plausible hypothesis is that molecules of a gas are in constant motion, colliding with one another and with the walls of their container. In collisions with the container walls, a force is imparted. It is this force that keeps a balloon distended. To measure the total force imparted by a gas is not a simple matter, but gas *pressure* can be measured rather easily. Pressure is a force per unit area, that is, a force divided by the total area over which the force is exerted.

$$P = \frac{F}{A} \tag{5.1}$$

LIQUID PRESSURE. The pressure of a gas is most often measured *indirectly* by comparison with a liquid pressure. The concept of liquid pressure is illustrated in Figure 5-1 for a liquid with density, d, contained in a cylinder with cross-sectional area, A, filled to a height, h. Equation (5.2) shows that *the pressure exerted by a liquid depends only on the height of the liquid column and the density of the liquid*. To establish this fact, recall the following: Weight (W) is a force. Weight (W) and mass (m) are proportional. The mass of a liquid is equal to the product of its density and volume ($m = d \cdot V$). The volume (V) of a cylinder is equal to the product of its height (h) and cross-sectional area (A).

$$P = \frac{F}{A} = \frac{W}{A} = \frac{mg}{A} = \frac{gdV}{A} = \frac{gdAh}{A} = ghd \tag{5.2}$$

MEASUREMENT OF GAS PRESSURE. The most familiar gas is air. Actually, air is a mixture of several gases—principally nitrogen (78.08%), oxygen (20.95%), argon (0.93%), and carbon dioxide (0.03%). Life on the surface of earth exists at the bottom of a "sea" of air called the atmosphere. The atmosphere extends for hundreds of kilometers above the earth's surface and has a mass of about 5×10^{18} kg, although about one half of this mass is concentrated in the lowest 6 km. All objects on earth are subjected to a pressure produced by this blanket of air. In 1643, Torricelli constructed the device pictured in Figure 5-2 to measure the pressure of the atmosphere. This device is called a **mercury barometer.**

As pictured in Figure 5-2a, if a long glass tube having *both ends open* is placed upright in a container of mercury, the mercury levels inside and outside the tube are the same. To create the situation in Figure 5-2b, first a long glass tube (say, about 1 m long) is *sealed at one end* and filled with liquid mercury. Next, the open end is kept closed while the tube is inverted

FIGURE 5-1
The concept of liquid pressure.

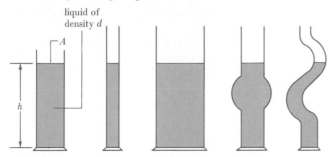

The pressure exerted by the liquid in the cylinder on the left is calculated in the text and shown, through equation (5.2), to depend only on the height of the liquid column (h) and the density of the liquid (d). If all of the vessels pictured above are filled to the same height with the same liquid, the liquid pressures are the same despite the fact that the container shapes and volumes are different.

The constant g, the acceleration due to gravity, has the same significance here as in equation (1.1). It is used to relate a force (weight) and a mass.

FIGURE 5-2
Measurement of atmospheric pressure with a mercury barometer.

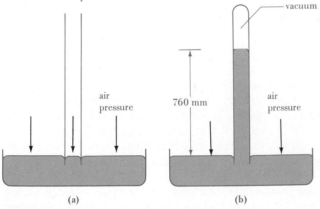

Black arrows represent the pressure exerted by the atmosphere.
(a) The mercury levels are equal inside and outside the open-end tube.
(b) A column of mercury 760 mm high is maintained in the closed-end tube.

One early hypothesis was that the top of the mercury column was attached to the top of the tube by an invisible thread.

into a container of mercury. Then this end is reopened. The mercury level in the tube does not drop to that in the outside container. Instead, it falls to a certain height and remains there. Something must maintain the mercury inside the tube at a greater height than that of the mercury outside the tube. Some of the original hypotheses used to explain this phenomenon involved forces within the tube. We now understand that these forces arise from *outside* the tube.

In the open-end tube (Figure 5-2a), the atmospheric blanket exerts a pressure on the surface of the mercury, both inside and outside the tube. These pressures are equal and the liquid levels are equal. With the closed-end tube (Figure 5-2b), there is no air inside the tube above the mercury. (Only a trace of mercury vapor is present.) The atmosphere exerts a force on the surface of the liquid. This force is transmitted through the liquid mercury, pushing it up into the tube. The column of liquid mercury in the tube exerts a downward pressure that depends on its height (and the density of liquid mercury). With a particular height of mercury in the tube, the pressure at the bottom of the mercury column and that of the atmosphere are equal and the column is maintained.

If a barometer is carried to a mountaintop, the mercury level falls. The level also falls during wet, stormy weather. It rises during clear, dry weather.

The height of mercury in a barometer is not constant but varies with the location and atmospheric conditions. When conditions are such that the height of mercury in a barometer at sea level and 0°C is 760 mm, atmospheric pressure is said to be normal. This statement relates two useful units of pressure, the standard atmosphere (atm) and the millimeter of mercury (mmHg).

1 atm = 760 mmHg   (at sea level and 0°C)

To honor Torricelli the pressure unit **torr** is sometimes used. The torr is defined as exactly (1/760) of a standard atmosphere.

760 torr = 1 atm

Thus, a pressure of 1 torr is the same as 1 mmHg (at sea level and 0°C).

Mercury is a relatively rare, expensive, and rather poisonous liquid. Why use it rather than water as the liquid in a barometer? The answer lies in the extreme height necessary for a water barometer, as calculated in Example 5-1.

Example 5-1 What is the height of a water column, in meters, that could be maintained by normal atmospheric pressure?

If we recognize that 1 atm of pressure will support a column of mercury 76.0 cm high, we can rephrase the question: What is the height of a column of water that exerts the same pressure as a column of mercury 76.0 cm (760 mm) high?

$$\text{pressure of Hg column} = gh_{Hg}d_{Hg} = g \times 76.0 \text{ cm} \times 13.6 \text{ g/cm}^3$$

$$\text{pressure of H}_2\text{O column} = gh_{H_2O}d_{H_2O} = g \times h_{H_2O} \times 1.00 \text{ g/cm}^3$$

$$g \times h_{H_2O} \times 1.00 \text{ g/cm}^3 = g \times 76.0 \text{ cm} \times 13.6 \text{ g/cm}^3$$

$$h_{H_2O} = 76.0 \text{ cm} \times \frac{13.6}{1.00} = 1.03 \times 10^3 \text{ cm} = 10.3 \text{ m}$$

SIMILAR EXAMPLE: Exercise 3.

Example 5-1 helps us to understand the operation of an old-fashioned suction pump for drawing water from a well. As illustrated through Figure 5-3, the pump action is used to evacuate air from a cylindrical pipe in the well. Atmospheric pressure, acting on the surface of the water in the well, *pushes* a column of water up the evacuated pipe. Even if all the air in the pipe could be evacuated (which it cannot), the column of water could not be raised higher than 10.3 m. The use of a straw for drinking liquids is based on the same principle as the suction pump.

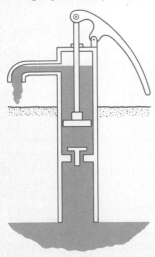

FIGURE 5-3
Pumping water by suction.

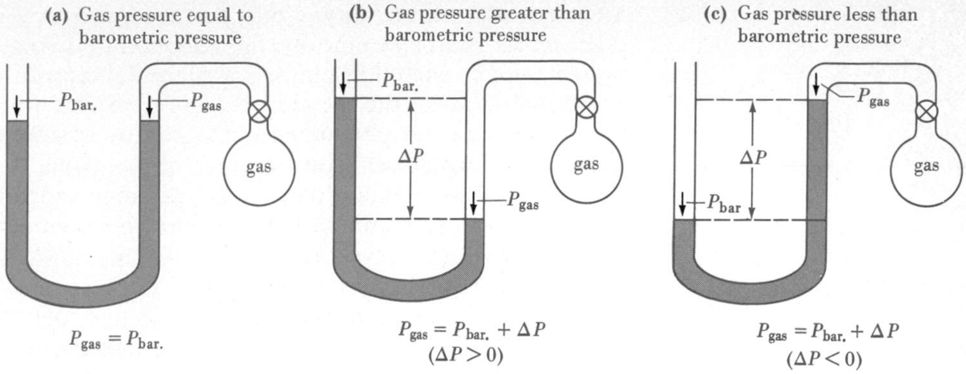

FIGURE 5-4
Measurement of gas pressure with an open-end manometer.

The possible relationships between a measured gas pressure and barometric pressure are pictured here.

(a) Gas pressure equal to barometric pressure

$P_{gas} = P_{bar.}$

(b) Gas pressure greater than barometric pressure

$P_{gas} = P_{bar.} + \Delta P$
$(\Delta P > 0)$

(c) Gas pressure less than barometric pressure

$P_{gas} = P_{bar.} + \Delta P$
$(\Delta P < 0)$

It is impractical to use a barometer to measure the pressures of gases other than air. One cannot usually introduce a mercury barometer directly into a container of a gas. A device that is commonly used to measure gas pressures in the laboratory is a **manometer.** The principle of an open-end manometer is illustrated in Figure 5-4. As long as the gas pressure being measured and the prevailing atmospheric (barometric) pressure are equal, the heights of the mercury columns in the two arms of the manometer are equal. A difference in height of the two arms means a difference between the gas pressure and barometric pressure. The device pictured in Figure 5-5, a closed-end manometer, is useful for measuring low gas pressures.

When the gas pressure to be measured is close to barometric pressure, greater accuracy is obtained with a manometer containing a nonvolatile liquid with a lower density than mercury. In Example 5-3 this means measuring a difference in liquid heights of 8.2 mm (for glycerol) instead of 0.76 mm (for mercury).

Example 5-2 What is the gas pressure, P_{gas}, if the conditions in Figure 5-4b are such that barometric pressure is 748.2 mmHg and the difference in mercury levels, $\Delta P = 25.0$ mm?

$$P_{gas} = P_{bar.} + \Delta P = 748.2 \text{ mmHg} + 25.0 \text{ mmHg} = 773.2 \text{ mmHg}$$

SIMILAR EXAMPLE: Exercise 4.

Example 5-3 What is the gas pressure, P_{gas}, if the conditions in Figure 5-4c are such that the manometer is filled with *glycerol* ($d = 1.26$ g/cm^3), the barometric pressure is 762.4 mmHg, and the difference in glycerol levels is 8.2 mm?

First, we must convert the difference in pressure, expressed as *8.2 mm of glycerol*, to an equivalent height of mercury. This can be done in exactly the same way as in Example 5-1.

$$gh_{Hg}d_{Hg} = g \times h_{Hg} \times 13.6 \text{ g/cm}^3 = gh_{glyc.}d_{glyc.} = g \times 8.2 \text{ mm} \times 1.26 \text{ g/cm}^3$$

$$h_{Hg} = 8.2 \text{ mm} \times \frac{1.26}{13.6} = 0.76 \text{ mmHg}$$

In the remaining calculation, $\Delta P = -0.76$ mmHg. (P_{gas} is less than $P_{bar.}$)

$$P_{gas} = P_{bar.} + \Delta P = 762.4 \text{ mmHg} - 0.76 \text{ mmHg} = 761.6 \text{ mmHg}$$

SIMILAR EXAMPLES: Exercises 4, 5.

FIGURE 5-5
Measurement of gas pressure with a closed-end manometer.

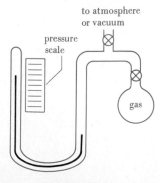

to atmosphere or vacuum

pressure scale

gas

When the manometer is open to the atmosphere, air pressure pushes the mercury level to the top of the closed end. When the manometer is connected to a source of high vacuum, the mercury level in the closed end falls until the levels in the two arms of the U-tube become equal. This is the condition of "zero pressure." When the manometer is connected to a container of gas, the mercury level in the closed end rests at a level that can be read on a scale. This is the measured pressure. Manometers of this type are generally used in the pressure range from about 5 to 300 mmHg. Of course, the longer the closed arm of the manometer, the higher the pressure that can be measured.

UNITS OF PRESSURE. Many different units are used currently to express pressure. This is one quantity for which the adoption of a single unit, as proposed by the SI system, might eliminate some confusion. Some units of pressure are based on the height of a liquid column, some on a mass per unit area, and some on an actual force (weight) per unit area. In the set of equalities that follows, several different ways of expressing normal atmospheric pressure are given. The first three, shown in color, are the units we have introduced to this point and will continue to use throughout this text. The fourth and fifth units are based on a mass (rather than a force) per unit area. The unit pounds per square inch (psi) is employed most commonly in engineering work. The sixth and seventh units, shown in boldface type, are the ones preferred in the SI system. The pascal is named after Blaise Pascal, a seventeenth-century scientist who made significant contributions to our understanding of pressure. In time, SI units may become the common ones for expressing pressure, but presently their use is limited. The unit millibar is commonly used by meteorologists.

normal atmospheric pressure

$$
\underset{(1)}{= 1 \text{ standard atmosphere (1 atm)}} = \underset{(2)}{760 \text{ mmHg}} = \underset{(3)}{760 \text{ torr}}
$$

$$
= \underset{(4)}{14.7 \text{ lb/in.}^2 \text{ (psi)}} = \underset{(5)}{1.033 \text{ kg/cm}^2}
$$

$$
= \underset{(6)}{\mathbf{101{,}325 \text{ newtons/m}^2 \ (N \cdot m^{-2})}} = \underset{(7)}{\mathbf{101{,}325 \text{ pascals (Pa)}}}
$$

$$
= \underset{(8)}{1.01325 \text{ bars}} = \underset{(9)}{1013.25 \text{ millibars (mb)}}
$$

(5.3)

5-3 The Simple Gas Laws

BOYLE'S LAW. Of the several relationships among gas variables, the first to be discovered was the one between gas pressure and volume. This was accomplished in 1662 by Robert Boyle. Boyle found that *the volume of a fixed amount of gas maintained at a constant temperature is inversely proportional to the gas pressure.* The meaning of this statement is suggested pictorially and graphically through Figure 5-6.

The gas pictured in Figure 5-6 is contained in a cylinder that is closed off by a freely moving "weightless" piston. The pressure of the gas is determined by the total weight placed on top of the piston. (This weight is a force that, divided by the area of the piston, yields the gas pressure directly.) If the weight on the piston is doubled, the pressure doubles and the gas volume decreases to one half of its original value; and so on.

The situation pictured in Figure 5-6 is like operating a football pump with the needle end plugged. The handle can be depressed to some extent, and the air in the pump is compressed. But then it becomes increasingly difficult to reduce the gas volume further (more pressure and thus more force is required).

FIGURE 5-6
Relationship between gas volume and pressure—Boyle's law.

When temperature and the amount of gas are held constant, a doubling of the pressure causes the volume to decrease to one half its original value.

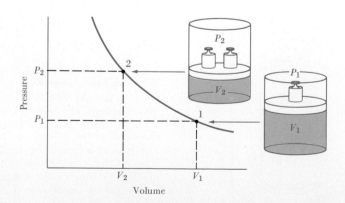

FIGURE 5-7
An application of Boyle's law—Example 5-4 visualized.

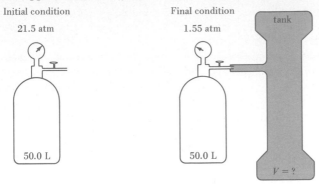

The inverse relationship between pressure and volume can be expressed mathematically.

$$P \propto \frac{1}{V}$$

or $P = \dfrac{a}{V}$ or $PV = a$ (a constant)   (5.4)

Equation (5.4) shows that the product of the pressure and volume of a fixed amount of gas at a constant temperature is a constant (a). The graph of the relationship $PV = a$ shown in Figure 5-6 is of a form called an equilateral (or rectangular) hyperbola.

Example 5-4   The volume of a large, irregularly shaped, closed tank is determined as follows. The tank is first evacuated, and then it is connected to a 50.0-L cylinder of compressed nitrogen gas. The gas pressure in the cylinder, originally at 21.5 atm, falls to 1.55 atm after it is connected to the evacuated tank. What is the volume of the tank? (See Figure 5-7.)

Equation (5.4) is written for the initial condition (i) and for the final condition (f).

$$P_i V_i = a = P_f V_f$$

Now solve for the final volume, V_f.

$$V_f = V_i \times \frac{P_i}{P_f} = 50.0 \text{ L} \times \frac{21.5 \text{ atm}}{1.55 \text{ atm}} = 694 \text{ L}$$

Of this volume, 50.0 L is that of the cylinder. The volume of the tank is 694 L − 50.0 L = 644 L.

SIMILAR EXAMPLES: Exercises 6, 7.

CHARLES' LAW.   The relationship between gas volume and temperature was discovered by the French physicist Charles in 1787 and restated by Gay-Lussac in 1802.

Figure 5-8 pictures a fixed amount of gas confined in a cylinder. The pressure is held constant while the temperature is varied. The volume of gas increases as the temperature is increased or decreases as the temperature is decreased; the relationship is linear (straight line). Three possibilities are indicated in the figure.

A common feature of the lines in Figure 5-8 is the point of intersection with the temperature axis. Although different at every other temperature, the gas volumes for the three cases shown all appear to reach a value of zero at some temperature below −270°C—actually at −273.15°C. The temperature −273.15°C corresponds to that at which the volume of a hypothetical gas would become zero. This temperature is called the **absolute zero of temperature.**

If the volume axis of Figure 5-8 is shifted 273.15°C

The final volume is the cylinder volume (50.0 L) plus that of the tank. The amount of gas remains constant when the cylinder is connected to the evacuated tank, but the pressure drops from 21.5 to 1.55 atm.

The following "common-sense" method can also be used. The final gas volume is equal to the initial volume multiplied by a ratio of pressures.

$$V_f = V_i \times \frac{\text{ratio of}}{\text{pressures}}$$

Only the ratio of pressures 21.5 atm/1.55 atm will produce a final volume larger than the initial volume.

$$V_f = 50.0 \text{ L} \times \frac{21.5 \text{ atm}}{1.55 \text{ atm}}$$
$$= 694 \text{ L}$$

All gases condense to liquids and solids before the temperature reaches the absolute zero; and when we speak of the volume of a gas we mean the free volume among the gas molecules, not the volume of the molecules themselves. Thus, the hypothetical gas referred to here is one whose molecules are point masses and which does not condense to a liquid or solid.

FIGURE 5-8

Gas volume as a function of Celsius temperature.

As the temperature of a fixed amount of gas at a constant pressure is lowered, the volume decreases. Three different starting conditions are pictured here: *A*, 10 cm^3 of gas at 1 atm and 100°C; *B*, 40 cm^3 of gas at 1 atm and 200°C; *C*, 100 cm^3 of gas at 1 atm and 300°C.

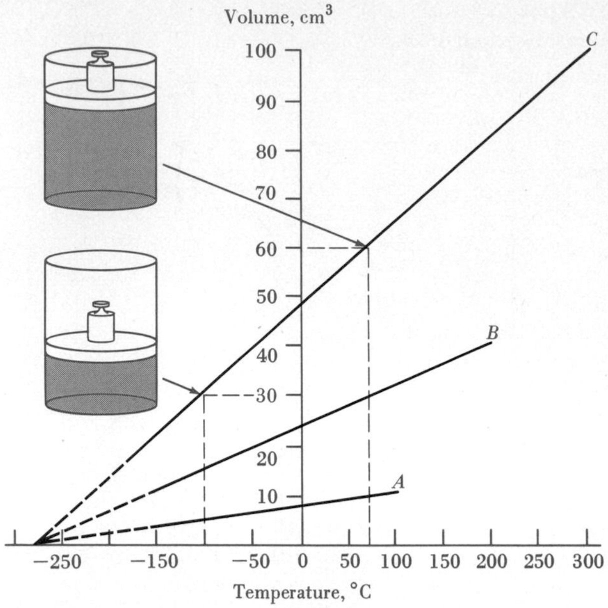

According to the SI convention, a kelvin temperature is referred to simply as kelvin and K, not degrees kelvin and °K. Thus, the kelvin temperature corresponding to 0°C is 273.15 K.

FIGURE 5-9

Gas volume as a function of kelvin temperature.

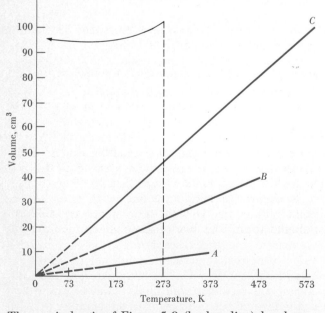

The vertical axis of Figure 5-8 (broken line) has been shifted 273.15° to the left. Note how the points *A*, *B*, and *C*, which were at 100, 200, and 300°C in Figure 5-8, now appear at 373, 473, and 573 K, respectively.

to the left, as shown in Figure 5-9, the straight lines then pass through the origin of the new axes. The origin corresponds to the hypothetical zero volume at the absolute zero of temperature. The further effect of shifting the volume axis in this way is that we must add 273.15 degrees to each temperature value. This leads to the following equation for converting from Celsius to **kelvin** or **absolute** temperature:

$$T \text{ (K)} = t \text{ (°C)} + 273.15 \qquad (5.5)$$

Thus, Charles' law may be stated in this way: *The volume of a fixed amount of gas at constant pressure is directly proportional to the kelvin (absolute) temperature.* Mathematically, this may be written as

$$V \propto T$$

or $\quad V = bT \quad$ (where *b* is a constant) $\qquad (5.6)$

From equation (5.6) we can see that doubling the kelvin (absolute) temperature of a gas causes its volume to double. (Explain why increasing the temperature of a gas from 1°C to 2°C would not cause its volume to double.)

Example 5-5 A 75.0-cm^3 sample of a gas at 10.0°C is heated to 100.0°C while the pressure is held constant at 1 atm. What volume does the gas occupy at 100.0°C?

Equation (5.6) is written for the initial condition (*i*) and for the final condition (*f*).

$$\frac{V_i}{T_i} = b = \frac{V_f}{T_f}$$

Now solve for the final volume, V_f.

$$V_f = V_i \times \frac{T_f}{T_i} = 75.0 \text{ cm}^3 \times \frac{(273 + 100) \text{ K}}{(273 + 10) \text{ K}}$$

The "commonsense" approach suggested in the marginal note on page 93 works here as well. That is, the final gas volume is the initial volume multiplied by a ratio of kelvin temperatures.

$$V_f = V_i \times \frac{\text{ratio of kelvin}}{\text{temperatures}}$$

Only the ratio 373 K/283 K will produce a final volume larger than the initial volume.

$$V_f = 75.0 \text{ cm}^3 \times \frac{373 \text{ K}}{283 \text{ K}}$$
$$= 98.9 \text{ cm}^3$$

$$= 75.0 \text{ cm}^3 \times \frac{373 \text{ K}}{283 \text{ K}} = 98.9 \text{ cm}^3$$

SIMILAR EXAMPLES: Exercises 8, 9.

STANDARD CONDITIONS OF TEMPERATURE AND PRESSURE. Because gas properties depend on the temperature and pressure, it is convenient to specify a particular temperature and pressure at which comparisons can be made. The standard temperature for gases is defined as $0°C = 273.15$ K and the standard pressure as 1 atm = 760 mmHg. Standard conditions are sometimes abbreviated as STP (or SC).

5-4 The Gas Laws and Development of the Atomic Theory

AVOGADRO'S HYPOTHESIS. An important early verification of Dalton's atomic theory seemed to come in 1808 when Gay-Lussac published his studies on the combining volumes of gases. Gay-Lussac reported that when gases react with one another they do so by volumes that are in the ratios of *small whole numbers*. For example, nitrogen and oxygen were known to form three different compounds, with the combining ratios of nitrogen to oxygen, by volume, being 2 : 1, 1 : 1, and 1 : 2. These simple ratios do not exist for reactions involving solids and liquids, nor do such ratios exist, even for gases, if masses are compared rather than volumes.

One explanation of the law of combining volumes was that equal volumes of different gases, under identical conditions of temperature and pressure, contain equal numbers of particles (atoms). If chemical combination involves the union of atoms in simple numerical ratios, the combining volumes should also be in simple numerical ratios. There were some valid objections to this line of reasoning, however. Dalton argued that in the reaction of hydrogen and oxygen to form water the number of particles of water (OH) formed should be the same as the number of atoms of hydrogen (H) and of oxygen (O) reacting. If the "equal volumes–equal numbers" hypothesis were correct, the ratio of volumes of reactants and products should have been 1 : 1 : 1. By experiment the ratio proved to be *two* volumes of hydrogen to *one* of oxygen and *two* of steam—2 : 1 : 2.

Avogadro published a paper in 1811 demonstrating that Gay-Lussac's law and Dalton's theory could be reconciled if one made *two* assumptions.

1. Equal volumes of different gases, under identical conditions of temperature and pressure, contain equal numbers of particles.
2. In many gases the ultimate particles are *molecules* consisting of a number of atoms joined together.

Avogadro proposed that hydrogen and oxygen both exist as *diatomic molecules,* that is, as H_2 and O_2, and that water has *two* H atoms for every O—H_2O! In the reaction of hydrogen and oxygen, the O_2 molecules split into half-molecules (atoms). The H_2 molecules and O half-molecules produce the same number of water molecules (H_2O) as H_2 molecules reacted. From two volumes of hydrogen and one of oxygen, two volumes of steam should form. This line of reasoning is pictured in Figure 5-10.

CANNIZZARO'S WORK. The scientific community was not ready for such bold assumptions as Avogadro's. His hypothesis was little used until it was promoted by Cannizzaro a half-century later. Cannizzaro reasoned as follows.

Take the atomic weight of hydrogen to be exactly 1. Assume that hydrogen exists as diatomic molecules, H_2. The molecular weight of hydrogen becomes exactly 2. Next, determine the volume of hydrogen gas that, under certain conditions of tem-

FIGURE 5-10
Formation of water—actual observation and Avogadro's hypothesis.

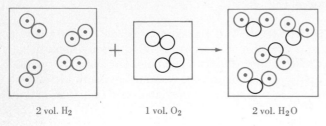

2 vol. H₂   1 vol. O₂   2 vol. H₂O

FIGURE 5-11
Cannizzaro's method illustrated.

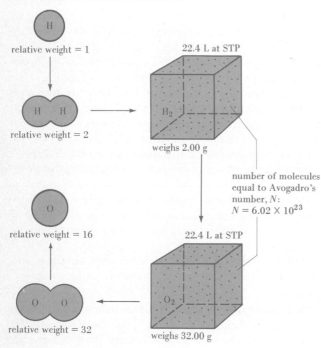

relative weight = 1

relative weight = 2

weighs 2.00 g

relative weight = 16

relative weight = 32

22.4 L at STP

weighs 32.00 g

number of molecules equal to Avogadro's number, N:
$N = 6.02 \times 10^{23}$

22.4 L at STP

weighs 2.00 g

TABLE 5-1
Cannizzaro's method—the atomic weight of nitrogen

Substance	Mol. wt. (relative to H = 1)	Nitrogen, % by mass	Relative mass of N per molecule
hydrogen	2	—	—
ammonia	17	82.5	14
nitrous oxide	44	63.7	28
nitric oxide	30	46.7	14
nitrogen dioxide	46	30.4	14
nitrogen gas	28	100.0	28

perature and pressure, weighs exactly 2 g. The conditions chosen were 0°C and 1 atm (STP), and the volume proved to be 22.4 L. Now, 22.4 L of some other gas at STP should contain the same number of molecules as does 22.4 L of hydrogen. The ratio of the mass of 22.4 L of this gas to the mass of 22.4 L of hydrogen should yield the molecular weight of the gas. This procedure is illustrated in Figure 5-11 for hydrogen and oxygen. By experiment, 22.4 L of oxygen at STP is found to weigh 32.00 g, and assuming the formula O₂, one arrives at an atomic weight of oxygen equal to 16.

This same procedure can be used to determine the atomic weights of other elements, as suggested through Table 5-1 for a series of gaseous nitrogen compounds. We may reason as follows.

1. 22.4 L of ammonia gas at STP weighs 17 g.
2. Ammonia consists of 82.5% N, by mass.
3. The relative mass of nitrogen in ammonia is 17 × 0.825 = 14.
4. 22.4 L of nitrous oxide at STP weighs 44 g.
5. Nitrous oxide consists of 63.7% N, by mass.
6. The relative mass of nitrogen in nitrous oxide is 44 × 0.637 = 28.
7. And so on.

The relative masses of nitrogen in all the molecules listed in Table 5-1 are either 14 or a multiple of 14. A plausible conclusion is that the atomic weight of nitrogen is 14, and that there is one N atom per molecule of ammonia, two per molecule of nitrous oxide, and so on.

It is customary to state **Avogadro's hypothesis** in two ways.

1. Equal volumes of different gases compared at the *same temperature and pressure* contain equal numbers of molecules.
2. Equal numbers of molecules of different gases compared at the *same temperature and pressure* occupy equal volumes.

Furthermore, the volume of a gas at a fixed temperature and pressure must be directly proportional to the amount of gas (i.e., to the number of molecules). If the amount of gas is doubled, the volume doubles; and so on. A simple mathematical statement of this fact is

$$V \propto n \quad \text{and} \quad V = cn \tag{5.7}$$

where c is a constant and n is the amount of gas (usually in moles).

At STP the number of molecules contained in 22.4 L of a gas is *1 mol* (see Figure 5-11). This quantity, 22.4 L of a gas at STP, is often referred to as the **molar volume of a gas.**

Students sometimes lose sight of the fact that Avogadro's law and statements derived from it, such as the molar volume of 22.4 L at STP, apply *only to gaseous substances*. There is no similar relationship dealing with liquids or solids.

5-5 The Ideal Gas Equation

The simple gas laws are stated again below.

Boyle's law: $V \propto \dfrac{1}{P}$ (*n* and *T* constant)

Charles' law: $V \propto T$ (*n* and *P* constant)

Avogadro's law: $V \propto n$ (*P* and *T* constant)

We may combine these three proportionalities into a single one and then replace that by an equality, where the proportionality constant is referred to as the **gas constant,** *R*.

$$V \propto \frac{nT}{P} \tag{5.8}$$

and

$$V = \frac{RnT}{P} \quad \text{or} \quad PV = nRT \tag{5.9}$$

Any gas that obeys all three of the gas laws will also obey equation (5.9). Such a gas is called an ideal gas, and equation (5.9) is called the **ideal gas equation.** [This is the equation of state of an idealized gas (recall Section 5-1).] Real gases can only approach the behavior implied by the ideal gas equation, as we shall learn shortly. Under suitable conditions, however, enough real gases do approach this behavior to make the equation very useful.

Before equation (5.9) can be applied to specific situations, a numerical value is needed for *R*. One of the simplest means of establishing this value is to substitute into equation (5.9) the molar volume at STP, 22.414 L.

We will generally express *R*, which has the same value for all gases, to three significant figures.

0.0821 L atm mol^{-1} K^{-1}

Alternative units for *R* are required in some of its applications.

$$R = \frac{PV}{nT} = \frac{1 \text{ atm} \times 22.414 \text{ L}}{1 \text{ mol} \times 273.15 \text{ K}} = 0.082056 \frac{\text{L atm}}{\text{mol K}} \tag{5.10}$$

In using the ideal gas equation, you should note the following:

1. There are five terms in the equation—*P*, *V*, *n*, *R*, and *T*. Four of them must be known; the equation is solved for the fifth. [In the examples that follow, all five terms are stated first (enclosed in brackets). This assists in identifying the unknown.]
2. Each term must be expressed in the proper units before substitution into the equation. The value of *R* serves as a guide in this. Since its units are liter atmospheres per mole per kelvin (L atm mol^{-1} K^{-1}), the unit for pressure must be atmospheres; for volume, liters; for amount of gas, moles; and for temperature, kelvins.

Example 5-6 What is the volume occupied by 3.50 g Cl_2(g) at 45°C and 745 mmHg?

$$P = 745 \text{ mmHg} \times \frac{1 \text{ atm}}{760 \text{ mmHg}} = \frac{745}{760} \text{ atm} = 0.980 \text{ atm}$$

$$V = ? \quad \text{(this is the unknown)}$$

$$n = 3.50 \text{ g Cl}_2 \times \frac{1 \text{ mol Cl}_2}{70.9 \text{ g Cl}_2} = \frac{3.50}{70.9} \text{ mol Cl}_2 = 0.0494 \text{ mol Cl}_2$$

$$R = 0.0821 \text{ L atm mol}^{-1} \text{ K}^{-1}$$

$$T = 45°\text{C} + 273 = 318 \text{ K}$$

$$PV = nRT$$

Divide both sides by P.

$$P\frac{V}{P} = \frac{nRT}{P}$$

$$V = \frac{nRT}{P}$$

To check the cancellation of units we have generally looked for the same unit in the numerator and the denominator of a setup, for example, atm/atm = 1. Here we need to note that a unit such as mol^{-1} is the same as 1/mol. Thus, $mol \times mol^{-1} = 1$. Also, $K^{-1} \times K = 1$.

$$V = \frac{0.0494\ \text{mol} \times 0.0821\ \text{L atm mol}^{-1}\ \text{K}^{-1} \times 318\ \text{K}}{0.980\ \text{atm}}$$

$$= 1.32\ \text{L}$$

SIMILAR EXAMPLES: Exercises 13 through 17.

Example 5-7   A gas cylinder has a volume of 65.0 L. The cylinder is filled with nitrogen gas to a pressure of 5.1 atm at 20°C. What mass of N_2 is contained in the cylinder?

$$\begin{cases} P = 5.1\ \text{atm} \\ V = 65.0\ \text{L} \\ n = ?\quad \text{(this is the unknown)} \\ R = 0.0821\ \text{L atm mol}^{-1}\ \text{K}^{-1} \\ T = 20°\text{C} + 273 = 293\ \text{K} \end{cases}$$

$$n = \frac{PV}{RT}$$

$$n = \frac{5.1\ \text{atm} \times 65.0\ \text{L}}{0.0821\ \text{L atm mol}^{-1}\text{K}^{-1} \times 293\ \text{K}} = 14\ \text{mol}$$

$$\text{no. g } N_2 = 14\ \text{mol } N_2 \times \frac{28.0\ \text{g } N_2}{1\ \text{mol } N_2}$$

$$= 3.9 \times 10^2\ \text{g } N_2$$

SIMILAR EXAMPLES: Exercises 13 through 17.

In Examples 5-6 and 5-7 a single set of conditions—a single state—was involved. A single application of the ideal gas equation was required. At times a gas is described under *two* different sets of conditions—the initial and final states. Here two applications of the ideal gas equation are required. However, if one or more of the gas variables is held constant, the solution generally takes a simplified form.

FIGURE 5-12
Pressure of a fixed amount of gas in a fixed volume as a function of temperature—Example 5-8 visualized.

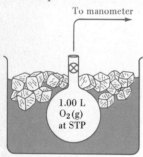

(a) Ice bath

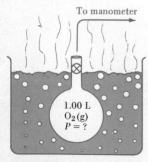

(b) Boiling water

(a) 1.00 L $O_2(g)$ at STP
(b) 1.00 L $O_2(g)$ at 100°C

Example 5-8   Pictured in Figure 5-12 is a 1.00-L flask of $O_2(g)$, first at STP and then at 100°C. What is the pressure of the gas at 100°C?

Since the volume (V) and amount of gas (n) remain constant, we can write the ideal gas equation in the form

$$\frac{n}{V} = \frac{P}{RT} = \frac{P_i}{RT_i} = \frac{P_f}{RT_f}$$

Then we can solve for P_f. (Note that P_i is standard pressure = 1.00 atm.)

$$P_f = P_i \times \frac{T_f}{T_i} = 1.00\ \text{atm} \times \frac{(100 + 273)\ \text{K}}{273\ \text{K}} = 1.00\ \text{atm} \times \frac{373\ \text{K}}{273\ \text{K}} = 1.37\ \text{atm}$$

Actually, what was done in Example 5-8 was the derivation of another simple gas law—Gay-Lussac's law. The pressure of a fixed amount of gas confined to a fixed volume is directly proportional to the absolute temperature. Moreover, this law also permits a "commonsense" solution based on the expression

$$P_f = P_i \times \frac{\text{ratio of kelvin}}{\text{temperatures}}$$

To establish that equation (5.11) is of the proper form, simply substitute the appropriate units for each term and check the cancellation of units.

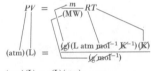

$$(atm)(L) = (L)(atm)$$

This technique is known as *dimensional analysis*. It is especially useful in deriving and memorizing equations.

Note that in this calculation we did not need to determine the actual amount of gas present, nor did we need to use the gas volume.

SIMILAR EXAMPLES: Exercises 11, 16.

5-6 Molecular Weight Determination

We have seen how Avogadro's hypothesis, as applied by Cannizzaro, can be used to establish molecular weights of gases (see Figure 5-11). A more direct approach is to use the ideal gas equation. For this purpose it is helpful to alter the equation slightly. The number of moles of gas, usually expressed as n, is also equal to the mass of gas, m, divided by its molecular weight, (MW), that is, $n = m/(\text{MW})$, and

$$PV = \frac{mRT}{(\text{MW})} \tag{5.11}$$

To determine the molecular weight of a gas using equation (5.11) requires measuring the volume (V) occupied by a known mass of gas (m) at a certain temperature (T) and pressure (P). The form of the ideal gas equation shown in equation (5.11) is not limited to the determination of molecular weights. It can be used in any application in which the quantity of gas is given or sought in grams rather than moles.

Example 5-9 What is the molecular weight of a gas if 1.81 g of the gas occupies a volume of 1.52 L at 25°C and 737 mmHg?

$$P = 737 \text{ mmHg} \times \frac{1 \text{ atm}}{760 \text{ mmHg}} = 0.970 \text{ atm}$$

$$V = 1.52 \text{ L}$$

$$m = 1.81 \text{ g}$$

$$(\text{MW}) = ?$$

$$R = 0.0821 \text{ L atm mol}^{-1} \text{ K}^{-1}$$

$$T = 25 + 273 = 298 \text{ K}$$

Rearrange equation (5.11) as follows.

$$(\text{MW}) = \frac{mRT}{PV}$$

and substitute the known values.

$$(\text{MW}) = \frac{1.81 \text{ g} \times 0.0821 \text{ L atm mol}^{-1} \text{ K}^{-1} \times 298 \text{ K}}{0.970 \text{ atm} \times 1.52 \text{ L}} = 30.0 \text{ g/mol}$$

SIMILAR EXAMPLE: Exercise 19.

Example 5-10 A glass vessel weighs 40.1305 g when clean, dry, and evacuated; 138.2410 g when filled with water at 25.0°C (density of water = 0.9970 g/cm³); and 40.2959 g when filled with propylene gas at 740.4 mmHg and 24.1°C. What is the molecular weight of propylene?

Two minor calculations are required first to establish values for the terms in equation (5.11). These involve determining the volume of the glass vessel (and hence the volume of the gas) and the mass of the gas. The gas constant, R, is expressed to four significant figures in this calculation to correspond to the other measured quantities. This allows the result also to be stated to four significant figures.

mass of water to fill vessel = 138.2410 g − 40.1305 g = 98.1105 g

$$\text{volume of water (volume of vessel)} = 98.1105 \text{ g } H_2O \times \frac{1 \text{ cm}^3 \text{ } H_2O}{0.9970 \text{ g } H_2O}$$

$$= 98.41 \text{ cm}^3 = 0.09841 \text{ L}$$

This method of determining molecular weight (the Dumas method) can be combined with an elemental analysis to yield the molecular formula of a gas. That is, if propylene is found to be 85.63% C and 14.37% H, by mass, what is its molecular formula?

$$\text{mass of gas} = 40.2959 \text{ g} - 40.1305 \text{ g} = 0.1654 \text{ g}$$

$$\text{temperature} = 24.1°C + 273.15 = 297.2 \text{ K}$$

$$\text{pressure} = 740.4 \text{ mmHg} \times \frac{1 \text{ atm}}{760.0 \text{ mmHg}} = 0.9742 \text{ atm}$$

$$(MW) = \frac{mRT}{PV} = \frac{0.1654 \text{ g} \times 0.08206 \text{ L atm mol}^{-1} \text{ K}^{-1} \times 297.2 \text{ K}}{0.9742 \text{ atm} \times 0.09841 \text{ L}} = 42.08 \text{ g/mol}$$

SIMILAR EXAMPLES: Exercises 20, 21.

5-7 Gas Densities

In Examples 5-9 and 5-10 we rearranged equation (5.11) to solve for molecular weight, (MW). A different rearrangement of the equation yields

$$\frac{m}{V} = \frac{(MW)P}{RT} \tag{5.12}$$

The term m/V is the mass of a gas divided by its volume—the **gas density.** Gas densities differ from those of solids and liquids in some important ways:

1. Because gas densities are so low, they are generally stated in g/L instead of g/cm^3.

2. Gas densities are strongly dependent on pressure and temperature, increasing as the gas pressure increases and decreasing as the temperature increases (see equation 5.12). Densities of liquids and solids do depend somewhat on temperature, but they are far less dependent on pressure.

3. The density of a gas is directly proportional to its molecular weight. No simple relationship exists between densities of liquids and solids and their molecular weights.

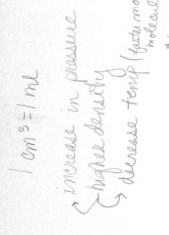

1 cm^3 = 1 ml

increase in pressure → higher density → decrease temp (faster moving molecules, being spread out more)

Example 5-11   What is the density of oxygen gas (O_2) at STP?
The terms for the right-hand side of equation (5.12) are readily obtainable for this example. The density is simply the left-hand side of the equation, m/V.

The density of a gas *at STP* can be easily calculated by dividing the molecular weight of the gas (in this case, 32.0 g/mol) by the molar volume at STP (22.4 L/mol). However, a calculation using equation (5.12) can be applied at various temperatures and pressures.

$$\frac{m}{V} = \frac{(MW)P}{RT} = \frac{32.0 \text{ g mol}^{-1} \times 1 \text{ atm}}{0.0821 \text{ L atm mol}^{-1} \text{ K}^{-1} \times 273 \text{ K}} = 1.43 \text{ g/L}$$

SIMILAR EXAMPLE: Exercise 22.

5-8 Gases in Chemical Reactions

We now have a new tool to apply to calculations dealing with gaseous reactants and/or products of a chemical reaction—the ideal gas equation. Specifically, information about gaseous species can be presented not only in grams and moles, but also in terms of gas volumes, temperatures, and pressures.

Example 5-12   What volume of $O_2(g)$, measured at 735 mmHg and 26°C, is produced when a 7.81-g sample of $KClO_3$ is decomposed? $2 \text{ KClO}_3(s) \xrightarrow{\Delta} 2 \text{ KCl}(s) + 3 \text{ O}_2(g)$

This problem is easiest to solve in two separate calculations. First, the number of moles of $O_2(g)$ is determined in the manner discussed in Chapter 4. Then the ideal gas equation is used to calculate the gas volume.

$$\text{no. mol } O_2(g) = 7.81 \text{ g KClO}_3 \times \frac{1 \text{ mol KClO}_3}{123 \text{ g KClO}_3} \times \frac{3 \text{ mol } O_2(g)}{2 \text{ mol KClO}_3} = 0.0952 \text{ mol } O_2$$

$$\begin{cases} P = 735 \text{ mmHg} \times \dfrac{1 \text{ atm}}{760 \text{ mmHg}} = 0.967 \text{ atm} \\[2mm] V = ? \\[2mm] n = 0.0952 \text{ mol} \\[2mm] R = 0.0821 \text{ L atm mol}^{-1} \text{ K}^{-1} \\[2mm] T = 26°C + 273 = 299 \text{ K} \end{cases}$$

$$V = \frac{nRT}{P} = \frac{0.0952 \text{ mol} \times 0.0821 \text{ L atm mol}^{-1} \text{ K}^{-1} \times 299 \text{ K}}{0.967 \text{ atm}} = 2.42 \text{ L}$$

SIMILAR EXAMPLES: Exercises 26, 27.

LAW OF COMBINING VOLUMES. The ideal gas equation, or more specifically Avogadro's hypothesis, is particularly useful in dealing with a second kind of calculation based on the balanced chemical equation. This relates to a situation in which either all the reactants and products are gases, or at least those involved in the particular calculation. We may reason as follows, taking as an example the formation of nitrogen dioxide from nitrogen oxide.

$$2 NO(g) + O_2(g) \longrightarrow 2 NO_2(g)$$

$$2 \text{ mol } NO(g) + 1 \text{ mol } O_2(g) \longrightarrow 2 \text{ mol } NO_2(g)$$

Now suppose that the gases are compared at the same T and P. Under these conditions 1 mol of gas, any gas, occupies a volume of V liters; 2 mol of a gas occupies $2V$ liters; and so on.

$$2V \text{ L } NO(g) + V \text{ L } O_2(g) \longrightarrow 2V \text{ L } NO_2(g)$$

Or, more simply, divide each coefficient by V:

$$2 \text{ L } NO(g) + 1 \text{ L } O_2(g) \longrightarrow 2 \text{ L } NO_2(g)$$

From this statement of the chemical equation, conversion factors of this kind may be written.

All we have really done in this development is to reaffirm Gay-Lussac's law of combining volumes.

$$\frac{2 \text{ L } NO_2(g)}{2 \text{ L } NO(g)} \backsimeq 1; \qquad \frac{2 \text{ L } NO_2(g)}{1 \text{ L } O_2(g)} \backsimeq 1; \qquad \frac{1 \text{ L } O_2(g)}{2 \text{ L } NO(g)} \backsimeq 1; \qquad \text{and so on.}$$

The use of such conversion factors in a calculation is illustrated in Example 5-13.

Example 5-13 Roasting of zinc sulfide is the first step in the preparation of zinc metal from its ore. The reaction may be represented by the equation

$$2 ZnS(s) + 3 O_2(g) \longrightarrow 2 ZnO(s) + 2 SO_2(g)$$

What volume of $SO_2(g)$ is produced in this reaction for every liter of $O_2(g)$ consumed. Both gases are measured at 740 mmHg and 25°C.

Since the reactant and product being compared are *both gases,* and are *both measured at the same temperature and pressure,* a ratio of combining volumes can be derived from the balanced equation and used as follows:

$$\text{no. L } SO_2(g) = 1.00 \text{ L } O_2(g) \times \frac{2 \text{ L } SO_2(g)}{3 \text{ L } O_2(g)} = 0.67 \text{ L } SO_2(g)$$

SIMILAR EXAMPLES: Exercises 28a, b.

You should note carefully the following points concerning a calculation of the kind performed in Example 5-13.

1. It was not necessary to use the specific information about temperature (25°C) and pressure (740 mmHg) at all. As long as the comparison is made at the *same T and P*, the relationship between volume and the number of molecules (or moles) of a gas is the same for all gases.

2. If a problem is stated in which temperature and pressure are *not* identical for the gases being compared, the method of Example 5-13 will not work. Then it is necessary to convert information about the gases to a mole basis and use conversion factors stated as mole ratios rather than volume ratios. (See Example 5-14.)

3. If the relationship sought in a problem is between a *solid* (or *liquid* in some cases) and a gas, it is again necessary to use a mole ratio as a conversion factor (as in Example 5-12).

Example 5-14 A 20.1-L sample of $H_2(g)$, measured at 0°C and 750 mmHg, is mixed with 11.2 L of $O_2(g)$ measured at 27°C and 720 mmHg. The mixture is ignited and reacts to produce water. What amount of water is formed?

$$2 H_2(g) + O_2(g) \longrightarrow 2 H_2O(l)$$

Although there appears to be more than enough $O_2(g)$ to combine with all the $H_2(g)$, the gas volumes are *not* at the same temperature and pressure. The law of combining volumes does *not* apply. To determine the limiting reagent, convert the quantities of the two reactants to moles and compare them.

$$n_{H_2} = \frac{PV}{RT} = \frac{750 \text{ mmHg} \times \dfrac{1 \text{ atm}}{760 \text{ mmHg}} \times 20.1 \text{ L}}{0.0821 \text{ L atm mol}^{-1} \text{ K}^{-1} \times 273 \text{ K}} = 0.885 \text{ mol}$$

$$n_{O_2} = \frac{PV}{RT} = \frac{720 \text{ mmHg} \times \dfrac{1 \text{ atm}}{760 \text{ mmHg}} \times 11.2 \text{ L}}{0.0821 \text{ L atm mol}^{-1} \text{ K}^{-1} \times 300 \text{ K}} = 0.431 \text{ mol}$$

The amount of O_2 required to react with the available H_2 is

$$\text{no. mol } O_2 \text{ required} = 0.885 \text{ mol } H_2 \times \frac{1 \text{ mol } O_2}{2 \text{ mol } H_2} = 0.442 \text{ mol } O_2$$

However, there is only 0.431 mol O_2 available. O_2 is the limiting reagent, and the amount of water produced is

$$\text{no. mol } H_2O = 0.431 \text{ mol } O_2 \times \frac{2 \text{ mol } H_2O}{1 \text{ mol } O_2} = 0.862 \text{ mol } H_2O$$

SIMILAR EXAMPLES: Exercises 28c, 29.

5-9 Mixtures of Gases

Except to establish the number of moles of gas, at no point in our previous use of the simple gas laws or the ideal gas equation has it been necessary to identify the gas. This is because, as a first approximation at least, all gases behave pretty much alike. The ideal gas equation is applicable to all gases under the appropriate conditions. As a result the ideal gas equation may be applied to a *mixture of gases* just as it is to a single gas. To do this it is only necessary to use for the value of *n* the *total* number of moles of molecules in the gaseous mixture.

Example 5-15   What is the pressure exerted by a mixture of 1.0 g H_2 and 5.0 g He when confined to a volume of 5.0 L at 20°C?

$$n_{tot.} = 1.0 \text{ g } H_2 \times \frac{1 \text{ mol } H_2}{2.0 \text{ g } H_2} + 5.0 \text{ g He} \times \frac{1 \text{ mol He}}{4.0 \text{ g He}}$$

$$= 0.50 \text{ mol } H_2 + 1.25 \text{ mol He} = 1.75 \text{ mol gas}$$

$$P = \frac{n_{tot.}RT}{V} \tag{5.13}$$

$$= \frac{1.75 \text{ mol} \times 0.0821 \text{ L atm mol}^{-1} \text{ K}^{-1} \times 293 \text{ K}}{5.0 \text{ L}}$$

$$= 8.4 \text{ atm}$$

SIMILAR EXAMPLES: Exercises 30, 31.

FIGURE 5-13
Dalton's law of partial pressures illustrated.

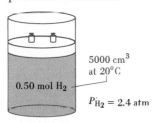

5000 cm^3
at 20°C

0.50 mol H_2

$P_{H_2} = 2.4$ atm

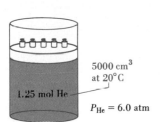

5000 cm^3
at 20°C

1.25 mol He

$P_{He} = 6.0$ atm

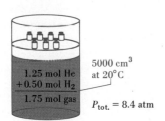

1.25 mol He
+0.50 mol H_2

1.75 mol gas

5000 cm^3
at 20°C

$P_{tot.} = 8.4$ atm

This figure indicates that the pressure of each gas is proportional to the number of moles of that gas. The total pressure of the mixture is the sum of the partial pressures of the individual gases.

In addition to his formulation of the atomic theory, John Dalton made an important contribution to the study of gaseous mixtures. Dalton considered that in a mixture of gases each gas expands to fill the container and exerts a **partial pressure** that is independent of the pressure of other gases. *The sum of these partial pressures is equal to the total pressure of the mixture* (see Figure 5-13). For a mixture of gases, A, B, . . .

$$P_{total} = P_A + P_B + \cdots \tag{5.14}$$

That Dalton's law of partial pressures leads to the same result as equation (5.13) in Example 5-15 can be shown in the following way:

$$P_{tot.} = P_A + P_B + \cdots = \frac{n_A RT}{V} + \frac{n_B RT}{V} + \cdots$$

$$= \frac{RT}{V}(n_A + n_B + \cdots)$$

$$= \frac{n_{tot.}RT}{V}$$

where $n_{tot.} = n_A + n_B + \cdots$.

An alternative approach to Dalton's law (known as Amagat's law) is useful in dealing with gaseous mixtures whose compositions are expressed in percent by volume. Here we begin with the expression

$$V_{tot.} = \frac{n_{tot.}RT}{P_{tot.}}$$

and again note that $n_{tot.} = n_A + n_B + \cdots$. This allows us to write

$$V_{tot.} = \frac{n_A RT}{P_{tot.}} + \frac{n_B RT}{P_{tot.}} + \cdots$$

$$= V_A + V_B + \cdots \tag{5.15}$$

The terms V_A, V_B, . . . are called partial volumes. The **partial volume** of a component in a gaseous mixture is the volume that would be occupied by that component if it were separated from the other components and maintained at the total pressure of the gaseous mixture. *The total volume of a gaseous mixture is the sum of the partial volumes of its components* (equation 5.15).

Still another useful expression for dealing with gaseous mixtures is obtained by taking the ratio of a partial pressure to a total pressure or a partial volume to a total volume.

$$\frac{P_A}{P_{tot.}} = \frac{n_A \frac{RT}{V_{tot.}}}{n_{tot.} \frac{RT}{V_{tot.}}} = \frac{n_A}{n_{tot.}}$$

$$\frac{V_A}{V_{tot.}} = \frac{n_A \frac{RT}{P_{tot.}}}{n_{tot.} \frac{RT}{P_{tot.}}} = \frac{n_A}{n_{tot.}}$$

$$\frac{n_A}{n_{tot.}} = \frac{P_A}{P_{tot.}} = \frac{V_A}{V_{tot.}} \tag{5.16}$$

Example 5-16 What are the partial pressures of H_2 and He in the gaseous mixture described in Example 5-15?

From the number of moles of each gas and the conditions stated in Example 5-15, we may calculate the partial pressures directly.

$$P_{H_2} = \frac{n_{H_2} \cdot RT}{V} = \frac{0.50 \text{ mol} \times 0.0821 \text{ L atm mol}^{-1}\text{ K}^{-1} \times 293 \text{ K}}{5.0 \text{ L}} = 2.4 \text{ atm}$$

$$P_{He} = \frac{n_{He} \cdot RT}{V} = \frac{1.25 \text{ mol} \times 0.0821 \text{ L atm mol}^{-1}\text{ K}^{-1} \times 293 \text{ K}}{5.0 \text{ L}} = 6.0 \text{ atm}$$

As expected, these partial pressures, when added together, yield the total pressure calculated in Example 5-15—8.4 atm.

A second method makes use of the total pressure first calculated in Example 5-15, together with expression (5.16).

$$P_{H_2} = \frac{n_{H_2}}{n_{tot.}} \times P_{tot.} = \frac{0.50}{1.75} \times 8.4 \text{ atm} = 2.4 \text{ atm}$$

$$P_{He} = \frac{n_{He}}{n_{tot.}} \times P_{tot.} = \frac{1.25}{1.75} \times 8.4 \text{ atm} = 6.0 \text{ atm}$$

SIMILAR EXAMPLE: Exercise 32.

Example 5-17 The major components of air are nitrogen, 78.08%; oxygen, 20.95%; argon, 0.93%; and carbon dioxide, 0.03%, by volume. What are the partial pressures of these four gases in a sample of air at normal atmospheric pressure, that is, 1.000 atm?

Volume percentages are directly related to ratios of partial to total volumes. In a total volume of 100.0 L of air, the partial volume of $N_2(g)$ is 78.08 L; $O_2(g)$, 20.95 L; and so on. We then substitute these values into equation (5.16).

$$P_{N_2} = \frac{V_{N_2}}{V_{tot.}} \times P_{tot.} = \frac{78.08 \text{ L}}{100.0 \text{ L}} \times 1.000 \text{ atm} = 0.7808 \text{ atm}$$

$$P_{O_2} = \frac{V_{O_2}}{V_{tot.}} \times P_{tot.} = \frac{20.95 \text{ L}}{100.0 \text{ L}} \times 1.000 \text{ atm} = 0.2095 \text{ atm}$$

$$P_{Ar} = \frac{V_{Ar}}{V_{tot.}} \times P_{tot.} = \frac{0.93 \text{ L}}{100.0 \text{ L}} \times 1.000 \text{ atm} = 0.0093 \text{ atm}$$

$$P_{CO_2} = \frac{V_{CO_2}}{V_{tot.}} \times P_{tot.} = \frac{0.03 \text{ L}}{100.0 \text{ L}} \times 1.000 \text{ atm} = 0.0003 \text{ atm}$$

SIMILAR EXAMPLE: Exercise 35.

A gaseous mixture is sometimes described by its **apparent molecular weight**—the mass of one mole of molecules of the gaseous mixture. The apparent molecular weight can be determined by adding together the contributions of each component to the mass of one mole of the mixture.

Example 5-18 Use data from Example 5-17 to calculate the apparent molecular weight of air.

The key to this calculation again lies in equation (5.16). The ratio of the number of moles of a gaseous component to the total number of moles of gas (i.e., $n_A/n_{tot.}$) is the same as the volume ratio (i.e., $V_A/V_{tot.}$). In "one mole of air," $n_{tot.} = 1.000$ and the numbers of moles of the individual gases are

0.7808 mol N_2;   0.2095 mol O_2;   0.0093 mol Ar;   0.0003 mol CO_2

The apparent molecular weight of air is

$$\left(0.7808 \text{ mol } N_2 \times \frac{28.01 \text{ g } N_2}{1 \text{ mol } N_2}\right) + \left(0.2095 \text{ mol } O_2 \times \frac{32.00 \text{ g } O_2}{1 \text{ mol } O_2}\right)$$

$$+ \left(0.0093 \text{ mol Ar} \times \frac{39.95 \text{ g Ar}}{1 \text{ mol Ar}}\right) + \left(0.0003 \text{ mol } CO_2 \times \frac{44.01 \text{ g } CO_2}{1 \text{ mol } CO_2}\right)$$

$$= 28.96 \text{ g/mol air}$$

SIMILAR EXAMPLE: Exercise 35.

FIGURE 5-14
Collection of a gas over water.

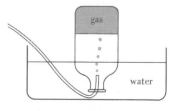

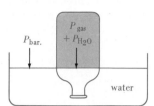

COLLECTION OF GASES OVER WATER. The simple device pictured in Figure 5-14, called a pneumatic trough, represented a great breakthrough in the study of gases in the seventeenth century. It afforded a means of isolating gaseous products of chemical reactions. Since it is based on displacing water from a container, the method works only for gases that are insoluble in water, such as nitrogen, oxygen, and hydrogen.

The gas that is collected is "wet." It is a mixture of the desired gas and water vapor. The gas being collected expands to fill the container and exerts its partial pressure: P_{gas}. Water vapor, produced by the evaporation of liquid water, also fills the container and exerts a partial pressure: P_{H_2O}. The pressure of the water vapor depends only on the temperature of the water. Water vapor pressure data are readily available in tabulated form (see Table 11-1). The concept of vapor pressure is explored more fully in Chapter 11.

According to Dalton's law the *total* pressure is the sum of the two partial pressures. Typically, the total pressure is measured simply by relating it to the prevailing pressure of the atmosphere (barometric pressure). If the container of gas is moved up or down until the water levels are equalized on the inside and outside, then the total gas pressure is made equal to the barometric pressure.

$$P_{bar.} = P_{tot.} = P_{gas} + P_{H_2O}$$

and

$$P_{gas} = P_{bar.} - P_{H_2O} \tag{5.17}$$

Example 5-19 A 1.49-g sample of Ag_2O(s) is decomposed by heating; the liberated oxygen is collected over water at 23°C and at a barometric pressure of 751 mmHg. The water vapor pressure at 23°C is 21.1 mmHg. What is the volume of "wet" $O_2(g)$ collected?

$$2\,Ag_2O(s) \longrightarrow 4\,Ag(s) + O_2(g)$$

This problem is best solved in two steps. First, we calculate the number of moles of O_2 produced in the chemical reaction. Then this information, together with other data about the gas collected, is used in the ideal gas equation.

$$\text{no. mol } O_2 = 1.49 \text{ g } Ag_2O \times \frac{1 \text{ mol } Ag_2O}{232 \text{ g } Ag_2O} \times \frac{1 \text{ mol } O_2}{2 \text{ mol } Ag_2O}$$

$$= 0.00321 \text{ mol } O_2$$

$$P_{O_2} = P_{\text{bar.}} - P_{H_2O} = 751 \text{ mmHg} - 21.1 \text{ mmHg}$$

$$= 730 \text{ mmHg} \times \frac{1 \text{ atm}}{760 \text{ mmHg}} = 0.961 \text{ atm}$$

$$V = ?$$

$$n \,\hat{=}\, 0.00321 \text{ mol}$$

$$R = 0.0821 \text{ L atm mol}^{-1}\text{ K}^{-1}$$

$$T = 23°C + 273 = 296 \text{ K}$$

$$V = \frac{nRT}{P} = \frac{0.00321 \text{ mol} \times 0.0821 \text{ L atm mol}^{-1}\text{ K}^{-1} \times 296 \text{ K}}{0.961 \text{ atm}}$$

$$= 0.0812 \text{ L} = 81.2 \text{ cm}^3$$

SIMILAR EXAMPLES: Exercises 36, 37.

5-10 Kinetic Molecular Theory of Gases

A theory of gases that was held in some regard at the turn of the nineteenth century was the caloric theory. A gas was thought of as comprised of stationary particles surrounded by a fluid caloric envelope. As temperature and pressure changed the caloric envelope was believed to increase or decrease in size, leading to changes in volume. This theory could be used to explain Boyle's law and Charles' law, but it could not explain all observed properties of gases, for example their ability to diffuse into one another.

The simple gas laws are empirical statements of the observed behavior of gases. These laws are reasonably accurate for most gases under normal conditions of temperature and pressure. As we have stated before, scientific laws express behavior through a correlation of observations or experiments. A scientific theory is an explanation of a law or a group of laws. A law is a statement of what will happen; a theory attempts to explain why this happens. A scientific theory is based on a model or concept from which various phenomena can be deduced logically.

The currently accepted theory for explaining gas behavior was developed during the middle nineteenth century. It is the kinetic molecular theory of gases, based on the following model.

1. A gas is comprised of extremely small particles called molecules (or atoms in some cases).
2. The molecules of a gas are usually separated by great distances. As a result they occupy only a very small fraction of the total gas volume. They are, in fact, assumed to be point masses.
3. There are assumed to be no intermolecular forces.
4. The molecules move constantly and randomly throughout the gas volume. As a result of their motion they undergo frequent collisions with one another and with the walls of their container.
5. Collisions between molecules are elastic. Individual molecules may gain or lose energy as a result of collisions; however, in a large collection of molecules at constant temperature, the total energy remains constant.

The basic equation of the kinetic molecular theory is obtained by totaling the forces exerted by the molecules of a gas as they collide with the walls of a container of volume V. The total force, divided by the area over which it is exerted, yields the gas pressure P. In terms of molecular properties, the force of molecular collisions depends on two factors. First is the frequency of molecular collisions with the container walls. In turn, this frequency depends on the speeds of the molecules; the faster they are moving, the more frequently they collide with the container walls. The second factor determining the forces of these collisions is the amount of translational kinetic energy (e_k) possessed by the gas molecules. Translational kinetic energy refers to the energy associated with the motion of a molecule through space: $e_k = \frac{1}{2}(mu^2)$, where m is the mass of the molecule and u is its speed (see also Appendix B). The derivation is greatly complicated, however, by the fact that molecules move in all directions and at different speeds. The result obtained, offered here without proof, is

$$PV = \frac{n'\overline{mu^2}}{3} \tag{5.18}$$

where n' represents the number of molecules in the volume V. We must pay special notice to the term $\overline{u^2}$. It represents the *average* of the *squares* of the molecular speeds (which is not the same as the square of the average speed). Based on this meaning of $\overline{u^2}$, we can represent the average translational kinetic energy of a collection of gas molecules as $\overline{e_k} = \frac{1}{2}(\overline{mu^2})$. By a slight rearrangement of equation (5.18), we obtain

$$PV = \tfrac{2}{3}n'(\tfrac{1}{2}\overline{mu^2}) \quad \text{and} \quad PV = \tfrac{2}{3}n'\overline{e_k} \tag{5.19}$$

Two additional relationships can be derived from equation (5.19) by (a) considering 1 mol of gas, $n' = N$, and (b) using the ideal gas equation for 1 mol of gas, $PV = RT$.

$$PV = \frac{2}{3}N\overline{e_k} = RT \tag{5.20}$$

$$\overline{e_k} = \frac{3}{2}\frac{R}{N}T = \frac{3}{2}kT \tag{5.21}$$

The constant k, which is the gas constant per *molecule*, is called the Boltzmann constant. The significance of equation (5.21) is that it provides a definition of temperature. The kelvin temperature of a gas is directly proportional to the average translational kinetic energy of its molecules. Now we have a new conception of what changes in temperature mean—changes in the intensity of molecular motion.

AVOGADRO'S LAW. Based on equation (5.19) we can write for two different gases (call them A and B)

$$P_A = \frac{2}{3}\frac{n'_A}{V_A}(\overline{e_k})_A \quad \text{and} \quad P_B = \frac{2}{3}\frac{n'_B}{V_B}(\overline{e_k})_B$$

If the two gases are compared at identical temperatures, $(\overline{e_k})_A = (\overline{e_k})_B$, and at identical pressures, $P_A = P_B$. Under these conditions the number of molecules per unit volume must be the same for the two gases.

$$\frac{n'_A}{V_A} = \frac{n'_B}{V_B}$$

Thus, if equal volumes of gases are compared ($V_A = V_B$), the number of molecules of the two gases must be equal, $n'_A = n'_B$. If equal numbers of molecules are compared ($n'_A = n'_B$), the volumes must be equal, $V_A = V_B$.

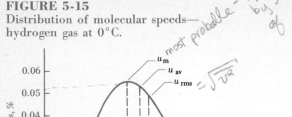

most probable - speed possessed
by largest %
of molecules

FIGURE 5-15
Distribution of molecular speeds—hydrogen gas at 0°C.

The ordinate values represent the percent of the molecules having a certain speed. The abscissas represent these speeds, based on an interval of 1 m/s. (For example, all molecules with speeds between 1499.5 and 1500.5 m/s are taken to have a speed of 1500 m/s.)

DISTRIBUTION OF MOLECULAR SPEEDS. The statements we have made about an average kinetic energy imply that in a collection of molecules there is a *distribution* of energies from very high to very low. And since $\overline{e_k} = \frac{1}{2}m\overline{u^2}$, there must be a distribution of speeds as well. Figure 5-15 shows a typical distribution.

Three different speeds are noted on the curve of Figure 5-15. These are the **most probable** or **modal speed, u_m**, the **average speed, $u_{av.} = \overline{u}$**, and the **root-mean-square speed, $u_{rms} = \sqrt{\overline{u^2}}$**. The root-mean-square speed is the square root of the average of the squares of the speeds of all the molecules in a sample. It can be derived from equation (5.21) by substituting for the value of $\overline{e_k}$.

$$\overline{e_k} = \frac{1}{2}m\overline{u^2} = \frac{3}{2}\frac{R}{N}T \qquad \overline{u^2} = \frac{3RT}{mN}$$

Since the product, mN, represents the mass of 1 mol of molecules, it can be replaced by the molecular weight of the gas, (MW).

$$u_{rms} = \sqrt{\overline{u^2}} = \sqrt{\frac{3RT}{(MW)}} \tag{5.22}$$

Derivations of u_m and $\overline{u}$ are more difficult than for u_{rms}, but these velocities prove to be proportional to u_{rms}.

$$u_m = 0.816\,u_{rms} \qquad u_{av.} = \overline{u} = 0.921\,u_{rms} \tag{5.23}$$

An analogy is suggested in Table 5-2 that may help you to appreciate the distinction among most probable, average, and root-mean-square speed.

To calculate molecular speeds with expressions (5.22) and (5.23) requires that the gas constant be expressed as

$$R = 8.314\ \text{J mol}^{-1}\,\text{K}^{-1}$$

Furthermore, to produce the proper cancellation of units, the joule (J) must be in terms of mass, length, and time. Since kinetic energy is expressed as K.E. $= \frac{1}{2}mv^2$, the joule must have the units of (mass) $\times$ (velocity)^2 = (kg)(m/s)^2. This leads to the value

$$R = 8.314\ \text{kg m}^2\,\text{s}^{-2}\,\text{mol}^{-1}\,\text{K}^{-1}$$

Example 5-20   What is the root-mean-square speed (u_{rms}) of H_2 molecules at 50°C?

Substitute into equation (5.22), noting that R must have the units described above and that the molecular weight, (MW), must be expressed in *kilograms* per mole.

This answer should seem reasonable when compared to Figure 5-15. At 0°C, u_{rms} for $H_2(g)$ is somewhat below 2000 m/s. Raising the temperature will increase u_{rms}.

$$u_{rms} = \sqrt{\frac{3 \times 8.314\ \text{kg m}^2\,\text{s}^{-2}\,\text{mol}^{-1}\,\text{K}^{-1} \times 323\ \text{K}}{2.016 \times 10^{-3}\ \text{kg/mol}}}$$
$$= \sqrt{4.00 \times 10^6\ \text{m}^2/\text{s}^2} = 2.00 \times 10^3\ \text{m/s}$$

TABLE 5-2
Analogy to the distribution of molecular speeds

Consider ten automobiles on a highway traveling at these speeds:

Speed, mi/h	(Speed)^2
40	1,600
42	1,764
45	2,025
48	2,304
50	2,500
50	2,500
55	3,025
57	3,249
58	3,364
60	3,600

most probable (modal) speed = 50

sum of speeds = $\sum$ speed = 505 $\sum$ (speed)^2 = 25,931

average speed = $\overline{\text{speed}}$ = $\dfrac{\sum \text{speed}}{10}$ = **50.5** $\overline{(\text{speed})^2}$ = $\dfrac{\sum (\text{speed})^2}{10}$ = 2593.1

root-mean-square speed

$= \sqrt{\overline{(\text{speed})^2}} = \sqrt{2593.1} =$ **50.9**

modal speed: 50   average speed: 50.5   rms speed: 50.9

FIGURE 5-16
Effusion through an orifice.

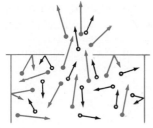

Average speeds of the two different types of molecules are suggested by the lengths of the arrows. The faster molecules (shown in color) effuse more rapidly.

A rubber balloon filled with hydrogen or helium gas gradually deflates, regardless of how tightly it is tied off. This is because gas molecules effuse through tiny, invisible holes in the rubber. As the hydrogen or helium effuses out of the balloon, air effuses in; but because the hydrogen or helium effuses more rapidly, the total number of molecules in the balloon decreases.

SIMILAR EXAMPLES: Exercises 42, 43.

GRAHAM'S LAW. Molecular speeds are quite high. A speed of 1500 m/s corresponds to about 1 mi/s or 3000 mi/h. However, when gases are allowed to mix or diffuse into one another, they do not do so at nearly the rate implied by the speeds of the gas molecules. This is because the molecules collide with one another with great frequency and may change direction as a result of collisions. They do not follow straight-line paths over long distances. Nevertheless, gases do diffuse or mix, and the rate at which this occurs is dependent on the speeds of the gas molecules.

A concept related to diffusion rates is pictured in Figure 5-16. Here molecules are allowed to escape from their container through a tiny orifice or pin hole. This escape through an orifice is called **effusion.** The rates of effusion of molecules are directly proportional to their speeds. Thus, in comparing two different gases at the same temperature, we may write

$$\frac{\text{rate of effusion of A}}{\text{rate of effusion of B}} = \frac{\overline{u_A}}{\overline{u_B}} = \frac{0.921(u_{rms})_A}{0.921(u_{rms})_B} = \frac{(u_{rms})_A}{(u_{rms})_B}$$

A further substitution is possible using equation (5.22):

$$\frac{\text{rate of effusion of A}}{\text{rate of effusion of B}} = \sqrt{\frac{3RT/(MW)_A}{3RT/(MW)_B}} = \sqrt{\frac{(MW)_B}{(MW)_A}} \tag{5.24}$$

The result shown in equation (5.24) is a kinetic-theory derivation of a nineteenth-century law called Graham's law: *The rates of effusion (or diffusion) of two different gases are inversely proportional to the square roots of their molecular weights.*

Equation (5.24) compares the *rates* of effusion of two different gases. Effusion *times* are inversely proportional to effusion rates. That is, the gas that effuses *fastest* (highest rate) takes the *shortest* time to do so.

$$\frac{\text{effusion time for B}}{\text{effusion time for A}} = \frac{\text{rate of effusion of A}}{\text{rate of effusion of B}} = \sqrt{\frac{(\text{MW})_B}{(\text{MW})_A}} \qquad (5.25)$$

Example 5-21   A sample of neon gas escapes through a tiny hole in 122 s. How long would it take an identical sample of hydrogen gas (i.e., one containing the same number of molecules per unit volume) to effuse under the same conditions of temperature and pressure?

Let H_2 be the gas represented as B in equation (5.25) and Ne, gas A.

$$\frac{\text{effusion time for } H_2}{\text{effusion time for Ne}} = \sqrt{\frac{2.02}{20.2}} = \sqrt{0.100} = 0.316$$

$$\text{effusion time for } H_2 = 0.316 \times \text{effusion time for Ne}$$

$$= 0.316 \times 122 \text{ s} = 38.6 \text{ s}$$

SIMILAR EXAMPLES: Exercises 45, 46.

A simple check is always possible in a calculation of this type. The lightest gas (smallest molecular weight) must effuse the fastest (the shortest effusion time).

APPLICATIONS OF DIFFUSION.   That gases diffuse into one another is a commonly experienced phenomenon. Natural gas is odorless. For commercial use small quantities of a gaseous organic sulfur compound are added to natural gas. The sulfur compound has an odor that can be detected in parts per billion (ppb) or less. When a natural gas leak occurs, we rely on the diffusion of this odorous gaseous compound for detection of the leak.

In the Manhattan Project during World War II one of the methods developed for separating the desired isotope $^{235}_{92}U$ from the predominant species $^{238}_{92}U$ involved gaseous diffusion. In this process uranium is obtained as the gaseous hexafluoride, $UF_6(g)$. Molecules of $^{235}UF_6$ diffuse a little faster than those of $^{238}UF_6$, producing a slight enrichment of the ^{235}U isotope. By repeating the process many times over, a separation of the isotopes can be achieved.

5-11 Nonideal Gases

We have stated on several occasions that real gases can be described by the ideal gas equation only under certain conditions. How serious are the departures from ideality displayed by real gases? An indication is given in Figure 5-17, where PV/RT is plotted as a function of P, for 1 mol of gas at a fixed temperature ($0°C$). If a gas is ideal, $PV/RT = 1$ for 1 mol of gas. The extent to which the measured value of PV/RT, called the compressibility factor, deviates from unity is a measure of the nonideality of a gas. Figure 5-17 suggests that all gases behave ideally at sufficiently low pressures, say below 1 atm, but that deviations become significant at increased pressures. At very high pressures the compressibility factor is always greater than 1.

That real gases display behavior that deviates from the ideal is not difficult to rationalize. For example, Boyle's law predicts that at very high pressures a gas volume becomes extremely small, approaching zero. This cannot be, however, because the molecules themselves occupy space and are practically incompressible.

Another source of failure stems from the kinetic theory assumption of no intermolecular forces. The force of collision of a gas molecule with a surface is reduced if the gas molecule is attracted by others within the gas (see Figure 5-18). If such forces of collision are reduced collectively, the pressure exerted by a real gas on the container walls is less than would be predicted for an ideal gas. Intermolecular attractive forces are responsible for compressibility factors less than unity, and they become increasingly important at low temperatures where molecular motion is diminished in intensity.

FIGURE 5-17
The behavior of real gases—compressibility factor for one mole of gas as a function of pressure at $0°C$.

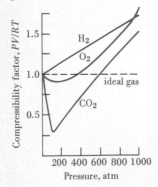

The subject of intermolecular forces is discussed more fully in Chapter 11.

FIGURE 5-18
Intermolecular
attractive forces.

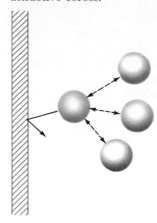

Attractive forces of other
molecules for the mole-
cule shown in color
cause that molecule to
exert less force when it
collides with the wall
than if these attractions
did not exist.

TABLE 5-3
Van der Waals constants
for several gases

Gas	a, L^2 atm mol^{-2}	b, L mol^{-1}
Ar	1.35	0.0322
Cl$_2$	6.49	0.0562
CO	1.49	0.0399
CO$_2$	3.59	0.0427
H$_2$	0.244	0.0266
He	0.034	0.0237
N$_2$	1.39	0.0391
O$_2$	1.36	0.0318
SO$_2$	6.71	0.0564

Calculated with the ideal
gas equation, the value of
P is 11.2 atm. If only the
b term is used in the van
der Waals equation, the
calculated pressure is
11.5 atm. Including the a
term in the van der
Waals equation reduces
the calculated pressure
by 1.62 atm. In-
termolecular forces of
attraction are the principal cause of the failure
of Cl$_2$(g) to behave ideally under the given con-
ditions.

To summarize, high pressures and low temperatures are the conditions that produce nonideal gas behavior. Conversely, gases tend to behave more ideally as the
pressure is lowered and/or the temperature is raised.

THE VAN DER WAALS EQUATION.   A number of equations of state have been proposed for real gases, equations that apply over a wider range of temperatures and
pressures than does the ideal gas equation. Such equations must account for the two
primary factors that cause deviations from ideal gas behavior: the volume associated
with the molecules themselves and the intermolecular forces of attraction. One equation that has found considerable application is the van der Waals equation.

$$\left(P + \frac{n^2 a}{V^2}\right)(V - nb) = nRT \tag{5.26}$$

In equation (5.26) V represents the volume of n moles of gas. The term $n^2 a/V^2$ is
related to the intermolecular forces of attraction. It is added to the pressure because
the measured pressure is lower than anticipated. The term b is related to the volume
of the gas molecules and must be subtracted from the measured volume. Thus,
$V - nb$ represents the *free* volume within the gas. The terms a and b have particular
values for particular gases, and they vary somewhat with temperature and pressure
(see Table 5-3). An equation of state for a nonideal gas is not as general as for an
ideal gas; the identity of the gas enters into the equation in some way. In Example
5-22 the pressure of a real gas is calculated with the van der Waals equation. Solving
equation (5.26) for either n or V is a more difficult matter and will not be considered
here.

Example 5-22   Use the van der Waals equation to calculate the pressure exerted by
1.00 mol Cl$_2$(g) when it is confined to a volume of 2.00 L at 273 K. For Cl$_2$(g):
$a = 6.49$ L^2 atm mol^{-2}; $b = 0.0562$ L/mol.
   Substitute the following values into equation (5.26).

$$P = ?$$
$$V = 2.00 \text{ L}$$
$$n = 1.00 \text{ mol}$$
$$R = 0.0821 \text{ L atm mol}^{-1} \text{ K}^{-1}$$
$$T = 273 \text{ K}$$
$$n^2 a = (1.00)^2 \text{ mol}^2 \times 6.49 \frac{\text{L}^2 \text{ atm}}{\text{mol}^2} = 6.49 \text{ L}^2 \text{ atm}$$
$$nb = 1.00 \text{ mol} \times 0.0562 \text{ L/mol} = 0.0562 \text{ L}$$

$$P = \left(\frac{nRT}{V - nb}\right) - \frac{n^2 a}{V^2}$$

$$= \frac{1.00 \text{ mol} \times 0.0821 \text{ L atm mol}^{-1} \text{ K}^{-1} \times 273 \text{ K}}{(2.00 - 0.0562) \text{ L}} - \frac{6.49 \text{ L}^2 \text{ atm}}{(2.00)^2 \text{ L}^2}$$

$$= 11.5 \text{ atm} - 1.62 \text{ atm} = 9.9 \text{ atm}$$ *note: only 2 significant figures.*

SIMILAR EXAMPLES: Exercises 47, 48.

5-12 Postscript: The Atmosphere

If the outer limit of the atmosphere is taken to be the distance at which its composition becomes the same as that of interplanetary space (very, very low density atomic
hydrogen gas, H), our atmosphere is about 10,000 km "thick." From another view-
point the atmospheric blanket is very "thin." The total mass of the atmosphere is

only about one-millionth that of the earth itself. Whether we think of the atmosphere as thick or thin, there is no question about the crucial role that it plays in the existence of life on earth.

Human beings and other animals depend on the oxygen content of the lower atmosphere to maintain their metabolic processes (see Section 25-3). Plants, through the process of photosynthesis, utilize carbon dioxide, a minor atmospheric component, and return oxygen to the atmosphere (see Section 25-2). Nitrogen, a vital element of life, is circulated among organisms through a complex cycle, called the nitrogen cycle, that originates with atmospheric nitrogen (see Section 26-2). Carbon dioxide plays a key role in maintaining the heat balance of the earth (see Section 26-3), as does gaseous ozone (O_3) found in the stratosphere. Ozone also serves to screen the earth from harmful ultraviolet radiation (see Section 20-4).

The first 80 km or so of the atmosphere is a region known as the homosphere, so called because the composition of the gaseous mixture is essentially uniform (homogeneous) throughout this region. The portion of the atmosphere beyond the 80-km limit is called the heterosphere. It consists of a succession of four layers of gases—molecular nitrogen (N_2), atomic oxygen (O), helium (He), and atomic hydrogen (H).

As suggested by Figure 5-19, the temperature of the atmosphere falls continuously for about the first 10 km above the earth's surface. This 10-km layer of air is that most familiar to us—the **troposphere.** Temperatures in the layer of air from about 10 to 40 km increase slowly from about 220 to 270 K. This is the region known as the **stratosphere.** (Supersonic transport aircraft fly in the lower regions of the stratosphere.) In the third atmospheric layer—the **mesosphere**—the temperature continues to rise to about 300 K and then falls to a minimum of about 180 K at 80 km. In the next layer of the atmosphere, the **thermosphere** (or **ionosphere**), temperatures rise continuously to 1500 K. In this region absorption of ultraviolet radiation from the sun causes gas molecules to ionize and/or dissociate. Thus, the atmosphere at these altitudes consists of positive and negative ions, free electrons, and neutral atoms and molecules. The term *ionosphere* is suggestive of this ionization process, and the term *thermosphere* suggests high temperatures associated with this ionized gas.

FIGURE 5-19
The atmosphere: structure, temperatures, composition, and other phenomena.

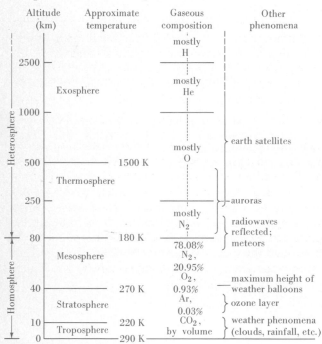

The values shown here are approximate. For example, the height of the troposphere varies from about 8 km at the poles to 16 km at the equator, and temperatures in the thermosphere vary greatly between day and night.

Example 5-23 At an altitude of 300 km the principal gaseous species is *atomic* oxygen, O. Oxygen atoms at this altitude have average speeds, $u_{av.} = 1.36 \times 10^3$ m/s. What is the approximate temperature corresponding to this molecular speed?

We need to use two equations from the kinetic molecular theory described earlier. Equation (5.23) provides a relationship between $u_{av.}$ and u_{rms}, from which we determine

$$u_{rms} = \frac{u_{av.}}{0.921} = \frac{1.36 \times 10^3 \text{ m/s}}{0.921} = 1.48 \times 10^3 \text{ m/s}$$

Equation (5.22) provides a relationship between u_{rms} and kelvin temperature. As in Example 5-20, the gas constant must be expressed as $8.314 \text{ J mol}^{-1} \text{ K}^{-1}$, and the molecular weight of the gas in *kilograms* per mole. (Recall also that 1 joule = $1 \text{ kg m}^2 \text{ s}^{-2}$.)

$$u_{rms} = \sqrt{\frac{3RT}{(MW)}} \quad \text{and} \quad (u_{rms})^2 = \frac{3RT}{(MW)} \quad \text{and} \quad T = \frac{(MW)(u_{rms})^2}{3R}$$

$$T = \frac{0.016 \text{ kg mol}^{-1} \times (1.48 \times 10^3)^2 \text{ m}^2 \text{ s}^{-2}}{3 \times 8.314 \text{ kg m}^2 \text{ s}^{-2} \text{ mol}^{-1} \text{ K}^{-1}} = 1.41 \times 10^3 \simeq 1400 \text{ K}$$

SIMILAR EXAMPLES: Exercises 42, 43.

AN ADDITIONAL INSIGHT INTO THE MEANING OF TEMPERATURE. The temperature of 1400 K calculated in Example 5-23, if achieved in bulk matter at the earth's surface, would be considered very high. (It would be high enough to cause a bright red glow in iron.) At high altitudes in the atmosphere there is a different significance to high temperature. The temperature is high in the sense that molecular speeds are high, but to experience the effects of high temperatures requires that heat be transferred through *frequent* molecular collisions. However, because the gas density at high altitudes is so low, molecular collisions occur only *infrequently*. A thermometer would give very low readings, even though in the midst of highly energetic, *but widely separated* molecules.

Meteors are extraterrestial chunks of matter that are trapped in the earth's gravitational field, disintegrate, and give off light as they fall through the atmosphere (hence the name "shooting stars"). The light emission from meteors is believed to be preceded by evaporation and ionization of atoms from the surface of the meteor. This evaporation of surface atoms in turn results from collisions with molecules of air. Meteors do not start to give up light until they fall to within 110 km of the earth's surface. The majority of them are completely vaporized in the range from about 80 to 110 km. Thus, meteors pass through the higher temperatures of the thermosphere without vaporizing. Instead, they vaporize in a lower-temperature region (about 220 K), but a region where gas densities are higher.

Summary

A gas is described *quantitatively* through four variables—pressure, temperature, volume, and amount of gas. Gas pressure is most readily measured by comparing it to the pressure exerted by a liquid column, usually mercury. This comparison, when applied to the atmosphere itself, is made with a mercury barometer. For gases other than the atmosphere a mercury manometer may be used. Once measured, pressure can be described in a number of different units—mmHg, atm, lb/in.², N/m², pascals, bars, and millibars.

Relationships between gas variables taken two at a time (with the remaining two held constant) are known as the simple gas laws. Most frequently encountered are Boyle's law, relating gas pressure and volume; Charles' law, relating gas volume and temperature; and Avogadro's law, relating volume and amount of gas. Other simple relationships, such as that between pressure and temperature for a fixed amount of gas in a fixed volume, can be readily deduced. A number of important ideas originate with the simple gas laws. Among these are the concept of an absolute zero of temperature, a temperature scale (kelvin) based on this absolute zero, a standard condition of temperature and pressure (STP), and the molar volume of a gas at STP—22.4 L/mol. Also, a variety of calculations is made possible by the simple gas laws.

By combining Boyle's, Charles', and Avogadro's laws, a more general statement of gas behavior is obtained that relates the four variables P, V, n, and T through the gas constant, R. This general statement is the ideal gas equation: $PV = nRT$. The equation can be solved for any one of the variables when values are known for the remaining variables. Simple modifications of the ideal gas equation can be written to deal with molecular weight and gas density determinations.

With a knowledge of the behavior of gases, chemical reactions involving gaseous reactants and/or products can be described more fully. Also, the ideal gas equation can be extended to mixtures of gases. At times this is done by simply using the total number of moles of gas in the ideal gas equation. In other cases partial pressures or partial volumes are employed. Particularly important applications of Dalton's law of partial pressures involve the collection of gases over water.

A theoretical basis for the ideal gas equation is provided by the kinetic molecular theory of gases. With this theory it is possible to establish a relationship between average molecular translational kinetic energy and kelvin temperature. Another relationship involves the root-mean-square speed of molecules, the temperature, and the molecular weight of a gas. Also, the rates and times of effusion of gases can be related to their molecular weights (Graham's law) through the kinetic molecular theory.

Real gases generally behave ideally only at high temperatures and low pressures. The causes of nonideal behavior are to be found in the presence of intermolecular attractions and in the finite volumes occupied by gas molecules. Alternative equations of state have been developed for real gases. The best known of these, perhaps, is the van der Waals equation.

Learning Objectives

As a result of studying Chapter 5, you should be able to

1. Explain the operation of a mercury barometer, an open-end manometer, and a closed-end manometer; and be able to use data obtained with these instruments.

2. Calculate, for a fixed amount of gas at a fixed temperature, how the volume changes with pressure, and vice versa.

3. Calculate, for a fixed amount of gas at a fixed pressure, how the volume changes with temperature, and vice versa.

4. Discuss the significance of the absolute zero of temperature; and calculate kelvin temperatures from Celsius temperatures, and vice versa.

5. State the standard conditions of temperature and pressure (STP) for a gas.

6. State the relationship between the volume and amount of gas at a fixed temperature and pressure; and use the molar volume of a gas at STP in calculations.

7. Use the ideal gas equation to calculate one gas variable—pressure, volume, temperature, amount of gas—when the other three are known.

8. Rearrange and use the ideal gas equation to calculate molecular weights and gas densities from experimental data.

9. Use the ideal gas equation, together with other data, in stoichiometric calculations for reactions in which gases are involved.

10. Calculate, for mixtures of gases, such quantities as partial pressures, total pressures, partial volumes, total volumes, compositions, and apparent molecular weights.

11. Perform calculations that deal with the collection of gases over water.

12. State the basic postulates of the kinetic molecular theory of gases.

13. Show how the results of the kinetic molecular theory can be used to deduce the simple gas laws.

14. Calculate the root-mean-square speed of the molecules of a gas of known molecular weight at a given temperature.

15. Relate the rates of effusion of gases to their molecular weights.

16. Describe the conditions under which a gas is most likely to behave as an ideal gas, and show how the van der Waals equation deals with nonideal behavior.

Some New Terms

One **atmosphere** is the pressure exerted by the atmosphere under normal conditions. This pressure will support a column of mercury 760 mm high at sea level at 0°C.

Avogadro's hypothesis states that equal volumes of different gases, compared under identical conditions of temperature and pressure, contain equal numbers of molecules.

A **barometer** is a device used to measure atmospheric pressure. In a mercury barometer this involves measuring the height of a column of liquid mercury that can be supported by the pressure of the atmosphere.

Boyle's law states that the volume of a fixed amount of gas at a constant temperature is inversely proportional to the gas pressure. That is, as the pressure is increased, the volume decreases.

Charles' law states that the volume of a fixed amount of gas at a constant pressure is directly proportional to the absolute (kelvin) temperature.

Dalton's law of partial pressures states that in a mixture of gases the total pressure is the sum of the partial pressures of the gases present.

The **gas constant, *R*,** is the numerical constant appearing in the ideal gas equation ($PV = nRT$).

Graham's law states that the rates of effusion (or diffusion) of two different gases are inversely proportional to the square roots of their molecular weights. **Effusion** refers to the escape of a gas from a container through a tiny hole. **Diffusion** refers to the spreading of a gas throughout a larger volume.

An **ideal gas** is one whose behavior can be predicted using the ideal gas equation. The behavior of a **nonideal gas** departs from that predicted by the ideal gas equation.

The **ideal gas equation** relates the pressure, volume, temperature, and number of moles of a gas (*n*) through the expression $PV = nRT$.

The **kinetic molecular theory of gases** is a model for describing gas behavior. It is based on a set of assumptions, and yields mathematical equations from which absolute temperature can be related to the average translational kinetic energy of molecules and from which the ideal gas equation can be deduced.

A **manometer** is a device used to measure the pressure of a gas. Sometimes this is done by comparing the gas pressure with barometric pressure (as in an open-end manometer), and sometimes a direct pressure measurement is involved (as in a closed-end manometer).

A **partial pressure** is the pressure exerted by an individual gas in a mixture, independently of the other gases present.

Pressure is a force per unit area. When applied to gases the concept of pressure is most easily understood in terms of the height of a liquid column that can be maintained by the gas.

The **van der Waals equation** is an equation of state for nonideal gases. It includes correction terms to account for intermolecular forces of attraction and for the volume occupied by the gas molecules themselves.

Exercises

Pressure and its measurement

1. Convert each of the following pressures to the equivalent pressure in atmospheres: **(a)** 738 mmHg; **(b)** 3.12 kg/cm²; **(c)** 70 psi; **(d)** 992 millibars (mb); **(e)** 1.67×10^5 N/m²

2. What height of mercury would you expect, under normal atmospheric conditions, in glass tubes filled in the manner of Figure 5-2b, for tubes with total lengths of **(a)** 10 cm; **(b)** 20 cm; **(c)** 100 cm; **(d)** 1000 cm?

3. Calculate the following quantities.
  (a) the height of a mercury column required to produce a pressure of 1.52 atm
  (b) the height of a column of liquid glycerol (density = 1.26 g/cm³) required to exert the same pressure as 2.40 m of carbon tetrachloride (density = 1.59 g/cm³)
  (c) the height of a column of liquid benzene (density = 0.879 g/cm³) required to exert a pressure of 1.10×10^4 N/m²
  (d) the density of a liquid if a 15.0-ft column is to exert a pressure of 10.0 lb/in.².

4. The mercury level in the arm of an open-end manometer open to the atmosphere is measured and found to be 385 mm above a certain reference point. In the arm of the manometer connected to a container of gas, the level is 195 mm above the same reference point. Measured barometric pressure is 752 mmHg. What is the pressure of the gas expressed in atmospheres?

5. In a laboratory experiment a gas is collected over water as suggested in Figure 5-14. Instead of equalizing the water levels, a difference of 2.4 cm exists, with the water level higher inside the gas container than outside. If barometric pressure is 748.7 mmHg, what is the pressure of the gas, also expressed in mmHg?

Boyle's law

6. A sample of $O_2(g)$ occupies a volume of 14.4 L at 748 mmHg. What is the new gas volume if, while the temperature and amount of gas are held constant, the pressure is **(a)** reduced to 615 mmHg; **(b)** increased to 1.72 atm?

7. A sample of nitrogen gas, which occupies a volume of 235 cm³ at a temperature of 25°C and a pressure of 755 mmHg, is expanded to a volume of 345 cm³. What is the new gas pressure?

Charles' law

8. A sample of neon gas occupies a volume of 125 cm³ at a pressure of 737 mmHg and 30°C. What will be the new gas volume if, while the pressure and amount of gas are held constant, the temperature is **(a)** increased to 60°C; **(b)** reduced to 0°C?

9. It is desired to increase the volume of a fixed amount of gas from 90.0 to 115 cm³ while holding the pressure constant. To what temperature must the gas be heated if the initial temperature is 23°C?

Additional applications of the simple gas laws

10. Indicate how the final volume v_f of a fixed amount of a gas is related to its initial volume v_i in each case.

(a) The gas pressure is decreased from 3 atm to 1 atm while the temperature is held constant at 25°C.

(b) The gas temperature is lowered from 400 to 100 K while the pressure is held constant at 1 atm.

(c) The gas temperature is raised from 200 to 300 K while the pressure is increased from 2 to 3 atm.

11. A fixed amount of gas, maintained in a volume of 148 cm³, exerts a pressure of 825 mmHg at 25°C. At what temperature will the gas pressure become exactly 1.00 atm?

12. A 10.00-g sample of a gas is introduced into an evacuated 2.50-L vessel at 0°C. It is desired to maintain the pressure of this gas constant as the temperature changes. How many grams of the gas must be allowed to escape if the temperature of the vessel is raised to 157°C?

Ideal gas equation

13. What is the volume occupied by 35.2 g $N_2(g)$ at 35°C and 741 mmHg?

14. A 10.0-L cylinder contains 52.0 g $O_2(g)$ at 25°C. What is the pressure exerted by this gas?

15. A sample of gas occupies a volume of 3.52 L at 30.0°C and 738 mmHg. What is the volume of this gas at 25.0°C and 758 mmHg?

16. A 25.0-L cylinder contains 128 g $N_2(g)$ at 10°C. How many grams of N_2 must be released to reduce the pressure in the cylinder to 1.75 atm?

17. A 725 g sample of Ne(g) is introduced into an evacuated 4.5-L cylinder, and the cylinder is heated until the gas pressure becomes 375 atm. What must be the gas temperature at this point?

★**18.** A 1.00-ft³ sample of He(g) at STP is introduced into a balloon and the balloon is released into the atmosphere. What is the pressure of He(g) inside this balloon when it has expanded to a volume of 75.0 L, assuming an air temperature of −20°C at this altitude?

Molecular weight determination

19. A 0.341-g sample of a gas occupies a volume of 355 cm³ at 98.7°C and 743 mmHg. What is the molecular weight of this gas?

20. A gaseous hydrocarbon weighing 0.185 g occupies a volume of 110 cm³ at 26°C and 743 mmHg. What is the molecular weight of this hydrocarbon? What conclusion can you draw about its molecular formula?

21. A 2.650-g sample of a gas occupies a volume of 428 cm³ at a pressure of 742.3 mmHg and 24.3°C. Analysis of this compound reveals a composition of 15.5% C, 23.0% Cl, and 61.5% F, by mass. What is the true molecular formula of the compound?

Gas densities

22. What is the density (in g/L) of $CO_2(g)$ at 30.3°C and 744 mmHg?

23. The density of phosphorus vapor at 310°C and 775 mmHg is 2.64 g/L. What is the true molecular formula of the phosphorus?

Cannizzaro's method

24. The following gaseous compounds all contain a common element, X. The data listed are the molecular weight and percent X for each compound. Construct a table similar to Table 5-1 and determine the atomic weight of X. What element do you think X is?

Compound	Molecular weight	X, %
nitryl fluoride	65.01	49.4
nitrosyl fluoride	49.01	32.7
thionyl fluoride	86.07	18.6
sulfuryl fluoride	102.07	31.4

Gases in chemical reactions

25. A particular coal sample contains 2.18% S, by mass. When the coal is burned, the sulfur is converted to $SO_2(g)$. What volume of $SO_2(g)$, measured at 25.0°C and 754 mmHg, is produced by burning 1.0×10^6 lb of this coal?

26. A method of removing $CO_2(g)$ from a spacecraft is to allow it to react with NaOH:

$$2 \, NaOH(s) + CO_2(g) \longrightarrow Na_2CO_3(s) + H_2O(l)$$

How many liters of $CO_2(g)$ at 26.0°C and 755 mmHg can be removed per kg NaOH?

27. A 2.15-g sample of a KCl–$KClO_3$ mixture is decomposed by heating and produces 90.2 cm³ $O_2(g)$, measured at 23.2°C and 741 mmHg. What is the percent $KClO_3$, by mass, in the mixture? (*Hint:* KCl in the mixture is unchanged.)

$$2 \, KClO_3(s) \longrightarrow 2 \, KCl(s) + 3 \, O_2(g)$$

28. The Haber process is the principal method for fixing atmospheric nitrogen (converting N_2 to nitrogen compounds).

$$N_2(g) + 3 \, H_2(g) \longrightarrow 2 \, NH_3(g)$$

(a) How many liters of $H_2(g)$ must be consumed to convert 4.0×10^3 L $N_2(g)$ to $NH_3(g)$ if all gases are measured at STP?

(b) How many liters of $NH_3(g)$ can be produced from 185 L $H_2(g)$ if the gases are measured at 525°C and 515 atm pressure?

(c) How many liters of $NH_3(g)$, *measured at STP*, can be produced from 185 L $H_2(g)$ measured at 525°C and 515 atm pressure?

29. 1.50 L $H_2S(g)$, measured at 23.0°C and 735 mmHg, is mixed with 4.45 L $O_2(g)$, measured at 26.1°C and 750 mmHg, and burned.

$$2 H_2S(g) + 3 O_2(g) \longrightarrow 2 SO_2(g) + 2 H_2O(g)$$

(a) How much $SO_2(g)$, in moles, is produced?
★ (b) If the excess reactant and the products of the reaction are collected at 748 mmHg and 120.0°C, what volume will they occupy?

Mixtures of gases

30. What is the volume occupied by a mixture of 15.0 g $Ne(g)$ and 30.1 g $Ar(g)$ at 10.0 atm pressure and 45.0°C?

31. A gas cylinder of 55.0 L volume contains $N_2(g)$ at a pressure of 20.0 atm and 23°C. How many grams of $Ne(g)$ must be introduced into this same cylinder to raise the total pressure to 75.0 atm?

32. A 1.50-L container of $H_2(g)$ at 765 mmHg and 25.0°C is connected to a 2.52-L container of $He(g)$ at 742 mmHg and 25.0°C. What is the *total* gas pressure after the gases have mixed, with the temperature remaining at 25.0°C?

33. A mixture of 4.0 g $H_2(g)$ and an unknown quantity of $He(g)$ is maintained at STP. If 10.0 g $H_2(g)$ is added to the mixture, while the conditions are maintained at STP, the volume of the gas doubles. How much He, in grams, is present?

34. The decomposition of ammonium nitrate, NH_4NO_3, occurs by the following reaction in the temperature range from 200 to 250°C.

$$NH_4NO_3(s) \longrightarrow N_2O(g) + 2 H_2O(g)$$

A 0.800-g sample of NH_4NO_3 is decomposed completely at 250°C in a closed evacuated vessel of 1.50-L volume.

(a) What is the partial pressure of $N_2O(g)$ in the vessel, in mmHg?
(b) What is the *total* gas pressure in the vessel, in mmHg?

35. When air is breathed by a human being, the exhaled (expired) air has a different composition from normal air. A typical analysis of expired air at 37°C and 760 mmHg yields the following composition, expressed as percent by volume: 75.1% N_2, 15.2% O_2, 3.8% CO_2, and 5.9% H_2O.

(a) What is the apparent molecular weight of this expired air? (Recall Example 5-18.)
(b) Would you expect the density of expired air to be greater or less than that of ordinary air at the same temperature and pressure? Explain.
(c) What is the ratio of the partial pressure of $CO_2(g)$ in expired air to that in ordinary air?

Collection of gases over water

36. A 2.65-g sample of Al is reacted with an excess of $HCl(aq)$ and the liberated $H_2(g)$ collected over water at 26°C at a barometric pressure of 746 mmHg. What is the volume of gas collected? Vapor pressure of H_2O at 26°C = 25.2 mmHg.

$$2 Al(s) + 6 HCl(aq) \longrightarrow 2 AlCl_3(aq) + 3 H_2(g)$$

37. A sample of $O_2(g)$ is collected over water at 22°C and 752 mmHg barometric pressure. The volume of gas collected is 84.8 cm³. How much O_2, in grams, is present in the gas? Vapor pressure of H_2O at 22°C = 19.8 mmHg.

38. A sample of $Ar(g)$ with a volume of 146 cm³ at 26°C and at a barometric pressure of 755 mmHg is passed through water at 26°C. What is the volume of the resulting gas when saturated with water vapor and again measured at 26°C and 755 mmHg barometric pressure? Water vapor pressure at 26°C = 25.2 mmHg.

39. A sample of $O_2(g)$ is collected over water at 25°C. The volume of the gas is 1.28 L. In a subsequent experiment it is determined that the mass of O_2 present is 1.58 g. What must have been the barometric pressure at the time the gas was collected? Vapor pressure of water at 25°C = 23.8 mmHg.

Kinetic molecular theory

40. Molecules of different gases compared at the same temperature have equal average translational kinetic energies. Do they also have equal speeds? Explain.

41. A kinetic theory verification of Avogadro's law was provided in the text. Verify Boyle's and Charles' laws using equations from Section 5-10.

42. The root-mean-square velocity, u_{rms}, of hydrogen molecules at 273 K is 1.84×10^3 m/s.

(a) At what temperature is u_{rms} for hydrogen equal to 3.68×10^3 m/s?
(b) What is u_{rms} for nitrogen at 273 K?

43. Calculate u_{rms}, in m/s, for $Cl_2(g)$ molecules at 25°C.

44. Following the method outlined in Table 5-2, determine $\bar{u}$ and u_{rms} for a group of six particles having the following speeds: 9.8×10^3, 9.0×10^3, 8.3×10^3, 6.5×10^3, 3.7×10^3, and 1.8×10^3 m/s.

Effusion of gases

45. What are the ratios of the diffusion rates for the following pairs of gases? (a) H_2 and O_2; (b) H_2 and D_2 (D = deuterium, i.e., ^2_1H); (c) $^{235}UF_6$ and $^{238}UF_6$

46. A sample of $N_2(g)$ effuses through a tiny hole in 38 s. What must be the molecular weight of a gas that requires 55 s to effuse under identical conditions?

Nonideal gases

47. The following data are given for three different gases. Indicate whether each behaves as an ideal or nonideal gas. (*Hint:* What is the value of PV/nRT for each gas?)
    (a) 1.00 mol $CO_2(g)$ confined to 1.20 L volume at 40°C exerts a pressure of 19.7 atm.
    (b) 0.113 g Ar(g) occupies a volume of 1.25 L at 0°C and 5.05×10^{-2} atm.

(c) 1.00 g $H_2(g)$ at 0°C and 200 atm occupies a volume of 63.06 cm³.

48. Calculate the pressure exerted by 1.00 mol $CO_2(g)$ confined to a volume of 855 cm³ at 30°C. Use (a) the ideal gas equation and (b) the van der Waals equation, where $a = 3.59$ L² atm mol^{-2} and $b = 0.0427$ L/mol. (c) Compare the results and explain.

Additional Exercises

1. Calculate the pressure, in N/m², exerted by a knife blade if one presses down with a force of 50 lb. Assume that the blade is 4 in. long and has a thickness along the edge of 0.01 in. What must be the mass of an object, in kilograms, that would exert the same pressure if its area of contact with the surface is 1.00 cm²?

2. A gas occupies a volume of 265 cm³ at 745 mmHg and 25°C. What *additional* pressure is required to reduce the gas volume to 215 cm³?

3. A sample of $N_2(g)$ occupies a volume of 58.0 cm³ under the existing barometric pressure. Increasing the pressure by 125 mmHg reduces the volume to 49.6 cm³. What is the prevailing barometric pressure?

4. Start with the conditions represented by points *A*, *B*, and *C* in Figure 5-8. Use Charles' law to calculate the volume occupied by each gas at 0, −100, −200, −250, and −270°C; and show that indeed the volume of each gas becomes zero at −273.15°C.

5. A 10.0-L cylinder contains 52.0 g $O_2(g)$ at 25°C. What is the pressure of this gas?

6. A 0.312-g sample of a gaseous hydrocarbon occupies a volume of 185 cm³ at 25.0°C and 745 mmHg. The hydrocarbon consists of 85.6% C and 14.4% H, by mass. What is the true molecular formula of this hydrocarbon?

7. Producer gas is a type of fuel gas made by passing air or steam through a bed of hot coal or coke. A typical producer gas has the following composition, in percent by volume: 8.0% CO_2, 23.2% CO, 17.7% H_2, 1.1% CH_4, and 50.0% N_2.
    (a) What is the apparent molecular weight of this gas?
    (b) What is the density of this gas at 25°C and 752 mmHg?
    (c) What is the partial pressure of CO in this gaseous mixture at STP?

8. A *mixture* of $H_2(g)$ and $O_2(g)$ is prepared by electrolyzing 1.32 g of water, and the mixture of gases is collected over water at 30°C when the barometric pressure is 748 mmHg. The volume of "wet" gas obtained is 2.90 L. What must be the vapor pressure of water at 30°C?

$$2\ H_2O(l) \xrightarrow{\text{electrolysis}} 2\ H_2(g) + O_2(g)$$

9. At what temperature will u_{rms} for Ne(g) be the same as u_{rms} for He(g) at 300 K?

*10. Recall the composition of ordinary air (Example 5-17). What volume of air, measured at STP, is required to complete the combustion of 1.00×10^3 L, at 23°C and 741 mmHg, of a natural gas of the following composition: 77.3% CH_4, 11.2% C_2H_6, 5.8% C_3H_8, 2.3% C_4H_{10} (and 3.4% noncombustible gases), by volume?

*11. A handbook states that mixtures of the anesthetic gas cyclopropane, $(CH_2)_3$, and air in which the volume percent of cyclopropane is between 2.4 and 10.3% are explosive. A sealed 1500-ml cylinder of $(CH_2)_3(g)$ at 2.50 atm and 25°C is placed inside a sealed fume hood of volume 72 ft³ containing air at 755 mmHg and 25°C. If the seal on the cylinder were to break and the $(CH_2)_3(g)$ to mix with the air inside the fume hood, would an explosive mixture be formed?

*12. A gaseous mixture of He and O_2 is found to have a density of 0.518 g/L at 25°C and 720 mmHg. What is the percent by mass of He in this mixture?

*13. If we replace the gas density, m/V, by the symbol, d, equation (5.12) can be rearranged to the form

$$\frac{m/V}{P} = \frac{d}{P} = \frac{(MW)}{RT}$$

This equation suggests that the ratio of gas density to gas pressure, at a constant temperature, should be a constant. The following gas density data were obtained for oxygen gas at various pressures at 273.15 K.

P, atm	d, g/L	d/P
1.000	1.428962	
0.750	1.071485	
0.500	0.714154	
0.250	0.356985	

(a) Calculate the values of d/P in the table, and with a graph or other means determine the best value of the term d/P for oxygen gas at 273.15 K. (This is the value corresponding to oxygen as an ideal gas.)

(b) Use the value of d/P determined in part (a) to calculate as precise a value of the atomic weight of oxygen as possible and compare it with that listed in an atomic weight table.

*14. A limestone sample is known to contain only magnesium and calcium carbonates. When a 0.4515-g sample of this limestone is decomposed by heating, 0.2398 g of mixed oxide (MgO and CaO) is obtained. How many liters of $CO_2(g)$ at 752.0 mmHg and 285.3 K will be produced by decomposing 50.0 lb of this limestone?

$$MCO_3(s) \longrightarrow MO(s) + CO_2(g) \quad \text{(where M = Ca or Mg)}$$

*15. A meteorological ("sounding") balloon is a synthetic rubber bag, filled with hydrogen gas, and carrying a set of instruments (the "payload"). Because this combination of bag, gas, and payload has a smaller mass than a corresponding volume of air, the balloon experiences a buoyant or lifting force that causes it to rise. As the balloon rises, it expands. From the following data, estimate the maximum height to which the balloon can rise.

mass of the balloon itself:	1200 g
payload:	1700 g
quantity of $H_2(g)$ introduced into balloon:	120 ft^3 at STP
diameter of balloon at maximum height:	25 ft
air pressures and temperatures as a function of altitude:	

km	mb	K
0	1.0×10^3	288
5	5.4×10^2	256
10	2.7×10^2	223
20	5.5×10^1	217
30	1.2×10^1	230
40	2.9×10^0	250
50	8.1×10^{-1}	250
60	2.3×10^{-1}	256

*16. Atmospheric pressure as a function of altitude can be calculated using an equation called the barometric formula.

$$P = P_0 \times 10^{-(MW)gh/2.303\,RT}$$

where P is the pressure, in atmospheres, at an altitude of h meters. P_0 is the pressure at sea level (usually taken to be 1 atm), g is the acceleration due to gravity (9.80 m/s^2); and (MW) is the molecular weight of air, expressed in *kilograms* per mole. R is expressed as 8.314 J mol^{-1} K^{-1}, and T is the kelvin temperature.

(a) Estimate barometric pressure at the top of Mt. Whitney in California. (Altitude: 14,494 ft; assume a temperature of 10°C.)

(b) Show that the observation that barometric pressure decreases by one-thirtieth in value for every 900-ft increase in altitude is consistent with the barometric formula.

Self-Test Questions

For questions 1 through 6 select the single item that best completes each statement.

1. The greatest pressure of the following is that exerted by **(a)** a column of liquid mercury 75.0 cm high ($d = 13.6$ g/cm^3); **(b)** 10.0 g $H_2(g)$ at STP; **(c)** a column of air 10 mi high; **(d)** a column of liquid CCl_4 60.0 cm high ($d = 1.59$ g/cm^3).

2. For a fixed amount of gas at a fixed pressure, changing the temperature from *100°C* to *200 K* causes **(a)** the gas volume to decrease; **(b)** the gas volume to double; **(c)** the gas volume to increase, but not to twice its original value; **(d)** no change in the gas volume.

3. A sample of $O_2(g)$ is collected over water at 23°C at a barometric pressure of 751 mmHg (vapor pressure of water at 23°C = 21 mmHg). The *partial* pressure of $O_2(g)$ in the sample collected is **(a)** 21 mmHg; **(b)** 751 mmHg; **(c)** 0.96 atm; **(d)** 1.02 atm.

4. A comparison is made at standard temperature and pressure (STP) of 0.50 mol $H_2(g)$ and 1.0 mol He(g). The two gases will **(a)** have equal average molecular kinetic energies; **(b)** have equal average molecular speeds; **(c)** occupy equal volumes; **(d)** have equal effusion rates.

5. A mixture of 0.50 mol $H_2(g)$ and 0.50 mol $SO_2(g)$ is introduced into a 10.0-L container at 25°C. The container has a "pinhole" leak. After a period of time **(a)** The partial pressure of $H_2(g)$ exceeds that of $SO_2(g)$ in the remaining gas. **(b)** The partial pressure of $SO_2(g)$ exceeds that of $H_2(g)$ in the remaining gas. **(c)** The partial pressures of the two gases remain equal throughout this time. **(d)** The partial pressures of both gases increase above their initial values.

6. To establish a pressure of 2.00 atm in a 2.24-L cylinder containing 1.60 g $O_2(g)$ at 0°C, **(a)** add 1.60 g $O_2(g)$; **(b)** release 0.80 g $O_2(g)$; **(c)** add 2.00 g He; **(d)** add 0.60 g He(g).

7. 0.10 mol He(g) is added to 2.24 L $H_2(g)$ at standard temperature and pressure (STP). This is followed by an increase in temperature to 100°C while the pressure and amount of gas are held constant. What is the final gas volume?

8. Explain briefly why
(a) The height of the mercury column in a barometer is

independent of the diameter of the barometer tube (i.e., whether the diameter is 1 mm, 1 cm, 10 cm, . . .).

(b) An open-end mercury manometer is useful for measuring pressures of the order of a few tenths of an atmosphere, but not for measuring either very low or very high pressures.

9. Calculate the number of L $H_2(g)$ (measured at 22°C and 745 mmHg) required to react with 30.0 L $CO(g)$ (measured at 0°C and 760 mmHg) in the reaction

$$3\ CO(g) + 7\ H_2(g) \longrightarrow C_3H_8(g) + 3\ H_2O(l)$$

10. A gaseous mixture consists of 50.0% O_2, 25.0% N_2, and 25.0% Cl_2, by mass, and is maintained at STP. Calculate the partial pressure of $Cl_2(g)$ in this mixture.

6 Thermochemistry

Our study in Chapter 4 focused on how chemical reactions can be used to transform matter from one chemical form (the reactants) to another (the products). In particular, we learned how to calculate the masses of reactants consumed and products formed in a chemical reaction. These calculations were expanded in scope in Chapter 5 following a discussion of the behavior of gases. Now, we shift our interest to quantities of energy exchanged between the reaction mixture and its surroundings: *Chemical reactions can be used as sources of energy or as means of storing energy.*

Chemical reactions in which energy exchanges occur principally as heat are called *thermochemical;* as light energy, *photochemical;* as electricity, *electrochemical.* Our attention in this chapter is on thermochemical reactions. We will continue to use the balanced chemical equation and its underlying basis—the law of conservation of mass. Equally important will be another great natural law—the law of conservation of energy (the first law of thermodynamics).

6-1 Work and Heat

Thermodynamics, of which thermochemistry is one important aspect, deals with relationships between heat energy and other energy forms known as work. In order to consider these relationships quantitatively, we need to develop clearer definitions of heat and work than those used by nonscientists.

Relevant portions of Appendix B should be reviewed along with this discussion.

WORK. In Appendix B a succession of physical quantities is described leading to the notion of work: Mechanical work is performed when a force acts through a distance. The quantity of work is the product of the force and the distance. The SI unit of work is the joule (J); it corresponds to a force of one newton (N) acting through a distance of one meter (m). That is,

1 joule = 1 newton × 1 meter

One way to think of work is as any form of energy transfer that can be expressed through the lifting or lowering of weights. Figure 6-1 suggests how work is performed when a gas is allowed to expand. Work associated with the expansion or compression of gases—**pressure–volume work**—is the only type that we consider explicitly in this chapter.

FIGURE 6-1
Pressure-volume work.

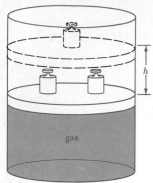

gas

When one of the weights is removed from the piston confining the gas, the remaining weight is lifted through the distance *h*. Work is performed.

It is unfortunate that in nutritional studies the unit called a Calorie is actually a kilocalorie. Thus, a "1 Calorie" soft drink actually has a food value of 1000 cal. Throughout this text, reference will always be to the defined calorie.

HEAT. Heat is energy that is transferred as a result of a temperature difference. Heat energy always flows from a warmer body (higher temperature) to a colder body (lower temperature). At the molecular level this means that molecules of the warmer body lose kinetic energy to those in the colder body when the two bodies are brought into contact. As a result, the average translational kinetic energy of the molecules in the warmer body is lowered—its temperature is lowered. In the cooler body the temperature is raised. Heat flows between two bodies until they reach the same temperature.

The quantity of heat, *q*, required to change the temperature of a substance depends on how much the temperature is to be changed, the quantity of substance, and its identity (type of molecules). **Heat capacity** is the quantity of heat required to raise the temperature of a substance by 1°C. Heat capacity, of course, depends on the quantity of substance. **Specific heat capacity,** or more simply **specific heat,** is the quantity of heat required to raise the temperature of one *gram* of substance by 1°C. **Molar heat capacity** is the quantity of heat required to raise the temperature of one *mole* of substance by 1°C.

Historically, heat energy has been defined by the unit the calorie. One **calorie (cal)** is the quantity of heat required to raise the temperature of 1 g of water from 14.5 to 15.5°C. Thus, we can say that, at 15°C, the specific heat of water is 1.000 cal g^{-1} °C^{-1} (or 1.000 cal per g per °C). The molar heat capacity of water at the same temperature is 18.02 cal mol^{-1} °C^{-1} since there is 18.02 g H_2O in 1 mol. Specific heat is itself a function of temperature, and this is why a particular temperature interval is chosen in the definition of the calorie. At 25°C, for example, the specific heat of water is 0.999 cal g^{-1} °C^{-1}. Throughout the temperature interval from 0 to 100°C the specific heat of water generally can be taken as 1.00 cal g^{-1} °C^{-1}.

The calorie is a rather small quantity of energy, and in many applications the larger unit, the **kilocalorie (kcal),** is used. 1 kcal = 1000 cal. Also, it should be noted that the calorie is *not* an SI unit. We will introduce the appropriate SI unit shortly, and we will then redefine the calorie in terms of the SI unit.

Example 6-1 How much heat is required to raise the temperature of 735 g of water from 21.0 to 98.0°C? (Assume that the specific heat of water remains at 1.00 cal g^{-1} °C^{-1} throughout this temperature range.)

The quantity of heat required to change the temperature of 735 g of water is 735 times as great as to change the temperature of 1 g of water. And the heat required to raise the temperature by 77.0°C (from 21.0 to 98.0°C) is 77 times as great as to raise the temperature 1°C. Thus, the specific heat of water must be multiplied by 735 and by 77.

$$\text{no. cal} = \frac{1.00 \text{ cal}}{\text{g water °C}} \times 735 \text{ g water} \times (98.0 - 21.0)°C = 5.66 \times 10^4 \text{ cal}$$

SIMILAR EXAMPLES: Exercises 1, 2.

Calculations involving the concept of specific heat can usually be performed by the line of reasoning used in Example 6-1, but it is perhaps simpler to think in terms of the following expression:

quantity of heat = *q*

$$= \underbrace{\text{mass of substance} \times \text{specific heat}}_{\text{heat capacity}} \times \text{temperature change} \quad (6.1)$$

The temperature change in equation (6.1) can be expressed as

$$\Delta t = t_f - t_i \quad (6.2)$$

where t_f is the final temperature, t_i is the initial temperature, and Δt (called "delta *t*") is the temperature change. By this convention, if the temperature of a substance is increased, the final temperature is greater than the initial temperature (noted symbolically as $t_f > t_i$) and Δt is positive (i.e., $\Delta t > 0$). The quantity of heat calculated is also *positive* and signifies that heat is *absorbed* or *gained* when the temperature of a substance is increased. If the temperature of a substance is lowered, the final temperature is less than the initial temperature (noted symbolically as $t_f < t_i$). In this case Δt is negative (i.e., $\Delta t < 0$). The quantity of heat is also *negative,* signifying that heat is *evolved* or *lost* when a substance is cooled.

An additional idea that enters into heat energy calculations is the **law of conservation of energy.** In interactions among objects or substances, the total energy remains constant. Thus, heat lost by one object must be gained by another. The simple laboratory method of determining the specific heat of a metal illustrated in Figure 6-2 is based on the law of conservation of energy. To apply this law in the present case we proceed as follows. Let q_{lead} represent the quantity of heat exchanged by the lead, and q_{water}, the quantity of heat exchanged by the water. Since the total energy must remain constant, the total quantity of heat must be zero.

$$q_{lead} + q_{water} = 0 \tag{6.3}$$

Moreover, the two terms must be equal in magnitude and opposite in sign. That is,

$$q_{lead} = -q_{water} \tag{6.4}$$

Example 6-2   Use data presented in Figure 6-2 to calculate the specific heat of lead. First, let us use equation (6.1) to calculate q_{water}.

$$q_{water} = 50.0 \text{ g water} \times \frac{1.00 \text{ cal}}{\text{g water }^\circ\text{C}} \times (28.8 - 22.0)^\circ\text{C} = 340 \text{ cal}$$

From equation (6.4) we may write

$$q_{lead} = -q_{water} = -340 \text{ cal}$$

Now, from equation (6.1) again, we obtain

$$q_{lead} = 150.0 \text{ g lead} \times \text{sp. ht. lead} \times (28.8 - 100.0)^\circ\text{C}$$
$$= -340 \text{ cal}$$

$$\text{sp. ht. lead} = \frac{-340 \text{ cal}}{150.0 \text{ g lead} \times (28.8 - 100.0)^\circ\text{C}}$$

$$= \frac{-340 \text{ cal}}{150 \text{ g lead} \times (-71.2)^\circ\text{C}} = 0.032 \frac{\text{cal}}{\text{g lead }^\circ\text{C}}$$

FIGURE 6-2
Determination of specific heat of lead— Example 6-2 illustrated.

(a) 150.0 g of lead at the temperature of boiling water (100.0°C).
(b) 50.0 g of water in a thermally insulated beaker at 22.0°C.
(c) Final lead–water mixture at a temperature of 28.8°C.

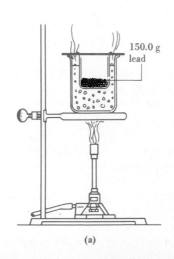

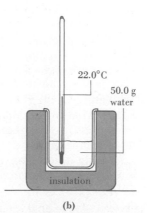

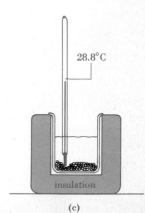

150.0 g lead

22.0°C

50.0 g water

insulation

28.8°C

insulation

(a)

(b)

(c)

FIGURE 6-3
Establishing the mechanical equivalent of heat.

As the two weights, *M*, fall under the influence of gravity, *g*, through the distance *h*, a quantity of work is done equal to $2Mgh$. This work is first expressed in the rotational motion of the paddles, P. Then this rotational kinetic energy is converted to heat as the paddles are slowed down and stopped by the water. The process can be repeated many times by raising the weights by winding the crank. The quantity of heat absorbed by the water is calculated by knowing the mass of water, its specific heat, and the temperature change (measured with the thermometer, T).

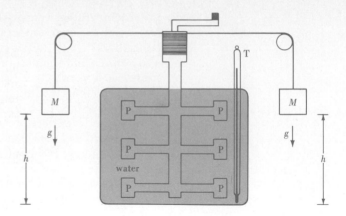

SIMILAR EXAMPLES: Exercises 3, 5.

A useful thought in distinguishing between work and heat is that work produces organized molecular motion and heat produces random or chaotic motion. In the work performed in Figure 6-1, all the atoms of the piston and weights move together, in the same direction and by the same distance. When heat energy is absorbed by a substance, the average translational kinetic energy of its molecules increases but the molecules continue their random, disorganized motion.

MECHANICAL EQUIVALENT OF HEAT. Figure 6-3 illustrates a significant experiment performed by Joule in 1847. The work associated with falling weights was converted to heat that was used to raise the temperature of water or other liquids. The number of joules of work done by the falling weights always turned out to be about 4.15 times greater than the number of calories of heat gained by the water. Joule's experiment established that mechanical work can be completely converted to heat, and suggests that heat and work can be measured in the same units—joules or calories. Since the joule is the basic energy unit of the SI system, we will henceforth express both heat and work in joules. Based on modern measurement, the *defined* calorie and kilocalorie are

$$1 \text{ cal} \equiv 4.184 \text{ J} \qquad \text{and} \qquad 1 \text{ kcal} \equiv 4.184 \text{ kJ} \tag{6.5}$$

6-2 Some Terminology

In the study of thermodynamics certain terms are defined and used in a rather precise way. We will not emphasize terminology, but you should acquire some familiarity with the following terms for use in this and later chapters.

By a thermodynamic **system** we mean that portion of the universe selected for investigation. A thermodynamic system may be as simple as a beaker of water or as complex as the contents of a blast furnace or a polluted lake. The **surroundings** represent that portion of the universe with which the system interacts. **Interactions** between a system and surroundings refer to the exchange of one or more of three basic entities—heat, work, and matter. Heat and work are different manifestations of energy discussed in the preceding section.

An **isothermal** system is one that is maintained at a constant temperature. If a system is insulated from its surroundings so that no heat may flow, it is said to be **adiabatic.** As an approximation at least, the contents of a Dewar flask (vacuum bottle) may be considered an adiabatic system. A **closed** system is one in which the total mass of substances is held constant. An **open** system is one that is permitted to exchange matter with its surroundings. A chemical reaction in an open beaker producing a gas is an example of an open system. An **isolated** system is one that is not permitted to interact, that is, not to exchange heat, work, or matter with its surroundings.

A mammal is essentially an isothermal system, and definitely not adiabatic.

6-3 The First Law of Thermodynamics

The first law of thermodynamics is simply another way of stating the law of conservation of energy. In an isolated system the total energy remains constant. If a system exchanges heat and/or work with its surroundings, this must occur in such a way that the total energy of the system and its surroundings remains constant. The first law of thermodynamics can be expressed most simply through a mathematical equation:

$$\Delta E = q - w \tag{6.6}$$

The convention that applies to equation (6.6) is that

q is a *positive* quantity if heat is *absorbed* by a system

q is a *negative* quantity if heat is *lost* by a system

w is a *positive* quantity if work is done *by* a system

w is a *negative* quantity if work is done *on* a system

E is called the internal energy of the system

ΔE is the change in internal energy for some process

Thermodynamics is independent of any particular theory of the structure of matter. It was, in fact, fully developed as a science before modern atomic theory. We now recognize, however, that internal energy represents the total energy possessed by the ultimate particles of matter. This includes chemical bond energy, kinetic energy of translational motion, vibrational energy associated with chemical bonds, rotational energy of polyatomic molecules, energy of intermolecular attractions, and so on.

Suppose that a system absorbs some heat energy ($q > 0$) but does work on the surroundings at the same time ($w > 0$). The change in internal energy, ΔE, is simply equal to the heat absorbed less the work done ($q - w$). If the quantity of heat absorbed is greater than the work done, ΔE is positive. If more work is done than heat absorbed, ΔE is negative.

Example 6-3 The expansion of a gas pictured in Figure 6-1 is conducted in such a way that 1000 J of heat is absorbed while the system performs 1850 J of work on its surroundings. What is ΔE for the system?

Heat absorbed is a positive quantity: $q = +1000$ J. Work done by the system is also a positive quantity: $w = +1850$ J.

$$\Delta E = q - w = +1000 \text{ J} - (+1850 \text{ J})$$
$$\Delta E = +1000 \text{ J} - 1850 \text{ J} = -850 \text{ J}$$

SIMILAR EXAMPLES: Exercises 7, 8.

Internal energy depends only on the conditions that characterize a system at the given time and not on how those conditions were achieved. The condition of a system is referred to as its **state** and any property that depends only on the state of a system is called a **function of state** (or a state function).

An analogy to a state function is provided in Figure 6-4, which relates to climbing a mountain. Path (a) is shorter but steeper; path (b) is longer and more gradual. The length of time required to climb the mountain depends on the path chosen, but the total elevation gain is fixed. The climbing time is analogous to q and w; the elevation gain, to ΔE. Furthermore, the loss of elevation in climbing back down the mountain is analogous to $-\Delta E$. That is, following a round trip to the top of the mountain (ΔE)

Authors vary in their thermodynamic notation. Sometimes the symbol U is used instead of E; different sign conventions may be adopted for q and w, and a different form used for equation (6.6).

FIGURE 6-4
An analogy to a thermodynamic function of state.

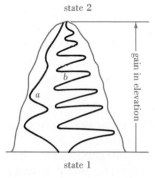

state 2

gain in elevation

state 1

The gain in elevation in climbing from the base to the summit of the mountain is independent of the path chosen. This elevation gain is analogous to ΔE in a thermodynamic system.

and back down $(-\Delta E)$, the elevation gain is zero. To state this same conclusion for a thermodynamic system, following a transition from state 1 to state 2 and back to the initial condition (state 1), the internal energy of a system must regain its original value. If it did not, energy would have been created or destroyed in the process, and this would be in violation of the first law of thermodynamics.

$$\text{state 1} \xrightarrow{\Delta E} \text{state 2} \xrightarrow{-\Delta E} \text{state 1} \qquad (6.7)$$

6-4 Application of the First Law of Thermodynamics to Chemical Reactions

In the symbolic representation of a chemical reaction, we can think of the reactants as representing one state of a thermodynamic system, state 1, with an internal energy of E_1. The products represent a different state, state 2, with an internal energy of E_2.

$$\text{reactants} \longrightarrow \text{products}$$
$$\underset{E_1}{\text{(state 1)}} \qquad \underset{E_2}{\text{(state 2)}}$$

Accompanying the reaction there occurs a change in internal energy,

$$\Delta E = E_2 - E_1$$

which according to the first law of thermodynamics can also be represented as

$$\Delta E = q - w \qquad (6.6)$$

The symbol q represents the quantity of heat and w is the quantity of work involved in the reaction. Because internal energy is a function of state, ΔE has a unique value for a given chemical reaction (recall Figure 6-4). But q and w do not have unique values; they depend on how the reaction is carried out. Thus, q and w must be measured, and ΔE calculated through equation (6.6).

DETERMINING ΔE BY BOMB CALORIMETRY. The laboratory determination of heat associated with a chemical reaction is carried out in a device called a **calorimeter.** The type shown in Figure 6-5 is called a bomb calorimeter.

The thermodynamic system in Figure 6-5 is the *contents* of the bomb. The steel bomb itself, the water in which the bomb is immersed, the thermometer, the stirrer, and so on, constitute the surroundings. After initiation of the reaction, the internal energy of the contents of the bomb changes by ΔE, which by the first law of thermodynamics is equal to $q - w$. The bomb

FIGURE 6-5
Experimental measurement of a heat of reaction.

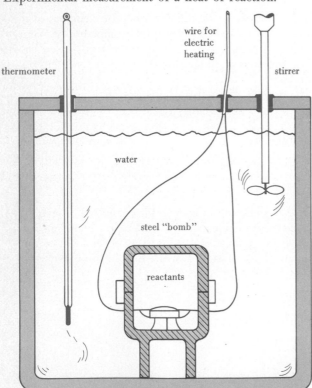

The heat of reaction is determined by measuring the total quantity of heat absorbed by the surroundings. The heat absorbed by the water is the product of its mass, specific heat, and temperature increase. The heat absorbed by the rest of the surroundings—bomb, stirrer, thermometer, etc.—is the product of their combined heat capacity and the temperature increase. A separate experiment is required to establish this combined heat capacity, the so-called heat capacity of the bomb.

confines the system to *constant volume: There is no opportunity for the system to do work or to have work done on it.* As a result, $w = 0$ and $\Delta E = q_V$ (where the subscript V stands for constant volume). The change in internal energy of the system is simply the quantity of heat exchanged with the surroundings, a quantity called the **heat of reaction at constant volume, q_V.**

Heat evolved in the reaction goes largely toward raising the temperature of the water surrounding the bomb. Small quantities, however, are also required to raise the temperature of the bomb itself, of the stirrer, and of other parts of the calorimeter. Thus, we need to deal with three quantities of heat—q_V, q_{water}, and q_{bomb}. The heat of reaction is q_V, and the heat effect in the surroundings is separated into that affecting the water, q_{water}, and that affecting the rest of the surroundings, q_{bomb}. If we follow the reasoning used in developing equations (6.3) and (6.4), we conclude that

$$q_V + q_{water} + q_{bomb} = 0 \tag{6.8}$$

and

$$q_V = -(q_{water} + q_{bomb}) \tag{6.9}$$

To evaluate q_{water} and q_{bomb} we use equation (6.1). When equations (6.1) and (6.9) are combined, we obtain the final expression

$$q_V = -[(\text{mass of water} \times \text{sp. ht. water} \times \text{temp. change})$$
$$+ (\text{heat capacity of bomb} \times \text{temp. change})] \tag{6.10}$$

Example 6-4 An 0.505-g sample of the solid hydrocarbon naphthalene, $C_{10}H_8$, is burned completely in an excess of $O_2(g)$ (about 25 atm) in a bomb calorimeter. The mass of water in the calorimeter is 1215 g, and the temperature of the water increases from 25.62 to 29.06°C as a result of the reaction. The heat capacity of the bomb is 826 J/°C. What is the heat of reaction at constant volume, q_V, expressed in **(a)** J/g $C_{10}H_8$; **(b)** kJ/mol $C_{10}H_8$; **(c)** kcal/mol $C_{10}H_8$?

(a) Solve equation (6.10) for q_V. This will be the heat of reaction based on an 0.505-g sample of $C_{10}H_8$. Note that the specific heat of water must now be expressed in its SI equivalent. That is, $1.00 \text{ cal g}^{-1} \text{ °C}^{-1} = 4.184 \text{ J g}^{-1} \text{ °C}^{-1}$.

$$q_V = -\left\{ \left[1215 \text{ g water} \times \frac{4.184 \text{ J}}{\text{g water °C}} \times (29.06 - 25.62)\text{°C} \right] \right.$$
$$\left. + [826 \text{ J/°C} \times (29.06 - 25.62)\text{°C}] \right\}$$
$$= -\{1.75 \times 10^4 \text{ J} + 2.84 \times 10^3 \text{ J}\} = -2.03 \times 10^4 \text{ J}$$

per gram of $C_{10}H_8$:

$$q_V = \frac{-2.03 \times 10^4 \text{ J}}{0.505 \text{ g } C_{10}H_8} = -4.02 \times 10^4 \text{ J/g } C_{10}H_8$$

(b) q_V *expressed in kJ/mol $C_{10}H_8$:*

$$q_V = \frac{-4.02 \times 10^4 \text{ J}}{\text{g } C_{10}H_8} \times \frac{128 \text{ g } C_{10}H_8}{1 \text{ mol } C_{10}H_8} \times \frac{1 \text{ kJ}}{1000 \text{ J}} = -5.15 \times 10^3 \text{ kJ/mol } C_{10}H_8$$

(c) q_V *expressed in kcal/mol $C_{10}H_8$:*

$$q_V = \frac{-5.15 \times 10^3 \text{ kJ}}{\text{mol } C_{10}H_8} \times \frac{1 \text{ kcal}}{4.184 \text{ kJ}} = -1.23 \times 10^3 \text{ kcal/mol } C_{10}H_8$$

Every bomb calorimeter assembly has its own distinctive heat capacity, which can only be determined by experiment. We should not expect to find the same heat capacities in Examples 6-4 and 6-5.

SIMILAR EXAMPLES: Exercises 19, 20.

Example 6-5 The combustion of benzoic acid is often used in experiments designed to establish the heat capacity of a bomb calorimetry assembly. If the combustion of a

1.000-g sample of benzoic acid ($C_7H_6O_2$) causes a temperature increase of 4.96°C when a bomb calorimeter contains 1085 g of water, what is the heat capacity of the bomb? (A handbook lists the heat of combustion of benzoic acid as 26.42 kJ/g.)

All the data required in equation (6.10) are known except the heat capacity of the bomb. We substitute these data and solve for this heat capacity.

$$q_V = -26.42 \text{ kJ} = -2.642 \times 10^4 \text{ J}$$
$$= -\left[\left(1085 \text{ g water} \times \frac{4.184 \text{ J}}{\text{g water } °C} \times 4.96°C \right) + (\text{ht. cap.} \times 4.96°C) \right]$$
$$= -2.642 \times 10^4 \text{ J} = -2.25 \times 10^4 \text{ J} - (\text{ht. cap.} \times 4.96°C)$$

$$\text{heat capacity of bomb} = \frac{-2.25 \times 10^4 \text{ J} + 2.642 \times 10^4 \text{ J}}{4.96°C} = \frac{0.39 \times 10^4 \text{ J}}{4.96°C}$$
$$= 7.9 \times 10^2 \text{ J/}°C$$

SIMILAR EXAMPLE: Exercise 18.

REPRESENTING ΔE IN A CHEMICAL EQUATION.   A chemical equation for the combustion of naphthalene (a hydrocarbon) can be written in the manner introduced in Chapter 4. From the result of Example 6-4 we can now add this important bit of thermochemical information.

The complete combustion of one mole of solid naphthalene, producing gaseous carbon dioxide and liquid water as the sole products, is accompanied by a decrease of internal energy equal to 5.15 × 10³ kJ.

$$C_{10}H_8(s) + 12 \text{ O}_2(g) \longrightarrow 10 \text{ CO}_2(g) + 4 \text{ H}_2O(l) \qquad \Delta \bar{E} = -5.15 \times 10^3 \text{ kJ/mol}$$
$$(6.11)$$

The unit kJ/mol appears in equation (6.11). We might ask, "per mole of what?" The /mol does not necessarily mean that exactly one mole of any reactant or product is involved. Rather, it signifies that the numbers of moles of reactants and products correspond to the stoichiometric coefficients in the balanced equation. Consequently, a value of ΔE must always be linked with a particular equation; and because it is so linked, the overbar symbol and the /mol unit are frequently omitted.

A line may be drawn above the symbol E (called an overbar) to signify that molar amounts of reactants and products are involved: that is, 1 mol $C_{10}H_8$, 12 mol O_2, 10 mol CO_2, and 4 mol H_2O. The negative sign for $\Delta \bar{E}$ in equation (6.11) means that the reaction is *exothermic*. That is, 5.15×10^3 kJ of heat is lost by the system: q_V is negative. Since $\Delta \bar{E} = q_V$, $\Delta \bar{E}$ is also negative. Thermochemical equations are sometimes written with fractional coefficients, as in the formation of one mole of NO(g) from its elements.

$$\tfrac{1}{2} \text{N}_2(g) + \tfrac{1}{2} \text{O}_2(g) \longrightarrow \text{NO}(g) \qquad \Delta \bar{E} = +90.37 \text{ kJ/mol}$$

A NEW THERMODYNAMIC FUNCTION—ENTHALPY, H.   If naphthalene is burned completely in a container *open to the atmosphere,* the amount of heat liberated is slightly different from that calculated in Example 6-4 and shown in equation (6.11). This is because the process proceeds in a different way. Rather than the volume of the system remaining constant, the pressure is constant (at 1 atm). The **heat of a reaction at constant pressure** is designated as q_P. If the only type of work considered is pressure–volume work (recall Figure 6-1), and if this work is performed at constant pressure, the quantity of work is the pressure multiplied by the volume change: $w = P \Delta V$. Now we can write

$$\Delta E = q_P - P \Delta V \quad \text{(at constant pressure)}$$

and

$$q_P = \Delta E + P \Delta V \quad \text{(at constant pressure)}$$

At this point we introduce a new thermodynamic property, called enthalpy, H, and define it as the sum of the internal energy of a system and its pressure–volume product. That is,

$$H = E + PV$$

TABLE 6-1
Internal energy (E) and enthalpy (H) compared

	E	$H = E + PV$
nature	fundamental	invented for convenience
most useful at:	constant volume (e.g., reaction in a bomb calorimeter)	constant pressure (e.g., reaction in an open beaker)
first-law statement under these conditions	$q_V = \Delta E$	$q_P = \Delta H$

For a change occurring at *constant pressure,*

$$\Delta H = \Delta E + P\,\Delta V = q_P \tag{6.12}$$

We will refer to ΔH as the enthalpy change for a reaction, or, more commonly, as the heat of reaction. The relationship between internal energy and enthalpy is summarized in Table 6-1.

RELATIONSHIP BETWEEN ΔH AND ΔE.   If only liquids and solids are involved in a reaction, very little volume change occurs. According to equation (6.12), in such reactions ΔH and ΔE should have essentially the same value (since $P\,\Delta V \simeq 0$). A greater change in volume occurs when reaction (6.11) is carried out at constant pressure. This results principally from the replacement of 12 mol of *gaseous* oxygen by 10 mol of *gaseous* carbon dioxide.

If we treat $O_2(g)$ and $CO_2(g)$ as ideal gases and compare them at the same T and P, we can write

$$V_{CO_2} = \frac{n_{CO_2}RT}{P} \qquad \text{and} \qquad V_{O_2} = \frac{n_{O_2}RT}{P}$$

The change in volume in the reaction is thus

$$\Delta V = V_{CO_2} - V_{O_2} = (n_{CO_2} - n_{O_2})\frac{RT}{P}$$

or

$$\Delta V = \Delta n_g \left(\frac{RT}{P}\right) \qquad \text{where } \Delta n_g = \text{number of moles of } \textit{gaseous} \tag{6.13}$$
$$\textit{products} \text{ minus the number of}$$
$$\text{moles of } \textit{gaseous reactants}$$

By substituting equation (6.13) into equation (6.12), we obtain

$$\Delta H = \Delta E + P\left(\frac{\Delta n_g RT}{P}\right)$$

and

$$\Delta H = \Delta E + \Delta n_g RT \tag{6.14}$$

Equation (6.14) is generally applied at 25°C (298 K), with values of ΔH and ΔE given in kJ. This requires that the gas constant R be expressed as 8.314×10^{-3} kJ mol^{-1} K^{-1}. The RT product becomes 8.314×10^{-3} kJ mol^{-1} K$^{-1} \times$ 298 K = 2.48 kJ/mol. Combining these features into equation (6.14), we obtain the useful expression

$$at\ 298\ K:\quad \Delta H\ (\text{in kJ}) = \Delta E\ (\text{in kJ}) + 2.48\ \text{kJ/mol} \times \Delta n_g\ \text{mol} \qquad (6.15)$$

Example 6-6 Use data presented in equation (6.11) to calculate ΔH for the complete combustion of 1 mol of naphthalene, $C_{10}H_8$, at 298 K.

Equation (6.11) provides us with a value of ΔE for the combustion reaction which can be substituted into equation (6.15). The change in number of moles of gaseous species, Δn_g, is 10 mol CO_2 − 12 mol O_2 = −2 mol gas.

$$\Delta H = -5.15 \times 10^3\ \text{kJ} + 2.48 \times (-2)\ \text{kJ} = -5.15 \times 10^3\ \text{kJ}$$

The essential difference between ΔH and ΔE for a reaction is the pressure–volume work associated with the expansion or compression of gases. In this case 12 mol of gas is compressed into 10 mol. Work is done *by* the surroundings *on* the system.

The magnitude of the heat of combustion of naphthalene at constant pressure ($\Delta H = q_P$) is just slightly greater (by 4.96 kJ) than that at constant volume ($\Delta E = q_V$). However, because of the rule that applies to significant figures when adding numbers, to three significant figures there is no difference between ΔH and ΔE.

SIMILAR EXAMPLES: Exercises 16b, 17.

EXPERIMENTAL DETERMINATION OF ΔH.   Combustion reactions are generally carried out at constant volume in a bomb calorimeter, yielding $q_V = \Delta E$. For these reactions a value of ΔH can be obtained with equation (6.15). This is what was done in Example 6-6. However, the simple calorimeter pictured in Figure 6-6 is much more commonly encountered in the general chemistry laboratory than is a bomb calorimeter. A chemical reaction is carried out in solution in a Styrofoam cup (generally in aqueous solution), and the temperature change is measured. Styrofoam is a good heat insulator, so that there is very little heat transfer with the surroundings. The system is essentially *adiabatic*. If a chemical reaction is exothermic, the heat released is retained within the solution and raises the temperature. If the reaction is endothermic, heat must be absorbed from the solution and its temperature falls. Because the reaction mixture is maintained under atmospheric pressure, the quantity of heat measured is at constant pressure: $q_P = \Delta H$.

Example 6-7   A 1.50-g sample of NH_4NO_3 is added to 35.0 g H_2O in a Styrofoam coffee cup and stirred until it dissolves. The temperature of the solution drops from 22.7 to 19.4°C. **(a)** Is the process endothermic or exothermic? **(b)** What is the heat of solution of NH_4NO_3 in water, expressed as kJ/mol NH_4NO_3?
(a) Since the water temperature decreases, the water must lose heat. This heat is absorbed by the NH_4NO_3 in order to dissolve. The process is endothermic.
(b) The two heat effects here can be designated, $q_{NH_4NO_3}$ and q_{water}. In the usual fashion we can write, $q_{NH_4NO_3} + q_{water} = 0$ and $q_{NH_4NO_3} = -q_{water}$.

$$q_{NH_4NO_3} = -\left[35.0\ \text{g H}_2\text{O} \times \frac{4.184\ \text{J}}{\text{g H}_2\text{O °C}} \times (19.4 - 22.7)°\text{C} \right] = 4.8 \times 10^2\ \text{J}$$

The quantity of heat just calculated is for a 1.50-g sample. For 1.00 mol NH_4NO_3, that is, for

$$NH_4NO_3(s) \longrightarrow NH_4NO_3(aq) \qquad \Delta \bar{H} = ?$$

$$\Delta \bar{H} = \frac{4.8 \times 10^2\ \text{J}}{1.50\ \text{g NH}_4\text{NO}_3} \times \frac{80\ \text{g NH}_4\text{NO}_3}{1\ \text{mol NH}_4\text{NO}_3} \times \frac{1\ \text{kJ}}{1000\ \text{J}} = 26\ \text{kJ/mol NH}_4\text{NO}_3$$

SIMILAR EXAMPLES: Exercises 14, 15.

6-5 Relationships Involving ΔH

If every thermodynamic property had to be determined by experiment, the value of thermodynamics would be limited. The power of thermodynamics is in permitting a large number of predictions to be made from a small number of measurements. Three important relationships involving enthalpy changes make this possible.

FIGURE 6-6
A calorimeter constructed from Styrofoam coffee cups.

The reaction mixture is in the inner cup. The outer cup provides additional thermal insulation from the surroundings. The cup is closed off with a cork stopper through which a thermometer and a stirrer are immersed into the reaction mixture.

1. ΔH Is an Extensive Property. Enthalpy change is directly proportional to the amounts of substances involved in a chemical reaction. For example, we can calculate the heat of combustion of 0.10 mol $C_{10}H_8$ from the result of Example 6-6. It is simply 0.10 mol $C_{10}H_8 \times (-5.15 \times 10^3$ kJ/mol $C_{10}H_8) = -5.15 \times 10^2$ kJ.

2. ΔH Changes Sign When a Process Is Reversed. Enthalpy (H), like internal energy (E), is a function of state. As in the mountain climbing analogy in Figure 6-4, if the direction of a process is reversed, the change in property $(\Delta E$ or $\Delta H)$ also reverses sign $(-\Delta E$ or $-\Delta H)$. Thus, if for the formation of nitrogen oxide from its elements

$$\tfrac{1}{2}N_2(g) + \tfrac{1}{2}O_2(g) \longrightarrow NO(g) \qquad \Delta\bar{H} = +90.37 \text{ kJ/mol}$$

then for the decomposition of nitrogen oxide into its elements

$$NO(g) \longrightarrow \tfrac{1}{2}N_2(g) + \tfrac{1}{2}O_2(g) \qquad \Delta\bar{H} = -90.37 \text{ kJ/mol}$$

3. Hess' Law of Constant Heat Summation.

If a process can be considered to occur in stages or steps, the enthalpy change for the overall process can be obtained by summing the enthalpy changes for the individual steps.

This statement is again a consequence of the fact that enthalpy is a function of state. It can perhaps best be understood by returning to the mountain climbing analogy of Figure 6-4. Imagine that a trip from the base to the summit of the mountain is made in stages. The elevation gain (or loss) can be determined for each stage, and the total elevation gain is the sum of the changes for each stage (e.g., $+1000$ m, -200 m, $+400$ m, etc.).

Suppose that we wish to determine the enthalpy change (ΔH) for the reaction

$$3\,C(\text{graphite}) + 4\,H_2(g) \longrightarrow C_3H_8(g) \qquad \Delta\bar{H} = ? \tag{6.16}$$

How should we proceed? If graphite and $H_2(g)$ are introduced into a reaction vessel, some reaction may occur but it will not go to completion. Neither will the products be limited to $C_3H_8(g)$: ΔH *for reaction* (6.16) *cannot be measured directly.* Instead, we must resort to an *indirect calculation* of the desired ΔH from values of ΔH that can be established by experiment, that is, from the heats of combustion of $C_3H_8(g)$, $C(\text{graphite})$, and $H_2(g)$.

(a) $C_3H_8(g) + 5\,O_2(g) \longrightarrow 3\,CO_2(g) + 4\,H_2O(l)$ $\Delta\bar{H} = -2220.1$ kJ/mol

(b) $C(\text{graphite}) + O_2(g) \longrightarrow CO_2(g)$ $\Delta\bar{H} = -393.5$ kJ/mol

(c) $H_2(g) + \tfrac{1}{2}O_2(g) \longrightarrow H_2O(l)$ $\Delta\bar{H} = -285.9$ kJ/mol

Example 6-8 Calculate ΔH for reaction (6.16) using equations (a), (b), and (c).

Our basic problem is this: What set of equations can we write so that when these equations are added together the result is equation (6.16)? The magnitude and sign of ΔH for each equation in the set must conform to the three rules stated at the beginning of this section.

Since equation (6.16) has $C_3H_8(g)$ appearing on the right side whereas equation (a) has $C_3H_8(g)$ on the left side, equation (a) must be reversed. Equation (6.16) has $C(\text{graphite})$ and $H_2(g)$ appearing on the left side and so do equations (b) and (c); these equations are *not* to be reversed. However, to yield the correct coefficients for $C(\text{graphite})$ and $H_2(g)$ in the final equation, equation (b) must be multiplied by 3 and equation (c) by 4.

In the summation below, any species that would appear on both sides of the final equation is canceled out.

$-$(a): $3\,CO_2(g) + 4\,H_2O(l) \longrightarrow C_3H_8(g) + 5\,O_2(g)$
$$\Delta\bar{H} = -(-2220.1 \text{ kJ/mol}) = +2220.1 \text{ kJ/mol}$$

The procedure for obtaining an overall (net) equation by the summation of separate equations was first illustrated in Figure 4-6.

$$3 \times \text{(b):} \quad 3\,\text{C(graphite)} + 3\,\text{O}_2(g) \longrightarrow 3\,\text{CO}_2(g) \qquad \Delta\bar{H} = 3 \times (-393.5\ \text{kJ/mol})$$
$$= -1180.5\ \text{kJ/mol}$$

$$4 \times \text{(c):} \quad 4\,\text{H}_2(g) + 2\,\text{O}_2(g) \longrightarrow 4\,\text{H}_2\text{O}(l) \qquad \Delta\bar{H} = 4 \times (-285.9\ \text{kJ/mol})$$
$$= -1143.6\ \text{kJ/mol}$$

$$3\,\text{C(graphite)} + 4\,\text{H}_2(g) \longrightarrow \text{C}_3\text{H}_8(g)$$
$$\Delta\bar{H} = (+2220.1 - 1180.5 - 1143.6)\ \text{kJ/mol}$$
$$\Delta\bar{H} = -104.0\ \text{kJ/mol}$$

SIMILAR EXAMPLES: Exercises 26, 27, 28.

6-6 Standard Enthalpies of Formation

Have you noticed that nowhere have we written an actual numerical value for either *E* or *H*? The reason is rather simple: *Absolute values of E and H do not exist.* Nevertheless, we have been successful in dealing with *changes* in these properties alone, that is, ΔE and ΔH.

We can return to our mountain analogy (Figure 6-4) for still another comparable situation. The difference in elevation between the summit and some fixed point at the base of the mountain can be determined very precisely, but what is the *absolute* elevation of the mountain? Do we mean by this the vertical distance between the mountaintop and the center of the earth? Do we mean the vertical distance between the mountaintop and the deepest trench in the ocean? No, by common agreement we mean the vertical distance between the mountaintop and mean sea level. If we arbitrarily assign to mean sea level an elevation of 0, all other points on earth can be assigned an elevation relative to this zero. The elevation of Mt. Everest is +8848 m; that of Badwater, Death Valley, California, is −86 m.

Whenever a thermodynamic quantity is written with a superscript °, this designates that the substances involved in the process are in their standard states. The overbar signifies molar quantities.

We can do the same thing with enthalpies. Here, *by convention,* we assign a value of zero to the enthalpies of the elements in their most stable forms at 1 atm pressure at the specified temperature, a condition referred to as the **standard state.** The enthalpies of compounds can then be related to this arbitrary zero. The difference in enthalpy between one mole of a compound in its standard state and its elements in their standard states is called the **standard molar enthalpy of formation** (or simply the **molar heat of formation**) and denoted as $\Delta\bar{H}_f^\circ$.

Extensive tables of standard enthalpies of formation are available that permit a variety of thermodynamic calculations. A few typical data are presented in Table 6-2. A more extensive tabulation is provided in Appendix D. Tabulated data are given most frequently for 25°C (298 K).

Let us apply the method of Example 6-8 to *calculate* the enthalpy change ($\Delta\bar{H}_{\text{rx}}^\circ$) for the combustion of 1 mol of ethane, $\text{C}_2\text{H}_6(g)$, with all reactants and products in their standard states.

$$\text{C}_2\text{H}_6(g) + \tfrac{7}{2}\,\text{O}_2(g) \longrightarrow 2\,\text{CO}_2(g) + 3\,\text{H}_2\text{O}(l) \qquad \Delta\bar{H}_{\text{rx}}^\circ = ? \tag{6.17}$$

The three equations that can be added together to yield equation (6.17) are

(a) $$\text{C}_2\text{H}_6(g) \longrightarrow 2\,\text{C(graphite)} + 3\,\text{H}_2(g)$$
$$\Delta\bar{H} = -\Delta\bar{H}_f^\circ[\text{C}_2\text{H}_6(g)]$$

(b) $$2\,\text{C(graphite)} + 2\,\text{O}_2(g) \longrightarrow 2\,\text{CO}_2(g)$$
$$\Delta\bar{H} = 2 \times \Delta\bar{H}_f^\circ[\text{CO}_2(g)]$$

(c) $$3\,\text{H}_2(g) + \tfrac{3}{2}\,\text{O}_2(g) \longrightarrow 3\,\text{H}_2\text{O}(l)$$
$$\Delta\bar{H} = 3 \times \Delta\bar{H}_f^\circ[\text{H}_2\text{O}(l)]$$

The subscript rx stands for reaction. It is often used to emphasize that a ΔH value is for a chemical reaction. Thus, the symbol $\Delta\bar{H}_{\text{rx}}^\circ$ signifies a standard molar heat of reaction.

$$\text{C}_2\text{H}_6(g) + \tfrac{7}{2}\,\text{O}_2(g) \longrightarrow 2\,\text{CO}_2(g) + 3\,\text{H}_2\text{O}(l) \qquad \Delta\bar{H}_{\text{rx}}^\circ = ? \tag{6.17}$$

Having just introduced the concept of enthalpies (heats) of formation, we should

TABLE 6-2
Some standard molar enthalpies (heats) of formation at 298 K

Substance	$\Delta\bar{H}^\circ_{f,\,298}$, kJ/mol
$CH_4(g)$	-74.85
$C_2H_2(g)$	226.73
$C_2H_4(g)$	52.30
$C_2H_6(g)$	-84.68
$C_3H_8(g)$	-103.85
$CO(g)$	-110.54
$CO_2(g)$	-393.51
$HCl(g)$	-92.30
$H_2O(l)$	-285.85
$NH_3(g)$	-46.19
$NO(g)$	90.37
$SO_2(g)$	-296.90

recognize that equation (a) is the *reverse* of the equation representing the formation of one mole of $C_2H_6(g)$ from its elements. $\Delta\bar{H}$ for equation (a) is the *negative* of the heat of formation of $C_2H_6(g)$. For equations (b) and (c) the $\Delta\bar{H}$ values are two and three times the heats of formation of $CO_2(g)$ and $H_2O(l)$, respectively. For reaction (6.17), then

$$\Delta\bar{H}^\circ_{rx} = \{2 \times \Delta\bar{H}^\circ_f[CO_2(g)] + 3 \times \Delta\bar{H}^\circ_f[H_2O(l)]\} - \{\Delta\bar{H}^\circ_f[C_2H_6(g)]\} \qquad (6.18)$$

Equation (6.18) is simply a specific application of a more general relationship that is expressed as

$$\Delta\bar{H}^\circ_{rx} = \left[\sum \nu_p \Delta\bar{H}^\circ_f(\text{products})\right] - \left[\sum \nu_r \Delta\bar{H}^\circ_f(\text{reactants})\right] \qquad (6.19)$$

The symbol Σ (Greek, sigma) means "the sum of." The terms that are added together are the products of the standard molar heats of formation ($\Delta\bar{H}^\circ_f$) and their stoichiometric coefficients, ν. One summation is required for the reaction products and another for the initial reactants. The heat of reaction is the summation of terms for the products minus the summation of terms for the reactants.

Example 6-9   Complete the calculation of $\Delta\bar{H}^\circ_{rx}$ for reaction (6.17).

The relationship of $\Delta\bar{H}^\circ_{rx}$ to heats of formation is expressed through equation (6.18). All that is required is the substitution of tabulated heat of formation data (see Table 6-2) into this equation.

$$\Delta\bar{H}^\circ_{rx} = 2 \times \Delta\bar{H}^\circ_f[CO_2(g)] + 3 \times \Delta\bar{H}^\circ_f[H_2O(l)] - \Delta\bar{H}^\circ_f[C_2H_6(g)]$$

$$= 2 \times (-393.5\ \text{kJ/mol}) + 3 \times (-285.8\ \text{kJ/mol}) - (-84.7\ \text{kJ/mol})$$

$$\Delta\bar{H}^\circ_{rx} = -787.0\ \text{kJ/mol} - 857.4\ \text{kJ/mol} + 84.7\ \text{kJ/mol} = -1559.7\ \text{kJ/mol}$$

SIMILAR EXAMPLES: Exercises 31, 32.

Example 6-10   The combustion of cyclopropane (used as an anesthetic) is represented as

$$(CH_2)_3(g) + \tfrac{9}{2}O_2(g) \longrightarrow 3\,CO_2(g) + 3\,H_2O(l) \qquad \Delta\bar{H}^\circ_{rx} = -2091.4\ \text{kJ/mol}$$

Use this value of $\Delta\bar{H}^\circ_{rx}$ and other data from Appendix D to calculate the standard enthalpy (heat) of formation of cyclopropane.

This question requires an application of equation (6.19), but we solve for an unknown heat of formation rather than for a heat of reaction. [Note that no $\Delta\bar{H}^\circ_f$ term appears for the $\tfrac{9}{2}O_2(g)$, since the heat of formation of a free element in its standard state is zero.]

$$\Delta\bar{H}^\circ_{rx} = 3\,\Delta\bar{H}^\circ_f[CO_2(g)] + 3\,\Delta\bar{H}^\circ_f[H_2O(l)] - \Delta\bar{H}^\circ_f[(CH_2)_3(g)]$$

$$= 3 \times (-393.5\ \text{kJ/mol}) + 3 \times (-285.8\ \text{kJ/mol}) - \Delta\bar{H}^\circ_f[(CH_2)_3(g)]$$

$$= -2091.4\ \text{kJ/mol}$$

$$\Delta\bar{H}^\circ_f[(CH_2)_3(g)] = -1180.5\ \text{kJ/mol} - 857.4\ \text{kJ/mol} + 2091.4\ \text{kJ/mol}$$

$$= +53.5\ \text{kJ/mol}$$

SIMILAR EXAMPLE: Exercise 33.

Since the heat of formation data used in Examples 6-9 and 6-10 were for 298 K, the result of each calculation also applies only at 298 K. Generally, ΔH values are not significantly temperature-dependent. This means that results obtained for 298 K are often applicable over a range of temperatures. (We discuss the temperature dependence of ΔH again in Chapter 15.)

6-7 Sources and Uses of Energy

In the United States energy consumption far exceeds the 10,000 to 12,000 kJ daily requirement to sustain human life. With 6% of the world's population, the United States consumes about 30% of the world's energy production, for agriculture, industry, transportation, and material comforts. The current annual rate of energy consumption in the United States is

$$7.9 \times 10^{16} \text{ kJ} = 1.9 \times 10^{16} \text{ kcal}$$
$$= 7.5 \times 10^{16} \text{ Btu} = 75 \text{ Quad}$$

The sources of this energy and the manner in which it is consumed are depicted in Figure 6-7.

FOSSIL FUELS.   As seen from Figure 6-7, the primary energy sources in use today are the so-called fossil fuels—coal, petroleum, and natural gas. These are carbon-containing materials that were formed from organisms that lived millions of years ago.

A direct consequence of the use of coal was the invention of the steam engine and the advent of the industrial revolution. The introduction of petroleum products as fuels in the past century ushered in the internal combustion engine and the jet age. Fossil fuels have served our needs well over the past few centuries, but what are the prospects for the future? Generally, the outlook for continued use of fossil fuels is bleak.

It is impossible to make accurate predictions regarding the available fossil fuel resources of the world and the rate at which they will be consumed. The prospects of new discoveries, both major and minor, are quite uncertain. Also uncertain are the rate at which energy requirements will increase and the extent to which fossil fuels will be supplanted by other energy sources. Perhaps improvements in technology will allow for the extraction of fuels from lower-grade materials (e.g., oil from tar sands and oil shales). Nevertheless, it is possible to produce some very rough estimates of how much longer fossil fuels will be available, as in Figure 6-8. Here the production of a fossil fuel is shown as a function of time, from the approximate date of its first general use to the approximate date anticipated for its exhaustion. As can be seen from this figure, the reserves of coal may last for several more centuries, but those of petroleum (including natural gas) seem likely to become unavailable much sooner.

SYNTHETIC FUELS FROM COAL.   Coal, as we have seen, is in considerably greater supply than petroleum, yet there has not been a significant increase in the production or use of coal in recent years. The difficulties with coal are twofold. First, there are considerable hazards

FIGURE 6-7
Sources and uses of energy in the United States.

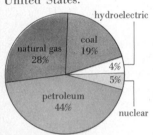

(a) Energy sources

(b) Energy uses

Current energy consumption in the United States is about 7.9×10^{16} kJ/yr.

One British thermal unit (Btu) is the amount of heat required to raise the temperature of one pound of water by one degree Fahrenheit. One quadrillion (10^{15}) Btu is often described as a Quad of energy.

The United States has reserves of coal variously estimated at 4900 to 21,400 Quads, probably about 30% of the world's reserves. Estimates of the U.S. petroleum reserves range from 245 to 950 Quads (about 6% of the world's reserves), and of natural gas, 244 to 950 Quads (about 8% of the world's reserves). Additional large energy reserves exist in the form of oil shales (western United States) and tar sands (western Canada). In a sense, reduction of the rate at which energy is consumed (conservation) is itself an energy resource.

FIGURE 6-8

Estimated world production of fossil fuels.

The exact shape and time of extinction for each curve depend on the estimate used for the total quantity of recoverable fuel and on the rate of production. The general shapes of the curves and the fact that total reserves of coal greatly exceed those of petroleum are rather well established.

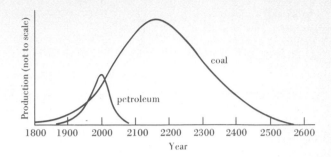

More aptly put: "There are two things wrong with coal today. We can't mine it and we can't burn it."—S. David Freeman, Director, Ford Foundation Energy Project, 1973

and expense in the deep mining of coal. Strip mining, which is less hazardous and expensive, is also much more damaging to the environment (although deep mines produce environmental damage as well). The second difficulty is in the burning of coal. The sulfur content of most coal is much too high to allow for its direct combustion without exceeding environmental limits for atmospheric SO_2. It is necessary either to remove the $SO_2(g)$ from the flue gases or to remove the sulfur before the coal is burned.

One promising possibility for greater utilization of coal reserves is to convert the coal to gaseous or liquid fuels, either in surface installations or while the coal is still underground. The necessary conversions are from carbon (in coal) to carbon monoxide, methane, and other hydrocarbons; chemical reactions are involved.

GASIFICATION OF COAL. Before the advent of cheap natural gas in the 1940s, gas produced from coal (variously called producer gas, town gas, or city gas) was widely used in the United States. This gas is manufactured by passing steam and air through heated coal, leading to reactions such as

$$C(s) + H_2O(g) \longrightarrow CO(g) + H_2(g) \tag{6.20}$$

$$CO(g) + H_2O(g) \longrightarrow CO_2(g) + H_2(g) \tag{6.21}$$

$$2\,C(s) + O_2(g) \longrightarrow 2\,CO(g) \tag{6.22}$$

$$C(s) + 2\,H_2(g) \longrightarrow CH_4(g) \tag{6.23}$$

A typical producer gas consists of about 23% CO, 18% H_2, 8% CO_2, and 1% CH_4, by volume; but it also contains about 50% N_2, since air is used in the process. Because so much of the gas is noncombustible (i.e., the nitrogen and carbon dioxide), producer gas has only about 10 to 15% of the heat value of natural gas.

Modern gasification processes

1. Use oxygen gas instead of air (thereby reducing the amount of nitrogen in the product to mere traces).
2. Provide for the removal of noncombustible CO_2 and of sulfur impurities.
3. Include a step in which CO and H_2, in the presence of a catalyst, are converted to methane:

$$CO(g) + 3\,H_2(g) \longrightarrow CH_4(g) + H_2O(g) \tag{6.24}$$

With these modifications it is possible to obtain a substitute or **synthetic natural gas (SNG),** a gaseous mixture with composition and heat value similar to natural gas.

LIQUEFACTION OF COAL. Processes are currently under development for the production of liquid fuels from coal. These generally involve first the gasification of coal to produce water gas—a mixture of CO and H_2—by reaction (6.20). This is followed by

FIGURE 6-9
Relative importance of different energy sources —present and future (speculative).

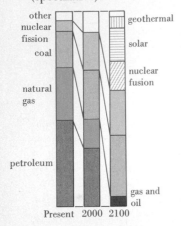

The projections for A.D. 2000 and 2100 are highly speculative. The actual situation might be much different. Solar energy might become an important energy source by 2000, or nuclear fusion may not be developed as an energy source even by 2100. Or perhaps biological sources may become increasingly important. One fact that seems quite certain for the time period shown is that natural gas and petroleum (oil) will decrease in importance.

The prospect of hydrogen becoming a major resource of the future and the various ways in which it might be used are referred to collectively as "the hydrogen economy."

catalytic reactions (the Fischer-Tropsch process) in which liquid hydrocarbons are formed.

$$n\,CO + (2n + 1)\,H_2 \longrightarrow C_nH_{2n+2} + n\,H_2O \tag{6.25}$$

In still another process water gas is converted to liquid methanol.

$$CO + 2\,H_2 \longrightarrow CH_3OH \tag{6.26}$$

In 1942, some 32 million gallons of aviation fuel were made from coal in Germany; and currently, in South Africa, the SASOL process for coal liquefaction is used to produce gasoline and a variety of other petroleum products and chemicals.

METHANOL, ETHANOL, AND HYDROGEN.   Methanol, CH_3OH, can be obtained from coal by reaction (6.26). It can also be produced by the thermal decomposition (pyrolysis) of wood, manure, sewage, or municipal waste. The heat of combustion of methanol is only about one half as great as a typical gasoline on a mass basis, but methanol has a high octane number—106. It has been tested and used as a fuel in internal combustion engines and has been found to burn cleaner than gasoline. Methanol can also be used for space heating, electric power generation, fuel cells, and organic synthesis.

Ethanol, C_2H_5OH, is produced mostly from ethylene, C_2H_4, which in turn is derived from petroleum. Current interest centers on the production of ethanol by the fermentation of organic matter, a process known throughout recorded history. Ethanol production by fermentation is probably in its most advanced state in Brazil, where sugar cane and cassava (manioc) are the plant matter ("biomass") principally employed. Brazil has set a goal of about 1.3 billion gallons of ethanol per year by 1985. By contrast, ethanol production in the United States currently is only about 200 million gallons per year.

Another potentially useful fuel currently derived mostly from petroleum is hydrogen. Its heat of combustion, per gram, is twice that of methane and about three times that of gasoline. There are, of course, problems attendant to using hydrogen as a fuel. It can form explosive mixtures with air. It is bulky to transport as a gas because of its very low density. It is difficult to transport as a liquid because of its very low boiling point (20 K). It dissolves in metals, causing them to become brittle. The greatest problem, however, is the large quantity of energy required to produce hydrogen. If a cheap energy source can be developed (solar energy, nuclear fusion), abundant hydrogen can be obtained by the decomposition of water, either by electrolysis or thermally. Hydrogen might then become the principal fuel for transportation, supplanting gasoline. It could also become the principal metallurgical reducing agent, supplanting carbon (coke) in such processes as the manufacture of iron and steel (see Section 26-7). A particular advantage of hydrogen over fossil fuels is that the only product of its combustion is H_2O. Increasingly, the buildup of carbon dioxide in the atmosphere that accompanies the burning of fossil fuels is being recognized as an environmental problem (see Section 26-3).

CONCLUSION.   Although there is considerable uncertainty as to what the profile of future energy sources will be, there is no doubt that it will differ greatly from that to which we are accustomed today. One possibility is suggested by Figure 6-9. (Nuclear fission and fusion are discussed in Chapter 23.)

Summary

Thermochemistry is concerned primarily with heat effects accompanying chemical reactions. To deal with this subject, however, it is first necessary to establish the meaning of heat and work. These are the two basic forms of energy transfer in chemical reactions. The relationship among heat, work, and the energy stored within a system (called internal energy, E) is embodied in the first law of thermodynamics. To apply this law in numerical calculations requires also that certain conventions be adopted for the signs (i.e., positive or negative) for heat, q, and work, w.

A chemical reaction can be treated as a thermodynamic system. The difference in internal energy between products and reactants, ΔE, is the heat of reaction at constant volume, q_V. This particular heat of reaction is what is measured when a reaction is carried out in a bomb calorimeter. In the bomb calorimeter, heat evolved in a combustion reaction (the system) is used to heat a quantity of water and other objects in the surroundings. The quantity of heat absorbed by the surroundings is easily calculated from the mass of water, its specific heat, the heat capacity of the rest of the surroundings, and the temperature change.

Chemical reactions are generally carried out at constant pressure, not constant volume. For such reactions it proves convenient to establish a new thermodynamic function called enthalpy, H. Enthalpy is defined so that for a process at constant pressure, $\Delta H = q_P$, where q_P is the heat of reaction at constant pressure. The experimental determination of q_P can be carried out in a calorimeter, and generally one of much simpler design than the bomb calorimeter. For most chemical reactions it is sufficient to say that $\Delta H = \Delta E$. Instances where this may not be the case are those in which processes are accompanied by large changes in gas volumes. In these cases ΔH and ΔE can be related through a simple mathematical equation.

By establishing an *arbitrary* zero of enthalpy for the elements in their most stable forms at 1 atm, it is possible to derive standard molar enthalpies (heats) of formation for compounds. These data can be compiled into extensive listings (see Appendix D). One use of heat of formation data is in combination with certain simple relationships involving ΔH values (principally Hess's law of constant heat summation). The result is that many heats of reaction can be determined *indirectly*, that is, by calculation rather than by experimentation.

Learning Objectives

As a result of studying Chapter 6, you should be able to

1. Distinguish between the energy forms heat and work.

2. Relate quantities of heat, temperature changes, and specific heats of substances through numerical calculations (i.e., calculate one of these quantities from known values of the other two).

3. Describe the meaning of the mechanical equivalent of heat, and write a conversion factor between calories and joules.

4. State the first law of thermodynamics and the sign conventions used for heat, q, and work, w.

5. Calculate a value of one of the following quantities from known values of the other two: ΔE, q, and w.

6. Explain the special feature of a function of state.

7. Calculate the heat of a reaction at constant volume, q_V, from bomb calorimetry data.

8. Explain the purposes served by the thermodynamic property of enthalpy (H), describe how ΔH is related to ΔE, and calculate one from the other for reactions involving gases.

9. Use data obtained with a "Styrofoam coffee cup" calorimeter to determine heat effects at constant pressure, q_p.

10. Combine known chemical equations and ΔH values to obtain the net equation for a desired reaction and the heat of that reaction (ΔH).

11. Explain how standard molar enthalpies of formation are established.

12. Calculate heats of reaction from tabulated standard molar enthalpies (heats) of formation, or heats of formation from experimentally determined heats of reaction.

Some New Terms

An **adiabatic** process is one in which no heat is exchanged between a system and its surroundings.

A **bomb calorimeter** is a device used to measure the heat of a combustion reaction at constant volume.

The **calorie (cal)** is the quantity of heat required to raise the temperature of one gram of water by one degree Celsius (specifically, from 14.5 to 15.5°C); 1 cal = 4.184 J.

A **calorimeter** is a device used to measure the quantity of heat exchanged between a thermodynamic system and its surroundings.

Endothermic processes absorb heat from the surroundings as they occur. The quantity of heat carries a *positive* sign.

Enthalpy (*H*) is a thermodynamic function, related to internal energy, and useful in describing constant-pressure processes; $H \equiv E + PV$.

Enthalpy change (ΔH) is the difference in enthalpy between two states of a system. If the process by which the change occurs is a chemical reaction, the enthalpy change is called the **heat of reaction.** @ constant pressure

Enthalpy (heat) of formation is the enthalpy change that accompanies the formation of a compound from the most stable forms of its elements.

Exothermic processes give off heat to the surroundings. The quantity of heat carries a *negative* sign.

The **first law of thermodynamics** states that in interactions between a system and its surroundings energy is neither created nor destroyed. The difference in the quantities of heat and work exchanged must be reflected as a change in internal energy of the system: $\Delta E = q - w$.

A **function of state (state function)** is a property that depends only on the state or present condition of a system and not on how this state is attained.

Heat is a form of energy that is most readily recognized by its ability to produce temperature changes. Heat is energy in transit due to a temperature difference.

Hess' law states that the enthalpy change for an overall or net process is simply the sum of the enthalpy changes for individual steps in the process.

The **internal energy (*E*)** of a system is the total energy attributed to the particles of matter and their interactions within the system.

An **isothermal** process is one that is conducted at a constant temperature.

The **joule (J)** is the basic SI unit of energy. It is the quantity of work done when a force of one newton acts through a distance of one meter.

The **law of conservation of energy** states that energy can neither be created nor destroyed in processes.

Specific heat is the quantity of heat required to raise the temperature of one gram of substance by one degree Celsius.

The **standard state** of a substance refers to the most stable form of that substance at 1 atm pressure.

The **surroundings** represents that portion of the universe with which a system interacts.

A **system** is the portion of the universe selected for a thermodynamic study.

Work is a form of energy that can be expressed as a force acting through a distance.

Exercises

Specific heat

1. How much heat, in calories, is required to raise the temperature of 415 g of chloroform from 21.4 to 51.0°C? The specific heat of chloroform ($CHCl_3$) is 0.23 cal g^{-1} °C^{-1}.

2. An electric range burner weighing 612 g is turned off after reaching a temperature of 515°C. Assume an average specific heat of 0.4 J g^{-1} °C^{-1} for the burner.
   (a) Approximately how much heat will the burner exchange with the surroundings in cooling back to room temperature (20°C)?
   (b) Approximately how much water, in grams, could be heated from room temperature to the boiling point if all this heat could be transferred to the water?

3. The experiment described in Example 6-2 is repeated using zinc instead of lead. The final water temperature is found to be 39.0°C. What is the specific heat of the zinc?

4. Magnesium metal has a specific heat of 1.04 J g^{-1} °C^{-1}. A 70.0-g sample of this metal, at a temperature of 99.8°C, is added to a beaker containing 50.0 g water at 30.0°C. The final water temperature is found to be 47.2°C. Is this result consistent with the law of conservation of energy? Explain.

5. A 74.8-g sample of copper metal at 143.2°C is added to an insulated vessel containing 165 cm³ of glycerol,

$C_3H_8O_3$(l) ($d = 1.26$ g/cm³), at 24.8°C. The final temperature is 31.1°C. The specific heat of copper is given as 0.393 J g^{-1} °C^{-1}. What is the *molar* heat capacity of glycerol?

*6. Brass has a density of 8.40 g/cm³ and a specific heat of 0.385 J g^{-1} °C^{-1}. A cube of brass 7.00 mm on an edge, initially at a temperature of 85.6°C, is immersed into 15.0 g of water at 25.2°C in an insulated container. What is the final temperature of the mixture?

First law of thermodynamics

7. What are the changes in internal energy of a system, ΔE, if the system
   (a) absorbs 50 J of heat and does 50 J of work?
   (b) absorbs 100 cal of heat and does 75 cal of work?
   (c) absorbs 150 cal of heat and does 675 J of work?
   (d) loses 20 J of heat and has 415 J of work done on it?
   (e) absorbs no heat and does 125 J of work?

8. *The internal energy of an ideal gas depends only on its temperature. A sample of an ideal gas is allowed to expand isothermally.*
   (a) Does the gas do work? (Recall Figure 6-1.)
   (b) Does the system exchange heat with the surroundings?
   (c) What happens to the temperature of the gas?

9. A famous experiment, reported to have been performed by the British physicist Joule while on his honeymoon in Switzerland, involved measuring the temperature of water at the top of a waterfall and again at the bottom. The temperature of the water at the bottom was found to be slightly higher than at the top. Explain this observation.

Heats of reaction

10. Only one of the following expressions can be used to describe the heat of a chemical reaction *regardless of how the reaction is carried out*. Which is the correct expression and why? **(a)** q_V; **(b)** q_P; **(c)** $\Delta E + w$; **(d)** ΔE; **(e)** ΔH

11. The reaction of quicklime (CaO) with water produces slaked lime [$Ca(OH)_2$], a substance widely used in the construction industry to make mortar and plaster. The reaction of quicklime and water is highly exothermic.

$$CaO(s) + H_2O(l) \longrightarrow Ca(OH)_2(s) \qquad \Delta \bar{H} = -350 \text{ kJ/mol}$$

(a) What is the heat of reaction per gram of calcium oxide (CaO) reacted?

(b) How much heat, in kilojoules, is associated with the production of 50.0 kg of slaked lime?

★**(c)** If 10.0 g CaO is added to 100.0 g H_2O at 25 °C in an insulated container, will the temperature of the resulting mixture reach the boiling point of water? [Use a value of 1.09 J g^{-1} °C^{-1} for the specific heat of $Ca(OH)_2(s)$.]

12. The combustion of methane gas in air is represented by the equation

$$CH_4(g) + 2 O_2(g) \longrightarrow CO_2(g) + 2 H_2O(l)$$
$$\Delta \bar{H} = -890 \text{ kJ/mol}$$

(a) If the heat of this combustion reaction could be utilized with 100% efficiency to heat water, how many liters of water could be heated from 22.5 °C to 48.2 °C by burning 1.50×10^3 g $CH_4(g)$?

(b) How many liters of $CH_4(g)$ at 22.5 °C and 748 mmHg must be burned to liberate 1.00×10^6 kJ of heat?

13. A particular natural gas consists, on a molar basis, of 83.0% CH_4, 11.2% C_2H_6, and 5.8% C_3H_8. The heats of combustion (ΔH) of these gases are -890 kJ/mol CH_4, -1559 kJ/mol C_2H_6, and -2219 kJ/mol C_3H_8. A 150.0-L sample of this gas, measured at 23.2 °C and 738 mmHg, is burned at constant pressure in an excess of oxygen gas. How much heat is given off to the surroundings?

14. A pellet of potassium hydroxide, KOH, weighing 0.150 g is added to 45.0 g of water in a Styrofoam coffee cup. The water temperature rises from 24.1 to 24.9 °C.

(a) What is the approximate heat of solution of KOH, expressed in kJ/mol KOH?

(b) How could the precision of the result be improved without modifying the apparatus in any way?

15. Care must be taken in preparing solutions of solutes that liberate heat on dissolving. The heat of solution of NaOH is -10 kcal/mol NaOH. To what approximate temperature will a sample of water, originally at 21 °C, be raised in the preparation of 250 cm^3 of 6 M NaOH? Assume that no effort is made to remove heat from the solution during its preparation.

16. An experimental determination of ΔE for the combustion of isopropyl alcohol yields a value of -33.41 kJ/g C_3H_7OH.

$$C_3H_7OH(l) + \tfrac{9}{2}O_2(g) \longrightarrow 3 CO_2(g) + 4 H_2O(l)$$

(a) What is $\Delta \bar{E}$ for this reaction, in kJ/mol C_3H_7OH?

(b) What is $\Delta \bar{H}$ for the reaction, in kJ/mol C_3H_7OH?

17. For each of the reactions listed below, indicate whether ΔH is equal to, greater than, or less than ΔE. (*Hint:* Recall that "greater than" means more positive, or less negative.)

(a) $C(\text{graphite}) + H_2O(g) \longrightarrow CO(g) + H_2(g)$;
$\Delta H = +130$ kJ

(b) $3 CO(g) + 7 H_2(g) \longrightarrow C_3H_8(g) + 3 H_2O(l)$;
$\Delta H = -628$ kJ

(c) the complete combustion of butanol, $C_4H_7OH(l)$, in excess $O_2(g)$

(d) the decomposition of $NH_4NO_3(s)$ into liquid water and gaseous dinitrogen oxide

Bomb calorimetry

18. A sample of substance that is burned in a bomb calorimeter is known to give off 20.9 kJ of heat. The quantity of water in the calorimeter is 1155 g, and the water temperature increases by 3.68 °C as a result of the combustion reaction. Calculate the heat capacity of the bomb.

19. What increase in temperature would you expect to occur in 1025 g of water in a bomb calorimeter if a 0.242-g sample of naphthalene is burned in an excess of $O_2(g)$ in the bomb? Assume a heat capacity of 802 J/°C for the bomb, and the heat of combustion of naphthalene calculated in Example 6-4.

20. The burning of 1.010 g of cane sugar, $C_{12}H_{22}O_{11}$, in a bomb calorimeter causes the temperature of the water to increase from 24.92 to 28.33 °C. The calorimeter contains 980.0 g of water. Assume a heat capacity of 785 J/°C for the bomb.

(a) What is the heat of combustion of the cane sugar (sucrose) expressed in kJ/mol $C_{12}H_{22}O_{11}$?

(b) Write a balanced equation for the combustion reaction, assuming that $CO_2(g)$ and $H_2O(l)$ are the sole products.

(c) Is the value obtained in part (a) ΔE or ΔH for the combustion reaction? Derive values for both ΔE and ΔH for the reaction.

(d) Verify the claim of sugar manufacturers that one teaspoon of sugar (about 4.8 g) "contains only 18 Cal." (*Note:* Reference here is to "large" Calories, that is, kilocalories.)

21. A bomb calorimetry experiment is carried out in two parts. First, a 1.148-g sample of benzoic acid is burned in an excess of $O_2(g)$ in a bomb immersed in 1215 g of water. The temperature of the water rises from 25.12 to 30.26°C. In a second experiment, a 0.895-g powdered coal sample is burned in the same calorimeter assembly. The temperature of 1187 g of water rises from 24.98 to 29.71°C as a result. How many metric tons (1 metric ton = 1000 kg) of this coal would have to be burned in a power plant to release 1.00×10^9 kJ of heat? (Use the value given in Example 6-5 for the heat of combustion of benzoic acid.)

22. Suppose that in the experiment described in Example 6-5 the combustion of the benzoic acid were carried out, in error, *before* the bomb was immersed in water. What approximate final temperature would the bomb reach?

Functions of state

23. Why is a thermodynamic function such as enthalpy (H) so useful, even though its absolute value cannot be determined?

24. Both graphite and diamond are pure forms of carbon. The heat of combustion of graphite, yielding $CO_2(g)$ as the only product and with all reactants and the product being at 25°C and 1 atm pressure, is -393.51 kJ/mol. If diamond is substituted for graphite, would you expect the heat of combustion to be exactly the same or different? Explain.

*25. Two identical steel springs are provided. Each is dissolved in 500 ml of 6 M HCl. One is dissolved in its relaxed condition; the other is compressed to one half its normal length and dissolved. How would you expect the magnitudes of the heat of reaction to compare in these two cases? Explain. (The reaction is exothermic.)

Hess's law

26. Given the heats of reaction

$$\tfrac{1}{2} N_2(g) + \tfrac{1}{2} O_2(g) \longrightarrow NO(g) \qquad \Delta \bar{H} = +90.37 \text{ kJ/mol}$$

$$\tfrac{1}{2} N_2(g) + O_2(g) \longrightarrow NO_2(g) \qquad \Delta \bar{H} = +33.85 \text{ kJ/mol}$$

determine the heat of the reaction

$$NO(g) + \tfrac{1}{2} O_2(g) \longrightarrow NO_2(g) \qquad \Delta \bar{H} = ?$$

27. The heats of combustion ($\Delta \bar{H}$) per mole of 1,3-butadiene [$C_4H_6(g)$], normal butane [$C_4H_{10}(g)$], and $H_2(g)$ are -2543.5, -2878.6, and -285.85 kJ/mol, respectively. Use these data to calculate the heat of hydrogenation of 1,3-butadiene to normal butane.

$$C_4H_6(g) + 2 H_2(g) \longrightarrow C_4H_{10}(g) \qquad \Delta \bar{H} = ?$$

28. Synthetic natural gas (SNG) is a gaseous mixture containing $CH_4(g)$ that is used in the synthesis of organic compounds. One reaction for the production of SNG is

$$4 CO(g) + 8 H_2(g) \longrightarrow 3 CH_4(g) + CO_2(g) + 2 H_2O(l)$$
$$\Delta \bar{H} = ?$$

Use the following data, as necessary, to determine $\Delta \bar{H}$ for this SNG reaction.

$$C(\text{graphite}) + \tfrac{1}{2} O_2(g) \longrightarrow CO(g) \quad \Delta \bar{H} = -110.54 \text{ kJ/mol}$$

$$CO(g) + \tfrac{1}{2} O_2(g) \longrightarrow CO_2(g) \Delta \bar{H} = -282.97 \text{ kJ/mol}$$

$$H_2(g) + \tfrac{1}{2} O_2(g) \longrightarrow H_2O(l) \Delta \bar{H} = -285.85 \text{ kJ/mol}$$

$$C(\text{graphite}) + 2 H_2(g) \longrightarrow CH_4(g) \Delta \bar{H} = -74.85 \text{ kJ/mol}$$

29. Methanol, a potential fuel source, can be prepared synthetically by heating carbon monoxide and hydrogen gases under pressure in the presence of a catalyst. The reaction is

$$CO(g) + 2 H_2(g) \longrightarrow CH_3OH(l) \qquad \Delta \bar{H} = ?$$

Use the fact that the heat of combustion of $CH_3OH(l)$ is $\Delta \bar{H} = -726.6$ kJ/mol $CH_3OH(l)$ and the following data to determine the heat of this reaction.

$$C(\text{graphite}) + \tfrac{1}{2} O_2(g) \longrightarrow CO(g) \quad \Delta \bar{H} = -110.54 \text{ kJ/mol}$$

$$C(\text{graphite}) + O_2(g) \longrightarrow CO_2(g) \Delta \bar{H} = -393.51 \text{ kJ/mol}$$

$$H_2(g) + \tfrac{1}{2} O_2(g) \longrightarrow H_2O(l) \quad \Delta \bar{H} = -285.85 \text{ kJ/mol}$$

30. A net reaction for a coal gasification process is given as

$$2 C(s) + 2 H_2O(g) \longrightarrow CH_4(g) + CO_2(g)$$

Show that this net equation can be established by an appropriate combination of equations (6.20), (6.21), and (6.24).

Enthalpies (heats) of formation

31. Use standard enthalpies of formation from Table 6-2 to determine the heat of the reaction

$$2 Cl_2(g) + 2 H_2O(l) \longrightarrow 4 HCl(g) + O_2(g) \qquad \Delta \bar{H}° = ?$$

32. What is the heat of combustion of $H_2S(g)$ if the reactants [$H_2S(g)$ and $O_2(g)$] and the products [$H_2O(l)$ and $SO_2(g)$] are measured at 25°C and 1 atm pressure. (*Hint:* Use data from Appendix D.)

33. Given the heat of the following reaction, determine the heat of formation of $CCl_4(g)$ at 25°C and 1 atm.

$$CH_4(g) + 4 Cl_2(g) \longrightarrow CCl_4(g) + 4 HCl(g)$$
$$\Delta \bar{H}° = -402 \text{ kJ/mol}$$

34. What is the enthalpy change ($\Delta \bar{H}°_{rx}$) for the net reaction if all reactants and products of the coal gasification process in Exercise 30 are measured at 25°C and 1 atm?

35. For the reaction

$$C_2H_4(g) + 3 O_2(g) \longrightarrow 2 CO_2(g) + 2 H_2O(l)$$
$$\Delta \bar{H}° = -1410.8 \text{ kJ/mol}$$

If the H_2O were obtained as a gas rather than a liquid, would the heat of reaction be greater (more negative) or smaller (less negative) than that indicated in the equation? Calculate the value of $\Delta \bar{H}$ in this case. (*Hint:* Refer to Appendix D.)

36. Use data from Appendix D, together with the fact that $\Delta \bar{H}°$ for the complete combustion of 1 mol $C_5H_{12}(l)$ is

−3534 kJ/mol, to calculate $\Delta\bar{H}°$ for the synthesis of 1 mol $C_5H_{12}(l)$ from $CO(g)$ and $H_2(g)$.

$$5\ CO(g) + 11\ H_2(g) \longrightarrow C_5H_{12}(l) + 5\ H_2O(l) \quad \Delta\bar{H}° = ?$$

37. The decomposition of limestone, $CaCO_3(s)$, into quicklime, $CaO(s)$, and $CO_2(g)$ is accomplished at a temperature of about 900°C in a gas-fired kiln. (Assume that heats of reaction under these conditions are essentially the same as at 25°C and 1 atm pressure.)

   (a) How much heat, in kilojoules, is required to cause the decomposition of 1000 kg $CaCO_3(s)$?

   (b) If the heat energy calculated in part (a) is supplied by the combustion of methane, $CH_4(g)$, what volume of the gas, measured at 24.5°C and 752 mmHg, is required?

38. Under the entry "H_2SO_4" a handbook lists several different values for the enthalpy of formation, $\Delta\bar{H}_f$. For example, for pure $H_2SO_4(l)$, $\Delta\bar{H}_f = -814$ kJ/mol H_2SO_4; for an aqueous solution that is 1.0 M H_2SO_4, $\Delta\bar{H}_f = -888$ kJ/mol H_2SO_4; for 0.25 M H_2SO_4, $\Delta\bar{H}_f = -890$ kJ/mol H_2SO_4; for 0.02 M H_2SO_4, $\Delta\bar{H}_f = -897$ kJ/mol H_2SO_4.

   (a) Explain why these values are not all the same.

Additional Exercises

In 1818, Dulong and Petit made the observation that the molar heat capacities of elements in their solid states are approximately constant from one element to another. They derived the expression: atomic weight × specific heat (in cal g^{-1} °C^{-1}) = 6.4 (approximately). The specific heats of silicon, phosphorus, and lead are 0.17, 0.19, and 0.031 cal g^{-1} °C^{-1}, respectively. Use this information in the following three exercises.

1. Comment on the validity of the law of Dulong and Petit when applied to silicon, phosphorus, and lead.

2. To raise the temperature of 50.0 g of a particular metal by 10°C requires 47.5 cal of heat. What is the approximate atomic weight of the metal? What might the metal be?

★3. A sample of tin weighing 248 g and at a temperature of 50.0°C is added to 100.0 cm³ of water at 25.0°C. Estimate the final water temperature.

★4. An old "trick" for cooling down a hot beverage quickly is to immerse a cold spoon into the liquid. A silver spoon weighing 3.20 oz and at a temperature of 70°F is placed in a cup of hot coffee (8 oz) at 160°F. What will be the resulting liquid temperature? Assume that the specific heat of coffee is 1 cal g^{-1} °C^{-1} and that of silver, 0.056 cal g^{-1} °C^{-1}. Comment on the effectiveness of this method of cooling a hot drink.

★5. A British thermal unit (Btu) is defined as the quantity of heat required to change the temperature of 1 lb of water by 1°F. Assuming the specific heat of water to be independent

   (b) When a concentrated aqueous solution of H_2SO_4 is diluted, does the solution temperature increase or decrease? Explain.

   ★(c) If 250 ml of 0.02 M H_2SO_4 is prepared by diluting pure $H_2SO_4(l)$ with water, estimate the change in temperature that occurs. [Assume that the $H_2SO_4(l)$ and the water used for its dilution are at the same temperature initially.]

★39. A 1.00-L sample (at STP) of a particular natural gas yields, upon complete combustion at constant pressure, $CO_2(g)$, $H_2O(l)$, and 43.6 kJ of evolved heat. If the gas is a mixture of $CH_4(g)$ and $C_2H_6(g)$, what is its composition, expressed as percent by volume?

★40. An alkane hydrocarbon has the formula C_nH_{2n+2}. Whatever the subscript for carbon, n, the subscript for hydrogen is "twice n plus 2." The heats of formation of the alkanes decrease (become more negative) as the numbers of carbon atoms increase. Starting with propane, C_3H_8, for each additional CH_2 group in the formula the heat of formation, $\Delta\bar{H}_f°$, changes by about −21 kJ/mol. Use this fact, together with data from Appendix D, to estimate the heat of combustion of normal heptane, $C_7H_{16}(l)$.

of temperature, how much heat is required to raise the temperature of the water in a 40-gal water heater from 70 to 150°F? **(a)** in Btu; **(b)** in kcal; **(c)** in kJ. (1 L = 1.06 qt.)

6. What are the changes in internal energy, ΔE, of a system if it

   (a) absorbs 112 cal of heat and does 255 J of work?

   (b) loses 185 cal of heat and has 825 J of work done on it?

   (c) loses 155 cal of heat and does 236 J of work?

7. An ideal gas is allowed to expand *adiabatically*.

   (a) Does the gas do work? (Recall Figure 6-1.)

   (b) Does the system exchange heat with the surroundings?

   (c) Does the internal energy of the gas increase or decrease?

8. A mixture of 125 g of copper and 375 g of water is heated in an open beaker from 24.0°C to 85.2°C. What is ΔH (in kJ) for the copper–water mixture? (Specific heats: water, 4.18 J g^{-1} °C^{-1}; copper, 0.393 J g^{-1} °C^{-1}.)

9. The combustion of normal octane in an excess of oxygen is represented by the equation

$$C_8H_{18}(l) + \tfrac{25}{2} O_2(g) \longrightarrow 8\ CO_2(g) + 9\ H_2O(l)$$
$$\Delta\bar{H} = -5.48 \times 10^3 \text{ kJ/mol}$$

How much heat, in kilojoules, is liberated to the surroundings per gallon of normal octane that is burned at constant pressure? (1 gal = 3.78 L; density of octane = 0.703 g/cm³.)

10. The heat of solution of potassium iodide in water is listed as $\Delta \bar{H} = +21.3$ kJ/mol KI. If a quantity of KI is added to a quantity of water at 23.5°C in a Styrofoam cup sufficient to produce 150.0 cm³ of 2.50 *M* KI, what will the final temperature be? (Assume a density of 1.30 g/cm³ and a specific heat of 4.18 J g⁻¹ °C⁻¹ for 2.50 *M* KI.)

11. A calorimeter that measures an exothermic heat of reaction by the quantity of ice that can be melted is called an **ice calorimeter.** Now consider that 0.100 L of methane gas, $CH_4(g)$, at 25.0°C and 744 mmHg is burned completely at constant pressure in an excess of air. The heat liberated is captured and used to melt 10.7 g of ice at 0°C. (The heat required to melt ice, called the heat of fusion, is 333.5 J/g.)
   (a) Write a chemical equation for the combustion reaction.
   (b) What is ΔH for this reaction, expressed as kJ/mol $CH_4(g)$?

12. Determine which of the following gases has the greater fuel value, on a per liter (STP) basis. That is, which gas has the greater heat of combustion? (*Hint:* Base your calculation on the combustible gases only— CH_4, C_3H_8, CO, and H_2.)
   (a) a coal gas consisting of 49.7% H_2, 29.9% CH_4, 8.2% N_2, 6.9% CO, 3.1% C_3H_8, 1.7% CO_2, and 0.5% O_2, by volume
   (b) sewage gas consisting of 66.0% CH_4, 30.0% CO_2, and 4.0% N_2, by volume

*13. Joule published his definitive results on the mechanical equivalent of heat in 1850. One of these results was that

"The quantity of heat capable of increasing the temperature of one pound of water by 1° Fahrenheit requires for its evolution the expenditure of a mechanical force represented by the fall of 772 lb through the space of one foot." Show that this result is equivalent to that presented in equation (6.5).

*14. Some of the butane, $C_4H_{10}(g)$, contained in a 200.0-L cylinder at 26.0°C is withdrawn and burned completely at constant pressure in an excess of air. As a result the pressure of the gas in the cylinder falls from 2.35 atm to 1.10 atm. The liberated heat is used to raise the temperature of 35.0 gal of water (1 gal = 3.78 L) from 26.0°C to 62.2°C. Assuming that the combustion products are $CO_2(g)$ and $H_2O(l)$ exclusively, what is the efficiency of the water heater? (That is, what percent of the heat of combustion was absorbed by the water?)

*15. One of the advantages of modern coal gasification processes over earlier ones is said to be that some of the heat required to bring about gasification (equation 6.20) is supplied by the methanation reaction (equation 6.24). Use data from Appendix D to show that this is indeed possible.

*16. The metabolism of glucose, $C_6H_{12}O_6$, yields $CO_2(g)$ and $H_2O(l)$ as products. Heat energy released in the process is converted to useful work with about 42% efficiency. Calculate the number of grams of glucose metabolized by a 58.0-kg person in climbing a mountain with an elevation gain of 1450 m. Assume that the work performed in the climb is about four times that required simply to lift 58.0 kg by 1450 m. The heat of formation of $C_6H_{12}O_6(s)$ is −1274 kJ/mol.

Self-Test Questions

For questions 1 through 6 select the single item that best completes each statement.

1. 1.00 *kcal* of heat is
   (a) absorbed when 1.00 cm³ water is heated from 14.5 to 15.5°C.
   (b) absorbed when 1.00 L water is heated from 20.0 to 30.0°C.
   (c) given off when 100.0 cm³ water is cooled from 20.0 to 10.0°C.
   (d) equal to 1.0×10^6 cal.

2. $\Delta E = +100$ J for a system that *gives off* 100 J of heat and **(a)** does 200 J of work; **(b)** has 200 J of work done on it; **(c)** does no work; **(d)** has 100 J of work done on it.

3. For *any* chemical reaction carried out in *any* manner, the quantity of heat associated with the reaction (the heat of reaction) is **(a)** $\Delta E + w$; **(b)** $\Delta E - w$; **(c)** q_V; **(d)** q_P.

4. For the reaction

$N_2(g) + O_2(g) \longrightarrow 2\ NO(g)$

(a) ΔE is less than ΔH; **(b)** ΔE is greater than ΔH; **(c)** $\Delta E = 0$; **(d)** $\Delta E = \Delta H$.

5. A handbook lists the heat of formation of $NH_3(g)$ as −46 kJ/mol. For the reaction

$2\ NH_3(g) \longrightarrow N_2(g) + 3\ H_2(g) \qquad \Delta H = ?$

(a) $\Delta H = -46$ kJ; **(b)** $\Delta H = +46$ kJ; **(c)** $\Delta H = +92$ kJ; **(d)** $\Delta H = +138$ kJ.

6. The heat of formation of $CO_2(g)$ is −394 kJ/mol, and that of $H_2O(l)$ is −286 kJ/mol. The heat of combustion of $C_5H_{12}(l)$ is $\Delta \bar{H} = -3534$ kJ/mol.

$C_5H_{12}(l) + 8\ O_2(g) \longrightarrow 5\ CO_2(g) + 6\ H_2O(l)$
$$\Delta \bar{H} = -3534 \text{ kJ/mol}$$

The heat of formation of $C_5H_{12}(l)$, in kJ/mol, is
   (a) +3534
   (b) $[-3534 - 5 \times (-394) - 6 \times (-286)]$
   (c) $[5 \times (-394) + 6 \times (-286) - 3534]$
   (d) $[5 \times (-394) + 6 \times (-286) + 3534]$

7. Explain briefly the difference in meaning of each pair of terms.

    (a) specific heat and molar heat capacity of a substance

    (b) endothermic and exothermic reaction

    (c) ΔH and ΔE for a reaction

    (d) heat of formation and heat of combustion of the hydrocarbon, $C_4H_{10}(g)$

8. A 1.50-kg piece of iron (specific heat $= 0.59$ J g^{-1} $^{\circ}C^{-1}$) is dropped into 0.755 kg of water and the water temperature is observed to rise from 21.3 to 38.6°C. What must have been the initial temperature of the iron?

9. The heat of combustion of phenol, $C_6H_5OH(s)$, is deter- mined in a bomb calorimeter and found to be -32.55 kJ/g C_6H_5OH.

    (a) Write a balanced equation for the combustion reac- tion.

    (b) What is $\Delta \bar{E}$ for this reaction, in kJ/mol C_6H_5OH?

    (c) What is $\Delta \bar{H}$ for this reaction, in kJ/mol C_6H_5OH?

10. The heats of combustion ($\Delta \bar{H}$) per mole of C(graphite) and CO(g) are -393.51 and -282.97 kJ/mol, respectively. In both cases $CO_2(g)$ is the sole product. For the formation of the poisonous gas, phosgene,

$$CO(g) + Cl_2(g) \longrightarrow COCl_2(g) \qquad \Delta \bar{H} = -108 \text{ kJ/mol}$$

Calculate the heat of formation of $COCl_2(g)$.

7 Electrons in Atoms

For several reasons we need to acquire a more detailed knowledge of atomic and molecular structure. A knowledge of atomic structure helps us to understand the forces between atoms that lead to the formation of molecules. With a knowledge of molecular structure we will be able to establish, among other things, whether a collection of molecules exists as a solid, liquid, or gas and whether substances will form a solution on mixing. In this chapter we study the electronic structures of atoms. By the electronic structure of an atom we mean, basically, a description of where the electrons in an atom are most likely to be found.

Rutherford's nuclear atomic model (Section 2-6) became possible only after basic principles had been established in two fundamental areas of physics—electricity and magnetism. To understand the behavior of electrons in atoms requires some appreciation of the nature of electromagnetic radiation. Our study will begin with this topic. Again, we will see how experiments and discoveries followed one another, for the most part in a logical fashion. But there were a few dramatic breakthroughs as well.

7-1 Electromagnetic Radiation

Electric charges and magnetic poles exert forces over a distance, through electric and magnetic fields. Furthermore, these fields are complementary. A changing electric field induces a magnetic field, and vice versa. If electrically charged particles move with respect to one another, alternating electric and magnetic fields are produced and propagated through the space or medium surrounding the particles. The mode of propagation is called a **wave.** Energy is associated with the electric and magnetic fields, and the wave becomes a means of transmitting energy over distances. This energy transfer is referred to as **electromagnetic radiation.**

FIGURE 7-1
The simplest wave motion—traveling wave in a string.

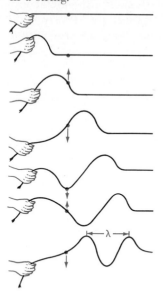

The hypothetical string pictured here is infinitely long. Waves pass along the string only in the left-to-right direction. The wave is called a traveling wave.

Wave motion is complex, but certain characteristics of waves can be thought of rather simply in terms of vibrations in a string, pictured in Figure 7-1. The motion which is propagated through the string is the up-and-down motion of the hand holding the end of the string, the motion starts at time, $t = 0$ (top), and continues through several intervals of time (bottom). The wave travels from left to right, but the vibrating medium (the string) moves up and down, that is, perpendicular to the direction of the wave itself. The position of a typical point on the string as a function of time is denoted by the colored dot. The arrows indicate the direction in which the dot is moving. At any instant of time the wave in the string consists of a number of regions in which the string is at a maximum or high point. These are called wave crests. There are a corresponding number of regions at a minimum or low point, called wave troughs. The distance between two successive crests (or troughs) is called the **wavelength,** usually designated by the Greek letter lambda, λ.

Another characteristic property of a wave is its **frequency,** designated by the Greek letter nu, ν. This is the number of wave crests or troughs that pass through a given point in a unit of time. Frequency can be expressed by the unit s^{-1} (i.e., per second), meaning the number of occurrences or events per second. For the vibrating string in Figure 7-1, this is simply the number of times per second that the hand driving the wave goes through its up-and-down motion.

The product of the length of a wave (λ) and the number of cycles produced per second (ν) indicates how far the wave front has traveled down the string in 1 s. This is *the velocity of the wave, c.*

$$c = \nu\lambda \tag{7.1}$$

Now let us return to a discussion of electromagnetic waves. Propagation of these waves results from the oscillations of charged particles. This is analogous to the up-and-down hand motion in Figure 7-1. The electromagnetic waves are complicated by the fact that they travel in all directions simultaneously, that is, they are three-dimensional waves. An additional complication is that the waves actually represent two kinds of vibrations occurring simultaneously. These are an oscillating electric field and, perpendicular to this (at a 90° angle), an oscillating magnetic field. Still another interesting difference between electromagnetic radiation and other kinds of waves is that electromagnetic radiation requires no medium for its transmission. It can travel through vacuum or empty space. Despite these complications, an electromagnetic wave is often represented in the simplified manner shown in Figure 7-2. This is adequate for our purposes.

FIGURE 7-2
An electromagnetic wave.

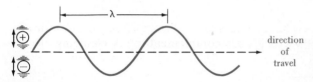

Although oversimplified, this representation does suggest that electromagnetic radiation results from the relative motion of electrically charged objects and denotes one characteristic of the radiation—the wavelength λ.

A radio station that broadcasts at 91.8 megahertz (MHz) transmits radio waves, a form of electromagnetic radiation, at a frequency of $91.8 \times 10^6 = 91,800,000$ cycles per second.

Also in use are the micron (μ), or micrometer (μm), which is 1×10^{-6} m, and the millimicron (mμ), which is 1×10^{-9} m. The SI convention calls for both the millimicron and the angstrom unit to be phased out.

FREQUENCY, WAVELENGTH, AND VELOCITY. Several different units are used to describe electromagnetic radiation. Frequency is often expressed by the unit s^{-1}, referred to as cycles per second. The SI unit for cycles per second is the **hertz (Hz).** Wavelength must have a unit of length, and logically this should be the meter (m). However, because so many kinds of electromagnetic radiation are of very short wavelength, some smaller units are necessary. All of the following are in common use.

1 centimeter (cm) $= 1 \times 10^{-2}$ m

1 nanometer (nm) $= 1 \times 10^{-9}$ m $= 1 \times 10^{-7}$ cm $= 10$ Å

1 angstrom (Å) $= 1 \times 10^{-10}$ m $= 1 \times 10^{-8}$ cm

A distinctive feature of electromagnetic radiation is that its velocity has a constant value of 2.997925×10^8 m/s in vacuum (usually rounded off to 3.00×10^8 m/s). Since ordinary light is a form of electromagnetic radiation, this characteristic velocity is often called the **speed of light.** The frequencies and wavelengths of a number of different kinds of electromagnetic radiation are compared in Figure 7-3.

Example 7-1   *A microwave device produces electromagnetic radiation of wavelength 0.85 cm. What is this wavelength expressed in meters and in nanometers?*

$$\text{no. m} = 0.85 \text{ cm} \times \frac{1 \text{ m}}{100 \text{ cm}} = 8.5 \times 10^{-3} \text{ m}$$

$$\text{no. nm} = 0.85 \text{ cm} \times \frac{1 \text{ nm}}{1 \times 10^{-7} \text{ cm}} = 8.5 \times 10^6 \text{ nm}$$

SIMILAR EXAMPLE: Exercise 1.

Example 7-2   *An FM radio station broadcasts on a frequency of 91.5×10^6 s^{-1}. What is the wavelength of these radio waves, in meters?*

FIGURE 7-3
The electromagnetic spectrum.

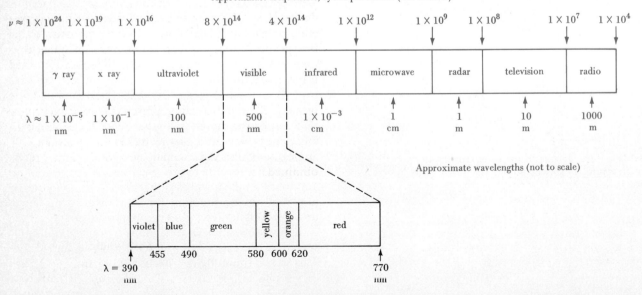

Since radio waves are a form of electromagnetic radiation, $c = 3.00 \times 10^8$ m/s. Solve equation (7.1) for λ.

$$\lambda = \frac{c}{\nu} = \frac{3.00 \times 10^8 \text{ m/s}}{91.5 \times 10^6 \text{ s}^{-1}} = 3.28 \text{ m}$$

SIMILAR EXAMPLES: Exercises 2, 4.

Example 7-3   Most of the light emitted by a sodium vapor lamp has a wavelength of 589 nm. What is the frequency of this radiation?

$$\begin{cases} c = 3.00 \times 10^8 \text{ m/s} \\ \lambda = 589 \text{ nm} \times \dfrac{1 \times 10^{-9} \text{ m}}{1 \text{ nm}} = 5.89 \times 10^{-7} \text{ m} \\ \nu = ? \end{cases}$$

Equation (7.1) must be rearranged to the form $\nu = c/\lambda$.

$$\nu = \frac{c}{\lambda} = \frac{3.00 \times 10^8 \text{ m/s}}{5.89 \times 10^{-7} \text{ m}} = 5.09 \times 10^{14} \text{ s}^{-1}$$

SIMILAR EXAMPLE: Exercise 3.

THE VISIBLE SPECTRUM.   The speed of light depends on the medium through which it travels. As a result, a light ray is bent or refracted as it passes from one medium to another. Because light waves of differing wavelengths are refracted differently, a light ray consisting of different wavelength components is dispersed into a spectrum of colors as the ray passes through a medium. The shortest wavelength light that the human eye can detect corresponds to the color violet; the longest, red.

These points are illustrated in Figure 7-4. Here a beam of "white" light is rendered parallel by a lens, passed through a narrow opening or slit, and then dispersed into its colored components by a glass prism. A device in which all these effects are achieved, together with a measurement of the intensity of each wavelength component, is called a **spectrometer.** If the spectrum is recorded photographically, the device is called a **spectrograph.**

White light is polychromatic (many-colored). If desired, after polychromatic light has been dispersed into a spectrum in a spectrometer, all the wavelength components except a narrow band can be blocked off. In this way monochromatic (one-color) light can be obtained.

FIGURE 7-4
The spectrum of white light.

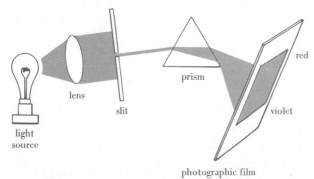

The wavelength components of white light have different speeds in a material medium. Because red light has the lowest frequency of the visible colors, it does not interact strongly with the medium. Its speed is not greatly reduced, and it is refracted least. Violet light, on the other hand, has the highest frequency of the visible colors. It has more opportunity to interact with the medium, is slowed down the most, and hence refracted the most.

7-2 Atomic Spectra

In the spectrum illustrated in Figure 7-4 the light source is "white" light. This could be sunlight or certain artificial light sources, for example, the heated fil-

FIGURE 7-5
The production of an atomic or line spectrum.

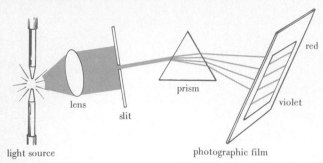

light source

lens

slit

prism

photographic film

red

violet

The light source depicted here is an electric arc between a pair of graphite electrodes. The substance whose spectrum is to be investigated is vaporized at the high temperature of the arc.

FIGURE 7-6
The Balmer series for hydrogen—a line spectrum.

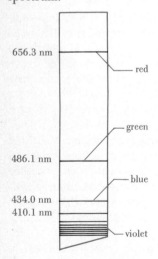

656.3 nm — red

green

486.1 nm — blue

434.0 nm
410.1 nm — violet

One of the first investigators to make extensive use of atomic spectra to identify chemical elements was the German chemist Bunsen (1811–1899). During the latter half of the nineteenth century the spectra of most of the elements had been investigated, including even that of the element helium, which was discovered to exist on the sun before it was discovered on earth.

When the Balmer equation is written in the form of (7.3), we see that a straight line is obtained if ν is plotted as a function of $1/n^2$. (See also Appendix 1.)

$$\nu = Rc\left(\frac{1}{2^2} - \frac{1}{n^2}\right)$$

$$\underbrace{\nu}_{y\,=} = \underbrace{-(Rc)}_{m\,\cdot} \underbrace{\left(\frac{1}{n^2}\right)}_{x} + \underbrace{\frac{Rc}{4}}_{b}$$

A common scientific procedure is to graph experimental data and then find a mathematical equation to describe the graph. Balmer, however, seems to have deduced his formula simply by manipulating numbers.

ament of an ordinary electric light bulb. Each wavelength component of the white light, after passing through the slit and prism, produces an image of the slit on the photographic film in the form of a line. White light consists of so many wavelength components that its spectrum is a continuum of these lines. The spectrum of white light is continuous, displaying a gradual blending of colors from red, through orange, yellow, green, and blue, to violet.

On the other hand, with light emitted by most heated substances, only a discrete number of images of the entrance slit—a series of colored lines—is observed. These spectra are discontinuous, or line, spectra. The production of a line spectrum is illustrated in Figure 7-5.

Among the most extensively studied atomic spectra during the nineteenth century was that of the element hydrogen. The visible spectrum of hydrogen is rather simple, consisting of a red line, a green line, and a number of blue and violet lines that appear to converge to a limit in the ultraviolet region. The first four lines in the spectrum (starting with the red line at 656.3 nm) were investigated very carefully by Ångstrom. They correspond to the wavelengths listed in Figure 7-6. In 1885, the Swiss physicist–school teacher, Johann Balmer, using just these four wavelength values, deduced, apparently by trial-and-error, the following formula for them:

$$\lambda \text{ (in Å)} = 3645.6\left(\frac{n^2}{n^2 - 4}\right)$$
$$\text{where } n = 3, 4, 5, \ldots \quad (7.2)$$

If $n = 3$ is substituted into Balmer's formula, the wavelength obtained is 656.2 nm, in excellent agreement with Ångstrom's measured value. If $n = 4$ the wavelength of the green line is obtained, and so on. A more commonly encountered form of the Balmer equation is shown below, written in terms of the frequencies of the spectral lines.

$$\nu = Rc\left(\frac{1}{2^2} - \frac{1}{n^2}\right)$$

$$= 3.2881 \times 10^{15} \text{ s}^{-1}\left(\frac{1}{2^2} - \frac{1}{n^2}\right) \quad (7.3)$$

R is a numerical constant, called the Rydberg constant, having a value of 10,967,800 m^{-1}; c is the velocity of light, 2.997925×10^8 m/s. To simplify calculations, the product $R \times c$ is also given in equation (7.3). To five significant figures, it is 3.2881×10^{15} s^{-1}.

Example 7-4 Use equation (7.3) to calculate the wavelength of the fourth line ($n = 6$) in the Balmer series of hydrogen. Compare this result with the value given in Figure 7-6.

$$\nu = 3.2881 \times 10^{15} \text{ s}^{-1} \times \left(\frac{1}{2^2} - \frac{1}{6^2}\right) = 3.2881 \times 10^{15} \text{ s}^{-1} \times \left(\frac{1}{4} - \frac{1}{36}\right)$$

$$= 3.2881 \times 10^{15} \text{ s}^{-1} \times (0.25000 - 0.02778) = 7.3068 \times 10^{14} \text{ s}^{-1}$$

$$\lambda = \frac{c}{\nu} = \frac{2.9979 \times 10^8 \text{ m/s}}{7.3068 \times 10^{14} \text{ s}^{-1}} = 4.103 \times 10^{-7} \text{ m} = 410.3 \text{ nm}$$

The calculated value (410.3 nm) and the measured value (410.1 nm) agree within 0.2 nm. This is good agreement.

SIMILAR EXAMPLES: Exercises 7, 8, 9.

By the early nineteenth century a wave theory of light had been firmly established and was successful in explaining continuous spectra (like the rainbow). But the existence of atomic or line spectra could not be explained by this wave theory. Not even the electromagnetic theory of radiation, introduced by James Maxwell in the 1860s, was successful in explaining atomic spectra. That such a simple equation as Balmer's could be used to correlate atomic spectral data suggested that there was some basic principle underlying all atomic spectra. Yet this principle was never discovered by the methods of nineteenth-century physics known as classical physics.

7-3 Quantum Theory

The key to the growing number of unsolved problems of nineteenth-century physics lay in a great breakthrough of modern science—the quantum theory. This theory was originated in 1900 by Max Planck (1858–1947) to explain a phenomenon known as blackbody radiation. Planck's revolutionary hypothesis was that energy (like matter) is *discontinuous,* and consists of large numbers of tiny *discrete* units called **quanta.** Whether a system gains or loses energy, it must do so in terms of these quanta. The energy associated with a quantum of electromagnetic radiation is proportional to the frequency of the radiation and is expressed by

$$E = h\nu \tag{7.4}$$

The proportionality constant, **h,** is called **Planck's constant** and has a value of 6.626×10^{-34} J s.

According to the laws of classical physics, the energy of a system should be able to assume any value and to change by any amount. By the principles of quantum theory, however, we find that the energy of a system can have only a unique set of values. This means that energy can change only by certain discrete amounts, called quantum jumps. A system may gain one quantum of energy, or two quanta, or three, and so on; but it cannot gain or lose $\frac{1}{3}$, $\frac{1}{2}$, $1\frac{1}{3}$, or $2\frac{1}{2}$ quanta. Figure 7-7 suggests an analogy between classical and quantum theory that may prove helpful in understanding the essential difference between the two.

There have been instances in the history of science when a new hypothesis proved useful in explaining one phenomenon but was not generally applicable to any others. It was only in the discovery of other applications of the quantum hypothesis that it acquired status as a significant new theory of science. The first notable new success came in 1905 with Albert Einstein's (1879–1955) quantum explanation of the photoelectric effect.

THE PHOTOELECTRIC EFFECT. An interesting phenomenon, first observed by H. Hertz in 1887, was the photoelectric effect, depicted in Figure 7-8. A beam of light is shown striking a particular metal surface. This causes the emission of electrons and

FIGURE 7-7
An analogy between quantum mechanics and classical mechanics.

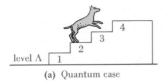

(a) Quantum case

(b) Classical case

To climb from level A to level B the dog in (a) proceeds through 4 steps. It may change its position (and hence its potential energy) only in these 4 discrete steps of fixed magnitude. The dog in (b) can change its position and potential energy from level A to level B in any number of steps of any size whatsoever. Case (a) corresponds to quantum changes in energy, case (b) to classical changes in energy.

FIGURE 7-8
The photoelectric effect.

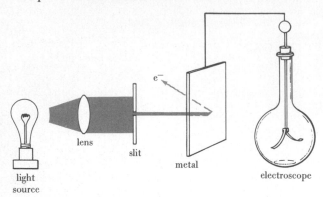

By losing electrons from its surface, the photoelectric metal acquires a positive charge, and the metal-foil leaves of the electroscope acquire the same type of charge. Having like charges, the leaves repel one another. (See Appendix B for a further description of an electroscope.)

FIGURE 7-9
Dependence of photoelectric effect on the frequency of light.

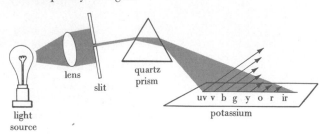

The lengths of the arrows represent the kinetic energies of the photoelectrons ejected by different wavelength components of light. Color code: uv, ultraviolet; v, violet; b, blue; g, green; y, yellow; o, orange; r, red; ir, infrared. Light with wavelength greater than 710 nm (red) causes no photoelectric effect on potassium.

The photoelectric effect has many practical applications, ranging from automatic door openers to light meters to light-sensitive elements in television cameras.

FIGURE 7-10
Photons of light visualized.

A light beam having this appearance

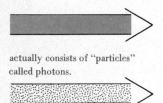

actually consists of "particles" called photons.

TABLE 7-1
Alternative expressions of energy associated with ultraviolet radiation with $\lambda = 250$ nm

	Energy of	
	Single photon	Mole of photons
joule	7.95×10^{-19}	4.79×10^{5}
kilojoule	7.95×10^{-22}	4.79×10^{2}
calorie	1.90×10^{-19}	1.14×10^{5}
kilocalorie	1.90×10^{-22}	1.14×10^{2}

leaves the surface positively charged. According to classical physics, both the *number* of electrons ejected from the surface and their *energies* should depend on the *intensity* or *brightness* of the incident light. The number of ejected electrons does depend on the intensity of the incident light, but the electron energies do not! These energies depend only on the frequency (or wavelength) of the light. Thus, the kinetic energies of electrons ejected by a feeble blue light are greater than those ejected by a bright red light (see Figure 7-9). As with atomic spectra, here was another phenomenon that defied explanation by classical physics.

Einstein proposed that electromagnetic radiation has particlelike characteristics and that "particles" of light, called **photons,** possess a characteristic energy, given by Planck's equation, $E = h\nu$. We can think of the energy of a light wave as being concentrated into photons. In the photoelectric effect these photons transfer energy during collisions with electrons. In each collision a photon gives up its entire energy—a quantum of energy—to an electron. As a result, the more energetic the photon, the more energy it transfers to an electron and the greater the kinetic energy of the ejected electron. Thus, the kinetic energy of the electron should depend on the light frequency (see Figure 7-9). The particlelike nature of light is suggested by Figure 7-10.

The product of h and ν yields the energy of a single photon of electromagnetic radiation in the unit joules. The energy of a typical photon is only a tiny fraction of a joule. Often we deal with the energy associated with a mole of photons, that is, Avogadro's number

(6.0225×10^{23}) of photons. This, of course, is a much larger energy. Several alternative expressions for the energy associated with ultraviolet radiation of wavelength 250 nm are presented in Table 7-1.

Example 7-5 The lowest-frequency light that can produce a photoelectric effect on potassium metal is $4.2 \times 10^{14}\,s^{-1}$. What is the energy of photons of this light?

$$E = h\nu = (6.626 \times 10^{-34}\,J\,s)(4.2 \times 10^{14}\,s^{-1}) = 2.8 \times 10^{-19}\,J$$

SIMILAR EXAMPLES: Exercises 10, 12.

Example 7-6 What is the energy, in kJ/mol, associated with a monochromatic radiation with a wavelength of 250 nm?

First we must determine the frequency of the radiation using equation (7.1). Then we apply the Planck equation (7.4) to determine the energy per photon. The following conversions are also required: *in the numerator*, joule → kilojoule; *in the denominator*, photon → mole photons.

$$\nu = \frac{c}{\lambda} = \frac{3.00 \times 10^8\,m/s}{2.50 \times 10^{-7}\,m} = 1.20 \times 10^{15}\,s^{-1}$$

$$\text{no. kJ/mol} = 6.626 \times 10^{-34}\frac{J\,s}{photon} \times 1.20 \times 10^{15}\,s^{-1}$$

$$\times \frac{1\,kJ}{1000\,J} \times \frac{6.02 \times 10^{23}\,photons}{1\,mol}$$

$$= 479\,kJ/mol$$

SIMILAR EXAMPLES: Exercises 10, 11.

7-4 The Bohr Atom

FIGURE 7-11
An unsatisfactory atomic model.

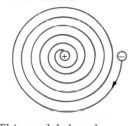

This model, based on classical physics, explains how an atom can emit light, though not why the light should show distinctive wavelength components (see again Figure 7-6). Furthermore, as energy is lost through light emission, the electron is drawn ever closer to the nucleus, eventually spiraling into it. This collapse of the atom would occur in a time interval much shorter than one second.

Even by the methods of classical physics, certain conclusions were possible concerning the behavior of electrons in atoms. It was obvious from the laws of electrostatics, for example, that negatively charged electrons could not remain at rest; otherwise, they would be attracted into the positively charged nucleus. Furthermore, the movement of electrons around the nucleus was a necessary condition to explain the emission of light. (Maxwell's theory stated that electromagnetic radiation resulted from the relative motion of electrically charged objects.) But a dilemma was posed for classical physics. According to classical physics, an electron must accelerate constantly in revolving about the nucleus of an atom. In doing so it would give off energy as light. Having lost energy, the electron should then be drawn closer to the nucleus. With each circuit the electron should lose more energy and be drawn still closer to the nucleus. This type of behavior suggests a spiraling motion in which the electron "falls" into the nucleus. But if electrons all were to suffer this fate, there could be no accounting for the fact that stable atoms consisting of electrically charged particles exist at all. The situation described here is pictured in Figure 7-11.

In 1913, a Danish physicist, Niels Bohr (1885–1962), produced a solution to this dilemma through an interesting combination of classical and quantum theory. His description of the hydrogen atom depicted the electron as orbiting around the nucleus, much as the earth revolves about the sun. It has been called the solar system atomic model. The basic assumptions of Bohr's theory were these:

1. There is *only a certain set of allowable orbits* for an electron in a hydrogen atom. These orbits, referred to as stationary states of motion, are circular paths about the nucleus. The motion of an electron within a stationary state can be described by

FIGURE 7-12
Bohr model of the hydrogen atom.

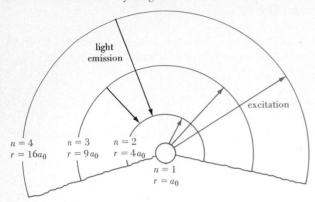

A portion of the hydrogen atom is pictured. The nucleus occupies a position at the center of the atom, and the electron is found in one of the discrete orbits $n = 1, 2, 3, \ldots$. Electron transitions corresponding to excitation of the atom are shown in color, those to light emission in black.

ordinary mechanics. However, even though classical theory would predict otherwise, *as long as an electron remains in a stationary state its energy remains constant and it does not emit light.*

2. The allowable stationary states are those in which certain properties of the electron have unique values. In particular, a property called the angular momentum must be an integral multiple of $h/2\pi$; that is, the angular momentum must be $nh/2\pi$ (where n is an integer and h is Planck's constant).

3. An electron can pass only from one stationary state to another. In such transitions fixed, discrete amounts of energy (quanta) are involved.

The atomic model for hydrogen based on these assumptions is pictured in Figure 7-12. The allowable states for the electron are numbered, $n = 1$, $n = 2$, $n = 3$, and so on. These numbers are called **quantum numbers.** The letters, K, L, M, N, . . . , are also used to designate the first few states. These designations by letter are based on terminology derived from spectroscopy.

Bohr was able to calculate several properties of the electron in a hydrogen atom using his theory. For one, he could calculate the radii of the allowable orbits. These could be expressed as the multiples, $2^2, 3^2$, $4^2, \ldots, n^2$, of a certain orbit of minimum radius, designated as $a_0 = 0.53$ Å. It was possible to calculate the velocities associated with the electron in each of these orbits, and, most important of all, the energy. The energy of the hydrogen atom, when its electron is in the orbit n, is

$$E_n = \frac{-B}{n^2} \tag{7.5}$$

The negative sign appears in equation (7.5) because the energy of interaction of the electron and the nucleus (proton) is attractive (see Appendix B).

B is a numerical constant with a value of 2.179×10^{-18} J.

One of the assumptions in Bohr's theory (assumption 3) is that discrete quantities of energy are involved when an electron in a hydrogen atom passes from one allowable state to another. Let us use equation (7.5) to calculate the *difference in energy* between two states. As a case of special interest, we choose the states: $n = 3$ and $n = 2$.

$$\Delta E = E_3 - E_2 = \left(\frac{-B}{3^2}\right) - \left(\frac{-B}{2^2}\right) = \left(\frac{B}{2^2}\right) - \left(\frac{B}{3^2}\right)$$

$$= B\left(\frac{1}{2^2} - \frac{1}{3^2}\right) \tag{7.6}$$

Equation (7.6) bears a striking resemblance to the Balmer equation (7.3)! Let us pursue this matter further. Normally, the electron in a hydrogen atom is believed to exist in the orbit closest to the nucleus ($n = 1$

or K state). Upon excitation a quantum of energy is absorbed and the electron jumps to a higher quantum state. Light emission occurs as the electron drops from a higher to a lower quantum state, and a unique quantity of energy is involved. Planck's equation (7.4) allows us to calculate the unique frequency (or wavelength) corresponding to this unique quantity of energy. Now let us return to our discussion of the transition of an electron from $n = 3$ to $n = 2$. The difference in energy between these two states, ΔE, is the energy of the emitted photon of light. We can write two equations for ΔE.

$$\Delta E = h\nu \tag{7.4}$$

$$\Delta E = B\left(\frac{1}{2^2} - \frac{1}{3^2}\right) \tag{7.6}$$

The frequency of the photon of light is

$$\nu = \frac{\Delta E}{h} = \frac{B}{h}\left(\frac{1}{2^2} - \frac{1}{3^2}\right) \tag{7.7}$$

The numerical value of the constant B/h can be determined readily.

$$\frac{B}{h} = \frac{2.179 \times 10^{-18}\ \text{J}}{6.626 \times 10^{-34}\ \text{J s}} = 3.289 \times 10^{15}\ \text{s}^{-1}$$

The constant B/h is practically identical to the product $R \times c$ in the Balmer equation (7.3)! We have succeeded in using results of Bohr's theory to derive the Balmer equation. If equation (7.7) is solved for a numerical result, the frequency obtained is that of the red line in the Balmer series. Every transition in which an electron moves from a higher-energy state to the state $n = 2$ produces a line in the Balmer series.

In Figure 7-13 the energies associated with the different allowable states for the electron in a hydrogen atom are shown as a group of lines. This representation is known as an **energy-level diagram.** In this diagram, by convention, the zero of energy is taken for the condition where the electron is completely separated (ionized) from the atom ($n = \infty$). Since a bound electron must absorb energy to become ionized and reach an energy state of zero, the energy of a bound electron must be negative. The smaller the value of n, the more negative the energy.

FIGURE 7-13
Energy-level diagram for the hydrogen atom.

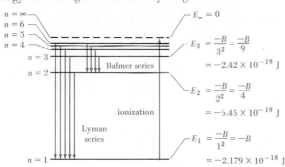

Among the features brought out by this diagram are

- Electron transitions from higher quantum levels to the level $n = 2$ produce lines in the Balmer series; to the level $n = 1$ lines in the Lyman series.
- Ionization of the normal hydrogen atom requires moving the electron from the level $n = 1$ to the level $n = \infty$, a process requiring 2.179×10^{-18} J.
- Energy differences between successive levels are smaller, the higher the values of n.

Example 7-7   What is the wavelength of the spectral line associated with the transition in which an electron in a hydrogen atom drops from the level $n = 2$ to the level $n = 1$?

This line is in the Lyman series, which is in the ultraviolet region of the spectrum. First, we need to calculate the difference in energy between the levels, $n = 2$ and $n = 1$, using equation (7.5).

$$E_1 = \frac{-B}{1^2} = -B \quad \text{and} \quad E_2 = \frac{-B}{2^2} = \frac{-B}{4}$$

$$\Delta E = E_2 - E_1 = \frac{-B}{4} - (-B) = \frac{3B}{4} = 0.75B$$

$$= 0.75 \times 2.179 \times 10^{-18}\,\text{J} = 1.634 \times 10^{-18}\,\text{J}$$

Now we can substitute this energy into Planck's equation (7.4) to obtain the frequency of the radiation.

$$\nu = \frac{E}{h} = \frac{1.634 \times 10^{-18}\,\text{J}}{6.626 \times 10^{-34}\,\text{J s}} = 2.466 \times 10^{15}\,\text{s}^{-1}$$

Finally, we use equation (7.1) to solve for the wavelength.

$$\lambda = \frac{c}{\nu} = \frac{2.998 \times 10^8\,\text{m/s}}{2.466 \times 10^{15}\,\text{s}^{-1}} \times \frac{1\,\text{nm}}{1 \times 10^{-9}\,\text{m}} = 121.6\,\text{nm}$$

SIMILAR EXAMPLES: Exercises 18, 19.

SHORTCOMINGS OF THE BOHR THEORY. The great success of the Bohr theory was in its ability to predict lines in the hydrogen atom spectrum. However, one of the discoveries of the time was that spectral lines have *fine structure,* especially in cases where excited atoms are placed in a magnetic field. That is, some principal lines were found actually to consist of a small number of very closed spaced lines. Fine structure in hydrogen spectra was explained through modification of the Bohr theory. However, the theory was never very successful in describing atomic spectra other than those of hydrogen, nor could it account for the ability of atoms to form molecules through chemical bonds.

7-5 Wave–Particle Duality

A demonstration that electrons have wavelike properties is provided by the electron microscope. In this device magnetic and electric fields are used to focus and direct electron beams, much as lenses and prisms are used in optical microscopes. But because of the shorter wavelengths associated with electrons, the resolving power of an electron microscope is much greater, thousands of times greater, than that of a light microscope. With the recently developed, high-resolution scanning electron microscope, it has become possible to obtain images of individual atoms, as shown in Figure 2-17.

In the year 1905, Einstein set to rest a centuries-old dispute concerning the nature of light. Newton had advanced the proposition that light has a corpuscular or *particle* nature, that is, that it consists of a stream of energetic particles. Another theory was that of Huygens, who proposed that light consists of waves of energy.

To choose between the two theories required that accurate measurements be made of the speed of light in vacuum and in various media. Newton's view required that light travel *faster* in denser media, while Huygen's required that light travel *more slowly* in denser media. Accurate measurements of the speed of light showed that light does indeed travel *more slowly* in a denser medium. Thus, the wave model became firmly established. And along with this was also established the view that matter and energy are two distinctly different natural qualities governed by different laws. However, to explain the photoelectric effect Einstein was required to think of photons of light as if they were particles. Thus, there emerged the idea that light has a *dual* nature—in some instances its behavior is better understood in terms of waves and in other cases, particles.

In 1924, the French physicist Louis De Broglie, considering the nature of light and matter, offered a startling proposition: *Not only does light display particlelike characteristics, but small particles may at times display wavelike properties.* De Broglie's proposal received experimental verification in 1927—through experiments that led directly to the development of the electron microscope. De Broglie's description of matter waves was in mathematical terms. The De Broglie wavelength associated with a particle is related to the particle momentum, p, and Planck's constant, h. (Momentum is the product of mass, m, and velocity, v.)

$$\lambda = \frac{h}{p} = \frac{h}{mv} \tag{7.8}$$

In equation (7.8) mass is in kilograms, velocity in m/s, and wavelength in meters.

Example 7-8 What is the wavelength associated with electrons traveling at one-hundredth the speed of light?

The electron mass, expressed in kilograms, is 9.11×10^{-31} kg. The electron velocity, v, is $0.01 \times 3.00 \times 10^8$ m/s $= 3.00 \times 10^6$ m/s. Planck's constant, $h = 6.626 \times 10^{-34}$ J s, and 1 J $= 1$ kg m^2 s^{-2}. Substituting these values into equation (7.8), we obtain

$$\lambda = \frac{6.626 \times 10^{-34} \text{ kg m}^2 \text{ s}^{-2} \text{ s}}{(9.11 \times 10^{-31} \text{ kg})(3.00 \times 10^6 \text{ m/s})} = 2.42 \times 10^{-10} \text{ m} = 0.242 \text{ nm}$$

SIMILAR EXAMPLE: Exercise 26.

7-6 The Uncertainty Principle

The laws of classical physics are often thought of as universal truths. They tell us what physical behavior is permitted and what future events will follow from the present state of a system. When a rocket is fired, the exact point of impact can be calculated. Errors in this calculation may arise from inaccuracies in measuring certain of the variables that affect the rocket trajectory. In principle, however, these variables can be determined with the highest precision, leading to a result of any desired degree of accuracy. In classical physics nothing is left to chance—physical behavior can be predicted with certainty.

During the 1920s Niels Bohr and Werner Heisenberg thought about hypothetical experiments designed to establish how precisely the behavior of subatomic particles could be determined. The two variables that determine this behavior are the position of a particle, x, and its momentum, p. [Recall that momentum is the product of mass (m) and velocity (v); that is, $p = mv$.] The conclusion they reached was that there must always be uncertainties in measurement such that the product of the uncertainty in position, Δx, and in momentum, Δp, is

The symbol $\geqslant$ stands for equal to or greater than.

$$\Delta x \, \Delta p \geqslant \frac{h}{2\pi} \tag{7.9}$$

The significance of this expression, referred to as the uncertainty principle, is that position and momentum cannot both be measured with great precision simultaneously. If an experiment is designed to locate the position of a particle with great precision, it is not possible to measure its momentum precisely. Its future actions (trajectory) cannot be predicted with certainty. Similarly, if the momentum is measured precisely, the position of the particle is not known with certainty. The uncertainty principle places a restriction on the precision with which we can make measurements and the certainty with which we can predict atomic events. But why should this be so?

Suppose that we wish to learn something of the behavior of an electron in a hydrogen atom by using a microscope with which to "see" the electron. What kind of microscope should this be? The resolving power of a microscope is limited to objects that are about the size of the wavelength of the light used. In an ordinary optical microscope using visible light, the resolving power is about 1000 nm. With an electron microscope the resolving power is about 1 nm.

The radius of the first Bohr orbit in a hydrogen atom is calculated to be 0.053 nm; thus, the diameter of a hydrogen atom is about 0.1 nm (10^{-10} m). Rutherford's experiments on the scattering of α particles by metal foils suggests that electrons are much smaller than the atoms containing them.

FIGURE 7-14
The uncertainty principle.

The uncertainty principle is not easy for most people to accept philosophically. Einstein spent a good deal of time from the middle 1920s until his death in 1955 attempting, unsuccessfully, to disprove it.

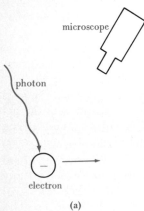

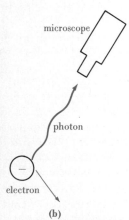

(a)

(b)

A free electron moves into focus of a hypothetical microscope (a). A photon of light strikes the electron and is reflected. In the collision the photon transfers momentum to the electron. The reflected photon is seen through the microscope, but the electron has moved out of focus (b). The exact position of the electron cannot be determined.

Suppose that the diameter of an electron is about 10^{-14} m. Light of this wavelength would have a frequency of 3×10^{22} s^{-1} ($c = \nu\lambda$) and an energy per photon of 2×10^{-11} J ($E = h\nu$). But from Figure 7-13 we can see that this energy is far, far in excess of that required to ionize the electron in a hydrogen atom completely. Thus, in our attempt to "see" an electron in an atom, the measuring system (the light used) would interfere greatly with the measurement. We could not hope to determine the electron position and momentum with any precision. This point is illustrated further in Figure 7-14.

7-7 Wave Mechanics

The concepts introduced in the two preceding sections carry a number of implications regarding atomic structure. With the Bohr theory it is possible to calculate both the radius of an orbit and the velocity of an electron in this orbit. The former represents a precise definition of electron position and the latter of electron momentum. But, according to the uncertainty principle, it is impossible to measure both these quantities

FIGURE 7-15
The electron as a matter wave.

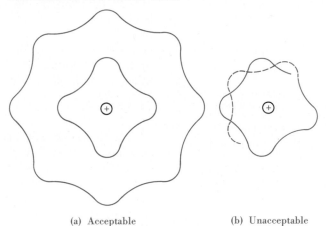

(a) Acceptable   (b) Unacceptable

The wave patterns in (a) are called standing waves. They have an integral number of wavelengths and are acceptable representations because the maxima and minima of the waves match exactly for consecutive circuits about the nucleus. The characteristic wavelength that will produce this condition is related to the electron mass and momentum through the De Broglie equation (7.8).

The pattern in (b) is unacceptable; the number of wavelengths is nonintegral. A maximum in one circuit (solid line) is cancelled by a minimum in the next (dashed), a condition called destructive interference.

The waves shown above are greatly oversimplified; they are actually three-dimensional waves.

precisely. We should not adopt a theory that allows for the precise prediction of that which cannot be measured.

A second implication, explored with great success by Schrödinger in 1927, is that electrons in atoms can be treated as matter waves. Their motion can be likened to wave motion. How are we to picture an electron in an atom according to this wave motion? Does it cease to exist as a particle? Is it literally smeared out into a wave? One answer to all these questions, of course, is that we do not know and cannot know. The Heisenberg uncertainty principle makes us ever aware of our inability to describe subatomic particles precisely.

The wave motion associated with electrons in atoms must correspond to certain allowable patterns (see Figure 7-15). These patterns can be described by mathematical equations, but this is a task well beyond the scope of our study. Let us say simply that the acceptable solutions of these wave equations are called **wave functions** or **orbitals** (denoted by the Greek letter psi, ψ). The orbital description of an electron in an atom permits us to establish its energy state and also to think about the electron in either of two ways. First, we can think of the electron as being a cloud of negative electric charge, with the density of the charge varying from point to point. Or we can continue to think of an electron as a particle, with the probability of finding the electron varying from point to point. The wave functions can be used to calculate charge densities and electron probabilities.

Perhaps the simplest wave function or orbital to describe is of the type known as $1s$. Figure 7-16 suggests two ways to view the electron described by a $1s$ orbital. The pattern of dots in Figure 7-16a represents the distribution of electronic charge. Where the dots are closely spaced there is a higher charge density. Alternatively, adhering to the view of electrons as particles, the same pattern of dots represents a probability distribution. Where the dots are closely spaced there is a higher probability of finding an electron than where the spacing is greater. The pattern of dots in Figure 7-16a extends in all directions and to all distances from the nucleus, with the spacing between dots gradually increasing. Because of this fact it is not possible to draw a picture that encompasses all of these dots. Instead, we must settle for a representation in which some portion (say, 90%) of all the dots are found. This means a region containing 90% of the charge density or in which there is a 90% probability of an electron being found. The spherical envelope in Figure 7-16b portrays the $1s$ orbital in this way. Figure 7-17 offers an analogy to the distribution of electron probabilities for the $1s$ orbital that may prove helpful.

FIGURE 7-16
The $1s$ orbital.

(a)

(b)

The orbital is represented in two ways here: (a) the distribution of electron charge density or the distribution of probabilities of finding an electron. (b) A spherical envelope enclosing 90% of the dots of representation (a). Also, the probabilities of finding an electron are identical for all points on the surface of the sphere.

The term *orbit* used in the Bohr theory suggests a definite pathway that electrons follow. The term *orbital* is intended to suggest that electrons are found in a much less definite, three-dimensional region. For example, we may speak of the region within which an electron is found 90% of the time.

FIGURE 7-17
Dart board analogy to a 1s orbital.

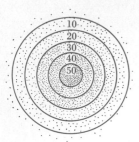

Summary of scoring		
200 darts score	"50"	
300	"40"	
400	"30"	
250	"20"	
200	"10"	
150		off the board
1500 darts	total	

Imagine that a single dart (analogous to the single electron in a hydrogen atom) is thrown at a dart board 1500 times. The board itself contains 90% of all the dart holes (1350 out of 1500). It is analogous to the region described by a 1s orbital. Where is the dart most likely to be found?

The density of dart holes (number of holes per unit area) is greatest in the "50" region. But if we ask what is the most likely score for a dart throw, it is "30" not "50" (400 throws out of 1500). Even though the density of dart holes is less in the "30" ring, the total area of the "30" ring is much greater than that of the "50" ring.

Similarly, although the probability of finding an electron is greatest in a small unit of volume at the nucleus, if we add up the probabilities for all the volume units equidistant from the nucleus, the greatest *total* probability is in a spherical shell of radius 0.053 nm (0.53 Å). This proves to be the same as the radius of the first Bohr orbit!

7-8 Electron Orbitals and Quantum Numbers

To produce acceptable solutions to the Schrödinger wave equation, it is necessary to assign *integral* values to three different parameters—three quantum numbers. Moreover, the allowable values for these quantum numbers are interrelated.

The first of these three is the **principal quantum number, n.** This quantum number may have only a *positive, nonzero integral* value.

$$n = 1, 2, 3, 4, \ldots \tag{7.10}$$

The second quantum number is the **orbital quantum number, l,** which may be zero or a positive integer. It cannot be negative and it cannot be any larger than $n - 1$ (where n is the principal quantum number).

$$l = 0, 1, 2, 3, \ldots, n - 1 \tag{7.11}$$

The third quantum number is called the **magnetic quantum number, m_l.** Its value may be positive or negative, may include zero, and may range from $-l$ to $+l$ (where l is the orbital quantum number.)

$$m_l = -l, -l + 1, -l + 2, \ldots, 0, 1, 2, \ldots, +l \tag{7.12}$$

The *orbital quantum number* is also commonly called the *azimuthal quantum number.*

FIGURE 7-18
The three *p* orbitals.

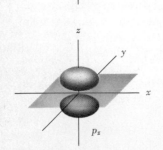

The electron clouds for *p* orbitals are not spherically symmetric. Each orbital consists of two lobes distributed along one of the three perpendicular axes through the nucleus. The two lobes are separated by a plane called a nodal plane. The probability of an electron being found in the nodal plane is zero.

Example 7-9   What are the possible values of l and m_l for an electron with the principal quantum number, $n = 3$?

From expression (7.11) we see that the allowable values of l are 0, 1, and 2. The allowable values of m_l depend on the value of l (expression 7.12).

If $l = 0$, there can be but a single value of m_l: 0.

If $l = 1$, there are three allowable values of m_l: $-1, 0, +1$.

If $l = 2$, there are five allowable values of m_l: $-2, -1, 0, +1, +2$.

SIMILAR EXAMPLES: Exercises 29, 31.

Example 7-10   Can an electron have the quantum numbers $n = 2$, $l = 2$, and $m_l = 2$?

No. The l quantum number cannot be greater than $n - 1$. Thus, if $n = 2$, l can be only 0 or 1. And if l can be only 0 or 1, m_l cannot be 2, because m_l can never be greater than l (expression 7.12).

SIMILAR EXAMPLE: Exercise 31.

Every combination of the three quantum numbers n, l, and m_l corresponds to a different electron orbital. All orbitals having the same value of the quantum number n are said to be in the same **principal electronic shell or principal level,** and all orbitals having the same l value are in the same **subshell** or **sublevel.**

The principal shells are numbered in accordance with the value of n, but they may also be denoted by letter. The *first* principal electronic shell or the K shell consists of orbitals with $n - 1$; the *second* principal shell or L shell, of orbitals with $n = 2$; and so on. The value of the quantum number n relates to the energies of electrons and their probable distances from the nucleus.

The primary significance of the l quantum number is that its value determines the geometrical shape of the electron cloud or electron probability distribution. All orbitals with the value $l = 0$ are s orbitals. If the s orbital is in the first principal shell ($n = 1$), it is a $1s$ orbital, if in the second principal shell, $2s$, and so on. The electron cloud or electron probability distribution for an s orbital has the shape of a sphere, with the atomic nucleus at its center. Because, when $l = 0$, m_l must also be 0, there can be only one orbital of the s type for each principal shell.

The orbital type corresponding to $l = 1$ is the p orbital. Because, when $l = 1$, m_l can have one of three values: $-1, 0, +1$, p orbitals always occur in sets of three. That is, there are three p orbitals in the p subshell. The electron clouds or electron probability distributions for orbitals of the p type are *not* spherically symmetric. The p orbitals are generally represented as dumbbell-shaped regions centered on the three perpendicular axes with the nucleus at the origin. Three p orbitals are pictured in Figure 7-18.

There is a set of five orbitals that have $l = 2$; these are the d orbitals, comprising the d subshell. The geometrical shapes corresponding to d orbitals, which are more complex than for s and p orbitals, are shown in Figure 7-19.

Some of the points discussed in the preceding paragraphs are illustrated through Table 7-2 and Example 7-11.

Example 7-11   Write an orbital designation for an electron with the quantum numbers $n = 4$, $l = 2$, and $m_l = 0$.

The type of orbital is determined by the l quantum number. If $l = 2$, the orbital is of the d type. Because $n = 4$, the designation is $4d$.

SIMILAR EXAMPLES: Exercises 32, 33.

FIGURE 7-19
The five *d* orbitals.

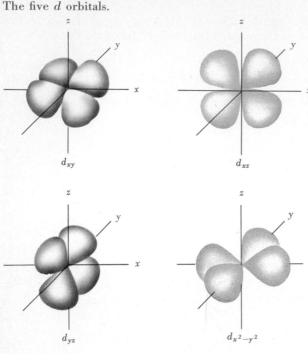

d_{xy}

d_{xz}

d_{yz}

$d_{x^2-y^2}$

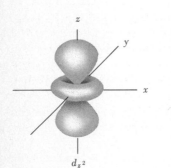

d_{z^2}

The designations *xy, xz, yz,* and so on are related to the values of the quantum number m_l ($l = 2$ for all *d* orbitals), but this is a detail not pursued further in the text.

7-9 Electron Spin— A Fourth Quantum Number

In 1925, Uhlenbeck and Goudsmit proposed that some unexplained features of fine structure in the hydrogen spectra could be understood if it were assumed that electrons possess a fourth quantum number. The property of electrons associated with this fourth quantum number has become known as electron spin. The electron is pictured as spinning on its axis as it moves about the nucleus, much as the earth spins on its axis as it revolves about the sun. There appear to be two possibilities for electron spin. The quantum number describing electron spin, m_s, may have a value of $+\frac{1}{2}$ or $-\frac{1}{2}$. Unlike the quantum numbers n, l, m_l, which have interrelated values, the value of m_s does not depend on the other three quantum numbers.

Although the concept of electron spin is useful, what proof is there that such a property exists, especially in view of the uncertainty principle? An experiment reported by Stern and Gerlach in 1922, although designed for a different purpose, seems to yield this proof. In the Stern-Gerlach experiment silver metal was vaporized in a furnace and a beam of silver atoms passed through an inhomogeneous (nonuniform) magnetic field. The beam was found to split in two (see Figure 7-20). A simplified explanation is based on these points:

1. An electron, by virtue of its spin, has associated with it a magnetic field.
2. A pair of electrons with opposing spins has no associated magnetic field.

TABLE 7-2
Electronic shells, orbitals, and quantum numbers

Principal shell	K		L				M							
$n =$	1	2	2	2	2	3	3	3	3	3	3	3	3	3
$l =$	0	0	1	1	1	0	1	1	1	2	2	2	2	2
$m_l =$	0	0	−1	0	+1	0	−1	0	+1	−2	−1	0	+1	+2
orbital designation	1s	2s	2p	2p	2p	3s	3p	3p	3p	3d	3d	3d	3d	3d
number of orbitals in subshell	1	1		3		1		3				5		
total number of orbitals $= n^2$	1			4								9		

FIGURE 7-20
The Stern–Gerlach experiment.

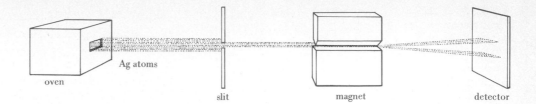

oven

Ag atoms

slit   magnet   detector

3. In a silver atom, 23 electrons have a spin of one type and 24 of the opposite type. The direction of deflection of a silver atom in a magnetic field depends on the type of spin on the "odd" electron.

4. In a collection of silver atoms there is an equal chance that the odd electron will have a spin of $+\frac{1}{2}$ or $-\frac{1}{2}$. Thus, the beam of atoms is split into two beams.

7-10 Many-Electron Atoms

To this point our description of electron orbitals has applied only to a hydrogen atom or hydrogenlike atoms—a species containing a single electron: H, He$^+$, Li^{2+}, and so on. Exact solutions of the necessary mathematical equations for many-electron atoms are almost impossibly difficult. Nevertheless, the results obtained for the hydrogen atom are approximately correct for more complex atoms. By this we mean that the same types of quantum numbers and orbitals are assumed to exist for many-electron atoms as for the hydrogen atom. As we discover in subsequent chapters how this particular concept of electron orbitals allows us to explain so many chemical phenomena, our confidence in this assumption should grow.

Through equation (7.5) and Figure 7-13 we demonstrated that the energy levels in the Bohr hydrogen atom are a function only of the principal quantum number, n. The same result is obtained with wave mechanics. That is, all the subshells in a given principal shell of the hydrogen atom ($s, p, d, f, \ldots$) are at the same energy. But there are some important differences between the hydrogen atom and many-electron atoms with respect to orbital energies.

FIGURE 7-21
The shielding effect.

The atom pictured is sodium. The charge on the nucleus is $+11$. The charge on the first electronic shell is -2, and on the second -8. If the nucleus and the two inner electronic shells are considered as a single unit (contained within the broken circle), the net charge on this unit is $+1$. For the time that the outer-shell electron spends outside the sphere of the inner electrons, the force on it is much the same as if the nucleus contained only a single unit of positive charge and there were only one electron in the atom.

NUCLEAR CHARGE EFFECT. The attractive force of the nucleus for a given electron increases as the nuclear charge increases. As a result, we should expect the energy of interaction between the electron and the nucleus—the orbital energy—to decrease (i.e., become more negative) with increasing atomic number.

SHIELDING EFFECT. In describing the attraction between a given electron and the nucleus, the presence of all the other electrons must also be considered. If the electron in question is the one farthest from the nucleus, the inner electrons reduce the effectiveness of the nucleus in attracting this outermost electron. They screen or "shield" the outermost electron from the full effects of the nucleus by partially neutralizing the nuclear charge. This shielding effect is illustrated in Figure 7-21.

The effectiveness of the shielding by inner electrons depends on the l value of the particular orbital for the outermost electron. For example, the p orbitals of an electronic shell extend to a greater distance from the nucleus than does the corresponding s orbital. Compared to being in an s orbital, an electron in a p orbital spends more time at distances from the nucleus where the shielding effect is strong (outside the dashed circle in Figure 7-21). The attractive force of the nucleus for an electron in a p orbital is less than for an electron in an s orbital, and the energy of the p orbital is higher than of the s orbital; in a d orbital the energy is higher still. The result of

FIGURE 7-22
Orbital energy diagram
for first three
electronic shells.

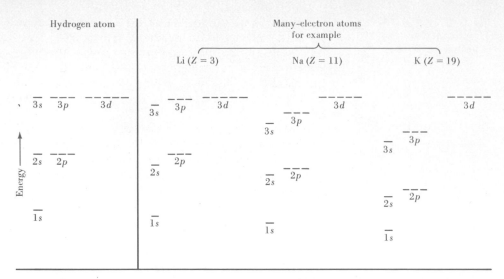

FIGURE 7-23
The order of
filling of electronic
subshells.

Follow the arrows from
top to bottom, and the
order of filling of sub-
shells is the same as that
in expression (7.13).
Another useful rule is
that for any given pair of
orbitals the one with the
lowest total of the n and
l quantum numbers fills
first. For example, the $3s$
orbital $(3 + 0 = 3)$ fills
before the $3p$
$(3 + 1 = 4)$. If the
$n + l$ sum is the same
for two orbitals, the one
with the lower n value
fills first. Thus, the $3d$
orbital $(3 + 2 = 5)$ fills
before the $4p$
$(4 + 1 = 5)$.

The most useful aid in
writing electron configu-
rations is the periodic
table of the elements.
The relationship between
this table and electron
configurations is estab-
lished in Chapter 8.

Energy levels for the subshells of the first three principal quantum levels are shown for a hydrogen atom and three typical many-electron atoms. Each many-electron atom has its own energy-level diagram. One of the distinctive features of the energy-level diagrams for many-electron atoms is the splitting of each quantum level into different energies for each subshell. Another feature is a steady decrease in the energies of these levels with increasing atomic number.

shielding is to cause a splitting of the energy level of each principal shell into separate levels for each subshell.

The ideas presented in this section are illustrated through Figure 7-22, which represents orbital energies for the first three shells of the hydrogen atom and some typical many-electron atoms.

7-11 Electron Configurations

In a normal hydrogen atom the electron is found in the lowest energy state possible, the $1s$ orbital. But what is the situation in a complex atom such as gold, with 79 electrons? We shall see that these electrons are distributed in a particular way among the orbitals in several electronic shells. A designation of the most likely distribution is called the **electron configuration** of the atom. We consider first three rules or principles that help in the assignment of a probable electron configuration to an atom, and then apply these principles to the various elements.

1. Electrons Occupy Orbitals in Such A Way as to Minimize the Energy of the Atom. Figure 7-22 implies an order in which electrons occupy orbitals, first the $1s$, then $2s$, $2p$, and so on. Actually, the energy of an atom is not minimized in most cases simply by filling the principal electronic shells in succession. One of the consequences of the splitting of energy levels for many-electron atoms is that at higher quantum numbers an "overlapping" occurs, for example, with $4s$ filling before $3d$. As a result, the order of filling of orbitals must be established by experiment. The order obtained is listed below and also depicted in Figure 7-23.

$$1s, 2s, 2p, 3s, 3p, 4s, 3d, 4p, 5s, 4d, 5p, 6s, 4f, 5d, 6p, 7s, 5f, 6d, 7p \qquad (7.13)$$

2. No Two Electrons in an Atom May Have All Four Quantum Numbers Alike— the Pauli Exclusion Principle. In 1926, Wolfgang Pauli studied the *absence* of lines in emission spectra which should have been present according to existing theory. His results led him to propose that any state or condition in which two electrons would have all four quantum numbers alike is *not* allowable. The first three quantum numbers, *n*, *l*, and m_l, determine a specific orbital. Two electrons may have these three quantum numbers alike, that is, they may have the same orbital designation; but if they do, they must have different values of m_s, the spin quantum number. *Only two electrons may exist in the same orbital and these electrons must have opposite spins.*

Because of this limit of two electrons per orbital, the maximum occupancy of a subshell can be obtained simply by *doubling* the number of orbitals in the subshell. That is, the *s* subshell consists of *one* orbital and has a capacity of *2* electrons; the *p* subshell consists of *three* orbitals and has a capacity of *6* electrons; and so on. The maximum occupancy of a principal shell is also *twice* the number of orbitals it contains (recall Table 7-2), leading to the expression

$$\text{maximum number of electrons in the electronic shell with principal quantum number } n = 2n^2$$

3. The Principle of Maximum Multiplicity—Hund's Rule. When orbitals of identical energy are available, electrons occupy these singly rather than in pairs. As a result, an atom tends to have as many unpaired electrons as possible. This behavior can be rationalized by saying that electrons, because they all carry the same electrical charge, seek out empty orbitals of nearly the same energy in preference to pairing up with electrons in half-filled orbitals.

7-12 Electron Configurations of the Elements

To apply the principles of the preceding section, we first need a shorthand notation to designate electron configurations. That of an atom of oxygen is represented below in two different ways.

spdf notation: O $1s^2 2s^2 2p^4$

orbital diagram: O

	1s	2s		2p	
	↓↑	↓↑	↓↑	↓	↓

The total number of electrons to be assigned is eight, the atomic number of oxygen. Two of these electrons are in the 1*s* subshell, two in the 2*s*, and four in the 2*p*. The *spdf* notation simply denotes the total number of electrons in each subshell. The orbital diagram breaks down each subshell into the individual orbitals and indicates the number of electrons for each orbital. This is done through the use of arrows. An arrow pointing upward corresponds to an electron with one type of spin and an arrow pointing down, to the other.

The orbital diagram indicates the number of unpaired electrons in an atom. This can also be done with *spdf* notation if the subshells are broken down into individual orbitals, such as

O $1s^2 2s^2 2p_x^2 2p_y^1 2p_z^1$

THE AUFBAU PROCESS. Let us consider the following hypothetical process—the building up of a more complex atom starting with the simplest atom, hydrogen. This hypothetical process is called the **Aufbau** process (meaning "building up" in German). In this process we proceed from one element to the next by adding a proton and the requisite number of neutrons to the nucleus and one electron to the appropriate orbital. It is to this "differentiating" electron that we pay particular attention.

Hydrogen, *Z* = 1. The lowest energy state available to the electron in a hydrogen atom is the 1*s* orbital. The electron configuration is $1s^1$.

Helium, *Z* = 2. In a helium atom a second electron goes into the 1*s* orbital. The two electrons must have opposing spins, $1s^2$.

Lithium, *Z* = 3. The differentiating electron cannot be accommodated in the 1*s*

orbital (Pauli exclusion principle). It must be placed in the next available orbital, the 2*s*. For lithium we may write, $1s^22s^1$.

Beryllium, $Z = 4$.   The configuration is $1s^22s^2$.

Boron, $Z = 5$.   The differentiating electron goes into the next available orbital, 2*p*. Boron has the configuration $1s^22s^22p^1$.

Carbon, $Z = 6$.   Here the rule of maximum multiplicity (Hund's rule) applies. The differentiating electron goes into a new 2*p* orbital.

Nitrogen, $Z = 7$.   The atom has three unpaired electrons in its 2*p* orbitals.

Oxygen, $Z = 8$.   The differentiating electron must go into an orbital already singly occupied. This reduces the number of unpaired electrons to two.

Fluorine, $Z = 9$, and Neon, $Z = 10$.   In the fluorine atom there is one unpaired electron; in the neon atom all electrons are paired and the second principal shell is filled.

Sodium, $Z = 11$, and Magnesium, $Z = 12$.   In the sodium atom the differentiating electron must go into a new orbital in the third principal shell, the 3*s*. In the magnesium atom there are two electrons in this orbital.

Na $1s^22s^22p^63s^1$ Mg $1s^22s^22p^63s^2$

Aluminum, $Z = 13$, to Argon, $Z = 18$.   In this series of six elements the 3*p* orbitals fill, leading to

Ar $1s^22s^22p^63s^23p^6$

Potassium, $Z = 19$, and Calcium, $Z = 20$.   We now encounter the first "irregularity" in the order of filling of orbitals. According to expression (7.13) and Figure 7-23, the 4*s* orbital fills *before* the 3*d*. The differentiating electrons for potassium and calcium go into the *s* orbital of the fourth electronic shell.

K $1s^22s^22p^63s^23p^64s^1$ Ca $1s^22s^22p^63s^23p^64s^2$

Since the first 18 electrons correspond to the configuration for argon, we may also write

K   [Ar]$4s^1$ and   Ca   [Ar]$4s^2$

Scandium, $Z = 21$ to Zinc, $Z = 30$.   This next series of elements is characterized by electrons filling the *d* orbitals of the third shell. The *d* subshell has a total capacity of 10 electrons, hence 10 elements are involved. The configuration for scandium is

Sc $1s^22s^22p^63s^23p^63d^14s^2$ or $[Ar]3d^14s^2$

and for zinc,

Zn $1s^22s^22p^63s^23p^63d^{10}4s^2$ or $[Ar]3d^{10}4s^2$

Gallium, $Z = 31$, to Krypton, $Z = 36$. In this series of six elements the $4p$ subshell is filled. The configuration for krypton is

Kr $1s^22s^22p^63s^23p^63d^{10}4s^24p^6$

Rubidium, $Z = 37$, to Xenon, $Z = 54$. In this series of 18 elements the subshells filled in succession are $5s$, $4d$, and $5p$.

Xe $1s^22s^22p^63s^23p^63d^{10}4s^24p^64d^{10}5s^25p^6$

Cesium, $Z = 55$, to Radon, $Z = 86$. In this series of 32 elements the subshells filled in succession are $6s$, $4f$, $5d$, and $6p$.

Rn $1s^22s^22p^63s^23p^63d^{10}4s^24p^64d^{10}4f^{14}5s^25p^65d^{10}6s^26p^6$

Francium, $Z = 87$ to ?. Francium starts a series of elements in which the subshells filled are $7s$, $5f$, $6d$, and presumably $7p$, although elements in which the $7p$ subshell is filled are not yet known.

The Aufbau principle outlined here is summarized in Appendix E, which is a complete listing of the probable electron configurations of the elements. Some slight discrepancies from the general order of filling of orbitals described here are noted with elements such as Cr ($Z = 24$), Cu ($Z = 29$), and La ($Z = 57$).

Example 7-12 Complete each of the following electron configurations using *spdf* notation.

(a) ———— $1s^22s^22p^63s^23p^5$

(b) Fe ($Z = 26$) $1s^22s^22p^63s^23p^63d(?)4s^2$

(c) As ($Z = 33$)   ?

(a) All electrons must be accounted for in an electron configuration. Add up the superscript numerals ($2 + 2 + 6 + 2 + 5$) to obtain the atomic number (17). Look up the correct element in a listing such as on the inside back cover or the periodic table on the inside front cover.

Cl ($Z = 17$) $1s^22s^22p^63s^23p^5$

(b) The total number of electrons is 26. The number accounted for in the portion of the electron configuration given is $2 + 2 + 6 + 2 + 6 + 2 = 20$. There must be six $3d$ electrons.

Fe ($Z = 26$) $1s^22s^22p^63s^23p^63d^64s^2$

(c) The first 18 electrons correspond to the configuration of Ar. The next two go into the $4s$ subshell; this accounts for 20 electrons. Following the $4s$ the next subshell to fill is the $3d$ (see Figure 7-23). The 10 $3d$ electrons bring the total to 30. The remaining 3 electrons go into the $4p$ subshell, resulting in the configuration

As ($Z = 33$) $1s^22s^22p^63s^23p^63d^{10}4s^24p^3$

SIMILAR EXAMPLES: Exercises 38, 39.

Sometimes the orbital designation of an atom is written in the order in which the orbitals fill, that is, (for Zn) $[Ar]4s^23d^{10}$. In this text we will group all subshells of the same principal shell together, regardless of the order in which orbitals fill, that is, (for Zn) $[Ar]3d^{10}4s^2$.

Summary

Introduced first in this chapter were several ideas concerning the nature of electromagnetic radiation, a useful starting point in the discussion of atomic structure. The dispersion of white light produces a continuous spectrum—a rainbow. However, most light originating from excited atoms produces a discontinuous or line spectrum—a series of colored lines. Perhaps simplest of the line spectra is that of hydrogen. The hydrogen spectrum can be described through an empirical equation (the Balmer equation).

Theoretical explanations of atomic spectra required a breakthrough in our thinking about energy. Planck proposed that energy exists as tiny discrete units called quanta. Einstein used Planck's quantum theory to explain the photoelectric effect, and Bohr applied classical and quantum mechanics to develop a model of the hydrogen atom. Bohr's model permits a calculation of the permissible energy levels for the electron in a hydrogen atom. Energies of photons emitted by excited hydrogen atoms correspond to differences in these electron energy levels. Frequencies of lines in the hydrogen spectrum can be predicted by combining Bohr's theory and the Planck equation.

Bohr's theory of the hydrogen atom, even when modified, fails to provide explanations of such phenomena as the formation of chemical bonds between atoms. For this a model of atomic structure must be based on a new form of quantum theory—wave mechanics. The essential ideas contributing to this newer quantum mechanics are De Broglie's concept of wave–particle duality (i.e., the existence of matter waves) and Heisenberg's uncertainty principle. These ideas were used by Schrödinger to provide a new picture of the hydrogen atom.

The essential feature of the Schrödinger atom is that an electron can no longer be described as a discrete particle in a discrete location. It may be thought of as a cloud of negative electrical charge having a certain geometrical shape. Also, it can be viewed as a particle whose probability of being found extends throughout space, though the probability is highest in certain three-dimensional regions. The variation of charge density or electron probability from point to point can be established through mathematical equations known as wave functions or orbitals. The key parameters that distinguish among orbitals are a set of three quantum numbers, n, l, and m_l. The specific orbital types described in this chapter are known as s, p, and d.

By considering the existence of a fourth quantum number (the spin quantum number) and a set of rules, the assignment of electrons to orbitals can be made. These assignments, called electron configurations, are made for the various elements as a conclusion to Chapter 7.

Learning Objectives

As a result of studying Chapter 7, you should be able to

1. Apply the basic expression relating the frequency, wavelength, and velocity of electromagnetic radiation, using appropriate units.

2. List the various kinds of radiation and their approximate location in the electromagnetic spectrum.

3. Explain the essential difference between a continuous and a line spectrum.

4. Apply the Balmer equation to calculate the wavelengths of lines in the spectrum of hydrogen.

5. Use Planck's equation to relate the frequency and energy content of electromagnetic radiation.

6. State Bohr's assumptions regarding the hydrogen atom and describe the picture of the hydrogen atom that results from them.

7. Calculate the energy of an electron in a hydrogen atom as a function of its principal quantum number, n.

8. Calculate the De Broglie wavelength of a particle.

9. Explain the basic differences between Bohr's and Schrödinger's descriptions of the hydrogen atom.

10. Apply the relationships among the three quantum numbers n, l, and m_l that result from wave mechanics.

11. Describe the shapes of the charge clouds or electron probability distributions corresponding to s, p, and d orbitals.

12. Explain how orbital energies for many-electron atoms differ from those of hydrogen.

13. Apply the three basic principles governing the assignment of electron configurations: the order of filling of electronic subshells, the Pauli exclusion principle, and Hund's rule of maximum multiplicity.

14. Use *spdf* notation or orbital diagrams to represent the electron configurations of different atoms.

Some New Terms

An **atomic spectrum (line spectrum)** is produced by dispersing the light emitted by excited atoms. Only a discrete set of wavelength components (seen as colored lines) is present.

A **continuous spectrum** is one in which all wavelength components of the visible portion of the electromagnetic spectrum are present.

Electromagnetic radiation is a form of energy propagated through mutually perpendicular electric and magnetic fields.

An **electron configuration** is a representation showing the orbital designations of all the electrons in an atom.

Hund's rule (of maximum multiplicity) states that whenever orbitals of equal energies are available, electrons are assigned to these orbitals singly before any pairing of electrons occurs.

An **orbital** describes the electron charge density or the probability of finding an electron in an atom. Wave mechanics requires the existence of several kinds of orbitals (s, p, d, f, . . .) which differ from one another in the geometrical shapes of the electron clouds they describe.

An **orbital diagram** is a representation of an electron configuration in which the most probable orbital designation and the spin of each electron in the atom are indicated.

The **Pauli exclusion principle** states that no two electrons may have all four quantum numbers alike. This limits the number of electrons with the same orbital designation to two. These two electrons must have opposing spins.

A **photon** is a "particle" of light. The energy of a beam of light is concentrated into these photons.

A **principal shell** refers to the collection of all orbitals having the same value of the principal quantum number, n. For example, the $3s$, $3p$, and $3d$ orbitals comprise the third principal shell ($n = 3$).

Quantum numbers are integral numbers whose values must be specified in order to solve the equations of wave mechanics. Three different quantum numbers are required: the *principal quantum number, n;* the *orbital quantum number, l;* and the *magnetic quantum number* m_l. The permitted values of these numbers are interrelated.

The **quantum theory** is based on the proposition that energy exists in the form of tiny, discrete units called quanta. Whenever an energy transfer occurs, it must involve an entire quantum or several quanta. The quantum of energy is described by the equation $E = h\nu$, where h, known as Planck's constant, has a value of 6.626×10^{-34} J s.

spdf notation is a method of describing the electron configuration of an atom in which the numbers of electrons assigned to each orbital are denoted by superscript numerals. For example, the electron configuration of chlorine, in *spdf* notation, is $1s^2 2s^2 2p^6 3s^2 3p^5$.

A **subshell** refers to a collection of orbitals of the same type. For example, the three, $2p$ orbitals constitute the $2p$ subshell. All electrons in a subshell have the same values of n and l.

The **uncertainty principle** states that, when measuring the position and momentum of fundamental particles of matter, uncertainties in measurement are inevitable. Submicroscopic particles cannot be described with certainty.

Wave mechanics is a form of quantum theory based on the concepts of wave–particle duality, the uncertainty principle, and the treatment of the electron in a hydrogen atom as a matter wave. Mathematical solutions of the equations of wave mechanics are known as **wave functions (ψ)**.

Exercises

Electromagnetic radiation

1. Restate each of the following wavelengths in the unit indicated.
   (a) 3000 Å = _____ nm
   (b) 1.56 μm = _____ cm
   (c) 3.6 cm = _____ nm
   (d) 2.18 μm = _____ Å
   (e) 0.62 μm = _____ m
   (f) 470 nm = _____ m

2. What are the wavelengths, in meters, associated with radiation of the following frequencies? (a) 1.00×10^{14} s^{-1}; (b) 8.6×10^{12} s^{-1}; (c) 2.0×10^9 s^{-1}

3. What is the frequency associated with radiation of each of the following wavelengths? (a) 1.8×10^{-3} cm; (b) 12.6 μm; (c) 480 Å; (d) 305 nm

4. The current international standard of time, the second, is defined as 9,192,631,770 cycles of a particular radiation emitted by $^{133}_{55}$Cs atoms. What is the wavelength of this radiation?

5. A certain radiation emitted by magnesium has a wavelength of 285.2 nm. Which of the following statements is (are) correct concerning this radiation?
   (a) It has a higher frequency than radiation with wavelength 315 nm.
   (b) It is in the visible region of the electromagnetic spectrum.
   (c) It has a greater speed in vacuum than does red light of wavelength, 7100 Å.
   (d) Its wavelength is longer than that of x rays.

6. How long does it take light from the sun, 93 million miles away, to reach the earth?

Atomic spectra

7. Use equation (7.3) to calculate the first four lines of the Balmer series of the hydrogen spectrum, starting with the *longest*-wavelength component.

8. A line is detected in the hydrogen spectrum at 1880 nm. Is this line in the Balmer series? Explain.

9. What is the wavelength limit to which the Balmer series for hydrogen converges; that is, what is the *shortest* possible wavelength in the series?

Quantum theory

10. A certain radiation has a frequency of 1.15×10^{15} s^{-1}. What is the energy of **(a)** a single photon; **(b)** a mole of photons of this radiation?

11. What is the wavelength, in nm, of light that has an energy content of exactly 100 kcal/mol? In what portion of the electromagnetic spectrum is this light?

12. Figure 7-3 establishes regions of the electromagnetic spectrum in terms of frequency limits. What are the energies, in joules per photon, corresponding to visible light?

The photoelectric effect

13. The lowest-frequency light that will produce the photoelectric effect on a material is called the *threshold frequency*.
   (a) The threshold frequency for platinum is 1.3×10^{15} s^{-1}. What is the energy of a quantum of this radiation?
   (b) Will platinum display the photoelectric effect when exposed to ultraviolet light? infrared light? Explain. (Refer to Figure 7-9.)

14. The *work function* of a photoelectric material is the energy that a photon of light must possess to just secure the release of an electron from the surface of a material. The corresponding frequency of the light is the threshold frequency. The higher the energy of the incident radiation, the more kinetic energy the ejected electrons have in moving away from the surface. The work function for the element mercury is equivalent to 435 kJ/mol of photons.
   (a) What is the threshold frequency?
   (b) What is the wavelength of light of this frequency?
   (c) Can the photoelectric effect be obtained with mercury using visible light?

15. In describing Einstein's quantum explanation of the photoelectric effect, Sir James Jeans made the following remark: "It not only prohibits killing two birds with one stone, but also the killing of one bird with two stones." Comment on the appropriateness of this analogy.

The Bohr atom

16. Use the description of the Bohr atom given in the text to determine **(a)** the radius, in nm, of the fifth Bohr orbit; **(b)** the energy of the atom when the electron is in this orbit.

17. Calculate the increase in **(a)** distance from the nucleus and **(b)** energy when an electron in a hydrogen atom is excited from the first to the third Bohr orbit.

18. What are **(a)** the frequency and **(b)** the wavelength, in μm, of the light emitted when the electron in a hydrogen atom drops from the energy level $n = 5$ to $n = 4$? **(c)** In what portion of the electromagnetic spectrum is this radiation?

19. Which of the following electron transitions requires that the greatest quantity of energy be *absorbed* by a hydrogen atom? **(a)** from $n = 1$ to $n = 2$; **(b)** from $n = 2$ to $n = 4$; **(c)** from $n = 3$ to $n = 6$; **(d)** from $n = \infty$ to $n = 1$. Explain.

$\star$**20.** The Bohr theory can be extended to one-electron species other than the hydrogen atom, for example He$^+$, Li^{2+}, and Be^{3+}. In these cases the energies are related to the quantum number, n, through the expression

$$E_n = \frac{-Z^2 B}{n^2}$$

where Z is the atomic number of the species and $B = 2.179 \times 10^{-18}$ J.
   (a) What is the energy of the lowest level ($n = 1$) of a He$^+$ ion?
   (b) What is the energy of the level $n = 3$ of a Li^{2+} ion?

$\star$**21.** Both the Balmer series of the hydrogen spectrum (Figure 7-6) and the energy-level diagram of Figure 7-13 feature lines that become very closely spaced in a certain region. There is a relationship between the two phenomena. Explain what this is.

The uncertainty principle

22. Describe the ways in which the Bohr model of the hydrogen atom violates the Heisenberg uncertainty principle.

23. Although Einstein himself made some early contributions to quantum theory, he was never able to accept the Heisenberg uncertainty principle. His objections were stated in many forms, one of the most famous being, "God does not play dice with the world." What do you suppose that Einstein meant by this remark?

$\star$**24.** Show that the uncertainty principle has little significance when applied to a macroscopic object like a moving automobile. (*Hint:* Assume that m is precisely known; assign a reasonable value to either Δx or Δv and estimate a value of the other.)

25. A proton is accelerated to one-tenth the velocity of light. Suppose that its velocity can be measured with a precision of $\pm 1\%$. What must be the uncertainty in its position?

Wave–particle duality

26. What must be the velocity of a beam of electrons if they are to display a De Broglie wavelength of 1 nm?

Wave mechanics

27. Describe briefly the several differences between the orbits of the Bohr atom and the orbitals of the wave mechanical atom. Are there any similarities?

28. The greatest probability of finding the electron in a small volume element of the $1s$ orbital of the hydrogen atom is at the nucleus. Yet the most probable distance of the electron from the nucleus is 0.53 Å. How can you reconcile these two statements?

Quantum numbers and electron orbitals

29. Select the correct answer: An electron that has the quantum numbers $n = 3$ and $m_l = 2$ **(a)** must have the quantum number $m_s = +\frac{1}{2}$; **(b)** must have the quantum number $l = 1$; **(c)** may have the quantum number $l = 0$, 1, or 2; **(d)** must have the quantum number $l = 2$.

30. With reference to Table 7-2, complete the entry for $n = 4$. The new subshell that arises is the f subshell. How many f orbitals are present in this subshell?

31. Which of the following sets of quantum numbers is not allowable? Why not?
  (a) $n = 2$, $l = 1$, $m_l = 0$ **(b)** $n = 2$, $l = 2$, $m_l = -1$
  (c) $n = 3$, $l = 0$, $m_l = 0$ **(d)** $n = 3$, $l = 1$, $m_l = -1$
  (e) $n = 2$, $l = 0$, $m_l = -1$ **(f)** $n = 2$, $l = 3$, $m_l = 2$

32. What type of electron orbital (i.e., s, p, d, or f) is designated: **(a)** $n = 2$, $l = 1$, $m_l = -1$; **(b)** $n = 4$, $l = 0$, $m_l = 0$; **(c)** $n = 5$, $l = 2$, $m_l = 0$.

33. What are the n and l quantum number designations for the subshells $3s$, $4p$, and $5d$?

34. How many orbitals can there be of each of the following types? Explain. **(a)** $2s$; **(b)** $3f$; **(c)** $4p$; **(d)** $5d$.

35. Which of the following statements is (are) correct for an electron that has the quantum numbers $n = 4$ and $m_l = -2$?
  (a) The electron is in the fourth principal shell.
  (b) The electron may be in a d orbital.
  (c) The electron may be in a p orbital.
  (d) The electron must have a spin quantum number, $m_s = +\frac{1}{2}$.

Electron configurations

36. Arrange the following group of orbitals in the order in which they fill with electrons: $5s$, $3p$, $3d$, $4p$, $5f$, $6p$, $6s$.

37. Five electrons in an atom have the quantum numbers given below. Arrange these electrons in order of increasing energy. If any two have the same energy, so indicate.
  (a) $n = 4$, $l = 0$, $m_l = 0$, $m_s = +\frac{1}{2}$
  (b) $n = 3$, $l = 1$, $m_l = -1$, $m_s = -\frac{1}{2}$
  (c) $n = 3$, $l = 2$, $m_l = 0$, $m_s = +\frac{1}{2}$

  (d) $n = 3$, $l = 2$, $m_l = -2$, $m_s = -\frac{1}{2}$
  (e) $n = 3$, $l = 0$, $m_l = 0$, $m_s = -\frac{1}{2}$

38. What element has the electron configuration represented in each example?
  (a) $1s^2 2s^2 2p^1$
  (b) $[Ar]3d^3 4s^2$

39. Complete the following using part (a) as an example.
  (a) Na ($Z = 11$) $1s^2 2s^2 2p^6 3s^1$
  (b) _____ $1s^2 2s^2 2p^6 3s^2 3p^3$
  (c) Zr ($Z = 40$) $[Kr]4d^{(?)}5s^2$
  (d) _____ $[Kr]4d^{(?)}5s^2 5p^4$
  (e) _____ $[Kr]4d^{(?)}5s^{(?)}5p^5$
  (f) Bi ($Z = 83$) $[Xe]4f^{(?)}5d^{(?)}6s^{(?)}6p^{(?)}$

40. Which of the following electron configurations is correct for phosphorus ($Z = 15$). What is wrong with each of the others?

41. On the basis of rules for electron configurations, indicate the number of **(a)** unpaired electrons in an atom of S; **(b)** $3d$ electrons in an atom of Cl; **(c)** $4p$ electrons in an atom of Ge; **(d)** $3s$ electrons in an atom of Rb; **(e)** $4f$ electrons in an atom of Pb.

42. The electron configurations described in the text are all for normal atoms in their ground states. An atom may absorb a quantum of energy and promote one or more electrons to a higher energy level; it becomes an "excited" atom. The following configurations represent excited states. Indicate why this is so. **(a)** $1s^2 2s^1 2p^1$; **(b)** $[Ne]3s^2 3p^2 3d^2$; **(c)** $[Ar]3d^{10}4s^1 4p^3$.

*__**43.**__ What would be the electron configuration of the element Cs in each case?
  (a) If there were *three* possibilities for electron spin.
  (b) If the quantum number, l, could have the value, n, and if all the rules governing electron configurations were otherwise valid.

Additional Exercises

1. The current international standard of length, the meter, is defined as 1,650,763.73 wavelengths of a certain radiation emitted by $^{86}_{36}Kr$ atoms.
   (a) What is this wavelength, in nm?
   (b) What is the frequency of this radiation?
   (c) Why is this a more exact standard than the distance between two scratches on a metal bar?

2. In what region of the electromagnetic spectrum would you expect to find radiation emitted by a hydrogen atom when the electron in the atom falls from the orbit $n = 6$ to the orbit $n = 4$?

3. The Lyman series of the hydrogen spectrum can be represented by the equation

$$\nu = 3.2881 \times 10^{15} \text{ s}^{-1}\left(\frac{1}{1^2} - \frac{1}{n^2}\right) \qquad \text{where } n = 2, 3, \ldots$$

   (a) Calculate the maximum and minimum wavelengths of lines in this series.
   (b) In what portion of the electromagnetic spectrum will this series be found?

4. What is the energy, in kJ/mol, of light having a wavelength of 715 nm?

5. Determine (a) the energy of an H atom when its electron is in the orbit $n = 5$; (b) the total energy required to ionize 1 mol of *normal* H atoms.

6. The subshell that arises after f is called the g subshell (i.e., s, p, d, f, g).
   (a) How many g orbitals are present in the g subshell?
   (b) In what principal electronic shell would the g subshell first occur, and what is the total number of orbitals in this principal shell?

7. Complete the following assignments by writing an acceptable value for the missing quantum number. What type of orbital is described by each set?
   (a) $n = ?, l = 2, m_l = 0, m_s = +\frac{1}{2}$
   (b) $n = 2, l = ?, m_l = -1, m_s = -\frac{1}{2}$
   (c) $n = 4, l = 2, m_l = 0, m_s = ?$
   (d) $n = ?, l = 0, m_l = ?, m_s = ?$

8. Which of the following electron configurations is correct for molybdenum ($Z = 42$)? Comment on the errors in each of the others.

(a) $[Ar]3d^{10}3f^{14}$
(b) $[Kr]4d^5 5s^1$
(c) $[Kr]4d^5 5s^2$
(d) $[Ar]3d^{14}4s^2 4p^8$
(e) $[Ar]3d^{10}4s^2 4p^6 4d^6$

9. A light year is defined as the distance that electromagnetic radiation can travel in space in 1 year.
   (a) What is this distance, in km?
   (b) What is the distance, in km, to Alpha Centauri, the star closest to our solar system, if this distance is listed as 4.3 light years?

★10. Use the equation given in Exercise 20 to calculate the wavelength, in nm, of the spectral line resulting from the transition of an electron from the orbit $n = 3$ to $n = 2$ in a Li^{2+} ion.

★11. The angular momentum of an electron in the Bohr hydrogen atom is mvr, where m is the mass of the electron, v, its velocity, and r, the radius of the Bohr orbit. Combine this fact with the second of the Bohr assumptions and other data given in the text to obtain for an electron in the third orbit ($n = 3$) of a hydrogen atom: (a) its velocity; (b) the number of revolutions about the nucleus it makes per second.

★12. Radio signals from Voyager 1 spacecraft on its trip to Jupiter in the late 1970s were broadcast at a frequency of 8.4 gigahertz. On earth this radiation was received by a 64-m antenna which was capable of detecting signals as weak as 4×10^{-21} watt (1 watt = 1 J/s). Approximately how many photons per second did the antenna intercept from this weak signal?

★13. Certain metallic compounds, when heated in open flames, impart characteristic colors to the flames: for example, sodium compounds, yellow; lithium, red; barium, green. "Flame tests" can be used to detect the presence of these metallic elements. For light to be emitted by atoms in the flame, these atoms must first absorb energy as a result of favorable collisions with other atoms or molecules in the flame.
   (a) If the flame temperature is 800°C, can collisions with other gaseous atoms or molecules possessing an average amount of kinetic energy supply the required energy to excite an atom to the point that visible light is emitted?
   (b) If not, how do you account for the excitation energy?

Self-Test Questions

For questions 1 through 6 select the single item that best completes each statement.

1. The *shortest* wavelength radiation of the following is
(a) 735 nm;   (b) 6.3×10^{-5} cm;   (c) 1.05 μm;
(d) 3.5×10^{-6} m.

2. A particular electromagnetic radiation with wavelength 200 nm,
   (a) has a higher frequency than radiation with wavelength 400 nm
   (b) is in visible region of electromagnetic spectrum

(c) has a higher velocity in vacuum than does radiation of wavelength 400 nm

(d) has a greater energy per photon than does radiation with wavelength 100 nm

3. The set of quantum numbers, $n = 2$, $l = 2$, $m_l = 0$ (a) describes an electron in a $2d$ orbital;   (b) describes an electron in a $2p$ orbital;   (c) describes one of five orbitals of a similar type;   (d) is not allowed.

4. The m_l quantum number for an electron in a $5d$ orbital (a) can have any value less than 5;   (b) may be zero;   (c) may be $+\frac{1}{2}$ or $-\frac{1}{2}$;   (d) is three.

5. The number of $2p$ electrons in an atom of Cl is   (a) 0;   (b) 2;   (c) 5;   (d) 6.

6. The number of unpaired electrons in an atom of scandium ($Z = 21$) is   (a) 3;   (b) 2;   (c) 1;   (d) 0.

7. What is the energy content, in kJ/mol of photons, of a red light with frequency 4.00×10^{14} s^{-1}?

8. The atomic spectrum of sodium contains two bright yellow lines, one at 589.0 nm and the other at 589.6 nm. Which of the two lines represents the greater energy per photon? What is the *difference* in energy per photon between the two?

9. The line at 434 nm in the Balmer series of the hydrogen spectrum corresponds to a transition of an electron from the nth to the second Bohr orbit. What is the value of n?

$$\nu = 3.2881 \times 10^{15} \text{ s}^{-1} \left(\frac{1}{2^2} - \frac{1}{n^2} \right)$$

10. Write out the complete electron configuration of (a) selenium ($Z = 34$), using *spdf* notation;   (b) iodine ($Z = 53$), using an orbital diagram.

8 Atomic Properties and the Periodic Table

The properties of matter in bulk are determined by the properties of individual atoms. This is why chemists study atomic structure. One atomic property introduced in Chapter 7 was the electron configuration. We begin this chapter by describing the periodic table of the elements. This discovery was, in many ways, the crowning achievement of nineteenth-century chemistry. Next, we turn to a discussion of the relationship between electron configurations and the periodic table. We will also consider a number of other properties of individual atoms that provide a basis for understanding chemical bonding.

One point that should be noted at the outset, however, is that individual atoms cannot be obtained in an isolated condition. Most of the atomic properties discussed here are actually derived from measurements on aggregations of atoms. In some cases the properties cannot be measured at all but only arrived at indirectly.

8-1 On the Idea of Order

One of the most important scientific activities is the search for order. If a large number of observations or objects can be arranged into categories according to some common features, it becomes easier to describe them. Moreover, it is often possible to discover an underlying cause for a particular order, and this discovery may lead, in turn, to a significant theory. Unfortunately, there is no way of knowing for certain what features to look for or the number of observations necessary to arrive at a classification scheme. Botanical observations, for example, were sufficiently numerous so that the task of ordering was completed in the eighteenth century. The field of chemistry was not ready for the establishment of order at the same time. The laws of chemical combination were not understood; the assignment of atomic weights was quite uncertain; too many elements remained undiscovered. Chemistry's turn came in the nineteenth century.

8-2 Periodic Law and the Periodic Table

A classification scheme of the elements similar to that used today was discovered independently and almost simultaneously by Dimitri Mendeleev and Lothar Meyer, in 1869. Their classifications were based on an early version of the periodic law.

If the elements are arranged in order of increasing atomic weight, certain sets of properties are found to recur periodically.

ATOMIC VOLUME.   A simple example of the periodic law involves the property, **atomic volume.** This is the volume occupied by one mole of atoms of an element. Atomic volume is the atomic weight of an element divided by its density.

$$\text{at. vol. (cm}^3/\text{mol)} = \text{at. wt. (g/mol)} \times \frac{1}{d}(\text{cm}^3/\text{g}) \tag{8.1}$$

Meyer's plot of atomic volumes, published in 1870, displayed a periodicity based on atomic weights. Figure 8-1 is a later version based on atomic numbers. Very striking is the fact that atomic volumes rise to a maximum periodically, with the alkali metals Li, Na, K, Rb, and Cs. Several other properties, such as electrical conductivity, thermal conductivity, and hardness, when plotted as a function of atomic number (or atomic weight), yield similar curves.

This result can be expressed only to two significant figures, since the graph (Figure 8-1) cannot be read any more exactly than this.

Example 8-1   Use data from Figure 8-1 to estimate the density of silver. Silver has an atomic number of 47 and an atomic weight of 108. From Figure 8-1, the atomic volume of the element with atomic number, 47, is 10 cm^3/mol.

$$\text{no. g/cm}^3 = \frac{108 \text{ g}}{1 \text{ mol}} \times \frac{1 \text{ mol}}{10 \text{ cm}^3} = 11 \text{ g/cm}^3$$

SIMILAR EXAMPLES: Exercises 1, 2.

MENDELEEV'S PERIODIC TABLE.   A tabular arrangement of the elements based on the periodic law is called a **periodic table.** In Mendeleev's 1871 periodic table the elements were arranged in 12 horizontal rows and eight vertical columns or groups. The eight groups were further divided into subgroups. To achieve the objective of bringing similar elements into appropriate groups and subgroups, it was necessary to leave blank spaces for elements undiscovered at the time and to make assumptions

FIGURE 8-1
An illustration of the periodic law—atomic volume as a function of atomic number.

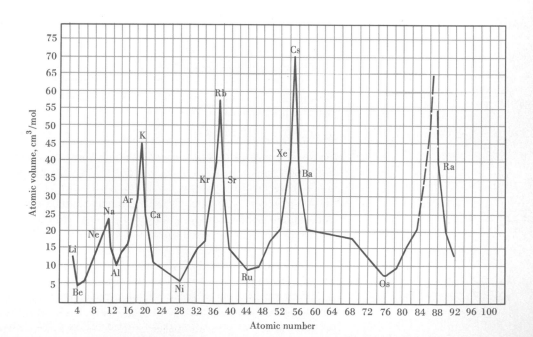

TABLE 8-1
Mendeleev's periodic table of 1871

R	Group I	Group II	Group III	Group IV	Group V	Group VI	Group VII	Group VIII
O	R_2O	RO	R_2O_3	RO_2	R_2O_5	RO_3	R_2O_7	RO_4
W	RCl	RCl_2	RCl_3	RCl_4 RH_4	RH_3	RH_2	RH	
1	H = 1							
2	Li = 7	Be = 9.4	B = 11	C = 12	N = 14	O = 16	F = 19	
3	Na = 23	Mg = 24	Al = 27.3	Si = 28	P = 31	S = 32	Cl = 35.5	
4	K = 39	Ca = 40	___ = 44	Ti = 48	V = 51	Cr = 52	Mn = 55	Fe = 56, Co = 59, Ni = 59, Cu = 63
5	(Cu = 63)	Zn = 65	___ = 68	___ = 72	As = 75	Se = 78	Br = 80	
6	Rb = 85	Sr = 87	?Yt = 88	Zr = 90	Nb = 94	Mo = 96	___ = 100	Ru = 104, Rh = 104, Pd = 106, Ag = 108
7	(Ag = 108)	Cd = 112	In = 113	Sn = 118	Sb = 122	Te = 125	I = 127	
8	Cs = 133	Ba = 137	?Di = 138	?Ce = 140	. . .	. . .	. . .	
9	. . .	. . .	. . .	. . .	. . .	. . .	. . .	
10	. . .	. . .	?Er = 178	?La = 180	Ta = 182	W = 184	. . .	Os = 195, Ir = 197, Pt = 198, Au = 199
11	(Au = 199)	Hg = 200	Tl = 204	Pb = 207	Bi = 208	. . .	. . .	
12	. . .	. . .	. . .	Th = 231	. . .	U = 240	. . .	

about atomic weights that were not known with certainty. Mendeleev's periodic table is reproduced in Table 8-1.

The elements within a particular subgroup of Mendeleev's table have similar physical and chemical properties, and these properties change gradually from top to bottom in the group. For instance, the alkali metals in group I have low melting points that decrease in the order

Li (174°C) > Na (97.8°C) > K (63.7°C) > Rb (38.9°C) > Cs (28.5°C)

These elements also have high atomic volumes, as illustrated in Figure 8-1. In moving across one of the rows in Mendeleev's table, the properties change rather dramatically from group to group.

At the top of Mendeleev's table are listed the formulas of the chlorides, hydrides, and oxides of the elements (R) in each group. Mendeleev was able to correlate these formulas with the group numerals.

CORRECTION OF ATOMIC WEIGHTS. To place them properly in his periodic table, Mendeleev made adjustments in the previously accepted atomic weights of a number of elements. One of these was indium. Indium was known to occur naturally in zinc ores and was assumed to form an oxide, InO, similar to that of zinc, ZnO. Based on the percent composition of this oxide (82.5% In) indium had been assigned an atomic weight of approximately 76. This atomic weight would have placed indium, a metal, between arsenic and selenium, both nonmetals. Mendeleev proposed that indium formed the oxide, In_2O_3. From this formula the atomic weight obtained for indium was 113. As a result, Mendeleev placed indium in the space between cadmium and tin, both metals.

Other atomic weights corrected by Mendeleev were those of beryllium (from 13.5 to 9) and uranium (from 120 to 240).

PREDICTION OF NEW ELEMENTS. Mendeleev deliberately left blank spaces in his periodic table for elements yet to be discovered. Not only did he predict the existence of these elements, but he predicted what their properties would be. The blank space at atomic weight 72 was for an element in the same group as silicon. Mendeleev called this element, eka-silicon. The remarkable agreement between some measured properties of germanium and Mendeleev's predicted values of these properties is brought out in Table 8-2.

In 1894, William Ramsay discovered helium and argon. He observed these gases to be unlike any other elements and assigned them to a new group of the periodic table, group 0. The remaining members of group 0—neon, krypton, xenon, and radon—were discovered soon thereafter.

ATOMIC NUMBER AS THE BASIS FOR THE PERIODIC LAW. In the early periodic table it was necessary to place certain pairs of elements out of order. For example, argon (at. wt. 39.9) was placed ahead of potassium (at. wt. 39.1). If this were not done, potassium, an active metal, would appear among the inert gases and argon, an inert gas, among a group of active metals. If the elements are arranged according to increasing atomic number, argon ($Z = 18$) naturally precedes potassium ($Z = 19$).

8-3 A Modern Periodic Table—The Long Form

Mendeleev's periodic table was a "short" form. Each main vertical group consisted of two subgroups. Most modern periodic tables are of the "long" form. The subgroups are separated from one another. Following are some of the features of the long form of the periodic table found on the inside front cover.

The horizontal rows of the table, which are arranged in order of increasing atomic number, are called **periods.** The vertical columns, which bring together similar elements, are called **groups** or **families.** The first period of the table consists of only two elements, hydrogen and helium. This is followed by two periods of eight elements each, lithium to neon, and sodium to argon. The fourth and fifth periods comprise 18 elements each, ranging from potassium to krypton and from rubidium

The term "eka" is derived from Sanskrit and means "first." That is, eka-silicon means literally, first comes silicon (and then comes the unknown element).

You will recall from Section 2-8 that the discovery of atomic numbers was made by Moseley in 1913. Reordering of the periodic table according to atomic number became possible at that time.

TABLE 8-2
Properties of germanium: predicted and observed

Property	Predicted: eka-silicon (1871)	Observed: germanium (1886)
atomic weight	72	72.6
density, g/cm^3	5.5	5.47
color	dirty gray	grayish white
density of oxide, g/cm^3	EsO_2: 4.7	GeO_2: 4.703
boiling point of chloride	$EsCl_4$: below 100°C	$GeCl_4$: 86°C
density of chloride, g/cm^3	$EsCl_4$: 1.9	$GeCl_4$: 1.887

The "oid" ending means like or similar to (as in humanoid). The lanthanoids are like lanthanum, and the actinoids like actinium. The names lanthanide and actinide are also commonly used in place of lanthanoid and actinoid.

to xenon. The sixth period is a long one of 32 members. To fit this period to a table which is held to a maximum width of 18 members requires that 14 members be extracted and placed at the bottom of the table. This series of 14 elements, which fits between lanthanum ($Z = 57$) and hafnium ($Z = 72$), is called the **lanthanoid** or rare earth series. The seventh and final period is incomplete but is believed to be a long one. A 14-member series, extracted from the seventh period and placed at the bottom of the table, is called the **actinoid** series.

The groups in the periodic table are designated by Roman numerals and letters. The A group elements are known as **representative elements.** The B group elements, together with those in group VIII and the lanthanoid and actinoid series, comprise the **transition elements.** The group labeled 0 contains the noble (inert) gases.

Example 8-2   Refer to the periodic table on the inside front cover and indicate:
(a) An element that is in group IVA and the fourth period.
(b) Two elements with properties similar to molybdenum.
(c) The known element that the unknown and undiscovered element, $Z = 114$, is most likely to resemble.

(a) The elements in the fourth period range from K ($Z = 19$) to Kr ($Z = 36$); those in group IVA, from C to Pb. The only element that is common to both of these groupings of elements is Ge ($Z = 32$).
(b) Molybdenum is in group VIB. Two other members of this group that it should resemble are chromium (Cr) and tungsten (W).
(c) Assuming that the seventh period has 32 members, the element $Z = 114$ should resemble lead, Pb ($Z = 82$), the element that it follows in group IVA.

SIMILAR EXAMPLES: Exercises 5, 6.

8-4 Electron Configurations and the Periodic Table

In Table 8-3 three typical groups of elements have been extracted from the periodic table and their electron configurations noted. The similarity in configurations among

TABLE 8-3
Electron configurations of some groups of elements

Group	Element	Configuration
IA	H	$1s^1$
	Li	$1s^22s^1$
	Na	$1s^22s^22p^63s^1$
	K	$1s^22s^22p^63s^23p^64s^1$
	Rb	$1s^22s^22p^63s^23p^63d^{10}4s^24p^65s^1$
	Cs	$1s^22s^22p^63s^23p^63d^{10}4s^24p^64d^{10}5s^25p^66s^1$
VIIA	F	$1s^22s^22p^5$
	Cl	$1s^22s^22p^63s^23p^5$
	Br	$1s^22s^22p^63s^23p^63d^{10}4s^24p^5$
	I	$1s^22s^22p^63s^23p^63d^{10}4s^24p^64d^{10}5s^25p^5$
0	He	$1s^2$
	Ne	$1s^22s^22p^6$
	Ar	$1s^22s^22p^63s^23p^6$
	Kr	$1s^22s^22p^63s^23p^63d^{10}4s^24p^6$
	Xe	$1s^22s^22p^63s^23p^63d^{10}4s^24p^64d^{10}5s^25p^6$

FIGURE 8-2
Electron configurations
and the periodic table.

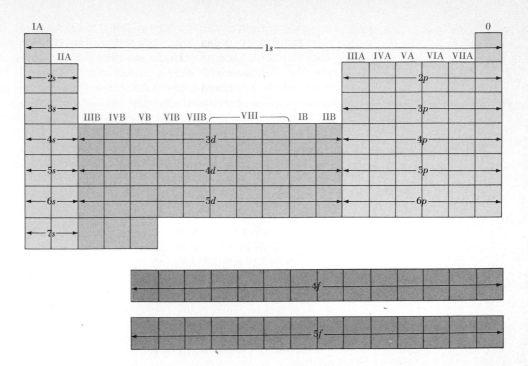

We must not lose sight of
the historical fact that
elements were first
grouped according to
similar properties, and
then the similarity of
electron configurations—
the theoretical basis—was
established.

Can you discover in Fig-
ure 8-2 the same order
of filling of electron orbitals that was presented
in Chapter 7, that is, $1s$,
$2s$, $2p$, $3s$, $3p$, $4s$, $3d$, and
so on?

Most atoms of the transi-
tions elements have two
outer-shell electrons in
the configuration ns^2.
Some, for example Cr,
Cu, Mo, Ag, and Au,
have only one—ns^1.

the elements in a group is quite apparent. All the atoms of group IA possess a single outer-shell electron in a s orbital. The halogen atoms (VIIA) have seven outer-shell electrons. These electrons have the configuration ns^2np^5, where n is the outermost shell or the shell of highest principal quantum number. The noble gases, with the exception of helium, which has only two electrons, have outermost shells with eight electrons in the configuration ns^2np^6.

Elements in the same group have similar physical and chemical properties and also similar electron configurations. We should begin to suspect that it is electron configurations that are principally responsible for the characteristic properties of the elements. Especially important is the electron configuration of the electronic shell of highest principal quantum number, that is, the *outermost* electronic shell.

Although it is not correct in all details, Figure 8-2 shows a periodic table in which the order of filling of orbitals is summarized. Here it can be seen that atoms of *representative* elements are characterized by the filling of s or p subshells of the electronic shell of highest principal quantum number. For the *transition* elements it is the d or f subshells of an *inner* electronic shell (not of the highest principal quantum number) that are partially filled. A d subshell fills for transition elements in the main body of the table, and an f subshell for the lanthanoid and actinoid elements.

For the representative elements the group numeral (IA, IIA, . . .) is identical to the number of electrons in s and p orbitals of the outermost electronic shell. In the case of transition elements, with the exception of groups IB and IIB, the group numeral is not the same as the number of outer-shell electrons. However, the B-group numerals do indicate the maximum number of electrons available for the formation of chemical compounds.

The properties of an element are determined largely by the electron configuration of the *outermost* electronic shell. Adjacent members of a series of *representative* elements in the same period (such as P, S, and Cl) have rather different properties because of differences in their outer-shell electron configurations. Within a transition series differences in electron configurations are found mostly in *inner* shells. As a result, within a transition series there are similarities among adjacent members of the same period (e.g., Fe, Co, and Ni) *as well as* within the same vertical group.

Although every other element has a definite place in the periodic table, the placement of hydrogen presents problems. Its electron configuration, $1s^1$, suggests that it be placed in group IA, as has been done in this text. However, it really does not resemble the alkali metals. Sometimes hydrogen is placed with the other elements whose outermost electronic shells contain just one electron less than that of a noble gas. This placement in group VIIA is not entirely satisfactory, however, because hydrogen does not particularly resemble the halogen elements. Still another alternative is to place hydrogen by itself at the top of the periodic table above carbon, an element that it does resemble in some respects. The uniqueness of hydrogen regarding its location in the periodic table stems from the fact that the hydrogen atom has only one electron.

Example 8-3 Based on the relationship between electron configurations and the periodic table, indicate how many **(a)** outer-shell electrons in an atom of bromine; **(b)** shells of electrons in an atom of strontium; **(c)** $5p$ electrons in an atom of tellurium; **(d)** $3d$ electrons in an atom of zirconium; **(e)** Unpaired electrons in an atom of indium. Determine the atomic number of each element and the location of the element in the periodic table. Then establish the significance of each location.

(a) Bromine ($Z = 35$) is a representative element in group VIIA. There are *seven* outer-shell electrons in all the atoms in this group.

(b) The highest principal quantum number for atoms in the fifth period is $n = 5$. There are *five* electronic shells in the Sr atom.

(c) Tellurium ($Z = 52$) follows the *second* transition series of elements, in which the $4d$ subshell is filled. The next subshell to receive electrons after the $4d$ is the $5p$. In the Aufbau process, Te is the fourth atom following Cd. The Te atom has *four* $5p$ electrons. (Alternatively, Te is in group VIA. All group VIA atoms have six outer-shell electrons, two s and four p. The outer-shell configuration of Te is $5s^2 5p^4$.)

(d) Zirconium ($Z = 40$) is in the *second* transition series, in which $4d$ orbitals fill. The filling of the $3d$ subshell occurs in the *first* transition series (from Sc to Zn). Thus, the $3d$ subshell of the Zr atom is filled. Zr has *ten* $3d$ electrons.

(e) Indium ($Z = 49$) is in group IIIA. The In atom has three outer-shell electrons with principal quantum number, $n = 5$, that is, $5s^2 5p^1$. The two $5s$ electrons are paired and the $5p$ electron is unpaired. The number of unpaired electrons in an In atom is one.

SIMILAR EXAMPLES: Exercises 13, 14.

8-5 Metals and Nonmetals

We have said much about the electron configurations of isolated atoms and their relation to the periodic table. But we must soon turn our attention to how these electron configurations may be altered, especially as atoms enter into chemical combination (the subject of Chapter 9). In this connection it proves useful to divide elements into four categories: metals, nonmetals, metalloids, and noble gases. In a way, the noble gases provide the clue by which this new classification may be understood.

The noble gases are all found in group 0 of the periodic table. Helium has the electron configuration, $1s^2$. The other noble gases all have eight electrons in the outermost shell, that is, the configuration $ns^2 np^6$. These prove to be very stable electron configurations. They can be altered only with great difficulty.

The electron configurations of the elements in groups IA and IIA differ from those of a noble gas by only one or two electrons in the s orbital of a new electronic shell. This fact is brought out especially clearly when we write electron configurations such as

K: $[Ar]4s^1$ and Ca: $[Ar]4s^2$

If a K atom is stripped of its outer-shell electron, it becomes the *ion* K$^+$, with the electron configuration [Ar]. A Ca atom acquires the [Ar] configuration following the removal of two electrons.

$$K([Ar]4s^1) \longrightarrow K^+([Ar]) + e^-$$

$$Ca([Ar]4s^2) \longrightarrow Ca^{2+}([Ar]) + 2e^-$$

Elements whose atoms have only a small number of electrons in the *s* and *p* orbitals of the electronic shell of highest principal quantum number are **metals.** The characteristic chemical properties of the metallic elements are based on the ease of removal of one or more electrons from their atoms to produce positive ions. Characteristic physical properties of metals—ability to conduct heat and electricity, ductility, malleability—also stem from these distinctive electron configurations.

Elements in groups VIIA and VIA have electron configurations with one and two electrons less than those of the corresponding noble gas. Atoms of these elements can acquire the electron configuration of a noble gas atom by *gaining* the appropriate number of electrons. For example, the electron configuration of Cl becomes that of Ar when *one* electron is gained; the Cl atom, of course, becomes the *negative ion* Cl$^-$. The S atom becomes the S^{2-} ion by gaining *two* electrons.

$$Cl([Ne]3s^23p^5) + e^- \longrightarrow Cl^-([Ar])$$

$$S([Ne]3s^23p^4) + 2e^- \longrightarrow S^{2-}([Ar])$$

Nonmetals are elements whose atoms can acquire a noble gas electron configuration by gaining a small number of electrons.

In our discussion of metals and nonmetals we have specifically mentioned only elements in groups IA, IIA, VIA, and VIIA. How shall we classify other elements in the periodic table, the transition elements, for instance? Most of the transition element atoms have two electrons in the *s* orbital of the outermost electronic shell; a few have only one. Thus, *they are all metals*. When transition metal atoms lose electrons, however, they do not generally acquire the electron configurations of noble gas atoms. Furthermore, under the appropriate conditions transition metal atoms may lose different numbers of electrons. Thus, an atom of iron may lose two electrons to form the ion, Fe^{2+},

$$Fe([Ar]3d^64s^2) \longrightarrow Fe^{2+}([Ar]3d^6) + 2e^-$$

or it may lose three electrons to form the ion, Fe^{3+},

$$Fe([Ar]3d^64s^2) \longrightarrow Fe^{3+}([Ar]3d^5) + 3e^-$$

There is hardly a classification scheme that does not have its borderline cases, and the metal–nonmetal scheme is no exception. Some of the remaining elements in groups IIIA, IVA, and VA are clear-cut nonmetals and some are predominantly metallic. A few of these elements, although having the appearance of metals, display nonmetallic properties as well. These elements are called **metalloids.**

The location of the metals, nonmetals, metalloids, and noble gases in the periodic table is illustrated through Figure 8-3. It is customary to separate the metals and nonmetals in a periodic table with a stepwise diagonal line. It is further understood that elements adjacent to the line exhibit some of the properties of both metals and nonmetals (i.e., are metalloids).

FIGURE 8-3
Metals, nonmetals,
metalloids, and
noble gases.

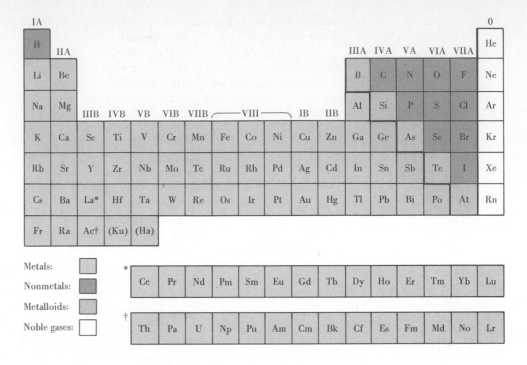

Metals:
Nonmetals:
Metalloids:
Noble gases:

8-6 Atomic Radius

A number of physical and chemical properties are related to the sizes of atoms, but atomic size is somewhat difficult to define. We have seen that the electron charge density falls off with increasing distance from the atomic nucleus, but nowhere does it reach zero. There is no precise outer boundary of an atom. Suppose that we try simply to describe the most probable distance from the nucleus to the outer-shell electron(s) and call this the atomic radius. Now let us consider the factors that affect this distance, that is, the factors that generally affect atomic size.

1. Variation of Atomic Sizes Within a Group of the Periodic Table. We consider the distance of an electron from the nucleus of an atom to depend primarily on the principal quantum number, n, of the electron. As a result we should expect that the higher the quantum number of the outermost electronic shell, the larger the atom. This generalization holds well for the group members of lower atomic numbers, where the percent increase in size from one period of elements to the next is large (e.g., from Li to Na to K in group IA). The percent increase in size from one period to the next is much smaller for the elements of higher atomic number (e.g., from K to Rb to Cs). In these elements the outer-shell electrons are held more tightly by the nucleus than would otherwise be expected. This is because inner-shell electrons in d and f subshells are not too effective in screening outer-shell electrons from the nucleus (recall Figure 7-21). Nevertheless, we still conclude that, in general, *the more electronic shells in an atom* (*the farther down a group of the periodic table*), *the larger the atom.*

2. Variation of Atomic Sizes Within a Period of the Periodic Table. Take the third period as an example. To build up each atom successively, starting with sodium, requires adding one unit of positive charge to the nucleus and one outer-shell electron. The number of inner-shell electrons remains constant at 10 (in the configuration $1s^2 2s^2 2p^6$). Thus, in sodium, with an atomic number of 11, the effective charge on the combination of nucleus and inner-shell electron core is $+1$. In magnesium, with atomic number 12, the effective charge is $+2$, and so on. The outer-shell elec-

The effect of filling an f subshell on atomic sizes, an effect called the lanthanoid contraction, is discussed in Section 21-1.

FIGURE 8-4
Covalent, ionic, and
metallic radii compared.

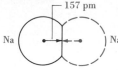

Covalent radius:

```
        ┌─ 157 pm
   Na  ( • • ) Na
```

Ionic radius:

```
        ┌─ 95 pm
  Na⁺ ( • • ) Cl⁻
```

Metallic radius:

```
        ┌─ 181 pm
   Na  ( • • ) Na
```

The **covalent radius** is
one half the distance be-
tween the centers of two
Na atoms in the gaseous
molecule $Na_2(g)$. The
ionic radius is based on
the distance between cen-
ters of ions in an ionic
compound, for example,
NaCl. Here, of course,
the cation and anions are
of different sizes. The
metallic radius is taken
as one half the distance
between the centers of
adjacent atoms in solid
metallic sodium.

FIGURE 8-5
Covalent radii of atoms.

The values plotted here
are for the bond type
known as the single cova-
lent bond. Radii are
given in picometers.

trons are attracted more strongly by the core and *the atomic radius decreases from left to right through a period of elements.*

3. Variation of Atomic Sizes Within a Transition Series. For the fourth and higher periods the situation is somewhat different from that described in generalization 2 for the portions of the periods that include transition elements. In a series of transition elements, additional electrons go into an *inner* electron shell, whereas the number of electrons in the *outer* shell tends to remain constant. Thus, the outer-shell electrons experience similar forces of attraction in these atoms. There is an initial sharp decrease in size for the first two or three members, but following that *atomic sizes change little in a transition series.* Consider Fe, Co, and Ni, for instance. Fe has 26 protons in the nucleus and 24 inner-shell electrons. In Co ($Z = 27$) there are 25 inner-shell electrons, and in Ni ($Z = 28$) there are 26. In each case the two outer shell electrons are under the influence of an effective charge of $+2$.

We should be more specific about what we mean by an atomic radius, since the radius of an atom depends on the environment in which it is found. For *bonded* atoms we customarily speak of a covalent radius, ionic radius, and, in the case of metals, a metallic radius. For atoms that are *not* bonded together, the radius is known as the van der Waals radius. Three different radii for a sodium atom are compared in Figure 8-4.

The unit that has been used for a long time to describe atomic dimensions, is the angstrom unit, Å. However, the angstrom is not a recognized SI unit. The appropriate SI unit is either the nanometer (nm) or the picometer (pm).

$$1 \text{ Å} = 1 \times 10^{-10} \text{ m} = 1 \times 10^{-8} \text{ cm} = 0.10 \text{ nm} = 100 \text{ pm}$$

We will use both the angstrom and the picometer in discussing atomic and molecular dimensions.

COVALENT RADIUS. In Chapter 10 we will picture covalent bonds as arising from the overlap of electron orbitals in the region between the centers of two atoms. The result is that the nuclei of bonded atoms approach each other more closely than do the nuclei of nonbonded ones. *The covalent radius is one half the distance between the nuclei of two identical atoms bonded together covalently.*

Actually covalent radii themselves are not constant but vary with the character of the covalent bond between atoms, as we shall learn in Chapter 9. Figure 8-5 com-

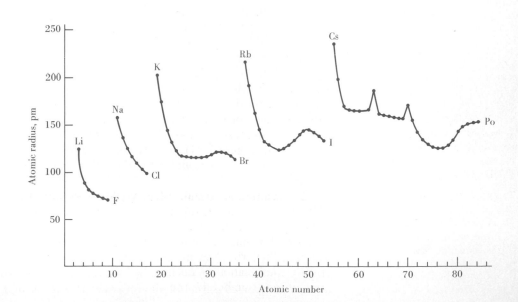

pares covalent radii based on the bond type called a single covalent bond. It illustrates the three generalizations about atomic sizes presented earlier in this section.

In the absence of statements to the contrary, the term "atomic radius" will be taken to mean *covalent radius*.

IONIC RADIUS.   When electrons are removed from a metal atom to form a *positive* ion (cation), a significant reduction in size occurs. Usually the electrons lost are those of the shell of highest principal quantum number, and the resulting ion has one shell less than the metal atom.

Figure 8-6 compares five species: a Na atom, a Mg atom, a Na^+ ion, a Mg^{2+} ion, and a Ne atom. The Na atom is larger than Mg for the reason stated in generalization 2 on page 180. The ions are considerably smaller than the corresponding metal atoms, for the reason stated above. Na^+, Mg^{2+}, and Ne are said to be **isoelectronic**—they have the same number of electrons (10) in identical configurations ($1s^2 2s^2 2p^6$). Ne has a nuclear charge of $+10$. Na^+ is smaller than Ne because Na has a nuclear charge of $+11$. Mg^{2+} is still smaller because it has a nuclear charge of $+12$.

When a nonmetal atom gains one or more electrons to form a *negative* ion (anion), there is an *increase* in size. The addition of electrons to an atom causes an increase in repulsions among the electrons. The electrons spread out more, and the size of the atom increases.

Figure 8-7 lists the radii of a number of ions, both negative and positive.

Example 8-4   The following species are isoelectronic with the noble gas argon. Without reference to figures or tables in the text, arrange them in order of increasing size: Ar, K^+, Cl^-, S^{2-}, Ca^{2+}.

The electron configuration that all five species have in common is $1s^2 2s^2 2p^6 3s^2 3p^6$. The greater the nuclear charge, the more tightly these electrons are held and the smaller the species. On this basis, Ca^{2+} is smallest, followed by K^+. If a species has an excess of electrons over protons, it becomes larger than the corresponding noble gas atom, the greater the negative charge the larger the size. Thus, S^{2-} is the largest species. The order is

$$Ca^{2+} < K^+ < Ar < Cl^- < S^{2-}$$

SIMILAR EXAMPLES: Exercises 20, 21.

FIGURE 8-7
Some representative ionic radii in picometers (pm).

Li⁺ 68	Be²⁺ 31														B³⁺ 20	C –	N³⁻ 171	O²⁻ 140	F⁻ 136
Na⁺ 95	Mg²⁺ 65														Al³⁺ 50	Si –	P³⁻ 212	S²⁻ 184	Cl⁻ 181
K⁺ 133	Ca²⁺ 99	Sc³⁺ 81	Ti²⁺ 90	V²⁺ 88	Cr²⁺ 84	Mn²⁺ 80	Fe²⁺ 76	Co²⁺ 74	Ni²⁺ 72	Cu²⁺ 72	Zn²⁺ 74							Se²⁻ 198	Br⁻ 196
Rb⁺ 148	Sr²⁺ 113																	Te²⁻ 225	I⁻ 216
Cs⁺ 169	Ba²⁺ 135																		

Many of the elements form more than a single ion, and these different ions have different sizes. The data listed here are meant only to be representative.

The observation of a large increase in size in the formation of a negative ion from a nonmetal atom applies only when *covalent* and ionic radii are being compared. There is no significant change in size when an electron is gained by a nonbonded atom. Here is a graphic illustration of the need to specify the environment of an atom when referring to its size.

FIGURE 8-6
A comparison of atomic and ionic sizes.

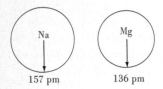

157 pm   136 pm

131 pm

95 pm   65 pm

The radii shown for Na and Mg are covalent radii; for Na^+ and Mg^{2+}, ionic radii; and for Ne, the van der Waals radius. The variations in size can be explained by the factors discussed in the text.

8-7 Ionization Energy (Ionization Potential)

We have described several ways in which atoms can be caused to lose electrons. This may occur through shining light of an appropriate frequency on a suitable material (photoelectric effect), through heating certain materials (thermionic effect), and through collisions between an electron beam and gaseous atoms. In any event atoms do not lose electrons spontaneously under normal conditions. They must absorb energy in order for ionization to occur. The **ionization energy (ionization potential),** *I*, of an atom is the energy that the gaseous atom must absorb in order that its most loosely held electron may become completely separated from it.

Ionization energies can be measured in cathode ray tubes in which the atoms of interest are present as a gas under low pressure. Some typical values are

$$Mg(g) \longrightarrow Mg^+(g) + e^- \qquad I_1 = 7.64 \text{ eV} \tag{8.2}$$

$$Mg^+(g) \longrightarrow Mg^{2+}(g) + e^- \qquad I_2 = 15.03 \text{ eV} \tag{8.3}$$

The symbol I_1 stands for first ionization energy, I_2, for second ionization energy, and so on. The loss of a second electron (as measured by I_2) occurs with greater difficulty than the first (as measured by I_1). This is because the electron being ionized would have to move away from an ion with a charge of $+2$ (Mg^{2+}) rather than from an ion with a charge of $+1$ (Mg^+).

The energy unit used in equations (8.2) and (8.3) is the **electron volt (eV).** One electron volt is the energy acquired by an electron as it falls through an electrical potential difference of 1 volt (V). The electron volt is a very small unit of energy and quite suitable to describe processes involving single atoms. When we consider atoms in large numbers, particularly in molar quantities, it proves convenient to express ionization energies in terms of ionizing 1 mol of gaseous atoms. The conversion relationship between electron volts per atom and kilojoules per mole (and kcal/mol) is given in equation (8.4).

$$1 \text{ eV/atom} = 96.49 \text{ kJ/mol} \ (=23.06 \text{ kcal/mol}) \tag{8.4}$$

TABLE 8-4
Ionization energies of the alkali metal (group IA) elements

Element	I_1, eV
Li	5.39
Na	5.14
K	4.34
Rb	4.18
Cs	3.89
Fr	3.83

The ease with which electrons can be removed from an atom requires consideration of several factors. It is reasonable to expect, however, that the farther a given electron is from the nucleus, the smaller the force by which it is attracted to the nucleus and the more easily it should be extracted: *Ionization energies decrease as the sizes of atoms increase.* This relationship is further illustrated through Table 8-4 for the alkali metal atoms and through Figure 8-8, in which ionization energies are plotted as a function of atomic number. The minima in Figure 8-8 come at the same atomic numbers as do the maxima in atomic volumes in Figure 8-1—the atomic numbers of the alkali metals. Another feature brought out by Figure 8-8 is the maxima in ionization energies that occur with the noble gases. The noble gas electron configuration is an exceptionally stable one that can be disrupted only with the expenditure of considerable energy.

If we consider the degree of metallic character of an element to be measured by the ease with which electrons can be removed from its atoms, then *the lower the ionization energy, the more metallic the element.* By this measure the atoms at the bottom of a group (larger atoms) are more metallic than those at the top (smaller atoms).

Table 8-5 lists ionization energies for the third period elements. With minor exceptions the trend in moving across a period (follow the colored stripe) is that ionization energies increase from group IA to group 0. (Recall that atoms get smaller and less metallic through the period.) Table 8-5 also lists stepwise ionization energies (I_1, I_2, . . .). Note particularly the large breaks that occur as indicated by the zigzag diagonal line. Consider magnesium as an example. Although the second electron is

FIGURE 8-8
First ionization energies
as a function of
atomic number.

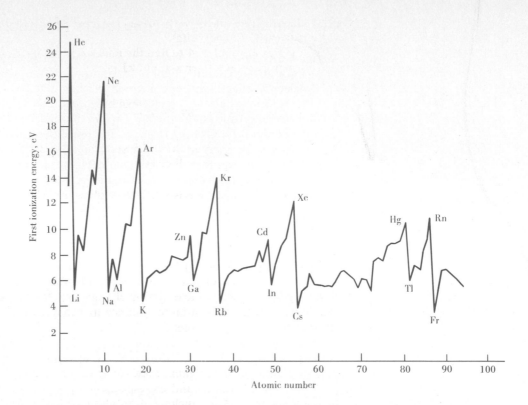

removed with greater difficulty than the first, when two electrons are removed from a
Mg atom it acquires the electron configuration of the noble gas Ne. Removal of a
third electron requires breaking into the especially stable octet of electrons characteristic of the outer shell of a noble gas atom (ns^2np^6). This is an event that is not likely
to occur in ordinary chemical processes. Similar reasoning would indicate that Na is
not likely to occur as an ion with a charge greater than $+1$, nor Al with a charge
greater than $+3$.

There are some exceptions to the regular increase of ionization energies in moving from left to right across the periodic table. For example, despite the fact that the
Al atom is smaller than Mg, the first ionization energy is *lower* for Al (6.0 eV) than
for Mg (7.6 eV). This is because the electron to be ionized in Al is in a higher energy
orbital ($3p$) than is the electron ($3s$) to be ionized in Mg.

TABLE 8-5
Ionization energies for the third-row elements (all values are in electron volts)

	Na	Mg	Al	Si	P	S	Cl	Ar
I_1	5.1	7.6	6.0	8.2	10.5	10.4	13.0	15.8
I_2	47.3	15.0	18.8	16.3	19.7	23.3	23.8	27.6
I_3		80.1	28.5	33.5	30.2	34.8	39.6	40.7
I_4			120.0	45.1	51.4	47.3	53.5	59.8
I_5				166.8	65.0	72.7	67.8	75.0
I_6					220.4	88.0	97.0	91.0
I_7						280.9	114.2	124.3

Example 8-5   How many joules of energy must be absorbed to convert to Li^+ all the atoms present in 1.00 mg Li(g)?

We need to use Table 8-4 to find the ionization energy of Li (per atom) and equation (8.4) to convert from eV per atom to kJ per mole. We must also convert from mg Li to mol Li.

$$\text{ionization energy } (I_1) = 5.39 \text{ eV} \times \frac{96.49 \text{ kJ/mol}}{1 \text{ eV}} = 520 \text{ kJ/mol}$$

$$\text{no. J} = 1.00 \times 10^{-3} \text{ g Li} \times \frac{1 \text{ mol Li}}{6.94 \text{ g Li}} \times \frac{520 \text{ kJ}}{1 \text{ mol Li}} \times \frac{1000 \text{ J}}{1 \text{ kJ}} = 74.9 \text{ J}$$

SIMILAR EXAMPLES: Exercises 25, 26.

8-8 Electron Affinity

Electron affinity (EA) is the enthalpy change, ΔH, for the process in which an electron is brought from an infinite distance away up to a neutral gaseous atom and absorbed by it. For example,

$$Cl(g) + e^- \longrightarrow Cl^-(g) \qquad EA = -3.7 \text{ eV} \tag{8.5}$$

A few other values for the gain of one electron are -3.57, -3.53, -3.06, -1.47, and -2.07 eV for F, Br, I, O, and S, respectively.

The attraction of the nucleus of an atom for an additional electron results in a release of energy when a gaseous atom gains a single electron (EA < 0). The gain of a second electron requires the absorption of energy to overcome electron–electron repulsions (EA > 0). The affinities of O and S for two electrons (to form the ions O^{2-} and S^{2-}) are $+7.3$ and $+4.0$ eV, respectively.

Electron affinity is a property that cannot be measured easily by experiment. This has been accomplished in a few cases, but most electron affinities have been derived indirectly from other measurements.

8-9 Electronegativity

Two criteria have been presented for expressing metallic and nonmetallic tendencies: ionization energy and electron affinity. These criteria are really quite adequate; yet it is useful to have a single criterion, especially when describing the bond type that results when atoms combine. The quantity we are seeking is called electronegativity. **Electronegativity** describes the ability of an atom to compete for electrons with another atom to which it is bonded. The electronegativity is related to ionization energy (I) and electron affinity (EA), since these quantities reflect the ability of an atom to lose or gain an electron. The most widely used electronegativity scale is based on an evaluation of bond energies and was devised by Linus Pauling. The basis of this scale is explored further in Chapter 9. Pauling's electronegativities are dimensionless (no units) numbers ranging from about 1 for very active metals to 4.0 for fluorine, the most active nonmetal. A few electronegativities are presented in Table 8-6.

As a rough rule most metals have electronegativities of about 1.7 or less; metalloids, about 2; and nonmetals, greater than 2.

TABLE 8-6
Electronegativities of selected elements

H 2.20						
Li 0.98	Be 1.57	B 2.04	C 2.55	N 3.04	O 3.44	F 3.98
Na 0.93	Mg 1.31	Al 1.61	Si 1.90	P 2.19	S 2.58	Cl 3.16
K 0.82	Ca 1.00	Ga 1.81		As 2.18		Br 2.96
Rb 0.82	Sr 0.95	In 1.78		Sb 2.05		I 2.66
Cs 0.79	Ba 0.89					

FIGURE 8-9
Paramagnetism illustrated.

(a) A sample is weighed in the absence of a magnetic field. **(b)** When the field is turned on, the balanced condition is upset. The sample gains weight because it is attracted into the magnetic field.

8-10 Magnetic Properties

On several occasions we have noted how atoms may display magnetic properties when placed in a magnetic field (recall, for example, the Stern–Gerlach experiment, Figure 7-20). One basic interaction between paired electrons in atoms and a magnetic field results in atoms being repelled by the field. This phenomenon is called **diamagnetism.** Although all substances have diamagnetic properties, the forces associated with diamagnetism are quite weak and easily overcome by another type of magnetic property that exists in some substances. This is the property of **paramagnetism.** A paramagnetic substance is attracted into a magnetic field.

An electron in motion, whether it be orbital motion or by virtue of its spin, can be likened to a tiny electric current. By the laws of electromagnetism, an electric current is expected to induce a magnetic field around it. In a species having only filled electronic orbitals all electrons are paired, and these individual magnetic effects cancel out. Such a species is repelled by a magnetic field (diamagnetic). If a species has unpaired electrons, however, the individual magnetic effects do not cancel out and the species displays paramagnetism. The more unpaired electrons in a species the stronger the attractive force it experiences when it is placed in a magnetic field. A straightforward method for measuring magnetic properties of a substance, illustrated in Figure 8-9, involves weighing the substance "in" and "out" of a magnetic field. If the substance is diamagnetic, it weighs less in the magnetic field; if paramagnetic, it weighs more.

The measurement of magnetic properties is an experimental method that can assist in establishing electron configurations of atoms and ions. For example, an atom of iron is assigned the electron configuration

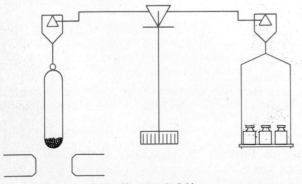

(a) No magnetic field

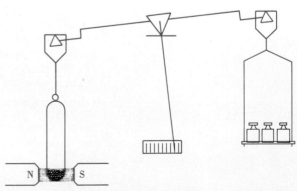

(b) Magnetic field turned on

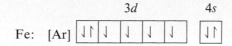

If an iron atom loses its two 4s electrons to form the ion Fe^{2+}, we should expect a species with four unpaired electrons.

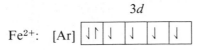

The experimentally determined paramagnetism of solids containing Fe^{2+} does indeed correspond to four unpaired electrons. For the species Fe^{3+} the expectation, confirmed by experiment, is *five* unpaired electrons.

Example 8-6 Which of the following species would you expect to be diamagnetic and which paramagnetic? **(a)** a Na atom; **(b)** a Mg atom; **(c)** a Cl^- ion; **(d)** a Ca^{2+} ion; **(e)** an Ag atom.

(a) Paramagnetic. The Na atom has a single 3s electron outside the Ne core. This electron is unpaired.

(b) Diamagnetic. The Mg atom has *two* 3s electrons outside the Ne core. They must be paired, as are all the other electrons.

(c) Diamagnetic. The Ar atom has all electrons paired ($1s^2 2s^2 2p^6 3s^2 3p^6$) and Cl^- is isoelectronic with Ar.

(d) Diamagnetic. Ca^{2+} is also isoelectronic with Ar.

(e) Paramagnetic. Even without knowing the exact electron configuration of Ag, because the atom has 47 electrons—an odd number—at least one of the electrons must be unpaired.

SIMILAR EXAMPLES: Exercises 35, 36, 37.

8-11 An Introduction to Descriptive Chemistry

Descriptive chemistry refers to an application of basic principles to a study of the elements and their compounds. For any given element this study might include describing the naturally occurring forms of the element; methods used to prepare the pure element from these naturally occurring forms; physical properties and uses of the pure element; typical reactions of the element; and the structures, properties, reactions, and uses of important compounds of the element. The basic principles that are essential to an understanding of these matters are drawn from a number of fundamental areas of chemistry. Included among these areas are atomic and molecular structure, thermodynamics, kinetics, equilibrium, acid-base theory, and oxidation-reduction. Inasmuch as our discussion of these aspects of chemistry will extend over a number of chapters, a systematic treatment of descriptive chemistry will have to come closer to the end of the text. However, we can use our present knowledge of atomic properties and periodic relationships as a basis upon which to preview the subject.

USING PERIODIC RELATIONSHIPS TO PREDICT PHYSICAL AND CHEMICAL PROPERTIES. Table 8-7 lists several properties of the halogen (VIIA) elements. Let us see if we can use our understanding of periodic relationships among the elements to fill in

The attraction of iron objects by a magnetic field is far stronger than can be accounted for by unpaired electrons alone. The special magnetic property possessed by iron and a few other metals and alloys is referred to as ferromagnetism (see Section 21-4).

TABLE 8-7
Some properties of the halogen (VIIA) elements

Element	Atomic number	Atomic weight	Melting point, K	Boiling point, K	Physical form at room temperature	Reaction with hydrogen: $H_2 + X_2 \longrightarrow 2\,HX$
F	9	19.00	53	85	yellow-green gas	?
Cl	17	35.45	172	239	greenish-yellow gas	rapid when exposed to light
Br	35	79.90	266	?	?	slow reaction
I	53	126.90	387	458	grayish-black solid	very slow reaction

the three blanks left in the table. The basic principle we will employ is that where a variation in some property is found within a group of elements, the variation is likely to be continuous from top to bottom in the group.

Example 8-7   Estimate the boiling point temperature of bromine.

Note that the atomic number of Br (35) is intermediate to those of Cl (17) and I (53). Also, the atomic weight of Br (79.90) is about intermediate to those of Cl and I [their average being $(35.45 + 126.90)/2 = 81.18$]. It is reasonable to expect that the boiling point of bromine, in kelvins, is the average of those of chlorine and iodine.

$$\text{bp of } Br_2 \cong \frac{239\text{ K} + 458\text{ K}}{2} = \frac{697\text{ K}}{2} = 348\text{ K}$$

(The actual boiling point of bromine is 332 K.)

SIMILAR EXAMPLES: Exercises 39, 40.

Example 8-8   What is the physical form of bromine at room temperature?

The melting point of bromine (266 K) is below 0°C, and the boiling point is well above room temperature. Bromine is a liquid at room temperature. Here is another way to arrive at this conclusion: The lighter halogens, fluorine and chlorine, are gases at room temperature, and iodine is a solid. It is reasonable to expect that the intermediate member of the group—bromine—will be in an intermediate state of matter—a liquid.

SIMILAR EXAMPLE: Exercise 45.

Example 8-9   What do you predict to be the nature of the reaction between fluorine gas and hydrogen?

From the data given in Table 8-7 it appears that the lighter the halogen, the more vigorous its reaction with hydrogen. We might predict a reaction of an explosive nature between fluorine and hydrogen.

SIMILAR EXAMPLE: Exercise 45.

USING PERIODIC RELATIONSHIPS TO WRITE CHEMICAL EQUATIONS.   A typical reaction in which all of the *alkali metal* (IA) and *halogen* (VIIA) elements participate is the formation of an ionic compound ("salt"). This occurs through the transfer of an electron from the metal to the nonmetal atom. We can write the following.

specific equations:

$$2\,Li(s) + F_2(g) \longrightarrow 2\,LiF(s)$$

$$2\,Li(s) + Cl_2(g) \longrightarrow 2\,LiCl(s)$$

and so on

$$2\,Na(s) + I_2(s) \longrightarrow 2\,NaI(s)$$

$$2\,Na(s) + Br_2(l) \longrightarrow 2\,NaBr(s)$$

and so on

or more simply

the general equation: $2\,M(s) + X_2 \longrightarrow 2\,MX(s)$ (8.6)

[where M is an alkali metal (IA) and X_2, a halogen element (VIIA)]

In a similar manner we can represent the reaction of a *halogen* with an *alkaline earth metal* by

the general equation: $M(s) + X_2 \longrightarrow MX_2(s)$ (8.7)
[where M is an alkaline earth (IIA) and X_2, a halogen element]

If we apply this scheme of writing generalized equations based on the periodic table to the *oxides* of the *alkaline earth (IIA) metals,* we obtain

the general equation: $2\,M(s) + O_2(g) \longrightarrow 2\,MO(s)$ (8.8)
[where M is an alkaline earth (IIA) metal]

But when we attempt to extend this scheme to reactions between the *alkali (IA) metals* and *oxygen,* we encounter a difficulty. The principal reactions are

$$4\,Li(s) + O_2(g) \longrightarrow 2\,Li_2O(s) \tag{8.9}$$

$$2\,Na(s) + O_2(g) \longrightarrow Na_2O_2(s) \tag{8.10}$$

$$M(s) + O_2(g) \longrightarrow MO_2(s) \qquad (\text{where } M = K,\ Rb,\ \text{or } Cs) \tag{8.11}$$

These three ions of oxygen are described in detail in Section 20-4.

The difficulty is that only with lithium does oxygen form the normal oxide ion, O^{2-}. With sodium it forms the peroxide ion, O_2^{2-}, and with the remaining alkali metals it forms the superoxide ion, O_2^{-}. We cannot write a *single* generalized equation to represent the reaction of the alkali metals with oxygen. Despite this "failure" that we have just illustrated, a reasonable starting assumption when writing chemical equations is still that periodic relationships do apply.

PREDICTING CHEMICAL COMPOUNDS. It is also reasonable to assume that if one element in a group of the periodic table forms a particular compound, other elements in the group are likely to form analogous compounds.

Example 8-10 Given the following names and formulas of known compounds, predict the formulas of the "unknown" compounds indicated: sodium periodate, $NaIO_4$; sodium hypochlorite, $NaClO$; phosphoric acid, H_3PO_4. **(a)** potassium perchlorate, ? **(b)** calcium hypochlorite, ? **(c)** arsenic acid, ?

(a) Potassium is in the same group (IA) as sodium, and chlorine is in the same group (VIIA) as iodine. We can substitute K for Na and Cl for I, leading to the formula $KClO_4$.

(b) The hypochlorite ion is the same species, ClO^-, whether combined with sodium or calcium ion. However, the calcium ion carries a charge of $+2$ (instead of $+1$ like sodium), leading to the formula $Ca(ClO)_2$.

(c) Arsenic and phosphorus are both in group VA. If arsenic forms similar compounds to phosphorus, we should expect the formula H_3AsO_4.

SIMILAR EXAMPLES: Exercises 42, 43.

Our flush of success over the predictions just made must now be tempered by these observations, which are based on the same data that we used in Example 8-10: (1) we would predict the formula, KFO_4, for potassium perfluorate, yet such a compound *does not exist;* and (2) *nitric* acid, which we would expect to be similar to *phosphoric* acid, has the formula HNO_3, not H_3NO_4.

As in predicting chemical reactions, there are simply times when the formulas of chemical compounds do not conform to our expectations based on periodic relationships. A complete understanding of descriptive chemistry requires the application of additional principles that we have not yet studied.

RELATIONSHIP OF PHYSICAL PROPERTIES TO ATOMIC PROPERTIES—GROUPS IA AND IIA. Table 8-8 lists three atomic properties and two physical properties of the elements of groups IA and IIA. Of the atomic properties, the atomic weight requires no special comment. The increase in atomic size from top to bottom of a group has been discussed previously, although we should note that the atomic radius shown in Table 8-8 is the metallic radius and not the covalent radius (recall Figure 8-4). The ionization energies in Table 8-8 follow the expected trends—decreasing from top to bottom in a group, and increasing in moving through a period from left to right (i.e., from group IA to group IIA).

Let us think of solid metals as consisting of aggregations of very large numbers of spherical atoms. Three factors should enter into establishing the density of a metal: the masses of the individual atoms, their volumes, and the manner in which the atoms are packed together. For the sake of discussion, let us assume that the packing arrangements are essentially the same for all the metals in Table 8-8 (although, in fact, they are not). This means that we should be able pretty much to correlate the physical property—density—with two atomic properties—atomic weight and atomic radius. An increase in atomic weight tends to produce an *increase* in density, and an increase in atomic radius (because of the corresponding increase in atomic volume), a *decrease* in density. When comparing an alkaline earth metal atom (e.g., Mg) with the immediately preceding alkali metal atom (e.g., Na), the atomic weight is seen to be slightly larger and the atomic radius, smaller. These factors both suggest an increase in density, which is borne out in every case in Table 8-8.

Relating density to atomic weight and atomic radius among elements within the same group of the periodic table is more difficult. Here, in progressing from top to

TABLE 8-8
Some properties of the alkali (IA) and alkaline earth (IIA) metals

Group	Element	Atomic weight	Atomic (metallic) radius, pm	First ionization energy, eV	Density, g/cm^3	Flame color
IA	Li	6.94	152	5.39	0.53	carmine
	Na	22.99	181	5.14	0.97	yellow
	K	39.10	227	4.34	0.86	violet
	Rb	85.47	248	4.18	1.53	bluish red
	Cs	137.34	266	3.89	1.87	blue
IIA	Be	9.01	112	9.31	1.85	—
	Mg	24.30	160	7.64	1.74	—
	Ca	40.08	197	6.11	1.55	orange-red
	Sr	87.62	215	5.69	2.54	scarlet
	Ba	137.34	222	5.40	3.51	green

We must not expect a direct correlation between flame color and ionization energy. Ionization energies correspond to the complete removal of an electron from an atom. Flame colors are related to energy differences between pairs of orbitals at lower quantum levels.

bottom, atomic weights increase, suggesting an increase in density. However, atomic radii also increase, suggesting a decrease in density. And the atomic radius factor is an especially important one since the volume of a sphere depends on the *third* power of its radius (i.e., $V = \frac{4}{3}\pi r^3$). The atomic weight of Na (23.0) is more than *three* times as great as that of Li (6.9). The corresponding increase in volume of the atom between Na and Li is less than *twofold*. The atomic weight factor is more important than the atomic radius factor, and we should expect the density of Na to be greater than that of Li, which it is. A similar comparison between Ca and Mg shows the proportionate increase in volume of the atom to be slightly greater than the increase in mass of the atom—the density of Ca is expected to be less than that of Mg.

The relatively low ionization energies of the group IA atoms suggest that the outer shell electrons in these atoms can be promoted to higher energy levels relatively easily. This might occur, for example, as a result of collisions between atoms in a gas flame. When these excited atoms revert to their normal (ground) states, characteristic amounts of energy are emitted in the form of light, imparting to the flame a characteristic color.

Summary

The periodic law is the historical basis of the tabular arrangement of the elements known as the periodic table. Mendeleev produced the first successful periodic table and used it to correct atomic weights and to predict the properties of undiscovered elements. Several modifications have been made to Mendeleev's original table, resulting in the currently used long form of the periodic table. Although the groupings of similar elements in Mendeleev's original table were based on *experimental* observations, a *theoretical* basis has also been established: Characteristic properties of the elements stem from their electron configurations. In the periodic table elements with similar electron configurations are grouped together.

To understand the physical and chemical properties of the elements, it is helpful to establish four categories: metals, nonmetals, metalloids, and noble gases. To compare the relative tendencies of atoms to behave as metals or nonmetals, it is necessary to study certain properties of the individual atoms—atomic properties. The atomic properties introduced in this chapter are atomic radius, ionization energy, electron affinity, and electronegativity. Magnetic properties of atoms and ions are helpful in establishing electron configurations.

Periodic relationships and atomic properties can be used to describe several aspects of descriptive chemistry, as illustrated in the concluding section of this chapter.

Learning Objectives

As a result of studying Chapter 8, you should be able to

1. Illustrate the periodic law by drawing graphs of selected properties of the elements as a function of atomic number.

2. Describe individual elements and groupings of elements in the periodic table by using the terms *periods, groups, families, representative elements,* and *transition elements.*

3. Describe the filling of electron orbitals (Aufbau process) by using the periodic table as a guide.

4. Cite the basic features of the electron configurations for representative, transition, and inner transition elements.

5. Distinguish among metals, nonmetals, metalloids, and noble gases, and locate them in the periodic table.

6. State the factors that affect atomic size; distinguish among the terms, covalent, ionic, metallic, and van der Waals radius; and describe the relationship between atomic size and location in the periodic table.

7. Use the properties of ionization energy, electron affinity, and electronegativity to evaluate the metallic/nonmetallic character of elements.

8. Relate the magnetic properties (diamagnetic or paramagnetic) of an atom or ion to its electron configuration.

9. Use periodic relationships to predict certain properties of the elements (e.g., melting point, boiling point).

10. Use the periodic table as an aid in writing chemical formulas and equations.

11. Relate certain observed physical properties of the elements (e.g., density) to atomic properties (e.g., atomic weight, atomic radius).

Some New Terms

The **actinoids (actinides)** are a series of elements ($Z = 90$ to 103) characterized by partially filled $5f$ orbitals in their atoms. All the actinoid elements are radioactive.

Covalent radius is one half the distance between the centers of two atoms that are bonded together covalently. It is the atomic radius associated with an element in its covalent compounds.

Diamagnetism refers to the repulsion by a magnetic field of a species in which all electrons are paired.

Electron affinity is the energy associated with the gain of an electron by a *gaseous* atom.

Electronegativity is a measure of the electron attracting power of an atom; metals have low electronegativities, nonmetals, high.

A **family** of elements is a numbered group from the periodic table, usually carrying a distinctive name: for example, group VIIA, the halogen family.

A **group** is a vertical column of elements in the periodic table; members of the group have similar properties.

Ionic radius is the radius of a spherical ion. It is the atomic radius associated with an element in its ionic compounds.

Ionization energy (ionization potential), *I,* is the energy required to remove the most loosely held electron from a *gaseous* atom.

Isoelectronic species have the same number of electrons (usually in the same configuration). Na^+ and Ne are isoelectronic.

The **lanthanoids (lanthanides)** are the series of elements ($Z = 58$ to 71) characterized by partially filled $4f$ orbitals in their atoms.

Metallic radius is one-half the distance between the centers of adjacent atoms in a metallic solid.

A **metalloid** is an element that may display both metallic and nonmetallic properties under the appropriate conditions (e.g., Si, Ge, As).

Metals are elements whose atoms have small numbers of electrons in the electronic shell of highest principal quantum number. Removal of an electron(s) from a metal atom occurs without great difficulty, producing a cation.

Noble gases are elements whose atoms have the electron configuration ns^2np^6 in the electronic shell of highest principal quantum number. (The noble gas, He, has the configuration, $1s^2$.)

Nonmetals are elements whose atoms tend to gain small numbers of electrons to form anions with the electron configuration of a noble gas.

Paramagnetism refers to the attraction of a magnetic field for a species containing unpaired electrons.

A **period** is a horizontal row of the periodic table. All members of a period have atoms with the same highest principal quantum number.

The **periodic law** refers to the periodic recurrence of certain physical and chemical properties when the elements are considered in terms of increasing atomic number.

The **periodic table** is an arrangement of the elements in which elements with similar physical and chemical properties are grouped together.

Representative elements are those whose atoms feature the filling of s or p orbitals of the electronic shell of highest principal quantum number.

Transition elements are those whose atoms feature the filling of d or f orbitals of an inner electronic shell.

Exercises

The periodic law

1. The element francium occurs naturally in uranium ores, but it is estimated that there is less than 30 g present in the earth's crust at any one time. Because it is extremely rare, very little is known of its properties. Estimate its density using Figure 8-1 and equation (8.1).

2. Use data from Figure 8-1 and equation (8.1) to estimate the density that will be exhibited by element 114, if and when it is discovered. Assume a mass number of 298.

3. The following melting points are given in degrees Celsius. Show that melting point is a periodic property of these elements: aluminum, 660; argon, −189; beryllium, 1278; boron, 2300; carbon, 3350; chlorine, −101; fluorine, −220; lithium, 179; magnesium, 651; neon, −249; nitrogen, −210; oxygen, −218; phosphorus, 590; silicon, 1410; sodium, 98; sulfur, 119.

The periodic table

4. There is every indication that the periodic table can be extended to elements of higher atomic numbers. What are the prospects of finding new elements within the existing periodic table at lower atomic numbers?

5. Assuming that the seventh period of elements is 32 members long, what would be the atomic number of the noble gas following radon (Rn)? of the alkali metal following francium (Fr)? What would you expect their approximate atomic weights to be?

6. With reference to the periodic table, identify
    (a) an element that is both in group IIIA and in the fifth period
    (b) an element similar to, and one unlike, sulfur
    (c) a highly reactive metal in the sixth period
    (d) the halogen element in the fifth period

(e) an element with atomic number greater than 50 that is similar chemically to the element with atomic number 18.

7. Find the several pairs of elements that are "out of order" in the periodic table in terms of increasing atomic weights and explain why it is necessary to arrange them in inverse order by atomic weight.

Periodic table and electron configurations

8. Explain why the several periods in the periodic table do not all have the same number of members.

9. Sketch a periodic table that would permit inclusion of all the members of each period into the main body of the table. How many "members" wide would the table have to be? Explain.

10. In what group of the periodic table is each of the following elements found?
   (a) $1s^2 2s^2 2p^6 3s^2 3p^6$ (b) $[Ar]3d^{10}4s^2 4p^2$
   (c) $1s^2 2s^2 2p^6 3s^1$ (d) $[Ar]3d^{10}4s^1$
   (e) $[Xe]4f^{14}5d^4 6s^2$

11. Use the periodic table as a guide to write electron configurations for (a) Ga; (b) Y; (c) Sn; (d) Ag.

12. In Figure 8-2 no designation is made about three of the elements in the seventh period. (a) What are these three elements? (b) What distinctive features would you expect for their electron configurations?

13. Based on the relationship between electron configurations and the periodic table, give the number of (a) outer shell electrons in an atom of gallium; (b) shells of electrons in an atom of tungsten; (c) elements whose atoms have six outer shell electrons; (d) unpaired electrons in an atom of arsenic; (e) transition elements in the sixth period of elements.

14. Use Figure 8-2 as a guide to indicate the number of (a) $4s$ electrons in a K atom; (b) $5p$ electrons in an I atom; (c) $3d$ electrons in an atom of Zn; (d) $2p$ electrons in an atom of S; (e) $4f$ electrons in an atom of Pb; (f) $3d$ electrons in an atom of Ni.

15. Write electron configurations of the following ions: (a) Rb^+; (b) Br^-; (c) O^{2-}; (d) Ba^{2+}; (e) Zn^{2+}; (f) Ag^+; (g) Bi^{3+}.

16. In Example 8-2 we concluded that the known element that the unknown and undiscovered element 114 would most closely resemble is Pb.
   (a) Write the electron configuration of Pb.
   (b) Propose a plausible electron configuration for element 114.

Atomic sizes

17. Explain why the sizes of atoms do not simply increase uniformly with increasing atomic number.

18. (a) Which is the smallest atom in group IIIA?
   (b) Which is the smallest of the following: Te, In, Sr, Po, Sb?

19. How would you expect the sizes of the hydrogen ion, H^+, and the hydride ion, H^-, to compare with that of the He atom? Explain.

20. The following species are isoelectronic with the noble gas krypton. Arrange them in order of increasing size and comment on the principles involved in doing so: Rb^+; Y^{3+}; Br^-; Kr; Sr^{2+}; Se^{2-}.

21. Arrange the following species in expected order of increasing size: Y; Li^+; Se; Br^-.

Ionization energies, electron affinities

22. Are there any atoms for which the second ionization energy (I_2) is smaller than the first (I_1)? Explain.

23. Although the first ionization energy (I_1) of Na is smaller than for Mg, the second ionization energy (I_2) of Na is much greater than for Mg. Why is this so?

24. The ion Na^+ and the atom Ne are isoelectronic. The ease of loss of an electron by a gaseous Ne atom is measured by I_1 and has a value of 21.6 eV. The ease of loss of an electron from a gaseous Na^+ ion is measured by I_2 for sodium and has a value of 47.3 eV. Why are these values not the same?

25. How much energy, in joules, must be absorbed to produce 3.50×10^{-5} mol Mg^{2+} ions from gaseous Mg atoms?

26. How much energy, in eV, must be absorbed to ionize completely all the third-shell electrons in a chlorine atom?

27. How much energy, in kJ, is involved when 1.00 g of chlorine, in the form of Cl atoms, is converted completely to Cl^- ions in the gaseous state?

28. Use data from the energy level diagram in Figure 7-13 to determine the ionization energy, in eV, of a hydrogen atom.

29. Use principles established in this chapter to arrange the following atoms in order of *increasing* value of the first ionization energy: Sr; Cs; S; F; As.

Electronegativities, metals and nonmetals

30. Use information from Table 8-6 and your knowledge of the periodic law to suggest plausible values of the electronegativities of the following: Se; Te; Ge; Sn.

31. Data for the transition elements are not included in Table 8-6. What range of values would you expect these elements to have? Explain.

32. Without referring to tables or figures in the text, indicate which of the atoms Bi, S, Ba, As, and Mg (a) is most metallic; (b) is most nonmetallic; (c) has the intermediate value when the five are arranged in order of increasing electronegativity.

33. Arrange the following elements in order of *decreasing* metallic character: Sc; Fe; Rb; Br; O; Ca; F; Te.

34. Table 8-6 lists approximate electronegativity values for metals, nonmetals, and metalloids. Use relevant data from tables and figures in this chapter to establish rough ranges of first ionization energies for metals, nonmetals, and metalloids. Is the first ionization energy a satisfactory criterion for metallic/nonmetallic behavior? Explain.

Magnetic properties

35. Unpaired electrons are found in only one of the following species. Indicate which is that one and explain why: F^-; Ca^{2+}; Fe^{2+}; S^{2-}.

36. Which of the following species would you expect to be diamagnetic and which paramagnetic: K^+; Cr^{3+}; Zn^{2+}; Co^{3+}; Sn^{2+}?

37. Write electron configurations consistent with the following data on number of unpaired electrons: V^{3+}, two; Cu^{2+}, one; Cr^{3+}, three.

38. Must all atoms with an odd atomic number be paramagnetic? Must all atoms with an even atomic number be diamagnetic? Explain.

Predictions based on periodic relationships

39. In 1829, Dobereiner found that when certain similar elements are arranged in groups of three the atomic weight of the middle member of the group is roughly the average of the other two. Explain why Dobereiner's method works in the first two and fails in the other three of the following cases: **(a)** the atomic weight of Na from those of Li and K; **(b)** the atomic weight of Br from those of Cl and I; **(c)** the atomic weight of Si from those of C and Ge; **(d)** the atomic weight of Sb from those of As and Bi; **(e)** the atomic weight of Ga from those of B and Tl.

40. Estimate the missing boiling point in the following series of compounds.
  (a) CH_4, $-164°C$; SiH_4, $-112°C$; GeH_4, $-90°C$; SnH_4, ?
  (b) H_2O, ?; H_2S, $-61°C$; H_2Se, $-41°C$; H_2Te, $-2°C$.
Does your estimate of the boiling point of water, based on the data given here, agree with the known value? (An explanation is presented in Chapter 11.)

41. The element gallium (eka-aluminum) was unknown in Mendeleev's time, and he predicted properties of this element much as he did for eka-silicon. Predict the following for gallium: **(a)** its density; **(b)** the formula and percent composition of its oxide. (*Hint*: Use Figure 8-1, equation (8.1), and Table 8-1.)

42. From the following formulas: boron trichloride, BCl_3; silver sulfate, Ag_2SO_4; calcium oxide, CaO; potassium nitrate, KNO_3; sodium sulfate, Na_2SO_4; predict plausible formulas for **(a)** strontium nitrate: **(b)** barium bromide; **(c)** silver oxide; **(d)** copper(I) telluride; **(e)** aluminum sulfate; **(f)** gallium oxide; **(g)** lithium nitride.

43. Each of the following formulas represents an actual compound, but one formula is inconsistent with the others given (i.e., not predictable from the others). Which is that one and what is the inconsistency? SiO_2; CO; H_2S; CCl_4; HCl; H_2O.

44. The alkali metal and alkaline earth metal elements react with hydrochloric acid [$HCl(aq)$] producing hydrogen gas and the metal chloride. The alkali metal elements and the heavier alkaline earth metals (Ca, Sr, Ba, and Ra) react with cold water to produce hydrogen gas and the metal hydroxide. Predict the probable outcome of each of the following. If no reaction occurs or if insufficient information is given with which to reach a conclusion, so state.
  (a) $Mg(s) + HCl(aq) \longrightarrow$
  (b) $Cs(s) + H_2O(l) \longrightarrow$
  (c) $Be(s) + H_2O(l) \longrightarrow$
  (d) $Na(s) + HI(aq) \longrightarrow$
  (e) $Si(s) + HCl(aq) \longrightarrow$
  (f) $Ba(s) + H_2O(l) \longrightarrow$

45. Use ideas presented in this chapter to indicate
  (a) three metals that you would expect to exhibit the photoelectric effect with visible light, and three that you would not expect to
  (b) the noble gas element that should have the highest density when in the liquid state
  (c) the approximate first ionization energy of fermium ($Z = 100$)
  (d) the approximate melting point of francium ($Z = 87$)
  (e) the approximate density of solid radium ($Z = 88$)
  (f) the approximate electronegativity of polonium ($Z = 84$)

Additional Exercises

1. Given that the density of tellurium ($Z = 52$) is 6.24 g/cm³, estimate its atomic weight using Figure 8-1 and equation (8.1).

2. Verify the statements made on page 174 concerning Mendeleev's correction of the atomic weight of indium. That is, show that if indium oxide is assumed to have the formula, InO, its atomic weight must be 76; and if In_2O_3, 113. (Recall that the oxide is 82.5% In, by mass.)

*★***3.** Studies conducted in 1880 showed that a volatile chloride of uranium had an approximate formula weight of 382 and a percent by mass of Cl of 37.34%. Pure uranium has a specific heat of 0.0276 cal g⁻¹ °C⁻¹. Use these data, together with the

statement of the law of Dulong and Petit given in the Additional Exercises of Chapter 6, to calculate the atomic weight of uranium and compare it with the value assigned by Mendeleev.

4. On the basis of the periodic table and rules for electron configurations, indicate the number of **(a)** $2p$ electrons in an atom of F; **(b)** $3s$ electrons in an atom of K; **(c)** $4d$ electrons in an atom of Se; **(d)** $4f$ electrons in an atom of Bi; **(e)** unpaired electrons in an atom of Sb; **(f)** elements in group IVA of the periodic table; **(g)** elements in the sixth period of the periodic table.

5. With reference to the periodic table, indicate **(a)** the most active nonmetal; **(b)** the transition metal with lowest atomic number; **(c)** a metalloid element whose atomic number places it exactly midway between two noble gas elements.

6. With reference to the periodic table, explain why
   (a) All nonmetals are representative elements, whereas among metals, some are representative and some are transition elements.
   (b) Metalloids are found only among the representative elements and not among the transition elements.

7. Use principles established in the text, but without reference to tables or figures, to arrange the following atoms in terms of
   (a) increasing first ionization energies: O; Rb; Br; Ca; Sc; Se; F; Cs; He.

   (b) decreasing metallic character: I; O; Cs; K; Te; F; Mg; Al.

8. Show that the trend in densities of the metals throughout group IIA listed in Table 8-8 is consistent with the atomic weights and atomic radii given.

9. Which of the following species has the greatest number of unpaired electrons: **(a)** Si; **(b)** I; **(c)** Cr^{3+}; **(d)** Br^-? Explain.

10. For the following groups of elements select the one that has the property noted.
   (a) The largest atom: H; Ar; Ag; Ba; Te; Au.
   (b) The lowest first ionization energy: B; Sr; Al; Br; Mg; Pb.
   (c) The greatest electron affinity: Na; I; Ba; Se; Cl; P.
   (d) The highest electronegativity: As; Ca; I; P; Ga; Se; Sn.
   (e) The largest number of unpaired electrons: F; N; S^{2-}; Mg^{2+}; Sc^{3+}; Ti^{3+}.

★**11.** Refer to Exercise 20 in Chapter 7 and calculate the *second* ionization energy (I_2) for the helium atom. Compare your result with the tabulated value of 54.416 eV.

★**12.** Various values are given for the radius of a hydrogen atom, including 37, 53, 120, and 210 pm. Why do you suppose there is such variability in the values given?

Self-Test Questions

For questions 1 through 6 select the single item that best completes each statement.

1. The element whose atoms have the electron configuration $[Kr]4d^{10}5s^25p^2$ **(a)** is in group IIA of the periodic table; **(b)** bears a similarity to the element Pb; **(c)** is similar to the element Se; **(d)** is a transition element.

2. An atom of As has **(a)** 5 electrons in the $4p$ subshell; **(b)** 10 electrons in the $4d$ subshell; **(c)** 6 electrons in the $3p$ subshell; **(d)** 3 electrons in the $4s$ subshell.

3. The largest of the following species is **(a)** an Ar atom; **(b)** a K^+ ion; **(c)** a Ca^{2+} ion; **(d)** a Cl^- ion.

4. The *highest* (first) ionization energy of the following elements is that of **(a)** Cs; **(b)** Cl; **(c)** I; **(d)** Li.

5. The *most* metallic of the following elements is **(a)** Mg; **(b)** Li; **(c)** K; **(d)** Ca.

6. The number of *unpaired* electrons in an atom of I is **(a)** 0; **(b)** 1; **(c)** 2; **(d)** 5.

7. For the atom $^{79}_{34}Se$ indicate the number of **(a)** protons in the nucleus; **(b)** neutrons in the nucleus; **(c)** electrons in the third principal electronic shell; **(d)** electrons in the $2s$ orbital; **(e)** $4p$ electrons; **(f)** electrons in the shell of highest principal quantum number.

8. Which of the following elements would you expect to have the highest electronegativity: Pb; Sn; Br; As; Al? Explain.

9. Give the symbol of the element **(a)** in group IVA that has the smallest atoms; **(b)** in the fifth period that has the largest atoms; **(c)** in group VIIA that has the lowest electronegativity.

10. With reference to the periodic table, briefly explain why
   (a) There are 2 elements in the first period, 8 in the third, 18 in the fifth, and 32 in the seventh.
   (b) Argon (Ar), with an atomic weight of 39.948, is placed ahead of potassium (K), with an atomic weight of 39.102.

9 Chemical Bonding I: Basic Concepts

While the atomic theory was being developed various ideas were also entertained about the *combinations* of atoms that lead to chemical compounds. In compounds, atoms are held together by forces known as chemical bonds. Electrons play a key role in chemical bonding.

In Chapter 7 we proposed that within individual atoms regions exist where the probability of finding electrons is high. A logical extension of this view is to consider that within *combinations* of atoms there also exist regions in which electron probabilities are high. That is, electrons in atoms are described by atomic orbitals and in molecules by molecular orbitals. The molecular orbital approach to chemical bonding is presented in Chapter 10. However, there are a number of simpler concepts concerning chemical bonds that we will consider first.

9-1 Importance of Electrons in Chemical Bonding

It is interesting that one of the first real clues about chemical bonding came from a group of elements that show little tendency to form chemical compounds at all. These are the noble (inert) gases, whose discovery was described in Chapter 8.

Beginning in 1916 several proposals about chemical bonding were made by two American chemists, Lewis and Langmuir, and a German, Kossel. Their basic reasoning was of the following sort. If the inert gases do not combine with other elements, perhaps there is something unique about their electron configurations that prevents their doing so. And if this is the case, perhaps atoms that do unite to form compounds experience changes in their electron configurations to make them more like inert gases. The theory developed around this model is commonly referred to as the Lewis theory. It is based on these essential propositions.

1. Electrons, especially those of the outermost (valence) electronic shell, play a fundamental role in chemical bonding.
2. In some cases chemical bonding results from the *transfer* of one or more electrons from one atom to another. This leads to the formation of positive and negative ions and a bond type known as **ionic.**
3. In other cases chemical bonding results from a mutual *sharing* of electrons between atoms. This leads to the formation of molecules having a bond type called **covalent.**

Until 1962, despite several attempts, no compounds of the inert gases had been synthesized. In fact, the strong belief in the uniqueness of the inert gas electron configurations tended to retard rather than stimulate research in this direction. Now, a considerable number of compounds of Kr, Xe, and Rn have been synthesized. These recent discoveries do not alter appreciably the applicability of the ideas presented here, although the term noble gases is now preferred to inert gases.

196

4. The transfer or sharing of electrons occurs to the extent that each atom involved acquires an especially stable electron configuration. Often this configuration is that of a noble (inert) gas, that is, involving eight outer-shell electrons, an **octet.**

Lewis Symbols.   To apply the Lewis theory, it is helpful to develop a special symbolism. The Lewis symbol of an element consists of the common chemical symbol, surrounded by a number of dots. The chemical symbol represents the kernel of the atom, consisting of the nucleus and *inner-shell* electrons. The dots represent the *outer-shell* or *valence* electrons. For example, an aluminum atom has a total of 13 electrons—2 in the first principal electronic shell, 8 in the second, and 3 in the outermost (third). The Lewis symbol, then, should consist of the chemical symbol Al with dots on *three* sides of the symbol. The formulation of this Lewis symbol is thus

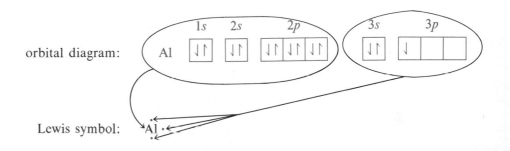

orbital diagram:

Lewis symbol:

The Lewis symbol we have written for aluminum differs from our expectation based on its electron configuration in one significant way. We have represented the three valence electrons as three widely separated dots, even though in the isolated atom two of these electrons are paired and one is unpaired. Lewis symbols lead to more accurate predictions of chemical formulas when the number of "unpaired" dots is maximized (usually to a limit of four) before dots are shown as pairs. This customary manner of writing Lewis symbols is illustrated in Figure 9-1 for the second row elements. This figure also illustrates that the number of outer-shell (valence) electrons equals the A group number.

When Lewis proposed his theory of chemical bonding, the concept of electron spin had not yet been established.

Example 9-1   Refer to Figure 9-1 and write Lewis symbols for the following groups of elements: **(a)** N, P, As, Sb, Bi; **(b)** Ca, Si, I.
(a)   These are the elements of group VA of the periodic table. Their atoms all have five valence electrons (ns^2np^3). The Lewis symbols feature five dots.

$\cdot \ddot{N} \cdot$ $\cdot \ddot{P} \cdot$ $\cdot \ddot{As} \cdot$ $\cdot \ddot{Sb} \cdot$ $\cdot \ddot{Bi} \cdot$

(b)   Ca is in group IIA, Si is in group IVA, and I is in group VIIA.

$\dot{Ca}$ $\cdot \dot{Si} \cdot$ $: \ddot{I} :$

SIMILAR EXAMPLE: Exercise 1.

LEWIS STRUCTURES.   Although Lewis's work dealt primarily with covalent bonding, we will depict both ionic and covalent bonds using his ideas. **A Lewis structure** is a

FIGURE 9-1
Lewis symbols of the
second row elements.

Group							
IA	IIA	IIIA	IVA	VA	VIA	VIIA	0
Li	Be·	·B·	·C·	·N·	:O·	:F:	:Ne:

FIGURE 9-2
Formation of
an ionic bond.

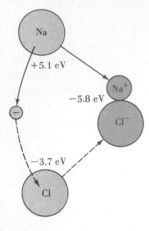

The processes depicted here have already been described in part in Chapter 8. The loss of an outer-shell electron by a sodium atom is accompanied by a decrease in size. There is essentially no change in size when an *isolated, nonbonded* chlorine atom gains an electron (recall marginal note on page 182). The energy data refer to the loss of an electron by Na ($+5.1$ eV), the gain of an electron by Cl(-3.7 eV), and the energy of interaction of the two ions (-5.8 eV). The energy requirements to produce gaseous Na from solid Na and Cl atoms from Cl_2 molecules are not described here but are discussed in Section 11-12.

combination of Lewis symbols representing the transfer or sharing of electrons in a chemical bond.

ionic bonding: $\text{Na}_\times + \cdot \ddot{\underset{\cdot\cdot}{\text{Cl}}} : \longrightarrow [\text{Na}]^+ [\times \ddot{\underset{\cdot\cdot}{\text{Cl}}} :]^-$ (9.1)

                              Lewis symbols   Lewis structure

covalent bonding: $\text{H}_\times + \cdot \ddot{\underset{\cdot\cdot}{\text{Cl}}} : \longrightarrow \text{H} \underset{\cdot\cdot}{\overset{\cdot\cdot}{\times}} \text{Cl} :$ (9.2)

                              Lewis symbols   Lewis structure

In these two examples the electrons from one atom are denoted as $\times$ and from the other atom, as $\cdot$. However, it is impossible to distinguish among the electrons in bonded atoms. In all subsequent Lewis structures we will use only the dot symbol $\cdot$. Lewis structures are discussed more fully in the sections that follow.

9-2 Ionic Bonding

Figure 9-2 suggests the formation of a pair of ions, Na^+ and Cl^-, from the free, isolated atoms. Energy is required to remove the outer-shell ($3s^1$) electron from an Na atom. This quantity is simply the first ionization energy—$I_1 = +5.1$ eV. Energy is released when a Cl atom absorbs an outer-shell electron. This quantity is measured by the electron affinity—$EA = -3.7$ eV. Once the ions have formed, they attract each other and move into close proximity. The energy of this interaction is -5.8 eV. The overall process in Figure 9-2 is feasible energetically. The net energy change is $5.1 - 3.7 - 5.8 = -4.4$ eV. (The energy requirement to convert solid to gaseous sodium and to dissociate Cl_2 molecules into Cl atoms has not been included in this analysis. This requirement is considerably less than 4.4 eV, however.)

The additional energy
associated with the clus-
tering of these ion pairs
into a crystal is about
-2.4 eV per ion pair.

The ion pair $(Na^+)(Cl^-)$ pictured in Figure 9-2 exerts attractive forces on additional ion pairs. The result of such attractive forces is the clustering of *large* numbers of Na^+ and Cl^- ions into a solid crystal. Thus, the formation of an ionic crystal is an integral part of the total process of ionic bonding. A portion of an ionic crystal of NaCl is pictured in Figure 9-3. Ionic crystal structures and the energetics of ionic compound formation are discussed further in Chapter 11. For the present, let us simply note the following facts about ionic bonds.

1. An ionic bond results from the transfer of electrons between a *metal* and a *nonmetal* atom. In this transfer the metal atom becomes a positively charged ion (cation) and the nonmetal, a negatively charged ion (anion).
2. The nonmetal atom gains a sufficient number of electrons to produce an anion with a noble gas electron configuration. Several different types of electron configurations are found for metal ions; some are listed in Table 9-1.
3. Except in the gaseous state, ionic compounds are not composed of simple ion pairs or small clusters of ions. In the solid state each ion surrounds itself with ions of the opposite charge, producing an orderly array called a crystal.

FIGURE 9-3
Formation of an ionic crystal.

Each Na^+ ion (small sphere) is surrounded by six Cl^- ions (large spheres); in turn, each Cl^- is surrounded by six Na^+. An enormous number of ions cluster together into a crystalline solid.

To distinguish clearly between ionic and covalent compounds, it is essential that Lewis structures for ionic compounds show that positive and negative ions are present, as illustrated here.

TABLE 9-1
Electron configurations for metal ions

"Octet"		"18"		"18 + 2"	"Various"	
Na^+	Mg^{2+}	Cu^+	Zn^{2+}	In^+	Cr^{2+}:	$[Ar]3d^4$
K^+	Ca^{2+}	Ag^+	Cd^{2+}	Tl^+	Cr^{3+}:	$[Ar]3d^3$
Rb^+	Sr^{2+}	Au^+	Hg^{2+}	Sn^{2+}	Mn^{2+}:	$[Ar]3d^5$
Cs^+	Ba^{2+}		Ga^{3+}	Pb^{2+}	Mn^{3+}:	$[Ar]3d^4$
Fr^+	Ra^{2+}		In^{3+}	Sb^{3+}	Fe^{2+}:	$[Ar]3d^6$
	Al^{3+}		Tl^{3+}	Bi^{3+}	Fe^{3+}:	$[Ar]3d^5$
	Sc^{3+}				Co^{2+}:	$[Ar]3d^7$
	Y^{3+}				Co^{3+}:	$[Ar]3d^6$
	La^{3+}				Ni^{2+}:	$[Ar]3d^8$
					Ni^{3+}:	$[Ar]3d^7$

Li^+ and Be^{2+} also have a noble gas electron configuration but not an octet of outer-shell electrons. Theirs is the configuration of helium: $1s^2$.

4. A formula unit of an ionic compound is the smallest collection of ions that is electrically neutral. The formula unit is obtained automatically when the Lewis structure is written.

In Table 9-1, the octet configuration is that of a noble gas (ns^2np^6). In the configuration labeled "18" all outer-shell electrons of the atom are removed. In the ion produced the new outer shell has 18 electrons, all paired, in the configuration $ns^2np^6nd^{10}$. In some cases all but the two s electrons of the outermost shell are removed from an atom. This produces ions with the "18 + 2" configuration. Since the filling of electronic shells to a total of 18 occurs in the transition series, the "18 + 2" configuration is limited to a few "post-transition" elements. In metal ion formation among the transition elements, the outer-shell s electron(s) and some number of inner-shell d electrons are removed. This produces a variety of possible electron configurations; several are listed in Table 9-1.

Example 9-2 Write Lewis structures for the following ionic compounds: **(a)** BaO; **(b)** $MgCl_2$; **(c)** potassium sulfide.

(a) Write the Lewis symbol and determine the number of electrons to be lost or gained by each atom in acquiring a noble gas electron configuration. The barium atom must lose two electrons and the oxygen atom must gain two.

$$Ba \quad + \quad \ddot{\underset{..}{O}}{:} \quad \longrightarrow \quad [Ba]^{2+}[\ddot{\underset{..}{:O}}{:}]^{2-}$$

Lewis structure

(b) A Cl atom can accept but a single electron, whereas a Mg atom must lose two electrons. Two Cl atoms are required for each Mg atom.

$$Mg \quad + \quad \begin{matrix} \ddot{\underset{..}{Cl}}{:} \\ \\ \ddot{\underset{..}{Cl}}{:} \end{matrix} \quad \longrightarrow \quad [\ddot{\underset{..}{:Cl}}{:}]^- [Mg]^{2+} [\ddot{\underset{..}{:Cl}}{:}]^-$$

Lewis structure

(c) We are not given the formula of potassium sulfide, but as noted earlier, the formula follows directly from the principles of writing Lewis structures.

$$\begin{matrix} K \\ \\ K \end{matrix} \quad + \quad \ddot{\underset{..}{:S}}{:} \quad \longrightarrow \quad [K]^+[\ddot{\underset{..}{:S}}{:}]^{2-}[K]^+$$

Lewis structure

SIMILAR EXAMPLES: Exercises 3, 4.

9-3 Covalent Bonding

In the Lewis structures (9.1) and (9.2) for NaCl and HCl, the Cl atom acquires the electron configuration of a noble gas atom. The tendency for the Cl atom to gain an electron is equally strong in either case, but from what atom, Na or H, can an electron be extracted most readily? To begin, we note that neither an Na atom nor an H atom will give up an electron freely. However, the energy necessary to extract the valence electron from Na (the first ionization energy) is much smaller than from H—5.14 eV compared to 13.6 eV. *Sodium is much more metallic than hydrogen.* In fact, hydrogen is a *nonmetal* under normal conditions; it does not give up an electron to another nonmetal. Bonding between a hydrogen and a chlorine atom involves the *sharing* of electrons. This leads to a bond type known as **covalent.**

The notion that a covalent bond involves the sharing of electrons between atoms needs to be explored much more fully. An in-depth treatment of the covalent bond is presented in Chapter 10. For the present let us see the kind of understanding of covalent bonding that is possible using the Lewis theory. First, we rewrite the Lewis structure for HCl to emphasize how the sharing of electrons and the attainment of noble gas electron configurations can be represented.

$$\text{H} \quad \text{Cl} \qquad\qquad (9.3)$$

The dashed circles are meant to represent the outermost electronic shells of the bonded atoms. The effective configuration of the outermost shell is established by counting the number of electrons lying on each circle. For the H atom this is two, corresponding to the configuration of He. For the Cl atom it is eight, corresponding to the configuration of the Ar atom. Note that the two electrons between H and Cl (:) have been counted twice, a consequence of the fact that these are the shared electrons. Some additional simple structures are shown in Figure 9-4.

We have already established and used the fact that several of the gaseous elements exist not as collections of isolated atoms but in *molecular* form. For example, we have noted these formulas: H_2, Cl_2, N_2, and O_2. Let us try now to establish these formulas through Lewis structures. The situation with H_2 and Cl_2 is relatively simple. Each atom acquires a noble gas electron configuration by sharing *one* pair of electrons with the atom to which it is bonded. The sharing of a single pair of electrons results in a **single** covalent bond, often represented by a dash sign (—). Electron pairs not involved in bond formation, **nonbonding pairs,** are sometimes also represented by dashes.

FIGURE 9-4
Some examples of covalent bonds.

Elements	H·	·Ö:	:Cl·	·N̈·
Compounds	H:Cl:	H:Ö: H	:Cl:Ö: :Cl:	H:N̈:H H
Names	hydrogen chloride	hydrogen oxide (water)	chlorine oxide	hydrogen nitride (ammonia)
Molecules	HCl	H_2O	Cl_2O	NH_3
Molecular weights	36.46	18.02	86.91	17.03

$$H\cdot \; + \; \cdot H \longrightarrow H:H \quad \text{or} \quad H\text{—}H \tag{9.4}$$

$$:\!\ddot{C}\!l\cdot \; + \; \cdot\ddot{C}\!l\!: \; \longrightarrow \; :\!\ddot{C}\!l\!:\!\ddot{C}\!l\!: \quad \text{or} \quad :\!\ddot{C}\!l\text{—}\ddot{C}\!l\!: \quad \text{or} \quad |\overline{C}\!l\text{—}\overline{C}\!l| \tag{9.5}$$

MULTIPLE COVALENT BONDS. If we attempt to extend what we have just written for H_2 and Cl_2 to the molecule N_2, this is what we obtain.

$$\cdot\ddot{N}\cdot \; + \; \cdot\ddot{N}\cdot \longrightarrow \cdot\ddot{N}:\ddot{N}\cdot \quad \text{(incorrect)}$$

This structure violates the octet rule. The N atoms appear to have only six outer-shell electrons. The situation can be improved if we consider that more than a single pair of electrons may be shared between atoms. The Lewis structure we write for N_2 is

$$\overset{\frown}{\underset{\smile}{:N ::: N:}} \quad \text{or} \quad :N\!\equiv\!N: \quad \text{or} \quad |N\!\equiv\!N| \tag{9.6}$$

The sharing of *three* pairs of electrons between two atoms, as in the N_2 molecule, is referred to as a **triple** covalent bond ($\equiv$).

In following this approach for O_2, we find a need to represent the sharing of *two* pairs of electrons between the O atoms. This sharing leads to a double covalent bond ($=$).

$$\overset{\frown}{\underset{\smile}{:O :: O:}} \quad \text{or} \quad :\ddot{O}\!=\!\ddot{O}: \quad \text{or} \quad |\overline{O}\!=\!\overline{O}| \quad \text{(incorrect)} \tag{9.7}$$

But why have we labeled the structures in (9.7) "incorrect"? We have done so because these structures fail to account for certain *experimentally determined* properties of O_2. We will return to this and other "failures" of the octet rule shortly. For now we simply make this point: *The fact that a plausible Lewis structure can be written for a species is not proof that this structure is the true electronic structure.* Experimental verification is always required.

9-4 Covalent Lewis Structures—Some Examples

Despite the warning with which we closed the preceding section, the subject of Lewis structures is worth pursuing still further. In considering some specific examples in this section, the following ideas will prove useful.

1. *All* the valence (outer-shell) electrons of the atoms in a Lewis structure must be accounted for.

2. *Usually,* each atom in a Lewis structure acquires an electron configuration with an outer-shell octet. [A few, notably hydrogen, are limited to an outer-shell duet (two electrons).]

3. *Usually,* all the electrons in a Lewis structure are paired.

4. *Usually,* both atoms in a bonded pair contribute equal numbers of electrons to the covalent bond, but sometimes both electrons in a bonded pair are derived from a single atom. (Such a bond is referred to as a coordinate covalent bond.)

5. *Sometimes* it is necessary to represent double or triple covalent bonds in a Lewis structure.

6. *Sometimes* it is impossible to draw a single Lewis structure that is consistent with all the available data. In these instances the true structure can only be represented as a *composite* or *hybrid* of two or more plausible structures. This situation, called resonance, is discussed in Section 9-6.

Other useful ideas in writing a Lewis structure are to

1. Start with a skeleton structure (a representation of the order in which atoms are bonded together).
2. Add up the total number of valence electrons and shift them around as necessary in an attempt to satisfy the requirements listed above.
3. Employ the concept of formal charge (described on page 203).

Example 9-3   Write a plausible Lewis structure for the molecule hydrazine, N_2H_4. Several skeleton structures for N_2H_4 are shown. Which is correct?

N—N—H—H—H—H   (a)

H—H—N—N—H—H   (b)

$$\begin{array}{cc} \text{H} & \text{H} \\ | & | \\ \text{H—N—N—H} \end{array}$$ (c)

There are times when the choice of skeleton structure is difficult, requiring perhaps that actual experimental evidence be provided. In the case under consideration we can reject structures (a) and (b) for this simple reason: If an H atom were to be covalently bonded to two other atoms simultaneously, this would place four electrons in the outer shell of the H atom. This is more electrons than the first shell can accommodate and certainly does not lead to the configuration of a noble gas. Or, to express this thought in another way, *a hydrogen atom can form only one single covalent bond.*

Our next step is to see how the stated requirements can be fulfilled by bringing together Lewis symbols for the atoms shown in structure (c).

$$\text{H} \cdot \rightarrow \cdot \overset{..}{\underset{..}{N}} \cdot \rightarrow \quad \leftarrow \cdot \overset{..}{\underset{..}{N}} \cdot \leftarrow \cdot \text{H} \longrightarrow \text{H} : \overset{H}{\underset{..}{N}} : \overset{H}{\underset{..}{N}} : \text{H} \quad \text{or} \quad \overset{H\ \ H}{\underset{..\ \ ..}{\text{H—N—N—H}}} \quad (9.8)$$

Structure (9.8) meets all the criteria for a plausible Lewis structure.

SIMILAR EXAMPLES: Exercises 8, 9, 11.

Example 9-4   Write a plausible Lewis structure for the molecule hydrogen cyanide, HCN.

When we attempt to assemble the Lewis symbols of H, C, and N into a Lewis structure, we encounter two difficulties: (1) neither the C nor the N atom acquires an outer-shell octet of electrons, and (2) two electrons each of the N and C atoms are left unpaired.

$$\text{H} \cdot \rightarrow \cdot \overset{..}{C} \cdot \leftarrow \cdot \overset{..}{N} : \longrightarrow \text{H} : \overset{..}{C} : \overset{..}{N} : \quad \text{(incorrect)}$$

Both difficulties can be overcome if the electrons designated below are shifted into the region between the C and N atoms, producing a triple covalent bond.

$$\text{H} : \overset{..}{C} : \overset{..}{N} : \longrightarrow \text{H} : C : : : N : \quad \text{or} \quad \text{H—C} \equiv \text{N} : \quad (9.9)$$

SIMILAR EXAMPLE: Exercise 10.

Example 9-5   Write a plausible Lewis structure for thionyl fluoride, SOF_2. (Experimental evidence indicates that the O and F atoms are bonded directly to the S atom.)

To establish the skeleton structure of a molecule, we generally need to identify the central atom—the atom to which the others are bonded. In the present case we eliminate the F atoms because, having seven outer-shell electrons, they normally form only one single covalent bond apiece. But the choice between S and O is not so easy. The outer-shell electron configurations of S and O are similar (both elements are in group VIA of the periodic table). It is for this reason that we need to have additional *experimental* information of the type given in the statement of the problem. *Sulfur is the central atom.*

To achieve a satisfactory Lewis structure, we must

In the margin:
Once formed, a coordinate covalent bond is no different from an ordinary covalent bond. Usually, no attempt is made to differentiate between them in a Lewis structure. If it is desired to do so, the usual dashed symbol for a bond can be replaced by an arrow pointing from the atom donating the electrons to the one receiving them.

1. Pair up two dots on the O atom.
2. Indicate a bond between the S and O atom in which *both electrons are donated by the S atom*—a **coordinate covalent bond.**

$$:\!\ddot{F}\!\cdot \quad \cdot\ddot{S}: \quad \overset{\cdot\cdot}{\underset{\cdot\cdot}{O}}: \longrightarrow :\!\ddot{F}\!:\!\ddot{S}\!:\!\ddot{O}: \text{ or } :\!\ddot{F}\!-\!\ddot{S}\!-\!\ddot{O}: \text{ or } :\!\ddot{F}\!-\!\ddot{S}\!\rightarrow\!\ddot{O}: \qquad (9.10)$$

$$\underset{\uparrow}{:\!\ddot{F}\!:} \qquad\qquad :\!\ddot{F}\!: \qquad\qquad :\!\ddot{F}\!:$$

SIMILAR EXAMPLE: Exercise 16.

THE CONCEPT OF FORMAL CHARGE. In Example 9-5 we had to use experimental evidence to establish the correctness of a skeleton structure. However, there is a method of "electron bookkeeping" that also can help us to make intelligent choices. In Example 9-4 all the ordinary rules for Lewis structures would have been met equally well by either of the following.

(a) (b)

$$H\!-\!C\!\equiv\!N: \qquad H\!-\!N\!\equiv\!C: \qquad\qquad (9.11)$$

Suppose that in addition to the counting scheme that lets us establish that atoms acquire outer-shell octets, we count electron dots in Lewis structures in the following way: Count all *nonbonding* electrons as belonging entirely to the atom in which they are found. Count *bonding* electrons by dividing them equally between the bonded atoms. Figure 9-5 illustrates how this counting procedure is applied to structures (a)

FIGURE 9-5
Selecting a plausible Lewis structure—the concept of formal charge illustrated.

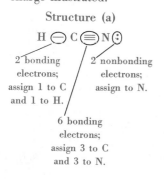

Structure (a)

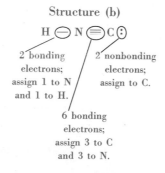

Structure (b)

Structure (a) is the more plausible because there are no formal charges on any of the atoms represented in the structure.

	Valence electrons	Electrons assigned in (a)	Formal charge
H	1	1	0
C	4	4	0
N	5	5	0

	Valence electrons	Electrons assigned in (b)	Formal charge
H	1	1	0
C	4	5	−1
N	5	4	+1

and (b) in (9.11). After having counted electrons in the manner indicated, determine whether any atoms in the Lewis structure have a **formal charge.**

The formal charge is the number of outer-shell (valence) electrons in an isolated atom minus the number of electrons assigned to that atom in a Lewis structure.

The rule that is used to establish structure (a) in (9.11) as the more plausible is this.

A Lewis structure in which there are no formal charges (i.e., where all formal charges are zero) is more plausible than one where formal charges are found.

A reminder about formal charges is necessary. Even if we choose to denote formal charges, as in structure (b),

$$H-\overset{+}{N}\equiv\overset{-}{C}: \tag{9.12}$$

individual atoms within a covalent molecule do not carry actual net charges.

Example 9-6 Show that, even in the absence of the experimental data given in Example 9-5, structure (1) is more plausible than structure (2).

$$:\overset{..}{F}-\overset{+}{\underset{..}{S}}-\overset{-}{\underset{..}{O}}: \qquad :\overset{..}{F}-\overset{+}{\underset{..}{O}}-\overset{-}{\underset{..}{S}}: \tag{9.13}$$

$$\qquad :\overset{..}{\underset{..}{F}}: \qquad\qquad :\overset{..}{\underset{..}{F}}:$$

$$\quad (1) \qquad\qquad (2)$$

In both structures (1) and (2), the F atoms would be assigned seven electrons and would have formal charges of $7 - 7 = 0$. In structure (1) the S atom is assigned five electrons, corresponding to an apparent loss of one electron from its normal complement of six valence electrons and a formal charge of $+1$. The formal charge of the O atom in structure (1) is $6 - 7 = -1$.

In structure (2) the S atom is assigned seven electrons and would have a formal charge of $6 - 7 = -1$. In turn, the O atom would have a formal charge of $6 - 5 = +1$.

Our choice between structures (1) and (2) cannot be made on the simple basis of the one that has no formal charges; both structures do. However, recall the discussion of atomic properties in Chapter 8. Which atom, S or O, has the greater electron affinity; the greater electronegativity; the more nonmetallic character? The answer to all these questions is oxygen. We should choose the Lewis structure in which O, not S, *appears* to gain an electron and acquire a formal charge of -1. Structure (1) is the more plausible of the two.

SIMILAR EXAMPLES: Exercises 13, 14.

Example 9-6 suggests a second rule on formal charges.

In choosing among alternatives having similar distributions of formal charges, the most plausible Lewis structure is that in which negative formal charges are placed on the more electronegative atoms.

9-5 Covalent Bonding in Polyatomic Ions

As we learned in Section 3-4, a polyatomic ion consists of two or more atoms. The forces that operate *within* such ions to hold atoms together are covalent bonds. We are now in a position to represent these bonds through Lewis structures. Consider, for example, the simple hydroxide ion, OH^-. Its Lewis structure must involve *eight*

The "extra" electron is shown in color in structure (9.14), but of course there is no way of distinguishing among the electrons in a species.

electrons: six from O, one from H, and one additional electron to account for the net charge of -1.

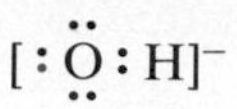

$$[:\overset{..}{\underset{..}{O}}:H]^-$$ (9.14)

Where does the OH^- ion gain its electron? Perhaps from a sodium atom that loses an electron to become a sodium ion. Thus, the *ionic* compound, sodium hydroxide, consists of simple Na^+ ions and the *polyatomic* anions, OH^-. Within the OH^- ions the O and H atoms are joined by single covalent bonds.

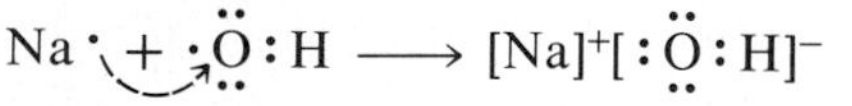

$$Na + \overset{..}{\underset{..}{O}}:H \longrightarrow [Na]^+[:\overset{..}{\underset{..}{O}}:H]^-$$ (9.15)

Example 9-7 Write a plausible Lewis structure for the chlorate ion, ClO_3^-.
Three points need to be established in writing this structure.

1. A number of different skeleton structures could be written for this ion, but it is generally true for oxyanions (polyatomic anions containing oxygen) that the O atoms are all bonded directly to the central nonmetal atom. The expected arrangement is

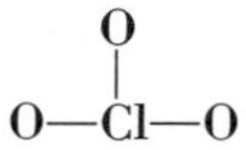

$$\begin{array}{c} O \\ | \\ O-Cl-O \end{array}$$

2. The number of electrons available for the structure is 7 (from Cl) $+3 \times 6$ (from O) $+1$ (to account for the net charge) $= 26$.
3. The Lewis symbols of two of the O atoms must be altered to pair up all the electron dots. The concept of coordinate covalency must also be used.

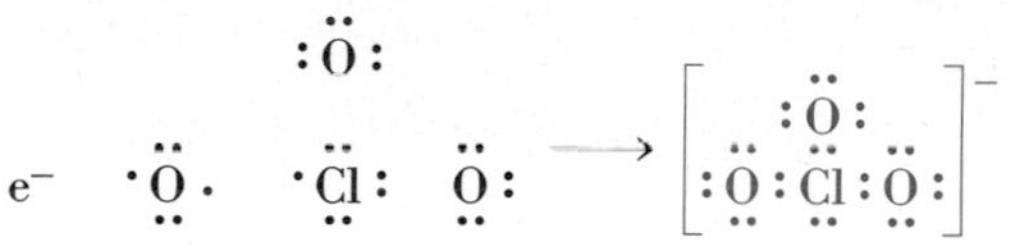

$$e^- \quad \cdot\overset{..}{\underset{..}{O}}\cdot \quad \cdot\overset{..}{\underset{..}{Cl}}: \quad \overset{..}{\underset{..}{O}}: \quad \overset{\overset{\textstyle :\overset{..}{O}:}{}}{\longrightarrow} \left[\begin{array}{c} :\overset{..}{O}: \\ :\overset{..}{\underset{..}{O}}:\overset{..}{\underset{..}{Cl}}:\overset{..}{\underset{..}{O}}: \end{array}\right]^-$$ (9.16)

SIMILAR EXAMPLE: Exercise 16.

We discover still another feature of the concept of formal charge when it is applied to the chlorate ion of Example 9-7. Each O atom has a formal charge of -1 and the Cl atom, $+2$. The sum of these formal charges is $-1-1-1+2 = -1$.

Where formal charges are assigned in a Lewis structure, these formal charges must total to zero for a neutral molecule, or to the net charge for a polyatomic ion.

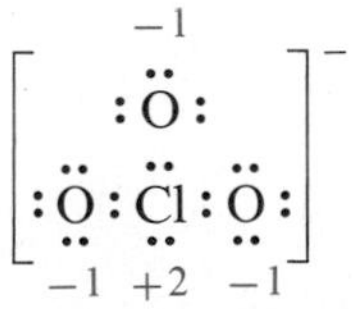

$$\left[\begin{array}{c} \overset{-1}{:\overset{..}{O}:} \\ \underset{-1\ +2\ -1}{:\overset{..}{\underset{..}{O}}:\overset{..}{\underset{..}{Cl}}:\overset{..}{\underset{..}{O}}:} \end{array}\right]^-$$ (9.17)

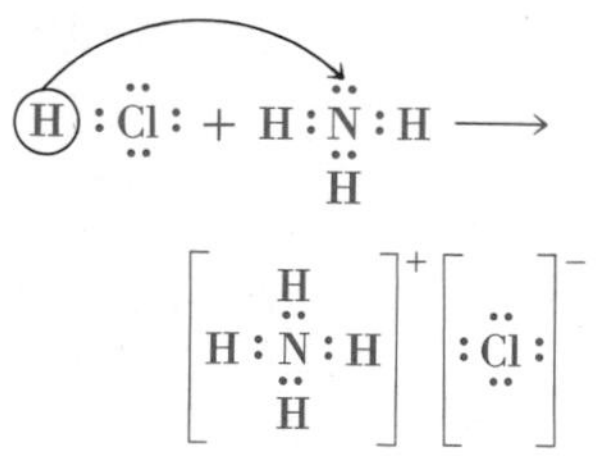

Figure 9-6 illustrates bonding in the polyatomic *cation* NH_4^+ and the formation of the ionic compound ammonium chloride by the reaction of hydrogen chloride and ammonia.

FIGURE 9-6
Formation of the ammonium ion, NH_4^+.

$$\overset{\frown}{(H)}:\overset{..}{\underset{..}{Cl}}: + H:\overset{..}{\underset{H}{N}}:H \longrightarrow$$

$$\left[\begin{array}{c} H \\ H:\overset{..}{\underset{H}{N}}:H \end{array}\right]^+ \left[:\overset{..}{\underset{..}{Cl}}:\right]^-$$

9-6 Resonance

We seem to be reaching the point of saying that in writing the Lewis structure of a species, within limits, "anything goes." In general, we should use only the valence electrons of the bonded atoms and adhere to as many of the rules as possible. That is, we should strive for electrons in pairs, stable octets for all bonded atoms, and a minimum of formal charges.

Let us try out this notion in writing a structure for sulfur dioxide, SO_2. The total number of dots that must appear in the Lewis structure is as follows: 6 (from S) + 2 × 6 (from O) = 18. Six attempts are illustrated in Figure 9-7.

The first four structures seem unsatisfactory because they fail to obey one or more of the rules. The structures obtained in trials 5 and 6 look equally plausible. Which one is correct? Actually, neither one is. Both structures suggest that one sulfur-to-oxygen bond is single and the other double. The properties of the SO_2 molecule determined by experiment indicate, however, that the two bonds are the same. How shall we represent this fact? One method is suggested through structure (9.18).

$$:\ddot{O}=\ddot{S}-\ddot{O}: \quad \longleftrightarrow \quad :\ddot{O}-\ddot{S}=\ddot{O}: \qquad (9.18)$$

The phenomenon we have been describing is called **resonance.** This is a situation in which more than one plausible structure can be written for a species and in which the true structure cannot be written at all. The true structure is considered to be a *hybrid* of the different plausible structures.

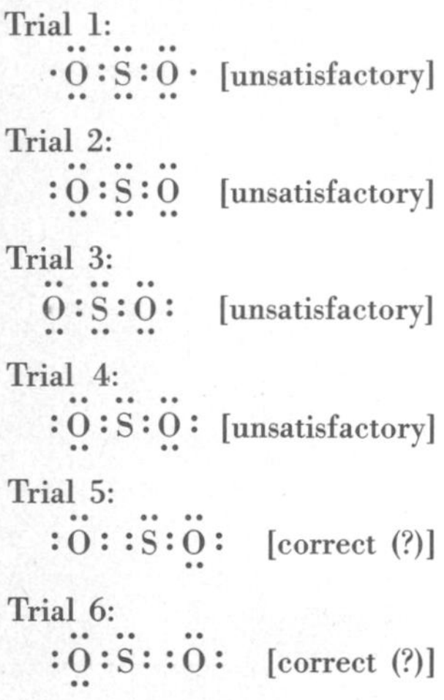

Example 9-8 Write the Lewis structure for the nitrate ion, NO_3^-. The O atoms are all bonded *in the same way* to the central N atom.

The structure we are seeking must make use of 24 electrons (23 valence electrons, plus the extra electron that conveys the charge of −1).

trial 1: $\left[\begin{array}{c} :\ddot{O}: \\ :\ddot{O}:\ddot{N}:\ddot{O}: \end{array} \right]^-$ (incorrect)

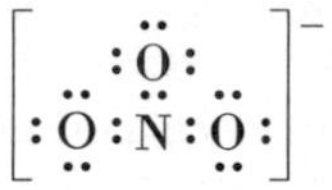

The difficulty with trial 1 is that the N atom appears to have only six outer-shell electrons.

trial 2: $\left[\begin{array}{c} :\ddot{O}: \\ :\ddot{O}:\ddot{N}::\ddot{O}: \end{array} \right]^-$

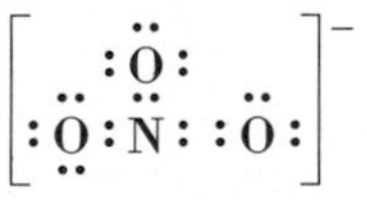

Trial 2 certainly produces a plausible structure, but it does not conform to the statement that all the nitrogen-to-oxygen bonds are equivalent. No single structure can be drawn to do this. Three equivalent structures can be written, and the true structure is a resonance hybrid of these.

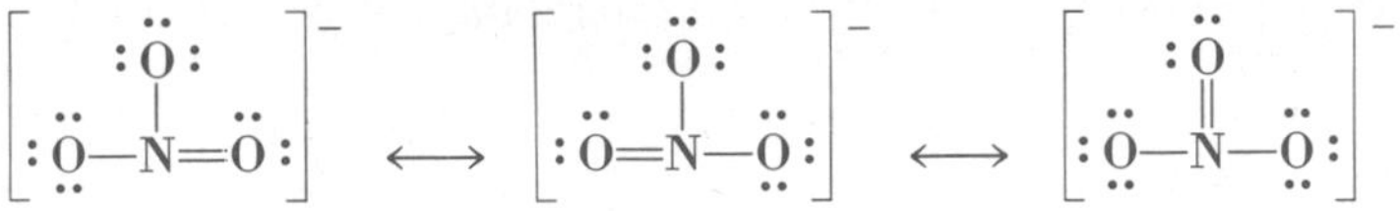

$$\left[\begin{array}{c} :\ddot{O}: \\ | \\ :\ddot{O}-N=\ddot{O}: \end{array} \right]^- \longleftrightarrow \left[\begin{array}{c} :\ddot{O}: \\ | \\ :\ddot{O}=N-\ddot{O}: \end{array} \right]^- \longleftrightarrow \left[\begin{array}{c} :\ddot{O} \\ \| \\ :\ddot{O}-N-\ddot{O}: \end{array} \right]^- \qquad (9.19)$$

SIMILAR EXAMPLES: Exercises 18, 19, 20.

9-7 Failure of the Octet Rule

ODD-ELECTRON SPECIES. If the total number of valence electrons in a Lewis structure is an *odd* number, two immediate conclusions follow about the structure. There must be

1. At least one unpaired electron somewhere.
2. At least one atom lacking a completed octet of electrons.

A mundane analogy that has often been used is to consider a mule to be a resonance hybrid of a horse and a donkey. The mule has some of the qualities of both, but it is a very distinctive animal. It is neither a horse part of the time and a donkey the rest, nor is it half-horse and half-donkey.

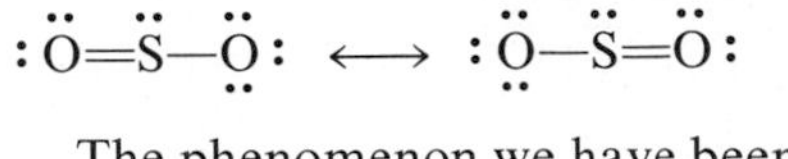

FIGURE 9-7
Attempts at writing a Lewis structure for SO_2.

Trial 1:
$\cdot\ddot{O}:\ddot{S}:\ddot{O}\cdot$ [unsatisfactory]

Trial 2:
$:\ddot{O}:\ddot{S}:\ddot{O}$ [unsatisfactory]

Trial 3:
$\ddot{O}:\ddot{S}:\ddot{O}:$ [unsatisfactory]

Trial 4:
$:\ddot{O}:\ddot{S}:\ddot{O}:$ [unsatisfactory]

Trial 5:
$:\ddot{O}::\ddot{S}:\ddot{O}:$ [correct (?)]

Trial 6:
$:\ddot{O}:\ddot{S}::\ddot{O}:$ [correct (?)]

Take the case of the NO_2 molecule. The total number of valence electrons is 17. The Lewis structure must display the features cited above. Actually, there are two plausible structures and the true structure is a resonance hybrid of the two.

$$:\ddot{O}-N=\ddot{O}: \longleftrightarrow :\ddot{O}=N-\ddot{O}: \qquad (9.20)$$

Because of the presence of an unpaired electron, an odd-electron species must be paramagnetic. NO_2 is paramagnetic, and so is NO. Molecules with an *even* number of electrons are expected to have all electrons paired and to be diamagnetic. Yet the molecule O_2, with 12 valence electrons, is *paramagnetic. The O_2 molecule must have unpaired electrons.* This is why structure (9.7), which seemed to meet all the criteria of an acceptable Lewis structure for O_2, was said to be incorrect. It contained no unpaired electrons. The measured distance between the O atoms in the O_2 molecule is less than that expected for a single covalent bond; the bond has some multiple-bond character. No single Lewis structure can be written to represent O_2. The best representation possible is a resonance hybrid with contributions from the following.

$$:\ddot{O}:\ddot{O}: \longleftrightarrow :\ddot{O}::\ddot{O}: \longleftrightarrow :\dot{O}:::\dot{O}: \qquad (9.21)$$

The seemingly simple oxygen molecule, O_2, is one of the most difficult to describe in terms of bonding theory. We make another, more successful, attempt in Section 10-4.

INCOMPLETE OCTETS.   We have encountered Lewis structures in which one or more atoms has not had an outer-shell octet of electrons and have used this as a basis for rejecting a structure. But occasionally, situations arise where such structures appear, in fact, to be correct. This is the case with the molecule BF_3. (For emphasis the valence electrons of the B atom are shown in color.)

$$:\ddot{F}:B:\ddot{F}: \\ :\ddot{F}: \qquad (9.22)$$

It might be reasoned, though, that the B atom in BF_3 will acquire an extra pair of electrons to complete its octet if at all possible. Such is the result when BF_3 and NH_3 react to produce the compound $H_3N \cdot BF_3$.

The lack of an octet of electrons around the B atom can be partially resolved if one considers the possibility of structures such as the following contributing to a resonance hybrid.

$$:\ddot{F}=B-\ddot{F}: \\ :\ddot{F}: \\ \uparrow \\ \downarrow \\ :\ddot{F}-B-\ddot{F}: \\ F: \\ \uparrow \\ \downarrow \\ :\ddot{F}-B=\ddot{F}: \\ :\ddot{F}:$$

There does, in fact, appear to be some multiple bond character to the B—F bonds in BF_3.

$$\begin{array}{ccc} & :\ddot{F}: & H \\ :\ddot{F}:B & :N:H & \text{or} \\ & :\ddot{F}: & H \end{array} \qquad \begin{array}{ccc} :\ddot{F}: & H \\ :\ddot{F}-B-N-H \\ :\ddot{F}: & H \end{array} \qquad (9.23)$$

The N atom donates both electrons to the boron–nitrogen bond; the bond is coordinate covalent.

THE EXPANDED OCTET.   Phosphorus forms two chlorides, PCl_3 and PCl_5. Covalent bonding in PCl_3 fulfills the basic criterion of Lewis structures—all atoms acquire an outer-shell octet. In PCl_5, because the five Cl atoms are bonded directly to the central P atom, ten electrons are found in the outer shell of the P atom. The "octet" has been expanded to ten electrons. In the molecule SF_6 it is expanded to 12. The situation may be depicted as follows.

$$(9.24)$$

octet   expanded octet   expanded octet

In Section 10-1 we shall see that when an atom acquires an octet of electrons the *s* and *p* subshells of the outermost electronic shell are filled. When the octet is expanded to 10 or 12 electrons, additional orbitals must be involved. The energy difference between the 2*p* and 3*s* sublevels is too great to provide this additional bonding possibility to the nonmetals in the second row of the periodic table. Once the *d* subshell becomes available to participate in bonding, expanded octets become possible. Thus, the phenomena described here are encountered with nonmetals in the third and higher periods, beginning with phosphorus.

At times, by combining the concepts of resonance and expanded octets, we can write Lewis structures that correspond even more closely to experimental observations. For example, if we attempt to write the Lewis structure of sulfate ion, SO_4^{2-}, *without* using expanded octets, we obtain the structure

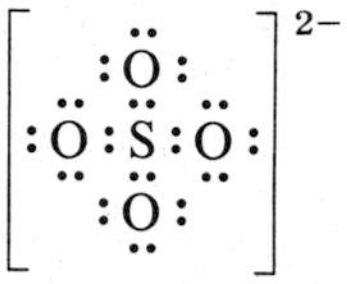

$$\left[\begin{array}{c} :\ddot{O}: \\ :\ddot{O}:S:\ddot{O}: \\ :\ddot{O}: \end{array} \right]^{2-} \tag{9.25}$$

Actually, the experimentally determined sulfur-to-oxygen bond distances in the sulfate ion appear to correspond to double bonds. This fact suggests that even the representation of Figure 9-8 has shortcomings. The Lewis theory of bonding is useful, but it is not perfect.

All the atoms in structure (9.25) carry a formal charge: -1 for each O atom and $+2$ for the S atom. We can write a structure with fewer formal charges by employing some sulfur-to-oxygen double bonds. However, since the sulfur-to-oxygen bonds are all equivalent, it is necessary to represent the resonance hybrid of the *six* Lewis structures shown in Figure 9-8. (Can you show that in each of these structures there is no formal charge on the S atom, and only *two* of the O atoms carry formal charges of -1?)

Measured bond distances in the sulfate ion indicate that in fact there is multiple-bond character to all the sulfur-to-oxygen bonds. This gives us still another reason to prefer the representation of Figure 9-8 over that of structure (9.25). Each structure in Figure 9-8 has two single and two double bonds, suggesting an average of a "$1\frac{1}{2}$" sulfur-to-oxygen bond.

The Lewis structure for SOF_2 presented in (9.10) used only single covalent bonds. Actually, the sulfur-to-oxygen bond has some double-bond character. The concept of the expanded octet allows us to write the following structure.

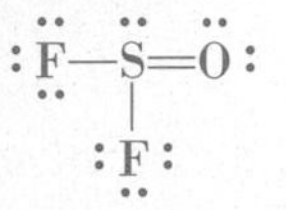

$$:\ddot{F}-\overset{\displaystyle ..}{S}=\ddot{O}: \\ \quad | \\ \quad :\ddot{F}:$$

Example 9-9 Experimental evidence indicates that the sulfur-to-oxygen bonds in SO_2 are essentially double bonds. What modification can be made to the structures shown in (9.18) to represent this fact?

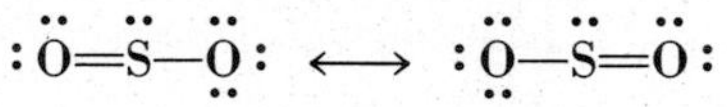

$$:\ddot{O}=\overset{\displaystyle ..}{S}-\ddot{O}: \longleftrightarrow :\ddot{O}-\overset{\displaystyle ..}{S}=\ddot{O}: \tag{9.18}$$

We need to add to the two possibilities shown in (9.18) a structure in which both of the sulfur-to-oxygen bonds are double bonds. This is easily done with an expanded octet for the S atom (ten valence electrons).

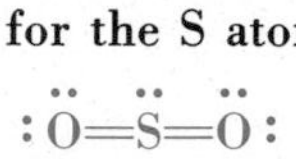

$$:\ddot{O}=\overset{\displaystyle ..}{S}=\ddot{O}: \tag{9.26}$$

SIMILAR EXAMPLES: Exercises 26, 27, 28.

FIGURE 9-8
Representation of bonding in the sulfate ion.

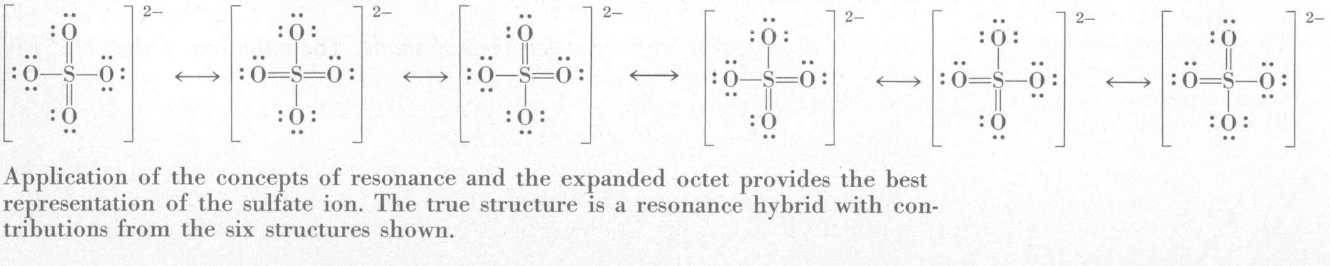

Application of the concepts of resonance and the expanded octet provides the best representation of the sulfate ion. The true structure is a resonance hybrid with contributions from the six structures shown.

FIGURE 9-9
Geometrical shape
of a molecule.

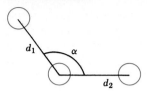

A hypothetical triatomic molecule is represented here. Although the atoms are actually in contact, for clarity only the centers of the atoms are shown (producing a so-called ball-and-stick model). To establish the shape of this molecule we must determine the distances between the centers of the bonded atoms (d_1 and d_2) and the angle between the adjacent bonds (α). Our primary concern is with the bond angle. For molecules with more than three atoms, additional bond distances and angles must be established, usually for a three-dimensional figure.

9-8 Molecular Shapes

When we speak of the shape of a molecule we mean the geometrical figure that results if the nuclei of the bonded atoms are joined by straight lines (see Figure 9-9). Because two points determine a straight line, all diatomic molecules are linear. Three points determine a plane, and all triatomic molecules are planar. For molecules with more than three atoms (polyatomic molecules), planar and even linear shapes are sometimes encountered. Usually, however, the atoms define a three-dimensional figure. Molecular shapes must be determined experimentally and cannot generally be predicted from empirical formulas.

FIGURE 9-10
Balloon analogy to
valence-shell electron-pair
repulsion.

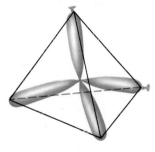

When two elongated balloons are twisted together at their centers, separation into four lobes occurs. To minimize mutual interferences, the lobes spread out into a tetrahedral shape. (A tetrahedron has four faces, each an equilateral triangle.) The lobes are analogous to valence-shell electron pairs. The distribution of the four pairs of outer-shell electrons of a neon atom is also shown.

VALENCE-SHELL ELECTRON-PAIR REPULSION (VSEPR) THEORY. Here is the basic idea of this theory of bonding: *Electrons in chemical bonds (single, double, and triple) and unshared electron pairs are negative charge centers that repel one another. They tend to remain as far apart as possible.* VSEPR (pronounced "vesper") theory pictures *pairs* of electrons assuming a certain orientation with respect to the nucleus of an atom.

Consider a noble gas atom such as Ne, for example. What orientation will the four pairs of valence electrons ($2s^2 2p^6$) assume? As suggested by the "balloon" analogy in Figure 9-10, the electron pairs are farthest apart when they occupy the corners of a regular tetrahedron with the atomic nucleus at its center. Now consider the methane molecule, CH_4, in which the central C atom has acquired the Ne electron configuration by forming covalent bonds with four H atoms.

$$H : \overset{\displaystyle H}{\underset{\displaystyle H}{\overset{..}{\underset{..}{C}}}} : H$$

The method predicts that CH_4 should be a *tetrahedral* molecule, with a C atom at the center of the tetrahedron and H atoms at the corners. This structure agrees with that established by experiment.

In NH_3 and H_2O the central atom is also surrounded by four pairs of electrons,

$$H : \overset{\displaystyle H}{\overset{..}{N}} : H \qquad \text{and} \qquad \overset{\displaystyle H}{: \overset{..}{\underset{..}{O}} : H}$$

but these molecules do not have a tetrahedral shape. The situation is this: VSEPR theory describes the distribution of electron pairs. The geometrical shape of a molecule, however, is described in terms of the geometrical figure that results by *joining the appropriate atomic nuclei by straight lines.*

In the NH_3 molecule only *three* of the electron pairs are *bonding* pairs and the fourth pair is a *nonbonding* or **lone pair.** The geometrical figure obtained by joining the nuclei of the H atoms with the nucleus of the N atom is not a tetrahedron but a

FIGURE 9-11
Geometric shapes based on the tetrahedral distribution of four electron pairs—CH_4, NH_3, and H_2O.

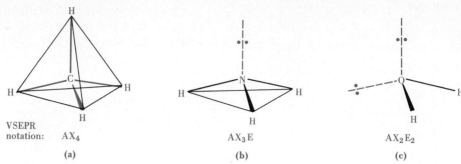

VSEPR notation: AX_4 AX_3E AX_2E_2

(a)   (b)   (c)

The geometric shape of the molecule is established by the lines shown in black. Lone pair electrons are shown as bold dots along broken lines originating at the central atom. Lone pair electrons are not involved in establishing the shape of a molecule.
(a) All electron pairs about the central carbon atom are bond pairs. (The valence of the carbon atom is said to be *saturated*.) The lines that establish the shape of the CH_4 molecule (in black) are different from those representing carbon-to-hydrogen bonds (in color).
(b) The lone pair of electrons is directed toward the "missing" corner of the tetrahedron. The figure remaining is a trigonal pyramid. The nitrogen-to-hydrogen bonds form three of the edges of this pyramid.
(c) The H_2O molecule is a V-shaped planar molecule outlined by the two oxygen-to-hydrogen bonds.

In a tetrahedral structure the central atom is *within* the tetrahedron and the atoms bonded to it occupy the four corners. In a trigonal pyramidal structure the central atom occupies one of the corners of the pyramid (which no longer has four identical faces).

pyramid (called a trigonal pyramid). The pyramid has H atoms at the base and an N atom at the apex. In the H_2O molecule two of the four pairs about the O atom are *bonding* pairs and two are *lone* pairs. The figure obtained by joining the nuclei of the two H atoms with that of the O atom is a *planar* figure. We can say that the molecule is "bent" or "V-shaped." The geometrical shapes of CH_4, NH_3, and H_2O are shown in Figure 9-11. Also introduced in Figure 9-11 is a special notation in which the number of lone-pair electrons in a structure is specified. For example, the notation AX_2E_2 signifies that two atoms of X and two lone pairs of electrons (E) are distributed about the central atom, A.

The bond angles expected for the tetrahedral distribution of electron pairs are 109.5°—called the tetrahedral bond angle. In CH_4 the H—C—H bond angles are, in fact, 109.5°. The bond angles in NH_3 and H_2O are slightly smaller: 107° for the H—N—H bond angle and 104.5° for H—O—H. These less-than-tetrahedral bond angles can be explained by assuming that the charge cloud of the lone-pair electrons spreads out and forces the bonding electrons closer together, reducing the bond angles. In H_2O, where there are two lone pairs, the effect is greatest and the bond angle smallest.

We continue our discussion of the VSEPR theory by turning to Table 9-2. Close inspection of some of the examples there suggests the need for an additional rule to govern cases where different alternatives appear to exist. For example, consider the case of five valence-shell electron pairs, with four bond pairs and one lone pair (i.e., AX_4E). Why is the lone pair shown in the triangular plane at the center of the bipyramid rather than at the top or bottom. The closer two pair of electrons are forced together, the stronger the repulsion between them. Repulsive forces become especially strong when the angle formed by lines joining two pair of electrons to the central atom is of the order of 90° (a right angle). Moreover, repulsions involving lone-pair electrons are stronger than those involving bond pairs, leading to the following order for repulsive forces.

lone pair–lone pair > lone pair–bond pair > bond pair–bond pair   (9.27)

Returning to the AX_4E case, placing the lone pair of electrons as indicated in Table

9-2 results in *two* interactions between a lone pair and a bond pair at 90°. If the lone pair were at the top or bottom of the bipyramid, there would be *three* such interactions. A similar situation is described in Example 9-12.

Example 9-10 Predict the geometrical shape of the molecule OCl_2.

Let us begin with a Lewis structure of this molecule. Since the O atom has two unpaired electrons, we should expect it, and not one of the Cl atoms, to be the central atom.

$$:\ddot{\ddot{Cl}}:\ddot{\ddot{O}}:\ddot{\ddot{Cl}}:$$

This Lewis structure helps us to count valence-shell electrons around the central atom as follows.

from the O atom = 6
from the Cl atoms, 2 × 1 = 2
total valence electrons = 8
valence-shell electron pairs = 4
number of bond pairs = 2
number of lone pairs = 2

For the geometrical shape corresponding to the distribution of two bond pairs and two lone pairs, that is, for AX_2E_2,

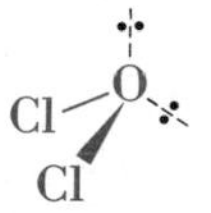

Conclusion: The molecule is V-shaped.

SIMILAR EXAMPLE: Exercise 29.

Example 9-11 The bonding scheme in hydrogen peroxide, H_2O_2, is H—O—O—H. Is this molecule linear in shape?

This molecule appears to have *two* central atoms, that is, both O atoms.

$$H:\ddot{\ddot{O}}:\ddot{\ddot{O}}:H$$

Whichever O atom we look at, we see four pairs of valence-shell electrons—two bond pairs and two lone pairs. This means that each H—O—O bond angle must correspond to the following electron-pair distributions.

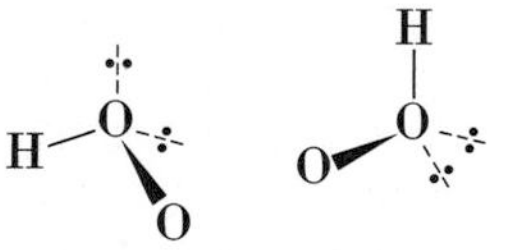

A plausible geometrical shape of the molecule H_2O_2 can be obtained by placing one of these structures over the other in such a way that the O—O bonds are superimposed.

For the molecule to be linear, both H—O—O bond angles would have to be 180°, but we see that they are more nearly tetrahedral.
Conclusion: H_2O_2 is not linear.

SIMILAR EXAMPLE: Exercise 29.

Example 9-12 Predict the shape of the polyatomic anion ICl_4^-.

The I atom has seven outer-shell electrons, as do all the Cl atoms. To bond the four Cl atoms to the central I atom (and to accommodate an extra electron to produce the ionic

TABLE 9-2
Molecular geometry as a function of geometrical distribution of valence-shell electron pairs and number of lone pairs

Number of electron pairs	Geometrical distribution of electron pairs	Number of lone pairs	VSEPR notation	Molecular geometry	Predicted bond angles	Example
2	linear	0	AX_2	X—A—X (linear)	180°	$BeCl_2$
3	trigonal planar	0	AX_3	(trigonal planar)	120°	BF_3
	trigonal planar	1	AX_2E	(V-shaped)	120°	SO_2[a]
4	tetrahedral	0	AX_4	(tetrahedral)	109.5°	CH_4
	tetrahedral	1	AX_3E	(trigonal pyramidal)	109.5°	NH_3
	tetrahedral	2	AX_2E_2	(V-shaped)	109.5°	OH_2
5	trigonal bipyramidal	0	AX_5	(trigonal bipyramidal)	90°, 120°	PCl_5

TABLE 9-2   (Continued)

Number of electron pairs	Geometrical distribution of electron pairs	Number of lone pairs	VSEPR notation	Molecular geometry	Predicted bond angles	Example
	trigonal bipyramidal	1	AX_4E[b]	(irregular tetrahedral)	90°, 120°	SF_4
	trigonal bipyramidal	2	AX_3E_2	(T-shaped)	90°	ClF_3
	trigonal bipyramidal	3	AX_2E_3	(linear)	180°	XeF_2
6	octahedral	0	AX_6	(octahedral)	90°	SF_6
	octahedral	1	AX_5E	(square pyramidal)	90°	BrF_5
	octahedral	2	AX_4E_2	(square planar)	90°	XeF_4

[a] For a discussion of the structure of SO_2, see page 214.
[b] For a discussion of the placement of the lone-pair electrons in this structure, see page 210.

charge of -1) requires an expanded octet for the I atom. A total of 36 electrons must appear in the Lewis structure.

$$\ddot{\underset{\cdot\cdot}{:}\overset{\cdot\cdot}{Cl}}\diagdown\quad\overset{\cdot\cdot}{\underset{\cdot\cdot}{Cl}:}$$

From the Lewis structure we determine that there are four bond pairs and two lone pairs: AX_4E_2. The corresponding molecular geometry is square planar.

Figure 9-12 suggests two possible structures for ICl_4^-, but by applying relationship (9.27) we see that the square planar structure is indeed the correct one. Repulsions between the two lone pairs of electrons would be much greater if they were located at adjacent corners of the central square than when they are located above and below the square.

SIMILAR EXAMPLES: Exercises 29, 34.

STRUCTURES WITH MULTIPLE COVALENT BONDS. VSEPR theory treats all chemical bonds as negative charge centers and does not distinguish between single and multiple bonds. That is, it treats multiple bonds *as if they were single bonds* and contained just a single electron pair in the bond. Let us apply this idea to the Lewis structures shown for SO_2 in (9.18). We count the electron pairs about the central S atom as if there were *three*. Two of these three pairs are bond pairs and one is a lone pair. The molecular geometry for this situation (i.e., corresponding to AX_2E) is a "bent" or V-shaped molecule. (See Table 9-2.)

$$O\diagup\!\!\!\!=\overset{\cdot\cdot}{S}\diagdown O$$

Example 9-13 What is the geometrical shape of the phosgene molecule, $COCl_2$?

We need first to draw a Lewis structure for this molecule. Although we might consider other possibilities, a structure with C as the central atom is most plausible. This structure has no formal charges.

$$\begin{array}{ccc} :\overset{\cdot\cdot}{\underset{\cdot\cdot}{Cl}}: & & :\overset{\cdot\cdot}{Cl}: \\ \underset{\cdot\cdot}{C}::\overset{\cdot\cdot}{\underset{\cdot\cdot}{O}}: & \text{or} & C\!=\!\overset{\cdot\cdot}{\underset{\cdot\cdot}{O}}: \\ :\overset{\cdot\cdot}{\underset{\cdot\cdot}{Cl}}: & & :\overset{\cdot\cdot}{\underset{\cdot\cdot}{Cl}}: \end{array}$$

FIGURE 9-12
Two predictions of the structure of ICl_4^- — Example 9-12 illustrated.

The observed structure is the square planar structure shown on the right.

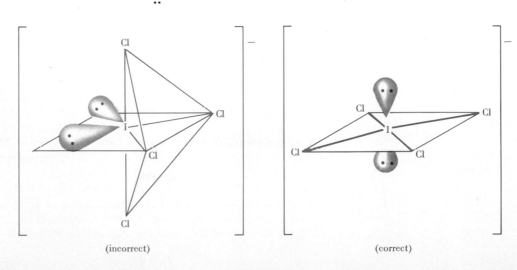

(incorrect)　　　　　　　　　(correct)

TABLE 9-3
Some representative bond energies and bond distances (lengths)[a]

Bond	Bond energy, kJ/mol	Bond distance		Bond	Bond energy, kJ/mol	Bond distance	
		Å	pm			Å	pm
H—H	435	0.74	74	C—O	360	1.43	143
H—C	414	1.10	110	C=O	736	1.23	123
H—N	389	1.00	100	C—Cl	326	1.77	177
H—O	464	0.97	97	N—N	163	1.45	145
H—F	565	1.01	101	N=N	418	1.23	123
H—Cl	431	1.36	136	N≡N	946	1.09	109
H—Br	364	1.51	151	F—F	155	1.28	128
H—I	297	1.70	170	Cl—Cl	243	1.99	199
C—C	347	1.54	154	Br—Br	192	2.28	228
C=C	611	1.34	134	I—I	151	2.66	266
C≡C	837	1.20	120				
C—N	305	1.47	147				
C=N	615	1.28	128				
C≡N	891	1.16	116				

[a] Bond energies and bond distances both depend on the environment of the given bond—the portion of the molecule close to the bond. The energy required to dissociate one H atom from an H_2O molecule, that is, to break one O—H bond, is about 502 kJ/mol. The value listed here, 464 kJ/mol, is the average for breaking both O—H bonds in water. Other values in the table are also averages, but they are generally adequate for making comparisons.

Counting the carbon-to-oxygen double bond as if it were a single bond, the central atom has three pairs of electrons distributed around it. All are bond pairs. The geometrical distribution of these electron pairs and the geometrical shape of the molecule are both trigonal *planar*. The structure can be represented as

$$\begin{array}{c} Cl \\ \diagdown \\ C{=}O \\ \diagup \\ Cl \end{array}$$

SIMILAR EXAMPLE: Exercise 30.

9-9 Bond Energies and Bond Distances

Energy is *released* when atoms join together through a chemical bond and must be *absorbed* if bonded atoms are to be separated. Let us define **bond energy** as the quantity of energy required to break one mole of chemical bonds in a *gaseous* species. In the SI system, bond energies are expressed in kilojoules per mole of bonds (kJ/mol).

It is reasonable to expect the strength of a chemical bond, as measured by bond energy, to depend on the nature of the bond between atoms. That is, a double bond is stronger than a single bond, and a triple bond, in turn, is stronger than a double bond. Furthermore, some correlation should be expected between bond energy and **bond distance (bond length)**—the distance between the nuclei of bonded atoms. The *stronger* a chemical bond, the *shorter* the bond distance. Some experimentally determined bond energies and bond distances that bring out this relationship are listed in Table 9-3.

The term that was described in Chapter 8 as the single covalent radius is simply one half the distance between the nuclei of identical atoms joined by a single covalent bond. Thus, the single covalent radii of hydrogen and fluorine are 37 pm (one half of 74 pm) and 64 pm (one half of 128 pm), respectively.

BOND ENERGIES AND ENTHALPY CHANGES. Bond energies are sometimes called **bond enthalpies** and expressed in the manner introduced in Chapter 6. For example,

bond breakage:
$$H_2(g) \longrightarrow 2\,H(g) \qquad \Delta H = +435 \text{ kJ/mol}$$

bond formation:
$$2\,H(g) \longrightarrow H_2(g) \qquad \Delta H = -435 \text{ kJ/mol}$$

Moreover, the heat of reaction, ΔH_{rx}, for a reaction involving *gaseous* species is simply the sum of the enthalpy changes for bond breakage and bond formation. Figure 9-13 depicts bond breakage and formation for the reaction

$$CH_4(g) + 4\,Cl_2(g) \longrightarrow CCl_4(g) + 4\,HCl(g)$$

> The method outlined here is limited to reactions involving gaseous species because tabulated bond energies are based on bond breakage or bond formation in gaseous molecules.

Example 9-14   Calculate ΔH for the reaction

$$CH_4(g) + 4\,Cl_2(g) \longrightarrow CCl_4(g) + 4\,HCl(g) \quad \Delta H_{rx} = ?$$

ΔH for bond breakage:

4 mol C—H bonds = 4 mol × (+414 kJ/mol)
$$= +1656 \text{ kJ}$$

4 mol Cl—Cl bonds = 4 mol × (+243 kJ/mol)
$$= +972 \text{ kJ}$$

ΔH for bond formation:

4 mol C—Cl bonds = 4 mol × (−326 kJ/mol)
$$= -1304 \text{ kJ}$$

4 mol H—Cl bonds = 4 mol × (−431 kJ/mol)
$$= -1724 \text{ kJ}$$

Heat of reaction:

$$\Delta H_{rx} = \Delta H_{\text{bond breakage}} + \Delta H_{\text{bond formation}}$$
$$= +1656 \text{ kJ} + 972 \text{ kJ} - 1304 \text{ kJ} - 1724 \text{ kJ}$$
$$= -400 \text{ kJ}$$

SIMILAR EXAMPLES: Exercises 39, 41.

To calculate the heat of a reaction by the use of tabulated bond energies generally offers no advantage over calculating this quantity from heat of formation data and expression (6.19). This is because heats of formation are usually known rather precisely, whereas bond energies are only average values. But there are

FIGURE 9-13
Bond breakage and formation in a chemical reaction—Example 9-14 illustrated.

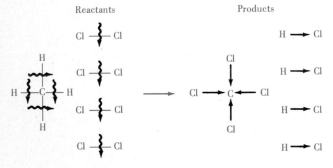

FIGURE 9-14
Behavior of polar molecules in an electric field.

In the absence of an electric field the molecules are oriented randomly. However, when one plate is made negative and the other positive, polar molecules tend to orient themselves in the manner shown here.

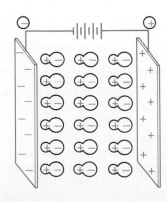

times when heat-of-formation data are either not known or not available. Here bond energies can prove particularly useful.

The relationship between bond energies and enthalpy changes also permits a simple prediction of whether a reaction will be endothermic or exothermic: In general,

if weak bonds $\longrightarrow$ strong bonds $\Delta H < 0$ (*exothermic*)

and   (9.28)

if strong bonds $\longrightarrow$ weak bonds $\Delta H > 0$ (*endothermic*)

Example 9-15 applies this idea to a reaction involving some highly reactive, unstable species for which heats of formation are not normally tabulated.

Example 9-15   One of the reactions leading to photochemical smog involves the attack of a hydrocarbon molecule, RH, by a hydroxyl radical, OH. The products are a water molecule, H_2O, and a hydrocarbon radical, R· (which undergoes further reaction). Is this reaction endothermic or exothermic?

$$RH(g) + OH(g) \longrightarrow H_2O(g) + R\cdot(g)$$

According to expression (9.28), we would say that a "weak" C—H bond is replaced by a "strong" O—H bond. The reaction is exothermic.

All the bonds in a hydrocarbon are either carbon-to-carbon or carbon-to-hydrogen. In the reaction in question, for every molecule of RH that reacts, one carbon-to-hydrogen bond is broken, requiring the absorption of 414 kJ/mol. An additional hydrogen-to-oxygen bond is formed, converting OH(g) to $H_2O(g)$ and releasing 464 kJ/mol. Since more energy is released in forming new bonds than is absorbed in breaking old bonds, the reaction is exothermic.

SIMILAR EXAMPLE: Exercise 40.

9-10 Partial Ionic Character of Covalent Bonds

We have established two basic bond types, ionic and covalent, and we have treated all bonds as if they were one or the other. But classification schemes generally involve borderline cases, and this is very much so with bond type.

If the medium between the electrodes in Figure 9-14 conducts electricity, a continuous flow of electric charge—an electric current—results. In this case the device does not function as a condenser.

POLAR MOLECULES AND DIPOLE MOMENTS.   The device depicted in Figure 9-14, consisting of a pair of electrodes separated by a *nonconducting* medium, is called a condenser or capacitor. Small quantities of electric charge can be stored on the electrodes, positive on one and negative on the other. The quantity of charge that can be stored depends on the medium between the electrodes. There is one class of substances that enhances this ability significantly. These are substances in whose molecules a separation of electric charge exists. Figure 9-14 suggests how such molecules line up in an electric field.

A separation into charge centers in a covalent molecule involves a shift of electrons toward one atom in a bond—the more electronegative (nonmetallic) atom. A molecule in which this charge separation exists is said to be **polar,** or it is termed a **dipole** because of the existence of two charge centers. For example, H_2 is a *nonpolar* molecule and HCl is *polar*.

$$H:H \qquad {}^{\delta+}H:\overset{\displaystyle ..}{\underset{\displaystyle ..}{Cl}}:{}^{\delta-}$$

The magnitude of the effect described here is denoted through the **dipole moment, μ.** The dipole moment is the product of the magnitude of the charges (δ) and the distance separating them (d). (The δ symbols suggest a small magnitude of charge, less than the charge of an electron.)

$$\mu = \delta d \tag{9.29}$$

If the product of the charge and the distance of separation has a magnitude of 3.34×10^{-30} coulomb $\cdot$ meter (C $\cdot$ m), the dipole moment, μ, has a value of 1 debye (D). How significant is charge separation in a molecule like HCl? The measured dipole moment of HCl is 1.03 D, and the bond distance listed in Table 9-3 is 136 pm. That is,

magnitude of charge $\times$ 1.36×10^{-10} m $= 1.03 \times 3.34 \times 10^{-30}$ C $\cdot$ m

magnitude of charge $= 3.44 \times 10^{-30}$ C $\cdot$ m$/1.36 \times 10^{-10}$ m $= 2.53 \times 10^{-20}$ C

This charge is about 16% of the charge on an electron (1.60×10^{-19} C), suggesting that HCl is about 16% ionic in character.

How does this new information change our outlook on the HCl molecule? Perhaps it is best simply to refer to HCl as a polar covalent molecule which is reasonably well represented by the simple Lewis structure

$$\text{H} \!:\! \overset{..}{\underset{..}{\text{Cl}}} \!: \tag{9.30}$$

but for which there is an ionic contribution

$$[\text{H}^+][:\overset{..}{\underset{..}{\text{Cl}}}:^-] \tag{9.31}$$

leading to a resonance hybrid

$$^{\delta+}\text{H}:\overset{..}{\underset{..}{\text{Cl}}}:^{\delta-} \tag{9.32}$$

MOLECULAR SHAPES AND DIPOLE MOMENTS. There is an electronegativity difference between C and O atoms, and, as expected, the molecule CO is polar. In the representation below the existence of a dipole moment in the carbon-to-oxygen bond is represented by a cross-base arrow ($\mapsto$). The arrow points to the atom that attracts electrons more strongly—the more electronegative atom.

C $\longmapsto$ O $\mu = 0.11$ D

The result described here for CO_2 is analogous to a tug-of-war contest between equally matched teams. Although there is a strong pull in each direction, the knot at the center of the rope does not move at all.

Even though there is an electronegativity difference and hence a **bond moment** along each carbon-to-oxygen bond, *the CO_2 molecule is nonpolar.* This can only mean that the effects of the two O atoms in attracting electrons cancel each other out. The O atoms lie along the same straight line through the central C atom.

O $\longleftmapsto$ C $\longmapsto$ O $\mu = 0$

There is also an electronegativity difference between H and O atoms, leading to an O—H bond dipole moment of 1.51 D. The resultant dipole moment in H_2O depends on how the two bond moments are directed. If H_2O were a linear molecule, the bond moments would be in opposite directions and there would be no resultant dipole moment. But the measured dipole moment of water is 1.84 D. The molecule cannot be linear. The two bond moments combine to yield this resultant dipole moment for a particular bond angle—104°.

O $\longleftmapsto$ H $\mu = 1.84$ D   (9.33)

104°

H

Verification of the tetrahedral shape of a molecule such as CCl_4 is possible through dipole moment measurements. Although the C—Cl bond moment is large (2.05 D), there is no resultant dipole moment in the molecule CCl_4. The chlorine

FIGURE 9-15
Dipole moments and
molecular shapes.

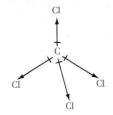

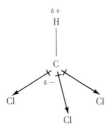

CCl_4: a nonpolar molecule

$\mu = 0$

CHCl$_3$: a polar molecule

$\mu = 1.92$ debye

The symmetrical distribution of the four C—Cl bonds in CCl_4 causes a cancellation of all bond dipole moments; there is no resultant dipole moment in the molecule. CCl_4 is nonpolar. The C—H bond has a dipole moment of essentially zero because the electronegativities of C and H are quite similar. The three C—Cl bond dipole moments cause the chlorine end of the molecule to develop a slight negative charge. The hydrogen end carries a slight positive charge. CHCl$_3$ is polar.

atoms must be arranged symmetrically about the carbon. Substituting a less electronegative atom for one of the Cl atoms, H for instance, leads to an imbalance of attractive forces for electrons and a resulting dipole moment. Figure 9-15 illustrates this point.

Example 9-16 Which of the molecules listed would you expect to be polar and which nonpolar? Cl_2, ICl, BeF_2, NO, SO_2, XeF_4.

Polar: ICl, NO, SO_2. ICl and NO are diatomic molecules with an electronegativity difference between the bonded atoms. SO_2 is a V-shaped molecule (nonlinear) with an electronegativity difference between S and O.

Nonpolar: Cl_2, BeF_2, XeF_4. Cl_2 is a homonuclear diatomic molecule. BeF_2 is linear. XeF_4 is a square planar molecule with the F atoms arranged symmetrically about the central Xe atom (recall Table 9-2).

SIMILAR EXAMPLES: Exercises 44, 45.

IONIC RESONANCE ENERGY. If the polarity of the bond A—B is about the same as the bonds A—A and B—B, in particular if all the bonds are nonpolar, the bond energy of A—B may be taken as the average of A—A and B—B.

Example 9-17 Use data from Table 9-3 to calculate the energy of the Br—Cl bond.

$$E_{Br—Cl} = \tfrac{1}{2}(E_{Br—Br} + E_{Cl—Cl})$$

$$E_{Br—Cl} = \tfrac{1}{2}(192 + 243) = 218 \text{ kJ/mol}$$

(The experimental value = 218 kJ/mol.)

SIMILAR EXAMPLE: Exercise 47.

When the method of Example 9-17 is applied to the HCl molecule, it fails. The average of the bond energies for H—H and Cl—Cl is about 339 kJ/mol, but the experimentally determined value is 431 kJ/mol. The H—Cl bond is stronger by 92 kJ/mol than would be the case if the three types of bonds (H—H, Cl—Cl, H—Cl) all were of equal polarity. We have already seen one method of assessing the relative ionic character of the HCl molecule based on dipole moment; here is another. The energy difference of 92 kJ/mol is the additional stabilization contributed by the ionic structure (9.31) to the true structure (9.32). It is called the **ionic resonance energy.**

ELECTRONEGATIVITY. Pauling's scale of electronegativity values presented in Section 8-9 was obtained through an extensive correlation of bond energies. Particular attention was given to ionic resonance energy. The equation developed for this purpose was

$$(\Delta EN)^2 = \frac{IRE}{96} \tag{9.34}$$

where ΔEN is the electronegativity difference between the bonded atoms and IRE is the ionic resonance energy in kilojoules per mole.

From electronegativity values Pauling was able to assign a "percent ionic character" to a bond, as illustrated in Figure 9-16. If the electronegativity difference is very small, a bond is essentially covalent; and if large, essentially ionic. There is no particular electronegativity difference at which the bond type changes from covalent to ionic. As a very rough rule, if the electronegativity difference exceeds about 1.7, a bond is greater than 50% ionic.

FIGURE 9-16
Percent ionic character of a chemical bond as a function of electronegativity difference.

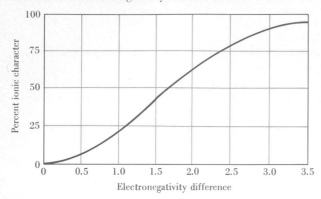

Example 9-18   Calculate the electronegativity difference between H and Cl using equation (9.34) and compare with data from Table 8-6. Estimate the percent ionic character in the H—Cl bond using Figure 9-16 and compare with the estimate based on dipole moment measurements.

The ionic resonance energy for HCl stated above is 92 kJ/mol.

$$(\Delta EN)^2 = \frac{92}{96} = 0.96 \qquad \Delta EN = (0.96)^{1/2} = 0.98$$

From Table 8-6,

$$\Delta EN = EN_{Cl} - EN_H = 3.16 - 2.20 = 0.96$$

From Figure 9-16,

$$\Delta EN \approx 0.9 - 1.0 \qquad \% \text{ ionic character} \approx 20\%$$

From dipole moment measurement (page 218), the H—Cl bond is about 16% ionic.

SIMILAR EXAMPLES: Exercises 49, 50.

9-11 Oxidation States

Because oxidation state refers to a number, the term *oxidation number* is often used synonymously. We will use the two terms interchangeably.

The concept of oxidation state (oxidation number) is a useful bookkeeping system to assess the extent to which electrons are lost, gained, or shared in chemical bond formation. It can be applied to the full range of bond types, from essentially ionic to essentially covalent. In NaCl a sodium atom transfers *one* electron to a chlorine atom; both atoms are in the oxidation state 1. Because the sodium atom becomes a positively charged ion, its oxidation state is taken to be $+1$. That of chlorine (in the form of chloride ion) is -1. *The oxidation states of ions are equal to their charges.* In $MgCl_2$ the oxidation state of Mg is $+2$ and that of Cl is -1.

For covalent compounds, although electrons are shared, we *arbitrarily* consider them to be transferred. Thus, in H_2O it is reasonable that the more electronegative atom—O—be assigned electrons from the H atoms. The O atom is considered to have gained two electrons and is assigned the oxidation state of -2. Each H atom is treated as if it had transferred one electron to the O atom; each H atom is assigned an oxidation state of $+1$.

Another arbitrary convention regarding the assignment of oxidation states is that the total of the oxidation states of all the atoms in a neutral formula unit or neutral molecule must be *zero*. In Cl_2 the two Cl atoms could be assigned oxidation states $+1$ and -1, leading to a total of 0 for the molecule. However, this assignment suggests an essential difference between the two atoms in Cl_2, and the two atoms are identical. The arbitrary assignment of 0 for the oxidation state of an atom in a free element resolves this difficulty.

From the examples just given it is obvious that some conventions or rules must be followed in assigning oxidation states. The six rules listed below are sufficient in dealing with all cases in this text, with the following important qualification: Whenever two rules contradict one another, follow the rule that appears *first* in the list.

1. The oxidation state of an atom in a free (uncombined) element is 0.
2. The total of the oxidation states of all the atoms in a neutral molecule or formula unit is 0. For an ion this total is equal to the charge on the ion, both in magnitude and sign.
3. In their compounds the alkali metals of group IA have an oxidation state of $+1$ and alkaline earth metals (IIA), $+2$.
4. In its compounds the oxidation state of hydrogen is $+1$; that of fluorine is -1.
5. In its compounds oxygen has an oxidation state of -2.
6. In their binary compounds with metals, the elements of group VIIA have an oxidation state of -1; those of group VIA, -2; and those of group VA, -3.

Example 9-19   What is the oxidation state of the underlined atom in each of the following? **(a)** $\underline{P}_4$; **(b)** $\underline{Al}_2O_3$; **(c)** $\underline{Mn}O_4^-$; **(d)** $Na\underline{H}$; **(e)** $H_2\underline{O}_2$; **(f)** $K\underline{O}_2$; **(g)** $\underline{Fe}_3O_4$.

(a) P_4: This formula represents a molecule of the element phosphorus. For an atom of a free element the oxidation state is 0 (rule 1). The oxidation state of P in P_4 is 0.
(b) Al_2O_3: The total of the oxidation numbers of all the atoms in a formula unit is 0 (rule 2). The oxidation state of O is -2 (rule 5). The total for three O atoms is -6. The total for two Al atoms is $+6$. The oxidation state of Al is $+3$.
(c) MnO_4^-: The total of the oxidation numbers of all the atoms in this ion must be -1 (rule 2). The total of the oxidation numbers of the four O atoms is -8. The oxidation state of Mn is $+7$.
(d) NaH: Rule 3 states that Na should have an oxidation state of $+1$. However, rule 4 indicates that the oxidation state of H should also be $+1$. If both atoms had an oxidation number of $+1$, the total for the formula unit would be $+2$. This would violate rule 2. *Rules 2 and 3 take precedence over rule 4.* The oxidation state of Na is $+1$; the total for the formula unit is 0; and the oxidation state of H in NaH must be -1.
(e) H_2O_2: Rule 4, stating that H has an oxidation state of $+1$, takes precedence over rule 5. The sum of the oxidation numbers of the two H atoms is $+2$ and that of the two O atoms is -2. The oxidation state of O in H_2O_2 is -1.
(f) KO_2: Rule 3 (requiring that the oxidation state of $K = +1$) takes precedence over rule 5. The sum of the oxidation numbers of the two O atoms is -1. The oxidation state of each O atom in KO_2 is $-\frac{1}{2}$.
(g) Fe_3O_4: The total of the oxidation numbers of the four O atoms is -8. For three Fe atoms the total is $+8$. This requires an assignment of $+2\frac{2}{3}$ for each Fe atom.

NaH is actually an ionic compound. An electron is transferred from Na to H, producing the ions Na^+ and H^-.

Compounds in which oxygen exhibits an oxidation state other than -2, such as H_2O_2 (hydrogen peroxide) and KO_2 (potassium superoxide), are discussed more fully in Section 20-4.

Examples (f) and (g) bring out the artificiality of the concept of oxidation state. What is the significance of a nonintegral oxidation state? In (f), a molecule of O_2 can be considered to gain an electron from a K atom, forming K^+ and O_2^-. In (g) two of the Fe atoms can be considered to have the oxidation state $+3$ and the third, $+2$. (This results in an average of $+2\frac{2}{3}$.)

SIMILAR EXAMPLES: Exercises 52, 53.

NOMENCLATURE.   Like formal charge, there is an arbitrariness about the concept of oxidation state, but it is useful. Its principal applications are to oxidation–reduction phenomena (see Chapter 19). For the present we will simply explore the relationship between oxidation state and nomenclature.

In Section 3-4 we noted that the names of a number of common polyatomic anions have endings other than "ide." In particular we discussed the naming of oxyanions of nonmetal atoms such as P, S, and Cl. The sodium compounds (salts) of

TABLE 9-4
Nomenclature of some ternary compounds

Oxidation state	Formula of acid	Name of acid	Formula of salt	Name of salt
+1	HClO	*hypochlorous* acid	NaClO	sodium *hypochlorite*
+3	HClO$_2$	*chlorous* acid	NaClO$_2$	sodium chlor*ite*
+5	HClO$_3$	chlor*ic* acid	NaClO$_3$	sodium chlor*ate*
+7	HClO$_4$	*perchloric* acid	NaClO$_4$	sodium *perchlorate*
+3	HNO$_2$	*nitrous* acid	NaNO$_2$	sodium nitr*ite*
+5	HNO$_3$	nitr*ic* acid	NaNO$_3$	sodium nitr*ate*
+4	H$_2$SO$_3$	sulf*urous* acid	Na$_2$SO$_3$	sodium sulf*ite*
+6	H$_2$SO$_4$	sulfur*ic* acid	Na$_2$SO$_4$	sodium sulf*ate*

several common polyatomic anions first described in Table 3-3 are shown again in Table 9-4. This table also lists the names and formulas of the covalent, hydrogen-containing compounds (called acids) from which these anions may be derived. We first described the naming of these compounds in terms of the number of oxygen atoms in the anion. We now see that this is actually a scheme based on the oxidation state of the central nonmetal atom.

——**Increasing oxidation state of central atom**——→

acid: hypo_____ous   _____ous   _____ic   per_____ic

anion: hypo_____ite   _____ite   _____ate   per_____ate

——**Increasing number of oxygen atoms**——→

The system of nomenclature whereby the following names were assigned in Section 3-4 is also based on oxidation states. The Roman numeral indicates the oxidation state of the metal.

In general, the "ic" and "ate" names are assigned to compounds in which the central atom has an oxidation state equal to the periodic group number. Halogen compounds are exceptional in that the "ic" and "ate" names are assigned to compounds in which the halogen atom has an oxidation state of +5.

iron(II) chloride = FeCl$_2$

iron(III) chloride = FeCl$_3$

Example 9-20 Supply appropriate names for the following compounds: **(a)** HIO$_4$; **(b)** SnCl$_4$.
(a) The oxidation state of iodine in this compound is +7. By analogy to the corresponding chlorine-containing compound in Table 9-4, we should name it periodic acid.
(b) The oxidation state of tin in this compound is +4. The compound is called tin(IV) chloride.

SIMILAR EXAMPLES: Exercises 54, 55.

Summary

The Lewis theory of chemical bonding is based on the behavior of electrons in the atoms being joined, especially the outer-shell or valence electrons. Although the Lewis theory can be used to predict formula units of ionic compounds, its greatest value is in describing covalently bonded species. This description takes the form of a framework or skeleton structure showing the order in which atoms are bonded together. The valence electrons of the bonded atoms are distributed through-

out the structure in accordance with a set of rules.

Sometimes experimental data must be used to establish the correct skeleton structure of a molecule. Lacking such data, the concept of formal charge may be used. At times, even though all the usual rules are followed, it is impossible to establish a single Lewis structure for a species. In these cases two or more plausible structures are obtained, and the true structure is said to be a resonance hybrid of these plausible structures. One of the basic assumptions of the Lewis theory is that all atoms in a Lewis structure acquire eight electrons in their outermost shells (an octet). This assumption fails for odd-electron species, for electron-deficient compounds, and for compounds in which the central atom can accommodate an "expanded octet," that is, 10 or 12 electrons.

Lewis theory does not address specifically the question of the shapes of molecules. A powerful method for doing so is the valence-shell electron-pair repulsion (VSEPR) theory. This method requires identifying the central atom of a structure, the number of pairs of valence electrons surrounding this atom, and the geometrical distribution of these electron pairs. The geometrical shape of a molecule depends both on the distribution of all valence-shell electron pairs and on whether these pairs are bonding pairs or lone (non-bonding) pairs.

Molecular properties such as bond energy and bond distance are not required in writing Lewis structures, but they are sometimes helpful in judging whether a structure is plausible. For example, they can be used to establish whether a covalent bond has multiple bond character. In addition, bond energies can be used to calculate enthalpy changes (heats) of reactions involving gaseous species.

In most of this chapter ionic and covalent bonds are treated as if they are distinctly different. In fact, most bonds have both partial ionic and partial covalent character. One indication of the partial ionic character of a covalent bond is the degree of separation of electrical charge that exists in the bond. This is measured through the dipole moment. Another measure of the partial ionic character of a bond is achieved through the use of electronegativities.

The chapter closes with the concept of oxidation states, which provides a system of counting electrons and assigning them to the atoms in a species. Several applications of this concept to the naming of compounds are considered in this chapter. The most important applications of the concept are to the study of oxidation-reduction, considered in Chapter 19.

Learning Objectives

As a result of studying Chapter 9, you should be able to

1. State the basic assumptions of the Lewis theory of chemical bonding.

2. Relate the Lewis symbol of an element to its position in the periodic table.

3. Write Lewis structures for simple ionic compounds.

4. Propose a plausible skeleton structure for a molecule, and assign valence electrons to this structure, using the basic rules of Lewis theory.

5. Assign formal charges to the atoms in a Lewis structure.

6. Select the most plausible Lewis structure among alternative structures by applying rules pertaining to formal charges.

7. Recognize situations in which resonance occurs and write plausible structures contributing to a resonance hybrid.

8. Identify odd-electron species and write Lewis structures for them.

9. Recognize molecules in which expanded octets are required and write Lewis structures based on these expanded octets of valence electrons.

10. Relate the geometrical shape of a molecule to the distribution of valence-shell electron pairs around the central atom.

11. Use bond energies to calculate ΔH of a reaction involving gases.

12. Relate electronegativity data, molecular geometry, and the dipole moments of molecules.

13. Describe the relationship between electronegativity difference and the percent ionic character of a bond.

14. State the basic oxidation state conventions and apply them in the assignment of oxidation states.

15. Use the oxidation state concept to assign names and formulas to chemical substances.

Some New Terms

Bond distance (bond length) is the distance between the nuclei of atoms joined by a chemical bond.

Bond energy (bond enthalpy) is the quantity of energy (usually expressed in J/mol) required to break one mole of chemical bonds in a gaseous species.

A **bonding pair** is a pair of electrons involved in bond formation.

In a **coordinate covalent bond** the electrons shared between atoms are contributed by just one of the atoms.

A **covalent bond** results from the sharing of electrons between atoms.

Dipole moment (μ) is a measure of the extent to which a separation of charges exists within a molecule. It is the product of the magnitude of the charge and the distance separating the charge centers. The unit used to measure dipole moment is the **debye**, 3.34×10^{-30} C m.

In a **double covalent bond** *two pairs* of electrons are shared between bonded atoms. The bond is represented by a double-dash sign ($=$).

Expanded octet is a term used to describe situations in which P, S, and certain other atoms in the third period or beyond are able to use 10 or 12 electrons in forming bonds.

Formal charge is the number of outer-shell (valence) electrons in an isolated atom minus the number of electrons assigned to that atom in a Lewis structure.

Incomplete octet is a term used to describe situations in which an atom fails to acquire eight outer-shell electrons when it forms a bond.

An **ionic bond** results from the transfer of electrons between metal and nonmetal atoms. Positive and negative ions are formed and held together by electrostatic attraction.

A **Lewis structure** is a combination of Lewis symbols that depicts the transfer or sharing of electrons in a chemical bond.

Lewis symbols are representations of the elements in which dots are placed around the chemical symbol to represent valence electrons.

A **lone pair** is a pair of electrons found in the valence shell of an atom and *not* involved in bond formation.

A **multiple covalent bond** is a bond in which more than two electrons are shared between the bonded atoms.

In a **nonpolar molecule** the centers of positive and negative charge coincide. That is, there is no separation of charge within the molecule.

An **octet** refers to the presence of *eight* electrons in the outermost (valence) electronic shell of an atom.

An **odd-electron species** is one in which the total number of valence electrons is an *odd* number. At least one unpaired electron is present in the species.

Oxidation state (oxidation number) is a measure of the number of electrons an atom gains, loses, or shares in becoming bonded to others.

In a **polar molecule** there exists a separation of electrical charge into positive and negative centers.

Resonance occurs when two or more plausible Lewis structures can be written for a species. The true structure is a composite or hybrid of these.

A **single covalent bond** results from the sharing of *one pair* of electrons between bonded atoms. It is represented by a single dash sign ($-$).

In a **triple covalent bond** *three pairs* of electrons are shared between the bonded atoms. The bond is represented by a triple-dash sign ($\equiv$).

The **valence-shell electron-pair repulsion (VSEPR) theory** relates the geometrical shape of a species to the geometrical distribution of electron pairs in the valence shell of the central atom.

Exercises

Lewis theory

1. Write Lewis symbols for the following species: **(a)** Xe; **(b)** Sn; **(c)** Sc^{3+}; **(d)** Br^-; **(e)** Ga; **(f)** Rb; **(g)** Ca^{2+}; **(h)** S^{2-}; **(i)** H.

2. What are some of the essential differences in the way in which Lewis structures are written for ionic and covalent bonds?

Ionic bonding

3. Write Lewis structures for the following ionic compounds: **(a)** BaS; **(b)** $CaCl_2$; **(c)** KI.

4. Derive the correct formulas for the following ionic compounds by writing Lewis structures: **(a)** lithium oxide; **(b)** sodium bromide; **(c)** strontium fluoride; **(d)** scandium chloride.

5. In all simple binary ionic compounds, the nonmetal atoms acquire the electron configurations of noble gas atoms. This is not always the case with the metal atoms. Explain why this is so.

6. Why is it inappropriate to use the term "molecules of NaCl" when describing *solid* sodium chloride? What would the term signify if one were describing *gaseous* sodium chloride?

Lewis structures

7. With reference to Lewis structures, what is meant by each of the following terms: **(a)** valence electrons; **(b)** octet; **(c)** unshared electron pairs; **(d)** multiple bonds; **(e)** coordinate covalent bonds; **(f)** resonance; **(g)** odd-electron species; **(h)** expanded octet?

8. By means of Lewis structures represent bonding between the following pairs of elements. Your structures should show clearly whether the bonding is essentially ionic or covalent. Give the name, formula, and formula weight of each. **(a)** Rb and Cl; **(b)** H and Se; **(c)** B and Cl; **(d)** Cs and S; **(e)** Sr and O; **(f)** F and O.

9. Write plausible Lewis structures for the following molecules, which contain only single covalent bonds: **(a)** Br_2; **(b)** ICl; **(c)** OF_2; **(d)** NI_3; **(e)** H_2Te.

10. The following molecules contain multiple covalent bonds. Give a plausible Lewis structure for each. **(a)** CO; **(b)** CS_2; **(c)** O_3; **(d)** H_2CO.

11. Indicate what is wrong with each of the following Lewis structures. Replace each by a more acceptable structure.

(a) H:H:N̈:Ö:H   **(b)** :Ö:N̈l:Ö:

(c) :B̈r:P::B̈r: **(d)** Ca
          :B̈r:

(e) $[\cdot \ddot{C} :: \ddot{N} :]^-$ **(f)** $[: \ddot{S} : C :: \ddot{N} :]^-$

(g) $[: \ddot{C}l]^+ [: \ddot{O} :]^{2-} [: \ddot{C}l]^+$

12. Suggest reasons why the following do not exist as stable molecules: **(a)** H_3; **(b)** HHe; **(c)** He_2; **(d)** H_3O.

Formal charge

13. Assign formal charges to the species represented below. If there are no formal charges present for certain of these species, so indicate.

(a) $: \ddot{I} - \ddot{I} :$

(b) (structure: S double bonded to two O)
$: \ddot{O} :: S :: \ddot{O} :$

(c) (structure: S bonded to two O)
$: \ddot{O} \overset{\ddot{S}}{} \ddot{O} :$

(d) (structure: C double bonded to O, bonded to two Cl)
$: \ddot{C}l \overset{C}{} \ddot{C}l :$ with $: \ddot{O} :$ above

(e) $[H - \ddot{O} - \ddot{O} :]^-$

(f) (structure: N bonded to two O)
$: \ddot{O} \overset{\dot{N}}{} \ddot{O} :$

(g) $: \ddot{F} - \ddot{S} = \ddot{O} :$ with $: \ddot{F} :$ below

14. Use the concept of formal charge to select the more likely skeleton structure for each of the following molecules: **(a)** H_2NOH or H_2ONH; **(b)** SCS or CSS; **(c)** NOCl or ONCl; **(d)** NNO or NON.

Polyatomic ions

15. Propose Lewis structures for the following ionic species containing sulfur-to-sulfur bonds: **(a)** S_2^{2-}; **(b)** S_3^{2-}; **(c)** S_4^{2-}; **(d)** S_5^{2-}.

16. The polyatomic anions below involve covalent bonds between O atoms and the central nonmetal atom. Propose plausible Lewis structures for each. **(a)** BrO_3^-; **(b)** ClO_2^-; **(c)** NO_2^-.

17. Represent each of the following compounds by an appropriate Lewis structure: **(a)** NH_4I; **(b)** NaOH; **(c)** HOCl; **(d)** $Ca(ClO_2)_2$.

Resonance

18. In the manner used to establish the structures for SO_2 shown in (9.18), demonstrate that there are *three* equivalent structures that can be written for SO_3.

19. In Example 9-8 the phenomenon of resonance was illustrated for the nitrate ion. Resonance is also involved in the nitrite ion, NO_2^-. Represent this fact through appropriate Lewis structures.

20. With reference to the ozone molecule, O_3, show that no single structure can be written for this molecule if it is assumed that the two O—O bonds are equivalent.

21. Dinitrogen oxide, N_2O, can be represented as a resonance hybrid of the structures shown below. Which structure(s) seems most plausible? What added experimental evidence would be helpful in resolving this question?

(a) $: N \equiv N - \ddot{O} :$ **(b)** $: \ddot{N} = N = \ddot{O} :$ **(c)** $: \ddot{N} - N \equiv O :$

Odd-electron species

22. As with the case of NO_2 described in the text, the molecule NO is paramagnetic. Represent this molecule through a Lewis structure(s).

23. NO_2 may dimerize (two molecules join) to the molecule N_2O_4. Write a plausible Lewis structure for N_2O_4. Do you think that N_2O_4 is diamagnetic or paramagnetic?

24. Which of the following species would you expect to be diamagnetic and which paramagnetic? Explain. (*Note:* Some of these species are not especially stable.) **(a)** OH^-; **(b)** OH; **(c)** NO_3; **(d)** SO_3; **(e)** SO_3^{2-}; **(f)** HO_2.

Expanded octets

25. Phosphorus and sulfur atoms make use of expanded octets in many of their compounds. Nitrogen and oxygen never do. Would you expect As and Se to resemble more nearly P and S or N and O? Explain.

26. Draw plausible Lewis structures for the following species, using the notion of expanded octets where necessary: **(a)** BrF_5; **(b)** PF_3; **(c)** ICl_3; **(d)** SF_4.

27. Indicate the nature of the sulfur-to-nitrogen bond in F_3SN (i.e., single, double, triple). (*Hint:* Use the notion of an expanded octet and ideas about formal charges.)

28. Exercise 18 refers to three *equivalent* structures for SO_3. Add to these *four* plausible structures based on an expanded octet for the sulfur atom. Of all seven structures (three from Exercise 18 and four here), which is the most plausible from the standpoint of formal charges? What experimental evidence would be helpful in assessing the relative importance of the various possible structures?

Molecular shapes

29. Use the valence-shell electron-pair repulsion theory to predict the geometrical shapes of the following species: **(a)** CO; **(b)** $SiCl_4$; **(c)** $SbCl_5$; **(d)** H_2Se; **(e)** ICl_3; **(f)** AlF_6^{3-}.

30. Each of the following molecules contains one or more multiple covalent bonds. Draw plausible Lewis structures to represent this fact, and predict the geometrical shape of each molecule. **(a)** CO_2; **(b)** N_2O; **(c)** NSF; **(d)** $ClNO_2$.

31. Can you think of an example of a molecule in which the central atom has *one bonding pair* and *three lone pairs* of electrons? What must be the shape of this molecule?

32. The structure of BF_3 is shown in Table 9-2 to be planar. If a fluoride ion is attached to the B atom of BF_3 through a coordinate covalent bond, the ion BF_4^- results. What is the geometrical shape of this ion?

33. Three possible Lewis structures for SO_3 were described in Exercise 18, and another four (based on an expanded octet for S) in Exercise 28. Why is the geometrical shape predicted for SO_3 *independent* of whichever of these Lewis structures is used?

34. Use the VSEPR theory to predict the geometrical shape of the triiodide ion, I_3^-.

Bond distances

35. In the gaseous state, HNO_3 molecules have two nitrogen-to-oxygen bond distances of 121 pm and one of 140 pm. Draw a plausible Lewis structure(s) to represent this fact.

36. A relationship between bond distances and single covalent radii of atoms is suggested in Section 9-9. Use this relationship and appropriate data from Table 9-3 to calculate the bond distances: **(a)** H—Cl; **(b)** C—N; **(c)** C—Cl; **(d)** C—F; **(e)** N—I.

37. Draw a sketch of the hydroxylamine molecule, H_2NOH, representing the geometrical shape of the molecule and, where possible, bond angles and distances.

Bond energies

38. Assuming that bond energies are additive, show that the total energy associated with the bonds in 1 mol of ethane, C_2H_6, is 2831 kJ. Would you expect the total bond energy in 1 mol of ethylene, C_2H_4, to be greater or less than for 1 mol of ethane? Explain.

39. Estimate the enthalpy change (ΔH) for the following reactions by using bond energies from Table 9-3.
 (a) $C_2H_6(g) + Cl_2(g) \longrightarrow C_2H_5Cl(g) + HCl(g);$
$$\Delta H = ?$$
 (b) $C_2H_4(g) + H_2(g) \longrightarrow C_2H_6(g) \qquad \Delta H = ?$

40. Without performing detailed calculations, indicate whether each of the following reactions is endothermic or exothermic.
 (a) $CH_4(g) + I(g) \longrightarrow CH_3(g) + HI(g)$
 (b) $H_2(g) + I_2(g) \longrightarrow 2\,HI(g)$
 (c) $C_2H_4(g) + Cl_2(g) \longrightarrow C_2H_4Cl_2(g)$

41. Use bond energies to calculate the heat of formation of $NH_3(g)$ and compare your result with the value listed in Appendix D.

$$N_2(g) + 3\,H_2(g) \longrightarrow 2\,NH_3(g)$$

*42. Use a value of 497 kJ/mol for the bond energy in $O_2(g)$ and other necessary data from the text to estimate the bond energy in NO(g).

Polar molecules

43. Estimate the percent ionic character of the HBr molecule, given that the dipole moment is 0.79 D.

44. Arrange the following in their expected order of increasing dipole moments: AsH_3; AsF_3; $AsCl_3$; $AsBr_3$; AsI_3.

45. Predict the shapes of the following molecules, and then predict which you would expect to have resultant dipole moments: **(a)** SO_2; **(b)** NO; **(c)** HBr; **(d)** NH_3; **(e)** H_2S; **(f)** C_2H_4; **(g)** BF_3; **(h)** SF_6; **(i)** CH_2Cl_2.

46. The molecule H_2O_2 has a dipole moment of 2.13 D. The bonding is H—O—O—H. Which of these bonds have bond dipole moments? Can the molecule be linear? Explain.

Partial ionic character of covalent bonds

47. Calculate the ionic resonance energies of HF and HBr. Do these values compare with that for HCl in the way you would expect? Explain.

48. Use electronegativity data to arrange the following bonds in terms of *increasing* ionic character: C—H, F—H; Na—Cl; Br—H; K—F.

49. Use the results of Exercise 47 and equation (9.34) to calculate the electronegativity differences between **(a)** H and F; **(b)** H and Br. Compare your results with values obtained from Table 8-6.

*50. Use the methods of Examples 9-17 and 9-18, together with appropriate data from the text, to estimate the O—O single bond energy. The measured N—O single bond energy is 201 kJ/mol.

Oxidation states

51. What is the relationship, if any, between the concepts of oxidation state and formal charge? Explain.

52. Indicate the oxidation state of the underlined element in each of the following species: **(a)** $\underline{Al}^{3+}$; **(b)** Ba$\underline{S}$; **(c)** Mg$\underline{S}O_3$; **(d)** Na$\underline{O}$H; **(e)** H$\underline{N}O_3$; **(f)** $\underline{I}O_4^-$; **(g)** $\underline{S}_2O_3^{2-}$; **(h)** $\underline{Cr}_2O_7^{2-}$; **(i)** K$\underline{Mn}O_4$.

53. Arrange the following sulfur-containing anions in order of *increasing* oxidation state of the sulfur atom: SO_3^{2-}; $S_2O_3^{2-}$; $S_2O_8^{2-}$; HSO_4^-; HS^-; $S_4O_6^{2-}$.

Nomenclature

54. Use ideas introduced in Section 9-11 to name the following compounds: **(a)** $FeBr_3$; **(b)** CrI_3; **(c)** $Ca(ClO)_2$; **(d)** $NaBrO_3$; **(e)** KIO_4; **(f)** $Na_2S_2O_8$.

55. Following the scheme outlined in Table 9-4, together with other ideas on nomenclature, supply appropriate formulas for the following compounds: **(a)** periodic acid; **(b)** tin(IV) oxide; **(c)** sodium selenate; **(d)** magnesium perchlorate; **(e)** gold(III) cyanide; **(f)** potassium iodide; **(g)** barium telluride.

Additional Exercises

1. Refer to Table 9-1 and write the complete electron configurations for the following ions: (a) Y^{3+}; (b) Cd^{2+}; (c) Sb^{3+}.

2. What are the principal assumptions made in writing Lewis structures for covalent molecules? Cite some exceptions to these assumptions.

3. A compound is found to consist of 47.5% S and 52.5% Cl, by mass. Write a Lewis structure for this compound and comment on its deficiencies. Write a different structure with the same ratio of S to Cl that is more plausible?

4. What is the formal charge on the indicated atom in each of the following?
   (a) oxygen in OH^- (structure 9.14)
   (b) nitrogen in NH_4^+ (Figure 9-6)
   (c) sulfur in SO_2 (structure 9.18)
   (d) nitrogen in NO_3^- (structure 9.19)
   (e) boron in $H_3N \cdot BF_3$ (structure 9.23)
   (f) phosphorus in PCl_5 (structure 9.24)
   (g) sulfur in SO_4^{2-} (Figure 9-8)
   (h) sulfur in SO_2 (structure 9.26)
   (i) iodine in ICl_4^- (Figure 9-12)

5. Comment on the irregularities that arise in assigning oxidation states to the elements in the following species: (a) Na_2O_2; (b) I_3^-; (c) $S_3O_6^{2-}$; (d) H_2CO.

6. What is the relationship between the geometrical shapes of (a) the ammonia molecule, NH_3, and the ammonium ion, NH_4^+; (b) sulfur trioxide, SO_3, and the sulfate ion, SO_4^{2-}?

7. One each of the following species is linear, V-shaped, planar, tetrahedral, and octahedral. Indicate the correct structure for each: (a) H_2Te; (b) C_2Cl_4; (c) CO_2; (d) $SbCl_6^-$; (e) SO_4^{2-}.

8. The bond energy in the carbon monoxide molecule, CO, is about 1070 kJ/mol. Use this fact, together with data from Table 9-3, to propose a plausible Lewis structure(s) for CO.

9. Indicate which of the following molecules you would expect to have a resultant dipole moment, and give reasons for your conclusions: (a) HCN; (b) SO_3; (c) CS_2; (d) COS; (e) $SOCl_2$; (f) SiF_4; (g) POF_3; (h) XeF_2.

*10. Carbon suboxide has the formula C_3O_2. The carbon-to-carbon distances are found to be 130 pm, and the carbon-to-oxygen distances are 120 pm. Propose plausible Lewis structures to account for these bond distances, and predict the geometrical shape of the molecule.

*11. The aldehyde propynal has the formula HCCCHO. Draw a sketch that represents bonding in the molecule and its shape. Include bond distances and bond angles.

*12. The total bond energy associated with all the bonds in

$$\overset{O}{\overset{\|}{}}$$

thiourethane, $H_2NCSCH_2CH_3$, is 4780 kJ/mol. Use this value, together with data from Table 9-3, to estimate the energy of the C—S bond. How would you expect this energy to compare in value with the bond energies of the carbon-to-sulfur bonds in CS_2?

*13. Estimate the enthalpies (heats) of formation of the following species at 25°C and 1 atm: (a) $N_2H_4(g)$; (b) $OH(g)$; (c) $CH_3(g)$. Use data from the text as necessary.

*14. The bond energy in the $O_2(g)$ molecule is 497 kJ/mol; and the heat of formation of $H_2O_2(g)$ is −136 kJ/mol. Use these values, together with other appropriate data from the text, to estimate the oxygen-to-oxygen single bond energy.

*15. The text states that the *bond* dipole moment of the O—H bond is 1.51 D; the H—O—H bond angle is 104°; and the resultant dipole moment of the H_2O molecule is 1.84 D. (See expression 9.33.)
   (a) Show by an appropriate geometric calculation that the three statements made above for the H_2O molecule are mutually consistent.
   (b) Use the same method as developed in part (a) to estimate the bond angle in H_2S, given that the H—S bond moment is 0.67 D and the resultant dipole moment of the H_2S molecule is 0.93 D.

Self-Test Questions

For questions 1 through 5 select the single item that best completes each statement.

1. Of the following species, the one containing a triple covalent bond is (a) NO_3^-; (b) CN^-; (c) CO_2; (d) $AlCl_3$.

2. In the ammonium ion, NH_4^+; (a) the four H atoms are situated at the corners of a square; (b) all bonds are ionic; (c) all bonds are coordinate covalent; (d) the N atom carries a formal charge.

3. The oxidation state of I in the ion $H_4IO_6^-$ is (a) −1; (b) +1; (c) +7; (d) +8.

4. All of the following molecules are linear except one. That one is (a) SO_2; (b) CO_2; (c) HCN; (d) C_2H_2.

5. Of the following molecules, all are polar except one. That one is (a) BCl_3; (b) CH_2Cl_2; (c) NO; (d) PCl_3.

6. A chemical compound is found to have the following per-

cent composition, by mass: 24.3% C, 71.6% Cl, and 4.1% H.

(a) What is the *empirical* formula of this compound?

(b) Draw a Lewis structure based on this empirical formula and comment on its inadequacies.

(c) Propose a *molecular* formula for the compound that results in a more plausible Lewis structure.

7. All of the species indicated below exist. Of the Lewis structures shown, one is plausible but the other three are much less so. Identify the plausible one and comment on the inadequacies of the other three.

(a) cyanate ion, $[:\overset{..}{O}:C::\overset{..}{N}:]^-$

(b) carbide ion, $[C\overset{..}{:}::\overset{..}{C}:]^{2-}$

(c) hypochlorite ion, $[:\overset{..}{\underset{..}{Cl}}:\overset{..}{\underset{..}{O}}:]^-$

(d) nitrogen oxide, $:\overset{..}{N}{=}\overset{..}{O}\cdot$

8. Draw Lewis structures for two different molecules having the formula C_3H_4. Is either of these molecules linear? Explain.

9. In which of the following molecules is the nitrogen-to-nitrogen bond distance expected to be the shortest: (a) N_2H_4; (b) N_2; (c) N_2O_4; (d) N_2O? Explain.

10. Predict the geometrical shapes of the following sulfur-containing species: (a) SO_2; (b) SO_3; (c) SO_4^{2-}.

11. Given the bond energies N-to-O bond in NO, 628 kJ/mol; H—H, 435 kJ/mol; N—H, 389 kJ/mol; O—H, 464 kJ/mol, calculate ΔH for the reaction

$$2\,NO(g) + 5\,H_2(g) \longrightarrow 2\,NH_3(g) + 2\,H_2O(g)$$

12. The following statements are not made as carefully as they might be. Criticize each one.

(a) Triatomic molecules have a planar shape.

(b) Molecules in which there is an electronegativity difference between the bonded atoms are polar.

(c) Lewis structures in which atoms carry formal charges are incorrect.

10 Chemical Bonding II: Additional Aspects

Ionic bonding is basically understandable in terms of the Coulombic force of attraction between oppositely charged objects—positive and negative ions. Our description of covalent bonding to this point has been less satisfactory. We have been successful in deriving formulas for a large number of covalent molecules by writing Lewis structures. Moreover, when combined with VSEPR theory, Lewis structures allow for a prediction of the shapes of molecules. However, they do not permit a prediction of bond energies or, in some cases, magnetic properties.

More adequate descriptions of the covalent bond require the methods of wave mechanics. As in Chapter 7 on the electronic structures of atoms, the discussion in this chapter is limited to a few basic ideas applied in a qualitative way. The basic requirement of a covalent chemical bond is that it correspond to a region between bonded atoms where the probability of finding electrons or the electron charge density is high. In this chapter we consider two basic approaches to describe these regions. These are known as the valence bond method and molecular orbital theory. Also considered

FIGURE 10-1
Bonding in H_2 represented
by atomic orbital overlap.

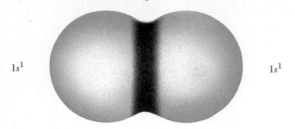

$1s^1$ $1s^1$

Each atomic orbital contains one unpaired electron. As
a result of the overlap of the two orbitals, the electrons become paired and a region of high electron
probability results—a covalent bond. Note how the
characteristic features of the 1s atomic orbital are retained, except in the region of overlap (shown in
black). (That is, compare this figure with Figure 7-16.)

FIGURE 10-2
Covalent bonding in H_2S represented
by atomic orbital overlap.

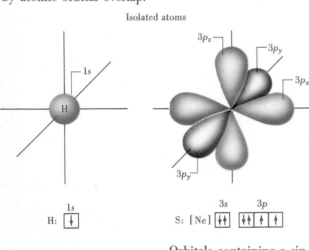

Isolated atoms

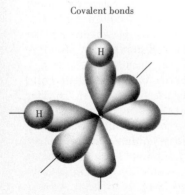

Covalent bonds

Orbitals containing a single electron are in color;
those with an electron
pair are grey. For S, only
3p orbitals are shown.
The 1s orbitals of two
hydrogen atoms overlap
with the $3p_y$ and $3p_z$
orbitals of the sulfur
atom. Here and else-
where in this chapter the
shapes of p orbitals have
been elongated along the
axes through the nucleus.
This permits a better visualization of geometric
structures and other
aspects of covalent
bonding.

in this chapter are bonding in metals and in semiconductors.

10-1 The Valence Bond Approach to Chemical Bonding

This method of describing the covalent bond originated within a year or two of Schrödinger's work on
the hydrogen atom. It views atoms involved in covalent
bond formation as being largely unchanged from their
isolated conditions. All that is considered to happen in
bond formation is that electron orbitals centered on the
individual atoms (called atomic orbitals) overlap. A
covalent bond, then, arises from the high electron
charge density (high electron probability) in the region
of atomic orbital overlap between bonded atoms. The
overlap of two 1s orbitals in a hydrogen molecule is
suggested in Figure 10-1.

The overlap of atomic orbitals involved in the formation of hydrogen-to-sulfur bonds in hydrogen sulfide is depicted in Figure 10-2. From this figure we
should note the following.

1. The number of covalent bonds between atoms is
such that, normally, all the unpaired electrons in the
original atoms become paired.
2. If valence electrons are counted in the same way as
in Lewis structures, each atom normally acquires a
noble gas electron configuration.
3. The geometrical shape of the molecule is determined by the orientation of the overlapping atomic
orbitals of the bonded atoms.

Example 10-1   Describe the structure of the NH_3 molecule by the valence bond method.
Step 1. Draw orbital diagrams for the separate atoms
that are to be bonded.

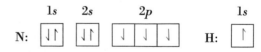

Step 2. Sketch the orbitals of the central atom (N)
that will be involved in the orbital overlap (see Figure
10-3).
Step 3. Complete the structure by bringing together
the bonded atoms and representing the orbital overlap.
Step 4. Describe the structure. NH_3 is a trigonal pyramidal molecule. The three H atoms lie in the same
plane. The N atom is situated at the apex of the pyramid
above the plane of the H atoms. The three H—N—H
bond angles are predicted to be 90°.

SIMILAR EXAMPLE: Exercise 1.

FIGURE 10-3
Bonding and structure
of NH_3 molecule—
Example 10-1
illustrated.

bonding orbitals of N atom

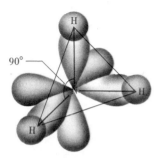

90°

covalent bonds formed

Orbitals with single electrons are shown in color;
those with an electron
pair are grey. Only bond-
ing orbitals are shown.
The 1s orbitals of three
hydrogen atoms overlap
with the three 2p orbit-
als of the nitrogen atoms.
The bond angles are pic-
tured to be 90°.

The number of hybrid
orbitals generated in a
hybridization scheme is
always equal to the total
number of pure atomic
orbitals involved.

If bonding in a molecule involves two or all three *p* orbitals of the valence shell of the central atom, we should expect 90° bond angles. This is because *p* orbitals are mutually perpendicular. The experimentally determined bond angle in H_2S, for example, is 92°. But the measured bond angles are 104.5° in H_2O and 107° in NH_3, not the predicted 90°. Taking H_2O as an example, the greater-than-90° bond angles can be rationalized as follows. Because O is more electronegative than H, there is a displacement of electrons toward the O atom in the O—H bonds in H_2O. This leaves the H atoms with a slight positive charge. The H atoms repel one another and cause an increase in the bond angle. The effect in NH_3 is similar. The effect is not as significant in H_2S because sulfur is not as electronegative as N and O.

In addition to its inadequacies in dealing with H_2O and NH_3, the unmodified valence bond method fails for other simple molecules, such as CH_4 and CO_2. We need to develop a modification of the simple valence bond method.

10-2 Hybridization of Orbitals

In the examples cited in the preceding section we were able to represent covalent bonding by starting with the normal or ground state electron configurations of the separated atoms. If we start with the ground state electron configurations of carbon and hydrogen, we would predict CH_2 rather than CH_4 as the simplest hydrocarbon molecule. To write an acceptable Lewis structure for CH_4 requires *four* unpaired electrons in the Lewis symbol of C. In terms of atomic orbitals this requirement can be met by postulating that a C atom acquires an "excited" electron configuration. One of the 2s electrons is "promoted" to the 2p subshell.

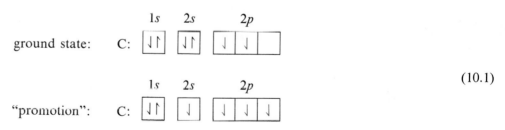

(10.1)

The molecular geometry predicted from the orbital diagram (10.1) would be a molecule with three mutually perpendicular C—H bonds (i.e., with bond angles of 90°). The fourth C—H bond would have no particular orientation with respect to the other three. However, we have already described the structure of CH_4 (see Table 9-2): The C—H bonds are directed in a tetrahedral fashion from the C atom; that is, the H—C—H bond angles are 109.5°. The orbital diagram (10.1) accounts for the correct number of bonds but not the correct orientation of these bonds.

The essential difficulty here is that the description of pure atomic orbitals (*s*, *p*, *d*, and *f*) is based on *isolated* atoms. There is no reason to assume, though, that the electron configurations of normal *isolated* atoms are applicable to *bonded* atoms. The fact that they are adequate in explaining some covalent bonds is fortunate. That they fail to apply in a large number of other cases is not an unreasonable finding either.

If the 2s and 2p orbitals are combined in an appropriate way, they generate a new set of orbitals. The set consists of four *identical* orbitals at exactly the tetrahedral bond angles, 109.5°. Orbitals obtained by this type of combination of pure atomic orbitals are called **hybrid orbitals.** The particular hybridization scheme described here and pictured in Figure 10-4 is called *sp*³ hybridization. The symbol *sp*³ signifies that one *s* and three *p* orbitals have been combined to produce a set of four new hybrid orbitals. A suitable orbital diagram for this hybridization scheme is

FIGURE 10-4
The sp^3 hybridization scheme.

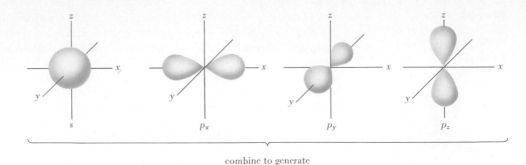

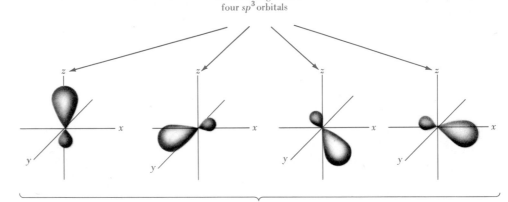

combine to generate
four sp^3 orbitals

FIGURE 10-5
Bonding and structure of CH_4.

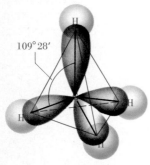

109°28′

geometric structure

which are represented
as the set

space-filling model

The four carbon orbitals involved in the bonding scheme (in grey) are $2sp^3$ hybrid orbitals. Those of the hydrogen atoms (in color) are $1s$. The structure is tetrahedral, with bond angles of about 109.5° (more exactly, 109°28′).

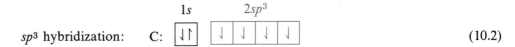

sp^3 hybridization: C: $\boxed{\downarrow\uparrow}$ $\boxed{\downarrow\;|\;\downarrow\;|\;\downarrow\;|\;\downarrow}$ (10.2)

The use of sp^3 hybrid orbitals in bond formation in CH_4 is pictured in Figure 10-5.

The term "scheme" (a systematic plan for attaining some objective) is an appropriate one for describing hybridization. The objective is an after-the-fact attempt to account for the geometrical shape that is experimentally observed for a molecule. Hybridization is not an actual physical phenomenon. There is no way during bond formation to see electron charge distributions rearranging from those described by pure atomic orbitals to those described by hybrid orbitals. Moreover, we should be prepared to accept the fact that for some covalent bonds no single hybridization scheme works well.

WATER (H_2O) AND AMMONIA (NH_3).   An sp^3 hybridization scheme produces good agreement with experiment and with VSEPR theory for the bond angles in H_2O and NH_3. In the orbital diagrams (10.3) and (10.4) the $2s$ and three $2p$ orbitals have been hybridized and the correct number of electrons placed in each orbital.

FIGURE 10-6
sp^3 hybrid orbitals and
bonding in H_2O and NH_3.

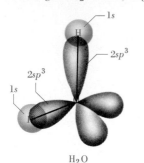

H_2O

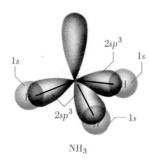

NH_3

$$
\begin{array}{cc}
1s & 2sp^3 \\
\end{array}
$$

N: $\boxed{\downarrow\uparrow}$ $\boxed{\downarrow\uparrow\,|\,\downarrow\,|\,\downarrow\,|\,\downarrow}$ $\qquad (10.3)$

$$
\begin{array}{cc}
1s & 2sp^3 \\
\end{array}
$$

O: $\boxed{\downarrow\uparrow}$ $\boxed{\downarrow\uparrow\,|\,\downarrow\uparrow\,|\,\downarrow\,|\,\downarrow}$ $\qquad (10.4)$

As shown in Figure 10-6, in H_2O there are two bonds and two unshared (lone) pairs of electrons; the molecule is V-shaped. In NH_3 there are three bonds and one unshared (lone) pair of electrons; the molecule has a trigonal pyramidal shape.

sp AND sp^2 HYBRID ORBITALS.   Two additional schemes for hybridizing s and p orbitals are shown in orbital diagrams (10.5) and (10.6), one for the boron atom and one for beryllium.

FIGURE 10-7
The sp^2 and sp hybridization schemes.

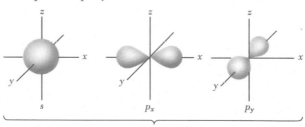

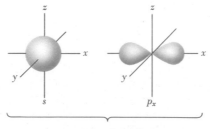

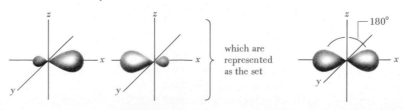

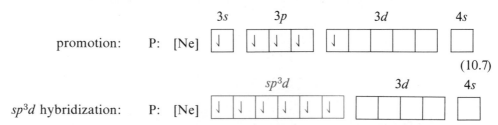

$$\text{(10.5)}$$

$$\text{(10.6)}$$

As pictured in Figure 10-7, the *three sp^2* hybrid orbitals are directed in the same plane at angles of 120°. The *two sp* hybrid orbitals are directed along a straight line, at a 180° angle. Thus, a molecule like BF_3 is *trigonal planar,* whereas $BeCl_2$ is *linear.*

d **HYBRID ORBITALS.** The concept of hybridization, if extended to include *d* orbitals, helps to explain the occurrence of expanded octets. Consider the molecule PCl_5, for example. The ground state electron configuration of phosphorus is

But to describe the formation of PCl_5 the following *hypothetical* process is proposed: (1) All the valence electrons are unpaired. (2) Electrons are placed successively and individually into $3s$, $3p$, and $3d$ orbitals. (3) All the orbitals that are occupied as a result are then hybridized.

$$\text{(10.7)}$$

A set of sp^3d hybrid orbitals is pictured in Figure 10-8 together with the trigonal bipyramidal molecule resulting from the use of these five orbitals in bond formation. In the molecule SF_6 there must be a set of *six* bonding orbitals. These are the sp^3d^2 hybrid orbitals.

Note again that the symbolism used to describe this scheme, sp^3d^2, indicates which atomic orbitals are involved.

$$\text{(10.8)}$$

The geometrical distribution of these orbitals, as shown in Figure 10-8, is octahedral.

HYBRID ORBITALS AND THE VALENCE-SHELL ELECTRON-PAIR REPULSION THEORY. In most cases the geometric structures of molecules predicted by the use of hybrid orbitals agree with predictions made by the valence-shell electron-pair repulsion (VSEPR) method. There is, of course, a connection between the two.

One hybrid orbital is produced for every pure atomic orbital that enters into a hybridization scheme. In a molecule each of the hybrid orbitals of a central atom normally acquires an electron pair, either a bond pair or a lone pair. Thus, the number of hybrid orbitals equals the number of electron pairs. Furthermore, in most

FIGURE 10-8
d hybrid orbitals—sp^3d and sp^3d^2.

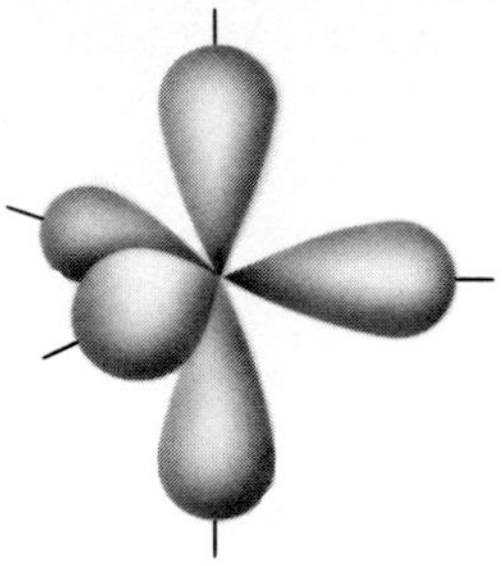

FIGURE 10-9
sp^2 hybridization and bonding in C_2H_4.

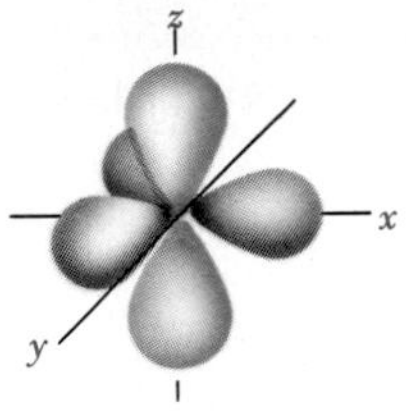

sp^3d orbitals

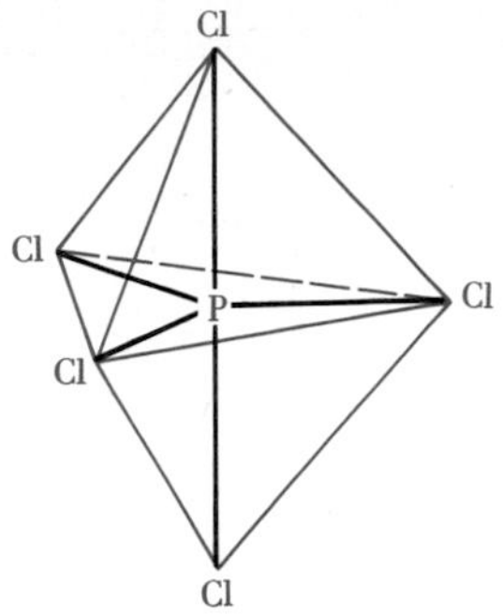

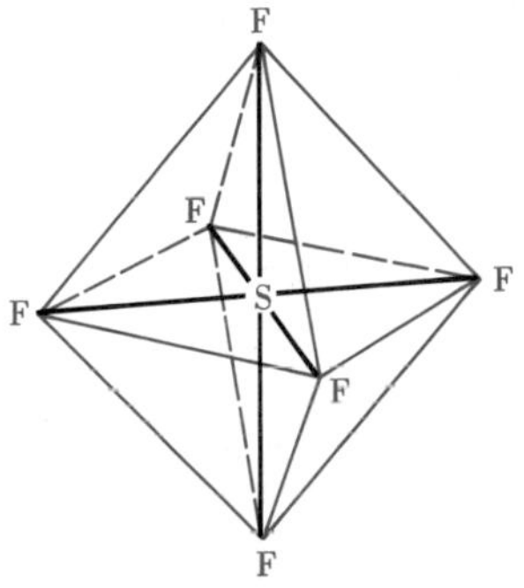

trigonal bipyramidal structure

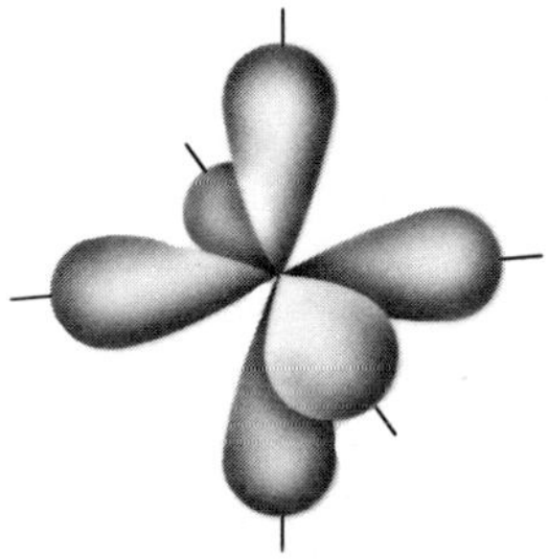

the set of orbitals $sp^2 + p$

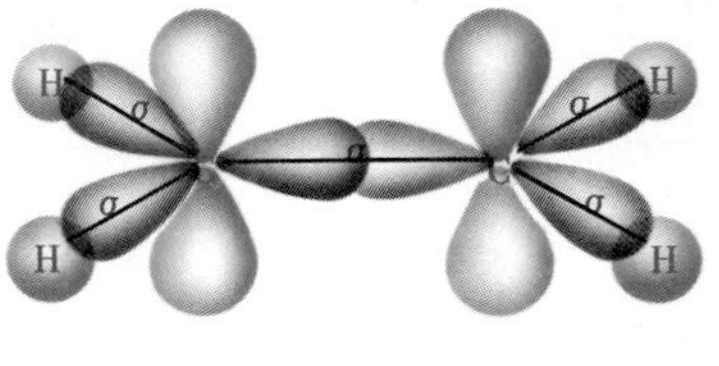

sigma (σ) bonds

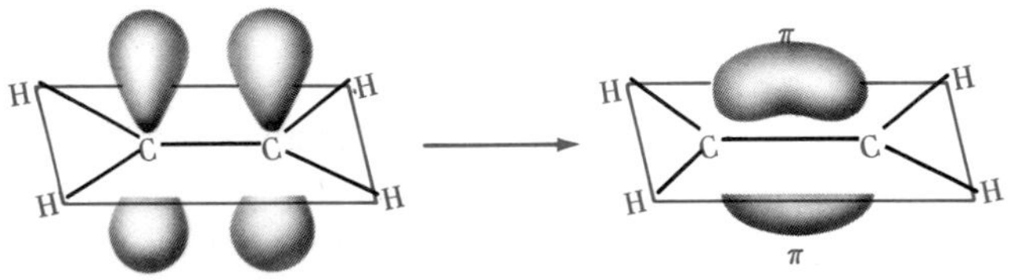

sp^3d^2 orbitals

octahedral structure

overlap of *p* orbitals leading to pi (π) bond

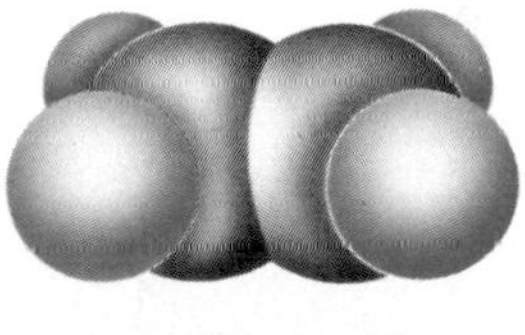

space–filling model

cases the orientation of hybrid orbitals is the same as that of electron pairs predicted by the VSEPR method. This fact is brought out in Table 10-1. There is one notable case where the valence bond method is superior to the VSEPR method. This is in structures where four pairs of valence electrons appear, but with the hybridization scheme dsp^2 rather than sp^3. These structures have a square planar geometry, a geometry not predicted by the VSEPR method. Examples of complex ions with this geometrical shape are described in Chapter 22.

TABLE 10-1
Hybrid orbitals and their geometric orientation

Pure orbitals	Hybrid orbitals	Orientation	Example	Bond angle
$s + p$	sp	linear	$BeCl_2$	180°
$s + p + p$	sp^2	trigonal planar	BF_3	120°
$s + p + p + p$	sp^3	tetrahedral	CH_4	109.5°
$d + s + p + p$	[a]dsp^2	square planar	$[Cu(NH_3)_4]^{2+}$	90°
$s + p + p + p + d$	[b]sp^3d	trigonal bipyramidal	PCl_5	120, 90°
$s + p + p + p + d + d$	[b]sp^3d^2	octahedral	SF_6	90°
$d + d + s + p + p + p$	[a]d^2sp^3	octahedral	$[Co(NH_3)_6]^{2+}$	90°

[a] These hybrid orbitals involve *d* orbital(s) from the next-to-outermost shell, together with *s* and *p* orbitals of the outermost shell. They are encountered commonly in the structures of complex ions (see Chapter 22).

[b] These hybrid orbitals involve *s*, *p*, and *d* orbitals, all from the outermost electronic shell. They are encountered in structures with an expanded octet that have nonmetals such as P, As, S, Cl, Br, and I as the central atom (see Chapter 20).

10-3 Multiple Covalent Bonds

To represent bonding in ethylene, C_2H_4, the following hybridization must be used. It results in both a set of hybrid orbitals (sp^2) *and a pure p orbital.*

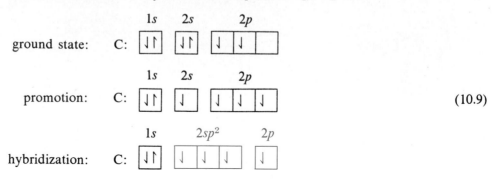

(10.9)

The three sp^2 hybrid orbitals are directed in a plane with an angle of 120° between each pair of orbitals. Bonding through these orbitals leads to a molecule in which all six atoms lie in the same plane. The molecule C_2H_4 is *planar.* The hybridization scheme, the nature of the orbital overlap, and a space-filling model of the molecule are illustrated in Figure 10-9.

In the double covalent bond in Figure 10-9 we picture one of the bonds as resulting from the overlap of sp^2 hybrid orbitals along the line joining the nuclei of the two carbon atoms. Orbitals that overlap in this "end-to-end" fashion produce a **sigma bond,** designated **σ bond.** Another bond arises from the overlap of the pure *p* orbitals and is characterized by regions of high electron density above and below the plane of the carbon and hydrogen atoms. This "side-by-side" overlap of *p* orbitals produces a bond known as a **pi bond,** designated **π bond.** The σ and π bonds are labeled in Figure 10-9.

Several aspects of the hybridization scheme outlined for C_2H_4 need further mention. First, the σ bond involves a greater overlap than does the π bond. As a result we should expect a carbon-to-carbon double bond (σ + π) to be stronger than a single bond (σ), but not twice as strong. The data given in Table 9-3 were as follows: C—C, 347 kJ/mol; C=C, 611 kJ/mol. Second, the shape of a molecule is determined only by the distribution of the orbitals leading to σ bonds (the "σ-bond framework"). This corresponds to the VSEPR method of treating multiple bonds as if they were single bonds. Finally, rotation about a double bond is restricted; that is, the double bond is quite rigid. Consider the space-filling model of C_2H_4 in Figure 10-9, for example. To twist one —CH$_2$ group out of the plane of the other would reduce the amount of overlap of the *p* orbitals. Rotation about a σ bond occurs more freely since the end-to-end overlap of orbitals is not disturbed by this type of motion.

In the molecule C_2H_2 (acetylene) the two carbon atoms are bonded through a triple covalent bond. The σ bonding scheme involves *sp* hybrid orbitals.

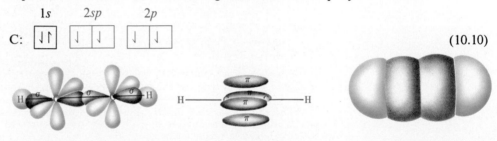

(10.10)

FIGURE 10-10
sp hybridization and bonding in C_2H_2.

formation of σ bonds   formation of π bonds   space–filling model

In the triple bond in C_2H_2 one of the $C\equiv C$ bonds is a σ bond and the other two are π bonds. These points are illustrated through Figure 10-10.

Example 10-2 Describe bonding in the molecule formaldehyde, H_2CO, in terms of atomic orbital overlap.

Generally, the following steps will lead to a plausible bonding scheme.

Step 1. Write a Lewis structure of the molecule. The total number of valence electrons is 2×1 (from H) + 4 (from C) + 6 (from O) = 12.

$$\begin{matrix} & H & \\ H &:\!\overset{\cdot\cdot}{C}\!:\!:\!\overset{\cdot\cdot}{O}\!: & \end{matrix}$$

Step 2. Identify multiple bonds in the Lewis structure. The carbon-to-oxygen double bond is a combination of a σ and a π bond. (A triple bond would signify one σ and two π bonds.)

Step 3. Establish the shape of the molecule using the VSEPR method. Since the central carbon atom forms three σ bonds, three electron pairs must be distributed. The molecule is *trigonal planar* (120° bond angles).

Step 4. Determine the pure and/or hybrid atomic orbitals that will produce the predicted geometry. A trigonal planar structure is based on sp^2 *hybrid orbitals* of the central atom.

Step 5. Sketch the orbitals and orbital overlap. The C atom uses two of its sp^2 hybrid orbitals to form σ bonds with the two H atoms. The remaining sp^2 hybrid orbital of the C atom is used to form a σ bond with oxygen. The p orbital of the carbon is used to form a π bond with oxygen. The orbitals employed by the H atoms are simply their $1s$ orbitals. The situation with oxygen is less clear because the bond angles in H_2CO are independent of whether pure $2p$ orbitals or hybridized orbitals are used. As shown in Figure 10-11, we have chosen an sp^2 hybridization scheme to place the lone pair valence electrons of oxygen farthest away from each other and from the bonding pairs.

FIGURE 10-11
Bonding and structure of the H_2CO molecule—Example 10-2 illustrated.

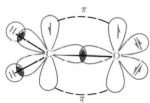

	$1s$	$2sp^2$	$2p$
sp^2 hybridization: O:	$\uparrow\downarrow$	$\uparrow\downarrow$ $\uparrow\downarrow$ $\uparrow$ $\downarrow$	$\downarrow$

SIMILAR EXAMPLE: Exercise 5.

Example 10-3 The following bond angles are given for the formic acid molecule, HCOOH: O—C—O, 118°; C—O—H, 108°. Propose a bonding scheme to account for the observed structure of formic acid.

A plausible Lewis structure for HCOOH is

$$\begin{matrix} :\!\overset{\cdot\cdot}{O}\!: & \\ H:\!\overset{\cdot\cdot}{C}\!:\!\overset{\cdot\cdot}{\underset{\cdot\cdot}{O}}\!:\!H & \end{matrix}$$

To form one double and two single covalent bonds, the central C atom must employ sp^2 hybrid orbitals. Use of these orbitals will also be consistent with the observed O—C—O bond angle (118° compared to a predicted 120°). The observed C—O—H bond angle (108°) is very close to the tetrahedral angle (109.5°). This suggests that the O atom employs sp^3 hybrid orbitals rather than pure p orbitals in forming bonds to carbon and hydrogen (recall the hybridization scheme presented in equation 10.4).

In Figure 10-12 bonds between atoms are drawn as straight lines; bonds are labeled σ or π; and the orbital overlap leading to each bond is indicated.

SIMILAR EXAMPLES: Exercises 8, 9, 10.

FIGURE 10-12
Bonding and structure of the HCOOH molecule—
Example 10-3 illustrated.

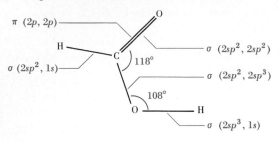

FIGURE 10-13
Energy of interaction of two hydrogen atoms as a
function of internuclear distance.

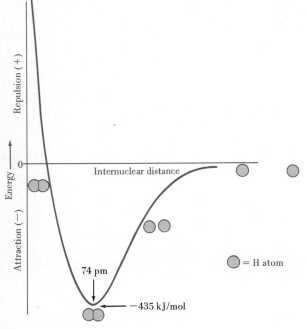

Two H atoms form the molecule H_2 at a particular internuclear distance (74 pm).

10-4 Molecular Orbital Theory

Imagine a process in which two hydrogen atoms are
brought together from an initial condition in which
they are "infinitely" far apart. The energy of interaction between the pair of atoms can be plotted as a
function of the distance between the nuclei of the two
atoms—the **internuclear distance.** Positive energies
indicate a net repulsive force between atoms; negative
energies, an attractive force.

As illustrated in Figure 10-13, the energy of interaction is zero for large internuclear distances: the
atoms neither attract nor repel one another. At very
small internuclear distances the atoms repel one another strongly. And over an intermediate range of internuclear distances an attractive force can develop
between the atoms. This force reaches a maximum at
one particular distance—74 pm (0.74 Å). This distance
is the H—H bond distance listed in Table 9-3. The energy of interaction at this distance is 435 kJ/mol—the
H—H bond energy. The challenge in developing a theoretical model of molecular structure is in being able to
predict molecular properties that are in close agreement with experiment.

The repulsive force between H atoms in very close
proximity is readily understood. It results from the
mutual repulsion of two positively charged nuclei. But
what is the source of the attractive forces at intermediate distances?

Figure 10-14 shows two different arrangements of
the protons and electrons in two hydrogen atoms that
are brought into close proximity. In the arrangement
where the electrons are located away from the internuclear region, the repulsion between protons is
strong. Energy is high and the arrangement is unstable.
The arrangement where the two electrons are located
between the atomic nuclei is of *lower* energy than in the
separated atoms. Thus, even classical theory (Coulomb's law) predicts that two H atoms should combine
to form an H_2 molecule. However, the energy of interaction predicted by classical theory is much less than
the experimental value, and the internuclear distance
is also in error. Success in these predictions is only possible using wave mechanics.

Let us think of electrons in terms of charge densities or probabilities extending over an entire molecule,
that is, in terms of **molecular orbitals.** One method of
deriving molecular orbitals is by an appropriate combination of atomic orbitals of the atoms being united
into a molecule. Wave mechanics allows two possibilities. One combination of two $1s$ orbitals represented in
Figure 10-14 produces a **bonding molecular orbital,**
designated σ_{1s}^{b}. Another combination produces an **anti-**

FIGURE 10-14
The interaction of two hydrogen atoms.

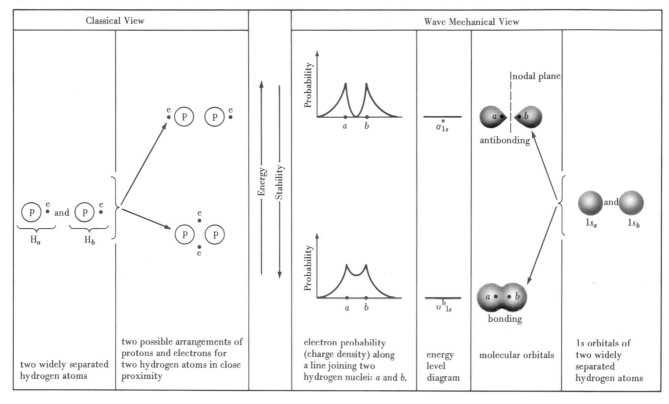

| Classical View | Wave Mechanical View |

two widely separated hydrogen atoms

two possible arrangements of protons and electrons for two hydrogen atoms in close proximity

electron probability (charge density) along a line joining two hydrogen nuclei: *a* and *b*.

energy level diagram

molecular orbitals

1s orbitals of two widely separated hydrogen atoms

bonding orbital, σ_{1s}^{*}. Also depicted in Figure 10-14 is an energy level diagram for these two molecular orbitals: *The bonding molecular orbital is at a lower energy than the separate atomic orbitals, and the antibonding molecular orbital at a higher energy.*

From the standpoint of electron probability or electron charge density, Figure 10-14 shows the bonding molecular orbital to correspond to a high electron probability or charge density *between* the atomic nuclei. Electron charge density concentrated in the internuclear region reduces the repulsive force between the positively charged nuclei. This permits bonding between the atoms—hence the term *bonding molecular orbital.* In the antibonding orbital the electron probability or charge density is much lower in the internuclear region. In fact, it falls to zero midway between the nuclei in a region known as the nodal plane. Electron charge density in the antibonding orbital is concentrated in regions away from the internuclear region, where it is ineffective in reducing internuclear repulsion—hence the term *antibonding molecular orbital.* The probability distributions for the bonding and antibonding molecular orbitals correspond roughly to the electron positions shown for the corresponding classical situations in Figure 10-14.

BASIC IDEAS CONCERNING MOLECULAR ORBITALS. To use molecular orbital theory to describe chemical bonding requires that we first establish some rules. These rules pertain to the particular molecular orbitals that arise when atomic orbitals are combined and the manner in which electrons are assigned to these molecular orbitals.

FIGURE 10-15
Molecular orbital diagrams for the diatomic molecules (or ions) formed from first period elements.

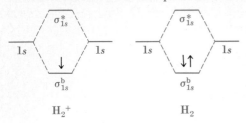

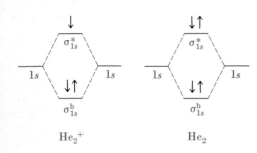

Note that both bonding schemes considered to this point (Lewis theory, valence bond method) have required a *pair* of electrons for a covalent bond. But one-electron bonds do exist and the molecular orbital theory accounts for them quite nicely.

1. The number of molecular orbitals produced is equal to the number of atomic orbitals combined.

2. Of the two molecular orbitals produced when two atomic orbitals are combined, one is a *bonding* molecular orbital at a *lower* energy than the original atomic orbitals. The other is an *antibonding* orbital at a *higher* energy.

3. Electrons normally seek the lowest energy molecular orbitals available to them in a molecule.

4. The maximum number of electrons that can be assigned to a given molecular orbital is *two* (Pauli exclusion principle).

5. Electrons enter molecular orbitals of identical energies *singly* before they pair up (Hund's rule).

6. Formation of a bond between atoms requires that the number of electrons in bonding molecular orbitals exceed the number of electrons in antibonding orbitals.

FIRST PERIOD ELEMENTS. Figure 10-15 suggests four possibilities for assigning electrons to the molecular orbitals depicted in Figure 10-14. Let us consider these possibilities.

H_2^+: This species has a single electron. The electron enters the σ_{1s}^b orbital and produces a bond between the two H atoms. We might call the bond a one-electron or "half" bond.

H_2: This molecule has two electrons, both of which enter the σ_{1s}^b orbital. A regular single covalent bond is formed.

He_2^+: This ion has a total of 3 electrons. Two are in the σ_{1s}^b orbital, and one in the σ_{1s}^*. The net number of bonding electrons is $2 - 1 = 1$. This species should exist as a stable ion, but with only a one-electron or "half" bond.

He_2: Two electrons are in the σ_{1s}^b orbital and two in the σ_{1s}^*. The net number of bonding electrons is $2 - 2 = 0$. No bond is produced. We should not expect to encounter He_2 as a stable species.

An important idea used in the preceding examples is the following.

bond order $= \frac{1}{2}$(no. e$^-$ in bonding M.O. $-$ no. e$^-$ in antibonding M.O.)   (10.11)

Bond order indicates whether a bond is single, double, triple (or one half, three halves, five halves).

Example 10-4 The bond energy of H_2 is 435 kJ/mol. Estimate the bond energies of H_2^+ and He_2^+.

The bond order in H_2 is one, that is, a single bond. In H_2^+ and He_2^+ the bond order is $\frac{1}{2}$. We should expect the bonds in these two species to be only about one-half as strong as in H_2, that is, about 220 kJ/mol. (Actual values: H_2^+, 255 kJ/mol; He_2^+, 251 kJ/mol.)

SIMILAR EXAMPLE: Exercise 17.

Example 10-5 Which of the four species described in Figure 10-15 are paramagnetic?

Paramagnetism requires the presence of unpaired electrons. One unpaired electron is found in H_2^+ and in He_2^+; these ions are paramagnetic.

SIMILAR EXAMPLE: Exercise 18.

SECOND PERIOD ELEMENTS. To apply the molecular orbital method to elements of the second period requires that molecular orbitals be formed from atomic orbitals of the second principal electronic shell. We will limit our discussion to *diatomic* molecules. Also, we note again that two molecular orbitals—one bonding and one anti-bonding—are produced for every pair of atomic orbitals in the separated atoms. Because there are *four* orbitals in each atom ($2s$, $2p_x$, $2p_y$, $2p_z$) we need to deal with *eight* new molecular orbitals—four bonding and four antibonding. We must also have an energy level diagram for these orbitals.

The molecular orbitals formed from $2s$ orbitals have the same characteristics as those derived from $1s$ orbitals, but they are at a higher energy. As illustrated in Figure 10-16, there are *two* possibilities for combinations of p orbitals. Those that overlap along the same straight line (i.e., end to end) combine to produce σ orbitals: σ_{2p}^b and σ_{2p}^*. Those that overlap in a parallel or sidewise fashion produce π orbitals: π_{2p}^b and π_{2p}^*. There are *two* molecular orbitals each of the π type because there are *two* pairs of p orbitals that are arranged in a parallel fashion. A distinction between bonding and antibonding molecular orbitals, first illustrated in Figure 10-14, is emphasized in Figure 10-16. In bonding molecular orbitals there is a high electron charge density between the atomic nuclei. In an antibonding orbital the electron

FIGURE 10-16

Representation of the molecular orbitals formed by a combination of $2p$ atomic orbitals.

These diagrams are meant simply to suggest the nature of the electron charge distribution for the several molecular orbitals. They are not exact in all details. Nodal planes for the antibonding orbitals are represented by the broken lines.

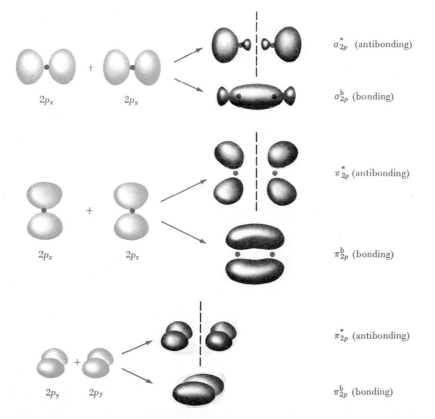

charge density falls to zero in a plane perpendicular to the line joining the atomic nuclei, at a point midway between them. An energy level diagram for these new orbitals is illustrated in Figure 10-17.

We can describe bonding in diatomic molecules of the second period elements (i.e., in Li_2, Be_2, B_2, . . .) in much the same way as for the first period elements. In this process we start with the σ_{1s}^b and the σ_{1s}^* orbitals filled. Then we add electrons successively to the available molecular orbitals formed from the second principal shells. Alternatively, we can think of the first shell (K shell) electrons as not involved in the bonding (i.e., as *nonbonding* electrons). This allows us simply to consider the assignment of the valence shell electrons of the two atoms. This assignment is depicted in Figure 10-17 (where the filled first electronic shells are denoted by the symbol KK).

Some of the previously unexplained features of the O_2 molecule can now be understood in terms of molecular orbitals. Each O atom brings *six* valence electrons to the diatomic molecule, O_2. There are *12* electrons to be assigned to molecular orbitals. The filling of these orbitals is depicted in Figure 10-17. The following features are brought out by the molecular orbital diagram for O_2.

1. The molecule has *two unpaired electrons,* even though the total number of electrons is even. Thus, the paramagnetism of O_2 is explained.

2. The total number of valence electrons in bonding orbitals is *eight,* and the number of electrons in antibonding orbitals is *four.* The excess of bonding over antibonding electrons is $8 - 4 = 4$. Counting two electrons per bond (i.e., using expression 10.11), this corresponds to a *double* covalent bond between oxygen atoms.

Example 10-6 Which of the species indicated in Figure 10-17 would you expect **(a)** *not* to exist as a stable molecule; **(b)** to have the highest bond energy?

(a) A stable diatomic molecule must have an excess of bonding over antibonding electrons. All the species in Figure 10-17 meet this criterion except two. These two—Be_2 and Ne_2—have equal numbers of bonding and antibonding electrons and a bond order of *zero* (recall expression 10.11). They do *not* exist as stable molecules.

(b) We would expect the greatest bond energy to exist in the species with the highest

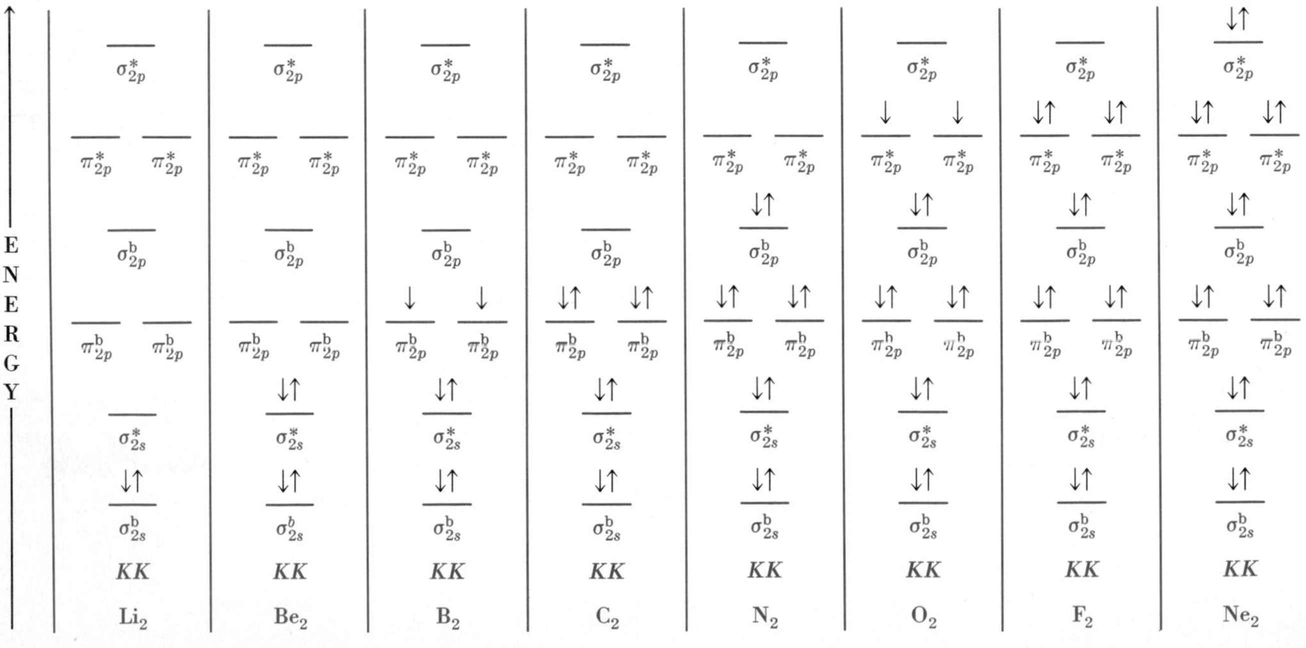

FIGURE 10-17
Molecular orbital diagrams for actual and hypothetical diatomic molecules of the second period elements.

The energy level diagram shown here is approximately correct for the second period elements. The principal exceptions are for O_2, F_2, and Ne_2, where the σ_{2p}^b orbital is at a lower energy than the π_{2p}^b orbitals. However, since all of these orbitals are filled with electrons, this variation in the energy level diagram has no effect on the matters discussed here.

bond order. Inspection of Figure 10-17 shows that the species with the largest excess of bonding over antibonding electrons is N_2.

SIMILAR EXAMPLES: Exercises 15, 16, 18.

Example 10-7 Represent bonding in the O_2^+ ion by means of a molecular orbital diagram.

The O_2 molecule has 12 valence electrons. In the *ion* O_2^+ there are 11 valence electrons. These are assigned to the available molecular orbitals in accordance with the principles established on page 240. The representation below is meant to resemble the orbital diagrams first developed for the electron configurations of atoms. The symbol *KK* means that electrons in the first electronic shells (*K* shells) are not involved in the bonding.

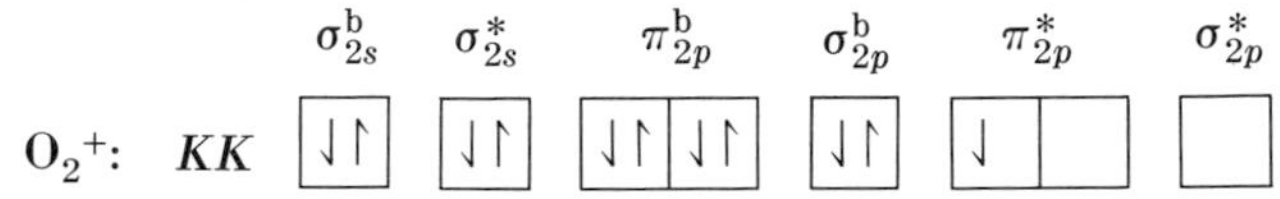

$$\sigma_{2s}^b \quad \sigma_{2s}^* \quad \pi_{2p}^b \quad \sigma_{2p}^b \quad \pi_{2p}^* \quad \sigma_{2p}^*$$

O_2^+: *KK* $\boxed{\downarrow\uparrow}$ $\boxed{\downarrow\uparrow}$ $\boxed{\downarrow\uparrow}\boxed{\downarrow\uparrow}$ $\boxed{\downarrow\uparrow}$ $\boxed{\downarrow}$ $\boxed{}$

SIMILAR EXAMPLES: Exercises 20, 23.

10-5 Bonding in the Benzene Molecule

Earlier in this chapter we described some simple hydrocarbons, specifically CH_4, C_2H_4, and C_2H_2, in terms of valence bond theory. These molecules can also be described through Lewis structures and by VSEPR theory. But there are some organic compounds, notably aromatic hydrocarbons, that cannot be described adequately by any one of these approaches. They require instead the use of molecular orbital theory. Aromatic hydrocarbons have structures based on the molecule benzene, C_6H_6. The term "aromatic" originally referred to the fragrant aromas associated with many, but not all, compounds of this type.

Michael Faraday discovered benzene in 1825 in the gas lines of London. Benzene presented the field of organic chemistry with a problem that was to consume the attention of researchers for 40 years: What is the structure of benzene? In 1834, it was shown to have the molecular formula C_6H_6. Thirty-one years later Friedrich Kekulé of the University of Bonn offered a structure.

Kekulé's hypothesis was that the benzene molecule consists of a flat, cyclic, hexagonal structure of alternate carbon-to-carbon single and double bonds. Each carbon atom is bonded to two other carbon atoms and to only one hydrogen atom. Kekulé accounted for the equivalence of the six carbon-to-carbon bonds by suggesting that the double bonds are not static, but instead oscillate from one position to another. To some chemists this suggested two discrete Kekulé forms of benzene that were in equilibrium with one another. A more correct view is not of discrete Kekulé structures but of a resonance hybrid structure toward which the Kekulé forms are the two principal contributing structures. This view is suggested by Figure 10-18.

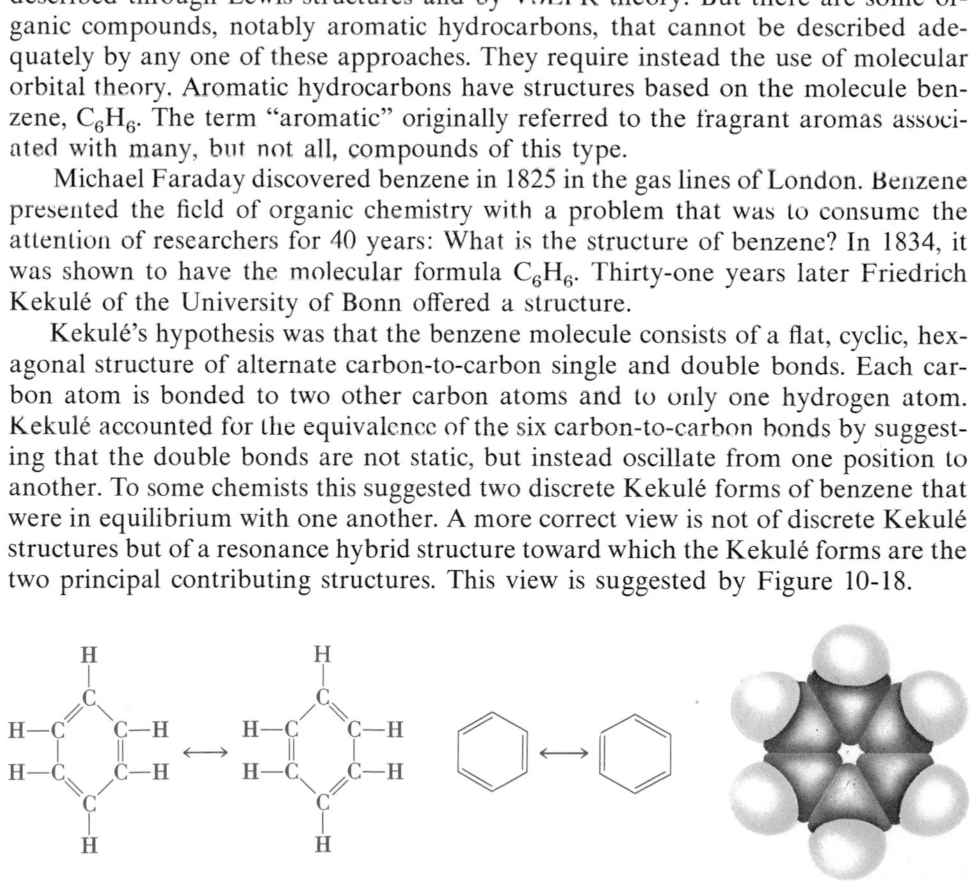

An interesting story is told of how Kekulé arrived at the structure of benzene. Supposedly, one evening Kekulé fell asleep sitting in front of the fireplace. He dreamt of organic molecules containing chains of carbon atoms taking the form of snakes. Suddenly, one of the snakes caught hold of its own tail, forming a whirling ring. Kekulé awakened, refreshed and enthused, and spent the remainder of the night working out his famous hypothesis. Kekulé is said to have written: "Let us learn to dream, gentlemen, and then perhaps we shall learn the truth."

FIGURE 10-18
Resonance in the benzene molecule and the Kekulé structures.

(a) Lewis structures for C_6H_6, showing alternate carbon-to-carbon single and double covalent bonds.
(b) Two equivalent Kekulé structures for benzene. A carbon atom is at each corner of the hexagonal structure and a hydrogen atom is bonded to each carbon. (The symbols for carbon and hydrogen are customarily *not* written in these structures.)
(c) A space-filling model.

(a)   (b)   (c)

FIGURE 10-19
Bonding in benzene,
C_6H_6, by the valence
bond method.

(a) Carbon atoms utilize
sp^2 and p orbitals (recall
Figure 10-9). Each car-
bon atom forms three σ
bonds, two with neigh-
boring C atoms in the
hexagonal ring and a
third with an H atom.
(b) The overlap in sidewise fashion of 2p orbit-
als produces three π
bonds. Thus, there are
three double bonds
(σ + π) between carbon
atoms in the hexagonal
ring. Two equivalent
structures are possible.

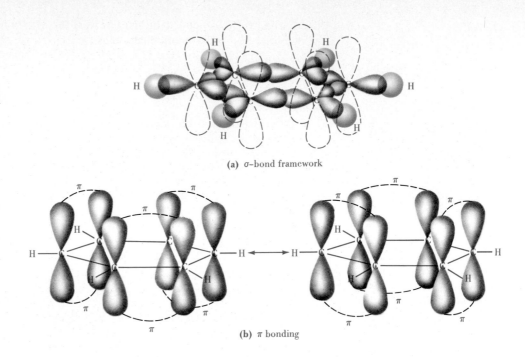

(a) σ–bond framework

(b) π bonding

FIGURE 10-20
Molecular orbital
representation of bonding
in benzene, C_6H_6.

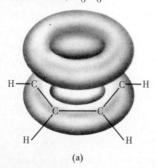

(a)

(b)

(a) A representation incorporating the σ-bond framework of Figure 10-19a and delocalized molecular orbitals (the doughnut-shaped regions) for the π bonds.
(b) A symbolic representation suggesting the delocalized nature of the π bonds (circle inscribed in the hexagon) that is often used in place of the Kekulé structures of Figure 10-18b.

As depicted in Figure 10-19, the valence bond
method of describing bonding in the benzene molecule
requires the use of sp^2 and *p* orbitals by the carbon
atoms. Overlap involving the sp^2 orbitals produces the
σ bond framework and is consistent with the formula
C_6H_6 and the hexagonal planar geometry (bond angles
of 120°). The valence bond method also accounts for
double-bond formation, through the sidewise overlap
of 2p orbitals yielding π bonds. However, the valence
bond method fails to account for the *equivalence* of the
carbon-to-carbon bonds (all are intermediate to single
and double bonds) without invoking the phenomenon
of resonance.

DELOCALIZED MOLECULAR ORBITALS. The key to a
better explanation of bonding in benzene requires introduction of the concept of delocalized molecular orbitals. In a **delocalized molecular orbital** high electron
probability or electron charge density extends over
three or more atoms instead of being limited (localized) to the internuclear region between two atoms.

The σ bond framework of Figure 10-19 describes
adequately the bonds formed within the plane of the
hexagonal ring of carbon and hydrogen atoms in C_6H_6.
However, let us consider combining the six 2p atomic
orbitals into molecular orbitals, a process that yields
three bonding and three antibonding molecular orbitals of the π type. The three bonding orbitals fill with six
electrons (one 2p electron from each C atom), and the
three antibonding orbitals remain empty. The sum of

the three bonding molecular orbitals describes the distribution of π electron charge in the molecule. This can be represented as two doughnut-shaped regions, one above and one below the plane of the C and H atoms (see Figure 10-20). Since they are spread out among six C atoms, these molecular orbitals are *delocalized*. The concept of delocalized electrons is carried over into the symbolic representation of the benzene molecule shown in Figure 10-20b. The circle inscribed within the hexagon represents the multiple bond character displayed by all six carbon atoms.

10-6 Other Structures with Delocalized Orbitals

The concept of delocalized orbitals that arises in molecular orbital theory is not limited to aromatic hydrocarbons. We should always look for this possibility in instances where a resonance hybrid is based on contributing structures in which multiple bonds appear. Consider, for example, the anion, NO_3^-, described in Chapter 9. In place of the resonance hybrid based on structures (9.19),

$$\left[\begin{array}{c} :\overset{\cdot\cdot}{O} \\ \| \\ N \\ :\overset{\cdot\cdot}{O}\overset{\cdot}{\cdot} \quad \overset{\cdot}{\cdot}\overset{\cdot\cdot}{O}: \end{array} \right]^- \longleftrightarrow \left[\begin{array}{c} :\overset{\cdot\cdot}{O}: \\ | \\ N \\ :\overset{\cdot\cdot}{O}\overset{\cdot}{\cdot} \quad \overset{\cdot}{O}: \end{array} \right]^- \longleftrightarrow \left[\begin{array}{c} :\overset{\cdot\cdot}{O}: \\ | \\ N \\ :\overset{\cdot\cdot}{O}: \quad \overset{\cdot}{O}: \end{array} \right]^-$$

FIGURE 10-21
Structure of the nitrate anion, NO_3^-.

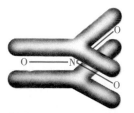

(a) σ-bond framework

(b) Delocalized π molecular orbital

we can write a single structure based on a σ bond framework and delocalized electrons in π molecular orbitals. The situation is pictured in Figure 10-21.

In the σ bond framework we assume sp^2 hybridization for each atom. Of the 24 valence shell electrons, 18 are assigned to the sp^2 hybrid orbitals. Six of these 18 are shared in the regions of orbital overlap. The other 12 are found as six lone pairs, two pairs each on the O atoms. The sp^2 hybridization scheme leaves each atom with a pure p orbital. The combination of these four p orbitals produces four molecular orbitals of the π type. Two of the π orbitals are bonding molecular orbitals, and two are antibonding. The remaining six of the 24 valence shell electrons are now assigned to these π orbitals. Four electrons go into the bonding molecular orbitals and two into the antibonding. With four bonding and two antibonding electrons, the total number of bonds arising from the π orbitals is $\frac{2}{2} = 1$. This π bond is apportioned among the three nitrogen-to-oxygen σ bonds. Thus, each N—O bond is a $1\frac{1}{3}$ bond, exactly the conclusion we reach by averaging the three Lewis structures! With molecular orbital theory it is unnecessary to invoke the phenomenon of resonance.

10-7 Bonding in Metals

In nonmetal atoms the valence electronic shells generally contain more electrons than they do vacant or partially filled orbitals. To illustrate, in the valence shell ($n = 2$) of an atom of F, there are four orbitals ($2s, 2p_x, 2p_y, 2p_z$) and seven electrons. By contrast, in sodium metal each atom has only one valence shell electron ($3s^1$) and four valence shell orbitals ($3s, 3p_x, 3p_y, 3p_z$). In solid metallic sodium, each Na atom is bonded, somehow, to eight nearest neighbors. There appear to be too few electrons to hold these atoms together. This is a feature shared by all metals—more valence shell orbitals than electrons. A bonding scheme for metals must also account for these distinctive properties that all metals share, more or less.

1. Ability to conduct electricity.
2. Ability to conduct heat.

FIGURE 10-22
Electron-sea model of metallic bonding.

The network of Na$^+$ ions is immersed in a sea of electrons, which are derived from the valence shells of all the Na atoms in the crystal. Electrons in this sea are not permanently attached to any particular Na$^+$ ion. They may move randomly, and they belong to the crystal as a whole.

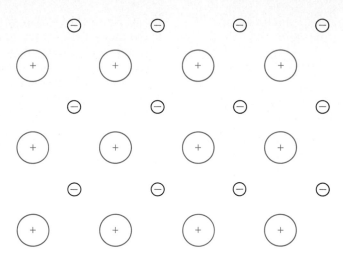

3. Ease of deformation [i.e., ability to be flattened into sheets (malleability) and to be drawn into wires (ductility)].
4. Lustrous appearance.

One simple model that can account for these properties is the **electron-sea model.** The metal is pictured as a network of positively charged ions immersed in a "sea of electrons" (see Figure 10-22). In sodium, for example, the ions would be Na$^+$ and one electron per atom would be contributed to the sea. These free electrons are responsible for the characteristic metallic properties. For example, if the ends of a bar of metal are connected to a source of electric current, electrons from the external source enter the bar at one end. Free electrons pass through the metal and leave the

FIGURE 10-23
Deformation of a metal compared to that of an ionic solid.

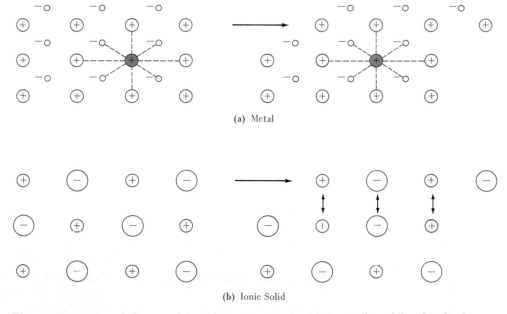

(a) Metal

(b) Ionic Solid

The environment of the metal ion shown in grey in (a) is unaffected by the displacement of a layer of ions above it. The displacement of a layer of ions in the ionic crystal (b) brings like-charged ions into proximity. A repulsive force is set up that could rupture the crystal.

Graphic illustration of the property of thermal conductivity is provided through simple operations in the chemical laboratory. One does not hesitate to stir a boiling mixture with a glass stirring rod, but if a metal spatula is used, a quick burning of the fingers results. In glass, a thermal insulator, heat energy is transferred only through atomic vibrations in the silicate structure. This process occurs quite slowly.

other end at the same rate. In thermal conductivity no electrons enter or leave the metal, but those in the region that is being heated gain kinetic energy and transfer this to other electrons. Both thermal and electrical conductivity can be thought of as a wave motion in the sea of electrons.

The ease of deformation of metals, say relative to an ionic crystal, is suggested in Figure 10-23. In the metal all the ions are positively charged, but their mutual repulsions are offset by the sea of electrons that flows around them. If one layer of metal ions is forced across another, perhaps by hammering, the internal structure remains essentially unchanged because the sea of electrons adjusts rapidly to the new situation. However, in an ionic crystal displacement of one layer of ions with respect to another brings like-charged ions into proximity. The strong repulsive forces set up between them can cause the ionic crystal to cleave or shatter. Ionic crystals are brittle.

The term *electron sea* is not very specific about the region occupied by free electrons in a metal. A more exact description results when molecular orbital theory is applied. For example, think of a metallic crystal of sodium as being a single giant molecule composed of all the atoms contained in the crystal. Imagine that each Na atom contributes one orbital and one electron to the creation of a set of molecular orbitals belonging to the entire crystal (molecule). This set consists of the same number of molecular orbitals as there are atoms in the crystal, and the set of molecular orbitals is half filled with electrons. The critical feature of the molecular orbitals in this set is that their energy levels are so closely spaced that electrons may move freely from one level to another.

This collection of closely spaced energy levels is called a band, more specifically a conduction band. This conduction band corresponds to the sea of electrons in the electron-sea model. Molecular orbital theory applied to metals is referred to as **band theory.**

FIGURE 10-24
A comparison of electron energy levels in insulators, metallic conductors, and semiconductors.

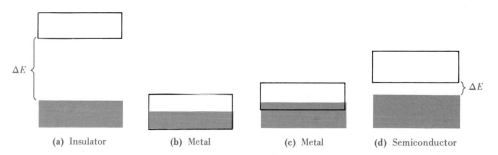

ΔE

$\}\Delta E$

(a) Insulator (b) Metal (c) Metal (d) Semiconductor

(a) In an insulator the valence band is filled with electrons and a large energy gap, ΔE, separates the valence band and conduction band (outlined in black). Few electrons can make the transition when an electric field is applied, and the insulator does not conduct electric current.
(b) In some metals the valence band is only partially filled with electrons, and the valence band also serves as a conduction band (for example, the half-filled 3s band in sodium).
(c) In other metals the valence band is filled, but a conduction band overlaps it. In an electric field electrons from the valence band can move through the conduction band. (For example, the empty 3p band of magnesium overlaps with the filled 3s valence band.)
(d) In a semiconductor the valence band is filled and the conduction band is empty. The energy gap between the two, ΔE, is small enough, however, that some electrons make the transition simply by acquiring extra thermal energy.

10-8 Semiconductors

In a nonconductor or electrical insulator, electrons are held tightly to particular atoms; that is, they are localized. In metallic conductors, as we have just seen, some of the electrons are delocalized. This accounts for the ability of metals to conduct heat and electricity. A third situation is that which exists in a semiconductor. Here a lower energy band (a valence band) is completely filled with electrons, but the next higher band, a conduction band, is separated from it by only a small energy gap. Electrons in the lower band may acquire enough energy, for example, thermal energy, to jump to a level in the conduction band. The electrical conductivity of a semiconductor, in contrast to metals, increases rapidly with temperature. This is because the number of electrons able to jump from the valence band to the conduction band increases as the temperature is raised. A simple comparison of insulators, conductors, and semiconductors is made in Figure 10-24.

Summary

The valence bond method views a covalent bond as resulting from the overlap of atomic orbitals of the bonded atoms. This produces a high electron probability (electron charge density) in the region of overlap. Some simple covalent molecules (e.g., H_2, HCl, H_2S) can be described adequately in terms of the overlap of pure *s* and/or *p* orbitals. In the majority of cases, however, these pure atomic orbitals must be hybridized. That is, they must be replaced by a new set of orbitals whose properties depend on the number and types of pure atomic orbitals used to form them. The geometrical shape of a molecule is determined by the spatial distribution of the orbitals involved in bond formation. For the most part the spatial distributions of hybrid atomic orbitals correspond to the distributions of electron pairs in VSEPR theory.

Two distinctive types of atomic orbital overlap are encountered in the valence bond method. One type (σ) involves end-to-end overlap along the line joining the nuclei of the bonded atoms. The other (π) requires a "sidewise" overlap of two *p* orbitals. Single covalent bonds are σ bonds: a double covalent bond consists of one σ and one π bond; and a triple covalent bond, one σ and two π bonds. The geometrical shape of a molecule is determined by its σ-bond framework.

In molecular orbital theory, when atoms join to form a molecule new regions of high electron probability—molecular orbitals—are established for the molecule as a whole. *Bonding* molecular orbitals correspond to high electron probability or electron charge density in the internuclear region between atoms. *Antibonding* molecular orbitals concentrate electron probability or charge density in regions away from the internuclear region. The numbers and kinds of molecular orbitals in a molecule are related to the corresponding atomic orbitals from which they arise. A procedure similar to the Aufbau process for the electron configurations of atoms can be employed to derive the electronic structure of a molecule: Electrons are assigned to the lowest-energy molecular orbitals available, and singly before electron pairing occurs. *Bond order* follows directly from the assignment of electrons to molecular orbitals: It is one-half the difference between the numbers of electrons in bonding molecular orbitals and in antibonding molecular orbitals. From the assignment of electrons to molecular orbitals one can also deduce whether a species is diamagnetic or paramagnetic.

A description of bonding in the benzene molecule (C_6H_6) requires the concept of delocalized molecular orbitals. These are regions of high electron probability that extend over *three or more* atoms in a molecule. Delocalized molecular orbitals also provide an alternative to the concept of resonance in dealing with species such as SO_2, SO_3, and NO_3^-. Finally, molecular orbital theory in the form called band theory can be applied to metals and semiconductors.

Learning Objectives

As a result of studying Chapter 10, you should be able to

1. Explain the fundamental basis of the valence bond method.

2. Write hybridization schemes for the formation of *sp*, *sp*^{2}, *sp*^{3}, *sp*^{3}*d*, and *sp*^{3}*d*^{2} hybrid orbitals.

3. Predict geometrical shapes of molecules in terms of the pure and hybrid orbitals employed in bonding.

4. Discuss the relationship between VSEPR theory and the valence bond method of predicting molecular geometry.

5. Describe the conditions leading to σ and to π bond formation and the different characteristics of these two types of bonds.

6. Propose plausible bonding schemes from experimental information about molecules (i.e., bond lengths, bond angles, and so on).

7. Explain the fundamental basis of molecular orbital theory.

8. Describe the differences between bonding and antibonding molecular orbitals.

9. Assign probable electron configurations, determine bond

orders, and predict magnetic properties of the diatomic molecules and ions of the first and second period elements.

10. Describe bonding in the benzene molecule (C_6H_6) through Lewis structures, valence bond theory, and molecular orbital theory.

11. Use the concept of delocalized orbitals to write a *single* structure to represent a resonance hybrid (i.e., in cases such as NO_3^-, SO_2, SO_3, etc.).

12. Discuss some of the distinctive properties of metals and the way in which these properties can be explained with the electron-sea model and the band theory of metals.

13. Use the idea of electron bands to contrast insulators, metals, and semiconductors.

Some New Terms

An **antibonding molecular orbital** describes zones of high electron charge density located away from the internuclear region between two atoms. This charge density detracts from bond formation between the atoms.

Band theory is a form of molecular orbital theory applied to metals.

Bond order is one-half the difference between the numbers of electrons in bonding and in antibonding molecular orbitals in a molecule.

A **bonding molecular orbital** describes zones of high electron probability or charge density in the internuclear region between two atoms.

A **delocalized molecular orbital** describes regions of high electron probability or charge density that extend over three or more atoms.

A **hybrid orbital** is one of a set of equivalent orbitals used to replace pure atomic orbitals in describing certain covalent bonds.

Hybridization refers to the combining of pure atomic orbitals to generate new (hybrid) orbitals.

A **pi (π) bond** results from the "sidewise" overlap of pure p orbitals. It consists of regions of high electron charge density above and below the line joining a pair of bonded atoms.

An *sp* **hybrid orbital** is one of two identical orbitals that result from the hybridization of one s and one p orbital of an atom. The angle between the two orbitals is $180°$.

An *sp²* **hybrid orbital** is one of three identical orbitals that

result from the hybridization of one s and two p orbitals of an atom. The angle between any two of the orbitals is $120°$.

An *sp³* **hybrid orbital** is one of four identical orbitals that result from the hybridization of one s and three p orbitals of an atom. The angle between any two of the four orbitals is the tetrahedral angle—$109.5°$.

An *sp³d* **hybrid orbital** is one of five orbitals that result from the hybridization of one s, three p, and one d orbital of an atom. The five orbitals are directed to the corners of a trigonal bipyramid.

An *sp³d²* **hybrid orbital** is one of six orbitals that result from the hybridization of one s, three p, and two d orbitals of an atom. The six orbitals are directed to the corners of a regular octahedron.

A **semiconductor** is a substance that is neither as good an electrical conductor as a metal nor as poor a conductor as an insulator. According to band theory, a semiconductor is characterized by a small energy gap between a filled valence band and an empty conduction band. Electrical conduction must be preceded by a transition of electrons from the valence to the conduction band.

A **sigma (σ) bond** results from the end-to-end overlap of pure or hybridized atomic orbitals along the straight line joining the nuclei of the bonded atoms.

The **valence bond method** treats a covalent bond in terms of the overlap of atomic orbitals. Electron probability (or charge density) is concentrated in the region of overlap.

Exercises

Valence bond method

1. In the manner employed in Example 10-1, describe the structure and bonding in **(a)** HCl; **(b)** ICl; **(c)** H_2Se; **(d)** NI_3.

2. Which of the following statements best describes the bond angle in H_2Se? Explain. **(a)** Greater than in H_2S; **(b)** less than in H_2S; **(c)** less than in H_2S, but not less than $90°$; **(d)** less than $90°$.

3. The Lewis structure of N_2 suggests a triple covalent bond. Describe this bonding in terms of atomic orbital overlap.

4. Indicate ways in which the valence bond method (atomic orbital overlap) is superior to Lewis structures in describing covalent bonds.

Hybridized atomic orbitals

5. In the manner depicted in Figure 10-11, indicate the structures of the following simple molecules in terms of the overlap of pure and hybridized atomic orbitals: **(a)** H_2CCl_2; **(b)** $BeCl_2$; **(c)** BF_3; **(d)** HCN.

6. Match each of the following species with one of these hybridization schemes: sp, sp^2, sp^3, sp^3d, sp^3d^2. **(a)** BrF_5; **(b)** CS_2; **(c)** SiF_4; **(d)** NO_3^-; **(e)** AsF_5. (*Hint:* You may find it helpful to write Lewis structures and to refer to the VSEPR theory.)

7. Predict the geometric shape of the ammonium ion, NH_4^+, and describe a bonding scheme that is consistent with this structure.

8. Describe a bonding scheme, based on pure and hybridized atomic orbitals, to account for the structure of the hydroxylamine molecule.

9. Acetic acid is a very common organic acid (5% by mass, in vinegar). Sketch a three-dimensional structure of the molecule that is consistent with this bonding scheme.

10. Use the method of Figure 10-12 to represent bonding in the molecule dimethyl ether, H_3COCH_3.

11. Describe a hybridization scheme for the central Cl atom in the molecule ClF_3 that is consistent with the geometrical structure pictured in Table 9-2. Which orbitals of the Cl atom are involved in overlap, and which are occupied by lone-pair electrons?

12. The molecule CO_2 features two carbon-to-oxygen double bonds and CO, a carbon-to-oxygen triple bond. Describe bonding schemes for these molecules that are consistent with these facts.

13. Which of the following molecules are linear? Which are planar? **(a)** $HC\equiv N$; **(b)** $N\equiv C-C\equiv N$; **(c)** $F_3C-C\equiv N$; **(d)** $H_2C=C=O$. (*Hint:* What type of hybrid orbitals are involved in the bonding?)

14. Based on the distinction between σ and π bonds and the data given, estimate the bond strength of a carbon-to-carbon triple bond. Compare your result with the value listed in

Table 9-3. Bond energies: C—C, 347 kJ/mol; C=C, 611 kJ/mol.

Molecular orbital method

15. With reference to the molecular orbital diagrams of the second period elements shown in Figure 10-17, which of the *stable* molecules are diamagnetic and which are paramagnetic?

16. Would you expect N_2^- and/or N_2^{2-} to be stable ionic species in the gaseous state? Explain.

17. The molecular orbital diagram of O_2 is shown in Figure 10-17, and for O_2^+ in Example 10-7. Which species, O_2 or O_2^+, has the stronger bond? Explain.

18. For each of the following indicate whether the species is diamagnetic or paramagnetic; if paramagnetic, indicate the number of unpaired electrons. **(a)** F_2; **(b)** N_2^+; **(c)** O_2^-.

19. The paramagnetism of gaseous B_2 has been established. Explain how this observation confirms that the π_{2p}^b orbitals are at a lower energy than the σ_{2p}^b orbital for B_2.

20. In the manner of Example 10-7, indicate the molecular orbital diagrams of **(a)** H_2^-; **(b)** N_2^+; **(c)** F_2^-; **(d)** Ne_2^+.

21. Describe the bond order of diatomic carbon, C_2, in terms of Lewis theory and molecular orbital theory, and explain why the results are different.

22. In all of our discussion of bonding we have not encountered a bond order higher than triple. Use the energy level diagram of Figure 10-17 to show why this is to be expected.

23. The molecular orbital method can be applied to heteronuclear (different nuclei) diatomic molecules as well as to homonuclear (same nuclei) diatomic molecules. Based on the energy level diagram of Figure 10-17, suggest suitable molecular orbital diagrams for **(a)** NO; **(b)** NO^+; **(c)** CO; **(d)** CN; **(e)** CN^-; **(f)** CN^+; **(g)** BN.

24. We have used the term isoelectronic to refer to atoms with identical electron configurations. In the molecular orbital theory this term can be applied to molecules as well. Which of the species of Exercise 23 are isoelectronic?

25. One of the characteristics of an antibonding molecular orbital is the presence of a nodal plane. Which of the *bonding* molecular orbitals considered in this chapter have nodal planes? Explain how a molecular orbital can have a nodal plane and still be a bonding molecular orbital.

Delocalized molecular orbitals

26. Explain how it is possible to get around the concept of resonance by using molecular orbital theory.

27. Represent chemical bonding in the molecule SO_3 **(a)** by writing a Lewis structure(s); **(b)** by using a combination of localized and delocalized orbitals.

28. In which of the following species would you expect to find delocalized molecular orbitals? Explain. **(a)** C_2H_4; **(b)** CO_3^{2-}; **(c)** NO_2; **(d)** H_2CO.

★29. In a manner similar to that outlined in Section 10-6, propose a bonding scheme for SO_2 that is consistent with structures (9.18). To do so requires introducing the concept of a *nonbonding* molecular orbital, one in which the electron charge density neither adds nor detracts from bond formation. Explain why this is necessary.

Metallic bonding

30. Which of the following factors are especially important in determining whether a substance has metallic properties? Explain. **(a)** Atomic number; **(b)** atomic weight; **(c)** number of valence electrons; **(d)** number of empty orbitals; **(e)** total number of electronic shells in the atom.

31. Based simply on the ground-state electron configurations of the isolated atoms, how would you expect the melting points and hardnesses of sodium, iron, and zinc to compare?

32. How many energy levels are present in the conduction band of a single crystal of sodium weighing 2.30 g? How many conduction electrons are present in the crystal?

Additional Exercises

1. Figure 10-13 represents the energy of interaction of two H atoms when the two electrons enter a bonding molecular orbital. Sketch a graph of energy vs. internuclear distance to represent the situation you would expect if the two electrons were to enter an antibonding orbital.

2. Show that both the valence bond method and molecular orbital theory provide an explanation for the existence of the covalent molecule Na_2 in the gaseous state.

3. The poisonous gas phosgene has the formula $COCl_2$. Propose an appropriate scheme of atomic orbital overlap for bonding in this molecule. What is its shape?

4. Lewis theory, which is based on the electron configurations of atoms, is satisfactory for explaining bonding in the ionic compound K_2O. However, it does not readily explain formation of the ionic compounds potassium superoxide, KO_2, and potassium peroxide, K_2O_2. **(a)** Show that molecular orbital theory can help provide this explanation. **(b)** Write Lewis structures consistent with this explanation.

5. The geometric structure of the molecule allene, $H_2C=C=CH_2$, is indicated below. Propose hybridization schemes for the three C atoms that are consistent with this structure.

★6. Suppose that the σ_{2p}^b orbitals were to fill before the π_{2p}^b orbitals. How would the number of unpaired electrons and the bond order in **(a)** B_2 and **(b)** C_2 compare to what is shown in Figure 10-17?

★7. Use the valence bond method (including hybridized orbitals as necessary) to propose a bonding scheme for **(a)** N_2O; **(b)** $ClNO_2$.

★8. Based on the plausible Lewis structure for HNO_3 in Exercise 35 of Chapter 9, propose a bonding scheme in the manner shown in Figure 10-12.

★9. It has been noted that He_2 does not exist as a stable molecule, but there is evidence that such a molecule can be formed between electronically excited He atoms. Suggest a bonding scheme based on molecular orbitals to account for this.

★10. The molecule formamide, $HCONH_2$, has the following approximate bond angles: H—C—O, 123°; H—C—N, 113°; N—C—O, 124°; C—N—H, 119°; H—N—H, 119°. The C—N bond length is 138 pm (1.38 Å). Two Lewis structures can be written for this molecule, with the true structure being a resonance hybrid of the two. Use the data given here (and refer to Table 9-3) to establish a bonding scheme—based on pure and hybridized atomic orbitals—for each structure.

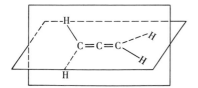

Self-Test Questions

For questions 1 through 4 select the single item that best completes each statement.

1. A molecule in which sp^2 hybrid orbitals are employed for bond formation by the central atom is **(a)** NH_3; **(b)** CO; **(c)** SCl_2; **(d)** H_2CO.

2. A molecule containing π bonds is **(a)** PCl_5; **(b)** N_2; **(c)** OF_2; **(d)** He_2.

3. In carbon-hydrogen-oxygen compounds **(a)** all oxygen-to-hydrogen bonds are π bonds; **(b)** all carbon-to-hydrogen bonds are σ bonds; **(c)** all carbon-to-carbon bonds are π bonds; **(d)** all carbon-to-carbon bonds consist of a σ and one or more π bonds.

4. Of the following, the species with a bond order of 1 is **(a)** H_2^+; **(b)** H_2^-; **(c)** He_2; **(d)** Li_2.

5. Propose a hybridization scheme to account for bonds formed by the carbon atom in each of the following molecules: **(a)** hydrogen cyanide, HCN; **(b)** chloroform, $CHCl_3$; **(c)** methyl alcohol, H_3COH; **(d)** carbamic acid,

$$\overset{\displaystyle O}{\underset{\displaystyle \|}{}}$$
H_2NCOH.

6. What is the total number of **(a)** σ bonds and **(b)** π bonds in the molecule H_3CNCO? (*Hint:* Draw a plausible Lewis structure.)

7. Explain how well each of the following methods describes the geometrical shape of the water molecule. **(a)** Lewis theory; **(b)** valence bond method using pure atomic orbitals; **(c)** valence bond method using hybridized atomic orbitals; **(d)** VSEPR theory.

8. Is it correct to say that when a diatomic molecule loses an electron the bond energy decreases, that is, that the bond is weakened? Explain.

9. Explain why the concept of delocalized molecular orbitals is essential to an understanding of bonding in the benzene molecule (C_6H_6).

10. Why does the hybridization scheme sp^3d *not* account for bonding in the molecule BrF_5? What hybridization scheme does work? Explain.

Liquids, Solids, and Intermolecular Forces

The covalent bond is an *intra*molecular force—a force that operates between atoms *within* a molecule. Such forces influence molecular shapes, bond energies, and many aspects of chemical behavior. Physical properties of the condensed states of matter—liquids and solids—stem from *inter*molecular forces, that is, forces *between* molecules. Intermolecular forces are themselves very closely related to intramolecular forces (i.e., to bond type). Thus, the present chapter follows naturally the material presented in the two preceding chapters.

We begin with an overview of some common properties of liquids and solids, and then show how these properties are related to intermolecular forces.

11-1 Comparison of the States of Matter

The atoms, ions, or molecules of a solid are in close contact. In many solids these structural units exist in a highly ordered network called a **crystal.** Crystals may have geometric shapes that are characterized by plane surfaces intersecting at fixed angles.

Not all solids are crystalline. If order among the structural units does not prevail over long distances, the solid is said to be **amorphous.** Whether a solid is crystalline or amorphous, it occupies a definite volume and maintains a definite shape. Solids are practically incompressible. This observation supports the idea that the atoms, ions, or molecules of a solid are in close contact.

The structural units of a liquid exist in close proximity, although they are usually not thought of as being in contact. Liquids are more compressible than solids because of the free volume that exists among the structural units of a liquid. The intermolecular forces in a liquid are strong enough to maintain a liquid sample in a fixed volume, but not strong enough to maintain a fixed shape. A liquid tends to flow; it covers the bottom and assumes the shape of its container. Liquids are fluid.

Gases are also fluid. Gases expand to fill their containers and thus possess neither a definite shape nor volume. Also, because their atoms or molecules are ordinarily far apart, gases are highly compressible. (In a typical gas at STP the molecules themselves occupy less than 1% of the total gas volume.)

The ability to flow—fluidity—is a property by which it is usually possible to distinguish a liquid from a solid, but there are exceptions. Glass, which does not flow readily at normal temperatures, is actually a supercooled liquid, not a solid. And some solids, for example, sodium and aluminum, can be made to flow when subjected to sufficient pressures.

A completely satisfactory distinction between solids and liquids must always be based on the structural arrangement of their atoms, ions, or molecules.

FIGURE 11-1
Comparison of the states of matter.

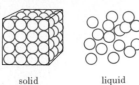

solid   liquid

gas

The structural units in a solid are constrained to fixed points. They may vibrate about these points but, ordinarily, may not move from them permanently. In a liquid there is some free volume among the structural units; motions are more vigorous; and the structure is more random. In a gas there is a great deal of free volume; motion is chaotic; disorder is at a maximum.

FIGURE 11-2
Imbalance of forces at the surface of a liquid.

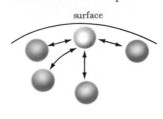

surface

interior

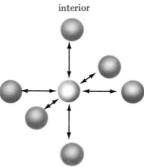

The molecule in the interior experiences equal attractive forces from all directions. The one at the surface is attracted only by other molecules in the surface and by molecules below the surface. The result is a net force toward the interior of the liquid.

A **wetting agent** is a substance that lowers the surface tension of water. It reduces the tendency for droplet formation and allows water to flow more easily over surfaces. Wetting agents are used in a variety of applications, ranging from industrial processes to dishwashing mixtures.

An additional comparison of the three states of matter is provided in Figure 11-1.

11-2 Surface Tension

A property that results directly from intermolecular forces in liquids is surface tension. Figure 11-2 suggests that molecules at the surface of a drop of liquid experience a net force drawing them toward the interior. This net force creates a tension in the liquid surface—almost as if the liquid were covered with a tight skin. For a given volume of liquid, the geometric shape having the minimum surface area is a sphere. The sphere also has the minimum ratio of surface molecules to bulk molecules. So liquids form spherical drops when they are allowed to fall freely.

Whether a drop of liquid can be maintained on a surface depends on the comparative strengths of two types of intermolecular forces—cohesive and adhesive forces. **Cohesive forces** are the intermolecular forces between like molecules in the drop. **Adhesive forces** are intermolecular forces between *unlike* molecules, that is, molecules of the liquid and molecules of the underlying surface. If cohesive forces are strong compared to adhesive forces, a liquid drop is maintained. If adhesive forces are strong, the drop spreads into a film; the liquid is said to wet the surface.

Cohesive forces in liquid mercury are strong. Mercury does not wet other surfaces. Its tendency is to form distinct drops, not a film. Water, on the other hand, wets many surfaces, such as glass and certain fabrics. This ability is essential if water is to be used as a cleaning agent. If a glass surface is coated with a film of oil or grease, water is no longer able to wet the glass and water droplets stand on the surface rather than form a film. Adding a detergent to the water produces two effects: The detergent solution dissolves the grease, exposing the clean glass surface, and the detergent lowers the surface tension of the water. The net result is that water is able to wet the glass once again.

Several phenomena related to surface tension are pictured in Figure 11-3.

11-3 Vaporization

Molecules with kinetic energies sufficiently above average may overcome the attractive forces of neighboring molecules and escape from a liquid surface into the gaseous or vapor state. This phenomenon is called vaporization or evaporation.

FIGURE 11-3
Some phenomena related to surface tension.

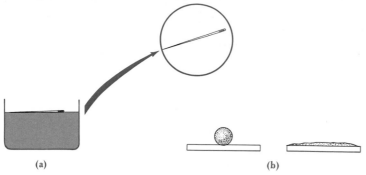

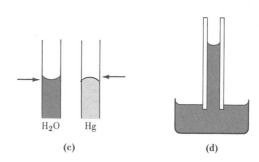

H_2O Hg

(a)   (b)   (c)   (d)

(a) *Surface tension illustrated.* That the surface of a liquid is under tension (as if covered by a skin) can be illustrated by carefully floating a clean steel needle on water. Steel is much more dense than water and the needle should sink, but the surface tension of the water supports the needle.
(b) *Wetting of a surface.* Whether a drop of liquid stands on a surface or spreads into a thin film depends on the ability of the liquid to wet the surface.
(c) *Meniscus formation.* If a liquid (such as water) wets glass, the meniscus of a column of the liquid in a glass tube is concave. If the liquid does not wet glass (as is the case with mercury), the meniscus is convex.
(d) *Capillary action.* Water wets the inside walls of a glass capillary and a thin film rises up the tube. This is accompanied by a rise in the bulk liquid level itself. The column of water is supported by its surface tension counteracting the force of gravity. The smaller the diameter of the tube, the higher the column of water that can be supported. The action of a sponge in soaking up water or of a blotter in soaking up ink depends on the rise of water into capillary pores of a fibrous material, such as cellulose. The penetration of water into soils also depends in part on capillary action.

In contrast to the effect with water, the level of liquid mercury is depressed in a glass capillary tube.

ENTHALPY (HEAT) OF VAPORIZATION. The loss of more energetic molecules by vaporization reduces the average kinetic energy of the remaining molecules in a liquid. The liquid temperature falls. This is why a cooling effect is observed when a volatile liquid such as ether or acetone is allowed to evaporate from one's skin. Alternatively, the temperature of a liquid can be kept constant during vaporization. But this requires the absorption of heat from the surroundings to replace the energy carried away by the vaporized molecules.

The quantity of heat associated with the vaporization of a fixed amount of liquid at a fixed temperature is the enthalpy (heat) of vaporization, denoted by ΔH_{vap}. The appropriate SI units for expressing ΔH_{vap} are J/mol and kJ/mol, but the units cal/g and kcal/mol are also commonly encountered.

Joules

Example 11-1 How much heat is required to convert 135 g of diethyl ether, $C_4H_{10}O$, from the liquid state at 20.0°C to the gaseous state at 30.0°C? The specific heat of liquid diethyl ether in the range 20 to 30°C, is 2.30 J g^{-1} °C^{-1} and its molar heat of vaporization at 30.0°C is 27.8 kJ/mol.

The heat required is the sum of two quantities: the heat required to raise the temperature of the liquid from 20.0°C to 30.0°C (calculated in the manner of Example 6-1) and that needed to vaporize the liquid at 30.0°C.

Energy to raise the temperature of the liquid ether:

$$\text{no. kJ} = 135 \text{ g ether} \times \frac{2.30 \text{ J}}{\text{g ether °C}} \times \frac{1 \text{ kJ}}{1000 \text{ J}}$$
$$\times (30.0 - 20.0)\text{°C}$$
$$= 3.10 \text{ kJ}$$

Specific Heat — heat associated w/ temp change

FIGURE 11-4
Establishing liquid–vapor equilibrium at constant temperature.

○→ molecules undergoing vaporization
○→ molecules undergoing condensation

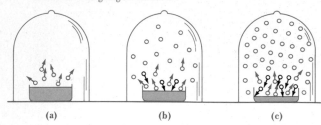

(a)   (b)   (c)

(a) A liquid is allowed to evaporate into a closed vapor volume. Initially only vaporization occurs.
(b) Condensation begins. However, because molecules are evaporating at a faster rate than they are condensing, the number of molecules in the vapor state continues to increase.
(c) The rate of condensation has become equal to the rate of vaporization. Dynamic equilibrium is established. The number of molecules present in the vapor state remains constant over time, as does the pressure exerted by the vapor.

Energy to vaporize the ether:

$$\text{no. kJ} = 135 \text{ g } C_4H_{10}O \times \frac{1 \text{ mol } C_4H_{10}O}{74.1 \text{ g } C_4H_{10}O}$$
$$\times \frac{27.8 \text{ kJ}}{1 \text{ mol } C_4H_{10}O}$$
$$= 50.6 \text{ kJ}$$

Total energy required:

$$\text{no. kJ} = 3.10 \text{ kJ} + 50.6 \text{ kJ}$$
$$= 53.7 \text{ kJ}$$

SIMILAR EXAMPLES: Exercises 5, 6.

VAPOR PRESSURE. The tendency for a liquid to vaporize *increases* as its temperature increases, but *decreases* with increased strength of intermolecular forces. In any event, if a liquid sample is allowed to evaporate into an unconfined vapor volume, the process will proceed until all the liquid has vaporized. On the other hand, if vaporization occurs into a *closed* vapor volume, as pictured in Figure 11-4, a different condition may result.

When a vapor is maintained in contact with a liquid, some molecules may return from the vapor to the liquid. This process, which is the reverse of vaporization, is called **condensation.** The extent of condensation depends on the concentration of vapor molecules (number of molecules per unit volume) and on the area of contact between the liquid and its vapor. In a container with both liquid and vapor, vaporization and condensation occur simultaneously. Although molecules continue to pass back and forth between liquid and vapor, if sufficient liquid is present, eventually a condition is reached in which no *additional* vapor is formed. This condition is one of **dynamic equilibrium.** The term *dynamic equilibrium* always implies that two opposing processes are occurring simultaneously in a closed system, and in such a way as to offset one another. As a result there is no net change with time once equilibrium has been established.

The vapor in equilibrium with a liquid, like any gas, exerts a pressure. A special name is given to the characteristic pressure exerted by this vapor; it is called the **vapor pressure.** Magnitudes of vapor pressures, like so many other properties, vary widely. Liquids with high vapor pressures are said to be volatile. Those with very low vapor pressures are nonvolatile. Diethyl ether and acetone are highly volatile liquids. Water at ordinary temperatures is a moderately volatile liquid; at 25°C its vapor pressure is 23.8 mmHg. Gasoline is a mixture of hydrocarbons, most of which are somewhat more volatile than water. Whether a liquid is volatile or nonvolatile at a given temperature is determined primarily by intermolecular forces.

Higher temp ⇒ more molecules w/sufficient energy to escape ⇒ higher vap. rate

When a vapor condenses, heat energy is evolved. The heat of condensation is the negative of the heat of vaporization.

same amt for a substance
$$\Delta H_{vap} = + \text{Energy}$$
$$\Delta H_{cond} = - \text{Energy}$$

High vapor pressure ⇒ exert alot of pressure ⇒ alot of molecules in gaseous state ⇒ Vaporize easily ⇒ volatile

Hydrocarbons derived from gasoline are important constituents of photochemical smog. The main source of these hydrocarbons is the unburned gasoline from internal combustion engines. A minor but significant source is vaporized gasoline from automobile gas tanks, filling-station operations, and so on.

FIGURE 11-5
Measurement of vapor pressure.

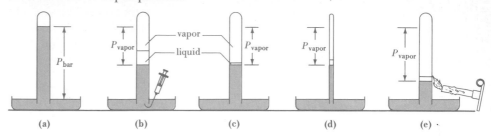

(a)   (b)   (c)   (d)   (e) ·

(a) A mercury barometer.
(b) A small volume of liquid is introduced to the top of the mercury column. The pressure of the vapor in equilibrium with the liquid depresses the level of the mercury in the barometer tube.
(c) By comparing this with situation (b) we see that vapor pressure is independent of the volume of liquid used to establish liquid-vapor equilibrium.
(d) Vapor pressure is also independent of the volume of vapor involved in the liquid–vapor equilibrium.
(e) An increase in temperature causes an increase in vapor pressure.

FIGURE 11-6
Vapor pressure curves of several liquids.

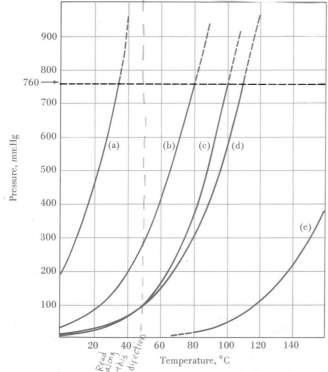

(a) Diethyl ether, $C_4H_{10}O$; **(b)** benzene, C_6H_6;
(c) water, H_2O; **(d)** toluene, C_7H_8; **(e)** aniline, C_6H_7N.

As an excellent first approximation, the vapor pressure of a liquid depends only on its temperature. The measurement of vapor pressure and its dependence on temperature are suggested by Figure 11-5.

A plot of vapor pressure as a function of temperature is known as a **vapor pressure curve.** Because vapor pressure increases with temperature, vapor pressure curves are always of the form shown in Figure 11-6. Vapor pressure data are also encountered in tabular form, as in Table 11-1 for water.

TABLE 11-1
Vapor pressure of water at various temperatures

Temperature, °C	Pressure, mmHg	Temperature, °C	Pressure, mmHg
0	4.6	60	149.4
10	9.2	70	233.7
20	17.5	80	355.1
21	18.7	90	525.8
22	19.8	91	546.0
23	21.1	92	567.0
24	22.4	93	588.6
25	23.8	94	610.9
26	25.2	95	633.9
27	26.7	96	657.6
28	28.3	97	682.1
29	30.0	98	707.3
30	31.8	99	733.2
40	55.3	100	760.0
50	92.5	110	1074.6

Boiling

vaporization occurs thru-out liquid; pockets
of vapor form; if vap press = atm press
than these pockets won't collapse but
instead push the liquid up & out of
way and then surface.

lower atm pressure = lower boiling pt.

BOILING AND THE BOILING POINT. The temperature at which the vapor pressure of a liquid is equal to normal atmospheric pressure (1 atm = 760 mmHg) has special significance. It is called the **normal boiling point.** The condition of boiling results when a liquid is heated in a container *open to the atmosphere* and vaporization occurs throughout the liquid rather than simply at the surface. Pockets of vapor form within the bulk of the liquid, rise to the surface, and escape. Once boiling begins the temperature remains constant until all the liquid has boiled away. Heat added to the liquid simply supplies the necessary heat of vaporization. The phenomenon of boiling is compared to ordinary vaporization in Figure 11-7.

From Figure 11-6 we see that the boiling point of a liquid varies with atmospheric pressure. (Simply shift the line shown at $P = 760$ mmHg to higher or lower pressures and determine the temperatures of the new points of intersection with the vapor pressure curves.) Atmospheric pressures less than 1 atm are encountered quite commonly, notably at high altitudes. At an altitude of 1500 m atmospheric pressure is 630 mmHg. The boiling point of water at this pressure is 95°C (203°F). To cook foods under these conditions of lower boiling temperatures, it is necessary to use longer cooking times. A "three-minute" boiled egg takes longer than 3 min to cook. The effect of high altitudes can be counteracted, however, by using a pressure cooker. In a pressure cooker the cooking water is maintained under higher pressure and its boiling point increases.

The elevation of Denver, Colorado, is 1609 m; that of Santa Fe, New Mexico, 2120 m; and that of Leadville, Colorado, 3170 m.

FIGURE 11-7
The phenomenon of boiling.

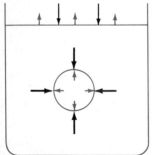

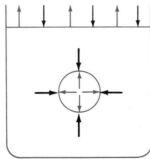

(a) Temperatures below the boiling point

(b) At the boiling point

Vapor pressure of the liquid is represented by colored arrows, atmospheric pressure by black arrows.
(a) At temperatures below the boiling point the vaporization of molecules from the surface of the liquid is unhindered by the pressure of the atmosphere. *If a large number of molecules were to form a gaseous pocket or bubble within the liquid, however, the bubble would collapse. This is because the pressure inside the bubble (vapor pressure) would be less than atmospheric. Thus bubbles do not form at all.
(b) At the boiling point the pressure to sustain the bubbles becomes equal to atmospheric pressure. The phenomenon of boiling involves vaporization into vapor pockets or bubbles. The bubbles, because they are less dense than the liquid, rise to the surface and escape.

THE CRITICAL POINT. The qualification "in a container open to the atmosphere" is an important one in describing the phenomenon of boiling. If a liquid is heated in a *closed* container, boiling does not occur. Instead, the temperature and vapor pressure rise continuously. In some cases pressures many times greater than normal atmospheric may be attained.

If sufficient liquid is enclosed in a sealed tube and heated, as suggested by Figure 11-8, the following observations are made.

1. The density of the liquid decreases; that of the vapor increases; and eventually they become equal.
2. The meniscus between the liquid and vapor becomes less distinct and eventually disappears.

The point at which these two conditions are reached is called the **critical point** of the liquid. The temperature at this point is called the **critical temperature** and the pressure, the **critical pressure.** The critical point is the highest temperature–pressure point on a vapor pressure curve.

no more liq, vapor pressure
cannot increase ⇒ all existance past
this point is of same state.

FIGURE 11-8
Attainment of the critical point.

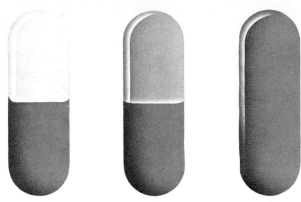

about 10°C below T_c about 1°C below T_c critical temp. T_c

The meniscus separating a liquid (bottom) from its vapor (top) disappears at the critical point. The liquid and vapor become indistinguishable.

TABLE 11-2
Some critical temperatures and pressures

Substance	Critical temperature, K	Critical pressure, atm
H_2	33.2	12.8
N_2	126.0	33.5
O_2	154.3	49.7
CH_4	191.1	45.8
CO_2	304.1	72.9
HCl	324.5	81.6
NH_3	405.5	111.5
SO_2	430.3	77.7
H_2O	647.3	218.2

critical pt

heat sufficient liquid → turns to vapor → as heating, molecules of liquid move quicker & spread apart ∴ liquid e decrease while vap e increases ⇒ density eventually equals & they (liq & vap) are essential the same & undistinguishable from each other.

If the vapor pressure of the Ag were appreciable, it would not have been possible to assume that the total gas volume remained constant at 113 L.

Several critical temperatures and pressures are listed in Table 11-2. Substances having a critical temperature above room temperature can be liquefied at room temperature by applying sufficient pressure to the gaseous state. Those with critical temperatures below room temperature require both the application of pressure *and* a lowering of temperature.

Pressure increase liquifies a vapor. If past crit. pt, decrease in temp must accompany press, increase for liquification to occur.

11-4 Some Calculations Involving Vapor Pressure and Related Concepts

EXPERIMENTAL DETERMINATION OF VAPOR PRESSURE. The method suggested by Figure 11-5 makes an effective demonstration but it is not widely used. For one thing, the method would not work for vapor pressures that are either very low or very high. Also, the results obtained are not very accurate. A somewhat more useful method is suggested in Example 11-2 (called the transpiration method). An inert gas is saturated with the vapor under study and the vapor pressure is calculated using the ideal gas equation.

Example 11-2   113 L of He gas at 1360°C and atmospheric pressure is passed through molten silver at 1360°C. The gas becomes saturated with silver vapor. As a result, a loss of mass of 0.120 g is recorded in the liquid silver. What is the vapor pressure, in mmHg, of liquid silver at 1360°C? Neglect any change in gas volume due to the vaporization of the silver.

Although the saturated gas is actually a mixture of He and Ag, we can deal with the Ag as if it were a single gas occupying a volume of 113 L. The necessary data follow.

$P_{Ag} = ?$ (this is the unknown)

$V_{Ag} = 113$ L   *remember: gas expands to fill whole volume so $V_{Ag} = V_{He} = V_{container}$*

$n_{Ag} = 0.120$ g Ag $\times \dfrac{1 \text{ mol Ag}}{108 \text{ g Ag}} = 0.00111$ mol Ag

$R = 0.0821$ L atm mol^{-1} K^{-1}

$T = 1360°C + 273 = 1633$ K

$PV = nRT$ $P = \dfrac{nRT}{V}$

$P = \dfrac{(0.00111 \text{ mol})(0.0821 \text{ L atm mol}^{-1} \text{ K}^{-1})(1633 \text{ K})}{113 \text{ L}}$

$= 1.32 \times 10^{-3}$ atm

$P = 1.32 \times 10^{-3}$ atm $\times \dfrac{760 \text{ mmHg}}{1 \text{ atm}} = 1.00$ mmHg

SIMILAR EXAMPLES: Exercises 11, 12.

AN APPLICATION OF VAPOR PRESSURE DATA—PREDICTION OF STATES OF MATTER

Example 11-3   As a result of a chemical reaction, 0.105 g H_2O is produced at a temperature of 50°C in a closed vessel of 482 cm³ volume. How much of this water is in the form of liquid and how much, vapor?

If both liquid and vapor are present, the pressure in the vessel will be the vapor pressure of water at 50°C, obtainable from Table 11-1 (92.5 mmHg). To determine the mass of vapor at this pressure requires an application of the ideal gas equation. The mass of liquid water may then be obtained by difference.

Mass of water vapor at 50°C:

$$PV = nRT \qquad PV = \frac{mRT}{(MW)} \qquad m = \frac{(MW)PV}{RT}$$

$m = ?$ (this is the unknown)

$(MW) = 18.0$ g H_2O/mol H_2O

$$P = 92.5 \text{ mmHg} \times \frac{1 \text{ atm}}{760 \text{ mmHg}} = 0.122 \text{ atm}$$

$$V = 482 \text{ cm}^3 \times \frac{1 \text{ L}}{1000 \text{ cm}^3} = 0.482 \text{ L}$$

$R = 0.0821$ L atm mol⁻¹ K⁻¹

$T = 50°C + 273 = 323$ K

$$m = \frac{18.0 \text{ g mol}^{-1} \times 0.122 \text{ atm} \times 0.482 \text{ L}}{0.0821 \text{ L atm mol}^{-1} \text{ K}^{-1} \times 323 \text{ K}} = 0.0399 \text{ g}$$

Mass of liquid and vapor:

mass $H_2O(g) = 0.0399$ g

mass $H_2O(l) = 0.105$ g $- 0.040$ g $= 0.065$ g

SIMILAR EXAMPLES: Exercises 22, 23, 24.

AN APPLICATION OF VAPOR PRESSURE DATA—COLLECTION OF GASES OVER WATER.

We have already considered another situation where some knowledge of vapor pressures is required—the collection of gases over water. This subject can be reviewed by rereading Section 5-9, in particular, Example 5-19.

AN EQUATION FOR CALCULATING VAPOR PRESSURES.

Interpolation (estimation of values) between points in a table or graph is difficult unless a linear relationship is involved. The vapor pressure curve is not linear—it becomes ever steeper as the temperature is increased. It is possible at times to convert a nonlinear relationship into a linear one by introducing a new function of the variables. Vapor pressure data yield a straight line when *logarithm of the vapor pressure (log P)* is plotted against *reciprocal of the kelvin temperature (1/T)*. As described in Appendix A-4, a straight line can always be expressed through a simple mathematical equation. The equations for the lines in Figure 11-9 are of the form

$$\log P = -A \left(\frac{1}{T}\right) + B \tag{11.1}$$

equation of straight line: $\underbrace{y}_{} = \underbrace{m \cdot x}_{} + \underbrace{b}_{}$

Margin notes:

Remember! @ any temp you know vapor pressure

What if the mass of water required to produce vapor at the equilibrium vapor pressure proved to be *greater* than the total mass of water present? In this case we would conclude that the vessel contained only vapor and no liquid. Moreover, the pressure of the vapor would be less than the equilibrium vapor pressure. This situation is encountered in some of the exercises.

The constant $-A$ corresponds to the slope (m), and the constant B to the intercept (b) of a straight line.

FIGURE 11-9

Vapor pressure data of Figure 11-6 replotted— log P versus $1/T$.

Data from Figure 11-6 have been recalculated and replotted as in the following example.

For benzene, at 60°C, the vapor pressure is 400 mmHg.
$\log P = \log 400 = 2.60$.
$T = 60°C = 333$ K;
$1/T = 1/333 = 0.00300 = 3.00 \times 10^{-3}$;
$1/T \times 10^3 = 3.00 \times 10^{-3} \times 10^3 = 3.00$.

The point corresponding to these data is marked by the arrow ($\rightarrow$).

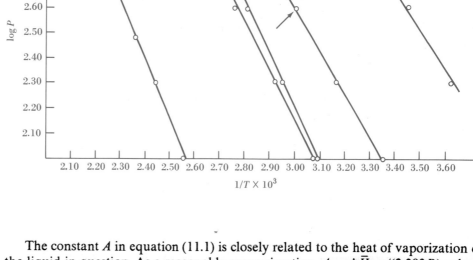

Many derivations yield natural logarithmic functions. For example, equation (11.2) can also be expressed as

$$\ln P = \frac{-\Delta \bar{H}_{vap}}{RT} + B'$$

Natural and common (base-10) logarithms are related through the factor 2.303 (see Appendix A-2).

To obtain equation (11.3) from (11.2), write equation (11.2) twice—once with the values P_2 and T_2 and once with P_1 and T_1. Subtract one equation from the other to eliminate B. Then simplify the resulting expression.

The constant A in equation (11.1) is closely related to the heat of vaporization of the liquid in question. As a reasonable approximation, $A = \Delta \bar{H}_{vap}/(2.303R)$, where R is the gas constant. Substituting this value of A into equation (11.1) yields an expression known as the Clausius–Clapeyron equation.

$$\log P = \frac{-\Delta \bar{H}_{vap}}{2.303RT} + B \tag{11.2}$$

To calculate the vapor pressure (P) at some given temperature (T) using equation (11.2) requires that numerical values be available for $\Delta \bar{H}_{vap}$ and B. Such data are often tabulated in handbooks, but a more usual approach is to eliminate the constant B by writing an expression based on two temperatures and the corresponding vapor pressures.

$$\log \frac{P_2}{P_1} = \frac{\Delta \bar{H}_{vap}}{2.303R}\left(\frac{T_2 - T_1}{T_1 T_2}\right) = \log P_2 - \log P_1 \tag{11.3}$$

In equation (11.3) the same units of pressure must be used for P_1 and P_2, and the same units of energy for $\Delta \bar{H}_{vap}$ and R. Also, temperatures must be expressed in kelvins.

$R = 8.314 \dfrac{J}{mol \cdot K}$

Example 11-4 Calculate the vapor pressure of water at 35°C with data from Table 11-1 and a value of 43.9 kJ/mol for $\Delta \bar{H}_{vap}$ of water.

Let P_2 be the "unknown" vapor pressure at the temperature, $T_2 = 35°C = 308$ K. Choose for P_1 and T_1 known data (Table 11-1) at a temperature close to 35°C. At $T_1 = 30°C = 303$ K, $P_1 = 31.8$ mmHg. Substituting these values into equation (11.3), we obtain

$$\log \frac{P_2 \text{ (mmHg)}}{31.8 \text{ mmHg}} = \frac{43.9 \times 10^3 \text{ J mol}^{-1}}{2.303 \times 8.314 \text{ J mol}^{-1} \text{ K}^{-1}}\left[\frac{308 \text{ K} - 303 \text{ K}}{(303 \text{ K})(308 \text{ K})}\right]$$

$$= 2.29 \times 10^3 \text{ K}\left(\frac{5.36 \times 10^{-5}}{\text{K}}\right) = 1.23 \times 10^{-1} = 0.123$$

$\log P_2 - \log P_1 = \log \dfrac{P_2}{P_1} =$

$\left[\dfrac{-\Delta \bar{H}_{vap}}{2.303RT_2} + B_2\right] - \left[\dfrac{-\Delta \bar{H}_{vap}}{2.303RT_1} + B_1\right]$

put into here

$= \dfrac{-\Delta \bar{H}_{vap}}{2.303R}\left(\dfrac{1}{T_2} - \dfrac{1}{T_1}\right) = \dfrac{-\Delta \bar{H}_{vap}}{2.303R}\left(\dfrac{T_1 - T_2}{T_1 T_2}\right) = \dfrac{\Delta \bar{H}_{vap}}{2.303R}\left(\dfrac{T_2 - T_1}{T_1 T_2}\right)$

Next, determine the *antilogarithm* of 0.123. (What is the number whose logarithm is 0.123?) The antilogarithm is 1.33 (see Appendix A). Thus,

$$\frac{P_2}{31.8 \text{ mmHg}} = 1.33$$
$$P_2 = 1.33 \times 31.8 \text{ mmHg} = 42.3 \text{ mmHg}$$

(The measured vapor pressure of water at 35°C is 42.18 mmHg.)

SIMILAR EXAMPLES: Exercises 16, 17.

11-5 Transitions Involving Solids

MELTING, MELTING POINT, AND HEAT OF FUSION. As the temperature of a crystalline solid is raised, its atoms, ions, or molecules undergo increasingly vigorous vibrations. Eventually, a temperature is reached at which the crystalline lattice is destroyed by these vibrations. The solid is converted to a liquid. This process is called **melting.** The reverse process, the conversion of a liquid to a solid, is called **freezing.** The temperature at which a pure liquid freezes and that at which the corresponding pure solid melts are identical. At this temperature, called either the **melting point** of the solid or the **freezing point** of the liquid, solid and liquid can coexist in dynamic equilibrium.

If heat is added to a mixture of solid and liquid at equilibrium, the solid is gradually converted to liquid *while the temperature remains constant.* When all the solid has melted, the temperature begins to rise. Conversely, the removal of heat from a mixture of solid and liquid at equilibrium results in the complete conversion of liquid to solid. The quantity of energy required to convert a given amount of solid to liquid is called the **enthalpy (heat) of fusion.** When a liquid freezes it *gives off* the heat of fusion.

A well-known example of a melting point is that of ice, 0°C.* At this temperature ice and liquid water, in contact with air under normal atmospheric pressure, are in equilibrium. The heat of fusion of ice is 6.02 kJ/mol.

A simple method for determining the freezing point of a liquid is to measure the temperature of a liquid sample continuously as it is allowed to cool. The temperature decreases with time until the freezing point is reached. As solid begins to form, the temperature remains constant with time. When all the liquid has frozen, the temperature is again free to fall. A graphic representation of temperature versus time is called a **cooling curve.** An example is shown in Figure 11-10 for liquid water.

The behavior represented by the solid-line portions of Figure 11-10 is idealized. At times a liquid can be cooled *below* its normal freezing point before freezing occurs—a condition called **supercooling.** This occurs when there are few nuclei present (such as suspended dust particles) on which solid crystals can form. When a supercooled liquid begins to freeze its temperature rises back to the normal freezing point, where freezing is completed. In these cases there is a dip in the cooling curve just before the straight-line portion.

FIGURE 11-10
Cooling curve for water.

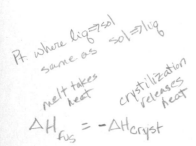

The broken line portion represents the condition of supercooling that occasionally occurs. (l) = liquid; (s) = solid.

Example 11-5 A 25.0-g cube of ice at 0.0°C is added to 100.0 g of liquid water at 22.0°C in a thermally insulated container (e.g., a Styrofoam cup). The heat of fusion of ice is 6.02 kJ/mol and the specific heat of liquid water is 4.18 J g⁻¹ °C⁻¹. What is the final condition reached—all liquid, or ice and liquid?

---

*If air is excluded, and solid and liquid water are in equilibrium with their own vapor (at a pressure of 4.58 mmHg), the equilibrium temperature is slightly different, +0.01°C.

Call the heat associated with melting the ice, q_{ice}, and that for cooling the original liquid water, q_{water}.

$$q_{ice} = 25.0 \text{ g ice} \times \frac{1 \text{ mol ice}}{18.0 \text{ g ice}} \times \frac{6.02 \text{ kJ}}{1 \text{ mol ice}}$$
$$= 8.36 \text{ kJ}$$

The maximum quantity of heat associated with cooling the liquid water would be for the temperature interval from 22.0 to 0.0°C.

$$q_{water} = 100 \text{ g water} \times \frac{4.18 \text{ J}}{\text{g water }°C} \times (0.0 - 22.0)°C$$
$$= -9196 \text{ J} = -9.20 \text{ kJ}$$

The maximum quantity of heat that could be liberated in cooling the original liquid water is greater than that required to melt all the ice. The final condition is one of "all liquid" at a temperature somewhat above 0°C.

SIMILAR EXAMPLE: Exercise 25.

SUBLIMATION. Vaporization of solids is also encountered, although in general the volatilities of solids are not as great as of liquids. The direct passage of molecules from a solid to a gas is called **sublimation.** The pressure exerted by the vapor in equilibrium with a solid is called the vapor pressure of the solid or the sublimation pressure. A plot of vapor pressure of a solid as a function of temperature can be called a sublimation curve. The enthalpy (heat) of sublimation is the quantity of heat required to convert a certain quantity of solid to vapor. It is related to the enthalpies of fusion and vaporization in a simple way:

$$\Delta H_{sub} = \Delta H_{fus} + \Delta H_{vap} \tag{11.4}$$

Two very common solids with significant volatilities are ice and dry ice (solid carbon dioxide).

11-6 Phase Diagrams

We have noted some general conditions under which a substance exists in three different states (phases) of matter. For instance, at high temperatures and low pressures we expect the gaseous state; at very low temperatures, the solid state. Moreover, we have described situations in which two states of matter coexist in equilibrium—liquid–vapor, solid–vapor, solid–liquid. All of this information can be summarized in a single graphic form called a phase diagram.

The phase diagram shown in Figure 11-11 for iodine is one of the simplest possible. Pressures and temperatures at which iodine exists as solid, liquid and gas are denoted by areas in the diagram. Equilibrium conditions between two states of matter are represented by curves along which areas adjoin. The curve OC is the vapor pressure curve of liquid iodine (C is the critical point). OB is the sublimation curve of iodine. The effect of pressure on the melting point of iodine is represented by the curve OD; it is called the fusion curve. The point O has a special significance. It gives the unique temperature and pressure at which solid, liquid, and vapor coexist in equilibrium. It is called a **triple point;** for iodine this is at 114°C and 91 mmHg. The normal melting point (114°C) and boiling point (184°C) are the temperatures at which a line at $P = 1$ atm intersects the fusion and vapor pressure curves, respectively.

↳ those @ 1atm pressure

In this calculation,

$q_{ice} + q_{water}$
$= 8.36 \text{ kJ} - 9.20 \text{ kJ}$
$= -0.84 \text{ kJ} = -840 \text{ J}.$

Consider that 840 J of heat is absorbed by 125.0 g of liquid water at 0°C, and the final temperature can be easily calculated. It is 1.6°C.

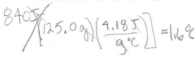

$$840 J / \left[(125.0 g) \left(\frac{4.18 J}{g°C} \right) \right] = 1.6°C$$

Residents of regions with cold climates are familiar with the fact that snow may disappear from the ground even though the temperature may fail to rise above 0°C. Under these conditions the snow does not melt; it sublimes.

The difference in melting point temperature and triple-point temperature is usually quite small. Generally, large pressure changes are required to produce appreciable changes in the solid–liquid equilibrium temperature.

FIGURE 11-11
Phase diagram for iodine.

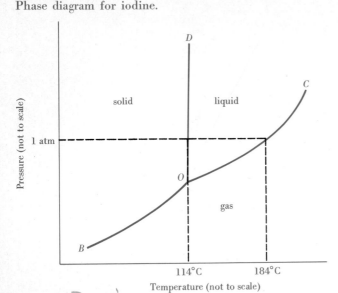

FIGURE 11-12
Phase diagram for carbon dioxide.

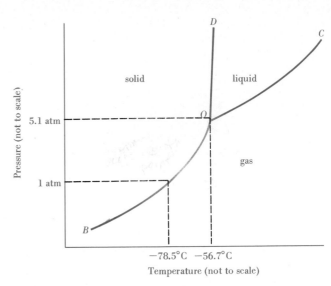

normally OD slants to right because!

@ given P, increase in T. allows more molecules to melt and substance now has greater P since liq is less dense

heat of surroundings doesn't raise Temp but is used @ ΔHsublimation

But for H₂O liq more dense higher T ⇒ more melt ⇒ more liq which is less dense ∴ decrease in P

FIGURE 11-13
Phase diagram for water.

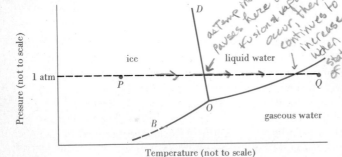

as Temp increased, P rises here while fusion & vaporization occur, then continues to increase when state of substance has completed change

Point *O*, the triple point, is at $+0.0098°C$ and 4.58 mmHg. The critical point, *C*, is at $374.1°C$ and 218.2 atm. The negative slope of the fusion curve, *OD*, is exaggerated in this diagram. An increase in pressure of about 100 atm is required to produce a decrease of $1°C$ in the melting point of ice.

The behavior of carbon dioxide, shown in Figure 11-12, differs from that of iodine in one important respect—the pressure at the triple point is greater than 1 atm. A line at $P = 1$ atm intersects the sublimation curve, not the vapor pressure curve. If solid carbon dioxide (dry ice) is heated in a container open to the atmosphere, it sublimes away at a constant temperature of $-78.5°C$ rather than undergoing melting. Liquid carbon dioxide can exist only under pressures in excess of 5.1 atm.

The phase diagram for water (Figure 11-13) presents another new feature: The fusion curve *OD* slopes *toward* the pressure axis. The melting point of ice *decreases* with increased pressure. This is unusual behavior for a solid. Example 11-6 illustrates how a phase diagram may be used to explain the physical behavior of a substance.

Example 11-6 A sample of ice is heated slowly in a closed cylinder under a constant pressure of 1 atm exerted by a freely moving piston (see Figure 11-14). Describe what happens as the temperature is raised from point *P* to *Q* in the phase diagram of Figure 11-13.

As long as the temperature remains in the region labeled ice only the solid phase exists. When the temperature reaches a point on the fusion curve *OD* (0°C) the ice begins to melt. The temperature remains constant as ice is converted to liquid. When melting is complete, the temperature increases again. No vapor appears in the cylinder until the temperature reaches 100°C, the temperature at which the vapor pressure is 1 atm. When all the liquid has vaporized, the temperature starts to rise again. The vol-

FIGURE 11-14
Predicting phase changes—Example 11-6 illustrated.

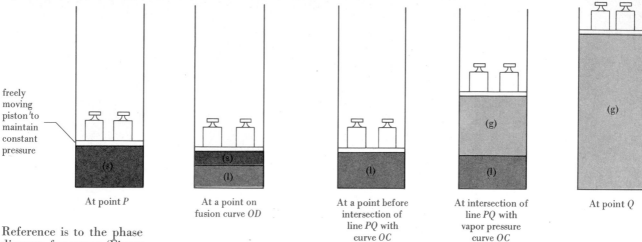

freely
moving
piston to
maintain
constant
pressure

(s)

At point *P*

(s)
(l)

At a point on
fusion curve *OD*

(l)

At a point before
intersection of
line *PQ* with
curve *OC*

(g)
(l)

At intersection of
line *PQ* with
vapor pressure
curve *OC*

(g)

At point *Q*

Reference is to the phase
diagram for water (Figure
11-13).

ume of the gas is directly proportional to the absolute temperature since the pressure and
amount of gas remain constant.

SIMILAR EXAMPLES: Exercises 29, 30.

11-7 Evidence of Intermolecular Forces— Condensed States of the Noble Gases

The simplest of the group 0 elements, helium, forms no stable chemical bonds. Given
this fact we might reason that helium would be a gas at all temperatures, right down
to absolute zero. In fact, helium does condense to a liquid at 4 K and freeze to a solid
(at 25 atm pressure) at 1 K. These data suggest the presence of weak intermolecular
forces that overcome thermal agitation and cause the gas to condense if the temperature is lowered sufficiently. But what kind of force can this be?

11-8 Van der Waals Forces

The process of induction
is described further in
Appendix B. A common-
place analogy is the at-
traction of a balloon to a
wall. The balloon is
charged by rubbing, and
the charged balloon in-
duces an opposite charge
on the wall.

INSTANTANEOUS AND INDUCED DIPOLES. In speaking of the electronic structure of
an atom or molecule we refer to the *probablity* of an electron being in a certain
region at a given instant of time. One event that may occur is suggested by Figure
11-15—an instantaneous displacement of electrons toward one region of an atom or
molecule. This displacement causes a normally nonpolar species to become polar; an
instantaneous dipole is formed. Following this, electrons in a neighboring atom or
molecule may be displaced, also leading to a dipole. This is a process of induction,
and the newly formed dipole is called an induced dipole.

FIGURE 11-15
Instantaneous and
induced dipoles.

$\delta+$ $\delta-$

(a)   (b)

$\delta+$ $\delta-$ $\delta+$ $\delta-$

(c)

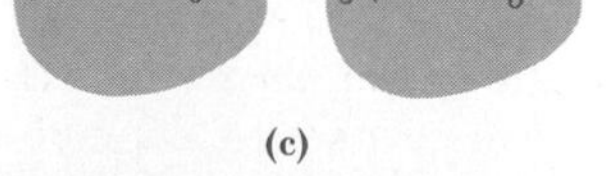

(a) *Normal condition.* In a nonpolar species there is a symmetrical charge distribution.
(b) *Instantaneous condition.* A displacement of the electronic charge produces an instantaneous dipole, with charges of $\delta+$ and $\delta-$.
(c) *Induced dipole.* The instantaneous dipole on the left induces a charge separation
in the species on the right. The result is a dipole–dipole interaction; the two dipoles
are attracted to each other.

FIGURE 11-16
Molecular shapes and polarizability.

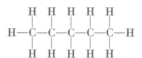

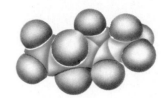

(a) Neopentane
bp = 9.5°C
T_c = 160.6°C

(b) Normal pentane
bp = 36.1°C
T_c = 196.5°C

The compact and symmetrical neopentane (a) is not as easily polarized as the elongated normal pentane (b). Neopentane is somewhat more difficult to liquefy than normal pentane; hence its lower critical temperature and normal boiling point. The two compounds have the same formula—C_5H_{12}.

Taken together these two events lead to an intermolecular force of attraction. Although it is proper to call this an instantaneous dipole–induced dipole interaction, the terms **dispersion force** and **London force** are generally used. (Fritz London offered a theoretical explanation of these forces in 1928.)

The ease with which an electron cloud is distorted by an external electric field (and hence the ease with which a dipole is induced) is called **polarizability.** In general, polarizability increases with the total number of electrons in a molecule. Since molecular weight is related in a general way to number of electrons, the polarizability of molecules and the strength of dispersion (London) forces increase with increased molecular weight. For example, radon (atomic weight, 222) has a much higher boiling point than does helium (atomic weight, 4)—211 K for Rn compared to 4 K for He. Another important factor in determining the strength of dispersion forces is molecular shape. Electrons in elongated molecules are more easily displaced than are those in small, compact, symmetrical molecules. Some physical properties of two substances with the identical numbers and kinds of atoms (isomers) but different molecular shapes are compared in Figure 11-16.

FIGURE 11-17
Dipole–dipole interactions.

The arrangement shown here is an idealized case. Normally, thermal motion upsets this orderly array, cancelling out most of the dipole–dipole attractions. But this tendency for dipoles to align themselves can be of considerable importance in affecting the properties of substances, especially solids.

Example 11-7 Arrange the following substances in the order in which you would expect their boiling points to increase: Cl_2, H_2, N_2, F_2.

Assuming that all intermolecular forces are of the London type, we would expect boiling points to increase in the same order as molecular weights.

$$H_2 < N_2 < F_2 < Cl_2$$

The observed boiling points are −253, −196, −188, and −34°C, respectively.

SIMILAR EXAMPLES: Exercises 31, 34.

DIPOLE–DIPOLE INTERACTIONS. In a *polar* substance molecules tend to become oriented with the positive end of one dipole directed toward the negative ends of neighboring dipoles. (An idealized situation is pictured in Figure 11-17.) This additional partial ordering of molecules can cause a substance to persist as a solid or liquid at temperatures higher than otherwise expected. For example, compare normal butane, C_4H_{10}, and acetone (dimethyl ketone), $(CH_3)_2CO$.

nonpolar: C_4H_{10} mol. wt. = 58   m.p. = $-138.3°C$ b.p. = $-0.5°C$

polar: $(CH_3)_2C{=}O$ mol. wt. = 58   m.p. = $-94.8°C$ b.p. = $56.2°C$

Example 11-8   Arrange the following compounds in expected order of normal boiling point.

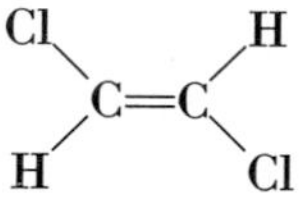

The compounds have the same molecular weight. A comparison must be made of their polarities. Compound A is nonpolar, since the electronegative Cl atoms are attached symmetrically with respect to the carbon–carbon double bond. Compound B has a nonsymmetrical arrangement of the Cl atoms; both are on the same side of the molecule. The molecule is polar. The nonpolar liquid should vaporize more readily and thus have the lower boiling point.

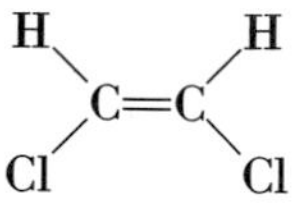

expected order of boiling points: $A < B$

observed boiling points: $47.7 < 60.3°C$

less suceptible to dipole/london bonds, more easily separated from neighbors.

SIMILAR EXAMPLES: Exercises 33, 35.

The intermolecular forces described in this section are the forces that cause a gas to depart from ideal gas behavior. The van der Waals equation of state takes these forces into account, and, collectively, these forces are called **van der Waals forces.** Surprisingly, perhaps, the greatest contribution for most substances, even those that have permanent dipole moments, is made by London forces. For one thing London forces involve an entire molecule, not just the dipolar portion. Also, whereas dipole–dipole interactions fall off sharply with increased temperature (see Figure 11-17), dispersion forces are unaffected by temperature. If a molecule is knocked out of position by thermal motion, a new dipole is instantaneously induced in the proper direction for an attractive force. There are some substances, however, in which dipole–dipole interactions are as, or more, important than London forces. These unusual dipole–dipole interactions are described in the next section.

11-9 Hydrogen Bonds

Normal boiling points of a series of similar compounds are plotted as a function of molecular weight in Figure 11-18. Normal behavior is that displayed by the group IVA hydrides—boiling point increases regularly with increasing molecular weight. Three striking exceptions are noted in the figure—NH_3, H_2O, and HF.

Figure 11-19 illustrates for hydrogen fluoride three important ideas about the dipole–dipole interactions known as **hydrogen bonds.**

1. The H atom in HF is a center of positive charge; the F atom, of negative charge. Dipoles tend to align in the usual fashion, the positive end of one dipole directed toward the negative end of another. This alignment places an H atom between two F atoms. Because of the very small size of the H atom, the dipoles come into close proximity, producing a strong dipole–dipole interaction.

A and B are different molecules because of the rigidity of the double bond described on page 236. That is, one molecule cannot be obtained from the other by rotation of a group about the double bond. Molecules of this type are described further in Chapter 24.

The importance of London forces relative to dipole–dipole interactions can be illustrated as follows: The nonpolar substance CCl_4, because of its much greater molecular weight (and polarizability), has a higher boiling point ($76.7°C$) than does CH_3Cl (b.p. $-24°C$). This is true despite the fact that CH_3Cl is a polar molecule.

FIGURE 11-18
Comparison of boiling points of some hydrides of the elements of groups IVA, VA, VIA, and VIIA.

The values for NH_3, H_2O, and HF are unusually high compared to those of other members of their groups.

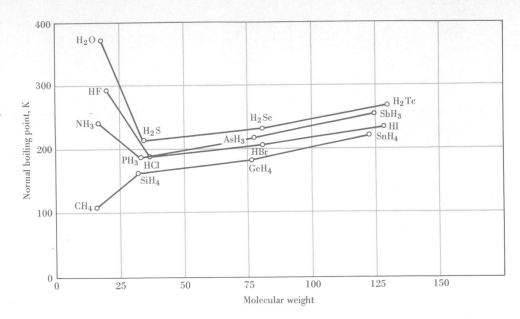

FIGURE 11-19
The hydrogen bond.

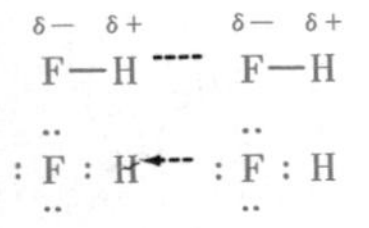

$$\overset{\delta-}{F}\!-\!\overset{\delta+}{H} \cdots \overset{\delta-}{F}\!-\!\overset{\delta+}{H}$$

$$:\!\overset{..}{F}\!:\!H \longleftarrow :\!\overset{..}{F}\!:\!H$$

two views of a hydrogen bond

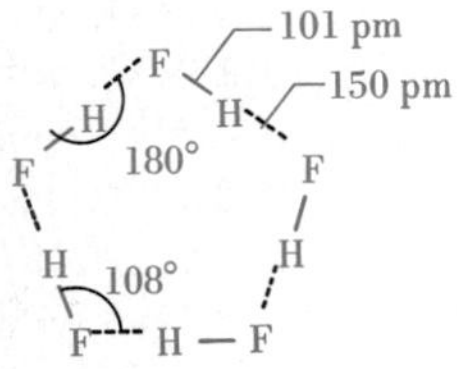

the species $(HF)_5$

In the gaseous state, several polymeric forms of the HF molecule exist in which the individual molecules (monomers) are held together through hydrogen bonds. A pentagonal arrangement of five HF molecules is shown here.

The type of hydrogen bonding described here is *inter*molecular. Another possibility is for a hydrogen atom to bridge two nonmetal atoms within the same molecule. This is called *intra*molecular hydrogen bonding (see Exercise 42).

2. Along with being strongly bonded to one F atom, an H atom acts as if it is weakly bonded to the F atom of a nearby HF molecule. This occurs through the lone pair of electrons on that F atom. In hydrogen bonding an H atom acts as a bridge between two nonmetal atoms of different molecules.

3. An essential part of hydrogen bond formation is the tendency for hydrogen bonds to form throughout a cluster of molecules, producing so-called polymeric structures. The bond angle between two nonmetal atoms bridged by an H atom (i.e., the bond angle, X—H⋯X) is usually about 180°. However, there are preferred orientations for hydrogen-bonded molecules, such as in the pentagonal structure, $(HF)_5$.

To summarize, the hydrogen bond is a rather strong intermolecular force, with energies of the order of 15 to 40 kJ/mol. (Van der Waals interactions correspond to energies of about 2 to 20 kJ/mol.) Hydrogen bonding is most likely to occur when an H atom in a molecule can be simultaneously attracted to a highly electronegative atom—F, O, or N—in a neighboring molecule. Weak hydrogen bonding may occur between an H atom of one molecule and a Cl or S atom of a neighboring molecule.

The most common substance in which hydrogen bonding occurs is ordinary water. Figure 11-20 indicates how one water molecule is held to four neighbors in a tetrahedral arrangement by hydrogen bonds. This is the structural arrangement in crystalline water or ice. Hydrogen bonds hold the water molecules in a rigid but rather open structure. As ice melts only a fraction

[handwritten margin notes at top: "ice structure has so many 'open' spaces that just as it begins to melt some of the structures breaks into pieces, allowing some 'bonded chunks' to come close together ∴ liquid more dense"]

FIGURE 11-20
Hydrogen bonding in water.

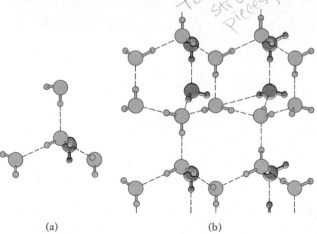

(a)   (b)

(a) Each water molecule is linked to four others through hydrogen bonds. The arrangement is tetrahedral. Each hydrogen atom is situated along a line joining two oxygen atoms, but somewhat closer to one oxygen atom (100 pm) than to the other (180 pm).
(b) The crystal structure of ice. Oxygen atoms are located in hexagonal rings arranged in layers. Positions available to the hydrogen atoms lie between pairs of oxygen atoms, again closer to one than to the other. This characteristic pattern is revealed at the macroscopic level in the hexagonal shapes of snowflakes.

Through hydrogen bonding, we can understand why a lake freezes from the top down and why the unfrozen water at the bottom of the lake is at a temperature somewhat above freezing (4°C = 39°F). It is interesting to speculate on the consequences for the macroscopic (real) world if there were no such phenomenon as hydrogen bonding in the molecular world. For one thing, would aquatic life be possible in cold climates?

of the hydrogen bonds are broken. Extensive hydrogen bonding persists in liquid water just above the melting point; liquid water retains an icelike structure. Evidence to this effect is provided by the low heat of fusion of water (6.02 kJ/mol), much less than would be expected if all the hydrogen bonds were to be broken.

The packing of molecules in liquid water at the melting point is closer than in ice; therefore, the liquid is more dense than ice. This is very unusual behavior for a substance. In all but a few cases, a liquid is less dense than the solid from which it is formed at the melting point. As liquid water is heated above the melting point, its density continues to *increase* as more hydrogen bonds are broken and further packing of the molecules occurs in the liquid state. Liquid water attains its maximum density at 3.98°C. Above this temperature the liquid behaves in a normal fashion; its density decreases with temperature.

It is not an overstatement to say that hydrogen bonding literally makes life possible. Living organisms are maintained through a series of chemical reactions involving complex structures, such as DNA and proteins. Certain bonds in these structures must be capable of breaking and reforming with relative ease. Only hydrogen bonds have just the right energies to permit this. The importance of hydrogen bonding in molecules of biological significance is explored further in Chapter 25.

Liquids in which hydrogen bonding occurs exhibit stronger than usual intermolecular forces. These liquids generally have high heats of vaporization. In acetic acid, H_3CCOOH, hydrogen bonding leads to the formation of dimers (double molecules) both in the liquid and *in the vapor state*. Not all the hydrogen bonds between molecules need to be broken to vaporize acetic acid and the heat of vaporization is abnormally *low*. Dimerization of acetic acid molecules is illustrated in Figure 11-21.

[handwritten margin notes: "held together tightly — need to add a lot of heat before you can pull them apart"]

11-10 Network Covalent Solids

[handwritten: "→ sharing of a pair of electrons"]

In some substances covalent bonds are not limited to individual small molecules. They extend throughout a single giant molecule in the form of a crystal. In such cases inter- and intramolecular forces are indistinguishable since the entire crystal is held together by strong covalent bonds.

FIGURE 11-21
Dimerization of gaseous acetic acid.

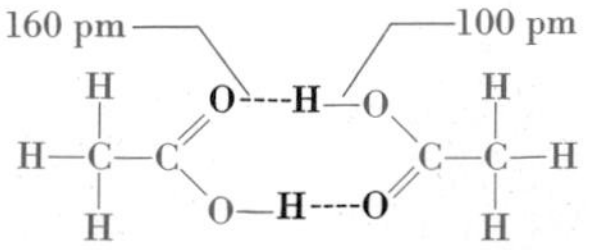

160 pm ——————— 100 pm

$$\text{H}-\overset{\displaystyle \text{H}}{\underset{\displaystyle \text{H}}{\text{C}}}-\text{C}\begin{matrix} \text{O}\text{---}\text{H}\text{---}\text{O} \\ \\ \text{O}\text{---}\text{H}\text{---}\text{O} \end{matrix}\text{C}-\overset{\displaystyle \text{H}}{\underset{\displaystyle \text{H}}{\text{C}}}-\text{H}$$

Hydrogen bonding permits molecules to exist in stable pairs (dimers).

The Lewis structure does suggest that an unsatisfied bonding capacity exists at the surface of a "macromolecule" such as diamond. This unsatisfied bonding capacity is responsible for some interesting properties of surfaces (see Section 12-8).

THE DIAMOND STRUCTURE. A scheme to describe how carbon atoms can bond, one to another is suggested in Figure 11-22. The two-dimensional Lewis structure (Figure 11-22a) is adequate only to establish the fact that this bonding involves ever increasing numbers of carbon atoms and leads to a giant covalent molecule. It provides no information about the three-dimensional structure. The entire crystal structure can be inferred from the portion shown in Figure 11-22b. Each atom is bonded to four others. Atoms 1, 2, and 3 lie in a plane with atom 4 above the plane. Atoms 1, 2, 3, and 5 define a tetrahedron with atom 4 inscribed in its center. When viewed from a particular direction, a nonplanar hexagonal arrangement of carbon atoms is seen (shown in grey).

The substance we have just described is diamond. If silicon atoms are substituted for one half the carbon atoms, the resulting structure is that of silicon carbide (carborundum). These two substances are both extremely hard. To break or scratch a crystal of diamond or silicon carbide requires that covalent bonds be broken. Both substances are widely used as abrasives, and diamond is the hardest substance known. They are nonconductors of heat and electricity and do not melt or sublime until very high temperatures are reached. SiC sublimes at 2700°C, and diamond melts above 3500°C.

Unlike most solids, silica does not have a sharp melting point. Instead, it softens and becomes liquid in the same way that glass does. Silica is a basic constituent of glass (see Section 26-8).

SILICA AND SILICATES. Silica, SiO_2, and silicate minerals comprise the bulk of the earth's crust. The bonding and structure of these materials is of great interest to chemists, but the subject is complex. There is an almost limitless number of structures to consider. Some of these structures are described in Section 20-9, but for the present we note this basic fact: A silicon atom, like a carbon atom, can form four bonds simultaneously, arranged in a tetrahedral fashion. In silica each Si atom is bonded to *four* O atoms and each O atom to *two* Si atoms. This structural arrangement extends throughout a very large network, as suggested by Figure 11-23. Certain properties of silica resemble those of diamond. The melting point is high (about 1700°C); silica, say in the form of quartz, is very hard; and silica is a nonconductor of electric current.

A further fact must be noted to explain why silicon dioxide is not composed of discrete molecules (SiO_2) as is carbon dioxide (CO_2). The silicon atom cannot form a π bond through the parallel overlap of one of its $3p$ orbitals with a $2p$ orbital of an O atom (recall Figure 10-9). Presumably, this is because the silicon atom is too large.

FIGURE 11-22
The diamond structure.

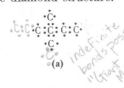

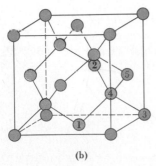

(b)

THE GRAPHITE STRUCTURE. Carbon atoms can be bonded together to produce a solid having properties very much different from diamond. This bonding involves sp^2 and p orbitals. The three orbitals of the sp^2 type are directed in a plane at angles of 120°; the p orbital is directed perpendicular to the plane, above and below. These are the same orbitals used by carbon atoms in C_6H_6 (described in Figure 10-19).

The crystal structure that results from this type of bonding is pictured in Figure 11-24; it is the graphite structure. Each carbon atom forms strong covalent bonds with three neighboring atoms in the same plane, giving rise to layers of carbon atoms in a hexagonal arrangement. The p electrons of the carbon atoms are *delocalized* (recall the discussion in Section 10-5). Bonding within layers is strong but between layers it is much weaker. Evidence of this is provided by bond distances. The C—C

(a) A portion of the Lewis structure. **(b)** Crystal structure. Each carbon atom is bonded to four others in a tetrahedral fashion. The segment of the entire crystal shown here is called a unit cell.

FIGURE 11-23
Bonding in silica—SiO₂.

● = silicon ⬤ = oxygen

FIGURE 11-24
The graphite structure.

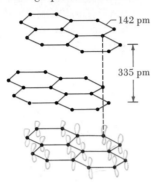

142 pm

335 pm

(handwritten notes: all these π bonds delocalize & form giant e⁻ cloud above & below layers (making conductive))

(handwritten notes: (high charged small sized ions) more attractive forces ⇒ come together easier ⇒ release more lattice energy)

FIGURE 11-25
Interionic forces of attraction.

Relative attractive force:

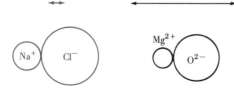

Mg²⁺

O²⁻

Na⁺ Cl⁻

Radius: Na⁺ = 95 pm
Cl⁻ = 181 pm

Mg²⁺ = 65 pm
O²⁻ = 140 pm

Radius sum = [distance between center of ions:] 276 pm *(handwritten: larger so less interionic attractive force)* 205 pm

Because of the higher charges on the ions and the closer proximity of their centers, the interionic attractive force between Mg²⁺ and O²⁻ is about seven times as great as between Na⁺ and Cl⁻.

bond distance within a layer is 142 pm (1.42 Å); between layers it is 335 pm (3.35 Å).

The distinctive properties of graphite are derived from its unique crystal structure. Because bonding between layers is weak, the layers can be made to glide over one another readily. As a result graphite may be used as a lubricant, either in pure dry form or suspended in oil. Layers flake off from a piece of graphite easily, making it valuable as the basic writing material in pencils. Because the *p* valence electrons in graphite are delocalized, they can be made to migrate through the planes of carbon atoms under the influence of an electric field. Graphite is a conductor of electric current and is widely used for electrodes in batteries and in industrial electrolysis processes. Diamond is not an electrical conductor because in its structure all of the valence electrons of the carbon atoms are localized or fixed permanently into single covalent bonds.

11-11 The Ionic Bond as an Intermolecular Force

Many of the physical properties of an ionic solid are determined by this one factor: How difficult is it to break up an ionic crystal lattice and separate the ions it contains? In an ionic crystal each ion exerts attractive forces on several neighboring ions. In addition, ions with like charge exert a repulsive force on one another. Determining the net strength of the forces within an ionic crystal is difficult, but this quantity can be calculated or determined by experiment. The quantity involved is called the lattice energy. **Lattice energy** is the quantity of energy liberated when separated ions, positive and negative, are allowed to come together to form an ionic crystal composed of 1 mol of formula units of a compound. Alternatively, lattice energy may be defined as the quantity of energy required to break up an ionic crystal, causing the complete separation of the ions in 1 mol of compound. The stronger the forces among ions, the more difficult it is to disrupt a crystal.

Any factor that contributes to strong attractive forces among ions will contribute to a high lattice energy. *The attractive force between a pair of oppositely charged ions increases with increased charge on the ions or with decreased ionic sizes,* as illustrated in Figure 11-25. The type of crystal structure is also a factor in establishing the magnitude of the lattice energy.

Lattice energies for most ionic compounds are sufficiently great that ions do not detach themselves readily from the crystal and pass into the gaseous state.

Ionic solids do not sublime at ordinary temperatures. All ionic solids can be melted by supplying enough thermal energy to destroy the crystalline lattice. In general the higher the lattice energy the higher the melting point.

— the more energy they release when coming together, the more they're attracted to each other & hold together

Example 11-9 Would you expect KI or CaO to have the higher melting point? *tighter*

The higher lattice energy is expected for the combination of small, highly charged ions—Ca^{2+} and O^{2-}. The expected order of melting points is $KI < CaO$. The observed melting points are 677°C for KI and 2590°C for CaO.

SIMILAR EXAMPLES: Exercises 47, 48.

Two conditions are necessary for electrical conductivity: (1) Charged particles must be present and (2) the particles must be able to migrate in an electric field. In a solid ionic compound only the first requirement is met. However, whenever ions enter the liquid state, by melting an ionic crystal or by dissolving it in a suitable solvent, the liquid becomes a good electrical conductor. The energy required to break up an ionic crystal in the process of dissolving comes as a result of the interaction of ions with the solvent. Thus, the extent to which an ionic solid will dissolve is again determined, at least in part, by the lattice energy of the solid. That is, the lower the lattice energy the greater the quantity of an ionic solid that can be dissolved in a given quantity of solvent (see Section 12-1).

solid ionic

when melted e⁻'s may move freely & conduct

lower U ⇒ held looser ⇒ less +/- pull between them ⇒ more easily dissolved

11-12 Calculation of Lattice Energy—The Born–Haber Cycle

The actual lattice energy of an ionic crystal need not be known to make qualitative predictions of the sort described in the preceding section. However, calculation of lattice energy does offer interesting illustrations of Hess's law and the atomic properties of ionization energy and electron affinity. The familiar reaction used to describe the formation of a mole of an ionic compound from its elements, for example NaCl(s), is

$$Na(s) + \tfrac{1}{2}Cl_2(g) \longrightarrow NaCl(s) \qquad \Delta\bar{H}_f^\circ = -411 \text{ kJ/mol} \qquad (11.5)$$

HESS
*Total Enthalpy change for a reaction is equal to sum of enthalpy changes in "subreactions"/stages.

But this is *not* the reaction that corresponds to the definition of lattice energy. Lattice energy, *U*, is based on the formation of a crystal from *gaseous* ions, a process that can be visualized in five steps:

*Ion. Potential
Energy needed to remove (loosely held) e⁻'s from gaseos state
Na(g) → Na⁺(g) + e⁻

1. Sublimation of solid sodium: $Na(s) \to Na(g)$; $\Delta\bar{H}_1 = +108$ kJ/mol
2. Ionization of gaseous atomic sodium: $Na(g) \to Na^+(g) + e^-$; $\Delta\bar{H}_2 = +496$ kJ/mol
3. Dissociation of gaseous chlorine: $\tfrac{1}{2}Cl_2(g) \to Cl(g)$; $\Delta\bar{H}_3 = +121$ kJ/mol
4. Ionization of gaseous atomic chlorine: $Cl(g) + e^- \to Cl^-(g)$; $\Delta\bar{H}_4 = -357$ kJ/mol
5. Combination of gaseous ions: $Na^+(g) + Cl^-(g) \to NaCl(s)$; $\Delta\bar{H}_5 = U = ?$

*Electron affinity
Energy associatd w/an e⁻ gain by a gaseous atom.
Cl(g) + e⁻ → Cl⁻(g)

The value of $\Delta\bar{H}_2$ is based on the ionization energy of Na, +5.14 eV/atom, and that of $\Delta\bar{H}_4$ on the electron affinity of Cl, −3.7 eV/atom. Both values are converted to kJ/mol by using the conversion factor from expression (8.4): 1 eV/atom = 96.49 kJ/mol. $\Delta\bar{H}_3$ is simply one half of the Cl—Cl bond energy (recall Table 9-3).

By an appropriate combination of the first four steps listed above with equation (11.5) we can obtain the lattice energy, *U*, that is, $\Delta\bar{H}_5$.

$$Na^+(g) + e^- \longrightarrow Na(g) \qquad -\Delta\bar{H}_2 = -496\,kJ/mol$$
$$Na(g) \longrightarrow Na(s) \qquad -\Delta\bar{H}_1 = -108\,kJ/mol$$
$$Cl^-(g) \longrightarrow Cl(g) + e^- \qquad -\Delta\bar{H}_4 = +357\,kJ/mol$$
$$Cl(g) \longrightarrow \tfrac{1}{2}Cl_2(g) \qquad -\Delta\bar{H}_3 = -121\,kJ/mol$$
$$\underline{Na(s) + \tfrac{1}{2}Cl_2(g) \longrightarrow NaCl(s) \qquad \Delta\bar{H}_f^\circ = -411\,kJ/mol}$$
$$Na^+(g) + Cl^-(g) \longrightarrow NaCl(s) \qquad \Delta\bar{H}_5 = U = -779\,kJ/mol \qquad (11.6)$$

The method illustrated here was developed by the German physicist Max Born and chemist Fritz Haber. Although we have applied the method to calculating the lattice energy of NaCl(s), a more common application of the Born–Haber cycle is in calculating electron affinities (see Additional Exercise 15).

11-13 Crystal Structures

Considerable insight into the structures of crystals can be gained by comparing them to the way in which spheres (marbles, cannonballs, or atoms) can be stacked.

CLOSEST PACKING OF SPHERES.   Unlike the case of cubes, there is no way of stacking spheres to fill all space; holes or voids remain among them. But there are certain arrangements in which the voids are kept to a minimum. These are known as **closest packed structures.** Two such structures, called **hexagonal closest packed** and **cubic closest packed,** are presented in Figure 11-26.

Imagine starting with a layer of identical spheres; this is layer (a) in Figure 11-26. Each sphere is in contact with six others arranged around it in a hexagonal fashion. Among the spheres there exist voids or holes. If the spheres are placed on a flat surface, as suggested for layer (a), these voids are all equivalent. Once the first sphere is placed into the next layer, the entire pattern for that layer is established; this is layer (b) in Figure 11-26. Again voids or holes appear within layer (b). But now the holes are of *two* different types. **Tetrahedral holes** in layer (b) fall directly over spheres in layer (a) and have this shape: △ . **Octahedral holes** in layer (b) fall directly over holes in layer (a) and have this shape: ✳ .

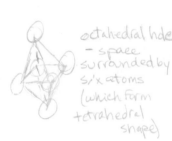

[handwritten notes:] tetrahedral hole — surrounded by 5 atoms

[handwritten notes:] octahedral hole — space surrounded by six atoms (which form tetrahedral shape)

FIGURE 11-26
Closest-packed structures.

Spheres in layer (a) are outlined in color. Those added as layer (b) are outlined in black.

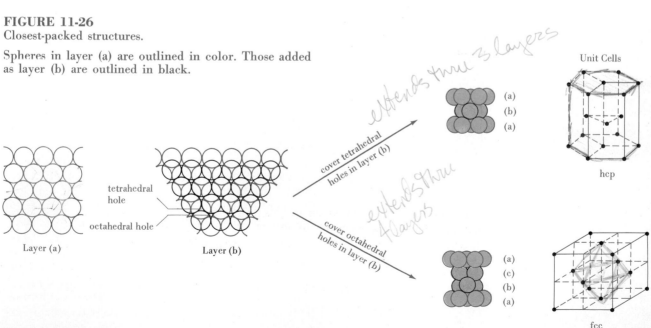

Layer (a)

tetrahedral hole

octahedral hole

Layer (b)

cover tetrahedral holes in layer (b)   *[handwritten:]* extends thru 3 layers

cover octahedral holes in layer (b)   *[handwritten:]* extends thru 4 layers

(a)
(b)
(a)

Unit Cells

hcp

(a)
(c)
(b)
(a)

fcc

FIGURE 11-27
Body-centered cubic
(bcc) unit cell.

The eight corner spheres
are each in contact with
an identical sphere at the
center of the cube.

A parallelepiped is a
three-dimensional struc-
ture having six faces ar-
ranged in three parallel
pairs. Each face is a par-
allelogram; in some cases
the parallelograms are
rectangles or squares.
Thus, a rectangular box
is a parallelepiped. So is
a cube.

TABLE 11-3
Crystal structures of some common metals

Hexagonal closest packed	Face-centered cubic	Body-centered cubic
cadmium	aluminum	iron
magnesium	copper	potassium
titanium	lead	sodium
zinc	silver	tungsten

Two possibilities now exist for the placement of the
third layer. In one arrangement, the **hexagonal closest
packed (hcp),** all the *tetrahedral* holes are covered. The
third layer is identical to layer (a) and the structure
begins to repeat itself. In the other arrangement, the
cubic closest packed, all the octahedral holes are covered. The spheres in layer (c) are out of line with those
in layer (a), so the two layers are not identical. In this
arrangement, when a fourth layer is added it is identical to layer (a) and the structure begins to repeat itself.

Also shown in Figure 11-26 are unit cells of the two
structures. The **unit cell** can be thought of as a parallelepiped formed by joining the centers of a group of
spheres. An important property of the unit cell is that
by shifting its position in three mutually perpendicular
directions (a process called translation) the entire
structure can be generated. To assist in visualizing the
hexagonal packing arrangement, three unit cells of the
hcp type are shown in Figure 11-26. The unit cell chosen for the cubic closest packed arrangement is called
face-centered cubic (fcc). A sphere appears in the center of each face of the unit cell.

In both the hcp and fcc arrangements, voids or
holes account for only 25.96% of the total volume.
Another arrangement for identical spheres in which
the packing is not quite so close results in the **body-
centered cubic structure (bcc).** In this structure voids
account for 31.98% of the total volume. A bcc unit cell
is pictured in Figure 11-27.

The best examples of crystal structures based on
the close packing of spheres are found among the metallic elements. Some examples are given in Table 11-3.

FIGURE 11-28
Diffraction of x rays by a crystal.

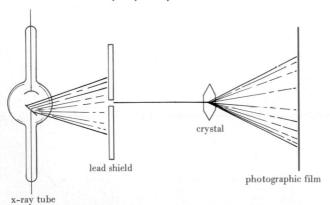

crystal

lead shield

photographic film

x-ray tube

X-RAY DIFFRACTION. The experimental determination of crystal structures is accomplished through the
technique of x-ray diffraction. When a beam of x rays
encounters atoms of a substance, the x rays interact
with the electrons and the original beam is reradiated,
diffracted or scattered in all directions. The scattering
pattern is related to the distribution and density of
electronic charge in the atoms encountered by the
x-ray beam. The scattered x rays must be made to produce a visible pattern, as on a photographic film, and
the microscopic structure of the object can only be inferred from this pattern.

Figure 11-28 suggests a method of scattering x rays
from a crystal, and Figure 11-29 represents a photographic record of the scattered x rays called a Laue
pattern (after Max von Laue, who pioneered in the
application of x rays to crystal-structure determination). Some aspects of scattering patterns can be ex-

FIGURE 11-29
Laue x-ray photograph
for crystalline sodium
chloride.

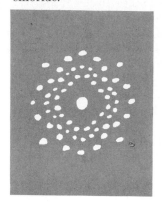

plained by a geometric analysis proposed by W. H. Bragg and W. L. Bragg in 1912, illustrated in Figure 11-30.

Figure 11-30 pictures two rays in a monochromatic x-ray beam, labeled a and b. Wave a is reflected from one plane of atoms or ions in a crystal and wave b from the next plane below. Wave b travels a greater distance than wave a. The additional distance is $2d \sin \theta$. In order to reinforce one another, the crests and troughs of the two waves must be in phase (line up with one another) as they approach the detector. To satisfy this requirement, the additional distance traveled by wave b must be an integral multiple of the wavelength of the x rays.

$$n\lambda = 2d \sin \theta \tag{11.7}$$

where $n = 1, 2, 3, \ldots$.

The spacing between atomic planes can thus be determined by knowing λ and measuring θ. Different orientations of the crystal allow for atomic spacings and electron densities to be determined along different directions through the crystal. With this type of information, the crystalline structure of a solid can be established. Altogether, this is generally a complex task which is greatly simplified by high-speed electronic computers.

Once a crystal structure has been established, certain properties can be determined by calculation. In Example 11-10 a metallic radius is calculated, and in Example 11-11, the density of a metal. In both of these calculations it is necessary to visualize a unit cell of the crystal structure. In addition, in Example 11-11 a method of counting the atoms in a unit cell is required.

The right triangle shown in Figure 11-31 must conform to the Pythagorean formula: $a^2 + b^2 = c^2$.

Example 11-10 At room temperature iron crystallizes in a bcc structure. By x-ray diffraction, the edge of the cubic cell corresponding to Figure 11-31 is found to be 286 pm. What is the radius of an iron atom?

Nine atoms are associated with the cube pictured in Figure 11-31. One atom is located at each of the eight corners of the cube and one at the center of the cube. The three atoms along a cube diagonal are in contact. The length of the cube diagonal (the distance from the farthest upper-right corner to the nearest lower-left corner) is four times the atomic radius. But also shown in Figure 11-31 is that the diagonal of a cube is equal to $\sqrt{3}$ times the length. The length, l, is what is given.

FIGURE 11-30
Determination of crystal structure by x-ray diffraction.

The two triangles outlined by the dotted lines are identical right triangles. The hypotenuse of each triangle is equal to the interatomic distance d. The side opposite the angle θ thus has a length of $d \sin \theta$. Wave b travels farther than wave a by the distance $2d \sin \theta$.

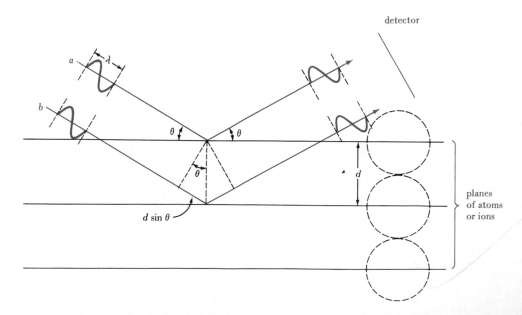

FIGURE 11-31
Determination of the atomic radius of iron—Example 11-10 illustrated.

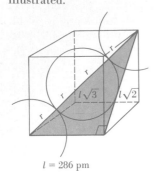

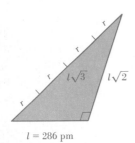

$l = 286$ pm

$l = 286$ pm

$$4r = l\sqrt{3} \qquad r = \frac{\sqrt{3} \times 286 \text{ pm}}{4} = \frac{1.73 \times 286 \text{ pm}}{4} = 124 \text{ pm} \ (1.24 \text{ Å})$$

SIMILAR EXAMPLES: Exercises 56, 57.

Example 11-11 Use data from Example 11-10 to calculate the density of iron.

Density is defined as mass per unit volume. Suppose we take as our unit of mass that of a unit cell, pictured in Figure 11-27 and again in Figure 11-31.

$$\text{volume} = l^3 = (2.86 \times 10^{-8} \text{ cm})^3 = 2.34 \times 10^{-23} \text{ cm}^3$$

What shall we use as the mass of a unit cell? It *cannot* be the mass of nine Fe atoms! Of the nine atoms associated with a bcc unit cell, only one belongs entirely to that cell. This is the atom at the center of the cube. As illustrated through Figure 11-32, the atoms at the corners of the unit cell are each shared among eight unit cells. "One eighth" of each corner atom belongs to the given unit cell. There are eight corner atoms, contributing a mass equivalent to $8 \times \frac{1}{8} = 1$ atom. The mass of a unit cell, then, is the mass of two Fe atoms.

$$\text{mass} = 2 \text{ atoms Fe} \times \frac{1 \text{ mol Fe}}{6.02 \times 10^{23} \text{ atoms Fe}} \times \frac{55.8 \text{ g Fe}}{1 \text{ mol Fe}} = 1.85 \times 10^{-22} \text{ g Fe}$$

$$\text{density of iron} = \frac{1.85 \times 10^{-22} \text{ g}}{2.34 \times 10^{-23} \text{ cm}^3} = 7.91 \text{ g/cm}^3$$

(Measured density of iron $= 7.86$ g/cm^3.)

SIMILAR EXAMPLES: Exercises 56, 57.

11-14 Ionic Crystal Structures

FIGURE 11-32
Apportioning atoms among bcc unit cells.

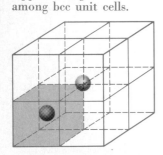

Eight unit cells are outlined. For clarity, only the centers of two atoms are pictured. Our attention is focused on the unit cell in color. The atom in the center of the cell belongs entirely to that cell. The corner atom is seen to be shared by all eight unit cells. Only one-eighth of the corner atom can be said to belong to any given unit cell.

When applied to ionic compounds, the packing-of-spheres model of crystal structures is complicated by two factors: (1) The structural units are not all of the same kind, and (2) they are not of the same size. In defining a unit cell of an ionic crystal structure alternative possibilities exist. Generally, a unit cell is chosen that:

1. By simple translation or displacement in three dimensions, generates the entire crystal.
2. Indicates the crystal coordination numbers.
3. Is consistent with the formula of the compound.

Unit cells of crystalline NaCl and CsCl are pictured in Figures 11-33 and 11-34.

The **crystal coordination number** is the number of nearest-neighboring ions of opposite charge to any given ion in a crystal. Note the Na$^+$ ion at the center of the unit cell in Figure 11-33. It is surrounded by *six* Cl$^-$ ions. The crystal coordination numbers of both Na$^+$ and Cl$^-$ are *six*. By contrast, the crystal coordination number of Cs$^+$ and of Cl$^-$ in Figure 11-34 is *eight*.

The number of ions pictured in the unit cell of NaCl is 27, but the unit cell does not contain 27 ions. The ions must be apportioned among the given and neighboring unit cells in the following way. Each Cl$^-$ ion in a corner position is shared by eight unit cells, and each Cl$^-$ in the center of a face, by two unit cells. This leads to a total number of Cl$^-$ ions for a unit cell of $(8 \times \frac{1}{8}) + (6 \times \frac{1}{2}) = 1 + 3 = 4$. There are 12 Na$^+$ ions along edges of the unit cell, and each edge is shared by four unit cells. The Na$^+$ in the very center of the unit cell belongs entirely to that cell: $(12 \times \frac{1}{4}) + (1 \times 1) = 3 + 1 = 4$. The unit cell has the equivalent of 4 Na$^+$ and 4 Cl$^-$ ions. The ratio of Na$^+$ to Cl$^-$ is 1:1, corresponding to the formula NaCl.

FIGURE 11-33
The unit cell of sodium chloride.

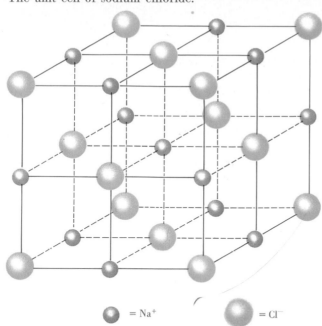

$\bigcirc$ = Na$^+$ $\bigcirc$ = Cl$^-$

For clarity, only the centers of the ions are shown. Oppositely charged ions are actually in contact.

FIGURE 11-34
The unit cell of cesium chloride.

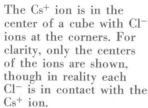

$\bigcirc$ = Cs$^+$ $\bigcirc$ = Cl$^-$

The Cs$^+$ ion is in the center of a cube with Cl$^-$ ions at the corners. For clarity, only the centers of the ions are shown, though in reality each Cl$^-$ is in contact with the Cs$^+$ ion.

FIGURE 11-35
Some unit cells of greater complexity.

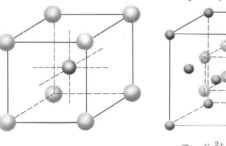

$\bigcirc$ = Ca^{2+} $\bigcirc$ = F$^-$

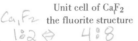
Unit cell of CaF$_2$ the fluorite structure

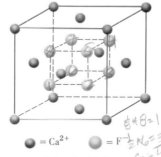

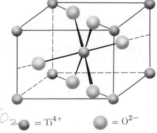

$\bigcirc$ = Ti^{4+} $\bigcirc$ = O^{2-}

Unit cell of TiO$_2$ the rutile structure

Example 11-12   The ionic radii of Na$^+$ and Cl$^-$ are 95 and 181 pm, respectively. What is the length of the unit cell of NaCl?

Again the key to solving this problem lies in understanding geometrical relationships in the unit cell. Along each edge of the unit cell (Figure 11-33) two Cl$^-$ ions are in contact with one Na$^+$. The edge length is equal to the radius of one Cl$^-$, plus the diameter of Na$^+$, plus the radius of another Cl$^-$, that is,

$$\text{edge length} = (r_{Cl^-}) + (r_{Na^+}) + (r_{Na^+}) + (r_{Cl^-})$$

$$= 2(r_{Na^+}) + 2(r_{Cl^-})$$

$$= [(2 \times 95) + (2 \times 181)]\ \text{pm}$$

$$= 552\ \text{pm}\ (5.52\ \text{Å})$$

SIMILAR EXAMPLE: Exercise 62.

Ionic compounds of the type M^{2+}X^{2-} (e.g., MgO, BaS, CaO) may form crystals of the NaCl type. For substances with formulas MX$_2$ or M$_2$X, the crystal structures are more complex. Because the cations and anions occur in unequal numbers, the crystals have two coordination numbers, one for the cation and another for the anion. Two typical structures of a more complex type are shown in Figure 11-35.

In CaF$_2$ (the fluorite structure) there are twice as many fluoride as calcium ions. The crystal coordination number of Ca^{2+} is eight, that of F$^-$ is four. In TiO$_2$ (the rutile structure) Ti^{4+} has a crystal coordination number of six and O^{2-}, three. In this structure two of the O^{2-} ions are within the interior of the cell, two are in the top face, and two in the bottom face of the cell. Ti^{4+} ions are found at the corners and the center of the cell.

RADIUS RATIO AND CRYSTAL COORDINATION NUMBER.   One approach to the structures of ionic crystals is to consider that anions come nearly into contact and cations fit into the holes or voids among the anions. Three different close packed arrangements of anions are shown in Figure 11-36. To determine which arrangement is applicable for a given ionic crystal requires a comparison of the sizes of the cations and anions. Will the cations fit into the voids among the anions?

In Figure 11-37 four anions are pictured lying in a plane. A fifth anion could rest in the hollow among the four, above the plane. A sixth could assume a corresponding position below the plane. In this octahedral packing arrangement we picture a cation as fitting into the hole among the six anions. As suggested in Figure 11-37, a right triangle can be constructed having legs equal to twice the radius of the anion, $2r_a$, and a hypotenuse equal to twice the radius of the anion plus twice

FIGURE 11-36
Some close packing arrangements of anions.

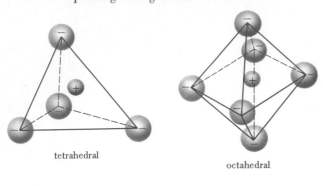

tetrahedral

octahedral

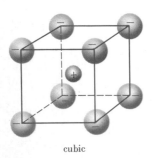

cubic

For clarity only the centers of the ions are shown; in reality the anions are considered to be in contact.

TABLE 11-4
Crystal coordination numbers and radius ratios[a]

Crystal coordination number	Type of arrangement	Radius ratio range (r_c/r_a)
4	tetrahedral	0.225–0.414
6	octahedral	0.414–0.732
8	cubic *body centered*	>0.732

[a] For ratios of $r_c/r_a > 1$, the cation is larger than the anion. In these cases cations form the close-packed arrangement of spheres into which anions fit. The structures are the same as those listed in the table, but based on r_a/r_c rather than r_c/r_a.

FIGURE 11-37
Calculation of radius ratio for octahedral packing.

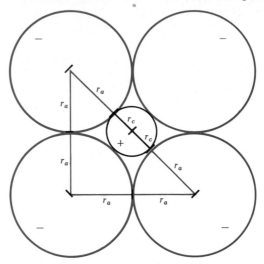

the radius of the cation, $2r_a + 2r_c$. By the Pythagorean formula, the sum of the squares of the sides of a right triangle is equal to the square of the hypotenuse.

$$(2r_a)^2 + (2r_a)^2 = (2r_a + 2r_c)^2 \tag{11.8}$$

Since both r_a and r_c are unknowns, this equation cannot be solved for unique values of r_a and r_c. However, it can be solved for the ratio, r_c/r_a.

$$\frac{r_c}{r_a} = 0.414 \tag{11.9}$$

The quantity expressed in equation (11.9) is called the **radius ratio.** It is the ratio of the radius of the cation to that of the anion. In ionic crystals the anions are not in direct contact. As a result the radius ratios corresponding to different types of structures are somewhat larger than calculated values and also somewhat variable. Table 11-4 lists radius ratio limits for three different types of ionic arrangements.

Summary

One result of the imbalance of forces between surface and bulk molecules in a liquid is surface tension and properties related to it. Another result is the tendency of energetic molecules at the surface to leave a liquid and pass into the vapor state—vaporization.

When vaporization and condensation occur simultaneously in a closed container, a condition of dynamic equilibrium is established. The vapor exerts a characteristic pressure called the vapor pressure of the liquid.

A graph of vapor pressure as a function of temperature is known as a vapor pressure curve. A point of special interest on this curve is the temperature at which the vapor pressure of a liquid becomes equal to the prevailing barometric pressure—the boiling point. Another point of interest is the high-temperature terminus of the vapor pressure curve—the critical point. At the critical point the liquid and gaseous states of a substance become indistinguishable.

The phase diagram of a substance represents the phases or states of matter that will exist at various temperatures and pressures. Characteristic points that can be located on a phase diagram are the normal melting and boiling points, the critical point, and the triple point. The triple point of a substance is the unique condition of temperature and pressure at which the solid, liquid, and gaseous state coexist. Phase diagrams can be used to describe the changes that occur when a sample of matter is heated, cooled, or subjected to a change in pressure.

Electrostatic attractions between instantaneous and induced dipoles are the most common intermolecular forces. The magnitude of these forces generally increases with increased molecular weight. For polar substances dipole–dipole attractions may also be significant. A hydrogen atom that is covalently bonded to one nonmetal atom may be attracted simultaneously to another nonmetal atom of high electronegativity in the same or a neighboring molecule. The concept of hydrogen bonding is crucial to an understanding of the properties of solid and liquid water and of the living state of matter.

In network covalent solids bonds extend throughout a crystalline structure. Network covalent solids have high melting points compared to other covalent substances. If all the electrons in a network covalent solid are localized into bonds, the solid is an electrical insulator. However, if some of the electrons are *delocalized,* the solid is an electrical conductor.

The strength of intermolecular forces in an ionic solid is expressed through the lattice energy. Lattice energy is largely a function of ionic size and charge. Lattice energies can be calculated from thermochemical data and certain atomic and molecular properties.

To describe the crystalline structure of a solid, for example a metal, it is useful to think in terms of a closely packed structure of spheres. The two closest-packed arrangements are the hexagonal and the cubic closest packed. A third arrangement, the body-centered cubic structure, is not so tightly packed. Experimental data required for a crystal-structure determination are acquired through the interaction of x rays with a crystal. This experimental method is known as x-ray diffraction.

Ionic crystal structures can be described through a packing-of-spheres model, but the matter is complicated by the fact that the spheres are not all of the same size. The ratio of the radius of the cation to that of the anion—the radius ratio—is a useful quantity for predicting the probable arrangement of ions in an ionic crystal. Another useful concept in describing crystals of all types is the unit cell. Knowledge of the unit cell of a crystal makes possible calculations involving densities, atomic radii, and other properties of substances.

Learning Objectives

As a result of studying Chapter 11, you should be able to

1. Explain surface tension and phenomena based on surface tension—drop formation, the wetting of surfaces, and capillary action.

2. Describe the condition of dynamic equilibrium between a liquid and its vapor and state the factors that affect the vapor pressure of a liquid.

3. Plot vapor pressure or log of vapor pressure as a function of temperature, and use such plots to relate vapor pressures and temperatures.

4. Calculate vapor pressures from experimental data, and use vapor pressure data to predict conditions for the existence of the vapor and/or liquid states.

5. Use the Clausius–Clapeyron equation to relate log of vapor pressure, temperature, and heat of vaporization of a liquid.

6. Describe the significance of the critical point of a substance.

7. Interpret the phase diagrams of a few simple substances, and use phase diagrams to predict changes that will occur as a substance is heated, cooled, or subjected to a change in pressure.

8. Describe the common types of intermolecular forces and how these forces influence the physical properties of a substance.

9. State the conditions that lead to hydrogen bond formation and describe some properties that result from hydrogen bond formation.

10. Give examples of network covalent solids and describe the bonding and characteristic properties of these solids.

11. Explain how lattice energy is related to ionic sizes and charges, and how physical properties of ionic solids are related to lattice energy.

12. Calculate lattice energies from thermochemical, atomic, and molecular data using the Born–Haber cycle.

13. Describe how crystal structures can be related to the close packing of spheres.

14. Explain the significance of the term, unit cell, and apply this concept to calculations involving quantities such as atomic radii and densities.

15. Apply the concept of the unit cell to ionic crystal structures to establish crystal coordination numbers and chemical formulas.

Some New Terms

Adhesive forces are intermolecular forces between unlike molecules, such as between molecules of a liquid and the surface with which it is in contact.

bcc is a symbol representing the body-centered cubic crystal structure.

Boiling is a process in which vaporization takes place throughout a liquid. It occurs at the temperature where the vapor pressure of a liquid is equal to the prevailing barometric pressure.

The **Born–Haber cycle** relates the lattice energy of an ionic solid to other quantities such as ionization energies, electron affinities, and heats of sublimation, dissociation, and formation.

Capillary action refers to the rise of a liquid in the pores of thin capillary tubes, a phenomenon related to adhesive and cohesive forces.

Cohesive forces are intermolecular forces between like molecules, such as the forces that hold a liquid in a droplike shape.

Condensation results from the passage of molecules from the vapor state to the liquid state.

The **critical point** refers to the condition of temperature and pressure where a liquid and its vapor become identical; it is the highest temperature point on the vapor pressure curve.

The **crystal coordination number** signifies the number of nearest neighboring ions of opposite charge to any given ion in a crystal.

fcc is an abbreviation denoting the cubic closest-packed crystal structure.

Freezing is the conversion of a liquid to a solid. Freezing of a liquid occurs at the same temperature as melting of the corresponding solid.

hcp is an abbreviation used to describe the hexagonal closest-packed crystal structure.

A **hydrogen bond** in an intermolecular attraction in which an H atom covalently bonded to one atom is attracted simultaneously to another electronegative atom of the same or a nearby molecule.

An **induced dipole** is an atom or molecule in which a separation of charge is produced by another dipole in its vicinity.

An **instantaneous dipole** is an atom or molecule in which a separation of charge is produced by a momentary displacement of electrons from their normal geometrical distribution.

An **intermolecular force** is a force of attraction *between* molecules.

Lattice energy is the quantity of energy released in the formation of one mole of an ionic solid from its separated gaseous ions.

London forces (dispersion forces) are the type of intermolecular forces associated with instantaneous and induced dipoles.

A **network covalent solid** is a substance in which covalent bonds extend throughout a crystal.

Normal boiling point is the temperature at which the vapor pressure of a liquid is 1 atm.

Normal melting point is the temperature at which the melting of a solid occurs at 1 atm pressure.

A **phase diagram** is a graphical representation of the phases or states of matter of a substance that will exist at various temperatures and pressures.

Radius ratio refers to the ratio of the radius of a cation to that of an anion in an ionic crystal. Its value is important in determining the structure of the crystal.

Sublimation refers to the direct passage of molecules from the solid to the gaseous state.

Surface tension results from an imbalance of intermolecular forces at the surface of a liquid, causing molecules to be drawn into the bulk of the liquid.

The **triple point** is the condition of temperature and pressure under which the solid, liquid, and gaseous states of a substance coexist at equilibrium.

A **unit cell** is a small collection of structural units of a crystal from which the entire crystal structure can be inferred.

van der Waals forces is a term used to describe, collectively, intermolecular forces of the London type and interactions between permanent dipoles.

Vaporization refers to the passage of molecules from the liquid to the gaseous state.

Vapor pressure is the pressure exerted by a vapor when it is in dynamic equilibrium with its liquid.

A **vapor pressure curve** is a graph of vapor pressure as a function of temperature.

X-ray diffraction is a method of crystal structure determination based on the interaction of a crystal with x rays.

Exercises

Surface tension and related properties

1. Describe briefly the meaning of the following terms: **(a)** surface tension; **(b)** adhesive force; **(c)** capillary action; **(d)** wetting agent; **(e)** meniscus.

2. A test for clean glassware in the laboratory involves observing the behavior of water on the surface. If droplets of water stand on the surface, the glass is dirty. If a continuous liquid film is produced, the glass is considered clean. Discuss the basis of this laboratory test.

3. Silicone oils are used in water repellents for treating tents, hiking boots, and similar items. Explain how they function.

*__4.__ When a candle is burned, the actual fuel is a mixture of gaseous hydrocarbons that appear at the end of the candle wick. Describe a series of steps by which the solid wax of the candle is consumed.

Vaporization

5. A sample of liquid benzene absorbs 1.00 kJ of heat at its normal boiling point of 80.1°C. 0.94 L $C_6H_6(g)$ at 80.1°C and 1 atm pressure is produced. What is ΔH_{vap} for C_6H_6, expressed in kJ/mol?

6. The molar heat of combustion of methane gas (CH_4) is -890 kJ/mol. How many liters of the gas, measured at 25.0°C and 748 mmHg, must be burned to provide for the evaporation of 1.00 L of liquid water at 100°C? The heat of vaporization of water at 100°C is 40.6 kJ/mol. The density of liquid water at 100°C = 0.958 g/cm³.

7. A double boiler is used in cooking sauces and other foods where a careful control of a low temperature is required. The operating principle is that water is boiled in an outside container to produce steam. The steam condenses on the walls of an inside container in which cooking occurs. (A related device used in the chemical laboratory is called a steam bath.)
  (a) By what means is heat energy conveyed to the food to be cooked?
  (b) What is the maximum temperature that can be achieved in the inside container?

8. When a volatile liquid is allowed to vaporize into the atmosphere from an ordinary glass container, the liquid temperature remains the same as that of the surroundings. If the same liquid is allowed to vaporize into the atmosphere from a thermally insulated container (a vacuum bottle or Dewar flask), its temperature falls below that of the surroundings. Explain this difference in behavior.

Vapor pressure and boiling point

9. From Figure 11-6 estimate **(a)** the vapor pressure of benzene at 50°C; **(b)** the normal boiling point of diethyl ether.

10. Use data from Table 11-1 to estimate **(a)** the boiling point of water in Santa Fe, New Mexico, if the prevailing atmospheric pressure is 600 mmHg; **(b)** the prevailing atmospheric pressure in Leadville, Colorado, if the observed boiling point of water is 89°C.

11. A sample of liquid bromine is allowed to establish equilibrium with its vapor at 25.0°C. A 250.0-cm³ sample of the saturated vapor is withdrawn and found to have a mass of 0.486 g. What is the vapor pressure of bromine at 25.0°C, expressed in mmHg?

12. A 25.0-L volume of He gas at 740 mmHg and 30.0°C is passed through 8.050 g of liquid aniline (mol. wt. 93.1) at 30.0°C. After the experiment the liquid is found to weigh 7.938 g. Assume that the He gas becomes saturated with aniline vapor and that the total gas volume and temperature remain constant. What is the vapor pressure of aniline at 30.0°C?

13. 10.0 L of $N_2(g)$ at 755 mmHg and 45.0°C is bubbled through $CCl_4(l)$ at 45.0°C. Assuming the gas becomes saturated with $CCl_4(g)$, what is the volume of the resulting gaseous mixture if the total pressure remains at 755 mmHg and the temperature, 45.0°C? The vapor pressure of CCl_4 at 45.0°C is 261 mmHg.

The Clausius–Clapeyron equation

14. Use the data plotted in Figure 11-9 to estimate **(a)** the normal boiling point of aniline; **(b)** the vapor pressure of toluene at 75°C.

15. By the method used to establish Figure 11-9, plot a graph of log P vs. $1/T$ for liquid yellow phosphorus. From the graph obtained, estimate its normal boiling point. Vapor pressure data: 76.6°C, 1; 128.0°C, 10; 166.7°C, 40; 197.3°C, 100; 251.0°C, 400 mmHg.

16. Isopropyl alcohol ("rubbing alcohol") has a vapor pressure of 10.0 mmHg at 2.4°C and 100.0 mmHg at 39.5°C. Calculate **(a)** the heat of vaporization and **(b)** the normal boiling point of isopropyl alcohol.

17. A handbook lists the normal boiling point of normal octane, a constituent of gasoline, as 125.8°C and its heat of vaporization as 33.9 kJ/mol. At what temperature does normal octane have a vapor pressure of 100.0 mmHg?

Critical point

18. Which substances listed in Table 11-2 can exist as liquids at room temperature (about 20°C)? Explain.

19. Can SO_2 be maintained as a liquid under a pressure of 100 atm at 0°C? Can liquid methane be obtained under the same conditions? (Refer to Table 11-2.)

Fusion

20. How much heat is required to melt a block of ice that measures 10.0 cm on an edge? The density of ice is 0.92 g/cm^3 and the heat of fusion is 6.02 kJ/mol.

21. It is desired to melt a 0.680-kg piece of Pb, starting with the sample at 21.0°C. What is the total quantity of heat required to do this? The melting point of Pb is 327.4°C; its heat of fusion is 4.774 kJ/mol; its average specific heat in the temperature range from 21 to 327.4°C is 0.134 J g^{-1} °C^{-1}.

States of matter and phase diagrams

22. 0.180 g H$_2$O(l) is sealed into an evacuated 2.50-L flask. What is the pressure of the vapor in the flask if the temperature is **(a)** 30°C; **(b)** 50°C; **(c)** 70°C? (*Hint:* Does any of the water remain as liquid or does it vaporize completely?)

23. A sample of water weighing 2.50 g is sealed in a 5.00-L flask at 120°C.
   (a) Show that the sample exists completely as vapor.
   (b) Estimate the temperature to which the flask must be cooled before liquid water condenses.

24. A 20.0-L vessel contains 0.100 mol H$_2$(g) and 0.050 mol O$_2$(g). The mixture is ignited with a spark and the reaction

$$2 H_2(g) + O_2(g) \longrightarrow 2 H_2O$$

goes to completion. The system is then cooled to 27°C. What is the final pressure in the vessel? (*Hint:* Is the H$_2$O formed present as a gas, a liquid, or a mixture of the two?)

25. The average specific heat of ice is 2.01 J g^{-1} °C^{-1}. The heat of fusion at 0°C is 6.02 kJ/mol. The average specific heat of liquid water is 4.18 J g^{-1} °C^{-1}. The average densities of ice and liquid water are 0.917 and 0.998 g/cm^3, respectively. Two ice "cubes," each with dimensions 4.0 cm × 2.5 cm × 2.7 cm, are taken from a freezer at −25.0°C and added to 400.0 cm^3 of liquid water at 32.0°C. Assuming that the container is perfectly insulated from the surroundings, what will be the final temperature of the contents of the container? What state(s) of matter will be present?

26. Explain why dry ice can be used to maintain frozen foods much more effectively than ordinary ice.

27. Why is the triple point of water (ice–liquid–vapor) a better fixed point for establishing a thermometric scale than either the melting point of ice or the boiling point of water?

28. Is it likely that any of the following states of matter will occur naturally at or near the surface of the earth, anywhere on earth? Explain. **(a)** Solid CO$_2$; **(b)** liquid CH$_4$; **(c)** gaseous SO$_2$; **(d)** liquid iodine; **(e)** liquid oxygen. (*Hint:* Use the appropriate phase diagrams and data from Table 11-2.)

29. Trace the changes that occur in the following samples in a device similar to that pictured in Figure 11-14. Be as specific as you can about the temperatures at which changes occur.
   (a) A sample of ice is heated from −20°C to 200°C at 600 mmHg.
   (b) A sample of iodine is heated from 0°C to 150°C at 100 mmHg.
   (c) A sample of carbon dioxide gas at 35°C is *cooled* at a constant pressure of 10 atm to −100°C.

30. Trace the changes that take place as a sample of water vapor originally at a pressure of 1.00 mmHg and at a temperature of −0.10°C is compressed at constant temperature until the pressure reaches 100 atm.

Van der Waals forces

31. One of the substances is out of order in the following list based on *increasing* boiling point. Identify it and put it in the proper place: N$_2$; O$_3$; F$_2$; Ar; Cl$_2$. Explain your reasoning.

32. Normal octane and isooctane are two of a large number of components of gasoline. Both have the formula C$_8$H$_{18}$. In normal octane all eight C atoms are arranged in a straight chain. The chain length in isooctane is five, with three of the C atoms attached as side chains. Which of the two liquids has the higher boiling point? Explain.

33. One of the following substances is a liquid at room temperature, whereas all the others are gaseous. Which do you think is the liquid? Explain. CH$_3$OH; C$_3$H$_8$; N$_2$; CO.

34. A handbook lists the following normal boiling points for a series of normal (straight-chain) alkanes: propane, C$_3$H$_8$, −42.1°C; butane, C$_4$H$_{10}$, −0.5°C; pentane, C$_5$H$_{12}$, 36.1°C; hexane, C$_6$H$_{14}$, 68.7°C; heptane, C$_7$H$_{16}$, 98.4°C; octane, C$_8$H$_{18}$, 125.6°C. Estimate the normal boiling point of the normal alkane nonane, C$_9$H$_{20}$.

35. Which member of each of the following pairs would you expect to have the highest boiling point? Explain. **(a)** Normal C$_7$H$_{16}$ or normal C$_{10}$H$_{22}$; **(b)** C$_3$H$_8$ or H$_3$C—O—CH$_3$; **(c)** H$_3$C—CH$_2$—S—H (ethyl mercaptan) or H$_3$C—CH$_2$—O—H (ethyl alcohol). (The term "normal" means that all C atoms are in a straight chain; that is, there is no branching of the chain.)

36. The boiling point of benzene, C$_6$H$_6$, is 80°C. When another atom or group of atoms is substituted for one of the H atoms, the boiling point changes. Explain the order of the following boiling points: C$_6$H$_5$Cl, 132°C; C$_6$H$_5$Br, 156°C; C$_6$H$_5$OH, 182°C.

Hydrogen bonding

37. Describe the conditions necessary for the formation of a hydrogen bond and how this bond differs from other intermolecular forces.

38. For each of the following substances indicate whether you would expect the principal intermolecular forces to be of the London type, dipole–dipole interactions, or hydrogen bonds: **(a)** HCl; **(b)** Br$_2$; **(c)** ICl; **(d)** HF.

39. Why doesn't hydrogen bonding occur between molecules of CH$_4$?

40. Figure 11-20 shows that one H$_2$O molecule can be bonded to four others through hydrogen bonds. What would you expect the situation to be with NH$_3$?

41. If water were a normal liquid, what would you expect to find for its **(a)** boiling point; **(b)** freezing point; **(c)** temperature of maximum density of the liquid; **(d)** relative densities of the solid and liquid states?

★**42.** In some cases a hydrogen bond can be formed *within* a single molecule. This is called an *intra*molecular hydrogen bond. Do you think this type of hydrogen bonding is an important factor in **(a)** C$_2$H$_6$; **(b)** H$_3$CCH$_2$OH; **(c)** H$_3$CCOOH; **(d)** *o*-phthalic acid, C$_6$H$_4$(COOH)$_2$? (The structure of *o*-phthalic acid is pictured on page 616.)

Network covalent solids

43. It is stated in the text that all the electrons in diamond are localized, but that in graphite certain electrons are delocalized.
   (a) Explain what is meant by the terms *localized* and *delocalized*.
   (b) Which electrons in graphite are delocalized?

44. Silicon carbide, SiC, crystallizes in a form similar to diamond, whereas the compound boron nitride, BN, crystallizes in a form similar to graphite.
   (a) Sketch the SiC structure in the manner of Figure 11-22.
   (b) Propose a bonding scheme to account for the structure of BN.
   (c) BN can be obtained in a diamondlike modification under high pressure, but SiC cannot be obtained in a graphitelike modification. Why do you suppose that this is so?

45. Based on data presented in the text, which would you expect to have the greater density, diamond or graphite? Explain.

46. Diamond is often used as a cutting medium in glass cutters. What property of diamond makes this possible? Could graphite function as well?

Ionic properties and bonding

47. Arrange the following ionic substances in the expected order of increasing lattice energy: CaO; MgBr$_2$; CsI.

48. The melting points of NaF, NaCl, NaBr, and NaI are 988, 801, 755, and 651°C, respectively. Are these data consistent with ideas developed in Section 11-11? Explain.

49. Which compound in each of the following pairs would you expect to be the more water soluble? **(a)** MgF$_2$ or BaF$_2$; **(b)** MgF$_2$ or MgCl$_2$.

★**50.** Use Coulomb's law to verify the conclusion concerning the relative strengths of the attractive forces in the ion pairs, Na$^+$Cl$^-$ and Mg^{2+}O^{2-}, presented in Figure 11-25.

Born–Haber cycle

51. The heat of formation of KF is -563 kJ/mol, and the heat of sublimation of K(s) is 90.0 kJ/mol. Use these data, together with other values from the text, to calculate the lattice energy of KF(s).

★**52.** The heat of formation of NaI(s) is -288 kJ/mol. Use this value, together with other data in the text, to calculate the lattice energy of NaI(s). (*Hint:* An extra step is required in addition to those used in establishing equation 11.6.)

Crystal structures

53. Define what is meant by **(a)** closest packing of spheres; **(b)** tetrahedral holes; **(c)** octahedral holes.

54. Explain why there are two arrangements for the closest packing of spheres rather than a single one.

55. Refer to appropriate figures in the text and verify that the number of nearest neighbors in the hcp and fcc structures is 12, and in the bcc, 8.

56. The unit cell of a fcc metallic structure is a cube with metal atoms at each of the eight corners and in the center of each of the six faces. Each atom in the center of a face is in contact with corner atoms as shown.

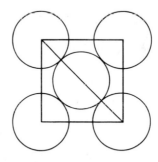

Assuming an atomic radius of 128 pm (1.28 Å) for a copper atom,
   (a) What is the length of the unit cell of copper?
   (b) What is the volume of the unit cell?
   (c) How many atoms belong to the unit cell?
   (d) What is the mass contained in the unit cell?
   (e) What is the calculated density of copper metal?

★**57.** Magnesium crystallizes in the hcp arrangement shown in Figure 11-26. The dimensions of the unit cell are height, 520 pm (5.20 Å); length of an edge, 320 pm (3.20 Å). Calculate the density of solid magnesium and compare with the measured value of 1.738 g/cm^3.

Ionic crystal structures

58. Solve equation (11.8) to obtain the radius ratio given in equation (11.9).

59. Use data from Chapter 8 and Table 11-4 to predict the crystal coordination number of the positive ions in the compounds **(a)** MgO; **(b)** CsBr.

60. In the manner illustrated in the text for NaCl, show that the formula of CsCl is consistent with its unit cell pictured in Figure 11-34.

61. Show that the unit cells for CaF_2 and TiO_2 shown in Figure 11-35 are consistent with their formulas.

62. Extend Example 11-12 by calculating **(a)** the volume of a unit cell; **(b)** the mass of a unit cell; **(c)** the density of NaCl.

Additional Exercises

1. The normal boiling point of acetone (a common solvent) is 56.5°C, and its heat of vaporization is 30.3 kJ/mol. What is the vapor pressure of acetone at 20.0°C?

★2. A quantity of steam (water vapor), measuring 100.0 L in volume at 100°C and 1 atm pressure, is passed into 1.00 kg of liquid water. The liquid water is maintained in a thermally insulated container (such as a vacuum bottle or Dewar flask) and has an initial temperature of 18.0°C. What will be **(a)** the final mass of liquid; **(b)** the final temperature? The heat of vaporization of water at 100°C is 40.6 kJ/mol.

3. A sample of pure, dry $N_2(g)$, measuring 150 cm³ in volume at 25.0°C and 750 mmHg, is passed through liquid benzene until the gas becomes saturated with benzene vapor. The volume of the gas is found to be 172 cm³ at a total pressure of 750 mmHg. What is the vapor pressure of benzene at 25.0°C?

4. A tray measuring 20.0 cm × 5.0 cm × 2.5 cm is filled with water at 18.0°C and placed in a freezer. The water freezes and the resulting ice is cooled to a temperature of −15.0°C. How much heat energy must be removed from the water to accomplish this? (Neglect the loss of heat by the tray itself.) Heat of fusion of ice, 6.02 kJ/mol; specific heat of ice, 2.01 J g^{-1} °C^{-1}; density of water, 1.00 g/cm³; specific heat of water, 4.18 J g^{-1} °C^{-1}.

★5. A more refined equation for the vapor pressure of $NH_3(l)$ than equation (11.3) is

$$\log P \text{ (mmHg)} = 9.95028 - 0.003863T - \frac{1473.17}{T}$$

What is the normal boiling point of $NH_3(l)$, expressed in °C?

6. The following data are given for CCl_4: normal melting point, −23°C; normal boiling point, 77°C; density of liquid, 1.59 g/cm³; heat of fusion, 30.5 kJ/mol; vapor pressure at 25°C, 110 mmHg.

   (a) How much heat must be absorbed to convert 10.0 g of solid CCl_4 to liquid at −23°C?

   (b) How much heat is required to vaporize 20.0 L of $CCl_4(l)$ at its normal boiling point?

   (c) What is the volume occupied by 1.00 mol of the saturated vapor of CCl_4 at 77°C?

   (d) What phases—solid, liquid, and/or vapor—are present if 3.5 g CCl_4 is maintained in a volume of 8.21 L at 25°C?

★7. A chemical catalog lists a cylinder of chlorine as having these approximate dimensions: 10 in. diameter × 45 in. height. The gas pressure is 100 psi (1 atm = 14.7 psi) at a temperature of 20°C. The contents of the cylinder weigh 150 lb. The melting point of chlorine is −103°C; its normal boiling point is −35°C; and its critical point is at 144°C and 76 atm. In what state(s) of matter does the chlorine exist in the cylinder?

8. To cool a bottle of pop quickly, it was placed in the freezer compartment of a refrigerator. Some time later the bottle was taken out; the pop was still liquid. When the cap was removed, however, the contents of the bottle froze instantaneously. Explain why this happened.

9. All three structures, diamond, graphite, and ice, feature a hexagonal arrangement of structural units when viewed along certain directions. How can this be reconciled with the facts that diamond is extremely hard and has a high melting point, graphite flakes easily and is a good electrical conductor, and ice is a low density solid with a low melting point?

10. Following are values of heats of vaporization of some typical liquids at their normal boiling points: H_2, 0.92 kJ/mol; CH_4, 8.16 kJ/mol; C_6H_6, 31.0 kJ/mol; H_2O, 40.7 kJ/mol. Explain the differences among these values.

11. Which of the following pure materials would you expect to conduct electric current in the state of matter indicated? Explain.   **(a)** $CO_2(g)$;   **(b)** NaCl(s);   **(c)** KI(l); **(d)** $SiO_2(s)$;   **(e)** C(s, graphite);   **(f)** C(s, diamond); **(g)** Al(s);   **(h)** $P_4(s)$;   **(i)** Hg(l).

★12. In a similar fashion to Exercise 62, calculate the density of CsCl.

★13. Establish an approximate radius ratio corresponding to a crystal coordination number of eight (cubic). Refer to Figure 11-34 (the CsCl structure) and note that along a diagonal of the cube two anions are in contact with a cation at the center. Also, assume that along an edge of the cube two anions are in contact (which, strictly speaking, they are not).

★14. In acetic acid vapor some molecules exist as monomers and some as dimers (as pictured in Figure 11-21). If the measured density of acetic acid vapor at 350 K and 1 atm pressure is 3.23 g/L, what percent of the molecules must exist as dimers? Would you expect this percent to increase or decrease with increasing temperature?

*15. Because the gain of an electron by the species $O^-(g)$ requires that energy be absorbed, the free oxide ion O^{2-} cannot be obtained in the gaseous state. Therefore, the second electron affinity of oxygen cannot be measured directly, that is,

$$O^-(g) + e^- \longrightarrow O^{2-}(g) \qquad EA_2 = ?$$

The O^{2-} ion can exist in the solid state, however, where the high energy requirement for its formation is offset by the large lattice energies of ionic oxides.

(a) Show that EA_2 can be calculated indirectly from the following data: heat of formation and lattice energy of MgO(s), heat of sublimation of Mg(s), ionization energies of Mg, bond energy of O_2, and EA_1 for O(g).

(b) The heat of sublimation of Mg(s) is 150 kJ/mol, the bond energy of $O_2(g)$ is 497 kJ/mol and the lattice energy of MgO is -3925 kJ/mol. Combine these data with other values in the text to calculate a value of EA_2 for oxygen.

Self-Test Questions

For questions 1 through 5 select the single item that best completes each statement.

1. Of the following liquids, the one with the highest normal boiling point is (a) $O_2(l)$; (b) $Ne(l)$; (c) $SO_3(l)$; (d) $Br_2(l)$.

2. The best electrical conductor of the following is (a) $SiO_2(s)$; (b) $Si(s)$; (c) $NaCl(s)$; (d) $Br_2(l)$.

3. Of the compounds HF, CH_4, CH_3OH, and N_2H_4, hydrogen bonding as an important intermolecular force is expected in (a) none of these; (b) two of these; (c) all but one of these; (d) all of these.

4. Of the following properties, the magnitude of one must always increase with temperature. That property is (a) vapor pressure; (b) density; (c) ΔH_{vap}; (d) surface tension.

5. Compared to diamond, the form of carbon known as graphite (a) is harder; (b) contains a higher percentage of carbon; (c) is a better electrical conductor; (d) has equal carbon-to-carbon bond distances in all directions.

6. A television commercial claims that a certain product makes water "wetter." Is there any basis to this claim? Explain.

7. Which of the following factors would you expect to affect the vapor pressure of a liquid? Explain.
   (a) intermolecular forces in the liquid
   (b) volume of liquid in the liquid–vapor equilibrium
   (c) volume of vapor in the liquid–vapor equilibrium
   (d) the size of the container in which the liquid–vapor equilibrium is established

(e) the temperature of the liquid

8. 10.0 g of steam at 100°C and 100.0 g of ice at 0°C are added to 100.0 g of liquid water at 20°C in a perfectly insulated container. (For water: heat of fusion = 6.02 kJ/mol, heat of vaporization = 40.7 kJ/mol.) Which of the following conditions will result? (a) The mixture will start to boil; (b) all of the liquid water will freeze; (c) all of the ice will melt; (d) a mixture of ice and liquid water will remain.

9. A dramatic lecture demonstration consists of continuously evacuating the water vapor produced from an open container of water in a bell jar. If the vacuum pump is sufficiently powerful, the water can be made to freeze. Indicate the principles involved in achieving this effect.

10. Explain what is wrong with the following approach used by a student to calculate the vapor pressure of a liquid at 50°C from a measured vapor pressure of 80 mmHg at 20°C.

$$\text{v.p. (at } 50°C) = 80 \text{ mmHg} \times \frac{323 \text{ K}}{293 \text{ K}} = 88 \text{ mmHg}$$

11. Argon, copper, sodium chloride, and carbon dioxide all crystallize in a fcc structure. How can this be and still have such a difference in their physical properties?

12. Summarize the characteristics of these four types of crystalline materials—ionic, molecular, network covalent, and metallic—in terms of intra- and intermolecular forces, structural units of the crystalline lattice, and expected physical properties, such as melting point, boiling point, and ability to conduct electricity.

12

Mixtures

The discussion of intermolecular forces in Chapter 11 was limited to *single* substances in the three states of matter. But intermolecular forces can exist among unlike molecules as well, and depending on the relative strengths of these forces a **heterogeneous** or **homogeneous** mixture results. The main body of this chapter is concerned with the properties of homogeneous mixtures or **solutions.**

12-1 Homogeneous and Heterogeneous Mixtures

A sample of matter having a fixed composition and uniform properties throughout is called a **phase.** For example, a sample of water at 25°C and 1 atm pressure exists as a single liquid phase. All properties of the water are uniform throughout this liquid phase. If a *small* quantity of salt (NaCl) is added to the water, the salt dissolves and the entire sample remains as a single liquid phase. The composition and properties of this new liquid phase, a salt solution, are different from those of pure water. This solution is a mixture, because it contains two different substances. It is **homogeneous,** because its properties are uniform throughout the liquid. If some sand (SiO_2) is added to water, the sand settles to the bottom of the liquid and remains there as an undissolved solid. This water–sand mixture is a *two*-phase mixture (liquid + solid) and it is **heterogeneous.** Its composition and properties are not uniform throughout. The composition and properties of the liquid phase are those of pure water; the composition and properties of the solid phase are those of sand.

FIGURE 12-1
Representation of intermolecular forces in a solution.

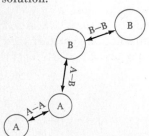

INTERMOLECULAR FORCES IN MIXTURES. The results obtained on mixing two substances are generally understandable in terms of intermolecular forces. However, in many cases our knowledge of these forces is inadequate for an accurate prediction. For the substances A and B let us represent the magnitude of the intermolecular forces between like molecules A $\leftrightarrow$ A and B $\leftrightarrow$ B and between unlike molecules as A $\leftrightarrow$ B (see Figure 12-1). Four possibilities are described below. In this description comparative forces are denoted by: $\simeq$ approximately equal to, $>$ greater than, $<$ smaller than, $\ll$ much smaller than.

The heat effects noted for solution formation in cases (1), (2), and (3) below can be understood in terms of this hypothetical process: First, particles of the solvent and of the solute are separated; this requires energy to be absorbed. Then the solute and solvent particles are allowed to move into close proximity, that is, to mix. Energy is released. The net energy change, ΔH, depends on the magnitudes of the energy changes associated with each step.

1. A ↔ B ≃ A ↔ A ≃ B ↔ B. If intermolecular forces between like and unlike molecules are of about the same strength, a random intermingling of molecules occurs and a homogeneous mixture or solution results. Most mixtures of liquid hydrocarbons (e.g., benzene–toluene) belong to this category. Properties of solutions of this type can generally be predicted from a knowledge of properties of the component substances: These solutions are said to be **ideal.** Specifically, the volume of an ideal solution is the sum of the volumes of the individual components ($\Delta V = 0$). The energies of interaction between like and unlike molecules are the same. There is no enthalpy change or heat effect on mixing the components ($\Delta H = 0$). Temperature remains constant as an ideal solution is formed from its components.

2. A ↔ B > A ↔ A, B ↔ B. If intermolecular forces between unlike molecules exceed those between like molecules, solution formation also occurs. However, the properties of solutions of this type cannot be predicted from those of the component substances. These solutions are **nonideal** (see Figure 12-2). The energy released in interactions between unlike molecules exceeds that required to separate like molecules. Energy is given off to the surroundings and the solution process is *exothermic* ($\Delta H < 0$).

3. A ↔ B < A ↔ A, B ↔ B. If forces of attraction between unlike molecules are smaller than between like molecules, complete mixing may occur, but again solutions of this type are **nonideal.** The solution process is endothermic ($\Delta H > 0$).

4. A ↔ B ≪ A ↔ A, B ↔ B. If intermolecular forces between unlike molecules are much smaller than those between like molecules, the substances remain segregated into a heterogeneous mixture. For example, in a mixture of water and octane (a component of gasoline) strong hydrogen bonds hold water molecules together in clusters. The nonpolar octane molecules cannot exert strong attractive forces on polar water molecules, and the two types of molecules do not mix.

FIGURE 12-2
Intermolecular force between unlike molecules leading to nonideal solution.

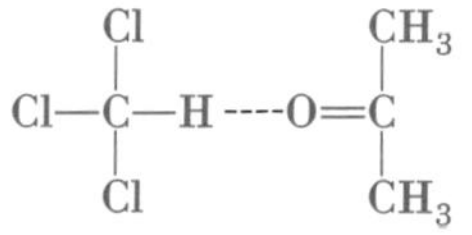

Hydrogen bonding occurs between the H atom of a $CHCl_3$ (chloroform) molecule and the O atom of a $(CH_3)_2CO$ (acetone) molecule. This intermolecular force between unlike molecules causes acetone–chloroform solutions to be nonideal.

As a further example of the process of solution formation, consider the dissolving of an ionic solid in water. In Figure 12-3 water dipoles are pictured as clustering around ions at the surface of a crystal. The negative ends of these dipoles are oriented toward the positive ions, and the positive ends of dipoles, toward negative ions. If these ion–dipole forces are sufficiently strong to overcome the interionic attractions within the crystal, dissolving will occur. Moreover, these ion–dipole forces persist in the solution. An ion surrounded by a cluster of water molecules is said to be *hydrated.* Energy is released when ions become hydrated. The greater the hydration energy in relation to the energy investment to separate ions from a crystal, the more energetically favorable is the solution process.

TYPES OF SOLUTIONS. By convention the component present in greatest quantity, or that component which determines the state of matter in which the solution exists, is called the **solvent.** The component(s) present in lesser quantity is (are) called the **solute(s).** A solution in which water is the solvent is called an **aqueous** solution. A solution containing a relatively large quantity of solute is said to be **concentrated.** If the quantity of solute is small, the solution is **dilute.** The term solution generally calls to mind a liquid solvent with either another liquid, a solid, or a gas as a solute. Three examples of solutions existing in the liquid state are

Gasoline: a mixture of a number of liquid hydrocarbons.
Seawater: an aqueous solution of sodium chloride and several other ionic solids.
Carbonated water: an aqueous solution of $CO_2(g)$.

FIGURE 12-3
Dissolving of an ionic crystal in water.

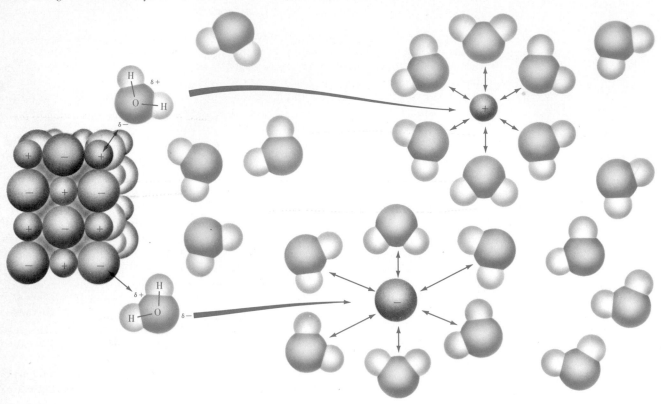

Not all mixtures of two or more metals (alloys) are solid solutions; some are heterogeneous, as is the lead-tin alloy known as solder. Also, in some mixtures a new intermetallic compound is formed.

Hydrogen, stored and transported either as a gas or liquid, is under serious study as a potential fuel to replace gasoline in many of its applications. One of the problems hydrogen presents though is its tendency to dissolve in metals and make them brittle.

Solutions may also exist in either the gaseous or solid state. Since the molecules of a gas are separated by great distances, molecules in a mixture of gases diffuse among one another randomly. *All gaseous mixtures are solutions.* The best known example of a gaseous solution is air, which consists of nitrogen, oxygen, argon, and trace amounts of other gases.

In a solid solution the solvent is a *solid* substance. The ability to form solid solutions is particularly common among metals, and such solid solutions are called **alloys.** In certain solid solutions, atoms of solute substitute for some of the solvent atoms in the crystalline lattice. These are called substitutional solid solutions and require that the solute and solvent atoms be of approximately the same size. Thus, copper (128 pm) and nickel (125 pm) form solid solutions in all proportions. In other solid solutions the solute atoms may take up positions in the interstices or holes in the solvent lattice. The formation of interstitial solid solutions requires that solute atoms be small enough to fit into the holes among the solvent atoms. Among the elements that often meet these requirements are carbon and hydrogen. Ordinary steels are alloys of iron and carbon.

12-2 Solution Concentration

To state the components present in a solution does not describe the solution fully. The additional information needed is the *concentration* of the solution. There are many ways to describe solution concentration, but all of them must express the quantity of solute present in a given quantity of solvent (or solution). That is, in every system of concentration we must establish the following points:

1. The units used to measure the solute.
2. Whether the second quantity measured is the solvent or the total solution.
3. The units used to measure this second quantity.

PERCENT BY MASS, PERCENT BY VOLUME, AND RELATED QUANTITIES. The statement, "5.00 g NaCl per 100.0 g of aqueous solution," has the following meaning: A solution is prepared by weighing out 5.00 g NaCl and dissolving it in 95.0 g H_2O, that is, a mass of water sufficient to produce 100.0 g of solution. The solution can be termed a 5.00% NaCl solution, *by mass.* This concentration unit, in which quantities of solute and solution are both measured by mass, is also referred to as **mass/mass percent** or as **% (mass/mass).**

The terms weight/weight percent and % (wt/wt) are also commonly encountered.

[handwritten: 5.00 g per 100 g TOTAL]

When a liquid solute is used it is often convenient to prepare a solution on a volume basis, such as dissolving 5.00 ml of ethanol in a sufficient volume of water to produce 100.0 ml of solution. This ethanol–water solution is 5.00% ethanol, *by volume;* or, because both quantities are measured in volume units, the term **volume/volume percent** or **% (vol/vol)** can be used.

Still another possibility is that mass and volume units are mixed. For example, if the solute is measured by mass and the quantity of solution by volume, the term **mass/volume percent** or **% (mass/vol)** can be used. If a solution concentration is given on a percent basis, but with no further specification as to whether it is mass/mass, vol/vol, or mass/vol, one should assume that percent by mass is intended.

Solution concentrations expressed as percents have no theoretical significance, but they are commonly encountered and it is necessary to be familiar with them. Mass/volume percent is widely used in biological and medical laboratories and mass/mass percent is the solution concentration most often used in industrial chemistry.

MOLAR CONCENTRATION (MOLARITY). Among the important ideas presented in Chapter 4 were

1. The stoichiometry of chemical reactions is based on relative *numbers* of reacting atoms, ions, or molecules.
2. Many chemical reactions are conducted in solution.

For these reasons we introduced at that time a solution concentration unit based on *numbers of solute particles*—the molar concentration.

Here we need simply note that molarity is the preferred SI concentration unit and repeat the defining equation.

$$\text{molar concentration } (M) = \frac{\text{number of moles solute}}{\text{number of liters soln}} \tag{12.1}$$

[handwritten: dependent on Temp, vol. increase, molarity decrease]

MOLAL CONCENTRATION (MOLALITY). Molarity is a function of temperature. This is because the quantity of solution is based on volume, and volume is a function of temperature. Suppose that a solution is prepared at 20°C using a volumetric flask calibrated at this same temperature but then the solution is used at 25°C. As the temperature increases from 20°C to 25°C, the amount of solute remains constant but the solution volume increases slightly. The number of moles of solute per liter (i.e., the molarity) *decreases* slightly.

Another situation in which molality concentration is useful is one in which the solvent, because it is a solid at room temperature, can be measured only by its mass not its volume.

For a variety of applications it is necessary to have a solution unit which is independent of temperature. The obvious unit is one in which *both* quantities, solute and solvent, are stated by mass. The mass of a substance is independent of temperature. A particularly useful unit is molality, in which the amount of solute is given in moles and the quantity of solvent (not solution) in kilograms. The units of molality then are *moles solute per kilogram solvent.* A solution in which 1 mol of NaCl is dissolved in 1000 g of water is described as a 1 molal solution and designated by the symbol 1 m NaCl.

Molal concentration is defined by equation (12.2).

$$\text{molal concentration } (m) = \frac{\text{number of moles solute}}{\text{number of kilograms solvent}} \qquad (12.2)$$

MOLE FRACTION. The concentration units molality and molarity have the amount of solute expressed on a number basis (in moles), but the quantity of solvent or solution is on a mass or volume basis. To relate physical properties of solutions to solution concentration, it is sometimes necessary to use a concentration unit in which all solution components are described on a mole basis. This can be done through the mole fraction. The mole fraction of component, i, designated χ_i, is the fraction of all the molecules in a solution which are of type, i. The mole fraction of component, j, is χ_j, and so on. The sum of the mole fractions of all the solution components is 1.

The mole fraction of a solution component is defined by equation (12.3).

$$\chi_i = \frac{\text{moles of component, } i}{\text{total moles of all soln components}} \qquad (12.3)$$

Another concentration unit related to the mole fraction is the **mole percent.** The mole percent of a solution component is the percent of all the molecules in a solution that are of a given type. Mole percents are simply mole fractions multiplied by 100.

12-3 Some Illustrative Examples Based on Solution Concentrations

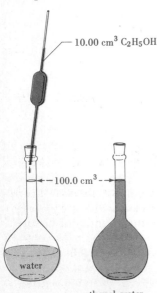

FIGURE 12-4
Preparation of an ethanol–water solution— Example 12-1 illustrated.

— 10.00 cm³ C₂H₅OH

←100.0 cm³—

water

ethanol–water solution: $d = 0.982$ g/cm³

Example 12-1 An ethanol–water solution is prepared by dissolving 10.00 cm³ of ethanol, C_2H_5OH ($d = 0.789$ g/cm³) in a sufficient volume of water to produce 100.0 cm³ of a solution with a density of 0.982 g/cm³ (see Figure 12-4). What is the concentration of this solution, expressed as **(a)** percent by volume; **(b)** percent by mass; **(c)** percent (mass/vol); **(d)** molarity; **(e)** molality; **(f)** mole fraction; and **(g)** mole percent of ethanol?

(a) Percent ethanol, by volume:

$$\% \text{ ethanol, by volume} = \frac{10.00 \text{ cm}^3 \text{ ethanol}}{100.0 \text{ cm}^3 \text{ soln}} \times 100 = 10.00\%$$

This concentration can also be expressed as 10.00% ethanol (vol/vol).

(b) Percent ethanol, by mass:

$$\text{no. g ethanol} = 10.00 \text{ cm}^3 \text{ ethanol} \times \frac{0.789 \text{ g ethanol}}{1.00 \text{ cm}^3 \text{ ethanol}} = 7.89 \text{ g ethanol}$$

$$\text{no. g soln} = 100.0 \text{ cm}^3 \text{ soln} \times \frac{0.982 \text{ g soln}}{1.00 \text{ cm}^3 \text{ soln}} = 98.2 \text{ g soln}$$

$$\% \text{ ethanol, by mass} = \frac{7.89 \text{ g ethanol}}{98.2 \text{ g soln}} \times 100 = 8.03\%$$

This concentration can also be expressed as 8.03% ethanol (mass/mass).

(c) Mass/volume percent ethanol:

$$\% \text{ ethanol (mass/vol)} = \frac{7.89 \text{ g ethanol}}{100.0 \text{ cm}^3 \text{ soln}} \times 100 = 7.89\%$$

(d) Molarity of ethanol:

To establish the various forms of percentage composition in parts (a), (b), and (c) did not require knowledge of the formula of ethanol. To express concentration on a molar basis, however, requires that a mole of ethanol be identified. For this the formula is required.

$$\text{no. mol } C_2H_5OH = 10.00 \text{ cm}^3 \text{ ethanol} \times \frac{0.789 \text{ g ethanol}}{1.00 \text{ cm}^3 \text{ ethanol}} \times \frac{1 \text{ mol } C_2H_5OH}{46.1 \text{ g } C_2H_5OH}$$

$$= 0.171 \text{ mol } C_2H_5OH$$

$$\text{no. L soln} = 100.0 \text{ cm}^3 \text{ soln} \times \frac{1 \text{ L soln}}{1000 \text{ cm}^3 \text{ soln}} = 0.1000 \text{ L soln}$$

$$\text{molarity} = \frac{0.171 \text{ mol } C_2H_5OH}{0.1000 \text{ L soln}} = 1.71\ M\ C_2H_5OH$$

(e) Molality of ethanol:
The key to determining molal concentration usually turns out to be establishing the mass of solvent (in kg) in a solution.

$$\text{mass soln} = 98.2 \text{ g soln [see part (b)]}$$

$$\text{mass ethanol} = 7.89 \text{ g ethanol [see part (b)]}$$

★ volumes are NOT additive! go to mass →

$$\text{mass } H_2O = \text{mass soln} - \text{mass ethanol} = 98.2 \text{ g} - 7.89 \text{ g} = 90.3 \text{ g } H_2O$$

$$\text{no. kg } H_2O = 90.3 \text{ g } H_2O \times \frac{1 \text{ kg } H_2O}{1000 \text{ g } H_2O} = 0.0903 \text{ kg } H_2O$$

$$\text{molality} = \frac{0.171 \text{ mol } C_2H_5OH}{0.0903 \text{ kg } H_2O} = 1.89\ m\ C_2H_5OH$$

(f) Mole fraction of ethanol:
The total moles of ethanol in the solution has already been calculated [see part (d)]. Now we must calculate the total moles of water present, based on the mass of water determined in part (e).

$$\text{no. mol } H_2O = 90.3 \text{ g } H_2O \times \frac{1 \text{ mol } H_2O}{18.0 \text{ g } H_2O} = 5.02 \text{ mol } H_2O$$

$$\chi_{C_2H_5OH} = \frac{0.171 \text{ mol } C_2H_5OH}{0.171 \text{ mol } C_2H_5OH + 5.02 \text{ mol } H_2O} = \frac{0.171}{5.19} = 0.0329$$

(g) Mole percent ethanol:

$$\text{mole percent } C_2H_5OH = 100 \cdot \chi_{C_2H_5OH} = 100 \times 0.0329 = 3.29\%$$

SIMILAR EXAMPLES: Exercises 6, 10, 13, 14, 18, 19.

saturated- holds as much dissolved solute while in contact w/ excess solute

12-4 Solubility Equilibrium

In some solutions the solute and solvent are miscible in all proportions. In these cases the solution never becomes saturated. Such is the case with ethanol–water solutions, for example.

One must be careful not to confuse solubility with the rate of dissolving. A saturated solution is always formed more rapidly as temperature is increased, but solubility is determined only by the concentration of the saturated solution.

When a sufficiently large quantity of solute is maintained in contact with a limited quantity of solvent, dissolving occurs continuously. After a time, however, the reverse process becomes increasingly important. This is the return of dissolved species (atoms, ions or molecules) to the undissolved state, a process called **precipitation.** When dissolving and precipitation occur at the same rate, the quantity of dissolved solute present in a given quantity of solvent remains constant with time. The process is one of dynamic equilibrium and the solution is said to be **saturated.** The formation of a saturated solution is suggested by Figure 12-5. The concentration of the saturated solution is referred to as the **solubility** of the solute in the given solvent. Solubility is generally a function of temperature, as suggested by the **solubility curves** in Figure 12-6.

solubility - amt. required to achieve saturation in given amt. of solvent

FIGURE 12-5
Formation of a saturated solution.

The lengths of the arrows represent the rate of dissolving (⟶) and the rate of precipitation (⟵----).
(a) When solute and solvent are first brought together only the process of dissolving occurs.

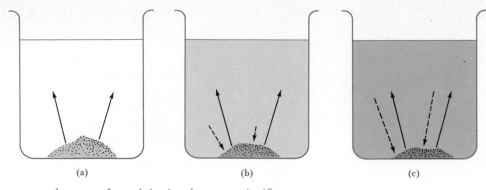

(a)   (b)   (c)

(b) After a time, though dissolving continues at the same rate, the rate of precipitation becomes significant.
(c) When the solution is saturated, dissolving and precipitation occur at the same rate. There is no further change in solution concentration with time.

FIGURE 12-6
Water solubility of several salts as a function of temperature.

For each curve, as illustrated here for $KClO_4$, regions above the curves (1) represent supersaturated solutions; points on the curves, S, saturated solutions; and regions below the curves (2), unsaturated solutions.

usually but not always

raise in temp shifts equilibrium to absorb energy. Dissolving more solute absorbs the added energy of Temp increase

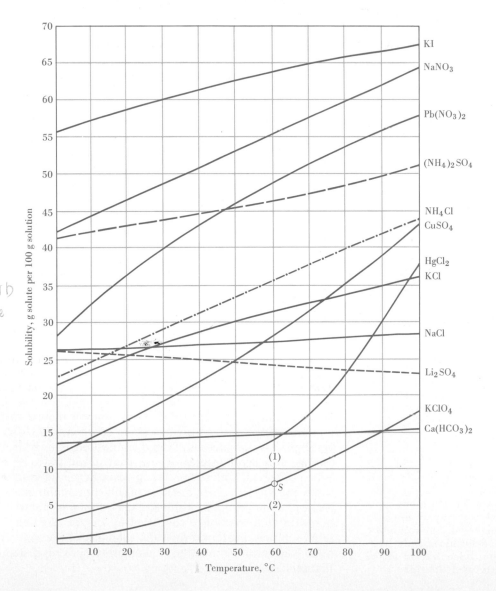

FIGURE 12-7
Effect of pressure on the solubility of a gas.

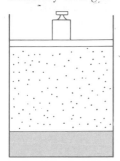

Suppose that a saturated solution is prepared at one temperature and then the temperature is changed to a value at which the solubility is lower (generally this means lowering the temperature). The usual result is that the excess solute precipitates from the solution. In some cases, however, all the solute may remain in solution. Because the quantity of solute in these cases is greater than in a normal saturated solution at the given temperature, such a solution is said to be **supersaturated.** If a few crystals of solute are brought into contact with a supersaturated solution, the excess solute will usually precipitate. A solution that contains less solute than required for saturation is **unsaturated.** The relationship of these terms to a solubility curve is illustrated in Figure 12-6.

With but a few exceptions the water solubilities of ionic compounds increase with temperature. Predicting those few exceptions is not so easily done. A useful generalization for our purposes is that about 95% of all ionic compounds exhibit increased water solubility with temperature. Exceptions are found primarily among compounds containing the anions SO_3^{2-}, SO_4^{2-}, SeO_4^{2-}, AsO_4^{3-}, and PO_4^{3-}.

The increased solubility with temperature that is characteristic of most compounds can serve as a basis for purifying them. A saturated solution of the impure compound is prepared at an elevated temperature. Then the solution is cooled to a lower temperature where the excess solute crystallizes from solution. For this method to be most effective, it is necessary that the impurities not form a solid solution with the substance being crystallized. Usually they do not. Sometimes it is necessary to recrystallize the desired solute several times. This may be the case if the original solution is saturated in one or more of the impurities.

EFFECT OF PRESSURE ON SOLUBILITY. The effect of pressure on the solubility of a gas is generally much more significant than the effect of temperature. The effect, as noted in Figure 12-7, is always the same: The solubility increases as the gas pressure is increased. Henry's law states that the concentration of a dissolved gas is proportional to the gas pressure above the liquid.

$$C = k \cdot P_{gas} \tag{12.4}$$

The proportionality constant, k, has a value depending on the units chosen for C and P.

Equilibrium between the gas above and the dissolved gas within a liquid is reached when the rates of evaporation and condensation of gas molecules become equal. The rate of condensation depends on the number of molecules per unit volume in the gaseous state. The rate of evaporation depends on the number of molecules of the dissolved gas per unit volume of solution. Thus, as the number of molecules per unit volume increases in the gaseous state (through an increase in pressure) the number per unit volume must also increase in the liquid state (through an increase in the solution concentration). An assumption in Henry's law is that solute and solvent molecules do not interact appreciably, that the gas is *nonreactive*.

A practical application of Henry's law is encountered in soft drinks. The dissolved gas is CO_2 and the higher the gas pressure maintained above the soda pop, the more CO_2 that can be kept dissolved. When a bottle is opened, the excess gas pressure is released and dissolved CO_2 escapes readily, usually rapidly enough to cause fizzing.

Unless otherwise specified, tabulated solubilities of gases are based on a gas pressure of 1 atm above a liquid.

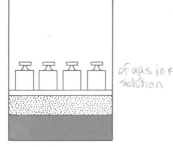

The concentration of dissolved gas is proportional to the pressure on the gas above the solution.

Example 12-2 100.0 g H_2O dissolves 437 cm³ H_2S(g) measured at STP. What is the molal concentration of a saturated solution at 10.0 atm pressure?

At STP:

$$\text{no. mol } H_2S = 437 \text{ cm}^3 \, H_2S \times \frac{1 \text{ L } H_2S}{1000 \text{ cm}^3 \, H_2S} \times \frac{1 \text{ mol } H_2S}{22.4 \text{ L } H_2S}$$

FBMEMBER

$$= 0.0195 \text{ mol } H_2S$$

$$\text{no. kg } H_2O = 100.0 \text{ g } H_2O \times \frac{1 \text{ kg } H_2O}{1000 \text{ g } H_2O} = 0.1000 \text{ kg } H_2O$$

$$\text{molality of } H_2S(aq) = \frac{0.0195 \text{ mol } H_2S}{0.1000 \text{ kg } H_2O} = 0.195 \, m$$

We may now solve equation (12.4) for k.

$$k = \frac{\text{conc.}}{P_{\text{gas}}} = \frac{0.195 \, m}{1 \text{ atm}}$$

Concentration $= k \cdot P_{gas}$

$k = \dfrac{c}{P_{gas}}$

At 10 atm: To determine the solubility at 10 atm pressure, we use equation (12.4) again, substituting the value of k just established.

$$\text{conc.} = k \cdot P_{\text{gas}} = \frac{0.195 \, m}{1 \text{ atm}} \times 10 \text{ atm} = 1.95 \, m$$

SIMILAR EXAMPLES: Exercises 25, 26.

The proportionality between concentration and pressure is direct. When the pressure increases tenfold (from 1 to 10 atm), the molality concentration must also increase by a factor of 10.

12-5 Colligative Properties

There exist four related properties whose values, to a first approximation at least, depend only on the number of solute particles in solution. That is, these properties do not depend on the identity of the solute. These four properties—vapor pressure lowering, boiling point elevation, freezing point depression, and osmotic pressure—are called colligative properties. Practical applications of colligative properties are numerous and varied. Also, the study of colligative properties has provided important methods of molecular weight determination and has contributed significantly to the development of solution theory.

VAPOR PRESSURE LOWERING. In the 1880s the French chemist F. M. Raoult conducted a systematic study of the vapor pressures of solutions and achieved this important result:

The vapor pressure of the solvent above a solution (P_A) is equal to the product of the vapor pressure of the pure solvent (P_A°) and its mole fraction in solution (χ_A).

Raoult found that the *lowering* of the vapor pressure of the solvent (ΔP) is equal to the product of the mole fraction of the *solute* (χ_B) and the vapor pressure of the pure solvent (P_A°). That is,

$$\Delta P = \chi_B P_A^\circ$$

In a binary solution (two components),
$\chi_A + \chi_B = 1$, and
$\chi_B = 1 - \chi_A$.

$$\Delta P = P_A^\circ - P_A$$
$$= (1 - \chi_A)P_A^\circ$$
$$P_A^\circ - P_A = P_A^\circ - \chi_A P_A^\circ$$
$$P_A = \chi_A P_A^\circ$$
$$\text{(12.5)}$$

$$P_A = \chi_A P_A^\circ \qquad \text{(12.5)}$$

if only one is volatile, $\chi_A P_A^\circ = P_{soln}$ also

If the solute(s) in a solution is also volatile, a similar expression can be written, that is,

$$P_B = \chi_B P_B^\circ \qquad \text{(12.6)}$$

In an ideal solution all components—solvent and solute(s) alike—adhere to Raoult's law over the entire concentration range. Solutions of benzene and toluene are essentially ideal. In all *dilute* solutions in which there is no chemical interaction among the components, Raoult's law will apply to the *solvent*. This is true whether the solution is ideal or nonideal. However, Raoult's law does not apply to the solute(s) in a dilute nonideal solution. This difference in behavior stems basically from this fact: Solvent molecules predominate in dilute solutions, and the solvent does not behave much differently than it would in the pure state. On the other hand, in dilute solutions solute molecules are surrounded by an overwhelming number of solvent molecules. This produces an environment for these solute molecules which is very much different from that in the pure solute. Although Raoult's law does not apply to the solute(s) in a dilute nonideal solution, Henry's law does.

Example 12-3 What are the partial and total vapor pressures at 25°C above a solution having equal numbers of molecules of benzene (C_6H_6) and toluene (C_7H_8)? The vapor pressures of benzene and toluene at this temperature are 95.1 and 28.4 mmHg, respectively.

If the solution contains equal numbers of molecules of each component, the mole fractions must both be 0.500.

Partial pressures:

$$P_{benz.} = \chi_{benz.} P^\circ_{benz.} = 0.500 \times 95.1 \text{ mmHg} = 47.6 \text{ mmHg}$$

$$P_{tol.} = \chi_{tol.} P^\circ_{tol.} = 0.500 \times 28.4 \text{ mmHg} = 14.2 \text{ mmHg}$$

Total vapor pressure:

$$P_{tot.} = P_{benz.} + P_{tol.} = 47.6 \text{ mmHg} + 14.2 \text{ mmHg} = 61.8 \text{ mmHg}$$

SIMILAR EXAMPLE: Exercise 29.

Example 12-4 What is the composition of the vapor in equilibrium with the benzene-toluene solution of Example 12-3?

The ratio of each partial pressure to the total pressure yields the mole fraction of that component in the vapor. This conclusion follows from equation (5.16). (That is, $n_A/n_{tot.} = \chi_A = P_A/P_{tot.}$.)

Vapor composition:

As is the case with liquid solutions, the sum of the mole fractions of the components in a gaseous or vapor mixture is 1. That is, $\chi_{benz.} + \chi_{tol.} = 0.770 + 0.230 = 1.000$.

$$\chi_{benz.} = \frac{P_{benz.}}{P_{tot.}} = \frac{47.6 \text{ mmHg}}{61.8 \text{ mmHg}} = 0.770$$

$$\chi_{tol.} = \frac{P_{tol.}}{P_{tot.}} = \frac{14.2 \text{ mmHg}}{61.8 \text{ mmHg}} = 0.230$$

SIMILAR EXAMPLE: Exercise 30.

REMEMBER (chapter 5)

$$\frac{n_A}{n_{tot}} = \frac{P_A}{P_{tot}} = \chi_A$$

TABLE 12-1
Vapor pressures and liquid–vapor compositions in benzene–toluene mixtures

Liquid composition, expressed as mole fraction benzene	Vapor pressures, mmHg			Vapor composition, expressed as mole fraction benzene
	$P_{benzene}$	$P_{toluene}$	P_{total}	
0.000	0.0	28.4	28.4	0.000
0.100	9.5	25.6	35.1	0.271
0.200	19.0	22.7	41.7	0.456
0.300	28.5	19.9	48.4	0.589
0.400	38.0	17.0	55.0	0.691
0.500	47.6	14.2	61.8	0.770
0.600	57.1	11.4	68.5	0.834
0.700	66.6	8.5	75.1	0.887
0.800	76.1	5.7	81.8	0.930
0.900	85.6	2.8	88.4	0.968
1.000	95.1	0.0	95.1	1.000

○ = nonvolatile solute
● = solvent

if you begin w/all solvent, then add a nonvolatile solute the surface area from which the solvent may vaporize is decreased & less of the solvent molecules w/sufficient k.e. are in a position to escape the liquid, so v.p. decrease

FIGURE 12-8

Liquid–vapor equilibrium for benzene–toluene mixtures at 25°C.

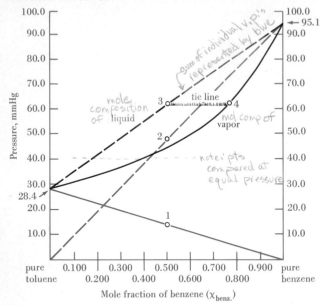

handwritten annotations on figure:
- sum of individual v.p's (represented by blue)
- mole composition of liquid
- tie line
- md comp of vapor
- note: pts compared at equal pressure

x-axis: Mole fraction of benzene ($\chi_{benz.}$)

pure toluene ... 0.100 0.200 0.300 0.400 0.500 0.600 0.700 0.800 0.900 ... pure benzene

y-axis: Pressure, mmHg

– – – – – partial vapor pressure of benzene
———— partial vapor pressure of toluene
– – – – total vapor pressure (and liquid composition)
———— vapor composition
o   data points from Examples 12–3 and 12–4

handwritten note:
purpose of tie line is to compare the diff. mol compositions of the liquid sdn. & its vapor @ a definite equilibrium pt.

LIQUID–VAPOR EQUILIBRIUM—IDEAL SOLUTIONS. Table 12-1 summarizes the data calculated in Examples 12-3 and 12-4 (shown in black) together with similar data for other benzene–toluene solutions. These data are presented graphically in Figure 12-8. This figure consists of four lines—three straight and one curved—spanning the entire concentration range. One straight line originates at $P = 0$ and increases to $P = 95.1$ mmHg at $\chi_{benz.} = 1$. This straight line represents the partial vapor pressure of benzene as a function of solution composition. It has the equation $P_{benz.} = \chi \cdot P^{\circ}_{benz.}$, signifying that benzene follows Raoult's law. Another straight line originates at $P = 28.4$ mmHg and falls to $P = 0$ when $\chi_{benz.} = 1$. This line represents the partial vapor pressure of toluene, which also follows Raoult's law. The third straight line ranges from $P = 28.4$ mmHg when $\chi_{benz.} = 0$ to $P = 95.1$ mmHg when $\chi_{benz.} = 1$. This line shows how the *total* vapor pressure of benzene–toluene solutions varies with solution composition. It is obtained simply by adding together points from the two straight lines that lie below it. For example, the pressure at point 3 is the sum of the pressures at points 1 and 2. Point 3 then represents the total vapor pressure of a benzene–toluene solution in which $\chi_{benz.} = 0.500$. As shown in Example 12-4, the vapor in equilibrium with this solution is richer still in benzene; the vapor has $\chi_{benz.} = 0.770$ (point 4). We can think of a line segment joining points 3 and 4, called a **tie line,** in this way: The tie line is plotted at a constant pressure equal to the total vapor pressure of a solution. One end of the tie line represents the composition of the liquid solution and the other end, the composition of the vapor. Imagine establishing a series of tie lines throughout the composition range. The vapor ends of these tie lines can be joined by a smooth curve, the fourth curve in Figure 12-8 (shown in black). From the relative placement of the liquid and vapor curves we see that for ideal solutions of two components, *the vapor phase is richer in the more volatile component.*

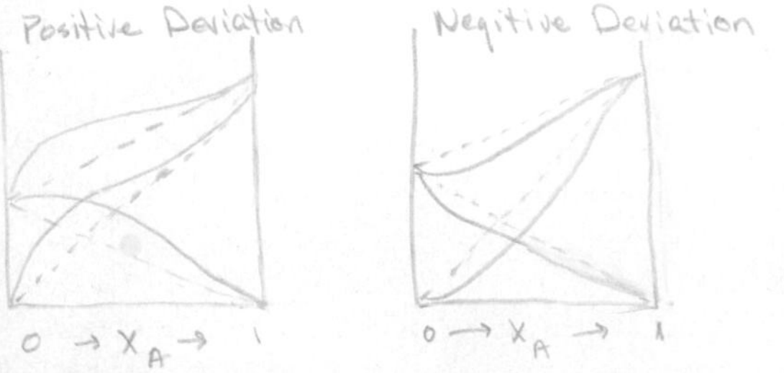

LIQUID–VAPOR EQUILIBRIUM—NONIDEAL SOLUTIONS. For a *nonideal* binary solution, the curves representing the partial vapor pressures of the two components and the total vapor pressure of the solution are *not straight lines.* Neither is it always the case that the vapor phase above nonideal solutions is richer in the more volatile component. Raoult's law simply does not apply throughout the concentration range for nonideal solutions. Unfortunately, further study of liquid–vapor equilibrium in nonideal solutions is beyond the scope of this text.

handwritten diagrams and notes:

Positive Deviation   Negitive Deviation

$0 \rightarrow X_A \rightarrow 1$ $0 \rightarrow X_A \rightarrow 1$

In Positive Deviation, vap. press. are greater than expected because the A∩B attractions are weaker than the A∩A & B∩B attractions, thus in molecules are not held as tightly in sdn as they are when in pure substance

In Neg. Deviation, same idea, but A∩B bounds hold soln molecules tighter. stronger than pure A∩A & B∩B

FIGURE 12-9

Vapor pressure lowering by a nonvolatile solute.

The phase diagram of the pure solvent is shown in color, and for the solvent containing a nonvolatile solute, in black. The freezing point and boiling point of the pure solvent are f.p.$_0$ and b.p.$_0$. The corresponding points for the solution are f.p. and b.p. The freezing point depression, ΔT_f, and boiling point elevation, ΔT_b, are indicated.

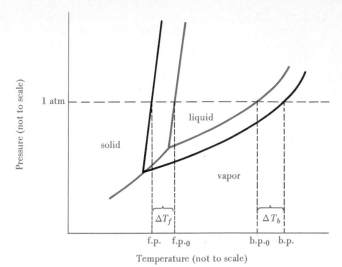

B.Pt. Elevation

if there exists a non volatile solute, O, in the solvent; s. area is decreased & fewer molecules may evaporate. ∴ Lower v.p. @ a given temp so higher degrees must be reached before v.p.= atm p, boiling pt.

F.pt. depression— solute particles get in way of crystallization process of solvent, need colder temp to overcome obstacle

The effects described here require that, in addition to being nonvolatile, the solute must be *insoluble* in the *solid* solvent. There are many solutions in which this requirement is met.

TABLE 12-2

Cryoscopic and ebullioscopic constants

Solvent	$K_f{}^a$	$K_b{}^a$
acetic acid	3.90	3.07
benzene	4.90	2.53
nitrobenzene	7.00	5.24
phenol	7.40	3.56
water	1.86	0.512

^{a} Values correspond to freezing point depressions and boiling point elevations, in degrees Celsius, due to 1 mol of solute particles dissolved in 1 kg of solvent. Units: °C kg solvent (mol solute)$^{-1}$.

UPON ADDITION OF A SOLUTE TO A SUBSTANCE, THERE EXISTS

FREEZING POINT DEPRESSION AND BOILING POINT ELEVATION. An assumption implicit throughout this section has been that both the solvent and the solute(s) are volatile. However, a very important class of solutions is that in which the solutes are *nonvolatile*. For such solutions the nonvolatile solute still lowers the vapor pressure of the solvent; the higher its concentration the greater the vapor pressure lowering. This effect is pictured in Figure 12-9. Here the vapor pressure curve and fusion curve for the solvent in a solution are superimposed on the phase diagram of the pure solvent.

The intersection of the vapor pressure and sublimation curves for a solvent containing a nonvolatile solute comes at a lower temperature than for the pure solvent. Also displaced to lower temperatures is the fusion curve. Now recall how freezing points and boiling points are established in a phase diagram. They are the temperatures at which a constant-pressure line at $P = 1$ atm intersects the fusion curve and the vapor pressure curve, respectively. Four points of intersection are indicated in Figure 12-9—the freezing points and the boiling points of the pure solvent and of the solvent in a solution. The freezing point of the solvent is *depressed* and the boiling point is *elevated*.

Freezing point depression and boiling point elevation are proportional to vapor pressure lowering, and hence to mole fraction concentration. For *dilute* solutions this proportionality can be extended to molality. Two simple mathematical relationships involving molality are

m= moles solute / kg solvent

$$\Delta T_f = K_f m \tag{12.7}$$

$$\Delta T_b = K_b m \tag{12.8}$$

In these equations ΔT_f and ΔT_b are the freezing point depression and boiling point elevation, respectively; m is the molality; K_f and K_b are proportionality constants. K_f is the **cryoscopic** or freezing point depression constant, and K_b the **ebullioscopic** or boiling point elevation constant. These constants, which are characteristic of the solvent, may be interpreted as the freezing point depression and the boiling point elevation for a 1 m solution. As a matter of fact, however, equations (12.7) and (12.8) often do not hold for solutions as concentrated as 1 m. Some typical values are listed in Table 12-2. Cooling curves for a pure solvent and a solution are compared in Figure 12-10.

Historically, freezing point measurements have been used to establish molecular

FIGURE 12-10
Cooling curves of a pure
solvent and a solution
compared.

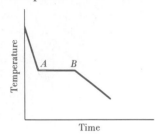

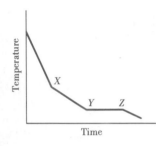

The cooling curve for the pure solvent has one horizontal break from A to B where complete freezing occurs. The cooling curve for the solution has a break at X, where the solvent begins to freeze from the solution (the freezing point). A second horizontal break from Y to Z represents the freezing of both components from solution as a mixture of solids (the eutectic temperature). The freezing points of solutions referred to in this section correspond to the point X. (The behavior described here assumes that the solute is insoluble in the solid solvent. Also, some supercooling is likely to occur at points A, X, and Y.)

formulas. The required calculations can be thought of in terms of three questions, as illustrated through Example 12-5.

solvent = H_2O : $F_p = 0°C$, $\Delta T_f = 1.86 \frac{kg\ °C}{mol}$

Example 12-5   (a) What is the molality of solute in an aqueous solution with a freezing point of $-0.450°C$? (b) If this solution was obtained by dissolving 2.12 g of an unknown compound in 48.92 g H_2O, what must be the molecular weight of the compound? (c) What is the true molecular formula of the compound if its analysis is 40.0% C, 53.3% O, and 6.7% H?

(a)   The molality of solute is readily established by using equation (12.7) with the value of K_f listed for water in Table 12-2.

$$m = \frac{\Delta T_f}{K_f} \qquad m = \frac{0.450°C}{1.86°C\ kg\ water\ (mol\ solute)^{-1}} = 0.242 \frac{mol\ solute}{kg\ water}$$

(b)   Here we use the defining equation for molal concentration (equation 12.2), but with a known molality (0.242) and an unknown molecular weight, (MW), of solute. The number of moles of solute is simply 2.12/(MW).

$$m = \frac{[2.12/(MW)]\ mol\ solute}{0.04892\ kg\ water} = 0.242 \frac{mol\ solute}{kg\ water}$$

$$(MW) = \frac{2.12}{0.04892 \times 0.242} = 179$$

(c)   The empirical formula of the compound can be determined from its percent composition by the method of Example 3-13. (The details of this calculation are left as an exercise for the student.) The result obtained is CH_2O. This empirical formula leads to a formula weight of 30. The experimentally determined molecular weight—179—is almost exactly six times as large. The true molecular formula is $C_6H_{12}O_6$.

SIMILAR EXAMPLES: Exercises 33, 34.

Example 12-5 illustrates how the measurement of a colligative property leads to a determination of molecular weight. There are limitations to this method, however, which must be understood. Remember that the boiling point of a liquid depends on atmospheric pressure. If boiling point elevation is to be used for molecular weight determination, it is necessary to maintain a constant barometric pressure. This is not particularly easy to accomplish and, as a result, boiling point elevation is not commonly employed. Because equation (12.7) is applicable only in dilute solutions (usually much less than $1\ m$), freezing points must be determined with considerable precision if water is the solvent ($K_f = 1.86$). In Example 12-5 the temperature measurement was made to $\pm0.001°C$. Temperature measurements of this precision are not possible with ordinary laboratory thermometers. Greater precision is possible when a solvent with a larger value of K_f is used, such as cyclohexane ($K_f = 20$) or, better still, camphor ($K_f = 40$). If a solute has a high molecular weight, the number of moles in a sample may be too small to affect the freezing point appreciably. For these solutes the measurement of osmotic pressure is a better method.

Whatever solvent is used, it must be of high purity and have a freezing point that is conveniently measured. The melting point of camphor is rather high for usual laboratory operations, but camphor is still desirable as a solvent because of its large K_f.

net flow of osmosis (desire to achieve concentration equil.) from more dilute → to more concentrated

concentrate   dilute

membrane on dilute side in surface contact with more solvent, so more will flow thru from that side.

FIGURE 12-11
An illustration of osmosis.

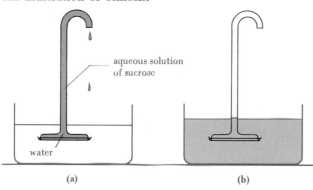

aqueous solution of sucrose

water

(a)   (b)

(a) Water molecules pass through the membrane creating a pressure within the funnel. This pressure causes the liquid level to rise and the solution to overflow. As this process continues, the solution inside the funnel becomes more dilute and the pure water outside the funnel becomes a dilute sucrose solution.
(b) As the concentrations of the solutions separated by the membrane become more nearly equal, the osmotic pressure difference falls and liquid flow stops.

There is a striking resemblance between equation (12.9) and the ideal gas equation, $PV = nRT$. Think of π as being equivalent to a gas pressure exerted by n moles of gas confined to a volume of V liters.

The phenomenon of freezing point depression also has practical applications. Perhaps best known are methods used to lower the freezing point of water. An antifreeze (usually ethylene glycol), when added to the cooling system of an automobile, protects the coolant from freezing in cold weather. The use of NaCl to lower the melting point of ice is also widely encountered, whether to de-ice roads or to prepare a freezing mixture for use in a home ice cream freezer.

OSMOTIC PRESSURE.   Certain membranes, though they appear to be continuous sheets or films, actually contain a network of submicroscopic holes or pores. Small solvent molecules may pass through these pores, but the passage of dissolved solute molecules is severely restricted. Membranes having this property are said to be **semipermeable.** They may be of animal or vegetable origin and occur naturally, such as pig's bladder and parchment, or they may be synthetic materials, such as cellophane.

Figure 12-11 pictures an aqueous sucrose (sugar) solution in a long glass tube separated from pure water by a semipermeable membrane (permeable to water only). Water molecules can pass through the membrane from either direction, and they do. But because the concentration of water molecules is *greater* in the pure water than in the solution, there is a net flow of water *from* the pure solvent *into* the solution. This net flow, called **osmosis,** causes the solution level in the tube to rise. The more concentrated the sucrose solution the higher the solution level rises. A 20% solution would be raised to about 150 m!

less solvent traveling from sdn allows more to pass from pure soly to soln.

The net flow of water into the sucrose solution can be reduced by applying a pressure to the solution. This increases the flow of water in the reverse direction. The pressure required to just stop the flow of water into the sucrose solution is known as the **osmotic pressure** of the solution. For the 20% sucrose solution this pressure is about 15 atm.

Osmotic pressure is included among the colligative properties because its magnitude depends only on the *number* of solute particles per unit volume of solution. It does not depend on the identity of the solute. The expression written below (known as the van't Hoff equation) works quite well for calculating osmotic pressures of *dilute* solutions. The osmotic pressure is represented by the symbol π; R is the gas constant (0.0821 L atm mol^{-1} K^{-1}); and T is the kelvin temperature. The term n represents the moles of *solute* and V is the volume (in liters) of *solution;* the ratio, n/V, then, is the *molarity* of the solution, M.

$$\pi = \left(\frac{n}{V}\right)RT = M \cdot RT \qquad (12.9)$$

Example 12-6   What is the osmotic pressure at 25°C of an aqueous solution that is 0.0010 M $C_{12}H_{22}O_{11}$ (sucrose)?

Direct substitution into equation (12.9) leads to the result

$$\pi = \frac{0.0010 \text{ mol} \times 0.0821 \text{ L atm mol}^{-1} \text{ K}^{-1} \times 298 \text{ K}}{L} = 0.024 \text{ atm } (= 18 \text{ mmHg})$$

SIMILAR EXAMPLES: Exercises 41, 42.

The 0.0010 M sucrose solution in Example 12-6 would have a molality of about 0.001 m. (In *dilute aqueous* solutions molarity and molality are essentially equal.) According to equation (12.7), we should expect a freezing point depression of about 0.00186°C for this solution. Such a small temperature difference is extremely difficult to measure with any precision. On the other hand, a pressure difference of 18 mmHg is rather easily measured. It corresponds to a solution height of about 0.25 m! This comparison suggests that measurement of osmotic pressure can be an important method of molecular weight determination when dealing with (a) very dilute solutions or (b) solutes of very high molecular weight.

Example 12-7   Polyvinyl chloride (PVC) is a plastic widely used in the manufacture of food wrap and phonograph records. An 0.61-g sample of PVC is dissolved in 250.0 cm³ of a suitable solvent at 25°C. The resulting solution has an osmotic pressure of 0.79 mmHg. What is the molecular weight of the PVC?

First we need to express the osmotic pressure in atm.

$$\text{no. atm} = 0.79 \text{ mmHg} \times \frac{1 \text{ atm}}{760 \text{ mmHg}} = 1.04 \times 10^{-3} \text{ atm}$$

Now we can apply equation (12.9) in a slightly modified form [i.e., with the number of moles of solute represented by mass of solute/(MW)].

$$\pi = \frac{[m/(MW)] \text{ RT}}{V}$$

$$(MW) = \frac{m \cdot R \cdot T}{\pi \cdot V} = \frac{0.61 \text{ g} \times 0.0821 \text{ L atm mol}^{-1} \text{ K}^{-1} \times 298 \text{ K}}{1.04 \times 10^{-3} \text{ atm} \times 0.250 \text{ L}} = 5.7 \times 10^4 \text{ g/mol}$$

SIMILAR EXAMPLES: Exercises 42, 43.

Polymeric substances are mixtures of molecules, and the molecules differ in size and mass. The value calculated here is an average molecular weight (see Section 26-10).

The walls (membranes) of red blood cells are approximately 10 nm thick and have pores (holes) about 0.8 nm in diameter. Water molecules are less than half this diameter and pass through easily. Potassium ions, which are found inside the cells, are also smaller than the pore diameters. But because the pore walls carry a positive electrical charge, potassium ions are repelled. Thus, factors other than simple size may be involved in determining what species can pass through the pores of a semipermeable membrane.

Perhaps the most important examples of osmosis are those found in living organisms. Consider red blood cells, for instance. If red blood cells are placed in pure water, the cells expand and eventually rupture as a result of water entering the cells through osmosis. The osmotic pressure associated with the fluid inside the cell is equivalent to that of an 0.95% sodium chloride solution. Thus, if the cells are placed in a sodium chloride solution (saline solution) of this concentration, there is no net flow of water through the cell walls and the cells remain stable. This solution is said to be *iso*tonic. If the salt solution has a higher concentration than about 0.95%, water flows out of the cells and the cells shrink. The solution is *hyper*tonic. If the salt concentration is less than 0.95%, water flows into the cells and the solution is said to be *hypo*tonic.

An interesting practical application of the idea that an external pressure can be used to stop the osmotic flow of water is found in the process of **reverse osmosis.** It is currently being tested as a method of desalinizing seawater or brackish water. As suggested by Figure 12-12, if a sufficiently high pressure is applied to a solution, the solvent can actually be forced to flow in the reverse direction, *from a solution into a pure solvent.* One of the problems associated with developing this scheme commercially is finding durable membrane materials with the required pore sizes and permeability properties.

FIGURE 12-12
Desalinization of seawater by reverse osmosis.

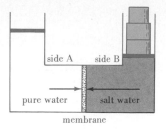

The membrane pictured here is permeable to water but not to sodium or chloride ions. The normal flow of water through the membrane, in the absence of external pressures, is from side A (pure water) to side B (salt water). If a pressure is exerted on side B that exceeds the osmotic pressure of the salt water, a net flow of water can be created in the *reverse* direction, that is, from the salt water (side B) to the pure water (side A). The magnitudes of the rates of flow of water molecules from each side are suggested by the lengths of the arrows.

12-6 Theory of Electrolytic Dissociation

Early investigators of the electrical properties of matter recognized that the ability to conduct electric current is not limited to metals. Some liquids and liquid solutions also conduct electric current. Pure liquid water is a very poor conductor of electric current, being essentially a nonconductor. The addition of certain solutes to water results in aqueous solutions that are excellent electrical conductors. Yet there are some solutes that do not enhance the electrical conductivity of water and still others that render it only weakly conducting. These three groups of solutes are termed **strong electrolytes, nonelectrolytes,** and **weak electrolytes,** respectively. Some representative examples are cited in Table 12-3 and illustrated in Figure 12-13.

For his doctoral dissertation (in 1884), a young Swedish chemist, Svante Arrhenius, undertook a careful investigation of the electrolytic conductivities of a variety of aqueous solutions. Prevailing opinion concerning ions in solution was that they form only as a result of the passage of electric current. Arrhenius, however, reached the conclusion that ions may exist in a solute and be dissociated from one another simply by dissolving the solute in water. The degree to which solute molecules are dissociated into ions he denoted by α, the **degree of dissociation.**

For a nonelectrolyte the electrolytic conductivity is extremely low; practically no ions exist in solution: $\alpha = 0$. For a weak electrolyte α is a small fractional number because in aqueous solution these solutes exist partly in ionic form and partly as

TABLE 12-3
Electrolytic properties of some aqueous solutions

| Nonelectrolytes | Strong electrolytes | | Weak electrolytes |
	Ionic compounds	Covalent compounds	
H_2O (water)	NaCl	HCl	$HCHO_2$ (formic acid)
C_2H_5OH (ethanol)	$MgCl_2$	HBr	$HC_2H_3O_2$ (acetic acid)
$C_6H_{12}O_6$ (glucose)	KBr	HI	HClO (hypochlorous acid)
$C_{12}H_{22}O_{11}$ (sucrose)	$KClO_4$	HNO_3	HNO_2 (nitrous acid)
$CO(NH_2)_2$ (urea)	KOH	H_2SO_4	H_2SO_3 (sulfurous acid)
$C_2H_6O_2$ (ethylene glycol)	$Al_2(SO_4)_3$	$HClO_4$	NH_3 (ammonia)
$C_3H_8O_3$ (glycerol)	$CuSO_4$		$C_6H_5NH_2$ (aniline)
	$LiNO_3$		
plus		**plus**	**plus**
many		**a few**	**many**
others	**plus**	**others**	**others**
	many		
	others		

[handwritten annotations: "forms good ion-dipole or H-bond", "dissociates well"]

FIGURE 12-13
Electrical conductivity of aqueous solutions.

For electric current to flow requires that electrical contact be made between the two metal rods immersed in solution. This contact through the solution is possible only if the solution contains ions. The magnitude of the current is estimated by the brilliance of the incandescent lamp.
(a) The ionic concentration is essentially zero; no current flows.
(b) A solution having a high conductivity, even though the solute concentration may be low, is a strong electrolyte.
(c) If even at fairly high concentrations a solute imparts only low electrical conductivity to a solution, it is termed a weak electrolyte.

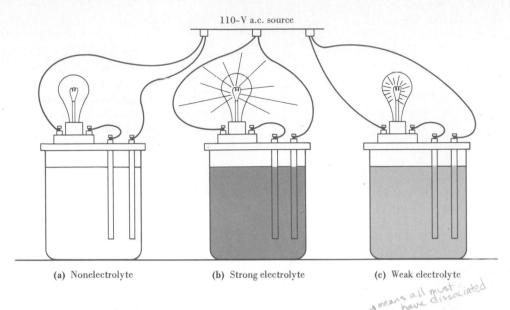

110-V a.c. source

(a) Nonelectrolyte   (b) Strong electrolyte   (c) Weak electrolyte

means all must have dissociated

undissociated molecules. In a strong electrolyte solution, especially at low concentrations, $\alpha = 1$. This value signifies that essentially complete dissociation of a solute into ions occurs in a strong electrolyte solution. Furthermore, from the measured electrolytic conductance it is possible to calculate the number of ions produced per mole of solute. For example, in NaCl, $MgCl_2$, and K_2SO_4 the numbers of ions per mole of substance are found to be *two, three,* and *three moles,* respectively.

Example 12-8 200.0 ml of 0.250 M Na_2SO_4 and 100.0 ml of 0.400 M NaCl are mixed. Assuming the solution volumes are additive, what is the molarity of Na^+ in the resulting solution?

Both Na_2SO_4 and NaCl are strong electrolytes, completely dissociated in aqueous solutions. In $Na_2SO_4(aq)$ there are twice as many moles of Na^+ ions as moles of Na_2SO_4. In NaCl(aq) the number of moles of Na^+ ions is the same as the number of moles of NaCl. We must determine the total number of moles of Na^+ ions present in a final solution of 300.0 ml volume.

$[\ ]$ - molar concentration

Concentrations are often expressed through bracket symbols, $[\ ]$. The solutions being mixed in this example can be represented as $[Na_2SO_4] = 0.250\ M$ and $[NaCl] = 0.400\ M$. The resulting molarity of Na^+ is $[Na^+] = 0.467\ M$. If a symbol to denote the concentration unit is omitted in the bracket notation, molarity is understood.

no. mol Na^+ in $Na_2SO_4(aq)$ = 200.0 ml $\times \dfrac{1\ L}{1000\ ml}$

$$\times \frac{0.250\ mol\ Na_2SO_4}{L} \times \frac{2\ mol\ Na^+}{1\ mol\ Na_2SO_4}$$

$$= 0.100\ mol\ Na^+$$

no. mol Na^+ in NaCl(aq) = 100.0 ml $\times \dfrac{1\ L}{1000\ ml}$

$$\times \frac{0.400\ mol\ NaCl}{L} \times \frac{1\ mol\ Na^+}{1\ mol\ NaCl}$$

$$= 0.0400\ mol\ Na^+$$

total mol Na^+ = 0.100 + 0.0400 = 0.140 mol Na^+

total soln volume = (200 + 100) ml $\times \dfrac{1\ L}{1000\ ml}$ = 0.300 L

molarity of Na^+ = $\dfrac{0.140\ mol\ Na^+}{0.300\ L}$ = 0.467 M

SIMILAR EXAMPLES: Exercises 44, 45.

The value of a scientific theory lies in its ability to provide explanations of a variety of seemingly unrelated phenomena. Arrhenius's theory, though developed to explain electrolytic conductivity, also provided a basis for understanding chemical reactions and chemical equilibria in solutions. The relationship of his theory to these other phenomena is explored in later chapters. One of the immediate successes of the theory was in explaining certain anomalous values of colligative properties first investigated by van't Hoff.

ANOMALOUS BEHAVIOR. Certain solutes produce a greater effect on colligative properties than expected. The van't Hoff factor, *i,* is defined as

$$i = \frac{\text{measured value}}{\text{expected value}} \quad \frac{\Delta T_f \text{ measured}}{\Delta T_f \text{ calculated as nonelectrolyte}} \tag{12.10}$$

> Arrhenius's theory of electrolytic dissociation was at first not accepted by his professors. Later, however, it was championed by such eminent chemists as van't Hoff and Ostwald. Arrhenius's theory marked the beginnings of the discipline of physical chemistry.

For a large group of solutes, such as urea, glycerol, and sucrose, *i* has a value of 1. For another equally large group of solutes, *i* has values greater than 1.

Example 12-9   The following freezing points are observed for 0.010 *m* aqueous solutions. Calculate the van't Hoff factor for each solute and account for its value: urea, $-0.0186°C$; acetic acid, $-0.0193°C$; magnesium chloride, $-0.054°C$.

To determine the van't Hoff factor, *i*, we must start with an expected value for the freezing point depression. According to equation (12.7), for an 0.010 *m* aqueous solution of a solute we normally expect $\Delta T_f = 0.0186°C$. $\Delta T_f = K_f m = (1.86)(0.010m)$

urea $(CO(NH_2)_2)$: $i = \dfrac{0.0186°C}{0.0186°C} = 1.00$

Urea is a nonelectrolyte and remains undissociated in aqueous solution.

acetic acid $(HC_2H_3O_2)$: $i = \dfrac{0.0193°C}{0.0186°C} = 1.04$

Acetic acid is a weak electrolyte. About 4% of the molecules are dissociated, producing two ions (H^+ and $C_2H_3O_2^-$) per molecule. ($\alpha = 0.04$.)

magnesium chloride $(MgCl_2)$: $i = \dfrac{0.054°C}{0.0186°C} = 2.9 \simeq 3$

A van't Hoff factor of $i \simeq 3$ suggests that $MgCl_2$ is dissociated in aqueous solution, producing three moles of ions per mole of compound. The reason why *i* is somewhat less than 3 is explained in the next section.

SIMILAR EXAMPLES: Exercises 46, 48.

12-7 Interionic Attractions

Once it had gained acceptance Arrhenius's theory stimulated great progress in physical chemistry. However, some observations made over the 40-year period from about 1880 to 1920 pointed to a need for refinements in this theory. Electrolytic conductances of concentrated solutions of strong electrolytes were found to be less than expected for complete dissociation into ions. These results suggest incomplete dissociation of a strong electrolyte in solution. Yet x-ray diffraction studies indicate that salts exist in 100% ionic form in the solid state. Should they not also be completely ionized in solution?

The modern view of electrolyte solutions is based on a theory proposed by Debye

FIGURE 12-14
Interionic attractions in aqueous solution.

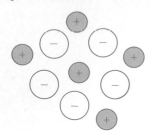

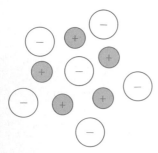

A positive ion in aqueous solution (top) is surrounded by a shell of negative ions. A negative ion (bottom) attracts positive ions to its immediate surroundings.

magnitudes of colligative properties are reduced due to interionic attractions. Though solute may be totally dissociated in a solvent, the # ions expected might not equal "i". Apparently, the more concentrated a soln the less the interionic distances. As a result, there still exists some attraction between the ions & they effect the properties less as independent particles.

ie/ you'd expect $MgCl_2$ to effect a H_2O soln as if you'd put 3 mols solute in ($Mg^{++}, 2Cl^-$), but if concentrated, ions can't get far enough apart & as result yield 2.9 effect.

Colloidal mixtures of silica in water can be prepared by acidifying aqueous solutions of sodium silicate (see Section 20-9). The production of colloidal particles may occur, as in this case, by the aggregation of large numbers of molecules through a process called *condensation*. A contrasting method that may sometimes be employed is that of *dispersion*. This involves breaking down larger particles mechanically, for example by grinding, until the particles are sufficiently small to remain suspended.

and Hückel in 1923. The theory states that in aqueous solutions salts do exist in completely ionized form. The ions, however, do not behave independently of one another. Instead, each positive ion is surrounded by a cluster of predominantly negative ions, and each negative ion by a cluster in which positive ions predominate. That is, each ion is enveloped by an ionic atmosphere with a net charge opposite in sign to the central ion (see Figure 12-14).

In an electric field the mobility of each ion is reduced because of the attraction or drag of its neighbors in the ionic atmosphere. Similarly, the magnitudes of colligative properties are reduced. Thus, each type of ion in an aqueous solution has a total concentration based on the amount of solute dissolved, called the stoichiometric concentration. But the ion also has an "effective" concentration, called the **activity,** which takes into account interionic attractions. If the activity is used in place of stoichiometric concentration, solution properties can be predicted quite well, especially for dilute solutions. To relate activity to concentration requires the use of an **activity coefficient.** The Debye–Hückel theory provides a theoretical basis for calculating activity coefficients and activities.

12-8 Colloidal Mixtures

In the opening section of this chapter we chose sand in water as an example of a heterogeneous mixture. From common experience we expect sand to settle to the bottom of such a mixture, even if the quantity of sand (silica, SiO_2) is very small. SiO_2 is very insoluble in water. Yet it is possible to prepare mixtures in which large quantities of silica, up to 30% by mass, are dispersed in water and remain dispersed for years! Such mixtures are clear, although with a faintly opalescent or milky cast. Obviously, these dispersions do not involve ordinary grains of sand. Neither do they consist of dissolved ions or molecules. They are called colloidal mixtures.

The freezing points of colloidal mixtures of silica in water are only slightly below 0°C. We conclude that these mixtures contain *small* numbers of particles in comparison to true solutions with comparable solute concentrations. But if the numbers of particles are small, their masses and physical dimensions must be huge compared to typical solute particles. The molecular weights, or more correctly, particle weights, of colloids range into the hundreds of thousands or millions. Figure 12-15 compares colloidal particles of different sizes and shapes with the more familiar particles of chemistry—atoms, ions, and molecules.

FIGURE 12-15
A comparison of colloidal, molecular, and atomic dimensions.

Some approximate sizes and shapes of colloidal particles are represented here. For comparison, particles with typical atomic and molecular dimensions are also shown.

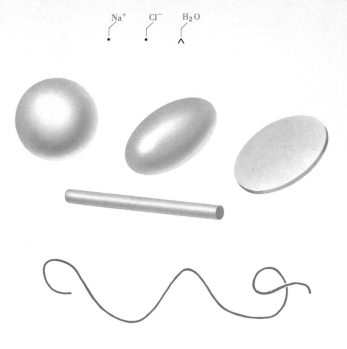

Figure 12-15 suggests that in order for a material to be classified as colloidal, one or more of its dimensions (length, width, or thickness) must fall in the approximate range 1 to 100 nm. If all the dimensions are less than 1 nm, the particles are in the molecular size range. If all the dimensions exceed 100 nm, the particles become of ordinary or macroscopic size (even if they are only visible under a microscope).

The colloidal particles in silica–water mixtures have a spherical shape; so do particles of bushy stunt virus. Some colloidal particles are rod shaped, for example, tobacco mosaic virus. Some have a disclike shape, like the gamma globulin in human blood plasma. Thin films, such as oil on water, are colloidal. And some colloids have the appearance of filaments or random coils, for example, cellulose fibers.

Determining whether a mixture is a true solution or colloidal is often possible by the method of Figure 12-16. When light is passed through a true solution, an observer viewing from a direction perpendicular to the beam sees no light. But in a colloidal suspension light is scattered in many directions and can be seen easily. This behavior, first studied by Tyndall in 1869, is known as the Tyndall effect. A common example of the Tyndall effect is the scattering of light by dust particles in the light beam of a movie projector in a darkened room.

One of the important characteristics of colloidal particles is their high ratio of surface area to volume. It is a well established fact that the atoms, ions, or mole-

As so aptly put years ago by Wilder Bancroft, "Colloid chemistry is the chemistry of bubbles, drops, grains, filaments, and films."

FIGURE 12-16
Light scattering by colloidal suspensions—the Tyndall effect.

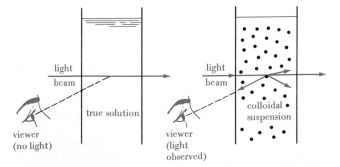

FIGURE 12-17
The phenomenon of electrophoresis.

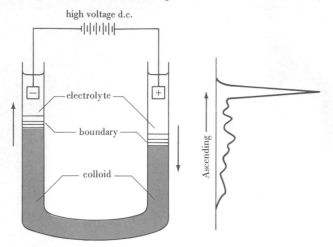

high voltage d.c.

(a) Simple electrophoresis cell

(b) Electrophoretic pattern
for human plasma

(a) A colloidal suspension in water is covered by an electrolyte solution. Electrodes are immersed in the electrolyte and connected to a high-voltage direct current source. Colloidal particles, because of their adsorbed ions, carry electrical charges. As a result they are attracted to one electrode and repelled by the other; this causes movement of the boundaries. In the illustration here, the particles are positively charged. If colloidal particles of differing types are present, they migrate at different rates in an electric field; the initial sharp boundary separates into several boundaries.
(b) This is a representation of an electrophoresis boundary in which the separation of several proteins in human blood plasma occurs. The large peak corresponds to albumin, the smaller peaks to several globulins.

cules at the surface of a substance behave somewhat differently than do those in the interior. This is because species at the surface are subject to forces different from those in the interior (recall the discussion of surface tension in Section 11-2). For ordinary materials the proportion of structural units at the surface is very small compared to the interior, and the distinctive phenomena associated with surfaces are masked. In colloidal materials these surface phenomena are often quite pronounced.

STABILITY OF COLLOIDS. One of the properties of surfaces is that of being able to attach species to themselves, a phenomenon called **adsorption.** In their formation, some colloidal particles adsorb large numbers of ions from solution and become electrically charged. Silica particles in the colloidal silica solutions mentioned previously adsorb hydroxide ions (OH^-) in preference to other ions. As a result the silica particles all acquire a negative charge. Having like charges, the particles repel one another. It is these mutual repulsions that overcome the force of gravity and keep the particles suspended.

Electrically charged colloidal particles can migrate in an electric field, just as do ions. The rates at which colloidal particles migrate depend on the magnitudes of the charges they carry, on their shapes, and on other factors. In the process of **electrophoresis,** illustrated in Figure 12-17, colloidal particles are separated according to these differences in mobility.

The electrostatic factor in stabilizing colloids is particularly important in a class referred to as lyophobic ("solvent fearing"). Another class of colloids called lyophilic ("solvent loving") owe their stabilities to an ability to swell in a solvent and remain suspended. If

TABLE 12-4
Some common types of colloids

Dispersed phase	Dispersion medium	Type	Examples
solid	liquid	sol	clay sols,[a] colloidal gold
liquid	liquid	emulsion	oil in water, milk, mayonnaise
gas	liquid	foam	soap and detergent suds, whipped cream, meringues
solid	gas	aerosol[b]	smoke, dust-laden air[c]
liquid	gas	aerosol[b]	fog, mist (as in aerosol products)
solid	solid	solid sol	ruby glass, certain natural and synthetic gems, blue rock salt, black diamond
liquid	solid	solid emulsion	opal, pearl
gas	solid	solid foam	pumice, lava, volcanic ash

[a] In water purification it is sometimes necessary to precipitate clay particles or other suspended colloidal materials. This is often done by treating the water with an aluminum compound, such as $Al_2(SO_4)_3$. The negatively charged clay particles are neutralized by Al^{3+} ions and coagulate or settle from solution. Clay sols are also suspected of adsorbing organic substances, such as pesticides, and distributing them in the environment.

[b] Smogs are complex materials that are at least partly colloidal. The suspended particles are both solid (smoke) and liquid (fog): smoke + fog = smog. Other constituents of smog are molecular, such as sulfur dioxide, carbon monoxide, nitric oxide, and ozone.

[c] The bluish haze of tobacco smoke and the brilliant sunsets in desert regions are both attributable to the scattering of light by colloidal particles suspended in air.

the suspending medium is water, the prefix "hydro" replaces "lyo" in these two terms (i.e., hydrophobic and hydrophilic). A large number of hydrophilic colloidal materials are of biochemical interest and are considered again in Chapter 25.

TYPES OF COLLOIDS. Colloidal mixtures can be categorized in part according to the phases of matter involved. A brief listing is provided in Table 12-4.

Summary

Whether the mixing of two substances produces a solution depends largely on the comparative strengths of intermolecular forces between like and unlike molecules. In the dissolving of an ionic solid in water, forces of attraction between ions and water dipoles must be compared to interionic forces of attraction within the ionic crystal.

In describing a solution it is necessary to indicate the relative proportions of solvent and solute. This can be done through a variety of expressions for solution concentration. The preferred SI unit of concentration is molarity.

In most instances the ability of a solvent to dissolve a solute is limited. A solution containing this limiting quantity of solute is saturated. Solubility of a solute is a function of temperature, often represented through a graph known as a solubility curve. Differences in solubility of a solute at different temperatures provide the basis of a method for purifying substances known as recrystallization. The solubilities of gases depend on pressure as well as temperature. A simple equation is available for relating the concentration of a gas in solution to the pressure of the gas above the solution (Henry's law).

Colligative properties depend primarily on the number of solute particles in a solution and not on the identity of the solute. These properties can all be related to the tendency of each solution component to lower the vapor pressure of other components in a solution. If this vapor pressure lowering conforms to Raoult's law, the solution is said to be ideal. Even if a solution is nonideal, Raoult's law still is obeyed by the solvent in a dilute solution.

An important class of solutions is that in which solutes are nonvolatile. For these solutions freezing point depression and boiling point elevation can be used to determine the molecular weight of a solute. When dealing with solutes of high molecular weight (e.g., polymers) the preferred colligative property for molecular weight determination is osmotic pressure. Osmosis and osmotic pressure are important phenomena encountered in a number of biological systems.

Solutes in aqueous solution belong to one of three classes. Either they exist almost exclusively in molecular form (nonelectrolyte), partly in molecular and partly in ionic form (weak electrolyte), or in a completely ionized form (strong electrolyte). Depending on which of the three cases prevails, an aqueous solution is either a nonconductor of electricity, a weak conductor, or a good electrical conductor. The expressions relating colligative properties to solution concentration must be modified if dissociation of a solute into ions occurs.

There are numerous instances, some of great practical importance, where a mixture does not fall neatly into the category of homogeneous (true solution) or heterogeneous. These intermediate cases are called colloidal mixtures. In general, if a material contains particles having one or more dimensions in the range 1 to 100 nm, it is called colloidal. Bubbles, droplets, grains, filaments, films, emulsions, foams, and aerosols, among others, fall into this category. Many distinctive features of colloidal materials stem from their high surface-to-volume ratios and their abilities to adsorb foreign molecules or ions on their surfaces.

Learning Objectives

As a result of studying Chapter 12, you should be able to

1. Explain the relationship between intermolecular forces of attraction and the formation of solutions, distinguishing between ideal and nonideal solutions.

2. Express solution concentration on a percent basis, or as molarity, molality, and mole fraction.

3. Distinguish among unsaturated, saturated, and supersaturated solutions, and describe how a solute may be purified by recrystallization.

4. Calculate the solubility of a gas in a liquid as a function of gas pressure.

5. Calculate partial vapor pressures of solution components, the total pressure, and the composition of vapor in equilibrium with an ideal solution.

6. Calculate, for dilute solutions of nonvolatile solutes, vapor pressures, freezing points, and boiling points.

7. Use the measured freezing points or boiling points of solutions as a basis for calculating molecular weights of solutes.

8. Describe the process of osmosis, and use osmotic pressure data in molecular weight determinations.

9. Distinguish among strong, weak, and nonelectrolytes, and calculate ionic concentrations in aqueous solutions of strong electrolytes.

10. Describe the modifications of Arrhenius's theory of electrolytic dissociation made necessary because of interionic attractions in aqueous solutions.

11. Describe some distinctive properties of colloidal mixtures, and state how colloids differ from true solutions and from heterogeneous mixtures.

Some New Terms

Adsorption refers to the attachment of ions or molecules to the surface of a material.

An **alloy** is a mixture of two or more metals. Some alloys are solid solutions, some are heterogeneous mixtures, and some are intermetallic compounds.

Colligative properties—vapor pressure lowering, freezing point depression, boiling point elevation and osmotic pressure—have values that depend only on the number of solute particles in solution.

A **colloidal mixture** contains particles that are intermediate in size to those of a true solution and an ordinary heterogeneous mixture.

Henry's law relates the solubility of a gas to the gas pressure maintained above a solution—$C = k \cdot P_{gas}$.

An **ideal solution** has certain properties (notably vapor pressure) that are predictable from the properties of the solution components.

Molality (**m**) is a solution concentration expressed as number of moles of solute per kilogram of solvent.

The **mole fraction** of solution component i (χ_i) is the fraction of all the molecules in a solution which are of that type (i.e., type i).

A **nonelectrolyte** is a substance that is essentially un-ionized, both in the pure state and in solution.

Osmosis is the net flow of solvent molecules through a semipermeable membrane, from a more dilute solution (or from a pure solvent) into a more concentrated solution.

Osmotic pressure is the pressure that would have to be applied to a solution to stop the passage of molecules from the pure solvent through a semipermeable membrane into the solution.

Precipitation refers to the separation of a solid from a liquid solution.

Raoult's law states that the vapor pressure of a solution component is equal to the product of the vapor pressure of the pure component and its mole fraction in solution. Raoult's law applies to all volatile components in an ideal solution and to the solvent in a dilute nonideal solution.

Reverse osmosis is the passage through a semipermeable membrane of solvent molecules *from a solution into a pure solvent*. It can be achieved by applying to the solution a pressure in excess of its osmotic pressure.

A **saturated solution** is one that contains the maximum quantity of solute that is normally possible.

A **semipermeable membrane** permits the passage of solvent molecules but restricts the flow of larger solute molecules (and ions in some cases). It is a film of material containing submicroscopic pores.

A **solute(s)** is (are) the solution component(s) present in lesser amount(s) (that is, in lesser amount than the solvent).

The **solvent** is the solution component present in greatest quantity or the component which determines the state of matter in which a solution exists.

A **strong electrolyte** is a substance that exists completely in the form of ions in solution.

A **supersaturated solution** contains, because of its manner of preparation, more solute than normally expected under the given condition.

An **unsaturated solution** contains less solute than the solution is capable of dissolving under the given conditions.

A **weak electrolyte** is a substance that is present in solution partly in the molecular form and partly as ions.

Exercises

Homogeneous and heterogeneous mixtures

1. For each of the following solutions, indicate which component is the solvent and which is the solute. Comment on any difficulties in applying these terms. **(a)** 10 g $C_2H_5OH(l)$ dissolved in 100 g $H_2O(l)$; **(b)** 50 g $CH_3OH(l)$ dissolved in 50 g $H_2O(l)$; **(c)** 10 g $CCl_4(l)$ dissolved in 50 g $C_6H_6(l)$ + 50 g $C_7H_8(l)$; **(d)** 1.0 M $Na_2SO_4(aq)$.

2. From the standpoint of intermolecular forces, comment on the common phrases "like dissolves like" and "oil and water don't mix."

3. When 50.0 ml of ethanol and 50.0 ml of water are mixed, heat is evolved and the resulting solution has a volume of 96.0 ml. Which of the four situations described on page 287 do you think applies?

4. Explain the observation that all metal nitrates are water soluble, whereas many metal sulfides are not. Among metal sulfides, which would you expect to be most soluble?

Percent concentration

5. A handbook lists the composition of a saturated aqueous solution of KI at 20°C as 144 g KI/100 g H_2O. Express this composition in the more conventional % (mass/mass)—that is, as g KI/100 g soln.

6. An aqueous solution having a density of 0.980 g/cm³ at 20°C is prepared by dissolving 11.3 ml CH_3OH ($d = 0.793$ g/cm³) in enough water to produce 75.0 ml of solution. What is the percent CH_3OH in this solution, expressed as **(a)** % (vol/vol); **(b)** % (mass/vol); **(c)** % (mass/mass)?

7. A sample of white vinegar is found to contain 6.10% acetic acid ($HC_2H_3O_2$), by mass. What mass of $HC_2H_3O_2$ is contained in a 0.500-L bottle of the vinegar? Assume a density of 1.01 g/cm³.

8. The calculations in Example 12-1 show that the percent ethanol, by mass, in a certain aqueous solution is less than the percent by volume in the same solution. Explain why you would expect this to be also true for all aqueous solutions of ethanol. Would it be true of all ethanol solutions, regardless of the other component? Explain.

9. Is either concentration term, percent by mass or percent by volume, independent of temperature? Explain.

Molar concentration

10. What is the molar concentration of methanol in the solution described in Exercise 6?

11. It is desired to prepare 250.0 ml of a standard solution having the molar concentration 0.0150 M $AgNO_3$. What mass of a sample known to be 99.68% $AgNO_3$, by mass, is required for this purpose?

12. How many ml of the ethanol–water solution described in Example 12-1 would have to be diluted with water to produce 500.0 ml of 0.250 M C_2H_5OH?

13. A 10.00% by mass solution of ethanol, C_2H_5OH, in water has a density of 0.9831 g/cm³ at 15°C and 0.9804 g/cm³ at 25°C. What is the molarity of C_2H_5OH in this solution at each temperature?

Molal concentration

14. What is the molal concentration of *p*-dichlorobenzene in a solution prepared by dissolving 1.50 g $C_6H_4Cl_2$ in 35.0 g of benzene, C_6H_6?

15. How many grams of iodine, I_2, must be dissolved in 250.0 ml of carbon tetrachloride, CCl_4 (density = 1.595 g/cm³) to produce a 0.175 m I_2 solution?

16. An aqueous solution of hydrofluoric acid is 30.0% HF, by mass, and has a density of 1.101 g/cm³. What are the molality and molarity of HF in this solution?

*17. A solution has a concentration described as 109.2 g KOH/L soln. The solution density is listed as 1.09 g/cm³. It is desired to convert 100.0 cm³ of this solution to one having a concentration of 0.250 m KOH. Which component, KOH or water, would you add to this solution? What mass of this component is necessary?

Mole fraction, mole percent

18. A solution is prepared by mixing the following numbers of moles of hydrocarbons: 1.15 mol C_7H_{16}, 1.48 mol C_8H_{18}, and 2.71 mol C_9H_{20}. What are **(a)** the mole fraction and **(b)** the mole percent of each component in this solution?

19. What is the mole fraction of **(a)** $C_6H_4Cl_2$ in the solution described in Exercise 14; **(b)** C_2H_5OH in the solution described in Exercise 13?

20. What mass of C_2H_5OH must be added to 100.0 ml of the solution described in Example 12-1(f) to increase the mole fraction of C_2H_5OH to 0.0500?

21. What volume of glycerol, $C_3H_8O_3$ ($d = 1.26$ g/cm³), must be added per kilogram of water to produce a solution with 10.0 mole percent $C_3H_8O_3$?

Solubility equilibrium

22. Refer to Figure 12-6 and estimate the temperature at which a saturated aqueous solution of $KClO_4$ is 1.00 m.

23. A solution prepared by dissolving 26.0 g $KClO_4$ in 500.0 g of water is brought to a temperature of 20°C.
  (a) Refer to Figure 12-6 and determine whether the solution is unsaturated or supersaturated at 20°C.

(b) Approximately what mass of $KClO_4$ must be added to make the solution saturated (if it is originally unsaturated) or what mass of $KClO_4$ can be crystallized from the solution (if it is originally supersaturated)?

24. A solid mixture consists of 95.0% NH_4Cl and 5.0% $(NH_4)_2SO_4$, by mass. A 50.0-g sample of this solid is added to 100.0 g of water at 90°C. With reference to Figure 12-6
  (a) Will all of the solid dissolve at 90°C?
  (b) If the resulting solution is cooled to 0°C, approximately what mass of NH_4Cl can be crystallized from solution?
  (c) Will $(NH_4)_2SO_4$ also crystallize at 0°C?

Solubility of gases

25. Certain natural waters contain dissolved $H_2S(g)$ [rotten-egg smell]. If such a water sample containing 0.5% by mass of dissolved H_2S is maintained under a pressure of $H_2S(g)$ of 740 mmHg, will the sample dissolve more H_2S or lose some that is already dissolved? (*Hint:* Use data from Example 12-2.)

26. Most natural gases consist of about 90% methane, CH_4. Assume that the water solubility of natural gas at 20°C and 1 atm gas pressure is about the same as that of CH_4, 0.02 g/kg water. If a sample of natural gas under a pressure of 15 atm is maintained in contact with 100.0 kg of water, how many grams of natural gas would you expect to dissolve?

27. Henry's law is often stated in this way: The mass of a gas dissolved by a given quantity of solvent at a fixed temperature is directly proportional to the pressure of the gas. Show how this statement is related to equation (12.4).

*28. Still another statement of Henry's law is this: A given quantity of liquid at a fixed temperature dissolves the same volume of gas at all pressures. What is the connection between this statement and the one given in Exercise 27? Under what conditions is this second statement not valid?

Raoult's law and liquid–vapor equilibrium

29. What are the partial and total vapor pressures above a solution obtained by mixing 50.0 g of benzene, C_6H_6, and 50.0 g of toluene, C_7H_8, at 25°C. The vapor pressures of pure benzene and toluene at this temperature are 95.1 and 28.4 mmHg, respectively.

30. With reference to Exercise 29, determine the composition of the vapor phase.

*31. Calculate the mole fraction of benzene in a benzene-toluene liquid solution that is in equilibrium at 25°C with a vapor phase that contains 62.0 mol percent C_6H_6. (Use data from Exercise 29.)

32. What would you expect to be the vapor pressure at 25°C above a solution in which 25.0 g of the *nonvolatile* solute, urea, $CO(NH_2)_2$, is dissolved in 525 g H_2O. The vapor pressure of water at 25°C is 23.8 mmHg.

Freezing point depression and boiling point elevation

33. The addition of 1.10 g of an unknown compound reduces the freezing point of 75.22 g of benzene from 5.51 to 4.90°C. What is the molecular weight of the unknown compound?

34. An unknown compound consists of 42.4% C, 2.4% H, 16.6% N, and 37.8% O. The addition of 6.45 g of this compound to 50.0 ml of benzene ($d = 0.879$ g/cm³) lowers the freezing point from 5.51 to 1.35°C. What is the true molecular formula of this substance?

35. The addition of 1.00 g of benzene, C_6H_6, to 80.00 g of cyclohexane, C_6H_{12}, reduces the freezing point of the cyclohexane from 6.5 to 3.3°C.
   (a) What is the value of K_f for cyclohexane?
   (b) Which do you think is the better solvent for molecular weight determinations by freezing point depression, benzene or cyclohexane? Explain.

36. What approximate proportions by volume of water ($d = 1.00$ g/cm³) and ethylene glycol, $C_2H_6O_2$ ($d = 1.12$ g/cm³), must be mixed to ensure protection of an automobile cooling system to -10°C? (Assume that equation 12.7 applies.)

37. Citrus growers know that it is not necessary to fire their smudge (smoke) pots even if the temperature is expected to drop several degrees (Fahrenheit) below the normal freezing point of water for several hours.
   (a) Why doesn't the citrus fruit freeze at the normal freezing point (32°F)?
   (b) Why do you suppose that lemons freeze at a higher temperature than do oranges?

38. The freezing point of an 0.01 m aqueous solution of a nonvolatile electrolyte is -0.072°C. What would you expect the normal boiling point of this same solution to be?

Osmotic pressure

39. When the stems of cut flowers are immersed in a concentrated salt solution, the flowers wilt. When a fresh cucumber is placed in a concentrated salt solution, it shrivels up (becomes pickled). Explain the basis of these phenomena.

40. Verify the statement in the text that a 20% sucrose solution would rise to a height of 150 m as a result of osmotic pressure.

41. At 25°C the average osmotic pressure of blood is 7.7 atm. What is the molar concentration of a glucose ($C_6H_{12}O_6$) solution that is isotonic with blood?

42. A solution containing 1.02 g of hemoglobin in 50.0 cm³

soln. has an osmotic pressure of 5.85 mmHg at 298 K. What is the molecular weight of hemoglobin?

43. An 0.50-g sample of polyisobutylene in 100.0 cm³ of benzene solution has an osmotic pressure at 25°C that is sufficient to support a 5.1-mm column of the solution ($d = 0.88$ g/cm³). What is the molecular weight of this sample of polyisobutylene?

Strong, weak, and nonelectrolytes

44. A solution is 0.10 M in KCl and 0.20 M in $MgCl_2$. What are the molarities of K^+, Mg^{2+}, and Cl^- in this solution?

45. Assuming no change in solution volume, what is the molarity of Cl^- in the solution obtained by adding 2.00 g $MgCl_2$ to 400.0 ml of 0.180 M $MgCl_2$?

46. Arrange the following aqueous solutions in order of increasing ability to conduct electric current. Comment on the reasons for this arrangement. 0.01 M NaCl; 1.0 M C_2H_5OH; 1.0 M $MgCl_2$; 0.01 M $HC_2H_3O_2$.

47. An aqueous solution of NH_3 conducts electric current only weakly. The same is true for an aqueous solution of $HC_2H_3O_2$. When these solutions are mixed, however, the resulting solution conducts electric current very well. Propose an explanation?

48. Predict the approximate freezing points of 0.10 m solutions of the following solutes dissolved in water: **(a)** urea; **(b)** NH_4NO_3; **(c)** $CaCl_2$; **(d)** $MgSO_4$; **(e)** ethanol; **(f)** HCl; **(g)** $HC_2H_3O_2$ (acetic acid).

Colloidal mixtures

49. Discuss some of the principal differences between colloidal mixtures and true solutions.

50. Describe what is meant by the terms **(a)** aerosol; **(b)** emulsion; **(c)** foam; **(d)** hydrophobic colloid; **(e)** electrophoresis

51. The particles of a particular arsenic trisulfide (As_2S_3) sol are negatively charged.
   (a) Sketch the results that would be obtained by the electrophoresis of this sol (see Figure 12-17).
   (b) Which 0.0005 M solution would be most effective in coagulating this sol: KCl, $MgCl_2$, $AlCl_3$, or Na_3PO_4? Explain.

*52. Suppose that 1.00 mg of gold is obtained in a colloidal dispersion in which the gold particles are assumed to be spherical, with a radius of 100 nm. (The density of gold is 19.3 g/cm³.)
   (a) What is the total surface area of the colloidal particles?
   (b) What is the surface area of a single cube of gold weighing 1.00 mg?

Additional Exercises

1. A solution of glycerol in water is 90.0% $C_3H_8O_3$ and 10.0% H_2O, by mass, and has a density of 1.235 g/cm^3. Determine **(a)** the molarity of glycerol (considering water as the solvent); **(b)** the molarity of water (considering glycerol as the solvent); **(c)** the molality of water in glycerol; **(d)** the mole fraction of glycerol; **(e)** the mole percent of water.

2. A certain brine solution contains 2.52% NaCl, by mass. If a 50.0 ml sample is found to weigh 51.1 g, how many liters of this brine would be required to extract 1 metric ton (1000 kg) of NaCl?

3. Calculate the molality of the ethanol–water solution described in Exercise 13. Does the molality differ at the two temperatures (i.e., 15 and 25°C)? Explain.

*4. Water and phenol are only partially miscible at temperatures up to 66.8°C. In a mixture prepared at 29.6°C from 50.0 g of water and 50.0 g of phenol, 32.8 g of a phase consisting of 92.5% water and 7.50% phenol is obtained. This can be considered a saturated solution of phenol in water. What is the percent by mass of water in the second phase—a saturated solution of water in phenol?

5. Assuming the volumes are additive, what is the molarity of NO_3^- in a solution obtained by mixing 325 ml of 0.231 M KNO_3, 625 ml of 0.510 M $Mg(NO_3)_2$, and 825 ml of H_2O?

6. In a molecular weight determination it is desired to achieve a freezing point depression of between 2 and 3°C. If a 50.0-g sample of benzene is used as the solvent, what mass of unknown must be taken if the estimated molecular weight of the unknown is **(a)** 50; **(b)** 75?

7. Use the concentration of an isotonic saline solution given in the text to determine the osmotic pressure of blood at body temperature, 37.0°C. (*Hint:* Recall that NaCl is completely dissociated in aqueous solutions.)

8. What pressure is required in the reverse osmosis depicted in Figure 12-12 if the salt water contains 3.0% NaCl, by mass. (*Hint:* Recall that NaCl is completely dissociated in aqueous solutions. Also, assume a temperature of 25°C.)

9. Solution A contains 0.515 g of urea, $CO(NH_2)_2$, dissolved in 85.0 g of water. Solution B contains 2.50 g of sucrose,

$C_{12}H_{22}O_{11}$, dissolved in 92.5 g of water. Above which solution is the water vapor pressure greater?

*10. The two solutions just described (Exercise 9) are placed in separate containers but in an enclosure in which their vapors may mix freely. Water evaporates from the solution of higher vapor pressure and condenses into the solution of lower vapor pressure. This process continues until both solutions have the same water vapor pressure. What are the compositions of solutions A and B when this vapor pressure equilibrium is reached?

*11. At 20°C liquid benzene has a density of 0.879 g/cm^3; liquid toluene, 0.867 g/cm^3. Assume that benzene–toluene solutions are ideal and
   (a) Calculate the densities of solutions containing 20, 40, 60, and 80 vol. % benzene.
   (b) Plot these data in a graph of density vs. volume percent composition.
   (c) Establish the equation

   $$d = \frac{1}{100}[0.879\ V + 0.867(100 - V)]$$

   where V = vol. % benzene.

*12. The following data are given for the densities of ethanol–water solutions at 15°C as a function of *volume* percent ethanol: 0%, 0.999 g/cm^3; 20.0%, 0.977 g/cm^3; 40.0%, 0.952 g/cm^3; 60.0%, 0.914 g/cm^3; 80.0%, 0.864 g/cm^3; 100%, 0.794 g/cm^3. Are ethanol–water mixtures ideal?

*13. Demonstrate that for a *dilute aqueous* solution the molality is essentially equal to the molar concentration.

*14. Show that for a dilute solution the mole fraction of solute is proportional to the molality, and that for a dilute *aqueous* solution the solute mole fraction is proportional to molarity.

*15. A saturated solution is prepared at 70°C containing 32.0 g $CuSO_4$ per 100.0 g soln. A 335-g sample of this solution is then cooled to 0°C and $CuSO_4 \cdot 5\ H_2O$ crystallizes out. If the concentration of a saturated solution at 0°C is 12.5 g $CuSO_4/100.0$ g soln, how many g $CuSO_4 \cdot 5\ H_2O$ would be obtained? (*Hint:* Note that the solution composition is stated in terms of $CuSO_4$ but that the solid that crystallizes is the hydrate, $CuSO_4 \cdot 5\ H_2O$.)

Self-Test Questions

For questions 1 through 6 select the single item that best completes each statement.

1. An *aqueous* solution is 0.01 M CH_3OH. The concentration of this solution is also very nearly **(a)** 0.01% CH_3OH (mass/vol); **(b)** 0.01 m CH_3OH; **(c)** $\chi_{CH_3OH} = 0.01$ (i.e., mole fraction $CH_3OH = 0.01$); **(d)** 0.99 M H_2O.

2. The most water soluble of the following compounds is **(a)** $C_6H_6(l)$; **(b)** $SiO_2(s)$; **(c)** $CH_3OH(l)$; **(d)** $C_{10}H_8(s)$.

3. The *best* electrical conductor of the following aqueous solutions is **(a)** 0.10 M NaCl; **(b)** 0.10 M C_2H_5OH (ethanol); **(c)** 0.10 M $HC_2H_3O_2$ (acetic acid); **(d)** 0.10 M $C_{12}H_{22}O_{11}$ (sucrose).

4. The aqueous solution with the *lowest* freezing point of the following group is **(a)** 0.01 m $MgSO_4$; **(b)** 0.01 m NaCl; **(c)** 0.01 m C_2H_5OH (ethanol); **(d)** 0.008 m MgI_2.

5. An ideal liquid solution is prepared having equal mole fractions of two volatile components, A and B. In the *vapor* above the solution
   (a) The mole fraction of A = 0.50.
   (b) The mole fractions of A and B are equal, but not necessarily 0.50.
   (c) The mole fractions of A and B are not likely to be equal.
   (d) There will be only one component present, whichever is the solvent.

6. The best method for determining the molecular weight of a polymeric substance generally involves measurement of **(a)** vapor density; **(b)** osmotic pressure; **(c)** freezing point depression; **(d)** boiling point elevation.

7. A solution is prepared by dissolving 1.00 g of naphthalene, $C_{10}H_8$, in 50.0 cm³ of benzene, C_6H_6 (density of C_6H_6 = 0.879 g/cm³).
   (a) What is the percent $C_{10}H_8$, by mass, in this solution?
   (b) What is the molal concentration of $C_{10}H_8$ in this solution?
   (c) What is the freezing point of the solution? [The freezing point of pure C_6H_6 is 5.51°C; K_f for C_6H_6 = 4.90°C kg solvent (mol solute)⁻¹.]

8. The boiling point of water at 735 mmHg is 99.07°C. What percent by mass of NaCl should be present in a water solution to raise the boiling point to 100.0°C? [K_b for water is 0.512°C kg solvent (mol solute)⁻¹.]

9. Pure liquid HCl is a poor electrical conductor. So is pure liquid water. When these two liquids are mixed, however, the resulting solution conducts electric current very well. How do you explain this?

10. How many ml of 0.25 M $MgCl_2$ must be added to 350 ml 0.25 M NaCl to produce a solution in which the concentration of chloride ion is 0.30 M Cl⁻?

13 Chemical Kinetics

The principles of stoichiometry permit us to calculate the amounts of substances that can be produced by a chemical reaction. They tell us nothing, however, about *how long* it takes for a reaction to occur. For an industrial process one might actually choose a reaction that gives a lower yield but proceeds *faster* than an alternative reaction yielding the same product. On the other hand, certain reactions that proceed extremely rapidly might not be desirable either—they might constitute explosions! And then there are circumstances where chemical reactions are unwanted. Here we prefer that whatever reaction does occur do so as slowly as possible. This is the objective of adding a rust inhibitor to the coolant in an automobile radiator or storing milk in a refrigerator.

The cases just cited suggest a need to be able to measure, control, and where possible *predict* the rates of chemical reactions. These topics are all part of the study of **chemical kinetics.** Also, chemical kinetics occasionally helps us to deduce something about the **mechanism** of a reaction. This is a detailed, step-by-step description of how initial reactants are converted to final products. Predictions of the rates of chemical reactions are based on mathematical equations called rate laws. Methods of deriving and using rate laws are the central topics of this chapter. Figure 13-1 outlines our task.

13-1 Rates of Chemical Reactions

By the **rate of a chemical reaction** we mean the rate or speed with which a reactant disappears or a product appears. More specifically we mean the rate at which the *concentration* of one of the reactants decreases or of one of the products increases with time. Commonly used units are mol/L for concentration and s, min, or h for time. A typical unit for a rate of reaction would be **mol L^{-1} s^{-1}** (mol per L per s).

For the purposes of illustration we will consider a simple classic study, the decomposition of dinitrogen pentoxide (N_2O_5) in an inert solvent (carbon tetrachloride).*

*H. Eyring and F. Daniels, *J. Amer. Chem. Soc.* **52**:1472 (1930).

Have soln, as Rt occurs the reactants w/in form products. So [ ] of reactants decrease as it "dissapears" while product is [ ] increases as it's formed.

FIGURE 13-1
The purpose of chemical kinetics.

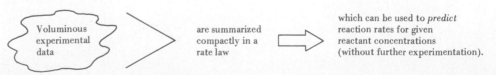

Voluminous experimental data   are summarized compactly in a rate law   which can be used to *predict* reaction rates for given reactant concentrations (without further experimentation).

FIGURE 13-2

Apparatus for determining the rate of decomposition of N_2O_5.

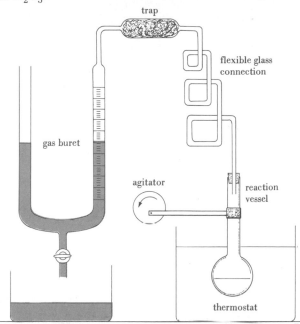

The reaction mixture is kept at a constant temperature and agitated continuously. The looped glass connection is flexible enough not to break during agitation. The oxygen gas evolved from the reaction mixture is passed through a trap and into a gas buret where its volume is measured.

TABLE 13-1

Decomposition of N_2O_5 (in CCl_4) at $45°C$—initial $[N_2O_5] = 1.40\ M$

Time, s	Total volume O_2, cm^3 (at STP)
0	0
423	1.32
753	2.18
1116	2.89
1582	3.63
1986	4.10
2343	4.46
⋮	⋮
∞[a]	5.93

[a] The symbol ∞ signifies that sufficient time has elapsed for the reaction to go to completion.

The bracket notation for concentration was introduced in the marginal note on page 302. That is, $[N_2O_5]$ represents the molar concentration of N_2O_5.

FIGURE 13-3

Decomposition of N_2O_5 (in CCl_4) at $45°C$.

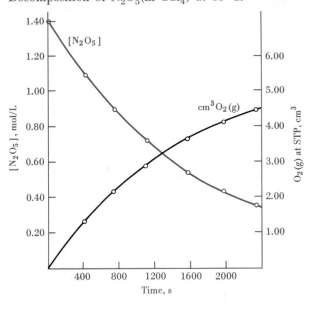

$$2\ N_2O_5(\text{in } CCl_4) \longrightarrow 2\ N_2O_4(\text{in } CCl_4) + O_2(g) \qquad (13.1)$$

One product, O_2, is a gas that is virtually insoluble in the reaction mixture. It escapes and can be collected. Figure 13-2 depicts the apparatus employed. The conditions under which the gas is collected are not important, since all gas volumes can be converted to a common standard, for example STP. Table 13-1 lists some actual data obtained in one experiment.

The data of Table 13-1 are plotted in two ways in Figure 13-3: in terms of the *formation* of $O_2(g)$ and the *disappearance* of N_2O_5. To obtain the data necessary for Figure 13-3 requires certain simple calculations. These are illustrated in Example 13-1.

Example 13-1 With reference to Table 13-1, what is $[N_2O_5]$ remaining at $t = 423$ s?

When the N_2O_5 in a solution having an initial concentration $[N_2O_5] = 1.40\ M$ decomposes completely (at time $t = \infty$, $[N_2O_5] = 0\ M$), $5.93\ cm^3\ O_2(g)$ is obtained at STP. After 423 s the volume of $O_2(g)$ collected is $1.32\ cm^3$ of a possible $5.93\ cm^3$. The fraction of the N_2O_5 decomposed is $1.32/5.93$. The *decrease* in concentration of N_2O_5 at this point is $(1.32/5.93) \times 1.40\ M = 0.312\ \text{mol } N_2O_5/L$. $[N_2O_5]$ remaining *undecomposed* after 423 s is $1.40 - 0.31 = 1.09\ M$.

SIMILAR EXAMPLES: Exercises 5a, 17a, b.

RATE OF REACTION: A VARIABLE QUANTITY. From Figure 13-3 we see that the concentration of N_2O_5 decreases with time at a rate that is initially rapid but then slows down; the reaction *decelerates*. For example, in the first 1000 s the concentration of N_2O_5 decreases from 1.40 M to about 0.80 M, a decrease of 0.60 mol N_2O_5/L. In the next 1000 s the decrease is only about 0.40 mol N_2O_5/L. Ordinarily, the reaction rate varies during the course of a reaction; we need to specify further exactly how the rate is to be expressed.

RATE OF REACTION EXPRESSED AS $-\Delta[N_2O_5]/\Delta t$. One approach to describing the rate of decomposition of N_2O_5 is to extract data from Figure 13-3 and tabulate these data (shown in color in Table 13-2). Column III of Table 13-2 lists the molar concentrations of N_2O_5 at the times shown in column I. Column II states the arbitrary time interval that we have chosen between data points—200 s. Column IV reports the changes in concentration that occur for each 200-s interval. The figures in column V represent the reaction rates; their values decrease continuously with time.

Starting with $[N_2O_5] = 1.40$ M at $t = 0$, with each succeeding time interval the concentration becomes smaller; $\Delta[N_2O_5]$ is a *negative* quantity. If the reaction rate is described as the rate of formation of a product, its concentration *increases* with time; and $\Delta[product]$, is a *positive* quantity. So that a reaction rate will always be a positive quantity we should define it as

$$\text{reaction rate} = \frac{-\Delta[\text{reactant}]}{\Delta t}$$

$$or \quad \text{reaction rate} = \frac{\Delta[\text{product}]}{\Delta t} \tag{13.2}$$

RATE OF REACTION EXPRESSED AS THE SLOPE OF A TANGENT LINE. When expressed as $-\Delta[N_2O_5]/\Delta t$, the reaction rate is simply an average value over the time interval chosen. For example, the rate of decomposition of N_2O_5 averages 2.5×10^{-4} mol L^{-1} s^{-1} in the interval from 1800 to 2000 s (see the last entry of Table 13-2). We might think of this as the reaction rate at the middle of the interval—1900. The reaction rate value would be slightly different if the concentration data were at 1850 and 1950 s, with $\Delta t = 100$ s. A unique value of the reaction rate is obtained only in the limit where the time interval is allowed to approach zero, that is, $\Delta t \to 0$. Under these circumstances the reaction rate becomes equal to the *negative of the slope of the tangent line* to the graph of $[N_2O_5]$ as a function of time. This curve has been replotted as Figure 13-4, where tangent lines are labeled.

It is better to extract points from a smooth curve drawn through experimental data points rather than to use raw data directly to construct Table 13-2. Individual data points are subject to experimental error.

TABLE 13-2
Decomposition of N_2O_5 (in CCl_4)—derived rate data

I Time, s	II Δt, s	III $[N_2O_5]$, mol/L	IV $\Delta[N_2O_5]$, mol/L	V Reaction rate $=$ $-\Delta[N_2O_5]/\Delta t$,[a] mol L^{-1} s^{-1} ($\times 10^4$)
0	200	1.40	−0.16	8.0
200	200	1.24	−0.14	7.0
400	200	1.10	−0.12	6.0
600	200	0.98	−0.11	5.5
800	200	0.87	−0.10	5.0
1000	200	0.77	−0.09	4.5
1200	200	0.68	−0.08	4.0
1400	200	0.60	−0.07	3.5
1600	200	0.53	−0.06	3.0
1800	200	0.47	−0.05	2.5
2000		0.42		

[a] The reaction rates in this column have been multiplied by 10^4. Thus, the reaction rate at time $= 0$ s is 8.0×10^{-4} mol L^{-1} s^{-1}.

In calculus notation, the ratio $-\Delta[N_2O_5]/\Delta t$ in the limit where $\Delta t \to 0$ can be replaced by the derivative $-d[N_2O_5]/dt$. That is,

$$\lim_{\Delta t \to 0} \frac{-\Delta[N_2O_5]}{\Delta t}$$

$$= \frac{-d[N_2O_5]}{dt}$$

Example 13-2 From Figure 13-4 determine the rate of decomposition of N_2O_5 at 1900 s.

FIGURE 13-4
Graphical determination of the rate of the
reaction $2 N_2O_5(\text{in } CCl_4) \rightarrow 2 N_2O_4(\text{in } CCl_4) + O_2(g)$.

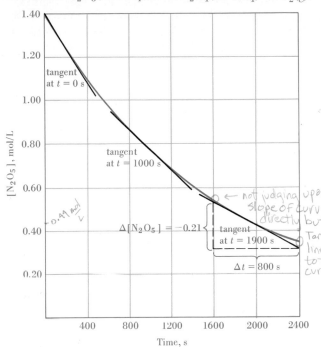

Reaction rates are determined from the slopes of the
tangent lines (see Examples 13-2 and 13-3).

[handwritten annotations:]

- $R_{rate} =$ derivative $=$ slope of tangent $= \frac{\Delta[\ ]}{\Delta t}$

For $[\ ]$ vs Time graphs $R_{rate} = \frac{-\Delta[\ ]}{\Delta t}$

using procedure in Table 13-2 to find R_{rate}, note this values can then vary according to the different intervals we base them on. The only time you get a definite, unique value for R_{rate} is if you base it as equal to the limit of the -slope of the tangent

like this you get a single unique value *another way to express R_{rate}*

If a graph of concentration of reactant vs. time is
available, a reaction rate is most readily established
through the slope of a tangent line, in this case at
$t = 1900$ s.

$$\text{reaction rate} = -\ \textit{slope of tangent} = \frac{-\Delta[N_2O_5]}{\Delta t}$$

$$= \frac{0.21 \text{ mol/L}}{800 \text{ s}} = 2.6 \times 10^{-4} \text{ mol L}^{-1}\text{ s}^{-1}$$

SIMILAR EXAMPLES: Exercises 7, 17e.

INITIAL RATE OF REACTION. To determine reaction
rates by the graphical method of Example 13-2 requires that data be collected over an extended period
of time. Occasionally, all that is required is the reaction
rate immediately after the reactants are brought together—the **initial rate of reaction.** For this one must
know the initial concentration of one reactant and its
concentration after a short time interval. In Figure
13-4 the initial reaction rate is the slope of the tangent
line at $t = 0$. The initial rate calculation works only if
the time interval chosen is so short that the tangent line
and the "concentration vs. time" curve practically coincide. In Figure 13-4 this corresponds to about the
first 200 s. Put in another way, the method usually
works over the time in which only a few percent of the
available reactant(s) are consumed.

[margin handwritten:] occasionally the only data needed is the Initial R_{rate}, The best values For the initial R_{rate} are obtained when judged over a time period in which few reactants react.

Example 13-3 For the reaction described by the data in
Table 13-2, what is the initial rate of reaction?
We need to use data for $t = 0$ and $t = 200$ s, the
shortest time interval available.

$$\text{initial rate} = \frac{-(1.24 - 1.40) \text{ mol } N_2O_5/L}{200 \text{ s}}$$

$$= 8.0 \times 10^{-4} \text{ mol } N_2O_5 \text{ L}^{-1}\text{ s}^{-1}$$

SIMILAR EXAMPLES: Exercises 5b, 17d.

13-2 The Rate Law for Chemical Reactions

For many reactions it is possible to express the reaction
rate through a mathematical equation known as the
rate law or **rate equation.** Consider the hypothetical
reaction

$$a\,A + b\,B + \cdots \longrightarrow g\,G + h\,H + \cdots \qquad (13.3)$$

where a, b, ... stand for coefficients in the balanced
equation. The reaction rate often can be expressed as

$$\text{rate} = k[A]^m[B]^n \cdots \qquad (13.4)$$

[handwritten:] $\leftarrow$ constant proportional to temp $\hookrightarrow$ molar concentration

In this expression the symbols [A], [B], ... represent

[handwritten margin notes:]

Rx rate $= \dfrac{mol}{L \cdot s}$

k is in whatever terms necessary to produce correct Rx rate terms

(careful zero-order)

"order of Rx" describe form of dependence of Rx rate upon [ ].

there are "orders of Rx" for each individual components Rx & there is an "order" of Rx for the overall Rx

The significance of the rate law expression seems to have first been recognized by the Norwegian mathematician Cato Guldberg and his brother-in-law chemist, Peter Waage. In 1865, they proposed that the force (rate) of a chemical reaction is equal to the product of the active masses (concentrations) of the reactants and an affinity coefficient (rate constant), with each active mass raised to some definite power. Moreover, they seemed to be fully aware that these definite powers were not necessarily integral numbers and not deducible from the balanced chemical equation. The Guldberg and Waage formulation is generally referred to as the **law of mass action.**

molar concentrations. The exponents m, n, . . . are generally small integral numbers, although in some cases they may be fractional or negative. It is important to note that there is *no* relationship between the exponents *m, n,* . . . and the corresponding coefficients in the balanced equation *a, b,* . . . . If in some cases they happen to be identical (e.g., $m = a$ or $n = b$), this is just a matter of chance; it is *not* to be expected.

The exponents in the rate equation are called the **order of the reaction.** If $m = 1$, the reaction is said to be *first order in A.* If $n = 2$, the reaction is *second order in B,* and so on. The total of the exponents $m + n + \cdots$ is the overall order of the reaction. The term k in equation (13.3) is called the **rate constant.** It is a proportionality constant that is characteristic of the particular reaction and is significantly dependent only on temperature. The reaction rate is usually expressed in the units *moles per liter per unit time,* for example, $mol\ L^{-1}\ s^{-1}$ or $mol\ L^{-1}\ min^{-1}$. The units of k depend on the order of the reaction.

[handwritten note:] → method to find $m, n, \ldots$ for rate equation by comparing initial Rx rates w/different concentration.

METHOD OF INITIAL RATES. This simple method of establishing the exponents in a rate equation involves measuring the initial rate of reaction for different sets of initial concentrations. The method is applied in Example 13-4 to the following reaction between peroxydisulfate and iodide ions.

$$S_2O_8{}^{2-}(aq) + 3\ I^-(aq) \longrightarrow$$
$$2\ SO_4{}^{2-}(aq) + I_3{}^-(aq) \qquad (13.5)$$

TABLE 13-3
Experimental data for the reaction
$S_2O_8{}^{2-} + 3\ I^- \longrightarrow 2\ SO_4{}^{2-} + I_3{}^-$

Experiment	Initial concentrations, M $[S_2O_8{}^{2-}]$	$[I^-]$	Initial reaction rate, mol $S_2O_8{}^{2-}$ $L^{-1}\ s^{-1}$
1	0.038	0.060	$R_1 = 1.4 \times 10^{-5}$
2	0.076	0.060	$R_2 = 2.8 \times 10^{-5}$
3	0.076	0.030	$R_3 = 1.4 \times 10^{-5}$

[handwritten notes:] each are initial Rx rates { based on S_2O_8 rate of disappearance ≠ rate of formation of SO_4

Example 13-4 The data in Table 13-3 were obtained for three reactions involving $S_2O_8{}^{2-}$ and I^-. Use these data to establish the order of reaction (13.5) with respect to $S_2O_8{}^{2-}$, the order with respect to I^-, and the overall order.

The rate equation for reaction (13.5) has the form

$$\text{reaction rate} = k[S_2O_8{}^{2-}]^m[I^-]^n \qquad (13.6)$$

Our task is to determine values of *m* and *n*. In experiments 1 and 2, $[I^-]$ is held constant and $[S_2O_8{}^{2-}]$ is increased by a factor of 2, from 0.038 to 0.076 M. (Note that we represent 0.076 as 2×0.038.) The reaction rate also increases by a factor of 2. As established in the ratios set up below, these observations require that $m = 1$.

$$R_2 = k(0.076)^m(0.060)^n = k(2 \times 0.038)^m(0.060)^n$$

$$= k(2)^m(0.038)^m(0.060)^n$$

$$= 2.8 \times 10^{-5}\ mol\ L^{-1}\ s^{-1}$$

$$R_1 = k(0.038)^m(0.060)^n = 1.4 \times 10^{-5} \text{ mol L}^{-1}\text{s}^{-1}$$

$$\frac{R_2}{R_1} = \frac{k(2)^m(0.038)^m(0.060)^n}{k(0.038)^m(0.060)^n} = 2^m = \left[\frac{2.8 \times 10^{-5}}{1.4 \times 10^{-5}} = 2\right]$$

[handwritten: 1st order]

[handwritten: say if $\frac{R_2}{R_1} = \frac{1.2 \times 10^{-5}}{1.4 \times 10^{-5}} = 8$, $\frac{R_2}{R_1}$, $m = 3$]

If $2^m = 2$, then $m = 1$. The reaction is first order in $S_2O_8^{2-}$.
    The data of experiments 2 and 3 are used to establish that $n = 1$ also.

$$R_2 = k(0.076)^m(0.060)^n = k(0.076)^m(2 \times 0.030)^n$$

$$= k(0.076)^m(2)^n(0.030)^n$$

$$= 2.8 \times 10^{-5} \text{ mol L}^{-1}\text{s}^{-1}$$

[handwritten: so... rate eightfolds upon double of concentration is 3rd order]

$$R_3 = k(0.076)^m(0.030)^n = 1.4 \times 10^{-5} \text{ mol L}^{-1}\text{s}^{-1}$$

*[margin note: If the initial reaction rate *doubles* with a doubling of initial concentration of a reactant, the reaction is first order in that reactant. If the initial rate *quadruples*, the reaction is second order in that reactant. For third-order reactions the initial rate would increase *eightfold* for a doubling of the reactant concentration.]*

$$\frac{R_2}{R_3} = \frac{k(0.076)^m(2)^n(0.030)^n}{k(0.076)^m(0.030)^n} = 2^n = \frac{2.8 \times 10^{-5}}{1.4 \times 10^{-5}} = 2$$

If $2^n = 2$, then $n = 1$. The reaction is first order in I^-. The overall order of the reaction is $m + n = 1 + 1 = 2$—second order.

SIMILAR EXAMPLES: Exercises 14, 15, 16.

Now that we have established the exponents in the peroxydisulfate–iodide rate equation (13.6), we can determine the value of the rate constant, k, as illustrated in Example 13-5.

Example 13-5 **(a)** Use the results of Example 13-4 and data from Table 13-3 to determine the value of k in the rate equation (13.6). **(b)** What is the initial rate of disappearance of $S_2O_8^{2-}$ in a reaction in which the initial concentrations are $[S_2O_8^{2-}] = 0.050\,M$ and $[I^-] = 0.025\,M$?
(a) We can use the data for any one of the three experiments of Table 13-3, together with the values $m = n = 1$. Equation (13.6) is solved for k.

[handwritten: $R_1 = k_1[A]^m[B]^n$]

$$k = \frac{R_1}{[S_2O_8^{2-}][I^-]} = \frac{1.4 \times 10^{-5} \text{ mol L}^{-1}\text{s}^{-1}}{0.038 \text{ mol/L} \times 0.060 \text{ mol/L}} = 6.1 \times 10^{-3} \text{ L mol}^{-1}\text{s}^{-1}$$

(b) Once a value of k has been established, as in part (a), the rate law can be used to predict the rate of reaction if the concentrations of reactants are known.

[handwritten: m,n,& k once found for a specific R_t, remain the same for all calculations of that R_t, even if concentrations vary, overall R_t rate same]

$$\text{rate} = k[S_2O_8^{2-}][I^-] = 6.1 \times 10^{-3} \text{ L mol}^{-1}\text{s}^{-1} \times 0.050 \text{ mol/L} \times 0.025 \text{ mol/L}$$

$$= 7.6 \times 10^{-6} \text{ mol L}^{-1}\text{s}^{-1}$$

SIMILAR EXAMPLES: Exercises 13, 14.

RATE OF REACTION BASED ON THE FORMATION OF PRODUCTS. At times the rate of a reaction is expressed in terms of the formation of a product rather than the disappearance of a reactant. If only a single reaction occurs, these two alternative forms are related through the chemical equation.

Example 13-6 What is the rate of formation of SO_4^{2-} in Experiment 1 of Table 13-3?
    From equation (13.5) we see that 2 mol SO_4^{2-} is produced for every mole of $S_2O_8^{2-}$ consumed. Thus,

$$\text{no. mol } SO_4^{2-} \text{ L}^{-1}\text{s}^{-1} = 1.4 \times 10^{-5} \text{ mol } S_2O_8^{2-} \text{ L}^{-1}\text{s}^{-1} \times \frac{2 \text{ mol } SO_4^{2-}}{1 \text{ mol } S_2O_8^{2-}}$$

$$= 2.8 \times 10^{-5} \text{ mol } SO_4^{2-} \text{ L}^{-1}\text{s}^{-1}$$

SIMILAR EXAMPLES: Exercises 6b, 21c.

13-3   Zero-Order Reactions

FIGURE 13-5
A straight-line plot for the zero-order reaction A → products.

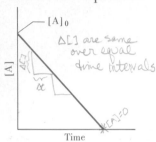

Δ[ ] are same over equal time intervals

[A] decreases from a maximum value of $[A]_0$ at time $t = 0$ to $[A] = 0$ at a time $t = [A]_0/k$. The rate constant $k = -$ (slope).

$[A]_t = -kt + [A]_0$

@ final t, $[A] = 0$ ∴ $+kt = [A]_0$

$t = [A]_0/k$

A classical case of first-order kinetics is radioactive decay, which is considered in Section 23-5.

obtained from "time/conc" graph @ pt where [ ] is a 4th mol/L

Occasionally, the rate of a reaction appears not to depend on the concentration of the reactant(s) at all. This situation is generally encountered whenever some other variable controls the rate of the reaction. This might be light intensity in a photo-chemical reaction or availability of enzyme in an enzyme-catalyzed reaction. In such cases the reaction proceeds at a *constant rate*.

$$\text{rate} = k = \text{constant} \quad =k[A]^m \quad m=0 \tag{13.7}$$

Equation (13.7) results from (13.4) if each of the exponents $m, n, \ldots$ equals zero. The reaction is **zero order**. The units of k in a zero-order reaction must be the same as those of the rate itself. Because the rate of the reaction is constant, a plot of concentration of a reactant as a function of time for a zero-order reaction is a *straight line* (see Figure 13-5). The line has a negative slope; the value of $k = -$(slope); the equation of the line is

$k = R$, rate = –slope of tang's = $\dfrac{-\Delta[A]}{\Delta t}$ for 0-order tangents ⟷ graph itself

$$[A]_t = -kt + [A]_0 \tag{13.8}$$

$mx + $ y-intercept

$[A]_0$ is the initial concentration of A and $[A]_t$ is its concentration at time t.

13-4   First-Order Reactions

The decomposition of N_2O_5 in CCl_4 is a first-order reaction in N_2O_5. This means that $[N_2O_5]$ appears in the rate equation to the first power. *another 1st order rxt could be $R = k[A]^m[B]^n$ w/ m/n = 1 fractions*

$$\text{rate of disappearance of } N_2O_5 = k[N_2O_5]$$

From the measured rate of reaction at a particular $[N_2O_5]$, it is possible to calculate the rate constant, k.

Example 13-7   When $[N_2O_5] = 0.44\,M$, the rate of decomposition of N_2O_5 is 2.6×10^{-4} mol L^{-1} s^{-1}. What is the value of k for this first-order reaction?

We rearrange the rate equation and solve for k.

$$k = \frac{\text{rate of reaction}}{[N_2O_5]} = \frac{2.6 \times 10^{-4} \text{ mol L}^{-1}\text{ s}^{-1}}{0.44 \text{ mol/L}} = 5.9 \times 10^{-4}\text{ s}^{-1}$$

SIMILAR EXAMPLE: Exercise 17f.

The method used to determine k in Example 13-7 is deceptively simple. The problem is twofold: How do we know that the reaction is first order? How do we determine the rate of the reaction at any given point without making additional measurements at other points? You may recognize the data in Example 13-7 as being based on the tangent line in Figure 13-4. This line can be drawn only after a significant portion of the concentration vs. time graph is plotted. This requires a number of experimental data points.

Consider a hypothetical reaction that is first order in A: A → B + C. The rate law for this reaction is

$$\text{rate of disappearance of } A = k[A]$$

This rate law can be put into a more useful form using the integration concept from calculus. We will forgo showing how this is done, for we are interested only in the result, which is

variables

$$\log [A]_t - \log [A]_0 = \log \frac{[A]_t}{[A]_0} = \frac{-kt}{2.303} \tag{13.9}$$

Remember: Rx rate changes thru-out
a single rx - you can even get
varying Rx rates for a
single conc if you vary
the intervals of Δ[] &
Δt. You can, however,
can a single unique
value for the Rx rate
@ a certain pt if you
let Rx rate = derivative
@ certain   slope of tangent
conc. pt.

since [] changes thru-out on Rx

TABLE 13-4

Decomposition of N_2O_5 (in CCl_4) at 45°C—data to test for first-order reaction

Time, s	$[N_2O_5]$, mol/L	$\log [N_2O_5]$
0	1.40	0.146[a]
400	1.10	0.041
800	0.87	−0.060
1200	0.68	−0.17
1600	0.53	−0.28
2000	0.42	−0.38

[a]This is the value of $\log [N_2O_5]_0$.

FIGURE 13-6

Test for a first-order reaction: decomposition of N_2O_5 (in CCl_4) at 45°C.

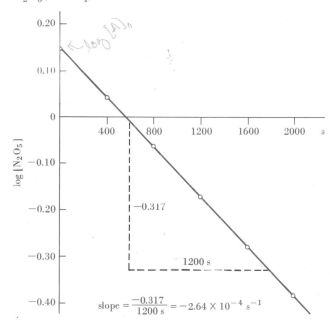

slope $= \dfrac{-0.317}{1200\ s} = -2.64 \times 10^{-4}\ s^{-1}$

In equation (13.9) the term $[A]_t$ represents the concentration of A at some time t. $[A]_0$ represents the initial concentration of A (at time $t = 0$). k is the rate constant for the reaction. Only the time t and the corresponding concentration, $[A]_t$, are variables in equation (13.9). Rearranging (13.9) slightly shows that it is the equation of a straight line if $\log [A]_t$ is plotted as a function of t.

$$\log [A]_t = -\left(\frac{k}{2.303}\right) t + \log [A]_0 \quad (13.10)$$

equation of straight line: $y = m \cdot x + b$

The rate data in Table 13-4 are derived from Table 13-2, and these data in turn are plotted in Figure 13-6 in the form $\log [N_2O_5]$ versus time. The result is a straight line! This is how we establish that the decomposition of N_2O_5 is a first-order reaction. Moreover, the value of the rate constant k can be derived from the slope of the line.

decreasing slope → looking @ decomposition

$$-\left(\frac{k}{2.303}\right) = m = \text{slope} = -2.64 \times 10^{-4}\ s^{-1}$$

(if were looking @ increase of products, then + m)

$$k = 2.303(2.64 \times 10^{-4}\ s^{-1}) = 6.08 \times 10^{-4}\ s^{-1}$$

The definitive way to describe the rate of a reaction is through the rate constant k and the order of the reaction. Calculations of the type shown in Example 13-8 then become possible.

Example 13-8 N_2O_5, initially at a concentration of 1.0 mol/L in CCl_4, is allowed to decompose at 45°C. At what time will $[N_2O_5]$ be reduced to 0.50 M?

The appropriate values must be substituted into equation (13.10).

$$\log [A]_0 = \log [N_2O_5]_0 = \log 1.0 = 0$$
$$\log [A]_t = \log [N_2O_5]_t = \log 0.50 = \log (5.0 \times 10^{-1})$$
$$= \log 5.0 + \log 10^{-1} = 0.6990 - 1.0000$$
$$= -0.30$$
$$k = 6.1 \times 10^{-4}\ s^{-1}$$
$$t = ?$$

$$\log [A]_t = -\left(\frac{k}{2.303}\right) t + \log [A]_0$$

$$-0.30 = -\frac{(6.1 \times 10^{-4}\ s^{-1})\, t}{2.303}$$

$$t = \frac{2.303 \times 0.30}{6.1 \times 10^{-4}\ s^{-1}} = 1.1 \times 10^3\ s$$

SIMILAR EXAMPLES: Exercises 20, 22.

The time calculated in Example 13-8 is known as the **half-life** for the reaction. This is the time required for the concentration of N_2O_5 to decrease to one half of some previous value. *If a reaction is first order, the half-life depends only on k; therefore, it is a constant.* If the values $[A]_t = \frac{1}{2}[A]_0$ and $t = t_{1/2}$ are substituted into equation (13.9), expressions (13.11) and (13.12) are obtained.

(handwritten: inverse)

(handwritten: $\log \frac{1}{2} = \frac{\log \frac{1}{2}[A]_0}{[A]_0} = \frac{-kt}{2.303}$ ← means sign change)

$$\log 2 = \frac{+kt_{1/2}}{2.303} = 0.3010 \tag{13.11}$$

$$t_{1/2} = \frac{0.693}{k} \tag{13.12}$$

In Example 13-8 the time required for $[N_2O_5]$ to be reduced to 0.25 M would be $(1100 + 1100)$ s; to 0.125 M, $(1100 + 1100 + 1100)$ s; and so on. The constancy of a half-life can be used as a test for a first-order reaction. (The half-life is not constant for reactions of other orders.) Furthermore, this test can be applied on a simple plot of concentration against time. (Try this with Figure 13-3. That is, starting with $[N_2O_5] = 1.40$ M at $t = 0$ s, at what time is $[N_2O_5] \simeq 0.70$ M? $[N_2O_5] \simeq 0.35$ M?)

(handwritten: to check if Rt is 1st order)

(handwritten: → solve for k using 2 dif $[A]_t$ & $[A]_0$, then $t_{1/2} = \frac{0.693}{k}$)

REACTIONS INVOLVING GASES. In chemical reactions involving gases, the reaction rate is sometimes measured in terms of increases or decreases in gas pressure. It is not difficult to derive equations similar to (13.9) and (13.10) for first-order gas-phase reactions. Consider a reaction $A(g) \rightarrow$ products that is known to be first order. The concentration of A on a mol/L basis can be expressed in terms of the partial pressure of A, as follows.

(handwritten: as gas A reacts, a lessening amt exists in mixture & thus pressure changes)

$$P_A V = n_A RT \qquad [A] = \frac{n_A}{V} = \frac{P_A}{RT}$$

By substituting this expression in equation (13.9) we obtain

(handwritten: $t_{1/2} = \frac{0.693}{k}$ applies for gases also)

$$\log \frac{[A]_t}{[A]_0} = \log \frac{(P_A)_t/RT}{(P_A)_0/RT} = \log \frac{(P_A)_t}{(P_A)_0} = \frac{-kt}{2.303} \tag{13.13}$$

and

$$\log (P_A)_t = \frac{-k}{2.303}t + \log (P_A)_0 \tag{13.14}$$

Equation (13.12), which relates the rate constant and the half-life of a first-order reaction, is independent of how the reaction rate is described. That is, the rate can be in terms of molar concentration or gas pressure (or simply in terms of the mass of a reactant, for that matter).

The first-order decomposition of di-*t*-butyl peroxide (DTBP) to acetone and ethane is represented by the equation

$$C_8H_{18}O_2(g) \longrightarrow 2\,C_3H_6O(g) + C_2H_6(g) \tag{13.15}$$

The partial pressure of DTBP as a function of time is plotted in Figure 13-7.

Example 13-9 Reaction (13.15) has a half-life of 80 min at 147°C. A reaction is started with pure DTBP at 147°C and 800 mmHg in a flask of constant volume. **(a)** At what time will the partial pressure of DTBP be 100 mmHg? **(b)** At what time will the partial pressure of DTBP be 700 mmHg **(c)** What will be the *total* gas pressure when the partial pressure of DTBP is 700 mmHg?

FIGURE 13-7
Decomposition of di-*t*-butyl peroxide (DTBP).

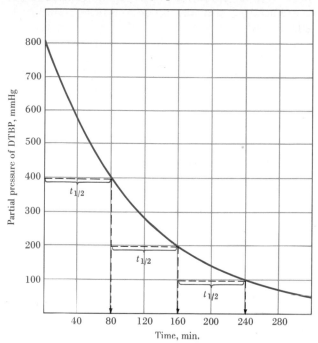

The rate of a gas-phase reaction can be followed by measuring the partial pressure of a gaseous species. The decomposition of di-*t*-butyl peroxide (DTBP) is described through equation (13.15). Three successive half-life intervals of 80 min each are indicated.

(a) In one half-life period the number of moles of DTBP in the reaction flask must decrease to one half the initial amount. This means that the partial pressure of DTBP also must decrease to one half its original value—to 400 mmHg. In a second half-life period the partial pressure falls to 200 mmHg, and in a third, to 100 mmHg. The time required for the partial pressure of DTBP to decrease from 800 to 100 mmHg is $3 \times t_{1/2} = 3 \times 80 \text{ min} = 240 \text{ min}$ (see Figure 13-7).

(b) Although we could estimate an answer from Figure 13-7, a more exact result can be obtained by using equation (13.14). Substitute partial pressures of DTBP for P_0 and P_t. The rate constant k has a value of $k = 0.693/t_{1/2} = 0.693/80 \text{ min} = 8.66 \times 10^{-3} \text{ min}^{-1}$.

$$\log 700 = \frac{-(8.66 \times 10^{-3}) \text{ min}^{-1} t}{2.303} + \log 800$$

$$\log 700 - \log 800 = \frac{-(8.66 \times 10^{-3}) \text{ min}^{-1} t}{2.303}$$

$$2.845 - 2.903 = -0.058 = \frac{-(8.66 \times 10^{-3}) \text{ min}^{-1} t}{2.303}$$

$$t = \frac{2.303 \times 0.058}{8.66 \times 10^{-3} \text{ min}^{-1}} = 15 \text{ min}$$

(c) Three moles of gaseous products appear for every mole of reactant consumed. A sufficient amount of reactant is consumed to cause a decrease in its partial pressure of 100 mmHg (i.e., from 800 mmHg to 700 mmHg). Products are formed in a sufficient amount to produce a partial pressure of 300 mmHg.

$$\text{total pressure} = P_{\text{DTBP}} + P_{\text{prod.}}$$
$$= 700 + 300 = 1000 \text{ mmHg}$$

SIMILAR EXAMPLES: Exercises 24, 25.

13-5 Second-Order Reactions

If the hypothetical reaction

$$A \longrightarrow B + C \tag{13.16}$$

is second order in A, this means that the rate law is

$$\text{rate of disappearance of A} = k[A]^2 \tag{13.17}$$

If the hypothetical reaction

$$A + B \longrightarrow C + D \tag{13.18}$$

is first order in A and first order in B, the overall order is second, and

$$\begin{array}{c}\text{rate of disappearance of A} \\ \text{(or of B)}\end{array} = k[A][B] \tag{13.19}$$

However, these facts about the reaction order *cannot* be deduced from the balanced equation. Reaction order can be determined *only from experimental rate data.* How must these data be treated to indicate if a reaction is second order? The situation is much more complex than with first-order reactions, and we limit ourselves to a single case: a reaction involving a *single* reactant and following the rate equation (13.17).

As in the case of first-order reactions, converting equation (13.17) to a more useful form requires the use of integration, a calculus procedure that is beyond the scope of this text. The result obtained has the form

$$\frac{1}{[A]_t} - \frac{1}{[A]_0} = kt$$

[handwritten: $R = k[A]^2$]

This can be rearranged into a more readily recognizable expression of a straight-line graph.

$$\frac{1}{[A]_t} = k \cdot t + \frac{1}{[A]_0}$$

(13.20)

equation of straight line:

$$y = mx + b$$

FIGURE 13-8
A straight-line plot for the second-order reaction   A → products.

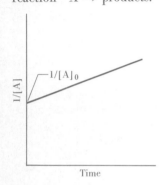

The reciprocal of the concentration, $1/[A]$, has its lowest value at the start of the reaction. As the reaction proceeds, $[A]$ decreases and $1/[A]$ increases, in a straight-line fashion. The slope of the line is the rate constant k.

A plot of $1/[A]_t$ as a function of time yields a straight line with a slope of k (see Figure 13-8). Each of the three terms in equation (13.20) must have the same units—L/mol. Since the product kt has the units L/mol, the units of k must be $L\,mol^{-1}\,(time)^{-1}$, that is, $L\,mol^{-1}\,s^{-1}$, $L\,mol^{-1}\,min^{-1}$, and so on. To establish the half-life for a second-order reaction of the type we have been describing, we substitute $t = t_{1/2}$ and $[A]_t = \frac{1}{2}[A]_0$ into equation (13.20).

$$\frac{1}{\frac{1}{2}[A]_0} = \frac{2}{[A]_0} = kt_{1/2} + \frac{1}{[A]_0}$$

$$kt_{1/2} = \frac{2}{[A]_0} - \frac{1}{[A]_0} = \frac{1}{[A]_0}$$

$$t_{1/2} = \frac{1}{k[A]_0}$$

(13.21)

The half-life for the reaction is not a constant. Its value depends on the initial concentration of reactant for every half-life interval.

Example 13-10 The data listed in Table 13-5 were obtained for the decomposition reaction: A → 2 B + C. **(a)** Establish the order of the reaction. **(b)** What is the rate constant k? **(c)** What is the half-life, $t_{1/2}$, if the initial $[A] = 1.00\,M$?
(a) The order of the reaction can be established by a simple graphical method. Plot the following three graphs.
　1. $[A]$ vs. time. (If a straight line, reaction is zero order.)
　2. $\log [A]$ vs. time. (If a straight line, reaction is first order.)
　3. $1/[A]$ vs. time. (If a straight line, reaction is second order.)
　These graphs are shown in Figure 13-9; the reaction is second order.
(b) The slope of graph 3 in Figure 13-9 is

$$k = \frac{(4.00 - 1.00)\ \text{L/mol}}{25\ \text{min}} = 0.12\ \text{L mol}^{-1}\ \text{min}^{-1}$$

(c) According to equation (13.21)

$$t_{1/2} = \frac{1}{k[A]_0} = \frac{1}{0.12\ \text{L mol}^{-1}\ \text{min}^{-1} \times 1.00\ \text{mol/L}} = 8.3\ \text{min}$$

SIMILAR EXAMPLES: Exercises 9, 10, 11.

TABLE 13-5
Kinetic data for Example 13-10

Time, min	[A], M	log [A]	1/[A]
0	1.00	0.00	1.00
5	0.63	−0.20	1.59
10	0.46	−0.34	2.17
15	0.36	−0.44	2.78
25	0.25	−0.60	4.00

FIGURE 13-9
Testing for the order of a reaction—Example 13-10 illustrated.

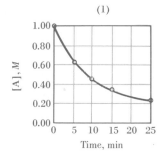

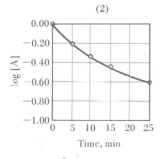

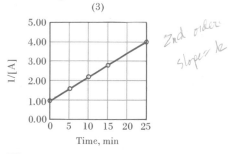

Really 2nd order; so much of one of reactants that you notice no concentration change & Rate appears dependent upon a single concentration

2nd order; slope = k

The straight-line plot is that shown in (3). The reaction is second order.

PSEUDO-FIRST-ORDER REACTIONS. The following reaction, the hydrolysis of ethyl acetate ($CH_3COOC_2H_5$) to acetic acid and ethanol, is second order:

$$CH_3COOC_2H_5 + H_2O \longrightarrow$$
$$CH_3COOH + C_2H_5OH \qquad (13.22)$$

rate of reaction $= k[CH_3COOC_2H_5][H_2O]$

However, suppose that we follow the hydrolysis of 1 L of an aqueous 0.01 M ethyl acetate solution to completion. The concentration of the ethyl acetate decreases from 0.01 M to essentially zero; 0.01 mol of ethyl acetate is consumed, and so is 0.01 mol H_2O. But now consider the molarity of water in this solution. One liter of the original solution contains approximately 1000 g H_2O, which is $1000/18 = 55$ mol H_2O. The original $[H_2O] \simeq 55$ M. After completion of the reaction, because only 0.01 mol H_2O is consumed, $[H_2O]$ is still about 55 M. During the reaction $[H_2O]$ remains essentially constant, so the rate of reaction appears independent of $[H_2O]$. This is the same result that would be observed if the reaction were zero order in H_2O. We can say that the reaction is pseudo-zero order in H_2O, first order in $CH_3COOC_2H_5$, and pseudo-first order overall. The reaction can be treated by the same methods presented for true first-order reactions. Thus, some second-order reactions can be converted to pseudo-first-order reactions if one of the reactants is present in overwhelming excess.

13-6 Collision Theory of Chemical Reactions

The statement that *chemical reactions occur as a result of collisions between molecules* may seem so obvious as not to require special mention. Yet a theory to explain chemical reactions based on molecular collisions did not become firmly established until the early decades of the twentieth century. The kinetic-molecular theory of gases had to be developed first.

Recall for a moment a number of ideas established in our discussion of the kinetic-molecular theory in Section 5-10. We noted that there is a distribution of kinetic energies and velocities of the molecules of a gaseous substance. Also, we were able to write an expression for the average molecular speed. Although beyond the scope of this text to do so, the number of collisions between molecules per unit time can be derived from kinetic-molecular theory. This is called the **collision frequency.**

For example, it can be shown that in a 1-L vessel containing 0.01 mol of a gaseous substance of molecu-

FIGURE 13-10
Distribution of molecular energies.

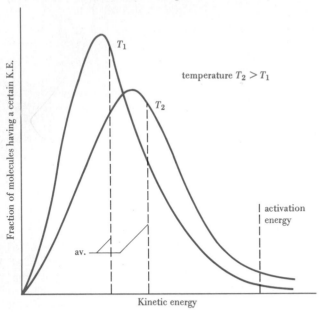

At the higher temperature, T_2, the distribution of energies is broadened; the average molecular kinetic energy increases; and many more molecules possess energies greater than the activation energy.

FIGURE 13-11
Molecular collisions and chemical reactions.

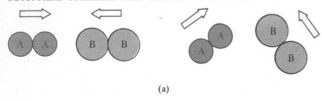

(a)

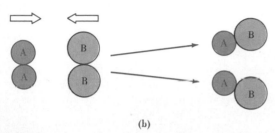

(b)

(a) Unfavorable collisions for chemical reaction.
(b) A favorable collision for chemical reaction.

lar weight 100 at approximately room temperature, the frequency of molecular collisions is of the order of 10^{30} collisions s^{-1}. If each of these collisions led to chemical reaction, the rate of reaction would be of the order of 10^6 mol L^{-1} s^{-1}.

$$\frac{10^{30} \text{ collisions } L^{-1} s^{-1}}{6.02 \times 10^{23} \text{ collisions/mol}} \simeq 10^6 \text{ mol } L^{-1} s^{-1} \quad (13.23)$$

The rate of reaction suggested by equation (13.23) is orders of magnitude greater than what is normally observed for chemical reactions. It is greater than the rate of decomposition of N_2O_5 by a factor of 10^{10}! Clearly, not all collisions between molecules can lead to chemical reaction (which is fortunate; otherwise, reactions of the living state would take place with astonishing speed and life would be very short). The observation that only a small fraction of molecular collisions are effective in producing chemical reaction is based on two factors: (1) Only the more energetic molecules in a mixture undergo reaction as a result of collisions. (2) The probability of a particular collision resulting in chemical reaction depends on the orientation of the colliding molecules.

The energy that molecules must possess in order to react is called the **activation energy.** With the kinetic-molecular theory it is possible to establish what fraction of all the molecules in a collection possess energies in excess of any particular value. A hypothetical activation energy and the fraction of molecules possessing energies in excess of this value are indicated on a distribution curve of molecular energies in Figure 13-10. Think of the rate of a chemical reaction as depending on the product of the collision frequency *and* the fraction of molecules with energies equal to or exceeding the activation energy. Because this fraction is usually so small, the rate of a reaction is much smaller than just the collision frequency itself. Moreover, the *higher* the activation energy, the *smaller* the fraction of activated molecules and the more slowly a reaction proceeds.

To visualize the reaction

$$A_2(g) + B_2(g) \longrightarrow 2 \, AB(g) \quad (13.24)$$

in terms of collision theory, assume that during a collision between a molecule of A_2 and B_2 the bonds A—A and B—B break and the bonds A—B form. The result is the conversion of the reactants A_2 and B_2 to the product AB. But as pictured in Figure 13-11, this assumption cannot hold for every collision. Molecules must have a particular orientation if the collision is to be effective in producing chemical reaction. As suggested in Figure 13-11, the number of unfavorable collision orientations for reaction generally exceeds the number of favorable ones. This means that the probability of a particular collision being favorable to reaction is usually small (i.e., the probability is much less than 1).

If we denote collision frequency as Z, the fraction of activated molecules as f, and the probability factor as p, the rate of a chemical reaction has the form

$$\text{rate of reaction} = p \cdot f \cdot Z \qquad (13.25)$$

Collision frequency is proportional to the concentrations of the molecular species involved in collisions (say that these species are A and B). That is, Z can be replaced by $[A] \times [B]$, and this more familiar rate expression can be written

$$\text{rate of reaction} \propto pf[A][B] = k[A][B] \qquad (13.26)$$

The collision theory thus seems to lead to a general rate equation for chemical reactions, but there are some serious shortcomings to the result we have just stated. Equation (13.26) describes a reaction that is second-order overall, yet we know that other reaction orders are possible. At times, the probability factor is much smaller than can be explained just in terms of molecular orientations. In a few cases it is very large, that is, much greater than 1. The theoretical basis of chemical kinetics must involve more than just the concepts associated with simple collision theory.

13-7 Transition State Theory

An important alternative to collision theory has been developed by the American chemist Henry Eyring (1901–   ), and others. It focuses on a hypothesized intermediate species, called an **activated complex,** that forms during an energetic collision. This species exists very briefly, and then dissociates—either back to the original reactants (in which case there is no reaction) or to product molecules. The activated complex for reaction (13.24) might be represented as follows.

$$
\begin{array}{ccccc}
\text{A} & \text{B} & \text{A} \cdots \text{B} & & \text{A}\!-\!\text{B} \\
| & + \; | & \vdots \quad \; \vdots & \longrightarrow & \\
\text{A} & \text{B} & \text{A} \cdots \text{B} & & \text{A}\!-\!\text{B}
\end{array}
\qquad (13.27)
$$

$$\underset{\text{reactants}}{} \qquad \underset{\substack{\text{activated}\\\text{complex}}}{} \qquad \underset{\text{products}}{}$$

An activated complex has old bonds stretched to the breaking point and new bonds only partially formed. Only if colliding molecules possess a large quantity of kinetic energy to invest in producing this strained species can an activated complex form. The energy required is the activation energy.

Another way of looking at activation energy is presented in Figure 13-12, which is a **reaction profile** for the reaction of $H_2(g)$ and $I_2(g)$ to form $HI(g)$. In Figure

Greater concentration more molecules higher frequency

$E_a = \dfrac{kJ}{mol}$

FIGURE 13-12
Energy profile for the reaction $H_2(g) + I_2(g) \rightarrow 2\,HI(g)$.

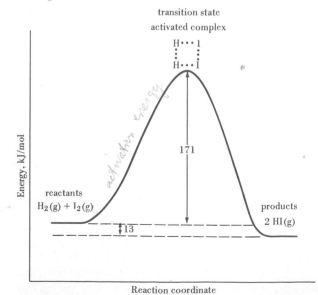

It may be helpful to think of activation energy in terms of this analogy: Imagine that a hike is being planned between two mountain valleys separated by a ridge through which there is a pass. The elevations of the valleys correspond to the energies of the reactants and products and the elevation of the pass corresponds to the activation energy. The ease or difficulty of the hike is not so much a matter of the difference in the elevation of the two valleys as it is of the elevation of the pass above the starting point. If the climb from the starting point to the pass is very long and steep, not too many individuals may care to take the hike, no matter how much "downhill" there is on the other side.

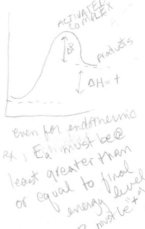

(handwritten annotations): ACTIVATED COMPLEX — products — ΔH = + — even for endothermic rx, Ea must be @ least greater than or equal to final energy level (ΔP must be "+")

TABLE 13-6
Specific rate constant, k, at several temperatures for the reaction
$2\,N_2O_5(\text{in } CCl_4) \longrightarrow 2\,N_2O_4(\text{in } CCl_4) + O_2(g)$

t, °C	T, K	$1/T$, K^{-1}	k, s^{-1}	$\log k$
0	273.2	3.66×10^{-3}	7.87×10^{-7}	-6.10
25	298.2	3.35×10^{-3}	3.46×10^{-5}	-4.46
35	308.2	3.24×10^{-3}	1.35×10^{-4}	-3.87
45	318.2	3.14×10^{-3}	4.98×10^{-4}	-3.30
55	328.2	3.05×10^{-3}	1.50×10^{-3}	-2.82
65	338.2	2.96×10^{-3}	4.87×10^{-3}	-2.31

13-12 energies of the species involved are plotted on the vertical axis and a quantity called the reaction coordinate, on the horizontal axis. The reaction coordinate can be thought of as representing the extent of the reaction. That is, the reaction starts with reactants on the left, passes through a transition state (activated complex), and ends with products on the right. The difference in energies between the reactants and products is ΔH for the reaction. The formation of HI(g) is a slightly exothermic reaction. The energy of the activated complex—the activation energy—is 171 kJ/mol greater than that of the reactants. Thus, a large energy barrier separates the reactants from the products. Only especially energetic reactant molecules can cross this barrier (by forming an activated complex which then dissociates into product molecules).

Figure 13-12 can also be used to describe the reverse process, the decomposition of HI(g) into $H_2(g)$ and $I_2(g)$. The activation energy for the reverse reaction is 184 kJ/mol. Figure 13-12 suggests the following important relationship between activation energy, E_a, and enthalpy (heat) of reaction, ΔH.

$$\Delta H = E_a(\text{forward}) - E_a(\text{reverse}) \qquad (13.28)$$

for HI formation: $\Delta H = 171\,\text{kJ/mol} - 184\,\text{kJ/mol} = -13\,\text{kJ/mol}$

Also, it should be apparent that for an *endothermic* reaction the activation energy must be equal to or exceed the endothermic heat of reaction (and usually it is greater).

Attempts at purely theoretical predictions of reaction rate constants using the collision or transition state theories have not been very successful. The principal value of these theories is in providing concepts to facilitate the discussion of experimentally observed reaction rate data. In the next section we see how the concept of activation energy enters into a discussion of the effect of temperature on reaction rates.

13-8 The Effect of Temperature on Reaction Rates— The Arrhenius Equation

As a practical matter we know that chemical reactions tend to go faster at higher temperatures. We speed up certain biochemical reactions by raising the temperature, for example in cooking foods. On the other hand, we slow down some reactions by lowering the temperature, such as in refrigerating or freezing cooked foods to prevent spoilage. But now we have an explanation for the profound effect of temperature on reaction rates: *Increasing the temperature increases the fraction*

Collision frequency increases with temperature and we might expect this also to be a factor in speeding up a chemical reaction. However, collision frequency (which is proportional to $\sqrt{T}$) accounts for less than 1% of the increase described here.

Remember, R.Rate $= \left(\%\ \text{molecules w}\right)\left(E > E_a\right)\left(\begin{smallmatrix}\text{collision}\\ \text{frequency}\end{smallmatrix}\right)$

of the molecules that have energies in excess of the activation energy (recall Figure 13-10). This factor is so important that for many chemical reactions it can lead to a doubling or tripling of the reaction rate for a temperature increase of only 10°C. Data to illustrate the effect of temperature on the decomposition of N_2O_5 are presented in Table 13-6.

Even without plotting the data in Table 13-6, we see that a graph of rate constant k against temperature would increase sharply with temperature. The graph would not be linear. This situation is reminiscent of that encountered with vapor pressure. In that case a steeply rising curve (Figure 11-6) was converted to a straight line (Figure 11-9) by plotting log P vs. $1/T$. Let us try a similar plot in this instance, that is, log k vs. $1/T$. The necessary data are given in Table 13-6 and plotted in Figure 13-13. The graph is indeed linear! Its equation has the form

equation of straight line:
$$\log k = \underbrace{\left(\frac{-E_a}{2.303 R}\right)}_{m}\ \underbrace{\frac{1}{T}}_{x}\ +\ \underbrace{A}_{b} \quad (13.29)$$
$$\underbrace{}_{y} = m \cdot x + b$$

Equation (13.29) can be used for a variety of purposes. One is to establish the activation energy of a reaction graphically, as shown in Figure 13-13. Variations of this equation are also frequently encountered. For example, in the manner outlined in the marginal note on page 261, the constant A can be eliminated from equation (13.29) to obtain the following useful equation.

$$\log \frac{k_2}{k_1} = \frac{E_a}{2.303 R}\left(\frac{T_2 - T_1}{T_2 T_1}\right) \quad (13.30)$$

$= \frac{-E_a}{2.303R}\left(\frac{1}{T_2} - \frac{1}{T_1}\right)$

$= \frac{+E_a}{2.303R}\left(\frac{1}{T_1} - \frac{1}{T_2}\right)$

In equation (13.30) T_2 and T_1 are two kelvin temperatures. k_2 and k_1 are the rate constants at these temperatures. E_a is the activation energy in J/mol. R is the gas constant, expressed as 8.314 J mol⁻¹ K⁻¹.

Example 13-11 Use data from Table 13-6 and Figure 13-13 to determine the temperature at which the half-life for the decomposition of N_2O_5 is 2 h.

Since both Table 13-6 and Figure 13-13 are based on rate constants, k, our first step must be to convert the half-life of 2 h to a corresponding value of k. For a first-order reaction this is done by using equation (13.12).

$$k = \frac{0.693}{t_{1/2}} = \frac{0.693}{2\ \text{h}} = \frac{0.693}{7200\ \text{s}} = 9.62 \times 10^{-5}\ \text{s}^{-1}$$

We may now proceed in either of two ways.
Graphical method:
We are seeking the temperature at which $k = 9.62 \times 10^{-5}$ and $\log k = \log 9.62 \times 10^{-5} = -4.02$. This point can be located directly on the straight-line

FIGURE 13-13
Temperature dependence of the rate constant k for the reaction $2\ N_2O_5(\text{in } CCl_4) \rightarrow 2\ N_2O_4(\text{in } CCl_4) + O_2(g)$.

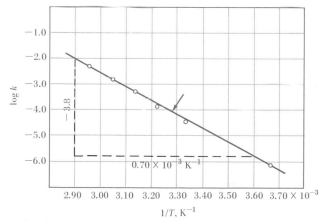

This graph can be used to establish the activation energy E_a for the reaction (see equation 13.29).

$$\text{slope of line} = \frac{-E_a}{2.303\ R} = \frac{-3.8}{0.70 \times 10^{-3}\ \text{K}^{-1}}$$
$$= -5.4 \times 10^3\ \text{K}$$

$E_a = 2.303 \times 8.314\ \text{J mol}^{-1}\ \text{K}^{-1} \times 5.4 \times 10^3\ \text{K}$
$= 1.0 \times 10^5\ \text{J/mol} = 1.0 \times 10^2\ \text{kJ/mol}$

The colored arrow is referred to in Example 13-11.

FIGURE 13-14
An example of homogeneous catalysis.

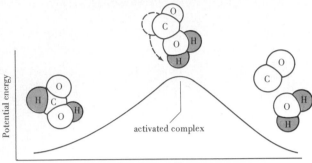

(a) Uncatalyzed Reaction

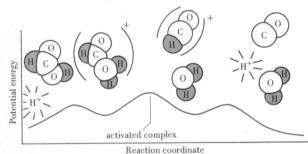

(b) Catalyzed Reaction

The potential energy of the activated complex—the activation energy—is lowered in the presence of a catalyst. [From: George C. Pimentel, ed., *Chemistry: An Experimental Science.* Freeman, San Francisco, 1963. Reproduced by permission of the Chemical Education Material Study.]

so that more molecules have the sufficient activation energy.

To continue the analogy introduced in the marginal note on page 328, a catalyst is like a guide who can ease the climb for a party of hikers by showing them an easier route (less steep) to their objective.

graph of Figure 13-13 (marked by the colored arrow). Corresponding to $\log k = -4.02$, $1/T = 3.28 \times 10^{-3}\,\text{K}^{-1}$; $T = (1/3.28 \times 10^{-3})\,\text{K} = 305\,\text{K} = 32°\text{C}$.
With equation (13.30):

Suppose that we denote as T_2 the temperature at which $k = 9.62 \times 10^{-5}\,\text{s}^{-1}$. For T_1, we must use some other temperature at which k is known. Suppose that we take $T_1 = 298.2\,\text{K}$ and $k_1 = 3.46 \times 10^{-5}\,\text{s}^{-1}$. The activation energy is $1.0 \times 10^5\,\text{J/mol}$. Solve equation (13.30) for T_2. (For simplicity units have been omitted below. The temperature is obtained in kelvins.)

$$\log \frac{k_2}{k_1} = \log \frac{9.62 \times 10^{-5}}{3.46 \times 10^{-5}} = \log 2.78 = 0.444$$

$$= \frac{1.0 \times 10^5}{2.303 \times 8.314}\left(\frac{T_2 - 298.2}{298.2 \times T_2}\right)$$

$$0.444 = 5.2 \times 10^3 \left(\frac{T_2 - 298.2}{298.2 \times T_2}\right)$$

$$132\,T_2 = 5.2 \times 10^3\,T_2 - 1.6 \times 10^6$$

$$5.1 \times 10^3\,T_2 = 1.6 \times 10^6$$
$$T_2 = 3.1 \times 10^2\,\text{K}$$

SIMILAR EXAMPLES: Exercises 29, 30, 31.

13-9 Catalysis

A reaction can be speeded up by increasing the fraction of the molecules that have energies in excess of the activation energy. Raising the temperature is one way to increase this fraction. Another way not requiring an increase in temperature is to find an alternate reaction pathway of *lower* activation energy. The function of a catalyst in a chemical reaction is to provide this alternate pathway. A catalyst enters into a chemical reaction in such a way that it undergoes no permanent change. As a result its formula does not appear in the net chemical equation (although its presence is generally indicated by an appropriate notation above the arrow sign). The success or failure of a commercial process for producing a substance often hinges on finding appropriate catalysts for the reactions involved. The ranges of temperatures and pressures that can be employed in industrial processes is not possible with biochemical reactions. The availability of appropriate catalysts for these reactions is absolutely crucial to the existence of living matter.

MECHANISM OF CATALYSIS. Figure 13-14 represents the decomposition of formic acid (HCOOH). In the uncatalyzed reaction a hydrogen atom must be transferred from one part of the formic acid molecule to another before the breaking of a C—O bond can occur.

The energy requirement for this atom transfer is large, resulting in a high activation energy and a slow reaction.

The acid-catalyzed decomposition of formic acid can be represented by

$$H-\overset{\overset{\displaystyle O}{\parallel}}{C}-O-H \xrightarrow{H^+} H_2O + CO \qquad (13.31)$$

In this reaction a hydrogen ion from solution attaches itself to the oxygen atom that is singly bonded to the carbon atom. The activated complex $(HCOOH_2)^+$ is formed. The C—O bond ruptures, and a hydrogen atom attached to a *carbon* atom in the intermediate species $(HCO)^+$ is released to the solution as a hydrogen ion. This reaction pathway does not require an atom transfer within the activated complex. Thus, it has a substantially lower activation energy than does the uncatalyzed reaction; it proceeds at a faster rate.

In the acid-catalyzed decomposition of formic acid the reactant and catalyst are present in a single phase. This type of catalysis is known as **homogeneous catalysis.** If the reaction of ethylene and hydrogen to form ethane gas is attempted in the gaseous state, the reaction rate is extremely low. The probable activated complex for this reaction is a four-membered cyclic ring, a structure of very high energy.

$$\begin{matrix} H \\ | \\ H \end{matrix} + \begin{matrix} CH_2 \\ \| \\ CH_2 \end{matrix} \longrightarrow \begin{matrix} H\cdots CH_2 \\ \vdots \quad \vdots \\ H\cdots CH_2 \end{matrix} \longrightarrow \begin{matrix} H-CH_2 \\ | \\ H-CH_2 \end{matrix} \qquad (13.32)$$

There is no practical way to catalyze this reaction homogeneously in the gaseous state. A different method of catalysis is required.

HETEROGENEOUS CATALYSIS. Reaction (13.32) can be changed from a homogeneous gaseous reaction to a reaction occurring on a surface—a heterogeneous reaction. If this surface material is properly chosen, the reaction rate can be increased significantly; the catalytic action is referred to as **heterogeneous catalysis.** The precise mechanism of heterogeneous catalysis is imperfectly understood. It appears, though, that the availability of d electrons and d orbitals in surface atoms of the catalyst plays an important role. Catalytic activity is associated with a large number of transition elements and their compounds.

The key requirement in heterogeneous catalysis is that reactants be *adsorbed* from a gaseous or solution phase onto the surface of the catalyst. Not all surface atoms are equally effective as catalysts; those that are constitute the **active sites** of a catalyst. Basically, then, heterogeneous catalysis involves (1) adsorption of reactants, (2) diffusion of reactants along the surface, (3) reaction at an active site to form adsorbed product, and (4) *desorption* of the product. These steps are described for the catalytic hydrogenation of ethylene to ethane in Figure 13-15. The net reaction is simply that described in equation (13.32), and the catalyst is unchanged.

FIGURE 13-15
Heterogeneous catalysis and the reaction $C_2H_4(g) + H_2(g) \rightarrow C_2H_6(g)$.

(a) Molecules of C_2H_4 and H_2 from the gaseous state are adsorbed on the surface of a catalyst (e.g., a finely divided form of nickel). The adsorption of H_2 involves dissociation of the molecules and formation of metal–hydrogen (M—H) bonds. The adsorbed H atoms are able to skip about the surface from one site to another. One of the H atoms is shown about to move from its site to a neighboring one where a molecule of C_2H_4 is adsorbed at an active site ($\star$).
(b) A C—H bond is formed between a C_2H_4 molecule and an H atom. The ethyl group $(-C_2H_5)$ that is formed remains adsorbed. A second H atom is shown about to migrate from a nearby site to the active site of the $-C_2H_5$ group.
(c) Attachment of a second H atom results in the formation of a molecule of C_2H_6. This molecule, because its bonding capacity is saturated through C—H bonds, is easily desorbed and escapes into the gaseous state.

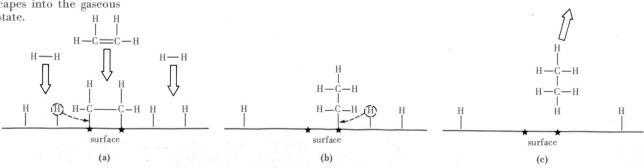

Heterogeneous catalysts must be carefully prepared and maintained, for they are easily "poisoned." Trace amounts of certain impurities may become bound to active sites and destroy their catalytic activity. This situation occurs if arsenic is present in platinum, for example, forming platinum arsenide at the active sites. It may also occur if gasoline containing lead compounds as additives is used in automobiles equipped with catalytic mufflers. [These catalysts are designed to promote the combustion of carbon monoxide and hydrocarbons (see Section 26-3).]

ENZYMES AS CATALYSTS. Some catalysts, for example platinum metal, can catalyze a wide variety of reactions. Unlike platinum, the catalysts associated with chemical reactions in a living organism must be very specific. These catalysts, called **enzymes,** are high-molecular-weight proteins. Many enzymes catalyze one particular reaction and no others. For example, in alcoholic fermentation, the six-carbon compound glucose is broken down into two molecules of ethanol and two of carbon dioxide.

$$C_6H_{12}O_6 \longrightarrow 2\,C_2H_5OH + 2\,CO_2 \qquad\qquad (13.33)$$

This process requires 12 enzymatic steps. In the last of these acetaldehyde is reduced to ethanol through the action of the enzyme alcohol dehydrogenase. (The necessary hydrogen atoms are furnished by other species in the reaction.)

FIGURE 13-16
The Michaelis–Menton mechanism of enzyme action.

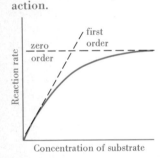

$$\mathrm{H-\underset{\underset{H}{|}}{\overset{\overset{H}{|}}{C}}-\overset{\overset{O}{\|}}{C}-H \xrightarrow[\text{dehydrogenase}]{\text{alcohol}} H-\underset{\underset{H}{|}}{\overset{\overset{H}{|}}{C}}-\underset{\underset{H}{|}}{\overset{\overset{H}{|}}{C}}-O-H} \qquad (13.34)$$

The simplest mechanism of enzyme action, known as the Michaelis–Menten mechanism, involves a reactant species, called the **substrate** (S), attaching itself to an active site on the enzyme (E). The result is an enzyme–substrate complex (ES). This complex dissociates to produce a product species (P) and the original enzyme (E). Thus, a two-step mechanism can be written, with each step being reversible (denoted by the double arrow $\rightleftharpoons$).

$$\mathrm{S + E \rightleftharpoons ES}$$

$$\mathrm{ES \rightleftharpoons E + P}$$

Figure 13-16, which shows how the reaction rate varies with substrate concentration, is consistent with this mechanism. Along the ascending portion of the curve the reaction is first order in S, because the rate at which the complex ES is formed is proportional to [S].

rate of disappearance of $S = k[S]$

At high concentrations of the substrate the reaction is zero order. The enzyme is saturated, and adding more substrate cannot accelerate the reaction.

rate of disappearance of $S = k'[S]^0 = k'$

Consider this analogy to a reaction mechanism. A person in Los Angeles meets an old acquaintance from New York City and asks, "When did you leave New York?" The answer: "This morning." The friend must have made the principal part of the journey by airplane. If the friend had answered, "Two weeks ago," the conclusion on the mode of travel used would not have been obvious. Further questioning (experimentation) would have been required.

13-10 Reaction Mechanisms

One important reason for studying chemical kinetics is the fact that knowledge of the the rate of a reaction can provide insight into the succession of steps by which the reaction proceeds—**the reaction mechanism.**

Each molecular event that significantly alters a molecule's energy or geometry is called an elementary process. The combined effect of all the elementary processes yields the net reaction. Unlike for the net reaction, concentration-term exponents in the rate law for an *elementary* process *are* the same as the corresponding coefficients in the balanced equation for the elementary process. Moreover, by writing rate equations for the elementary steps and combining them in the appropriate fashion, a rate equation can be determined for the net reaction. A test of the *plausibility* of a mechanism is that it yield the same rate equation as that determined experimentally. The word "plausible" is emphasized because there is no way to *prove* a reaction mechanism. It is often possible to propose several mechanisms that are consistent with the observed rate law.

Other ideas essential to developing a plausible reaction mechanism are

1. Elementary processes in which a single molecule dissociates—**unimolecular**—or two molecules collide—**bimolecular**—are much more probable than a process requiring the simultaneous collision of three bodies—**termolecular.**

2. All elementary processes are reversible and may reach a **steady-state condition.** In the steady state the rates of the forward and reverse processes become equal. The concentration of some intermediate becomes constant with time.

3. One elementary process may occur much more slowly than all the others. In this case it determines the rate at which the overall reaction proceeds and is called the **rate-determining step.**

THE HYDROGEN–IODINE REACTION. Is there a reaction for which the mechanism consists of a single elementary step? That is, is there a reaction for which the mechanism is that implied by the net equation? For the better part of the twentieth century the classic example of such a reaction had been that of gaseous hydrogen and iodine.

$$H_2(g) + I_2(g) \longrightarrow 2\,HI(g) \tag{13.35}$$

rate of formation of $HI = k[H_2][I_2]$

The rate equation for a simple one-step bimolecular process based on the collision of an H_2 and an I_2 molecule is the same as that for the net reaction.

However, as a result of a detailed experimental investigation of this reaction, it was concluded that the hydrogen–iodine reaction is probably more complex than the one-step mechanism of (13.35).* In particular the proposal was made of a two-step mechanism. In the first step, which occurs rapidly, iodine molecules are believed to dissociate into iodine atoms. The second step is believed to involve the simultaneous collision of *two* iodine atoms and a hydrogen molecule. We should expect this *termolecular* step to occur much more slowly and to be the *rate-determining step.* This step is illustrated in Figure 13-17.

(fast) $$I_2(g) \xrightleftharpoons[k_2]{k_1} 2\,I(g) \tag{13.36}$$

(slow) $$\underline{2\,I(g) + H_2(g) \xrightarrow{k_3} 2\,HI(g)} \tag{13.37}$$

net: $$I_2(g) + H_2(g) \longrightarrow 2\,HI(g) \tag{13.35}$$

The net equation obtained by adding together the two elementary processes is indeed that expected. If the reversible step (13.36) reaches a *steady-state condition,* we can write

rate of disappearance of I_2 = rate of formation of I_2

$$k_1[I_2] = k_2[I]^2$$

and $$[I]^2 = \frac{k_1}{k_2}[I_2] \tag{13.38}$$

FIGURE 13-17
Rate-determining step in the reaction $H_2(g) + I_2(g) \rightarrow 2\,HI(g)$.

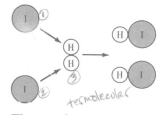

The rate-determining step for this reaction appears to involve the simultaneous collision of two I atoms with an H_2 molecule.

A k_4 arrow is omitted simply because the reverse of the k_3 step is believed to occur infrequently.

*J. Sullivan, *J. Chem. Phys.* **46:**73 (1967).

For the *rate-determining step* (13.37) we can write

$$\text{rate of formation of HI} = k_3[\text{I}]^2[\text{H}_2] \tag{13.39}$$

By substituting into the rate equation (13.39) the expression for $[\text{I}]^2$ from the steady-state condition (13.38), we obtain for the rate of the net reaction

$$\text{rate of formation of HI} = \frac{k_1 k_3}{k_2}[\text{H}_2][\text{I}_2] = k[\text{H}_2][\text{I}_2] \quad (\text{where } k = k_1 k_3/k_2) \tag{13.40}$$

Equation (13.40) agrees with the observed rate law. Whether the mechanism outlined here is the actual mechanism of the reaction we cannot say. All that we can say is that it is *plausible*.

THE HYDROGEN–BROMINE REACTION—A CHAIN REACTION.

Next we might ask which of the two mechanisms described for the hydrogen–iodine reaction is most likely to apply to the hydrogen–bromine reaction.

$$\text{H}_2(g) + \text{Br}_2(g) \longrightarrow 2\,\text{HBr}(g) \tag{13.41}$$

$$\text{observed rate of formation of HBr} = \frac{k'[\text{H}_2][\text{Br}_2]^{1/2}}{1 + k''([\text{HBr}]/[\text{Br}_2])}$$

The answer can only be neither! The mechanisms presented for the hydrogen–iodine reaction would lead to the predicted rate equation: rate $= k[\text{H}_2][\text{Br}_2]$. The observed rate equation is much more complex than this, suggesting also that the mechanism is very complex. One mechanism that leads to the observed rate equation is the following.

$$\text{Br}_2 \longrightarrow 2\,\text{Br} \tag{13.42a}$$

$$\text{Br} + \text{H}_2 \longrightarrow \text{HBr} + \text{H}$$
$$\text{H} + \text{Br}_2 \longrightarrow \text{HBr} + \text{Br}$$
$$\text{Br} + \text{H}_2 \longrightarrow \text{HBr} + \text{H} \tag{13.42b}$$
$$\text{H} + \text{Br}_2 \longrightarrow \text{HBr} + \text{Br}$$

$$\text{and so on}$$

$$\text{H} + \text{HBr} \longrightarrow \text{H}_2 + \text{Br}$$
$$\text{Br} + \text{Br} \longrightarrow \text{Br}_2 \tag{13.42c}$$

A Br atom formed in step (13.42a) reacts with an H_2 molecule to produce a molecule of HBr and an H atom. This H atom, in turn, reacts with a Br_2 molecule to produce a second HBr molecule and a Br atom. This Br atom reacts with an H_2 molecule; and so on. Such a cycle is called a **chain reaction.** A chain reaction may proceed through thousands of elementary steps before it is terminated by reactions such as (13.42c). The net change represented by all the elementary steps shown in equation (13.42b) is

$$\text{H}_2(g) + \text{Br}_2(g) \longrightarrow 2\,\text{HBr}(g) \tag{13.41}$$

Chain reactions are encountered again in connection with certain reactions between halogens and hydrocarbons (Section 24-2), in polymerization reactions (Sections 24-3 and 26-10) and in the formation of photochemical smog (Section 26-3).

Example 13-12 The thermal decomposition of ozone to oxygen, $2\,\text{O}_3(g) \rightarrow 3\,\text{O}_2(g)$, has the observed rate law

$$\text{rate of disappearance of O}_3 = k\frac{[\text{O}_3]^2}{[\text{O}_2]}$$

Show that the following mechanism is consistent with this experimental rate law.

(fast) $\qquad O_3 \underset{k_2}{\overset{k_1}{\rightleftharpoons}} O_2 + O$ $\qquad\qquad$ (13.43)

An "INTERMEDIATE" constant [ ] in steady state condition

(slow) $\quad O + O_3 \xrightarrow{k_3} 2\,O_2$ $\qquad\qquad$ (13.44)

Assume that the first step (13.43) reaches the steady-state condition.

rate formation O = rate disappearance O

Write expressions for these two rates based on the balanced equation (13.43) for the elementary process.

rate formation $O = k_1[O_3] = k_2[O_2][O]$ = rate disappearance O $\qquad$ (13.45)

Solve equation (13.45) for the steady-state concentration of the intermediate, that is, [O].

$$[O] = \frac{k_1[O_3]}{k_2[O_2]}$$ $\qquad\qquad$ (13.46)

Assume that the second step, (13.44), is the rate-determining step and write its rate equation.

rate disappearance of $O_3 = k_3[O][O_3]$ $\qquad\qquad$ (13.47)

Substitute the steady-state concentration of O from equation (13.46) into equation (13.47). Combine k_1, k_2, and k_3 into a single constant, k.

rate disappearance $O_3 = \dfrac{k_1 k_3[O_3][O_3]}{k_2[O_2]} = k\dfrac{[O_3]^2}{[O_2]}$ $\qquad$ (13.48)

SIMILAR EXAMPLES: Exercises 43, 44.

Summary

The rate of a chemical reaction describes how fast the concentration of a reactant decreases or that of a product increases with time. An initial reaction rate can be established simply by dividing the change in concentration of a reactant by the time interval over which this change occurs. The method works as long as no more than a few percent of a reactant has been consumed. Beyond this initial stage, the exact reaction rate is given by the slope of a tangent line to a concentration vs. time graph. One goal of the kinetic study of a reaction is to establish a rate law for the reaction, having the form

reaction rate = $k[A]^m[B]^n$. . . .

The order of a reaction is related to the concentration-term exponents in the rate law. A few reactions proceed at a constant rate, *independent* of the concentrations of the reactants. These are zero-order reactions. A first-order reaction usually has a single concentration term appearing in the rate law and that term is raised to the first power. The most common forms of a second-order rate law have either a single concentration term appearing to the second power or two concentration terms each appearing to the first power. One method of establishing the order of a reaction requires measuring the *initial* reaction rate in a series of experiments. A second method requires plotting appropriate functions of reactant concentration against time so as to obtain a straight-line graph. A third method involves substituting experimentally determined concentrations and times into appropriate mathematical equations.

An important characteristic associated with reaction rates is the half-life of a reaction. For a first-order reaction the half-life is *independent* of the concentration of the reactant. For other reaction orders the half-life is concentration-dependent.

The theoretical basis of chemical kinetics originates from these essential ideas: Chemical reactions occur as a result of collisions between molecules. Only collisions in which molecules possess sufficient energy and a proper geometrical orientation will be effective in yielding a product. Transition state theory depicts the course of a chemical reaction through an energy diagram, called a reaction profile. In this profile the energies of reactants and products are represented together with that of the activated complex. Such a profile permits a visualization of the enthalpy change and activation energy of a reaction.

a plot slope

methods of determining Rt order

Raising the temperature of a reaction mixture increases the fraction of the molecules possessing energies in excess of the activation energy—the reaction speeds up. A mathematical equation can generally be written to relate rate constants, temperatures, and activation energy for a reaction. Another method of speeding up a reaction is to employ a catalyst. Some catalyst molecules participate as intermediates in a homogeneous chemical reaction and are regenerated. In another type of catalysis—heterogeneous catalysis—the catalyst provides a surface on which the desired reaction proceeds at a faster rate. Biochemical reactions are promoted by catalysts called enzymes.

It is sometimes possible to propose a mechanism for a reaction. This is done by postulating a series of elementary processes, writing rate equations for them, and combining these elementary rate equations into a rate law for the net reaction.

Learning Objectives

As a result of studying Chapter 13, you should be able to

1. Establish the exact rate of a chemical reaction from the slope of a tangent line to a concentration vs. time graph.

2. Determine the *initial* rate of a chemical reaction, either from a graph or by calculation.

3. Apply the method of initial rates to derive the rate law for a reaction.

4. Use a rate law, together with rate data, to calculate a rate constant, k, or use the rate law and rate constant to calculate rate data.

5. Establish, through rate data, equations, and graphs, whether a reaction is zero order, first order, or second order.

6. Use the concept of the half-life of a reaction for zero-order, first-order, and second-order reactions.

7. Describe the collision theory of reactions, stating the factors that affect collision frequency and those that lead to favorable collisions.

8. Describe a reaction in terms of a reaction profile, activated complex, and activation energy.

9. Use the Arrhenius equation in calculations involving rate constants, temperatures, and activation energies.

10. Describe the role of a catalyst in a chemical reaction, and explain the essential difference between homogeneous and heterogeneous catalysis.

11. Explain what is meant by a reaction mechanism, and distinguish between elementary processes and a net chemical reaction.

12. Use the concepts of steady-state condition and rate-determining step in testing a plausible mechanism for a chemical reaction.

Some New Terms

An **activated complex** is an intermediate product in the mechanism of a chemical reaction.

Activation energy is the energy that molecules must possess so that collisions between them will lead to chemical reaction.

Active sites are the locations at which catalytic activity occurs, whether on the surface of a heterogeneous catalyst or an enzyme.

The **Arrhenius equation** mathematically relates reaction rate to temperature and activation energy for a reaction.

A **bimolecular process** is an elementary process in a reaction mechanism involving the collision of two molecules.

Catalysis is the speeding up of a reaction in the presence of an agent **(catalyst)** that changes the reaction mechanism to one of lower activation energy.

Collision frequency is the number of collisions occurring between molecules in a unit of time.

Collision theory describes reactions in terms of molecular collisions—the frequency of collisions and the probability that collisions will lead to chemical reaction.

A **first-order** reaction is one for which the sum of the concentration-term exponents in the rate law is 1.

The **half-life** is the time required for one half of a reactant to be consumed in a chemical reaction.

The **initial rate of a reaction** is the rate or speed with which a reaction proceeds immediately after the reactants are brought together.

The **method of initial rates** is an experimental method of determining the order of a chemical reaction based on the measurement of initial rates of reaction.

The **order of a reaction** relates to the exponents of the concentration terms in the rate law for the reaction.

The **rate constant, k,** is the proportionality constant in the rate law of a chemical reaction.

A **rate-determining step** in a reaction mechanism is an elementary process that proceeds much more slowly than other steps. It determines the rate of the overall reaction.

The **rate law (rate equation)** for a reaction relates the reaction rate to the concentrations of reactants. It has the form: reaction rate $= k[A]^m[B]^n. \ldots$

The **rate of a chemical reaction** is a measure of the rate or speed with which a reactant is consumed or a product is formed.

A **reaction mechanism** is a detailed description of a chemical reaction in terms of a set of elementary steps or processes by which the reaction is proposed to occur.

A **reaction profile** is a graphical representation of a chemical reaction in terms of the energies of the reactants, activated complex(es), and products.

A **second-order** reaction is one for which the sum of the concentration-term exponents in the rate law is 2.

A **steady-state condition** is reached in an elementary process when a species is formed and consumed at equal rates. The concentration of the species remains constant with time.

A **termolecular process** is an elementary process in a reaction mechanism in which three molecules collide simultaneously.

A **unimolecular process** is an elementary process in a reaction mechanism in which a single molecule dissociates.

A **zero-order** reaction proceeds at a rate that is *independent* of reactant concentrations.

Exercises

Rates of chemical reactions

1. Suppose that a reaction involves the decomposition of the substance A: A $\rightarrow$ B + C. What is the meaning of each of the following terms with respect to this reaction? **(a)** $[A]_0$; **(b)** $[A]_t$; **(c)** $\Delta[A]$; **(d)** Δt; **(e)** $-\Delta[A]/\Delta t$; **(f)** $\Delta[B]/\Delta t$; **(g)** $t_{1/2}$.

2. What is meant by the order of a chemical reaction? What are the differences among zero-, first-, and second-order reactions?

3. The decomposition of acetaldehyde in the gaseous phase is found to be a second-order reaction. Write the rate equation for this decomposition: $CH_3CHO \rightarrow CH_4 + CO$.

4. For the reaction A $\rightarrow$ products, what are the units of the rate constant, k, if the reaction is **(a)** zero order in A; **(b)** first order in A; **(c)** second order in A?

5. A small volume of a solution of N_2O_5 in an inert solvent has an initial concentration of $N_2O_5 = 0.550\ M$. The solution is allowed to decompose and the volume of $O_2(g)$ at STP is measured. After 85 s the volume of $O_2(g)$ is 6.2 cm³, and after a long period of time, 182 cm³.
   (a) What is $[N_2O_5]$ at 85 s?
   (b) What is the initial rate of reaction, expressed as mol $N_2O_5\ L^{-1}\ s^{-1}$?
   (c) What would you expect $[N_2O_5]$ to be after 2.0 min? (*Hint:* Assume that the initial rate of reaction remains constant for the first 2 min.)

6. Refer to Exercise 5 and equation (13.1).
   (a) Determine the initial *volume* of the N_2O_5 solution. [*Hint:* How many mol N_2O_5 must have been present in the 0.550 M solution to account for all of the liberated $O_2(g)$?]
   (b) Determine the initial rate of formation of $O_2(g)$. (*Hint:* Use the result of Exercise 5b.)
   (c) Calculate the total volume of O_2 (at STP) released at the end of 1.0 min.

7. From the tangent line in Figure 13-4, estimate the rate of decomposition of N_2O_5 at $t = 1000$ s. How would you ex-

pect this rate to compare with the initial rate of decomposition? with the rate at 1900 s? Explain.

8. The rate of a reaction in the gaseous state is generally established by measuring the *total* gas pressure as a function of time, not just the partial pressure of a reactant or product. Thus, in the hypothetical reaction A(g) $\rightarrow$ 2 B(g) + C(g), the total pressure *increases* while the partial pressure of A(g) decreases. If the initial pressure of A(g) is 1000 mmHg
   (a) What is the total pressure when the reaction has gone to completion?
   (b) What is the total gas pressure when the partial pressure of A(g) has fallen to 800 mmHg?

Three different sets of data of [A] vs. time are given in the table below for the reaction A $\rightarrow$ products. (*Hint:* There are several ways of arriving at answers for each of the following four exercises.)

I		II		III	
Time, s	[A], M	Time, s	[A], M	Time, s	[A], M
0	1.00	0	1.00	0	1.00
25	0.78	25	0.75	25	0.80
50	0.61	50	0.50	50	0.67
75	0.47	75	0.25	75	0.57
100	0.37	100	0.00	100	0.50
150	0.22			150	0.40
200	0.14			200	0.33
250	0.08			250	0.29

9. Which of these sets of data corresponds to a **(a)** zero-order, **(b)** first-order, **(c)** second-order reaction?

10. What is the approximate half-life for each reaction?

11. What is the approximate concentration of A remaining after 110 s in the **(a)** zero-order, **(b)** first-order, **(c)** second-order reaction?

12. What is the approximate rate of the reaction, expressed

as mol A $L^{-1} s^{-1}$, at $t = 75$ s for the **(a)** zero-order, **(b)** first-order, **(c)** second-order reaction?

Method of initial rates

13. What is the initial rate of reaction expected in the peroxydisulfate–iodide reaction (see Example 13-5) if the initial concentrations are $[S_2O_8^{2-}] = 0.15\ M$ and $[I^-] = 0.010\ M$?

14. The rate of the reaction $2\ HgCl_2 + C_2O_4^{2-} \rightarrow 2\ Cl^- + 2\ CO_2(g) + Hg_2Cl_2(s)$ is followed by measuring the number of moles of Hg_2Cl_2 that precipitate per liter of solution per minute.

Expt.	$[HgCl_2]$, M	$[C_2O_4^{2-}]$, M	Initial rate, mol L^{-1} min^{-1}
1	0.105	0.15	1.8×10^{-5}
2	0.105	0.30	7.1×10^{-5}
3	0.052	0.30	3.5×10^{-5}
4	0.052	0.15	8.9×10^{-6}

(a) From the data given, determine the order of the reaction with respect to $HgCl_2$, with respect to $C_2O_4^{2-}$, and overall.
(b) What is the value of the rate constant, k?
(c) What would be the initial rate of reaction if $[HgCl_2] = 0.020\ M$ and $[C_2O_4^{2-}] = 0.22\ M$?

15. Listed below are the *initial rates,* expressed as the rate of decrease of partial pressure of a reactant, for this reaction at 826°C.

$$2\ NO(g) + 2\ H_2(g) \longrightarrow N_2(g) + 2\ H_2O(g)$$

With initial $P_{H_2} =$ 400 mmHg		With initial $P_{NO} =$ 400 mmHg	
Initial P_{NO}, mmHg	Rate, mmHg/s	Initial P_{H_2}, mmHg	Rate, mmHg/s
359	0.750	289	0.800
300	0.515	205	0.550
152	0.125	147	0.395

(a) What is the order of the reaction with respect to NO, with respect to H_2, and overall?
(b) Write the rate equation for this reaction.

*16. Hydroxide ion is involved in the mechanism of the following reaction but is not consumed in the net reaction.

$$OCl^- + I^- \xrightarrow{\ OH^-\ } OI^- + Cl^-$$

(a) From the data given, determine the order of the reaction with respect to OCl^-, I^-, and OH^-.
(b) What is the overall reaction order?

(c) Write the rate equation and determine the value of the rate constant, k.

$[OCl^-]$, M	$[I^-]$, M	$[OH^-]$, M	Rate formation OI^-, mol $L^{-1} s^{-1}$
0.0040	0.0020	1.00	4.8×10^{-4}
0.0020	0.0040	1.00	5.0×10^{-4}
0.0020	0.0020	1.00	2.4×10^{-4}
0.0020	0.0020	0.50	4.6×10^{-4}
0.0020	0.0020	0.25	9.4×10^{-4}

First-order reactions

17. Benzenediazonium chloride decomposes by a first-order reaction in water, yielding $N_2(g)$ as a product.

$$C_6H_5N_2Cl \longrightarrow C_6H_5Cl + N_2(g)$$

The reaction can be followed by measuring the volume of $N_2(g)$ evolved as a function of time. The following data were obtained for the decomposition of an $0.071\ M$ solution at 50°C. ($t = \infty$ corresponds to the completed reaction.)

Time, min	$N_2(g)$, cm^3	Time, min	$N_2(g)$, cm^3
0	0	18	41.3
3	10.8	21	44.3
6	19.3	24	46.5
9	26.3	27	48.4
12	32.4	30	50.4
15	37.3	∞	58.3

(a) Use the method of Example 13-1 to determine $[C_6H_5N_2Cl]$ remaining after 21 min.
(b) Construct a table similar to Table 13-2, with an interval of time, $\Delta t = 3$ min. That is, determine $[C_6H_5N_2Cl]$ at 3, 6, 9, ... min; $\Delta[C_6H_5N_2Cl]$ over every 3-min interval; and $\Delta[C_6H_5N_2Cl]/\Delta t$ for each 3-min interval.
(c) Plot a graph similar to Figure 13-3 showing both the formation of $N_2(g)$ and the disappearance of $C_6H_5N_2Cl$ as a function of time.
(d) What is the initial rate of the reaction?
(e) From the graph of part (c), estimate the rate of the reaction through the slope of the tangent to the curve at $t = 21$ min. Compare with the reported value of 1.1×10^{-3} mol $C_6H_5N_2Cl$ L^{-1} min^{-1}.
(f) Write a rate-law expression for the first-order decomposition of $C_6H_5N_2Cl$ and use the method of Example 13-7 to estimate a value of k based on the rate determined in parts (d) and (e).
(g) Determine $t_{1/2}$ for this reaction by calculation (equation 13.12) and by estimation from the graph of the rate data.

(h) At what time should the decomposition of the sample be three fourths completed?

(i) Plot log $[C_6H_5N_2Cl]$ vs. time (as in Figure 13-6) and show that the reaction is indeed first order.

(j) Determine k from the slope of the log plot of part (i).

18. A given substance, A, decomposes by a first-order reaction. Starting initially with $[A] = 2.00\ M$, after 200 min $[A] = 0.250\ M$. For this reaction what is (a) $t_{1/2}$; (b) k?

19. An 80-g sample of N_2O_5 is dissolved in CCl_4 and allowed to decompose at 45°C. From the half-life listed in the text, determine (a) how long it will take for the quantity of N_2O_5 to be reduced to 5 g; (b) the volume of $O_2(g)$ produced at this point, measured at STP.

20. A certain first-order decomposition has a value of $k = 1.0 \times 10^{-3}\ s^{-1}$. Starting with a concentration of reactant of $0.50\ M$, how long must the reaction proceed for the reactant to be 80% decomposed?

21. For the reaction $2\ A \rightarrow B + C$, the following data are obtained for [A] as a function of time.

Time, min	[A], M
0	0.80
8	0.60
24	0.35
40	0.20

(a) By suitable means establish the order of the reaction.

(b) What is the value of the rate constant k?

(c) Calculate the rate of formation of B at $t = 30$ min.

22. In the first-order reaction A → products, it is found that 99% of the original amount of reactant A decomposes in 132 min. What is the half-life, $t_{1/2}$, of this decomposition reaction?

23. The half-life for the first-order decomposition of nitramide is 123 min at 15°C: $NH_2NO_2(aq) \rightarrow N_2O(g) + H_2O(l)$. If 165 ml of a $0.105\ M\ NH_2NO_2$ solution is allowed to decompose, how long must the reaction proceed to produce 50.0 cm³ of "wet" $N_2O(g)$ measured at 15°C and a barometric pressure of 756 mmHg? (The vapor pressure of water at 15°C is 12.8 mmHg.)

24. Concerning the decomposition of di-*t*-butyl peroxide (DTBP) described in Example 13-9,

(a) Extract data from Figure 13-7; plot log P_{DTBP} vs. time; establish a value of k from this plot; and compare this value with that used in Example 13-9b.

★(b) Determine the time at which the total gas pressure is 2000 mmHg; 2100 mmHg.

25. The decomposition of gaseous dimethyl ether at 504°C is

$$(CH_3)_2O(g) \longrightarrow CH_4(g) + H_2(g) + CO(g)$$

The following data are partial pressures of dimethyl ether as a function of time.

Time, s	$P_{(CH_3)_2O}$, mmHg
0	312
390	264
777	224
1195	187
3155	78.5
∞	2.5

(a) Show that the reaction is first order.

(b) What is the rate constant, k?

(c) What is the total gas pressure at 390 s?

(d) What is the *maximum* gas pressure that can develop during the course of the reaction?

(e) What is the total gas pressure at $t = 1000$ s?

Collision theory; activation energy

26. The relationship of molecular collisions to reaction rates is explored in Section 13-6. From this discussion explain why

(a) The reaction rate cannot be calculated from the collision frequency alone.

(b) The rate of a chemical reaction may increase so dramatically with temperature while the collision frequency increases much more slowly.

(c) The addition of a catalyst to a reaction mixture can have such a pronounced effect on the rate of a reaction, even if the temperature is held constant.

27. For the reaction $A + B \rightarrow C + D$, the heat of reaction is $+21$ kJ/mol. The activation energy of the forward reaction is 84 kJ/mol.

(a) What is the activation energy of the reverse reaction?

(b) In the manner of Figure 13-12 sketch a graph of the energy profile of this reaction.

28. By an appropriate sketch or other means, verify the statement in the text that the activation energy of an *endothermic* reaction must be equal to or exceed the heat of reaction. Is there a similar relationship for an exothermic reaction?

Effect of temperature on reaction rate

29. Experiment 3 of Table 13-3 (the peroxydisulfate–iodide reaction) is repeated at several temperatures and the following rate constants are established: 3°C, $k = 1.4 \times 10^{-3}$ L mol⁻¹s⁻¹; 13°C, 2.9×10^{-3}; 24°C, 6.2×10^{-3}; 33°C, 1.20×10^{-2}.

(a) Construct a graph of log k vs. $1/T$.

(b) What is the activation energy, E_a, of the reaction?

(c) Calculate a value of the rate constant, k, at 40°C.

(d) What would be the *initial rate* of the reaction for Experiment 3 at 50°C?

30. The rate constant for the reaction $H_2(g) + I_2(g) \rightarrow 2\,HI(g)$ has been determined at the following temperatures: 556 K, $k = 1.2 \times 10^{-4}\,L\,mol^{-1}\,s^{-1}$; 666 K, $k = 3.8 \times 10^{-2}\,L\,mol^{-1}\,s^{-1}$. Estimate the activation energy for the net reaction and compare with the value shown in Figure 13-12.

31. The reaction $2\,NO_2(g) \rightarrow 2\,NO(g) + O_2(g)$ has a rate constant $k = 1.0 \times 10^{-10}$ at 300 K and an activation energy of 111 kJ/mol. At what temperature will this reaction have a rate constant $k = 1.0 \times 10^{-5}$? (Assume no change in the reaction mechanism.)

32. A statement is made in the text that as a rule of thumb reaction rates double for a temperature increase of about 10°C.

(a) What must be the approximate activation energy for this statement to be true for a reaction at room temperature?

(b) Would you expect this rule of thumb to apply to the formation of HI from its elements at room temperature? (Recall Figure 13-12.) Explain.

33. With reference to the data in Table 13-6 and Figure 13-13, to what temperature must a solution in which $[N_2O_5] = 0.15\,M$ be heated to have an *initial rate* of decomposition equal to that of a 1.25 M solution at 0°C?

Catalysis

34. The following statements are sometimes encountered with reference to catalysis, but they are not stated as carefully as they might be. What slight modifications should be made in each of them?

(a) A catalyst is a substance that speeds up a chemical reaction but does not take part in the reaction.

(b) The function of a catalyst is to lower the activation energy for a chemical reaction.

35. What is the principal difference between the catalytic activity of platinum metal and an enzyme?

36. In a particular enzyme reaction the following data are obtained on the rate of disappearance of substrate S.

Time, min	[S], M
0	1.0
20	0.90
60	0.70
100	0.50
160	0.20

What is the order of this reaction with respect to S in the concentration range studied?

37. A statement is made in the text that in enzyme-catalyzed reactions the reaction is first order at low substrate concentrations and becomes zero order at high concentrations. Certain gas-phase reactions with a heterogeneous catalyst are found to be first order at low gas pressures and zero order at high pressures. What is the connection between these two situations?

Reaction mechanisms

38. We have used the terms order of a reaction and molecularity of an elementary process (i.e., unimolecular, bimolecular, termolecular). What is the relationship, if any, between these two terms?

39. The rate-determining step in a reaction mechanism is sometimes referred to as a bottleneck. Comment on the appropriateness of this analogy.

40. Show that by combining equations (13.43) and (13.44) the correct balanced net equation for the thermal decomposition of O_3 is obtained.

41. The mechanism proposed for the reaction $2\,ICl + H_2 \rightarrow 2\,HCl + I_2$ consists of two steps.

(slow) $\quad H_2 + ICl \longrightarrow HI + HCl$

(fast) $\quad ICl + HI \longrightarrow HCl + I_2$

Predict a plausible rate law for the net reaction and explain how it is consistent with the mechanism.

42. For this reversible elementary process, determine the steady-state concentration of N_2O_2, that is, $[N_2O_2]$.

$$2\,NO \underset{k_2}{\overset{k_1}{\rightleftharpoons}} N_2O_2$$

43. For the reaction $2\,NO(g) + O_2(g) \rightarrow 2\,NO_2(g)$, the rate law, expressed as the rate of formation of NO_2, is found to be rate $= k[NO]^2[O_2]$.

(a) Is the rate law consistent with the one-step mechanism $2\,NO + O_2 \rightarrow 2\,NO_2$?

(b) Why is this one-step mechanism not very plausible?

44. Show that the rate law in Exercise 43 is consistent with the following mechanism.

(fast) $\qquad 2\,NO \underset{k_2}{\overset{k_1}{\rightleftharpoons}} N_2O_2$

(slow) $\quad N_2O_2 + O_2 \overset{k_3}{\longrightarrow} 2\,NO_2$

45. The collision theory states that chemical reactions occur as a result of collisions between molecules. A unimolecular elementary process in a reaction mechanism involves dissociation of a *single* molecule. How can these two ideas be compatible? Explain.

Additional Exercises

1. Some of the following statements are true regarding the following *first-order* reaction $2 A \rightarrow B + C$, and some are not. Indicate which are true and which are false. Explain your reasoning.
 (a) The rate of the reaction decreases as more and more B and C are formed.
 (b) The time required for one half of the substance A to react is directly proportional to the quantity of A present.
 (c) A plot of [A] vs. time yields a straight line.
 (d) The rate of formation of C is one half the rate of disappearance of A.

2. The first-order reaction $A \rightarrow$ products has a half-life of 150 s.
 (a) What percent of a sample of A remains *unreacted* 600 s after a reaction has been started?
 (b) What is the rate of the reaction, in mol $A\,L^{-1}\,s^{-1}$, when [A] = 0.50 M?

*3. With reference to equation (13.1), Table 13-1, Example 13-2, and Figure 13-3,
 (a) What is the rate of formation of O_2, expressed in cm^3 $O_2(STP)\ s^{-1}$, 1900 s after the reaction described by the data in Table 13-1 has been started?
 (b) What is the rate of formation of O_2 in $cm^3\ O_2(STP)$ s^{-1} when $[N_2O_5] = 0.60\ M$?

4. The rate-determining step in the hydrogen–iodine reaction has been proposed to be $2\ I(g) + H_2(g) \rightarrow 2\ HI(g)$. The rate constant for this termolecular elementary step has been determined at 520 and 710 K, yielding values of 3.96×10^{-5} and $1.61 \times 10^{-4}\ L^2\ mol^{-2}\ s^{-1}$, respectively. What is the activation energy, E_a, for this elementary process?

*5. The decomposition of ethylene oxide at 690 K is followed by measuring the total gas pressure as a function of time.

$$(CH_2)_2O(g) \longrightarrow CH_4(g) + CO(g)$$

The data obtained are

Time, min	$P_{tot.}$, mmHg
10	139.14
20	151.67
40	172.65
60	189.15
100	212.34
200	238.66
∞	249.88

 (a) What must be the initial total pressure (i.e., the pressure of the pure ethylene oxide at $t = 0$)?
 (b) What is the order of the reaction?

*6. The following data were obtained for the reaction $2\ A + B \rightarrow$ products. Use appropriate methods introduced in this chapter to derive the rate law for this reaction.

Experiment 1, [B] = 1.00 M		Experiment 2, [B] = 0.50 M	
Time, min	[A], M	Time, min	[A], M
0	1.000×10^{-3}	0	1.000×10^{-3}
1	0.951×10^{-3}	1	0.975×10^{-3}
5	0.779×10^{-3}	5	0.883×10^{-3}
10	0.607×10^{-3}	10	0.779×10^{-3}
20	0.368×10^{-3}	20	0.607×10^{-3}

*7. The data listed below were obtained for the decomposition of acetaldehyde at 518°C: $CH_3CHO(g) \rightarrow CH_4(g) + CO(g)$.

Time, s	$P_{tot.}$, mmHg
0	363
42	397
73	417
105	437
190	477
310	517
480	557
665	587

 (a) By an appropriate graphical method, establish the order of the reaction.
 (b) Obtain a value of the rate constant, k.

*8. The peroxydisulfate–iodide reaction (13.5) can be followed by the series of reactions

$$S_2O_8{}^{2-}(aq) + 3\ I^-(aq) \longrightarrow 2\ SO_4{}^{2-}(aq) + I_3{}^-(aq)$$

$$2\ S_2O_3{}^{2-}(aq) + I_3{}^-(aq) \longrightarrow S_4O_6{}^{2-}(aq) + 3\ I^-(aq)$$

$$I_3{}^-(aq) + starch(aq) \longrightarrow blue\ complex$$

Solutions of the two reactants $S_2O_8{}^{2-}$ and I^- are mixed in the presence of a fixed amount of $S_2O_3{}^{2-}$ and starch indicator. The $I_3{}^-$ produced in the main reaction reacts very rapidly with $S_2O_3{}^{2-}$ until all of the $S_2O_3{}^{2-}$ is consumed. Then the $I_3{}^-$ combines with the starch to produce a deep blue color in the solution. The rate of reaction can be related to the appearance of this blue color. In an experiment, 25.0 ml of 0.20 M $(NH_4)_2S_2O_8$, 25.0 ml of 0.20 M KI, 10.0 ml of 0.010 M $Na_2S_2O_3$, and 5.0 ml of starch solution are mixed. How long after mixing will the blue color appear? (*Hint:* Use the rate constant established in Example 13-5.)

*9. Show that the following mechanism is consistent with the rate law established for the iodide–hypochlorite reaction in Exercise 16.

(fast) $OCl^- + H_2O \underset{k_2}{\overset{k_1}{\rightleftharpoons}} HOCl + OH^-$

(slow) $I^- + HOCl \overset{k_3}{\longrightarrow} HOI + Cl^-$

(fast) $HOI + OH^- \underset{k_5}{\overset{k_4}{\rightleftharpoons}} H_2O + OI^-$

*10. The accompanying table presents two possible mechanisms for the reaction

$$H_3AsO_3 + I_3^- + H_2O \longrightarrow products$$

Show that each of these mechanisms is consistent with the experimentally determined rate law

$$rate = k\frac{[H_3AsO_3][I_3^-]}{[H^+][I^-]^2}$$

Self-Test Questions

For questions 1 through 10 select the single item that best completes each statement.

1. For the reaction $A \rightarrow$ products, a plot of $[A]$ vs. time is found to be a straight line. The order of this reaction is (a) zero; (b) first; (c) second; (d) impossible to determine from this graph.

2. A kinetic study of the reaction $A \rightarrow$ products yields the following data.

Time, s	[A], M
0	2.00
500	1.00
1500	0.50
3500	0.25

The order of the reaction must be   (a) zero;   (b) first; (c) second;   (d) either first or second, but not zero.

3. A *first-order* reaction $A \rightarrow$ products has a half-life of 100 s. Whatever the quantity of substance A involved in a particular reaction, (a) the reaction goes to completion in 200 s; (b) the quantity of A remaining after 200 s is half of what remains after 100 s; (c) the same quantity of A is consumed for every 100 s of the reaction; (d) 100 s elapses before the reaction begins.

4. In the *first-order* decomposition of substance A the following concentrations are found to exist at the indicated times following the start of the reaction.

Mechanism 1

$$H_3AsO_3 \underset{k_2}{\overset{k_1}{\rightleftharpoons}} H_2AsO_3^- + H^+$$

$$H_2O + I_3^- \underset{k_4}{\overset{k_3}{\rightleftharpoons}} H_2OI^+ + 2\,I^-$$

rate-detg. $H_2OI^+ + H_2AsO_3^- \overset{k_5}{\longrightarrow} H_2AsO_3I + H_2O$

$$H_2AsO_3I \overset{k_6}{\longrightarrow} products$$

Mechanism 2

$$H_3AsO_3 \underset{k_2}{\overset{k_1}{\rightleftharpoons}} H_2AsO_3^- + H^+$$

$$H_2O + I_3^- \underset{k_4}{\overset{k_3}{\rightleftharpoons}} H_2OI^+ + 2\,I^-$$

$$H_2OI^+ + H_2AsO_3^- \underset{k_6}{\overset{k_5}{\rightleftharpoons}} H_2AsO_3I + H_2O$$

rate-detg. $H_2AsO_3I \overset{k_7}{\longrightarrow} products$

Time, s	[A], M
0	1.00
50	0.61
100	0.37
150	0.22

The exact rate of the reaction at $t = 100$ s is   (a) 0.0048 mol $L^{-1} s^{-1}$;   (b) 0.0030 mol L^{-1} s^{-1};   (c) greater than 0.0030 but less than 0.0048 mol $L^{-1} s^{-1}$;   (d) equal to the initial rate of reaction, 0.0078 mol $L^{-1} s^{-1}$.

5. The rate equation for the reaction $2\,A + B \rightarrow C$ is found to be: *rate of appearance of C* $= k[A][B]$. (a) The unit of k must be s^{-1};   (b) $t_{1/2}$ is a constant;   (c) the value of k depends on the initial concentrations of A and B;   (d) the rate of formation of C is one half the rate of disappearance of A.

6. The decomposition of substance A is *second order*: $A \rightarrow$ products. The initial rate of decomposition when $[A]_0 = 0.50\ M$ is   (a) the same as the initial rate of decomposition for any other value of $[A]_0$;   (b) half as great as when $[A]_0 = 1.00\ M$;   (c) five times as great as when $[A]_0 = 0.10\ M$;   (d) four times as great as when $[A]_0 = 0.25\ M$.

7. The rate of a chemical reaction generally increases rapidly, even for small temperature increases, because of a rapid increase with temperature in   (a) the collision frequency;   (b) the fraction of molecules with energies in excess of the activation energy;   (c) the activation energy; (d) the average kinetic energy of gas molecules.

8. The reaction $A + B \rightarrow C + D$ has $\Delta \bar{H} = +25$ kJ/mol, and an activation energy, $E_a =$ **(a)** -25 kJ/mol; **(b)** less than $+25$ kJ/mol; **(c)** more than $+25$ kJ/mol; **(d)** either less than $+25$ kJ/mol or more than $+25$ kJ/mol, which can only be determined by experiment.

9. A catalyst speeds up a chemical reaction by increasing **(a)** the average kinetic energy of molecules; **(b)** the frequency of molecular collisions; **(c)** the activation energy of the reaction; **(d)** the proportion of molecules with energies in excess of the activation energy.

10. For the net reaction $A + B \rightarrow 2\,C$, which proceeds by a *single-step bimolecular mechanism,* the following equation is applicable: **(a)** $t_{1/2} = 0.693/k$; **(b)** rate of disappearance of $A = k[A][B]$; **(c)** rate of appearance of $C =$ rate disappearance A; **(d)** $\log [A] = (-k/2.303)t + \log [A]_0$.

11. The reaction $A \rightarrow$ products is first order in A.
  (a) If 1.60 g A is allowed to decompose for 20 min, the

mass of A remaining undecomposed is found to be 0.40 g. What is the half-life, $t_{1/2}$, of this reaction?
  (b) Starting with 1.60 g A, what is the mass of A remaining undecomposed after 33.2 min?

12. The decomposition of acetaldehyde, CH_3CHO, can be catalyzed by $I_2(g)$.
uncatalyzed reaction: $E_a = 190$ kJ/mol

$$CH_3CHO(g) \longrightarrow CH_4(g) + CO(g)$$

catalyzed reaction: $E_a = 136$ kJ/mol

$$CH_3CHO(g) + I_2(g) \longrightarrow CH_3I(g) + HI(g) + CO(g)$$

$$CH_3I(g) + HI(g) \longrightarrow CH_4(g) + I_2(g)$$

The following enthalpies of formation are also given: $\Delta \bar{H}_f^\circ[CH_3CHO(g)] = -166$ kJ/mol; $\Delta \bar{H}_f^\circ[CH_4(g)] = -74.9$ kJ/mol; $\Delta \bar{H}_f^\circ[CO(g)] = -110.5$ kJ/mol. Sketch the reaction profiles for the catalyzed and uncatalyzed reactions, representing ΔH and E_a.

14

Principles of Chemical Equilibrium

Equilibrium is a condition in which two opposing processes occur at equal rates. As a result, no further *net* change occurs in a system at equilibrium. We have already encountered two equilibrium situations. Let us review them briefly.

1. When a liquid vaporizes into a closed container, there comes a time when molecules return to the liquid state at the same rate at which they leave it. That is, vapor *condenses* at the same rate at which liquid *vaporizes*. Even though molecules continue to pass back and forth between the liquid and vapor, at equilibrium the pressure exerted by the vapor remains constant with time.
2. When a solute dissolves in a solvent, a point is reached where the rate at which additional solute particles dissolve is just matched by the rate at which solute particles precipitate. The solution becomes saturated and its concentration remains constant with time.

One of the characteristics of a system at equilibrium, then, is that certain properties acquire values that remain unchanged with time. In this chapter we turn our attention to dynamic equilibrium in chemical reactions and focus on the constant property known as the equilibrium constant.

14-1 The Condition of Chemical Equilibrium

The situation pictured by the three graphs in Figure 14-1 is unlike anything encountered in our discussion of stoichiometry in Chapter 4. In Experiment 1, 0.00150 mol each of H_2 and I_2 are allowed to react. In the usual fashion we can write

forward reaction: $H_2(g) + I_2(g) \longrightarrow 2 HI(g)$ (14.1)

However, as soon as some HI has formed it begins to dissociate back to H_2 and I_2.

reverse reaction: $2 HI(g) \longrightarrow H_2(g) + I_2(g)$ (14.2)

Thus, two reactions occur simultaneously—a forward and a reverse reaction. When the amount of HI reaches 0.00234 mol, it ceases to increase. The amounts of HI, H_2, and I_2 present all remain constant with time. The two opposing reactions continue to occur but now at equal rates. A condition of dynamic equilibrium is achieved. It is customary to write the forward and reverse reactions together by using a double arrow ($\rightleftharpoons$).

344

FIGURE 14-1

Three approaches to equilibrium in the reaction $H_2(g) + I_2(g) \rightleftharpoons 2\,HI(g)$.

The data plotted here are from Table 14-1.

—— mol H_2 = mol I_2
—— mol HI

equal ratio of moles HI to moles H_2 & I_2 in all instances

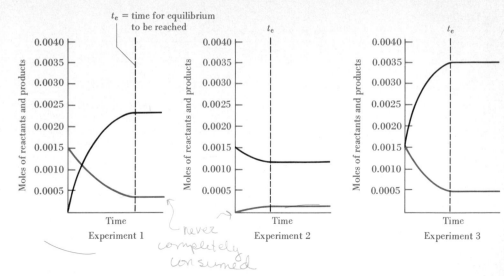

never completely consumed

The equilibrium amounts of reactants for reaction (14.3) can be determined as follows. Weighed quantities of the initial reactants are sealed into glass vessels and maintained at a constant temperature until equilibrium has been established. Then the vessels are quickly chilled and the contents transferred to aqueous solution at room temperature. These solutions are titrated with sodium thiosulfate using starch as an indicator. The titration data are used to establish the amount of I_2 present at equilibrium. Other equilibrium amounts can be related to the amount of I_2 (see Exercise 11).

$$H_2(g) + I_2(g) \rightleftharpoons 2\,HI(g) \tag{14.3}$$

Experiment 2 represents a different approach to equilibrium in the same reaction, this time starting with pure HI and forming H_2 and I_2. Once more a time is reached following which there is no further net change, this time because the rate of re-formation of HI from H_2 and I_2 has become equal to the rate of dissociation of HI. Experiment 3 represents still another situation. Here all three reactants are present initially.

Two points should be noted in Figure 14-1: (1) In no case is any reacting species consumed completely, and (2) based only on the amounts of reactants and products at equilibrium, there is no apparent common feature in the three situations.

THE HYDROGEN–IODINE–HYDROGEN IODIDE EQUILIBRIUM.   Let us explore the equilibrium condition in reaction (14.3) a little more fully, using data from Table 14-1. In particular, let us seek the constant property of a chemical reaction at equilibrium referred to in the introduction to this chapter.

In Table 14-2 three different attempts are summarized for the three experiments of Table 14-1. One of these expressions does, in fact, give almost identical numerical values in all three cases. The expression has the form

$$K_c = \frac{[HI]^2}{[H_2][I_2]} = 50.2 \quad \text{(at } 445\,°C) \tag{14.4}$$

The symbol K_c denotes an expression based on molar concentrations at equilibrium.

Three experiments are not enough to establish the constant value of expression (14.4), but repeated experimentation at 445°C would yield the same value. The significance of expression (14.4) is that whenever equilibrium is established among $H_2(g)$, $I_2(g)$, and HI(g) at 445°C, the particular ratio of molar concentrations must have a value of 50.2. We need to say a great deal more about the quantity K_c, called an **equilibrium constant.** Before we do, let us illustrate its significance in the following example.

Example 14-1   If the equilibrium concentrations of HI and I_2 in reaction (14.3) at 445°C are found to be [HI] = 2.02×10^{-3} M and [I_2] = 1.68×10^{-3} M, what must be the equilibrium concentration of H_2?

The three equilibrium concentrations must be related through expression (14.4). We

TABLE 14-1
Three approaches to equilibrium in the reaction $H_2(g) + I_2(g) \rightleftharpoons 2\ HI(g)$[a]

Exper- iment	Initial amounts, mol			Equilibrium amounts, mol		
	$H_2(g)$	$I_2(g)$	$HI(g)$	$H_2(g)$	$I_2(g)$	$HI(g)$
1	1.50×10^{-3}	1.50×10^{-3}	—	3.30×10^{-4}	3.30×10^{-4}	2.34×10^{-3}
2	—	—	1.50×10^{-3}	1.65×10^{-4}	1.65×10^{-4}	1.17×10^{-3}
3	1.50×10^{-3}	1.50×10^{-3}	1.50×10^{-3}	4.96×10^{-4}	4.96×10^{-4}	3.51×10^{-3}

[a] Temperature $= 445°C$; volume of reaction mixture $= 0.8000$ L.

need simply to solve (14.4) for $[H_2]$ and substitute the known molar concentrations.

$$[H_2] = \frac{[HI]^2}{K_c[I_2]} = \frac{(2.02 \times 10^{-3})^2}{50.2 \times 1.68 \times 10^{-3}}$$

$$= 4.84 \times 10^{-5}\ M$$

SIMILAR EXAMPLES: Exercises 14, 15.

THE EQUILIBRIUM CONSTANT, K_c. From the specific example of the hydrogen–iodine–hydrogen iodide reaction, we now turn to the general case of a reversible reaction at a condition of equilibrium.

For the generalized reaction

$$a\,A + b\,B + \cdots \rightleftharpoons g\,G + h\,H + \cdots$$

The equilibrium constant expression has the form

$$\frac{[G]^g[H]^h \cdots}{[A]^a[B]^b \cdots} = K_c \tag{14.5}$$

The numerator is the product of the concentrations of the species written on the right side of the equation ([G], [H], . . .), each concentration being raised to a power given by the coefficient in the balanced equation (g, h, . . .). The denominator is the product of the concentrations of the species written on the left side of the equation ([A], [B], . . .), again with each concentration raised to a power given by the coefficient in the balanced equation (a, b, . . .).

A more exact formulation of the equilibrium constant based on activities (effective concentrations) is presented in Chapter 15. For present purposes, however, the approach described here is quite satisfactory.

The numerical value of the **equilibrium constant, K_c**, depends uniquely on the particular reaction and on the temperature.

TABLE 14-2
In search of a constant ratio of concentrations to describe equilibrium in the reaction $H_2(g) + I_2(g) \rightleftharpoons 2\ HI(g)$[a]

	Experiment 1	Experiment 2
Try: $\dfrac{[HI]}{[H_2][I_2]}$	$\dfrac{2.92 \times 10^{-3}}{(4.12 \times 10^{-4})(4.12 \times 10^{-4})} = 1.72 \times 10^4$	$\dfrac{1.46 \times 10^{-3}}{(2.06 \times 10^{-4})(2.06 \times 10^{-4})} = 3.44 \times 10^4$
Try: $\dfrac{2 \times [HI]}{[H_2][I_2]}$	$\dfrac{2 \times 2.92 \times 10^{-3}}{(4.12 \times 10^{-4})(4.12 \times 10^{-4})} = 3.44 \times 10^4$	$\dfrac{2 \times 1.46 \times 10^{-3}}{(2.06 \times 10^{-4})(2.06 \times 10^{-4})} = 6.88 \times 10^4$
Try: $\dfrac{[HI]^2}{[H_2][I_2]}$	$\dfrac{(2.92 \times 10^{-3})^2}{(4.12 \times 10^{-4})(4.12 \times 10^{-4})} = 50.2$	$\dfrac{(1.46 \times 10^{-3})^2}{(2.06 \times 10^{-4})(2.06 \times 10^{-4})} = 50.2$

[a] Equilibrium concentrations are from Table 14-1; temperature $= 445°C$.

Equilibrium concentrations, mol/L		
$[H_2(g)]$	$[I_2(g)]$	$[HI(g)]$
4.12×10^{-4}	4.12×10^{-4}	2.92×10^{-3}
2.06×10^{-4}	2.06×10^{-4}	1.46×10^{-3}
6.20×10^{-4}	6.20×10^{-4}	4.39×10^{-3}

14-2 Additional Relationships Involving Equilibrium Constants

RELATIONSHIP OF K_c TO THE BALANCED CHEMICAL EQUATION. The reversible reaction involving $SO_2(g)$, $O_2(g)$, and $SO_3(g)$ is described in three different ways below.

$$2 SO_2(g) + O_2(g) \rightleftharpoons 2 SO_3(g) \qquad K_c(a) = 2.8 \times 10^2 \text{ at } 1000 \text{ K} \quad (14.6a)$$

$$2 SO_3(g) \rightleftharpoons 2 SO_2(g) + O_2(g) \qquad K_c(b) = ? \quad (14.6b)$$

$$SO_2(g) + \tfrac{1}{2} O_2(g) \rightleftharpoons SO_3(g) \qquad K_c(c) = ? \quad (14.6c)$$

The expressions obtained for the equilibrium constants are

$$K_c(a) = \frac{[SO_3]^2}{[SO_2]^2[O_2]} = 2.8 \times 10^2 \text{ at } 1000 \text{ K} \quad (14.7a)$$

$$K_c(b) = \frac{[SO_2]^2[O_2]}{[SO_3]^2} = ? \quad (14.7b)$$

$$K_c(c) = \frac{[SO_3]}{[SO_2][O_2]^{1/2}} = ? \quad (14.7c)$$

For a given set of initial conditions, the equilibrium concentrations of SO_2, O_2, and SO_3 acquire a unique set of values. This is true regardless of which of the three expressions in (14.7) we choose to represent equilibrium. The K_c values for the expressions in equations (14.7a–c) must be related in some way. Since

$$\frac{[SO_2]^2[O_2]}{[SO_3]^2} = \frac{1}{[SO_3]^2/([SO_2]^2[O_2])}$$

Experiment 3

$$\frac{4.39 \times 10^{-3}}{(6.20 \times 10^{-4})(6.20 \times 10^{-4})} = 1.14 \times 10^4$$

$$\frac{2 \times 4.39 \times 10^{-3}}{(6.20 \times 10^{-4})(6.20 \times 10^{-4})} = 2.28 \times 10^4$$

$$\frac{(4.39 \times 10^{-3})^2}{(6.20 \times 10^{-4})(6.20 \times 10^{-4})} = 50.1$$

then

$$K_c(b) = \frac{1}{K_c(a)} = \frac{1}{2.8 \times 10^2} = 3.6 \times 10^{-3}$$

and since

$$\frac{[SO_3]}{[SO_2][O_2]^{1/2}} = \left\{ \frac{[SO_3]^2}{[SO_2]^2[O_2]} \right\}^{1/2}$$

then

$$K_c(c) = \{K_c(a)\}^{1/2} = \sqrt{2.8 \times 10^2} = 1.7 \times 10^1$$

To summarize this discussion we note simply that

1. Whatever expression is used for K_c, it must be matched to the corresponding balanced chemical equation.
2. If an equation is reversed, the value of K_c is inverted; that is, the new equilibrium constant is the reciprocal of the old one.
3. If the coefficients in a balanced equation are multiplied by a common factor (2, 3, . . .), the new equilibrium constant will be the old one raised to the corresponding power (2, 3, . . .).
4. If the coefficients in a balanced equation are divided by a common factor (2, 3, . . .), the new equilibrium constant will be the corresponding root of the old one (square root, cube root, . . .).

Careful! "raised to power of", don't simply multiply numerical k_c value by 2 if equation coefficients multiplied by 2; you're to take it to the 2nd power

Example 14-2 For the reaction $NH_3 \rightleftharpoons \frac{1}{2}N_2 + \frac{3}{2}H_2$, $K_c = 5.2 \times 10^{-5}$ at 298 K. What is the value of K_c at 298 K for the reaction $N_2 + 3H_2 \rightleftharpoons 2NH_3$?
To obtain the desired equation, the original equation must be (1) reversed and (2) doubled. Thus,

(1) $\frac{1}{2}N_2 + \frac{3}{2}H_2 \rightleftharpoons NH_3$ $K_c(1) = \dfrac{[NH_3]}{[N_2]^{1/2}[H_2]^{3/2}} = \dfrac{1}{5.2 \times 10^{-5}} = 1.9 \times 10^4$

(2) $N_2 + 3H_2 \rightleftharpoons 2NH_3$ $K_c(2) = ?$

$$K_c(2) = \frac{[NH_3]^2}{[N_2][H_2]^3} = \left\{ \frac{[NH_3]}{[N_2]^{1/2}[H_2]^{3/2}} \right\}^2 = \{K_c(1)\}^2 = (1.9 \times 10^4)^2 = 3.6 \times 10^8$$

SIMILAR EXAMPLES: Exercises 4, 5, 6.   *not* $2*(K_c(1))$

COMBINING EQUILIBRIUM CONSTANT EXPRESSIONS. Suppose that we are given the following equilibrium constant data at 25°C.

$$N_2(g) + O_2(g) \rightleftharpoons 2NO(g) \qquad K_c = 4.1 \times 10^{-31} \tag{14.8}$$

$$N_2(g) + \tfrac{1}{2}O_2(g) \rightleftharpoons N_2O(g) \qquad K_c = 2.4 \times 10^{-18} \tag{14.9}$$

and that we wish to establish K_c for the reaction

$$N_2O(g) + \tfrac{1}{2}O_2(g) \rightleftharpoons 2NO(g) \qquad K_c = ? \tag{14.10}$$

We may obtain (14.10) by an appropriate combination of equations (14.8) and (14.9), as follows.

(1)	$N_2(g) + O_2(g) \rightleftharpoons 2NO(g)$	$K_c(1) = 4.1 \times 10^{-31}$
(2)	$N_2O(g) \rightleftharpoons N_2(g) + \tfrac{1}{2}O_2(g)$	$K_c(2) = 1/(2.4 \times 10^{-18})$
		$= 4.2 \times 10^{17}$

net: $N_2O(g) + \tfrac{1}{2}O_2(g) \rightleftharpoons 2NO(g)$ $K_c = ?$

We can write several expressions for K_c for reaction (14.10).

$$K_c(\text{net}) = \frac{[NO]^2}{[N_2O][O_2]^{1/2}} = \underbrace{\frac{[NO]^2}{[N_2][O_2]}}_{K_c(1)} \times \underbrace{\frac{[N_2][O_2]^{1/2}}{[N_2O]}}_{K_c(2)} = K_c(1) \times K_c(2)$$

$$K_c(\text{net}) = K_c(1) \times K_c(2) = 4.1 \times 10^{-31} \times 4.2 \times 10^{17} = 1.7 \times 10^{-13} \qquad (14.11)$$

The important generalization established by (14.11) is that

The equilibrium constant for a net reaction is the product of the equilibrium constants for the individual reactions being added.

THE EQUILIBRIUM CONSTANT EXPRESSED AS K_p. Equilibrium constants for gaseous systems can be based on the partial pressures of gases rather than on molar concentrations. An equilibrium constant written in this way is called a **partial pressure equilibrium constant** and is denoted by the symbol K_p. To illustrate the relationship between K_p and K_c for a reaction, let us consider again reaction (14.6a).

$$2\,SO_2(g) + O_2(g) \rightleftharpoons 2\,SO_3(g) \qquad K_c = 2.8 \times 10^2 \text{ at 1000 K} \qquad (14.6a)$$

$$K_c = \frac{[SO_3]^2}{[SO_2]^2[O_2]}$$

but also according to the ideal gas law, $PV = nRT$ and

$$[SO_3] = \frac{n_{SO_3}}{V} = \left(\frac{P_{SO_3}}{RT}\right) \qquad [SO_2] = \frac{n_{SO_2}}{V} = \left(\frac{P_{SO_2}}{RT}\right)$$

$$[O_2] = \frac{n_{O_2}}{V} = \left(\frac{P_{O_2}}{RT}\right)$$

Substituting the circled terms for concentrations in K_c, we obtain the expression

$$K_c = \frac{(P_{SO_3}/RT)^2}{(P_{SO_2}/RT)^2(P_{O_2}/RT)} = \frac{(P_{SO_3})^2}{(P_{SO_2})^2(P_{O_2})} \times RT \qquad (14.12)$$

The ratio of partial pressures shown in color in (14.12) is the equilibrium constant, K_p. The relationship between K_p and K_c for reaction (14.6a) is

$$K_c = K_p \times RT \qquad \text{and} \qquad K_p = \frac{K_c}{RT} = K_c(RT)^{-1} \qquad (14.13)$$

If a similar derivation were carried out for the general reaction

$$a\,A(g) + b\,B(g) + \cdots \rightleftharpoons g\,G(g) + h\,H(g) + \cdots \qquad (14.14)$$

the result would be

$$K_p = K_c(RT)^{\Delta n} \qquad (14.15)$$

where Δn is the difference in the stoichiometric coefficients of *gaseous* products and reactants; that is, $\Delta n = (g + h + \cdots) - (a + b + \cdots)$. In reaction (14.6a), $\Delta n = 2 - (2 + 1) = -1$, just as noted in equation (14.13).

Example 14-3 Complete the calculation of K_p for reaction (14.6a) from the data given.

If we agree always to express concentrations in K_c on a molar basis and partial pressures in K_p as atm, the appropriate value of R to use in equation (14.15) is 0.0821 L atm mol^{-1} K^{-1}. The solution to equation (14.13) then becomes

Handwritten margin notes:

constant factor

$PV = nRT$

$\frac{P}{RT} = \frac{n}{V} = [\ ]$

$K_c = \frac{[SO_3]^2}{[SO_2]^2[O_2]}$

$= \frac{\left(\frac{n_{SO_3}}{V}\right)^{n_1}}{\left(\frac{n_{SO_2}}{V}\right)^{n_2}\left(\frac{n_{O_2}}{V}\right)^{n_3}} = \frac{\left(\frac{P_{SO_3}}{RT}\right)^{n_1}}{\left(\frac{P_{SO_2}}{RT}\right)^{n_2}\left(\frac{P_{O_2}}{RT}\right)^{n_3}}$

these move to numerator

$= \frac{P_{SO_3}}{(P_{SO_2})(P_{O_2})} * (RT)^{(n_2+n_3) - n_1}$

K_p Unless otherwise noted, the partial pressures in a K_p expression are in atm.

$= K_p * (RT)^{\Delta n}$

K_p & K_c have no definite unit values, vary for each case

We have defined K_c and K_p so as to eliminate the need to attach units to their numerical values. If units were assigned to K_c for equation (14.6a), they would be L mol^{-1}. The unit for K_p would be atm^{-1}. The units for $(RT)^{-1}$ are L^{-1} atm^{-1} mol. For the product $K_c \times (RT)^{-1}$, the units are L mol^{-1} L^{-1} atm^{-1} mol = atm^{-1}, the same as the unit for K_p.

$$K_p = K_c(RT)^{-1} = 2.8 \times 10^2(0.0821 \times 1000)^{-1} = \frac{2.8 \times 10^2}{0.0821 \times 1000} = 3.4$$

SIMILAR EXAMPLES: Exercises 25, 26.

Example 14-4   What is the value of K_p for the hydrogen–iodine–hydrogen iodide reaction at 445°C?

For the reaction $H_2(g) + I_2(g) \rightleftharpoons 2\,HI(g)$, $\Delta n = 0$. This means that in the expression $K_p = K_c(RT)^{\Delta n}$, $K_p = K_c$ (since any number raised to the "0" power has a value of 1). Referring to expression (14.4), we conclude that

$$K_p = K_c = 50.2$$

SIMILAR EXAMPLES: Exercises 25, 26.

EQUILIBRIA INVOLVING PURE LIQUIDS AND SOLIDS (HETEROGENEOUS REACTIONS).
For reactions occurring in a homogeneous gaseous or liquid solution a concentration term (or pressure term) appears in the equilibrium constant expression for each reacting species. For a heterogeneous reaction, concentration terms for solids and liquids will not normally appear because the concentrations (activities) of pure solids and liquids do not change during the course of a chemical reaction. For example, the equilibrium constant expression for the decomposition of calcium carbonate

As discussed in Section 15-6, equilibrium constant expressions should be written in terms of activities, and the activities of pure solids and liquids are defined to be 1.000.

$$CaCO_3(s) \rightleftharpoons CaO(s) + CO_2(g) \qquad (14.16)$$

is written as

$$K_c = \frac{[CaO(s)][CO_2(g)]}{[CaCO_3(s)]} = [CO_2(g)] \qquad (14.17)$$

K_p can be written in a similar fashion, and the relationship between K_p and K_c is that derived from equation (14.15), with $\Delta n = 1$.

$$K_c = [CO_2(g)] \qquad K_p = P_{CO_2} \qquad K_p = K_c(RT) \qquad (14.18)$$

According to (14.18), a simple statement of the equilibrium pressure of $CO_2(g)$ in contact with $CaO(s)$ and $CaCO_3(s)$ is in itself a value of the equilibrium constant, K_p. And the pressure of the CO_2 does not depend on the quantities of $CaO(s)$ and $CaCO_3(s)$ present.

A liquid–vapor equilibrium is a *physical* equilibrium (no chemical reaction is involved), but the principles just presented also apply. For the vaporization equilibrium of water we can write

$$H_2O(l) \rightleftharpoons H_2O(g)$$

$$K_c = [H_2O(g)] \qquad K_p = P_{H_2O} \qquad K_p = K_c(RT) \qquad (14.19)$$

Thus, equilibrium vapor pressures of water can be viewed simply as equilibrium constants, K_p, at different temperatures.

Example 14-5   When equilibrium is established in the following reaction at 60°C, the partial gas pressures are found to be $P_{HI} = 3.65 \times 10^{-3}$ atm and $P_{H_2S} = 9.96 \times 10^{-1}$ atm. What is the value of K_p for the reaction?

$$H_2S(g) + I_2(s) \rightleftharpoons 2\,HI(g) + S(s) \qquad K_p = ?$$

Recall that terms for pure solids do not appear in an equilibrium constant expression. The value of K_p is

$$K_p = \frac{(P_{HI})^2}{(P_{H_2S})} = \frac{(3.65 \times 10^{-3})^2}{9.96 \times 10^{-1}} = 1.34 \times 10^{-5}$$

SIMILAR EXAMPLE: Exercise 3.

14-3 Predicting the Direction and Extent of a Reaction

At each point in the progress of a reaction it is possible to formulate a ratio of concentrations having the same form as the equilibrium constant expression. This generalized ratio is called the **reaction quotient,** often designated by the symbol **Q.** For the generalized reversible reaction (14.14) the reaction quotient is

$$Q = \frac{[G]^g[H]^h \cdots}{[A]^a[B]^b \cdots} \tag{14.20}$$

If the values substituted into the reaction quotient, Q, are a valid set of equilibrium concentrations, then Q will be found to equal K_c.

What is the likelihood that any chosen set of *initial* concentrations for the reactants and products in a reversible reaction will in fact be equilibrium concentrations? The likelihood is exceedingly small! Further reaction must occur in which *all* of the reactant and product concentrations change until the reaction quotient, Q, becomes equal to K_c. Depending on the relationship of Q to K_c, reaction proceeds either in the forward direction (to the right) or in the reverse direction (to the left).

The three experiments pertaining to reaction (14.21) were described through Figure 14-1 and Tables 14-1 and 14-2. They are presented again in Table 14-3.

$$H_2(g) + I_2(g) \rightleftharpoons 2\,HI(g) \qquad K_c = 50.2 \text{ at } 445\,^\circ C \tag{14.21}$$

Let us focus on the *initial* concentrations of the reactants. In experiment 1 only $H_2(g)$ and $I_2(g)$ are present initially. This means that $[HI(g)] = 0$ and the reaction quotient, $Q = 0$; whereas K_c for the reaction is 50.2. We know that in order for equilibrium to be established in experiment 1 some $HI(g)$ must be produced. The reaction proceeds in the forward direction or to the right. As $[HI(g)]$ increases, $[H_2(g)]$ and $[I_2(g)]$ decrease. The reaction quotient, Q, increases in value until it becomes equal to the equilibrium constant, K_c.

A net reaction proceeds from left to right (the forward reaction) if $\qquad Q < K_c$ $\tag{14.22}$

In experiment 2 of Table 14-3, only $HI(g)$ is present initially, no $H_2(g)$ and $I_2(g)$. If $[H_2(g)] = [I_2(g)] = 0$, the reaction quotient is infinitely large, $Q = \infty$. Again, the value of $K_c = 50.2$. In this situation we know that in order for equilibrium to be established, reaction must proceed in the *reverse* direction, that is, to the left. In this way the concentrations of $H_2(g)$ and $I_2(g)$ increase while $[HI(g)]$ decreases. Ultimately, the value of Q becomes equal to K_c and equilibrium is established.

A net reaction proceeds from right to left (the reverse direction) if $\qquad Q > K_c$ $\tag{14.23}$

In experiment 3 of Table 14-3, all three reactants are present initially and the direction of the reaction is not immediately obvious. However, consideration of the criteria just established shows that (14.22) applies; that is, $Q = 1$ is less than $K_c = 50.2$. The reaction proceeds to the right.

The criteria for predicting the *direction* of chemical change in a reversible reaction are illustrated in Figure 14-2 and applied in Example 14-6. Predicting the *extent* of reaction (i.e., actual equilibrium concentrations from initial concentrations) requires additional algebraic calculations and is the subject of Example 14-11.

TABLE 14-3
Predicting the direction of change in a reversible chemical reaction $H_2(g) + I_2(g) \rightleftharpoons 2\,HI(g)$; $K_c = 50.2$[a]

Experiment	Initial concentrations, mol/L[b]			Initial reaction quotient $Q = \dfrac{[HI(g)]^2}{[H_2(g)][I_2(g)]}$
	$[H_2(g)]$	$[I_2(g)]$	$[HI(g)]$	
1	1.88×10^{-3}	1.88×10^{-3}	0	$Q = \dfrac{0}{(1.88 \times 10^{-3})(1.88 \times 10^{-3})} = 0$
2	0	0	1.88×10^{-3}	$Q = \dfrac{(1.88 \times 10^{-3})^2}{0 \times 0} = \infty$
3	1.88×10^{-3}	1.88×10^{-3}	1.88×10^{-3}	$Q = \dfrac{(1.88 \times 10^{-3})^2}{(1.88 \times 10^{-3})(1.88 \times 10^{-3})} = 1.00$

[a] Temperature = 445°C; reaction volume = 0.800 L.
[b] Initial concentrations are obtained by dividing initial numbers of moles (Table 14-1) by the reaction volume, 0.800 L.

Volume terms will cancel from a reaction quotient or an equilibrium constant expression only if the total of the exponents of the concentration terms in the numerator equals that in the denominator.

FIGURE 14-2
Predicting the direction of change in a reversible reaction.

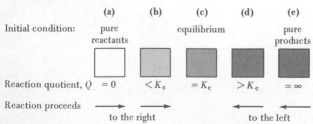

	(a)	(b)	(c)	(d)	(e)
Initial condition:	pure reactants		equilibrium		pure products

| Reaction quotient, Q | $= 0$ | $< K_c$ | $= K_c$ | $> K_c$ | $= \infty$ |

Reaction proceeds   → → to the right   ← ← to the left

From Table 14-3 Experiment 1 corresponds to initial condition (a), Experiment 2 to condition (e), and Experiment 3 to (b). The situation in Example 14-6 corresponds to condition (d).

Example 14-6 For the reaction $CO(g) + H_2O(g) \rightleftharpoons CO_2(g) + H_2(g)$, $K_c = 1.00$ at about 1100 K. The following amounts of substances are brought together at this temperature and allowed to react: 1.00 mol CO 1.00 mol H_2O, 2.00 mol CO_2, and 2.00 mol H_2. Relative to their initial amounts, which of the reactants will be present in greater amount and which, in lesser amount, when equilibrium is established?

Basically, all that is required here is to determine the direction in which the reaction proceeds; and for this we use the criteria (14.22) and (14.23). To substitute concentrations into the reaction quotient, we assume an arbitrary reaction volume, V. Its value proves to be immaterial since in this case volume cancels out.

$$Q = \frac{[CO_2][H_2]}{[CO][H_2O]} = \frac{(2.00/V)(2.00/V)}{(1.00/V)(1.00/V)} = 4.00$$
$$4.00 > K_c = 1.00$$

Because $Q > K_c$ reaction proceeds to the left. When equilibrium is established, the amounts of CO_2 and H_2 will have *decreased* from their initial values and the amounts of CO and H_2O will have *increased*.

SIMILAR EXAMPLES: Exercises 16, 17, 21.

14-4 Equilibrium Calculations— Some Illustrative Examples

Calculations relating to the condition of chemical equilibrium are among the most important encountered in chemistry. Such calculations are the subject of the next several chapters. In this section we consider several examples that employ the general equilibrium principles established in the first three sections of the chapter. Each example is followed by a brief section labeled

Comparison of Q and K_c	Direction of net chemical reaction
$Q < K_c$	to the right
$Q > K_c$	to the left
$Q < K_c$	to the right

"comments," which describes the special features illustrated. The collection of these "comments" constitutes the basic methodology of equilibrium calculations.

Example 14-7   Equilibrium is established in the reaction $N_2O_4(g) \rightleftharpoons 2\,NO_2(g)$ at 25°C. The quantities of reactant and product present in a 3.00-L vessel are 7.64 g N_2O_4 and 1.56 g NO_2. What is the value of K_c for this reaction?

Equilibrium amounts, mol	Equilibrium concentrations, mol/L
$7.64 \text{ g N}_2\text{O}_4 \times \dfrac{1 \text{ mol N}_2\text{O}_4}{92.0 \text{ g N}_2\text{O}_4} = 0.0830 \text{ mol N}_2\text{O}_4$	$[\text{N}_2\text{O}_4] = \dfrac{0.0830 \text{ mol N}_2\text{O}_4}{3.00 \text{ L}} = 0.0277\,M$
$1.56 \text{ g NO}_2 \times \dfrac{1 \text{ mol NO}_2}{46.0 \text{ g NO}_2} = 0.0339 \text{ mol NO}_2$	$[\text{NO}_2] = \dfrac{0.0339 \text{ mol NO}_2}{3.00 \text{ L}} = 0.0113\,M$

$$K_c = \frac{[\text{NO}_2]^2}{[\text{N}_2\text{O}_4]} = \frac{(1.13 \times 10^{-2})^2}{2.77 \times 10^{-2}} = 4.61 \times 10^{-3}$$

SIMILAR EXAMPLES: Exercises 7, 13.

Comments:
Correct substitutions must be made into an equilibrium constant expression, K_c. To ensure that this is done, it is helpful to tabulate the data and label each item carefully. Equilibrium concentrations in mol/L must be used, *not* simply equilibrium amounts in moles.

Example 14-8   A 0.0200-mol sample of SO_3 is introduced into an evacuated 1.52-L vessel and heated to 900 K, where equilibrium is established. The amount of SO_3 present at equilibrium is found to be 0.0142 mol. What are the values of **(a)** K_c and **(b)** K_p at 900 K for the reaction

$$2\,SO_3(g) \rightleftharpoons 2\,SO_2(g) + O_2(g)?$$

In the table of data below, the key item is the amount of SO_3 that undergoes decomposition: $(0.0142 - 0.0200)$ mol $SO_3 = -0.0058$ mol SO_3. (The negative sign signifies that this amount of reactant is consumed.) In the row labeled "change" we must relate the changes in amounts of SO_2 and O_2 to this change in amount of SO_3. For this we use the balanced equation, in particular the stoichiometric coefficients 2, 2, and 1. That is, 1 mol SO_2 and $\frac{1}{2}$ mol O_2 are produced for every mole of SO_3 consumed.

the reaction:	$2SO_3(g)$	$\rightleftharpoons$	$2SO_2(g)$	$+$	$O_2(g)$
initial amounts:	0.0200 mol		0.00 mol		0.00 mol
change:	-0.0058 mol		0.0058 mol		0.0029 mol
equilibrium amounts:	0.0142 mol		0.0058 mol		0.0029 mol
equilibrium concentrations:	$[SO_3]$ $= 0.0142$ mol/1.52 L $= 9.34 \times 10^{-3}\ M$		$[SO_2]$ $= 0.0058$ mol/1.52 L $= 3.8 \times 10^{-3}\ M$		$[O_2]$ $= 0.0029$ mol/1.52 L $= 1.9 \times 10^{-3}\ M$

(a) $K_c = \dfrac{[SO_2]^2[O_2]}{[SO_3]^2} = \dfrac{(3.8 \times 10^{-3})^2(1.9 \times 10^{-3})}{(9.34 \times 10^{-3})^2} = 3.1 \times 10^{-4}$

(b) $K_p = K_c(RT)^{\Delta n} = 3.1 \times 10^{-4}(0.0821 \times 900)^{(2+1)-2}$
$\qquad = 3.1 \times 10^{-4}(0.0821 \times 900)^1 = 2.3 \times 10^{-2}$

SIMILAR EXAMPLES: Exercises 8 through 11.

Comments:
1. The chemical equation for a reversible reaction can serve *both* to establish the form of the equilibrium constant expression *and* to provide the conversion factors to relate the equilibrium amounts and concentrations to the specified initial conditions.
2. Whether working with K_c or K_p or the relationship between them, these expressions must be based on the chemical equation that is given, not on what may have been used in other situations: K_p and K_c were related for the sulfur dioxide–oxygen–sulfur trioxide reaction in Example 14-3, but the result here is different because the form of the chemical equation is different.

Example 14-9 Ammonium hydrogen sulfide, NH_4HS, dissociates appreciably, even at room temperature.

$$NH_4HS(s) \rightleftharpoons NH_3(g) + H_2S(g) \qquad K_p = 1.08 \times 10^{-1} \text{ at } 25°C$$

Just as we know

$n_{NH_3} = n_{H_2S}$

we know partial p's

$P_{NH_3} = P_{H_2S}$

A sample of $NH_4HS(s)$ is introduced into an evacuated flask and allowed to establish equilibrium at 25°C. What is the total gas pressure at equilibrium?

K_p for this reaction is simply the product of the partial pressures of $NH_3(g)$ and $H_2S(g)$, each stated in atm. Moreover, since these gases are produced in equimolar amounts by the dissociation of NH_4HS, $P_{NH_3} = P_{H_2S}$; and $P_{tot.} = P_{NH_3} + P_{H_2S} = 2 \times P_{NH_3}$:

$$K_p = (P_{NH_3})(P_{H_2S}) = (P_{NH_3})(P_{NH_3}) = (P_{NH_3})^2 = 1.08 \times 10^{-1} = 10.8 \times 10^{-2}$$

$$P_{NH_3} = \sqrt{10.8 \times 10^{-2}} = 3.29 \times 10^{-1}\text{ atm}$$

$$P_{tot.} = 2 \times P_{NH_3} = 2 \times 3.29 \times 10^{-1}\text{ atm} = 0.658\text{ atm}$$

SIMILAR EXAMPLES: Exercises 27, 28, 29.

Comments:
When writing a K_p equilibrium constant expression look for relationships among partial pressures of the reactants. If the total gas pressure needs to be related to reactant partial pressures, this can be done through equations presented in Chapter 5 (e.g., equations 5.13, 5.14, or 5.16).

this is initial value

Example 14-10 An 0.0240-mol sample of $N_2O_4(g)$ is allowed to dissociate and come to equilibrium with $NO_2(g)$ in an 0.372-L flask at 25°C. What is the percent dissociation of the N_2O_4?

FIGURE 14-3
Equilibrium in the reaction $N_2O_4(g) \rightleftharpoons 2 NO_2(g)$ at 25°C—Example 14-10 illustrated.

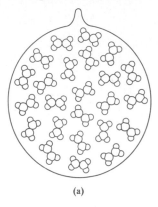

(a)

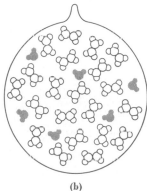

(b)

$\text{\small◯◯}$ – N_2O_4 $\text{\small●}$ = NO_2

Each "molecule" illustrated represents 0.001 mol.
(a) Initially, pure N_2O_4 is introduced into an evacuated glass bulb and the bulb is sealed. The illustration shows 24 "molecules," corresponding to 0.024 mol N_2O_4.
(b) At equilibrium, some molecules of N_2O_4 have dissociated to NO_2 (shown in grey). The illustration contains 21 N_2O_4 and 6 NO_2 "molecules," corresponding to 0.021 mol N_2O_4 and 0.006 mol NO_2.

$$N_2O_4(g) \rightleftharpoons 2 NO_2(g) \qquad K_c = 4.61 \times 10^{-3} \text{ at } 25°C$$

By percent dissociation we mean the percent of the N_2O_4 molecules present initially that are converted to NO_2 (see Figure 14-3). This requires a determination of the number of moles of reactant and product at equilibrium. But now for the first time we must introduce an algebraic unknown, x. Suppose that we let x = the number of moles of N_2O_4 that dissociate. We enter this value into the row labeled "change" in the table below. The amount of NO_2 produced is $2x$.

the reaction:	$N_2O_4(g)$	$\rightleftharpoons$	$2 NO_2(g)$
initial amounts:	0.0240 mol		0.00 mol
change:	$-x$ mol		$+2x$ mol
equilibrium amounts:	$(0.0240 - x)$ mol		$2x$ mol
equilibrium concentrations:	$[N_2O_4]$ = $(0.0240 - x \text{ mol})/0.372$ L		$[NO_2]$ = $2x$ mol/0.372 L

The equation that must be solved here is a quadratic equation of the form $ax^2 + bx + c = 0$. For such an equation the general solution is

$$x = \frac{-b \pm \sqrt{b^2 - 4ac}}{2a}$$

The symbol $\pm$ in this equation signifies that there are two possible roots. Of the two, only one can have physical meaning. In this problem x is defined to be a *positive* quantity and it must be smaller than 0.0240.

$$K_c = \frac{[NO_2]^2}{[N_2O_4]} = \frac{\left(\frac{2x}{0.372}\right)^2}{\frac{0.0240 - x}{0.372}} = \frac{4x^2}{0.372(0.0240 - x)}$$

$$= 4.61 \times 10^{-3}$$

$$4x^2 = 4.12 \times 10^{-5} - (1.71 \times 10^{-3})x$$

$$x^2 + (4.28 \times 10^{-4})x - 1.03 \times 10^{-5} = 0$$

$$x = \frac{-4.28 \times 10^{-4} \pm \sqrt{(4.28 \times 10^{-4})^2 + 4 \times 1.03 \times 10^{-5}}}{2}$$

$$x = \frac{-4.28 \times 10^{-4} \pm \sqrt{(1.83 \times 10^{-7}) + (4.12 \times 10^{-5})}}{2}$$

$$x = \frac{-4.28 \times 10^{-4} \pm \sqrt{4.14 \times 10^{-5}}}{2}$$

$$x = \frac{-4.28 \times 10^{-4} \pm 6.43 \times 10^{-3}}{2}$$

$$= \frac{-4.28 \times 10^{-4} + 6.43 \times 10^{-3}}{2}$$

$$= \frac{6.00 \times 10^{-3}}{2}$$

$$x = 3.00 \times 10^{-3} \text{ mol } N_2O_4 \quad \text{(dissociated)}$$

The percent dissociation of the N_2O_4 is given by the expression

$$\% \text{ dissoc. of } N_2O_4 = \frac{3.00 \times 10^{-3} \text{ mol } N_2O_4 \text{ dissoc.}}{0.0240 \text{ mol } N_2O_4 \text{ initially}} \times 100$$

$$= 12.5\%$$

SIMILAR EXAMPLES: Exercises 30, 31, 33.

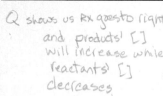

Remember:

[Products]
────────
[reactants]

Comments:
When one or more of the quantities in an equilibrium constant expression must be stated in terms of the algebraic unknown, x, there are usually several ways in which x can be defined. No hard-and-fast rules need be given, except to note that it is necessary to establish through this definition whether x is:

1. An *amount* of substance (mol) or a *concentration* (mol/L).
2. Stated in terms of a reactant *consumed* or a *product* formed.
3. A positive or a negative quantity. (A definition that makes x a positive quantity is generally preferred.)

Example 14-11   A solution at 25°C is prepared having the initial concentrations: $[Fe^{3+}] = 0.5000\,M$; $[Hg_2^{2+}] = 0.5000\,M$; $[Fe^{2+}] = 0.03000\,M$; $[Hg^{2+}] = 0.03000\,M$. The following reaction occurs among these ions.

$$2\,Fe^{3+}(aq) + Hg_2^{2+}(aq) \rightleftharpoons 2\,Fe^{2+}(aq) + 2\,Hg^{2+}(aq) \qquad K_c = 9.14 \times 10^{-6} \text{ at } 25°C$$

What will be the ionic concentrations when equilibrium is established?
Although it is not necessary to do so, a comparison of Q and K_c may help us to visualize the solution to this problem.

$$Q = \frac{[Fe^{2+}]^2[Hg^{2+}]^2}{[Fe^{3+}]^2[Hg_2^{2+}]} = \frac{(0.03000)^2(0.03000)^2}{(0.5000)^2(0.5000)} = \frac{8.10 \times 10^{-7}}{1.25 \times 10^{-1}}$$

$$= 6.48 \times 10^{-6} < K_c = 9.14 \times 10^{-6}$$

Since Q is smaller than K_c, a reaction must proceed to the right (recall criterion 14.22). Let us define x as the number of moles per liter of Fe^{3+} that are converted to Fe^{2+}. The several equilibrium concentrations can then be expressed in terms of x, as shown below.

the reaction:	$2\,Fe^{3+}(aq)$	$+$	$Hg_2^{2+}(aq)$	$\rightleftharpoons$	$2\,Fe^{2+}(aq)$	$+$	$2\,Hg^{2+}(aq)$
initial concentrations:	$0.5000\,M$		$0.5000\,M$		$0.03000\,M$		$0.0300\,M$
change:	$-x\,M$		$-x/2\,M$		$+x\,M$		$+x\,M$
equilibrium concentrations:	$(0.5000 - x)\,M$		$(0.5000 - x/2)\,M$		$(0.03000 + x)\,M$		$(0.03000 + x)\,M$

Q shows us Rx goes to right and products' [ ] will increase while reactants' [ ] decreases

$$K_c = \frac{(0.03000 + x)^2(0.03000 + x)^2}{(0.5000 - x)^2(0.5000 - x/2)} = 9.14 \times 10^{-6}$$

Solving this equation can be simplified greatly if the following assumption is made, and *if the assumption proves valid.* If x is much smaller than 0.5000, then $(0.5000 - x) \simeq 0.5000$, and $(0.5000 - x/2) \simeq 0.5000$. This assumption leads to the expression

assume "x" is so small the difference in subtraction is insignificant.

$$K_c = \frac{(0.03000 + x)^2(0.03000 + x)^2}{(0.5000)^2(0.5000)} = 9.14 \times 10^{-6}$$

$$(0.03000 + x)^4 = 1.14 \times 10^{-6} = 114 \times 10^{-8}$$

Take the *fourth* root of each side of this equation (i.e., take the square root twice).

$$(0.03000 + x)^2 = 10.7 \times 10^{-4}$$

$$(0.03000 + x) = 3.27 \times 10^{-2}$$

$$x = 3.27 \times 10^{-2} - 0.03000 = 2.7 \times 10^{-3}$$

We see that the simplifying assumption is indeed valid: 2.7×10^{-3} is much smaller than 0.5000.

TABLE 14-4
Equilibrium constants, K_p, for the reaction
$2\,SO_2(g) + O_2(g) \rightleftharpoons 2\,SO_3(g)$
at several temperatures

T, K	$1/T$, K^{-1}	K_p	$\log K_p$
800	12.5×10^{-4}	9.1×10^2	2.96
850	11.8×10^{-4}	1.7×10^2	2.23
900	11.1×10^{-4}	4.2×10^1	1.62
950	10.5×10^{-4}	1.0×10^1	1.00
1000	10.0×10^{-4}	3.2×10^0	0.51
1050	9.5×10^{-4}	1.0×10^0	0.00
1100	9.1×10^{-4}	3.9×10^{-1}	−0.41
1170	8.5×10^{-4}	1.2×10^{-1}	−0.92

FIGURE 14-4
Temperature dependence of the equilibrium constant
K_p for the reaction $2\,SO_2(g) + O_2(g) \rightleftharpoons 2\,SO_3(g)$.

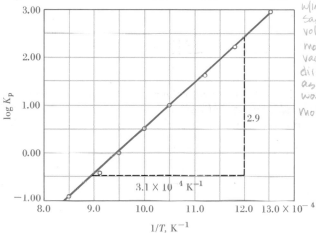

all are w/in the same volume, moles vary directly as would molarity.

This graph can be used to establish the heat of reaction, $\Delta \bar{H}^\circ$ (see equation 14.24).

$$\text{slope} = \frac{-\Delta \bar{H}^\circ}{2.303R} = \frac{2.9}{3.1 \times 10^{-4}\ \text{K}^{-1}} = 9.4 \times 10^3\ \text{K}$$

$$\Delta \bar{H}^\circ = -2.303 \times 8.314\ \text{J mol}^{-1}\,\text{K}^{-1} \times 9.4 \times 10^3\ \text{K}$$
$$= -1.8 \times 10^5\ \text{J/mol} = -180\ \text{kJ/mol}$$

Note the similarity of this equation (14.24) to equation (11.2), which describes the variation of the vapor pressure of a liquid with temperature. And in this chapter we have demonstrated that the vapor pressure of a substance is itself a K_p value (equation 14.19).

$A(\ell) \rightleftharpoons A(g)$

$K_p = P_{A(g)}$

Equilibrium concentrations:

$[\text{Fe}^{2+}] = 0.03000 + x = 0.03000 + 2.7 \times 10^{-3}$
$= 3.27 \times 10^{-2}\ M$

$[\text{Hg}^{2+}] = 0.03000 + x = 0.03000 + 2.7 \times 10^{-3}$
$= 3.27 \times 10^{-2}\ M$

$[\text{Fe}^{3+}] = 0.5000 - x = 0.5000 - 2.7 \times 10^{-3}$
$= 4.97 \times 10^{-1}\ M$

$[\text{Hg}_2^{2+}] = 0.5000 - x/2 = 0.5000 - 1.4 \times 10^{-3}$
$= 4.99 \times 10^{-1}\ M$

SIMILAR EXAMPLES: Exercises 19, 22, 23, 24.

Comments:
1. It is sometimes useful to compare the reaction quotient, Q, to the equilibrium constant, K_c, to determine the direction in which a reaction will proceed.
2. In some equilibrium calculations—often those involving species in aqueous solution—one can work with molar concentrations exclusively. No specific reference is made to moles of reactants and solution volumes.
3. Algebraic solutions can often be greatly simplified if one recognizes certain relationships among the terms in the equilibrium constant expression. Usually, these simplifications take the form of x being much smaller than some other numerical value to which it is added or from which it is subtracted.

14-5 The Effect of Temperature on Equilibrium

Kp & Kc

In general, the equilibrium constant for a reaction is temperature-dependent. Values of K_p for the sulfur dioxide–oxygen–sulfur trioxide reaction at several temperatures are tabulated in Table 14-4, together with familiar functions of these data—$\log K$ and $1/T$. Figure 14-4 shows that a plot of $\log K$ vs. $1/T$ yields a straight line. The equation of this straight line is

equation of straight line:

$$\underbrace{\log K}_{y} = \underbrace{\frac{-\Delta \bar{H}^\circ}{2.303R}}_{m} \cdot \underbrace{\frac{1}{T}}_{x} + \underbrace{\text{constant}}_{b} \quad (14.24)$$

Moreover, as illustrated twice previously (in establishing equations 11.3 and 13.30), the constant term can be eliminated from equation (14.24) to yield a result with a familiar form (called the van't Hoff equation).

$$\log \frac{K_2}{K_1} = \frac{\Delta \bar{H}^\circ}{2.303R}\left(\frac{T_2 - T_1}{T_2 T_1}\right) = \frac{\Delta H}{2.303R}\left(\frac{1}{T_1} - \frac{1}{T_2}\right) \quad (14.25)$$

K_2 and K_1 are the equilibrium constants at the kelvin temperatures T_2 and T_1. $\Delta \bar{H}^\circ$ is the standard molar heat of reaction. Both positive and negative values of $\Delta \bar{H}^\circ$ are possible. Equilibrium constant values may increase with temperature (if $\Delta \bar{H}^\circ > 0$) or decrease with temperature (if $\Delta \bar{H}^\circ < 0$). The assumption is made in equation (14.25) that $\Delta \bar{H}^\circ$ is independent of temperature; and in most cases it is reasonably so.

Example 14-12 Use data from Table 14-4 and Figure 14-4 to estimate the temperature at which $K_p = 1.0 \times 10^6$ for the reaction

$$2 \, SO_2(g) + O_2(g) \rightleftharpoons 2 \, SO_3(g)$$

For substitution into equation (14.25), use a known value from Table 14-4 and the heat of reaction established through Figure 14-4. That is,

$$\begin{cases} T_2 = 800 \text{ K} & K_2 = 9.1 \times 10^2 \\ T_1 = ? & K_1 = 1.0 \times 10^6 \end{cases} \quad \Delta \bar{H}^\circ = -1.8 \times 10^5 \, J/mol$$

$$\log \frac{9.1 \times 10^2}{1.0 \times 10^6} = \log 9.1 \times 10^{-4} = -3.04 = \frac{-1.8 \times 10^5}{2.303 \times 8.314} \left(\frac{800 - T_1}{800 T_1} \right)$$

$$\frac{2.303 \times 8.314 \times 3.04 \times 800}{1.8 \times 10^5} T_1 = 800 - T_1$$

$$0.26 T_1 = 800 - T_1$$

$$1.26 T_1 = 800 \qquad T_1 = \frac{800}{1.26} = 635 \text{ K}$$

SIMILAR EXAMPLES: Exercises 36, 37.

14-6 Altering Equilibrium Conditions— Le Châtelier's Principle

To this point we have emphasized quantitative solutions to questions concerning equilibrium conditions. Yet there are times when *qualitative* statements about equilibrium are sufficient. Moreover, in cases where necessary data are not available, qualitative descriptions are all that can be given.

A useful qualitative statement about the influence that different factors produce on the condition of equilibrium is attributed to the French chemist Henri Le Châtelier (1884). Le Châtelier's principle is easier to illustrate than to state. Nevertheless, let us begin with this statement.

If a system at equilibrium is disturbed, the equilibrium condition is upset. A reaction proceeds in that direction that tends to relieve the disturbing influence.

Systems at equilibrium can be disturbed by adding or removing reactants or products, or by changing the temperature, pressure, or volume of the system.

EFFECT OF CHANGING CONCENTRATIONS. Let us return to the equilibrium at 1000 K for the reaction

$$2 \, SO_2(g) + O_2(g) \rightleftharpoons 2 \, SO_3(g) \qquad K_c = 2.8 \times 10^2 \text{ at } 1000 \text{ K} \qquad (14.6a)$$

Figure 14-5a pictures a particular equilibrium mixture for this reaction, and Figure 14-5b, a disturbance in the form of adding 1.00 mol SO_3 to the mixture. How will the amounts of the reactants change to reestablish equilibrium? One approach is to calculate the reaction quotient, Q, immediately after the addition of the 1.00 mol SO_3. The addition of any quantity of SO_3 to an equilibrium mixture makes the value of Q larger than that of K_c. The reaction proceeds to the left, that is, in the reverse direction. The amount of each species present in the new equilibrium can be calcu-

FIGURE 14-5
Changing equilibrium conditions by increasing the amount of one of the reactants in the reaction $2 SO_2(g) + O_2(g) \rightleftharpoons 2 SO_3(g)$; $K_c = 280$ at 1000 K.

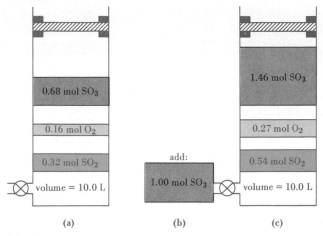

(a) The original equilibrium condition.
(b) Disturbance caused by the addition of 1.00 mol SO_3.
(c) The new equilibrium condition.

*Remember!
Prod & react never
completely consumed.*

FIGURE 14-6
Effect of volume change on equilibrium condition in the reaction $2 SO_2(g) + O_2(g) \rightleftharpoons 2 SO_3(g)$; $K_c = 280$ at 1000 K.

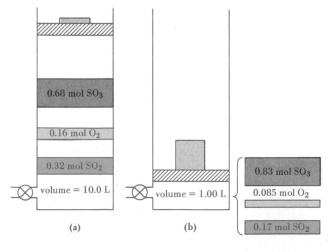

*decrease in Volume shifts
to side w/smallest #
moles reacting species*

*For this
case only,
not always
will products
increase, Depends on ✳*

lated by the methods of Section 14-4. These conditions are represented in Figure 14-5c.

An alternative, *qualitative* approach is simply to note that the addition of SO_3 to the original equilibrium mixture represents a disturbance that can only be relieved by the removal of some of the added SO_3. This is the result when reaction proceeds to the left—some of the added SO_3 is converted to SO_2 and O_2. In the new equilibrium there should be greater amounts of all the reactants than in the original equilibrium (although obviously the additional amount of SO_3 must be less than 1.00 mol).

Example 14-13 Predict the effect of introducing additional $H_2(g)$ into an equilibrium mixture of $N_2(g)$, $H_2(g)$, and $NH_3(g)$.

$$N_2(g) + 3 H_2(g) \rightleftharpoons 2 NH_3(g)$$

The disturbance here is relieved as reaction proceeds from left to right. However, only a portion of the added $H_2(g)$ is consumed in this reaction. When equilibrium is reestablished, there will still be more $H_2(g)$ than present originally. The amount of $NH_3(g)$ will also be greater, but the amount of $N_2(g)$ will have *decreased*. [$N_2(g)$ must be consumed along with $H_2(g)$ if a reaction is to occur from left to right. Note that additional $H_2(g)$ was added to the original equilibrium mixture but no additional $N_2(g)$.]

SIMILAR EXAMPLES: Exercises 41, 42.

CHANGE OF VOLUME. The equilibrium mixture of Figure 14-6a is subjected to a different kind of stress in Figure 14-6b. Here, nothing is added or removed from the reaction mixture. Instead, the reaction volume is decreased to one tenth of its original value by increasing the external pressure on the mixture. Once more a readjustment of equilibrium amounts of the reactants must occur in accord with the expression for K_c.

$$K_c = \frac{[SO_3]^2}{[SO_2]^2[O_2]} = \frac{(n_{SO_3}/V)^2}{(n_{SO_2}/V)^2(n_{O_2}/V)} \quad \left[= \frac{(n_{SO_3})^2 \left(\frac{1}{V^2}\right)}{(n_{SO_2})^2(n_{O_2}) \left(\frac{1}{V^3}\right)} \right]$$

$$\left[= \frac{(n_{SO_3})}{(n_{SO_2})(n_{O_2})} \left(\frac{V^3}{V^2}\right) \right] = \frac{(n_{SO_3})^2}{(n_{SO_2})^2(n_{O_2})} \cdot V = 2.8 \times 10^2$$

$$(14.26)$$

From equation (14.26) it follows that, if V is reduced by a factor of 10, the ratio

$$\frac{(n_{SO_3})^2}{(n_{SO_2})^2(n_{O_2})}$$

*since K_c remains constant,
however Volume Varies, $\frac{n_{SO_2}}{n_{SO_2}n_{O_2}}$
must vary to counteract.*

must increase by a factor of 10. The equilibrium amount of SO_3 must increase and the amounts of SO_2 and O_2 must decrease. A quantitative description of the new equilibrium amounts is given in Figure 14-6b.

In the reaction $2 SO_2(g) + O_2(g) \rightleftharpoons 2 SO_3(g)$, 3 mol of *gases* on the left produces 2 mol of *gases* on the right. The product of the reaction, SO_3, occupies a smaller volume than the reactants from which it is formed. Thus, forcing an equilibrium system into a smaller volume results in the production of additional SO_3.

*When the volume of a reaction mixture involving **gases** is decreased, a net reaction proceeds in the direction in which the number of moles of **gases** becomes smaller. If the reaction volume is increased, the net reaction proceeds in the direction producing a larger number of moles of **gases**.*

CHANGE OF PRESSURE. The disturbance pictured in Figure 14-6 can be described as a change in external pressure as well as a change in volume. After all, the reduction in gas volume was produced by an increase in external pressure. However, increasing the total gas pressure in Figure 14-6 by adding an inert gas would *not* affect the equilibrium condition. This would leave the number of moles per liter of each reacting species unchanged. In short, a change in pressure will alter an equilibrium condition only if this pressure change produces changes in concentrations.

The effect of pressure changes on equilibria involving condensed phases is generally insignificant, because solids and liquids are practically incompressible. (See Exercise 48 for a notable exception, however.)

EFFECT OF TEMPERATURE. We have already considered the effect of temperature on the equilibrium constant, K. From the standpoint of Le Châtelier's principle, changing the temperature of an equilibrium mixture is equivalent to adding heat to or removing heat from the system. The disturbance produced by adding heat is minimized when reaction proceeds in the direction of the heat absorbing (endothermic) reaction. Lowering the temperature requires removal of heat from a system. The heat evolving (exothermic) reaction is favored in this case. (The system attempts to replace the heat that is being removed.) In summary:

Raising the temperature of an equilibrium mixture causes the equilibrium condition to shift in the direction of the endothermic reaction. Lowering the temperature causes a shift in the direction of the exothermic reaction.

Example 14-14 Is the conversion of $SO_2(g)$ to $SO_3(g)$ favored at high or low temperatures?

$$2 SO_2(g) + O_2(g) \rightleftharpoons 2 SO_3(g) \qquad \Delta \bar{H}° = -180 \text{ kJ/mol}$$

Raising the temperature favors the endothermic reaction, the reverse reaction above. To favor the forward (exothermic) reaction requires that the temperature be lowered. Therefore, conversion of $SO_2(g)$ to $SO_3(g)$ is favored at *low* temperatures.

SIMILAR EXAMPLES: Exercises 46, 47.

14-7 Kinetic Basis of the Equilibrium Constant

Several times we have referred to the fact that when equilibrium is established, the rates of a forward and a reverse reaction become equal. Can this statement serve as the basis for a theoretical derivation of the equilibrium constant expression? The first attempts to relate reaction rates and the equilibrium constant are generally attributed to Guldberg and Waage (see again the marginal note on page 318). They proposed that the equilibrium condition could be described simply by equating the rate laws for the forward and reverse reactions. However, this formulation did not always lead to the equilibrium constant expression. Guldberg and Waage did not always use coefficients from the balanced equation as the exponents in their rate law expressions. Other investigators, including van't Hoff, seem to have arrived at the correct form of the equilibrium constant expression (14.5) before Guldberg and Waage.

Let us pursue Guldberg and Waage's lead a bit further. For this we use as an illustrative example the hydrogen–iodine–hydrogen iodide reaction

$$H_2(g) + I_2(g) \underset{k_2}{\overset{k_1}{\rightleftharpoons}} 2\,HI(g) \tag{14.27}$$

Suppose that the *mechanism* of this reaction involves simple one-step bimolecular processes of the type first described in Section 13-10. We would write the following simple rate laws and equate them for the equilibrium condition. Rearrangement of equation (14.28) then leads to an expression that has exactly the same form as the equilibrium constant established in (14.4).

$$\text{rate forward} = k_1[H_2][I_2] = k_2[HI]^2 = \text{rate reverse} \tag{14.28}$$

$$K_c = \frac{k_1}{k_2} = \frac{[HI]^2}{[H_2][I_2]} \tag{14.29}$$

But there is a serious objection to this derivation! An alternate mechanism was presented in Section 13-10 for reaction (14.27). This was the two-step mechanism outlined through equations (14.30) and (14.31). What is the equilibrium condition when described by this alternate mechanism?

$$\text{(fast)} \qquad\qquad I_2(g) \underset{k_2}{\overset{k_1}{\rightleftharpoons}} 2\,I(g) \tag{14.30}$$

$$\text{(slow)} \quad 2\,I(g) + H_2(g) \underset{k_4}{\overset{k_3}{\rightleftharpoons}} 2\,HI(g) \tag{14.31}$$

Let us first establish the *steady-state condition* for each of these steps.

$$\text{(fast)} \qquad k_1[I_2] = k_2[I]^2 \quad \textit{and} \quad \frac{k_1}{k_2} = \frac{[I]^2}{[I_2]} \tag{14.32}$$

$$\text{(slow)} \quad k_3[I]^2[H_2] = k_4[HI]^2 \quad \textit{and} \quad \frac{k_3}{k_4} = \frac{[HI]^2}{[H_2][I]^2} \tag{14.33}$$

rate governing step

Now, solve equation (14.32) for $[I]^2$ and substitute into equation (14.33) to obtain

$$\frac{k_3}{k_4} = \frac{[HI]^2}{\dfrac{k_1}{k_2}[I_2][H_2]} \quad \textit{and} \quad \frac{k_1 k_3}{k_2 k_4} = \frac{[HI]^2}{[I_2][H_2]} = K_c \tag{14.34}$$

Again we obtain an expression having the same form as the equilibrium constant expression first written in equation (14.4)! Actually, <u>any plausible mechanism of a reversible reaction can be used to derive an equilibrium constant expression of the expected form</u> (i.e., as written in equation 14.5).

EFFECT OF A CATALYST ON EQUILIBRIUM.   The presence of a catalyst in a reversible reaction has the effect of speeding up *both* the forward and reverse reactions. The condition of equilibrium is reached more quickly, but the equilibrium amounts of the reacting species are *unchanged* by the catalyst. Thus, for a given set of reaction conditions the equilibrium amounts of $SO_2(g)$, $O_2(g)$, and $SO_3(g)$ have fixed values. This is true whether the reaction is carried out as a slow homogeneous gas-phase

[handwritten margin notes: catalyst just changes rx rate so that pt. of equilibrium is reached either sooner or later than normal, but same equilibrium conditions are reached.]

reaction or a faster heterogeneous reaction on the surface of a catalyst. Or, stated in another way, the presence of a catalyst does not change the numerical value of the equilibrium constant. *because concentrations aren't changing.*

$$2 SO_2(g) + O_2(g) \rightleftharpoons 2 SO_3(g) \qquad K_c = 2.8 \times 10^2 \text{ at } 1000 \text{ K} \qquad (14.6a)$$

The role of a catalyst is to change the mechanism of a chemical reaction to one involving a lower activation energy (recall Figure 13-14). Also, a catalyst has no affect on the condition of equilibrium in a reversible reaction. Taken together these facts mean that an equilibrium condition must be *independent* of the reaction mechanism. Thus, as stated above, the kinetic derivation of the equilibrium constant should not depend on the particular mechanism chosen. Moreover, we could always conceive of a *hypothetical* catalyst that changes the mechanism of a reversible reaction to the simple one-step process suggested by the balanced chemical equation. The equilibrium constant expression derived from such a mechanism would then always be of the form written in equation (14.5).

14-8 The Ammonia Synthesis Reaction

Elemental nitrogen occurs in the atmosphere to the extent of 78% by volume; but because of nitrogen's relative inertness, nitrogen-containing compounds do not occur extensively in nature. Artificial methods of synthesizing nitrogen compounds, referred to as nitrogen fixation, are important industrial processes. The principal method used involves the reaction of nitrogen and hydrogen to form ammonia. The ammonia can be converted to oxides of nitrogen, which in turn can be used to produce nitric acid and nitrate salts. The ammonia synthesis reaction was perfected by the German chemist Fritz Haber in 1908. The primary interest in ammonia synthesis at that time was for the production of high explosives. Now, the principal use is in the manufacture of fertilizers. Current annual production in the United States is about 17 million tons, making ammonia about the third most important manufactured chemical.

Some relevant data on the ammonia synthesis reaction are

$$N_2(g) + 3 H_2(g) \rightleftharpoons 2 NH_3(g) \qquad (14.35)$$

At 298 K: $K_p = 6.2 \times 10^5 \qquad \Delta \bar{H}^\circ = -92.38 \text{ kJ/mol}$

For every 4 mol of gases reacting [1 mol $N_2(g)$ and 3 mol $H_2(g)$], 2 mol NH_3 is produced. Increasing the pressure forces the reaction mixture into a smaller volume and favors the reaction producing the smaller number of moles of gas—the production of $NH_3(g)$. The forward reaction is *exothermic*. The exothermic reaction is favored if the temperature is *lowered*. Thus, the optimum conditions for the production of NH_3 appear to be *high pressures* and *low temperatures*.

These "optimum" conditions, however, do not take into account the importance of reaction rate. Although the equilibrium production of NH_3 is favored at low temperatures, the rate of its formation is so slow as to make the method unfeasible. One way of speeding up the reaction is to raise the temperature (even though the equilibrium concentration of NH_3 decreases in doing so). Another way is to use a catalyst. The usual operating conditions for the Haber process are about 550°C, pressures ranging from 100 to 1000 atm, and a catalyst—usually Fe_3O_4 with small added amounts of other metal oxides. The dramatic difference between the theoretical optimum conditions and the actual operating conditions is suggested through Figure 14-7.

Another way to increase the rate of production of NH_3 is to remove NH_3 continuously as it is formed. This is done by liquefying the NH_3 and recycling the N_2 and

[handwritten margin notes:] $H_2 \, \& \, N_2$ ↓ NH_3 ↓ N_xO_y ↙ ↘ *nitrate salts* *nitric acid*

FIGURE 14-7
Equilibrium conversion of $N_2(g)$ and $H_2(g)$ to $NH_3(g)$ as a function of temperature and pressure.

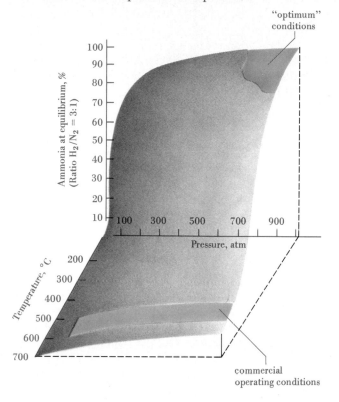

FIGURE 14-8
Ammonia synthesis reaction.

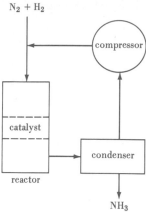

The gaseous N_2–H_2 mixture is introduced into a reactor at high temperature and pressure in the presence of a catalyst. The gaseous N_2–H_2–NH_3 mixture leaves the reactor and is cooled as it passes through a condenser. Liquefied NH_3 is removed, and the remaining N_2 H_2 mixture is compressed and returned to the reactor.

Reasons why a substance like NH_3 is easily liquefied and N_2 and H_2 are not were discussed in Sections 11-8 and 11-9.

H_2, which do not liquefy. The disturbance caused by the continuous removal of NH_3 displaces the equilibrium toward the production of more NH_3. In fact, the mixture need not be allowed to come to equilibrium at all. In this way practically 100% conversion of N_2 and H_2 to NH_3 is possible. A schematic representation of the ammonia synthesis reaction is provided in Figure 14-8.

Summary

When two competing reactions—a forward and a reverse reaction—occur simultaneously, reaction does not go to completion. The extent to which such a *reversible* reaction proceeds is to the point where the rates of the forward and reverse reactions become equal. At this point of equilibrium no further net change occurs and the amounts of the reacting species remain constant with time. The chemical equation describes the proportions in which reactants participate in a reversible reaction. A quantitative description of the equilibrium condition is provided through the equilibrium constant expression.

The *form* of the equilibrium constant expression, K_c, is established through the balanced chemical equation. The *numerical value* of the constant is determined by experiment. If a reaction involves gases, an equilibrium constant expression, K_p, can be written based on partial pressures of gases. K_p and K_c can be interrelated through a simple mathematical equation. The equilibrium constant expression is inverted when a chemical equation is written in the reverse direction. If two or more equations are added together, the equilibrium constant for the resulting net reaction is the product of the constants for the individual reactions.

A ratio of initial reactant concentrations can be formed in the same manner as the equilibrium constant expression. By comparing the numerical value of this ratio, called the reaction quotient, Q, to K_c, one can determine the *direction* in which a reaction proceeds. An algebraic equation can be written and solved to determine the *extent* of reaction, that is, to relate the final equilibrium to the initial conditions. A variety of possibilities exists for equilibrium calculations, but the basic principles and algebraic techniques involved are few in number.

One variable that generally has a profound effect on the value of an equilibrium constant is temperature. The van't Hoff equation relates log K to the standard molar enthalpy change for a reaction and to kelvin temperature. The qualitative effect of temperature on a reversible reaction is that raising the temperature favors the endothermic reaction and lowering the temperature, the exothermic reaction. The addition of a catalyst to a reversible reaction speeds up the forward and reverse reactions equally. Equilibrium is attained more rapidly but without any change in the equilibrium concentrations. Le Châtelier's principle states that an equilibrium condition undergoes modification or "shifts" whenever an equilibrium mixture is disturbed. The shift is always in the direction to relieve the disturbance. These disturbances may take the form of the addition or removal of reacting species or of changes in volumes, temperatures, or pressures.

Chemical equilibrium has theoretical links to other areas of chemistry. In this chapter the relationship of the equilibrium constant expression to the rates of forward and reverse reactions is established.

Learning Objectives

As a result of studying Chapter 14, you should be able to

1. Describe the condition of equilibrium in a reversible reaction.

2. Write the equilibrium constant expression, K_c, from the chemical equation for a reversible reaction.

3. Derive K_c values for situations where chemical equations are reversed, multiplied through by constant coefficients or added together.

4. Write an equilibrium constant expression in terms of partial pressures of gases, K_p; and relate a value of K_p to the corresponding value of K_c.

5. Calculate a numerical value of an equilibrium constant if equilibrium conditions are given.

6. Predict the direction in which a reaction proceeds toward equilibrium by comparing the reaction quotient, Q, to K_c.

7. Calculate the final equilibrium condition in a reversible reaction from a given set of initial conditions.

8. Relate the equilibrium constant to the standard molar enthalpy of reaction, $\Delta \bar{H}°$, and to kelvin temperature, both graphically and algebraically.

9. Make qualitative predictions of how equilibrium conditions change when an equilibrium mixture is disturbed.

10. Demonstrate that the form of an equilibrium constant expression is consistent with the rate laws and mechanisms of chemical reactions.

Some New Terms

Equilibrium refers to a condition where a forward and reverse process proceed at equal rates and no further net change occurs (e.g., amounts of reactants and products remain constant with time).

An **equilibrium constant** describes the relationship among the concentrations (or partial pressures in some cases) of the substances within an equilibrium system. The numerical value of the constant does not depend on how the equilibrium condition is attained.

K_c is a relationship that exists among the equilibrium concentrations of reactants and products in a reversible reaction at a given temperature.

K_p, the partial pressure equilibrium constant, is a relationship that exists among the partial pressures of gaseous reactants and products in a reversible reaction at equilibrium at a given temperature.

Le Châtelier's principle states that if a system at equilibrium is disturbed, the equilibrium condition is upset, and a reaction proceeds in that direction which tends to relieve the disturbing influence.

The **reaction quotient, Q,** is a ratio of concentration terms having the same form as the equilibrium constant expression, but usually applied to *nonequilibrium* conditions. It is used to establish the direction in which a reaction occurs to establish equilibrium.

Exercises

Writing equilibrium constant expressions

1. Write an equilibrium constant expression, K_c, for each of the following reactions.
   (a) $CO(g) + H_2O(g) \rightleftharpoons CO_2(g) + H_2(g)$
   (b) $2 NO(g) + O_2(g) \rightleftharpoons 2 NO_2(g)$
   (c) $CS_2(g) + 4 H_2(g) \rightleftharpoons CH_4(g) + 2 H_2S(g)$
   (d) $NH_3(g) + O_2(g) \rightleftharpoons NO(g) + H_2O(g)$
   (not balanced)
   (e) $NO_2(g) + H_2(g) \rightleftharpoons NH_3(g) + H_2O(g)$
   (not balanced)

2. Write an equilibrium constant expression, K_c, for the formation of *1 mol* of each of the following *gaseous* compounds from its *gaseous elements:* (a) NO; (b) HCl; (c) NH_3; (d) ClF_3; (e) NOCl.

3. Each of the following reversible reactions involves one or more condensed states of matter. Write an equilibrium constant expression, K_c, for each.
   (a) $CaCO_3(s) \rightleftharpoons CaO(s) + CO_2(g)$
   (b) $2 NaHCO_3(s) \rightleftharpoons Na_2CO_3(s) + CO_2(g) + H_2O(g)$
   (c) $CO_2(s) \rightleftharpoons CO_2(g)$
   (d) $(CH_3)_2CO(l) \rightleftharpoons (CH_3)_2CO(g)$
   (e) $CS_2(l) + Cl_2(g) \rightleftharpoons CCl_4(l) + S_2Cl_2(l)$
   (not balanced)
   (f) $Na_2CO_3(s) + C(s) + N_2(g) \rightleftharpoons NaCN(s) + CO(g)$
   (not balanced)

4. From the values of K_c given

$$CO(g) + H_2O(g) \rightleftharpoons CO_2(g) + H_2(g)$$
$$K_c = 23.2 \text{ at } 600 \text{ K}$$

$$SO_2(g) + \tfrac{1}{2}O_2(g) \rightleftharpoons SO_3(g) \quad K_c = 56 \text{ at } 900 \text{ K}$$

$$2 H_2S(g) \rightleftharpoons 2 H_2(g) + S_2(g)$$
$$K_c = 2.3 \times 10^{-4} \text{ at } 1405 \text{ K}$$

$$2 NO_2(g) \rightleftharpoons 2 NO(g) + O_2(g)$$
$$K_c = 1.8 \times 10^{-6} \text{ at } 457 \text{ K}$$

determine values of K_c for the following reactions.
   (a) $CO_2(g) + H_2(g) \rightleftharpoons CO(g) + H_2O(g)$
   (b) $2 SO_2(g) + O_2(g) \rightleftharpoons 2 SO_3(g)$
   (c) $H_2S(g) \rightleftharpoons H_2(g) + \tfrac{1}{2}S_2(g)$
   (d) $NO(g) + \tfrac{1}{2}O_2(g) \rightleftharpoons NO_2(g)$

5. Determine K_c for the reaction $\tfrac{1}{2}N_2(g) + \tfrac{1}{2}O_2(g) + \tfrac{1}{2}Br_2(g) \rightleftharpoons NOBr(g)$ from the following information (at 298 K).

$$2 NO(g) \rightleftharpoons N_2(g) + O_2(g) \quad K_c = 2.4 \times 10^{30}$$
$$NO(g) + \tfrac{1}{2}Br_2(g) \rightleftharpoons NOBr(g) \quad K_c = 1.4$$

6. Given the equilibrium constant values

$$N_2(g) + \tfrac{1}{2}O_2(g) \rightleftharpoons N_2O(g) \quad K_c = 3.4 \times 10^{-18}$$

$$N_2O_4(g) \rightleftharpoons 2 NO_2(g) \quad K_c = 4.6 \times 10^{-3}$$
$$\tfrac{1}{2}N_2(g) + O_2(g) \rightleftharpoons NO_2(g) \quad K_c = 4.1 \times 10^{-9}$$

Determine a value of K_c for the reaction

$$2 N_2O(g) + 3 O_2(g) \rightleftharpoons 2 N_2O_4(g)$$

Experimental determination of equilibrium constants

7. Equilibrium is established in the reversible reaction $A(g) + B(g) \rightleftharpoons 2 C(g)$. Following are the equilibrium concentrations: [A] = 0.52, [B] = 0.60, [C] = 0.40. What is the value of K_c for this reaction?

8. The following reaction was allowed to come to equilibrium: $2 A(g) + B(g) \rightleftharpoons C(g)$. The *initial* amounts of the reactants present in a 2.00-L vessel were 1.20 mol A and 0.80 mol B. The amount of A, at equilibrium, was found to be 0.90 mol. What is the value of K_c for this reaction?

9. 1.00 g PCl_5 is introduced into a 250-ml flask and the flask heated to 250°C, where dissociation of PCl_5 is allowed to reach equilibrium: $PCl_5(g) \rightleftharpoons PCl_3(g) + Cl_2(g)$. The quantity of $Cl_2(g)$ present at equilibrium is found to be 0.25 g. What is the value of K_c for the dissociation reaction at 250°C?
3.8 $\times 10^{-2}$

10. A classical experiment used to establish principles of chemical equilibrium was the reaction of ethanol (C_2H_5OH) and acetic acid (CH_3COOH) to produce ethyl acetate and water.

$$C_2H_5OH(l) + CH_3CO_2H(l) \rightleftharpoons$$
$$CH_3CO_2C_2H_5(l) + H_2O(l) \quad (14.36)$$

The reaction can be followed by analyzing the equilibrium mixture for its acetic acid content.

$$2 CH_3CO_2H(aq) + Ba(OH)_2(aq) \longrightarrow$$
$$Ba(CH_3CO_2)_2(aq) + 2 H_2O \quad (14.37)$$

An experiment is performed in which 1.000 mol of acetic acid and 0.500 mol of ethanol are mixed and allowed to come to equilibrium. A sample representing exactly one-hundredth of the total equilibrium mixture requires 28.85 ml of 0.1000 M $Ba(OH)_2$ for its titration. Show that the equilibrium constant, K_c, for reaction (14.36) is 4.0. (*Hint:* It is not necessary to know the volume of the reaction mixture.)

*★11.** The decomposition of HI(g) is represented by the equation $2 HI(g) \rightleftharpoons H_2(g) + I_2(g)$. HI(g) is introduced into five identical 400-cm^3 glass bulbs, and the five bulbs are maintained at 623 K. Each bulb is opened after a period of time and analyzed for I_2 by titration with 0.0150 M $Na_2S_2O_3(aq)$.

$$I_2(aq) + 2 Na_2S_2O_3(aq) \longrightarrow Na_2S_4O_6(aq) + 2 NaI(aq)$$

What is the value of K_c at 623 K? (Data are presented on page 366.)

Bulb number	Original amount of $HI(g)$, g	Bulb opened after, h	Volume of 0.0150 M $Na_2S_2O_3$ required for titration, ml
1	0.300	2	20.96
2	0.320	4	27.90
3	0.315	12	32.31
4	0.406	20	41.50
5	0.280	40	28.68

Equilibrium relationships

12. For the reaction $CO(g) + H_2O(g) \rightleftharpoons CO_2(g) + H_2(g)$, $K_c = 1.00$ at about 1100 K. Which one of the following statements must always be true concerning this reaction at equilibrium? Explain.

   (a) $[CO] = [H_2O] = [CO_2] = [H_2]$
   (b) $[CO] \times [H_2O] = [CO_2] \times [H_2]$
   (c) $[CO] = [H_2O]$ and $[CO_2] = [H_2]$
   (d) $[CO] \times [H_2O] = [CO_2] \times [H_2] = 1.00$

13. An equilibrium mixture at 1000 K contains 0.276 mol H_2, 0.276 mol CO_2, 0.224 mol CO, and 0.224 mol H_2O.

$CO_2(g) + H_2(g) \rightleftharpoons CO(g) + H_2O(g)$

   (a) Show that for this reaction, K_c is independent of the reaction volume, V.
   (b) Determine the value of K_c.

14. A mixture of SO_2, SO_3, and O_2 gases is maintained in a 10.0-L flask at a temperature at which $K_c = 100$ for the reaction $2 SO_2(g) + O_2(g) \rightleftharpoons 2 SO_3(g)$.
   (a) If the number of moles of SO_2 and SO_3 in the flask are equal, how much O_2 is present?
   (b) If the number of moles of SO_3 in the flask is twice the number of moles of SO_2, how much O_2 is present?

15. For the dissociation of $I_2(g)$ at about 1200 K, $I_2(g) \rightleftharpoons 2 I(g)$, $K_c = 1 \times 10^{-2}$. What volume vessel is required if it is desired that 1.00 mol I_2 and 0.50 mol I be present at equilibrium?

Direction and extent of chemical change

16. Can a mixture of 2 mol O_2, 1 mol SO_2, and 8 mol SO_3 be maintained indefinitely in a 10.0-L flask at a temperature at which $K_c = 100$ in the reaction $2 SO_2(g) + O_2(g) \rightleftharpoons 2 SO_3(g)$? If not, in what direction will a reaction occur?

17. For the reaction $3 A(g) + B(g) \rightleftharpoons 2 C(g)$,
   (a) Write the equilibrium constant expression for K_c.
   (b) If, at a given temperature, $K_c = 9.0$, can a mixture of 2.00 mol each of A, B, and C exist in equilibrium in a 1.00-L flask?
   (c) What must be the volume of the flask if the mixture described in part (b) is to exist at equilibrium?

18. Starting with 1.00 mol each of $CO(g)$ and $COCl_2(g)$ in a

1.50-L reaction vessel at 668 K, what is the number of moles of $Cl_2(g)$ produced at equilibrium?

$CO(g) + Cl_2(g) \rightleftharpoons COCl_2(g)$ $K_c = 1.2 \times 10^3$ at 668 K

19. With reference to Example 14-6, what will be the amounts of $CO(g)$, $H_2O(g)$, $CO_2(g)$, and $H_2(g)$ when equilibrium is established?

20. 3.00 mol $SbCl_3$ and 1.00 mol Cl_2 are introduced into an evacuated 5.00-L vessel and equilibrium is established at 248°C. How many moles of $SbCl_5$, $SbCl_3$, and Cl_2 are present at equilibrium?

$SbCl_5(g) \rightleftharpoons SbCl_3(g) + Cl_2(g)$
$$K_c = 2.5 \times 10^{-2} \text{ at } 248°C$$

21. If 0.390 mol SO_2, 0.156 mol O_2, and 0.657 mol SO_3 are introduced simultaneously into a 1.90-L reaction vessel at 1000 K,
   (a) Is this mixture at equilibrium?
   (b) If not, in what direction must a reaction proceed to establish equilibrium?

$2 SO_2(g) + O_2(g) \rightleftharpoons 2 SO_3(g)$
$$K_c = 2.8 \times 10^2 \text{ at } 1000 \text{ K}$$

★22. Calculate the actual equilibrium amounts of SO_2, O_2, and SO_3 in Exercise 21.

23. An aqueous solution is made 1.00 M in $AgNO_3$ and 1.00 M in $Fe(NO_3)_2$ and allowed to come to equilibrium. What are the values of $[Ag^+]$, $[Fe^{2+}]$, and $[Fe^{3+}]$ when equilibrium is established?

$Ag^+(aq) + Fe^{2+}(aq) \rightleftharpoons Fe^{3+}(aq) + Ag(s)$ $K_c = 2.98$

★24. Solid iron metal is added to a solution having the concentrations $[Cr^{3+}] = 0.250\ M$, $[Cr^{2+}] = 0.0500\ M$, and $[Fe^{2+}] = 0.00100\ M$. What are the concentrations of these ions when equilibrium is established?

$2 Cr^{3+}(aq) + Fe(s) \rightleftharpoons 2 Cr^{2+}(aq) + Fe^{2+}(aq)$
$$K_c = 10.34$$

Partial pressure equilibrium constant, K_p

25. Determine values of K_p for the first four reversible reactions listed in Exercise 4.

26. Determine values of K_c for each of the following reversible reactions.
   (a) $SO_2Cl_2(g) \rightleftharpoons SO_2(g) + Cl_2(g)$
$$K_p = 2.9 \times 10^{-2} \text{ at } 303 \text{ K}$$
   (b) $2 NO(g) + O_2(g) \rightleftharpoons 2 NO_2(g)$
$$K_p = 1.48 \times 10^4 \text{ at } 184°C$$
   (c) $Sb_2S_3(s) + 3 H_2(g) \rightleftharpoons 2 Sb(s) + 3 H_2S(g)$
$$K_p = 0.429 \text{ at } 713 \text{ K}$$

27. A sample of air with an original mole ratio of nitrogen to oxygen of 79:21 is heated to 2500 K. When equilibrium is established, the mole percent NO present is found to be 1.8%. Calculate K_p for the reaction

$$N_2(g) + O_2(g) \rightleftharpoons 2\,NO(g) \qquad K_p \text{ at } 2500\,K = ?$$

(*Hint:* The result is independent of both volume and total pressure. The presence of other gases in air can be neglected.)

28. A sample of $NH_4HS(s)$ is introduced into a 1.60-L flask containing 0.170 g NH_3. What is the total gas pressure when equilibrium is established in the flask at 25°C?

$$NH_4HS(s) \rightleftharpoons NH_3(g) + H_2S(g) \qquad K_p = 0.108 \text{ at } 25°C$$

29. In the manufacture of sodium carbonate by the Solvay process, $NaHCO_3(s)$ is decomposed by heating.

$$2\,NaHCO_3(s) \rightleftharpoons Na_2CO_3(s) + CO_2(g) + H_2O(g)$$
$$K_p = 0.23 \text{ at } 100°C$$

(a) If a sample of $NaHCO_3(s)$ is brought to a temperature of 100°C in a closed container, what will be the total gas pressure (in atm) at equilibrium?
(b) A mixture of 1.00 mol each of $NaHCO_3(s)$ and $Na_2CO_3(s)$ is introduced into a 2.50-L flask in which $P_{CO_2} = 2.10$ atm and $P_{H_2O} = 715$ mmHg. When equilibrium is established (at 100°C), will the partial pressures of $CO_2(g)$ and $H_2O(g)$ be greater or less than their initial partial pressures? Explain.
*(c) Starting with the initial conditions of part (b), what will be the partial pressures of $CO_2(g)$ and $H_2O(g)$ when equilibrium is established at 100°C?

Dissociation reactions

30. If the reaction mixture described in Example 14-10 were transferred to a 10.0-L vessel, would the percent dissociation increase, decrease, or remain the same? Explain.

31. Calculate the actual percent dissociation referred to in Exercise 30.

32. When 1.00 mol $I_2(g)$ is introduced into an evacuated 1.00-L flask at 1200°C, it is 5% dissociated into I atoms. For the reaction $I_2(g) \rightleftharpoons 2\,I(g)$ at 1200°C, what is the value of (a) K_c; (b) K_p?

33. What is the percent dissociation of HI(g) into its gaseous elements at 350°C?

$$H_2(g) + I_2(g) \rightleftharpoons 2\,HI(g) \qquad K_p = 7 \times 10^1$$

*34. A sample of pure $PCl_5(g)$ is introduced into an evacuated flask and allowed to dissociate. If the fraction of the molecules originally present that dissociate is denoted by α, and if the total gas pressure at equilibrium is P, show that for the reaction

$$PCl_5(g) \rightleftharpoons PCl_3(g) + Cl_2(g) \qquad K_p = \frac{\alpha^2 P}{1 - \alpha^2}$$

35. With reference to the equation established in Exercise 34, if $K_p = 1.78$ at 250°C,
(a) What is the percent dissociation of PCl_5 at 250°C and 1 atm total pressure?

(b) Under what total pressure must the gaseous mixture be maintained to limit dissociation of PCl_5 to 10.0%?

Effect of temperature on equilibrium constants

36. Estimate the value of K_p for the reaction $2\,SO_2(g) + O_2(g) \rightleftharpoons 2\,SO_3(g)$ at 25°C. Use data from Table 14-4 and Figure 14-4.

37. The following equilibrium constants have been determined for the reaction $H_2(g) + I_2(g) \rightleftharpoons 2\,HI(g)$: $K_c = 50.0$ at 448°C and 66.9 at 350°C. Use these data to estimate $\Delta \bar{H}°$ for the reaction and compare your result with that given in the reaction profile of Figure 13-12.

38. Use data from Example 14-7 and Appendix D to estimate the value of K_p at 100°C for the reaction $N_2O_4(g) \rightleftharpoons 2\,NO_2(g)$.

39. The following data are given for the temperature variation of K_p (partial pressures expressed in atm) for the reaction

$$2\,NaHCO_3(s) \rightleftharpoons Na_2CO_3(s) + CO_2(g) + H_2O(g)$$

t, °C	K_p
30	1.66×10^{-5}
50	3.90×10^{-4}
70	6.27×10^{-3}
100	2.31×10^{-1}

(a) Plot a graph similar to Figure 14-4 and determine $\Delta \bar{H}°$ for the reaction.
(b) Calculate the temperature at which the total gas pressure above a mixture of $NaHCO_3(s)$ and $Na_2CO_3(s)$ is 2.00 atm.

Le Châtelier's principle

40. What effect would decreasing the volume of the reaction mixture have on the equilibrium condition in each of the following reactions?
(a) $C(s) + H_2O(g) \rightleftharpoons CO(g) + H_2(g)$
(b) $CO(g) + H_2O(g) \rightleftharpoons CO_2(g) + H_2(g)$
(c) $4\,HCl(g) + O_2(g) \rightleftharpoons 2\,Cl_2(g) + 2\,H_2O(g)$

41. Continuous removal of one of the products of a chemical reaction has the effect of causing the reaction to go to completion. Explain.

42. A mixture of $HCl(g)$, $O_2(g)$, $H_2O(g)$, and $Cl_2(g)$ is brought to equilibrium at 200°C.

$$4\,HCl(g) + O_2(g) \rightleftharpoons 2\,H_2O(g) + 2\,Cl_2(g)$$

What would be the effect on the equilibrium amount of $HCl(g)$ if
(a) additional $O_2(g)$ were added to the mixture?

(b) $Cl_2(g)$ were removed from the reaction mixture?

(c) the volume of the reaction mixture were increased to twice its original value?

(d) a catalyst were added to the reaction mixture?

'43. The *exothermic* reaction $A(g) + B(g) \rightleftharpoons 2\,C(g)$ proceeds to an equilibrium condition at 200°C. Which of the following statements is true? Explain. (*Hint:* There may be more than one correct statement.)

(a) If the mixture is transferred to a reaction vessel of twice the volume, the amounts of reactants and products will remain unchanged.

(b) Addition of an appropriate catalyst will result in the formation of a greater amount of $C(g)$.

(c) Lowering the reaction temperature to 100°C will not alter the equilibrium condition.

' (d) Addition of an inert gas, such as helium, will have little or no effect on the equilibrium.

44. Show that the percent dissociation in reaction (1) depends on the volume of the reaction vessel and in reaction (2) it does not. Explain this difference from the standpoint of Le Châtelier's principle.

(1) $SO_2Cl_2(g) \rightleftharpoons SO_2(g) + Cl_2(g)$

(2) $CS_2(g) \rightleftharpoons C(s) + S_2(g)$

'45. For which of the following reactions would you expect the percent dissociation to increase with increasing temperature? Explain.

(a) $NO(g) \rightleftharpoons \frac{1}{2}N_2(g) + \frac{1}{2}O_2(g)$
$$\Delta\bar{H}° = -90.2\text{ kJ/mol}$$

(b) $SO_3(g) \rightleftharpoons SO_2(g) + \frac{1}{2}O_2(g)$
$$\Delta\bar{H}° = +98.9\text{ kJ/mol}$$

(c) $N_2H_4(g) \rightleftharpoons N_2(g) + 2\,H_2(g)$
$$\Delta\bar{H}° = -95.4\text{ kJ/mol}$$

(d) $COCl_2(g) \rightleftharpoons CO(g) + Cl_2(g)$
$$\Delta\bar{H}° = +108.3\text{ kJ/mol}$$

46. Explain why all dissociation reactions of the type $A_2(g) \rightleftharpoons 2\,A(g)$ proceed to a greater extent at higher temperatures [e.g., $I_2(g) \rightleftharpoons I(g)$].

'47. The reaction $N_2(g) + O_2(g) \rightleftharpoons 2\,NO(g)$, $\Delta\bar{H}° = +181$ kJ/mol, occurs whenever a substance is burned in air. This reaction occurs in internal combustion engines, leading to the formation of oxides of nitrogen that are involved in the production of photochemical smog. High-compression engines, characteristic of large automobiles, operate at high temperatures.

(a) What effect do these high temperatures have on the equilibrium production of $NO(g)$?

(b) What effect does high temperature have on the rate of this reaction?

48. The freezing of $H_2O(l)$ at 0°C can be represented as

$$H_2O(l,\ d = 1.00\text{ g/cm}^3) = H_2O(s,\ d = 0.92\text{ g/cm}^3)$$

Explain why the application of pressure to ice at 0°C causes the ice to melt. Is this behavior to be expected of solids in general?

Additional Exercises

1. Assume that when equilibrium is established in a certain mixture of sulfur dioxide, oxygen, and sulfur trioxide, $[SO_2] = [SO_3]$. Show that $[O_2]$ has the same value regardless of which of the three expressions (14.7a), (14.7b), or (14.7c) is used to describe the equilibrium.

2. Use data from Exercise 5 to determine K_p at 298 K for the reaction

$$\tfrac{1}{2}N_2(g) + \tfrac{1}{2}O_2(g) + \tfrac{1}{2}Br_2(g) \rightleftharpoons NOBr(g) \qquad K_p = ?$$

3. The high-temperature dissociation of salicylic acid is represented by the equation

$$C_7H_6O_3(g) \rightleftharpoons C_6H_6O(g) + CO_2(g)$$

As a result of an experiment carried out at 200°C, an initial sample of 0.300 g $C_7H_6O_3$ in a 50.0-cm³ vessel yielded an equilibrium mixture in which the partial pressure of $CO_2(g)$ was found to 1.50 atm. What are (a) K_c and (b) K_p for this reaction at 200°C?

4. Formamide is used as an intermediate and solvent in the manufacture of pharmaceuticals, dyes, and agricultural chemicals. At elevated temperatures it decomposes to $NH_3(g)$ and $CO(g)$.

$$HCONH_2(g) \rightleftharpoons NH_3(g) + CO(g) \qquad K_c = 4.84 \text{ at } 400 \text{ K}$$

If 0.100 mol $HCONH_2(g)$ is allowed to dissociate in a 1.50-L flask at 400 K, what will be the *total* pressure at equilibrium?

5. With reference to Exercise 10 and the reaction described there,

$$C_2H_5OH(l) + CH_3CO_2H(l) \rightleftharpoons$$
$$CH_3CO_2C_2H_5(l) + H_2O(l)\ .\ K_c = 4.0$$

15.5 g C_2H_5OH, 25.0 g CH_3CO_2H, 45.5 g $CH_3CO_2C_2H_5$, and 52.0 g H_2O are mixed and allowed to react.

(a) In what direction will a net reaction occur?

(b) What will be the equilibrium quantities of each of the reacting species?

6. A solution is prepared having $[Fe^{3+}] = 0.4000\ M$ and $[Hg_2^{2+}] = 0.2500\ M$. What are the values of $[Fe^{3+}]$, $[Fe^{2+}]$, $[Hg_2^{2+}]$, and $[Hg^{2+}]$ when equilibrium is established?

$$2\,Fe^{3+}(aq) + Hg_2^{2+}(aq) \rightleftharpoons 2\,Fe^{2+}(aq) + 2\,Hg^{2+}(aq)$$
$$K_c = 9.14 \times 10^{-6}$$

7. Use data from Appendix D to establish if the forward reaction is favored by high or low temperatures.

$$2 NO(g) + O_2(g) \rightleftharpoons 2 NO_2(g)$$

8. An equilibrium condition is attained in the reaction

$$Fe_3O_4(s) + 4 H_2(g) \rightleftharpoons 3 Fe(s) + 4 H_2O(g)$$

at 150°C. What would be the effect on $[H_2O(g)]$ if
   (a) additional $H_2(g)$ were introduced into the mixture?
   (b) more $Fe(s)$ were added to the reaction mixture?
   (c) a catalyst were employed?

9. Use Le Châtelier's principle to make qualitative predictions about
   (a) the effect on the amount of $Cl_2(g)$ at equilibrium if the volume of the reaction vessel in Exercise 18 is increased from 1.50 L to 2.50 L
   (b) the effect on the equilibrium amounts of $SbCl_5$, $SbCl_3$, and Cl_2 if a catalyst is used to speed up the reaction in Exercise 20
   (c) the effect on the percent dissociation of $HI(g)$ in Exercise 33 if the total pressure exerted by an equilibrium mixture is increased from 1.0 atm to 10.0 atm
   (d) the effect on the total pressure exerted by $CO_2(g)$ and $H_2O(g)$ in equilibrium with $NaHCO_3(s)$ and $Na_2CO_3(s)$ if the temperature is raised from 25°C to 200°C (see Exercise 39)

10. What is the percent dissociation of $H_2S(g)$ if 1.00 mol H_2S is introduced into an evacuated 1.10-L vessel at 1000 K?

$$2 H_2S(g) \rightleftharpoons 2 H_2(g) + S_2(g) \qquad K_c = 1.0 \times 10^{-6}$$

*11. A mixture of $H_2S(g)$ and $CH_4(g)$ in the mole ratio 2:1 was allowed to come to equilibrium at 700°C and a total gas pressure of 1 atm. The *equilibrium* mixture was analyzed for the amount of H_2S present; 9.54×10^{-3} mol H_2S was found.

The CS_2 present at equilibrium was converted successively to H_2SO_4 and then to $BaSO_4$; 1.42×10^{-3} mol $BaSO_4$ was obtained. Use these data to determine K_p at 700°C for the reaction

$$2 H_2S(g) + CH_4(g) \rightleftharpoons CS_2(g) + 4 H_2(g)$$
$$K_p \text{ at } 700°C = ?$$

*12. From the data given, calculate the equilibrium amounts of the reactants and products shown in   (a) Figure 14-5c; (b) Figure 14-6b.

*13. For the loss of water by the trihydrate, $CuSO_4 \cdot 3 H_2O$, the following data are given at 298 K.

$$CuSO \cdot 3 H_2O(s) \rightleftharpoons CuSO_4 \cdot H_2O(s) + 2 H_2O(g)$$
$$K_p = 5.43 \times 10^{-5} \text{ (pressures in atm) } \Delta \overline{H}° = +113 \text{ kJ/mol}$$
   (a) Write an equation similar to equation (14.25) to show how the partial pressure of water vapor varies with temperature above a mixture of the two hydrates.
   (b) At what temperature is the partial pressure of the water vapor above the hydrates equal to 0.100 atm?

*14. The decomposition of the poisonous gas phosgene is represented by the equation $COCl_2(g) \rightleftharpoons CO(g) + Cl_2(g)$. The values of K_p for this reaction are listed as $K_p = 6.7 \times 10^{-9}$ at 100°C and $K_p = 4.44 \times 10^{-2}$ at 395°C. At what temperature is $COCl_2$ 15% dissociated when the total gas pressure is maintained at 1.00 atm?

*15. What is the apparent molecular weight of the gaseous mixture that results when $COCl_2(g)$ is allowed to dissociate at 395°C and a total pressure of 3.00 atm?

$$COCl_2(g) \rightleftharpoons CO(g) + Cl_2(g) \ K_p = 4.44 \times 10^{-2} \text{ at } 395°C$$

Self-Test Questions

For questions 1 through 8 select the single item that best completes each statement.

1. In the reversible reaction $H_2(g) + I_2(g) \rightleftharpoons 2 HI(g)$, a mixture that initially contains 2 mol H_2 and 1 mol I_2 produces, at equilibrium,   (a) 1 mol HI;   (b) 2 mol HI;   (c) more than 2 mol HI but less than 4 mol HI;   (d) less than 2 mol HI.

2. Equilibrium is established in the reaction $2 SO_2(g) + O_2(g) \rightleftharpoons 2 SO_3(g)$ at a temperature at which $K_c = 100$. If the number of moles of SO_3 in the equilibrium mixture is equal to the number of moles of SO_2,
   (a) the number of moles of O_2 is also equal to the number of moles of SO_2;
   (b) the number of moles of O_2 is half the number of moles of SO_2;
   (c) $[O_2] = 0.01 \ M$;
   (d) $[O_2]$ may have any of several different values.

3. The volume of the reaction vessel containing an equilibrium mixture in the reaction $SO_2Cl_2(g) \rightleftharpoons SO_2(g) + Cl_2(g)$ is increased. When equilibrium is reestablished,
   (a) the amount of $Cl_2(g)$ will have increased;
   (b) the amount of $SO_2(g)$ will have decreased;
   (c) the amount of $Cl_2(g)$ will have remained unchanged;
   (d) the amount of $SO_2Cl_2(g)$ will have increased.

4. Which of the following statements is true when equilibrium is established in the reaction $A + B \rightleftharpoons C + D$; $K_c = 10.0$?   (a) $[C][D] = [A][B]$;   (b) $[C] = [A]$ and $[B] = [D]$;   (c) $[A][B] = 0.10 \times [C][D]$;   (d) $[A] = [B] = [C] = [D] = 10.0 \ M$.

5. Equilibrium in a mixture of $CO(g)$, $H_2O(g)$, $CO_2(g)$, and $H_2(g)$ is established in a 1.00-L container at 1000 K. The following data are given.

$$CO(g) + H_2O(g) \rightleftharpoons CO_2(g) + H_2(g)$$
$$\Delta \overline{H}° = -42 \text{ kJ/mol} \qquad K_c = 0.66$$

The equilibrium amount of $H_2(g)$ can be increased by **(a)** adding a catalyst; **(b)** increasing the temperature; **(c)** transferring the mixture to a 10.0-L container; **(d)** none of the methods described in (a), (b), or (c).

6. 1.00 mol *each* of $CO(g)$, $H_2O(g)$, and $CO_2(g)$ are introduced into a 10.0-L flask at a temperature at which $K_c = 10.0$ for the reaction

$$CO(g) + H_2O(g) \rightleftharpoons CO_2(g) + H_2(g) \qquad K_c = 10.0$$

When the reaction reaches a state of equilibrium,
- **(a)** the amount of H_2 will be 1.00 mol.
- **(b)** the amount of $CO_2(g)$ will be greater than 1.00 mol, and the amounts of $CO(g)$, $H_2O(g)$, and $H_2(g)$ will each be less than 1.00 mol.
- **(c)** the amounts of all reactants and products will be greater than 1.00 mol.
- **(d)** the amounts of $CO_2(g)$ and $H_2(g)$ will each be greater than 1.00 mol, and the amounts of $CO(g)$ and $H_2O(g)$ will each be less than 1.00 mol.

7. For the reaction $2 NO_2(g) \rightleftharpoons 2 NO(g) + O_2(g)$, $K_c = 1.8 \times 10^{-6}$ at 184°C. At 184°C, the value of K_c for the reaction $NO(g) + \frac{1}{2} O_2(g) \rightleftharpoons NO_2(g)$ is **(a)** 0.9×10^6; **(b)** 7.5×10^2; **(c)** 5.6×10^5; **(d)** 2.8×10^5.

8. For the dissociation reaction $2 H_2S(g) \rightleftharpoons 2 H_2(g) + S_2(g)$, $K_p = 1.2 \times 10^{-2}$ at 1065°C. For this same reaction at 298 K **(a)** K_c is less than K_p; **(b)** K_c is greater than K_p; **(c)** $K_c = K_p$; **(d)** whether K_c is less than, equal to, or greater than K_p depends on the total gas pressure.

9. Describe several ways in which the balanced chemical equation for a reversible reaction is used in equilibrium calculations. Explain why the balanced equation *in itself* is insufficient for determining the composition of an equilibrium mixture.

10. Explain briefly the relationship between
- **(a)** the rates of chemical reactions and the condition of equilibrium
- **(b)** the reaction quotient, Q, and the equilibrium constant expression, K_c
- **(c)** the equilibrium constants K_c and K_p

11. A 0.0010-mol sample of $S_2(g)$ is allowed to dissociate in an 0.500-L flask at 1000 K. When equilibrium is reached, 1.0×10^{-11} mol $S(g)$ is present. What is K_c for the reaction $S_2(g) \rightleftharpoons 2 S(g)$ at 1000 K?

12. Into a 1.00-L vessel at 1000 K are introduced 0.100 mol each of $NO(g)$ and $Br_2(g)$ and 0.0100 mol $NOBr(g)$.

$$2 NO(g) + Br_2(g) \rightleftharpoons 2 NOBr(g)$$
$$K_c = 1.32 \times 10^{-2} \text{ at } 1000 \text{ K}$$

- **(a)** In what direction must a net reaction occur?
- **★(b)** What is the partial pressure of $NOBr(g)$ in the vessel when equilibrium is established?

15 Thermodynamics, Spontaneous Change, and Equilibrium

Thermodynamics deals with the relationship between heat and other energy forms (work). Its development was one of the most significant scientific achievements of the nineteenth century. This development was largely the product of physicists and engineers seeking ways to improve the efficiencies of heat engines. The need to improve heat engines has again come into prominence because of current worldwide energy problems. But for the past 75 years the most important applications have been in chemistry. This is because the laws of thermodynamics provide powerful tools for studying chemical reactions. The first law of thermodynamics, presented in Chapter 6, serves as a basis for describing quantities of heat energy associated with chemical reactions. The second law of thermodynamics deals with the *direction* and *extent* of a chemical reaction in a more satisfactory manner than was possible in Chapter 14. Specifically, it allows us to derive values of equilibrium constants from certain thermodynamic properties of substances, properties established in this chapter. The third law of thermodynamics provides a starting point for the experimental evaluation of these thermodynamic properties.

15-1 In Search of a Criterion for Spontaneous Change

In a way "spontaneous" is not an ideal term for what we are attempting to describe here, because the word has a practical implication of something that occurs rapidly. Spontaneous processes may in fact occur quite slowly (such as the rusting of iron). What is intended by the term spontaneous is more along the lines of a *precise* dictionary definition: "acting in accordance with or resulting from natural feeling, temperament, or disposition, or from a native internal proneness, readiness, or tendency, without compulsion, constraint, or premeditation."

On occasion in Chapter 14 we used the phrase "the direction of chemical change." By this we meant whether a reaction would proceed in the forward or the reverse direction, based on a comparison of the reaction quotient, Q, and the equilibrium constant, K. To be more precise we should add one word to this phrase and speak of "the direction of *spontaneous* chemical change."

A **spontaneous or natural process** is one that occurs in a system left to itself; no external action is required to make it happen. A spontaneous change continues until a system reaches a state of equilibrium; then no further net change occurs. For example, the running down of a tightly wound clock occurs spontaneously, but the clock cannot rewind itself. The winding of a clock is a *nonspontaneous* process: External action (winding by human hands) is required to make this process happen. A more interesting example from a chemical standpoint is the corrosion ("rusting") of an iron pipe exposed to the atmosphere. Although the process may occur quite slowly, it does so *continuously* and in the *same* direction. The amount of iron decreases and the amount of rust increases until a final state of equilibrium is reached where practically all of the iron has been consumed. To reverse this process, that is, to convert the rust back into pure iron (and to fabricate the iron into a pipe essentially identical to the original one) is *not* impossible. However, the process is certainly not spontaneous. In fact, this nonspontaneous reverse process is essentially what occurs in the manufacture of iron from iron ore. From these examples we might draw the following conclusions.

371

FIGURE 15-1
Direction of spontaneous change in a mechanical system.

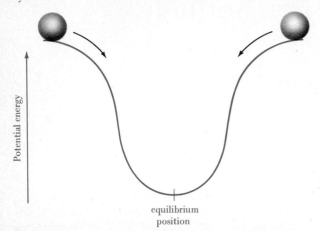

Whether we consider the ball on the left or the one on the right, the direction of spontaneous change is downhill. The ball reaches a position of equilibrium when it comes to rest at the bottom, the point of lowest potential energy.

In a scientific paper written in 1875, Berthelot stated this conclusion in the form of a *principle of maximum work:* "All chemical changes occurring without the intervention of outside energy tend toward the production of bodies, or a system of bodies, which liberate more heat."

FIGURE 15-2
Search for a criterion for spontaneous chemical change.

We are looking for some thermodynamic property, here called the "chemical potential," that has a minimum value at a point in the reaction between the pure reactants and the pure products. At this point the reaction is at equilibrium, with the reaction quotient Q equal to the equilibrium constant K. For a condition on either side of the equilibrium point, spontaneous reaction will occur in the direction of the equilibrium point.

1. If a process is found to be spontaneous, the reverse process is nonspontaneous.
2. Both spontaneous and nonspontaneous processes are *possible,* but only spontaneous processes will occur *naturally.* Nonspontaneous processes require the system to be acted upon in some way.

Useful as these two conclusions are, there is a third one that will prove to be most important of all. This is a conclusion that permits us to predict whether the forward or the reverse direction is the direction of spontaneous change. We will refer to this as a criterion for spontaneous change. To begin, we might look to mechanical systems for a clue: A ball rolls downhill and water flows to a lower level. Figure 15-1 illustrates a common feature of these processes—a decrease in potential energy.

For a chemical system the property analogous to the potential energy of a mechanical system is the internal energy or the closely related property of enthalpy. In the 1870s, the French chemist Berthelot and the Danish chemist J. Thomsen proposed that the direction of spontaneous change was that in which the enthalpy of a system decreases. An enthalpy decrease means that heat is given off by the system to the surroundings. Their conclusion then was that *exothermic* reactions should be spontaneous. In fact, exothermic processes generally are spontaneous, but so are some endothermic ones! The melting of ice at room temperature, the evaporation of liquid ether from an open beaker, and the dissolving of ammonium nitrate in water are all examples of *spontaneous, endothermic* processes. In abandoning enthalpy as our criterion for spontaneous change, let us agree that it would be use-

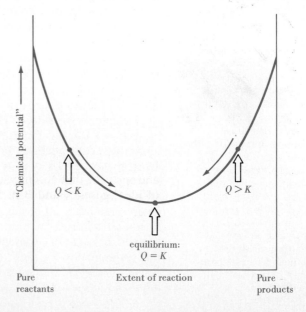

so if an enthalpy decrease/heat release is not basis for spontaneanty, what is?

ful to find some other thermodynamic function having the properties implied by Figure 15-2: We seek a function whose value falls to a minimum at the point where a chemical reaction has reached equilibrium.

FIGURE 15-3
The mixing of ideal gases.

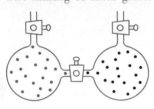

(a) Before mixing

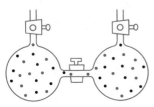

(b) After mixing

• gas A   • gas B

The total volume of the system and the total gas pressure are the same in each case pictured above. The net change that occurs is this.
Before mixing:
Each gas is confined to one half of the total volume (a single bulb) at a pressure of 1.00 atm.
After mixing:
Each gas has expanded *not enthalpy* into the total volume (both bulbs) yielding a partial pressure of 0.500 atm.

how about ENTROPY

The concept of entropy originated from studies on the efficiencies of heat engines (see Section 15-8). The change in entropy of a system is the quantity of heat it exchanges reversibly with its surroundings divided by the kelvin temperature. If the temperature is held constant,

$$\Delta S = \frac{q_{rev}}{T}$$

Later work by Boltzmann and others established that entropy is related to the microscopic order in a system.

15-2 Entropy and Disorder

In continuing our search for a criterion for spontaneous change, perhaps we should focus on endothermic processes. After all, these are the ones that occur in contradiction to the Berthelot–Thomsen principle of maximizing the quantity of heat evolved in a process. The three spontaneous, endothermic processes mentioned in the closing paragraph of Section 15-1 share a common characteristic. To see what this is, let us first consider a simpler case—the mixing of ideal gases.

Figure 15-3 depicts a situation in which one ideal gas, labeled A, is introduced into a glass bulb at a pressure of 1.00 atm. A second ideal gas, B, is introduced into a second bulb, which is identical to the first. Again the pressure is brought to 1.00 atm. The two bulbs are joined by a stopcock valve. This initial situation is pictured in Figure 15-3a. Assume that no chemical reaction is possible between the two gases, and now imagine that the valve between the two bulbs is opened. Intuitively, what would we expect to happen?

We know that the molecules of a gas are in constant motion and that they will move into whatever space is available to them—gases expand to fill their containers. In this case each gas expands into the bulb containing the other—the gases mix. The mixing will continue until the partial pressure of each gas becomes 0.500 atm in each bulb, as illustrated through Figure 15-3b. Since the mixing of ideal gases is a spontaneous process, we might next inquire as to what property of the system has undergone a change.

The internal energy and enthalpy of an ideal gas depend only on temperature, not on the gas pressure or volume. For the mixing of ideal gases at constant temperature, $\Delta E = \Delta H = 0$. One characteristic of the system, however, is greatly altered by the mixing process: the degree of *order* that prevails. In the initial condition of Figure 15-3a there is some degree of order, at least to the extent that all the molecules of A are found on one side of the valve and all those of B on the other side. After mixing, half of the molecules of A and half of those of B are found in each bulb. The molecules have reached the maximum state of mixing or *disorder* possible. A thermodynamic property that relates to the degree of disorder in a system is called **entropy** and denoted by the symbol **S**.

The higher the degree of randomness or disorder in a system, the greater its entropy.

Entropy (like enthalpy in Chapter 6) is defined in such a way as to be a function of state. When the state or condition of a system is defined, corresponding unique values exist for all state functions, including entropy. Moreover, as with other thermodynamic properties, we can speak of the **entropy, S,** of a system and the **change in entropy, ΔS,** that accompanies a process involving the system. Thus, for the mixing of ideal gases A and B in Figure 15-3,

$$A(g) + B(g) \longrightarrow \text{mixture of A(g) and B(g)}$$

$$\Delta S = S_{\text{final state}} - S_{\text{initial state}}$$

$$\Delta S = S_{\text{mixt. of gases}} - [S_{A(g)} + S_{B(g)}]$$

$$\Delta S > 0$$

goes for more disorder, thus leads toward a higher entropy (15.1)

We will forgo numerical computations involving entropy for the moment and simply note that since the degree of disorder *increases* in the mixing of ideal gases, entropy *increases*, and ΔS is a *positive* quantity, that is, $\Delta S > 0$. Now let us see what we can conclude about the entropy changes for the three endothermic processes described earlier.

In the melting of ice, a crystalline solid (recall Figure 11-20) is replaced by a less-structured liquid. Disorder and entropy increase in the process of melting. Molecules in the gaseous state, because of the large free volume in which they can move, have a much higher entropy than in the corresponding liquid state: The process of evaporation is accompanied by an increase in entropy. In the dissolving of an ionic solid such as ammonium nitrate in water, a crystalline solid and a pure liquid are replaced by a mixture of ions and water molecules in the liquid (solution) state (recall Figure 12-3). There is a tendency for some ordering of water molecules to occur around the ions in solution (referred to as hydration of the ions). However, this ordering tendency is not as great as the disorder produced by destroying the original crystalline solid. Again, disorder and entropy increase in the dissolving process. The increased disorder accompanying the types of processes described here is pictured in Figure 15-4. In general, entropy *increases* are expected to accompany processes in which

pure liquids or liquid solutions are formed from solids;

gases are formed, either from solids or liquids;

the number of molecules of gases increases in   (15.2)
the course of a chemical reaction;

the temperature of a substance is increased. (Increased temperature means increased molecular motion, whether it be vibrational motion of atoms or ions in a solid or translational motion of molecules in a liquid or gas.)

FIGURE 15-4
Entropy and disorder—three processes in which entropy increases.

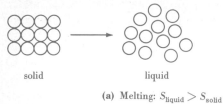

solid   liquid

(a) Melting: $S_{\text{liquid}} > S_{\text{solid}}$

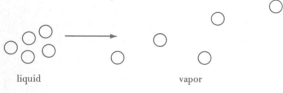

liquid   vapor

(b) Vaporization: $S_{\text{vapor}} > S_{\text{liquid}}$

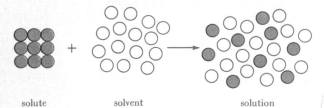

solute   solvent   solution

(c) Dissolving: $S_{\text{soln}} > (S_{\text{solvent}} + S_{\text{solute}})$

Each of the processes pictured here results in greater disorder and an increased entropy.

guidelines for predicting entropies

Example 15-1 Predict whether each of the following processes involves an increase or decrease in entropy.
(a) The explosive decomposition of NH_4NO_3:

$$2\,NH_4NO_3(s) \longrightarrow 2\,N_2(g) + 4\,H_2O(g) + O_2(g)$$

(b) The conversion of SO_2 to SO_3:

$$2\,SO_2(g) + O_2(g) \longrightarrow 2\,SO_3(g)$$

(c) The extraction of NaCl(s) from seawater:

$$NaCl(aq) \longrightarrow NaCl(s) + aq$$

(d) The "water gas" reaction:

$$CO(g) + H_2O(g) \longrightarrow CO_2(g) + H_2(g)$$

(a) Here a solid yields a large quantity of gas. An *increase* in entropy is expected.

(b) Three moles of gaseous reactants produce 2 mol of gaseous product. The *loss* of 1 mol of gas in the course of the reaction suggests a greater degree of order in the $SO_3(g)$ than in the gases from which it is formed. A *decrease* in entropy is expected.

(c) The ions, Na^+ and Cl^-, achieve a high degree of order when they leave the solution state and arrange themselves into the crystalline state. A *decrease* in entropy is expected.

(d) The entropies of the four gases are all likely to be different because of differences in their molecular structures. There should be an entropy change in the reaction. However, because no change occurs in the number of gaseous molecules in the course of the reaction, the entropy change is likely to be small. Moreover, there is no way for us to conclude from the relationships stated in (15.2) whether the entropy increases or decreases.

SIMILAR EXAMPLES: Exercises 1, 2, 3.

15-3 Free Energy and Spontaneous Change

THE MEANING OF $\Delta S_{universe}$.   A moment's reflection on the conclusions of the preceding section tells us that we still have not found a suitable, *single* criterion for spontaneous change. For example, based on the entropy change of the water alone, how do we explain the *spontaneous* freezing of water at $-10\,°C$? If the entropy of the system increases when ice melts, it must *decrease* when water freezes. The answer to this puzzle lies in the fact that there are always *three* entropy changes that must be assessed. These are the entropy change of the system, of the surroundings, and the total of the two—the so-called entropy change of the "universe." The relationship among the three is

$$\Delta S_{total} = \Delta S_{universe} - \Delta S_{system} + \Delta S_{surroundings} \tag{15.3}$$

It is beyond the scope of this text to pursue this matter, but for all *spontaneous* processes it can be shown that

$$\Delta S_{univ} = \Delta S_{syst} + \Delta S_{surr} > 0 \tag{15.4}$$

Equation (15.4) is in fact one important statement of the second law of thermodynamics.

All spontaneous or natural processes produce an increase in the entropy of the universe.

The freezing of water is accompanied by a decrease in entropy of the *system*. But as long as the temperature is below $0\,°C$, the entropy of the *surroundings* increases to a greater extent. The total entropy change is *positive* and the process is indeed spontaneous.

FREE ENERGY.   Total entropy change is a valid criterion for spontaniety (spontaneous change), but is it practical? Assessment of the entropy change of the surroundings can be a very tedious process, even impossible at times if the interactions between a system and its surroundings are not completely known. What we need is a criterion that can be applied *just to the system itself.* This criterion is based on a new thermodynamic property called the **free energy,** represented by the symbol *G*. *G* (like enthalpy, *H*) is designed solely for convenience. By defining it in terms of state functions, it too is a state function. The following equations should seem reasonable to you. Both of the properties that we have previously attempted to use as criteria for spontaneous change—enthalpy and entropy—appear in them. Unfortunately, proof of equation (15.7) is beyond the scope of the present discussion.

The symbol G is used to honor the American mathematical physicist J. Willard Gibbs, who formulated the free energy function (1876). The function itself is often called Gibbs free energy. Although discovered by Helmholtz later than by Gibbs, equation (15.6) is often called the Gibbs–Helmholtz equation.

Defining equation for free energy, G:

$$G = H - TS \qquad \text{enthalpy} - (\text{temp})(\text{order}) \qquad (15.5)$$

For a change occurring at constant temperature:

$$\Delta G = \Delta H - T\Delta S \qquad (15.6)$$

For a process occurring at constant temperature and pressure in a closed system, spontaneous change is accompanied by a decrease *in free energy, that is,*

$$\boxed{\Delta G < 0 \quad \text{then rxn is spontaneous}} \qquad (15.7)$$

If free energy is to carry an energy unit, then we see from equation (15.6) that the **free energy change, ΔG,** must represent a difference in two energy terms. The first of the two, ΔH, is familiar to us. The term, $T\Delta S$, is less obviously recognizable as an energy term. However, if the $T\Delta S$ product is to carry a unit of energy (J or kJ), then entropy itself must have the units of energy divided by kelvin temperature (J/K). The product of the two terms, $T\Delta S$, is sometimes called the "organizational energy" of a system.

APPLICATION OF THE NEW CRITERION FOR SPONTANEOUS CHANGE. In considering H and S separately we concluded that decreases in H (ΔH negative) and increases in S (ΔS positive) both tend to favor *spontaneous* change. For a process in which ΔH is negative and ΔS is positive, equation (15.6) indicates a *negative* value of ΔG. It should come as no surprise, then, that our new criterion (15.7) also states that for a spontaneous process at constant temperature and pressure, $\Delta G < 0$.

Altogether there are four possibilities for ΔG, based on the signs of ΔH and ΔS. These possibilities are presented in Table 15-1. An important conclusion can be drawn from the data in this table: At low temperatures the enthalpy change ΔH is more significant in determining the sign of ΔG, whereas at high temperatures ΔS assumes this role.

Example 15-2 Is a dissociation reaction of the type $AB(g) \rightarrow A(g) + B(g)$, favored at high or low temperatures?

This reaction involves the breaking of a bond between A and B. Energy must be absorbed; $\Delta H > 0$. Because 2 mol of gas is produced from 1 mol, increased disorder results from the reaction; $\Delta S > 0$.

$$\Delta G = \underbrace{\Delta H}_{+} - \underbrace{T\Delta S}_{-} = \underbrace{?}_{+ \text{ or } -}$$

Whether the dissociation process is spontaneous at some given temperature depends on the relative values of the ΔH term and the $T\Delta S$ term, and this we do not know. But what

TABLE 15-1
Criterion for spontaneous change: $\Delta G = \Delta H - T\Delta S$

Case	ΔH	ΔS	ΔG	Result	Example
1	−	+	−	spontaneous at all temp.	$2\,N_2O(g) \longrightarrow 2\,N_2(g) + O_2(g)$
2	−	−	− / +	spontaneous at low temp. / nonspontaneous at high temp.	$H_2O(l) \longrightarrow H_2O(s)$
3	+	+	+ / −	nonspontaneous at low temp. / spontaneous at high temp.	$2\,NH_3(g) \longrightarrow N_2(g) + 3\,H_2(g)$
4	+	−	+	nonspontaneous at all temp.	$3\,O_2(g) \longrightarrow 2\,O_3(g)$

Handwritten margin notes:

ΔH negative, then exothermic rxn

ΔS positive, then rxn increases in disorder

$\Delta H - T\Delta S = \Delta G < 0$ — constant conditions

TABLE 15-1

Case II: $\Delta H - T\Delta S$
$\Delta H < 0$ & $\Delta S < 0$
then for $\Delta H - T\Delta S$ to be < 0 (spontaneous rxn requires $\Delta G < 0$), then ΔH must be larger (more neg) than ΔS, so want small T

Case III: $\Delta H - T\Delta S$
if this is > 0 & this is > 0 then the ΔS part must be more than ΔH so difference = ΔG is < 0 thus want constant multiple T to be large.

THINK: endothermic rxn will be spontaneous if there is more heat around to absorb

we can say is that the higher the temperature, the more negative the term $-T \Delta S$ becomes. At some temperature the magnitude of $T \Delta S$ will just exceed that of ΔH. At this temperature, $\Delta G < 0$. Dissociation reactions of this type are favored at *high* temperatures. An example would be $I_2(g) \rightarrow I(g) + I(g)$.

SIMILAR EXAMPLES: Exercises 6, 7.

An interesting conclusion can be drawn from the result of Example 15-2: There should exist an upper temperature limit for the stabilities of chemical compounds. No matter how positive the value of ΔH for dissociation of a compound into its atoms, the term $T \Delta S$ will eventually exceed it as temperature is increased. If we consider the complete range of temperatures from absolute zero to the interior temperatures of the stars (about 3×10^7 K) we find that molecules exist over only a very small portion of this range (up to about 1×10^4 K).

15-4 Standard Free Energy Change, $\Delta G°$

As with other thermodynamic functions, the free energies of chemical substances depend on their state or condition. To facilitate the use of this function in calculations, we need to establish standard state conditions, as was done with enthalpy in Chapter 6. The standard state conventions we shall use are

for a solid: the pure substance at 1 atm pressure.

for a liquid: the pure substance at 1 atm pressure.

for a gas: an ideal gas at 1 atm partial pressure.   (15.8)

for a solute: an ideal solution at 1 M concentration.

The free energy change that results when reactants and products are all in their standard states is referred to as the **standard free energy change, $\Delta G°$.** For a reaction in which a compound is formed from its elements, $\Delta G°_{rx}$ is referred to as the **standard free energy of formation** of the compound, represented by $\Delta G°_f$. If the amount of compound formed is *one mole,* as is usually the case in tabulated data, it is customary to use an overbar on the symbol, that is, $\Delta \overline{G}°_f$. And according to convention, standard free energies of formation of the elements in their most stable forms at 1 atm pressure are defined as *zero* at the specified temperature. Selected values of $\Delta \overline{G}°_f$ at 298 K are tabulated in Appendix D.

Additional relationships needed in applying the free energy function are similar to the three presented in Section 6-5 for enthalpy.

1. ΔG is an extensive property.
2. ΔG changes sign when a process is reversed.
3. ΔG for a net or overall process can be obtained by summing the ΔG values for the individual steps.

$\Delta \overline{H}°_f = -30.59$ KJ/mole

Example 15-3   The standard molar heat of formation of silver oxide at 298 K is -30.59 kJ/mol. The standard molar free energy change, $\Delta \overline{G}°$, for the dissociation of silver oxide at 298 K is given below. What is $\Delta \overline{S}°$ for this reaction?

$$2 \, Ag_2O(s) \longrightarrow 4 \, Ag(s) + O_2(g) \qquad \Delta \overline{G}° = +22.43 \text{ kJ/mol}$$

In applying equation (15.6) we note that the reactants and products are in their standard states. Therefore, $\Delta \overline{G}° = \Delta \overline{H}° - T \Delta \overline{S}°$, and

$$\Delta \overline{S}° = \frac{\Delta \overline{H}° - \Delta \overline{G}°}{T}$$

The concept of a "mole of reaction" has the same meaning here as that introduced in Section 6-4. One mole of reaction signifies that the amounts of reactants and products involved in the reaction, in moles, are given by the stoichiometric coefficients in the chemical equation. The overbar on the symbol $\Delta \overline{G}°$ indicates that a mole of reaction is being considered.

Before we can solve for $\Delta \bar{S}^\circ$, it is necessary to have a value of $\Delta \bar{H}^\circ$. This can be obtained in the familiar manner employed in Section 6-6.

$$\Delta \bar{H}^\circ = 4\{\Delta \bar{H}_f^\circ [Ag(s)]\} + \Delta \bar{H}_f^\circ [O_2(g)] - 2\{\Delta \bar{H}_f^\circ [Ag_2O(s)]\}$$

$$\Delta \bar{H}^\circ = \quad\quad 0 \quad\quad + \quad\quad 0 \quad\quad - 2(-30.59 \text{ kJ/mol})$$

$$\Delta \bar{H}^\circ = +61.18 \text{ kJ/mol}$$

$$\Delta \bar{S}^\circ = \frac{61.18 \text{ kJ/mol} - 22.43 \text{ kJ/mol}}{298 \text{ K}} = 0.130 \text{ kJ mol}^{-1} \text{ K}^{-1} = 130 \text{ J mol}^{-1} \text{ K}^{-1}$$

SIMILAR EXAMPLES: Exercises 10, 12.

Example 15-4 What is $\Delta \bar{G}^\circ$ at 298 K for the reaction $C(s) + CO_2(g) \rightleftharpoons 2\ CO(g)$? Will the reaction proceed spontaneously in the forward direction at 298 K?

Here we can use tabulated data from Appendix D to determine $\Delta \bar{G}^\circ$.

$$\Delta \bar{G}^\circ = 2\{\Delta \bar{G}_f^\circ [CO(g)]\} - \Delta \bar{G}_f^\circ [C(s)] - \Delta \bar{G}_f^\circ [CO_2(g)]$$

$$\Delta \bar{G}^\circ = 2(-137.28 \text{ kJ/mol}) - 0 - (-394.38 \text{ kJ/mol})$$

$$= -274.56 + 394.38 = +119.82 \text{ kJ/mol}$$

The large positive value of $\Delta \bar{G}^\circ$ indicates that the reaction will *not* occur spontaneously to the extent that $CO(g)$ would be produced at a partial pressure of 1 atm (its standard state). At best, only a trace amount of $CO(g)$ would be produced before equilibrium is established. [After we consider the relationship between $\Delta \bar{G}^\circ$ and K_p in Section 15-6, a calculation of the equilibrium partial pressure of $CO(g)$ will be possible.]

SIMILAR EXAMPLE: Exercise 11.

15-5 Free Energy and Equilibrium

We have established that for a spontaneous process $\Delta G < 0$, and for a nonspontaneous process $\Delta G > 0$. Some mention should now be made of the situation in which $\Delta G = 0$. This is the condition of equilibrium. When a system is at equilibrium, there is an equal tendency for a process to proceed in either the forward or the reverse direction. Even an infinitesimal change in one of the experimental variables (e.g., temperature or pressure) will cause a net change to occur. As long as a system at equilibrium is left undisturbed, however, there will be no net change with time.

A hypothetical condition of equilibrium is presented in Figure 15-5 as the intersection of two graph lines. One line represents the temperature variation of the enthalpy change and the other, the product $T \Delta S$. The ΔH line may have a positive slope in some cases and negative in others, but the slope is generally quite gradual—ΔH is not especially temperature sensitive. The slope of the $T \Delta S$ line is expected to be rather steep, because of the presence of the temperature factor, T, in the product, $T \Delta S$. The vertical distance from the $T \Delta S$ line to the ΔH line represents $\Delta G = \Delta H - T \Delta S$. Starting at the left side of the figure, ΔG is positive and large. The magnitude of ΔG decreases with increasing temperature. At the right side of the figure, $T \Delta S$ exceeds ΔH in value and ΔG is negative. At the point of intersection of the two lines, $\Delta G = 0$ and the system is at equilibrium.

PHASE TRANSITIONS. Consider for a moment the application of Figure 15-5 to a description of the melting of ice. If we deal with the solid and liquid in their standard states (i.e., under 1 atm pressure), the intersection of the two lines comes at 273.15 K. This is the normal melting point of ice. At temperatures above 273.15 K the melting of ice is a spontaneous process. Below 273.15 K it is nonspontaneous. At 273.15 K ice and liquid water are at equilibrium. This equilibrium condition provides us with a method for determining the entropy change, ΔS°, for the melting of ice. (The molar heat of fusion of ice at 273.15 K is 6.02 kJ/mol.)

FIGURE 15-5

Free energy change as a function of temperature.

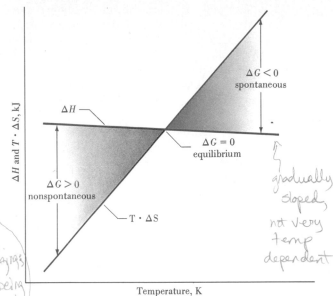

the ones changing, not being formed →

when assuming all positive values on graph.

$$\Delta G = \Delta H - T\Delta S$$

we see if $T\Delta S$ is much smaller than ΔH then ΔG will be positive & ∴ nonspon.

If $T\Delta S > \Delta H$ then $\Delta G = \Delta H - T\Delta S < 0$ & ∴ spontaniety

$\Delta H = \Delta H_{fusion} = 6.02 \frac{kJ}{mol}$

For the process $H_2O(s, 1\ atm) \rightleftharpoons H_2O(l, 1\ atm)$ *at* 273.15 K:

$$\Delta \overline{G}^\circ = \Delta \overline{H}^\circ - T\Delta \overline{S}^\circ = 0 \qquad \Delta \overline{H}^\circ = T\Delta \overline{S}^\circ$$

$$\Delta \overline{S}^\circ = \frac{\Delta \overline{H}^\circ}{T}$$

$$\Delta \overline{S}^\circ = \frac{6.02\ kJ/mol}{273.15\ K} = 2.20 \times 10^{-2}\ kJ\ mol^{-1}\ K^{-1}$$
$$= 22.0\ J\ mol^{-1}\ K^{-1}$$

The calculation just performed is a specific application of the general equation that can be written to describe any process at equilibrium.

$$\left.\begin{array}{l} \Delta G = \Delta H - T\Delta S = 0 \\[2mm] \Delta S = \dfrac{\Delta H}{T} \end{array}\right\} \text{ at equilibrium} \qquad (15.9)$$

When the process involves a transition between phases at constant temperature—melting, freezing, vaporization, condensation, and so on—descriptive subscripts may be used. That is, the subscripts "tr," "fus," "vap," and so on, may be attached to ΔG, ΔH, ΔS, and T. If we wish to describe a transition involving molar quantities of substances in their standard states, equation (15.9) is modified in the usual fashion: overbars and superscript degree signs are placed on the thermodynamic symbols.

$$\Delta \overline{S}^\circ_{tr} = \frac{\Delta \overline{H}^\circ_{tr}}{T_{tr}} \qquad (15.10)$$

✱ any phase transition situations use this eqn because such a situation represents equilibr.

Example 15-5 What is the standard molar entropy of vaporization of water at 100°C? The standard molar enthalpy of vaporization at 100°C is 40.7 kJ/mol.

Let us first translate this verbal description of the process into a chemical equation, *since @ vaporiecation gas & liq @ equil.*

$$H_2O(l, 1\ atm) \rightleftharpoons H_2O(g, 1\ atm)$$
$$\Delta \overline{H}^\circ_{vap} = 40.7\ kJ/mol \qquad \Delta \overline{S}^\circ_{vap} = ? \quad \frac{\Delta g}{= \Delta x10}$$

and then apply equation (15.10).

$$\Delta \overline{S}^\circ_{vap} = \frac{\Delta \overline{H}^\circ_{vap}}{T_{bp}} = \frac{40.7\ kJ/mol}{373.15\ K}$$
$$= 0.109\ kJ\ mol^{-1}\ K^{-1} = 109\ J\ mol^{-1}\ K^{-1}$$

equil temp for a vap rxn is the b. pt.

SIMILAR EXAMPLES: Exercises 21, 22.

A useful generalization, known as Trouton's rule, is that for many liquids at their normal boiling points the standard molar *entropy of vaporization* has a value of about 88 J mol⁻¹ K⁻¹.

$$\Delta \overline{S}^\circ_{vap} = \frac{\Delta \overline{H}^\circ_{vap}}{T_{bp}} \simeq 88\ J\ mol^{-1}\ K^{-1} \qquad (15.11)$$

If the degree of disorder produced in transferring 1 mol of molecules from liquid to vapor at 1 atm pres-

TABLE 15-2
Some standard molar entropy changes for phase transitions

Substance	T_{tr}, K	$\Delta \bar{H}^\circ_{tr}$, kJ/mol	$\Delta \bar{S}^\circ_{tr}$, J mol^{-1} K^{-1}
Vaporization			
acetone	329	30.25	91.9
ethanol	351	39.33	112.1
n-hexane	342	28.58	83.6
lead	2024	179.41	88.6
Fusion			
benzene	279	9.83	35.2
camphor	452	6.86	15.2
lead	601	4.77	7.9
mercury	234	2.30	9.8

FIGURE 15-6
Liquid-vapor equilibrium and the direction of spontaneous change.

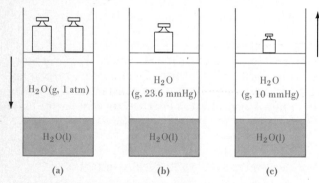

(a)   (b)   (c)

(a) At 25°C and 1 atm, the direction of spontaneous change is the condensation of $H_2O(g)$. For the vaporization process $H_2O(l, 1 \text{ atm}) \rightarrow H_2O(g, 1 \text{ atm})$, $\Delta G = \Delta G^\circ = +8.58$ kJ/mol.
(b) At 25°C and 23.6 mmHg, the equilibrium $H_2O(l) \rightleftharpoons H_2O(g)$ is established; $\Delta G = 0$.
(c) At 25°C and 10 mmHg, the vaporization of water is spontaneous; for $H_2O(l, 10 \text{ mmHg}) \rightarrow H_2O(g, 10 \text{ mmHg})$, $\Delta G < 0$.
In each case the black arrow suggests the direction of spontaneous change.

In equation (15.14) R is the gas constant [8.314 J mol^{-1} K^{-1}] and T is the kelvin temperature.

sure is roughly comparable for different liquids, then we should expect similar values of $\Delta \bar{S}^\circ_{vap}$. Instances where Trouton's rule fails are also understandable. In water and ethanol, hydrogen bonding among molecules produces a greater degree of order than otherwise expected in the liquid state. The degree of disorder produced in the vaporization process is greater than normal, and $\Delta \bar{S}^\circ_{vap} > 88$ J mol^{-1} K^{-1}. There is no regularity in the values of entropies of fusion, other than that they are generally smaller for metals than for most other substances. Selected values for a few phase transitions are presented in Table 15-2.

15-6 Relationship of $\Delta \bar{G}^\circ$ to K

For the vaporization of water, $H_2O(l) \rightleftharpoons H_2O(g)$, the standard molar free energy change at 25°C is $\Delta \bar{G}^\circ = +8.58$ kJ/mol. The positive sign of $\Delta \bar{G}^\circ$ cannot mean that water is incapable of vaporizing at 25°C, for practical experience tells us that it can. The difficulty here is that $\Delta \bar{G}^\circ$ refers to liquid water and water vapor in their *standard states*. Its positive value simply tells us that $H_2O(l)$ at 1 atm pressure will not spontaneously produce $H_2O(g)$ at *1 atm pressure* at 25°C, that is,

$$H_2O(l, 1 \text{ atm}) \rightleftharpoons H_2O(g, 1 \text{ atm})$$
$$\Delta \bar{G}^\circ = +8.58 \text{ kJ/mol} \quad (15.12)$$

The equilibrium vapor pressure of water at 25°C is found experimentally to be 23.76 mmHg = 0.0313 atm. The condition of equilibrium is

$$H_2O(l, 0.0313 \text{ atm}) \rightleftharpoons H_2O(g, 0.0313 \text{ atm})$$
$$\Delta \bar{G} = 0 \quad (15.13)$$

Figure 15-6 suggests the difference between the $\Delta \bar{G}^\circ$ of (15.12) and $\Delta \bar{G}$ of (15.13).

THE EXPRESSION $\Delta \bar{G}^\circ = -2.303RT \log K$. $\Delta \bar{G}^\circ$ alone can be used to determine the direction of spontaneous change only if the reactants and products are in their standard states. Because reaction conditions of interest to us will often be nonstandard ones, we should pursue the criterion for spontaneous change further. To do so we need to relate the molar free energy change for nonstandard conditions $\Delta \bar{G}$, to the corresponding molar free energy change for a process in which all reactants and products are in their standard states, $\Delta \bar{G}^\circ$. The key term for doing this is the reaction quotient, Q, written for the specified states of reactants and products. Unfortunately, a derivation of this relationship is beyond the scope of the present discussion, and the equation below is given without proof.

$$\Delta \bar{G} = \Delta \bar{G}^\circ + 2.303RT \log Q \quad (15.14)$$

FIGURE 15-7
Free energy change, equilibrium and the direction of spontaneous change.

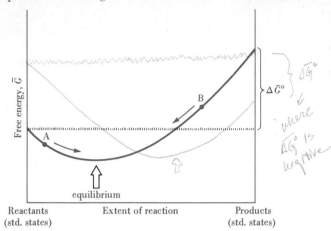

Free energy is plotted as a function of the extent of reaction. The difference between the standard state molar free energies of reactants and products is the standard molar free energy change, $\Delta \bar{G}°$. The equilibrium point lies somewhere between pure reactants and pure products. The free energy of the equilibrium mixture is lower than that of pure reactants, pure products, and any other mixture of the two. Mixtures, A and B, will each undergo spontaneous change in the direction of the equilibrium mixture. If $\Delta \bar{G}°$ is large and positive, the minimum in free energy lies very close to the pure reactant side and very little reaction occurs before equilibrium is reached. If $\Delta \bar{G}°$ is large and negative, the minimum in free energy lies very close to the product side, and the reaction goes essentially to completion. Only if $\Delta \bar{G}°$ has a small magnitude, either positive or negative, will the equilibrium mixture have appreciable amounts of both reactants and products.

The value of equation (15.14) lies in applying it to situations in which the specified states of the reactants and products are *equilibrium* states. That is, if we apply it to a system at equilibrium, the value of ΔG is *zero* and the reaction quotient, Q, is the *equilibrium constant*, K. This leads to the following expression.

At equilibrium:

$$\Delta \bar{G} = \Delta \bar{G}° + 2.303 RT \log K = 0$$

$$and \qquad \Delta \bar{G}° = -2.303 RT \log K \qquad (15.15)$$

The most important implication of equation (15.15) is that from tabulated thermodynamic data (as in Appendix D) we can derive $\Delta \bar{G}°$ values, and from these, in turn, equilibrium constants, K. As a result, many of the calculations considered in Chapter 14 now become possible without the need for direct experimental equilibrium measurements. We will attempt such calculations shortly, but first we need to consider two other matters briefly.

The first of these matters is that we have completed our search for a criterion for spontaneous change. We can replace the hypothetical graphical representation of Figure 15-2 with our final result, as is done in Figure 15-7. The second matter is discussed in the following section.

THE THERMODYNAMIC EQUILIBRIUM CONSTANT. To be used in equation (15.15), equilibrium constants must be modified to a form called the thermodynamic equilibrium constant. This is a constant that is written in terms of activities or "effective concentrations" (see Section 12-7). For the general reaction

$$a\,A + b\,B + \cdots \rightleftharpoons g\,G + h\,H + \cdots \qquad (15.16)$$

the thermodynamic equilibrium constant is

$$K = \frac{(a_G)^g (a_H)^h \cdots}{(a_A)^a (a_B)^b \cdots} \qquad (15.17)$$

In applying equation (15.17), only numerical values of the activities are substituted for the a symbols: Each a symbol is actually a ratio of the equilibrium activity of a reactant to its activity in its standard state. However, standard states have been defined to have unit activity—the denominator in each ratio is 1. Thus, as long as the units used to express equilibrium activities are those used to define standard states, activity units in the a ratios cancel. This leaves a unitless but numerically equal to the equilibrium activity. The following are useful approximations in establishing activities.

For pure solids and liquids: The activity is taken as unity, $a = 1$.

In cases where these assumptions of ideal conditions cannot be made, it is necessary to include a corrective factor known as the activity coefficient, γ, to relate gas pressures and solute concentrations to activities. That is, $a_{gas} = \gamma \cdot P_{gas}$ and $a_{solute} = \gamma \cdot [solute]$. However, this added complication is beyond the scope of the present discussion.

For gases: Assume ideal gas behavior. Then the activity of a gas is equal to its pressure in atm.

For components in solution: Assume ideal behavior (e.g., no interionic attractions). Then if the standard state is taken to be 1 M, the activity is equal to the molar concentration.

Whether K_c or K_p can be used in equation (15.15) depends on whether either of these expressions results from writing a thermodynamic equilibrium constant expression. One of the cases cited in Example 15-6 results in a value of K_p because partial pressures are substituted for activities. Another yields a value of K_c, with dilute concentrations substituting for activities. In the remaining case, because both concentrations and partial pressures appear in the equilibrium constant expression, the symbol K is used.

Example 15-6 Write thermodynamic equilibrium constant expressions for the following reversible processes, making appropriate substitutions for activities.

(a) $C(s) + H_2O(g) \rightleftharpoons CO(g) + H_2(g)$
(b) $PbI_2(s) \rightleftharpoons Pb^{2+}(aq) + 2\,I^-(aq)$
(c) $O_2(g) + 2\,S^{2-}(aq) + 2\,H_2O(l) \rightleftharpoons 4\,OH^-(aq) + 2\,S(s)$

(a) $K = \dfrac{(a_{CO(g)})(a_{H_2(g)})}{(a_{C(s)})(a_{H_2O(g)})} = \dfrac{(P_{CO})(P_{H_2})}{(1)(P_{H_2O})} = \dfrac{(P_{CO})(P_{H_2})}{(P_{H_2O})} = K_p$

(b) $K = \dfrac{(a_{Pb^{2+}})(a_{I^-})^2}{a_{PbI_2(s)}} = \dfrac{[Pb^{2+}][I^-]^2}{1} = [Pb^{2+}][I^-]^2 = K_c$

(c) $K = \dfrac{(a_{S(s)})^2(a_{OH^-})^4}{(a_{O_2(g)})(a_{S^{2-}})^2(a_{H_2O})^2} = \dfrac{(1)^2[OH^-]^4}{(P_{O_2})[S^{2-}]^2(1)^2} = \dfrac{[OH^-]^4}{(P_{O_2})[S^{2-}]^2}$

SIMILAR EXAMPLE: Exercise 27.

ILLUSTRATIVE EXAMPLES. The following two examples make use of equation (15.15) to provide information about equilibrium in the reaction described in Example 15-3.

$$2\,Ag_2O(s) \rightleftharpoons 4\,Ag(s) + O_2(g) \tag{15.18}$$

Example 15-7 Use data from Example 15-3, as needed, to determine the equilibrium partial pressure of $O_2(g)$ above a mixture of $Ag(s)$ and $Ag_2O(s)$ at 298 K. Treat the $O_2(g)$ as an ideal gas.

The equilibrium in question is described by the expression

Here we discover a reason for specifying "a mole of reaction." The "mol^{-1}" term in the value of $\Delta \overline{G}°$ provides for the cancellation of "mol^{-1}" in the units of R, making $\log K$ a dimensionless quantity. A balanced chemical equation must always be given for a reaction. From this, the "mole of reaction," the $\Delta \overline{G}°$ value, and the equilibrium constant expression, K, can all be established in a consistent way.

$$K = \frac{(a_{Ag(s)})^4(a_{O_2(g)})}{(a_{Ag_2O(s)})^2} = \frac{(1)^4(P_{O_2})}{(1)^2} = P_{O_2} = K_p$$

To determine the value of $K = K_p$ at 298 K we use $\Delta \overline{G}°$ from Example 15-3 in equation (15.15).

$$\Delta \overline{G}° = -2.303 RT \log K = +22.43 \text{ kJ/mol}$$

$$= -2.303 \times 8.314\,\text{J mol}^{-1}\text{K}^{-1} \times 298\,\text{K} \times \log K = 22.43 \times 10^3\,\text{J/mol}$$

$$\log K = \frac{-22.43 \times 10^3}{2.303 \times 8.314 \times 298} = -3.931 = 0.069 - 4.00$$

$$K = 1.17 \times 10^{-4}$$

But we have already seen that $K = K_p = P_{O_2}$. Therefore, $P_{O_2} = 1.17 \times 10^{-4}$ atm

SIMILAR EXAMPLES: Exercises 29, 30, 31.

We could have arrived at this conclusion directly from the statement of the question. Since the *equilibrium condition* described has reactants and products in their *standard states* (pure solids and a partial pressure of 1 atm for an ideal gas), $\Delta \bar{G} = \Delta \bar{G}^\circ = 0$.

Equation (14.25), which describes the effect of temperature on the equilibrium constant, K, can be derived from the Gibbs–Helmholtz equation, as follows.

$$\Delta \bar{G}^\circ = \Delta \bar{H}^\circ - T \Delta \bar{S}^\circ$$

$$\frac{\Delta \bar{G}^\circ}{T} = \frac{\Delta \bar{H}^\circ}{T} - \Delta \bar{S}^\circ$$

$$\frac{-2.303 \cdot R \cdot T \log K}{T}$$

$$= \frac{\Delta \bar{H}^\circ}{T} - \Delta \bar{S}^\circ$$

$$\log K = \frac{-\Delta \bar{H}^\circ}{2.303 \cdot R \cdot T}$$

$$+ \frac{\Delta \bar{S}^\circ}{2.303 \cdot R}$$

By assuming that $\Delta \bar{H}^\circ$ and $\Delta \bar{S}^\circ$ are independent of temperature, the equation above can be written for two different temperatures, the constant term, $\Delta \bar{S}^\circ / 2.303 \cdot R$, eliminated, and the result obtained.

$$\log \frac{K_2}{K_1}$$

$$= \frac{\Delta \bar{H}^\circ}{2.303 \cdot R} \left(\frac{T_2 - T_1}{T_1 T_2} \right)$$

$$(14.25)$$

Example 15-8   At what approximate temperature will the dissociation pressure of $Ag_2O(s)$ become equal to 1 atm? Use data from Example 15-3, and indicate any assumptions that are made in this estimation.

Step 1. If the dissociation pressure is to be 1 atm, this means that $P_{O_2} = 1$ atm, and $K = K_p = P_{O_2} = 1$ (see Example 15-7).

Step 2. If we know a value of K, we can determine the corresponding value of $\Delta \bar{G}^\circ$ with equation (15.15).

$$\Delta \bar{G}^\circ = -2.303 RT \log K = -2.303 RT \log 1 = 0$$

We are looking for the temperature where $\Delta \bar{G}^\circ = 0$ for reaction (15.18).

Step 3. The temperature at which we do have some data for this reaction is 298 K. From Example 15-3 we obtain the values

$$\Delta \bar{G}^\circ = +22.43 \text{ kJ/mol} \qquad \Delta \bar{H}^\circ = +61.18 \text{ kJ/mol} \qquad \Delta \bar{S}^\circ = 0.130 \text{ kJ mol}^{-1} \text{ K}^{-1}$$

In discussing Figure 15-5 we noted that ΔH for a reaction often does not change appreciably with temperature. Let us assume that at the temperature we are seeking, $\Delta \bar{H}^\circ$ has the same value as at 298 K, that is, $+61.18$ kJ/mol. Let us assume further that at this same temperature $\Delta \bar{S}^\circ$ has approximately the same value as it has at 298 K, that is, 0.130 kJ mol^{-1} K^{-1}. Now we need to solve equation (15.6) for temperature.

$$\Delta \bar{G}^\circ = \Delta \bar{H}^\circ - T \Delta \bar{S}^\circ = 61.18 \text{ kJ/mol} - T \times 0.130 \text{ kJ mol}^{-1} \text{ K}^{-1} = 0$$

$$T \simeq \frac{61.18 \text{ kJ/mol}}{0.130 \text{ kJ mol}^{-1} \text{ K}^{-1}} \simeq 470 \text{ K} \ (\simeq 200°C)$$

SIMILAR EXAMPLES: Exercises 35, 36, 37.

$\Delta \bar{G}^\circ$ **AS A FUNCTION OF TEMPERATURE.**   Quite often, values of $\Delta \bar{G}^\circ$ and K are desired at temperatures other than those at which values of $\Delta \bar{H}^\circ$ and $\Delta \bar{S}^\circ$ are available. Exact expressions that incorporate the temperature dependence of ΔH and ΔS are available to relate $\Delta \bar{G}^\circ$ and T. However, it is beyond the scope of this text to consider these. Fortunately, the *approximation* introduced in Example 15-8 often works: ΔH and ΔS are taken to be temperature-independent. The Gibbs–Helmholz equation can then be written as follows.

$$\Delta \bar{G}^\circ \simeq \Delta \bar{H}^\circ - T \times \Delta \bar{S}^\circ \qquad \begin{array}{l} \text{Regardless of the temperature in question, assume that} \\ \text{these quantities have the same values as at some other} \\ \text{temperature at which they are known or can be determined} \end{array}$$

15-7   The Third Law of Thermodynamics

Although absolute values of most thermodynamic properties cannot be determined, it is possible to obtain *absolute* values of the entropy for many substances. This is made possible by the third law of thermodynamics.

The entropy of a pure perfect crystal at the absolute zero of temperature is zero.

The perfect crystal at the absolute zero of temperature represents the greatest order possible for a thermodynamic system. As the temperature is raised from values near 0 K, the entropy increases. Absolute entropies are always positive quantities.

If more than one form of a solid substance exists, there will be a small entropy of transition at the temperature at which the low-temperature form is converted to a new solid phase. Another sharp increase in entropy comes at the melting point of the solid, and a still larger increase at the boiling point of the liquid. Entropies of transition can be calculated with equation (15.9). The determination of entropy changes in temperature ranges where there are no phase transitions requires the calorimetric determination of specific heats.

The variation of absolute entropy with temperature is illustrated through Figure 15-8. Typical entropy data are included in Appendix D; and such data are used in Example 15-9.

Example 15-9   Does the conclusion of Example 15-2 apply to the reaction $2\,NO(g) \rightleftharpoons N_2(g) + O_2(g)$? That is, is this dissociation favored at high temperatures?

The reaction in Example 15-2 conformed to case 3 in Table 15-1, that is, dissociation was favored at high temperatures. We reached that conclusion easily because we had no difficulty in assessing signs of both ΔH and ΔS. What are the signs of ΔH and ΔS for the dissociation of $NO(g)$? The answer cannot be found so simply: Chemical bonds are formed as well as broken, and none of the notions about entropy change offered through (15.2) applies here. Let us take the following approach: (1) determine the standard entropy change at 298 K from tabulated *absolute* entropies (see Appendix D); (2) determine the standard enthalpy change for the reaction at 298 K; and (3) refer to Table 15-1.

At 298 K:

$$\Delta \overline{S}^\circ = \overline{S}^\circ[N_2(g)] + \overline{S}^\circ[O_2(g)] - 2\{\overline{S}^\circ[NO(g)]\}$$

$$\Delta \overline{S}^\circ = \quad 191.50 \quad + \quad 205.02 \quad - \quad 2(210.62) \quad = -24.72\ \text{J mol}^{-1}\,\text{K}^{-1}$$

At 298 K:

$$\Delta \overline{H}^\circ = \Delta \overline{H}_f^\circ[N_2(g)] + \Delta \overline{H}_f^\circ[O_2(g)] - 2\{\Delta \overline{H}_f^\circ[NO(g)]\}$$

$$\Delta \overline{H}^\circ = \quad\quad 0 \quad\quad + \quad\quad 0 \quad\quad - 2(+90.37) \quad\quad = -180.74\ \text{kJ/mol}$$

A process with $\Delta H < 0$ and $\Delta S < 0$ corresponds to case 2 in Table 15-1. The forward process is spontaneous at low temperatures ($\Delta G < 0$). Thus, the dissociation of $NO(g)$ is favored at *low* temperatures.

SIMILAR EXAMPLES: Exercises 35, 36, 37.

Since the dissociation of $NO(g)$ is favored at low temperatures, its formation is favored at high temperatures. As a result, the production of $NO(g)$ from $N_2(g)$ and $O_2(g)$ occurs in any combustion process in air. The higher the combustion temperature, the more important the production of $NO(g)$ becomes, and the more serious the attendant air pollution problems that are created.

FIGURE 15-8
Absolute entropy as a function of temperature.

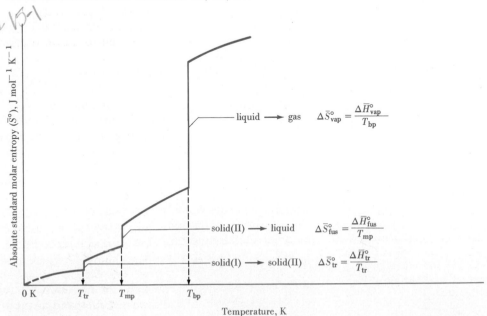

The absolute standard molar entropy of a hypothetical substance is represented here. According to the third law of thermodynamics, an absolute entropy of zero is expected at 0 K. However, experimental measurements cannot be carried to this temperature—extrapolation techniques are required (broken-line portion of graph). A transition at T_{tr} is noted from the solid(I) to solid(II) form of the substance. Melting occurs at the normal melting point, T_{mp}, and vaporization at the normal boiling point, T_{bp}.

SKIP

15-8 Heat Engines

EFFICIENCY OF HEAT ENGINES. The underlying principle behind the second law of thermodynamics was deduced by a French engineer, Sadi Carnot, in 1820 in a study of the efficiencies of heat engines. William Thomson (Lord Kelvin) recognized the significance of Carnot's work and saw in it the basis of the second law of thermodynamics and an absolute temperature scale.

The basic principle of a heat engine is that heat (q_h) is absorbed by the working substance of the engine at a high temperature (T_h). This heat is converted partly to work (w), and the remainder (q_l) is rejected to the surroundings at a lower temperature (T_l). This process is pictured in Figure 15-9.

The efficiency of the engine is governed by the ratio, w/q_h. If all the heat absorbed could be converted to work, the engine would be 100% efficient. The second law of thermodynamics places an absolute limit on the efficiency of a heat engine, *and it is never 100%.* The expression obtained is

$$\text{efficiency} = \frac{T_h - T_l}{T_h} \qquad (15.19)$$

where temperatures are in kelvins.

If a steam engine were operated between 100°C (the boiler temperature) and 25°C (the condenser temperature), it would have an efficiency of only 0.20, that is, 20%. 80% of the heat supplied to the boiler would be rejected to the surroundings.

$$\text{efficiency} = \frac{373 - 298}{373} = \frac{75}{373} = 0.20$$

Conventional electric power plants are based on the burning of a fossil fuel, which provides the heat energy to convert water to steam. The steam powers a turbine, which in turn drives an electric generator. The process is not very efficient. The efficiency can be improved by using a higher working temperature for the turbine (T_h). This can be accomplished by operating the system under high pressure, at temperatures well in excess of 100°C, using superheated steam. Still, however, most electric power plants operate at efficiencies less than 40%. Thus, more than half the heat required to produce electric power is waste heat. The implications of this fact are two fold: wasted heat means wasted fuel, and as the waste heat enters the environment it leads to thermal pollution.

THERMAL POLLUTION. The origin of thermal pollution in a power plant operation is suggested by Figure 15-10. The best known effects of thermal pollution are those produced on fish and other aquatic life. The differential growth rate of algae with changing tempera-

FIGURE 15-9
Schematic representation of a heat engine.

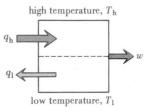

high temperature, T_h

q_h

w

q_l

low temperature, T_l

The efficiency of the engine is determined by the quantity of heat q_h that is converted to work w. The smaller the quantity of heat rejected to the surroundings at the lower temperature q_l, the more efficient the engine. The second law of thermodynamics places a theoretical maximum efficiency on every heat engine; it is always less than 100% (see equation 15.19).

By comparison, a diesel engine is about 35% efficient, and an internal combustion gasoline engine 25% efficient, in converting heat to work.

FIGURE 15-10
Efficiency of a fossil-fuel electric power plant and the origin of thermal pollution.

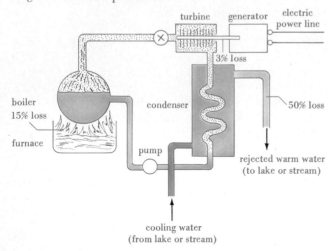

Water is heated in a boiler and converted to super-heated steam. The steam is allowed to expand in the turbine where it does work by turning the rotor blades. The turbine shaft drives the electric generator, which produces electric power. Conversion of mechanical to electrical energy in the turbine-generator combination is nearly 100% efficient.

If the steam were to be rejected to the surroundings, all the makeup water in the boiler would have to be cold water. This would require more fuel for heating than if the steam is returned to the boiler. However, because of its greatly expanded volume after performing work in the turbine, the steam must be condensed back to liquid before being returned to the boiler. Condensation of the steam occurs in the condenser. The water that is used for cooling carries away up to 50% of the heat energy that was released in burning the fuel. It is this waste heat that produces thermal pollution.

tures can cause one type of algae, which is ideal food for fish, to be displaced by another, which is a poor food or even toxic to fish. Also, because the metabolic rate goes up with temperature, fish need more food and oxygen as the temperature of their water rises. Temperature changes affect other physiological processes too. Some fish are killed by the thermal shock of even relatively small temperature changes.

Even if all other pollution sources could be controlled, the loss of heat as one energy form is converted to another would still be unavoidable. Means must be found to minimize the amount of waste heat or put the waste heat to useful purposes. The first objective can be met through more efficient heat engines and through the direct conversion of other energy forms to electricity (such as with fuel cells and solar energy devices). The second can be met by using waste heat from power plants to supply hot water for industrial, commercial or residential purposes, or using this water for agricultural purposes. None of these proposals to deal with thermal pollution is well developed currently, and the problem is likely to grow significantly before solutions are found.

Summary

One goal of this chapter is to develop a criterion for the direction of spontaneous change. Another is to establish a means of calculating equilibrium constants from thermodynamic properties.

Although most exothermic reactions are spontaneous, many *endothermic* reactions are spontaneous as well. This means that enthalpy change alone cannot serve as the criterion for spontaneous change. A common feature shared by systems in which spontaneous endothermic processes occur is that they undergo an increase in disorder or randomness. The thermodynamic property that relates to the degree of disorder or randomness in a system is *entropy, S*. At times, a quali-

tative assessment of the entropy change in a system can be made simply by inspecting the situation. For example, the conversions solid → liquid, solid → gas, and liquid → gas all produce an increase in disorder and in entropy. If entropy change alone is to be used as the criterion of spontaneous change, it is necessary to assess the *total* entropy change accompanying a process. This is the sum of the entropy changes of the system *and* its surroundings. Such an assessment is inconvenient and sometimes very difficult to make.

The thermodynamic property devised to provide a criterion for spontaneous change based just on the system itself is the *free energy* (or Gibbs free energy), *G*.

Both enthalpy and entropy appear in the basic free energy equations. For example, for a process at constant temperature, the Gibbs–Helmholz equation can be written: $\Delta G = \Delta H - T\Delta S$. The criterion for spontaneous change in a closed system at constant temperature and pressure is that there be a *decrease in free energy:* $\Delta G < 0$.

The free energy criterion for *equilibrium* is that $\Delta G = 0$. At equilibrium, then, $\Delta H = T\Delta S$; and this expression can be used to calculate any one of the three terms when the other two are known. This expression is particularly useful in dealing with phase transitions. For liquid–vapor equilibria at the normal boiling point, Trouton's rule is often applicable as well: $\Delta\bar{S}^{\circ}_{vap} = \Delta\bar{H}^{\circ}_{vap}/T_{bp} \simeq 88\ \text{J mol}^{-1}\ \text{K}^{-1}$.

The standard free energy change for a reaction, ΔG°, accompanies the conversion of reactants in their standard states to products in their standard states.

Tabulated free energy data are usually standard molar free energies of formation $\Delta\bar{G}^{\circ}_f$. A most useful relationship exists between standard molar free energy change and the equilibrium constant, K: $\Delta\bar{G}^{\circ} = -2.303RT \log K$. If a value of ΔG° can be obtained, say by the appropriate combination of tabulated thermodynamic data, a value of K can be calculated. The K values obtained from $\Delta\bar{G}^{\circ}$ data are thermodynamic equilibrium constants. These equilibrium constant expressions are based on the *activities* of reactants and products. Activities can be related to the more familiar terms concentration and pressure through a few simple conventions.

A key to successful thermodynamic calculations is in being able to determine entropy changes from tabulated *absolute* molar entropies. The third law of thermodynamics provides a basis upon which absolute entropies may be established experimentally.

Learning Objectives

As a result of studying Chapter 15, you should be able to

1. Explain the meaning of the term "spontaneous change" as it applies to chemical reactions.

2. Discuss the significance of entropy and its relationship to the degree of disorder within a system.

3. Predict for certain processes whether entropy increases or decreases.

4. Write the principal equations that define free energy and free energy change and state the basic criterion for spontaneous change, $\Delta G < 0$.

5. Relate $\Delta\bar{G}^{\circ}$, $\Delta\bar{H}^{\circ}$, and $\Delta\bar{S}^{\circ}$ through the Gibbs–Helmholtz equation: $\Delta\bar{G}^{\circ} = \Delta\bar{H}^{\circ} - T\Delta\bar{S}^{\circ}$.

6. Calculate entropy and enthalpy changes for phase transitions, using the condition of equilibrium ($\Delta G = 0$).

7. Write thermodynamic equilibrium constant expressions for chemical reactions and relate these to K_c and K_p.

8. Calculate equilibrium constants using tabulated data and the relationship between $\Delta\bar{G}^{\circ}$ and K.

9. Explain how the third law of thermodynamics makes possible an evaluation of *absolute* entropies of substances.

10. Describe how a heat engine functions and the factors that limit its efficiency.

Some New Terms

Entropy, S, is a measure of the degree of *disorder* in a system; the higher the degree of *disorder,* the greater the entropy.

Entropy change, ΔS, expresses the extent to which the degree of order changes as the result of some process. A positive ΔS means an increase in *disorder.*

Entropy change of the universe, ΔS_{univ}, refers to the total entropy change (system and surroundings) that accompanies a process. For every spontaneous change, $\Delta S_{univ} > 0$.

Free energy, G, is a thermodynamic function designed to provide a criterion for spontaneous changes. It is defined in terms of enthalpy and entropy through the equation: $G = H - TS$.

Free energy change, ΔG, indicates the direction of spontaneous change. For a spontaneous process at constant temperature and pressure in a closed system, free energy decreases, $\Delta G < 0$.

A heat engine is a device for converting heat into work. The engine absorbs heat at a high temperature, converts part of it to work, and rejects the remaining heat to the surroundings at a lower temperature.

The second law of thermodynamics relates to the direction of spontaneous change. All spontaneous processes produce an increase in the entropy of the universe.

A spontaneous or natural process is one that is able to take place in a system left to itself. No external action is required to make the process go, although in some cases the process may take a very long time to occur.

Standard free energy change, ΔG°, is the free energy change that accompanies a process when reactants and products are all in their standard states. Of special significance is the equation relating the standard molar free energy change and the equilibrium constant: $\Delta \bar{G}^\circ = -2.303RT \log K$.

The **standard free energy of formation, $\Delta \bar{G}_f^\circ$**, is the standard free energy change associated with the formation of 1 mol of compound from its elements.

The **thermodynamic equilibrium constant** is an equilibrium constant expression written with activities. In dilute solutions activities can be replaced by molar concentrations. In ideal gases partial pressures in atm can be substituted for activities.

The **third law of thermodynamics** postulates that *the entropy of a pure perfect crystal at the absolute zero of temperature is zero.*

Trouton's rule states that at their normal boiling points the entropies of vaporization of many liquids are about 88 J mol^{-1} K^{-1}.   *notice the terms*

Exercises

Entropy and disorder

1. Based on the relationship of entropy to the degree of order in a system, indicate whether each of the following changes represents an increase or decrease in entropy in a system: **(a)** the freezing of ethanol; **(b)** the sublimation of dry ice; **(c)** the burning of a rocket fuel.

2. Indicate whether you would expect the entropy to increase or decrease in each of the following reactions. If a determination cannot be made simply by inspecting the equation, state why this is so.
   (a) $(CH_3CH_2)_2O(l) \longrightarrow (CH_3CH_2)_2O(g)$
   (b) $CuSO \cdot 3H_2O(s) + 2 H_2O(g) \longrightarrow CuSO_4 \cdot 5H_2O(s)$
   (c) $SO_3(g) + H_2(g) \longrightarrow SO_2(g) + H_2O(g)$
   (d) $H_2S(g) + O_2(g) \longrightarrow H_2O(g) + SO_2(g)$
   [not balanced]
   (e) $H_2(g) + I_2(g) \longrightarrow HI(g)$ [not balanced]

3. Which one in each pair of substances would you expect to have the greater entropy? Explain the basis of your reasoning.
   (a) 1 mol $H_2O(l, 1$ atm, $50°C)$ *or* 1 mol $H_2O(g, 1$ atm, $50°C)$
   (b) 50.0 g Fe(s, 1 atm, 20°C) *or* 0.80 mol Fe(s, 1 atm, 20°C)
   (c) 1 mol $Br_2(l, 1$ atm, $58°C)$ *or* 1 mol $Br_2(s, 1$ atm, $-10°C)$
   (d) 0.10 mol $O_2(g, 0.10$ atm, $25°C)$ *or* 0.10 mol $O_2(g, 10.0$ atm, $25°C)$

4. Explain why
   (a) Some exothermic reactions do not occur spontaneously.
   (b) Some reactions in which the entropy of the system increases also do not occur spontaneously.

* 5. Use ideas from this chapter and Chapter 6 to comment on the meaning of a famous remark attributed to Rudolf Clausius (1865): "Die Energie der Welt is konstant; die Entropie der Welt strebt einem Maximum zu. [The energy of the world is constant; the entropy of the world increases toward a maximum.]"

Free energy and spontaneous change

6. The decomposition of nitrosyl bromide is represented by the equation

$$NOBr(g) \longrightarrow NO(g) + \tfrac{1}{2} Br_2(g)$$
$$\text{at 298 K, } \Delta \bar{H}^\circ = +8.54 \text{ kJ/mol}$$

Do you expect this decomposition to occur to a greater extent at low or high temperatures?

7. From the data given, indicate which of the four cases in Table 15-1 applies for each reaction.
   (a) $H_2(g) \longrightarrow 2 H(g)$
   (b) $2 SO_2(g) + O_2(g) \longrightarrow 2 SO_3(g)$
   at 298 K, $\Delta \bar{H}^\circ = -197.8$ kJ/mol
   (c) $N_2H_4(g) \longrightarrow N_2(g) + 2 H_2(g)$
   at 298 K, $\Delta \bar{H}^\circ = -95.4$ kJ/mol
   (d) $N_2(g) + 3 Cl_2(g) \longrightarrow 2 NCl_3(l)$
   at 298 K, $\Delta \bar{H}^\circ = +230$ kJ/mol

8. For the process pictured in Figure 15-3, what are the values (positive, negative, or zero) for ΔH, ΔS, and ΔG? Explain your reasoning.

9. What values of ΔH, ΔS, and ΔG would you expect for the formation of an ideal solution of liquid components (i.e., are these values positive, negative, or zero)? Explain.

Standard free energy change

10. At what temperature will the following reaction have the values of $\Delta \bar{G}^\circ$, $\Delta \bar{H}^\circ$, and $\Delta \bar{S}^\circ$ given?

$$2 PbS(s) + 3 O_2(g) \longrightarrow 2 PbO(s) + 2 SO_2(g)$$

$\Delta \bar{G}^\circ = -777.8$ kJ/mol $\Delta \bar{H}^\circ = -843.7$ kJ/mol
$\Delta \bar{S}^\circ = -0.165$ kJ mol^{-1} K^{-1}

11. Use data from Appendix D to determine values at 298 K of $\Delta \bar{G}^\circ$ for the following reactions.
   (a) $N_2(g) + 3 H_2(g) \longrightarrow 2 NH_3(g)$
   (b) $C_2H_2(g) + 2 H_2(g) \longrightarrow C_2H_6(g)$
   (c) $Fe_3O_4(s) + 4 H_2(g) \longrightarrow 3 Fe(s) + 4 H_2O(g)$
   (d) $MgO(s) + 2 HCl(g) \longrightarrow MgCl_2(s) + H_2O(g)$

12. A particular handbook lists values of $\Delta \bar{G}_f^\circ$ and $\Delta \bar{H}_f^\circ$ but

not molar entropies. From the data given below, determine $\Delta \bar{S}^\circ$ for the following reaction. All data are for 298 K.

$$NH_3(g) + HCl(g) \longrightarrow NH_4Cl(s)$$

$\Delta \bar{H}_f^\circ$: $NH_3(g) = -46.2 \text{ kJ/mol}$
$HCl(g) = -92.3 \text{ kJ/mol}$
$NH_4Cl(s) = -315.4 \text{ kJ/mol}$

$\Delta \bar{G}_f^\circ$: $NH_3(g) = -16.6 \text{ kJ/mol}$
$HCl(g) = -95.3 \text{ kJ/mol}$
$NH_4Cl(s) = -203.9 \text{ kJ/mol}$

13. The absolute entropies of $F_2(g)$ and $F(g)$ at 298 K are given as 187.61 and 158.66 J mol^{-1} K^{-1}, respectively. Use these data, together with that listed below, to estimate the bond energy of the F_2 molecule. Compare your result with that listed in Table 9-3.

$$F_2(g) \longrightarrow 2 \, F(g) \qquad \Delta \bar{G}^\circ = 123.85 \text{ kJ/mol at 298 K}$$

*14. The free energy of combustion of *n*-octane is indicated below. What would be the free energy change if the water were produced as a gas instead of as a liquid? (*Hint:* Use data from Appendix D.)

$$C_8H_{18}(l) + \tfrac{25}{2} O_2(g) \longrightarrow 8 \, CO_2(g) + 9 \, H_2O(l)$$
$$\text{at 298 K, } \Delta \bar{G}^\circ = -5.28 \times 10^3 \text{ kJ/mol}$$

Free energy and equilibrium

15. If a graph similar to Figure 15-5 were drawn for the process $H_2O(l, 1 \text{ atm}) \rightarrow H_2O(g, 1 \text{ atm})$
(a) At what temperature would the two lines intersect?
(b) What would be the value of ΔG at this point?

16. At how many different temperatures and within what temperature range can the following equilibrium be established?

$$H_2O(l, 0.50 \text{ atm}) \Longleftrightarrow H_2O(g, 0.50 \text{ atm})$$

17. Refer to Figures 11-11 and 15-5. Does solid or liquid iodine have the lower free energy at 110°C and 1 atm pressure?

18. Refer to Figures 11-12 and 15-5. Which has the lowest free energy at 1 atm and $-60°C$, solid, liquid, or gaseous carbon dioxide? Explain.

Phase transitions

19. Example 15-5 dealt with the standard molar enthalpy and entropy of vaporization of water at 100°C. Use data from Appendix D to determine corresponding values at 25°C.

20. Which of the following substances would you expect to obey Trouton's rule most closely? Explain your reasoning.
(a) HF; (b) $C_6H_5CH_3$; (c) CH_3OH.

21. Estimate the normal boiling point of bromine, Br_2, in the following two-step procedure, and compare your result with the measured value of 58.8°C.

(a) Determine $\Delta \bar{H}_{vap}^\circ$ for Br_2 with data from Appendix D.
(b) Estimate the normal boiling point, assuming that Trouton's rule is obeyed. (You will also be assuming that $\Delta \bar{H}_{vap}^\circ$ is independent of temperature.)

22. Use the method outlined in Exercise 21 to estimate the normal boiling point of mercury.

Relationship of $\Delta \bar{G}$, $\Delta \bar{G}^\circ$, Q, and K

23. Refer to situation (c) in Figure 15-6. Calculate $\Delta \bar{G}$ at 298 K for the vaporization process described, that is,

$$H_2O(l, 10 \text{ mmHg}) \longrightarrow H_2O(g, 10 \text{ mmHg}) \qquad \Delta \bar{G} = ?$$

(*Hint:* Recall equation 15.14.)

24. The following data are given for the vaporization of CCl_4 at 298 K.

$$CCl_4(l, 1 \text{ atm}) \longrightarrow CCl_4(g, 1 \text{ atm})$$

$$\Delta \bar{S}^\circ = 94.98 \text{ J mol}^{-1} \text{ K}^{-1}$$
$$\Delta \bar{H}_f^\circ[CCl_4(l)] = -139.3 \text{ kJ/mol}$$
$$\Delta \bar{H}_f^\circ[CCl_4(g)] = -106.7 \text{ kJ/mol}$$

(a) Show that the vaporization of $CCl_4(l)$ to produce $CCl_4(g)$ at *1 atm pressure* does not occur spontaneously.
(b) Calculate the equilibrium vapor pressure of CCl_4 at 298 K.

25. For the reaction $2 \, SO_2(g) + O_2(g) \rightleftharpoons 2 \, SO_3(g)$, $K_c = 2.8 \times 10^2$ at 1000 K.
(a) What is K_p for this reaction?
(b) What is $\Delta \bar{G}^\circ$ at 1000 K?
(c) In what direction will a reaction occur if one mixes in a 2.50-L vessel at 1000 K 0.40 mol SO_2, 0.18 mol O_2, and 0.72 mol SO_3?

The thermodynamic equilibrium constant

26. Why must the thermodynamic equilibrium constant, K, be used in the expression $\Delta \bar{G}^\circ = -2.303RT \log K$, rather than simply K_c? What are the circumstances under which K_c and/or K_p may be used in place of K?

27. Use the principles established in Example 15-6 to write thermodynamic equilibrium constant expressions for the following reactions. Do any of these expressions correspond to K_c or to K_p?
(a) $2 \, NO(g) + O_2(g) \rightleftharpoons 2 \, NO_2(g)$
(b) $MgSO_3(s) \rightleftharpoons MgO(s) + SO_2(g)$
(c) $HC_2H_3O_2(aq) \rightleftharpoons H^+(aq) + C_2H_3O_2^-(aq)$
(d) $2 \, NaHCO_3(s) \rightleftharpoons Na_2CO_3(s) + H_2O(g) + CO_2(g)$
(e) $MnO_2(s) + 4 \, H^+(aq) + 2 \, Cl^-(aq) \rightleftharpoons$
$$Mn^{2+}(aq) + 2 \, H_2O(l) + Cl_2(g)$$

28. A laboratory method of preparing $H_2(g)$ involves passing steam over hot iron: $3 \, Fe(s) + 4 \, H_2O(g) \rightleftharpoons Fe_3O_4(s) + 4 \, H_2(g)$. For this reaction we can write the thermodynamic equilibrium constant (assuming that H_2O and H_2 behave as ideal gases)

$$K = K_p = \frac{(P_{H_2})^4}{(P_{H_2O})^4}$$

The partial pressure of $H_2(g)$ is independent of the amounts of Fe(s) and Fe_3O_4(s) present. Can we conclude that the production of $H_2(g)$ from $H_2O(g)$ could be accomplished regardless of what proportions of Fe(s) and Fe_3O_4(s) are used? Explain.

$\Delta\bar{G}°$ and K

29. For the reaction $Cl_2(g) \rightleftharpoons 2\ Cl(g)$, $K_p = 2.45 \times 10^{-7}$ at 1000 K. What is the value of $\Delta\bar{G}°$ for this reaction at 1000 K?

30. Use data from Appendix D to determine K_p at 298 K for the reaction $2\ NO(g) + O_2(g) \rightleftharpoons 2\ NO_2(g)$.

31. For the decomposition of $CaCO_3$(s) to CaO(s) and $CO_2(g)$ (a basic reaction in the manufacture of portland cement) the standard molar free energy and enthalpy changes at 298 K are

$$CaCO_3(s) \rightleftharpoons CaO(s) + CO_2(g)$$
$$\Delta\bar{H}° = +178.45 \text{ kJ/mol}$$
$$\Delta\bar{G}° = +130.30 \text{ kJ/mol}$$

   (a) Does the decomposition of $CaCO_3$(s) occur to any appreciable extent at room temperature?
   (b) Does $CaCO_3$(s) decompose more readily by raising or lowering the temperature from 298 K?

32. The equilibrium constant at 298 K for the reaction $CO(g) + Cl_2(g) \rightleftharpoons COCl_2(g)$ is $K_p = 5.64 \times 10^{35}$. What is the standard free energy of formation of $COCl_2(g)$ at 298 K, that is, $\Delta\bar{G}_f°[COCl_2(g)]$? (*Hint:* You will need to use some data from Appendix D.)

33. In Example 15-4 the statement was made that only a trace of CO(g) would be produced in the reaction C(s) + $CO_2(g) \rightleftharpoons 2\ CO(g)$ at 298 K. Show by calculation that this is in fact the case.

*34. A statement is made in the caption to Figure 15-7 that only if $\Delta\bar{G}°$ has a small magnitude, either positive or negative, will an equilibrium mixture contain appreciable amounts of all reactants and products. Illustrate this statement using hypothetical data.

Additional Exercises

1. Would you expect the reaction $Cl_2(g) \rightleftharpoons 2\ Cl(g)$ to occur to a greater extent in the forward direction by raising or lowering the temperature? Explain your reasoning.

2. Use data from Appendix D to complete the discussion of Example 15-1(d); that is, does entropy increase or decrease in this reaction?

$$CO(g) + H_2O(g) \longrightarrow CO_2(g) + H_2(g)$$

The variation of $\Delta\bar{G}°$ with temperature

35. Use data from Appendix D to determine for the reaction

$$CO(g) + H_2O(g) \rightleftharpoons CO_2(g) + H_2(g)$$

   (a) $\Delta\bar{H}°$, $\Delta\bar{S}°$, and $\Delta\bar{G}°$ at 298 K
   (b) K_p at 1100 K (*Hint:* Recall the assumptions made in Examples 15-8 and 15-9.)
   (c) Compare the result calculated in part (b) with that found in Example 14-6.

36. In Example 14-12 the approximate temperature at which $K_p = 1.0 \times 10^6$ for the reaction $2\ SO_2(g) + O_2(g) \rightleftharpoons 2\ SO_3(g)$ was determined by calculation. Obtain another estimate of this temperature with data from Appendix D and the Gibbs–Helmholtz equation (15.6). Compare your result to that of Example 14-12.

37. For the reaction $C(s) + CO_2(g) \rightleftharpoons 2\ CO(g)$ described in Example 15-4 and Exercise 33
   (a) Is conversion of $CO_2(g)$ to CO(g) favored at high or low temperatures?
   (b) If the equilibrium partial pressure of $CO_2(g)$ is maintained at 1.00 atm, at approximately what temperature does the equilibrium partial pressure of CO(g) become equal to that of $CO_2(g)$?

*38. At its normal boiling point of 78.4°C, the entropy of vaporization of ethanol is $+112.1$ J mol^{-1} K^{-1}. Use these data, together with values from Appendix D, to estimate the following at 298 K for $C_2H_5OH(g)$: (a) $\Delta\bar{H}_f°$, (b) $\Delta\bar{G}_f°$, (c) $\bar{S}°$.

Heat engines

39. If a steam electric power plant discharges condensate at 40°C and is found to operate at 36% efficiency,
   (a) What is the minimum temperature of the steam used in the plant?
   (b) Why is the actual steam temperature probably higher than that calculated in part (a)?

*40. Assume that the steam [i.e., $H_2O(g)$] in the plant referred to in Exercise 39 is in equilibrium with liquid water. What is the steam pressure corresponding to the minimum steam temperature calculated in part (a)?

3. Use data from Appendix D to determine values at 298 K of $\Delta\bar{G}°$ and K for the following reactions.
   (a) $NO(g) + O_2(g) \rightleftharpoons NO_2(g)$ (not balanced)
   (b) $HCl(g) + O_2(g) \rightleftharpoons H_2O(g) + Cl_2(g)$
       (not balanced)
   (c) $Fe_2O_3(s) + H_2(g) \rightleftharpoons Fe_3O_4(s) + H_2O(g)$
       (not balanced)

4. An *equilibrium* mixture at 1000 K contains 0.276 mol H_2, 0.276 mol CO_2, 0.224 mol CO, and 0.224 mol H_2O.

$$CO_2(g) + H_2(g) \rightleftharpoons CO(g) + H_2O(g)$$

(a) What is the equilibrium constant, K_p, for this reaction at 1000 K?

(b) Calculate $\Delta \bar{G}°$ at 1000 K.

(c) In what direction will a spontaneous reaction occur if one brings together in a vessel at 1000 K: 0.0500 mol CO_2, 0.070 mol H_2, 0.0400 mol CO, and 0.0850 mol H_2O?

5. Use data from Appendix D and other information from this chapter to determine the temperature at which the dissociation of $I_2(g)$ molecules becomes appreciable [e.g., with the $I_2(g)$ being 50% dissociated at 1 atm total pressure]. $I_2(g) \rightleftharpoons 2 I(g)$.

*6. Will these processes occur spontaneously as written at 298 K?

(a) $SO_2(g, 0.010 \text{ atm}) + \frac{1}{2} O_2(g, 0.0010 \text{ atm}) \longrightarrow$
$$SO_3(g, 2.0 \text{ atm})$$

(b) $C_2H_2(g, 1.2 \text{ atm}) + 2 H_2(g, 0.020 \text{ atm}) \longrightarrow$
$$C_2H_6(g, 3.0 \text{ atm})$$

(c) $Br_2(l, 1 \text{ atm}) \longrightarrow Br_2(g, 0.10 \text{ atm})$

*7. Many statements have been made about the first and second laws of thermodynamics that draw upon everyday life. One such statement, in gambler's parlance, is "You can't win and you can't even break even." Explain the basis of this statement.

*8. The normal boiling point of cyclohexane, C_6H_{12}, is 80.7 °C. Estimate the temperature at which the vapor pressure of cyclohexane is 100 mmHg.

*9. From the data given in Exercise 39 of Chapter 14, estimate a value of $\Delta \bar{S}°$ at 298 K for the reaction

$$2 NaHCO_3(s) \longrightarrow Na_2CO_3(s) + H_2O(g) + CO_2(g)$$

*10. One of the steps that appears to be involved in smog formation is the reaction of atomic oxygen, O(g), with molecular oxygen, $O_2(g)$. [The atomic oxygen is produced by the action of sunlight on $NO_2(g)$.]

$$O(g) + O_2(g) \rightleftharpoons O_3(g)$$

A sample of air is found to contain 10 parts per million of $O_3(g)$ by volume. If the $O_3(g)$ is assumed to be in equilibrium with $O_2(g)$ and O(g), what would have to be the approximate partial pressure of atomic oxygen present in the air, that is, $P_{O(g)}$?

*11. Use appropriate data from Appendix D to estimate the percent dissociation of $H_2O(g)$ at 2000 K if the total pressure is maintained at 1.00 atm.

$$H_2O(g) \rightleftharpoons H_2(g) + \frac{1}{2} O_2(g)$$

Self-Test Questions

For questions 1 through 6 select the single item that best completes each statement.

1. For a process to occur spontaneously,
   (a) the entropy of the system must increase.
   (b) the entropy of the surroundings must increase.
   (c) both the entropy of the system and of the surroundings must increase.
   (d) the entropy of the universe must increase.

2. The change in free energy accompanying a reaction is a measure of
   (a) the quantity of heat given off to the surroundings.
   (b) the direction in which a net reaction occurs.
   (c) the increased molecular disorder that occurs in the system.
   (d) how rapidly the reaction occurs.

3. If a reaction has $\Delta H < 0$ and $\Delta S < 0$, the reaction proceeds furthest in the forward direction at   (a) low temperatures;   (b) high temperatures;   (c) all temperatures;   (d) no temperature.

4. If it is necessary to employ electric current (electrolysis) to carry out a chemical reaction, then for that reaction
   (a) $\Delta H > 0$;   (b) $\Delta G = \Delta H$;   (c) $\Delta G > 0$;   (d) $\Delta S < 0$.

5. For the reaction $Br_2(g) \rightarrow 2 Br(g)$, we should expect that at all temperatures   (a) $\Delta H < 0$;   (b) $\Delta S > 0$;   (c) $\Delta G < 0$;   (d) $\Delta S < 0$.

6. If $\Delta \bar{G}° = 0$ for a reaction, then   (a) $\Delta \bar{H}° = 0$;   (b) $\Delta \bar{S}° = 0$;   (c) $K = 0$;   (d) $K = 1$.

7. Explain briefly why
   (a) The change in entropy in a system is not always a suitable criterion for spontaneous change.
   (b) $\Delta \bar{G}°$ is of such importance in dealing with the question of spontaneous change, even though the conditions employed in a reaction are often nonstandard state.

8. Explain the relationships among $\Delta \bar{G}$, $\Delta \bar{G}°$, and the equilibrium constant, K, for a reversible chemical reaction.

9. A handbook lists the following standard enthalpies of formation at 298 K for cyclopentane, C_5H_{10}: $\Delta \bar{H}_f°[C_5H_{10}(l)] = -105.9 \text{ kJ/mol}$ and $\Delta \bar{H}_f°[C_5H_{10}(g)] = -77.2 \text{ kJ/mol}$.
   (a) Estimate the normal boiling point of cyclopentane.
   (b) Estimate $\Delta \bar{G}°$ for the vaporization of cyclopentane at 298 K.
   (c) Comment on the significance of the sign of $\Delta \bar{G}°$.

10. The following data are given at 298 K.

	$\Delta \bar{H}_f^\circ$, kJ/mol	$\bar{S}^\circ$, J mol^{-1} K^{-1}
$NH_4NO_3(s)$	-365.56	151.08
$N_2O(g)$	$+81.55$	219.99
$H_2O(l)$	-285.85	69.96

For the reaction $NH_4NO_3(s) \rightarrow N_2O(g) + 2\ H_2O(l)$

(a) Is the reaction endothermic or exothermic?

(b) What is the value of $\Delta \bar{G}^\circ$ at 298 K?

(c) What is the value of the equilibrium constant, K, at 298 K?

(d) Does the reaction tend to be spontaneous at temperatures above 25°C, below 25°C, both, or neither? Explain.

16 Solubility Equilibria in Aqueous Solutions

In Chapters 14 and 15 we emphasized gas-phase equilibria. This permitted us to illustrate the basic concepts of equilibrium in a straightforward manner. Equally important are applications of equilibrium principles to aqueous solutions. These applications range from routine procedures of analytical chemistry to reactions occurring in living organisms. We begin our study of solution equilibria with the simple case of slightly soluble (sparingly soluble) solutes in water. Throughout this chapter and the several that follow, we will use the concepts of the preceding two chapters repeatedly.

16-1 The Solubility Product Constant, K_{sp}

Silver chromate is slightly soluble (sparingly soluble) in water. The equilibrium existing in a saturated solution is

$$Ag_2CrO_4(s) \overset{H_2O}{\rightleftharpoons} 2\, Ag^+(aq) + CrO_4^{2-}(aq) \qquad (16.1)$$

for which the thermodynamic equilibrium constant expression is

$$K = \frac{(a_{Ag^+})^2(a_{CrO_4^{2-}})}{(a_{Ag_2CrO_4(s)})} \qquad (16.2)$$

This expression can be simplified considerably by applying the conventions introduced in Section 15-6. The activity of a pure solid $= 1$; and in *dilute* solutions molar concentrations may be substituted for activities of solutes. The result obtained is

$$K = K_c = \frac{[Ag^+]^2[CrO_4^{2-}]}{(1)} = [Ag^+]^2[CrO_4^{2-}] \qquad (16.3)$$

Because only molar concentration terms appear in expression (16.3), it is appropriate to refer to this equilibrium constant by the symbol K_c. However, a special term and symbolism are used instead. An equilibrium constant expression representing equilibrium between a slightly soluble ionic compound and its ions in aqueous solution is called a **solubility product constant,** designated K_{sp}. For a saturated aqueous solution of Ag_2CrO_4 at 25°C,

$$K_{sp} = [Ag^+]^2[CrO_4^{2-}] = 2.4 \times 10^{-12} \qquad (16.4)$$

Some typical solubility product constants are listed in Table 16-1.

TABLE 16-1
Solubility product constants at 25°C

Solute	Solubility equilibrium	K_{sp}
aluminum hydroxide	$Al(OH)_3(s) \rightleftharpoons Al^{3+}(aq) + 3\,OH^-(aq)$	1.3×10^{-33}
barium carbonate	$BaCO_3(s) \rightleftharpoons Ba^{2+}(aq) + CO_3^{2-}(aq)$	5.1×10^{-9}
barium hydroxide	$Ba(OH)_2(s) \rightleftharpoons Ba^{2+}(aq) + 2\,OH^-(aq)$	5×10^{-3}
barium sulfate	$BaSO_4(s) \rightleftharpoons Ba^{2+}(aq) + SO_4^{2-}(aq)$	1.1×10^{-10}
bismuth(III) sulfide	$Bi_2S_3(s) \rightleftharpoons 2\,Bi^{3+}(aq) + 3\,S^{2-}(aq)$	1×10^{-97}
cadmium sulfide	$CdS(s) \rightleftharpoons Cd^{2+}(aq) + S^{2-}(aq)$	8.0×10^{-27}
calcium carbonate	$CaCO_3(s) \rightleftharpoons Ca^{2+}(aq) + CO_3^{2-}(aq)$	2.8×10^{-9}
calcium fluoride	$CaF_2(s) \rightleftharpoons Ca^{2+}(aq) + 2\,F^-(aq)$	2.7×10^{-11}
calcium hydroxide	$Ca(OH)_2(s) \rightleftharpoons Ca^{2+}(aq) + 2\,OH^-(aq)$	5.5×10^{-6}
calcium sulfate	$CaSO_4(s) \rightleftharpoons Ca^{2+}(aq) + SO_4^{2-}(aq)$	9.1×10^{-6}
chromium(III) hydroxide	$Cr(OH)_3(s) \rightleftharpoons Cr^{3+}(aq) + 3\,OH^-(aq)$	6.3×10^{-31}
cobalt(II) sulfide	$CoS(s) \rightleftharpoons Co^{2+}(aq) + S^{2-}(aq)$	4.0×10^{-21}
copper(II) sulfide	$CuS(s) \rightleftharpoons Cu^{2+}(aq) + S^{2-}(aq)$	6.3×10^{-36}
iron(II) sulfide	$FeS(s) \rightleftharpoons Fe^{2+}(aq) + S^{2-}(aq)$	6.3×10^{-18}
iron(III) hydroxide	$Fe(OH)_3(s) \rightleftharpoons Fe^{3+}(aq) + 3\,OH^-(aq)$	4×10^{-38}
lead(II) chloride	$PbCl_2(s) \rightleftharpoons Pb^{2+}(aq) + 2\,Cl^-(aq)$	1.6×10^{-5}
lead(II) chromate	$PbCrO_4(s) \rightleftharpoons Pb^{2+}(aq) + CrO_4^{2-}(aq)$	2.8×10^{-13}
lead(II) iodide	$PbI_2(s) \rightleftharpoons Pb^{2+}(aq) + 2\,I^-(aq)$	7.1×10^{-9}
lead(II) sulfate	$PbSO_4(s) \rightleftharpoons Pb^{2+}(aq) + SO_4^{2-}(aq)$	1.6×10^{-8}
lead(II) sulfide	$PbS(s) \rightleftharpoons Pb^{2+}(aq) + S^{2-}(aq)$	8.0×10^{-28}
lithium phosphate	$Li_3PO_4(s) \rightleftharpoons 3\,Li^+(aq) + PO_4^{3-}(aq)$	3.2×10^{-9}
magnesium carbonate	$MgCO_3(s) \rightleftharpoons Mg^{2+}(aq) + CO_3^{2-}(aq)$	3.5×10^{-8}
magnesium fluoride	$MgF_2(s) \rightleftharpoons Mg^{2+}(aq) + 2\,F^-(aq)$	3.7×10^{-8}
magnesium hydroxide	$Mg(OH)_2(s) \rightleftharpoons Mg^{2+}(aq) + 2\,OH^-(aq)$	2×10^{-11}
magnesium phosphate	$Mg_3(PO_4)_2(s) \rightleftharpoons 3\,Mg^{2+}(aq) + 2\,PO_4^{3-}(aq)$	1×10^{-25}
manganese(II) sulfide	$MnS(s) \rightleftharpoons Mn^{2+}(aq) + S^{2-}(aq)$	2.5×10^{-13}
mercury(I) chloride	$Hg_2Cl_2(s) \rightleftharpoons Hg_2^{2+}(aq) + 2\,Cl^-(aq)$	1.3×10^{-18}
mercury(II) sulfide	$HgS(s) \rightleftharpoons Hg^{2+}(aq) + S^{2-}(aq)$	1.6×10^{-52}
nickel(II) sulfide	$NiS(s) \rightleftharpoons Ni^{2+}(aq) + S^{2-}(aq)$	3.2×10^{-19}
silver bromide	$AgBr(s) \rightleftharpoons Ag^+(aq) + Br^-(aq)$	5.0×10^{-13}
silver carbonate	$Ag_2CO_3(s) \rightleftharpoons 2\,Ag^+(aq) + CO_3^{2-}(aq)$	8.1×10^{-12}
silver chloride	$AgCl(s) \rightleftharpoons Ag^+(aq) + Cl^-(aq)$	1.6×10^{-10}
silver chromate	$Ag_2CrO_4(s) \rightleftharpoons 2\,Ag^+(aq) + CrO_4^{2-}(aq)$	2.4×10^{-12}
silver iodide	$AgI(s) \rightleftharpoons Ag^+(aq) + I^-(aq)$	8.5×10^{-17}
silver sulfate	$Ag_2SO_4(s) \rightleftharpoons 2\,Ag^+(aq) + SO_4^{2-}(aq)$	1.4×10^{-5}
silver sulfide	$Ag_2S(s) \rightleftharpoons 2\,Ag^+(aq) + S^{2-}(aq)$	6.3×10^{-50}
strontium carbonate	$SrCO_3(s) \rightleftharpoons Sr^{2+}(aq) + CO_3^{2-}(aq)$	1.1×10^{-10}
strontium sulfate	$SrSO_4(s) \rightleftharpoons Sr^{2+}(aq) + SO_4^{2-}(aq)$	3.2×10^{-7}
tin(II) sulfide	$SnS(s) \rightleftharpoons Sn^{2+}(aq) + S^{2-}(aq)$	1.0×10^{-25}
zinc sulfide	$ZnS(s) \rightleftharpoons Zn^{2+}(aq) + S^{2-}(aq)$	1.0×10^{-21}

Margin notes (handwritten):

SO_4^{2-} sulfate
SO_3^{2-} sulfite
S^{2-} sulfide
CrO_4^{2-} chromate
I^- Iodide
PO_4^{2-} phosphate
PO_3^{2-} phosphite
Cl^- chloride
Br^- Bromide
Sn^{2+} tin
Ag^{1+} silver
IO_3^{2-} iodate
$C_2O_4^{2-}$ oxalate

Example 16-1 Write a solubility product constant expression for the formation of a saturated aqueous solution of Bi_2S_3.

Unless otherwise indicated, the solubility product constant expression is based on the balanced equation written per mole of the solid solute. The ionic concentration terms appearing in the expression are raised to powers equal to the coefficients on the right side of the balanced equation. No term appears in the denominator of a K_{sp} expression. (Why?)

$$Bi_2S_3(s) \rightleftharpoons 2\,Bi^{3+}(aq) + 3\,S^{2-}(aq)$$

$$K_{sp} = [Bi^{3+}]^2[S^{2-}]^3$$

SIMILAR EXAMPLES: Exercises 1, 2.

16-2 Relationship Between Solubility and K_{sp}

There are several ways of establishing the solubility product constant of a slightly soluble ionic compound, but one of the most direct is to relate K_{sp} to the experimentally determined solubility.

Example 16-2   A 100.0 ml-sample is removed from a water solution saturated in MgF_2 at $25°C$. The water is completely evaporated from the sample and a 13-mg deposit of $MgF_2(s)$ is obtained. What is K_{sp} for MgF_2 at $25°C$?

$$MgF_2(s) \rightleftharpoons Mg^{2+}(aq) + 2 F^-(aq) \qquad K_{sp} = ?$$

We must first determine the molar solubility of MgF_2 and then relate the ionic concentrations, $[Mg^{2+}]$ and $[F^-]$, to it.

$$\text{no. mol } MgF_2 \text{ per L satd. soln.} = \frac{13 \text{ mg } MgF_2}{100.0 \text{ ml soln.}} \times \frac{1 \text{ g } MgF_2}{1000 \text{ mg } MgF_2} \times \frac{1 \text{ mol } MgF_2}{62.3 \text{ g } MgF_2} \times \frac{1000 \text{ ml soln.}}{1 \text{ L soln.}}$$

$$= 2.1 \times 10^{-3} \text{ mol } MgF_2/L$$

Key factors in the expression below are those (shown in color) which relate the number of moles of ions in solution to the number of moles of solute dissolved.

$$[Mg^{2+}] = 2.1 \times 10^{-3} \frac{\text{mol } MgF_2}{L} \times \frac{1 \text{ mol } Mg^{2+}}{1 \text{ mol } MgF_2} = 2.1 \times 10^{-3} M$$

$$[F^-] = 2.1 \times 10^{-3} \frac{\text{mol } MgF_2}{L} \times \frac{2 \text{ mol } F^-}{1 \text{ mol } MgF_2} = 2 \times 2.1 \times 10^{-3} = 4.2 \times 10^{-3} M$$

$$K_{sp} = [Mg^{2+}][F^-]^2 = (2.1 \times 10^{-3})(4.2 \times 10^{-3})^2 = 3.7 \times 10^{-8}$$

SIMILAR EXAMPLES: Exercises 5, 6.

Tabulations of solubility product constants for slightly soluble ionic compounds amount to tabulations of solubilities. As illustrated in Example 16-3, if a K_{sp} value is known, the solubility of a solute is easily calculated.

Example 16-3   Calculate the molar solubility of Ag_2CrO_4 in water.
The formation of a saturated solution is represented by

$$Ag_2CrO_4(s) \rightleftharpoons 2 Ag^+(aq) + CrO_4{}^{2-}(aq) \qquad K_{sp} = 2.4 \times 10^{-12}$$

Two moles of Ag^+ and *one* mole of $CrO_4{}^{2-}$ ion appear in the saturated solution for every mole of Ag_2CrO_4 dissolved. Thus, if S represents the number of moles of Ag_2CrO_4 that have dissolved per liter of saturated solution, then at equilibrium

$$[Ag^+] = 2S \qquad [CrO_4{}^{2-}] = S$$

The solubility product relationship must be satisfied by these concentrations.

$$K_{sp} = [Ag^+]^2[CrO_4{}^{2-}] = (2S)^2(S) = 2.4 \times 10^{-12}$$

$$4S^3 = 2.4 \times 10^{-12} \qquad \text{doubled \& squared}$$

$$S^3 = 0.60 \times 10^{-12}$$

$$S = (0.60)^{1/3} \times 10^{-4} = 0.84 \times 10^{-4}$$

$$S = \text{molar solubility} = 8.4 \times 10^{-5} \text{ mol } Ag_2CrO_4/L$$

SIMILAR EXAMPLES: Exercises 8a, 9, 10.

Examples 16-2 and 16-3 demonstrate that molar solubilities and solubility product constants are related, but they are by no means identical. Each can be used as a basis for *calculating* the other. Their numerical values will never be equal.

LIMITATION OF K_{sp} TO SLIGHTLY SOLUBLE SOLUTES. We have used the term "slightly soluble solute" or "sparingly soluble solute" in our discussion of the solubility product constant. Cannot similar expressions be written for saturated solutions of moderately or highly soluble ionic compounds, for example, $NaCl$, KNO_3, or $NaOH$? The answer is that for these solutes we can indeed write solubility equilibrium equations similar to (16.1) and *thermodynamic* equilibrium constant expressions similar to (16.2). What we cannot do is substitute ionic concentrations for ionic activities, as we did in deriving equations (16.3) and (16.4) from (16.2). Saturated solutions of moderately or highly soluble ionic compounds are simply much too concentrated to permit the assumption that activities and molar concentrations are equal. Without this assumption much of the value of the solubility product concept is lost. Even if the qualifier, "slightly soluble," is not always used in describing solubility equilibrium, whenever a K_{sp} value is given it will be for a slightly soluble ionic compound.

THE COMMON ION EFFECT. In the situations discussed to this point, the saturated solution has contained ions derived from a *single* source, the pure solid solute. What will be the effect on equilibrium in the saturated solution if some of these ions are introduced from a *second* source as well? As an example, suppose that to the saturated solution of Ag_2CrO_4 described in Example 16-3 is added some CrO_4^{2-} ion—a *common ion*—from a source such as $K_2CrO_4(aq)$.

According to Le Châtelier's principle, the addition of a reactant to a system at equilibrium creates a disturbance. The equilibrium condition shifts in such a way as to relieve this disturbance. In this case,

to an original equilibrium mixture

$$Ag_2CrO_4(s) \rightleftharpoons 2\,Ag^+(aq) + CrO_4^{2-}(aq)$$

is added CrO_4^{2-}.

The reverse reaction is favored,

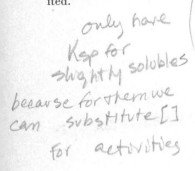

leading to a new equilibrium in which

additional $Ag_2CrO_4(s)$ precipitates,	$[Ag^+]$ is less than in the original equilibrium,	$[CrO_4^{2-}]$ is greater than in the original equilibrium.

The common ion effect is pictured in Figure 16-1.

In the case just considered, the net effect of adding a common ion was to *reduce* the amount of dissolved Ag_2CrO_4: *The solubility of a slightly soluble ionic compound is lowered by the presence of a second solute which furnishes a common ion.* This point can be illustrated by recalculating the molar solubility in the presence of the common ion, as is done in Example 16-4.

Example 16-4 What is the molar solubility of Ag_2CrO_4 in $0.10\,M\,K_2CrO_4(aq)$?

Many of the ionic compounds for which K_{sp} values are given are often termed "insoluble." Their solubilities are very limited.

only have Ksp for slightly solubles because for them we can substitute [ ] for activities

FIGURE 16-1
The common ion effect in solubility equilibrium.

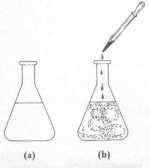

(a)   (b)

(a) A clear saturated solution. **(b)** Addition of a small volume of a solution containing a common ion. The common ion reduces the solubility of the solute and the excess solid precipitates.

The ion concentrations at equilibrium, *derived from the dissolved Ag_2CrO_4*, can be written as they were in Example 16-3. That is, if we let S = the molar solubility of Ag_2CrO_4, then $[Ag^+] = 2S$ and $[CrO_4^{2-}] = S$. However, an *additional* 0.10 mol CrO_4^{2-}/L is derived from the $K_2CrO_4(aq)$. Thus, the *total* ion concentrations at equilibrium are $[Ag^+] = 2S$ and $[CrO_4^{2-}] = 0.10 + S$. This kind of information can generally be summarized effectively in the following manner.

$$Ag_2CrO_4(s) \rightleftharpoons 2\,Ag^+(aq) + CrO_4^{2-}(aq)$$

from $Ag_2CrO_4(s)$:	$2S$	S mol/L
from 0.10 M $K_2CrO_4(aq)$:	—	0.10 mol/L
equilibrium concentrations:	$2S$	$(0.10 + S)$ mol/L

The usual K_{sp} relationship must be satisfied, that is,

$$K_{sp} = [Ag^+]^2[CrO_4^{2-}] = (2S)^2(0.10 + S) = 2.4 \times 10^{-12}$$

Because we know that S in this case will be even smaller than it was in Example 16-3, we can safely assume that $S \ll 0.10$ and $(0.10 + S) \simeq 0.10$.

$$(2S)^2(0.10) = 4S^2(0.10) = 2.4 \times 10^{-12}$$

$$S^2 = 6.0 \times 10^{-12}$$

$$S = 2.4 \times 10^{-6} = 2.4 \times 10^{-6} \text{ mol } CrO_4^{2-}/L$$

Since we defined S to be the number of moles of Ag_2CrO_4 dissolved per liter of 0.10 M $K_2CrO_4(aq)$, the value of S is, in fact, the molar solubility we are seeking: 2.4×10^{-6} mol Ag_2CrO_4/L.

SIMILAR EXAMPLES: Exercises 8b, c, 17, 20.

The molar solubility of Ag_2CrO_4 in the presence of 0.10 M CrO_4^{2-} calculated in Example 16-4 (2.4×10^{-6} mol Ag_2CrO_4/L) proves to be 35 times less than its value in pure water calculated in Example 16-3 (8.4×10^{-5} mol Ag_2CrO_4/L). The common ion effect is a pronounced one! Calculation of the effect of Ag^+ ion on the solubility of Ag_2CrO_4 would show an even more striking effect produced by this common ion.

THE DIVERSE ("UNCOMMON") ION EFFECT—THE SALT EFFECT. Having just established the effect of common ions, we might next ask: Do ions different from those involved in the solubility equilibrium ("uncommon" ions) have an effect on the solubilities of sparingly soluble ionic compounds? They do, but in a different way from common ions. The effect is significant but not as striking as the common ion effect. Moreover, the presence of "uncommon" ions tends to *increase* rather than decrease solubility. As the total ionic concentration of a solution increases, interionic attractions become more important (recall Section 12-7). Activities (effective concentrations) become smaller than the stoichiometric or measured concentrations. For the ions involved in the solution process this means that higher concentrations must appear in solution before equilibrium is established—*the solubility increases*. Figure 16-2 suggests the different effect of common and "uncommon" ions.

The "uncommon" or diverse ion effect is more commonly referred to as the **salt effect.** In all further discussion of solubility equilibrium, we will neglect the salt effect. This means that we must limit our discussion to very dilute solutions with all ions derived from a sparingly soluble solute, or to more concentrated solutions with a common ion present.

OTHER FACTORS AFFECTING THE SOLUBILITIES OF SPARINGLY SOLUBLE IONIC COMPOUNDS. In each calculation performed to this point we have assumed that all of the dissolved solute appears in solution as separated cations and anions. There are cir-

FIGURE 16-2
Comparison of the common ion effect and the salt effect.

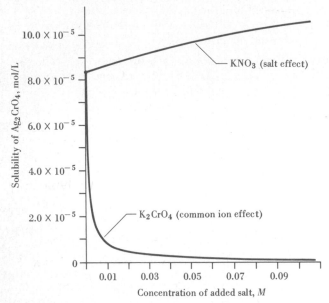

The presence of the common ion CrO_4^{2-}, derived from $K_2CrO_4(aq)$, reduces the solubility of Ag_2CrO_4 by a factor of about 35 over the concentration range shown (from 0 to 0.10 M added salt). Over this same range, the solubility of Ag_2CrO_4 is increased by the presence of the "uncommon" or diverse ions from KNO_3, but only by about 25% or so.

cumstances, however, in which ions in solution may join to form an **ion pair.** For example, in a saturated solution of magnesium fluoride ion pairs consisting of a Mg^{2+} and an F^- ion, that is, MgF^+, might be found. To the extent that ion-pair formation occurs in a solution, the free ion concentrations tend to be reduced. This means that the amount of solute that must dissolve to maintain the required *free* ion concentrations to satisfy the K_{sp} expression increases: *Solubility increases when ion-pair formation occurs in solution.* Although ion-pair formation can be significant in some cases (especially for a moderately soluble solute yielding ions with high charge), we will not consider its effect on solubility equilibrium.

A much more significant factor is the possibility of an ionic species being *simultaneously* involved in the solubility equilibrium and in some other equilibrium process. These other equilibrium processes generally involve acid-base reactions, oxidation-reduction reactions, or complex ion formation. We will consider each of these possibilities in some detail later.

CONSTANCY OF K_{sp}. When considering such phenomena as the salt effect, the value of K_{sp} based on molar concentrations varies depending on the ionic atmosphere. But for all the applications considered in this text we will assume that the form of the solubility product constant expression and the value of K_{sp} remain unchanged. As with other equilibrium constants, however, values of K_{sp} are temperature-dependent.

16-3 Precipitation Reactions

We have used the term *solubility equilibrium* to describe the phenomena encountered to this point. But we learned in Chapter 14 that an equilibrium condition can be approached from *either* direction. If the equilibria of the preceding sections are approached by starting with ions in solution and producing pure, undissolved solute, then the process involved is a *precipitation reaction*. And there is a great deal that we can say about precipitation reactions from the standpoint of solubility product constants.

CRITERION FOR PRECIPITATION FROM SOLUTION. The most basic question we can ask about a precipitation reaction is whether it will in fact occur for a given set of conditions. Suppose that a solution is made simultaneously 0.10 M in Ag^+ and 0.10 M in Cl^-. Should a precipitate of $AgCl(s)$ form? To answer this question we must begin with a balanced equation to represent equilibrium between the slightly soluble solute and its ions together with a value of K_{sp} for this equilibrium.

$$AgCl(s) \rightleftharpoons Ag^+(aq) + Cl^-(aq) \tag{16.5}$$

$$K_{sp} = [Ag^+][Cl^-] = 1.6 \times 10^{-10} \tag{16.6}$$

When applied to precipitation reactions, the reaction quotient, Q, is often called the ion product. This is because Q consists only of the product of ion concentration terms; there is no denominator in the expression.

Now recall how we dealt with the question of the direction of net change in Section 14-3. We formulated a quantity called the reaction quotient, Q, and compared its value with that of the equilibrium constant, K. In this case the reaction quotient is simply the product, $[Ag^+][Cl^-]$, based on the initial concentrations of these ions.

$$Q = (0.10)(0.10) = 1 \times 10^{-2} > K_{sp} = 1.6 \times 10^{-10}$$

We conclude that reaction should occur to the left or in the reverse direction in equation (16.5)—*precipitation should occur.*

More general conclusions about precipitation from solution are

Precipitation should occur if $Q > K_{sp}$.

Precipitation does not occur if $Q < K_{sp}$.

A solution is just saturated if $Q = K_{sp}$. (16.7)

(handwritten margin notes:)

when a s.s.s. is dissolved as much as possible, the product of [ions] = Ksp, the ion product Q is much larger so we have more ions present initially than should be for equilibrium. System works to reduce [ions]; shifts left.

will occur ← until [ ]'s of ions have decreased

ILLUSTRATIVE EXAMPLES. In each of the following three examples a different idea important to a successful quantitative description of precipitation reactions is presented.

The point illustrated by Example 16-5 is that the criterion for precipitation must be applied to ion concentrations *after* mixing of solutions. That is, any possible dilution effects must be taken into account.

Example 16-5 Should a precipitate form if 10.0 ml 0.0010 M $AgNO_3$(aq) is added to 500.0 ml 0.0020 M K_2CrO_4(aq)?

$$Ag_2CrO_4(s) \rightleftharpoons 2\,Ag^+(aq) + CrO_4{}^{2-}(aq) \qquad K_{sp} = 2.4 \times 10^{-12}$$

$$\text{no. mol } Ag^+ = 0.0100\text{ L} \times \frac{0.0010\text{ mol } AgNO_3}{1\text{ L}} \times \frac{1\text{ mol } Ag^+}{1\text{ mol } AgNO_3}$$

$$= 1.0 \times 10^{-5}\text{ mol } Ag^+$$

$$\text{no. mol } CrO_4{}^{2-} = 0.500\text{ L} \times \frac{0.0020\text{ mol } K_2CrO_4}{1\text{ L}} \times \frac{1\text{ mol } CrO_4{}^{2-}}{1\text{ mol } K_2CrO_4}$$

$$= 1.0 \times 10^{-3}\text{ mol } CrO_4{}^{2-}$$

total solution volume = 0.0100 L + 0.500 L = 0.510 L

After mixing:

$$[Ag^+] = \frac{1.0 \times 10^{-5}\text{ mol } Ag^+}{0.510\text{ L}} = 2.0 \times 10^{-5}\,M$$

$$[CrO_4{}^{2-}] = \frac{1.0 \times 10^{-3}\text{ mol } CrO_4{}^{2-}}{0.510\text{ L}} = 2.0 \times 10^{-3}\,M$$

Application of criterion:

$$Q = [Ag^+]^2[CrO_4{}^{2-}] = (2.0 \times 10^{-5})^2(2.0 \times 10^{-3}) = 8.0 \times 10^{-13}$$

$$= 8.0 \times 10^{-13} < K_{sp}\ (= 2.4 \times 10^{-12})$$

Note that if the ion product, Q, had been based on the ion concentrations as given, rather than after mixing, we would have concluded, *erroneously*, that precipitation should occur. That is,

$(1 \times 10^{-3})^2$

$\quad \times (2 \times 10^{-3}) > K_{sp}$

Precipitation of Ag_2CrO_4(s) will *not* occur.

SIMILAR EXAMPLES: Exercises 23, 24, 25.

In Example 16-6 our interest centers on whether precipitation goes to completion, that is, on the concentration of an ion (Mg^{2+}) *remaining* in solution after precipitation has occurred.

Here we're starting with soln of $[Mg^{2+}]$. OH^- is added, & thus $Mg(OH)_2$ ppts. The ppt occurs to form until $[Mg^{2+}][OH^-]^2 = K_{sp}$. Since K_{sp} & OH^- are constants, we know what $[Mg^{2+}]$ must remain in order to maintain K_{sp} equilibrium.

Example 16-6 The first step in the extraction of magnesium metal from seawater involves precipitating Mg^{2+} as $Mg(OH)_2(s)$. The magnesium ion concentration in seawater is about 0.059 *M*. If a seawater sample is treated so that its $[OH^-]$ is maintained at 2.0×10^{-3} *M*, **(a)** what will be the concentration of Mg^{2+} remaining in solution after precipitation has occurred? **(b)** Can we say that precipitation is complete?

$$Mg(OH)_2(s) \rightleftharpoons Mg^{2+}(aq) + 2\,OH^-(aq) \qquad K_{sp} = 2 \times 10^{-11}$$

(a) There is no question that precipitation will occur since the product, $[Mg^{2+}][OH^-]^2 = (0.059)(2.0 \times 10^{-3})^2 = 2.4 \times 10^{-7}$, exceeds K_{sp}. We need to determine the $[Mg^{2+}]$ remaining in a solution from which some solid $Mg(OH)_2$ has precipitated. For the saturated solution at equilibrium, we substitute $[OH^-]$ (maintained constant at 2.0×10^{-3} *M*) into the solubility product constant expression and solve for $[Mg^{2+}]$.

$$K_{sp} = [Mg^{2+}][OH^-]^2 = [Mg^{2+}] \times (2.0 \times 10^{-3})^2 = 2 \times 10^{-11}$$

$$[Mg^{2+}] = \frac{2 \times 10^{-11}}{4 \times 10^{-6}} = 5 \times 10^{-6}\,M$$

(b) $[Mg^{2+}]$ in the seawater is reduced from 0.059 *M* to 5×10^{-6} *M* as a result of the precipitation reaction. We can conclude that precipitation is complete.

A useful rule of thumb is that precipitation is complete if less than one part per thousand (0.1%) of the original solute is left unprecipitated.

SIMILAR EXAMPLES: Exercises 27, 28.

In the example just considered the concentration of one of the precipitating ions (OH^-) was kept constant while the other ion (Mg^{2+}) was removed from solution. In some precipitation reactions *both* ion concentrations change during the precipitation, as illustrated in Example 16-7.

Example 16-7 A 50.0-ml sample of 0.0152 *M* $Na_2SO_4(aq)$ is added to 50.0 ml of 0.0125 *M* $Ca(NO_3)_2(aq)$. **(a)** Should precipitation of $CaSO_4(s)$ occur? **(b)** Will precipitation of Ca^{2+} be complete?

$$CaSO_4(s) \rightleftharpoons Ca^{2+}(aq) + SO_4^{2-}(aq) \qquad K_{sp} = 9.1 \times 10^{-6}$$

(a) The concentrations of ions present *after* the solutions are mixed are

$$[Ca^{2+}] = 0.0500\text{ L} \times \frac{0.0125\text{ mol } Ca^{2+}}{\text{L}} \times \frac{1}{0.100\text{ L}} = 6.25 \times 10^{-3}\,M$$

← concentration; total volume

$$[SO_4^{2-}] = 0.0500\text{ L} \times \frac{0.0152\text{ mol } SO_4^{2-}}{\text{L}} \times \frac{1}{0.100\text{ L}} = 7.60 \times 10^{-3}\,M$$

$$Q = [Ca^{2+}][SO_4^{2-}] = (6.25 \times 10^{-3})(7.60 \times 10^{-3})$$
$$= 4.75 \times 10^{-5} > K_{sp} = 9.1 \times 10^{-6}$$

Precipitation of $CaSO_4(s)$ should occur.

(b) Let us outline the course of the precipitation reaction in the following way.

The essential difference between a problem of this type and the one considered in Example 16-6 is in the last column of figures. In Example 16-6, $[OH^-]$ was the same, initially and after precipitation was completed. Here, $[SO_4^{2-}]$ *decreases* from its initial to its equilibrium value.

	$CaSO_4(s) \rightleftharpoons$ $Ca^{2+}(aq)$ +	$SO_4^{2-}(aq)$
initial concentrations [from part (a)]:	0.00625 *M*	0.00760 *M*
consumed in the precipitation:	$0.00625 - x$	$0.00625 - x$
equilibrium concentrations:	x	$0.00760 - (0.00625 - x)$
		$0.00760 - 0.00625 + x$
		$0.00135 + x$

Calcium determines amt. ppt. It is present in limiting quantity *notice x is not the change in concentrations, but the new concentration.*

What we have indicated above is that if $[Ca^{2+}]$ falls from $0.00625\ M$ to x, then the precipitation of $CaSO_4(s)$ must have consumed a concentration of Ca^{2+} equal to $0.00625 - x$. One SO_4^{2-} ion is removed from solution for every Ca^{2+} ion. The decrease in $[SO_4^{2-}]$ must also be $0.00625 - x$. The $[SO_4^{2-}]$ *remaining* in solution at equilibrium is the initial concentration minus that which is consumed: $0.00760 - (0.00625 - x) = 0.00135 + x$.

We can now substitute the equilibrium concentrations into the K_{sp} expression and solve for x.

$$K_{sp} = [Ca^{2+}][SO_4^{2-}] = (x)(0.00135 + x) = 9.1 \times 10^{-6}$$

The result is a quadratic equation: $x^2 + 0.00135x - 9.1 \times 10^{-6} = 0$. Solution of this equation by the quadratic formula leads to the result

$$x = \frac{-0.00135 \pm \sqrt{(0.00135)^2 + 4 \times 9.1 \times 10^{-6}}}{2} = \frac{-0.00135 \pm 6.2 \times 10^{-3}}{2}$$

$$[Ca^{2+}] = x = \frac{4.8 \times 10^{-3}}{2} = 2.4 \times 10^{-3}\ M$$

The percentage of calcium ion left in solution can be expressed as

$$\%\ Ca^{2+}\ remaining = \frac{2.4 \times 10^{-3}\ M\ Ca^{2+}}{6.25 \times 10^{-3}\ M\ Ca^{2+}} \times 100 = 38\%$$

Precipitation of Ca^{2+} is incomplete.

SIMILAR EXAMPLES: Exercises 29, 30.

The usual simplifying assumption that x is very small compared to a number to which it is added or from which it is subtracted does not work here. The number 0.00135 is itself quite small. The calculated value of x turns out to be larger, not smaller, than 0.00135.

COMPLETENESS OF PRECIPITATION. Example 16-7 suggests the following generalization concerning the completeness of a precipitation reaction.

Value to consider	Complete precipitation is likely if value is	Precipitation may not be complete if value is
K_{sp} of precipitate	small	large
concentration of common ion in saturated solution	large	small

by "complete" they mean such that only 0.01% of original [ion] exists. → So if Ksp is large, [ions] must also be large @ equil. & thus most likely will be present greater than 0.1% of initial [].

Unfortunately, there is no simple definition of "small" and "large." We can say, though, that in Example 16-7, K_{sp} was large compared to most of the K_{sp} values in Table 16-1, and the concentration of common ion (SO_4^{2-}) in the saturated solution was small. These are both conditions which lead to *incompleteness* of precipitation.

16-4 Writing Net Ionic Equations

Suppose that we are asked to predict whether a chemical reaction occurs when the following solutions are mixed.

$$Pb(NO_3)_2(aq) + KI(aq) \longrightarrow ? \tag{16.8}$$

An appropriate start would be to write (16.8) in the *ionic* form, that is, to show the actual ionic species that are brought together.

$$Pb^{2+}(aq) + 2\,NO_3^-(aq) + K^+(aq) + I^-(aq) \longrightarrow ? \tag{16.9}$$

Now we can use our chemical knowledge. Is there any combination of cations and anions in (16.9) that might produce a precipitate?

TABLE 16-2
Some general rules for the water solubilities of common ionic compounds[a]

1. All common salts of the alkali (IA) metals—including ammonium (NH_4^+) salts—are **soluble.**
2. All nitrates (NO_3^-), chlorates (ClO_3^-), perchlorates (ClO_4^-), and acetates ($C_2H_3O_2^-$) are **soluble.** Silver acetate is only **moderately soluble.**
3. The chlorides (Cl^-), bromides (Br^-), and iodides (I^-) of most metals are **soluble.** The principal *exceptions* are those of Pb^{2+}, Ag^+, and Hg_2^{2+}.
4. All sulfates (SO_4^{2-}) are **soluble** *except* for those of Sr^{2+}, Ba^{2+}, Pb^{2+}, and Hg_2^{2+}. Sulfates of Ca^{2+} and Ag^+ are only **moderately soluble.**
5. All carbonates (CO_3^{2-}), chromates (CrO_4^{2-}), and phosphates (PO_4^{3-}) are **insoluble** *except* for those of the alkali metals (including ammonium).
6. The group IA metal hydroxides (OH^-) are **soluble.** The hydroxides of Ca^{2+}, Sr^{2+}, and Ba^{2+} are **moderately soluble.** The rest of the hydroxides are **insoluble.**
7. The sulfides of all metals are **insoluble** *except* for those of NH_4^+ and the IA and IIA metals.

[a] Generally speaking, if a saturated solution of an ionic solute is greater than about 0.10 M, the solute is said to be soluble; if less than about 0.01 M, insoluble; and if of an intermediate concentration (i.e., between about 0.01 M and 0.10 M), moderately soluble.

PbI_2 is the only one of the potential products [KNO_3, KI, and $Pb(NO_3)_2$ are the other possible ion combinations] that is listed in Table 16-1. We might conclude that it is "insoluble" and the other potential products are all soluble. As long as [Pb^{2+}] and [I^-] in the mixed solution are large enough that $[Pb^{2+}][I^-]^2 > K_{sp}(PbI_2)$, a precipitate should form.

$$Pb^{2+}(aq) + 2\,I^-(aq) \longrightarrow PbI_2(s) \tag{16.10}$$

The method used here is not entirely satisfactory, however. The available listing of solubility product constants may not be extensive enough to be certain that the absence of a substance from the list means that it is soluble. A better approach is to employ a set of **solubility rules,** as in Table 16-2. From Table 16-2 we note that all nitrates are soluble—neither $Pb(NO_3)_2$ nor KNO_3 is expected to precipitate. Among the common iodides, all are soluble *except* those of Ag^+, Hg_2^{2+}, and Pb^{2+}. A precipitate of PbI_2 should form.

Example 16-8 Predict whether a chemical reaction will occur in each of the following cases. If so, write a net ionic equation to represent the reaction.
(a) $NaOH(aq) + MgCl_2(aq) \longrightarrow$?
(b) $(NH_4)_2SO_4(aq) + CuCl_2(aq) \longrightarrow$?
(a) The ions present in the mixed solution are Na^+, Mg^{2+}, OH^-, and Cl^-. Since all common sodium compounds are soluble in water (see Table 16-2), the Na^+ remains in solution. Magnesium hydroxide is *insoluble*. The reaction is

$$Mg^{2+}(aq) + 2\,OH^-(aq) \longrightarrow Mg(OH)_2(s) \tag{16.11}$$

Note that it was possible to go directly to the net ionic equation without first having to write an equation with the "spectator" ions—Na^+ and Cl^-.
(b) An analysis of the solubility relationships for all the possible ionic combinations in this case indicates that there are no insoluble combinations. All ammonium compounds are soluble, and $CuCl_2$ and $CuSO_4$ are not among the few insoluble chlorides and sulfates cited in Table 16-2. No chemical reaction occurs, a fact that can be indicated as follows

$$(NH_4)_2SO_4(aq) + CuCl_2(aq) \longrightarrow \text{no reaction} \tag{16.12}$$

SIMILAR EXAMPLES: Exercises 31, 32.

We will return to the topic of net ionic equations on later occasions as we master additional concepts relating to solution equilibria.

16-5 Qualitative Analysis

Qualitative analysis refers to a set of laboratory procedures that can be used to *separate* and *test* for the presence of ions in solution. The analysis is said to be *qualitative* because it reveals only which ions are present in a mixture. The analysis does not necessarily indicate the compounds from which these ions are derived, or the actual amounts present (i.e., not the masses or concentrations of substances). Perhaps more so than any other set of common laboratory procedures, qualitative analysis illustrates the complete range of solution equilibrium concepts considered in this and subsequent chapters. We begin with a brief introduction to qualitative analysis and will add to this discussion in later chapters.

Qualitative analysis can be applied both to anions and cations. A particular set of procedures used to perform an analysis is called a **qualitative analysis scheme.** Our interest will center on the analysis of cations by the scheme outlined in Figure 16-3 (in a form called a flow chart or flow diagram).

As indicated in Figure 16-3, the basic approach is to separate the common cations into groups (usually five) by *precipitation*. For example, only the members of

FIGURE 16-3
A qualitative analysis scheme for cations.

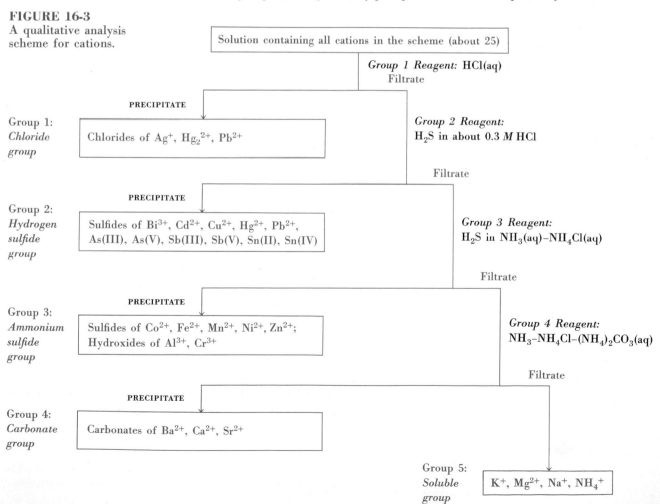

Other approaches to the qualitative analysis scheme are often encountered. For example, analysis can begin with the soluble group cations (group 5), followed by analysis of groups 1 through 4 on a separate sample.

group 1, that is, only Ag^+, Hg_2^{2+}, and Pb^{2+}, form insoluble chlorides. All other common metal chlorides are water soluble. Group 2 cations are precipitated as sulfides from acidic solution. Cations of group 3 form insoluble sulfides in basic solution. The group 4 cations are precipitated as carbonates. The cations of group 5 remain in solution throughout the separation of the other four groups.

Following the separation of cations into the five major groups, further separation and testing must be done within each group. The end result is to establish for an unknown mixture the presence or absence of each ion in the scheme. Detailed flow diagrams can be written for this further separation and testing; examples will be presented in later chapters. A valuable adjunct to the qualitative analysis scheme is a table of solubilities (see Table 16-2).

Example 16-9 You are given as an unknown a *colorless* aqueous solution and told that it contains *none, one, two,* or *all three* of the following ions: Ag^+, Ba^{2+}, Cu^{2+}. You decide to treat a sample of the unknown with $(NH_4)_2CO_3(aq)$, and when you do you obtain a *white* precipitate. What conclusions can you draw from this observation? (Use data from Figure 16-3 and Table 16-2, together with general laboratory knowledge.)

All three of the possible ions produce an insoluble carbonate (Table 16-2, item 5). The formation of a precipitate only allows you to conclude that at least one of the three ions is present. The general laboratory knowledge expected of you is some familiarity with the colors of ionic compounds and of ions in solution. Of the three ions, only Cu^{2+} displays a color in solution (blue). Compounds containing Cu^{2+} are also colored. The observation that neither the solution nor the precipitate is colored *suggests* that Cu^{2+} is *absent*. (It is still possible—although probably not likely—that Cu^{2+} is present in the solution. At very low concentrations its color cannot be detected.) Whether the white precipitate is $Ag_2CO_3(s)$, $BaCO_3(s)$, or a mixture of the two cannot be determined from this single observation.

SIMILAR EXAMPLES: Exercises 34, 36.

Can you see how an unambiguous answer to Example 16-9 could be obtained by following the qualitative analysis scheme of Figure 16-3? If the sample were first treated with HCl(aq), formation of a precipitate would prove the presence of Ag^+ (the only one of the three ions that forms an insoluble chloride). Lack of a precipitate would establish the *absence* of Ag^+. The filtrate—the solution remaining after removal of any AgCl(s)—could now be treated with H_2S. Here, formation of a precipitate would prove the presence of Cu^{2+}, and lack of a precipitate, the absence of Cu^{2+}. *After* the removal of any Ag^+ or Cu^{2+} present in the unknown, a test could be performed for Ba^{2+} using the group 4 reagent.

A reagent is a substance or mixture that is used to produce a chemical reaction.

16-6 Precipitation Reactions in Quantitative Analysis

In Chapter 3 we described how the precipitation of a solid can be used to determine the exact composition of a sample of matter (see, e.g., Example 3-15). What does it take to conduct a successful precipitation analysis? One factor that we assessed through Examples 16-6 and 16-7 is that precipitation must be as complete as possible (so that very little of the desired substance is left in solution). In purifying a precipitate by washing, it is sometimes necessary to do so with a solution containing a common ion rather than with pure water. This is because the precipitate is less soluble in the common ion solution than it is in pure water. Figure 16-4 outlines these ideas as applied to the gravimetric determination (determination by weighing) of calcium.

FIGURE 16-4
Outline of a gravimetric analysis for calcium.

Ca^{2+}(aq) [from sample being analyzed]

add excess
$(NH_4)_2C_2O_4$(aq)

$CaC_2O_4 \cdot H_2O$(s) [impure]

wash with dilute $(NH_4)_2C_2O_4$(aq);
filter and dry

$CaC_2O_4 \cdot H_2O$(s) [pure]

heat to 500°C

$CaCO_3$(s) [pure]

A weighed sample is dissolved [usually in HCl(aq)] to obtain Ca^{2+} (aq). The solution is treated with excess $(NH_4)_2C_2O_4$(aq) and the hydrate, $CaC_2O_4 \cdot H_2O$, precipitates. The precipitate is purified by washing with $(NH_4)_2C_2O_4$(aq). When heated strongly, $CaC_2O_4 \cdot H_2O$(s) is converted to $CaCO_3$(s), in which form the calcium is finally weighed.
In the washing process it is important to use a solution that provides a common ion to reduce the solubility of $CaC_2O_4 \cdot H_2O$(s) but does not leave a residue that would contaminate the final $CaCO_3$(s). Upon heating, $(NH_4)_2C_2O_4$ undergoes decomposition to gaseous products—NH_3, H_2O, CO, and CO_2.

FRACTIONAL PRECIPITATION. Another technique that can be better understood using principles of solubility equilibrium is that of fractional precipitation. This term refers to a situation in which two or more ions in solution, each capable of being precipitated by the same reagent, are *separated* by the use of that reagent: *One ion is precipitated and the other(s) remains in solution.* The primary condition for a successful fractional precipitation is that there be a significant difference in the solubilities of the substances to be separated. (Usually this means a significant difference in their K_{sp} values.)

Example 16-10 considers the separation of Cl^-(aq) and I^-(aq) through the use of Ag^+(aq). The data needed to describe this separation are the solubility equilibrium equations and solubility product constants for AgCl and AgI.

$$AgCl(s) \rightleftharpoons Ag^+(aq) + Cl^-(aq)$$
$$K_{sp} = 1.6 \times 10^{-10} \qquad (16.13)$$

$$AgI(s) \rightleftharpoons Ag^+(aq) + I^-(aq)$$
$$K_{sp} = 8.5 \times 10^{-17} \qquad (16.14)$$

Example 16-10   To a solution that has $[Cl^-] = 0.010\ M$ and $[I^-] = 0.010\ M$, is slowly added $0.10\ M$ $AgNO_3$(aq) (see Figure 16-5). **(a)** Show that AgI(s) precipitates before AgCl(s). **(b)** At the point AgCl(s) begins to precipitate, what is $[I^-]$ remaining in solution? **(c)** Is separation of I^-(aq) and Cl^-(aq) by fractional precipitation feasible?
(a) As a drop of the $AgNO_3$(aq) enters the solution, $[Ag^+]$ builds up from a value of zero to a point where one of the ion products, Q, exceeds the corresponding

FIGURE 16-5
Fractional precipitation—Example 16-10 illustrated.

(a) $0.10\ M$ $AgNO_3$(aq) is slowly added to a solution that is $0.010\ M$ in Cl^- and $0.010\ M$ in I^-.
(b) This is the condition at the point where AgCl(s) would just begin to precipitate. Essentially all of the I^- has precipitated as AgI(s), leaving $[I^-] = 5.3 \times 10^{-9}\ M$ in the solution. The Cl^- and I^- have been separated.

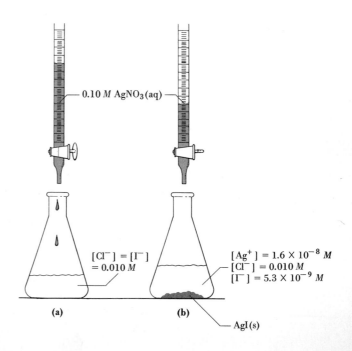

0.10 M AgNO₃(aq)

$[Cl^-] = [I^-]$
$= 0.010\ M$

$[Ag^+] = 1.6 \times 10^{-8}\ M$
$[Cl^-] = 0.010\ M$
$[I^-] = 5.3 \times 10^{-9}\ M$

(a)   (b)

AgI(s)

AgI(s) will ppt 1st because its Ksp is exceeded by Q before Ksp(AgCl) is exceeded.

K_{sp}. Then precipitation occurs. The required values of $[Ag^+]$ at which precipitation begins are

for AgI(s) to ppt: $Q = [Ag^+][I^-] = [Ag^+](0.010) = 8.5 \times 10^{-17} = K_{sp}$

$$[Ag^+] = 8.5 \times 10^{-15}\ M$$

for AgCl(s) to ppt: $Q = [Ag^+][Cl^-] = [Ag^+](0.010) = 1.6 \times 10^{-10} = K_{sp}$

$$[Ag^+] = 1.6 \times 10^{-8}\ M$$

At the point where the precipitation of AgI has been essentially completed and that of AgCl begins, newly formed precipitate changes in color from yellow (AgI) to white (AgCl). This color change is not as good an indicator that separation has been achieved as is the direct measurement of $[Ag^+]$. At the point in question, $[Ag^+] = 1.6 \times 10^{-8}\ M$. Methods of determining such ion concentrations are discussed in Section 19-6.

Since the required $[Ag^+]$ to start the precipitation of AgI(s) is less than that for AgCl(s), AgI(s) will be first to precipitate. As long as a significant quantity of AgI(s) is forming, the free silver ion concentration in solution is not able to attain the value required for the precipitation of AgCl(s).

(b) As more and more AgI(s) precipitates, the concentration of iodide ion gradually decreases; and this permits $[Ag^+]$ to increase. When $[Ag^+]$ reaches $1.6 \times 10^{-8}\ M$, precipitation of AgCl(s) begins. Next, we need to answer the question: What is $[I^-]$ at the point where $[Ag^+] = 1.6 \times 10^{-8}\ M$? For this we use the solubility product constant expression for AgI, and solve for $[I^-]$.

$$K_{sp} = [Ag^+][I^-] = (1.6 \times 10^{-8})[I^-] = 8.5 \times 10^{-17}$$

$$[I^-] = 5.3 \times 10^{-9}\ M$$

AgCl does not begin ppt until all AgI has ppt'ed 1st.

(c) Before AgCl(s) begins to precipitate, $[I^-]$ will have been reduced from 1.0×10^{-2} to $5.3 \times 10^{-9}\ M$. Essentially all of the I^- will have precipitated from solution as AgI(s) while the Cl^- remains in solution. Fractional precipitation is feasible for separating mixtures of Cl^- and I^-.

SIMILAR EXAMPLES: Exercises 38, 39.

Summary

Equilibrium between a slightly soluble ionic compound and its ions in aqueous solution is expressed through a *solubility product constant,* K_{sp}. The form of a K_{sp} expression is that of a product of ionic concentration terms. One concentration term is included for each ion produced by the slightly soluble solute. In the usual fashion, these terms are raised to powers based on the balanced chemical equation for the solubility equilibrium. In using the K_{sp} expression two situations are commonly encountered: (1) The ions in a saturated solution may be derived *solely* from the slightly soluble solute. (2) An additional salt(s) may be present in solution which contributes ions either *common to* or *different from* those of the slightly soluble solute.

The solubility of a slightly soluble ionic compound is greatly affected by the presence of *common ions* derived from a dissolved salt (known as the common ion effect). Qualitatively, that is, reasoning from Le Châtelier's principle, the expected effect of common ions is to *reduce* considerably the solubility of a slightly soluble ionic compound. Quantitatively, the effect can be calculated through the solubility product expression. A salt whose ions are *different* from those of a slightly soluble solute generally causes a *small increase* in the solubility of the solute. The pairing of solute ions in a saturated solution will also cause the solubility to be somewhat higher than predicted from the K_{sp} value.

Additional factors that may significantly affect the solubilities of slightly soluble solutes are considered in later chapters.

An "ion product," Q, having the same form as the K_{sp} expression is based on whatever ion concentrations happen to exist in a solution. A comparison of Q and K_{sp} provides a criterion for precipitation: If $Q > K_{sp}$, precipitation should occur; if $Q < K_{sp}$, the solution remains unsaturated. Another matter of interest concerns the completeness of a precipitation reaction. If a precipitation reaction is carried out in the presence of a high concentration of a common ion, it generally goes to completion. Combinations of factors, such as the lack of sufficient common ion and moderate solubility of the precipitate, can result in incomplete precipitation. At times, ions in solution can be separated by *fractional precipitation.* One type of ion is removed by precipitation while the others remain in solution.

Precipitation reactions find extensive application in *qualitative analysis.* In the qualitative analysis scheme for cations, the cations are first separated into groups based on differing solubilities of their compounds. Further separation and testing of ions is then done within each group. In addition to their employment in qualitative analysis, solubility relationships can be stated through *solubility rules* and expressed through *net ionic equations.*

Learning Objectives

As a result of studying Chapter 16, you should be able to

1. Write the solubility product constant expression, K_{sp}, for a slightly soluble ionic compound.

2. Calculate K_{sp} from the experimentally determined solubility of a slightly soluble ionic compound, or the solubility from a known value of K_{sp}.

3. Calculate the effect of common ions on the aqueous solubilities of slightly soluble ionic compounds.

4. Describe how the presence of "uncommon" ions or the formation of ion pairs affects solute solubilities.

5. State and apply the criterion for precipitation from solution.

6. Determine, by calculation, whether the precipitation of a slightly soluble solute will be complete for a given set of conditions.

7. Explain how fractional precipitation works and the conditions under which it can be used.

8. State the general rules that apply to the water solubilities of ionic compounds.

9. Write net ionic equations based on solubility rules.

10. Describe how precipitation reactions are employed in a qualitative analysis scheme, and draw conclusions about the presence or absence of ions in an unknown from experimental observations.

Some New Terms

The **common ion effect** describes the effect on a solution equilibrium of a substance which furnishes ions that can participate in the equilibrium. For example, Na_2SO_4, when added to a saturated solution of $BaSO_4$, furnishes common ions (SO_4^{2-}), which reduce the solubility of the $BaSO_4$.

Fractional precipitation is a technique in which ions in solution are separated by the addition of a precipitating reagent. One ion is selectively precipitated while the others remain in solution.

Ion-pair formation refers to the association of cations and anions in solution. Such combinations, when they occur to a significant extent, can have an effect on solution equilibria.

Molar solubility is the molar concentration of solute (mol/L) in a saturated solution.

A **net ionic equation** is a chemical equation in which only the species actually participating in a chemical reaction are shown. In addition, the form in which each reactant

occurs is also represented, that is, as ions in solution, insoluble precipitate, gas, and so on.

Qualitative analysis is a laboratory method for determining the presence or absence of certain cations or anions in a sample. The method is based on a variety of solution equilibrium concepts.

Quantitative analysis refers to the analysis of substances or mixtures to determine the *quantities* of the various components rather than their mere presence or absence.

The **salt effect** describes the effect of ions *different* from those directly involved in a solution equilibrium. These ions alter the general ionic atmosphere in which the equilibrium reaction occurs. The salt effect is also known as the diverse or "uncommon" ion effect.

The **solubility product constant, K_{sp},** is an equilibrium constant expression describing a saturated solution of a slightly soluble ionic compound. It is the product of ionic concentration terms, with each term raised to an appropriate power.

Exercises

The meaning of K_{sp}

1. Write solubility product constant expressions, K_{sp}, for the following equilibria. For example, $AgCl(s) \rightleftharpoons Ag^+(aq) + Cl^-(aq)$; $K_{sp} = [Ag^+][Cl^-]$.
   (a) $Ag_2SO_4(s) \rightleftharpoons 2\,Ag^+(aq) + SO_4^{2-}(aq)$; $K_{sp} = ?$
   (b) $Hg_2C_2O_4(s) \rightleftharpoons Hg_2^{2+}(aq) + C_2O_4^{2-}(aq)$; $K_{sp} = ?$
   (c) $Ra(IO_3)_2(s) \rightleftharpoons Ra^{2+}(aq) + 2\,IO_3^-(aq)$; $K_{sp} = ?$
   (d) $Ni_3(PO_4)_2(s) \rightleftharpoons 3\,Ni^{2+}(aq) + 2\,PO_4^{3-}(aq)$; $K_{sp} = ?$
   (e) $PuO_2CO_3(s) \rightleftharpoons PuO_2^{2+}(aq) + CO_3^{2-}(aq)$; $K_{sp} = ?$

2. Write the solubility equilibrium equations that are described by the following K_{sp} expressions. For example,
   $$K_{sp} = [Ag^+][Cl^-] \qquad AgCl(s) \rightleftharpoons Ag^+(aq) + Cl^-(aq)$$
   (a) $K_{sp} = [Fe^{3+}][OH^-]^3$ (b) $K_{sp} = [BiO^+][OH^-]$
   (c) $K_{sp} = [Hg_2^{2+}][I^-]^2$ (d) $K_{sp} = [Pb^{2+}]^3[AsO_4^{3-}]^2$
   (e) $K_{sp} = [Cu^{2+}]^2[Fe(CN)_6^{4-}]$
   (f) $K_{sp} = [Mg^{2+}][NH_4^+][PO_4^{3-}]$

3. Based on terminology from Section 12-4, how would you describe a clear solution (no precipitate) for which the ion product Q (a) $< K_{sp}$; (b) $> K_{sp}$?

4. A statement is made in the text that although K_{sp} and molar solubility of a slightly soluble ionic compound are related, they can never be equal. Demonstrate why this is so.

K_{sp} and solubility (Use data from Table 16-1 as necessary.)

5. A saturated aqueous solution of lead iodate is found to be 2.15×10^{-5} M $Pb(IO_3)_2$. What is the solubility product constant, K_{sp}, for lead iodate?

$$Pb(IO_3)_2(s) \rightleftharpoons Pb^{2+}(aq) + 2 IO_3^-(aq) \qquad K_{sp} = ?$$

6. The solubility of barium oxalate in water is found to be 22 mg/L. What is K_{sp} for barium oxalate?

$$BaC_2O_4(s) \rightleftharpoons Ba^{2+}(aq) + C_2O_4^{2-}(aq) \qquad K_{sp} = ?$$

7. A 0.200 M Na_2SO_4 solution is saturated with Ag_2SO_4 and is found to have $[Ag^+] = 9.2 \times 10^{-3}$ M. What is the value of K_{sp} for Ag_2SO_4 obtained from these data?

8. Calculate the molar solubility of PbI_2 in **(a)** pure water; **(b)** 0.010 M KI(aq); **(c)** 0.030 M $Pb(NO_3)_2$(aq).

9. Which of the following saturated aqueous solutions has the highest concentration of magnesium ion? **(a)** $MgCO_3$; **(b)** MgF_2; **(c)** $Mg_3(PO_4)_2$.

10. How many parts per million (ppm) of fluoride ion (i.e., g F^- per 10^6 g of solution) are present in a water solution that is saturated in CaF_2?

*11. One of the substances sometimes responsible for the "hardness" of water is $CaSO_4$. A particular water sample has 130 ppm of $CaSO_4$ (g $CaSO_4$ per 10^6 g of water). If this water is boiled in a tea kettle, approximately what fraction of the water must be evaporated before a deposit of $CaSO_4$(s) begins to form? Assume that the solubility of $CaSO_4$ does not change appreciably with temperature in the range 0 to 100°C.

12. A 50.00-ml sample of a saturated solution of PbI_2 requires 16.2 ml of a certain $AgNO_3$(aq) for its titration. What is the molarity of this $AgNO_3$(aq)?

$$\underset{\text{[from satd. } PbI_2(aq)]}{I^-(aq)} + \underset{\text{[from } AgNO_3(aq)]}{Ag^+(aq)} \longrightarrow AgI(s)$$

13. A saturated solution of CaC_2O_4 is prepared in pure water and a 100.0-ml sample of the solution is withdrawn and titrated with 2.1 ml 0.00122 M $KMnO_4$(aq). What is the value of K_{sp} for CaC_2O_4 obtained from these data?
$$CaC_2O_4(s) \rightleftharpoons Ca^{2+}(aq) + C_2O_4^{2-}(aq); \quad K_{sp} = ?$$

$$5 C_2O_4^{2-}(aq) + 2 MnO_4^-(aq) + 16 H^+(aq) \longrightarrow$$
$$2 Mn^{2+}(aq) + 8 H_2O + 10 CO_2(g)$$

*14. To precipitate as Ag_2S(s) all the Ag^+ present in 338 ml of a saturated solution of $AgBrO_3$ requires 30.4 cm³ of H_2S(g) measured at 23°C and 748 mmHg. What is K_{sp} for $AgBrO_3$?

$$2 Ag^+(aq) + H_2S(g) \longrightarrow Ag_2S(s) + 2 H^+(aq)$$

*15. S is the molar solubility of a slightly soluble ionic compound in pure water. Derive an algebraic relationship between S and K_{sp} in each of the following cases. Then calculate the molar solubility. For example, K_{sp} of AgCl = 1.6×10^{-10}; $S = \sqrt{K_{sp}} = 1.3 \times 10^{-5}$ mol AgCl/ L.
 (a) $Mg(OH)_2$: $K_{sp} = 2 \times 10^{-11}$
 (b) Ag_2CO_3: $K_{sp} = 8.1 \times 10^{-12}$
 (c) $Al(OH)_3$: $K_{sp} = 1.3 \times 10^{-33}$
 (d) Li_3PO_4: $K_{sp} = 3.2 \times 10^{-9}$
 (e) Bi_2S_3: $K_{sp} = 1 \times 10^{-97}$

The common ion effect (Use data from Table 16-1 as necessary.)

16. The two salts KI and KNO_3 are found to have different effects on the solubility of AgI in water. Describe the effect of each of these salts and explain why the effects are different.

17. Demonstrate that the effect of Ag^+ ion on reducing the water solubility of Ag_2CrO_4 is even greater than the effect of CrO_4^{2-} described in Example 16-4. [*Hint:* What is the solubility of Ag_2CrO_4 in 0.10 M $AgNO_3$(aq)?]

18. Plot a graph similar to Figure 16-2 to show how the solubility of lead iodate varies with concentration of KIO_3(aq) ranging from pure water to 0.10 M KIO_3(aq).

$$Pb(IO_3)_2(s) \rightleftharpoons Pb^{2+}(aq) + 2 IO_3^-(aq)$$
$$K_{sp} = 3.2 \times 10^{-13}$$

19. A large excess of $Mg(OH)_2$(s) is maintained in contact with 500.0 ml of its saturated solution. What will be the $[Mg^{2+}]$ at equilibrium in the final solution in each of these cases?
 (a) 500.0 ml of pure water is added to the mixture and equilibrium is reestablished between $Mg(OH)_2$(s) and the saturated solution.
 (b) 100.0 ml of the clear saturated solution is removed from the original mixture and added to 500.0 ml of pure water.
 (c) 25.00 ml of clear saturated solution is removed from the original mixture and added to 250.0 ml 0.065 M $MgCl_2$(aq).
 (d) 50.00 ml of clear saturated solution is removed from the original mixture and added to 150.0 ml 0.150 M KOH(aq).

20. A particular water sample is saturated in CaF_2 at the same time that it has a Ca^{2+} content of 130 ppm (i.e., 130 g Ca^{2+} per 10^6 g of water sample). What is the fluoride ion content of the water in ppm?

$$CaF_2(s) \rightleftharpoons Ca^{2+}(aq) + 2 F^-(aq) \qquad K_{sp} = 2.7 \times 10^{-11}$$

*21. Calculate the solubility of MgF_2 in 5.50×10^{-4} M $MgCl_2$(aq). How effective is the common ion (Mg^{2+}) in reducing the solubility of MgF_2 in this case? Explain.

Criterion for precipitation from solution (Use data from Table 16-1 as necessary.)

22. What must be the $[OH^-]$ in a solution that is 0.25 M in Fe^{3+} to just cause the precipitation of $Fe(OH)_3(s)$?

23. Should precipitation occur if 27.0 mg $MgCl_2 \cdot 6H_2O$ is added to 450 ml 0.050 M KF?

$$MgF_2(s) \rightleftharpoons Mg^{2+}(aq) + 2 F^-(aq) \qquad K_{sp} = 3.7 \times 10^{-8}$$

24. Should precipitation of $PbCl_2(s)$ occur when 155 ml 0.016 M KCl(aq) is added to 245 ml 0.153 M $Pb(NO_3)_2$(aq)?

25. Should precipitation occur in the following cases?
    (a) 1.0 mg NaCl is added to 1.00 L 0.10 M $AgNO_3$(aq).
    (b) One drop (0.05 ml) of 0.20 M KBr is added to 200 ml of a saturated solution of AgCl.
    (c) One drop (0.05 ml) of 0.0150 M NaOH(aq) is added to 5.0 L of a solution containing 2.0 mg Mg^{2+} per liter.

26. If 0.80 g KCl is added to 0.75 L of a solution saturated in Ag_2CO_3, will AgCl precipitate?

Completeness of precipitation (Use data from Table 16-1 as necessary.)

27. When 200 ml of 0.350 M K_2CrO_4(aq) is added to 200 ml of 0.100 M $AgNO_3$(aq),
    (a) Should a precipitate form?
    (b) What $[Ag^+]$ is left unprecipitated?

28. A certain sample of water is found to have magnesium present at a concentration, $[Mg^{2+}] = 1.0 \times 10^{-2} M$?
    (a) What $[OH^-]$ must be maintained in this water to remove essentially all of the magnesium by precipitation as $Mg(OH)_2(s)$? (*Hint:* Assume that all the Mg^{2+} has been removed when $[Mg^{2+}] = 1 \times 10^{-6} M$.)
    (b) What $[OH^-]$ should be maintained to effect the removal of 90% of the dissolved magnesium?

29. What percent of the Ba^{2+} present in solution is precipitated as $BaCO_3(s)$ if *equal volumes* of 0.0020 M Na_2CO_3(aq) and 0.0010 M $BaCl_2$(aq) are mixed? (*Hint:* Recall that dilutions occur.)

*30. What is $[Pb^{2+}]$ remaining in solution if 225 ml 0.15 M KCl(aq) is added to 135 ml 0.12 M $Pb(NO_3)_2$(aq)? (*Hint:* Look for a simplifying assumption but not the usual one.)

Net ionic equations (Use data from Tables 16-1 and 16-2 as necessary.)

31. Predict whether a reaction is likely to occur in each of the following cases. If so, write a net ionic equation.
    (a) NaI(aq) + $ZnSO_4$(aq) $\longrightarrow$
    (b) $CuSO_4$(aq) + Na_2CO_3(aq) $\longrightarrow$
    (c) $AgNO_3$(aq) + $CuCl_2$(aq) $\longrightarrow$
    (d) BaS(aq) + $CuSO_4$(aq) $\longrightarrow$
    (e) CuS(s) + H_2O(l) $\longrightarrow$
    (f) NaOH(aq) + $FeCl_3$(aq) $\longrightarrow$
    (g) $CaCl_2$(aq) + Na_3PO_4(aq) $\longrightarrow$
    (h) Na_2SO_4(aq) + $(NH_4)_2S_2O_8$(aq) $\longrightarrow$

32. You are provided with the following solutions and pure substances: NaOH(aq), K_2SO_4(aq), $Mg(NO_3)_2$(aq), $BaCl_2$(aq), NaCl(aq), $Ca(NO_3)_2$(aq), Ag_2SO_4(s), $BaSO_4$(s). Write net ionic equations to show how you would use these reagents to obtain the following substances: (a) $CaSO_4$(s); (b) $Mg(OH)_2$(s); (c) KCl(aq); (d) AgCl(s).

Qualitative analysis (Use data from Tables 16-1 and 16-2 and Figure 16-3 as necessary.)

33. Explain the following restrictions in the qualitative analysis scheme for cations.
    (a) An unknown solution cannot simply be divided into about two dozen samples and a test performed for a single different ion on each sample.
    (b) The separation of ions into groups cannot be done in just any order desired, such as groups 4, 3, 1, 2, and 5.

34. In Example 16-9 a statement was made that the distinctive blue color of Cu^{2+}(aq) might not be detectable if a solution were too dilute. Suppose that you had an unknown solution in which $[Cu^{2+}] = 0.02 M$.
    (a) Show that you would be able to precipitate CuS from 10 ml of this solution by adding a few drops of 0.50 M Na_2S. (1 drop = 0.05 ml.)
    (b) How much CuS would be obtained, and do you think that it would be visible to the unaided eye?

35. Why is Pb^{2+} found in qualitative analysis groups 1 *and* 2? Why do you suppose that none of the other common cations is found in more than one group?

36. What reagent solution might you use to separate the cations in the following pairs of solids, that is, with one ion appearing in solution and the other in a precipitate? (*Hint:* Refer to Figure 16-3; consider pure water also to be a reagent.)
    (a) $BaCl_2$(s) and NaCl(s)
    (b) $MgCO_3$(s) and Na_2CO_3(s)
    (c) $AgNO_3$(s) and KNO_3(s)
    (d) $PbSO_4$(s) and $Pb(NO_3)_2$(s)

37. In the qualitative analysis scheme for the silver group, $PbCl_2$(s) is separated from AgCl(s) by dissolving it in hot water. Given the K_{sp} values for $PbCl_2$ listed below, assess the water solubility of $PbCl_2$(s) at low and high temperatures. Use the definitions of soluble, insoluble, and moderately soluble given in the footnote to Table 16-2.

for $PbCl_2$: $K_{sp} = 1.6 \times 10^{-5}$ at 25°C;
$K_{sp} = 3.3 \times 10^{-3}$ at 80°C

Fractional precipitation (Use data from Table 16-1 as necessary.)

38. Assume that the seawater sample described in Example 16-6 contains approximately 440 g Ca^{2+} per metric ton of sea water (1 metric ton = 1000 kg; density of seawater $\simeq$ 1.03 g/cm^3).

(a) Should $Ca(OH)_2(s)$ precipitate from seawater under the conditions stated, that is, with $[OH^-]$ maintained at 2.0×10^{-3} M?

(b) Is the separation of Ca^{2+} and Mg^{2+} from seawater by fractional precipitation feasible?

39. To a solution that is 0.250 M in NaCl(aq) and 0.0022 M in KBr(aq) is slowly added $AgNO_3(aq)$.

(a) Which precipitate should form first, AgCl(s) or AgBr(s)?

(b) Can the Cl^- and Br^- be separated effectively by this fractional precipitation?

40. Which of the following reagents would work best to separate Ba^{2+} and Ca^{2+} from a solution in which both are present at a concentration of 0.05 M?   (a) 0.10 M NaCl(aq); (b) 0.50 M $Na_2SO_4(aq)$;   (c) 0.001 M NaOH(aq); (d) 0.50 M $Na_2CO_3(aq)$.

Additional Exercises

1. Equilibrium is established between excess $PbI_2(s)$ and the ions Pb^{2+} and I^- in 100 ml of a saturated aqueous solution. What would you expect to happen to $[Pb^{2+}]$ in the saturated solution (i.e., increase, decrease, . . .) if to the mixture of saturated solution and excess solute were added   (a) 50 ml H_2O;   (b) 50 ml KI(aq);   (c) 50 ml $KNO_3(aq)$?

2. A handbook lists the molar solubilities of $BaSO_4$ and $BaCO_3$ as 1.0×10^{-5} mol $BaSO_4$/L and 7.1×10^{-5} mol $BaCO_3$/L, respectively. When 0.50 M $Na_2CO_3(aq)$ is added to a saturated solution of $BaSO_4$, a precipitate of $BaCO_3(s)$ is observed to form. How do you account for this observation, given the fact that $BaCO_3$ is more soluble than is $BaSO_4$?

3. A solution is saturated with $PbSO_4$ at 50°C ($K_{sp} = 2.3 \times 10^{-8}$). How many mg $PbSO_4$ will precipitate from 425 ml of this solution if it is cooled to 25°C? ($K_{sp} = 1.6 \times 10^{-8}$.)

4. A 25.00-ml sample of a saturated solution of $SrCrO_4$ requires 30.6 ml of 0.0115 M $Fe(NO_3)_2(aq)$ for its titration according to the reaction

$$CrO_4^{2-}(aq) + 3\ Fe^{2+}(aq) + 8\ H^+(aq) \longrightarrow$$
$$Cr^{3+}(aq) + 3\ Fe^{3+}(aq) + 4\ H_2O$$

Calculate K_{sp} for $SrCrO_4$.

5. If 10.0 mg $CaCl_2$ is added to 250.0 ml 0.03500 M Na_2SO_4

(a) Should precipitation occur?

(b) Will precipitation of Ca^{2+} be complete?

6. The electrolysis of an aqueous solution of $MgCl_2$ can be represented by the equation

$$Mg^{2+}(aq) + 2\ Cl^-(aq) + 2\ H_2O \longrightarrow$$
$$Mg^{2+}(aq) + 2\ OH^-(aq) + H_2(g) + Cl_2(g)$$

The electrolysis of a 315-ml sample of 0.220 M $MgCl_2$ is

continued until 1.04 L $H_2(g)$ at 23°C and 748 mmHg has been collected. Will $Mg(OH)_2(s)$ precipitate when electrolysis is carried to this point?

★7. In the Mohr titration, chloride ion is determined by titration with silver nitrate. The solution also contains a small volume of dilute $K_2CrO_4(aq)$ as an indicator. After a sufficient volume of $AgNO_3(aq)$ has been added to a sample to precipitate all of the chloride ion, a red precipitate of Ag_2CrO_4 forms. Explain the basis of this titration. That is, why does the red color not appear while AgCl(s) is still precipitating? Why does the red precipitate appear immediately after the AgCl(s) has finished precipitating?

★8. A mixture of $PbSO_4(s)$ and $PbS_2O_3(s)$ is shaken with pure water until a saturated solution is formed. Both solids remain in excess. What is $[Pb^{2+}]$ in the saturated solution? (*Hint:* Both of the following equilibrium expressions are required, and a third equation as well.)

$$PbSO_4(s) \rightleftharpoons Pb^{2+}(aq) + SO_4^{2-}(aq) \quad K_{sp} = 1.6 \times 10^{-8}$$
$$PbS_2O_3(s) \rightleftharpoons Pb^{2+}(aq) + S_2O_3^{2-}(aq) \quad K_{sp} = 4.0 \times 10^{-7}$$

★9. Use the method of the preceding exercise to determine $[Pb^{2+}]$ in a saturated solution in contact with a mixture of $PbCl_2(s)$ and $PbBr_2(s)$.

$$PbCl_2(s) \rightleftharpoons Pb^{2+}(aq) + 2\ Cl^-(aq) \quad K_{sp} = 1.6 \times 10^{-5}$$
$$PbBr_2(s) \rightleftharpoons Pb^{2+}(aq) + 2\ Br^-(aq) \quad K_{sp} = 4.0 \times 10^{-5}$$

★10. 2.50 g $Ag_2SO_4(s)$ is added to a beaker containing 0.150 L 0.025 M $BaCl_2$.

(a) Write an equation for any reaction that occurs.

(b) Describe the final contents of the beaker, that is, the masses of any precipitates present and the concentrations of the ions in solution.

Self-Test Questions

For questions 1 through 6 select the single item that best completes each statement.

1. Pure water is saturated with the slightly soluble solute, PbI_2. In this saturated solution,   (a) $[Pb^{2+}] = [I^-]$;

(b) $[Pb^{2+}] = K_{sp}$ of PbI_2;   (c) $[Pb^{2+}] = \sqrt{K_{sp}}$ of PbI_2;   (d) $[Pb^{2+}] = 0.5\ [I^-]$.

2. The addition of 10.0 g Na_2SO_4 to 500.0 ml of saturated aqueous $BaSO_4$ has the following effect on the saturated solution: (a) reduces $[Ba^{2+}]$; (b) reduces $[SO_4^{2-}]$; (c) increases the solubility of $BaSO_4$; (d) has no effect.

3. The slightly soluble solute Ag_2CrO_4 is expected to be *most* soluble in (a) pure water; (b) 0.10 *M* K_2CrO_4; (c) 0.25 *M* KNO_3; (d) 0.40 *M* $AgNO_3$.

4. Cu^{2+} and Pb^{2+} are both present in an aqueous solution. To precipitate one of the ions and leave the other in solution, add (a) $H_2S(aq)$; (b) $H_2SO_4(aq)$; (c) $HNO_3(aq)$; (d) $NH_4NO_3(aq)$.

5. The addition of $K_2CO_3(aq)$ to the following solutions is expected to yield a precipitate in every case but one. That one is (a) $BaCl_2(aq)$; (b) $CaBr_2(aq)$; (c) $(NH_4)_2SO_4(aq)$; (d) $Pb(NO_3)_2(aq)$.

6. A *large* excess of $MgF_2(s)$ is maintained in contact with 1.00 L of pure water to produce a saturated solution of MgF_2. When an additional 1.00 L of pure water is added to the mixture and equilibrium reestablished, compared to its value in the original saturated solution, $[Mg^{2+}]$ will be (a) the same; (b) twice as large; (c) half as large; (d) some unknown fraction of the original $[Mg^{2+}]$.

7. The solubility product constant of PbI_2 is

$$PbI_2(s) \rightleftharpoons Pb^{2+}(aq) + 2\,I^-(aq) \qquad K_{sp} = 7.1 \times 10^{-9}$$

Determine the mg Pb^{2+} per milliliter in an 0.050 *M* KI solution that is also saturated with PbI_2.

8. Arrange the following *saturated* solutions according to increasing $[Ag^+]$.
   (a) Ag_2SO_4, $K_{sp} = 1.4 \times 10^{-5}$
   (b) Ag_2CO_3, $K_{sp} = 8.1 \times 10^{-12}$
   (c) $AgCl$, $K_{sp} = 1.6 \times 10^{-10}$
   (d) $AgNO_3$
   (e) $AgI(s)$, $K_{sp} = 8.5 \times 10^{-17}$

9. A solution is 0.010 *M* in K_2CrO_4 and 0.010 *M* in K_2SO_4. To this solution is slowly added 0.10 *M* $Pb(NO_3)_2(aq)$.
   (a) What should be the first substance to precipitate from the solution? $PbCrO_4$ or $PbSO_4$
   (b) What is $[Pb^{2+}]$ at the point where the second substance precipitates?
   (c) Are the two substances effectively separated by this fractional precipitation? Explain. (K_{sp} for $PbCrO_4 = 2.8 \times 10^{-13}$; K_{sp} for $PbSO_4 = 1.6 \times 10^{-8}$.)

10. Each of the following statements represents a misuse of the solubility product concept. Tell what is wrong with each statement.
   (a) Of two slightly soluble ionic compounds, the one with the larger K_{sp} is the more soluble.
   (b) The solubility product constant of $MgCO_3$ is $K_{sp} = 5.3 \times 10^{-8}$. This means that in *all* solutions containing $MgCO_3$, $[Mg^{2+}] = [CO_3^{2-}]$ and $[Mg^{2+}][CO_3^{2-}] = 5.3 \times 10^{-8}$.

9,b) $K_{sp}(PbSO_4) = [Pb^{2+}][SO_4^{2-}] = 1.6 \times 10^{-8}$

$[Pb^{2+}][0.010\,M\,SO_4^{2-}] = 1.6 \times 10^{-8}$

so $[Pb^{2+}] = 1.6 \times 10^{-6}$

@ time of $PbSO_4$ ppt.

9,c) $K_{sp} = [Pb^{2+}][CrO_4^-] = 2.8 \times 10^{-13}$

$[1.6 \times 10^{-6}][CrO_4^-] = 2.8 \times 10^{-13}$

$[CrO_4^-] =$

this is pt where $PbCrO_4$ stops ppt-ing (because $PbSO_4$ starts)

since this is less than 0.1% of original [ ], yes effectively separated

17 Acids and Bases

handwritten margin note: acid has H⁺ anywhere in equation

In this chapter we continue our treatment of aqueous solution equilibria by discussing two types of substances that figure prominently in many of these equilibria—acids and bases. We begin with an overview of acid-base theories; and then proceed to describe a number of topics of special interest, some qualitatively and some quantitatively.

17-1 Acid-Base Theories

A practical scheme has long existed for classifying a variety of substances into the three categories of acids, bases, and salts. Consider acids, for instance. In general, these substances have a sour taste, produce a prickling sensation on the skin, dissolve certain metals, and dissolve carbonate minerals to produce carbon dioxide gas. Also, they produce color changes in plant dyes (e.g., changing the color of litmus from blue to red). The most striking feature of bases (alkalis) is their ability to neutralize the characteristic properties of acids. Other properties of basic substances are a bitter taste, a slippery feel, and the ability to affect the color of plant dyes (changing the color of litmus from red to blue). A salt is produced when a base neutralizes an acid.

Once practical knowledge of acids and bases began to accumulate, theories to account for their characteristic behavior soon followed. Three important acid-base theories are considered next.

ARRHENIUS'S THEORY. In his theory of electrolytic dissociation Svante Arrhenius proposed that an electrolyte dissociates into ions upon dissolving in water: a strong electrolyte dissociates completely; a weak electrolyte, only partially. A substance that dissociates to produce hydrogen ions (H^+) is an acid. A base dissociates to produce hydroxide ions (OH^-). Equations for acid-base neutralization reactions can be written in complete, ionic, and net ionic forms.

complete:
$$\underset{\text{acid}}{HCl} + \underset{\text{base}}{NaOH} \longrightarrow \underset{\text{salt}}{NaCl} + \underset{\text{water}}{H_2O} \tag{17.1}$$

ionic:
$$\underset{\text{acid}}{H^+ + Cl^-} + \underset{\text{base}}{Na^+ + OH^-} \longrightarrow \underset{\text{salt}}{Na^+ + Cl^-} + \underset{\text{water}}{H_2O} \tag{17.2}$$

net ionic:
$$H^+ + OH^- \longrightarrow H_2O \tag{17.3}$$

---

margin text:

The term acid is derived from the Latin *acetum,* which means vinegar. The acid constituent in vinegar is acetic acid, H_3CCOOH. The term alkali is derived from the Arabic word for ashes. Until relatively recent times, the principal source of bases or alkalis was wood ashes.

Lavoisier considered oxygen to be present in all acids. In fact, he proposed the name oxygen from the Greek for "acid forming." According to Lavoisier's concept of an acid, it should have been possible to isolate oxygen from the salts of muriatic acid (hydrochloric acid). However, all such attempts failed. In 1810, Humphry Davy isolated the element chlorine from muriatic acid. In doing so he proved that only hydrogen and chlorine occur in this acid. As a result, Davy proposed that *hydrogen,* not oxygen, is the element common to all acids.

The net ionic equation is an especially appropriate representation of a neutralization reaction according to the Arrhenius theory. It brings out this essential point: *A neutralization reaction involves the combination of hydrogen and hydroxide ions to form water.*

There is another way in which the Arrhenius theory explains the process of neutralization better than any prior theory. The heats of neutralization of strong acids and bases are found to be essentially constant: -55.90 kJ/mol water formed. That ΔH_{neu} is independent of the identity of the strong acid and base is readily understood: The essential reaction is always $H^+ + OH^- \rightarrow H_2O$ (17.3).

The Arrhenius theory also met with success in explaining the catalytic activity of acids in certain reactions. The acids that prove to be the most effective catalysts are also those that have the best electrical conductivity, that is, strong acids. The stronger the acid, the higher the H^+ concentration in solution. It is the H^+ ion that is the actual catalyst in these reactions. An example of an acid-catalyzed reaction, the decomposition of formic acid, was considered in Section 13-9.

BRØNSTED–LOWRY THEORY.   Despite its successes and continued usefulness, the Arrhenius theory does have limitations. For one, it does not recognize any constituent other than OH^- as imparting basic properties to a substance. This requirement led to representing the ionization of aqueous solutions of ammonia as

$$NH_4OH \rightleftharpoons NH_4^+ + OH^- \qquad (17.4)$$

Yet the substance NH_4OH (ammonium hydroxide) appears not to exist. That is, it cannot be isolated in pure form as can sodium hydroxide ($NaOH$).

Moreover, even in Arrhenius's time, reactions were being conducted in *nonaqueous* solvents, such as liquid ammonia. Some of these reactions seemed to have the characteristics of acid-base reactions. Clearly, however, OH^- was not present, since there were no oxygen atoms in the system at all. For example, ammonium chloride and sodium amide react in liquid ammonia as follows.

complete: $\qquad\qquad NH_4Cl + NaNH_2 \longrightarrow NaCl + 2\,NH_3 \qquad (17.5)$

ionic: $\qquad NH_4^+ + Cl^- + Na^+ + NH_2^- \longrightarrow Na^+ + Cl^- + 2\,NH_3 \qquad (17.6)$

net ionic: $\qquad\qquad NH_4^+ + NH_2^- \longrightarrow 2\,NH_3 \qquad (17.7)$

Reaction (17.7) can be considered an acid-base reaction with NH_4^+ analogous to H^+ and NH_2^- to OH^-. This analogy is clearly possible by using a more general theory of acids and bases proposed independently by J. N. Brønsted in Denmark and T. M. Lowry in Great Britain in 1923. According to the Brønsted–Lowry theory, an acid is a **proton donor** and a base, a **proton acceptor,** as suggested in reaction (17.8).

$$NH_4^+ + NH_2^- \rightleftharpoons NH_3 + NH_3 \qquad (17.8)$$
$$\text{acid(1)}\quad\text{base(2)}\qquad\text{acid(2)}\quad\text{base(1)}$$

A number of ideas are implicit in equation (17.8). An acid, call it acid(1), loses a proton and becomes base(1). Similarly, base(2) gains a proton and becomes acid(2). In general proton transfer reactions are reversible. When base(1) regains a proton, it forms acid(1). Base(1) is said to be the **conjugate base** of acid(1). Similarly, acid(2) is the **conjugate acid** of base(2). Figure 17-1 may help you to visualize the proton transfer involved in reaction (17.8).

The direction in which an acid-base reaction tends to occur, that is, the direction of proton transfer, depends on the relative strengths of the species involved. If the acid is strong, its conjugate base is weak, and vice versa. The net reaction goes from strong acid and base to weak acid and base. We will explore some of these relationships between conjugate acids and bases more fully in Section 17-8.

In acid-base theory the term proton means the nucleus of a hydrogen atom, that is, H^+.

NH_3 (the solvent) is both the conjugate base of the acid NH_4^+ and the conjugate acid of the base NH_2^- in equation (17.8).

By Arrhenius Theory: NH_4^+ dissociates to produce H^+ ion ∴ NH_4^+ is acid

By Brønsted-Lowry: NH_4^+ donates proton(H^+) ∴ acid

NH_2^- accepts proton(H^+) ∴ base

FIGURE 17-1
Brønsted–Lowry acid-base reaction.

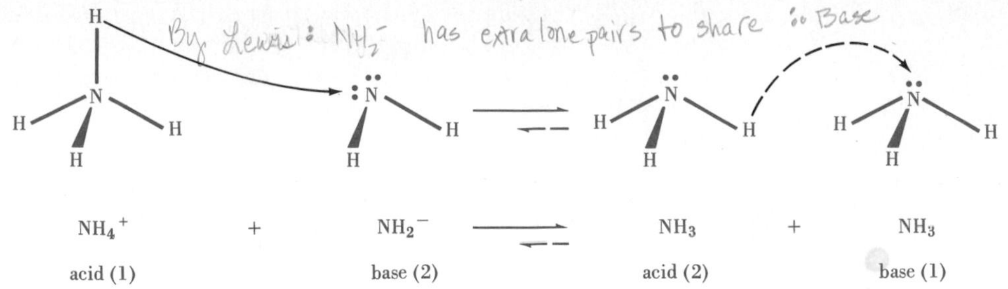

By Lewis: NH_2^- has extra lone pairs to share ∴ Base

$$NH_4^+ \quad + \quad NH_2^- \quad \longrightarrow \quad NH_3 \quad + \quad NH_3$$

acid (1)   base (2)   acid (2)   base (1)

This figure depicts the proton transfer in reaction (17.8). The solid arrows represent the forward reaction and the broken arrows, the reverse reaction. The NH_4^+/NH_2^- combination is a strong acid/strong base combination, whereas NH_3/NH_3 is a weak acid/weak base combination. The forward direction is favored in this acid-base reaction.

Two additional features of the Brønsted–Lowry theory are illustrated here: (1) Any species that is an acid by the Arrhenius theory remains an acid (similarly for bases). (2) Certain species that would not be classified as bases by the Arrhenius theory actually are, such as ClO^- and HPO_4^{2-}.

Example 17-1 For each of the following, identify the acids and bases involved in both the forward and reverse reactions.

(a) $HClO_2 + H_2O \rightleftharpoons H_3O^+ + ClO_2^-$
(b) $ClO^- + H_2O \rightleftharpoons HOCl + OH^-$
(c) $NH_3 + H_2PO_4^- \rightleftharpoons NH_4^+ + HPO_4^{2-}$
(d) $HCl + H_2PO_4^- \rightleftharpoons H_3PO_4 + Cl^-$

Consider $HClO_2$ in reaction (a). It is converted to ClO_2^- by losing a proton (H^+). $HClO_2$ must be an acid, and ClO_2^- is its conjugate base. H_2O accepts the proton lost by $HClO_2$. Then H_2O must be a base and H_3O^+, its conjugate acid. In reaction (b), H_2O acts as an acid and OH^- is its conjugate base. Reactions (c) and (d) illustrate another species that can either donate or accept a proton—an **amphiprotic** species. In this case the amphiprotic species is $H_2PO_4^-$.

(a) $HClO_2 + H_2O \rightleftharpoons H_3O^+ + ClO_2^-$
    acid(1)   base(2)   acid(2)   base(1)

(b) $H_2O + ClO^- \rightleftharpoons HOCl + OH^-$
    acid(1)   base(2)   acid(2)   base(1)

(c) $H_2PO_4^- + NH_3 \rightleftharpoons NH_4^+ + HPO_4^{2-}$
    acid(1)   base(2)   acid(2)   base(1)

(d) $HCl + H_2PO_4^- \rightleftharpoons H_3PO_4 + Cl^-$
    acid(1)   base(2)   acid(2)   base(1)

SIMILAR EXAMPLES: Exercises 4, 5.

Acid-base reactions in solvents containing neither H^+ nor OH^- can also be described by the "solvent system" method: Associated with every solvent there is a characteristic cation and anion, for example,

$$2 SO_2 \rightarrow SO^{2+} + SO_3^{2-}$$

Any substance that produces the characteristic cation (SO^{2+}) in the solvent is an acid, and any substance that produces the characteristic anion (SO_3^{2-}) is a base.

LEWIS THEORY. G. N. Lewis developed an alternative to the Arrhenius theory of acids and bases at about the same time as Brønsted and Lowry (1923). Lewis's theory has some additional advantages over the Brønsted–Lowry theory in that it permits the acid-base classification to be used for some reactions in which *neither* H^+ *nor* OH^- is present. Moreover, it allows reactions to be considered of the acid-base type even when there is no solvent at all, such as certain reactions in the gas phase.

In Lewis theory an acid is an **electron pair acceptor** and a base is an **electron pair donor.** From what we know about chemical bonding, then, we should expect acids to be species with *available orbitals and a lack of electrons.* Bases are species with *lone-pair electrons available for sharing.* In addition, an acid-base reaction leads to the formation of a covalent bond between the acid and the base.

With the definitions just given, we would classify H^+ as an acid because of the presence of an empty orbital (1s) that can accept a pair of electrons. OH^- and NH_3

are bases because of the presence in these species of available lone-pair electrons.

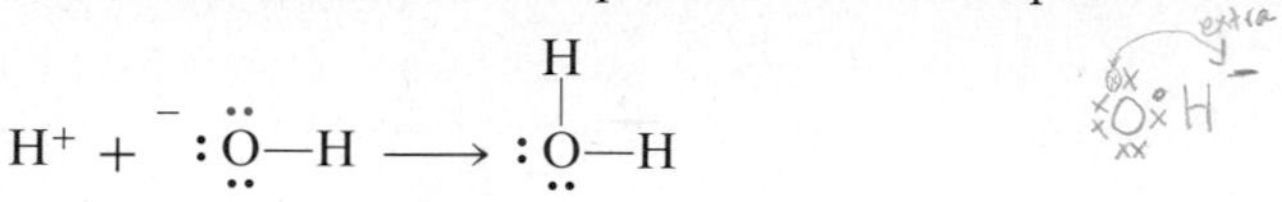

$$H^+ + ^-:\overset{..}{\underset{..}{O}}-H \longrightarrow :\overset{\overset{\textstyle H}{|}}{\underset{..}{O}}-H \tag{17.9}$$

$$H^+ + H-\overset{..}{\underset{\underset{\textstyle H}{|}}{N}}-H \longrightarrow \left[H-\overset{\overset{\textstyle H}{|}}{\underset{\underset{\textstyle H}{|}}{N}}-H \right]^+ \tag{17.10}$$

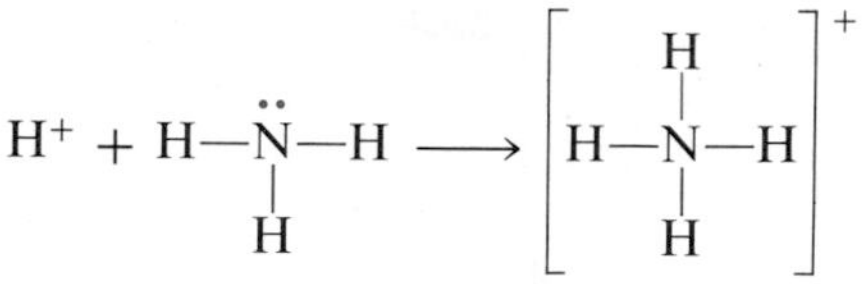

In the following reaction, which we first encountered in Section 9-7, BF_3 acts as a Lewis acid and NH_3 as a Lewis base. An electron-pair bond (a coordinate covalent bond) is formed between the B and the N atom.

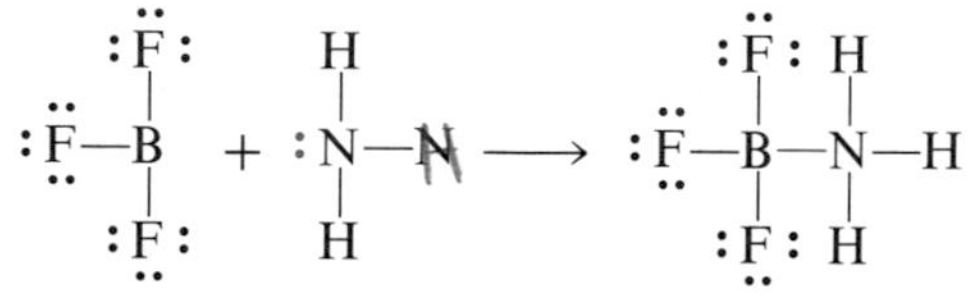

$$:\!\overset{..}{\underset{..}{F}}\!: \quad \text{...} \quad :\!\overset{..}{\underset{..}{F}}\!: \; H$$

$$:\overset{\overset{\textstyle :\overset{..}{F}:}{|}}{\underset{\underset{\textstyle :\overset{..}{\underset{..}{F}}:}{|}}{F}}-B \;+\; :\overset{\overset{\textstyle H}{|}}{\underset{\underset{\textstyle H}{|}}{N}}-H \longrightarrow :\overset{\overset{\textstyle :\overset{..}{F}:}{|}}{\underset{\underset{\textstyle :\overset{..}{\underset{..}{F}}:}{|}}{F}}-\overset{\overset{\textstyle H}{|}}{\underset{\underset{\textstyle H}{|}}{B}}-\overset{\overset{\textstyle H}{|}}{\underset{\underset{\textstyle H}{|}}{N}}-H \tag{17.11}$$

Example 17-2 Each of the following is an acid-base reaction in the Lewis sense. Which species is the acid and which is the base?

(a) $BF_3 + F^- \longrightarrow BF_4^-$

(b) $Zn^{2+}(aq) + 4\,NH_3(aq) \longrightarrow [Zn(NH_3)_4]^{2+}$

(c) (Lewis structures showing $SO_2 + H_2O \longrightarrow H_2SO_3$ with rearrangement of a pair of electrons and one hydrogen atom)

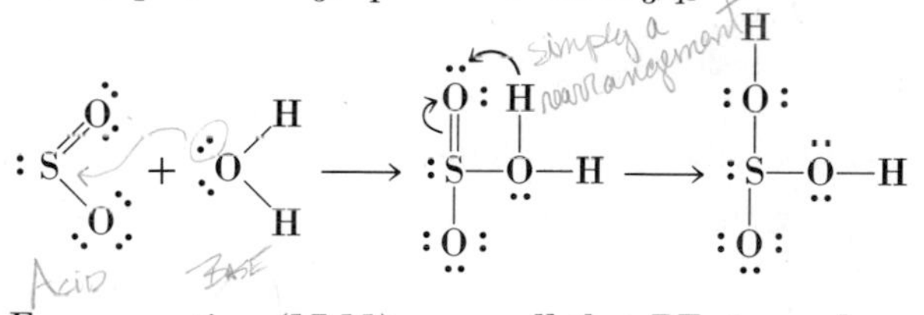

(a) From equation (17.11) we recall that BF_3 is an electron-deficient molecule, with a vacant orbital on the B atom. The fluoride ion has an outer-shell octet of electrons. BF_3 is the electron pair acceptor—the acid. F^- is the electron pair donor—the base.

(b) Again recalling equation (17.11), we identify NH_3 as an electron pair donor—a base. Although from the equation written it is not clear just how the Zn^{2+} ion accepts electrons from the NH_3, it must indeed do so. Altogether four pairs of electrons are accepted by the Zn^{2+} to form the *complex ion* $[Zn(NH_3)_4]^{2+}$. Complex ion formation is discussed in Chapter 22.

(c) Here the key to identifying the acid and the base is through the structure of the product molecule (sulfurous acid, H_2SO_3). This molecule has three oxygen atoms bonded to sulfur, whereas SO_2 has two. The third oxygen atom is the oxygen atom from H_2O, which supplies an electron pair. H_2O is the Lewis base and SO_2 is the Lewis acid. Note also that a rearrangement of a pair of electrons and one hydrogen atom occurs in this reaction.

SIMILAR EXAMPLES: Exercises 7, 8.

17-2 Self-ionization (Autoionization) of Water

Water is a very poor electrical conductor. Yet the fact that it does conduct electric current feebly indicates that some ions are present. According to the Arrhenius theory, these ions, which arise from the ionization of water molecules themselves, are H^+ and OH^-.

[handwritten: PROTON ACCEPTOR, PROTON DONOR]

[handwritten top: covalent bond "shared pair of e⁻'s"]

[handwritten: $[H:\overset{..}{\underset{..}{O}}:]^-$]

$$H_2O \rightleftharpoons H^+ + OH^- \tag{17.12}$$

The self-ionization of water is also an acid-base reaction in the Brønsted–Lowry sense (equation 17.13). One water molecule acts as an acid; it loses a proton. Another water molecule acts as a base. It accepts the proton, with which it forms a coordinate covalent bond through an unshared pair of electrons on the oxygen atom. The resulting ions are H_3O^+, called **hydronium ion,** and OH^-, **hydroxide ion.** The ionization reaction is reversible, and in the reverse reaction a hydronium ion loses a proton to a hydroxide ion. In fact, since acid(2) and base(1) are *much* stronger than acid(1) and base(2), the reverse reaction is much more significant than the forward reaction. *Equilibrium is displaced far to the left.*

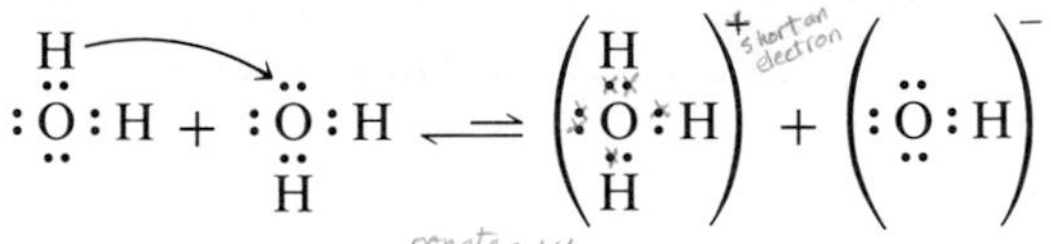

[handwritten margin: rxn displace from the STRONG acids/bases to the WEAK.]

$$\overset{\text{H}}{:\underset{..}{\overset{..}{O}}:H} + :\underset{..}{\overset{..}{O}}:H \rightleftharpoons \left(\overset{\text{H}}{:\underset{\text{H}}{\overset{..}{O}}:H}\right)^+ + \left(:\underset{..}{\overset{..}{O}}:H\right)^- \tag{17.13}$$

acid (1)   base (2)   *[handwritten: donates H⁺ (proton)]*   acid (2)   base (1)

[handwritten: accepts H⁺ (proton)]

Because the self-ionization of water is a reversible process, we can describe it through a thermodynamic equilibrium constant expression.

*[handwritten margin: *Review pp. 301-304]*

$$K = \frac{(a_{H_3O^+})(a_{OH^-})}{(a_{H_2O})^2}$$

Since the activity of the water—a pure liquid—is 1,

[handwritten margin: Stoichiometric concentration ⇒ total [ ] of ion in aqueous soln. based on amt of solute dissolved.]

$$K = (a_{H_3O^+})(a_{OH^-})$$

Furthermore, because the ion concentrations are very small, we can substitute molar concentrations for activities, leading to the final expression

[handwritten margin: Activity – "effective" concentration of ion which takes into account interionic attractions.]

$$K_w = [H_3O^+][OH^-]$$

There are several experimental methods of determining the concentrations of H_3O^+ and OH^- in pure water. All lead to the result that

at 25°C in pure water: $[H_3O^+] = [OH^-] = \boxed{1.0 \times 10^{-7}\ M}$

That $[H_3O^+]$ and $[OH^-]$ in pure water must be equal can be seen from the balanced equation (17.13). The equilibrium constant for the self-ionization of water is called the **ion product of water,** and represented as K_w. At 25°C,

$$K_w = [H_3O^+][OH^-] = 1.0 \times 10^{-14} \tag{17.14}$$

Like all other equilibrium constants, the ion product of water is temperature-dependent. At 25°C, $K_w = 1.01 \times 10^{-14}$; at 60°C, 9.62×10^{-14}; at 100°C, 5.5×10^{-13}.

[handwritten margin: higher temps, more dissociation, greater ion concentrations]

THE NATURE OF THE PROTON IN AQUEOUS SOLUTION. One reason for favoring equation (17.13) over (17.12) for the self-ionization of water is that it conforms to the Brønsted–Lowry theory, which is more general than the Arrhenius theory. There is another equally important reason. The Arrhenius theory (i.e., through equation 17.12) postulates the existence of H^+ ions in aqueous solution. Recall that the H^+ ion is simply a lone proton (the nucleus of a hydrogen atom). Because of their very small size and high positive charge density, we should expect H^+ ions to seek out centers of negative charge with which to form bonds. Thus, hydrogen ions, H^+, are not expected to exist in water solution, *[handwritten: instead the H⁺ attaches to H₂O ⇒ H₃O⁺]*

What is the situation with respect to the hydronium ion, H_3O^+, whose existence in water solution is postulated by the Brønsted–Lowry theory? Does it exist? For many years this seemed an unanswerable question, and hydronium ion was thought

Theoretical calculations of the concentration of free H^+ ions or lone protons in water yield results as low as $10^{-130}\ M$. As described by N. V. Sidgwick (1950), this corresponds to *one* free proton in 10^{70} universes filled with a $1\ M$ acid solution!

just to be the simplest of a series of species called *hydrated* protons: $[H(H_2O)_n]^+$. That is, if $n = 1$, the species obtained is $[H(H_2O)]^+ = H_3O^+$—the hydronium ion. However, there is experimental evidence, established through x-ray diffraction studies, for the existence of the hydronium ion in the solid state. What was once thought to be a monohydrate of perchloric acid, $HClO_4 \cdot H_2O$, is now known to be the ionic compound $H_3O^+ClO_4^-$. This salt, which we might call hydronium perchlorate, is quite similar structurally to ammonium perchlorate, $NH_4^+ClO_4^-$. In addition, recent experimental evidence has established the presence of H_3O^+ in aqueous solution. A current view is of the existence of structures such as the one depicted in Figure 17-2—a central hydronium ion hydrogen-bonded to three H_2O molecules.

The formula of the structure shown in Figure 17-2 could be written as $H_3O^+ \cdot 3\ H_2O$ or $H_9O_4^+$, but we will not do so. Generally, all ions are hydrated in aqueous solution, and we have found it satisfactory just to use the symbol (aq) to represent this fact. Throughout this chapter we refer to hydronium ion in solution as H_3O^+ or $H_3O^+(aq)$.

> Hydroxide ion appears to form a similar species: $OH^- \cdot 3\ H_2O$ or $H_7O_4^-$. We will continue simply to speak of OH^- or $OH^-(aq)$.

17-3 Strong Acids and Bases

FIGURE 17-2
The hydronium ion in aqueous solution.

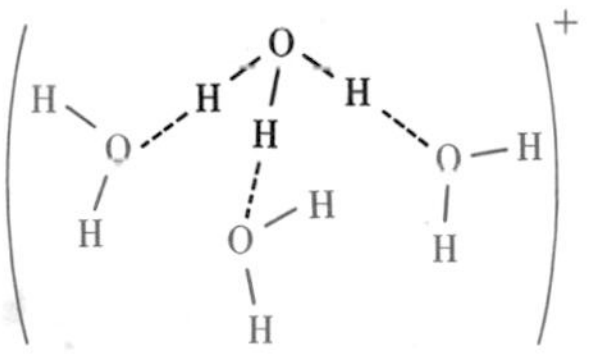

Represented here is a probable species that exists in aqueous solution. The central H_3O^+ is hydrogen-bonded to three H_2O molecules.

When an acid is added to water, as in aqueous solutions of hydrochloric acid, two acid-base reactions occur simultaneously.

$$H_2O + H_2O \rightleftharpoons H_3O^+ + OH^- \qquad (17.15)$$
$$\text{acid} \quad \text{base} \qquad\quad \text{acid} \quad \text{base}$$

$$HCl + H_2O \longrightarrow H_3O^+ + Cl^- \qquad (17.16)$$
$$\text{acid} \quad \text{base} \qquad\quad \text{acid} \quad \text{base}$$

The self-ionization of water, the forward reaction in equation (17.15), occurs only to a slight extent. By contrast, the ionization of HCl, a strong acid, goes essentially to completion (equation 17.16). In calculating the concentration of H_3O^+ in an aqueous HCl solution, it is generally acceptable to consider the ionization of HCl to be the sole source of H_3O^+.

Example 17-3 Calculate $[H_3O^+]$, $[Cl^-]$, and $[OH^-]$ in 0.015 M HCl(aq).

If we assume that HCl (a) is the sole source of H_3O^+ and (b) is completely ionized in aqueous solution,

$$[H_3O^+] = 0.015\ M$$

Furthermore, as indicated by equation (17.16), one Cl^- is produced for every H_3O^+. Therefore,

$$[Cl^-] = 0.015\ M$$

To calculate $[OH^-]$ we must recognize (a) that all the OH^- is derived from the self-ionization of water (equation 17.15) and (b) that in any aqueous solution $[H_3O^+]$ and $[OH^-]$ must have values consistent with the ion product of water.

$$K_w = [H_3O^+][OH^-] = 1.0 \times 10^{-14}$$
$$(0.015)[OH^-] = 1.0 \times 10^{-14}$$

$$[OH^-] = \frac{1.0 \times 10^{-14}}{1.5 \times 10^{-2}} = 0.67 \times 10^{-12}$$

$$= 6.7 \times 10^{-13}\ M$$

SIMILAR EXAMPLES: Exercises 10, 11.

TABLE 17-1
Some common strong
acids and bases

Acids	Bases
HCl	NaOH
HBr	KOH
HI	RbOH
$HClO_4$	CsOH
HNO_3	$Ca(OH)_2$
H_2SO_4	$Sr(OH)_2$
	$Ba(OH)_2$

The concentration of OH^- in 0.015 M HCl (Example 17-3) is much smaller than in pure water (6.7×10^{-13} M compared to 1.0×10^{-7} M). The addition of an acid represses the ionization of water. That is, it favors the reverse of reaction (17.15). The result is in accord with Le Châtelier's principle—disturbing a system at equilibrium causes the equilibrium condition to shift in that direction which minimizes the disturbance. If $[H_3O^+]$ in a water solution is increased, $[OH^-]$ must decrease so that the ion product remains constant.

To use the method of Example 17-3 requires knowledge of what compounds are strong acids and what compounds are strong bases. These substances are limited enough in number that a simple listing is possible, as in Table 17-1.

17-4 pH and pOH

The concentrations of H_3O^+ and OH^- encountered in aqueous solutions are variable over an extreme range. In 0.015 M HCl, $[H_3O^+]$ is 0.015 M. In saturated $Ca(OH)_2(aq)$, $[H_3O^+] = 4.2 \times 10^{-13}$ M. The advantage of writing the exponential form $[H_3O^+] = 4.2 \times 10^{-13}$ M over the decimal form $[H_3O^+] = 0.00000000000042$ M is obvious; but a further simplification is also possible. This is accomplished through the pH notation introduced in 1909 by Søren Sørensen, a Danish biochemist. (Sørensen meant the symbol pH to stand for "potential of hydrogen.") If a solution has

$$[H_3O^+] = 10^{-x} \tag{17.17}$$

$$pH = x \tag{17.18}$$

The mathematical operation in transforming equation (17.17) to equation (17.18) involves the use of logarithms. The pH is the *negative* of the logarithm of $[H_3O^+]$.

$$pH = -\log[H_3O^+] \tag{17.19}$$

In 0.01 M HCl

$$[H_3O^+] = 10^{-2}$$

$$pH = -\log[H_3O^+] = -\log(10^{-2}) = -(-2) = 2$$

And for 0.0050 M HCl

$$[H_3O^+] = 5.0 \times 10^{-3}$$

$$pH = -\log(5.0 \times 10^{-3}) = -(\log 5.0 + \log 10^{-3})$$
$$= -(0.70 - 3) = 2.30$$

To determine $[H_3O^+]$ from a known pH value requires the *inverse* operation from taking logarithms. An **antilogarithm** refers to a number having a given logarithm. For example, antilog (-5) refers to a number whose logarithm is -5; the number is 10^{-5}. For nonintegral values it is helpful to express pH as the sum of two numbers, one a positive decimal quantity, the other a negative integer. The antilog of each part is taken, and the result is expressed in exponential form. For example, if pH = 5.30,

$$-\log[H_3O^+] = pH = 5.30$$
$$\log[H_3O^+] = -5.30 = 0.70 - 6$$
$$[H_3O^+] = antilog\ 0.70 \times antilog\ -6$$
$$[H_3O^+] = 5.0 \times 10^{-6}$$

pOH is defined in an analogous fashion to pH.

$$pOH = -\log[OH^-] \tag{17.20}$$

Operations involving logarithms can now be performed on electronic calculators with hardly a thought being given to the matter. However, make it a point to keep in mind the ideas outlined here and in Appendix A. If you do, you are more likely to catch the type of errors that come from "punching the wrong button."

In this inverse operation it is helpful to note that the correct $[H_3O^+]$ can be bracketed between two powers of 10. A pH of 5.30 must correspond to $[H_3O^+]$ between 1×10^{-5} and 1×10^{-6} M.

Since in any dilute aqueous solution at 25°C $[H_3O^+][OH^-] = 1.0 \times 10^{-14}$, we obtain the useful expression (17.21) by taking negative logarithms.

$$-(\log [H_3O^+][OH^-]) = -\log (1.0 \times 10^{-14})$$
$$-\log [H_3O^+] - \log [OH^-] = -(-14.00)$$
$$pH + pOH = 14.00 \qquad \text{[handwritten: } pH = 14.00 - pOH\text{]} \tag{17.21}$$

The pH values associated with a number of common materials are listed in Figure 17-3. As indicated in the figure, pH = 7 represents a neutral solution; pH < 7, acidic; and pH > 7, basic.

Conversions between $[H_3O^+]$ and pH are generally encountered as part of a larger problem, as illustrated through Example 17-4.

Example 17-4 "Milk of magnesia" is a suspension of $Mg(OH)_2$ in water commonly used as an antacid and laxative. What is the pH of "milk of magnesia?"

$$Mg(OH)_2(s) \rightleftharpoons Mg^{2+}(aq) + 2\ OH^-(aq) \qquad {}^*K_{sp} = \underline{2 \times 10^{-11}}$$

Let us assume that $Mg(OH)_2$ is the sole source of OH^- in solution and solve for $[OH^-]$ in the manner presented in Chapter 16.

let S = no. mol $Mg(OH)_2$ per L of satd. soln; $[Mg^{2+}] = S$ and $[OH^-] = 2S$.

$$[Mg^{2+}][OH^-]^2 = (S)(2S)^2 = 4S^3 = 2 \times 10^{-11}$$

$$S^3 = 5 \times 10^{-12} \qquad S = 2 \times 10^{-4}$$

$$[OH^-] = 2S = 2 \times 2 \times 10^{-4} = 4 \times 10^{-4}\ M$$

$$pOH = -\log (4 \times 10^{-4}) = -(\log 4 + \log 10^{-4}) = -(0.6 - 4) = 3.4$$

$$pH = 14.0 - pOH = 14.0 - 3.4 = 10.6$$

[handwritten: CAREFUL! — ORIGINALLY ASKED FOR WAS pH]

SIMILAR EXAMPLES: Exercises 17, 18.

17-5 Weak Acids and Weak Bases

Most acids and bases are weak, and the situation that results when they are dissolved in water is more complex than that encountered in Example 17-3. Again two ionization reactions occur simultaneously.

$$\underset{\text{acid}}{H_2O} + \underset{\text{base}}{H_2O} \rightleftharpoons \underset{\text{acid}}{H_3O^+} + \underset{\text{base}}{OH^-} \tag{17.15}$$

[handwritten: if H_2O present in rx, this ionization equation will always occur; there will always be $1.0 \times 10^{-14} = K_w = [H_3O^+][OH^-]$.]

$$\underset{\text{acid}}{HOCl} + \underset{\text{base}}{H_2O} \rightleftharpoons \underset{\text{acid}}{H_3O^+} + \underset{\text{base}}{OCl^-} \tag{17.22}$$

The ionization of HOCl represented by equation (17.22) differs from that of HCl represented by (17.16) in one highly significant way: The ionization of HOCl is a truly reversible reaction and must be represented by an equilibrium constant expression.

$$K_a = \frac{[H_3O^+][OCl^-]}{[HOCl]} = 2.95 \times 10^{-8}$$

The term K_a is called the **ionization constant** of hypochlorous acid. Its value, 2.95×10^{-8}, must be determined by experiment. Ionization constants for a few typical weak acids and bases are listed in Table 17-2. The symbol K_a is commonly used to represent the ionization constant of a weak acid and K_b for a weak base. pK is a commonly used shorthand designation for an ionization constant: $pK = -\log K$. Thus, for hypochlorous acid, $pK_a = -\log (2.95 \times 10^{-8}) = -(-7.53) = 7.53$.

[handwritten left margin notes:]

$^*K_{sp}$ = Solubility Product constant.

Represents equilibrium between a slightly soluble ionic compd & its ions in aqueous soln.

$= \dfrac{[\text{ions / products}]}{[\text{solid / reactant}]}$

(activity of pure solid = 1)

The terms dissociation and ionization are often used synonomously, but there is some objection to this. Dissociation means "coming apart," as in the dissociation of $N_2O_4(g)$: $N_2O_4(g) \rightleftharpoons 2\ NO_2(g)$. It is also appropriate to speak of the dissociation of NaCl upon dissolving.

$$NaCl(s) \rightarrow Na^+(aq) + Cl^-(aq)$$

Ionization of Na and Cl atoms occurs in the formation of crystalline NaCl, not when it dissolves. With HOCl, however, dissociation and ionization (ion formation) occur simultaneously. In our discussion of weak acids and bases we will emphasize proton transfer (and hence ion formation). For this reason we will use the term ionization.

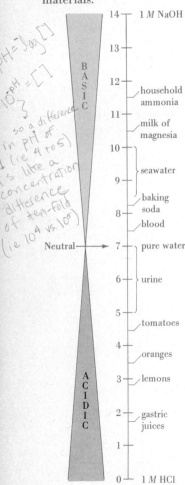

FIGURE 17-3
The pH scale and pH values of some common materials.

(handwritten notes in margin):
$pH = log[\ ]$
$10^{pH} = [\ ]$
so a difference in pH of 1 (ie 4 to 5) is like a concentration difference of ten-fold (ie 10^4 vs 10^5)

pH scale:
- 14 — 1 M NaOH
- 13 —
- 12 — household ammonia
- 11 — milk of magnesia
- 10 —
- 9 — seawater
- 8 — baking soda / blood
- 7 — pure water (Neutral)
- 6 — urine
- 5 — tomatoes
- 4 — oranges
- 3 — lemons
- 2 — gastric juices
- 1 —
- 0 — 1 M HCl

BASIC / ACIDIC

It is important to note that a change of pH of one unit represents a ten-fold change in $[H_3O^+]$. For example, orange juice is about 10 times more acidic than tomato juice.

TABLE 17-2
Ionization constants for some weak acids and weak bases in water at 25°C

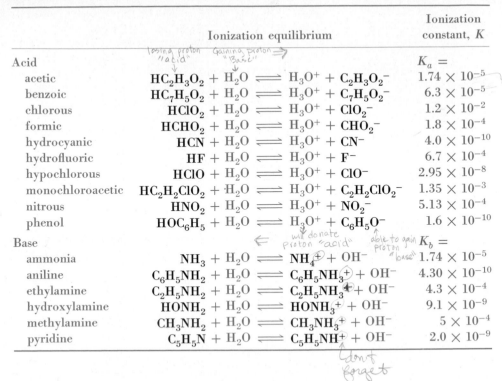

	Ionization equilibrium	Ionization constant, K
Acid	*(handwritten: losing proton "acid" → Gaining proton "Base")*	$K_a =$
acetic	$HC_2H_3O_2 + H_2O \rightleftharpoons H_3O^+ + C_2H_3O_2^-$	1.74×10^{-5}
benzoic	$HC_7H_5O_2 + H_2O \rightleftharpoons H_3O^+ + C_7H_5O_2^-$	6.3×10^{-5}
chlorous	$HClO_2 + H_2O \rightleftharpoons H_3O^+ + ClO_2^-$	1.2×10^{-2}
formic	$HCHO_2 + H_2O \rightleftharpoons H_3O^+ + CHO_2^-$	1.8×10^{-4}
hydrocyanic	$HCN + H_2O \rightleftharpoons H_3O^+ + CN^-$	4.0×10^{-10}
hydrofluoric	$HF + H_2O \rightleftharpoons H_3O^+ + F^-$	6.7×10^{-4}
hypochlorous	$HClO + H_2O \rightleftharpoons H_3O^+ + ClO^-$	2.95×10^{-8}
monochloroacetic	$HC_2H_2ClO_2 + H_2O \rightleftharpoons H_3O^+ + C_2H_2ClO_2^-$	1.35×10^{-3}
nitrous	$HNO_2 + H_2O \rightleftharpoons H_3O^+ + NO_2^-$	5.13×10^{-4}
phenol	$HOC_6H_5 + H_2O \rightleftharpoons H_3O^+ + C_6H_5O^-$	1.6×10^{-10}
Base	*(handwritten: will donate proton "acid" ← → able to gain proton "base")*	$K_b =$
ammonia	$NH_3 + H_2O \rightleftharpoons NH_4^+ + OH^-$	1.74×10^{-5}
aniline	$C_6H_5NH_2 + H_2O \rightleftharpoons C_6H_5NH_3^+ + OH^-$	4.30×10^{-10}
ethylamine	$C_2H_5NH_2 + H_2O \rightleftharpoons C_2H_5NH_3^+ + OH^-$	4.3×10^{-4}
hydroxylamine	$HONH_2 + H_2O \rightleftharpoons HONH_3^+ + OH^-$	9.1×10^{-9}
methylamine	$CH_3NH_2 + H_2O \rightleftharpoons CH_3NH_3^+ + OH^-$	5×10^{-4}
pyridine	$C_5H_5N + H_2O \rightleftharpoons C_5H_5NH^+ + OH^-$	2.0×10^{-9}

(handwritten: don't forget)

Example 17-5   The pH of an 0.100 M HCN solution is found to be 5.2. What is the value of K_a for HCN?

Suppose we let x be the number of moles per liter of HCN that ionize. The relevant data are summarized below the equation. Note that $[H_3O^+] = [CN^-]$ since these ions are formed in equal numbers in the reaction.

$$HCN + H_2O \rightleftharpoons H_3O^+ + CN^-$$

placed in solution:	0.100 M	—	—
changes:	$-x\,M$	$+x\,M$	$+x\,M$
at equilibrium:	$(0.100 - x)\,M$	$x\,M$	$x\,M$

But $x = [H_3O^+]$ and is directly related to the measured pH.

$$\log [H_3O^+] = -5.2 = 0.8 - 6.0$$

$$x = [H_3O^+] = 6 \times 10^{-6}$$

*(handwritten: pH = potential to ionize, to yield H⁺ = 6 * 10⁻⁶)*

The ionization constant, K_a, is

$$K_a = \frac{[H_3O^+][CN^-]}{[HCN]} = \frac{x \cdot x}{0.100 - x} = \frac{(6 \times 10^{-6})(6 \times 10^{-6})}{0.100 - (6 \times 10^{-6})}$$

$$= \frac{(6 \times 10^{-6})^2}{1.00 \times 10^{-1}} = 36 \times 10^{-11} = 4 \times 10^{-10}$$

SIMILAR EXAMPLES: Exercises 20, 22.

Example 17-6   A saturated aqueous solution of the weak base aniline contains 36.0 g $C_6H_5NH_2/L$. What is the pH of this solution?

$$C_6H_5NH_2 + H_2O \rightleftharpoons C_6H_5NH_3^+ + OH^- \qquad K_b = 4.30 \times 10^{-10}$$

Assume aniline is soul contributer of OH⁻

Even though it is a weak base, let us assume that the aniline produces practically all of the OH^- in solution. This is equivalent to saying that aniline is a stronger base than is H_2O. As in Example 17-5, the relevant data are summarized below the ionization equilibrium equation. A preliminary calculation is required in this case, though: the *molar* concentration of aniline in a saturated aqueous solution.

$$[C_6H_5NH_2] = \frac{36.0 \text{ g } C_6H_5NH_2}{L} \times \frac{1 \text{ mol } C_6H_5NH_2}{93.1 \text{ g } C_6H_5NH_2} = 0.387 \, M$$

$$C_6H_5NH_2 \; + \; H_2O \; \rightleftharpoons \; C_6H_5NH_3^+ \; + \; OH^-$$

placed in solution:	$0.387 \, M$	—	—
changes:	$-x \, M$	$+x \, M$	$+x \, M$
at equilibrium:	$(0.387 - x) \, M$	$x \, M$	$x \, M$

$$K_b = \frac{[C_6H_5NH_3^+][OH^-]}{[C_6H_5NH_2]} = \frac{x \cdot x}{0.387 - x} = 4.30 \times 10^{-10}$$

To solve this equation, let us assume that x is very small, that is, $x \ll 0.387$.

that is, assume very weak acid, little ionization, & ∴ little change in activity[ ] (x)

Our assumption proves to be valid: $1.29 \times 10^{-5} = 0.0000129 \ll 0.387$.

MAKE!! SURE!! answer can prove to be very off! otherwise.

$$\frac{x^2}{0.387} = 4.30 \times 10^{-10} \qquad x^2 = 1.66 \times 10^{-10} \qquad [OH^-] = x = 1.29 \times 10^{-5} \, M$$

$$pOH = -\log [OH^-] = -(\log 1.29 \times 10^{-5}) = 4.89$$

REALIZE WHAT "X" STANDS FOR.

$$pH = 14.00 - pOH = 14.00 - 4.89 = 9.11$$

SIMILAR EXAMPLES: Exercises 21, 24.

Example 17-7   What is $[H_3O^+]$ in $0.00250 \, M \; HNO_2(aq)$?

The initial steps to describe the ionization equilibrium are the same here as they were for HCN in Example 17-5 and $C_6H_5NH_2$ in Example 17-6.

$$HNO_2 \quad + \; H_2O \; \rightleftharpoons \; H_3O^+ + NO_2^- \qquad K_a = 5.13 \times 10^{-4}$$

placed in solution:	$0.00250 \, M$	—	—
changes:	$-x \, M$	$+x \, M$	$+x \, M$
at equilibrium:	$(0.00250 - x) \, M$	$x \, M$	$x \, M$

$$K_a = \frac{[H_3O^+][NO_2^-]}{[HNO_2]} = \frac{x \cdot x}{0.00250 - x} = 5.13 \times 10^{-4} \tag{17.23}$$

Let us make the usual assumption, that is, $x \ll 0.00250$ and $(0.00250 - x) \simeq 0.00250$.

$$\frac{x^2}{0.00250} = 5.13 \times 10^{-4} \qquad x^2 = 1.28 \times 10^{-6} \qquad [H_3O^+] = x = 1.13 \times 10^{-3} \, M$$

The value of x is about 40% as large as 0.00250. This is much too large to neglect. That is, if $x = 1.13 \times 10^{-3}$, then $0.00250 - x = 0.00250 - 0.00113 = 0.00137 \not\simeq 0.00250$.

Since our assumption failed, we must return to equation (17.23) and seek an exact solution. This means solving a quadratic equation.

Can see Good chance an assumption like this will fail; look how small [ ] init is.

$$\frac{x^2}{0.00250 - x} = 5.13 \times 10^{-4}$$

$$x^2 + 5.13 \times 10^{-4} \, x - 1.28 \times 10^{-6} = 0$$

−(4)(−1.28 ×10⁻⁶)

$$x = \frac{-5.13 \times 10^{-4} \pm \sqrt{(5.13 \times 10^{-4})^2 + 4 \times 1.28 \times 10^{-6}}}{2}$$

$$= \frac{-5.13 \times 10^{-4} \pm 2.32 \times 10^{-3}}{2}$$

use appropriate sign (+ or −) so that [ ] turns out to be positive #.

$$x = [H_3O^+] = 9.04 \times 10^{-4} \, M$$

SIMILAR EXAMPLES: Exercises 27, 28.

FIGURE 17-4
Percent ionization of an acid as a function of concentration.

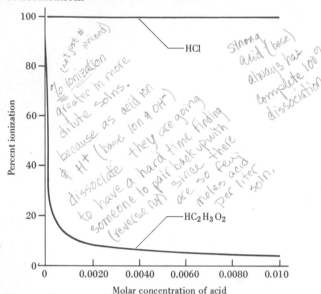

Over the concentration range shown, HCl(aq) is considered to be completely ionized—100% ionized. The percent ionization of $HC_2H_3O_2$(aq) increases from about 4% in 0.010 M $HC_2H_3O_2$ to essentially 100% when the solution becomes extremely dilute.

[handwritten notes]
$K_a/K_b = \dfrac{[ions]}{[acid/base]}$

so if K_a/K_b is large #, means large [ ] activity of ions, good dissociation ∴ can't make assumption "x" is small

if WEAK acid/base [ ] is large, its slight dissociation will be insignificant in comparison

[handwritten bottom left]
% ionization = $\dfrac{[x]}{total [ ]}$ * 100

= $\dfrac{\frac{mol}{L} H^+/OH^-}{M \text{ of Soln}}$ * 100 = ___ %

The simplifying assumption in Examples 17-5 and 17-6, which failed in Example 17-7, was this: treat the weak acid or weak base as if it is essentially nonionized. When can such an assumption be made? The best answer, of course, is to attempt the assumption, test its validity, and proceed accordingly. The following generalization is also useful.

Value to consider	The assumption that most of a weak acid or base remains essentially nonionized usually	
	Works if value is	Does not work if value is
K_a or K_b	less than about 1×10^{-5}	greater than about 1×10^{-5}
total concentration of weak acid or weak base	greater than about 0.01 M	less than about 0.01 M

In Example 17-7, the value of K_a for HNO_2 is seen to be greater than 1×10^{-5}, and the concentration of HNO_2 placed in solution, less than 0.01 M. The usual simplifying assumption failed.

DEGREE OF IONIZATION. In a similar manner to the percent dissociation of gases discussed in Chapter 14 (recall Example 14-10), we can establish a percent ionization of a weak acid or base. In Example 17-8 we calculate the percent ionization of acetic acid at three different concentrations and reach this conclusion: *The degree of ionization (percent ionization) of a weak electrolyte increases as the solution becomes more dilute.* This conclusion is pictured and contrasted to the behavior of a strong acid in Figure 17-4.

Example 17-8 Calculate the percent ionization of acetic acid in **(a)** 1.0 M; **(b)** 0.10 M; **(c)** 0.010 M $HC_2H_3O_2$(aq).

$$HC_2H_3O_2 + H_2O \rightleftharpoons H_3O^+ + C_2H_3O_2^-$$
$$K_a = 1.74 \times 10^{-5}$$

In each case let us make the assumption that the weak acid is essentially nonionized, so that we can write $[HC_2H_3O_2] = 1.0\ M$ in (a), 0.10 M in (b), and 0.010 M in (c).

(a) $K_a = \dfrac{[H_3O^+][C_2H_3O_2^-]}{[HC_2H_3O_2]} = \dfrac{x \cdot x}{1.0} = 1.74 \times 10^{-5}$

$x^2 = 1.74 \times 10^{-5}$ $x = [H_3O^+] = 4.2 \times 10^{-3}\ M$

% ionization of $1.0\ M\ HC_2H_3O_2$ = $\dfrac{4.2 \times 10^{-3}\ \text{mol}\ H_3O^+/L}{1.0\ \text{mol}\ HC_2H_3O_2/L} \times 100$
$= 0.42\%$

In this expression $[HC_2H_3O_2] = 1.0 - x \simeq 1.0\ M$ because x is assumed to be much smaller than 1.0.

To complete the calculations of parts (b) and (c), substitute the appropriate value of $[HC_2H_3O_2]$. Solve for $[H_3O^+]$ and then determine the percent ionization.

(b) $0.10\ M\ HC_2H_3O_2(aq)$ is 1.3% ionized.

(c) $0.010\ M\ HC_2H_3O_2(aq)$ is 4.2% ionized.

SIMILAR EXAMPLES: Exercises 29, 30.

17-6 Polyprotic Acids

All of the acids listed in Table 17-2 are weak *monoprotic* acids. They produce only one proton per acid molecule, even if at times there is more than one hydrogen atom in the molecule. For example, in acetic acid only the hydrogen atom shown in color is ionizable.

$$H-\underset{\underset{H}{|}}{\overset{\overset{H}{|}}{C}}-\overset{\overset{O}{\|}}{C}-O-H + H_2O \rightleftharpoons H_3O^+ + \left(H-\underset{\underset{H}{|}}{\overset{\overset{H}{|}}{C}}-\overset{\overset{O}{\|}}{C}-O\right)^- \qquad (17.24)$$

There are some acids, however, that contain *more than one* ionizable hydrogen atom per molecule. These are called **polyprotic acids.** Ionization constants for several polyprotic acids are listed in Table 17-3. A polyprotic acid important to the qualitative analysis scheme is hydrogen sulfide (hydrosulfuric acid), H_2S. Its ionization occurs in two steps.

first ionization step: $H_2S + H_2O \rightleftharpoons H_3O^+ + HS^-$ (17.25)

second ionization step: $HS^- + H_2O \rightleftharpoons H_3O^+ + S^{2-}$ (17.26)

An ionization constant expression can be written for each step in the usual manner.

$$K_{a_1} = \frac{[H_3O^+][HS^-]}{[H_2S]} = 1.1 \times 10^{-7} \qquad K_{a_2} = \frac{[H_3O^+][S^{2-}]}{[HS^-]} = 1.0 \times 10^{-14} \qquad (17.27)$$

TABLE 17-3
Ionization constants of some common polyprotic acids

Acid	Ionization equilibria	Ionization constants, K
carbonic	$H_2CO_3 + H_2O \rightleftharpoons H_3O^+ + HCO_3^-$	$K_{a_1} = 4.2 \times 10^{-7}$
	$HCO_3^- + H_2O \rightleftharpoons H_3O^+ + CO_3^{2-}$	$K_{a_2} = 5.6 \times 10^{-11}$
hydrosulfuric	$H_2S + H_2O \rightleftharpoons H_3O^+ + HS^-$	$K_{a_1} = 1.1 \times 10^{-7}$
	$HS^- + H_2O \rightleftharpoons H_3O^+ + S^{2-}$	$K_{a_2} = 1.0 \times 10^{-14}$
oxalic	$H_2C_2O_4 + H_2O \rightleftharpoons H_3O^+ + HC_2O_4^-$	$K_{a_1} = 5.4 \times 10^{-2}$
	$HC_2O_4^- + H_2O \rightleftharpoons H_3O^+ + C_2O_4^{2-}$	$K_{a_2} = 5.4 \times 10^{-5}$
phosphoric	$H_3PO_4 + H_2O \rightleftharpoons H_3O^+ + H_2PO_4^-$	$K_{a_1} = 5.9 \times 10^{-3}$
	$H_2PO_4^- + H_2O \rightleftharpoons H_3O^+ + HPO_4^{2-}$	$K_{a_2} = 6.2 \times 10^{-8}$
	$HPO_4^{2-} + H_2O \rightleftharpoons H_3O^+ + PO_4^{3-}$	$K_{a_3} = 4.8 \times 10^{-13}$
phosphorous	$H_3PO_3 + H_2O \rightleftharpoons H_3O^+ + H_2PO_3^-$	$K_{a_1} = 5.0 \times 10^{-2}$
	$H_2PO_3^- + H_2O \rightleftharpoons H_3O^+ + HPO_3^{2-}$	$K_{a_2} = 2.5 \times 10^{-7}$
sulfurous	$H_2SO_3 + H_2O \rightleftharpoons H_3O^+ + HSO_3^-$	$K_{a_1} = 1.3 \times 10^{-2}$
	$HSO_3^- + H_2O \rightleftharpoons H_3O^+ + SO_3^{2-}$	$K_{a_2} = 6.3 \times 10^{-8}$

(just as 2nd Ionization energy is higher)

harder to pull H⁺ from already negatively charged acid (too much attraction)

∴ more negatively charged the acid, the weaker it is.

One generalization about polyprotic acids is that the second ionization constant, K_{a_2}, is always *smaller* than the first, K_{a_1}. This is a reasonable finding: We would expect greater difficulty in separating a proton from a double negatively charged ion, S^{2-} (the second step) than from a single negatively charged ion, HS^- (the first step). These additional observations about the H_2S equilibria are generally applicable to polyprotic acids (with the qualification noted in item 3).

1. All species involved in the ionization equilibria exist together in a single solution phase. For any given equilibrium condition, each concentration—$[H_2S]$, $[H_3O^+]$, $[HS^-]$, $[S^{2-}]$—has a fixed value.

2. The concentrations of all species present in the solution must be consistent with both ionization constant expressions, K_{a_1} and K_{a_2}.

3. Because HS^- is much weaker an acid than H_2S, H_3O^+ is produced almost exclusively in the *first* ionization step. [This assumption is valid only if $K_{a_1} \gg K_{a_2}$ (say by a factor of 10^3 or 10^4). Where the difference between K_{a_1} and K_{a_2} is not so great, the second ionization may also be significant as a source of H_3O^+.]

4. Even though further ionization in the second step occurs to a very limited extent, this ionization is the *only* source of S^{2-}.

Example 17-9   A saturated aqueous solution of $H_2S(g)$ at $25°C$ and 1 atm is $0.10 M$ H_2S. Calculate $[H_2S]$, $[H_3O^+]$, $[HS^-]$, and $[S^{2-}]$ in this solution.

To describe initial and equilibrium concentrations in the usual manner, we need to write *two* expressions, one for each ionization step. The number of moles per liter of H_2S ionized in the first step is represented by x. The number of moles per liter of HS^- that undergo further ionization in the second step is represented by y. The final equilibrium concentrations are shown in color.

$$H_2S \; + \; H_2O \; \rightleftharpoons \; H_3O^+ \; + \; HS^-$$

placed in solution:	$0.10\,M$	—	—
changes:	$-x\,M$	$+x\,M$	$+x\,M$
after first ionization:	$(0.10 - x)\,M$	$x\,M$	$x\,M$

$$HS^- \; + \; H_2O \; \rightleftharpoons \; H_3O^+ \; + \; S^{2-}$$

from first ionization:	$x\,M$	$x\,M$	—
changes:	$-y\,M$	$+y\,M$	$+y\,M$
after second ionization:	$(x - y)\,M$	$(x + y)\,M$	$y\,M$

Equilibrium concentrations

$$[H_2S] = 0.10 - x \qquad [H_3O^+] = x + y \qquad [HS^-] = x - y \qquad [S^{2-}] = y$$

Equilibrium constant expressions

$$K_{a_1} = \frac{[H_3O^+][HS^-]}{[H_2S]} = \frac{(x + y)(x - y)}{0.10 - x} = 1.1 \times 10^{-7} \tag{17.28}$$

$$K_{a_2} = \frac{[H_3O^+][S^{2-}]}{[HS^-]} = \frac{(x + y)y}{x - y} = 1.0 \times 10^{-14} \tag{17.29}$$

To calculate the equilibrium concentration of each species is not nearly so difficult as first appearances might suggest. The key to a simple solution is in Statement 3 above. The concentration of H_3O^+ produced in the second ionization, y, is very much smaller than that produced in the first ionization, x. That is, $y \ll x$. This means that $(x + y) \simeq (x - y) \simeq x$. Furthermore, because H_2S is quite a weak acid, most of it remains nonionized. This means that $x \ll 0.10$ and $(0.10 - x) \simeq 0.10$. With these assumptions we obtain the much simpler algebraic expressions:

$$K_{a_1} = \frac{x \cdot x}{0.10} = 1.1 \times 10^{-7} \quad and \quad K_{a_2} = \frac{x \cdot y}{x} = 1.0 \times 10^{-14}$$

$$x^2 = 1.1 \times 10^{-8}$$

$$x = 1.0 \times 10^{-4} \qquad\qquad y = 1.0 \times 10^{-14}$$

Equilibrium concentrations

(handwritten left margin: $0.10 - x$ but assumed "x" so small that [ ] remains unchanged)

$$[H_2S] = 0.10\ M \qquad [H_3O^+] = 1.0 \times 10^{-4}\ M$$
$$[HS^-] = 1.0 \times 10^{-4}\ M \qquad [S^{2-}] = 1.0 \times 10^{-14}\ M$$

(handwritten right margin: these [ ] affected by "y" but assumed y ≪ x that significantly their [ ] is really only "x")

SIMILAR EXAMPLES: Exercises 33, 34.

As long as Statement 3 of page 424 is valid, the general ideas illustrated by Example 17-9 are that

1. $[H_3O^+]$ in a polyprotic acid solution can be calculated through the K_{a_1} expression alone.

2. The molar concentration of the anion produced in the second ionization step of a diprotic acid (e.g., $[S^{2-}]$) is numerically equal to K_{a_2}.

Where the assumption is not valid, calculation of the concentrations of species in solution is more difficult, because it requires the simultaneous solution of two or more algebraic equations. Additional aspects of equilibria involving polyprotic acids are considered in Chapter 18.

(handwritten: STOP! / PROG TEST UP TO ABOVE LINE)

17-7 Cations and Anions as Acids and Bases—Hydrolysis

The acids and bases chosen for discussion in Section 17-5 were all neutral molecular species. However, proton donors and acceptors are not limited to neutral molecules. Ions can sometimes act in these capacities as well. For example, the second and subsequent ionization steps of a polyprotic acid must involve an anion acting as an acid, as in the ionization

$$HS^- + H_2O \rightleftharpoons H_3O^+ + S^{2-} \tag{17.26}$$

Each of the following is also an acid-base reaction. Let us think about how they might be described.

$$NH_4^+ + H_2O \rightleftharpoons NH_3 + H_3O^+ \tag{17.30}$$

$$C_2H_3O_2^- + H_2O \rightleftharpoons HC_2H_3O_2 + OH^- \tag{17.31}$$

Reaction (17.30) suggests that NH_4^+ is an *acid*, able to donate a proton to water, a *base*. The equilibrium constant for this reaction could be referred to as the acid ionization constant, K_a, of the ammonium ion, NH_4^+. Reaction (17.31) shows $C_2H_3O_2^-$ acting as a *base* by accepting a proton from water, an *acid*. Here the applicable equilibrium constant would be a base ionization constant, K_b, for the acetate ion, $C_2H_3O_2^-$. Some tabulations of ionization constants do include K_a for NH_4^+ and K_b for $C_2H_3O_2^-$, but many do not. However, these quantities are related to other equilibrium constants, as we shall soon see.

HYDROLYSIS. In pure water at 25°C, $[H_3O^+] = [OH^-] = 1.0 \times 10^{-7}\ M$. When a salt such as NaCl is added to water, complete dissociation into Na^+ and Cl^- ions occurs, but these ions do not influence the ionization of water. The pH of the solution remains at 7.

$$Na^+ + Cl^- + H_2O \longrightarrow \text{no reaction}$$

When ammonium chloride is added to water, the pH falls below 7. This means that $[H_3O^+]$ in the solution increases and $[OH^-]$ decreases. A reaction producing H_3O^+ must occur between the added ions and water molecules. Chloride ion cannot act as an acid, and it is too weak a base to accept a proton from H_2O. However, as we have already seen, a reaction does occur between ammonium ion and water.

$$Cl^- + H_2O \longrightarrow \text{no reaction}$$

$$NH_4^+ + H_2O \rightleftharpoons NH_3 + H_3O^+ \tag{17.30}$$

A reaction between water and an ionic species in solution is called a **hydrolysis** reaction. The ammonium ion undergoes hydrolysis and chloride ion does not.

When sodium acetate is added to water, the pH rises above 7. This means that $[OH^-]$ in the solution increases and $[H_3O^+]$ decreases. Sodium ion has neither acidic nor basic properties, but acetate ion undergoes hydrolysis.

$$Na^+ + H_2O \rightleftharpoons \text{no reaction}$$

$$C_2H_3O_2^- + H_2O \rightleftharpoons HC_2H_3O_2 + OH^- \tag{17.31}$$

The equilibrium constant for reaction (17.31) can be called the **hydrolysis constant** and denoted by the symbol K_h. Its form is

$$K_h = \frac{[HC_2H_3O_2][OH^-]}{[C_2H_3O_2^-]} \tag{17.32}$$

Suppose that we think of the hydrolysis of acetate ion as the net reaction resulting from the sum of two other reactions. Then K_h may be derived from two more familiar equilibrium constants. These statements are simply applications to a hydrolysis reaction of the general remarks about equilibrium constants made in Section 14-2. Note that the first of the two reactions below is the *reverse* of the ionization of acetic acid.

$$
\begin{array}{ll}
C_2H_3O_2^- + H_3O^+ \rightleftharpoons H_2O + HC_2H_3O_2 & K = 1/K_a(\text{acetic acid}) \\
2\,H_2O \rightleftharpoons H_3O^+ + OH^- & K = K_w \\
\hline
C_2H_3O_2^- + H_2O \rightleftharpoons HC_2H_3O_2 + OH^- & K_h = K_w/K_a
\end{array}
$$

For the hydrolysis of acetate ion

$$K_h = \frac{[HC_2H_3O_2][OH^-]}{[C_2H_3O_2^-]} = \underbrace{\frac{[HC_2H_3O_2]}{[H_3O^+][C_2H_3O_2^-]}}_{1/K_a} \times \underbrace{[H_3O^+][OH^-]}_{K_w}$$

$$= \frac{1.0 \times 10^{-14}}{1.74 \times 10^{-5}} = 5.7 \times 10^{-10}$$

For the hydrolysis of ammonium ion

$$K_h = \frac{[H_3O^+][NH_3]}{[NH_4^+]} = \frac{K_w}{K_b} = \frac{1.0 \times 10^{-14}}{1.74 \times 10^{-5}} = 5.7 \times 10^{-10}$$

where K_b is the ionization constant of NH_3.

That the values of K_h are identical for acetate ion and ammonium ion is just a matter of chance. It happens because K_a for acetic acid and K_b for ammonia are equal numerically.

With numerical values of hydrolysis constants, it is possible to assess the extent to

[handwritten top margin: REASONS WHY MANY SALT SOLN'S AREN'T NEUTRAL:]

which a hydrolysis reaction occurs. However, a number of qualitative statements about hydrolysis can also prove helpful.

[handwritten: NaOH HCl]

- In general, salts are completely dissociated into ions in aqueous solutions.
- Salts of strong acids and strong bases (such as NaCl) do not undergo hydrolysis: pH = 7.

[handwritten: NaOH HC₂H₃O₂]

- Salts of *weak* acids and strong bases (such as $NaC_2H_3O_2$) undergo hydrolysis, producing a <u>basic</u> solution: pH > 7. It is the *anion* in such a salt which hydrolyzes, with $K_h = K_w/K_a$.

[handwritten: $H_2O + C_2H_3O_2^- \rightleftharpoons HC_2H_3O_2 + OH^-$]

- Salts of strong acids and *weak* bases (such as NH_4Cl) undergo hydrolysis to produce an acidic solution: pH < 7. In such a salt it is the *cation* that hydrolyzes, and $K_h = K_w/K_b$.

[handwritten: NH_3 HCl → $NH_4^+ + H_2O \rightleftharpoons NH_3 + H^+$]

- Salts of *weak* acids and *weak* bases (such as $NH_4C_2H_3O_2$) hydrolyze, but whether the resulting solution is neutral, acidic, or basic depends on the relative values of K_a and K_b.

[handwritten: whichever has smaller Ka & Kb, thus larger K_H, is one which will take soln priority]

Example 17-10 Predict whether you would expect each of the following solutions to be acidic, basic, or neutral: **(a)** $NaNO_2(aq)$; **(b)** $KCl(aq)$; **(c)** $NH_4CN(aq)$; **(d)** $NaHCO_3(aq)$.

(a) The ions in solution are Na^+, which does not hydrolyze, and NO_2^-, which does: $NO_2^- + H_2O \rightleftharpoons HNO_2 + OH^-$. Because the hydrolysis reaction produces OH^-, the solution is basic.

(b) Neither K^+ nor Cl^- undergoes hydrolysis. $KCl(aq)$ is neutral—pH = 7.

(c) *Both* NH_4^+ and CN^- hydrolyze in aqueous solution, one to produce H_3O^+ and the other, OH^-.

$$NH_4^+ + H_2O \rightleftharpoons NH_3 + H_3O^+ \qquad K_h = \frac{K_w}{K_b} = \frac{1.0 \times 10^{-14}}{1.74 \times 10^{-5}} = 5.7 \times 10^{-10}$$

$$CN^- + H_2O \rightleftharpoons HCN + OH^- \qquad K_h = \frac{K_w}{K_a} = \frac{1.0 \times 10^{-14}}{4 \times 10^{-10}} = 2.5 \times 10^{-5}$$

Because of the larger value of its K_h, we should expect CN^- to hydrolyze to a greater extent than NH_4^+. This means that $[OH^-] > [H_3O^+]$ and the solution will be basic.

(d) In this instance HCO_3^- can undergo further ionization as an *acid* or hydrolysis (acting as a *base*). We need to compare equilibrium constants for these two possibilities.

$$HCO_3^- + H_2O \rightleftharpoons H_3O^+ + CO_3^{2-} \qquad K_{a_2} = 5.6 \times 10^{-11}$$

$$HCO_3^- + H_2O \rightleftharpoons H_2CO_3 + OH^- \qquad K_h = \frac{K_w}{K_{a_1}} = \frac{1.0 \times 10^{-14}}{4.2 \times 10^{-7}} = 2.4 \times 10^{-8}$$

The hydrolysis constant K_h is larger than the second ionization constant, K_{a_2}. We expect $[OH^-] > [H_3O^+]$ and the solution to be basic

SIMILAR EXAMPLES: Exercises 38, 39.

Example 17-11 What is the pH of an 0.50 *M* NaCN solution?
The concentrations of the several species involved in the hydrolysis reaction are summarized below, where $x = [OH^-]$.

$$CN^- + H_2O \rightleftharpoons HCN + OH^-$$

placed in solution:	0.50 *M*	—	—
changes:	−*x M*	+*x M*	+*x M*
equilibrium concentrations:	(0.50 − *x*) *M*	*x M*	*x M*

[handwritten left margin notes:]
a salt is what forms when acid neutralizes base.

the strong ones dissociate. the weak ones achieve some equil. w/ H₂O

For ex. the salt Na·C₂H₃O₂. is formed from neutralization of NaOH & HC₂H₃O₂.

NaOH is STRONG base so its conjugate Na⁺ will be weak (& thus can't act as good)

HC₂H₃O₂ is WEAK acid so its conjugate C₂H₃O₂⁻ must be strong *thus can pull H⁺ from water. Good base.

[handwritten bottom right: as a weak acid it will reach equil w/ H₂O taking up some of the H⁺ & resulting in OH's (BASIC)]

[handwritten bottom left:]
$CN^- + H^+ \rightleftharpoons HCN \quad \frac{1}{K_a}$
$2H_2O \rightleftharpoons H^+ + OH^- \quad K_w$
$$\frac{[H^+][OH^-]}{[H^+][CN^-]} / [HCN] = \frac{[OH^-][HCN]}{[CN^-]}$$

[handwritten: polyprotic]

These concentrations may now be substituted into the hydrolysis equilibrium expression, where the hydrolysis constant is written as $K_h = K_w/K_a$.

$$K_h = \frac{[OH^-][HCN]}{[CN^-]} = \frac{K_w}{K_a} = \frac{1.00 \times 10^{-14}}{4.0 \times 10^{-10}} = 2.5 \times 10^{-5}$$

$$\frac{x \cdot x}{0.50 - x} = \frac{x^2}{0.50 - x} = 2.5 \times 10^{-5}$$

Next a familiar assumption can be made. If $x \ll 0.50$, then $0.50 - x \simeq 0.50$.

$$x^2 = (0.50)(2.5 \times 10^{-5}) = 1.2 \times 10^{-5} = 12 \times 10^{-6}$$

$$x = [OH^-] = (12 \times 10^{-6})^{1/2} = 3.5 \times 10^{-3}$$

$$[H_3O^+] = \frac{K_w}{[OH^-]} = \frac{1.00 \times 10^{-14}}{3.5 \times 10^{-3}} = 2.9 \times 10^{-12}$$

$$pH = -\log[H_3O^+] = -\log(2.9 \times 10^{-12}) = 11.5$$

SIMILAR EXAMPLES: Exercises 40, 41.

17-8 Molecular Structure and Acid Strength

For the quantitative descriptions of aqueous solutions that we have been emphasizing in this chapter, we have had to consider two questions primarily: Do the ionization processes in solution go to completion or do they reach a state of equilibrium? If an equilibrium condition is involved, what are the numerical values of the relevant equilibrium constants? If we had answers to these questions, we generally were able to calculate any concentrations of interest to us.

There are numerous instances, however, when we seek only some qualitative answers: Why is HClO a weak acid whereas $HClO_4$ is a strong acid? Why is ethanol, C_2H_5OH, so much weaker an acid than phenol, C_6H_5OH? The answers to questions of this type come through relating acid strength (or base strength) to molecular structure.

STRENGTHS OF BRØNSTED–LOWRY ACIDS AND BASES.   Before attempting to answer questions like the two just posed, we need to say a bit more about acid and base strength in relation to the Brønsted–Lowry theory. If we study the following two reactions, we find that the first goes essentially to completion and the other proceeds hardly at all in the forward direction.

$$HClO_4 \;+\; OH^- \;\rightleftharpoons\; H_2O \;+\; ClO_4^-$$

acid(1)	base(2)	acid(2)	base(1)
very strong	strong	weak	very weak

$$HCO_3^- \;+\; Br^- \;\rightleftharpoons\; HBr \;+\; CO_3^{2-}$$

acid(1)	base(2)	acid(2)	base(1)
weak	very weak	very strong	strong

Each reaction illustrates a statement first made on page 413: *A Brønsted–Lowry acid-base reaction is always favored in the direction of the stronger to the weaker acid/base combination.* This statement in turn suggests the need for a listing of Brønsted–Lowry acids and bases according to their relative strengths. Such a list is provided in Table 17-4. A point to note in Table 17-4 is that the stronger an acid, the weaker its conjugate base.

The relative placement of some of the entries in Table 17-4 follows directly from what we have learned elsewhere in this chapter. For example, acetic acid ($K_a = 1.74 \times 10^{-5}$) is stronger than carbonic acid ($K_{a_1} = 4.2 \times 10^{-7}$); and carbonic

TABLE 17-4
Relative strengths of some common Brønsted–Lowry acids and bases

	Acid		Conjugate base	
	perchloric acid	$HClO_4$	perchlorate ion	ClO_4^-
	hydroiodic acid	HI	iodide ion	I^-
	hydrobromic acid	HBr	bromide ion	Br^-
	hydrochloric acid	HCl	chloride ion	Cl^-
	nitric acid	HNO_3	nitrate ion	NO_3^-
	sulfuric acid	H_2SO_4	hydrogen sulfate ion	HSO_4^-
	hydronium ion[a]	H_3O^+	water[a]	H_2O
	hydrogen sulfate ion	HSO_4^-	sulfate ion	SO_4^{2-}
	nitrous acid	HNO_2	nitrite ion	NO_2^-
	acetic acid	$HC_2H_3O_2$	acetate ion	$C_2H_3O_2^-$
	carbonic acid	H_2CO_3	bicarbonate ion	HCO_3^-
	ammonium ion	NH_4^+	ammonia	NH_3
	bicarbonate ion	HCO_3^-	carbonate ion	CO_3^{2-}
	water	H_2O	hydroxide ion	OH^-
	methanol	CH_3OH	methoxide ion	CH_3O^-
	ammonia	NH_3	amide ion	NH_2^-

(left margin: increasing acid strength ↑) (right margin: increasing base strength ↓)

[a] The hydronium ion/water combination refers to the ease with which a proton is passed from one water molecule to another; that is, $H_3O^+ + H_2O \rightleftharpoons H_3O^+ + H_2O$.

acid is a stronger acid than its anion, HCO_3^- ($K_{a_2} = 5.6 \times 10^{-11}$). Why do we rank $HClO_4$ ahead of HCl, however? In water solution both of these acids are so strong that they are completely ionized. Water is said to have a **leveling effect** on these two acids. Water is a strong enough base to accept protons from either acid, blurring whatever differences may exist between them.

To distinguish between the strengths of $HClO_4$ and HCl, we need to use a **differentiating solvent.** This is a solvent that is itself such a weak base that it will accept protons from the stronger of the two acids more readily than from the weaker one. Diethyl ether, $C_2H_5OC_2H_5$, has less affinity for protons than does water. In diethyl ether, $HClO_4$ is still essentially completely ionized but HCl is only partially ionized. In considering the relative strengths of acids and bases, then, the solvent chosen for these comparisons plays an important role. Our attention, nevertheless, will continue to be focused on aqueous solutions.

BOND STRENGTH AND ACID STRENGTH.   The strength of an acid or base must relate ultimately to the ease with which a proton is lost or gained.

It is an oversimplification to relate the acidities of the binary acids HX to the mere breaking of the bond H—X. For one thing, bond energies are given for the dissociation of gaseous species, and here we are dealing with species in solution. Nevertheless, it would seem that the stronger the H—X bond, the weaker the acid. This generalization does hold true for the series of acids, HF, HCl, HBr, HI, for which the bond energies increase in the order

$$HI < HBr < HCl < HF$$

whereas the acid strengths decrease in the order

$$HI > HBr > HCl > HF$$

strong   weak

The fact that HF is a weak acid whereas the other hydrogen halides are very strong has always seemed an anomaly. One recent proposal involves ion pairs held together by strong hydrogen bonds. This keeps $[H_3O^+]$ from being as large as otherwise expected.

$$HF + H_2O \longrightarrow$$
$$(^-F\cdots H_3O^+) \rightleftharpoons$$
$$H_3O^+ + F^-$$

they have "oxygen bridges"

FIGURE 17-5
Formation and acidic behavior of an oxyacid.

$$E_xO_y + nH_2O \longrightarrow --E-O \{ H$$

bond breakage for acidic behavior

one free oxy pulling the e's

other groups—
either OH or O
(occasionally H)

$H_5IO_6 \sim HIO_2$

(handwritten structures)
H—O—I—O—H with O above and O—H, and H below

O—I—O—H

one free oxy doing the e⁻ pulling

electronegativity

An alternative view leading to the same ordering for these acids is that, as the anion decreases in size, its affinity for a proton increases. The anion is the conjugate base of the acid, and as its base strength increases, the acid strength decreases.

STRENGTHS OF OXYACIDS. With the exception of a few binary acids such as the hydrogen halides and H_2S, in most acids the proton that is lost is attached to an oxygen atom. In general a nonmetal oxide, E_xO_y, will undergo a reaction with water to produce an acidic solution. The oxide is called an **acid anhydride** (from the Greek, "without water"). The acid is an **oxyacid.** It has at least one E—O—H bond and other bonds that are either E—O—H or E—O (or occasionally, E—H). The formation and acidic behavior of an oxyacid are pictured in Figure 17-5.

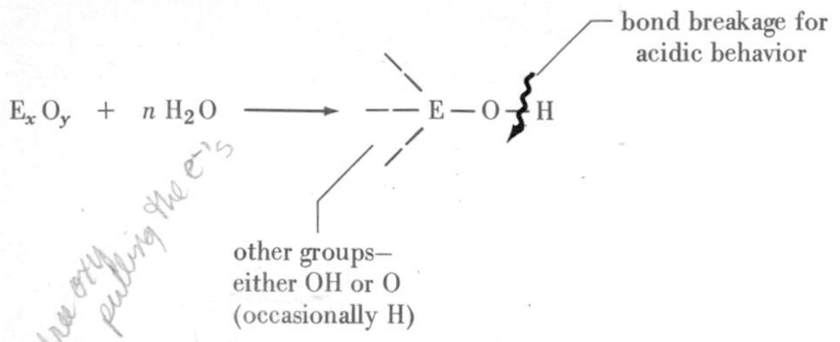

As indicated in Figure 17-5, in order for an oxyacid to display acidic properties, an E—O—H bond must break between the O and H atoms. The more readily this bond is broken, the stronger the oxyacid. What are the factors that will favor this bond breakage? In general, these are any factors that promote the withdrawal of electrons from the O—H bond toward the nonmetal atom, E. These factors include a high electronegativity of the central atom, E, and an increased number of oxygen atoms (not OH groups) bonded directly to the E atom. The fact that HNO_3 is a strong acid whereas H_3PO_4 is weak is consistent with these statements.

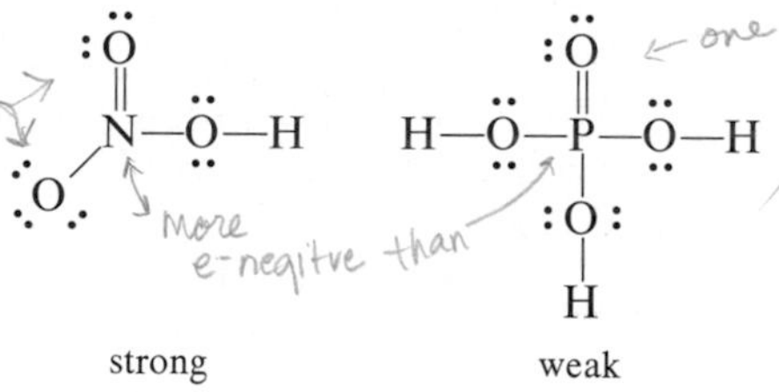

2 oxygens w/drawing electrons

more e-negitve than

one oxygen to w/draw e⁻

$$\text{(17.33)}$$

strong   weak

In the series of oxyacids of chlorine, as the oxidation state and formal charge on the Cl atom increase, this atom becomes more of a "positive charge" center. Electron density is drawn away from the O—H bond, leading to increased acid strength. The variation in acid strengths is from HOCl, quite a weak acid, to $HOClO_3$ (perchloric acid), one of the strongest known.

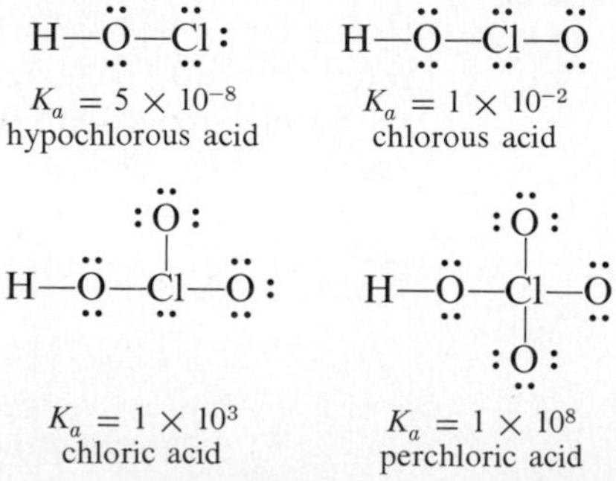

$K_a = 5 \times 10^{-8}$
hypochlorous acid

$K_a = 1 \times 10^{-2}$
chlorous acid

$$\text{(17.34)}$$

$K_a = 1 \times 10^3$
chloric acid

$K_a = 1 \times 10^8$
perchloric acid

The following general rule often can be applied to oxyacids with the formula, $EO_m(OH)_n$ (where E is the central atom): If
$m = 0$, $K_{a_1} \simeq 10^{-7}$;
$m = 1$, $K_{a_1} \simeq 10^{-2}$;
$m = 2$, K_{a_1} is large;
$m = 3$, K_{a_1} is very large.

no free oxy "very weak acid" TO

many free oxy much e⁻ w/drawal from other bonds of OH strong acid

STRENGTHS OF ORGANIC ACIDS.   Ethanol and acetic acid both have an O—H group bonded to a carbon atom, but acetic acid is a much stronger acid than is ethanol.

$$
\begin{array}{cc}
\underset{\text{acetic acid, } pK_a = 4.76}{
\begin{array}{c}
\text{H} \quad \overset{\displaystyle ..}{\underset{\displaystyle }{\text{O}}}: \\
\mid \qquad \parallel \\
\text{H} - \text{C} - \text{C} - \overset{..}{\underset{..}{\text{O}}} - \text{H} \\
\mid \\
\text{H}
\end{array}
}
&
\underset{\text{ethanol, } pK_a = 15.9}{
\begin{array}{c}
\text{H} \quad \text{H} \\
\mid \qquad \mid \\
\text{H} - \text{C} - \text{C} - \overset{..}{\underset{..}{\text{O}}} - \text{H} \\
\mid \qquad \mid \\
\text{H} \quad \text{H}
\end{array}
}
\end{array}
\tag{17.35}
$$

We might rationalize that the carbonyl oxygen atom in acetic acid ($>$C=O), being highly electronegative, withdraws electrons from the O—H bond. The result is that the proton can be lost more readily. Another reason for acetic acid being the stronger acid is based on the anions that are formed.

$$
\underset{\text{acetate ion}}{
\begin{array}{c}
\text{H} \quad \overset{..}{\text{O}}\overset{..}{} \\
\mid \qquad \diagup \\
\text{H} - \text{C} - \text{C} \\
\mid \qquad \diagdown \\
\text{H} \quad \overset{..}{\underset{..}{\text{O}}}{}^{-}
\end{array}
}
\longleftrightarrow
\begin{array}{c}
\text{H} \quad \overset{..}{\text{O}}{}^{-} \\
\mid \qquad \diagup \\
\text{H} - \text{C} - \text{C} \\
\mid \qquad \diagdown \\
\text{H} \qquad \text{O}
\end{array}
\qquad
\underset{\text{ethoxide ion}}{
\begin{array}{c}
\text{H} \quad \text{H} \\
\mid \qquad \mid \\
\text{H} - \text{C} - \text{C} - \overset{..}{\underset{..}{\text{O}}}: {}^{-} \\
\mid \qquad \mid \\
\text{H} \quad \text{H}
\end{array}
}
\tag{17.36}
$$

In acetate anion resonance occurs. Two plausible structures can be written in which the double bond is shifted from one O atom to the other. The net effect is that each carbon-to-oxygen bond is a "$\frac{3}{2}$" bond and each O atom carries "$\frac{1}{2}$" negative charge. In short, the excess unit of negative charge is spread out. This reduces the ability of either O atom to attract a proton and makes acetate ion a moderately weak Brønsted–Lowry base. Ethoxide ion, on the other hand, has no resonance possibilities. The unit of negative charge is centered on a single O atom. The ion is a much stronger base than acetate ion. If a conjugate base is strong, then the corresponding acid is weak (recall Table 17-4).

[handwritten margin note: ∴ thus weak base, so strong acid]

The substitution of groups can also have an effect on the strength of an organic acid. Replacement of one H atom by a Cl atom in acetic acid produces mono-chloroacetic acid.

$$
\underset{\text{monochloroacetic acid, } pK_a = 2.87}{
\begin{array}{c}
:\overset{..}{\text{C}}\text{l}: \quad \overset{..}{\text{O}}: \\
\mid \qquad \parallel \\
\text{H} - \text{C} - \text{C} - \overset{..}{\underset{..}{\text{O}}} - \text{H} \\
\mid \\
\text{H}
\end{array}
}
\tag{17.37}
$$

[handwritten margin note: farther away from carboxyl group, less w/drawing effect or weaker acid]

The highly electronegative Cl atom helps to draw electrons away from the O—H bond. The bond is weakened; the proton is lost more readily; and the acid is a stronger acid than acetic acid. Viewed from the standpoint of the anion, the electronegative Cl atom causes an additional spreading out of the unit of negative charge beyond that described in (17.36). Monochloroacetate ion has less attraction for a proton than does acetate ion; it is a weaker base.

The electron-withdrawing power of certain groups in molecules is called the **inductive effect.** The inductive effect is strongest when the substituent group is adja-

cent to the carboxyl group ($-\text{C}\overset{\displaystyle\diagup\text{O}}{\diagdown_{\text{O—H}}}$). The effect falls off rapidly as the substituent

is moved farther away on a hydrocarbon chain. The benzene ring system (phenyl group) exhibits an inductive effect and accounts for the greater acid strength of phenol (17.38) relative to ethanol.

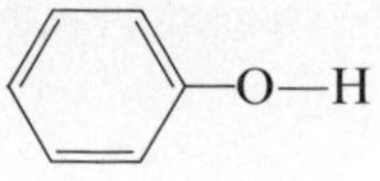

phenol, $pK_a = 10.0$

(17.38)

Electronegative substituents on the phenyl group may also exhibit an inductive effect, causing the acid to become somewhat stronger.

Example 17-12 Which member of each pair of acids is the stronger?

(a) 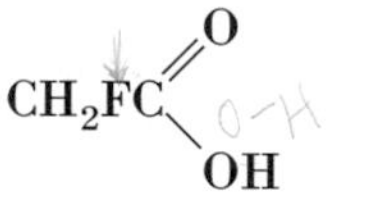 CH₂FC (I) or CH₂ClC (II)

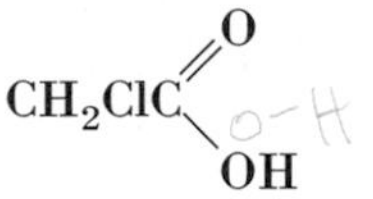

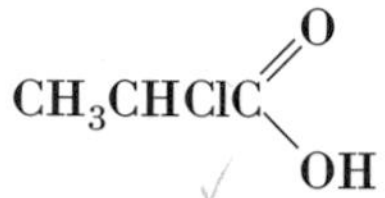

(b) 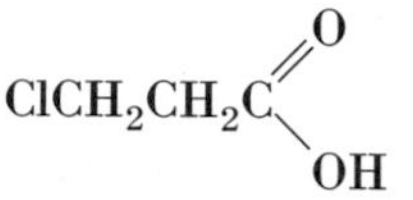 ClCH₂CH₂C (I) or CH₃CHClC (II)

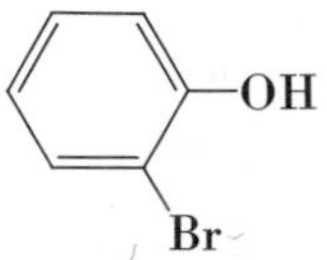

(c) 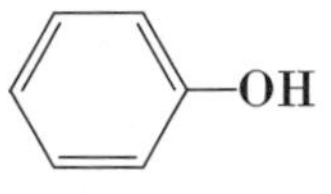 (I) or (II)

(a) Because F is more electronegative than Cl, it should exert a stronger pull on the electrons in the O—H bond than does Cl. Monofluoroacetic acid (I) is a stronger acid than monochloroacetic acid (II).

(b) The location from which the Cl atom exerts the strongest inductive effect is directly adjacent to the carboxyl group. Compound II is more acidic than compound I.

(c) The electronegative Br atom draws electrons away from the O—H group. *o*-Bromophenol (II) is a stronger acid than phenol (I).

SIMILAR EXAMPLES: Exercises 46, 47.

Summary

Much of solution chemistry can be explained by acid-base theories. Most useful is the Brønsted–Lowry theory, which describes acid-base reactions in terms of proton transfer. The tendencies for substances to lose or gain protons can be measured and the relative strengths of acids and bases tabulated. These relative acid and base strengths determine the direction in which an acid-base reaction will occur—from the stronger to the weaker acid/base combination. The Lewis acid-base theory views an acid-base reaction in terms of covalent bond formation between an electron pair acceptor (acid) and an electron pair donor (base). Its greatest use is in situations where reactions occur in the absence of a solvent or in which an acid contains no hydrogen atoms.

In the self-ionization of water a proton is trans-

ferred from one water molecule to another, producing the ions H_3O^+ and OH^-. The equilibrium constant used to describe this self-ionization is the ion product of water, K_w. Constancy of the ion product permits a calculation of $[OH^-]$ from a known value of $[H_3O^+]$, and vice versa. Useful shorthand designations for $[H_3O^+]$ and $[OH^-]$ introduced in this chapter are the symbols pH and pOH. In any aqueous solution at 25°C, pH + pOH = 14.00.

Strong acids ionize completely in aqueous solutions, producing H_3O^+. In an aqueous solution, a strong base ionizes completely, producing OH^-. With weak acids and weak bases ionization of the molecular acid or base does not go to completion. Equilibrium exists between the nonionized acid or base and its ions in solution, and is described through the ionization constant, K_a or K_b. Some weak acids ionize to produce more than one proton per acid molecule, and they do so in a stepwise fashion. That is, polyprotic acid molecules first lose one proton; the remaining acid anions then lose the second proton; and so on. Distinct ionization constants K_{a_1}, K_{a_2}, . . . apply to each step. Calculations involving ionization equilibria are similar to those introduced for gas-phase equilibria in Chapter 14, although some additional considerations are necessary for polyprotic acids.

In a reaction between an ion and water—a hydrolysis reaction—the ion can be treated as a weak acid or weak base and an appropriate K_a or K_b value estab-lished. An alternative approach is to describe hydrolysis of an ion through a hydrolysis constant K_h. This constant is related to K_w of water and the ionization constant of the conjugate acid or base of the ion. Hydrolysis accounts for the fact that many salt solutions are not neutral (pH $\neq$ 7).

Molecular structure establishes whether a substance will have acidic or basic properties. More than this, however, molecular structure affects whether an acid or base is strong or weak. Differences in molecular structure among similar types of compounds often can be used to predict relative values of K_a or K_b.

Learning Objectives

As a result of studying Chapter 17, you should be able to

1. Describe the similarities and differences among the Arrhenius, Brønsted–Lowry, and Lewis theories of acids and bases.

2. Identify Brønsted–Lowry conjugate acids and bases and write equations to represent acid-base reactions.

3. Identify Lewis acids and bases and write equations for acid-base reactions involving them.

4. Describe the nature of the proton in aqueous solution, with special attention to the hydronium ion, H_3O^+.

5. Name the common strong acids and bases.

6. Calculate ionic concentrations in aqueous solutions of strong electrolytes, and relate $[H_3O^+]$ and $[OH^-]$ through K_w.

7. Use the basic relationships among $[H_3O^+]$, $[OH^-]$, pH, and pOH in numerical calculations.

8. Identify a weak acid or weak base, write a chemical equation to represent its ionization, and set up its ionization constant expression.

9. Calculate, from appropriate data, ionization constants, concentrations of nonionized weak acids or bases, or concentrations of their ions in solution.

10. Describe the ionization of a polyprotic acid in aqueous solution and calculate the concentrations of the different species present in such a solution.

11. Predict which ions hydrolyze and whether salt solutions are acidic, basic, or neutral.

12. Write ionic equations to represent hydrolysis, and calculate the pH of salt solutions in which hydrolysis occurs.

13. Predict the direction of acid-base reactions from a knowledge of the relative strengths of Brønsted–Lowry acids and bases.

14. Predict the relative strengths of acids and bases from a knowledge of their molecular structures.

Some New Terms

Arrhenius acid-base theory views an acid as a substance that produces H^+ and a base as a substance that produces OH^- in aqueous solution.

Brønsted–Lowry theory describes acids as proton donors and bases as proton acceptors.

A **conjugate base** is the substance that remains after a Brønsted–Lowry acid has lost a proton. A **conjugate acid** is formed when a Brønsted–Lowry base gains a proton. Every acid has its conjugate base and every base, its conjugate acid.

Degree of ionization (or percent ionization) refers to the extent to which molecules of a weak acid or base have ionized. The degree of ionization increases as the weak electrolyte solution is diluted.

Hydrolysis refers to a reaction between water molecules and an ionic species in solution. Generally, as a result of hydrolysis, the solution pH $\neq$ 7.

A **hydrolysis constant, K_h,** describes a solution equilibrium in which hydrolysis occurs. For the hydrolysis of an anion, $K_h = K_w/K_a$ (where K_a is the ionization constant of the conjugate acid of the anion). For the hydrolysis of a cation, $K_h = K_w/K_b$ (where K_b is the ionization constant of the conjugate base of the cation).

Hydronium ion, H_3O^+, is the principal form in which protons are found in aqueous solution. The terms "hydrogen ion" and "hydronium ion" are often used synonomously.

An **ionization constant** is an equilibrium constant describing the ionization of a weak acid or weak base. The symbol

K_a is used for weak-acid ionizations and K_b for weak-base ionizations.

The **ion product of water**, K_w, is the product of $[H_3O^+]$ and $[OH^-]$ in pure water or an aqueous solution. This product must have a unique value which depends only on temperature. At 25°C, $K_w = 1.0 \times 10^{-14}$.

Lewis acid-base theory considers an acid to be an electron pair acceptor and a base, an electron pair donor. An acid-base reaction consists of the formation of a covalent bond between the acid and the base.

An **oxyacid** is an acid in which the ionizable hydrogen atom(s) is bonded through an oxygen atom to a central nonmetal atom, that is, E—O—H. Other groups bonded to the central atom are either additional —OH groups or O atoms (or occasionally, H atoms).

pH is a shorthand designation for $[H_3O^+]$ in a solution. It is defined as pH $= -\log[H_3O^+]$.

pK is a shorthand designation for an ionization constant; $pK = -\log K$. pK values are useful when comparing the relative strengths of acids or bases.

pOH relates to the $[OH^-]$ in an aqueous solution; pOH $= -\log[OH^-]$.

A **polyprotic acid** is an acid capable of losing more than a single proton in acid-base reactions. The protons are lost in a stepwise fashion, with the first proton being the most readily lost, then the second, and so on.

Self-ionization is an acid-base reaction in which one solvent molecule acts as an acid and donates a proton to another solvent molecule acting as a base.

Exercises

Brønsted–Lowry theory of acids and bases

1. According to the Brønsted–Lowry theory, which of the following would you expect to be acidic and which basic?
(a) HNO_2; (b) ClO^-; (c) NH_2^-; (d) NH_4^+; (e) $CH_3NH_3^+$.

2. What are the conjugate acids of the following bases?
(a) OH^-; (b) Cl^-; (c) ClO^-; (d) CN^-.

3. The following acids are all *monoprotic* (yielding a single proton). Write the formula of the conjugate base of each acid: (a) HIO_4; (b) $HC_3H_5O_2$; (c) C_6H_5COOH; (d) $C_6H_5NH_3^+$.

4. Substances that can either lose or gain protons are said to be *amphiprotic*. Which of the following are amphiprotic?
(a) OH^-; (b) NH_3; (c) H_2O; (d) HS^-; (e) NO_3^-; (f) HCO_3^-; (g) HSO_4^-; (h) HNO_3.

5. For each of the following identify the acids and bases involved in both the forward and reverse reactions.
(a) $HBrO + H_2O \rightleftharpoons H_3O^+ + BrO^-$
(b) $HSO_4^- + H_2O \rightleftharpoons H_3O^+ + SO_4^{2-}$
(c) $C_6H_5NH_2 + H_2O \rightleftharpoons C_6H_5NH_3^+ + OH^-$
(d) $S^{2-} + H_2O \rightleftharpoons HS^- + OH^-$

6. One of the advantages of the Brønsted–Lowry theory is that acid-base reactions can be described in nonaqueous solvents. Indicate whether each of the following would be an acid, a base, or either in pure liquid acetic acid, $HC_2H_3O_2$, as a solvent: (a) $C_2H_3O_2^-$; (b) H_2O; (c) $HC_2H_3O_2$; (d) $HClO_4$. (*Hint:* Think of analogous situations in water or ammonia.)

Lewis theory of acids and bases

7. Indicate whether each of the following is a Lewis acid or base: (a) OH^-; (b) $B(OH)_3$; (c) $AlCl_3$; (d) CH_3NH_2. (*Hint:* You may find it useful to draw Lewis electronic structures.)

8. Each of the following is an acid-base reaction in the Lewis sense. Which is the acid and which is the base? Explain.
(a) $SO_3 + H_2O \longrightarrow H_2SO_4$
(b) $Al(OH)_3(s) + OH^-(aq) \longrightarrow [Al(OH)_4]^-$

9. Show that in each of the following practical cases a Lewis acid-base reaction is involved. Identify the acid and the base. (*Hint:* Draw Lewis electronic structures as an aid to describing these reactions.)
(a) A method of removing sulfur dioxide from the exhaust gases of a power plant is to allow the $SO_2(g)$ to combine with lime, $CaO(s)$. The result is the ionic compound $CaSO_3(s)$.
(b) Carbon dioxide gas can be removed from confined quarters (such as spacecraft) by allowing it to combine with an alkali metal hydroxide, yielding an alkali metal bicarbonate. For example, $CO_2(g) + LiOH(s) \rightarrow LiHCO_3(s)$.

Strong acids, strong bases, and pH

10. What is $[H_3O^+]$ and $[OH^-]$ in each of the following solutions?
(a) 0.0010 M HCl
(b) 0.040 M NaOH
(c) 0.0020 M H_2SO_4
(d) 1.3 $\times$ 10^{-3} M $Ba(OH)_2$

11. What is $[H_3O^+]$ in a solution obtained by dissolving 24.8 cm³ HCl(g), measured at 30°C and 740 mmHg, in 2.50 L of water solution?

12. What is the pH of each of the following solutions?
(a) 1 $\times$ 10^{-4} M HCl
(b) 0.00020 M HI
(c) 3.50 $\times$ 10^{-3} M H_2SO_4
(d) 2.50 $\times$ 10^{-4} M NaOH

13. What is the pOH of each of the following solutions?
(a) 1 $\times$ 10^{-3} M KOH
(b) 0.050 M NaOH

(c) 4.0×10^{-4} M $Ba(OH)_2$

(d) 1.50×10^{-3} M HCl

14. What is the pH of a water solution containing 2.05 g $Ba(OH)_2 \cdot 8 H_2O$ in 450 ml?

15. What is the pH of a solution obtained by mixing 280 ml of 3.00×10^{-3} M H_2SO_4 and 220 ml of 6.40×10^{-2} M HCl?

16. What volume, in milliliters, of concentrated sulfuric acid (98.0% H_2SO_4, by mass; $d = 1.84$ g/ml) is required to produce 4.00 L of solution with pH = 1.65?

17. What $[Mg^{2+}]$ must be present in saturated $Mg(OH)_2(aq)$ so that the solution will have pH = 9.50?

★18. Use the method of Example 17-4 to determine $[OH^-]$ in a saturated aqueous solution of $Al(OH)_3$. Why cannot the answer obtained be correct? How must the calculation be modified to obtain a correct answer?

★19. Can a solution of pH 8 be prepared by dissolving HCl in water? If it is possible to do so, indicate how. If it is not possible, say why not.

Weak acids, weak bases, and pH (Use data from Table 17-2 as necessary.)

20. Normal caproic acid, $HC_6H_{11}O_2$, is found in small amounts in milk fat, coconut oil, and palm oil. It is used in making artificial flavors. A saturated aqueous solution of the acid contains 11 g/L and has pH = 2.94. Calculate K_a for n-caproic acid.

21. What is $[CHO_2^-]$ in a solution prepared by dissolving 25.0 g of formic acid, $HCHO_2$, in 1400 ml of water solution?

22. The pH of an 0.250 M aqueous solution of β-picoline, an organic solvent used in the chemical industry, is found to be 9.7. What is K_b for this base?

$$CH_3C_5H_4N + H_2O \rightleftharpoons CH_3C_5H_4NH^+ + OH^- \qquad K_b = ?$$

23. The compound o-nitrophenol, $HC_6H_4NO_3$, is slightly soluble in water and ionizes as a weak acid. A saturated solution of o-nitrophenol has pH = 4.53. What is the solubility of o-nitrophenol in water, expressed in g/L?

$$HC_6H_4NO_3 + H_2O \rightleftharpoons H_3O^+ + C_6H_4NO_3^-$$
$$K_a = 5.9 \times 10^{-8}$$

24. The active ingredient in aspirin is acetylsalicylic acid.

$$HC_9H_7O_4 + H_2O \rightleftharpoons H_3O^+ + C_9H_7O_4^-$$
$$K_a = 2.75 \times 10^{-5}$$

What is the pH of the solution obtained by dissolving two aspirin tablets in a glass of water. Assume that each tablet contains 0.32 g of acetylsalicylic acid and that the volume of water is 250 ml.

25. An aqueous solution of dichloroacetic acid, prepared by dissolving 0.10 mol $HC_2HCl_2O_2$ per liter of solution, has pH = 1.30. What is K_a for this acid?

26. A particular vinegar is found to contain 6.0% acetic acid by mass. How much of this vinegar, in milligrams, should be added to 1.00 L of water to produce a solution with pH = 4.50?

$$HC_2H_3O_2 + H_2O \rightleftharpoons H_3O^+ + C_2H_3O_2^-$$
$$K_a = 1.74 \times 10^{-5}$$

27. A solution is prepared by dissolving 5.0×10^{-3} mol of monochloroacetic acid per liter of aqueous solution. What is the pH of this solution?

$$HC_2H_2ClO_2 + H_2O \rightleftharpoons H_3O^+ + C_2H_2ClO_2^-$$
$$K_a = 1.35 \times 10^{-3}$$

28. The organic base piperidine is found in small amounts in black pepper. A water solution is prepared by dissolving 125 mg of piperidine in 0.250 L of water solution. What is the pH of this solution?

$$C_5H_{11}N + H_2O \rightleftharpoons C_5H_{11}NH^+ + OH^-$$
$$K_b = 1.6 \times 10^{-3}$$

Degree of ionization (Use data from Table 17-2 as necessary.)

29. A solution is prepared by dissolving 0.50 mol of formic acid, $HCHO_2$, per liter of aqueous solution. What is the percent ionization of the acid in this solution?

30. What must be the total molarity of acetic acid, $HC_2H_3O_2$, in an aqueous solution if the acid is to be 1.00% ionized?

31. The percent ionization found for acetic acid in Example 17-8 was 0.42% in 1.0 M $HC_2H_3O_2$, 1.3% in 0.10 M $HC_2H_3O_2$, and 4.2% in 0.010 M $HC_2H_3O_2$. Could we reasonably expect to find that the percent ionization is 13% in 0.0010 M $HC_2H_3O_2$ and 42% in 0.00010 M $HC_2H_3O_2$? Explain.

★32. An early method of determining the degree of ionization of a weak electrolyte was by measuring the depression of the freezing point of the solvent. What would you expect for the measured freezing point of 0.050 M $HC_2H_3O_2(aq)$?

Polyprotic acids (Use data from Table 17-3 as necessary.)

33. A water solution saturated with CO_2 is about 0.034 M $CO_2(aq)$. CO_2 combines with water to a very limited extent to produce H_2CO_3 (carbonic acid). H_2CO_3 ionizes as a weak diprotic acid. For purposes of calculation, the CO_2 can be treated as if all of it were converted to H_2CO_3. Calculate the pH and $[CO_3^{2-}]$ in a saturated aqueous solution of CO_2. (*Hint:* Take $[H_2CO_3] = 0.034$ M.)

34. For the following solutions of $H_2S(aq)$, determine $[H_3O^+]$, $[HS^-]$, and $[S^{2-}]$: **(a)** 0.075 M H_2S; **(b)** 0.0050 M H_2S; **(c)** 1.0×10^{-5} M H_2S.

35. The assumptions on page 424 refer to a *di*protic acid. Consider the *tri*protic acid H_3A that ionizes as follows.

$$H_3A + H_2O \rightleftharpoons H_3O^+ + H_2A^- \quad K_{a_1}$$

$$H_2A^- + H_2O \rightleftharpoons H_3O^+ + HA^{2-} \quad K_{a_2}$$

$$HA^{2-} + H_2O \rightleftharpoons H_3O^+ + A^{3-} \quad K_{a_3}$$

(a) What relationship would you expect among K_{a_1}, K_{a_2}, and K_{a_3}; that is, which is largest, smallest?
(b) Under what conditions will $[HA^{2-}] = K_{a_2}$?
(c) Will $[A^{3-}] = K_{a_3}$? Explain.

36. A common polyprotic acid is phosphoric acid, H_3PO_4, which ionizes in three steps. Use assumptions similar to those on page 424 to calculate the following concentrations in 0.100 M H_3PO_4(aq): $[H_3O^+]$, $[H_2PO_4^-]$, $[HPO_4^{2-}]$, and $[PO_4^{3-}]$. (*Hint:* Refer to Exercise 35.)

37. The antimalarial drug quinine, $C_{20}H_{24}O_2N_2$, has a water solubility of 1.00 g per 1900 ml of solution. Quinine is a weak diprotic *base* with $K_{b_1} = 1.08 \times 10^{-6}$ and $K_{b_2} = 1.5 \times 10^{-10}$.
(a) Write equations to represent the ionization equilibria corresponding to K_{b_1} and K_{b_2}.
(b) What is the pH of a saturated aqueous solution of quinine?

Hydrolysis (Use data from Tables 17-2 and 17-3 as necessary.)

38. Write net ionic equations to show which of the following ions hydrolyze in aqueous solution: (a) NO_3^-; (b) BrO^-; (c) NH_4^+; (d) I^-; (e) $C_6H_5NH_3^+$.

39. Which of the following aqueous solutions would you expect to be acidic, which basic, and which neutral? (a) KCl; (b) NH_4NO_3; (c) $NaNO_3$; (d) KI; (e) $Ca(ClO)_2$.

40. Sodium benzoate ($NaC_7H_5O_2$) is a commonly used food preservative, at a concentration of about 0.10%, by mass. Determine the pH of a 0.10%, by mass, solution of sodium benzoate in water.

41. It is desired to produce an aqueous solution of pH = 8.75 by dissolving one of the following salts in water. Which salt would you use and at what molar concentration? (a) NH_4Cl; (b) $KHSO_4$; (c) KNO_2; (d) $NaNO_3$.

42. Aniline, $C_6H_5NH_2$, forms a salt, anilinium hydrochloride, $C_6H_5NH_3Cl$, as a result of a reaction with HCl. Write an ionic equation to represent the hydrolysis of the anilinium ion, and calculate the pH of 0.050 M $C_6H_5NH_3Cl$.

43. In addition to hydrolysis, certain salts can affect the pH of water as a result of ionization as an acid or base. For each of the following ions, write equations to represent ionization and hydrolysis. Then use data from Tables 17-2 and 17-3 to predict whether each ion makes the solution acidic or basic. (a) HSO_3^-; (b) HS^-; (c) $HC_2O_4^-$.

Molecular structure and acid strength (Use data from Tables 17-2, 17-3, and 17-4 as necessary.)

44. Based on the Brønsted–Lowry theory, explain why
(a) Acetic acid is a strong acid in NH_3(l) whereas it is a weak acid in H_2O(l).
(b) Ammonia is a strong base in $HC_2H_3O_2$(l) whereas it is a weak base in H_2O(l).

45. Predict the direction favored in each of the following acid-base reactions. That is, does the reaction tend to go more in the forward or the reverse direction?
(a) $HBr + OH^- \rightleftharpoons H_2O + Br^-$
(b) $HSO_4^- + NO_3^- \rightleftharpoons HNO_3 + SO_4^{2-}$
(c) $CH_3OH + C_2H_3O_2^- \rightleftharpoons HC_2H_3O_2 + CH_3O^-$
(d) $HC_2H_3O_2 + CO_3^{2-} \rightleftharpoons HCO_3^- + C_2H_3O_2^-$
(e) $HNO_2 + ClO_4^- \rightleftharpoons HClO_4 + NO_2^-$
(f) $H_2CO_3 + CO_3^{2-} \rightleftharpoons HCO_3^- + HCO_3^-$

46. Arrange the following in the order of increasing acid strength.

(a) HI

(b) HCl

(c) H—C—C—O—H (with H O groups)

(d) Cl—C—C—O—H (with H O groups)

(e) F—C—C—O—H (with F O groups)

(f) I—C—C—O—H (with H O groups)

47. Which is the stronger base, propylamine, $CH_3CH_2CH_2NH_2$, or aniline, ⬡—NH_2? (*Hint:* Compare the conjugate acids and recall the situation involving ethanol and phenol.)

48. Our explanation of the varying strengths of oxyacids was given based on electronegativity, oxidation state, and formal charge of the central atom. In comparing the acid strengths of acetic acid and ethanol, we reasoned from the standpoint of the anions. That is, because of the increased spread of the negative charge, acetate ion is a weaker base than ethoxide ion. Show that this same line of reasoning can be applied in describing the strengths of oxyacids.

49. Phosphorous acid has the formula H_3PO_3, but it is listed in Table 17-3 as a *di*protic acid (two ionizable hydrogen atoms). Also, notice that K_{a_1} for phosphorous acid is about the same as K_{a_1} for phosphoric acid. Propose a Lewis structure for H_3PO_3 that is consistent with these facts.

50. Use the general rule in the marginal note on page 430 (a) to estimate the value of K_{a_1} for H_3AsO_4; (b) to write a Lewis structure for hypophosphorous acid, H_3PO_2, for which $pK_{a_1} = 1.1$.

Additional Exercises

1. Calculate $[H_3O^+]$ and pH in each of the following solutions: **(a)** 0.0015 M HI(aq); **(b)** 0.0030 M Ca(OH)$_2$(aq); **(c)** saturated Ba(OH)$_2$(aq) [containing 39 g Ba(OH)$_2 \cdot$ 8 H$_2$O per liter].

2. What volume of 0.500 M NaOH must be diluted to 1.00 L to produce a solution with pH = 11.35?

3. In a manner similar to that used in equation (17.13), represent the self-ionization of the following liquid solvents: **(a)** NH$_3$; **(b)** HF; **(c)** CH$_3$OH; **(d)** HC$_2$H$_3$O$_2$; **(e)** H$_2$SO$_4$.

4. Draw Lewis structures corresponding to the cations and anions formed in each of the self-ionization reactions of the preceding exercise.

5. What are the pH values of the following solutions of benzoic acid?
  (a) 0.10 M HC$_7$H$_5$O$_2$ **(b)** 1.0×10^{-3} M HC$_7$H$_5$O$_2$
  (c) 1.0×10^{-5} M HC$_7$H$_5$O$_2$

6. The pH of a saturated solution of phenol at room temperature is 4.90. What is the solubility of phenol in grams per liter of saturated solution?

$$HC_6H_5O + H_2O \rightleftharpoons H_3O^+ + C_6H_5O^-$$
$$K_a = 1.6 \times 10^{-10}$$

7. Calculate the pH of a solution that is 0.100 M NH$_4$Cl, 0.080 M NH$_4$ClO$_4$, and 0.200 M NH$_4$NO$_3$

8. With information from this chapter, but without performing calculations, arrange the following 0.10 M solutions in order of *increasing* pH: **(a)** KCl; **(b)** NH$_3$; **(c)** HCl; **(d)** HC$_2$H$_3$O$_2$; **(e)** NH$_4$NO$_3$; **(f)** LiOH; **(g)** NaNO$_2$.

9. For 0.100 M H$_2$C$_2$O$_4$(aq), calculate $[H_3O^+]$, $[HC_2O_4^-]$, and $[C_2O_4^{2-}]$. For oxalic acid, $K_{a_1} = 5.4 \times 10^{-2}$; $K_{a_2} = 5.4 \times 10^{-5}$.

10. Which compound is the more *acidic* in each of the following pairs? Explain your reasoning.
  (a) ClCH$_2$CH$_2$OH or CH$_3$CH$_2$OH
  (b) CH$_3$NH$_2$ or ClCH$_2$NH$_2$
  (c) ClCH$_2$CH$_2$COOH or CH$_2$FCOOH.

*11. Show that when $[H_3O^+]$ of a solution is reduced to one-half of its original value, the pH value increases by 0.30 unit, *regardless of the initial pH.* Can it also be said that when any solution is diluted to one-half of its original concentration its pH value increases by 0.30 unit? Explain.

*12. What total concentration of HC$_2$H$_3$O$_2$ must be placed in aqueous solution if the solution is to have the same freezing point as 0.150 M HC$_2$H$_2$ClO$_2$ (monochloroacetic acid)?

*13. What is the pH of 0.0010 M NaC$_2$H$_2$ClO$_2$ (sodium monochloroacetate)? (*Hint:* What assumption usually made in an equilibrium calculation is not applicable here?)

*14. What is the pH of a solution that is 0.250 M HC$_2$H$_3$O$_2$ and 0.250 M HCHO$_2$?

*15. Data are given in the chapter on K_w as a function of temperature. Use these data to verify that the heat of neutralization of strong acids by strong bases is about -56 kJ/mol H$_2$O produced.

Self-Test Questions

For questions 1 through 8 select the single item that best completes each statement.

1. The number of moles of hydronium ion, H$_3$O$^+$, in 300 ml of 0.0050 M H$_2$SO$_4$ is **(a)** 0.0075; **(b)** 0.0015; **(c)** 0.0030; **(d)** 0.0050.

2. A solution has a pH of 5.0. In this solution $[OH^-]$ must be **(a)** 1.0×10^{-9} M; **(b)** 1.0×10^{-7} M; **(c)** greater than 1.0×10^{-5} M; **(d)** 1.0×10^{-5} M.

3. $[H_3O^+]$ in 0.10 M HC$_3$H$_5$O$_2$ (propionic acid) is
  (a) equal to $[H_3O^+]$ in 0.10 M HNO$_2$ (nitrous acid)
  (b) less than $[H_3O^+]$ in 0.10 M HI (hydroiodic acid)
  (c) greater than $[H_3O^+]$ in 0.10 M HBr (hydrobromic acid)
  (d) equal to 0.10 mol H$_3$O$^+$/L

4. An aqueous solution is 0.10 M in the weak base, CH$_3$NH$_2$. In this solution **(a)** $[H_3O^+] = 0.10$ M; **(b)** $[OH^-] = 0.10$ M; **(c)** pH < 7; **(d)** pH < 13.

5. $[H_3O^+]$ in 0.010 M NaClO (sodium hypochlorite) is **(a)** 0.010 M; **(b)** 1.0×10^{-12} M; **(c)** greater than 1.0×10^{-7} M; **(d)** less than 1.0×10^{-7} M.

6. One of the following ions is amphiprotic, that is, has the ability to lose or gain a proton in aqueous solution. That one is **(a)** HCO$_3^-$; **(b)** CO$_3^{2-}$; **(c)** Cl$^-$; **(d)** NH$_4^+$.

7. Which of the following is a Lewis base? **(a)** Al^{3+}; **(b)** CO$_2$; **(c)** NH$_3$; **(d)** BF$_3$.

8. The reaction of acetic acid, HC$_2$H$_3$O$_2$, with a base will proceed furthest toward completion (to the right) when that base is **(a)** H$_2$O; **(b)** NH$_3$; **(c)** Cl$^-$; **(d)** HClO$_4$.

9. Arrange the following 10 solutions in order of decreasing $[H_3O^+]$ in aqueous solution: **(a)** 0.01 M NH$_3$; **(b)** 0.01 M H$_2$S; **(c)** 0.01 M HNO$_3$; **(d)** 0.01 M NaNO$_2$; **(e)** 0.01 M H$_2$SO$_4$; **(f)** 0.01 M NaOH; **(g)** 0.01 M NaCl; **(h)** 0.01 M Ba(OH)$_2$; **(i)** 0.01 M NH$_4$Cl; **(j)** 0.01 M HC$_2$H$_3$O$_2$.

10. How many grams of benzoic acid, $HC_7H_5O_2$, must be dissolved in 250.0 ml of water solution to produce a solution with pH = 2.60?

$$HC_7H_5O_2 + H_2O \rightleftharpoons H_3O^+ + C_7H_5O_2^-$$
$$K_a = 6.3 \times 10^{-5}$$

11. Explain why you would expect
(a) $HClO_4$ to be a stronger acid than HNO_3
(b) trifluoroacetic acid, $HC_2F_3O_2$, to be a stronger acid than acetic acid, $HC_2H_3O_2$
(c) *o*-chloroaniline to be a weaker base than aniline

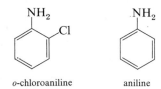

    o-chloroaniline   aniline

12. Explain the following statements.
(a) $[H_3O^+]$ in a strong acid solution doubles as the total concentration of the acid is doubled, whereas for a weak acid $[H_3O^+]$ increases only by $\sqrt{2}$.
(b) Even though two H_3O^+ ions are produced for every S^{2-} ion produced by the ionization of H_2S in aqueous solution, $[S^{2-}]$ is not simply $\frac{1}{2}[H_3O^+]$; it is *much* smaller than this.

Additional Aspects of Equilibria in Aqueous Solutions

Go See Otto!

In this chapter we combine a number of ideas presented previously. For example, we apply the common ion effect to equilibria involving weak acids and bases. This will help us to understand how certain solutions—buffer solutions—are able to resist changes in their pH. The common ion effect is also important in the action of acid-base indicators. An understanding of these indicators, together with other equilibrium principles, makes possible a detailed description of acid-base titrations.

The presence of acids or bases in aqueous solution can affect significantly the solubilities of slightly soluble ionic compounds. Adjusting solute solubilities through control of pH is crucial to the qualitative analysis scheme. Finally, as in the two preceding chapters, some conclusions will just be stated qualitatively, through net ionic equations.

18-1 The Common Ion Effect in Acid-Base Equilibria

In Chapter 16 we studied the effect of common ions on the solubilities of slightly soluble ionic compounds. Equilibria involving weak acids and bases are also greatly affected by common ions. For example, when calculating $[OH^-]$ in an aqueous solution of HCl we find $[OH^-]$ to be *very much smaller* than in pure water. This is so because the high concentration of H_3O^+ produced by the strong acid causes the self-ionization equilibrium of water to shift far to the left.

$$H_2O + H_2O \rightleftharpoons H_3O^+ + OH^- \qquad (18.1)$$

In the presence of an acid or base, equilibrium condition shifts

A strong base represses the self-ionization of water by increasing the concentration of the common ion, OH^-. Actually, the self-ionization of water occurs to such a slight extent that even weak acids and bases produce a common ion effect on this equilibrium.

SOLUTIONS OF WEAK ACIDS AND STRONG ACIDS. A strong acid represses the ionization of a weak acid just as it represses the self-ionization of water, through the common ion, H_3O^+. For example,

439

$$HC_2H_3O_2 + H_2O \rightleftharpoons H_3O^+ + C_2H_3O_2^- \qquad K_a = 1.74 \times 10^{-5} \qquad (18.2)$$

In the presence of
a strong acid,
equilibrium shifts

inventory:

$[H_2O]\ [OH^-]\ [H_3O^+]$

$[HC_2H_3O_2]\ [C_2H_3O_2^-]$

EN:

$[H_3O^+] = [C_2H_3O_2^-] + [OH^-]$

material
Balance:

$0.100 = [HA] + [A^-]$

since K_a is low
& dealing w/ fairly
concentrated amts
we neglect $[A^-]$;

$\therefore [0.100] = [HA]$

Example 18-1 **(a)** Determine $[H_3O^+]$ and $[C_2H_3O_2^-]$ in 0.100 *M* $HC_2H_3O_2$. **(b)** Then determine these same quantities in a solution that is 0.100 *M* in both $HC_2H_3O_2$ and HCl.

(a) This calculation is performed in the same way as those of Chapter 17. Let $x = [H_3O^+] = [C_2H_3O_2^-]$; $[HC_2H_3O_2] = 0.100 - x \simeq 0.100$.

$$K_a = \frac{[H_3O^+][C_2H_3O_2^-]}{[HC_2H_3O_2]} = \frac{x^2}{0.100} = 1.74 \times 10^{-5}$$

$$x^2 = 1.74 \times 10^{-6} \qquad x = [H_3O^+] = [C_2H_3O_2^-] = 1.32 \times 10^{-3}\ M$$

(b) Here we rewrite equation (18.2) and arrange the relevant data below the equation in the usual fashion.

$$HC_2H_3O_2 + H_2O \rightleftharpoons H_3O^+ + C_2H_3O_2^-$$

	$HC_2H_3O_2$	H_3O^+	$C_2H_3O_2^-$
from weak acid:	$(0.100 - x)\ M$	$x\ M$	$x\ M$
from 0.100 *M* HCl:	—	0.100 *M*	—
equilibrium condition:	$(0.100 - x)\ M$	$(0.100 + x)\ M$	$x\ M$

Because of the repression of the ionization of $HC_2H_3O_2$ by the strong acid, HCl, we should expect x to be very small. This means that $(0.100 - x) \simeq (0.100 + x) \simeq 0.100$, and leads to the expression

$$K_a = \frac{[H_3O^+][C_2H_3O_2^-]}{[HC_2H_3O_2]} = \frac{(0.100 + x)(x)}{0.100 - x} = \frac{0.100(x)}{0.100} = 1.74 \times 10^{-5}$$

$$x = [C_2H_3O_2^-] = 1.74 \times 10^{-5}\ M \qquad 0.100 + x = [H_3O^+] = 0.100\ M$$

$$0.100 - x = [HC_2H_3O_2] = 0.100\ M$$

SIMILAR EXAMPLE: Exercise 2.

To summarize the effect of HCl on the ionization of $HC_2H_3O_2$ calculated in Example 18-1,

This assumption will not be valid if the strong acid is very dilute and/or if K_a of the weak acid is large.

1. All of the H_3O^+ in a mixture of a strong acid and a weak acid is assumed to come from the strong acid.

2. In the *absence* of any additional solute, $[C_2H_3O_2^-] = [H_3O^+]$. In the *presence* of a strong acid, $[C_2H_3O_2^-] \ll [H_3O^+]$. In Example 18-1 the common ion from the strong acid, H_3O^+, caused nearly a 100-fold decrease in $[C_2H_3O_2^-]$—from 1.32×10^{-3} *M* to 1.74×10^{-5} *M*.

The ideas presented here about solutions of weak and strong acids are, of course, also applicable to mixtures of weak and strong bases.

SOLUTIONS OF WEAK ACIDS AND SALTS OF WEAK ACIDS. The salt of a weak acid is a strong electrolyte—it is completely dissociated into its ions in aqueous solution. One of the ions, the *anion,* is a common ion in the ionization equilibrium of the weak acid. The presence of this common ion represses the ionization of the weak acid. For example,

$$HC_2H_3O_2 + H_2O \rightleftharpoons H_3O^+ + C_2H_3O_2^- \qquad K_a = 1.74 \times 10^{-5} \qquad (18.2)$$

In the presence of
acetate salts,
equilibrium shifts

In mixtures of acetic acid and sodium acetate, $[C_2H_3O_2^-]$ has a "high" value and $[H_3O^+]$ is diminished by the presence of this common ion, as illustrated in Example 18-2.

Example 18-2   Calculate $[H_3O^+]$ and $[C_2H_3O_2^-]$ in a solution that is 0.100 M both in $HC_2H_3O_2$ and in $NaC_2H_3O_2$.

The setup below is very similar to that used in Example 18-1(b), except that $NaC_2H_3O_2$ provides the common ion.

$$HC_2H_3O_2 \; + \; H_2O \; \rightleftharpoons \; H_3O^+ \; + \; C_2H_3O_2^-$$

from weak acid:	$(0.100 - x)\,M$	$x\,M$	$x\,M$
from 0.100 M NaC₂H₃O₂:	—	—	0.100 M
equilibrium condition:	$(0.100 - x)\,M$	$x\,M$	$(0.100 + x)\,M$

Again we should expect x to be very small, because of the repression of the ionization of $HC_2H_3O_2$ caused by the anion, $C_2H_3O_2^-$. $(0.100 - x) \simeq (0.100 + x) \simeq 0.100$, and

$$K_a = \frac{[H_3O^+][C_2H_3O_2^-]}{[HC_2H_3O_2]} = \frac{(x)(0.100 + x)}{0.100 - x} = \frac{(x)\,0.100}{0.100} = 1.74 \times 10^{-5}$$

$$x = [H_3O^+] = 1.74 \times 10^{-5}\,M \qquad 0.100 + x = [C_2H_3O_2^-] = 0.100\,M$$

SIMILAR EXAMPLES: Exercises 2, 3.

Once more we should add that the same ideas illustrated here apply to mixtures of weak bases and their salts. (In this case the salt provides a common cation.)

18-2   Buffer Solutions

If even a very small quantity of either an acid or a base is added to pure water, the pH changes dramatically. Pure water, as outlined below and pictured in Figure 18-1, has no resistance to a change in pH—it has no buffer capacity.

Pure water has a pH = 7. Addition of 0.001 mol HCl (e.g., 1.00 ml of 1.0 M HCl) to 1.00 L of pure water produces $[H_3O^+] = 10^{-3}\,M$ and pH = 3. Addition of 0.001 mol NaOH (e.g., 40 mg NaOH) to 1.00 L of pure water produces $[OH^-] = 10^{-3}\,M$, pOH = 3, and pH = 11.   (18.3)

The acetic acid–sodium acetate solution of Example 18-2 *does* have a capacity to resist a change in pH—it is a **buffer** solution. Let us see how this fact can be established, first qualitatively and then quantitatively. In any solution of acetic acid and sodium acetate, the K_a expression for acetic acid must apply.

$$K_a = \frac{[H_3O^+][C_2H_3O_2^-]}{[HC_2H_3O_2]} = 1.74 \times 10^{-5} \qquad (18.4)$$

Suppose we take a solution that has equal concentrations of $HC_2H_3O_2$ and $NaC_2H_3O_2$ (as in Example 18-2). In such a solution $[H_3O^+] = 1.74 \times 10^{-5}\,M$. Now consider adding a small quantity of either an acid (H_3O^+) or a base (OH^-) to this solution. The following reactions will occur in the buffer solution.

$$C_2H_3O_2^- + \underset{\text{(small added amt.)}}{H_3O^+} \longrightarrow HC_2H_3O_2 + H_2O \qquad (18.5)$$

$$HC_2H_3O_2 + \underset{\text{(small added amt.)}}{OH^-} \longrightarrow C_2H_3O_2^- + H_2O \qquad (18.6)$$

In the first instance (18.5) a small amount of the salt ($C_2H_3O_2^-$) is converted to the acid ($HC_2H_3O_2$). In the second instance (18.6) a small amount of the acid is con-

FIGURE 18-1
Change in pH of water upon addition of a small quantity of acid or base.

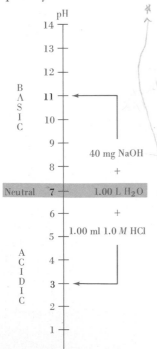

Handwritten annotations:

✱ $HC_2H_3O_2$ soln is prepared. In a buffer soln, there is however not just Acid initially, but also some of the conjugate base ion. Here salt → $NaC_2H_3O_2$ is used to supply the initial $C_2H_3O_2^-$.

amt. of $C_2H_3O_2^-$ from both supplies is equal to amt of $HC_2H_3O_2$

add OH⁻ → $H_2C_2H_3O_2 \longrightarrow C_2H_3O_2^- + H$ add H₃O⁺
shift right →   ← shifts left

verted to the salt. In both cases, basically all that has occurred is a very slight change in the ratio $[C_2H_3O_2^-]/[HC_2H_3O_2]$ in equation (18.4). This change is so slight that $[H_3O^+]$ remains essentially constant. The acetic acid–sodium acetate buffer maintains a constant pH at a value of about 5.

To obtain a buffer in basic solution, a weak base and its salt can be used, as illustrated in Example 18-3.

[handwritten margin note: my this is too weak base to rx w/water.]

Example 18-3 Show that an aqueous solution of NH_3 and NH_4Cl is a basic buffer solution.

The key to buffer action, as we have just seen, is the presence in solution of at least *two* components. One reacts with small amounts of added acid and the other with small amounts of added base. In the present case these components are NH_3 and NH_4^+, respectively.

[handwritten margin note: not good buffer soln (Since strong acids & bases completely dissociate rx goes to near completion & soln is virtually all ions.)]

$$NH_3 + \underset{\text{(small added amt.)}}{H_3O^+} \longrightarrow NH_4^+ + H_2O$$

$$NH_4^+ + \underset{\text{(small added amt.)}}{OH^-} \longrightarrow NH_3 + H_2O$$

[handwritten margin note: It is the forward rx of acid or Base to ions]

In any aqueous solution containing NH_3 and NH_4^+ the ionization equilibrium of NH_3 must be established ($NH_3 + H_2O \rightleftharpoons NH_4^+ + OH^-$) and the K_b expression must be satisfied.

[handwritten margin note: $NH_3 + H_2O \rightleftharpoons NH_4^+ + OH^-$; present initially (from NH_4Cl) ; present initially]

$$K_b = \frac{[NH_4^+][OH^-]}{[NH_3]} = 1.74 \times 10^{-5}$$

[handwritten margin note: $[NH_4^+] \sim [NH_3]$]

[handwritten margin note: NH_3 is Weak so it only rx w/small amts.]

If approximately equal concentrations of the base (NH_3) and its salt (NH_4^+) are maintained in solution, then $[OH^-] \simeq 1 \times 10^{-5}$, $pOH \simeq 5$, and $pH \simeq 9$. Thus, ammonia-ammonium chloride solutions are basic buffer solutions.

SIMILAR EXAMPLE: Exercise 6.

CALCULATING THE pH OF BUFFER SOLUTIONS. To calculate the pH of a buffer solution requires that equilibrium concentrations of the weak acid (or weak base) and its salt be substituted into an ionization constant expression, as illustrated in Example 18-4.

Example 18-4 Calculate the pH of a buffer solution that is 0.15 M $HC_2H_3O_2$–0.50 M $NaC_2H_3O_2$.

In the usual fashion we can write

$$HC_2H_3O_2 + H_2O \rightleftharpoons H_3O^+ + C_2H_3O_2^- \quad K_a = 1.74 \times 10^{-5}$$

placed in solution:	0.15 M	—	0.50 M
changes:	$-x\,M$	$+x\,M$	$+x\,M$
equilibrium condition:	$(0.15 - x)\,M$	$x\,M$	$(0.50 + x)\,M$

Assume that $x \ll 0.15$ (which means also that $x \ll 0.50$).

$$\frac{[H_3O][C_2H_3O_2^-]}{[HC_2H_3O_2]} = \frac{x(0.50)}{(0.15)} = 1.74 \times 10^{-5}$$

$$x = [H_3O^+] = 5.2 \times 10^{-6} \qquad pH = -\log(5.2 \times 10^{-6}) = 5.28$$

SIMILAR EXAMPLE: Exercise 8.

A USEFUL EQUATION FOR BUFFER SOLUTIONS. As simplifying assumptions in Example 18-4 we substituted the stoichiometric concentrations (the concentrations placed in solution) for the equilibrium concentrations of $HC_2H_3O_2$ and $C_2H_3O_2^-$.

That is, we substituted: $0.15 - x \simeq 0.15$ and $0.50 + x \simeq 0.50$. Such assumptions are generally valid so long as (a) neither buffer component is too dilute and (b) K_a (or K_b) is not too large. From this starting point let us return to equation (18.4), that is,

$$K_a = \frac{[H_3O^+][C_2H_3O_2^-]}{[HC_2H_3O_2]} = 1.74 \times 10^{-5} \qquad (18.4)$$

First, take the negative logarithm of both sides of equation (18.4),

$$-\log K_a = -\log \frac{[H_3O^+][C_2H_3O_2^-]}{[HC_2H_3O_2]}$$

and then rearrange this equation to

$$-\log K_a = -\log [H_3O^+] - \log \frac{[C_2H_3O_2^-]}{[HC_2H_3O_2]} \qquad (18.7)$$

Now recall the two designations we introduced in Chapter 17: $pH = -\log [H_3O^+]$ and $pK_a = -\log K_a$.

$$pK_a = pH - \log \frac{[C_2H_3O_2^-]}{[HC_2H_3O_2]} \qquad (18.8)$$

Finally, solve equation (18.8) for pH.

see page 422

because pH includes negative sign

An equation written in this form for the general weak acid, HA, is $pH = pK_a + \log [A^-]/[HA]$. It is sometimes called the **Henderson–Hasselbalch equation.**

$$pH = pK_a + \log \frac{[C_2H_3O_2^-]}{[HC_2H_3O_2]} \qquad = pK_a - \log \frac{[HA]}{[A^-]} \qquad (18.9)$$

For acetic acid, $K_a = 1.74 \times 10^{-5}$ and $pK_a = -\log (1.74 \times 10^{-5}) = 4.76$.

$$pH = 4.76 + \log \frac{[C_2H_3O_2^-]}{[HC_2H_3O_2]} \qquad (18.10)$$

A similar equation can be derived for a buffer consisting of a weak base and its salt. For an ammonia–ammonium ion buffer,

$$pOH = pK_b + \log \frac{[NH_4^+]}{[NH_3]} = 4.76 + \log \frac{[NH_4^+]}{[NH_3]} \qquad (18.11)$$

PREPARING BUFFER SOLUTIONS. According to equation (18.10), preparation of a buffer solution with $pH = 4.76$ should be an easy matter. Dissolve acetic acid and sodium acetate in water in equimolar concentrations (i.e., $[HC_2H_3O_2] = [C_2H_3O_2^-] = 1.0\ M$, $[HC_2H_3O_2] = [C_2H_3O_2^-] = 0.50\ M$, and so on). Then, $[C_2H_3O_2^-]/[HC_2H_3O_2] = 1.00$; $\log 1.00 = 0$; and $pH = pK_a = 4.76$.

What if we wish to prepare a buffer solution with $pH = 5.10$? One method would be to find a weak acid having $pK_a = 5.10$ (corresponding to $K_a = 7.9 \times 10^{-6}$). A solution with equimolar concentrations of this acid and its salt would have a $pH = 5.10$. Finding such an acid is not always possible and an alternative method is illustrated in Example 18-5.

want $\log \frac{[C_2H_3O_2^-]}{[HC_2H_3O_2]}$ to equal 0, so you want $\frac{[\ ]}{[\ ]} = 1$

Example 18-5   What mass of $NaC_2H_3O_2$ must be dissolved in 0.300 L of $0.250\ M$ $HC_2H_3O_2$ to produce a solution with $pH = 5.10$? (Assume that the solution volume remains constant.)

Basically what we must calculate is $[C_2H_3O_2^-]$ in the buffer solution. We can do this by converting pH to $[H_3O^+]$ and then substituting appropriate data (i.e., $[H_3O^+]$ and $[HC_2H_3O_2]$) into equation (18.4). Alternatively, we can use equation (18.10).

$$5.10 = 4.76 + \log \frac{[C_2H_3O_2^-]}{0.250} \qquad \log \frac{[C_2H_3O_2^-]}{0.250} = 5.10 - 4.76 = 0.34$$

$$\frac{[C_2H_3O_2^-]}{0.250} = \text{antilogarithm } 0.34 = 2.19$$

$$[C_2H_3O_2^-] = 0.250 \times 2.19 = 0.548\,M$$

$$\text{no. g } NaC_2H_3O_2 = 0.300\,L \times \frac{0.548\ \text{mol } NaC_2H_3O_2}{L} \times \frac{82.0\ \text{g } NaC_2H_3O_2}{1\ \text{mol } NaC_2H_3O_2}$$

$$= 13.5\ \text{g } NaC_2H_3O_2$$

SIMILAR EXAMPLES: Exercises 9, 10.

CALCULATING pH CHANGES IN BUFFER SOLUTIONS. To calculate the change in pH produced by adding a small amount of acid or base to a buffer solution requires the use of two expressions. Chemical equations representing buffer action (such as equations 18.5 and 18.6) are used to establish changes in concentration of the weak acid (or base) and its salt. The equilibrium constant expression, often in the form of equation (18.9), is used to determine pH. The method is illustrated in Example 18-6.

Example 18-6 What is the effect on the pH of a $0.100\,M$ $HC_2H_3O_2$–$0.100\,M$ $NaC_2H_3O_2$ buffer solution produced by adding 0.001 mol H_3O^+ to 1.00 L of solution?

For a $0.100\,M$ $HC_2H_3O_2$–$0.100\,M$ $NaC_2H_3O_2$ buffer solution, equation (18.10) indicates a pH = $pK_a = 4.76$.

Let us now describe solution concentrations before and after the addition of 0.001 mol H_3O^+.

$$C_2H_3O_2^- + \quad H_3O^+ \quad \longrightarrow \quad HC_2H_3O_2 + H_2O$$

	$C_2H_3O_2^-$	H_3O^+	$HC_2H_3O_2$
in original buffer:	$0.100\,M$	$1.74 \times 10^{-5}\,M$	$0.100\,M$
add:		$+0.001\,M$	
changes:	$-0.001\,M$	$-0.001\,M$	$+0.001\,M$
in final buffer:	$0.099\,M$	?	$0.101\,M$

The pH of the buffer solution after addition of the acid can now be calculated using equation (18.10).

$$pH = 4.76 + \log\frac{0.099}{0.101} = 4.76 + \log 0.98 = 4.76 - 0.01 = 4.75$$

The effect of adding 0.001 mol H_3O^+ to the original buffer solution is to lower its pH from 4.76 to 4.75.

SIMILAR EXAMPLES: Exercises 14, 15.

In Example 18-6 it could have been demonstrated just as readily that the addition of 0.001 mol OH^-/L of the buffer solution would result in a

$$pH = 4.76 + \log\frac{0.101}{0.099} = 4.76 + \log 1.02 = 4.76 + 0.01 = 4.77$$

When added to pure water these amounts of acid and base (i.e., 0.001 mol) produce pH changes of *4 units* (recall Figure 18-1). In the acetic acid–sodium acetate buffer, the corresponding change in pH is only *0.01 unit!*

If a buffer solution is diluted, both $[A^-]$ and $[HA]$ are reduced by the same factor. The *ratio* $[A^-]/[HA]$ remains constant, and so does the pH (pH = pK_a + $\log[A^-]/[HA]$). Resistance to pH changes upon dilution is another important property of buffer solutions.

BUFFER CAPACITY AND BUFFER RANGE. In Example 18-6 addition of more than 0.100 mol H_3O^+/L would cause essentially complete conversion of $C_2H_3O_2^-$ to

$$\frac{[C_2H_3O_2{}^-]}{[HC_2H_3O_2]}$$

one is 10 times larger than other

HC$_2$H$_3$O$_2$. The result would be a mixture of a weak acid (HC$_2$H$_3$O$_2$) and a strong acid (the unreacted H$_3$O$^+$). The solution would become considerably more acidic than the original buffer, and the buffering action would be destroyed. *If you added more A or B than original ion*

Buffer capacity refers to the amount of acid or base that may be added to a buffer solution before its pH changes appreciably. In general, the maximum capacity to resist pH changes exists when the concentrations of weak acid (or base) and its salt are kept large and approximately equal to one another. The buffer has its maximum capacity at pH = pK_a (or pOH = pK_b). Whenever the ratio of salt to weak electrolyte concentration is either less than about 0.10 or greater than about 10, the buffer loses its effectiveness. Since $\log 0.10 = -1$ and $\log 10 = +1$, this means that the effective buffer range is about one pH unit on either side of the value, pH = pK.

For acetic acid–sodium acetate buffers the effective range is from about pH 3.76 to 5.76. For ammonia–ammonium chloride, the range is about pH 8.24 to 10.24.

Among the factors that can lead to a condition of acidosis, in which there is a decrease in the pH of blood, are heart failure, kidney failure, diabetes mellitus, persistent diarrhea, or a long-term, high-protein diet. A temporary condition of acidosis may result from prolonged, intensive exercise. A temporary condition also occurs when young children, in a belligerent mood, hold their breaths. They pass out and then breathe normally again while unconscious. Alkalosis, characterized by an increase in the pH of blood, may occur as a result of hyperventilation (overbreathing caused by anxiety or hysteria), exposure to high altitudes (altitude sickness), or severe vomiting.

BLOOD AS A BUFFERED SOLUTION. A complex natural process for the control of pH occurs in human blood, which normally has a pH of 7.4. Serious illness or death can result from variations of a few tenths of a pH unit. Maintenance of proper pH values in processes occurring in living organisms is crucial because the functioning of enzymes—the catalysts for these processes—is pH-dependent.

Several factors are involved in the control of the pH of blood. A particularly important one is the ratio of dissolved HCO$_3{}^-$ (bicarbonate ion) to H$_2$CO$_3$ (carbonic acid). CO$_2$(g) is moderately soluble in water, and in aqueous solution reacts to a limited extent to produce H$_2$CO$_3$. Nevertheless, in using K_{a_1} we treat the dissolved CO$_2$ as if it were completely converted to H$_2$CO$_3$. Moreover, although H$_2$CO$_3$ is a weak diprotic acid, in the carbonic acid–bicarbonate ion buffer system we deal only with the first ionization step: H$_2$CO$_3$ is the weak acid and HCO$_3{}^-$ is the conjugate base (salt).

$$CO_2(aq) + H_2O \rightleftharpoons H_2CO_3(aq) \tag{18.12}$$

$$H_2CO_3 + H_2O \rightleftharpoons H_3O^+ + HCO_3{}^- \qquad K_{a_1} = 4.2 \times 10^{-7} \tag{18.13}$$

$$HCO_3{}^- + H_2O \rightleftharpoons H_3O^+ + CO_3{}^{2-} \qquad K_{a_2} = 5.6 \times 10^{-11} \tag{18.14}$$

Carbon dioxide enters the blood from tissues as the by-product of metabolic reactions. In the lungs CO$_2$(g) is exchanged for O$_2$(g), which is transported throughout the body by the blood.

To establish the pH of the buffer we use an expression similar to (18.9). For carbonic acid, p$K_{a_1} = -\log 4.2 \times 10^{-7} = 6.4$.

$$pH = pK_{a_1} + \log \frac{[HCO_3{}^-]}{[H_2CO_3]} = 6.4 + \log \frac{[HCO_3{}^-]}{[H_2CO_3]}$$

The [HCO$_3{}^-$]/[H$_2$CO$_3$] ratio required to maintain a pH of 7.4 is

$$7.4 = 6.4 + \log \frac{[HCO_3{}^-]}{[H_2CO_3]}$$

The *antilogarithm* of 1.0 is 10.

$$\log \frac{[HCO_3{}^-]}{[H_2CO_3]} = 7.4 - 6.4 = 1.0 \qquad and \qquad \frac{[HCO_3{}^-]}{[H_2CO_3]} = 10$$

We have stated that the maximum buffer capacity occurs when the concentrations of an acid and its salt are about equal. In blood the concentration of bicarbonate ion is about 20 times greater than that of H$_2$CO$_3$. How can this be an effective buffer? For one thing, the need to neutralize excess acid (lactic acid produced by exercise) is usually greater than the need to neutralize excess base. The high proportion of HCO$_3{}^-$ helps in this regard. Also, if additional carbonic acid is needed to neutralize excess alkalinity, CO$_2$(g) in the lungs can be reabsorbed to build up the H$_2$CO$_3$ content of the blood. Finally, important contributions to maintaining the pH

The 20:1 ratio of $[HCO_3^-]$ to $[H_2CO_3]$ in blood would correspond to a pH of about 7.7. This is offset by the $H_2PO_4^-/HPO_4^{2-}$ buffer system with a pH of 7.2.

of blood are made by other components in the blood, such as the phosphate buffer $H_2PO_4^-/HPO_4^{2-}$.

18-3 Acid-Base Indicators

We have discussed at some length the meaning of pH, how it can be calculated, and how it can be controlled; but we have not yet commented on how it may be measured experimentally. One simple method involves the use of indicators. An **acid-base indicator** is a *weak acid* for which the nonionized acid (HIn) has one color [color(1)] and the anion another [color(2)]. When a small amount of indicator is placed in a solution, depending on whether the ionization equilibrium of the indicator is displaced toward the acid or anion form, the solution acquires either color(1) or color(2).

An acid-base indicator is usually prepared as a solution (in water, ethanol, or some other solvent). In acid-base titrations a small volume (a few drops) of the indicator is added to the solution being titrated. In another form, porous paper is impregnated with an indicator solution and dried. When this paper is moistened with the solution being tested, it acquires a color determined by the pH of the solution. This paper is often called "pH paper."

$$HIn + H_2O \rightleftharpoons H_3O^+ + In^- \tag{18.15}$$

color(1)   color(2)

[handwritten: @ color 1, 90% acid @ color 2, 90% ions ⟹ neutral when acid ≈ ions w/inbetween color]

To assess the pH range in which an indicator will work, we can again write an expression similar to equation (18.9), that is,

$$pH = pK_a + \log \frac{[In^-]}{[HIn]} \tag{18.16}$$

[handwritten: @ least 90% of soln in one form or other   thus again ratio 0.10 or 10 a ±1 range to achieve color change from normal soln.]

In general, if 90% or more an indicator is in the form, HIn, the solution in which it is found will assume color(1). If 90% or more is in the form, In^-, the solution acquires color(2). These two conditions correspond roughly to the ratios: $[In^-]/[HIn] \simeq 0.10$ and $[In^-]/[HIn] \simeq 10$. The logarithms of these ratios are about -1 and $+1$, respectively. Thus, an indicator changes from color(1) to color(2) over a pH range of about *2 units*. At the middle of this range, that is, with $[HIn] = [In^-]$, $pH = pK_a$. At this midpoint the solution color is a "mixture" of color(1) and color(2). The indicator is said to be undergoing a color change. The colors and pH ranges for several common indicators are illustrated in Figure 18-2.

[handwritten: from the −1 of normal pH to +1 of normal]

18-4 Neutralization Reactions and Titration Curves

Suppose that a 0.10 M HCl solution is also made 0.10 M in NaOH. If there were no reaction, the solution would have $[H_3O^+]$ and $[OH^-]$ each equal to 0.10 M. The product of the two concentrations would be $(0.10)(0.10) = 0.010 = 1.0 \times 10^{-2}$. However, in any aqueous solution at 25°C the product of these ion concentrations must be equal to $K_w = 1.0 \times 10^{-14}$. *A solution cannot be simultaneously 0.10 M in HCl and 0.10 M in NaOH.* A chemical reaction must occur in which H_3O^+ and OH^- combine to form water.

$$H_3O^+ + OH^- \longrightarrow 2 H_2O \tag{18.17}$$

The method of carrying out the neutralization of an acid by a base is known as titration. Titration reactions were introduced and discussed in Section 4-4, but we left a critical question unanswered at that time: How does one select an indicator for a titration? We are now in a position to answer this question.

In a titration one of the solutions to be neutralized, say the acid, is placed in a flask or beaker. The other solution, the base, is contained in a buret and is added to the acid, first rapidly and then dropwise, until neutralization of the acid is accomplished. This occurs at the **equivalence point** of the titration. In the laboratory one attempts to locate the equivalence point through the color change of an acid-base

note difference in 2 terms

indicator. The titration is stopped at the precise point where the indicator changes color. This is the **end point** of the titration. What is needed is to match the end point of the titration with the equivalence point of the neutralization reaction. This can be done if we know both the pH value at the equivalence point and the pH range over which the indicator changes color.

FIGURE 18-2
pH ranges for some common indicators.

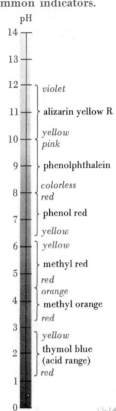

pH

14 —

13 —

12 — } *violet*

11 — → alizarin yellow R

10 — } *yellow* / *pink*

9 — → phenolphthalein

8 — } *colorless* / *red*

7 — → phenol red

— } *yellow*

6 — } *yellow*

5 — → methyl red
— } *red*
— } *orange*

4 — → methyl orange
— } *red*

3 —

2 — } *yellow* / thymol blue (acid range) / *red*

1 —

0 —

TITRATION OF A STRONG ACID BY A STRONG BASE. For the titration of 25.00 ml 0.1000 M HCl (a strong acid) by 0.1000 M NaOH (a strong base) we can calculate the pH of the solution at various points during the titration. These data can be plotted in the form of solution pH versus volume of added base, a form called a **titration curve.** From this curve we can establish the pH at the equivalence point and thus select a suitable indicator for the titration. The necessary calculations are of the type involved in Example 18-7.

Example 18-7 What is the pH at each of the following points in the titration of 25.00 ml 0.1000 M HCl by 0.1000 M NaOH? **(a)** Before the addition of any NaOH; **(b)** after the addition of 24.00 ml 0.1000 M NaOH; **(c)** at the equivalence point; **(d)** after the addition of 26.00 ml 0.1000 M NaOH.

(a) Before the addition of NaOH we are dealing only with 0.1000 M HCl. This solution has $[H_3O^+] = 0.1000$ M and a pH = 1.00.

(b) The total number of moles of H_3O^+ to be titrated is

$$\text{no. mol } H_3O^+ = 0.02500 \text{ L} \times \frac{0.1000 \text{ mol } H_3O^+}{L} = 2.500 \times 10^{-3} \text{ mol } H_3O^+$$

The number of moles of OH^- present in 24.00 ml 0.1000 M NaOH is

$$\text{no. mol } OH^- = 0.02400 \text{ L} \times \frac{0.1000 \text{ mol } OH^-}{L} = 2.400 \times 10^{-3} \text{ mol } OH^-$$

This point in the neutralization reaction can be represented by

$$H_3O^+ \quad + \quad OH^- \quad \longrightarrow \quad 2 H_2O$$

this is in moles

initial:	2.500×10^{-3} mol	—
add:		2.400×10^{-3} mol
after reaction:	0.100×10^{-3} mol	—

The 0.100×10^{-3} mol H_3O^+ is present in 49.00 ml of solution (the original 25.00 ml of acid plus the added 24.00 ml of base).

then change to Molarity

$$[H_3O^+] = \frac{0.100 \times 10^{-3} \text{ mol } H_3O^+}{0.04900 \text{ L}} = 2.04 \times 10^{-3} \text{ } M$$

$$pH = -\log [H_3O^+] = -\log (2.04 \times 10^{-3}) = 2.69$$

(c) At the equivalence point 2.500×10^{-3} mol NaCl has been produced and is present in 50.00 ml of solution. The solution is 0.0500 M NaCl. Since neither Na^+ nor Cl^- hydrolyzes, this solution has a pH = 7.00.

(d) Beyond the equivalence point excess OH^- is present. For example, following the addition of 26.00 ml 0.1000 M NaOH,

$$H_3O^+ \quad + \quad OH^- \quad \longrightarrow \quad 2 H_2O$$

initial:	2.500×10^{-3} mol	—
add:		2.600×10^{-3} mol
after reaction:	—	0.100×10^{-3} mol

change to [ ]

$$[OH^-] = \frac{0.100 \times 10^{-3} \text{ mol } OH^-}{0.05100 \text{ L}} = 1.96 \times 10^{-3} \text{ } M$$

careful!

$$pOH = -\log (1.93 \times 10^{-3}) = 2.71 \qquad pH = 14.00 - 2.71 = 11.29$$

SIMILAR EXAMPLE: Exercise 26a.

initially

M = $\dfrac{0.0250 L \times 0.100 \text{ mols ions}}{L}$

$\dfrac{}{0.050 L}$

quantitatively:

$H_3O^+ + OH^- \longrightarrow 2H_2O$

initial: 2.500 × 10⁻³

add: 2.500 × 10⁻³

after: 0 M

But still must remember H_2O always has self-ionization [ ] of $[H^+] = [OH^-] = 1.0 \times 10^{-7}$ ∴ pH=7 by ion product, K_w

FIGURE 18-3
Titration curve for strong acid by a strong base—25.00 ml of 0.1000 M HCl by 0.1000 M NaOH.

The indicators whose color-change ranges fall along the steep portion of the titration curve are all suitable for this titration. Thymol blue changes color too soon (that is, before 25.00 ml of base has been added); alizarin yellow-R, too late.

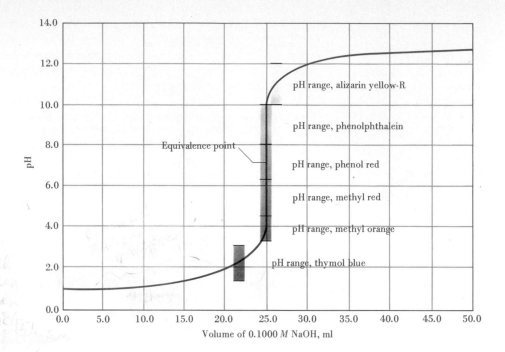

The titration curve for Example 18-7 is shown in Figure 18-3. The most interesting feature of this NaOH–HCl titration curve is that the pH changes quite slowly until just before the equivalence point is reached. At the equivalence point the pH rises very sharply, by approximately 6 units for an addition of only 0.10 ml of base (corresponding to 2 drops). Beyond the equivalence point the pH again changes quite slowly as excess NaOH is added. Any indicator whose color changes in the pH range from about 4 to 10 is suitable for this titration.

TITRATION OF A WEAK ACID BY A STRONG BASE. The titration of 0.1000 M $HC_2H_3O_2$ by 0.1000 M NaOH yields a titration curve somewhat different from the HCl–NaOH titration. The neutralization reaction is

$$HC_2H_3O_2 + OH^- \longrightarrow H_2O + C_2H_3O_2^- \tag{18.18}$$

Points on this titration curve are calculated in Example 18-8.

Example 18-8 What is the pH at each of the following points in the titration of 25.00 ml 0.1000 M $HC_2H_3O_2$ by 0.1000 M NaOH? **(a)** Before the addition of any NaOH; **(b)** after the addition of 15.00 ml 0.1000 M NaOH; **(c)** at the equivalence point; **(d)** after the addition of 26.00 ml 0.1000 M NaOH.

(a) The initial pH—that of 0.1000 M $HC_2H_3O_2$—can be obtained from the calculation in Example 18-1(a): $pH = -\log(1.32 \times 10^{-3}) = 2.88$.

(b) The total number of moles of acetic acid to be neutralized is

$$\text{no. mol } HC_2H_3O_2 = 0.02500 \text{ L} \times \frac{0.1000 \text{ mol } HC_2H_3O_2}{\text{L}}$$

$$= 2.500 \times 10^{-3} \text{ mol } HC_2H_3O_2$$

At this point in the titration the number of moles of OH^- added $= 0.01500 \text{ L} \times 0.1000 \text{ mol } OH^-/\text{L} = 1.500 \times 10^{-3} \text{ mol } OH^-$. Reaction has proceeded to the point where 60% of the acid has been neutralized:

Handwritten margin notes (top left): Don't forget to put mole into molarity

$$* \quad HC_2H_3O_2 \quad + \quad OH^- \quad \longrightarrow \quad C_2H_3O_2^- \quad + H_2O$$

(handwritten labels above: acid, base, conjugate)

initial:	2.500×10^{-3} mol	—	—
add:		1.500×10^{-3} mol	
after reaction:	1.000×10^{-3} mol	—	1.500×10^{-3} mol
concentrations:	$\dfrac{1.000 \times 10^{-3} \text{ mol}}{0.04000 \text{ L}}$	—	$\dfrac{1.500 \times 10^{-3} \text{ mol}}{0.04000 \text{ L}}$
	$= 2.500 \times 10^{-2}$ *M*		$= 3.750 \times 10^{-2}$ *M*

The concentrations of the buffer components can be substituted directly into equation (18.10) for a calculation of the pH of the solution.

Handwritten margin note (left): solve using pH equation for buffer soln. / since it is one @ this pt; consist partially of acid & its salt (ion)

$$pH = 4.76 + \log \frac{[C_2H_3O_2^-]}{[HC_2H_3O_2]} = 4.76 + \log \frac{3.750 \times 10^{-2}}{2.500 \times 10^{-2}}$$

Handwritten margin note (right): w/ strong acid just calculate directly

$$= 4.76 + \log 1.500 = 4.94$$

(c) At the equivalence point neutralization has just been completed, and the solution is the same as one would get by dissolving 2.500×10^{-3} mol $NaC_2H_3O_2$ in 50.00 ml. In this solution $[C_2H_3O_2^-] = 2.500 \times 10^{-3}/0.05000 = 0.05000$ *M*. The question is: What is the pH of 0.05000 *M* $NaC_2H_3O_2$? The answer lies in recognizing that $C_2H_3O_2^-$ hydrolyzes, and then using the method of Example 17-11.

Handwritten margin note (left): its like having added just $NaC_2H_3O_2$ because the H^+ from acid & OH^- from base have neutralized each other thus we're left w/ Na^+ & $C_2H_3O_2^-$ / Na^+ too weak to react, / so $C_2H_3O_2^-$ is only thing left to cause pH disturbance; / it "Hydrolyzes"

$$C_2H_3O_2^- + H_2O \rightleftharpoons HC_2H_3O_2 + OH^- \quad *$$

$$K_h = \frac{[HC_2H_3O_2][OH^-]}{[C_2H_3O_2^-]} = \frac{K_w}{K_a} = \frac{1.0 \times 10^{-14}}{1.74 \times 10^{-5}} = 5.7 \times 10^{-10}$$

$$[OH^-] = [HC_2H_3O_2] = x \qquad [C_2H_3O_2^-] = 0.05000 - x \simeq 0.05000$$

$$\frac{x \cdot x}{0.05000} = 5.7 \times 10^{-10} \qquad x^2 = 2.85 \times 10^{-11} \qquad x = [OH^-] = 5.3 \times 10^{-6} \text{ M}$$

$$pOH = 5.3 \qquad pH = 14.0 - 5.3 = 8.7$$

(d) After the addition of 26.00 ml 0.1000 *M* NaOH, the total number of moles of OH^- added $= 0.02600$ L $\times 0.1000$ mol $OH^-/L = 2.600 \times 10^{-3}$ mol OH^-. Of this total amount, 2.500×10^{-3} mol OH^- is consumed to neutralize the $HC_2H_3O_2$. This leaves 0.100×10^{-3} mol OH^- in 51.00 ml of solution.

$$[OH^-] = \frac{0.100 \times 10^{-3} \text{ mol } OH^-}{0.05100 \text{ L}} = 1.96 \times 10^{-3} \text{ M}$$

Handwritten note: note proper volume (circled 0.05100 L)

$$pOH = 2.71 \qquad pH = 14.00 - 2.71 = 11.29$$

SIMILAR EXAMPLES: Exercises 26b, 27.

The principal features to note in the titration curve of a weak acid by a strong base, illustrated in Figure 18-4, are these.

Handwritten margin note (left): not enough conjugate yet to achieve equil. Buffer

1. The initial pH is higher than in the titration curve of a strong acid by a strong base (because the weak acid is only partially ionized).
2. There is a fairly sharp increase in pH at the start of the titration. [The acetate ion resulting from the neutralization reaction (18.18) acts as a common ion and represses the ionization of acetic acid.]
3. Over a long section of the curve preceding the equivalence point, the pH changes quite gradually. (Solutions represented by this portion of the titration curve contain appreciable concentrations of both $HC_2H_3O_2$ and $C_2H_3O_2^-$. These are *buffer* solutions.)
4. The pH at the equivalence point is greater than 7. (This results from the hydrolysis of $C_2H_3O_2^-$.) *normally, hydrolysis doesn't contribute enough*
5. Beyond the equivalence point the titration curve for a weak acid by a strong base

Handwritten margin note (bottom left): all OH^- & H^+ cancelled out / only ion left Na^+ & / $C_2H_3O_2^-$ no effect / ∴ makes basic

Handwritten note (bottom): be appreciable- but not / the only H^+ contributor

FIGURE 18-4
Titration curve for weak acid by a strong base—25.00 ml of 0.1000 M HC$_2$H$_3$O$_2$ by 0.1000 M NaOH.

Phenolphthalein is a suitable indicator for this titration, but methyl red is not. When exactly one half of the acid is neutralized, [HC$_2$H$_3$O$_2$] = [C$_2$H$_3$O$_2^-$], and pH = pK_a = 4.76.

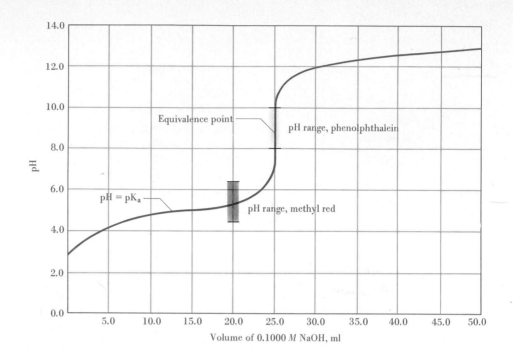

is identical to that of a strong acid by a strong base. (In this portion of the titration, the pH is determined only by the concentration of free OH$^-$.)

because initial pH is higher

← **6.** The steep portion of the titration curve at the equivalence point occurs over a narrower pH range (from about pH 7 to 11) than for the titration of a strong acid by a strong base (about pH 4 to 10).

7. The selection of indicators suitable for the titration of a weak acid by a strong base is more limited than for a strong acid by a strong base.

TITRATION OF A WEAK POLYPROTIC ACID.   The most striking visual evidence that a polyprotic acid ionizes in distinctive steps is provided in the neutralization of the acid. For example, in the neutralization of phosphoric acid, essentially all H$_3$PO$_4$ molecules are first converted to the salt, NaH$_2$PO$_4$; then NaH$_2$PO$_4$ is converted to Na$_2$HPO$_4$; and finally, Na$_2$HPO$_4$ is converted to Na$_3$PO$_4$. That is,

the one mol of each added completely neutralize each other so for further ionization must add 1 mol NaOH @ each step

$$\text{H}_3\text{PO}_4 + \text{OH}^- \longrightarrow \text{H}_2\text{PO}_4^- + \text{H}_2\text{O}$$

equivalence is then reached soln most all Na$^+$ & H$_2$PO$_4^-$ H$_2$PO$_4^-$ hydrolyzes

is followed by

Ⓑ $$\text{H}_2\text{PO}_4^- + \text{OH}^- \longrightarrow \text{HPO}_4^{2-} + \text{H}_2\text{O}$$

acidic $\text{H}_2\text{PO}_4^- + \text{H}_2\text{O} \rightleftarrows \text{H}_3\text{O}^+ + \text{HPO}_4^-$ @ $1/8$ α_i

is followed by   *another mol NaO$_4$l added*

Ⓒ $$\text{HPO}_4^{2-} + \text{OH}^- \longrightarrow \text{PO}_4^{3-} + \text{H}_2\text{O}$$

→ ep. reached

Corresponding to these three distinctive stages there are three equivalence points. For every mole of H$_3$PO$_4$, 1 mol NaOH is required to reach the first equivalence point. At this first equivalence point, the solution is essentially one of NaH$_2$PO$_4$, which, because of the further ionization of H$_2$PO$_4^-$, is an *acidic* solution (pH about 4.5).

$$\text{H}_2\text{PO}_4^- + \text{H}_2\text{O} \rightleftharpoons \text{H}_3\text{O}^+ + \text{HPO}_4^{2-}$$

An additional mole of NaOH is required to titrate the acid to its second equivalence point. At this second equivalence point, the solution is *basic* (pH about 9.5) because the hydrolysis of HPO$_4^{2-}$

@ this 2nd e.p., hydrolysis again occurs, this time ion (∴ soln) takes on more basic characteristics than acidic

$$HPO_4^{2-} + H_2O \rightleftharpoons H_2PO_4^- + OH^- \qquad K_h = \frac{K_w}{K_{a_2}} = \frac{1.0 \times 10^{-14}}{6.2 \times 10^{-8}} = 1.6 \times 10^{-7}$$

occurs to a greater extent than the further ionization of HPO_4^{2-},

$$HPO_4^{2-} + H_2O \rightleftharpoons H_3O^+ + PO_4^{3-} \qquad K_{a_3} = 4.8 \times 10^{-13}$$

The complete neutralization of H_3PO_4 to Na_3PO_4 cannot normally be accomplished in a titration, even though a third equivalence point might be expected. This *3rd e.p.t. not reached* is because the pH of the strongly hydrolyzed Na_3PO_4 solution at the third equivalence point is higher than can normally be attained in an acid-base titration, approaching pH = 13. An Na_3PO_4 solution is nearly as basic as the NaOH solution that would be used to form it in a titration. Thus, the titration curve (see Figure 18-5) has two breaks instead of three, with the breaks equally spaced in terms of volume of titrant (NaOH) used.

STOP

FINAL TO HERE. FOR THIS CHAPTER

18-5 Solubility and pH

If the ions derived from a slightly soluble solute are able to enter into acid-base reactions with H_3O^+ or OH^-, the solubility of the solute will be affected by pH. Consider $Mg(OH)_2$, for example. OH^- ions derived from the solubility equilibrium can react with H_3O^+ to form H_2O.

$$Mg(OH)_2(s) \rightleftharpoons Mg^{2+}(aq) + 2\,OH^-(aq) \qquad K_{sp} = 1.8 \times 10^{-11} \qquad (18.19)$$

$$OH^- + H_3O^+ \longrightarrow 2\,H_2O \qquad\qquad (18.20)$$

According to Le Châtelier's principle, reaction (18.20) creates a disturbance in the equilibrium represented by (18.19)—removal of OH^-. The equilibrium condition in (18.19) is displaced to the right as $Mg(OH)_2(s)$ dissolves to replace OH^- drawn off by reaction (18.20). In fairly acidic solutions reactions (18.19) and (18.20) go to completion and $Mg(OH)_2$ is highly soluble. The net reaction is

FIGURE 18-5
Titration of a weak polyprotic acid by a strong base—10.00 ml of 0.1000 *M* H_3PO_4 by 0.1000 *M* NaOH.

why acidic here?

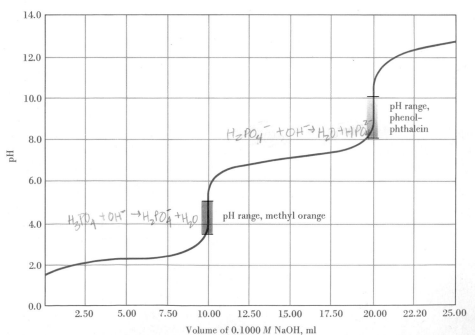

$$Mg(OH)_2(s) + 2\,H_3O^+(aq) \longrightarrow Mg^{2+}(aq) + 4\,H_2O \qquad (18.21)$$

These further ideas about writing net ionic equations are suggested by the way in which equation (18.27) is written: Ionic formulas are used only for strong electrolytes (salts and strong acids and bases). In equation (18.27) the salt, calcium acetate, is written in ionic form. Molecular formulas are used for nonelectrolytes (H_2O), weak electrolytes ($HC_2H_3O_2$), gases (CO_2), and solids ($CaCO_3$). If the equilibrium condition is displaced far to the right, a single arrow is used and the reaction is said to go to completion. If significant amounts of all reactants and products coexist at equilibrium, a double arrow is used.

Example 18-9   Write a net ionic equation to represent the dissolving of $CaCO_3(s)$ in $HC_2H_3O_2(aq)$.

Carbonate ions are produced by the reaction

$$CaCO_3(s) \rightleftharpoons Ca^{2+}(aq) + CO_3^{2-}(aq) \qquad (18.22)$$

Hydronium ion is furnished by the ionization of acetic acid.

$$HC_2H_3O_2(aq) + H_2O \rightleftharpoons H_3O^+(aq) + C_2H_3O_2^-(aq) \qquad (18.23)$$

Carbonate ion, CO_3^{2-}, is a stronger base than acetate ion, $C_2H_3O_2^-$ (recall Table 17-4) and accepts protons from H_3O^+.

$$CO_3^{2-}(aq) + H_3O^+(aq) \longrightarrow HCO_3^-(aq) + H_2O \qquad (18.24)$$

Reaction (18.24) promotes both the ionization of $HC_2H_3O_2$ (18.23) and the dissolving of $CaCO_3$ (18.22). Further reaction occurs between HCO_3^- and H_3O^+.

$$HCO_3^-(aq) + H_3O^+(aq) \longrightarrow H_2CO_3(aq) + H_2O \qquad (18.25)$$

Finally, H_2CO_3 dissociates.

$$H_2CO_3(aq) \longrightarrow H_2O + CO_2(g) \qquad (18.26)$$

The net reaction that occurs when calcium carbonate dissolves in acetic acid, obtained by combining equations (18.22) through (18.26), is

$$CaCO_3(s) + 2\,HC_2H_3O_2(aq) \longrightarrow Ca^{2+}(aq) + 2\,C_2H_3O_2^-(aq) + H_2O + CO_2(g) \qquad (18.27)$$

SIMILAR EXAMPLE: Exercise 43.

We have seen that the solubility of $Mg(OH)_2$ is pH-dependent. So are the conditions under which it precipitates from solution, as illustrated in the following examples.

Example 18-10   Should $Mg(OH)_2$ precipitate from a solution that is 0.010 M $MgCl_2$ if the solution is also made 0.10 M NH_3?

First consider the ionization equilibrium in $NH_3(aq)$.

$$NH_3 + H_2O \rightleftharpoons NH_4^+ + OH^- \qquad K_b = 1.74 \times 10^{-5}$$

If we let $x = [NH_4^+] = [OH^-]$ and $[NH_3] = (0.10 - x) \simeq 0.10$, we obtain

$$K_b = \frac{[NH_4^+][OH^-]}{[NH_3]} = \frac{x \cdot x}{0.10} = 1.74 \times 10^{-5}$$

$$x^2 = 1.74 \times 10^{-6} \qquad x = [OH^-] = 1.32 \times 10^{-3}\,M$$

Now we can rephrase the question as: Should $Mg(OH)_2(s)$ precipitate from a solution in which $[Mg^{2+}] = 1.0 \times 10^{-2}\,M$ and $[OH^-] = 1.32 \times 10^{-3}\,M$? We must compare the ion product with K_{sp}.

$$[Mg^{2+}][OH^-]^2 = (1.0 \times 10^{-2})(1.32 \times 10^{-3})^2$$
$$= 1.74 \times 10^{-8} > K_{sp} = 1.8 \times 10^{-11}$$

Precipitation should occur.

SIMILAR EXAMPLES: Exercises 32, 35.

Example 18-11   What $[NH_4^+]$ must be maintained to prevent the precipitation of $Mg(OH)_2$ from a solution that is 0.010 M $MgCl_2$ and 0.10 M NH_3?

The maximum value the ion product may have before precipitation occurs is 1.8×10^{-11}. This allows us to determine the maximum $[OH^-]$ that can be tolerated.

$$[Mg^{2+}][OH^-]^2 = (1.0 \times 10^{-2})[OH^-]^2 = 1.8 \times 10^{-11}$$

$$[OH^-]^2 = 1.8 \times 10^{-9} \qquad [OH^-] = 4.2 \times 10^{-5}\,M$$

Next we determine what $[NH_4^+]$ must be present in 0.10 M NH$_3$ to maintain $[OH^-] = 4.2 \times 10^{-5}\,M$.

$$K_b = \frac{[NH_4^+][OH^-]}{[NH_3]} = \frac{[NH_4^+](4.2 \times 10^{-5})}{0.10} = 1.74 \times 10^{-5}$$

$$[NH_4^+] = 0.041\,M$$

SIMILAR EXAMPLE: Exercise 34.

The solution described in Example 18-11, since it contains both NH$_3$ and its cation, NH$_4^+$, is a *buffer solution*. What we have illustrated is how a buffer solution can be used to control a precipitation reaction.

Note from the K_{sp} values in Table 16-1 that MgCO$_3$ is more soluble than BaCO$_3$, CaCO$_3$, and SrCO$_3$.

In the qualitative analysis scheme outlined in Figure 16-3, the cation group 4 reagent was listed as NH$_3$–NH$_4$Cl–(NH$_4$)$_2$CO$_3$. The cations precipitating as carbonates in group 4 are Ba^{2+}, Ca^{2+}, and Sr^{2+}. Mg^{2+} remains in solution and continues on to cation group 5 (together with K$^+$, Na$^+$, and NH$_4^+$). The pH of the cation group 4 precipitating reagent is critical. If the pH is too low, the carbonates will not precipitate completely. If the pH is too high, Mg(OH)$_2$ will precipitate along with the Group 4 carbonates. The NH$_3$–NH$_4^+$ buffer maintains just the proper pH to accomplish the desired separation.

18-6 H$_2$S Equilibria Revisited—Precipitation and Solubilities of Metal Sulfides

As we learned in Section 17-6, hydrogen sulfide in aqueous solutions (hydrosulfuric acid) is a weak *di*protic acid. One of the conclusions we reached in our previous discussion of H$_2$S is that in an aqueous solution of H$_2$S, $[S^{2-}] = K_{a_2} = 1.0 \times 10^{-14}\,M$. With this value of $[S^{2-}]$ we can determine which metal ions can be precipitated as sulfides—an important question in the qualitative analysis scheme.

Example 18-12 Which of the following ions will precipitate from a solution that is kept saturated in H$_2$S and 0.010 M in each: Pb^{2+}, Zn^{2+}, and Mn^{2+}?

$$K_{sp}(PbS) = 8 \times 10^{-28} \qquad K_{sp}(ZnS) = 1.0 \times 10^{-21} \qquad K_{sp}(MnS) = 2.5 \times 10^{-13}$$

In a saturated H$_2$S solution, $[S^{2-}] = K_{a_2} = 1.0 \times 10^{-14}\,M$.

The following comparisons must be made.

$$[Pb^{2+}][S^{2-}] = (1.0 \times 10^{-2})(1.0 \times 10^{-14}) = 1.0 \times 10^{-16} > K_{sp} = 8 \times 10^{-28}$$

PbS precipitates.

$$[Zn^{2+}][S^{2-}] = (1.0 \times 10^{-2})(1.0 \times 10^{-14}) = 1.0 \times 10^{-16} > K_{sp} = 1.0 \times 10^{-21}$$

ZnS precipitates.

$$[Mn^{2+}][S^{2-}] = (1.0 \times 10^{-2})(1.0 \times 10^{-14}) = 1.0 \times 10^{-16} < K_{sp} = 2.5 \times 10^{-13}$$

MnS does not precipitate.

THE COMMON ION EFFECT IN H$_2$S SOLUTIONS.   Selective precipitation of sulfides cannot be accomplished very effectively in a solution containing only H$_2$S. In the qualitative analysis scheme the H_2S equilibria are altered by the presence of other substances. As we learned in the opening section of this chapter, the ionization of a weak acid can be repressed by the presence of a strong acid. Adding a strong acid (HCl) to an aqueous solution of H_2S reduces $[S^{2-}]$ in solution.

When $[H_3O^+]$ in solution is determined by some species other than the H_2S itself, calculations pertaining to H_2S equilibria are more readily made by using a *combination* of ionization constants. In a manner we have employed before, we can write

$$\begin{array}{llr} H_2S + H_2O \rightleftharpoons H_3O^+ + HS^- & K_{a_1} = 1.1 \times 10^{-7} & (18.28) \\ HS^- + H_2O \rightleftharpoons H_3O^+ + S^{2-} & K_{a_2} = 1.0 \times 10^{-14} & (18.29) \\ \hline H_2S + 2\,H_2O \rightleftharpoons 2\,H_3O^+ + S^{2-} & K = K_{a_1} \times K_{a_2} & (18.30) \end{array}$$

and

$$K_{a_1} \times K_{a_2} = \frac{[H_3O^+]^2[S^{2-}]}{[H_2S]} = (1.1 \times 10^{-7})(1.0 \times 10^{-14}) = 1.1 \times 10^{-21} \qquad (18.31)$$

Remember, in aqueous solutions containing only H_2S, $[H_3O^+]$ is *very much larger* than $[S^{2-}]$. It is not just twice as large, as equation (18.30) might seem to suggest. In such solutions practically all the H_3O^+ is produced in the first ionization step, together with HS^-. Very little of the HS^- ionizes further to produce S^{2-}.

Equation (18.30) states that for every molecule of H_2S that undergoes *complete* ionization, two H_3O^+ ions and one S^{2-} ion are formed. Equation (18.31) indicates that if any *two* of the three concentration terms it relates are known, the remaining one can be calculated. For solutions of pure H_2S in water, the only concentration term known is $[H_2S]$. Here the ionization steps must be considered separately and the methods of Example 17-9 used.

SELECTIVE PRECIPITATION OF METAL SULFIDES. In the qualitative analysis scheme (recall Figure 16-3) the *least* soluble of the metal sulfides of cation group 3 is ZnS and the *most* soluble of those of cation group 2 is CdS (see Table 16-1). To achieve an effective separation of these two groups of cations, precipitation must be carried out under conditions where CdS precipitates and ZnS does not. The necessary conditions are established in Example 18-13.

Example 18-13 What $[H_3O^+]$ must be maintained in a saturated H_2S solution (0.10 *M* H_2S) to precipitate CdS, but not ZnS, if Cd^{2+} and Zn^{2+} are each present initially at a concentration of 0.10 *M*?

$$CdS(s) \rightleftharpoons Cd^{2+}(aq) + S^{2-}(aq) \qquad K_{sp} = 8.0 \times 10^{-27}$$

$$ZnS(s) \rightleftharpoons Zn^{2+}(aq) + S^{2-}(aq) \qquad K_{sp} = 1.0 \times 10^{-21}$$

If ZnS is *not* to precipitate,

$$[Zn^{2+}][S^{2-}] < K_{sp} = 1.0 \times 10^{21}$$

$$(0.10)[S^{2-}] < 1.0 \times 10^{-21}$$

$$[S^{2-}] < 1.0 \times 10^{-20}$$

The maximum value of $[S^{2-}]$ before ZnS will precipitate is 1.0×10^{-20} *M*. The $[H_3O^+]$ required to maintain this $[S^{2-}]$ is

$$K_{a_1} \times K_{a_2} = \frac{[H_3O^+]^2[S^{2-}]}{[H_2S]} = \frac{[H_3O^+]^2(1.0 \times 10^{-20})}{0.10} = 1.1 \times 10^{-21}$$

$$[H_3O^+]^2 = 1.1 \times 10^{-2} \qquad [H_3O^+] = 1.0 \times 10^{-1} = 0.10\ M$$

That CdS should precipitate under these conditions is easily demonstrated.

$$[Cd^{2+}][S^{2-}] = (0.10)(1.0 \times 10^{-20}) = 1.0 \times 10^{-21} > K_{sp} = 8.0 \times 10^{-27}$$

Any concentration of $[H_3O^+]$ greater than 0.10 *M* ensures that no ZnS precipitates. In actual practice the cation group 2 precipitating reagent is kept at about pH = 0.5 ($[H_3O^+] = 0.3\ M$).

SIMILAR EXAMPLES: Exercises 38, 39.

DISSOLVING METAL SULFIDES.   The qualitative analysis scheme requires that sulfides be both precipitated and redissolved. To discuss the dissolving of metal sulfides it is again helpful to combine two equilibrium reactions. Consider, for example, the dissolving of $PbS(s)$ in $HCl(aq)$.

$$PbS(s) \rightleftharpoons Pb^{2+}(aq) + S^{2-}(aq)$$
$$K_{sp} = 8 \times 10^{-28} \qquad (18.32)$$

$$\underset{\text{(from HCl)}}{2\,H_3O^+} + \underset{\text{(from PbS)}}{S^{2-}} \rightleftharpoons H_2S(aq) + 2\,H_2O$$
$$K = \frac{1}{K_{a_1} \times K_{a_2}} = 9.1 \times 10^{20} \qquad (18.33)$$

$$PbS(s) + \underset{\text{(from HCl)}}{2\,H_3O^+} \rightleftharpoons Pb^{2+}(aq) + H_2S(aq) + 2\,H_2O$$
$$K = \frac{K_{sp}}{K_{a_1} \times K_{a_2}} = 7.3 \times 10^{-7} \qquad (18.34)$$

Equation (18.32) describes the solubility equilibrium for PbS; K_{sp} is the applicable equilibrium constant. Sulfide ion from PbS combines with H_3O^+ from HCl to produce H_2S. This equilibrium reaction is described by the *reciprocal* of $K_{a_1} \times K_{a_2}$. The net equation (18.34) indicates that as S^{2-} from PbS is converted to $H_2S(aq)$, Pb^{2+} appears in solution—the PbS dissolves. Because of the small value of K for reaction (18.34), however, we should not expect very much PbS to dissolve. In fact, we know that the reverse of reaction (18.34) is strongly favored in 0.3 M HCl: Pb^{2+} precipitates with the group 2 cations in the qualitative analysis scheme. Nevertheless, high concentrations of H_3O^+ will promote some dissolving of PbS, as illustrated in Example 18-14.

Example 18-14   What is the concentration of Pb^{2+} in a 1.0 M HCl solution that is saturated in PbS?

The relevant data are listed below equation (18.34), which represents the dissolving reaction.

$$PbS(s) + 2\,H_3O^+ \rightleftharpoons Pb^{2+} + H_2S(aq) + 2\,H_2O \qquad (18.34)$$

initial concentrations:	1.0 M		
changes:	$-2x\,M$	$+x\,M$	$+x\,M$
at equilibrium:	$(1.0 - 2x)\,M$	$x\,M$	$x\,M$

$$K = \frac{[Pb^{2+}][H_2S]}{[H_3O^+]^2} = \frac{x \cdot x}{(1.0 - 2x)^2} = 7.3 \times 10^{-7} \qquad (18.35)$$

If, as we suspect, x is very small, then $(1.0 - 2x) \simeq 1.0$.

$$\frac{x^2}{1} = 7.3 \times 10^{-7} \qquad x = [Pb^{2+}] = 8.5 \times 10^{-4}\,M$$

SIMILAR EXAMPLES: Exercises 37, 41.

Note that in writing equilibrium constant expression (18.35), no terms appear for the pure solid, $PbS(s)$, or for water.

18-7   Postscript: Equivalent Weight and Normality

The concept of equivalent weight was used in place of molecular weight during the early history of chemistry. It is still used to some extent in analytical chemistry. In this section we give brief consideration to equivalent weight and to a concentration scale based on equivalent weight—normality.

Let us define an **equivalent weight (or equivalent),** say of substances A and B, such that

One equiv of A reacts completely with one equiv of B,

that is,

Recall the significance of
the symbol ⇌, described
on page 12.

1 equiv A ⇌ 1 equiv B (18.36)

This definition of an equivalent obviously requires that we know the kind of reaction in which substances participate. For the present we will limit our discussion to acid-base reactions.

EQUIVALENT WEIGHT IN ACID-BASE REACTIONS. *An equivalent is a quantity of substance that will liberate or react with one mole of hydrogen ions.* The equivalent weight of HCl is equal to its molecular weight, that is, 36.46 g HCl/equiv HCl. The reaction of HCl with NaOH can be represented by the ionic equation

$$\underbrace{H^+(aq) + Cl^-(aq)}_{\substack{1 \text{ mol} = \\ 1 \text{ equiv}}} + \underbrace{Na^+(aq) + OH^-(aq)}_{\substack{1 \text{ mol} = \\ 1 \text{ equiv}}} \longrightarrow$$

$$HOH + Na^+(aq) + Cl^-(aq) \qquad (18.37)$$

An equivalent of NaOH is identical to a mole; the equivalent weight of NaOH is 40.00 g NaOH/equiv NaOH.

When completely neutralized with NaOH, sulfuric acid (H_2SO_4) yields *two* H^+ ions per molecule. That is,

$$\underbrace{2\,H^+(aq) + SO_4{}^{2-}(aq)}_{\substack{1 \text{ mol} \\ \frac{1}{2} \text{ mol} = \\ 1 \text{ equiv}}} + \underbrace{2\,Na^+(aq) + 2\,OH^-(aq)}_{\substack{2 \text{ mol} \\ 1 \text{ mol} = \\ 1 \text{ equiv}}} \longrightarrow$$

$$2\,H_2O + 2\,Na^+(aq) + SO_4{}^{2-}(aq) \qquad (18.38)$$

There are *two* equivalents in 1 mol of H_2SO_4. The equivalent weight is *one half* the molecular weight—49.04 g H_2SO_4/equiv H_2SO_4.

The situation with phosphoric acid, H_3PO_4, is rather complicated. If H_3PO_4 participates in reaction (18.39) its equivalent weight is equal to its molecular weight. In reaction (18.40) the equivalent weight is *one half* the molecular weight; and in reaction (18.41), *one third* the molecular weight.

$$H_3PO_4(aq) + NaOH(aq) \longrightarrow NaH_2PO_4(aq) + H_2O \qquad (18.39)$$

$$H_3PO_4(aq) + 2\,NaOH(aq) \longrightarrow Na_2HPO_4(aq) + 2\,H_2O \qquad (18.40)$$

$$H_3PO_4(aq) + 3\,NaOH(aq) \longrightarrow Na_3PO_4(aq) + 3\,H_2O \qquad (18.41)$$

Example 18-15   Determine the equivalent weight of the following substances: **(a)** HNO_3; **(b)** $Ba(OH)_2$.
(a) Since only 1 mol H^+ can be produced per mole of HNO_3,

1 equiv HNO_3 = 1 mol HNO_3

equiv wt HNO_3 = mol wt HNO_3 = 63.0 g HNO_3

(b) In water solution $Ba(OH)_2$ produces *2* mol OH^- for every mole of compound, and this 2 mol OH^- is capable of reacting with 2 mol H^+. One mole of $Ba(OH)_2$ is *2* equivalents.

2 equiv $Ba(OH)_2$ = 1 mol $Ba(OH)_2$

1 equiv $Ba(OH)_2$ = $\frac{1}{2}$ mol $Ba(OH)_2$

equiv wt $Ba(OH)_2$ = $\frac{1}{2}$f wt $Ba(OH)_2$ = $\frac{1}{2} \times 171.4$ = 85.7 g $Ba(OH)_2$

SIMILAR EXAMPLES: Exercises 49, 50.

NORMALITY. Normality concentration is defined in a similar fashion to molarity.

$$\text{normality } (N) = \frac{\text{number equiv solute}}{\text{number liters soln}} \qquad (18.42)$$

A 1 *normal* solution of HCl can be prepared by dissolving 36.5 g HCl (1 equiv) in 1 L of a water solution. Since this mass of HCl is also 1 mol, the solution is also 1 *molar*.

$$1\,N\,\text{HCl} = \frac{36.5\text{ g HCl}}{1.00\text{ L soln.}} = 1\,M\,\text{HCl}$$

Normality and molarity are not equal for an aqueous solution of $Ba(OH)_2$, however.

$$0.01\,M\,Ba(OH)_2 = \frac{0.01\text{ mol }Ba(OH)_2}{1.00\text{ L soln.}} = \frac{0.02\text{ equiv }Ba(OH)_2}{1.00\text{ L soln.}} = 0.02\,N\,Ba(OH)_2$$

The relationship between normality and molarity concentration is

$$\text{normality} = n \times \text{molarity} \qquad (18.43)$$

where n represents the number of moles of hydrogen ions per mole of compound that a solute is capable of releasing (acid) or reacting with (base).

Example 18-16 What are the normalities of the following solutions? **(a)** 0.50 M NaOH; **(b)** 0.02 M H_2SO_4.
(a) One mole of OH^- is produced for every mole of NaOH. The value of n in equation (18.43) is $n = 1$. Normality and molarity are identical for NaOH solutions. 0.50 M NaOH = 0.50 N NaOH.
(b) Two moles of H^+ are produced for every mole of H_2SO_4 involved in an acid-base reaction; $n = 2$. Normality = 2 × molarity = 2 × 0.02 M = 0.04 N H_2SO_4.

SIMILAR EXAMPLES: Exercises 51, 52.

SOLUTION STOICHIOMETRY AND NORMALITY CONCENTRATION. Equation (18.42) can be recast in a slightly different form.

$$\text{no. equiv solute} = \text{normality} \times \text{no. L soln.} \qquad (18.44)$$

Another common expression is obtained by dividing both sides of this equation by 1000. One-thousandth of a liter is a milliliter (ml) and one-thousandth of an equivalent is a **milliequivalent (mequiv).** Thus, normality concentration can be expressed either as equiv/L or mequiv/ml.

$$\text{no. mequiv solute} = \text{normality} \times \text{no. ml soln.} \qquad (18.45)$$

According to the definition of the equivalent (equation 18.36), we can write

no. equiv A $\backsim$ no. equiv B

$$V_A \cdot N_A \backsim V_B \cdot N_B \qquad (18.46)$$

In equation (18.46) the symbols N refer to normality concentration, and the subscripts indicate whether the solute is reactant A or B. The symbols V refer to the corresponding solution volumes. These volumes can be expressed either in liters or in milliliters, as long as the same unit is used for each.

Example 18-17 The concentration of an HCl(aq) solution is established by titration with a standard solution of $Ba(OH)_2$(aq). The HCl(aq) is then used to determine the normality of an NaOH(aq) solution. A 25.00-ml sample of the HCl(aq) is titrated with 30.08 ml 0.1000 N $Ba(OH)_2$. A 10.00-ml sample of the NaOH(aq) requires 25.10 ml of the HCl(aq) for its titration. What is the normality of the NaOH(aq)?

Standardization of HCl(aq):

$$V_{HCl} \times N_{HCl} = V_{Ba(OH)_2} \times N_{Ba(OH)_2}$$

$$25.00 \text{ ml} \times N_{HCl} = 30.08 \text{ ml} \times \frac{0.1000 \text{ mequiv}}{\text{ml}}$$

$$N_{HCl} = 0.1203 \frac{\text{mequiv}}{\text{ml}}$$

Determination of NaOH(aq):

$$V_{NaOH} \times N_{NaOH} = V_{HCl} \times N_{HCl}$$

$$10.00 \text{ ml} \times N_{NaOH} = 25.10 \text{ ml} \times \frac{0.1203 \text{ mequiv}}{\text{ml}}$$

$$N_{NaOH} = 0.3020 \frac{\text{mequiv}}{\text{ml}}$$

SIMILAR EXAMPLES: Exercises 53, 54, 55.

The concepts of equivalent weight and normality do not offer any important new possibilities for problem solving. Example 18-17 could have been done just as well with moles and molar concentrations.

You will notice that in Example 18-17 we wrote no chemical equations. This represents one of the principal advantages of normality concentration. When expressed in equivalents, reactants always combine in a 1:1 ratio and equation (18.46) is applicable. Even if the chemical formula of a substance is not known, its equivalent weight and normality concentrations with respect to certain types of reactions can still be established. These advantages, however, are offset by this principal disadvantage: *The equivalent weight of a substance and the normality concentration of its solutions depend on the particular reaction in which the substance participates* (recall equations 18.39 through 18.41). Equivalent weight and normality often *do not* have unique values, as do formula (molecular) weight and molarity. Although equivalent weight and normality will continue to be used for some time, it is likely that their use will decrease in importance. This is especially true now that the mole has been established as the base unit for amount of substance in the SI system.

Summary

The degree of ionization of a weak acid in aqueous solution can be reduced by adding to the solution either a strong acid or a salt of the weak acid. In each case this occurs through the common ion effect. In a strong acid/weak acid solution, $[H_3O^+]$ and pH are established by the strong acid. The weak acid/salt combination is an especially significant one: A solution with these two components resists changes in its pH—it is a buffer solution. A weak base/salt combination is also a buffer.

The typical buffer solution has one component to react with small added amounts of acid and another to react with bases. The pH at which a buffer solution functions is determined by the pK value of the weak acid or base and the *ratio* of molar concentration of salt to that of weak acid or base: $\text{pH} = \text{p}K_a + \log([A^-]/[HA])$ and $\text{pOH} = \text{p}K_b + \log([BH^+]/[B])$ [where A is a weak acid (e.g., $HC_2H_3O_2$) and B is a weak base (e.g., NH_3)]. The molar concentrations of salt and weak acid or weak base also establish the buffer capacity—the amount of added acid or base that the buffer is capable of neutralizing. Buffer action plays a critical role in the functioning of blood and other fluids in living organisms. Buffers are widely used in the procedures of analytical chemistry, and they are used in industrial processes.

In a titration solutions of an acid and a base are allowed to react to the point where exact neutralization occurs. At this point (the equivalence point) there is neither excess acid nor base in solution but simply the salt produced by their neutralization. Whether this salt solution is acidic, basic, or neutral is determined by whether the salt can hydrolyze or undergo further ionization. A titration curve represents the pH of the solution being titrated as a function of the volume of titrating agent (titrant) added. Generally, a titration curve shows a sharp change in solution pH at the equivalence point. The object in choosing an indicator for a titration is to find one that changes color as close to the equivalence point as possible. An acid-base indicator is a weak acid, and the selection of an appropriate indicator for a titration requires matching its pK_a value to the pH at the equivalence point.

By using the common ion effect to alter equilibria of weak acids and bases, other solution equilibria can be affected as well. For example, by controlling pH through the use of buffers, solution concentrations of such ions as OH^-, $CO_3{}^{2-}$, and S^{2-} can be established and maintained over an extreme range. Such control allows for the selective precipitation or dissolving of ionic compounds—procedures that are at the very core of qualitative analysis.

All of the aqueous solution equilibria presented in this and the preceding two chapters can be described through mathematical equations and calculations. Where qualitative information is all that is required, net ionic equations describe matters very nicely.

Learning Objectives

As a result of studying Chapter 18, you should be able to

1. Describe the effect of common ions on the ionization of weak acids and bases, and calculate the concentrations of species present in solutions of weak acids or bases and their common ions.

2. Explain why pure water cannot resist changes in pH and the way in which buffer solutions work to maintain a constant pH.

3. Derive and use the basic equations that relate to the pH of buffer solutions.

4. Calculate the pH of a buffer solution from concentrations of the buffer components and a value of K_a or K_b, and describe how to prepare a buffer having a specific pH.

5. Determine the changes in pH of a buffer solution that result from the addition of acids or bases.

6. Define the terms "buffer range" and "buffer capacity" and establish numerical values for them.

7. Describe how the blood buffer system works.

8. Explain how an acid-base indicator works.

9. Calculate the pH values necessary to plot titration curves for various combinations of weak and strong acids and bases.

10. Plot titration curves and extract significant information from such curves—initial pH, buffer region, equivalence point, indicator selection.

11. Describe, through chemical equations and through calculations, the effect of pH on the precipitation and solubilities of slightly soluble substances.

12. Relate $[H_3O^+]$ and $[S^{2-}]$ in $H_2S(aq)$ solutions through a combined equilibrium constant, $K_{a_1} \times K_{a_2}$.

13. Predict conditions under which metal sulfides will precipitate from solution and conditions under which they will dissolve.

14. Apply the concepts of equivalent weight and normality to solution stoichiometry problems, especially those involving titrations.

Some New Terms

An **acid-base indicator** is a substance that can be used to measure the pH of a solution. The indicator takes on one color when in its nonionized acid form and a different color in its anion form. Its color in a particular solution depends on which form predominates.

A **buffer** solution resists a change in pH. It contains components capable of reacting with (neutralizing) small added amounts of acid or base.

Buffer capacity refers to the amount of acid and/or base that a buffer solution can neutralize and still maintain an essentially constant pH.

Buffer range refers to the range of pH values over which a particular buffer system will function.

The **end point of a titration** is the point in the titration where the indicator used undergoes a color change. A properly chosen indicator for a titration must have its end point correspond as closely as possible to the equivalence point of the titration reaction.

The **equivalence point of a neutralization reaction** refers to the condition in which an acid and a base neutralize one another.

Equivalent weight is a definition of a quantity of substance based on the extent to which it enters into certain kinds of reactions. For example, an equivalent weight of acid liberates 1 mol of H^+, and an equivalent weight of base reacts with 1 mol of H^+.

The **Henderson–Hasselbalch equation** is the equation relating pH, pK_a, and the concentrations of weak acid and salt in a buffer solution: $\mathbf{pH} = \mathbf{p}K_a + \log([A^-]/[HA])$. [For a buffer consisting of a weak base and its salt, $\mathbf{pOH} = \mathbf{p}K_b + \log([BH^+]/[B])$.]

Normality (N) is a concentration unit used in conjunction with the concept of equivalent weight. Normality is the number of equivalents of solute per liter of solution.

A **titration curve** is a graph of solution pH versus volume of titrant added. It outlines how pH changes during the course of an acid-base titration, and it can be used to establish such features as the equivalence point of the titration.

Exercises

The common ion effect in acid-base equilibria (Use data from Table 17-2, as necessary.)

1. Describe the effect on the pH of the solution produced by adding   **(a)** $NaNO_2$ to $HNO_2(aq)$;   **(b)** $NaNO_3$ to $HNO_3(aq)$. Why are the effects not the same? Explain.

2. Calculate the concentration of the ionic species indicated.
   (a) $[H_3O^+]$ in a solution that 0.100 *M* HCl and 0.100 *M* HClO
   (b) $[NO_2^-]$ in a solution that is 0.100 *M* $NaNO_2$ and 0.0100 *M* HNO_2
   (c) $[Cl^-]$ in a solution that is 0.200 *M* HCl and 0.0500 *M* $HC_2H_3O_2$
   (d) $[C_2H_3O_2^-]$ in a solution that is 0.100 *M* HCl and 0.300 *M* $HC_2H_3O_2$
   (e) $[OH^-]$ in a solution that is 0.200 *M* $(NH_4)_2SO_4$ and 0.500 *M* NH_3

3. Lactic acid, $HC_3H_5O_3$, is found in sour milk. A solution prepared by dissolving 10.0 g $NaC_3H_5O_3$ in 100.0 ml of 0.0500 *M* $HC_3H_5O_3$ has pH = 4.11. What is K_a of lactic acid?

4. What is the pOH of a solution obtained by adding 1.20 mg of aniline hydrochloride ($C_6H_5NH_3^+Cl^-$) to 2.00 L of 0.100 *M* aniline ($C_6H_5NH_2$)?

$$C_6H_5NH_2 + H_2O \rightleftharpoons C_6H_5NH_3^+ + OH^-$$
$$K_b = 4.30 \times 10^{-10}$$

*5. You are given 250.0 ml of 0.100 *M* $HC_3H_7O_2(aq)$ (propionic acid, $K_a = 1.35 \times 10^{-5}$). You wish to adjust the pH of this acid by adding an appropriate solution to it. What volume, in ml, would you add of   **(a)** 1.00 *M* HCl to lower its pH to 1.00;   **(b)** 1.00 *M* $NaC_3H_7O_2$ to raise its pH to 4.00;   **(c)** water to raise its pH by 0.15 unit?

Buffer solutions (Use data from Tables 17-2 and 17-3 as necessary.)

6. Indicate which of the following aqueous solutions are buffer solutions. Explain your reasoning. (*Hint:* You must also consider whether any reactions occur between the solution components listed.)
   (a) 0.100 *M* NaCl
   (b) 0.100 *M* NaCl–0.100 *M* NH_4Cl
   (c) 0.100 *M* CH_3NH_2 – 0.150 *M* $CH_3NH_3^+Cl^-$

   (d) 0.100 *M* HCl–0.050 *M* $NaNO_2$
   (e) 0.100 *M* HCl–0.200 *M* $NaC_2H_3O_2$

7. In the text the $H_2PO_4^-/HPO_4^{2-}$ combination is mentioned as playing a role in maintaining the pH of blood.
   (a) Write equations to show how a solution containing these ions functions as a buffer.
   (b) Verify that this buffer mixture is most effective at pH = 7.2.
   (c) Calculate the pH of a buffer solution in which $[H_2PO_4^-] = 0.050$ *M*   and $[HPO_4^{2-}] = 0.150$ *M*. (*Hint:* Recall that phosphoric acid ionizes in three distinct steps and focus on the second step.)

8. Calculate the pH values of the following buffer solutions.
   (a) 0.125 *M* $HCHO_2$–0.082 *M* $NaCHO_2$
   (b) 0.165 *M* HClO–0.216 *M* NaClO
   (c) 1.52 *M* NH_3–0.660 *M* $(NH_4)_2SO_4$

9. What mass of $(NH_4)_2SO_4$, in grams, must be added to 250.0 ml of 0.100 *M* NH_3 to yield a solution with pH = 9.35?

10. You are given the task of preparing a buffer solution with pH = 3.50 and have available the following solutions, all 0.100 *M*: $HCHO_2$, $HC_2H_3O_2$, H_3PO_4, $NaCHO_2$, $NaC_2H_3O_2$, and NaH_2PO_4. Describe how you would prepare this buffer solution. (*Hint:* What volumes of which solutions would you use?)

*11. Suppose that the task in the preceding exercise were modified to require exactly 1.00 L of the buffer solution with pH = 3.50 and that the solutions available were 0.10 *M* $NaCHO_2$, 0.10 *M* $NaC_2H_3O_2$, 0.10 *M* NaH_2PO_4, and 1.00 *M* HCl. Describe how you would prepare this buffer solution.

12. Verify the statement in the text that a ratio of $[HCO_3^-]/[H_2CO_3] \simeq 20$ should correspond to a pH of about 7.7 for a carbonic acid–bicarbonate ion buffer. Would a buffer of $HC_2H_3O_2$ and $NaC_2H_3O_2$ function well with $[C_2H_3O_2^-]/[HC_2H_3O_2] \simeq 20$? Explain.

13. Compare the following buffers with respect to their **(a)** pH and   **(b)** buffer capacities: 0.010 *M* $HC_2H_3O_2$–0.010 *M* $NaC_2H_3O_2$ and 1.0 *M* $HC_2H_3O_2$–0.50 *M* $NaC_2H_3O_2$.

14. With respect to the buffer solution described in Example 18-4, that is, 0.150 *M* $HC_2H_3O_2$–0.50 *M* $NaC_2H_3O_2$,

(a) What is the pH if 1.00 ml of 12 M HCl is added to 100.0 ml of the buffer?

(b) What is the pH if 5.00 ml of 12 M HCl is added instead?

15. A buffer solution is prepared by dissolving 1.00 g NH_3 and 5.00 g $(NH_4)_2SO_4$ in 0.500 L of water solution.

(a) What is the pH of this solution?

(b) If 1.00 g NaOH is added to this solution, what is the pH?

(c) How many drops of 12 M HCl must be added to 0.500 L of the original buffer solution to change its pH to 9.00? (1 drop $\simeq$ 0.05 ml.)

16. An acetic acid–sodium acetate buffer can be prepared by allowing an *excess* of $NaC_2H_3O_2$ to react with HCl. The strong acid HCl is converted to the weak acid $HC_2H_3O_2$.

$$C_2H_3O_2^- \quad + \quad H_3O^+ \quad \longrightarrow \quad HC_2H_3O_2 + H_2O$$
(from $NaC_2H_3O_2$)   (from HCl)

(a) If 10.0 g $NaC_2H_3O_2$ is added to 0.300 L of 0.200 M HCl, what is the pH of the resulting solution?

(b) If 1.00 g $Ba(OH)_2$ is added to the solution in part (a), what happens to the pH?

(c) What is the capacity of the buffer solution in part (a) toward $Ba(OH)_2$?

(d) What is the pH of the solution in part (a) following the addition of 5.2 g $Ba(OH)_2$?

Acid-base indicators (Use data from Tables 17-2 and 17-3 as necessary.)

17. In the use of acid-base indicators

(a) Why is it generally sufficient to use a *single* indicator in an acid-base titration but often necessary to use *several* indicators to establish the approximate pH of a solution?

(b) Why do you suppose the amount of indicator used in an acid-base titration must be kept as small as possible?

★(c) How would you account for the fact that thymol blue actually can be used in *two* different pH ranges? Its color changes from red to yellow in the range pH = 1.2 to 2.8, and from yellow to blue in the range pH = 8.0 to 9.6.

18. A handbook lists the following data for some acid-base indicators.

		Color change
bromphenol blue	$K_a = 1.41 \times 10^{-4}$	yellow → blue
		(acid) (anion)
bromcresol green	$K_a = 2.09 \times 10^{-5}$	yellow → blue
bromthymol blue	$K_a = 7.9 \times 10^{-8}$	yellow → blue
2,4-dinitrophenol	$K_a = 1.26 \times 10^{-4}$	colorless → yellow
chlorophenol red	$K_a = 1.0 \times 10^{-6}$	yellow → red
thymolphthalein	$K_a = 1.0 \times 10^{-10}$	colorless → blue

(a) Which of these indicators change color in acidic solution, which in basic solution, and which near the neutral point?

(b) What is the approximate pH of a solution if bromcresol green indicator assumes a green color; if chlorophenol red assumes an orange color?

19. With reference to the indicators listed in Exercise 18, what would be the color of each combination?

(a) 2,4-dinitrophenol when placed in 0.100 M HCl(aq)

(b) chlorophenol red in 1.00 M NaCl

(c) thymolphthalein in 1.00 M NH_3

(d) bromthymol blue in 1.00 M NH_4NO_3

(e) bromcresol green in seawater (recall Figure 17-3)

(f) bromphenol blue in saturated CO_2(aq) [Consider CO_2(aq) to be 0.034 M H_2CO_3.]

20. The indicator methyl red has a $pK_a = 4.95$. It changes color from red to yellow over the pH range 4.4 to 6.2.

(a) If the indicator is placed in a buffer solution that is 0.10 M $HC_2H_3O_2$–0.10 M $NaC_2H_3O_2$, what percent of the indicator will be in the acid form and what percent in the anion form?

(b) Which form of the indicator do you think has the "stronger" (more visible) color, the acid form (red) or the anion form (yellow)? Explain your reasoning.

Neutralization reactions

21. A 22.50-ml sample of 0.0500 M H_2SO_4 is required to titrate a 25.00-ml sample of an unknown strong base. What is [OH$^-$] in the base?

22. Excess $Ca(OH)_2$(s) is shaken with water to produce a saturated solution. A 50.00-ml sample of the clear, saturated solution is withdrawn and titrated. What volume of 0.1032 M HCl is required for this titration? (*Hint:* Refer to Table 16-1 for relevant data.)

23. Two solutions are mixed: (A) 100.0 ml of an HCl(aq) with pH = 2.50 and (B) 100.0 ml of an NaOH(aq) with pH = 11.00. What is the pH of the solution that results from mixing A and B?

★24. With reference to Exercise 23, if solution A were 100.0 ml of a weak acid solution with pH = 2.50 and solution B, 100.0 ml of a weak base solution with pH = 11.00

(a) Would the final pH on mixing the two solutions be greater or less than that calculated in Exercise 23? Explain.

(b) What additional information would be required to calculate the actual pH of the mixed solution?

Titration curves

25. In the text several differences were pointed out between the titration curves for a strong acid by a strong base and a weak acid by a strong base (i.e., between Figures 18-3 and 18-4). One point that is identical in the two curves, however, is the volume of 0.1000 M NaOH required to reach the equivalence point. Explain why this should be so.

26. Sketch the following titration curves. Indicate the initial pH and the pH corresponding to the equivalence point. Indicate the volume of titrant required to reach the equivalence point, and select a suitable indicator from Figure 18-2.
   (a) 25.0 ml 0.100 M KOH by 0.200 M HI
   (b) 10.0 ml 1.00 M NH_3 by 0.250 M HCl

27. Sketch a series of titration curves for the following three hypothetical weak acids when titrated with 0.1000 M NaOH. Select suitable indicators for the titrations. (*Hint:* Select a few key points at which to estimate the pH of the solution, for example, the initial pH, the condition pH = pK_a, and the pH at the equivalence point.)
   (a) 10.00 ml of 0.1000 M HX; $K_a = 1.0 \times 10^{-3}$
   (b) 10.00 ml of 0.1000 M HY; $K_a = 1.0 \times 10^{-5}$
   (c) 10.00 ml of 0.1000 M HZ; $K_a = 1.0 \times 10^{-7}$

28. A 10.00-ml solution that is 0.0400 M H_3PO_4 and 0.0150 M NaH_2PO_4 is titrated with 0.0200 M NaOH. Sketch the titration curve obtained. (*Hint:* How many equivalence points are there? Are they equally spaced along the volume axis as in Figure 18-5?)

29. Determine the pH at the point where the original acid is 90% neutralized in the titration of
   (a) 25.00 ml of 0.1000 M HCl by 0.1000 M NaOH (see Figure 18-3)
   (b) 25.00 ml of 0.1000 M $HC_2H_3O_2$ by 0.1000 M NaOH (see Figure 18-4)

*30. Thymol blue in its acid range is not a suitable indicator for the titration of HCl by NaOH. Suppose that a student uses thymol blue by mistake in the titration of Figure 18-3, and suppose that the indicator end point is taken to be at pH = 2.
   (a) Would there be a sharp color change [i.e., produced by a single drop of titrant solution (0.1000 M NaOH)]?
   (b) What percent of the original HCl remains unneutralized at this point?

*31. What volume of 0.1000 M NaOH must be added to just reach the following?
   (a) a pH of 3.0 in the titration of 25.00 ml 0.1000 M HCl (Figure 18-3)
   (b) a pH of 5.25 in the titration of 25.00 ml of 0.1000 M $HC_2H_3O_2$ (Figure 18-4)
   (c) a pH of 7.50 in the titration of 10.00 ml of 0.1000 M H_3PO_4 (Figure 18-5)

Solubility and pH (Use data from tables in Chapters 16 and 17 as necessary.)

32. Should the following precipitates form under the given conditions?
   (a) $PbI_2(s)$ from a solution that is 2×10^{-3} M HI, 2×10^{-3} M NaI, and 1×10^{-3} M $Pb(NO_3)_2$
   (b) $Mg(OH)_2(s)$ from 2.50 L of 0.0150 M $Mg(NO_3)_2$ to which is added 1 drop (0.05 ml) of 1.00 M NH_3

(c) $Al(OH)_3(s)$ from a solution that is 1.0×10^{-2} M in Al^{3+}, 0.01 M $HC_2H_3O_2$, and 0.01 M $NaC_2H_3O_2$

33. The solubility of $Mg(OH)_2$ in a particular buffer solution is found to be 1.00 g/L. What must be the pH of this buffer solution?

34. To 0.350 L of 0.100 M NH_3 is added 0.150 L of 0.100 M $MgCl_2$.
   (a) Show that $Mg(OH)_2$ should precipitate from this solution.
   (b) What mass of $(NH_4)_2SO_4$ must be added to cause the $Mg(OH)_2$ to redissolve?

35. Will $Fe(OH)_3(s)$ precipitate from a buffer solution that is 0.50 M $HC_2H_3O_2$ and 0.15 M $NaC_2H_3O_2$, if the solution is also made to be 0.25 M in Fe^{3+}?

36. A handbook lists the solubility of $CaHPO_4$ as 0.32 g $CaHPO_4 \cdot 2\ H_2O$/L. The same handbook gives the K_{sp} value as

$$CaHPO_4(s) \rightleftharpoons Ca^{2+}(aq) + HPO_4^{2-}(aq)$$
$$K_{sp} = 1 \times 10^{-7}$$

   (a) Are these data consistent with one another? (That is, are the molar solubilities the same when derived in two different ways?)
   (b) How do you account for the "discrepancy"? (*Hint:* Recall the nature of phosphate species in solution.)

37. Use the following data to calculate the molar solubility of $Mg(OH)_2$ in 1.00 M $NH_4Cl(aq)$. (*Hint:* Use the method of Example 18-14.)

$$Mg(OH)_2(s) \rightleftharpoons Mg^{2+}(aq) + 2\ OH^-(aq)$$
$$K_{sp} = 1.8 \times 10^{-11}$$

$$NH_3 + H_2O \rightleftharpoons NH_4^+ + OH^-$$
$$K_b = 1.74 \times 10^{-5}$$

Precipitation and solubilities of metal sulfides

38. Should FeS precipitate from a solution that is saturated in H_2S (0.10 M), 0.0020 M in Fe^{2+}, and at a pH = 3.5? [$K_{sp}(FeS) = 6.3 \times 10^{-18}$.]

39. What pH must be maintained in a solution saturated in H_2S (0.10 M H_2S) and 10^{-3} M in Zn^{2+} to prevent ZnS from precipitating? [$K_{sp}(ZnS) = 1 \times 10^{-21}$.]

40. In the qualitative analysis of a mixture of cations, Pb^{2+} is first precipitated as $PbCl_2$. Later, PbS is precipitated when the saturated $PbCl_2$ solution is saturated with H_2S and its pH adjusted to about 0.5. Show that in fact the precipitation of PbS should occur under these conditions.

$$PbCl_2(s) \rightleftharpoons Pb^{2+} + 2\ Cl^- \qquad K_{sp} = 1.6 \times 10^{-5}$$
$$PbS(s) \rightleftharpoons Pb^{2+} + S^{2-} \qquad K_{sp} = 8 \times 10^{-28}$$

41. What is the solubility of FeS, in g/L, in a buffer solution that is 0.500 M $HC_2H_3O_2$–0.250 M $NaC_2H_3O_2$? (*Hint:*

Assume that $[H_3O^+]$ remains constant in the dissolving reaction.) $[K_{sp}(FeS) = 6.3 \times 10^{-18}$; $K_a(HC_2H_3O_2) = 1.74 \times 10^{-5}$; $K_{a_1} \times K_{a_2}(H_2S) = 1.1 \times 10^{-21}.]$

*42. Calculate the molar solubility of CoS(s) in 0.100 M $HC_2H_3O_2$. (*Hint:* One approach involves combining three equilibrium equations.) $[K_a(HC_2H_3O_2) = 1.74 \times 10^{-5}$; $K_{sp}(CoS) = 4 \times 10^{-21}$; $K_{a_1} \times K_{a_2}(H_2S) = 1.1 \times 10^{-21}.]$

Net ionic equations

43. Predict whether a reaction will occur in each of the following cases. If so, write a balanced equation and indicate whether the reaction goes to completion or is reversible.
   (a) $NH_4^+ + C_2H_3O_2^- + H_3O^+ + I^- \longrightarrow$
   (b) $Na^+ + I^- + Zn^{2+} + SO_4^{2-} \longrightarrow$
   (c) $Na^+ + HPO_4^{2-} + H_2O \longrightarrow$
   (d) $NH_4^+ + NO_3^- + Ba^{2+} + OH^- \longrightarrow$
   (e) $Al(OH)_3(s) + HC_2H_3O_2(aq) \longrightarrow$
   (f) $Na^+ + C_3H_5O_2^- + H_3O^+ + Br^- \longrightarrow$
   (g) $HNO_2(aq) + H_3O^+ + Cl^- \longrightarrow$
   (h) $NH_3(aq) + H_2CO_3(aq) \longrightarrow$

44. The following pertain to the precipitation or dissolving of metal sulfides under the conditions indicated. In each case predict whether a reaction proceeds to any significant extent in the forward direction (to the right).
   (a) $Cu^{2+}(aq) + H_2S(\text{satd. aq}) \longrightarrow$
   (b) $Mg^{2+}(aq) + H_2S(\text{satd. aq}) \xrightarrow{0.3\,M\,HCl}$
   (c) $PbS(s) + HCl(0.3\,M) \longrightarrow$
   (d) $Ag^+(aq) + H_2S(\text{satd. aq}) \xrightarrow{0.3\,M\,HCl}$

*45. Write net ionic equations to represent the following observations.
   (a) When a concentrated aqueous solution of $CaCl_2$ is added to $Na_2HPO_4(aq)$, a white precipitate is formed that is 38.7% Ca, by mass.
   (b) When a piece of dry ice $[CO_2(s)]$ is placed into a clear saturated solution of "lime water" $[Ca(OH)_2(aq)]$, bubbles of gas are evolved. At first a white precipitate forms, but then it redissolves.

Applications of various equilibrium principles (Use tables from Chapters 16 and 17 as necessary.)

46. What aqueous concentrations of the following substances are required to produce the desired pH values?
   (a) $HC_2H_3O_2$ in 0.050 M $NaC_2H_3O_2$ to obtain a pH of 4.22
   (b) $Ba(OH)_2$ to produce a solution with pH = 12.65
   (c) Aniline, $C_6H_5NH_2$, to obtain a pH of 8.95
   (d) NH_4Cl to obtain a pH of 5.5

47. Use appropriate values of equilibrium constants to determine whether a solution can be simultaneously
   (a) 0.10 M NH_3 *and* 0.10 M NH_4Cl, with pH = 5.0

   (b) 0.10 M $NaC_2H_3O_2$ *and* 0.20 M HI
   (c) 0.10 M KCl *and* 0.50 M $NaNO_3$
   (d) 1×10^{-3} M $Pb(NO_3)_2$ *and* 1×10^{-4} M K_2CrO_4
   (e) 1.00 M $MgCl_2$ *and* pH = 8.0.

48. The single equilibrium equation written below can be applied to different phenomena described in this or the preceding chapter.

$$HC_2H_3O_2 + H_2O \rightleftharpoons H_3O^+ + C_2H_3O_2^-$$
$$K_a = 1.74 \times 10^{-5}$$

Indicate the phenomenon—ionization of pure acid, common ion effect, buffer action, hydrolysis—for each of the following combinations of concentrations.
   (a) $[H_3O^+]$ and $[HC_2H_3O_2]$ are high; $[C_2H_3O_2^-]$ is very low.
   (b) $[C_2H_3O_2^-]$ is high; $[HC_2H_3O_2]$ and $[H_3O^+]$ are very low.
   (c) $[HC_2H_3O_2]$ is high; $[H_3O^+]$ and $[C_2H_3O_2^-]$ are low.
   (d) $[HC_2H_3O_2]$ and $[C_2H_3O_2^-]$ are high; $[H_3O^+]$ is low.

Equivalent weight and normality concentration

49. Calculate the equivalent weights of the following substances for use in acid-base reactions: (a) $HClO_4$; (b) $Mg(OH)_2$;   (c) $HC_3H_5O_2$ (propionic acid).

50. In aqueous solutions carbonate ion, CO_3^{2-}, can react with *two* H^+ ions yielding one molecule of H_2O and one of CO_2. What mass of $Na_2CO_3 \cdot 10\ H_2O$, in grams, is required to produce 2.00 L of 0.175 N $Na_2CO_3(aq)$?

51. What are the normality concentrations corresponding to the following molarities? Indicate any cases in which more than one normality seems possible.   (a) 0.24 M KOH(aq); (b) 0.001 M HI;   (c) 2.0×10^{-3} M $Ca(OH)_2$;   (d) 0.15 M H_3PO_4; e (e) 0.01 M C_6H_5COOH (benzoic acid).

52. It is desired to prepare a standard solution of $Ba(OH)_2$ for use in acid-base titrations. What is the approximate maximum *normality* solution that can be prepared if the solubility of barium hydroxide is 3.89 g $Ba(OH)_2$ per 100 g of solution at 20°C?

53. What volume of 0.115 N NaOH is required for the complete neutralization of 10.00 ml of 0.188 N H_2SO_4?

54. A 25.00-ml sample of $H_3PO_4(aq)$ requires 31.15 ml of 0.242 N KOH for its titration in reaction (18.40). What is the normality of this $H_3PO_4(aq)$ if it is always to be used   (a) in reaction (18.39);   (b) in reaction (18.40);   (c) in reaction (18.41)?

55. A sample of battery acid is to be analyzed for its sulfuric acid content. A 1.00-ml sample weighs 1.239 g. This sample is diluted to 250.0 ml and 10.00 ml of the diluted acid requires 32.40 ml 0.0100 N $Ba(OH)_2$ for its titration. What is the mass percent H_2SO_4 in the battery acid?

Additional Exercises

1. In Example 17-8, the percent ionization of $HC_2H_3O_2$ was calculated for (a) 1.0 M, (b) 0.10 M, and (c) 0.010 M $HC_2H_3O_2$. Recalculate the percent ionization if each of the three solutions is also made 0.10 M in $NaC_2H_3O_2$. Explain why these results differ from those of Example 17-8.

2. Although we say that a buffered solution has a constant pH and $[H_3O^+]$, small changes in pH do translate into significant changes in $[H_3O^+]$. What is the *percent* increase in $[H_3O^+]$ when the pH of blood drops from 7.4 to 7.3?

3. A buffer solution is prepared by dissolving 1.00 g each of benzoic acid, $HC_7H_5O_2$, and sodium benzoate, $NaC_7H_5O_2$, in 150.0 ml of water solution.

$$HC_7H_5O_2 + H_2O \rightleftharpoons H_3O^+ + C_7H_5O_2^-$$
$$K_a = 6.3 \times 10^{-5}$$

(a) What is the pH of this buffer solution?
(b) Which buffer component, and in what quantity, must be added to the solution to change its pH to 4.00?

4. A solution is 0.350 M $HCHO_2$ (formic acid). What mass of sodium formate ($NaCHO_2$) must be added to 250.0 ml of this solution to produce a buffer solution of pH = 3.50? [$K_a(HCHO_2) = 1.8 \times 10^{-4}$.]

5. If to 100.0 ml of the buffer solution prepared in the preceding exercise is added 0.50 ml 15 M NH_3, what will be the pH of the resulting solution?

6. In what approximate pH range would you expect each of the following buffer solutions to be most effective?
(a) HNO_2–$NaNO_2$;   (b) CH_3NH_2–$CH_3NH_3^+Cl^-$;
(c) NH_3–$(NH_4)_2SO_4$.

7. It is desired to bring the pH of 0.500 L of 0.500 M $NH_4Cl(aq)$ to a value of 7.0. How many drops (1 drop = 0.05 ml) of which of the following solutions would you use: 10.0 M HCl, 10.0 M NH_3?

8. Use data from Table 17-3, together with the K_{sp} value below, to determine if 1.00 g $H_2C_2O_4$ (oxalic acid) can be dissolved in 0.200 L of 0.150 M $CaCl_2$ without the formation of a precipitate.

$$CaC_2O_4(s) \rightleftharpoons Ca^{2+}(aq) + C_2O_4^{2-}(aq)$$
$$K_{sp} = 1.3 \times 10^{-9}$$

9. It is desired to carry out the separation of Fe^{2+} and Mn^{2+} from an acetic acid–sodium acetate buffer by precipitating $FeS(s)$ and leaving Mn^{2+} in solution.
(a) Show that this can be accomplished in a solution that is 0.50 M $HC_2H_3O_2$ and 0.40 M $NaC_2H_3O_2$.
(b) What is the pH of the solution *after* precipitation of $FeS(s)$? [*Hint:* Why doesn't the pH remain constant?] The following conditions apply.
$[Fe^{2+}] = 0.10$ M; $[Mn^{2+}] = 0.10$ M; $[H_2S] = 0.10$ M;
$K_{sp}(FeS) = 6.3 \times 10^{-18}$; $K_{sp}(MnS) = 2.5 \times 10^{-13}$;

$K_a(HC_2H_3O_2) = 1.74 \times 10^{-5}$;
$K_{a_1} \times K_{a_2}(H_2S) = 1.1 \times 10^{-21}$.

*10. The titration of a weak acid by a weak base is not a particularly satisfactory procedure because the pH does not increase sharply at the equivalence point. Demonstrate this fact by sketching a titration curve for the neutralization of 10.00 ml of 0.100 M $HC_2H_3O_2$ by 0.100 M NH_3.

*11. At times a salt of a weak base can be titrated by a strong base. Use appropriate data from the text to sketch a titration curve for the titration of 20.00 ml of 0.0500 M $C_6H_5NH_3^+Cl^-$ by 0.1000 M NaOH.

*12. Carbonic acid is a weak diprotic acid (H_2CO_3), having $K_{a_1} = 4.2 \times 10^{-7}$ and $K_{a_2} = 5.6 \times 10^{-11}$. The equivalence points for the titration of this acid come at approximately pH 4 and pH 9. Suitable indicators for use in titrating carbonic acid or carbonate solutions are methyl orange and phenolphthalein.
(a) Sketch the titration curve that would be obtained in titrating a sample of $NaHCO_3(s)$ with 1.00 M HCl.
(b) Sketch the titration curve for $Na_2CO_3(s)$ by 1.00 M HCl.
(c) What volume of 0.100 M HCl is required for the neutralization of 1.00 g $NaHCO_3(s)$?
(d) What volume of 0.100 M HCl is required for the complete neutralization of 1.00 g $Na_2CO_3(s)$?
(e) A sample of NaOH contains a small amount of Na_2CO_3. For titration to the phenolphthalein end point, 0.1000 g of this sample requires 23.98 ml 0.1000 M HCl. An additional 0.78 ml is required to the methyl orange end point. What is the mass percent of Na_2CO_3 in the sample?

*13. Thymol blue indicator has *two* pH ranges. It changes color from red to yellow in the pH range 1.2 to 2.8, and from yellow to blue in the pH range 8.0 to 9.6. What is the color of the indicator in the following situations?
(a) The indicator is placed in 350 ml 0.205 M HCl.
(b) To the solution in part (a) is added 250 ml 0.500 M $NaNO_2$.
(c) To the solution in part (b) is added 150 ml 0.100 M NaOH.
(d) To the solution in part (c) is added 5.00 g $Ba(OH)_2$.

*14. Calculate the molar solubility of MnS in water (a) based only on the solubility product expression, K_{sp}, for MnS and (b) taking into account the hydrolysis of S^{2-} to HS^-. (c) Explain the difference in the results obtained.

*15. An expression that is sometimes used to calculate the pH of $NaH_2PO_4(aq)$ is pH = $\frac{1}{2}(pK_{a_1} + pK_{a_2})$. Verify this expression. Does it apply to all concentrations of $NaH_2PO_4(aq)$? What would be the pH of 0.10 M $NaHCO_3(aq)$ based on a similar expression?

Self-Test Questions

For questions 1 through 6 select the single item that best completes each statement.

1. To raise the pH of 1.00 L of 0.50 M HCl(aq) *significantly*, add   **(a)** 0.50 mol $HC_2H_3O_2$;   **(b)** 1.00 mol NaCl; **(c)** 0.60 mol $NaC_2H_3O_2$;   **(d)** 0.40 mol NaOH.

2. To convert NH_4^+ to NH_3 in aqueous solution,   **(a)** add H_3O^+;   **(b)** raise the pH;   **(c)** add KNO_3(aq);   **(d)** add $BaSO_4$(s).

3. If an indicator is to be used in an acid-base titration having an equivalence point in the pH range 8 to 10, the indicator must   **(a)** be a weak base;   **(b)** have an ionization constant of about $K_a = 1 \times 10^{-9}$;   **(c)** ionize in two steps; **(d)** be added to the solution only after the solution has become alkaline.

4. When a solution of a weak monoprotic acid has been *half*-neutralized by a strong base,   **(a)** the pH $= \frac{1}{2}pK_a$; **(b)** pH $= \frac{1}{2}$ of the pH value at the equivalence point; **(c)** pH $=$ twice the initial pH value;   **(d)** pH $= pK_a$.

5. To increase the molar solubility of $CaCO_3$(s) in a saturated aqueous solution, add   **(a)** $NaHSO_4$;   **(b)** Na_2CO_3; **(c)** NH_3;   **(d)** more water.

6. The best way to ensure the complete precipitation from a saturated aqueous solution of H_2S of some metal ion, M^{2+}, as its sulfide, MS(s), is to   **(a)** add an acid;   **(b)** increase $[H_2S]$ in the solution;   **(c)** raise the pH;   **(d)** heat the solution.

7. Which of the following solids are likely to be more soluble in acidic solution, and which in basic solution? Which are likely to have a solubility that is independent of pH? Explain.   **(a)** $H_2C_2O_4$;   **(b)** $MgCO_3$;   **(c)** CdS;   **(d)** KCl; **(e)** $NaNO_3$;   **(f)** $Ca(OH)_2$.

8. A $HCHO_2$–$NaCHO_2$ buffer solution is to be prepared; K_a($HCHO_2$, formic acid) $= 1.8 \times 10^{-4}$.
   (a) What mass of $NaCHO_2$ must be dissolved in 0.500 L 0.650 M $HCHO_2$ to produce a pH of 3.90?
   (b) If to the 0.500 L of buffer solution produced in part (a) is added one small pellet of NaOH (about 0.20 g), what will the new pH value be?

9. 25.0 ml 0.0100 M $HC_7H_5O_2$ (the monoprotic acid, benzoic acid, $K_a = 6.3 \times 10^{-5}$) is titrated by 0.0100 M $Ba(OH)_2$. Calculate the pH   **(a)** of the initial acid solution;   **(b)** after the addition of 6.25 ml of 0.0100 M $Ba(OH)_2$;   **(c)** at the equivalence point;   **(d)** after the addition of a total of 15.0 ml 0.0100 M $Ba(OH)_2$.

10. A buffer solution is 0.25 M $HC_2H_3O_2$–0.15 M $NaC_2H_3O_2$, saturated in H_2S (0.10 M) and has $[Mn^{2+}] = 0.015$ M.

$K_a(HC_2H_3O_2) = 1.74 \times 10^{-5}$

$K_{a_1} \times K_{a_2}(H_2S) = 1.1 \times 10^{-21}$

$K_{sp}(MnS) = 2.5 \times 10^{-13}$

   (a) Show that MnS will not precipitate from this solution.
   (b) Which buffer component would you increase in concentration, and to what minimum value, to ensure that precipitation of MnS would begin?

Oxidation-Reduction and Electrochemistry

In Chapters 17 and 18 we studied reactions involving the transfer of protons—acid-base reactions. In this chapter we consider reactions involving a transfer of electrons—oxidation-reduction reactions. We will reexamine the concepts of spontaneous change, equilibrium, and free energy change in relation to a new property—the electrode potential. Also explored will be the electrolytic basis of corrosion and some practical uses of electrochemistry.

19-1 Oxidation-Reduction: Some Definitions

When an iron object is exposed to the atmosphere, it rusts. In simplified fashion the process can be represented as

$$4 \, Fe(s) + 3 \, O_2(g) \longrightarrow 2 \, Fe_2O_3(s) \tag{19.1}$$

In this reaction iron combines with oxygen. Originally, the term "oxidation" was applied to reactions in which a substance combines with oxygen.

Iron rust is an oxide of iron and so are most iron ores. In simplified fashion, the production of iron from iron ore is described by

$$Fe_2O_3(s) + 3 \, CO(g) \longrightarrow 2 \, Fe(l) + 3 \, CO_2(g) \tag{19.2}$$

Reaction (19.2) involves the oxidation of $CO(g)$ to $CO_2(g)$. The oxygen atoms required for this oxidation come from the $Fe_2O_3(s)$, which is reduced. Originally, the term "reduction" was used to denote reactions in which oxygen is removed from a substance. Both oxidation and reduction occur in reactions (19.1) and (19.2). They are called oxidation-reduction reactions. A definition of oxidation-reduction based solely on a transfer of oxygen atoms is too restrictive. For one thing, it would limit us to reactions in which oxygen atoms are involved. We need broader definitions.

One type of reaction that we should include in the category "oxidation-reduction" is the displacement of silver ion from aqueous solution by copper metal (see Figure 19-1). In addition, it is useful to separate the net reaction into an **oxidation half-reaction** and a **reduction half-reaction.** Finally, our definitions should assign the

FIGURE 19-1
Displacement of Ag^+ from aqueous solution by copper metal.

$$Cu(s) + 2\,Ag^+(aq) \longrightarrow$$
$$Cu^{2+}(aq) + 2\,Ag(s)$$

Copper metal displaces Ag^+ ion from solution, producing a feathery deposit of metallic silver (a silver tree).

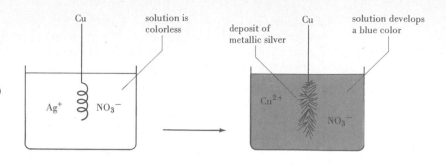

Cu — solution is colorless — deposit of metallic silver — Cu — solution develops a blue color

Ag^+ — NO_3^- — Cu^{2+} — NO_3^-

term "oxidation" to describe what happens to the copper, and "reduction" to the silver ion. The result of these requirements is

oxidation: $Cu(s) \longrightarrow Cu^{2+}(aq) + 2\,e^-$ (19.3)

reduction: $Ag^+(aq) + e^- \longrightarrow Ag(s)$ (19.4)

net reaction: $Cu(s) + 2\,Ag^+(aq) \longrightarrow Cu^{2+}(aq) + 2\,Ag(s)$ (19.5)

all free elements oxidation zero (unless ions)

The oxidation state of copper *increases* from 0 to +2 (corresponding to the loss of two electrons by each copper atom). The oxidation state of silver *decreases* from +1 to 0 (corresponding to the gain of one electron by each silver ion). What equations (19.3), (19.4), and (19.5) seem to require by way of definitions of oxidation and reduction is:

- Oxidation is a process in which the oxidation state of some element *increases* and in which electrons appear on the right-hand side of the oxidation half-equation.
- Reduction is a process in which the oxidation state of some element *decreases* and in which electrons appear on the left-hand side of the reduction half-equation.
- Oxidation and reduction half-reactions must always occur together. Moreover, the total number of electrons associated with the oxidation process must be equal to the total number associated with the reduction process.

also physical size of atom decreases More e⁻'s tighter hold, closer together

19-2 Balancing Oxidation-Reduction Equations

Before adding half-equations (19.3) and (19.4) to obtain the net equation (19.5), it was necessary to *multiply (19.4) by the factor "2."* This was done because equal numbers of electrons must appear in the oxidation and in the reduction half-equations. One method of balancing oxidation-reduction equations is based on this requirement. It is called the half-reaction or ion electron method. Another approach is based on the definitions of oxidation and reduction in terms of oxidation states. This is the oxidation state change method. These two methods are discussed next. The most versatile approach, however, is to use whichever method seems most appropriate to a given situation, including a combination of the two.

As noted in Section 17-2, protons in aqueous solutions are hydrated. In Chapters 17 and 18 we chose to represent them as hydronium ions— H_3O^+—because of the importance of this species in a discussion of acids and bases. Here we will represent them as H^+, to simplify the appearance of oxidation-reduction equations.

ION ELECTRON OR HALF-REACTION METHOD. In this method the oxidation and reduction half-equations are balanced separately and then combined into a balanced net equation. Consider the reaction of sulfite and permanganate ions in acidic solution.

$[H_3O \rightarrow H_2O + H^+]$

$$SO_3^{2-} + H^+ + MnO_4^- \longrightarrow SO_4^{2-} + Mn^{2+} + H_2O$$

Acid – dissociates to produce H⁺
Base – dissociates to produce OH⁻

Step 1. *Identify the species involved in oxidation state changes and write "skeleton" half-equations based on them.* In comparing the two oxyanions of sulfur, the

[handwritten top margin: -2·3=-6 3×6 S=+4 total   -2·4=-8 3×6 S=+6 total]

oxidation state of S in SO_3^{2-} is $+4$, and in SO_4^{2-} it is $+6$. The oxidation half-reaction involves the conversion of sulfite to sulfate ion. The oxidation state of Mn decreases from $+7$ to $+2$ in the net reaction. The conversion of MnO_4^- to Mn^{2+} occurs in the reduction half-reaction.

[handwritten left margin: I have an Ite-ous 3 lower oxy state Because I ate-ic 3 higher oxy state]

oxidation: $SO_3^{2-} \longrightarrow SO_4^{2-}$

reduction: $MnO_4^- \longrightarrow Mn^{2+}$

[handwritten: +7 -8 ... +2]

Step 2. *Balance each half-equation "atomically."* To show the same number of atoms of each type on both sides of the equation, it is often necessary to add H_2O and either H^+ (for acidic solutions) or OH^- (for basic solutions). For an acidic solution add *one* H_2O molecule for every O atom needed on the side that is deficient in O atoms. To the *opposite* side of the half-equation, add *two* H^+ for every H_2O molecule that was used.

[handwritten right margin: so need oxygen on left. Add by adding H₂O. Now counterbalance extra Hydrogen w/ H⁺ cations.]

[handwritten left margin: remember this is in Acid soln; add H₂O & H⁺ to sides]

oxidation: $SO_3^{2-} + H_2O \longrightarrow SO_4^{2-} + 2 H^+$

[handwritten: 3 oxygens ... 4 oxygens]

reduction: $MnO_4^- + 8 H^+ \longrightarrow Mn^{2+} + 4 H_2O$

[handwritten: 4 oxygens]

Step 3. *Balance each half-equation "electrically."* To the *right-hand* side of the *oxidation* half-equation, add the number of electrons necessary to achieve the same net charge on both sides of the half-equation. Do the same to the *reduction* half-equation by adding electrons to the *left-hand* side.

[handwritten right margin: add e⁻ to necessary side so charge balances]

oxidation: $SO_3^{2-} + H_2O \longrightarrow SO_4^{2-} + 2 H^+ + 2 e^-$

[handwritten: 2⁻ ... 0]   (net charge on each side, -2)

[handwritten left margin: remember: e⁻ is negative, reduce "+" charged side]

reduction: $MnO_4^- + 8 H^+ + 5 e^- \longrightarrow Mn^{2+} + 4 H_2O$

[handwritten: +7 ... +2]   (net charge on each side, $+2$)

Step 4. *Obtain the net oxidation-reduction equation by combining the half-equations.* Multiply through the oxidation half-equation by *5* and through the reduction half-equation by *2* This will result in 10 e^- on *each* side of the net equation. These terms will then cancel out. Electrons must not appear in the net equation.

$5 SO_3^{2-} + 5 H_2O \longrightarrow 5 SO_4^{2-} + 10 H^+ + \cancel{10 e^-}$

$2 MnO_4^- + 16 H^+ + \cancel{10 e^-} \longrightarrow 2 Mn^{2+} + 8 H_2O$

$5 SO_3^{2-} + 5 H_2O + 2 MnO_4^- + 16 H^+ \longrightarrow 5 SO_4^{2-} + 10 H^+ + 2 Mn^{2+} + 8 H_2O$

Step 5. *Simplify.* If the net equation contains the same species on both sides of the equation, cancel it from the side where it appears in lesser amount. Subtract *five* H_2O from each side of the net equation of step 4. This leaves *three* H_2O on the right. Also subtract *ten* H^+ from each side, leaving *six* on the left.

[handwritten: -10 -2 +6 = -6 = -10 +4 0]

$5 SO_3^{2-} + 2 MnO_4^- + 6 H^+ \longrightarrow 5 SO_4^{2-} + 2 Mn^{2+} + 3 H_2O$ (19.6)

Step 6. *Verify.* Check the final net equation to ensure that it is balanced both "atomically" and "electrically." For example, show that in equation (19.6) the net charge on each side of the equation is -6.

OXIDATION STATE CHANGE METHOD. This method focuses on the oxidation state changes that occur in the reaction. These changes can be taken as equivalent to the numbers of electrons appearing in the half-equation method. Let us illustrate by balancing the equation

[handwritten left margin: look @ specific element change]

$NO_2(g) + H_2(g) \longrightarrow NH_3(g) + H_2O(g)$

Step 1. *Assign oxidation states to all atoms on both sides of the equation.*

[handwritten circles: +4 -2   0   -3 +1   +1 -2]

$NO_2(g) + H_2(g) \longrightarrow NH_3(g) + H_2O(g)$

Step 2. *Identify the atoms that undergo changes in oxidation state and indicate the increase or decrease in oxidation state (O.S.) per atom.*

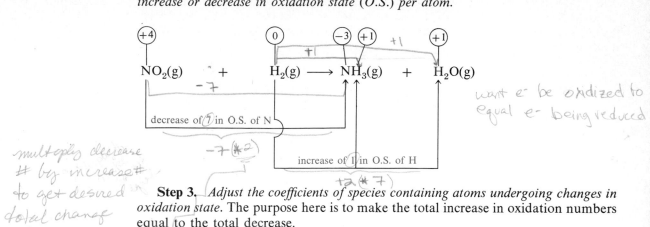

want e⁻ be oxidized to equal e⁻ being reduced

decrease of 7 in O.S. of N

increase of 1 in O.S. of H

multiply decrease # by increase # to get desired total change

Step 3. *Adjust the coefficients of species containing atoms undergoing changes in oxidation state.* The purpose here is to make the total increase in oxidation numbers equal to the total decrease.

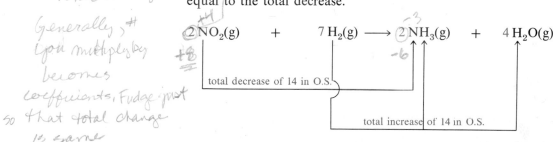

total decrease of 14 in O.S.

total increase of 14 in O.S.

(generally, # you multiply by becomes coefficients. Fudge just so that total change is same

Although the required ratio of H to N atoms on the left-hand side of the equation is 7:1, we use the ratio 14 H:2 N to avoid a fractional coefficient for H_2 (i.e., to avoid using $\frac{7}{2}$). The coefficient of NH_3 must be "2" because that of NO_2 is "2." This accounts for six of the 14 H atoms. The other eight must appear in the form of 4 H_2O.

Step 4. *Without changing coefficients previously established, balance the remainder of the equation by inspection.* Normally, there are species not involved in oxidation state changes whose coefficients must be established in this way. In the present case, however, the equation was balanced as a result of step 3.

$$2\,NO_2(g) + 7\,H_2(g) \longrightarrow 2\,NH_3(g) + 4\,H_2O(g) \tag{19.7}$$

OTHER EXAMPLES. At times in balancing oxidation-reduction equations:

1. More than a single oxidation and/or reduction process may be involved, as in the reaction

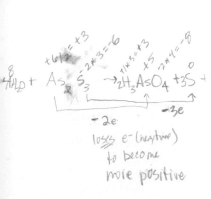

$$As_2S_3(s) + NO_3^- + H^+ \longrightarrow H_3AsO_4 + NO(g) + S(s) + H_2O$$

Note that both the As and S of $As_2S_3(s)$ undergo an increase in oxidation state. The balanced oxidation half-equation is

oxidation: $As_2S_3(s) + 8\,H_2O \longrightarrow 2\,H_3AsO_4 + 3\,S(s) + 10\,H^+ + 10\,e^-$

[The reduction involves converting NO_3^- to NO(g). Balancing of the net equation is considered in Exercise 3e.]

2. The same substance may undergo both oxidation and reduction, as in the reaction $Cl_2(g) + H_2O \longrightarrow ClO^- + H^+ + Cl^-$. Here $Cl_2(g)$ must appear in both skeleton half-equations.

oxidation: $Cl_2(g) \longrightarrow 2\,ClO^-$ (not balanced)
reduction: $Cl_2(g) \longrightarrow 2\,Cl^-$ (not balanced)

(The balancing of this equation is considered in Exercise 3f.)

loses e⁻ (negative) to become more positive

3. The reaction may occur in basic solution. Here we must deal with OH^- and H_2O, rather than H^+ and H_2O, as illustrated in Example 19-1.

That a reaction occurs in basic solution is established by the presence of $\dot{O}H^-$ in the equation to be balanced.

Example 19-1   Balance the oxidation-reduction equation

$$Cr(OH)_3(s) + ClO^- + OH^- \longrightarrow CrO_4^{2-} + Cl^- + H_2O$$

The oxidation state of Cr in $Cr(OH)_3$ is $+3$, and in CrO_4^{2-} it is $+6$. In ClO^- the oxidation state of Cl is $+1$, and in Cl^- it is -1.

oxidation: $Cr(OH)_3(s) \longrightarrow CrO_4^{2-}$
reduction: $ClO^- \longrightarrow Cl^-$

Whether we add H_2O or OH^-, and to whichever side of a half-equation, we find ourselves adding *both* H and O atoms. Moreover, since the H : O ratio is not the same for H_2O and for OH^-, we must be careful how we do this. Table 19-1 suggests how to proceed. In the skeleton oxidation half-equation there are *three* O atoms on the left and *four* on the right. According to Table 19-1, to gain *one* O atom on the left we should add *two* OH^- to the left and *one* H_2O to the right:

oxidation: $Cr(OH)_3(s) + 2\,OH^- \longrightarrow CrO_4^{2-} + H_2O$ (not balanced)

The H atoms are still out of balance—*five* on the left and *two* on the right. Again according to Table 19-1, we should add *three* H_2O to the right and *three* OH^- to the left:

oxidation: $Cr(OH)_3(s) + 5\,OH^- \longrightarrow CrO_4^{2-} + 4\,H_2O$

Completing the reduction half-equation is a bit simpler because only O atoms are out of balance in the skeleton half-equation.

reduction: $ClO^- + H_2O \longrightarrow Cl^- + 2\,OH^-$

Next we add electrons to balance the net charge in each half-equation.

oxidation: $Cr(OH)_3(s) + 5\,OH^- \longrightarrow CrO_4^{2-} + 4\,H_2O + 3\,e^-$
reduction: $ClO^- + H_2O + 2\,e^- \longrightarrow Cl^- + 2\,OH^-$

Finally, we must multiply through the oxidation half-equation by 2 and through the reduction half-equation by 3. Together with adding together the two half-equations below, we combine the step of simplifying the net equation by canceling the appropriate number of H_2O and OH^-.

oxidation: $2\,Cr(OH)_3 + 10\,OH^- \longrightarrow 2\,CrO_4^{2-} + 8\,H_2O + \cancel{6\,e^-}$
reduction: $3\,ClO^- + 3\,H_2O + \cancel{6\,e^-} \longrightarrow 3\,Cl^- + 6\,OH^-$

net: $2\,Cr(OH)_3(s) + 3\,ClO^- + 4\,OH^- \longrightarrow 2\,CrO_4^{2-} + 3\,Cl^- + 5\,H_2O$

$$\text{(19.8)}$$

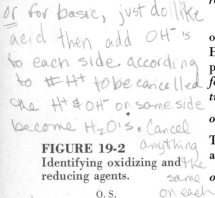

FIGURE 19-2
Identifying oxidizing and reducing agents.

O. S.

+7
+6
+5
+4
+3
+2
+1
0
−1
−2
−3

Oxidation process (Substance is a reducing agent.)

Reduction process (Substance is an oxidizing agent.)

TABLE 19-1
Achieving a balance of H_2O and OH^- in oxidation and reduction half-equations in basic solutions[a]

	To the side deficient in oxygen	To the other side
to balance O atoms:	for every O atom required, add two OH^-	add one H_2O
	To the side deficient in hydrogen	To the other side
to balance H atoms:	for every H atom required, add one H_2O	add one OH^-

[a]Another technique commonly used to balance an oxidation-reduction equation in basic solution is to treat the reaction *as if* it occurred in acidic solution (i.e., by using H^+ and H_2O to balance the half-equations). A number of OH^- ions equal to the number of H^+ ions in the net equation is added to *both* sides of the net equation. H^+ and OH^- are combined into H_2O and the net equation is simplified.

SIMILAR EXAMPLES: Exercises 3, 4.

The terms "oxidizing agent" and "reducing agent" sometimes cause confusion because the oxidizing agent is *not* oxidized, nor is the reducing agent reduced. But, by simple analogy, neither does a cleaning agent become cleaned nor a coloring agent become colored.

OXIDIZING AND REDUCING AGENTS. In an oxidation-reduction reaction an element in the substance that is oxidized undergoes an increase in oxidation state. The substance that is oxidized is called the **reducing agent,** because it *makes possible* the reduction process. Similarly, some element in the substance that is reduced undergoes a decrease in oxidation state. This substance is an **oxidizing agent,** because it *makes possible* the oxidation process. The oxidizing and reducing agents in an oxidation-reduction reaction can be identified readily by oxidation state changes, as outlined in Figure 19-2.

substance is oxidized, loses e⁻ & thus allows another to gain e⁻ & reduce

Example 19-2 What are the oxidizing and reducing agents in equations (19.6) and (19.7)?

$$5\,SO_3^{2-} + 2\,MnO_4^- + 6\,H^+ \longrightarrow 5\,SO_4^{2-} + 2\,Mn^{2+} + 3\,H_2O \tag{19.6}$$

$$2\,NO_2(g) + 7\,H_2(g) \longrightarrow 2\,NH_3(g) + 4\,H_2O(g) \tag{19.7}$$

(19.6): SO_3^{2-} is oxidized to SO_4^{2-}; SO_3^{2-} is the reducing agent, MnO_4^- is reduced to Mn^{2+}; MnO_4^- is the oxidizing agent.

(19.7): The oxidation state of N decreases from $+4$ in NO_2 to -3 in NH_3. NO_2 is reduced; and NO_2 is the oxidizing agent. The oxidation state of H increases from 0 in H_2 to $+1$ in NH_3 and in H_2O. H_2 is oxidized; and H_2 is the reducing agent.

SIMILAR EXAMPLE: Exercise 7.

19-3 Measurement of Oxidation and Reduction Tendencies

FIGURE 19-3
An electrochemical half-cell.

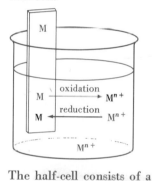

The half-cell consists of a metal electrode, M, immersed in an aqueous solution of its ions, M^{n+}. (The anions required to maintain electrical neutrality in the solution are not shown.)

If a solution of $Zn(NO_3)_2$ is substituted for the $AgNO_3$ solution in Figure 19-1, no reaction occurs. Why does copper behave differently toward Ag^+ and Zn^{2+}? To answer this and other basic questions, it is helpful to think in terms of a device in which an oxidation-reduction reaction can actually be separated into two distinct half-reactions.

MEASUREMENT OF ELECTROMOTIVE FORCE (EMF). Figure 19-3 pictures a metal strip, M, called an **electrode,** immersed in a solution containing the metal ions, M^{n+}. *cations - short on e⁻* The entire assembly is called a **half-cell.** Three kinds of interactions can take place between the metal atoms on the electrode and the metal ions in solution.

The situation illustrated in Figure 19-3 is limited to metals that do not react with water.

• A metal ion M^{n+} may collide with the electrode and undergo no change.
• A metal ion may collide with the electrode, gain n electrons and be converted to a metal atom M. *The ion is reduced.*
• A metal atom M on the electrode may lose n electrons and enter the solution as the ion M^{n+}. *The metal atom is oxidized.*

Equilibrium between the metal and its ions, which is quickly established, can be represented as

$$M(s) \underset{\text{reduction}}{\overset{\text{oxidation}}{\rightleftharpoons}} M^{n+}(aq) + n\,e^- \tag{19.9}$$

The net numbers of electrons on the electrode before and after equilibrium is established will be slightly different. As a result, the electrode acquires a very slight electrical charge; the solution acquires the opposite charge. One method of assessing these slight electrical charges is to construct *two* half-cells and connect them electri-

[handwritten note, top left: Two ½-cells connected; charge will flow TO the electrode w/ the smaller electrical charge.]

This process is analogous to water flowing from a higher to a lower level; its potential energy decreases in the process. Also, no matter how slight the difference in level between two points, water will always flow between them.

cally. This way the electric charge can flow from a region of high charge density or high electrical potential to a region of lower charge density or lower electrical potential. Even the slightest difference in electrical potential is enough to set up a flow of charged particles—an electric current. A combination of two half-cells is called an **electrochemical cell.** The reaction occurring in the electrochemical cell pictured in Figure 19-4 is between copper metal and silver ion, as we shall see shortly.

The electrical connection of the two half-cells must be done in a special way. *Both* the metal electrodes *and* the solutions have to be connected, so that a continuous circuit is formed through which charged particles flow. The electrodes can simply be connected by a metal wire which permits the flow of electrons.

The flow of electric current between the solutions must be in the form of a migration of ions. This cannot occur through a wire but only through another solution which "bridges" the two half-cells; this connection is called a **salt bridge.**

When these connections have been made, the following changes occur. Copper atoms lose electrons at the copper electrode and enter the solution as Cu^{2+} ions. The electrons lost by the copper atoms pass through the wire and the electrical measuring circuit to the silver electrode. Here Ag^+ ions from solution gain electrons and deposit as silver metal. Without a salt bridge, the solution in the copper half-cell would acquire excess Cu^{2+} and a net positive charge. In the silver half-cell there would be a deficiency of Ag^+ in solution, an excess of anions, and a negative charge buildup. Electric current could not continue to flow. The salt bridge allows for passage of electric current between the solutions. Consider the copper half-cell. Excess Cu^{2+} ions in this half-cell enter the salt bridge and migrate toward the silver half-cell. Also, anions from the salt solution (NO_3^-) migrate into the copper half-cell. In the silver half-cell, NO_3^- ions migrate out of the half-cell and NH_4^+ ions from the salt bridge migrate in. The net reaction that occurs is

FIGURE 19-4
Measurement of the electromotive force of a cell.

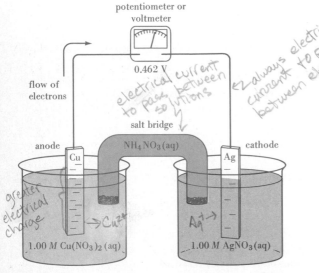

[handwritten annotations on figure: electrical current to pass between solutions; ← always electrical current to pass between electrodes; greater electrical charge]

The cell consists of two half-cells with electrodes joined by a wire and solutions by a salt bridge. (The ends of the salt bridge are plugged with a porous material that allows ions to migrate but prevents the bulk flow of liquid.) The potentiometer or voltmeter measures the difference in electrical potential between the two electrodes; this difference is 0.462 volt (V).

[handwritten note: The greater the electrical difference, the speedier, easier the flow will travel. "Electromotive force"]

oxidation: $Cu(s) \longrightarrow Cu^{2+}(aq) + 2\,e^-$
reduction: $2\{Ag^+(aq) + e^- \longrightarrow Ag(s)\}$

net: $Cu(s) + 2\,Ag^+(aq) \longrightarrow$
$$Cu^{2+}(aq) + 2\,Ag(s) \quad (19.5)$$

The reading on the meter in the electrical circuit (0.462 V) is also of significance. It represents the **potential difference** between the two half-cells. Since this potential difference is the "driving force" for electrons, it is often called the **electromotive force (emf)** of the cell or the **cell potential.** The unit used to measure electrical potential is the **volt,** so the cell potential is also

anode—"oxidation site"
atoms on anode lose
e⁻ & become oxidized.
Cations go into soln, &
e⁻ flow off electrode.

cathode — "reduction site"
e⁻'s are flowing on to
cathode from anode. Cations
in soln hit electrode pick
up these e⁻'s & become
reduced. (now are metal atoms)

FIGURE 19-5
The reaction $Zn(s) + Cu^{2+}(aq) \rightarrow Zn^{2+}(aq) + Cu(s)$
occurring in an electrochemical cell.

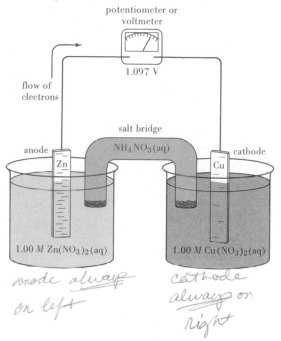

potentiometer or
voltmeter

1.097 V

flow of
electrons

salt bridge

$NH_4NO_3(aq)$

anode

Zn

cathode

Cu

$1.00\ M\ Zn(NO_3)_2(aq)$

$1.00\ M\ Cu(NO_3)_2(aq)$

anode always on left

cathode always on right

Luigi Galvani (1737–1798) is generally credited with the discovery of current electricity. Alessandro Volta (1745–1827) made the first cells of the type being described here. He was able to generate electric currents by stacking up, alternately, pieces of copper and zinc (i.e., Cu, Zn, Cu, . . .) separated by pieces of paper soaked in NaCl(aq). This particular device is called a voltaic pile.

referred to as the **cell voltage.** One definition of the unit, volt, helps to relate it to other units: The passage of one coulomb of electric charge through a potential difference of one volt produces a quantity of work equal to one joule.

$$1 \text{ joule (J)} = 1 \text{ volt (V)} \times 1 \text{ coulomb (C)} \qquad (19.10)$$

Returning to the question of why copper does not displace zinc ion from solution, the answer can be found in constructing an electrochemical cell based on this reaction. Such a cell is pictured in Figure 19-5, where we see that zinc shows a greater tendency to be oxidized than does copper. The reaction that occurs spontaneously in the electrochemical cell is

oxidation: $\quad Zn(s) \longrightarrow Zn^{2+}(aq) + 2\ e^-$
reduction: $\quad \underline{Cu^{2+}(aq) + 2\ e^- \longrightarrow Cu(s)}$
net: $\qquad\quad Zn(s) + Cu^{2+}(aq) \longrightarrow$
$$Zn^{2+}(aq) + Cu(s) \quad (19.11)$$

CELL DIAGRAMS AND TERMINOLOGY.   Sketching an electrochemical cell as in Figures 19-4 and 19-5 is somewhat tedious and cumbersome. A symbolic representation is often used to describe a cell. This representation is called a **cell diagram.** For the electrochemical cell of Figure 19-5 it takes the form

anode

salt
bridge

ions written directly on oppo
sides of salt bridge (!)

cathode

$$Zn(s)|Zn^{2+}(aq)||Cu^{2+}(aq)|Cu(s) \qquad (19.12)$$

half-cell   half-cell

The electrode shown *at the left* is the one at which *oxidation* occurs; it is called the **anode.** At the electrode shown *at the right, reduction* occurs; this electrode is called the **cathode.** A single vertical line represents the boundary between an electrode and another phase (e.g., a solution or a gas). A double vertical line signifies that solutions are joined by a salt bridge. The parenthetical expressions in equation (19.12) are simply the familiar (s) and (aq), but they can be used to specify electrode and solution conditions. For example, the symbol $Zn^{2+}(0.10\ M)$ would indicate that a solution is 0.10 molar in Zn^{2+}. Combinations such as $Zn|Zn^{2+}$ and $Cu^{2+}|Cu$ are often called couples. For example, $Zn|Zn^{2+}$ is the oxidation couple.

The electrochemical cells considered to this point are all of a type that *produce* electricity as a result of *spontaneous* chemical change. They are all called **galvanic** or **voltaic** cells. Another possibility that we will consider later (Section 19-9) is the production of a *non-spontaneous* chemical change through the *consumption* of electricity.

STANDARD ELECTRODE POTENTIALS. We have been describing the *measurement* of potential *differences*. However, if we could figure out a way to assign numerical values to metal–metal ion combinations or couples, we might then have a means of *calculating* cell potentials. The way to do this is to choose a particular couple and assign it a value of *zero*. Other couples can then be compared to this reference electrode.

The reference electrode for potential measurements is the **standard hydrogen electrode (S.H.E.),** pictured in Figure 19-6. The standard hydrogen electrode involves H^+ ions in solution at unit activity ($a = 1$); for simplicity we will take this to be essentially $1\,M\,H^+$. H_2 molecules in the gaseous state are at a pressure of 1 atm. The oxidized (H^+) and reduced (H_2) forms of hydrogen come into contact on an inert platinum metal surface and impart a characteristic potential to the surface. The temperature is taken to be exactly 25°C. The conditions specified here can be written in the form of an equation. Also, they can be represented through a half-cell couple.

Another symbol commonly used for standard electrode potentials is $\mathcal{E}°$.

$$2\,H^+(a = 1) + 2\,e^- \xrightleftharpoons{on\ Pt} H_2(g,\ 1\ atm) \qquad E° = 0.0000\ \text{volt (V)} \tag{19.13}$$

$$H^+(a = 1)\,|\,H_2(g,\ 1\ atm)\,|\,Pt \tag{19.14}$$

By international agreement, a **standard electrode potential, $E°$,** is based on the tendency for a **reduction** process to occur at the electrode. To represent other standard electrodes, we can write expressions of this sort.

By an earlier convention, standard electrode potentials were based on oxidation processes. One must be careful in using tabulated data to determine which convention applies.

$$Cu^{2+}(1\ M) + 2\,e^- \rightleftharpoons Cu(s) \qquad E° = ? \tag{19.15}$$

$$Cl_2(g,\ 1\ atm) + 2\,e^- \rightleftharpoons 2\,Cl^-(1\ M) \qquad E° = ? \tag{19.16}$$

In all cases the ionic species are present in aqueous solution at unit activity (approximately $1\,M$); gases are at 1 atm pressure. Where no solid substance is indicated, the potential is established on an inert platinum electrode.

To determine values of $E°$ for electrodes such as those of (19.15) and (19.16), we need to measure the potential difference between *two* electrodes. This can be done through an electrochemical cell with a S.H.E. as one electrode and the standard electrode in question as the other. In the following voltaic cell the measured potential difference is 0.337 V. Since this is the emf of the cell formed from two standard electrodes, it is referred to as the **standard cell potential, $E°_{cell}$.**

$$Pt\,|\,H_2(g,\ 1\ atm)\,|\,H^+(1\ M)\,||\,Cu^{2+}(1\ M)\,|\,Cu(s) \qquad E°_{cell} = 0.337\ V \tag{19.17}$$

The reaction that occurs in the voltaic cell of (19.17) is

oxidation: $\quad H_2(g,\ 1\ atm) \longrightarrow 2\,H^+(1\ M) + 2\,e^-$

reduction: $\quad Cu^{2+}(1\ M) + 2\,e^- \longrightarrow Cu(s)$

net: $\quad H_2(g,\ 1\ atm) + Cu^{2+}(1\ M) \longrightarrow 2\,H^+(1\ M) + Cu(s)$
$$E°_{cell} = 0.337\ V \tag{19.18}$$

According to reaction (19.18) $Cu^{2+}(1\ M)$ must be reduced more easily than is $H^+(1\ M)$. The standard electrode potential representing the reduction of $Cu^{2+}(aq)$ to $Cu(s)$ is $+0.337$ V.

$$Cu^{2+}(1\ M) + 2\,e^- \rightleftharpoons Cu(s) \qquad E° = +0.337\ V \tag{19.19}$$

When a standard hydrogen electrode is combined with a standard zinc electrode, the S.H.E. acts as the *cathode* and the standard zinc electrode as the *anode*. The measured value of $E°_{cell}$ is 0.760 V.

$$Zn(s)\,|\,Zn^{2+}(1\ M)\,||\,H^+(1\ M)\,|\,H_2(g,\ 1\ atm)\,|\,Pt \qquad E°_{cell} = 0.760\ V \tag{19.20}$$

The reaction that occurs in the voltaic cell (19.20) is

FIGURE 19-6
The standard hydrogen electrode (S.H.E.).

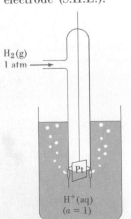

this has greater tendency to be oxidized so undergoes the reduction so H₂E.

oxidation: $\quad$ $Zn(s) \longrightarrow Zn^{2+}(1\ M) + 2\ e^-$

reduction: $\quad$ $2\ H^+(1\ M) + 2\ e^- \longrightarrow H_2(g,\ 1\ atm)$

net: $\qquad$ $Zn(s) + 2\ H^+(1\ M) \longrightarrow Zn^{2+}(1\ M) + H_2(g,\ 1\ atm)$
$$E^\circ_{cell} = 0.760\ V \qquad (19.21)$$

Here reduction of $Zn^{2+}(1\ M)$ must occur with *greater difficulty* than that of $H^+(1\ M)$, since oxidation, not reduction, occurs at the zinc electrode. E°_{cell} of equation (19.21) describes the tendency for zinc to become *oxidized*. *If we consider the reduction tendency to be the opposite of the oxidation tendency,* then

$$Zn^{2+}(1\ M) + 2\ e^- \rightleftharpoons Zn(s) \qquad E^\circ = -0.760\ V \qquad (19.22)$$

has smaller tendency to reduce than does H₂, so when put in cell with SHE, it will favor oxidation

In summary:

an elements electrode potential is based on TENDENCY TO REDUCE.

- The potential of the standard hydrogen electrode is set at zero.
- Any electrode at which a *reduction* half-reaction shows a *greater* tendency to occur than does $2\ H^+(1\ M) + 2\ e^- \rightarrow H_2(g,\ 1\ atm)$ has a *positive* electrode potential, E°.
- Any electrode at which a *reduction* half-reaction shows a *lesser* tendency to occur than does $2\ H^+(1\ M) + 2\ e^- \rightarrow H_2(g,\ 1\ atm)$ has a *negative* electrode potential, E°.
- If the tendency for a reduction process is given by E°, the oxidation tendency is simply the negative of this value, that is, $-(E^\circ)$.
- With these ideas it is possible to develop extensive tabulations of standard electrode potentials, as suggested by Table 19-2.

19-4 A New Criterion for Spontaneous Change

before using ΔG (free energy change)

The conclusions of the preceding section lead rather naturally to a new criterion for spontaneous change. To predict whether an oxidation-reduction reaction will proceed spontaneously when the reactants and products are present in their standard states, we proceed as follows.

1. Write the proposed reduction half-equation and a standard reduction potential, E°_{red}, to describe it. This will be a value of E° from Table 19-2.
2. Write the proposed oxidation half-equation. The standard oxidation potential, E°_{ox}, is the *negative* of the E° value listed in Table 19-2.
3. Combine the half-equations into a net oxidation-reduction equation. *Add* together the oxidation and reduction potentials to obtain E°_{cell}.
4. If E°_{cell} is *positive*, the reaction will occur spontaneously as written. If E°_{cell} is *negative*, the reaction will not occur spontaneously as written.
5. If a cell reaction is reversed, E°_{cell} changes sign.

Example 19-3 Will aluminum metal displace Cu^{2+} ion from aqueous solution? That is, will this reaction occur spontaneously as written?

$$2\ Al(s) + 3\ Cu^{2+}(1\ M) \longrightarrow 3\ Cu(s) + 2\ Al^{3+}(1\ M)$$

The net equation is obtained by adding together (19.23) and (19.24). Note that the coefficients must first be adjusted so as to eliminate electrons, e^-, from the net equation.

Since the potential of an electrode is an intensive property of the electrode system, it does not depend on the amounts of materials involved. E° values are *unaffected* by multiplying half-equations by a constant coefficient.

oxidation: $\quad 2\ \{Al(s) \longrightarrow Al^{3+}(1\ M) + 3\ e^-\}$
$$E^\circ_{ox} = -(-1.66) = +1.66\ V \qquad (19.23)$$

reduction: $\quad 3\ \{Cu^{2+}(1\ M) + 2\ e^- \longrightarrow Cu(s)\}$
$$E^\circ_{red} = +0.337\ V \qquad (19.24)$$

net: $\qquad 2\ Al(s) + 3\ Cu^{2+}(1\ M) \longrightarrow 3\ Cu(s) + 2\ Al^{3+}(1\ M)$
$$E^\circ_{cell} = +2.00\ V \qquad (19.25)$$

Since E°_{cell} is *positive*, the net reaction will occur spontaneously as written.

reverse for electrolytic cells

TABLE 19-2
Some selected standard electrode potentials

Reduction half-reaction	$E°$, V
Acidic solution	
$F_2(g) + 2\,e^- \longrightarrow 2\,F^-(aq)$	+2.87
$O_3(g) + 2\,H^+(aq) + 2\,e^- \longrightarrow O_2(g) + H_2O$	+2.07
$S_2O_8^{2-}(aq) + 2\,e^- \longrightarrow 2\,SO_4^{2-}(aq)$	+2.01
$H_2O_2(aq) + 2\,H^+(aq) + 2\,e^- \longrightarrow 2\,H_2O$	+1.77
$MnO_4^-(aq) + 8\,H^+(aq) + 5\,e^- \longrightarrow Mn^{2+}(aq) + 4\,H_2O$	+1.51
$PbO_2(s) + 4\,H^+(aq) + 2\,e^- \longrightarrow Pb^{2+}(aq) + 2\,H_2O$	+1.455
$Cl_2(g) + 2\,e^- \longrightarrow 2\,Cl^-(aq)$	+1.360
$Cr_2O_7^{2-}(aq) + 14\,H^+(aq) + 6\,e^- \longrightarrow 2\,Cr^{3+}(aq) + 7\,H_2O$	+1.33
$MnO_2(s) + 4\,H^+(aq) + 2\,e^- \longrightarrow Mn^{2+}(aq) + 2\,H_2O$	+1.23
$O_2(g) + 4\,H^+(aq) + 4\,e^- \longrightarrow 2\,H_2O$	+1.229
$2\,IO_3^-(aq) + 12\,H^+(aq) + 10\,e^- \longrightarrow I_2(s) + 6\,H_2O$	+1.195
$Br_2(l) + 2\,e^- \longrightarrow 2\,Br^-(aq)$	+1.065
$NO_3^-(aq) + 4\,H^+(aq) + 3\,e^- \longrightarrow NO(g) + 2\,H_2O$	+0.96
$Ag^+(aq) + e^- \longrightarrow Ag(s)$	+0.799
$Fe^{3+}(aq) + e^- \longrightarrow Fe^{2+}(aq)$	+0.771
$O_2(g) + 2\,H^+(aq) + 2\,e^- \longrightarrow H_2O_2(aq)$	+0.682
$I_2(s) + 2\,e^- \longrightarrow 2\,I^-(aq)$	+0.54
$Cu^+(aq) + e^- \longrightarrow Cu(s)$	+0.52
$H_2SO_3(aq) + 4\,H^+(aq) + 4\,e^- \longrightarrow S(s) + 3\,H_2O$	+0.45
$Cu^{2+}(aq) + 2\,e^- \longrightarrow Cu(s)$	+0.337
$SO_4^{2-}(aq) + 4\,H^+(aq) + 2\,e^- \longrightarrow H_2SO_3(aq) + H_2O$	+0.17
$Sn^{4+}(aq) + 2\,e^- \longrightarrow Sn^{2+}(aq)$	+0.15
$S(s) + 2\,H^+(aq) + 2\,e^- \longrightarrow H_2S(g)$	+0.141
$\mathbf{2\,H^+(aq) + 2\,e^- \longrightarrow H_2(g)}$	**0.0000**
$Pb^{2+}(aq) + 2\,e^- \longrightarrow Pb(s)$	−0.126
$Sn^{2+}(aq) + 2\,e^- \longrightarrow Sn(s)$	−0.136
$Fe^{2+}(aq) + 2\,e^- \longrightarrow Fe(s)$	−0.440
$Zn^{2+}(aq) + 2\,e^- \longrightarrow Zn(s)$	−0.763
$Al^{3+}(aq) + 3\,e^- \longrightarrow Al(s)$	−1.66
$Mg^{2+}(aq) + 2\,e^- \longrightarrow Mg(s)$	−2.37
$Na^+(aq) + e^- \longrightarrow Na(s)$	−2.714
$Ca^{2+}(aq) + 2\,e^- \longrightarrow Ca(s)$	−2.89
$K^+(aq) + e^- \longrightarrow K(s)$	−2.925
$Li^+(aq) + e^- \longrightarrow Li(s)$	−3.045
Basic solution	
$O_3(g) + H_2O + 2\,e^- \longrightarrow O_2(g) + 2\,OH^-$	+1.24
$ClO^-(aq) + H_2O + 2\,e^- \longrightarrow Cl^- + 2\,OH^-$	+0.89
$O_2(g) + 2\,H_2O + 4\,e^- \longrightarrow 4\,OH^-(aq)$	+0.401
$CrO_4^{2-}(aq) + 4\,H_2O + 3\,e^- \longrightarrow Cr(OH)_3(s) + 5\,OH^-$	−0.13
$S(s) + 2\,e^- \longrightarrow S^{2-}(aq)$	−0.48
$2\,H_2O + 2\,e^- \longrightarrow H_2(g) + 2\,OH^-(aq)$	−0.828
$SO_4^{2-}(aq) + H_2O + 2\,e^- \longrightarrow SO_3^{2-}(aq) + 2\,OH^-(aq)$	−0.93

Handwritten margin notes:

oxidizing agents → increasing

power (potential to reduce based on scale w/ 2H⁺ + 2e⁻ → H₂ equal to 0.0V)

E° becomes more negative

two diff. rxns

spontaneous

not spontaneous

Ca(s) 2H₃O⁺ → H₂(s) + Ca²⁺ + 2H₂O

FROM CHAP 20: should be obvious these have lower tendency to reduce for we know most of them when react w/acid ⇒ H₂(g)

↗ spontaneous system

SIMILAR EXAMPLES: Exercises 14, 15.

Example 19-4   Does the following cell, *as diagrammed*, function as a voltaic cell?

$$Pt\,|\,Cl_2(g,\ 1\ atm)\,|\,Cl^-(1\ M)\,||\,Cu^{2+}(1\ M)\,|\,Cu(s)$$

According to the convention for writing cell diagrams, the proposed reaction involves oxidation at the chlorine electrode (anode) and reduction at the copper (cathode).

oxidation: $2\ Cl^-(1\ M) \longrightarrow Cl_2(g,\ 1\ atm) + 2\ e^-$ $E^\circ_{ox} = -(+1.360) = -1.360\ V$

reduction: $Cu^{2+}(1\ M) + 2\ e^- \longrightarrow Cu(s)$ $E^\circ_{red} = 0.337\ V$

net: $Cu^{2+}(1\ M) + 2\ Cl^-(1\ M) \longrightarrow Cu(s) + Cl_2(g,\ 1\ atm)$

$$E^\circ_{cell} = -1.023\ V \qquad (19.26)$$

Since the proposed cell reaction has $E^\circ_{cell} < 0$, it is not a spontaneous reaction as written. The cell diagrammed does not function as a voltaic cell.

SIMILAR EXAMPLES: Exercises 20, 21.

A conclusion that a reaction is nonspontaneous under standard state conditions does not mean that it is nonspontaneous under all conditions, as we shall see later. Also, even though we use an electrochemical cell as the basis of predicting spontaneous change, the reaction can be carried out simply by bringing the reactants together. This was the case in the reaction pictured in Figure 19-1.

19-5   Electrical Work and Free Energy

In Chapter 15 we learned two things about *nonspontaneous* processes. One is that it takes work to make a nonspontaneous process occur, and the other is that for such a process, $\Delta G > 0$. Conversely, for a *spontaneous* process, $\Delta G < 0$. We did not stress this point at that time but a *spontaneous process is capable of doing work;* and $-\Delta G$ is the maximum amount of work that can be done.

$$-\Delta G = w_{max} \qquad (19.27)$$

When a reaction is carried out in a voltaic cell, it also does work. Moreover, this is the maximum amount of work obtainable for the process. The quantity of electrical work done in a voltaic cell is

Recall from equation (19.10) that 1 joule = 1 volt × 1 coulomb. The product, $n\mathfrak{F}E_{cell}$, in equation (19.28) has a unit of work or energy, the joule.

$$w_{elec.} = n\mathfrak{F}E_{cell} \qquad (mol\ e^-)\left(\frac{C}{mol\ e^-}\right)(volt) = C \times volt = \underline{Joule} \qquad (19.28)$$

where n is the number of moles of electrons. $\mathfrak{F}$, the **Faraday constant,** is the electrical charge per mole of electrons, 96,500 C/mol e^-. E_{cell} is the emf of the voltaic cell, in volts.

If we equate (19.27) and (19.28), we obtain the expression

$$\Delta G = -n\mathfrak{F}E_{cell} \qquad (19.29)$$

and if the reactants and products in the half-cells are in their standard states,

$$\Delta G^\circ = -n\mathfrak{F}E^\circ_{cell} \qquad (19.30)$$

One of the most significant applications of equation (19.30) comes in combining it with equation (15.15).

$$\Delta \overline{G}^\circ = -2.303RT \log K = -n\mathfrak{F}E^\circ_{cell} \qquad (19.31)$$

or $$E^\circ_{cell} = \frac{2.303RT}{n\mathfrak{F}} \log K \qquad (19.32)$$

FIGURE 19-7
Variation of E_{cell} with
ion concentrations for
the cell reaction
$Zn(s) + Cu^{2+}(aq) \rightarrow$
$Zn^{2+}(aq) + Cu(s)$.

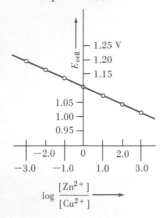

TABLE 19-3
Variation of E_{cell} with ion concentrations for the cell reaction
$Zn(s) + Cu^{2+}(aq) \longrightarrow Zn^{2+}(aq) + Cu(s)$

$[Zn^{2+}]$, M	$[Cu^{2+}]$, M	$\dfrac{[Zn^{2+}]}{[Cu^{2+}]}$	$\log\dfrac{[Zn^{2+}]}{[Cu^{2+}]}$	E_{cell}°, V
1.0	1.0×10^{-4}	1.0×10^{4}	4.0	0.98
1.0	1.0×10^{-3}	1.0×10^{3}	3.0	1.01
1.0	1.0×10^{-2}	1.0×10^{2}	2.0	1.04
1.0	1.0×10^{-1}	1.0×10^{1}	1.0	1.07
1.0	1.0	1.0	0	1.10
1.0×10^{-1}	1.0	1.0×10^{-1}	-1.0	1.13
1.0×10^{-2}	1.0	1.0×10^{-2}	-2.0	1.16
1.0×10^{-3}	1.0	1.0×10^{-3}	-3.0	1.19
1.0×10^{-4}	1.0	1.0×10^{-4}	-4.0	1.22

If we use a value of 8.314 J mol^{-1} K^{-1} for R and limit the temperature to 25°C (the temperature at which E° values are usually tabulated), the term $2.303RT/\mathcal{F}$ becomes 0.0592 V. This permits us to write

$$E_{cell}^{\circ} = \frac{0.0592}{n} \log K \tag{19.33}$$

Example 19-5   What is K_c for the following reaction at 25°C?

$$Cu^{2+}(aq) + Sn^{2+}(aq) \longrightarrow Sn^{4+}(aq) + Cu(s)$$

First we must determine E_{cell}° for this reaction.

oxidation: $Sn^{2+}(aq) \longrightarrow Sn^{4+}(aq) + 2\,e^{-}$ $E_{ox}^{\circ} = -(+0.15) = -0.15$ V
reduction: $\underline{Cu^{2+}(aq) + 2\,e^{-} \longrightarrow Cu(s)}$ $E_{red}^{\circ} = +0.337$ V
net: $Cu^{2+}(aq) + Sn^{2+}(aq) \longrightarrow Sn^{4+}(aq) + Cu(s)$
            $E_{cell}^{\circ} = 0.19$ V

The number of moles of electrons involved in the cell reaction is 2.

$$E_{cell}^{\circ} = \frac{0.0592}{2} \log K = 0.19 \qquad \log K = \frac{2 \times 0.19}{0.0592} = 6.4$$

$K_c = $ antilog $6.4 = 2.5 \times 10^{6}$

SIMILAR EXAMPLES: Exercises 33, 34.

19-6 E_{cell} as a Function of Concentrations

If the galvanic cell pictured in Figure 19-5 is operated at different concentrations of Zn^{2+} and Cu^{2+}, E_{cell} is found to vary in the manner suggested by Table 19-3 and Figure 19-7. The equation of the straight line in Figure 19-7 is

$$E_{cell} = 1.10 - 0.03 \log \frac{[Zn^{2+}]}{[Cu^{2+}]} \tag{19.34}$$

Relationships of this type were first studied by Walter Nernst (1864–1941). Equation (19.34) is a specific example of the general equation now known as the **Nernst equation.** We have shown how this equation may be established by experiment, but it can also be derived from thermodynamics. For the reaction

$$a\,A + b\,B + \cdots \rightleftharpoons g\,G + h\,H + \cdots \tag{19.35}$$

Usually, the value of n can be determined by inspection of the cell reaction. Sometimes it is necessary to separate a reaction into half-reactions to determine the number of moles of electrons involved.

If the values $[Zn^{2+}] = [Cu^{2+}] = 1.0$ M are substituted into equation (19.34), the log term becomes equal to zero. $E_{cell} = E_{cell}^{\circ} = 1.097 = 1.10$ V, just as shown in Figure 19-5. E_{cell} changes by about 0.03 V for every tenfold change in the ratio $[Zn^{2+}]/[Cu^{2+}]$.

we can write

$$\Delta \bar{G} = \Delta \bar{G}^{\circ} + 2.303RT \log Q \qquad (15.14)$$

Substituting equations (19.29) and (19.30) into (15.14), we obtain

$$-n\mathfrak{F}E_{cell} = -n\mathfrak{F}E_{cell}^{\circ} + 2.303RT \log Q$$

and
$$E_{cell} = E_{cell}^{\circ} - \frac{2.303RT}{n\mathfrak{F}} \log Q$$

Q is the reaction quotient and has the form established in Chapters 14 and 15. At 25°C, $2.303RT/\mathfrak{F}$ has the value 0.0592 V. The term n is the number of moles of electrons transferred in the cell reaction. At 25°C the Nernst equation becomes

$$E_{cell} = E_{cell}^{\circ} - \frac{0.0592}{n} \log \frac{(a_G)^g (a_H)^h \cdots}{(a_A)^a (a_B)^b \cdots} \qquad (19.36)$$

Example 19-6 Calculate E_{cell} for the voltaic cell pictured in Figure 19-8.

$$Pt | Fe^{2+}(0.10\ M),\ Fe^{3+}(0.20\ M) || Ag^+(1.0\ M) | Ag(s)$$

We begin by using data from Table 19-2 to determine E_{cell}°. Then we apply the Nernst equation.
Determining E_{cell}°:

oxidation: $Fe^{2+}(aq) \longrightarrow Fe^{3+}(aq) + e^-$
$\qquad\qquad E_{ox}^{\circ} = -(+0.771) = -0.771$ V
reduction: $Ag^+(aq) + e^- \longrightarrow Ag(s)$
$\qquad\qquad \underline{E_{red}^{\circ} = +0.799\ \text{V}}$
net: $\overline{Fe^{2+}(aq) + Ag^+(aq) \longrightarrow Fe^{3+}(aq) + Ag(s)}$
$\qquad\qquad E_{cell}^{\circ} = +0.028$ V $\qquad\qquad (19.37)$

Nernst equation:

$$E_{cell} = E_{cell}^{\circ} - \frac{0.0592}{n} \log \frac{[Fe^{3+}]}{[Fe^{2+}][Ag^+]}$$

Substitute: $E_{cell}^{\circ} = 0.028$ V; $n = 1$; $[Fe^{2+}] = 0.10\ M$; $[Fe^{3+}] = 0.20\ M$; $[Ag^+] = 1.0\ M$.

$$E_{cell} = 0.028 - \frac{0.0592}{1} \log \frac{0.20}{0.10 \times 1.0}$$
$$= 0.028 - (0.0592 \log 2.0)$$

$$E_{cell} = 0.028 - (0.0592 \times 0.301)$$
$$= 0.028 - 0.018 = 0.010\ \text{V}$$

SIMILAR EXAMPLES: Exercises 28, 29.

THE NERNST EQUATION AND LE CHÂTELIER'S PRINCIPLE. Le Châtelier's principle can provide qualitative results about oxidation-reduction reactions. For example, what conditions favor the *reverse* of reaction (19.37)?

$$Fe^{2+}(aq) + Ag^+(aq) \longrightarrow Fe^{3+}(aq) + Ag(s)$$
$$E_{cell}^{\circ} = +0.028\ \text{V} \qquad (19.37)$$

FIGURE 19-8
A voltaic cell with non-standard conditions— Example 19-6 illustrated.

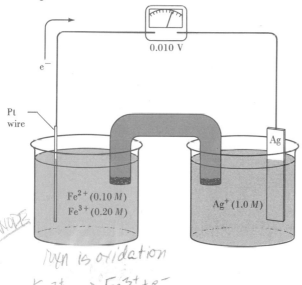

The terms $\Delta \bar{G}^{\circ}$ and E_{cell}° both refer to a reaction in which reactants and products are in their standard states. By contrast, $\Delta \bar{G}$ and E_{cell} refer to nonstandard conditions.

The fact that $E_{cell} > 0$ signifies that reaction (19.37) will occur spontaneously. As it does so, however, $[Fe^{3+}]$ increases and $[Fe^{2+}]$ and $[Ag^+]$ decrease. E_{cell} decreases as the reaction proceeds and becomes zero when equilibrium is reached.

The reverse reaction is favored by *reducing* the concentration of Fe^{2+} and/or Ag^+ and *increasing* the concentration of Fe^{3+}. The Nernst equation allows us to *calculate* the effect of these conditions on E_{cell}.

Example 19-7 What is the *minimum* Ag^+ concentration at which the cell of Figure 19-8 is able to function as a voltaic cell?

$$Pt|Fe^{2+}(0.10\ M),\ Fe^{3+}(0.20\ M)||Ag^+(?\ M)|Ag(s)$$

The cell will not function as a voltaic cell having Fe^{2+}/Fe^{3+} as the anode if $E_{cell} \leqslant 0$. A system with $E_{cell} = 0$ is at equilibrium.

$$E_{cell} = 0 = 0.028 - \frac{0.0592}{1} \log \frac{0.20}{0.10 \times [Ag^+]}$$
$$= 0.028 - 0.0592(\log 2.0 - \log[Ag^+])$$

$$0 = 0.028 - 0.0592(0.301 - \log[Ag^+])$$
$$0.0592 \log[Ag^+] = 0.0178 - 0.028$$

$$\log[Ag^+] = \frac{-0.010}{0.0592} = -0.17 \qquad [Ag^+] = 0.68\ M$$

SIMILAR EXAMPLE: Exercise 31.

MEASUREMENT OF pH. Suppose that we set up the electrochemical cell illustrated in Figure 19-9. This consists of *two* hydrogen electrodes. One is a *standard hydrogen electrode* (S.H.E.) and the other is a hydrogen electrode immersed in a solution of unknown $[H^+]$. The two half-cells are joined by a salt bridge, resulting in the voltaic cell,

$$Pt,\ H_2(g,\ 1\ atm)|H^+(x\ M)||H^+(1\ M)|H_2(g,\ 1\ atm),\ Pt \tag{19.38}$$

The reaction occurring in this cell is

oxidation: $H_2(g,\ 1\ atm) \longrightarrow 2\ H^+(x\ M) + 2e^-$
reduction: $2\ H^+(1\ M) + 2e^- \longrightarrow H_2(g,\ 1\ atm)$
net: $2\ H^+(1\ M) \longrightarrow 2\ H^+(x\ M) \tag{19.39}$

Any electrochemical cell in which the net cell reaction is simply the change in concentration of some species (here $[H^+]$) is called a concentration cell. A **concentration cell** consists of two half-cells with *identical electrodes* but differing ion concentrations. Because the electrodes are identical, $E°$ for the oxidation will be numerically equal and opposite in sign to $E°$ for the reduction. As a result, $E°_{cell} = 0$. However, because the ion concentrations do differ, there is a potential difference between the two half-cells. The Nernst equation for reaction (19.39) takes the form

$$E_{cell} = E°_{cell} - \frac{0.0592}{2} \log \frac{(x)^2}{(1)^2}$$

FIGURE 19-9
A concentration cell, consisting of two hydrogen electrodes, for measuring pH.

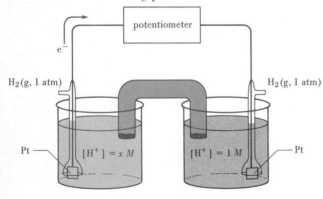

which simplifies to

$$E_{cell} = 0 - \frac{0.0592}{2} \times 2 \log \frac{x}{1} = -0.0592 \log x$$

Since x is simply $[H^+]$ in the unknown solution, and $-\log x = -\log[H^+] = pH$, our final result is

$$E_{cell} = 0.0592 \ pH \qquad (19.40)$$

If an "unknown" solution has a pH of 3.50, for example, the measured cell voltage in Figure 19-9 will be $E_{cell} = 0.0592 \times 3.50 = 0.207$ V.

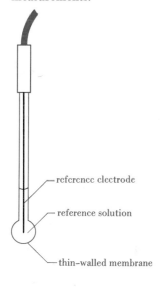

FIGURE 19-10
A glass electrode for pH measurements.

— reference electrode

— reference solution

— thin-walled membrane

THE GLASS ELECTRODE. Constructing and using a hydrogen electrode is difficult. The Pt metal surface must be specially prepared and maintained, gas pressures must be controlled, and the electrode cannot be used in the presence of strong oxidizing or reducing agents. A better approach to pH measurement than that of Figure 19-9 replaces the S.H.E. with some other reference electrode of precisely known $E°$ value. The second hydrogen electrode is replaced by a **glass electrode.** The basic feature of this electrode is a thin glass membrane of a carefully regulated chemical composition. When the electrode is dipped into a solution, depending on the type and concentration of ions present, a potential is established on the outer surface of the membrane. This potential is registered through a reference electrode immersed in a solution inside the membrane. The glass electrode, pictured in Figure 19-10, immersed in an "unknown" solution serves as a half-cell. When combined with another reference half-cell, the assembly functions as a voltaic cell.

The most commonly used glass electrodes are those whose potentials are determined by $[H^+]$ in solution. These are the glass electrodes used on common laboratory pH meters. Other glass electrodes have been developed that can function with Na^+, K^+, or other cations. These so-called **ion-selective electrodes** are especially valuable because ordinary half-cells cannot use very active metals as electrodes. These metals react with water to liberate $H_2(g)$.

MEASUREMENT OF K_{sp}. One of the best ways to measure K_{sp} of a slightly soluble salt is electrochemically. Consider this concentration cell.

$$Pb(s)|Pb^{2+}(sat'd. \ PbCl_2(aq))\|Pb^{2+}(0.10 \ M)|Pb(s) \qquad E_{cell} = 0.024 \ V \qquad (19.41)$$

In the anode half-cell lead is immersed in a saturated aqueous solution of $PbCl_2$. In the cathode half-cell a second lead strip is dipped into a solution with $[Pb^{2+}] = 0.10 \ M$. The two half-cells are connected by a salt bridge and the difference in potential between the two electrodes is measured. It is 0.024 V. The cell reaction occurring in this concentration cell is

oxidation: $\cancel{Pb(s)} \longrightarrow Pb^{2+}(sat'd. \ PbCl_2(aq)) + \cancel{2e^-}$
reduction: $Pb^{2+}(0.10 \ M) + \cancel{2e^-} \longrightarrow \cancel{Pb(s)}$
net: $Pb^{2+}(0.10 \ M) \longrightarrow Pb^{2+}(sat'd. \ PbCl_2(aq))$

$$E_{cell} = 0.024 \ V \qquad (19.42)$$

Example 19-8 With the data listed in (19.42), calculate K_{sp} for $PbCl_2$.

$$PbCl_2(s) \rightleftharpoons Pb^{2+}(aq) + 2 \ Cl^-(aq) \qquad K_{sp} = ?$$

Let us represent $[Pb^{2+}]$ in saturated $PbCl_2(aq)$ by x. Then we apply the Nernst equation to the cell reaction (19.42). (Recall that $E°_{cell} = 0$ for a concentration cell.)

$$E_{cell} = E°_{cell} - \frac{0.0592}{2} \log \frac{x}{0.10} = 0 - 0.0296(\log x - \log 0.10) = 0.024$$

$$-0.0296[(\log x) + 1] = 0.024 \qquad \log x = \frac{-0.0296 - 0.024}{0.0296} = -1.8$$

$$x = [Pb^{2+}] = 1.6 \times 10^{-2} M \qquad [Cl^-] = 2[Pb^{2+}] = 3.2 \times 10^{-2} M$$

$$K_{sp} = [Pb^{2+}][Cl^-]^2 = (1.6 \times 10^{-2})(3.2 \times 10^{-2})^2 = 1.6 \times 10^{-5}$$

SIMILAR EXAMPLES: Exercises 30, 31.

19-7 Production of Electric Energy by Chemical Change

An important use of galvanic cells is the production of electric energy by chemical change. Three rather different devices are considered in this section. Two have been in use for a long time; the third is a new type coming into use.

LECLANCHÉ (DRY) CELL. A familiar flashlight cell is pictured in Figure 19-11. Oxidation occurs at a zinc anode and reduction at an inert carbon cathode. The electrolyte is a moist paste of MnO_2, $ZnCl_2$, NH_4Cl, and carbon black. The difference in potential of the two electrodes is about 1.5 V. Because there is no free liquid in the cell, it is called a "dry" cell.

The anode reaction is simple—the oxidation of zinc atoms to Zn^{2+}.

oxidation: $Zn(s) \longrightarrow Zn^{2+}(aq) + 2 e^-$

The reduction is more complex. Essentially, it involves the reduction of MnO_2 to a series of compounds having Mn in a +3 oxidation state, for example, Mn_2O_3.

reduction: $2 MnO_2(s) + H_2O + 2 e^- \longrightarrow Mn_2O_3(s) + 2 OH^-(aq)$

An acid-base reaction occurs between the OH^- and NH_4^+.

acid-base reaction: $NH_4^+(aq) + OH^-(aq) \longrightarrow NH_3(g) + H_2O$

A buildup of a layer of $NH_3(g)$ around the cathode cannot be allowed; it would disrupt the electric current. This is prevented by a reaction between Zn^{2+} and $NH_3(g)$ leading to the formation of complex ions, such as $[Zn(NH_3)_4]^{2+}$. (Complex ion formation is discussed in Chapter 22.)

complex ion formation: $Zn^{2+}(aq) + 4 NH_3(aq) \longrightarrow [Zn(NH_3)_4]^{2+}(aq)$

The Leclanché cell is called a **primary** cell. The electrode reactions cannot be reversed. The cell is not rechargeable. Rechargeable cells, called **secondary** cells, use oxidation-reduction reactions that can be reversed by an external electric energy source. The most familiar one is probably the lead storage cell.

LEAD STORAGE CELL. The electrodes in this cell are plates of a lead–antimony alloy. The anodes are impregnated with spongy lead metal, the cathodes with red-brown lead dioxide. The electrolyte is a dilute sulfuric acid solution. When the cell pictured in Figure 19-12 is allowed to discharge, the following reactions occur. [Think of these as involving the oxidation of Pb^0 and the reduction of Pb^{4+}, both to Pb^{2+}, followed by the precipitation of $PbSO_4(s)$.]

oxidation: $Pb(s) + SO_4^{2-}(aq) \longrightarrow PbSO_4(s) + 2 e^-$

reduction: $\underline{PbO_2(s) + 4 H^+(aq) + SO_4^{2-} + 2 e^- \longrightarrow PbSO_4(s) + 2 H_2O}$

net: $Pb(s) + PbO_2(s) + 4 H^+ + 2 SO_4^{2-} \longrightarrow 2 PbSO_4(s) + 2 H_2O$

$$E_{cell} \simeq 2.0 \text{ V} \qquad (19.43)$$

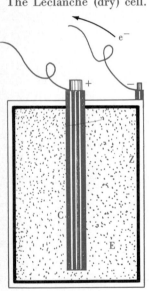

FIGURE 19-11
The Leclanché (dry) cell.

Code: C = carbon rod serving as cathode—reduction occurs.

Z = zinc container serving as anode—oxidation occurs.

E = electrolyte—a moist paste of MnO_2, $ZnCl_2$, NH_4Cl, and carbon black.

FIGURE 19-12
A lead storage cell.

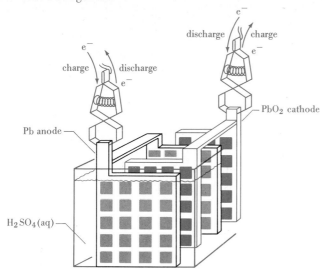

When the plates become coated with $PbSO_4(s)$ and the electrolyte has been diluted by the water produced, the cell is in a discharged condition. To recharge it, electrons are made to flow in the opposite direction using an external electric source. The net reaction that occurs is the reverse of (19.43).

To prevent short circuiting, alternating anode and cathode plates are separated by sheets of insulating material. A group of anodes is connected together electrically, and a group of cathodes is similarly connected. This "parallel" connection increases the electrode area in contact with the electrolyte solution and increases the current-delivering capacity of the cell. Cells are then joined together in "series" fashion, + to −, to produce a battery. In a 6-V battery there are three cells; in a 12-V battery, six cells.

FUEL CELLS. A galvanic cell produces electric energy with a high efficiency, as high as 90%. This is in contrast to the 30 to 40% efficiencies encountered in the combustion–steam turbine–electric generator method. Cannot fuels be consumed more efficiently by a direct conversion of chemical energy to electricity? The objective of a fuel cell is to achieve this conversion. Presently, fuel cells have achieved their most publicized successes as energy devices for space vehicles. Their uses will undoubtedly multiply in the future as conventional fuel sources dwindle and alternative means of energy production are explored more fully. The basic process involved in a fuel cell is

$$\text{fuel} + \text{oxygen} \longrightarrow \text{oxidation products}$$

The essential requirement is an electrode system in which this reaction can be carried out. The free energy change of the reaction is released as electric energy.

A fuel cell consists of a compartment containing electrolyte (a concentrated aqueous solution or a molten salt) and porous electrodes into which the gases and electrolyte may diffuse and enter into reaction. One of the simplest and most successful fuel cells, shown schematically in Figure 19-13, involves the reaction of $H_2(g)$ and $O_2(g)$ to form water. In an alkaline solution (e.g., 25% KOH) these reactions occur.

oxidation: $2\,H_2(g) + 4\,OH^-(aq) \longrightarrow 4\,H_2O + 4\,e^-$
reduction: $O_2(g) + 2H_2O + 4\,e^- \longrightarrow 4\,OH^-(aq)$
net $2\,H_2(g) + O_2(g) \longrightarrow 2\,H_2O(l)$

A fuel cell is more properly described as an energy conversion device rather than as an electrical battery. As long as fuel and $O_2(g)$ are available, it will produce electricity. It does not have a limited capacity, as does a primary cell. However, neither can it store electrical energy, as does a secondary cell.

FIGURE 19-13
Schematic representation of a hydrogen-oxygen fuel cell.

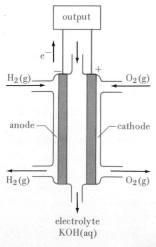

19-8 Electrochemical Mechanism of Corrosion

An important group of oxidation-reduction processes are those involved in corrosion. The fact that the combined cost of corrosion protection and the losses due to corrosion amount to billions of dollars annually lends practical as well as theoretical importance to this subject.

The processes involved in the corrosion of iron can be demonstrated in a particularly graphic manner as pictured in Figure 19-14. The object undergoing corrosion is an iron nail. The nail is embedded in a gel of agar in water. Incorporated in the gel are the acid-base indicator, phenolphthalein, and the substance $K_3[Fe(CN)_6]$ (potassium ferricyanide).

Following are the observations that can be made within hours of starting the experiment. At the head of the nail and at the tip a deep blue precipitate forms. Along the body of the nail the agar gel acquires a pink color. The blue precipitate, known as Turnbull's blue, establishes the presence of iron(II). The pink color, of course, is characteristic of phenolphthalein in a basic solution. From these observations we can write two half-equations.

oxidation: $2\,Fe(s) \longrightarrow 2\,Fe^{2+}(aq) + 4\,e^-$
reduction: $O_2 + 2\,H_2O + 4\,e^- \longrightarrow 4\,OH^-(aq)$

Thus, in the corrosion of the nail oxidation occurs at the two ends. Electrons given up in the oxidation pass along the body of the nail where they are used to reduce dissolved O_2. The reduction product, OH^-, is detected by the phenophthalein. The overall corrosion reaction is an electrochemical one. With a bent nail, oxidation occurs at three points: the head, the tip, and the bend. The nail is preferentially corroded at these points because the strained metal is more active (more anodic) than the unstrained metal.

With some metals, such as aluminum, the corrosion products (Al_2O_3) form a tough adherent coating that protects the underlying metal from further corrosion. But iron oxide (rust) flakes off an object, constantly exposing fresh surface which corrodes. A number of methods, of varying degrees of effectiveness, have been devised to protect a metal from corrosion. The simplest involves coating the surface with paint or some other protective coating. An iron surface is protected in this way only as long as the coating does not chip or peel off.

Another method of protecting an iron surface is to plate it with a thin layer of a second metal. Iron can be coated with copper by electroplating or with tin by dipping into the molten metal. In both cases protection of

FIGURE 19-14
Corrosion of an iron nail.

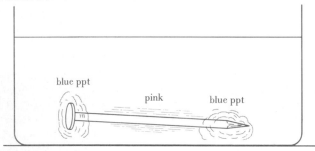

blue ppt

pink   blue ppt

The nail is imbedded in an agar gel that is impregnated with phenolphthalein and $K_3[Fe(CN)_6]$.

It is this very difference in behavior of the corrosion products that explains why cans made of iron deteriorate fairly rapidly under environmental conditions, whereas aluminum cans have an almost unlimited lifetime.

FIGURE 19-15
Protection of iron against electrolytic corrosion.

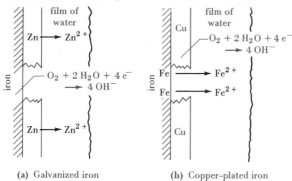

(a) Galvanized iron   (b) Copper-plated iron

In the anodic reaction (oxidation), the metal that is more easily oxidized loses electrons to produce metal ions. In case (a) this is zinc; in case (b), iron. In the cathodic reaction (reduction), oxygen gas, which is dissolved in a thin film of adsorbed water, is reduced to hydroxide ion. Rusting of iron does not occur in (a), but it does in (b).

$$Fe^{2+} + 2\,OH^- \longrightarrow Fe(OH)_2(s)$$

$$4\,Fe(OH)_2(s) + O_2 + 2\,H_2O \longrightarrow 4\,Fe(OH)_3(s)$$

$$2\,Fe(OH)_3(s) \longrightarrow Fe_2O_3 \cdot H_2O + 2\,H_2O$$
$$\text{rust}$$

FIGURE 19-16
Protection of an iron tank from corrosion with a sacrificial magnesium anode.

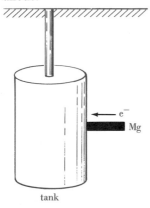

tank

Oxidation occurs at the Mg anode, and reduction occurs on the iron tank.

A lead storage battery could function as the external energy source for electrolysis reaction (19.44).

the underlying iron is achieved only as long as the coating remains intact. If the coating is cracked, as when a "tin" can is dented for example, the underlying iron is exposed and corrodes. Iron, being more active than copper and tin, undergoes oxidation; the reduction half-reaction occurs on the plating. When iron is coated with zinc (galvanized iron) the situation is different. Zinc is more active than iron. If a break occurs in the zinc plating, the iron is still protected. Zinc is oxidized in place of iron, and corrosion products protect zinc from further corrosion. The difference in these two types of protective action is brought out in Figure 19-15. Still another method can be used to protect iron and steel objects—ships, storage tanks, pipelines, plumbing systems. This involves connecting to the object, either directly or through a wire, a chunk of magnesium or other active metal. Oxidation occurs at the active metal and it gradually dissolves. The iron surface acquires electrons from the oxidation of the active metal; the iron acts as a cathode and supports a *reduction* half-reaction. As long as some of the active metal remains the iron is protected. This type of corrosion protection is called cathodic protection and the magnesium or other active metal is called, appropriately, a "sacrificial anode." The method is illustrated in Figure 19-16.

19-9 Electrolysis and Nonspontaneous Chemical Change

Let us return to a consideration of Figure 19-5, which already has provided us with insights into several electrochemical phenomena. If the cell is allowed to function spontaneously, electrons flow from zinc to copper, and the net chemical change is that of equation (19.11).

voltaic cell: $Zn(s) + Cu^{2+} \longrightarrow Cu(s) + Zn^{2+}$
$$E^\circ_{cell} = +1.10\ V$$

By connecting the electrodes to an external energy source—either a generator or a voltaic cell of sufficient emf—electrons can be made to flow in the opposite direction. The chemical reaction in this case is the reverse of (19.11). In an electrolysis reaction electric energy is used to produce a chemical change that will not occur spontaneously: E_{cell} is negative.

electrolysis: $Cu(s) + Zn^{2+} \longrightarrow Zn(s) + Cu^{2+}$
$$E^\circ_{cell} = -1.10\ V \qquad (19.44)$$

PREDICTING ELECTRODE REACTIONS. If a potential difference exceeding 1.10 V is applied to the cell of Figure 19-5, with zinc as the cathode and copper the anode, the electrolysis reaction (19.44) will occur. Similar calculations can be made regarding other electrolysis reactions. However, what actually happens may not always correspond to these calculations.

In many cases the voltage necessary to bring about a particular electrode reaction may exceed that which is calculated theoretically. Interactions may occur between the electrode surface and the species involved in an electrode reaction. This may require that an overpotential be applied in order for the electrode reaction to occur. Overpotentials are particularly common when gases are involved. For example, the overpotential for the discharge of $H_2(g)$ at a mercury cathode is approximately 1.5 V, whereas on a platinum cathode it is practically zero.

A second complicating factor is that if the material being electrolyzed contains several species capable of undergoing oxidation and reduction competing electrode reactions may occur. In the electrolysis of *molten* sodium chloride only one oxidation and one reduction are possible.

oxidation: $2\,Cl^- \longrightarrow Cl_2(g) + 2\,e^-$
reduction: $2\,Na^+ + 2\,e^- \longrightarrow 2\,Na(l)$

In the electrolysis of *aqueous* sodium chloride *two* oxidation and *two* reduction half-reactions must be considered.

oxidation: $2\,Cl^- \longrightarrow Cl_2(g) + 2\,e^-$ $E^\circ_{ox} = -1.36\ V$ (19.45)
            $2\,H_2O \longrightarrow O_2(g) + 4\,H^+ + 4\,e^-$ $E^\circ_{ox} = -1.23\ V$ (19.46)

reduction: $2\,Na^+ + 2\,e^- \longrightarrow 2\,Na(s)$ $E^\circ_{red} = -2.71\ V$ (19.47)
            $2\,H_2O + 2\,e^- \longrightarrow H_2(g) + 2\,OH^-$ $E^\circ_{red} = -0.83\ V$ (19.48)

The electrode potentials for half-reactions (19.45) and (19.46) are similar in magnitude. Exact values depend on $[Cl^-]$ in the one case and $[H^+]$ in the other. If the NaCl solution is concentrated, the oxidation half-reaction (19.45) is favored; if it is quite dilute, (19.46).

As far as the reduction half-reaction is concerned, the reduction of water occurs much more readily than that of Na^+. Generally, only the half-reaction (19.48) occurs. The principal exception is if liquid mercury is used as a cathode. In this case, because of the high overpotential of hydrogen on mercury and the solubility of sodium metal in liquid mercury, the half-reaction (19.47) is actually observed.

Example 19-9 With reference to Figure 19-17, predict the electrode reactions and the net electrolysis reaction that will occur when the anode is made of **(a)** copper and **(b)** platinum.

In both cases the cathode reaction is the same, the deposition of copper from the $CuSO_4$ solution.

reduction: $Cu^{2+}(aq) + 2\,e^- \longrightarrow Cu(s)$ $E^\circ_{red} = +0.34\ V$

(a) The half-reaction occurring at the anode is the oxidation of copper.

oxidation: $Cu(s) \longrightarrow Cu^{2+}(aq) + 2\,e^-$ $E^\circ_{ox} = -0.34\ V$

The net electrolysis reaction is

$Cu(s)[anode] \longrightarrow Cu(s)[cathode]$ (19.49)

Copper from the anode dissolves at the same rate that Cu^{2+} ions are deposited at the cathode. The solution concentration remains unchanged.

(b) Platinum metal is much too difficult to oxidize in an electrolysis process. Also, the tendency for SO_4^{2-} to be oxidized to $S_2O_8^{2-}$ is very low ($E^\circ_{ox} = -2.01\ V$). The oxidation that occurs most readily is that of water, as shown by equation (19.46).

oxidation: $2\,H_2O \longrightarrow O_2(g) + 4\,H^+(aq) + 4\,e^-$ $E^\circ_{ox} = -1.23\ V$

An overpotential is simply the potential difference in excess of that calculated theoretically required to produce electrolysis.

FIGURE 19-17
Predicting electrode reactions in electrolysis—Example 19-9 illustrated.

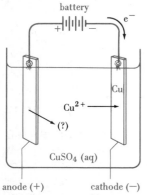

anode (+)   cathode (−)

Electrons are forced onto the copper cathode by the external source (battery). Cu^{2+} ions are attracted to the cathode and are reduced to $Cu(s)$. The oxidation half-reaction at the anode depends on the metal used for the anode.

In aqueous solutions, when other oxidation and reduction half-reactions are not feasible, the oxidation and/or reduction of water can always be brought about. These are the half-reactions (19.46) and (19.48), respectively.

The overall electrolysis reaction is

$$2\,Cu^{2+}(aq) + 2\,H_2O \longrightarrow$$
$$2\,Cu(s) + 4\,H^+(aq) + O_2(g)$$
$$E^\circ_{cell} = -0.89\,V \qquad (19.50)$$

SIMILAR EXAMPLE: Exercise 41.

Gold and silver are commonly found as impurities in copper. These metals are less active than copper, that is, less readily oxidized. They do not enter into the anode reaction, but simply deposit at the bottom of the electrolysis tank as sludge. Their value is usually more than enough to cover the cost of the electrolysis.

INDUSTRIAL ELECTROLYSIS PROCESSES. An important use of electrolysis is in refining metals. The usual metallurgical smelting process produces copper metal that is too impure for most of its intended uses. For example, the presence of arsenic lowers the electrical conductivity of copper, making it unfit for the manufacture of wire and other electrical conductors. The electrolysis process described by equation (19.49) in Example 19-9(a) is used to refine copper to purities of 99.95%. A large chunk of impure copper is taken as the anode and a strip of pure copper is taken as the cathode. During electrolysis copper is transported continuously through the solution (as Cu^{2+}) from anode to cathode.

FIGURE 19-18
Electrolysis of NaCl(aq) in a diaphragm cell.

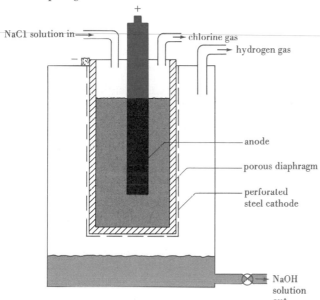

NaCl solution in

chlorine gas

hydrogen gas

anode

porous diaphragm

perforated steel cathode

NaOH solution out

In the normal operation of a liquid mercury chloralkali cell there is a small loss of mercury, as the metal and as inorganic salts. Although the percentage loss is small, the net loss of mercury to the surroundings can be considerable. Prior to 1970 this loss amounted to hundreds and perhaps thousands of kilograms per day. Mercury can be a serious environmental pollutant.

As described previously, electrolysis of concentrated NaCl(aq) involves the oxidation of Cl^- (equation 19.45) and the reduction of H_2O (equation 19.48).

$$2\,Na^+ + 2\,Cl^- + 2\,H_2O \xrightarrow{electrolysis}$$
$$2\,Na^+ + 2\,OH^- + H_2(g) + Cl_2(g) \qquad (19.51)$$

This is among the most important commercial electrolysis reactions. It can be used to produce both NaOH(aq) and $Cl_2(g)$ [and $H_2(g)$ for that matter]. One type of cell used in this electrolysis is shown in Figure 19-18. It is called the **Nelson** or **diaphragm cell.** The cell consists of two compartments. The inner compartment is made of a porous material supported by a steel mesh outer wall. A graphite rod serving as the anode is immersed in NaCl(aq) in the inner compartment; the steel mesh is the cathode. $Cl_2(g)$ is produced at the anode by the oxidation of $Cl^-(aq)$ and is drawn off. As the NaCl(aq) seeps through the porous diaphragm, the reduction half-reaction occurs at the outside wall. $H_2(g)$ is drawn off, and NaOH(aq) drops to the bottom of the outer compartment.

One of the disadvantages of the diaphragm cell is that the NaOH(aq) cannot be kept entirely free of chloride ion. Although acceptable for many applications, this alkali is not satisfactory for use in the textile industry. An alternative method involves using a mercury cathode. Instead of $H_2(g)$, sodium metal is produced at the cathode and dissolves in the mercury. [Recall that $H_2(g)$ has a very high overpotential on liquid mercury.] The Na–Hg amalgam is decomposed with water. $H_2(g)$ and pure NaOH(aq) are produced and the mercury is regenerated and returned to the electrolysis cell.

Among the common substances that are produced almost exclusively by electrolytic processes are the alkali metals, magnesium, aluminum, chlorine, fluorine, hydrogen peroxide, and sodium hydroxide. It is not an overstatement that modern industry and modern society in general could not function without the availability of electrolysis reactions.

FARADAY'S LAWS OF ELECTROLYSIS.   The relationship between quantities of electric energy consumed and chemical change produced in electrolysis is one of the many important questions to which Michael Faraday (1791–1867) sought answers. Faraday's first law of electrolysis notes that

the amount of chemical change produced is proportional to the quantity of electric charge that passes through an electrolytic cell.

Faraday's second law of electrolysis states that

a given quantity of electricity produces the same number of equivalents of any substance in an electrolysis.

An equivalent of substance is associated with *1 mol of electrons* in a half-reaction (see also, Section 19-10). Let us rewrite the electrolysis reaction (19.50) in terms of half-equations based on the transfer of *1 mol of electrons* between the anode and the cathode.

oxidation: $\frac{1}{2} H_2O \longrightarrow \frac{1}{4} O_2(g) + H^+(aq) + e^-$
reduction: $\frac{1}{2} Cu^{2+}(aq) + e^- \longrightarrow \frac{1}{2} Cu(s)$

From these half-equations we would define one (electrochemical) equivalent as equal to $\frac{1}{2}$ mol H_2O, $\frac{1}{4}$ mol O_2, 1 mol H^+, $\frac{1}{2}$ mol Cu^{2+}, and $\frac{1}{2}$ mol Cu(s). Thus, the passage of 1 mol of electrons through the electrolysis cell of Figure 19-17 is signaled by the dissolution of $\frac{1}{2}$ mol Cu (31.77 g) at the anode. Simultaneously, this same amount of Cu is deposited at the cathode. One ampere (A) of electric current represents the passage of 1 coulomb of charge per second (C/s). Thus, the product, current × time (s), yields the total quantity of charge transferred, in coulombs (C). The Faraday constant, 96,500 C/mol e, allows for a conversion between coulombs of charge and moles of electrons. Two types of calculation are possible with this kind of information. One may make electrical measurements and calculate the extent of chemical change (as in Example 19-10). Alternatively, by determining the extent of chemical change one can establish the quantity of electricity involved in an electrolysis. Methods of determining chemical change include weighing a deposit on an electrode or titrating a product of the electrolysis. An electrolytic cell designed for the purpose of determining quantities of electric charge is called a **coulometer** (see Exercise 46).

Example 19-10   What mass of copper, in grams, is deposited by a current of 1.50 A in 1.00 h in the electrolysis of a $CuSO_4$ solution?
   The electrode reaction is $Cu^{2+}(aq) + 2 e^- \rightarrow Cu(s)$, which yields the conversion factor 1 mol Cu $\backsimeq$ 2 mol e^-. The calculation is done in three steps, although the steps could be combined easily into a single setup.

no. C = $1.00\,h \times \dfrac{60\,min}{1\,h} \times \dfrac{60\,s}{1\,min} \times \dfrac{1.50\,C}{1\,s} = 5.40 \times 10^3\,C$

no. mol e^- = $5.40 \times 10^3\,C \times \dfrac{1\,mol\,e^-}{9.65 \times 10^4\,C} = 5.60 \times 10^{-2}\,mol\,e^-$

no. g Cu = $5.60 \times 10^{-2}\,mol\,e^- \times \dfrac{1\,mol\,Cu}{2\,mol\,e^-} \times \dfrac{63.55\,g\,Cu}{1\,mol\,Cu} = 1.78\,g\,Cu$

SIMILAR EXAMPLES: Exercises 43, 44.

19-10 Postscript: Equivalent Weight and Normality in Oxidation-Reduction Reactions

A balanced oxidation-reduction equation can be used in stoichiometric calculations just like any other balanced equation. Sometimes, however, oxidation-reduction reactions are treated from the standpoint of equivalent weight and normality rather than the mole and molar concentration. Let us explore briefly how this is done.

For oxidation-reduction reactions *an equivalent is the amount of substance associated with 1 mol of electrons in a half-reaction.* Consider the equation

$$5\ Fe^{2+} + MnO_4^- + 8\ H^+ \longrightarrow 5\ Fe^{3+} + Mn^{2+} + 4\ H_2O \qquad (19.52)$$

Expressed as balanced half-equations, it becomes

oxidation: $5\ Fe^{2+} \longrightarrow 5\ Fe^{3+} + 5\ e^-$

reduction: $MnO_4^- + 8\ H^+ + 5\ e^- \longrightarrow Mn^{2+} + 4\ H_2O$

We conclude that 5 mol Fe^{2+} is 5 equiv Fe^{2+}, or that 1 mol Fe^{2+} is 1 equiv Fe^{2+}. The situation with MnO_4^- is that 1 mol MnO_4^- is 5 equiv MnO_4^-, or $\frac{1}{5}$ mol MnO_4^- is 1 equiv MnO_4^-. Stated in another way:

$$1\text{ mol }Fe^{2+} \backsimeq \tfrac{1}{5}\text{ mol }MnO_4^- \qquad \text{and} \qquad 1\text{ equiv }Fe^{2+} \backsimeq 1\text{ equiv }MnO_4^-$$

> Fe²⁺ is obtained by dissolving iron metal (Fe) in an aqueous solution of an acid.

Example 19-11   A piece of pure iron wire weighing 0.1568 g is dissolved in acidic solution and titrated with 26.24 ml of a $KMnO_4(aq)$ solution. What is the normality of the $KMnO_4(aq)$?

$$\text{no. equiv }Fe^{2+} = \text{no. equiv Fe} = \text{no. mol Fe} = 0.1568\text{ g Fe} \times \frac{1\text{ mol Fe}}{55.85\text{ g Fe}}$$

$$= 2.808 \times 10^{-3}\text{ equiv }Fe^{2+}$$

$$\text{no. equiv }KMnO_4 = \text{no. equiv }MnO_4^- = \text{no. equiv }Fe^{2+} = 2.808 \times 10^{-3}$$

$$\text{normality }KMnO_4 = \frac{2.808 \times 10^{-3}\text{ equiv }KMnO_4}{0.02624\text{ L}} = 0.1070\ N\ KMnO_4$$

> The endpoint of a permanganate titration is signaled by the first lasting pink color in solution.

SIMILAR EXAMPLE: Exercise 49.

Example 19-12   25.8 ml of the 0.1070 N $KMnO_4$ solution described in Example 19-11 is used to titrate 50.0 ml of a saturated solution of sodium oxalate, $Na_2C_2O_4$. What is the solubility of $Na_2C_2O_4$ in g/L?

$$5\ C_2O_4^{2-} + 2\ MnO_4^- + 16\ H^+ \longrightarrow 2\ Mn^{2+} + 8\ H_2O + 10\ CO_2(g) \qquad (19.53)$$

The number of equivalents of MnO_4^- used in the titration is

$$\text{no. equiv }MnO_4^- = 0.0258\text{ L} \times \frac{0.1070\text{ equiv }MnO_4^-}{L} = 2.76 \times 10^{-3}\text{ equiv }MnO_4^-$$

Using the basic idea that 1 equiv $Na_2C_2O_4 \backsimeq 1$ equiv $C_2O_4^{2-} \backsimeq 1$ equiv MnO_4^-, we can express the solubility of $Na_2C_2O_4$ in equiv/L, that is, in *normality* concentration.

$$\frac{2.76 \times 10^{-3}\text{ equiv }C_2O_4^{2-}}{0.0500\text{ L}} = 5.52 \times 10^{-2}\ N\ Na_2C_2O_4$$

The final step is to convert from equiv $Na_2C_2O_4$ to g $Na_2C_2O_4$. For this we turn to equation (19.53). The reduction of 2 mol of MnO_4^- to Mn^{2+} involves 10 mol of electrons (recall the reduction half-equation in 19.52). Ten moles of electrons must also be associated with the oxidation of 5 mol $Na_2C_2O_4$ to $CO_2(g)$. The amount of $Na_2C_2O_4$ associated

$\left(5 \text{ mol } Na_2C_2O_4 \sim 10\,e^- \right) * \frac{1}{10}$

$\frac{1}{2} \text{ mol} \qquad \sim / e^- (\sim 1\,eq)$

$(\#\,eq)\left(\frac{1 \text{ mol}}{2\,eq} \right)\left(\frac{134\,g}{1\,mol} \right)$

with *1 mol of electrons* is 0.500 mol $Na_2C_2O_4$: The equivalent weight of $Na_2C_2O_4$ is one half its formula weight, or $0.500 \times 134 = 67.0$ g $Na_2C_2O_4$/equiv $Na_2C_2O_4$.

$$\text{solubility} = \frac{5.52 \times 10^{-2} \text{ equiv } Na_2C_2O_4}{L} \times \frac{67.0 \text{ g } Na_2C_2O_4}{1 \text{ equiv } Na_2C_2O_4} = 3.70 \text{ g } Na_2C_2O_4/L$$

SIMILAR EXAMPLES: Exercises 48, 50.

If the $KMnO_4$ solution of Examples 19-11 and 19-12 were used in a reaction in which MnO_4^- is reduced to $MnO_2(s)$ instead of Mn^{2+}, its normality concentration would *not* be 0.1070 *N*. This is because reduction of MnO_4^- to MnO_2 involves *3 mol* of electrons per mole of MnO_4^-, whereas reduction to Mn^{2+} involves *5 mol* of electrons per mole of MnO_4^-. *Molarity* is *independent* of the reaction in which a solution participates. At times, *normality* may not be. One of the drawbacks of equivalent weight and normality, then, is that one must have prior knowledge of the type of reaction in which a solution is to be used.

Summary

The basic changes that occur in an oxidation-reduction reaction are most readily seen by separating the net reaction into two half-reactions. In one of these certain atoms undergo an increase in oxidation state—this is the oxidation half-reaction. In the reduction half-reaction the oxidation states of certain atoms decrease. In an oxidation half-equation electrons appear on the right-hand side, and in a reduction half-equation, on the left-hand side. A net oxidation-reduction reaction must involve both oxidation and reduction, and the same number of electrons must appear in each half-equation. The latter requirement is the basis of balancing oxidation-reduction equations.

Even if all the conditions just stated are met, an oxidation-reduction reaction may still not occur spontaneously as written. In some instances the reverse reaction is favored. To establish which is the case, it is necessary to have a measure of the relative tendencies for oxidation and reduction. These tendencies are described through standard electrode potentials, $E°$. Standard electrode potentials are expressed relative to the standard hydrogen electrode (S.H.E.), which is assigned a value of $E° = 0.0000$ V. Extensive tabulations of standard electrode potentials are available. Such data can be used to design voltaic cells, which produce electricity from chemical change. If an electrochemical cell functions as a voltaic cell, that is, if it has $E_{cell} > 0$, the cell reaction proceeds spontaneously as written.

Cell voltages predicted from tabulations of *standard* electrode potentials are always $E°_{cell}$ values, based on standard state conditions for reactants and products. For non-standard-state conditions, the Nernst equation permits calculation of E_{cell} from $E°_{cell}$ and the activities of reactants and products. Among the applications of cell potential measurements made possible by the Nernst equation are the determination of pH and K_{sp} values. Another important application of cell potential measurements is in calculations of standard free energy changes ($\Delta\bar{G}°$) and equilibrium constants, K. Voltaic cells are encountered in such practical devices as flashlight and lead storage batteries. Less desirable are the voltaic cell reactions encountered in electrochemical corrosion.

Even though a particular oxidation-reduction reaction will not occur spontaneously as written, the reaction can be made to occur in an electrolytic cell. Here some external source of electricity is used to force electrons to flow in the direction opposite to that in which they would flow spontaneously. Numerous important industrial processes make use of electrolysis. The amount of chemical change produced in an electrolysis cell is directly related to the quantity of electrical charge passing through the cell, as stated through Faraday's laws of electrolysis.

Learning Objectives

As a result of studying Chapter 19, you should be able to

1. Identify an oxidation-reduction reaction in terms of changes in oxidation state and balance the net equation for the reaction.

2. Separate an oxidation-reduction equation into half-equations, complete and balance the half-equations, and combine them into a balanced net oxidation-reduction equation.

3. Identify oxidizing and reducing agents.

4. Describe a voltaic (galvanic) cell in terms of the electrodes, salt bridge, half-cell reactions, net cell reaction, and cell diagram.

5. Describe the standard hydrogen electrode (S.H.E.) and explain how other standard electrode potentials are related to it.

6. Use tabulated $E°$ values to determine $E°_{cell}$ for an oxidation-reduction reaction, and predict whether the reaction will occur spontaneously as written.

7. Describe the effect of varying conditions (concentration, gas pressures) on E_{cell} values, both qualitatively and quantitatively.

8. Use the relationships that exist among $\Delta \bar{G}°$, $E°_{cell}$, and K.

9. Describe some common voltaic cells—the dry cell, lead storage cell, and fuel cell.

10. Explain the corrosion of metals in electrochemical terms.

11. Describe the essential features of an electrolytic cell and the way in which it differs from a voltaic (galvanic) cell.

12. Identify oxidation and reduction processes that might be involved in an electrolysis reaction and choose among them to predict the most probable reaction.

13. Calculate relationships between amount of chemical change and quantity of electric charge involved in an electrolysis.

14. Apply the concepts of equivalent weight and normality to stoichiometric calculations involving oxidation-reduction reactions.

Some New Terms

The **anode** is an electrode at which an oxidation half-reaction occurs.

The **cathode** is an electrode at which a reduction half-reaction occurs.

Cell potential, E_{cell}, is the term used to describe the difference in potential between two electrodes in an electrochemical cell. If $E_{cell} > 0$, the cell reaction will occur spontaneously as written. If $E_{cell} < 0$, it will not. If $E_{cell} = E°_{cell}$, the reactants and products at each electrode are in their standard states.

An **electrochemical cell** is a device in which an oxidation-reduction reaction is carried out in the form of separate half-reactions for oxidation and reduction.

An **electrode potential** is the electrical potential that exists on an electrode in contact with the oxidized and reduced form of some substance.

An **electrolytic cell** is an electrochemical cell in which a *nonspontaneous* reaction is carried out by using an external source of electric energy.

The **Faraday, $\mathcal{F}$,** is the quantity of electrical charge associated with 1 mol of electrons, $96,500 \ C/mol \ e^-$.

Faraday's laws of electrolysis describe the relationship between the quantity of electrical charge that passes through an electrolytic cell and the amount of chemical change that occurs.

A **half-reaction** describes one portion of a net oxidation-reduction reaction, either the oxidation or the reduction.

The **Nernst equation** is used to relate E_{cell}, $E°_{cell}$, and the activities of the reactants and products involved in a cell reaction.

In an **oxidation** process certain atoms undergo an *increase* in oxidation state.

An **oxidation-reduction** reaction is the net reaction that results from oxidation and reduction processes occurring simultaneously.

An **oxidizing agent** is a substance that makes oxidation possible. It itself is reduced.

A **reducing agent** is a substance that makes reduction possible. It itself is oxidized.

In a **reduction** process certain atoms undergo a *decrease* in oxidation state.

A **voltaic cell** is an electrochemical cell in which the cell reaction occurs spontaneously. For a voltaic cell, $E_{cell} > 0$.

Exercises

Definitions and terminology

1. Indicate the essential *differences* in meanings of the following pairs of terms: **(a)** oxidation and reduction; **(b)** oxidizing agent and reducing agent; **(c)** half-reaction and net reaction; **(d)** voltaic (galvanic) and electrolytic cell; **(e)** anode and cathode; **(f)** E_{cell} and $E°_{cell}$.

Balancing oxidation-reduction equations

2. Complete and balance each of the following half-equations, and indicate whether the process is an oxidation or a reduction.
   - **(a)** $S_2O_8{}^{2-} \longrightarrow SO_4{}^{2-}$
   - **(b)** $HNO_3 \longrightarrow N_2O$ (acidic solution)
   - **(c)** $CH_4 \longrightarrow CO_2$ (acidic solution)
   - **(d)** $Br^- \longrightarrow BrO_3{}^-$ (basic solution)
   - **(e)** $NO_3{}^- \longrightarrow NH_3$ (basic solution)

3. Balance the following equations using the half-reaction (ion electron) method.
   - **(a)** $Cu(s) + H^+ + NO_3{}^- \longrightarrow Cu^{2+} + NO(g) + H_2O$

$E_{cell} = E_{cathode} - E_{anode}$

(b) $Zn(s) + H^+ + NO_3^- \longrightarrow Zn^{2+} + NH_4^+ + H_2O$

(c) $H_2O_2 + MnO_4^- + H^+ \longrightarrow Mn^{2+} + H_2O + O_2(g)$

(d) $C_8H_{16}O + Cr_2O_7^{2-} + H^+ \longrightarrow$
$$C_8H_{14}O + Cr^{3+} + H_2O$$

(e) $As_2S_3(s) + H^+ + NO_3^- + H_2O \longrightarrow$
$$H_3AsO_4 + NO(g) + S(s)$$

(f) $Cl_2(g) + H_2O \longrightarrow Cl^- + ClO^- + H^+$

4. Balance the following equations for oxidation-reduction reactions occurring in basic solution.

(a) $Br_2(l) + OH^- \longrightarrow Br^- + BrO_3^- + H_2O$

(b) $[Fe(CN)_6]^{3-} + N_2H_4(g) + OH^- \longrightarrow$
$$[Fe(CN)_6]^{4-} + N_2(g) + H_2O$$

(c) $As_2S_3(s) + OH^- + H_2O_2 \longrightarrow$
$$AsO_4^{3-} + H_2O + SO_4^{2-}$$

(d) $CrI_3(s) + H_2O_2 + OH^- \longrightarrow$
$$CrO_4^{2-} + IO_4^- + H_2O$$

(e) $P_4(s) + OH^- + H_2O \longrightarrow H_2PO_2^- + PH_3(g)$

5. Balance the following equations by the oxidation state change method.

(a) $Cl_2(g) + H_2O(g) \longrightarrow HCl(g) + O_2(g)$

(b) $PbO(s) + NH_3(g) \longrightarrow Pb(s) + N_2(g) + H_2O(g)$

(c) $Cu(s) + H^+ + NO_3^- \longrightarrow Cu^{2+} + H_2O + NO(g)$

(d) $Zn(s) + H^+ + NO_3^- \longrightarrow Zn^{2+} + H_2O + N_2O(g)$

(e) $MnO_4^- + NO_2^- + H^+ \longrightarrow Mn^{2+} + NO_3^- + H_2O$

6. Balance the following oxidation-reduction reactions by an appropriate method.

(a) $(NH_4)_2Cr_2O_7(s) \longrightarrow Cr_2O_3(s) + N_2(g) + H_2O(g)$

(b) $S_2O_3^{2-} + H^+ + Cl_2(g) \longrightarrow SO_4^{2-} + Cl^- + H_2O$

(c) $CN^- + MnO_4^- + OH^- \longrightarrow$
$$MnO_2(s) + CNO^- + H_2O$$

(d) $Fe_2S_3(s) + H_2O + O_2(g) \longrightarrow Fe(OH)_3(s) + S(s)$

(e) $C_2H_5OH(aq) + MnO_4^- + OH^- \longrightarrow$
$$C_2H_3O_2^- + MnO_2(s) + H_2O$$

(f) $CS_2(g) + H_2S(g) + Cu(s) \longrightarrow Cu_2S(s) + CH_4(g)$

(g) $P(s) + H^+ + NO_3^- \longrightarrow H_2PO_4^- + NO(g) + H_2O$

Oxidizing and reducing agents

7. Identify the oxidizing and reducing agents in the following equations encountered in this chapter: (19.5); (19.6); (19.11); (19.18); (19.25); (19.37); (19.43); (19.52); (19.53).

8. With reference to Table 19-2, arrange the following oxidizing agents in order of increasing power in acidic solution: $I_2(s)$, $IO_3^-(aq)$, $F_2(g)$, $Na^+(aq)$, $Zn^{2+}(aq)$, $PbO_2(s)$.

9. Certain substances can act only as oxidizing agents, others only as reducing agents. But there are some substances that can act in either capacity. Refer to Table 19-2 and indicate the situation for each of the following: **(a)** $Zn(s)$;   **(b)** $S^{2-}(aq)$;   **(c)** $H_2SO_3(aq)$;   **(d)** $Al^{3+}(aq)$;   **(e)** $I_2(s)$;   **(f)** $S(s)$;   **(g)** $Cr_2O_7^{2-}(aq)$.

Standard electrode potentials

10. E_{cell}° for the following reaction is found to be $+0.34$ V.

$$3 Pd(s) + Cr_2O_7^{2-} + 14 H^+ \longrightarrow 3 Pd^{2+} + 2 Cr^{3+} + 7 H_2O$$

What is the standard electrode potential for the half-reaction

$$Pd^{2+}(aq) + 2 e^- \longrightarrow Pd(s) \qquad E^\circ = ?$$

11. From the observations listed, estimate the approximate value of the standard electrode potential for the half-reaction $M^{2+}(aq) + 2 e^- \rightarrow M(s)$.

(a) The metal M dissolves in $HNO_3(aq)$ but not in $HCl(aq)$; it displaces $Ag^+(aq)$ but not $Cu^{2+}(aq)$.

(b) The metal M dissolves in $HCl(aq)$ producing $H_2(g)$, but displaces neither $Zn^{2+}(aq)$ nor $Fe^{2+}(aq)$.

12. You are given the task of estimating the standard electrode potential of indium.

$$In^{3+}(aq) + 3 e^- \longrightarrow In(s) \qquad E^\circ = ?$$

You do not have any electrical equipment but you do have all of the metals listed in Table 19-2 and water solutions of their ions. Describe the experiments you would perform and indicate the accuracy you would expect in your results.

★13. Two electrochemical cells are assembled in which these reactions occur.

$$V^{2+} + VO^{2+} + 2 H^+ \longrightarrow 2 V^{3+} + H_2O \qquad E_{cell}^\circ = 0.616 \text{ V}$$

$$V^{3+} + Ag^+ + H_2O \longrightarrow VO^{2+} + 2 H^+ + Ag(s)$$
$$E_{cell}^\circ = 0.439 \text{ V}$$

Use these data and other values from Table 19-2 to calculate the standard electrode potential for the half-reaction

$$V^{3+} + e^- \longrightarrow V^{2+}$$

Predicting oxidation-reduction reactions

14. Assume that all reactants and products are in their standard states and predict whether a reaction will occur in each case.

(a) $Sn(s) + Zn^{2+} \longrightarrow Sn^{2+} + Zn(s)$

(b) $2 Fe^{3+} + 2 I^- \longrightarrow 2 Fe^{2+} + I_2(s)$

(c) $4 NO_3^- + 4 H^+ \longrightarrow 3 O_2(g) + 4 NO(g) + 2 H_2O$

(d) $O_2(g) + 2 Cl^- \longrightarrow 2 ClO^-$ (in basic soln.)

(e) $2 H_2O_2(aq) \longrightarrow 2 H_2O + O_2(g)$

15. Use data from Table 19-2, assume that all reactants and products are in their standard states, and predict whether **(a)** $Mg(s)$ will displace Sn^{2+} from aqueous solution; **(b)** lead metal will dissolve in 1 M HCl;   **(c)** SO_4^{2-} will oxidize Fe^{2+} to Fe^{3+} in acidic solution;   **(d)** SO_4^{2-} will oxidize Fe^{2+} to Fe^{3+} in basic solution;   **(e)** $Cl_2(g)$ will displace I^- from aqueous solution to produce I_2.

16. According to standard electrode potentials, Na metal seemingly should displace Mg^{2+} from aqueous solution. Yet if this reaction is attempted, it is found not to occur. Explain. (*Hint:* What reaction does occur?)

17. Copper does not dissolve in $HCl(aq)$, but it does dissolve in $HNO_3(aq)$, producing $Cu^{2+}(aq)$.

(a) Explain this difference in behavior.

(b) Write an equation to represent the reaction of Cu and $HNO_3(aq)$.

18. Zinc will react with 1 M HCl to displace $H_2(g)$, but copper will not. If a piece of copper metal is joined to one of zinc and the pair of metals immersed in an HCl solution, bubbles of $H_2(g)$ appear at the copper metal.
   (a) Does this mean that copper reacts with HCl(aq)?
   (b) What is the reaction that occurs?
   (c) What is the function of the copper metal?

19. What observations would you expect to make if copper and silver metal are joined together in the manner described in Exercise 18 and the metals immersed in   (a) HCl(aq);   (b) $HNO_3(aq)$?

Voltaic (galvanic) cells

20. In each of the following examples, diagram a voltaic cell that utilizes the given reaction. Label the anode and cathode; indicate the direction of electron flow; write a balanced equation for each cell reaction; and calculate $E°_{cell}$ from data in Table 19-2.
   (a) $Cu(s) + Fe^{3+} \longrightarrow Cu^{2+} + Fe^{2+}$
   (b) $Sn^{2+} + Cr_2O_7^{2-} + H^+ \longrightarrow Sn^{4+} + Cr^{3+} + H_2O$
   (c) $Cl_2(g) + H_2O \longrightarrow Cl^- + O_2(g) + H^+$

21. For each of the following cell diagrams, write the equation for the cell reaction. Use data from Table 19-2 to establish whether the cell functions as a voltaic cell.
   (a) $Sn(s)|Sn^{2+}(aq)||Zn^{2+}(aq)|Zn(s)$
   (b) $Pt|O_2(g, 1 \text{ atm})|H_2O_2(aq)||Ag^+(aq)|Ag(s)$
   (c) $Pt|O_2(g, 1 \text{ atm})|H^+(aq)||Cu^{2+}(aq)|Cu(s)$

22. Use data from Table 19-2 to design a voltaic cell in which   (a) $Cl_2(g)$ is reduced to Cl^- and Fe(s) is oxidized to Fe^{2+};   (b) Zn^{2+} is displaced from solution as Zn(s);   (c) The net cell reaction is $2 Cu^+ \rightarrow Cu^{2+} + Cu(s)$.

Concentration dependence of E_{cell}—the Nernst equation

23. Write a Nernst-equation expression (19.36) for the following oxidation-reduction reactions in the text: (19.5); (19.25); (19.52).

24. Write an equation to represent the oxidation of Mn^{2+} to $MnO_2(s)$ by $O_2(g)$ in acidic solution. Will this reaction occur as written if all other reactants and products are in their standard states and   (a) $[H^+] = 10 M$;   (b) $[H^+] = 1.0 M$;   (c) $[H^+] = 0.10 M$;   (d) pH = 9.50?

25. Use data from Table 19-2 to show that the oxidation of Cl^- to $Cl_2(g)$ by $Cr_2O_7^{2-}$ in acidic solution will not occur with reactants and products in their standard states. Nevertheless, this method can be used to produce $Cl_2(g)$ in the laboratory. Explain why this is so. (*Hint:* What experimental conditions would you use?)

26. The following substances, all oxidizing agents, can be found in Table 19-2: $Cl_2(g)$, $O_2(g)$, $MnO_4^-(aq)$, $H_2O_2(aq)$, $F_2(g)$.
   (a) For which of them is the oxidizing power dependent on pH?

(b) For which is the oxidizing power independent of pH?
   (c) Where the oxidizing power is pH dependent, which are better oxidizing agents in acidic and which in basic solution?

27. If $[Zn^{2+}]$ is maintained at 1.0 M
   (a) What is the minimum $[Cu^{2+}]$ for which reaction (19.11) is still spontaneous?
   (b) Does the displacement of $Cu^{2+}(aq)$ by Zn(s) go to completion?

28. A galvanic cell is constructed of two hydrogen electrodes, one immersed in a solution with H^+ at 1.0 M and the other in 1.0 M KOH.
   (a) Determine E_{cell} for the reaction that occurs.
   (b) Compare your result with the standard electrode potential for the reduction of H_2O to $H_2(g)$ in basic solution.

29. If the 1.0 M KOH solution of Exercise 28 is replaced by 1.0 M NH_3
   (a) Will E_{cell} be higher or lower than in 1.0 M KOH?
   (b) What will be the value of E_{cell}?

30. A voltaic cell is constructed as follows.

$$Ag(s)|Ag^+(\text{sat'd. } Ag_2SO_4(aq))||Ag^+(0.10 \, M)|Ag(s)$$

What is the value of E_{cell}?

*31. It is desired to construct the following voltaic cell to have $E_{cell} = 0.050$ V. What $[Cl^-]$ must be present in the cathode half-cell to achieve this result?

$$Ag(s)|Ag^+(\text{sat'd. } AgI(aq))||Ag^+(\text{sat'd. } AgCl, \, x \, M \, Cl^-)|Ag(s)$$

$\Delta \overline{G}°$, $E°_{cell}$, and K

32. Use data from Table 19-2 and appropriate equations from the text to determine $\Delta \overline{G}°$ for the following reactions. (*Hint:* The equations are not balanced.)
   (a) $Al(s) + Zn^{2+} \longrightarrow Al^{3+} + Zn(s)$
   (b) $Pb^{2+} + MnO_4^- + H^+ \longrightarrow$
$$PbO_2(s) + Mn^{2+} + H_2O$$
   (c) $H^+ + Cl^- + MnO_2(s) \longrightarrow Mn^{2+} + H_2O + Cl_2(g)$

33. Determine the value of K at 298 K for reaction (a) in Exercise 32. Does the displacement of Zn^{2+} by Al(s) go to completion? (*Hint:* What is $[Zn^{2+}]$ remaining at equilibrium if Al metal is added to a solution with $[Zn^{2+}] = 1.0 \, M$?)

*34. The standard electrode potential corresponding to the reduction $Cr^{3+} + e^- \rightarrow Cr^{2+}$ is $E° = -0.407$ V. If excess Fe(s) is added to a solution in which $[Cr^{3+}] = 1.00 \, M$, what will be $[Fe^{2+}]$ when equilibrium is established at 298 K? $Fe(s) + 2 Cr^{3+} \rightleftharpoons Fe^{2+} + 2 Cr^{2+}$.

Batteries and fuel cells

35. The nickel–cadmium cell features one electrode of $Cd(OH)_2(s)$ on cadmium metal serving as the anode on discharge. The second electrode is coated with $Ni(OH)_2$ and $Ni(OH)_3$. The electrolyte is concentrated KOH(aq). Write

electrode reactions and the net cell reaction for the discharge of this cell.

36. One type of fuel cell (see Figure 19-13) uses a mixture of molten carbonates as an electrolyte. In this mixture some decomposition of carbonate ion occurs.

$$CO_3^{2-} \longrightarrow CO_2 + O^{2-}$$

Write the electrode reactions corresponding to the net cell reaction

$$CH_4(g) + 2\,O_2(g) \longrightarrow CO_2(g) + 2\,H_2O(g).$$

(*Hint:* Use O^{2-} in writing electrode reactions.)

Electrochemical mechanism of corrosion

37. Comment on the appropriateness of the statement "A corroding metal is like a galvanic cell."

38. How would the observations in Figure 19-14 differ if **(a)** a copper wire were wrapped around the iron nail; **(b)** the nail were driven through a piece of zinc?

39. Natural gas transmission pipes are sometimes protected against corrosion by maintaining a small potential difference between the pipe and an inert electrode buried in the ground. Describe how the method works. (*Hint:* Is the inert electrode maintained positively or negatively charged with respect to the pipe?)

Electrolysis reactions

40. The commercial production of magnesium involves the electrolysis of molten $MgCl_2$. Why cannot the simpler electrolysis of $MgCl_2(aq)$ be used instead?

41. Predict the probable products when the following substances are electrolyzed using platinum electrodes. Use data from Table 19-2 as appropriate. **(a)** $CuCl_2(aq)$; **(b)** $HCl(aq)$; **(c)** $H_2SO_4(aq)$; **(d)** $BaCl_2(l)$; **(e)** $KI(aq)$; **(f)** $KOH(aq)$.

42. If a lead storage battery is charged at too high a voltage, gases are produced at each electrode. (It is possible to recharge a lead storage battery only because of the high overpotential for gas formation on the electrodes.)
  (a) What are these gases?
  (b) Write a cell reaction to describe their formation.

Additional Exercises

1. Balance the following oxidation-reduction equations.
  (a) $H_2SO_3(aq) + MnO_4^- + H^+ \longrightarrow$
$$Mn^{2+} + SO_4^{2-} + H_2O$$
  (b) $Ag(s) + H^+ + NO_3^- \longrightarrow Ag^+ + H_2O + NO(g)$
  (c) $B_2Cl_4 + OH^- \longrightarrow BO_2^- + Cl^- + H_2O + H_2(g)$
  (d) $C_2H_5NO_3 + Sn + H^+ \longrightarrow$
$$NH_2OH + C_2H_5OH + Sn^{2+} + H_2O$$

Faraday's laws of electrolysis

43. What mass of metal would be deposited from a solution of each of the following ions by the passage of 1.20 A of electric current for 0.600 h in an electrolysis cell? **(a)** Cu^{2+}; **(b)** Ag^+; **(c)** Fe^{2+}; **(d)** Al^{3+}.

44. How long would it take to electrodeposit all the Cu^{2+} in 0.500 L of 0.100 M $CuSO_4$ with a current of 2.20 A?

45. A concentrated aqueous solution of Na_2SO_4 is electrolyzed with a current of 2.50 A for 0.283 h. Assuming that the anode reaction is exclusively that given in equation (19.46), what volume of $O_2(g)$ is produced at 25°C and 740 mmHg pressure if **(a)** the gas is dry; **(b)** the gas is saturated with water vapor?

46. In a silver coulometer, Ag^+ ions are reduced to $Ag(s)$ at a platinum cathode. If a quantity of electricity is passed through a silver coulometer such that in a period of 840 s 2.08 g of silver is deposited, **(a)** how many coulombs of electrical charge must have been passed and **(b)** what was the magnitude of the electric current?

Stoichiometry of oxidation-reduction reactions

47. Calculate the equivalent weights of the following substances, according to the half-reactions described in Exercise 2: **(a)** $(NH_4)_2S_2O_8$; **(b)** HNO_3; **(c)** CO_2; **(d)** $KBrO_3$; **(e)** KNO_3.

48. An iron ore sample weighing 0.8500 g is dissolved in $HCl(aq)$, treated with a reducing agent to convert all the iron to $Fe^{2+}(aq)$, and then titrated with exactly 36.10 ml of 0.0410 M $K_2Cr_2O_7$. What is the percent Fe, by mass, in the ore sample?

$$6\,Fe^{2+} + Cr_2O_7^{2-} + 14\,H^+ \longrightarrow 6\,Fe^{3+} + 2\,Cr^{3+} + 7\,H_2O$$

49. 40.10 ml of a $K_2Cr_2O_7$ solution is required for the titration of an 0.2050-g sample of $FeSO_4$ that is 99.72% pure. What are **(a)** the molarity; **(b)** the normality of the $K_2Cr_2O_7$ solution?

50. The $KMnO_4$ solution described in Example 19-11 is used in the following titration. What volume of the solution, in milliliters, will be required to react with 0.0525 g $Na_2S_2O_3$?

$$S_2O_3^{2-} + MnO_4^- + H_2O \longrightarrow MnO_2(s) + SO_4^{2-} + OH^-$$
$$\text{(not balanced)}$$

  (e) $Ag(s) + CN^- + O_2(g) + OH^- \longrightarrow$
$$[Ag(CN)_2]^- + H_2O$$

2. Use data from Table 19-2, assume that all species are in their standard states, and predict whether **(a)** $Zn(s)$ will displace $Al^{3+}(aq)$ from solution; **(b)** MnO_4^- will oxidize Cl^- to $Cl_2(g)$ in acidic solution; **(c)** $O_2(g)$ will oxidize S^{2-}

to free S in basic solution; **(d)** Cu(s) will dissolve in 1 M HCl; **(e)** the reaction $Mn^{2+} + O_2(g) + H_2O \rightarrow MnO_2(s) + H_2O_2 + H^+$ will occur as written.

3. The standard potential difference is measured for the reaction

$$3\,U^{4+} + 2\,NO_3^- + 2\,H_2O \longrightarrow$$
$$3\,UO_2^{2+} + 2\,NO(g) + 4\,H^+ \qquad E^\circ_{cell} = 0.63\ V$$

What must be the standard electrode potential for the reduction of UO^{2+} to U^{4+} in acidic solution?

4. A $KMnO_4(aq)$ solution is to be standardized by titration against $As_2O_3(s)$. A 0.1097-g sample of As_2O_3 requires 26.10 ml of the $KMnO_4(aq)$ for its titration. What are the molarity and normality of the $KMnO_4(aq)$?

$$As_2O_3 + MnO_4^- + H^+ \longrightarrow H_3AsO_4 + Mn^{2+} + H_2O$$
$$\text{(not balanced)}$$

5. $Mn^{2+}(aq)$ can be determined by titration with $MnO_4^-(aq)$.

$$Mn^{2+} + MnO_4^- + OH^- \longrightarrow MnO_2(s) + H_2O$$

A 15.00-ml sample of an aqueous solution of Mn^{2+} requires 20.06 ml of the $KMnO_4(aq)$ of the preceding exercise for its titration. What are the molarity and normality of the Mn^{2+} solution?

6. In a similar fashion to Exercise 27, would you say that the displacement of Pb^{2+} from a 1.0 M $Pb(NO_3)_2$ solution can be carried to completion by tin metal?

7. Why do you suppose that copper plumbing is preferred to cast iron in modern construction?

8. The silver–zinc cell utilizes a Zn/ZnO combination for one electrode, Ag/AgO for the other, and saturated KOH as the electrolyte. The Zn/ZnO electrode is the anode on discharge. Write balanced equations to represent the discharge and recharging of this cell.

9. Estimate the voltage of the fuel cell pictured in Figure 19-13.

*10. A piece of nickel and a piece of iron are placed in a solution with $[Ni^{2+}] = 1.0\ M$ and then connected to a battery in such a way that the nickel becomes the anode.
   (a) Write the equation for the electrolysis reaction.
   (b) The iron has a surface area of 165 cm². How long must electrolysis be continued with a current of 1.50 A to build a 0.050-mm-thick deposit of nickel on the iron? (Density of nickel = 8.90 g/cm³.)

*11. 100.0-ml solutions with ion concentrations of 1.000 M were placed in each of the half-cell compartments of the cell pictured in Figure 19-5. The cell was operated as an *electrolysis* cell, with copper as the anode, and zinc as the cathode. A current of 0.500 A was used. Assume that the only electrode reactions occurring were those involving Cu/Cu^{2+} and Zn/Zn^{2+}. Electrolysis was stopped after 10.00 h and the cell was allowed to function as a *galvanic* cell. What was E_{cell}?

*12. A common reference electrode is the silver–silver chloride electrode, which consists of a silver wire coated with AgCl(s) and immersed in 1 M KCl.

$$AgCl(s) + e^- \longrightarrow Ag(s) + Cl^-(1\ M) \quad E^\circ = +0.2223\ V$$

   (a) What is E°_{cell} when this electrode is a *cathode* in combination with a zinc electrode as an *anode?*
   (b) Cite several reasons why this electrode is easier to use than a standard hydrogen electrode.
   (c) By comparing the potential of the silver–silver chloride electrode with that of the standard silver electrode, determine K_{sp} for AgCl.

*13. Recovery as a by-product during the metallurgical extraction of lead is an important source of silver. The percentage of silver in lead was determined by the following procedure. A 1.050-g sample of lead was dissolved in a small quantity of nitric acid to produce an aqueous solution of Pb^{2+} and Ag^+. The solution was made up to a volume of 350 ml with water; a pure Ag electrode was immersed in the solution and the potential difference between this electrode and a S.H.E. was found to be 0.503 V. What was the percent Ag, by mass, in the lead metal?

*14. The course of a precipitation titration reaction can be followed by measuring the emf of a cell. The resulting curve of cell voltage vs. volume of titrant shows a sharp change in cell voltage at the equivalence point. The following cell is set up.

$$Hg(l)\,|\,Hg_2Cl_2(s)\,|\,Cl^-(0.10\ M)\,\|\,Ag^+(x\ M)\,|\,Ag(s)$$

The potential of the electrode in the anode half-cell remains constant throughout the titration. Its value is $E^\circ_{ox} = -0.2802\ V$.
   (a) Write an equation for the cell reaction.
   (b) Calculate E_{cell} at various points in the titration of 50.0 ml of 0.0100 M $AgNO_3$ carried out in the cathode half-cell compartment by adding 0.01000 M KI.
   (c) Sketch a titration curve for the titration.

Self-Test Questions

For questions 1 through 6 select the single item that best completes each statement.

1. In the reaction $Cu(s) + 2\,H_2SO_4(aq) \rightarrow CuSO_4(aq) + 2\,H_2O + SO_2(g)$ **(a)** $H_2SO_4(aq)$ is oxidized; **(b)** Cu(s) is reduced; **(c)** $SO_2(g)$ is the reducing agent; **(d)** $H_2SO_4(aq)$ is the oxidizing agent.

2. The process in which NpO_2^+ is converted to Np^{4+} **(a)** is an oxidation half-reaction; **(b)** can only occur in an elec-

trochemical cell;   **(c)** is a reduction half-reaction;   **(d)** is an oxidation-reduction reaction.

3. The reaction $Cu^{2+}(aq) + 2\,Cl^-(aq) \rightarrow Cu(s) + Cl_2(g)$ has $E^\circ_{cell} = -1.02$ V. This reaction   **(a)** can be made to produce electricity in a voltaic cell;   **(b)** can be made to occur in an electrolysis cell;   **(c)** occurs whenever Cu^{2+} and Cl^- are brought together in aqueous solution;   **(d)** can occur in acidic solution but not in basic solution.

4. The value of E°_{cell} for the oxidation-reduction reaction $Zn(s) + Pb^{2+}(1.0\ M) \rightarrow Zn^{2+}(1.0\ M) + Pb(s)$ is $+0.66$ V. For the reaction $Zn(s) + Pb^{2+}(0.10\ M) \rightarrow Zn^{2+}(0.10\ M) + Pb(s)$, $E_{cell} =$ **(a)** $+0.63$ V;   **(b)** $+0.66$ V;   **(c)** $+0.69$ V;   **(d)** $+0.72$ V.

5. For the reaction $Co(s) + Ni^{2+} \rightarrow Co^{2+} + Ni(s)$, $E^\circ_{cell} = +0.03$ V. If cobalt metal is added to a water solution in which $[Ni^{2+}] = 1.0\ M$

  (a) The reaction will not proceed in the forward direction at all.

  (b) The displacement of Ni^{2+} from solution by cobalt will go to completion.

  (c) The displacement of Ni^{2+} from solution by cobalt will proceed to a considerable extent, but the reaction will stop before the nickel ion is completely displaced.

  (d) Only the reverse reaction will occur.

6. A quantity of electrical charge that brings about the deposition of 4.5 g Al from Al^{3+} at a cathode will also produce the following volume (STP) of $H_2(g)$ from H^+ at a cathode: **(a)** 5.6 L;   **(b)** 11.2 L;   **(c)** 22.4 L;   **(d)** 44.8 L.

7. Write half-equations to represent the oxidation of PbO(s) to $PbO_2(s)$ and the reduction of $MnO_4^-(aq)$ to $MnO_2(s)$, both in basic solution. Then write a complete and balanced oxidation-reduction equation by combining the two half-equations.

8. Diagram a voltaic (galvanic) cell in which the following reaction occurs. Label the anode and cathode; use a table of standard electrode potentials to determine E°_{cell}; and balance the equation for the cell reaction.

$$Zn(s) + H^+ + NO_3^- \longrightarrow Zn^{2+} + H_2O + NO(g)$$

9. The voltaic (galvanic) cell indicated below registers an $E_{cell} = +0.108$ V. What is the pH of the unknown solution?

$$Pt\,|\,H_2(g,\ 1\ atm)|H^+(x\ M)||H^+(0.10\ M)|H_2(g,\ 1\ atm)|\,Pt$$

10. For the reaction $2\,Cu^+ + Sn^{4+} \rightarrow 2\,Cu^{2+} + Sn^{2+}$, $E^\circ_{cell} = -0.005$ V. A solution is prepared that is 1.00 M in each of the four ions. Estimate the concentration of each of the four ions when equilibrium is established at 298 K.

20 The Chemistry of Selected Representative Elements

Although numerous applications of chemical principles have been considered in earlier chapters, the remainder of the text emphasizes applications. We will be particularly interested in interrelating basic ideas of earlier chapters through a systematic study of the physical and chemical properties of selected chemical elements and their compounds. Such a study is called "descriptive chemistry."

The elements selected for study in this chapter are

Groups IA and IIA:	The most metallic elements.
Boron and aluminum:	A nonmetal and a metal in group IIIA.
Group VIIA:	The most nonmetallic group of elements.
Oxygen and sulfur:	Two important nonmetals.
Group VA:	Emphasis is placed on **nitrogen** and **phosphorus** and the variation of properties from nonmetallic to metallic within the group.
Silicon:	A nonmetal that offers an interesting contrast to carbon, which in turn is the subject of Chapters 24 and 25.

20-1 The Alkali (IA) and Alkaline Earth (IIA) Metals

There are two distinctive groups of properties of the metallic state. Chemical properties include the ability to form ionic compounds with nonmetals and reactivity with water and/or acids. These properties result from low ionization energies, low electronegativities, and negative electrode potentials. Physical properties of metals include electrical conductivity, thermal conductivity, high melting point, and hardness. These properties reflect the nature of the metallic bond. The elements of groups IA and IIA, although not as metallic as some in terms of physical properties, display the chemical properties of metals to the highest degree.

The relationship of these physical properties to the nature of the metallic bond was explored in Section 10-7.

The primary feature of the IA and IIA elements responsible for these metallic properties is their electron configurations. The IA atoms have a single s electron, and the IIA atoms a pair of s electrons, in an outermost electronic shell beyond a noble gas core. These electrons are transferred with relative ease to nonmetal atoms to produce ionic compounds. In the pure metals, however, the availability of electrons to form metallic bonds is limited, especially with the IA metals.

497

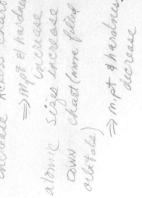

RELATIONSHIP OF PHYSICAL PROPERTIES TO ATOMIC PROPERTIES.   Table 20-1 summarizes some properties of the IA and IIA metals. With reference to this table, melting point and hardness are expected to increase as the number of bonding electrons increases and as atomic size decreases (both factors leading to stronger interatomic bonds). Some additional properties can best be understood in terms of the ionic properties listed in Table 20-1.

The more negative the electrode potential, the less the tendency for the metal ion to be reduced and the greater the tendency for the reverse process to occur—that is, for the metal atom to be oxidized. The case of lithium in Table 20-1 is worthy of special mention.

In general, ions in aqueous solution interact with neighboring water molecules. The energy of this interaction is called the **hydration energy.** The hydration energy of the Li^+ ion is unusually large because this small ion can bond strongly to water molecules. This large hydration energy stabilizes the Li^+ ion in solution, making it especially difficult to reduce. Thus, the standard reduction potential of lithium is more negative than those of the other alkali metals.

Recall the discussion of the hydrated proton in Section 17-2.

SOLUBILITIES AND FORMATION OF COMPLEXES.   Because of the large size and small charge of the IA metal ions, interionic attractive forces within a crystalline ionic compound are rather easily overcome by attractive forces between ions and water dipoles (see Figure 12-3). Most alkali metal compounds are highly water soluble. Among the alkaline earth compounds a considerable number are either only moderately soluble or "insoluble" (e.g., carbonates, fluorides, hydroxides, and phosphates). This is a consequence of the smaller size and higher charge of IIA metal ions compared to IA metal ions, leading to higher lattice energies. Alkali metal ions do not tend to form complex ions in solution. Again, this is because alkali metal ions have large size, small charge, and correspondingly low charge density. The alkaline earth metal ions, on the other hand, do form some important complexes. Ca^{2+} forms a complex with ethylenediaminetetraacetic acid (EDTA) that plays a role in water treatment. Mg^{2+} is a constituent of chlorophyll, a catalyst in photosynthesis.

Complex ion formation is discussed in Chapter 22.

TABLE 20-1
Some properties of the alkali and alkaline earth metals

Element	Melting point, °C	Hardness[a]	Electrical conductivity[b]	Ion	Ionic radius, pm	Hydration energy of ion, kJ/mol[c]	Standard reduction potential, V
Li	179	0.6	17.4	Li^+	68	515	−3.045
Na	97.8	0.4	35.2	Na^+	95	405	−2.714
K	63.7	0.5	23.1	K^+	133	322	−2.925
Rb	38.9	0.3	13.0	Rb^+	148	293	−2.925
Cs	28.5	0.3	8.1	Cs^+	169	264	−2.923
Be	1278	ca. 5	8.8	Be^{2+}	31	—	−1.85
Mg	651	2.0	36.3	Mg^{2+}	65	1925	−2.37
Ca	845	1.5	35.2	Ca^{2+}	99	1653	−2.87
Sr	769	1.8	7.0	Sr^{2+}	113	1485	−2.89
Ba	725	ca. 2	—	Ba^{2+}	135	1276	−2.90

[a] Hardness measures the ability of substances to scratch, abrade, or indent one another. On Mohs' scale, 10 minerals are selected as standards, ranging from the softest, talc, to the hardest, diamond. These are assigned values of 1 and 10, respectively. Other values: wax (0°C) 0.2; asphalt, 1–2; coal, 2.2; fingernail, 2.5; copper, 2.5–3; brass, 3–4; iron, 4–5; agate, 6–7; topaz, 8; chromium 9. Each substance is capable of scratching only others of hardness values less than its own.
[b] On a scale relative to silver, 100; copper, 95.9; gold, 67.5.
[c] For the process $M^{n+}(g) + aq = M^{n+}(aq)$.

FIGURE 20-1
Diagonal relationships.

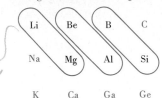

The encircled pairs of elements exhibit many similar properties.

DIAGONAL RELATIONSHIPS.   The element lithium has a number of properties that set it apart from the other alkali metals. For example, its carbonate, fluoride, and phosphate are only slightly soluble in water. In this respect, and in its chemical behavior, lithium bears a strong resemblance to magnesium. The similarity of lithium to magnesium is referred to as a "diagonal relationship." Such relationships also exist between Be and Al, and between B and Si (see Figure 20-1). The Li–Mg similarity is thought to result from the approximately equal sizes of the Li and Mg atoms and of the Li^+ and Mg^{2+} ions.

CHEMICAL PROPERTIES.   The chemistry of the IA and IIA metals is largely a reflection of the ease of oxidation of the metal atoms to the metal ions. The direct action of a halogen (X_2) on the metal (M) results in a **binary halide.**

$$\text{IA} \qquad\qquad\qquad\qquad \text{IIA}$$
$$2\,M(s) + X_2(g) \longrightarrow 2\,MX(s) \qquad M(s) + X_2(g) \longrightarrow MX_2(s)$$

Oxides result from the direct action of oxygen on the metals, but the products are varied. The principal reactions are

$$\text{IA} \qquad\qquad\qquad\qquad\qquad \text{IIA}$$
$$4\,Li(s) + O_2(g) \longrightarrow 2\,Li_2O(s)$$
$$2\,Na(s) + O_2(g) \longrightarrow Na_2O_2(s) \qquad 2\,M(s) + O_2(g) \longrightarrow 2\,MO(s)$$
$$M(s) + O_2(g) \longrightarrow MO_2(s) \qquad \text{(where M = any group IIA element;}$$
$$\text{(where M = K, Rb, Cs)} \qquad\qquad Ba \text{ also forms } BaO_2\text{.)}$$

In normal oxides (e.g., Li_2O) the oxide ion is that discussed in earlier chapters—O^{2-}. In peroxides (e.g., Na_2O_2) the anion is O_2^{2-}, and in superoxides (e.g., KO_2) the anion is O_2^-. The structures of these anions are discussed in Section 20-4.

Because the IIA metal ions and the oxide ion are small and carry a relatively high charge ($+2$ and -2, respectively), the IIA metal oxides have high lattice energies and high melting points. Typical data are listed in Table 20-2.

Ionic hydrides are produced by the reaction of the metals with hydrogen gas. In these compounds the hydrogen exists as hydride ion $[H:]^-$.

$$\text{IA} \qquad\qquad\qquad\qquad \text{IIA}$$
$$2\,M(s) + H_2(g) \longrightarrow 2\,MH(s) \qquad M(s) + H_2(g) \longrightarrow MH_2(s)$$
$$\text{(where M = any group IA element)} \qquad \text{(where M = Ca, Sr, Ba)}$$

Ionic hydrides react with water to liberate hydrogen gas. The hydrogen is produced both by the oxidation of H^- and the reduction of H_2O. CaH_2, a gray powdered solid, is used as a "portable" source of hydrogen, say for filling weather observation balloons.

$$H^- + H_2O \longrightarrow OH^-(aq) + H_2(g) \tag{20.1}$$

TABLE 20-2
Alkaline earth metal oxides

Oxide	Mp, °C	Lattice energy, kJ/mol
BeO	2530	—
MgO	2800	3934
CaO	2580	3524
SrO	2430	3308
BaO	1923	3127

Nitrides are formed by the direct union of $N_2(g)$ with the IIA metals at high temperature, but only lithium among the IA metals reacts in this way. Here is an additional resemblance of lithium and magnesium.

$$\text{IA} \qquad\qquad\qquad\qquad \text{IIA}$$
$$6\,Li(s) + N_2(g) \longrightarrow 2\,Li_3N(s) \qquad 3\,M(s) + N_2(g) \longrightarrow M_3N_2(s)$$

These metal nitrides react with water to form $NH_3(g)$ and the metal hydroxide. However, this method of "fixing" atmospheric nitrogen cannot compete commercially with the Haber process (Section 14-8).

The group IA metals react with cold water to produce hydrogen and an aqueous solution of the metal **hydroxide.** The lighter metals float on water and race along the surface as they react. Often the heat evolved is sufficient to melt the metal and ignite the liberated hydrogen.

$$2\,M(s) + 2\,H_2O \longrightarrow 2\,M^+(aq) + 2\,OH^-(aq) + H_2(g) \tag{20.2}$$

Of the group IIA metals only the heavier ones (Ca, Sr, Ba, and Ra) react with

[handwritten margin notes: IA rxt w/ cold H₂O — Mg - if oxide coating removed — Be - not @ all]

cold water to produce the metal hydroxide and $H_2(g)$. Magnesium reacts very slowly with cold water but at appreciable rates with hot water.

$$Mg(s) + H_2O \xrightarrow{\text{heat}} MgO(s) + H_2(g) \tag{20.3}$$

The lack of reactivity of magnesium with cold water is due to the protective action of an impervious coating of magnesium oxide on the metal. If magnesium is dissolved in liquid mercury, the protective oxide does not form and the metal displays its expected high reactivity toward water.

[margin: From equations (20.2) and (20.3) we can see that water is not a good agent for fighting fires involving active metals.]

[handwritten margin: MgO(s)+H₂O → true path ... contradictory?]

The reactions just described are impractical for producing the hydroxides in quantity. Alkaline earth metal hydroxides are more readily produced by the action of water on the metal oxide. The metal oxide in turn is obtained by decomposition of the metal carbonate.

$$MCO_3(s) \xrightarrow{\text{heat}} MO(s) + CO_2(g) \tag{20.4}$$

$$MO(s) + H_2O \longrightarrow M(OH)_2(s) \tag{20.5}$$

Alkali metal hydroxides are produced in quantity by electrolysis of the aqueous chlorides (recall Section 19-9).

The group IA and IIA metals react with acids such as HCl to produce $H_2(g)$. In these reactions the metal is a reducing agent and H^+ is an oxidizing agent.

$$\overset{\text{IA}}{2\,M(s) + 2\,H^+ \longrightarrow 2\,M^+ + H_2(g)} \qquad \overset{\text{IIA}}{M(s) + 2\,H^+ \longrightarrow M^{2+} + H_2(g)}$$

[handwritten: more tendency to lose e⁻ than H⁺]

If there is present in the solution NO_3^-, SO_4^{2-}, or some other oxidizing agent stronger than H^+, $H_2(g)$ may not be produced. Instead one obtains a reduction product such as $NO(g)$, $N_2O(g)$, NH_4^+, or $SO_2(g)$.

[handwritten: $2M(s) + 2H^+ + NO_3^- \longrightarrow NO + 2M^+ + 2OH^-$]

THE SPECIAL CASE OF BERYLLIUM.   Beryllium (and to some extent magnesium) is rather different from the heavier members of group IIA. It can be seen from Table 20-1 that beryllium has a considerably higher melting point and is much harder than the other group IIA elements. Also, its chemical properties differ significantly. Some of its distinctive chemical properties are

[margin: In general, the properties of the first member of a group of the periodic table differ somewhat from those of other elements in the group. Several examples are cited in this chapter.]

1. Be is quite unreactive toward air and water.
2. BeO does not react with water. [For the other oxides, $MO + H_2O \rightarrow M(OH)_2$.]
3. Be and BeO dissolve in strongly basic solutions to form the ion BeO_2^{2-}. [$BeO(s) + 2\,OH^-(aq) \rightarrow BeO_2^{2-}(aq) + H_2O$.]
4. $BeCl_2$ and BeF_2 in the molten state are poor conductors of electric current.

The chemical behavior of beryllium is best understood in terms of the small size and high ionization energy of the Be atom. The tendency to form the Be^{2+} ion is limited, and the ability of the Be atom to form covalent bonds is more pronounced. The reactivity of BeO with strongly basic solutions indicates that the oxide has acidic properties. This in turn is associated with small ionic size and high charge (see Section 20-4). Covalent bond formation by beryllium appears to involve hybridized atomic orbitals. Figure 20-2 depicts bonding through sp hybrid orbitals in gaseous $BeCl_2$ and through sp^3 hybrid orbitals in solid $BeCl_2$.

PREPARATION AND USES OF THE GROUP IA AND IIA METALS.   The metals occur only in ionic form. This may be as NaCl and $MgCl_2$ in seawater, $CaCO_3$ in limestone and marble, or as complex minerals such as $LiAlSi_2O_6$ (spodumene) and $Be_3Al_2(SiO_3)_6$ (beryl).

based on tendency to reduce others in aqueous solⁿ

FIGURE 20-2
Covalent bonds in $BeCl_2$.

$BeCl_2(g)$

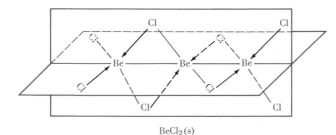

$BeCl_2(s)$

In gaseous $BeCl_2$ discrete molecules exist with the bonding scheme shown in the figure. In solid $BeCl_2$ two Cl atoms are bonded to a Be atom through normal covalent bonds. Two others are bonded by coordinate covalent bonds, using lone pair electrons of the Cl atoms (bonds shown as arrows). The arrangement is essentially tetrahedral. Of course, once formed, these two types of bonds cannot be distinguished from one another. $BeCl_2$ units are linked into long chain-like polymeric molecules—$(BeCl_2)_n$.

Beryllium is an extremely poisonous substance, and this fact has probably limited experimentation with it. The maximum permissible level of beryllium dust in air is only about 10^{-6} g/m^3.

Because the IA and IIA elements have large negative electrode potentials, reduction of the metal ions cannot be achieved in aqueous solution. Reduction cannot be carried out either with chemical reducing agents or electrolytically (with the exceptions noted in Section 19-9). Reduction can be accomplished with certain reducing agents in nonaqueous media at high temperatures, but for the most part these methods are not commercially feasible. Commercial production is generally by the electrolysis of a molten salt, as in the Dow process for magnesium (see Section 26-7).

Although compounds of the group IA and IIA elements are used in quite a number of ways, the metals themselves are not widely used except for sodium, magnesium, and beryllium. Magnesium has a lower density than any other structural metal and is used widely in fabricating lightweight objects, such as aircraft parts, generally as an alloy with aluminum and other metals. Magnesium metal is also used as a reducing agent in certain metallurgical processes. Sodium is used as a reducing agent in chemical syntheses and as a heat exchange medium in nuclear reactors.

Because of its ability to withstand metal fatigue, an alloy of copper containing about 2% beryllium is used in making springs, clips, and electrical contacts. Other beryllium alloys are used for structural purposes where light weight is a primary requirement. Because the beryllium atom has so little stopping power for x rays or neutrons, beryllium is used in fabricating "windows" for x ray tubes and for various components in nuclear reactors.

20-2 Boron and Aluminum

Because they are both members of group IIIA we expect similarities between boron and aluminum, and they do exist. But there are differences too. Furthermore, based on the "diagonal relationship," we expect similarities between boron and silicon and between aluminum and beryllium.

The tendency to lose electrons (three) to acquire a noble gas electron configuration is much less pronounced in B and Al than in the group IA and IIA elements. In fact, although the B^{3+} ion may be postulated when describing certain phenomena, the expenditure of energy required to produce it is much too high. The ion B^{3+} does not actually exist. Even the ion Al^{3+} is not commonly encountered, except perhaps in $AlF_3(s)$. It is attainable in aqueous solution, however, as the hydrated ion $[Al(H_2O)_6]^{3+}$ or in other complex forms.

Boric acid, $B(OH)_3$, is a weak acid. Furthermore, boric acid and borate compounds bear a number of similarities to silicic acid and silicates (see Section 20-9). Aluminum hydroxide, $Al(OH)_3$, is primarily basic, although it does exhibit acidic properties under certain circumstances (see Section 20-4). Whereas aluminum forms a single, ill-defined hydride, boron forms a series of hydrides. In this boron again resembles silicon, although the boron hydrides have more complex structures than those of silicon. Because the bonded pair B—N is isoelectronic with the bonded pair C—C, there exist a number of covalent boron–nitrogen compounds with properties similar to organic compounds (Chapter 24). Some naturally occurring compounds of boron and their uses are described in Chapter 26.

COVALENT BONDING. Compounds of B and Al are characterized by

1. Bonding through sp^2 hybrid orbitals.
2. A deficiency of electrons in the outermost shell of the bonded B or Al atom.

AlF_3 has considerable ionic character. Its melting point is 1040°C, compared to 194°C for $AlCl_3$ (at 5.2 atm), 97.5°C for $AlBr_3$, and 191°C for AlI_3.

Let us consider some of the consequences of these factors as they relate to the **halides** of boron and aluminum. The boron halides (e.g., BCl_3) exist as planar, monomeric species in the gaseous state or in nonpolar solvents. But in the case of $AlCl_3$, $AlBr_3$, and AlI_3, to overcome their electron deficiency, the Al atoms make use of the $3s$ and *all* $3p$ orbitals to form sp^3 hybrid orbitals. There would still be only six electrons available for these orbitals in the simple molecules (one each from three halogen atoms and three from the Al). However, by forming *dimers,* that is, Al_2Cl_6, Al_2Br_6, and Al_2I_6, a fourth pair of electrons is made available. This occurs through "back bonding" by a halogen atom, as pictured in Figure 20-3.

FIGURE 20-3
Bonding in Al_2Cl_6.

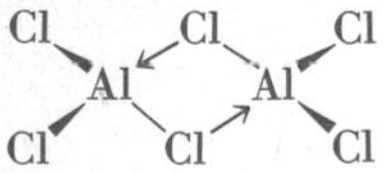

The halides of boron and aluminum are excellent examples of Lewis acids; they are electron pair acceptors. The B or Al atom in these species can accommodate a pair of electrons from a donor atom in another molecule to form an **adduct.** We have considered one of these adducts on previous occasions: $F_3B : NH_3$. The formation of adducts by aluminum halides requires that the dimer molecule split into two monomers. For example, aluminum chloride forms an adduct with diethyl ether— $(C_2H_5)_2O : AlCl_3$. The arrangement of the three Cl atoms and the O atom around the Al atom is tetrahedral.

The halides of boron and aluminum are commonly used reagents in organic chemistry where, through adduct formation, they can lower the activation energy of certain chemical reactions. In this respect they function as catalysts.

ADDITIONAL ASPECTS OF THE CHEMISTRY OF ALUMINUM. Aluminum metal is a good reducing agent; it is easily oxidized.

$$Al^{3+}(aq) + 3\,e^- \longrightarrow Al(s) \qquad E° = -1.66\text{ V} \tag{20.6}$$

In principle, the reverse of half-equation (20.6), the oxidation of Al(s), can be combined with any reduction process having a potential more positive than -1.66 V. Thus, Al(s) is able to displace a large number of metal ions from solution, including H^+.

$$2\,Al(s) + 6\,H^+ \longrightarrow 2\,Al^{3+} + 3\,H_2(g) \tag{20.7}$$

When dissolved in an oxidizing acid, such as H_2SO_4, the reduction product may be a substance other than $H_2(g)$.

$$2\,Al(s) + 12\,H^+(aq) + 3\,SO_4^{2-} \longrightarrow 2\,Al^{3+} + 6\,H_2O + 3\,SO_2(g) \tag{20.8}$$

When powdered aluminum is mixed with iron oxide and ignited, a reaction occurs with the evolution of a great deal of heat, sufficient to melt the iron produced. This process, called the **thermite** process, can be used as a "portable" welding source. It can also be used for the preparation of metals such as manganese and chromium.

$$Fe_2O_3(s) + 2\,Al(s) \longrightarrow Al_2O_3(s) + 2\,Fe(l) \tag{20.9}$$

A compound of aluminum that is a widely used reducing agent, especially in organic chemistry, is lithium aluminum hydride, $LiAlH_4$. The $[AlH_4]^-$ ion can be thought of as an adduct of AlH_3 and H^-. The reduction of water by $LiAlH_4$ is represented in equation (20.10).

$$AlH_4^- + 4 H_2O \longrightarrow Al(OH)_3(s) + OH^- + 4 H_2(g) \qquad (20.10)$$

Acid-base reactions of $Al(OH)_3$, complex ion formation by Al^{3+}, and the preparation, properties, and uses of Al metal are described later in the text.

20-3 The Halogen Elements (VIIA)

The term "halogen" meaning "salt former" is most appropriate in describing the group VIIA elements. As is typical of nonmetals, they form ionic compounds or salts when combined with metals. No matter which of several criteria are used, the halogen elements stand out as a group of very active nonmetals. Their outer-shell electrons are in the configuration ns^2np^5—one electron short of the stable octet of the noble gases. Their ionization energies are high—electrons are lost with great difficulty. Their electron affinities are large and negative—electrons are gained readily. They have high electronegativities. Their standard electrode potentials are large and positive, meaning that reduction of free elements to halide ions occurs readily. Some of these data are listed in Table 20-3.

PHYSICAL PROPERTIES. The physical properties of the halogens listed in Table 20-3 reflect a number of ideas presented in Chapter 11. Two halogen atoms, X, form a strong bond between them in the molecule, X_2. However, there is little residual attractive force between molecules. Intramolecular forces are strong and intermolecular forces are weak. The intermolecular forces among halogen molecules are of the instantaneous dipole–induced dipole type (London forces). Since the strengths of

The lower than expected bond energy of F_2 (see Table 20-3) can probably be attributed to the repulsions of lone pair electrons, made significant by the small size of the F atoms and the short bond distance. Also, there may be some d-orbital contributions to bonding in the higher halogens, leading to partial multiple-bond character, that would be lacking in F_2.

TABLE 20-3
Some properties of the halogen elements

| Element | Atomic properties | | | | | |
	First ioniz. energy, eV/atom	Electron affinity, eV/atom	Electro-negativity	Ionic radius, X^-, pm	Covalent radius (X), pm	Bond energy (X—X), kJ/mol
F	17.42	−3.57	3.98	136	72	155
Cl	12.96	−3.7	3.16	181	99	243
Br	11.81	−3.53	2.96	196	114	192
I	10.45	−3.06	2.66	216	133	151

| Element | Physical properties | | |
	Melting point, °C	Boiling point, °C	Physical form at room temperature
F	−220	−188	yellow-green gas
Cl	−101	−34.6	greenish yellow gas
Br	−7.2	+58.8	red-brown liquid
I	+114	+184.4	grayish black solid

these forces increase with increasing molecular weight, we find that fluorine and chlorine are gases at room temperature, bromine is a liquid, and iodine is a solid.

PREPARATION.   To see what types of reactions must be used to extract the halogens from natural sources, let us first consider their standard electrode potentials.

$$X_2 + 2\,e^- \longrightarrow 2\,X^-(aq) \qquad E° \text{ for} \begin{cases} F_2 = +2.87 \text{ V} \\ Cl_2 = +1.36 \text{ V} \\ Br_2 = +1.07 \text{ V} \\ I_2 = +0.54 \text{ V} \end{cases}$$

[handwritten: ease of being reduced in a rxn]

[handwritten: free element w/charge 0; halide ion; extra electron]

Fluorine has such a strong tendency to exist in the reduced form, F^-, that the reverse process, oxidation of F^- to F_2, cannot be achieved by a spontaneous chemical reaction. Electrolytic methods must be employed, generally utilizing HF in molten KHF_2 as the electrolyte.

[margin note: The hydrogen difluoride ion, HF_2^-, features a strong hydrogen bond, with an H^+ ion about midway between two F^- ions, $[F—H—F]^-$.]

$$2\,H^+ + 2\,F^- \xrightarrow{\text{electrolysis}} H_2(g) + F_2(g) \qquad (20.11)$$

[handwritten: conductor of electricity]

Chlorine can be prepared by oxidation of $Cl^-(aq)$ with a strong oxidizing agent, for example, potassium permanganate.

[handwritten: has a higher standard Electrode potential]

$$2\,MnO_4{}^- + 16\,H^+ + 10\,Cl^- \longrightarrow 2\,Mn^{2+} + 8\,H_2O + 5\,Cl_2(g) \qquad (20.12)$$

Most commercially produced chlorine, however, is obtained by electrolysis, either of a molten chloride ($NaCl$, $MgCl_2$) or an aqueous solution.

[margin note: Note that this is the same electrolysis reaction as (19.51).]

$$2\,Cl^- + 2\,H_2O \xrightarrow{\text{electrolysis}} 2\,OH^- + H_2(g) + Cl_2(g) \qquad (20.13)$$

The source of Cl^- is from brine wells, salt mines, or seawater. Another important commercial reaction is the Deacon process, which involves the oxidation of $HCl(g)$ in the presence of a catalyst (a mixture of metal chlorides)

$$4\,HCl(g) + O_2(g) \rightleftharpoons 2\,H_2O(g) + 2\,Cl_2(g) \qquad (20.14)$$

Bromine can be extracted from seawater, where it occurs to an extent of about 70 parts per million (ppm), by a simple oxidation-reduction reaction employing $Cl_2(g)$ as an oxidizing agent.

$$Cl_2(g) + 2\,Br^-(aq) \longrightarrow Br_2(l) + 2\,Cl^-(aq) \qquad (20.15)$$

TABLE 20-4
Selected reactions of the halogen elements

Reaction with	Reaction equation	Remarks
alkali metals	$2\,M + X_2 \longrightarrow 2\,MX$	
alkaline earth metals	$M + X_2 \longrightarrow MX_2$	
other metals (e.g., Fe)	$2\,Fe + 3\,X_2 \longrightarrow 2\,FeX_3$	
hydrogen	$H_2 + X_2 \longrightarrow 2\,HX$	extremely rapid with F_2; rapid with Cl_2 when exposed to light; slow with Br_2 and I_2
sulfur	$4\,X_2 + S_8 \longrightarrow 4\,S_2X_2$	with Cl_2 and Br_2; F_2 forms SF_6
phosphorus	$6\,X_2 + P_4 \longrightarrow 4\,PX_3$	PX_5 also forms (except with I_2)
other halogens	$I_2 + Cl_2 \longrightarrow 2\,ICl$	other known interhalogen compounds include ClF, ClF_3, $BrCl$, BrF, BrF_3, BrF_5, IBr, ICl_3, IF_5, and IF_7
water	(1) $2\,X_2 + 2\,H_2O \longrightarrow 4\,H^+ + 4\,X^- + O_2(g)$	reverse reaction with I_2
	(2) $X_2 + H_2O \longrightarrow H^+ + X^- + XOH$	with F_2, reaction (1) only
hydrogen sulfide	$8\,X_2 + 8\,H_2S \longrightarrow 16\,H^+ + 16\,X^- + S_8$	
ammonia	$3\,X_2 + 8\,NH_3 \longrightarrow 6\,NH_4X + N_2$	I_2 forms $NI_3 \cdot NH_3$

[handwritten annotations in table: ionic compds; reacts so halide has -1; $F_2 > Cl_2 > Br_2 > I_2$ rxn speed (or electronegativity)]

Figure 20-4
Structure of the I_3^- ion.

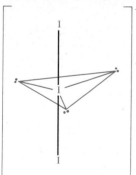

The bond-pair electrons
are shown as black lines.
The lone-pair electrons
of the central I atom are
represented by bold dots.

Iodine is obtainable in small amounts from dried seaweed, since certain marine plants absorb and concentrate I^- selectively in the presence of Cl^- and Br^-. From such sources, oxidation of I^- by a variety of oxidizing agents is possible. A more abundant natural source of iodine is sodium iodate, $NaIO_3$, found in large deposits in Chile. To release iodine from iodate ion requires the use of a *reducing* agent, for example, sodium hydrogen sulfite (bisulfite).

$$2\,IO_3^- + 5\,HSO_3^- \longrightarrow I_2 + 5\,SO_4^{2-} + 3\,H^+ + H_2O \tag{20.16}$$

CHEMICAL PROPERTIES. The halogen elements undergo a variety of chemical reactions, a few of which are summarized in Table 20-4. With active metals the halogens form ionic compounds. With nonmetals their compounds are covalent. All of the halogens react with hydrogen gas but at different rates and in some cases by different mechanisms (see Section 13-10). Two different oxidation-reduction reactions are noted for the reaction of the halogens with water.

INTERHALOGEN COMPOUNDS AND POLYHALIDE ANIONS. An interesting type of reaction involves the combination of two halogens to form interhalogen compounds or polyhalide ions. Several interhalogen compounds are listed in Table 20-4. The structures of these compounds correspond reasonably well to predictions based on the valence-shell electron-pair repulsion method described in Section 9-8. Figure 20-4 pictures the triiodide ion, I_3^-. In this polyatomic ion there are five pairs of electrons in a trigonal bipyramidal arrangement. The ion is linear. I_3^- is formed by the reaction of I_2 with I^-(aq).

$$I_2 + I^-(aq) \rightleftharpoons I_3^-(aq) \tag{20.17}$$

FIGURE 20-5
Structure of the H_5IO_6
molecule.

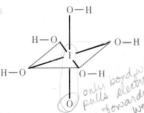

If we represent H_5IO_6 in
the form $IO_m(OH)_n$,
$m = 1$ and $n = 5$, that
is, $IO(OH)_5$. According
to the marginal note on
page 430, we should ex-
pect $K_{a_1} \simeq 10^{-2}$; the
measured value is
$K_{a_1} = 5.1 \times 10^{-4}$.

OXYACIDS OF THE HALOGENS. An important class of compounds of the halogens are the oxygen containing acids and their salts. No stable oxygen-containing acid of fluorine is known, but each of the other halogens forms oxyacids. A summary is presented in Table 20-5.

Chlorine forms a complete set of acids. For bromine and iodine several of the acids do not exist or, at least, have not been completely characterized. A system of nomenclature for these acids and their salts, based on the oxidation state of the halogen atom, has already been discussed (Section 9-11).

In our previous discussion of factors affecting the strengths of oxyacids (Section 17-8) we indicated why an acid such as HIO_4 is very strong. The case with H_5IO_6, **paraperiodic acid,** is different, even though I is in the oxidation state $+7$. As shown in Figure 20-5, in H_5IO_6 six O atoms are arranged octahedrally about a central I atom. Five of the O atoms have H atoms bonded to them. The relationship between acid strength and number of O atoms bonded to the central nonmetal atom applies only to the number of O atoms *with no attached H atoms*. In this case there is only

TABLE 20-5
Oxyacids of the halogens

Oxidation state of the halogen	Chlorine	Bromine	Iodine
+1	HClO	HBrO	HIO
+3	$HClO_2$	—	—
+5	$HClO_3$	$HBrO_3$	HIO_3
+7	$HClO_4$	—	HIO_4; H_5IO_6; $H_4I_2O_9$

one such O atom. Paraperiodic acid is a weak polyprotic acid, and only the first H atom ionizes with comparative ease: $pK_{a_1} = 3.29$; $pK_{a_2} = 8.3$; $pK_{a_3} = 11.6$.

ELECTRODE (REDUCTION) POTENTIAL DIAGRAMS. A number of diagrammatic and graphic methods have been developed to facilitate the discussion of oxidation-reduction reactions. We introduce a particularly simple form here. In the diagram in Figure 20-6, several species containing chlorine are represented. Various pairs of these species are joined by line segments. Written above each line is the standard electrode potential for the reduction of one species to the other. These two species, a couple, form the basis of a skeleton reduction half-equation.

Example 20-1 Use data from Figure 20-6 to predict whether the following reaction will occur. All reactants and products are in their standard states.

$$Cl_2(g) + H_2O \longrightarrow Cl^- + ClO_3^- + H^+$$

$E°$ values are taken from the line segments joining Cl_2 with Cl^- and with ClO_3^-.

oxidation: $\quad Cl_2(g) + 6 H_2O \longrightarrow 2 ClO_3^- + 12 H^+ + 10 e^-$
$$E°_{ox} = -(+1.47) = -1.47 \text{ V}$$

reduction: $\quad 5[Cl_2(g) + 2 e^- \longrightarrow 2 Cl^-] \qquad E°_{red} = +1.36 \text{ V}$

net: $\quad 3 Cl_2(g) + 3 H_2O \longrightarrow 5 Cl^- + ClO_3^- + 6 H^+$
$$E°_{cell} = -0.11 \text{ V} \qquad (20.18)$$

The reaction will not occur spontaneously.

SIMILAR EXAMPLES: Exercises 13, 14.

Suppose that the desired potential from Figure 20-6 is for a pair of species *not* joined by a line segment. What is the appropriate value to use? To determine the potential for an unknown *half-reaction* from known potentials requires that these potentials *first be converted to free energy changes,* which are additive. This requires multiple use of expression (19.30), $-\Delta \overline{G}° = n\mathscr{F}E°$, where n may have *different* values for the half-reactions being combined.

Example 20-2 Refer to Figure 20-6. Determine a value of the standard potential for the reduction of ClO_3^- to ClO^- in basic solution.

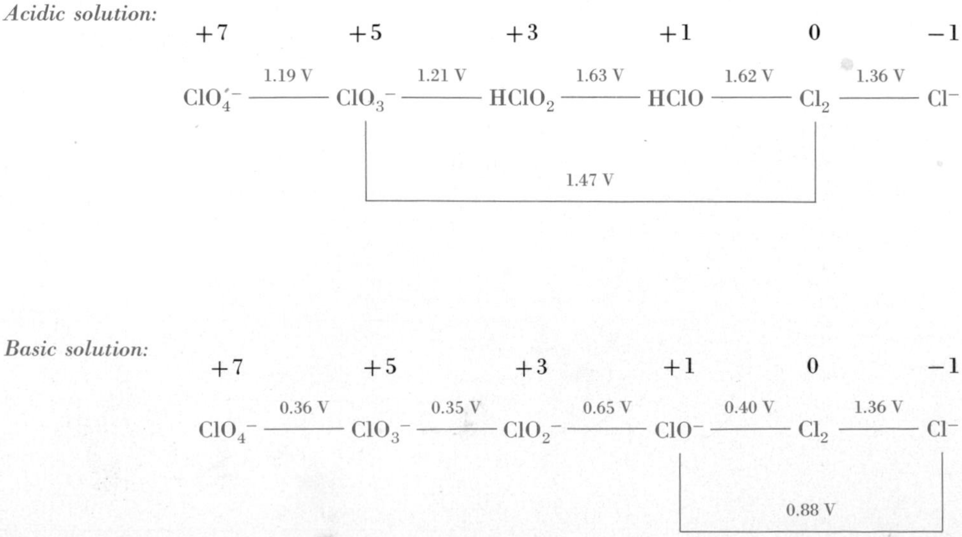

FIGURE 20-6
Standard electrode potential diagram for chlorine.

The numbers in black are the oxidation states of the chlorine. The potentials listed are for reduction processes with reactants and products at unit activity. The data for acidic solution are at $[H^+] = 1 M$; in basic solution, $[OH^-] = 1 M$. Because of the basic properties of ClO_2^- and ClO^-, the weak acids $HClO_2$ and $HClO$ are formed in acidic solution.

Acidic solution:

$+7$		$+5$		$+3$		$+1$		0		-1
ClO_4^-	—1.19 V—	ClO_3^-	—1.21 V—	$HClO_2$	—1.63 V—	$HClO$	—1.62 V—	Cl_2	—1.36 V—	Cl^-

1.47 V

Basic solution:

$+7$		$+5$		$+3$		$+1$		0		-1
ClO_4^-	—0.36 V—	ClO_3^-	—0.35 V—	ClO_2^-	—0.65 V—	ClO^-	—0.40 V—	Cl_2	—1.36 V—	Cl^-

0.88 V

[Marginal handwritten notes, left side:]
standard free energy change ("n" means energy released # is thus a result of SPONTANEOUS rx)
n ← moles of e⁻
𝓕 ← constant related to e⁻'s electrical charge
E° ← electromotive force "driving force" of e⁻'s to reduce
$\Delta \overline{G}° = n\mathscr{F}E°$

The desired half-reaction is the sum of the two half-reactions that follow. Electrode potentials are converted to free energy changes through the expression $-\Delta \overline{G}^\circ = n\mathfrak{F}E^\circ$.

$$\text{ClO}_3^- + \text{H}_2\text{O} + 2\,\text{e}^- \longrightarrow \text{ClO}_2^- + 2\,\text{OH}^- \qquad -\Delta\overline{G}^\circ = 2\,\mathfrak{F}\,(0.35)$$
$$\text{ClO}_2^- + \text{H}_2\text{O} + 2\,\text{e}^- \longrightarrow \text{ClO}^- + 2\,\text{OH}^- \qquad -\Delta\overline{G}^\circ = 2\,\mathfrak{F}\,(0.65)$$

$$\text{ClO}_3^- + 2\,\text{H}_2\text{O} + 4\,\text{e}^- \longrightarrow \text{ClO}^- + 4\,\text{OH}^- \qquad -\Delta\overline{G}^\circ = 2\,\mathfrak{F}\,(0.35 + 0.65)$$

But for the desired half-reaction we may also write that $-\Delta\overline{G}^\circ = n\mathfrak{F}E^\circ$, and in this half-reaction $n = 4$, whereas it was two in each of the others.

$$E^\circ = \frac{-\Delta\overline{G}^\circ}{n\mathfrak{F}} = \frac{2\,\mathfrak{F}\,(0.35 + 0.65)}{4\,\mathfrak{F}} = 0.50\ \text{V}$$

[handwritten] $-\Delta \overline{G}^\circ = 2\mathfrak{F}(1.00) = 4\mathfrak{F}E^\circ$

$E^\circ = \dfrac{2\mathfrak{F}(1.00)}{4\mathfrak{F}} = \dfrac{1}{2} = 0.5$

SIMILAR EXAMPLES: Exercises 15, 36.

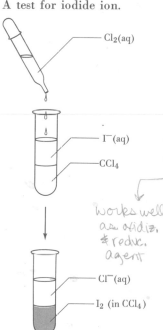

FIGURE 20-7
A test for iodide ion.

[handwritten labels on figure] Cl₂(aq); I⁻(aq); CCl₄; Cl⁻(aq); I₂ (in CCl₄)

[handwritten] works well as oxidiz. & reduc. agent

ANALYTICAL CHEMISTRY OF THE HALOGENS. Analysis of Cl^-, Br^-, and I^- in aqueous solutions is easily accomplished using $\text{AgNO}_3(\text{aq})$. The silver halides are all insoluble. Moreover, because of the differences in K_{sp} of AgCl, AgBr, and AgI, fractional precipitation is also possible. That is, first AgI, then AgBr, and finally AgCl precipitates from an aqueous solution containing Cl^-, Br^-, and I^- as $\text{AgNO}_3(\text{aq})$ is slowly added (recall Section 16-6).

Br^- and I^- can also be detected by reaction with $\text{Cl}_2(\text{aq})$. The Cl_2 is reduced to Cl^- and Br^- or I^- is oxidized to Br_2 or I_2. The liberated Br_2 or I_2 is extracted from aqueous solution with a solvent, such as CCl_4 (see Figure 20-7). In CCl_4 iodine has a violet color; bromine, red. To analyze the oxyanions of Cl, Br, and I, oxidation-reduction reactions can be used to reduce the oxyanion to the free halogen or the halide ion. Then the methods just outlined can be used. *[handwritten]* — above paragraph *[handwritten]* chart 20-2

[handwritten] Iodine and iodide ion are among the most versatile reagents used in oxidation-reduction reactions in analytical chemistry. For example, a standard solution of I_2 can be used to titrate certain reducing agents, such as sulfite ion.

$$\text{SO}_3^{2-} + \text{I}_2 + \text{H}_2\text{O} \longrightarrow \text{SO}_4^{2-} + 2\,\text{I}^- + 2\,\text{H}^+ \tag{20.19}$$

Iodide ion can be used in the titration of strong oxidizing agents.

$$2\,\text{MnO}_4^- + 10\,\text{I}^- + 16\,\text{H}^+ \longrightarrow 2\,\text{Mn}^{2+} + 5\,\text{I}_2 + 8\,\text{H}_2\text{O} \tag{20.20}$$

Sometimes, a reaction such as (20.20) is conducted by adding an *excess* of $\text{I}^-(\text{aq})$ to the oxidizing agent and then titrating the liberated I_2 with sodium thiosulfate. Even when iodine is present only in trace amounts, it forms a deep blue complex with starch. The end point of reaction (20.21) is marked by the disappearance of the blue color associated with the starch iodine complex.

$$\text{I}_2 + 2\,\text{S}_2\text{O}_3^{2-} \longrightarrow 2\,\text{I}^- + \text{S}_4\text{O}_6^{2-} \tag{20.21}$$

$$\text{I}_2 + \text{starch} \longrightarrow \text{blue complex} \tag{20.22}$$

In reactions in which iodine is used as an analytical reagent, it is usually in the presence of I^- and therefore exists mostly as triiodide ion, I_3^-, as noted in equation (20.17).

20-4 Oxygen

Oxygen, in group VIA, is not as nonmetallic as its neighbor fluorine in group VIIA. However, because of its small size it is about as electronegative as chlorine and other members of group VIIA. It is more electronegative than other members of group VIA. It is a typical nonmetal. With metals it forms ionic compounds and with nonmetals, covalent compounds.

The electron configuration of oxygen is $1s^2 2s^2 2p^4$. Two of the $2p$ electrons are unpaired. What might appear to be one of the simplest covalent bonds involving O atoms is actually one of the more complex—the bond between two O atoms in O_2. Bonding in this molecule was described in Chapters 9 and 10.

Another form of oxygen that can be described somewhat more successfully by Lewis structures is **O_3, ozone,** although even here a resonance hybrid must be proposed. The O—O bonds are equivalent and have a bond order halfway between a single and double bond.

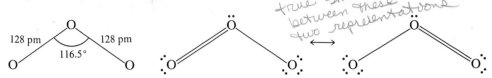

true structure between these two representations

128 pm   128 pm
116.5°

The natural abundance of O_3(g) at sea level is only 0.04 ppm (parts per million). In the stratosphere (11 to 50 km), the ozone content is appreciably greater, although still present only in trace amounts. It reaches about 10 ppm at altitudes of 25 to 30 km. Atmospheric ozone plays a vital role in maintaining life on earth because of the ability of O_3 molecules to absorb ultraviolet radiation in the range 210 to 290 nm. The effect is twofold. First, ultraviolet radiation of this wavelength is damaging to tissues. Many forms of life on earth would not be possible if this radiation were not screened out. Also, in absorbing ultraviolet radiation O_3 molecules undergo dissociation with the evolution of heat energy, thereby helping to maintain a heat balance in the atmosphere.

The principal chemical reactions in the upper atmospheric production of ozone are

$$O_2 + h\nu \longrightarrow O + O \tag{20.23}$$

$$O_2 + O + M \longrightarrow O_3 + M \tag{20.24}$$

Equation (20.23) describes the dissociation of an O_2 molecule following the absorption of ultraviolet radiation. The reaction of atomic and molecular oxygen to produce ozone is represented by equation (20.24). The need for a "third body" M [e.g., N_2(g)] is to carry off excess collision energy, else the O_3 produced would be so energetic as to decompose spontaneously.

The energy released in reaction (20.26) accounts for the heating effect associated with the absorption of ultraviolet radiation by ozone molecules.

$$O_3 + h\nu \longrightarrow O_2 + O \tag{20.25}$$

$$O_3 + O \longrightarrow 2 O_2 \tag{20.26}$$

The production of ozone directly from oxygen gas does not proceed spontaneously under normal conditions in the lower atmosphere.

$$3 O_2(g) \longrightarrow 2 O_3(g) \qquad \Delta \bar{H}° = +285 \text{ kJ/mol} \qquad \Delta \bar{G}° = +327 \text{ kJ/mol}$$

This reaction will proceed to some extent under high-energy conditions as in electrical discharges. Thus, some ozone is produced during electric storms, and ozone formation is the cause of the sharp pungent odor sometimes encountered around electric machinery. The principal method of preparing laboratory ozone is by passing a silent electric discharge through oxygen gas.

Both O_2(g) and O_3(g) are good oxidizing agents. Among ordinary chemical species, the oxidizing ability of O_3(g) is exceeded only by F_2(g) and F_2O(g).

$$O_2(g) + 4 H^+ + 4 e^- \longrightarrow 2 H_2O \qquad E° = +1.229 \text{ V} \tag{20.27}$$

$$O_3(g) + 2 H^+ + 2 e^- \longrightarrow O_2(g) + H_2O \qquad E° = +2.07 \text{ V} \tag{20.28}$$

An important oxidation-reduction reaction used to determine trace amounts of O_3 and other oxidants in polluted air involves oxidation of iodide ion.

$$2\,I^- + O_3(g) + H_2O \longrightarrow 2\,OH^- + I_2 + O_2(g) \tag{20.29}$$

After acidification of the solution, the iodine is titrated with sodium thiosulfate solution, using a starch indicator (see equations 20.21 and 20.22).

WATER AND HYDROGEN PEROXIDE. Certainly the most important and widespread compound of oxygen is water, H_2O. Probably more properties of water have been measured than of any other chemical substance, yet water is still the subject of considerable physical and chemical research.

Among the properties of water commented upon elsewhere in the text are its molecular structure; its crystal structure; its vapor pressure, heat of vaporization, specific heat, heat of fusion, melting point, triple point, boiling point, and critical point; its ionization constant; and its solvent properties.

Although not as important as water, **hydrogen peroxide, H_2O_2,** also has a number of interesting properties. The molecule features an O—O bond as indicated by the simple Lewis structure

$$H\!:\!\overset{\cdot\cdot}{\underset{\cdot\cdot}{O}}\!:\!\overset{\cdot\cdot}{\underset{\cdot\cdot}{O}}\!:\!H$$

FIGURE 20-8
Geometric structure of
H_2O_2.

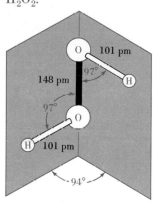

The geometric structure of the molecule is shown in Figure 20-8.

Hydrogen peroxide enters into a wide variety of oxidation-reduction reactions in which its role is summarized by the electrode potential diagrams:

acidic solution ($[H^+] = 1\,M$): $O_2 \xrightarrow{\;0.68\text{ V}\;} H_2O_2 \xrightarrow{\;1.77\text{ V}\;} H_2O$

basic solution ($[OH^-] = 1\,M$): $O_2 \xrightarrow{\;-0.6\text{ V}\;} O_2^- \xrightarrow{\;0.4\text{ V}\;} HO_2^- \xrightarrow{\;0.87\text{ V}\;} OH^-$

$\xrightarrow{\;-0.08\text{ V}\;}$

In reaction (20.30) H_2O_2 acts as an oxidizing agent (reduced to H_2O). In reaction (20.31) it assumes the role of a reducing agent (oxidized to O_2).

$$H_2O_2 + 2\,I^- + 2\,H^+ \longrightarrow 2\,H_2O + I_2 \tag{20.30}$$

$$H_2O_2 + Cl_2 \longrightarrow O_2 + 2\,H^+ + 2\,Cl^- \tag{20.31}$$

Hydrogen peroxide can be prepared by a variety of methods. For the laboratory preparation of small quantities, the addition of barium peroxide to cold, dilute, aqueous sulfuric acid is a convenient method.

$$BaO_2(s) + H_2SO_4(aq) \longrightarrow BaSO_4(s) + H_2O_2(aq)$$

Pure hydrogen peroxide is a pale-blue, syrupy liquid, with a freezing point of $-0.46\,°C$. The liquid is considerably more dense than water (1.47 g/cm^3 at $0\,°C$). The pure compound is very unstable. The decomposition of H_2O_2 is a spontaneous, exothermic reaction which is readily catalyzed by light and a variety of materials (such as iron and copper).

$$2\,H_2O_2(l) \longrightarrow 2\,H_2O(l) + O_2(g) \qquad \Delta \bar{H}° = -197\text{ kJ/mol}$$
$$\Delta \bar{G}° = -245\text{ kJ/mol}$$

H_2O_2 ionizes to a limited extent in aqueous solution. It is a very weak acid.

$$H_2O_2 + H_2O \rightleftharpoons H_3O^+ + HO_2^- \qquad K_a = 2 \times 10^{-12} \tag{20.32}$$

Aside from its industrial uses, such as in bleaching wood pulp, hydrogen peroxide has been used in the household as a mild antiseptic and as a bleaching agent. Be-

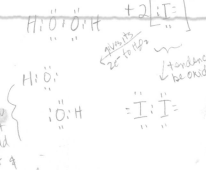

cause of its instability, its concentration for most uses is limited to a 3% aqueous solution. One desirable feature of hydrogen peroxide is that its reaction products are harmless—liquid water and oxygen gas.

OXIDES OF THE ELEMENTS.   With the exception of the lighter noble gases, oxygen forms compounds with all of the other elements. There are several ways in which oxides may be classified. For example, among the ionic oxides we may speak of normal oxides, peroxides, and superoxides. **Normal oxides,** such as Li_2O and CaO, involve the simple oxide anion. The **peroxide** ion, such as in Na_2O_2, features an O—O bond and oxidation state of -1 for oxygen. The **superoxide** ion, such as in KO_2, also features an O—O bond, an unpaired electron, and an oxidation state of $-\frac{1}{2}$ for oxygen. Lewis structures of these ions are

oxide ion   *peroxide ion*   *superoxide ion*

$$\left[:\ddot{O}:\right]^{2-}\qquad \left[:\ddot{O}:\ddot{O}:\right]^{2-}\qquad \left[:\ddot{O}:\ddot{O}:\right]^{-}$$

All three ions are unstable toward water and can only be found in the solid state. The reaction of O^{2-} with water is a simple hydrolysis. The decompositions of O_2^{2-} and O_2^- involve oxidation and reduction. Both of these ions, especially O_2^-, are good oxidizing agents.

oxide: $\qquad\qquad O^{2-} + H_2O \longrightarrow 2\,OH^-(aq)$ (20.33)

peroxide: $\qquad 2\,O_2^{2-} + 2\,H_2O \longrightarrow O_2(g) + 4\,OH^-(aq)$ (20.34)

superoxide: $\quad 4\,O_2^- + 2\,H_2O \longrightarrow 3\,O_2(g) + 4\,OH^-(aq)$ (20.35)

In one classification scheme oxides are viewed as being acidic, basic, or amphoteric. To see how this scheme works consider the hypothetical oxide formed by an element E and denoted by the formula E_xO_y. Consider that this oxide reacts with water to form a hydroxy compound containing one or more bonds, E—O—H.

$$E_xO_y + H_2O \longrightarrow {\diagdown}E\text{—}O\text{—}H \qquad (20.36)$$

Any factor that draws electrons toward the atom E strengthens the E—O bond, weakens the O—H bond, and causes the hydroxy compound to dissociate as an acid. The ability of the compound to donate protons is brought out even more strongly in a basic solution.

$$\diagdown E\text{—}O\text{—}(H + OH^-) \longrightarrow \left(\diagdown E\text{—}O\right)^- + H_2O \qquad (20.37)$$

Small size and high charge (oxidation state) on the atom E lead to the acidic properties described by equation (20.37). This type of behavior is expected if the element E is nonmetallic.

A large size and small charge on the atom E favor rupture of the E—O bond. The hydroxy compound dissociates to produce OH^- and the atom E becomes a cationic species. This tendency to dissociate as a base is accentuated when the hydroxy compound is placed in an acidic solution. Behavior of this type is associated with metallic character in the element E.

$$\diagdown E\text{—}(O\text{—}H + H^+) \longrightarrow \left(\diagdown E\right)^+ + H_2O$$

In some cases the hydroxy compound may act either as an acid or a base. This behavior is known as **amphoterism** and is associated with elements having both metallic and nonmetallic properties.

Another description of the effect of ionic size and charge is based on charge density—the ratio of cationic charge to cationic radius (in pm). The higher the charge density, the more acidic the hydroxy compound; the lower the charge density, the more basic. This relationship is illustrated by the charge densities listed below:

$$\begin{array}{cccc} Mg^{2+} & Be^{2+} & Al^{3+} & B^{3+} \\ 0.031 & 0.065 & 0.060 & 0.15 \end{array}$$

The factors described here are illustrated by considering the series of hydroxy compounds: $Be(OH)_2$, $Mg(OH)_2$, $B(OH)_3$, and $Al(OH)_3$. The order of basicity, from strongest to weakest, is

$$Mg(OH)_2 > Be(OH)_2 \simeq Al(OH)_3 > B(OH)_3$$
$$\text{basic} \qquad \text{amphoteric} \qquad \text{acidic}$$

The comparative radii and charges on the ions are

$$Mg^{2+},\ 65\ pm \quad Be^{2+},\ 31\ pm \quad Al^{3+},\ 50\ pm \quad B^{3+},\ 20\ pm$$

This classification scheme is further illustrated through Table 20-6.

TABLE 20-6
Classification of some oxides

Acidic		Basic		Amphoteric
Representative elements				
Cl_2O	P_4O_{10}	Na_2O	BeO	GeO
SO_2	CO_2	K_2O	Al_2O_3	Sb_2O_3
SO_3	SiO_2	MgO	SnO	
N_2O_5	B_2O_3	CaO	PbO	
Transition elements				
CrO_3		Sc_2O_3	ZnO	
MoO_3		TiO_2	Cr_2O_3	
WO_3		ZrO_2		
Mn_2O_7				

20-5 Sulfur

ALLOTROPY. The existence of an element in two or more forms is known as allotropy. For example, O_2 and O_3 are allotropic forms of oxygen. Sulfur displays allotropy to a remarkable degree, existing both in a variety of different molecular and physical forms. Figure 20-9 represents some of the molecular forms of sulfur. Among the different physical forms of sulfur are

1. rhombic sulfur (S_α), which has sixteen S_8 rings in a unit cell and converts at 96°C to
2. monoclinic sulfur (S_β). Monoclinic sulfur is thought to have six S_8 rings in its unit cell; it melts at 119°C to produce
3. a liquid comprised of S_8 molecules (S_λ). At about 160°C the S_8 rings open up and join together into long spiral-chain molecules resulting in
4. a thick viscous liquid sulfur (S_μ). If this liquid is poured into cold water
5. plastic sulfur is formed. Liquid sulfur boils at 445°C. Several different molecular species are found in the vapor phase.

The allotropy of sulfur as a function of temperature is summarized as follows:

$$S_\alpha \xrightarrow{96°C} S_\beta \xrightarrow{119} S_\lambda \xrightarrow{160} S_\mu \xrightarrow{445}$$
$$S_8(g) \longrightarrow S_6 \longrightarrow S_4 \xrightarrow{1000} S_2 \xrightarrow{2000} S$$

BONDING IN SULFUR COMPOUNDS. Based on periodic relationships and electron configurations, a similarity is to be expected between sulfur and oxygen. Both elements form ionic compounds with active metals and both form some similar covalent compounds, such as CO_2 and CS_2. However, there are factors that produce some differences between oxygen and sulfur compounds.

FIGURE 20-9
Different molecular forms of sulfur.

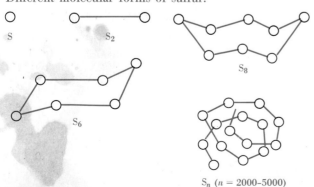

The O atom has a single covalent bond radius of 74 pm, whereas that of an S atom is 104 pm. The comparative electronegativities are 3.44 for O and 2.58 for S. Hydrogen bonding is not a significant feature in sulfur compounds as it is in some oxygen compounds. Although H_2O has a very high boiling point ($+100°C$) for a compound of such a low molecular weight (18), the boiling point of H_2S (molecular weight, 34) is more nearly normal ($-61°C$).

OXIDES OF SULFUR. A number of oxides of sulfur have been reported: S_2O, S_2O_3, SO_2, S_2O_7, SO_3, and SO_4. Of these only sulfur dioxide, SO_2, and sulfur trioxide, SO_3, are commonly encountered. Some interesting aspects of the structures of these compounds are outlined in Figure 20-10.

Sulfur dioxide gas can be prepared commercially by the direct oxidation (burning) of sulfur or by "roasting" metal sulfides.

$$S(s) + O_2(g) \longrightarrow SO_2(g) \tag{20.38}$$

$$2\,ZnS(s) + 3\,O_2(g) \longrightarrow 2\,ZnO(s) + 2\,SO_2(g) \tag{20.39}$$

The principal commercial source of sulfur trioxide is the catalytic oxidation of $SO_2(g)$. The catalyst employed is platinum metal or vanadium pentoxide.

$$2\,SO_2(g) + O_2(g) \rightleftharpoons 2\,SO_3(g) \tag{20.40}$$

OXYACIDS OF SULFUR. Sulfur dioxide is considered to be the acid anhydride of sulfurous acid, but this acid has never been isolated in pure form.

$$SO_2(g) + H_2O(l) \longrightarrow H_2SO_3(aq)$$

The acid is a weak diprotic acid: $K_{a_1} = 1.3 \times 10^{-2}$ and $K_{a_2} = 6.3 \times 10^{-8}$.

Sulfurous acid and its salts (sulfites) are good reducing agents, as in the following reaction with $Cl_2(g)$.

$$Cl_2(g) + SO_3{}^{2-}(aq) + H_2O \longrightarrow$$
$$2\,Cl^-(aq) + SO_4{}^{2-}(aq) + 2\,H^+(aq) \tag{20.41}$$

But they can also act as oxidizing agents, as in this reaction with $H_2S(g)$.

$$2\,H_2S(g) + 2\,H^+(aq) + SO_3{}^{2-}(aq) \longrightarrow$$
$$3\,H_2O + 3\,S(s) \tag{20.42}$$

Sulfur trioxide is the acid anhydride of sulfuric acid. In contrast to H_2SO_3, H_2SO_4 can be obtained in pure form. It is a liquid having a density of about 1.86 g/cm³ and a freezing point of 10°C.

Although the acid results from the action of $SO_3(g)$ on pure water, the product is in the form of a fog or mist. The usual commercial method involves dis-

FIGURE 20-10
Structures of some sulfur oxides.

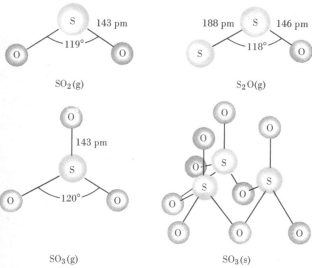

The SO_2 molecule owes its shape to bonding based on sp^2 hybrid orbitals about the S atom. The molecule S_2O has a similar structure but with an S atom substituted for one O atom. In the gaseous state SO_3 molecules are planar with 120° bond angles. In the solid state the basic structural unit has three SO_3 groups bonded together as shown. The S—O bonds in the ring appear to be single bonds, and those projecting above the ring double. Some *d* orbital contribution to the bonding must occur in $SO_3(s)$ to account for all the bonds to S atoms. (The use of expanded octets in writing Lewis structures of sulfur compounds was discussed in Section 9-7.)

solving $SO_3(g)$ in pure H_2SO_4, to produce pyrosulfuric acid, $H_2S_2O_7$. This is followed by dilution with water.

$$SO_3 + H_2SO_4 \longrightarrow H_2S_2O_7$$

$$H_2S_2O_7 + H_2O \longrightarrow 2\,H_2SO_4$$

Sulfuric acid is also a diprotic acid. Its dissociation goes to completion in the first step, but not in the second step except in dilute solutions.

$$H_2SO_4 + H_2O \longrightarrow H_3O^+ + HSO_4^- \qquad K_1 \text{ (very large)}$$

$$HSO_4^- + H_2O \rightleftharpoons H_3O^+ + SO_4^{2-} \qquad K_2 = 1.0 \times 10^{-2}$$

Dilute sulfuric acid enters into all the common reactions of a strong mineral acid, such as neutralizing bases and dissolving active metals to liberate $H_2(g)$.

Concentrated sulfuric acid has some distinctive properties. It has a very strong affinity for water, strong enough even to remove chemically bonded water from certain compounds. When concentrated sulfuric acid is dropped onto ordinary cane sugar, the sugar becomes charred or carbonized.

$$C_{12}H_{22}O_{11} \xrightarrow{H_2SO_4 \text{ (conc.)}} 12\,C(s) + 11\,H_2O \tag{20.43}$$

The concentrated acid is a moderately good oxidizing agent, which enables it to dissolve some metals that are actually less active than hydrogen.

$$Cu(s) + 2\,H_2SO_4 \longrightarrow Cu^{2+} + SO_4^{2-} + 2\,H_2O + SO_2(g) \tag{20.44}$$

The concentrated acid is also widely used as a reagent in organic chemistry and for the production of the fertilizer, "superphosphate."

THIO COMPOUNDS. In thio compounds a S atom replaces an O atom. Replacement of one of the O atoms in sulfate ion, SO_4^{2-}, leads to thiosulfate ion, $S_2O_3^{2-}$. The formal oxidation state of S in $S_2O_3^{2-}$ is $+2$ although, as can be seen from Figure 20-11, the two S atoms are not equivalent. Thiosulfate ion results when an alkaline solution of sodium sulfite is boiled with elemental sulfur.

$$SO_3^{2-} + S \longrightarrow S_2O_3^{2-}$$

Thiosulfate ion is decomposed in acidic solution. The product sulfur is obtained, first in a colloidal state, and then as an allotropic crystalline modification, **rho** sulfur, composed of the S_6 ring molecules pictured in Figure 20-9.

$$S_2O_3^{2-} + 2\,H^+ \longrightarrow H_2O + SO_2(g) + S(s) \tag{20.45}$$

Thiosulfate ion is a reducing agent widely used in analytical chemistry in conjunction with iodine. For example, in the analysis of Cu^{2+} an excess of iodide ion is added, and the liberated iodine is titrated with a standard solution of $Na_2S_2O_3$. The iodine–thiosulfate reaction was described through equations (20.21) and (20.22).

$$2\,Cu^{2+} + 4\,I^- \longrightarrow 2\,CuI(s) + I_2 \tag{20.46}$$

SULFIDES. All metals form sulfides, and most sulfides are insoluble in water. Most are soluble in acidic solution, however, and a few are soluble in basic solution. The differing solubilities of the sulfides allow metal ions to be separated into different groups. This, as we have already learned, is the basis of a qualitative analysis scheme for cations (see Figure 16-3).

Whether a given sulfide will precipitate from solution depends on the pH of the solution. Predictions require the use of the ionization constants for H_2S and solubility product constants for metal sulfides, as previously described in Section 18-6. For

Although all acids produce a stinging sensation on the skin, concentrated H_2SO_4 produces severe burns as a result of reactions like that shown in equation (20.43).

FIGURE 20-11
Structures of some thio ions.

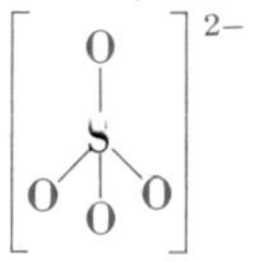

SO_4^{2-} sulfate ion

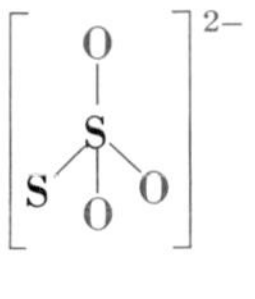

$S_2O_3^{2-}$ thiosulfate ion

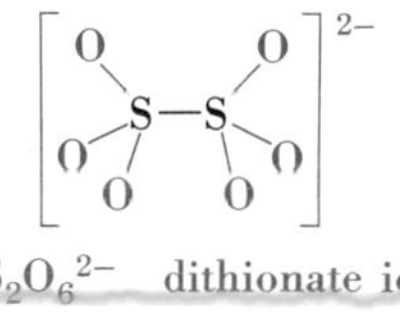

$S_2O_6^{2-}$ dithionate ion

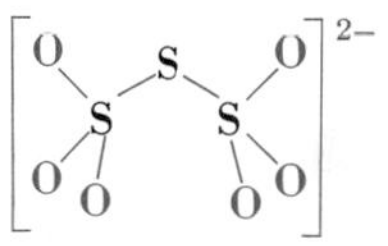

$S_3O_6^{2-}$ trithionate ion

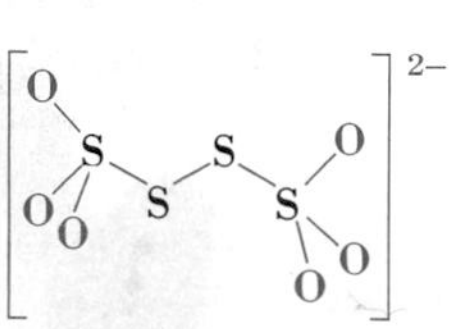

$S_4O_6^{2-}$ tetrathionate ion

the simple metal ion M^{2+}, which precipitates from an aqueous solution saturated in H_2S, we may write

$$H_2S(aq) + M^{2+}(aq) \longrightarrow MS(s) + 2\,H^+(aq) \tag{20.47}$$

Dissolution of a metal sulfide in an acid can be represented by the reverse of equation (20.47), that is,

$$MS(s) + 2\,H^+(aq) \longrightarrow M^{2+}(aq) + H_2S(g) \tag{20.48}$$

Some metal sulfides have such low values of K_{sp} that they will not dissolve in ordinary mineral acids. However, they can be partially dissolved in a strong oxidizing acid. The sulfide ion is oxidized to insoluble elemental sulfur, and the metal ion appears in solution.

$$3\,CuS(s) + 8\,H^+ + 2\,NO_3^- \longrightarrow 3\,Cu^{2+} + 4\,H_2O + 2\,NO(g) + 3\,S(s) \tag{20.49}$$

Mercuric sulfide, HgS, which has the smallest solubility of all metal sulfides ($K_{sp} = 1.6 \times 10^{-52}$), can be dissolved in a mixture of nitric and hydrochloric acids called *aqua regia*. Here dissolution involves oxidation of the sulfide ion and formation of the complex ion $HgCl_4^{2-}$.

$$HgS(s) + 4\,H^+ + 4\,Cl^- + 2\,NO_3^- \longrightarrow$$
$$HgCl_4^{2-}(aq) + 2\,H_2O + 2\,NO_2(g) + S(s) \tag{20.50}$$

Still another basis for dissolving certain metal sulfides is the acidic or amphoteric nature of certain metal sulfides. This property is exhibited in basic solutions containing high concentrations of S^{2-} or S_n^{2-} (polysulfide ion).

$$SnS(s) + S_2^{2-}(aq) \longrightarrow SnS_3^{2-}(aq) \tag{20.51}$$

In reaction (20.51) oxidation of tin(II) to tin(IV) occurs along with dissolution of the sulfide; SnS_3^{2-}, called thiostannate ion, is the sulfur analog of SnO_3^{2-}, stannate ion. Upon acidification of a solution containing SnS_3^{2-}, insoluble SnS_2 precipitates.

$$SnS_3^{2-}(aq) + 2\,H^+(aq) \longrightarrow SnS_2(s) + H_2S(g) \tag{20.52}$$

Some of the points discussed in the preceding few paragraphs are further illustrated through Table 20-7.

The precipitation of metal sulfides is generally not carried out directly with $H_2S(g)$. Other precipitating agents can be used instead. One commonly employed laboratory method is based on the hydrolysis of thioacetamide. This hydrolysis reaction is quite slow at room temperature, but it proceeds at an appreciable rate at higher temperatures.

Polysulfides, which can be prepared by boiling a solution of a soluble metal sulfide with free sulfur, have sulfur atoms bonded together in a chain.

$$\left[:\!\overset{..}{\underset{..}{S}}\!:\right]^{2-} + \,:\!\overset{..}{\underset{..}{S}}\!: \longrightarrow$$
$$\left[:\!\overset{..}{\underset{..}{S}}\!:\!\overset{..}{\underset{..}{S}}\!:\right]^{2-}$$
disulfide ion

$$\left[:\!\overset{..}{\underset{..}{S}}\!:\!\overset{..}{\underset{..}{S}}\!:\right]^{2-} + \,:\!\overset{..}{\underset{..}{S}}\!: \longrightarrow$$
$$\left[:\!\overset{..}{\underset{..}{S}}\!:\!\overset{..}{\underset{..}{S}}\!:\!\overset{..}{\underset{..}{S}}\!:\right]^{2-}$$
trisulfide ion

$$\left[:\!\overset{..}{\underset{..}{S}}\!:\!\overset{..}{\underset{..}{S}}\!:\!\overset{..}{\underset{..}{S}}\!:\right]^{2-} +$$
$$(n-3)\,\overset{..}{\underset{..}{S}}\!: \longrightarrow$$
$$\left[:\!\overset{..}{\underset{..}{S}}\!:\right]_n^{2-}$$
polysulfide ion

$$\underset{\text{thioacetamide}}{CH_3\overset{\overset{\displaystyle S}{\|}}{C}NH_2} + H_2O \longrightarrow \underset{\text{acetamide}}{CH_3\overset{\overset{\displaystyle O}{\|}}{C}NH_2} + H_2S(g) \tag{20.53}$$

TABLE 20-7
Solubilities of some metal sulfides

| H$_2$O | 0.3 M HCl (K_{sp}) | Soluble in | | KOH(aq) or (NH$_4$)$_2$S$_n$(aq) (K_{sp}) |
		3 M HNO$_3$ (K_{sp})	Aqua regia (K_{sp})	
K$_2$S	MnS (2.5×10^{-13})	CdS (8.0×10^{-27})	HgS (1.6×10^{-52})	SnS (1.0×10^{-25})
Na$_2$S	FeS (6.3×10^{-18})	PbS (8×10^{-28})		As$_2$S$_3$
CaS	CoS (4.0×10^{-21})	CuS (6.3×10^{-36})		Sb$_2$S$_3$
	ZnS (1.0×10^{-21})			

20-6 The Nitrogen Family (VA)

Perhaps the most interesting feature of the group VA elements is that properties of both nonmetals and metals are displayed within the group. (The stepwise diagonal line in the periodic table passes through group VA.) The electron configurations of the elements provide only a limited clue to their metallic–nonmetallic behavior. Their outer shell electron configurations are ns^2np^3. There are a number of ways in which this electron configuration may be altered when a group VA atom enters into compound formation. Several possibilities are worthy of special mention.

One possibility is the gain or, more likely, sharing of three electrons in the outer shell to produce the noble gas configuration ns^2np^6 and an oxidation state of -3. This is especially so for the smaller atoms N and P. For the larger atoms—As, Sb, and Bi—the p^3 set of electrons may be lost. This leads to an electron configuration involving a next-to-outermost shell of 18 and an outer shell of 2. This so-called "18 + 2" configuration is adopted by a number of ionic species derived from metals. In some cases, all five outer shell electrons may be involved in compound formation, leading to the oxidation state of $+5$. Finally, for nitrogen in particular, all oxidation states from -3 to $+5$ are possible.

ASSESSMENT OF METALLIC–NONMETALLIC CHARACTER.   To aid in this discussion a number of properties have been listed in Table 20-8. For group VA the usual decrease of ionization energy with increasing atomic number is noted. This establishes the order of metallic character within the group. Nitrogen is least metallic and bismuth is most metallic. Of course, all these ionization energies are high compared to the IA and IIA elements. The electronegativities indicate a high degree of nonmetallic character for nitrogen and less so for the remaining members of the group. None of the elements in group VA is highly metallic, however.

Three of the elements—phosphorus, arsenic, and antimony—exhibit allotropy. For phosphorus the stable form at room temperature is red phosphorus. Its physical properties are those of a nonmetal. For example, its triple point is at 590°C and 43 atm pressure; red phosphorus sublimes without melting. For arsenic and antimony the stable allotropic forms are the "metallic" ones. They have high densities, fair thermal conductivities, and limited abilities to conduct electricity. Bismuth has no nonmetallic allotropic forms.

THERMODYNAMIC CONSIDERATIONS.   The increase of metallic behavior from top to

TABLE 20-8

Selected properties of group VA elements

Element	Atomic radius, pm	Electro- negativity	Ionization energy,[a] eV/atom	Physical form (at room temperature)	Density of solid, g/cm^3	Comparative electrical conductivity[b]
N	150	3.04	14.53	gas	1.03 (-252°C)	—
P	190	2.19	10.49	waxlike white solid	1.82	
				red-colored solid	2.36	10^{-15}
As	200	2.18	9.81	yellow solid	2.03	
				gray solid with metallic luster	5.73	4.6
Sb	220	2.05	8.64	yellow solid	5.3	
				silvery white metallic solid	6.69	4.2
Bi	—	1.9	7.29	pinkish white metallic solid	9.75	1.4

[a] The energy required to ionize one electron from a gaseous atom.
[b] These values are relative to an assigned value of 100 for silver.

TABLE 20-9
Some thermodynamic properties of group VA hydrides

Hydride	Free energy of formation $\Delta \bar{G}_f^\circ$ (298 K), kJ/mol	Standard electrode potential, V
ammonia, NH_3	-16.6	$N_2 \xrightarrow{+0.27} NH_3$
phosphine, PH_3	$+8.8$	$P_4 \xrightarrow{-0.03} PH_3$
arsine, AsH_3	$+157.7$	$As \xrightarrow{-0.54} AsH_3$
stibine, SbH_3	$+147.7$	$Sb \xrightarrow{-0.51} SbH_3$
bismuthine, BiH_3	$+230(?)$	$Bi \xrightarrow{-0.80(?)} BiH_3$

bottom in group VA is reflected by the data listed in Table 20-9. Covalent bond formation with hydrogen is expected for nonmetallic elements. The negative value of the free energy of formation of $NH_3(g)$ suggests it is a stable molecule that forms spontaneously from its elements in their standard states. The values for the other hydrides are positive and increase in magnitude with increasing atomic number, suggesting decreasing stabilities. In fact, BiH_3 is so unstable that its properties have not been measured with any accuracy.

Another indication of the order of stability of the covalent hydrides is provided by electrode potentials. A decrease in electrode potential signifies a decreasing tendency toward reduction of the free element to the hydride.

ACID-BASE BEHAVIOR OF GROUP VA OXIDES. The group VA elements form a number of different oxides. Table 20-10 describes the solubilities of the oxides X_2O_3 or X_4O_6 in acidic and basic solutions. The oxides of nitrogen, phosphorus, and arsenic are acidic; antimony oxide is amphoteric; and the oxide of bismuth has only basic properties. The acid-base behavior of the oxides establishes the gradation of properties within group VA, from nonmetallic to metallic, about as well as any other criterion.

20-7 Additional Aspects of Nitrogen Chemistry

Figure 20-12 lists some important nitrogen-containing species through an electrode potential diagram. The variability of the oxidation state of nitrogen in its compounds suggests an unusually rich chemistry.

TABLE 20-10
Acid-base behavior of some group VA oxides

Oxide	Nature of oxide	Principal product(s) obtained when dissolved in	
		Acidic soln.	Basic soln.
N_2O_3	acidic	HNO_2	NO_2^-
P_4O_6	acidic	H_3PO_3	$H_2PO_3^-$, HPO_3^{2-}
As_4O_6	acidic	$H_3AsO_3(HAsO_2)$	AsO_3^{3-} (AsO_2^-)
Sb_2O_3	amphoteric	SbO^+	SbO_2^-
Bi_2O_3	basic	BiO^+ and Bi^{3+}	insoluble

FIGURE 20-12
Electrode potential diagram for nitrogen.

Acidic solution:

$$NO_3^- \xrightarrow{+0.81 \text{ V}} NO_2 \xrightarrow{+1.07 \text{ V}} HNO_2 \xrightarrow{+0.99 \text{ V}} NO \xrightarrow{+1.59 \text{ V}} N_2O \xrightarrow{+1.77 \text{ V}} N_2 \xrightarrow{-1.87 \text{ V}} NH_3OH^+ \xrightarrow{+1.46 \text{ V}} N_2H_5^+ \xrightarrow{+1.24 \text{ V}} NH_4^+$$

Basic solution:

$$NO_3^- \xrightarrow{-0.85 \text{ V}} NO_2 \xrightarrow{+0.88 \text{ V}} NO_2^- \xrightarrow{-0.46 \text{ V}} NO \xrightarrow{+0.76 \text{ V}} N_2O \xrightarrow{+0.94 \text{ V}} N_2 \xrightarrow{-3.04 \text{ V}} NH_2OH \xrightarrow{+0.74 \text{ V}} N_2H_4 \xrightarrow{+0.10 \text{ V}} NH_3$$

The values given are for reduction processes, with $[H^+] = 1 \, M$ in acidic solutions and $[OH^-] = 1 \, M$ in basic solutions.

AMMONIA AND AMMONIUM COMPOUNDS. With few exceptions, commonly used nitrogen compounds are produced from atmospheric $N_2(g)$. The principal chemical reaction, considered in detail in Section 14-8, involves that of $N_2(g)$ and $H_2(g)$ to form $NH_3(g)$.

In aqueous solution NH_3 is a weak base. The neutralization of $NH_3(aq)$ by an acid, such as HCl or H_2SO_4, results in an ammonium salt. The oxidation state of the nitrogen remains unchanged in this acid-base reaction. The properties of the ammonium ion are similar to those of the alkali metal ions. This means that all common ammonium salts are water soluble.

Liquid ammonia and aqueous solutions of ammonia have a number of important uses, chiefly in the formulation of liquid fertilizers. Under the name "anhydrous ammonia" liquid NH_3 can be applied directly to fields. Alternatively, aqueous solutions of NH_3, NH_4NO_3, and urea may be used. Aqueous ammonia is also found in the household in a variety of cleaning agents.

HYDRAZINE AND HYDROXYLAMINE. These two compounds are derivatives of NH_3. Hydrazine, N_2H_4, results from replacing an H atom of NH_3 by an $-NH_2$ group and hydroxylamine, NH_2OH, by replacing an H atom with an $-OH$ group. Both compounds are weak bases, still weaker than NH_3. Because of its two N atoms, N_2H_4 can accept two protons, in a stepwise fashion. The ion $N_2H_6^{2+}$ can be obtained in appreciable concentrations only in strongly acidic solutions, however.

$$NH_2OH(aq) + H_2O \rightleftharpoons NH_3OH^+ + OH^- \qquad K_b = 9.1 \times 10^{-9}$$

$$N_2H_4(aq) + H_2O \rightleftharpoons N_2H_5^+ + OH^- \qquad K_{b_1} = 8.5 \times 10^{-7}$$

$$N_2H_5^+(aq) + H_2O \rightleftharpoons N_2H_6^{2+} + OH^- \qquad K_{b_2} = 8.9 \times 10^{-16}$$

Hydrazine and hydroxylamine form salts analogous to ammonium salts, that is, salts such as $[NH_3OH]Cl$, $N_2H_5NO_3$, and $N_2H_6SO_4$. These salts all hydrolyze in water to yield acidic solutions.

Both hydrazine and hydroxylamine can act either as an oxidizing or reducing agent (usually the latter) depending on the pH and the substance with which they react (see Figure 20-12). Hydrazine and some of its derivatives, burn in air with the evolution of considerable heat and are used as rocket fuels.

$$N_2H_4(l) + O_2(g) \longrightarrow N_2(g) + 2 \, H_2O(l) \tag{20.54}$$

NITRIC ACID AND NITRATE SALTS. An important use of NH_3 is in the synthesis of nitric acid. The key reaction in this synthesis is the Ostwald reaction, in which $NH_3(g)$ is oxidized to $NO(g)$ on a platinum catalyst. The $NO(g)$ is then converted to $NO_2(g)$. The reaction of $NO_2(g)$ with water produces $HNO_3(aq)$ and $NO(g)$, which is recycled.

$$4\,NH_3(g) + 5\,O_2(g) \xrightleftharpoons{Pt} 4\,NO(g) + 6\,H_2O(g) \tag{20.55}$$

$$2\,NO(g) + O_2(g) \longrightarrow 2\,NO_2(g) \tag{20.56}$$

$$3\,NO_2(g) + H_2O \longrightarrow 2\,HNO_3(aq) + NO(g) \tag{20.57}$$

Pure nitric acid is a colorless liquid with a density of 1.50 g/cm^3. It dissociates at temperatures slightly above its melting point ($-42\,°C$) to dinitrogen pentoxide and water. Ordinary concentrated nitric acid is an aqueous solution with a density of 1.41 g/cm^3 and a concentration of about $15\,M\,HNO_3$. It generally has a yellow color due to the presence of dissolved oxides of nitrogen. Nitrate salts are produced by neutralizing nitric acid.

In addition to its acidic properties, nitric acid is a good oxidizing agent. It can dissolve some metals that are less active than hydrogen. The reaction of an active metal, such as magnesium, with very dilute HNO_3 yields some $H_2(g)$. In all other cases the product of the reduction half-reaction is $NO_2(g)$, $NO(g)$, $N_2O(g)$, or even NH_4^+.

In dilute $HNO_3(aq)$:

$$4\,Zn(s) + 10\,H^+ + 2\,NO_3^- \longrightarrow 4\,Zn^{2+} + 5\,H_2O + N_2O(g) \tag{20.58}$$

In cold dilute $HNO_3(aq)$:

$$3\,Cu(s) + 8\,H^+ + 2\,NO_3^- \longrightarrow 3\,Cu^{2+} + 4\,H_2O + 2\,NO(g) \tag{20.59}$$

In warm concentrated $HNO_3(aq)$:

$$Cu(s) + 4\,H^+ + 2\,NO_3^- \longrightarrow Cu^{2+} + 2\,H_2O + 2\,NO_2(g) \tag{20.60}$$

Nitric acid also reacts with nonmetallic elements, generally with the formation of an oxyacid and $NO(g)$.

In cold concentrated $HNO_3(aq)$:

$$S(s) + 2\,HNO_3(aq) \longrightarrow H_2SO_4(aq) + 2\,NO(g) \tag{20.61}$$

In hot concentrated $HNO_3(aq)$:

$$3\,I_2(s) + 10\,HNO_3(aq) \longrightarrow 6\,HIO_3(aq) + 2\,H_2O + 10\,NO(g) \tag{20.62}$$

OXIDES OF NITROGEN.   Nitrogen oxide (nitric oxide) is formed by the oxidation of $NH_3(g)$ or the reaction of a metal with dilute nitric acid. Air oxidation of $NO(g)$ or the reaction of a metal with warm concentrated nitric acid can be used to prepare nitrogen dioxide, NO_2. (NO_2 exists in equilibrium with dinitrogen tetroxide, N_2O_4.) Dinitrogen pentoxide, N_2O_5, can be prepared by the decomposition of pure nitric acid. Dinitrogen oxide (nitrous oxide), $N_2O(g)$, results from the action of dilute nitric acid on active metals or by the decomposition of ammonium nitrate.

$$NH_4NO_3(s) \longrightarrow 2\,H_2O(g) + N_2O(g) \tag{20.63}$$

Lewis structures of four oxides of nitrogen are shown below. Whether these oxides are diamagnetic or paramagnetic must be determined by experiment, but inferences are possible from the structures. (The schemes are simplified in that single structures are shown rather than true resonance hybrids.)

paramagnetic   *diamagnetic*

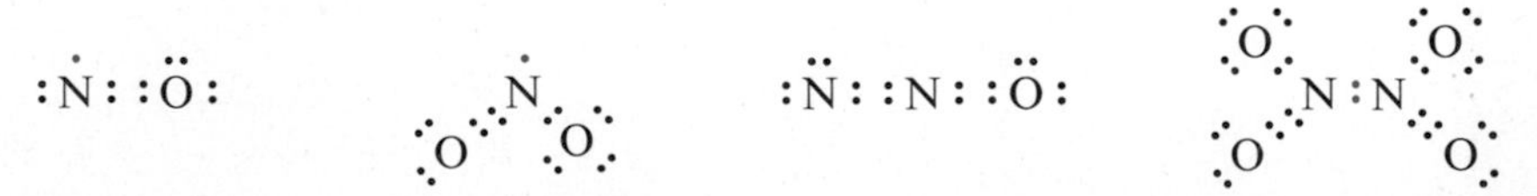

:N::Ö: N   :N::N::Ö: N:N

nitrogen oxide   nitrogen dioxide   dinitrogen oxide   dinitrogen tetroxide

20-8 Additional Aspects of Phosphorus Chemistry

ALLOTROPY.   Phosphorus is prepared by heating a phosphate containing mineral with sand and coke in an electric furnace. The overall reaction is

$$2\ Ca_3(PO_4)_2(s) + 10\ C(s) + 6\ SiO_2(s) \longrightarrow$$
$$6\ CaSiO_3(l) + 10\ CO(g) + P_4(g) \qquad (20.64)$$

The phosphorus vapor is condensed to solid white phosphorus, which, because of its inflammability, is stored under water.

The molecular form in solid white phosphorus is the same as in the vapor, P_4. The geometric shape of P_4 molecules is tetrahedral. Bonding between molecules is weak, and as a result white phosphorus has a low melting point and a fairly high vapor pressure. Even though it is formed directly by the condensation of $P_4(g)$, solid white phosphorus is metastable: Eventually it changes to solid red phosphorus. Bonding in red phosphorus is more complex than in white phosphorus and is still not completely understood. It appears to involve opening up and interlinking P_4 tetrahedra into chains. Because there is no simple mechanism for this molecular rearrangement, the conversion of solid white to solid red phosphorus occurs extremely slowly at room temperature. The triple point of red phosphorus is 590°C and 43 atm pressure. Thus, red phosphorus sublimes without melting (at about 420°C); it can only be melted by maintaining it under a pressure in excess of 43 atm.

OXIDES AND OXYACIDS OF PHOSPHORUS.   The two principal oxides of phosphorus, P_4O_6 and P_4O_{10}, can be considered as derivatives of the P_4 molecule. In P_4O_6 one O atom bridges each pair of P atoms, hence six O atoms in the molecule. In P_4O_{10} there is an additional O atom bonded to each P atom in the tetrahedron. These structural features are brought out in Figure 20-13.

P_4O_6 and P_4O_{10} are the acid anhydrides of H_3PO_3 and H_3PO_4, respectively. These and other oxyacids are listed in Table 20-11. A complication suggested in this table is explored more fully in Figure 20-14—the existence of ortho, meta, and pyro acids.

Orthophosphoric acid forms when P_4O_{10} reacts with an excess of H_2O.

$$P_4O_{10} + 6\ H_2O \longrightarrow 4\ H_3PO_4 \qquad (20.65)$$

When orthophosphoric acid is heated to temperatures in excess of 215°C, **pyrophosphoric** acid results.

$$2\ H_3PO_4 \longrightarrow H_4P_2O_7 + H_2O \qquad (20.66)$$

FIGURE 20-13
Molecular structures of P_4O_6 and P_4O_{10}.

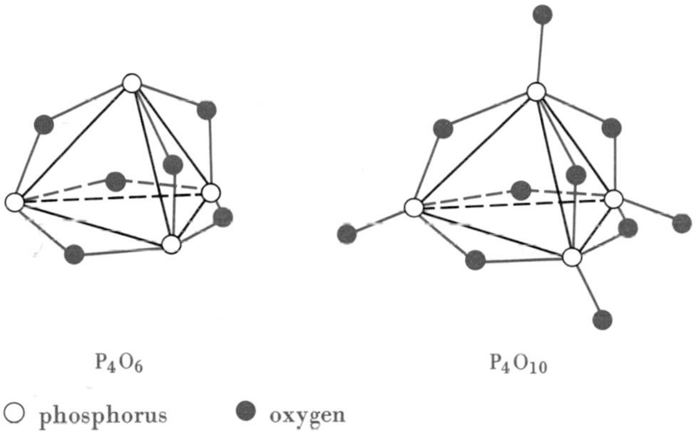

P_4O_6 P_4O_{10}

○ phosphorus   ● oxygen

TABLE 20-11
Oxyacids of phosphorus

Oxidation state	Formula	Name	
+1	H_3PO_2	hypophosphorous acid	
+3	H_3PO_3	ortho	
	HPO_2	meta	phosphorous acid
	$H_4P_2O_5$	pyro	
+4	$H_4P_2O_6$	hypophosphoric acid	
+5	H_3PO_4	ortho	
	HPO_3	meta	phosphoric acid
	$H_4P_2O_7$	pyro	

(handwritten note: "low oxy state" pointing to +1 row; "high oxy state" pointing to +5 row)

Although several oxidation states of phosphorus are represented in the table, the oxidation state of +5 is most common. Among naturally occurring phosphorus compounds, only this oxidation state is observed.

FIGURE 20-14
Some oxyacids of phosphorus(V).

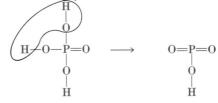

When either the ortho or pyro acid is heated to temperatures in excess of about 300°C, a glassy product is formed. This is probably a polymerized form of **meta-phosphoric** acid, $(HPO_3)_n$, where $n = 2, 3, 4, 6$. For example, the sodium salt, $Na_6P_6O_{18}$ is called sodium hexametaphosphate (used in water softening). Another complex phosphoric acid, whose sodium salts have been commonly used in detergents, is $H_5P_3O_{10}$—triphosphoric acid (see Section 26-6).

To apply the scheme of Figure 20-14 to other acids, proceed in this way.

1. Start with an oxide in which the element E, is in the oxidation state, $+n$.
2. Consider that by reacting the oxide with water the hydroxy compound $E(OH)_n$ is formed. (n is the oxidation number of E.)
3. If the hydroxy compound, $E(OH)_n$, actually exists, it is the *ortho* acid (i.e., H_nEO_n). When this hydroxy compound does not exist, the loss of one water molecule yields the ortho acid.
4. The *meta* acid is formed from the ortho acid by the loss of one water molecule.

This scheme does not work in all cases. For example, phosphorous acid, H_3PO_3, does not have the structure $P(OH)_3$. Instead, one of the H atoms is bonded directly to the P atom.

Example 20-3 Supply names and/or formulas for the following: **(a)** H_3AsO_4; **(b)** meta-arsenous acid; **(c)** magnesium pyroarsenate.

(a) Arsenic is in the oxidation state $+5$. The hypothetical hydroxy compound would be $As(OH)_5$. Loss of one molecule of water leads to H_3AsO_4. The acid is called **orthoarsenic acid;** it is analogous to H_3PO_4.

(b) This is an "ous" acid and so must have arsenic in the oxidation state $+3$. The hydroxy compound corresponding to this oxidation state is $As(OH)_3$ or H_3AsO_3 and does actually exist. It is the ortho acid. The loss of one molecule of water leads to $HAsO_2$; this is the meta acid.

(c) The pyroacid is formed by the loss of one water molecule by two H_3AsO_4 molecules and has the formula $H_4As_2O_7$. The magnesium ion is Mg^{2+}, so magnesium pyroarsenate is $Mg_2As_2O_7$.

SIMILAR EXAMPLE: Exercise 42.

IONIZATION AND NEUTRALIZATION OF H_3PO_4. Phosphoric acid is a polyprotic acid; the molecule of H_3PO_4 has three ionizable H atoms. Ionization equilibria for acids of this type were discussed in Section 17-6. The stepwise neutralization of H_3PO_4 and a resulting titration curve were discussed in Section 18-4.

20-9 Silicon

The element silicon crystallizes in the same type of fcc structure as diamond (see Figure 11-22). Whereas diamond is an electrical insulator and graphite is a moderately good electrical conductor, silicon is classified as a semiconductor. The structure of silicon dioxide (silica) was described in Figure 11-23. Quartz is a common form of silica, but there are more than a dozen forms altogether.

SILICATE MINERALS. If carbon is the key element of the living world, silicon is no less the key element of the inanimate or mineral world. Only the simple alkali metal silicates can be obtained as water-soluble compounds. The vast majority of silicates are highly insoluble. A common feature of silicate minerals is the complexity of the silicate anions. Nevertheless, within these complex anions the basic structural unit is a simple tetrahedral arrangement of four O atoms about a central Si atom. These tetrahedra may exist (a) as separate units; (b) joined into chains or rings in groups of 2, 3, 4, or 6; (c) joined together into long single chains or (d) double chains; (e) arranged in sheets; (f) linked into a three-dimensional network. Two of these possibilities are pictured in Figure 20-15 and several examples are listed in Table 20-12.

COLLOIDAL SILICA. CO_2 is an acidic oxide and dissolves in basic solutions, forming

> The triangular BO_3 unit in borates is analogous to the tetrahedral SiO_4 unit in silicates. The BO_3 units may be joined into rings, chains, or sheets. Crystalline boron oxide (B_2O_3) has a structure in which two-dimensional sheets are linked together. In sodium metaborate, $Na_3B_3O_6$, the anion has a ring structure.

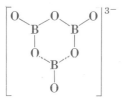

> To establish the charge on a silicate anion, it is convenient to think of each Si atom as carrying a charge of $+4$ and each O, -2. Thus, the charge on the anion, SiO_4, is -4, and on Si_2O_7, -6.

TABLE 20-12
Some representative silicate minerals

Mineral type	Anion type	Mineral	Composition
(a) orthosilicate	SiO_4^{4-}	zircon	$ZrSiO_4$
		olivine	Mg_2SiO_4
(b) polysilicate	$Si_2O_7^{6-}$	thortveitite	$Sc_2Si_2O_7$
	$Si_6O_{18}^{12-}$	beryl	$Be_3Al_2Si_6O_{18}$
(c) pyroxene	Si—O chains	diopside	$CaMg(SiO_3)_2$
		spodumene	$LiAl(SiO_3)_2$
(d) amphibole	Si—O double chains	tremolite	$Ca_2Mg_5(Si_4O_{11})_2(OH)_2$
		hornblende	$(Ca,Na,K)_{2-3}(Mg,Fe,Al)_5$ $(Si,Al)_2Si_6O_{22}(OH)_2$
(e) mica	Si—O sheets	muscovite	$KAl_2(AlSi_3O_{10})(OH)_2$
(f) zeolite	Si—O three-dimensional network	natrolite	$Na_2Al_2Si_3O_{10} \cdot 2\,H_2O$

carbonates. SiO_2 dissolves slowly in strong bases, forming a variety of silicates.

$$SiO_2(s) + 2\,NaOH(aq) \longrightarrow \underset{\text{sodium metasilicate}}{Na_2SiO_3(aq)} + H_2O \qquad (20.67)$$

$$SiO_2(s) + 4\,NaOH(aq) \longrightarrow \underset{\text{sodium orthosilicate}}{Na_4SiO_4(aq)} + 2\,H_2O \qquad (20.68)$$

Aqueous sodium silicate is often encountered under the common name of "water glass."

Acidification of an aqueous carbonate solution produces H_2CO_3, which decomposes to H_2O and $CO_2(g)$. The silicic acids resulting from the acidification of aqueous silicate solutions are also unstable; they tend to decompose to silica. Depending on the pH, the silica may be obtained as a colloidal dispersion, a gelatinous precipitate, or as a solidlike gel in which all of the liquid is entrapped. Water molecules are eliminated between neighboring silicic acid molecules, producing a polymer of ever increasing length.

This is the process referred to in Section 12-8 for the preparation of colloidal silica.

FIGURE 20-15
Silicate anions.

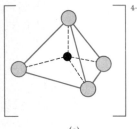

(a)

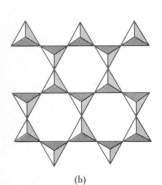

(b)

(a) The simple anion SiO_4^{4-} consists of a central Si atom surrounded by four O atoms in a tetrahedral arrangement.
(b) A portion of the anion structure of mica. The cations required to balance the electric charge of the sheet anions are located above and below each sheet. Usually these are Al^{3+} and K^+. Talc and clay have similar structures.

$$SiO_4{}^{4-} + 4\,H^+ \longrightarrow Si(OH)_4$$

ORGANOSILICON COMPOUNDS. The uniqueness of carbon chemistry stems from the ability of carbon atoms to link together into chains. This phenomenon is also displayed by silicon and germanium, but only to a very limited extent. The bonds Si—Si and Ge—Ge are simply not strong enough to permit long chains to exist. Nevertheless, a series of **silanes** can be prepared, up to a limit of six silicon atoms per chain.

monosilane   disilane   trisilane   hexasilane

A mixture of the silanes can be prepared by reacting magnesium silicide, Mg_2Si, with dilute HCl. Pure monosilane results from the reaction of lithium aluminum hydride and silicon tetrachloride.

$$LiAlH_4 + SiCl_4 \longrightarrow LiCl + AlCl_3 + SiH_4(g) \qquad (20.69)$$

The silanes are thermally unstable. Moderate heating of the higher silanes causes decomposition to the lower silanes, and above 500°C, to the elements. Like the hydrocarbons, the silanes are combustible. In fact, they ignite spontaneously in air.

$$SiH_4 + 2\,O_2 \longrightarrow SiO_2 + 2\,H_2O \qquad (20.70)$$

Other atoms can be substituted for hydrogen atoms in the silanes rather easily. For example, the series of compounds SiH_3Cl, SiH_2Cl_2, $SiHCl_3$, and $SiCl_4$ results from the successive substitution of Cl for H atoms in SiH_4.

The hydrolysis of $(CH_3)_2SiCl_2$ produces a compound, $(CH_3)_2Si(OH)_2$, dimethyl silanol, which undergoes a reaction (polymerization) by the successive elimination of water molecules from among large numbers of the silanol molecules. The result is a material consisting of long-chain molecules, belonging to a class of polymers called **silicones.**

$$(CH_3)_2SiCl_2 + 2\ H_2O \longrightarrow (CH_3)_2Si(OH)_2 + 2\ HCl \qquad (20.71)$$

$$
\begin{array}{ccc}
& CH_3 & & CH_3 \\
& | & & | \\
HO-Si-O & (H \quad HO) & -Si-OH & \xrightarrow{-H_2O} \quad \xrightarrow{-H_2O} \\
& | & & | \\
& CH_3 & & CH_3
\end{array}
$$

$$
\begin{array}{ccc}
CH_3 & \left(\!\! \begin{array}{c} CH_3 \\ | \end{array} \!\!\right) & CH_3 \\
| & & | \\
HO-Si-O & -Si-O & -Si-OH \\
| & & | \\
CH_3 & \left(\!\! \begin{array}{c} | \\ CH_3 \end{array} \!\!\right)_{\!n} & CH_3
\end{array} \qquad (20.72)
$$

Depending on the length of the chains and the degree of crosslinking between chains, the silicones are obtained either as oils or as rubberlike materials. Silicone oils are not volatile; they may be heated to high temperatures without decomposition. Also, they can be cooled to very low temperatures without becoming too viscous or solidifying. (Hydrocarbon oils become very viscous at low temperatures.) Silicone rubbers retain their elasticity at low temperatures; they are chemically resistant and thermally stable.

Summary

The alkali (IA) and alkaline earth (IIA) elements are very active metals. Atoms of these metals have low ionization energies, low electronegativities, and highly negative standard reduction potentials. These factors all indicate that the metal atoms are easily oxidized to the metal ions—M^+ for group IA and M^{2+} for group IIA. This oxidation occurs when the metals react with the halogens, oxygen, nitrogen, hydrogen, acids, and, in most cases, water. With few exceptions, alkali metal compounds are highly water soluble, but numerous compounds of the alkaline earth metals are not.

Boron and aluminum display some similarity to each other and a number of differences. Additional similarities, known as diagonal relationships, exist between boron and silicon and between beryllium and aluminum. Aluminum exhibits many of the reactions of an active metal. However, it also forms important covalent compounds and has an amphoteric hydroxide. Some boron and aluminum compounds are "electron deficient"; these compounds are strong Lewis acids.

The halogens (group VIIA) are the most active nonmetal elements. They form binary compounds with metals (halides), compounds among themselves (interhalogens), and a variety of oxyacids. The oxidation states of the halogen elements in their compounds can range from -1 to $+7$. As a result, many of the characteristic reactions involving halogen compounds are oxidation-reduction reactions. The standard reduction potential diagram introduced in this chapter is especially useful in summarizing the oxidation-reduction chemistry of the halogens.

Almost all of the elements form oxides. One classification scheme for oxides is based on the ionic form of the oxygen, that is, normal oxide, peroxide, and superoxide. Another scheme views oxides as the anhydrides of hydroxy compounds (oxyacids). Whether these hydroxy compounds ionize as acids or bases depends on such factors as the size and charge (oxidation state) of the central atom. The chemistry of ozone and hydrogen peroxide centers on oxidation-reduction reactions.

Sulfur, another important nonmetal, exists in a variety of molecular and physical forms. The oxidation state of sulfur in its compounds ranges from -2 to $+6$ ($+7$ in $S_2O_8^{2-}$), so again oxidation-reduction reactions are an important part of its chemistry. The differing solubilities of metal sulfides in water, acids, and bases are an important aspect of the qualitative analysis scheme for cations.

The elements of group VA display a variation of nonmetallic–metallic behavior ranging from the active nonmetal $N_2(g)$ to the moderately metallic $Bi(s)$. The chemistry of nitrogen centers on oxidation-reduction reactions in which the oxidation state of N can range from -3 to $+5$. Also, a number of acid-base reactions are encountered in which ammonia and its derivatives are weak bases. Phosphorus, like sulfur, exists in different physical forms. Much of the chemistry of phosphorus is based on its oxides and oxyacids.

Silicon, through its ability to form a vast number of different silicates, is a key element of the mineral world. Because silicon forms stable bonds with H, C, and O atoms, there is also an organic silicon chemistry.

Learning Objectives

As a result of studying Chapter 20, you should be able to

1. Explain physical properties of representative elements in terms of their atomic properties and their positions in the periodic table.

2. Describe methods that can be used to prepare several representative elements from their naturally occurring compounds.

3. Describe some important uses of certain of the more common representative elements and their compounds.

4. Write equations for reactions of the group IA and IIA metals with the halogens, oxygen, nitrogen, hydrogen, acids, and water.

5. Write equations for oxidation-reduction reactions involving the halogens, nitrogen compounds, sulfur compounds, O_2, O_3, and H_2O_2.

6. Use electrode (reduction) potential diagrams as a source of information about oxidation-reduction reactions.

7. Explain the important functions served by atmospheric ozone.

8. Describe oxides of the elements using the terms "normal oxides," "peroxides," "superoxides," "acidic," "basic," and "amphoteric."

9. Describe the different molecular and/or physical forms of oxygen, sulfur, and phosphorus.

10. Describe a number of thio compounds by name, formula, and structure.

11. Predict whether a metal sulfide will dissolve in water, acids, bases, or under special oxidizing conditions [i.e., in $HNO_3(aq)$ or aqua regia].

12. Name oxyacids—ortho, meta, pyro—according to a scheme based on the loss of H_2O molecules.

13. Describe the principal forms of silicate minerals in terms of the arrangements of Si and O atoms.

14. Write formulas and names for some organosilicon compounds and equations for their reactions.

Some New Terms

Allotropy refers to the existence of an element in two or more forms, such as O_2 and O_3 or rhombic and monoclinic sulfur.

Amphoterism is the ability of certain element oxides and hydroxy compounds to behave either as acids or bases.

Diagonal relationships refer to similarities that exist between certain pairs of elements in different groups and periods of the periodic table. Examples are Li/Mg, Be/Al, and B/Si.

An **electrode (reduction) potential diagram** tabulates the standard electrode (reduction) potentials for oxidation-reduction couples of an element and various of its ionic and compound forms.

An **interhalogen compound** is a binary covalent compound between two halogen elements, for example, ICl or BrF_3.

A **meta acid** is formed by the elimination of two H atoms and one O atom (i.e., H_2O) from the ortho acid.

An **ortho acid** is an oxyacid containing the maximum number of OH groups possible.

The **peroxide ion** has the structure $[:\overset{..}{\underset{..}{O}}:\overset{..}{\underset{..}{O}}:]^{2-}$.

A **polyhalide anion** consists of two or more halogen atoms covalently bonded into a polyatomic ion, for example, triiodide ion, I_3^-.

A **pyro acid** is formed by the elimination of two H atoms and one O atom (i.e., H_2O) from between two molecules of the ortho acid.

A **silicone** is an organosilicon polymer containing $-O-Si-O-$ bonds.

The **superoxide ion** has the structure $[:\overset{..}{\underset{}{O}}:\overset{..}{\underset{..}{O}}:]^-$.

A **thio** compound is one in which an S atom replaces an O atom. For example, in thiosulfate ion an O atom in SO_4^{2-} is replaced by an S atom, producing the ion $S_2O_3^{2-}$.

Exercises

Group IA and IIA elements

1. Write balanced equations to represent the reaction of (a) calcium metal with $Cl_2(g)$; (b) potassium metal with $O_2(g)$; (c) barium metal with cold water; (d) calcium oxide with water; (e) magnesium nitride with water.

2. Express equation (20.1) as two half-equations, one an oxidation and the other a reduction.

3. Be reacts with concentrated $NaOH(aq)$ to produce $H_2(g)$. Write a balanced equation for the reaction.

4. Magnesium oxide is produced by heating Mg in air. From 0.200 g Mg, 0.305 g of product is obtained.
   (a) Can the compound be pure MgO? Explain.
   (b) What other substance is probably present?
   (c) How would you test for the presence of the substance referred to in part (b)?

5. What mass of $CaH_2(s)$ would be required to produce sufficient $H_2(g)$ to fill a 200-L observation balloon at 755 mmHg and 18°C?

6. What is the pH of the resulting solution if 0.500 L $NaCl(aq)$ is electrolyzed for 123 s with a current of 1.40 A? Does the result depend on the concentration of NaCl present? Explain.

Boron and aluminum

7. Explain what is meant by the terms (a) electron-deficient compound; (b) dimer; (c) Lewis acid; (d) adduct.

8. Identify the oxidizing and reducing agents in equation (20.8).

9. Write half-equations for the oxidation and reduction half-reactions in the reduction of water by AlH_4^-.

10. The enthalpies of formation $(\Delta \bar{H}_f^\circ)$ of $Al_2O_3(s)$, $Fe_2O_3(s)$, $MnO_2(s)$, and $MgO(s)$ are -1670, -824, -519, and -602 kJ/mol, respectively.
   (a) Determine the heat of the thermite reaction (20.9) per mole of Fe produced.
   (b) Determine the heat of reaction per mole of Mn, if MnO_2 is substituted for Fe_2O_3.
   (c) Show that if MgO were substituted for Fe_2O_3 the reaction would be endothermic. (Al does not reduce MgO to Mg.)

The halogen elements

11. Use data from Table 20-3 to predict the melting points and boiling points of the interhalogen compounds BrCl and IBr. What principle is involved?

12. Which of the following reactions will occur as written? Make a general statement about which halogen elements will displace others from an aqueous solution of the halide ion.

$$Cl_2 + 2\,I^-(aq) \longrightarrow 2\,Cl^-(aq) + I_2$$
$$I_2 + 2\,Br^-(aq) \longrightarrow 2\,I^-(aq) + Br_2$$
$$Br_2 + 2\,Cl^-(aq) \longrightarrow 2\,Br^-(aq) + Cl_2$$

13. Write a half-equation for the half-reaction in basic solution represented by $ClO^- \xrightarrow{0.88\ V} Cl^-$.

14. In Example 20-1 it was concluded that the reaction represented by equation (20.18) would not occur spontaneously in acidic solution. Will this reaction occur spontaneously in basic solution? Explain.

15. Use the methods of Examples 20-1 and 20-2 to determine (a) the standard electrode potential for the reduction of HClO to Cl^-; (b) whether the reaction $2\,HClO \rightarrow HClO_2 + H^+ + Cl^-$ will occur spontaneously if the reactant and products are in their standard states.

16. Write a plausible equation to represent a reaction in which HClO acts as (a) an oxidizing agent; (b) a reducing agent; (c) an acid.

17. Is the forward reaction (20.14) favored by high or low (a) pressure; (b) temperature? Explain. Le Chatelier's

*18. The following data are given.
$$IO_3^- + 3\,H_2SO_3(aq) \longrightarrow I^- + 3\,SO_4^{2-} + 6\,H^+$$
$$E_{cell}^\circ = 0.92\ V$$
$$HIO(aq) + H^+ + I^- \longrightarrow I_2(s) + H_2O \qquad E_{cell}^\circ = 0.91\ V$$

Use these data together with values from Table 19-2 to complete the standard electrode potential diagram shown.

$$IO_3^- \xrightarrow{(?)} HIO \xrightarrow{(?)} I_2(s) \xrightarrow{0.54\ V} I^-$$
$$\underset{(?)}{\rule{8cm}{0.4pt}}$$

Oxygen

19. Use Lewis structures or other information from this chapter to explain the fact that
   (a) H_2S exists as a gas at room temperature while H_2O is a liquid.
   (b) O_3 is diamagnetic.
   (c) The O–O bond lengths in O_2, O_3, and H_2O_2 are 121, 128, and 148 pm, respectively.

20. Write oxidation and reduction half-equations for the reaction of (a) peroxide and (b) superoxide ion with water.

21. In a disproportionation reaction the same substance is both oxidized and reduced. Write an equation for a plausible disproportionation of H_2O_2.

22. Would you expect a water solution of sodium peroxide to be acidic, basic, or neutral? Explain your reasoning.

23. Ozone is a powerful oxidizing agent. Write equations to represent oxidation of **(a)** I^- to I_2 in acidic solution; **(b)** sulfur in the presence of moisture to sulfuric acid; **(c)** $[Fe(CN)_6]^{4-}$ to $[Fe(CN)_6]^{3-}$ in basic solution. In each case $O_3(g)$ is reduced to $O_2(g)$.

24. Show that the structure of the O_3 molecule given in the text is consistent with the prediction made with VSEPR theory.

***25.** It has been estimated that if all the ozone in the atmosphere were brought to sea level at STP, the gas would form a layer 0.3 cm thick. Estimate the number of O_3 molecules in the earth's atmosphere.

Sulfur

26. What is the oxidation state of S in each of these oxides: SO_2; SO_3; SO_4; S_2O; S_2O_3; S_2O_7?

27. Use Lewis structures or other information from this chapter to explain the fact that
  (a) S_2Cl_2 has a structure similar to H_2O_2.
  (b) SO_2 possesses a dipole moment but SO_3 does not.
  (c) A metastable purple sulfur, S_2, exists that is paramagnetic.

28. Sulfur is found in nature in elemental form as sulfides—for example, Cu_2S—and as sulfates—for example $CaSO_4 \cdot 2 H_2O$ (gypsum)—but not as sulfites. Why do you suppose this is so?

29. Complete and balance the following equations. If no reaction occurs, so state.

  (a) $FeS(s) + O_2 \xrightarrow{heat}$
  (b) $Zn(s) + H^+ + SO_4^{2-}$ (conc.) $\longrightarrow$
  (c) $HSO_4^-(aq) + OH^-(aq) \longrightarrow$
  (d) $HS^- + H_2O \longrightarrow$
  (e) $FeS(s) + H^+ + NO_3^- \longrightarrow$
  (f) $PbS(s) + H^+ + Cl^- \longrightarrow$

30. A statement is made in the text that HgS is the least soluble of all metal sulfides; yet K_{sp} for Bi_2S_3 is 1×10^{-96} compared to 1.6×10^{-52} for HgS. Explain this apparent discrepancy.

31. What mass of Na_2SO_3 must have been present in a sample that required 26.50 ml of 0.0510 M $KMnO_4$ for its titration in an acidic solution?

32. A 1.100-g sample of copper ore is dissolved and the $Cu^{2+}(aq)$ is treated with excess KI. The liberated iodine requires 11.24 ml of 0.1000 M $Na_2S_2O_3$ for its titration. What is the percent copper, by mass, in the ore? (*Hint:* Recall equation 20.46.)

Nitrogen

33. Write balanced equations to represent the following.
  (a) equilibrium between nitrogen dioxide and dinitrogen tetroxide in the gaseous state
  (b) the decomposition of HNO_3 by heating
  (c) the neutralization of $NH_3(aq)$ by $H_2SO_4(aq)$
  (d) the dissolving of silver metal in conc. $HNO_3(aq)$
  (e) the complete combustion of the rocket fuel, dimethyl hydrazine, $(CH_3)_2NNH_2$

34. What is the pH of an aqueous solution that is **(a)** 0.025 M in NH_2OH; **(b)** 0.015 M in $[NH_3OH]Cl$?

35. Indicate the oxidation state of N in each of the species shown in Figure 20-12.

36. NH_3OH^+ can act as an oxidizing agent in acidic solutions.
  (a) Write a half-equation to represent the reduction of NH_3OH^+ to NH_4^+.
  (b) Use data from Figure 20-12 to determine $E°$ for the half-reaction described in part (a).
  (c) Will the oxidation of Fe^{2+} to Fe^{3+} occur in acidic solution with NH_3OH^+ as the oxidizing agent?

37. The heat of formation of hydrazine $(\Delta \bar{H}_f°)$ is $+50.63$ kJ/mol. What is the heat of combustion of hydrazine as represented in equation (20.54)?

***38.** An active metal can be used to reduce NO_3^- to $NH_3(g)$ in basic solution. (The following equation is *not* balanced.)

$$NO_3^- + Zn(s) + OH^- + H_2O \longrightarrow$$
$$Zn(OH)_4^{2-} + NH_3(g) \quad (20.73)$$

The NH_3 can be neutralized by passing the gas into an excess of HCl(aq). The unreacted HCl can then be titrated with standard NaOH. In this way a quantitative estimate of NO_3^- is possible. In one analysis a 25.00-ml sample of a solution containing nitrate ion was treated according to (20.73). The liberated NH_3 was passed into 50.00 ml of 0.1500 M HCl. The excess HCl required 32.10 ml of 0.1000 M NaOH for its titration. What was $[NO_3^-]$ in the original sample?

Phosphorus

39. A certain phosphate rock contains 58.0% $Ca_3(PO_4)_2$. What mass of this rock is required to produce 100 kg of phosphorus, assuming no loss of material in reaction (20.64)?

40. Write chemical equations to show why
  (a) A solution of Na_3PO_4 is strongly basic.
  (b) The first equivalence point in the titration of H_3PO_4 is on the acid side of neutral.

41. Hypophosphorous acid, H_3PO_2, is a *monoprotic* acid (ionizes in a single step). Propose a structure for this acid.

42. Use the scheme of Figure 20-14 to supply plausible

names and/or formulas for the following: **(a)** calcium metaphosphate, **(b)** potassium pyrophosphate; **(c)** $NaSbO_2$; **(d)** $NaBiO_3$; **(e)** sodium orthobismuthate.

Silicon

43. With reference to Table 20-12, show that the formula given for mica is consistent with the usual oxidation states and ionic charges of its constituents.

44. Apply the scheme of page 520 **(a)** to justify the names and formulas of sodium orthosilicate and sodium metasilicate; **(b)** to write the formula of the pyrosilicate anion.

45. In a manner similar to equations (20.71) and (20.72),
(a) write equations to represent the hydrolysis of $(CH_3)_3SiCl$, followed by the elimination of water from the resulting silanol molecules.
(b) Does a silicone polymer form?
(c) What would be the corresponding product obtained from CH_3SiCl_3?

Additional Exercises

1. Describe briefly what is meant by each of the following terms: **(a)** diagonal relationship; **(b)** hydration energy; **(c)** thermite reaction; **(d)** interhalogen compound; **(e)** polyhalide anion; **(f)** oxyacid; **(g)** polyprotic acid; **(h)** electrode potential diagram; **(i)** amphoterism; **(j)** allotropy; **(k)** thio compound; **(l)** aqua regia; **(m)** silane; **(n)** silicone; **(o)** ortho acid.

2. Supply a name or formula for each of the following:
(a) Ag_2O; **(b)** lithium peroxide; **(c)** potassium superoxide; **(d)** calcium nitride; **(e)** $Ba(OH)_2$; **(f)** B_2H_6, **(g)** $LiAlH_4$; **(h)** hydrogen difluoride ion; **(i)** $NaHSO_3$; **(j)** BrF_5; **(k)** OCl^-; **(l)** periodic acid; **(m)** calcium bromate; **(n)** H_2O_2; **(o)** $H_2S_2O_7$; **(p)** $S_2O_3^{2-}$; **(q)** AsS_4^{3-}; **(r)** calcium metasilicate; **(s)** tetrasilane; **(t)** hydroxylamine sulfate; **(u)** BiH_3; **(v)** dinitrogen trioxide; **(w)** $H_2PO_4^-$; **(x)** $CaHPO_4$; **(y)** $Zn_2P_2O_7$.

3. Complete each of the following equations. If no reaction occurs, so state.
(a) $CaH_2(s) + H_2O \longrightarrow$ N.R. ?
(b) $Mg(s) + HC_2H_3O_2(aq) \longrightarrow$
(c) $BeO(s) + H_2O \longrightarrow$
(d) $Ra(s) + H_2O \longrightarrow$
(e) $NaCl(aq) \xrightarrow{electrolysis}$ $Na^+ + Cl^-$
(f) $2Cl_2(g) + 2H_2O \longrightarrow 4HCl + O_2$
(g) $H_2O_2 + MnO_4^- + H^+ \longrightarrow$
(h) $MgO(s) + 2H^+ + 2Cl^- \longrightarrow H_2O + Mg^+ + 2Cl^-$
(i) $CO_2(g) + Na^+ + OH^- \longrightarrow$
(j) $Na^+(aq) + H_2S(aq) \longrightarrow$
(k) $Al(s) + Cr_2O_3(s) \longrightarrow$
(l) $Zn^{2+}(aq) + H_2S(aq) \longrightarrow$
(m) $CuS(s) + H^+ + Cl^- \longrightarrow$

4. Explain why the compound SF_6 exists but OF_6 does not; PCl_5 but not NCl_5.

5. Show that the hypothetical process pictured in Figure 20-14 for developing the oxyacids of phosphorus(V) also leads to correct formulas for **(a)** carbonic acid (H_2CO_3); **(b)** nitrous acid (HNO_2); **(c)** nitric acid (HNO_3); **(d)** sulfuric acid (H_2SO_4); **(e)** periodic acid (HIO_4).

6. Write equations to show how H_2O_2 **(a)** oxidizes NO_2^- to NO_3^- in acidic solution; **(b)** oxidizes $SO_2(g)$ to SO_4^{2-} in basic solution; **(c)** reduces MnO_4^- to Mn^{2+} in acidic solution; **(d)** reduces $Cl_2(g)$ to Cl^- in basic solution.

7. Paraperiodate ion can be prepared by oxidizing an iodate with chlorine in basic solution.

$$IO_3^- + Cl_2 + OH^- \longrightarrow H_3IO_6^{2-} + Cl^-$$

A precipitate is formed with silver ion.

$$H_3IO_6^{2-} + Ag^+ \longrightarrow Ag_3IO_5(s) + H_2O + H^+$$

A solution of paraperiodic acid is formed when an aqueous suspension of Ag_3IO_5 is treated with chlorine gas.

$$Ag_3IO_5(s) + Cl_2(g) + H_2O \longrightarrow$$
$$H_5IO_6(aq) + AgCl(s) + O_2(g)$$

The paraperiodic acid can be crystallized from this solution.
(a) Balance all the equations given.
(b) What mass of paraperiodic acid can be prepared from 100 g $NaIO_3$?

*8. The electrolysis of 0.250 L of 0.220 M $MgCl_2$ is conducted until 104 ml of gas (a mixture of H_2 and water vapor) is collected at 23.0°C and 748 mmHg. Will $Mg(OH)_2$ precipitate if electrolysis is carried to this point? (Use 21 mmHg as the vapor pressure of the solution.)

*9. Propose a plausible bonding scheme for O_3 involving hybridized orbitals?

*10. The preparation of bromine from seawater utilizing reaction (20.15) is carried out at a pH of 3.5. The liberated Br_2 is swept from the seawater with a current of air and then passed into a basic solution where Br_2 undergoes both oxidation and reduction (disproportionation) to BrO_3^- and Br^-. When the solution is acidified, the reaction reverses itself.
(a) Write chemical equations to represent each step in the process.
(b) What mass of seawater must be processed to recover 1 metric ton (1000 kg) of Br_2, assuming no losses? (Seawater contains 70 ppm Br.)

*11. A handbook lists the concentration of a saturated solution of I_2 in water as 1.33×10^{-3} M. The handbook also gives the following expression.

$$I_2(aq) \rightleftharpoons I_2(CCl_4) \qquad K = \frac{[I_2]_{CCl_4}}{[I_2]_{aq}} = 85.5$$

A 10.0-ml sample of a saturated solution of I_2 in water is shaken with 10.0 ml CCl_4. After equilibrium is established, the two liquid layers are separated. (*Hint:* Recall Figure 20-7.)

(a) What mass of I_2, in mg, remains in the water layer?
(b) If the 10.0-ml water layer in (a) is extracted with a second 10.0-ml portion of CCl_4, what will be the number of mg I_2 remaining in the water?
(c) If the 10.0-ml sample of saturated $I_2(aq)$ had originally been extracted with 20.0 ml CCl_4, would the quantity of I_2 remaining in the aqueous solution have been less than, equal to, or greater than in part (b)? Explain.

★ 12. Estimate the percent dissociation of $Cl_2(g)$ into $Cl(g)$ at 1 atm total pressure and 1000 K. Use data from Appendix D and equations established elsewhere in the text, as necessary.

Self-Test Questions

For questions 1 through 5 select the single item that best completes each statement.

1. To displace Br_2 from an aqueous solution containing Br^-, add (a) $I_2(aq)$; (b) $Cl_2(aq)$; (c) $Cl^-(aq)$; (d) $I_3^-(aq)$.

2. All of the following molten compounds are good electrical conductors except (a) $BeCl_2$; (b) KF; (c) CsI; (d) $NaCl$.

3. Under the appropriate conditions each of the following can act as an oxidizing agent except (a) I_2; (b) Cl^-; (c) F_2; (d) BrO_3^-.

4. Of the following group of oxides, the one that is expected to be amphoteric is (a) SO_2; (b) CO_2; (c) K_2O; (d) Al_2O_3.

5. To dissolve mercuric sulfide, HgS ($K_{sp} = 1.6 \times 10^{-52}$), use (a) $HNO_3(aq)$; (b) $HCl(aq)$; (c) a mixture of $HNO_3(aq)$ and $HCl(aq)$; (d) $NaOH(aq)$.

6. Use principles from this chapter and elsewhere in the text to explain why
(a) The electrolysis of molten NaCl yields $Cl_2(g)$, whereas electrolysis of NaCl(aq) yields both $H_2(g)$ and $Cl_2(g)$.
(b) In the titration of boric acid, H_3BO_3, with NaOH(aq) only one equivalence point is noted.
(c) I_2 is much more soluble in KI(aq) than in pure H_2O.
(d) Certain metals, for example, copper, are soluble in nitric acid but not in hydrochloric acid.

(e) Labels on household bleaches usually carry a warning that the bleach should not be used in combination with cleaners containing NH_3. (Bleaches usually have hypochlorite ion, ClO^-, as their "active" ingredient.)

7. Supply a name or formula for each of the following: (a) magnesium nitride; (b) $Ca(OH)_2$; (c) RbO_2; (d) ClO_2^-; (e) bromic acid; (f) thiosulfate ion; (g) trisilane; (h) $CaSiO_3$; (i) $Mg_2P_2O_7$; (j) calcium dihydrogen phosphate.

8. Write balanced equations to represent (a) electrolysis of molten $MgCl_2$; (b) decomposition of $BaCO_3(s)$ by heating; (c) reaction of CaO(s) with water; (d) displacement of Cu(s) from a solution of Cu^{2+} by Al(s); (e) reduction of iodate ion to free iodine in acidic solution by hydrogen sulfide ion; (f) reduction of Cl_2 to Cl^- in acidic solution by hydrogen peroxide; (g) dissolution of ZnS(s) in concentrated $HNO_3(aq)$; (h) oxidation of $NH_3(g)$ to NO(g) by $O_2(g)$.

9. Discuss why iodine and iodide ion solutions are so useful in analytical chemistry.

10. A portion of a standard electrode (reduction) potential diagram is given below. What is the $E°$ value for the reduction of H_2SeO_3 to H_2Se?

$$SeO_4^{2-} \xrightarrow{\text{1.15 V}} H_2SeO_3 \xrightarrow{\text{0.74 V}} Se \xrightarrow{-0.35\text{ V}} H_2Se$$
$$\underbrace{\qquad\qquad\qquad\qquad}_{(?)}$$

21 The Chemistry of Transition Elements

Atoms or ions of the main transition series (sometimes called the "d block" elements) contain partially filled d orbitals. Partially filled f orbitals are characteristic of atoms of the inner-transition ("f block") elements. All the elements in the middle section of the periodic table fit one or another of these descriptions. Thus, more than half the elements belong either to a transition or inner-transition series. The chemistry of these elements has both theoretical and practical significance.

We begin with a brief survey of some properties of the first transition series. This is followed by a further consideration of chromium and the iron group as representative members of this series. We conclude with a further discussion of qualitative analysis by considering specifically the cation group containing a number of transition metal ions—the ammonium sulfide group.

One important characteristic of transition elements is the ability to form complex ions. This tendency is mentioned in several instances in this chapter and explored more fully in the next.

21-1 Some Properties of the Transition Elements

The properties listed in Table 21-1 for elements of the first transition series ($Z = 21$ to 30) are clearly those of a group of metallic elements. High melting points, good electrical conductivity, and moderate to extreme hardness result from the ready availability of electrons and orbitals for metallic bonding.

ATOMIC RADII. In a transition series the essential difference in atomic structure between successive elements involves one unit of positive charge on the nucleus and one electron in an orbital of an *inner* electronic shell. This is a rather small difference and does not produce significant changes in atomic size, especially in the middle of the series. As a result, in the first transition series one finds strong similarities among *horizontal* groupings of elements. For example, the three elements, Fe, Co, and Ni—the iron triad—are usually discussed as a single group (see Section 21-4).

LANTHANOID CONTRACTION. When an element in the first transition series is compared with those of the second and third series within the same group, some important differences are noted. Table 21-2 lists representative data for the members of group VIB—Cr, Mo, and W. The most notable feature of the table is that the atomic

The exact elements that comprise the transition series are the subject of some debate, especially regarding groups IB and IIB. One view is that a transition element must have partially filled d orbitals in either (a) the neutral atom or (b) the atom in certain oxidation states. Copper does not have d orbital vacancies in the neutral atom nor in Cu(I), but it does in Cu(II). Copper is a transition element. Zinc does not have d orbital vacancies in either the neutral atom or in Zn(II). Zinc does not fit this definition of a transition element. However, we have included both groups IB and IIB in the discussion that follows since this is so often done.

[handwritten margin notes at top: "have so few m d-orbital tend to lose all them & 4s shell ∴ exist in highest oxy state most often"]

[handwritten note above Cr column: "as w/ Cu here is but 4s shell is ½ being completed w/the removal of a 4s e⁻ → the normal addition of another e⁻"]

TABLE 21-1
Selected properties of elements of the first transition series

	Sc	Ti	V	Cr	Mn	Fe
atomic number	21	22	23	24	25	26
electron configuration[a]	$3d^14s^2$	$3d^24s^2$	$3d^34s^2$	$3d^54s^1$	$3d^54s^2$	$3d^64s^2$
atomic (metallic) radius, pm	161	145	132	127	124	124
ionization energies, eV/atom						
first	6.54	6.82	6.74	6.77	7.44	7.87
second	12.80	13.58	14.65	16.50	15.64	16.18
third	24.76	27.49	29.31	30.96	33.67	30.65
electrode potential,[b] V	-2.08	-1.63	-1.2	-0.91	-1.18	-0.44
oxidation states[c]	3	2, 3, 4	2, 3, 4, 5	**2, 3**, 6	**2**, 3, 4, 7	**2**, 3
melting point, °C	1397	1672	1710	1900	1244	1530
density, g/cm³	2.99	4.49	5.96	7.20	7.20	7.86
hardness[d]	—	—	—	9.0	5.0	4.5
electrical conductivity[e]	—	2	3	10	2	17

[handwritten note left of ionization energies: "energy absorbed to release MLB electron"]

[handwritten note: "to reduce @ cathode"]

[a] Each atom has an argon inner core configuration.
[b] For the reduction process $M^{2+}(aq) + 2 e^- \rightarrow M(s)$ [except for scandium, where the ion is $Sc^{3+}(aq)$].
[c] Common oxidation states; the most stable is shown in boldface.
[d] Hardness values are on the Mohs scale (see Table 20-1).
[e] Compared to an arbitrarily assigned value of 100 for silver.

radii of Mo and W are the same. Along with the usual filling of *s*, *p*, and *d* sublevels, in the interval of elements separating Mo and W the 4*f* sublevel is also filled. Electrons in an *f* subshell are not very effective in screening outershell electrons from the nucleus. As a result these outer-shell electrons are held more tightly than would otherwise be the case and atomic size does not increase as expected. The limited shielding ability of 4*f* electrons means that in the series of elements in which the 4*f* sublevel is being filled, an actual decrease in atomic size occurs. This filling occurs in the lanthanoid series ($Z = 58$ to 71), so the phenomenon is called the lanthanoid contraction. An example of ways in which Cr differs from Mo and W is that Cr can exist in aqueous solution as the simple hydrated ions $Cr^{2+}(aq)$ and $Cr^{3+}(aq)$. Ionic forms of Mo and W are polyatomic (e.g., oxyanions). The oxidation state $+3$ is very common for Cr, whereas higher oxidation states ($+5$, $+6$) are favored with Mo and W.

[margin note: The terms "lanthanoid" and "lanthanide" are often used interchangeably.]

[handwritten margin note: "their ions are polyatomic, e.g. they bond with oxygen to form anions"]

[handwritten note far left: "14 OH⁻ + NH₄"]

ELECTRON CONFIGURATIONS AND OXIDATION STATES. All the elements of the first transition series have an electron configuration with the following characteristics:

1. An inner core of electrons in the argon configuration.

TABLE 21-2
Some properties of group VIB—Cr, Mo, W

Transition series	Element	Atomic number	Electron configuration	Atomic radius, pm	Standard electrode potential,[a] V	Oxidation states[b]
first	Cr	24	$[Ar]3d^54s^1$	127	-0.744	2, **3**, 6
second	Mo	42	$[Kr]4d^55s^1$	139	-0.2	2, 3, 4, **5**, **6**
third	W	74	$[Xe]4f^{14}5d^46s^2$	139	—	2, 3, 4, 5, **6**

[a] For the reduction process $M^{3+} + 3 e^- \rightarrow M(s)$.
[b] Common oxidation states; the most stable is shown in boldface.

(handwritten margin notes, top) inner d-shell is finished w/new e⁻ & w/one from 4s shell; then the 4s is returned to & completed

Co	Ni	Cu	Zn
27	28	29	30
$3d^7 4s^2$	$3d^8 4s^2$	$3d^{10} 4s^1$	$3d^{10} 4s^2$
125	125	128	133
7.86	7.64	7.73	9.39
17.06	18.17	20.29	17.96
33.50	35.17	36.83	39.72
−0.28	−0.25	+0.34	−0.76
2, 3	2	1, 2	2
1495	1455	1083	419
8.9	8.90	8.92	7.14
—	—	2.5–3.0	—
24	24	97	27

(handwritten notes, right) → for all, the inner core is [Ar] ⇒ $1s^2 2s^2 p^6 3s^2 p^6$

(handwritten note) ← variable states because they may lose 4s & 3d orbitals

2. Two electrons in the 4s orbital for eight members and one 4s electron for the remaining two (Cr and Cu).

3. A number of electrons in 3d orbitals varying from one in scandium to ten in copper and zinc.

The 4s electrons of the first transition series atoms can be extracted without great difficulty, meaning that most of these atoms should exhibit an oxidation state of +2. In Cr and Cu the atoms must also lose a 3d electron to become dipositive ions, since they have but a single 4s electron. Sc loses its $3d^1$ and $4s^2$ electrons in forming the ion Sc^{3+}, and so exhibits only the oxidation state +3. Because electrons in 3d orbitals may be lost along with 4s electrons, the transition metals display variable oxidation states. The maximum oxidation state corresponds to the periodic table group number (such as +6 for the members of group VIB). Not all the possible oxidation states for a transition metal are equally stable, however. In general, the initial members of a transition series (e.g., Sc, Ti, V) tend to exist in their higher oxidation states and the later members (e.g., Fe, Co, Ni) in lower oxidation states. Reactions involving changes in oxidation state—oxidation-reduction reactions—are very common among the transition elements.

(handwritten margin note, left) since they have a completed shell in the d level (Cu's ½ shell has stability similar to a full shell)

IONIZATION ENERGIES AND ELECTRODE POTENTIALS.   The ionization energies are fairly constant across the first transition series of elements. The values of the first ionization energies are about the same as for the group IIA metals. Standard electrode potentials increase in value gradually across the transition series. However, with the exception of the oxidation of Cu to Cu^{2+}, all of these elements are more readily oxidized than hydrogen. This means that they dissolve in acidic solution with the production of $H_2(g)$.

(handwritten margin note, left) $M^{2+} + 2e^- \rightarrow M(s)$ $2H^+ \rightarrow H_2(g)$

21-2   An Overview of the First Transition Series

SCANDIUM.   Scandium is a rare metal. Its presence in the earth's crust has been estimated at from 5 to 30 ppm, and it is found only in a few mineral deposits. The commercial uses of scandium are very limited and its production is measured in gram or kilogram amounts, not tonnages. One recent use has been as a component in high-intensity lamps. The pure metal can be prepared by electrolysis of a fused mixture of $ScCl_3$ with other chlorides.

TITANIUM.   Titanium is one of the more abundant elements, comprising 0.6% of the earth's crust. Its low density and high structural strength make it desirable as a construction material for the aerospace industry. These properties, together with its corrosion resistance, contribute to its usefulness in the chemical industry for pipes, component parts of pumps, and reaction vessels. The maximum oxidation state of Ti in its compounds (+4) is also its principal oxidation state. Its most important compound, commercially, is titanium dioxide, TiO_2. Because of its whiteness, opacity, and inertness, TiO_2 is widely used as a pigment in paints and as a paper whitener. The metallurgy of Ti is described in Section 26-7.

VANADIUM.   Vanadium is used mainly in alloys with other metals. About 80% of the vanadium produced goes into the manufacture of steel; and its alloys with Ti for the aerospace industry are becoming increasingly important. The metal and its oxide, V_2O_5, are used as catalysts in the chemical industry. Pure V can be produced by the electrolysis of a mixture of salts containing VCl_2, but the pure metal is not always needed. For example, the reduction of V_2O_5 and Fe_2O_3 by Al yields a material, **ferrovanadium,** that can be added directly to iron in steelmaking. The principal oxidation state of V in its compounds is also its maximum oxidation state, +5. Various reducing agents can be used to convert vanadium from this to lower oxidation states, as outlined below:

This is another example of the thermite reaction discussed in Section 20-2.

$$VO_3^- \xrightarrow{\ Fe^{2+}\ } VO^{2+} \xrightarrow{\ Sn^{2+}\ } V^{3+} \xrightarrow{\ Zn\ } V^{2+} \tag{21.1}$$

CHROMIUM.   Although it is found only to the extent of 100 ppm in the earth's crust, chromium is one of the most important industrial metals. The principal ore is **chromite,** $Fe(CrO_2)_2$, from which a mixture of Fe and Cr called **ferrochrome** is obtained by reduction.

$$Fe(CrO_2)_2 + 4\,C \longrightarrow \underset{\text{ferrochrome}}{Fe + 2\,Cr} + 4\,CO(g) \tag{21.2}$$

Ferrochrome may be added directly to iron, together with other metals, to produce steel. Chromium metal is very hard and maintains a bright surface through the protective action of an invisible oxide coating; it is often used to plate other metals. The principal compounds of chromium are the chromates, which can be obtained from chromite ore through the following reaction at high temperature:

$$4\,Fe(CrO_2)_2 + 8\,Na_2CO_3 + 7\,O_2 \longrightarrow 2\,Fe_2O_3 + 8\,Na_2CrO_4 + 8\,CO_2 \tag{21.3}$$

Other chromates, dichromates, and oxides of chromium can be produced from Na_2CrO_4.

Important industrial uses of chromium compounds center on their oxidizing power and on their colors. In the chrome tanning process, hides are immersed in $Na_2Cr_2O_7(aq)$, which is then reduced by $SO_2(g)$ to soluble basic chromic sulfate, $Cr(OH)SO_4$. Collagen, a protein in hides, reacts to form an insoluble complex chromium compound. The hides become tough, pliable, and resistant to biological decay; they are converted to **leather.** Paint pigments employ $PbCrO_4$ for yellow and orange colors and Cr_2O_3 for green. Additional aspects of the chemistry of chromium are considered in Section 21-3.

MANGANESE.   Manganese is another metal required in the manufacture of steel (about 13 to 20 lb Mn per ton of steel). Manganese is the twelfth most abundant

element in the earth's crust (about 0.1%). Although plentiful sources of the metal exist in the world, no manganese ores are currently being mined in the United States. A possible future source of the metal may come from manganese nodules on the ocean floor (see Section 26-4).

Manganese can exist in all oxidation states ranging from $+2$ to $+7$. The important chemical reactions of manganese compounds then are oxidation-reduction reactions, which can be summarized through an electrode potential diagram.

Acidic solution, $[H^+] = 1\ M$:

$$MnO_4^- \xrightarrow{\ 0.56\ V\ } MnO_4^{2-} \xrightarrow{\ 2.26\ V\ } MnO_2 \xrightarrow{\ 0.95\ V\ } Mn^{3+} \xrightarrow{\ 1.49\ V\ } Mn^{2+} \xrightarrow{\ -1.18\ V\ } Mn \qquad (21.4)$$

with the bracketed potentials $1.70\ V$ (from MnO_4^- to MnO_2) and $1.23\ V$ (from MnO_2 to Mn^{2+}).

Basic solution, $[OH^-] = 1\ M$:

$$MnO_4^- \xrightarrow{\ 0.56\ V\ } MnO_4^{2-} \xrightarrow{\ 0.3\ V\ } MnO_3^- \xrightarrow{\ 0.8\ V\ } MnO_2 \xrightarrow{\ 0.2\ V\ } Mn(OH)_3 \xrightarrow{\ 0.1\ V\ } Mn(OH)_2 \xrightarrow{\ -1.55\ V\ } Mn \quad (21.5)$$

with the bracketed potentials $0.60\ V$ and $-0.04\ V$.

A number of important conclusions can be drawn from these data. For example,

1. $Mn^{3+}(aq)$ is unstable. Its decomposition is spontaneous.

$$2\ Mn^{3+} + 2\ H_2O \longrightarrow Mn^{2+} + MnO_2(s) + 4\ H^+ \qquad E^\circ_{cell} = 0.54\ V \qquad (21.6)$$

2. $MnO_4^{2-}(aq)$ is unstable in acidic solution. Reaction (21.7) is spontaneous.

$$3\ MnO_4^{2-} + 4\ H^+ \longrightarrow MnO_2(s) + 2\ MnO_4^- + 2\ H_2O \qquad E^\circ_{cell} = 1.70\ V \qquad (21.7)$$

3. But $MnO_4^{2-}(aq)$ can be obtained in alkaline solution.

$$3\ MnO_4^{2-} + 2\ H_2O \longrightarrow MnO_2(s) + 2\ MnO_4^- + 4\ OH^- \qquad E^\circ_{cell} = 0.04\ V \qquad (21.8)$$

That is, if $[OH^-]$ is kept sufficiently high, the reverse of reaction (21.8) becomes important. Significant concentrations of MnO_4^{2-} (manganate ion) may then be present in solution.

Aqueous solutions of permanganate ion (usually as $KMnO_4$) are widely used in the chemical laboratory. In quantitative analysis they are employed for a variety of titration reactions, such as the determination of nitrites, hydrogen peroxide, iron in iron ore, and calcium after precipitation as the oxalate (CaC_2O_4). In organic chemistry $MnO_4^-(aq)$ can be used to oxidize alcohols and unsaturated hydrocarbons (see Chapter 24).

IRON, COBALT, AND NICKEL. Iron, with annual worldwide production measured in the hundreds of millions of tons, is, of course, the most important metal to modern civilization. It is found widely distributed throughout the earth's crust in an abundance of about 4.7%. Cobalt is among the rarer metals; it constitutes but 20 ppm of the earth's crust. It occurs in sufficiently concentrated deposits (ores) so that its annual production runs into the millions of pounds. Cobalt is used primarily in alloys with other metals. Nickel ranks twenty-fourth in abundance among the elements in the earth's crust. Its ores are mainly the sulfides, oxides, silicates, and arsenides. Particularly large deposits are found in Canada. Of the 300 million pounds or so of nickel that are consumed annually in the United States, about 80% is used in the production of alloys. Another 15% is used for electroplating; and the remainder for miscellaneous purposes. Some of the properties and typical reactions of the iron triad elements are considered in Section 21-4.

"dissolve in acid..." 1e/ HCl

"dissolve in oxidizing acid..." 1e/ HNO₃
→ H₂(g)

COPPER. Copper is especially desirable as a metal because of its excellent electrical conductivity (exceeded only by that of silver). About one half of the copper used in the United States is for electrical applications. Because of its corrosion resistance, copper is also used in plumbing and in other construction applications (about 20%). Copper forms some well-known alloys: brasses are alloys of Cu and Zn; bronzes contain Cu with various amounts of Sn, Zn, and Pb; German silver is an alloy of Cu, Zn, and Ni. Of all the first transition series metals, only copper is less active than hydrogen. That is, copper is not oxidized by H⁺ in acidic solution.

as seen on chart page 530

$$Cu(s) + 2 H^+ \longrightarrow Cu^{2+}(aq) + H_2(g) \quad E°_{cell} = -0.34 \text{ V} \tag{21.9}$$

However, copper can be dissolved in oxidizing acids, such as $HNO_3(aq)$.

$$3 Cu(s) + 8 H^+ + 2 NO_3^- \longrightarrow 3 Cu^{2+} + 4 H_2O + 2 NO(g)$$
$$E°_{cell} = 0.62 \text{ V} \tag{21.10}$$

−3e−

Even though we say that copper is corrosion resistant, it does corrode in moist air to produce a *green* basic copper carbonate. This is the green color associated with copper roofing and gutters and with bronze statues. Fortunately, this corrosion product forms a tough adherent coating that protects the underlying metal. The corrosion reaction is complex but can be summarized as follows:

$$2 Cu + H_2O + CO_2 + O_2 \longrightarrow Cu_2(OH)_2CO_3 \tag{21.11}$$
moist air   basic copper carbonate

The principal oxidation state of copper in its compounds is +2, and chief among these compounds is $CuSO_4$. It can be made by dissolving copper in sulfuric acid. Copper can also exist in the oxidation state +1, but some copper(I) compounds are unstable in aqueous solution (i.e., they disproportionate to Cu^{2+} and Cu).

$$2 Cu^+(aq) \longrightarrow Cu^{2+}(aq) + Cu(s) \quad E°_{cell} = 0.37 \text{ V} \tag{21.12}$$

Trace amounts of copper are essential to life, but larger amounts are toxic, especially to bacteria, algae, and fungi. This explains the use of copper sulfate and other copper-containing pesticides in the form of the basic acetate, carbonate, chloride, or hydroxide. The metallurgy of copper is described in Section 26-7.

ZINC. As pointed out in the marginal note on page 529, zinc is not always considered among the transition elements. It has no *d*-orbital vacancies, neither in the metal atom nor in its ion, Zn^{2+}.

Zn: [Ar]$3d^{10}4s^2$ Zn^{2+}: [Ar]$3d^{10}$

Because of its electron configuration we should expect zinc to resemble the metals of group IIA and it does somewhat. However, because 3*d* electrons are not too effective in shielding 4*s* electrons from the nucleus, these outer-shell electrons are more tightly held in Zn than in Ca. Zinc is not as active as the group IIA metals. Nevertheless, it is active enough to dissolve in a nonoxidizing acid (such as HCl) to displace $H_2(g)$. An interesting property of Zn which also sets it apart from the group IIA metals is the amphoterism of its oxide and hydroxide.

...resembles because the 3d shell is not v.effective in shielding so much so that... so even though zinc has this completed shell more than the IIA elements, there is virtually no difference

$$ZnO + H_2O \longrightarrow Zn\diagup^{O-H}_{\diagdown O-H} \quad \xrightarrow[\text{base}]{2 H^+} Zn^{2+} + 2 H_2O$$
$$\xrightarrow{2 OH^-} ZnO_2^{2-} + 2 H_2O \tag{21.13}$$

presence of acid   → Zn(OH)₂

The principal uses of zinc (about 90%) are in galvanizing steel and in the manufacture of brass and other alloys. (The protective action of zinc on iron was discussed in Section 19-8.)

21-3 Additional Aspects of the Chemistry of Chromium

The name "chromium" is derived from the Greek word *chroma*, meaning color.

Chromium, with the electron configuration $[Ar]3d^54s^1$, is in group VIB of the periodic table. It exists in several different oxidation states, but principally $+2$, $+3$, and $+6$. Table 21-3 illustrates the variability of oxidation states, the colors of ions, and the behavior of the oxides and hydroxy compounds.

OXIDATION-REDUCTION.   Chromium is a moderately active metal.

That chromium is a moderately active metal means that it is oxidized readily.

$$Cr^{2+}(aq) + 2\,e^- \longrightarrow Cr(s) \qquad E° = -0.91 \text{ V} \tag{21.14}$$

$$Cr(s) \longrightarrow$$
$$Cr^{2+}(aq) + 2\,e^-$$
$$E°_{ox} = -(-0.91)$$
$$= +0.91 \text{ V}$$

It dissolves in HCl(aq), producing $H_2(g)$, and in $H_2SO_4(aq)$, producing $SO_2(g)$.

Chromous ion is oxidized readily to chromic ion, even by air; Cr(II) is a very good reducing agent.

That Cr(II) is a good reducing agent means that $Cr^{2+}(aq)$ is easily oxidized.

$$Cr^{3+}(aq) + e^- \longrightarrow Cr^{2+}(aq) \qquad E° = -0.41 \text{ V} \tag{21.15}$$

$$Cr^{2+}(aq) \longrightarrow$$
$$Cr^{3+}(aq) + e^-$$
$$E°_{ox} = -(-0.41)$$
$$= +0.41 \text{ V}$$

Dichromate ion is a good oxidizing agent,

$$Cr_2O_7^{2-} + 14\,H^+ + 6\,e^- \longrightarrow 2\,Cr^{3+} + 7\,H_2O \qquad E° = +1.33 \text{ V} \tag{21.16}$$

but its strength is noticeably dependent on pH. Its most powerful action occurs in strongly acidic solutions. (Explain why, using Le Châtelier's principle.)

In basic solution CrO_4^{2-} is much weaker an oxidizing agent than is $Cr_2O_7^{2-}$ in acidic solution.

$$CrO_4^{2-} + 4\,H_2O + 3\,e^- \longrightarrow Cr(OH)_3(s) + 5\,OH^- \qquad E° = -0.13 \text{ V} \tag{21.17}$$

As a result, CrO_4^{2-} itself can be obtained by oxidizing CrO_2^- in basic solution with a good oxidizing agent like H_2O_2. (Because of its acidic properties, H_2O_2 occurs in basic solutions as the anion HO_2^-.)

$$2\,CrO_2^- + 3\,HO_2^- \longrightarrow 2\,CrO_4^{2-} + H_2O + OH^- \tag{21.18}$$

Some Examples of Oxidation-Reduction Reactions Employing Chromium Compounds

1. Removal of $O_2(g)$ impurity from other gases. [The $O_2(g)$ oxidizes $Cr^{2+}(aq)$ to $Cr^{3+}(aq)$.]

$$4\,Cr^{2+} + O_2(g) + 4\,H^+ \longrightarrow 4\,Cr^{3+} + 2\,H_2O \tag{21.19}$$

2. Production of $Cr_2O_3(s)$ by the decomposition of ammonium dichromate.

$$(NH_4)_2Cr_2O_7(s) \xrightarrow{\text{heat}} Cr_2O_3(s) + N_2(g) + 4\,H_2O \tag{21.20}$$

3. Quantitative analysis of iron. [A solution with unknown $[Fe^{2+}]$ is titrated with standard $K_2Cr_2O_7(aq)$.]

TABLE 21-3
Oxidation states of chromium

Oxidation state	Oxide	Hydroxy compound	Behavior	Ion	Name	Color
+2	CrO	$Cr(OH)_2$	basic	Cr^{2+}	chromium(II)	light blue
+3	Cr_2O_3	$Cr(OH)_3$	amphoteric	Cr^{3+} / CrO_2^-	chromium(III) / chromite	violet or green / green
+6	CrO_3	H_2CrO_4 / $H_2Cr_2O_7$	acidic	CrO_4^{2-} / $Cr_2O_7^{2-}$	chromate / dichromate	yellow / orange

$$6 \, Fe^{2+} + Cr_2O_7^{2-} + 14 \, H^+ \longrightarrow 6 \, Fe^{3+} + 2 \, Cr^{3+} + 7 \, H_2O \qquad (21.21)$$

4. Chrome tanning of leather. [The effective agent is basic chromic sulfate, $Cr(OH)SO_4$.]

$$Cr_2O_7^{2-} + 3 \, SO_2(g) + H_2O \longrightarrow 3 \, SO_4^{2-} + 2 \, Cr(OH)^{2+} \qquad (21.22)$$

ACID-BASE PROPERTIES.   The oxides and hydroxy compounds of chromium afford good examples with which to test the general scheme established in Section 20-4. Recall that a hydroxy compound is considered to result from the action of water on an element oxide. Then, if the hydroxy compound possesses an acidic character, it reacts in a basic solution, producing an anion. If the hydroxy compound has basic properties, it reacts in an acidic solution, producing a cationic species. Finally, if it undergoes reaction in both acidic and basic solution, the compound is amphoteric. The observed reactions for the oxides and hydroxy compound of chromium listed in Table 21-3 are

Cr(II): $CrO + H_2O \longrightarrow HO-\overset{2+}{Cr}-OH$ $\;\xrightarrow{\;H^+\;} Cr^{2+} + 2 \, H_2O$

$\xrightarrow{\;OH^-\;}$ no reaction

Cr(III): $Cr_2O_3 + H_2O \longrightarrow HO-Cr\big\langle{}^{OH}_{OH}$ $\;\xrightarrow{\;H^+\;} Cr^{3+} + 3 \, H_2O$

$\xrightarrow{\;OH^-\;} CrO_2^- + 2 \, H_2O$

Cr(VI): $CrO_3 + H_2O \longrightarrow$ $O\!=\!\overset{+6}{Cr}\big\langle{}^{OH}_{OH}$ with $O\!=$ $\;\xrightarrow{\;H^+\;}$ no Cr(VI) cation

$\xrightarrow{\;OH^-\;} CrO_4^{2-} + 2 \, H_2O$

In the case of Cr(VI), consider that first a hypothetical hydroxy compound, $Cr(OH)_6$, is formed, which then loses two molecules of water to become $CrO_2(OH)_2$. This is the same scheme for formulating oxyacids that was illustrated in Figure 20-14.

From these observations we conclude that CrO and $Cr(OH)_2$ are basic; Cr_2O_3 and $Cr(OH)_3$ are amphoteric; CrO_3 and H_2CrO_4 are acidic.

THE CHROMATE–DICHROMATE EQUILIBRIUM.   The red oxide CrO_3 dissolves in water to produce a strongly acidic solution. Although chromic acid H_2CrO_4 might be postulated as a product of the reaction, such a compound has never been isolated in the pure state. The observed reaction is

$$2 \, CrO_3(s) + H_2O \longrightarrow 2 \, H^+ + Cr_2O_7^{2-}$$

It is possible to crystallize a dichromate salt, such as $Na_2Cr_2O_7$ or $K_2Cr_2O_7$, from a water solution of CrO_3. If the solution is made basic, the color turns from red-orange to yellow. From basic solutions only chromate salts can be crystallized, for example, Na_2CrO_4 or K_2CrO_4. CrO_3 can be obtained by the action of concentrated sulfuric acid on a chromate or dichromate. This red solid is a powerful oxidizing agent and, in conjunction with concentrated sulfuric acid, is commonly used as a cleaning solution for laboratory glassware. Its principal mode of action is to oxidize grease.

Whether a solution contains Cr(VI) as dichromate or chromate ion is a function of pH, since the equilibrium between these ions depends on the concentration of H^+.

(handwritten margin notes: "that does not exist in acidic soln", "Soln is made acidic result", "more acidic, right side eqn increases, ∴ more Cr₂O₇²⁻")

$$2\,CrO_4^{2-} + 2\,H^+ \rightleftharpoons Cr_2O_7^{2-} + H_2O \tag{21.23}$$

$$K_c = \frac{[Cr_2O_7^{2-}]}{[CrO_4^{2-}]^2[H^+]^2} = 3.2 \times 10^{14} \tag{21.24}$$

$$\frac{[Cr_2O_7^{2-}]}{[CrO_4^{2-}]^2} = 3.2 \times 10^{14}[H^+]^2 \tag{21.25}$$

Equation (21.23) is actually the sum of two equilibrium expressions. The first is an acid-base reaction.

$$H^+ + CrO_4^{2-} \rightleftharpoons HCrO_4^-$$

The second involves the elimination of a water molecule from between two $HCrO_4^-$ ions (a dehydration reaction).

$$2\,HCrO_4^- \rightleftharpoons Cr_2O_7^{2-} + H_2O$$

FIGURE 21-1
Structures of CrO_4^{2-} and $Cr_2O_7^{2-}$.

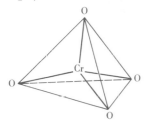

Quantitative calculations of the relative amounts of the two ions as a function of $[H^+]$ can be made with equation (21.25), but a qualitative prediction can be made as well by applying Le Châtelier's principle to equation (21.23). Clearly, $Cr_2O_7^{2-}$ is the predominant species in acidic solution and CrO_4^{2-} in basic solution. Control of the chromate–dichromate equilibrium through the control of pH is important in applications where dichromate solutions are used as oxidizing agents and chromate solutions to precipitate metal chromates.

The structures of the CrO_4^{2-} and $Cr_2O_7^{2-}$ ions are suggested by Figure 21-1. The Cr atom is at the center of a tetrahedron with O atoms at the corners. In $Cr_2O_7^{2-}$ two tetrahedra share an O atom. The Cr-O distance in the Cr—O—Cr link is somewhat greater than the other Cr-O distances.

CHROME PLATING. Chrome plating of steel is accomplished from an aqueous solution containing CrO_3 and H_2SO_4 in a mass ratio of about 100:1. The plating obtained is thin and porous and tends to develop cracks. Usual practice is first to plate the steel with copper or nickel, which is the true protective coating. Then chromium is plated over this for decorative purposes. The technical art of chrome plating is well understood, but the mechanism of the electrodeposition has not been established. The efficiency of chrome plating is limited by the fact that reduction of Cr(VI) to Cr(0) produces only $\frac{1}{6}$ mol Cr per Faraday: Large quantities of electric energy are required for chrome plating relative to other types of metal plating.

21-4 Additional Aspects of the Chemistry of the Iron Triad Elements

For a very long time, in its pure form and especially in various alloys, iron has been the most widely used metal. It is understandable, therefore, that a great deal of practical and scientific knowledge exists about the occurrence, production, refining, properties, and uses of iron. These topics are considered further in Chapter 26. Here we wish simply to consider some aspects of the physical and chemical behavior of iron, cobalt, and nickel.

As expected, there are some similarities between Fe, Co, and Ni and other elements in the first transition series, such as Ti, V, Cr, and Mn. Various combinations of these metals form alloys with one another, and all of the simple metal ions have orbitals available for complex ion formation. But there are also differences between the Fe-Co-Ni group and the elements preceding them in the first transition series. For example, Fe, Co, and Ni do not form stable oxyanions like VO_3^-, CrO_4^{2-}, and MnO_4^-. Also, they do not exhibit the same variability of oxidation state. d electrons are not as much involved in bonding in elements in which the d level is more than half filled (the $3d$ level is half filled with Mn).

(handwritten margin note: "still vary in oxy states, but not as much other transition elements")

FERROMAGNETISM. One unique property possessed only by iron, cobalt, and nickel among the common chemical elements is that of **ferromagnetism**. Although the ions Fe^{2+}, Co^{2+}, and Ni^{2+} all have unpaired electrons, the property of ferromagnetism cannot be accounted for by the paramagnetism of these species alone. In the solid

(handwritten note: "attraction to magnetic field by unpaired e⁻")

much too magnetic just to be explained by paramagnetism

state the metal ions are thought to be grouped together into small regions containing rather large numbers of metal ions. These regions are called **domains.** Instead of the individual magnetic moments of the ions within a domain being randomly oriented, their effects are all directed in the same way. In an unmagnetized piece of iron the domains are oriented in several directions and their magnetic effects cancel. When the metal is placed into a magnetic field, however, the domains are lined up and a strong resultant magnetic effect is produced. This ordering of the domains persists when the object is removed from the magnetic field, and thus permanent magnetism results (see Figure 21-2).

The key to ferromagnetism involves two basic factors: that the species involved have unpaired electrons (a property possessed by many species) and that interionic distances be of just the right magnitude to make possible the ordering of ions into domains. If atoms are too large, interactions among them are too weak to produce this ordering. With small atoms the tendency is for atoms to pair and their magnetic moments to cancel. This critical factor of atomic size is just met in Fe, Co, and Ni. However, it is possible to prepare alloys of metals other than these three in which this condition is met (e.g., Al-Cu-Mn, Ag-Al-Mn, and Bi-Mn). Also, certain rare earth elements, for example, gadolinium (Gd) and dysprosium (Dy), are ferromagnetic at low temperatures.

METAL CARBONYLS. The transition metals, with few exceptions, form compounds with carbon monoxide. These compounds are called metal carbonyls. In the simple metal carbonyls listed in Table 21-4

1. Each CO molecule contributes an electron pair to an empty orbital of the metal atom.
2. All electrons are paired (most metal carbonyls are diamagnetic).
3. The metal atom acquires the electron configuration of the noble gas, Kr.

The structures of the simple carbonyls in Figure 21-3 are those that one might predict from VSEPR theory (i.e., based on a number of electron pairs equal to the number of CO molecules).

The procedure outlined here for predicting the formula of a simple carbonyl will not work for transition elements with odd atomic numbers. For example, the species $Mn(CO)_5$ would have 35 electrons; this would make it an odd-electron (paramagnetic) species. The compound formed instead is the *binuclear* carbonyl shown in Figure 21-4, $Mn_2(CO)_{10}$. In this carbonyl all electrons are paired.

FIGURE 21-2
The phenomenon of ferromagnetism.

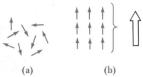

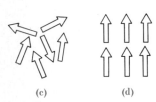

(a)   (b)

(c)   (d)

The existence of magnetic domains has been verified by observing, through the microscope, the behavior of magnetic powders suspended in a liquid film on a polished metal surface. The alignment of the domains referred to here probably actually involves the growth of domains with favorable orientations at the expense of those with unfavorable orientations when an object to be magnetized is placed in a magnetic field.

(a) In ordinary paramagnetism the magnetic moments of the atoms or ions are randomly distributed.
(b) In a ferromagnetic material the magnetic moments are aligned into domains.
(c) In an unmagnetized piece of the material the domains are randomly oriented.
(d) In a magnetic field the domains are oriented in the direction of the field and the material becomes magnetized.

TABLE 21-4
Three metal carbonyls

| | Number of e | | |
	From metal	From CO	Total
$Cr(CO)_6$	24	12	36
$Fe(CO)_5$	26	10	36
$Ni(CO)_4$	28	8	36

easily predicted when even # electrons (diamagnetic)

FIGURE 21-3
Structures of some simple carbonyls.

Metal carbonyls are produced in several ways. Nickel metal combines with CO(g) at ordinary temperatures and pressures in a reversible reaction.

$$Ni(s) + 4\,CO(g) \rightleftharpoons Ni(CO)_4(l) \qquad (21.26)$$

With iron it is necessary to use higher temperatures (200°C) and CO pressures (100 atm).

$$Fe(s) + 5\,CO(g) \longrightarrow Fe(CO)_5(g) \qquad (21.27)$$

In other cases the carbonyl is obtained by reducing a metal compound in the presence of CO(g).

OXIDATION STATES. Variability of oxidation states is a characteristic of transition elements. It is encountered within the iron triad elements, even if not to the same degree as with certain other transition elements like vanadium and manganese. The oxidation state +2 is commonly encountered with all three elements.

$$Fe^{2+}:\ [Ar]3d^6 \qquad Co^{2+}:\ [Ar]3d^7 \qquad Ni^{2+}:\ [Ar]3d^8$$

For cobalt and nickel the oxidation state +2 is the most stable, but for iron the most stable is the oxidation state +3. *Iron however is more stable @ +3*

tendency is to simply lose 4/s shell & none of d-electrons

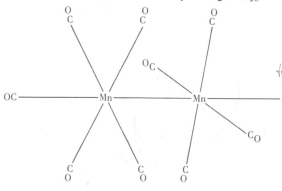

	3d
Fe²⁺: [Ar]	↓↑ ↓ ↓ ↓ ↓

	3d	
Fe³⁺: [Ar]	↓ ↓ ↓ ↓ ↓	← ½-filled shell ($3d^5$)

An electron configuration in which the d subshell is half-filled, with all electrons unpaired, has a special stability. This fact suggests that Fe(II) can be oxidized to Fe(III) without great difficulty; for example, in the presence of oxygen at 1 atm pressure and with $[H^+] = 1\ M$,

∴ favored rxn

$$4\,Fe^{2+} + O_2(g) + 4\,H^+ \longrightarrow 4\,Fe^{3+} + 2\,H_2O$$
$$E^\circ_{cell} = 0.44\ V \qquad (21.28)$$

But even at lower partial pressures of oxygen, as in the atmosphere, and in less acidic media, reaction (21.28) may still be spontaneous.

For Co(II) and Ni(II) the loss of an additional electron does not lead to an electron configuration with half-filled d orbitals.

	3d
Co³⁺: [Ar]	↓↑ ↓ ↓ ↓ ↓

	3d
Ni³⁺: [Ar]	↓↑ ↓↑ ↓ ↓ ↓

Carbon monoxide poisoning results from an action similar to carbonyl formation. CO molecules coordinate with Fe atoms in hemoglobin, displacing the oxygen molecules normally carried by the hemoglobin. The metal carbonyls themselves are also very poisonous.

FIGURE 21-4
Structure of a binuclear carbonyl: $Mn_2(CO)_{10}$.

As a result the conversion of Co(II) to Co(III) and Ni(II) to Ni(III) is accomplished not nearly so easily as in the case of iron. Consider, for example, this standard electrode potential for cobalt:

$$Co^{3+}(aq) + e^- \longrightarrow Co^{2+}(aq) \qquad E° = 1.82 \text{ V} \qquad (21.29)$$

The *reduction* of Co^{3+} to Co^{2+} occurs readily; Co(III) compounds tend to be very good oxidizing agents. For reasons explained in Section 22-12, the +3 oxidation state of cobalt can be stabilized, however, if the Co^{3+} is part of a complex ion. The oxidation state +3 is also difficult to achieve with nickel and is not commonly encountered. Nickel compounds in higher oxidation states do find use, however, as electrode materials in the nickel–cadmium cell (see Exercise 35, Chapter 19).

Because of the stability attributed to the electron configuration of Fe^{3+}, iron does not commonly occur in higher oxidation states. The oxidation state +6 is attainable, however, under very strong oxidizing conditions.

$$2 \, Fe(OH)_3(s) + 3 \, OCl^- + 4 \, OH^- \longrightarrow 2 \, FeO_4^{2-} + 3 \, Cl^- + 5 \, H_2O \qquad (21.30)$$

The ion FeO_4^{2-} is called **ferrate ion,** and a few ferrate salts have been prepared. For example, barium ferrate, $BaFeO_4$, is a purple colored solid. As might be expected the ferrates are unstable and are powerful oxidizing agents.

SOME REACTIONS OF THE IRON TRIAD ELEMENTS. The reactions of the iron triad elements are many and varied. The metals are all more active than hydrogen and liberate $H_2(g)$ from an acidic solution.

$$Ni(s) + 2 \, H^+ + 2 \, Cl^- \longrightarrow Ni^{2+} + 2 \, Cl^- + H_2(g) \qquad (21.31)$$

Hydrated, colored ions are characteristic of the iron triad elements. For example, Co^{2+} and Ni^{2+} are red and green, respectively. In aqueous solution Fe^{2+} is pale green and Fe^{3+} is colorless. Generally, solutions of Fe^{3+} are yellow to brown in color, but this color is probably due to the presence of the species $FeOH^{2+}$. This ion forms by the hydrolysis of Fe^{3+}.

$$Fe^{3+} + 2 \, H_2O \longrightarrow FeOH^{2+} + H_3O^+ \qquad (21.32)$$

Salts of the iron triad elements usually crystallize from solution as hydrates. The hydrate of Co(II) chloride has an interesting application. When exposed to atmospheric moisture, depending on the partial pressure of $H_2O(g)$, the hydrate assumes different colors. In dry air, water of hydration is lost and the solid acquires a blue color; as the humidity increases, the solid undergoes a gradual color change to pink. This reaction has been used as an inexpensive, if somewhat crude, method of moisture determination.

$$CoCl_2 \cdot 6 \, H_2O(s) \rightleftharpoons CoCl_2 \cdot 2 \, H_2O(s) + 4 \, H_2O(g) \qquad (21.33)$$
$$\text{(pink)} \qquad\qquad \text{(blue)}$$

The basis for writing formulas and systematic names of complex ions such as these is taken up in Chapter 22.

The **hexacyanoferrates** are compounds containing the complex ions $[Fe(CN)_6]^{4-}$ and $[Fe(CN)_6]^{3-}$. Their systematic names are hexacyanoferrate(II) and hexacyanoferrate(III), respectively, but they are also commonly called **ferrocyanide** and **ferricyanide**. Iron(III) in aqueous solution yields a dark blue precipitate, **Prussian blue,** when treated with potassium hexacyanoferrate(II). The exact structure of Prussian blue is not known, but it is probably quite complex. It has the empirical formula $Fe_7C_{18}N_{18} \cdot 10 \, H_2O$. A simplified representation of this reaction is

$$4 \, Fe^{3+} + 3[Fe(CN)_6]^{4-} \longrightarrow Fe_4[Fe(CN)_6]_3 \qquad (21.34)$$

Iron(II) compounds yield a blue precipitate, **Turnbull's blue,** when treated with potassium hexacyanoferrate(III). The reaction appears to proceed in two stages. The

TABLE 21-5
Some qualitative tests for iron(II) and iron(III)

Reagent	Iron(II)	Iron(III)
NaOH(aq)	green precipitate	red-brown precipitate
$K_4[Fe(CN)_6]$	white precipitate, turning blue rapidly	Prussian blue precipitate
$K_3[Fe(CN)_6]$	Turnbull's blue precipitate	red-brown coloration (no precipitate)
KSCN	no coloration	deep-red coloration

first is an oxidation-reduction reaction in which iron(II) is oxidized to iron(III) and hexacyanoferrate(III) is reduced to hexacyanoferrate(II).

$$Fe^{2+} + [Fe(CN)_6]^{3-} \longrightarrow Fe^{3+} + [Fe(CN)_6]^{4-} \tag{21.35}$$

This is followed by reaction (21.34). Turnbull's blue is a lighter shade than Prussian blue and this may be caused by the admixture of some $K_2\{Fe[Fe(CN)_6]\}$, which is white.

The qualitative tests for iron just described, together with two others, are summarized in Table 21-5. The dark-red coloration resulting from the reaction of thiocyanate ion, SCN^-, with Fe^{3+} is the basis of an extremely sensitive test for iron. The composition of the product appears to be $[Fe(H_2O)_5NCS]^{2+}$.

A very distinctive reaction of Ni^{2+} that can be used both for its qualitative detection and its quantitative determination is the formation of a neutral complex with dimethylglyoxime, which precipitates from solution as a brilliant scarlet precipitate. In addition to the coordination bonds between N atoms and the Ni^{2+} ion, this complex features hydrogen bonds.

$$\tag{21.36}$$

21-5 Qualitative Analysis Revisited

In this return to the subject of qualitative analysis, we will attempt to do two things. We will consider

1. How individual ions are separated and identified following precipitation of a cation group.
2. How the procedures of qualitative analysis can be used to illustrate the descriptive chemistry of the elements.

The aspects of qualitative analysis that have already been considered are (a) the solubility relationships used in cation group separations (Section 16-5), (b) the use of H_2S as a precipitating agent (Section 18-6), and (c) the importance of buffering action in qualitative analysis procedures (Section 18-5). For the present discussion we choose the group of eight cations that constitute the ammonium sulfide group (cation group 3 of Figure 16-3). The scheme is described in the numbered items listed below and through the flow chart in Figure 21-5. The notes provide additional detail on a few interesting points.

1. A solution containing the ammonium sulfide group cations, together with cations of groups 4 and 5 (see Figure 16-3), is treated with $(NH_4)_2S$ in an NH_3–NH_4Cl buffer solution. Under these conditions Fe^{3+}, Cr^{3+}, and Al^{3+} precipitate as hydroxides, and Fe^{2+}, Co^{2+}, Ni^{2+}, Mn^{2+}, and Zn^{2+} as sulfides. All the cations of the ammonium sulfide group are found in the mixed hydroxide–sulfide precipitate. The solution contains cations of groups 4 and 5 (see Figure 16-3).

Note: Since the sulfides of the ammonium sulfide group are more soluble than most, a basic solution is needed to obtain a sufficiently high $[S^{2-}]$ for their precipitation

FIGURE 21-5
Qualitative analysis of the ammonium sulfide group.

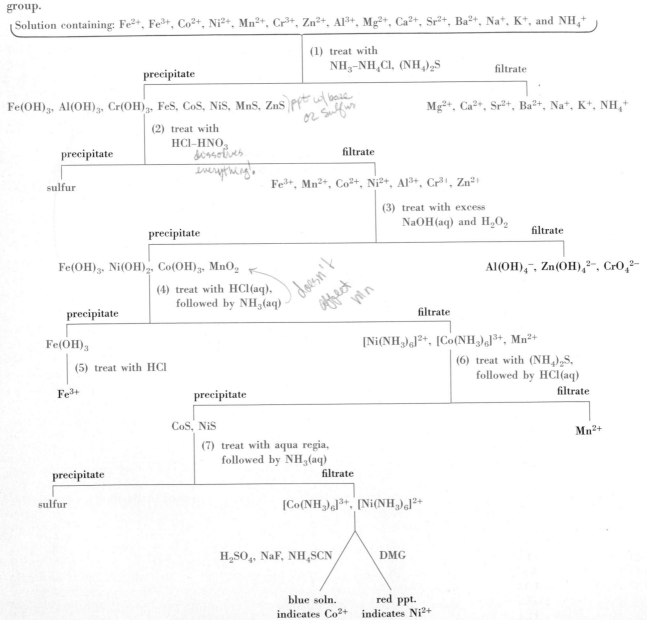

(recall equation 18.31). But the pH of the solution must not be too high or Mg^{2+} will precipitate as $Mg(OH)_2$. The use of a NH_3–NH_4Cl buffer solution to regulate the precipitation of $Mg(OH)_2$ was the subject of Example 18-11.

2. The precipitate from step 1 is treated with HCl–HNO_3 (aqua regia). Any precipitated sulfur is filtered off.

Note: All components of the precipitate in step 1 are readily soluble in HCl(aq) except NiS and CoS. These require aqua regia for their dissolution (recall equation 20.50).

3. The solution from step 2 is treated with an excess of NaOH. The hydroxides of Al, Zn, and Cr, which are *amphoteric,* dissolve. The hydroxides of Fe, Co, Ni, and Mn are basic and precipitate from the solution. The mixture is next treated with H_2O_2, which oxidizes CrO_2^- to CrO_4^{2-} in the solution and Co(II) to Co(III) and Mn(II) to Mn(IV) in the precipitate. The solution can be subjected to further separations and tests for Al, Zn, and Cr. A yellow color in the solution at this point (CrO_4^{2-}) indicates the presence of chromium. The precipitate consists of $Fe(OH)_3$, $Ni(OH)_2$, $Co(OH)_3$, and MnO_2.

Note: The oxidation of CrO_2^- to CrO_4^{2-} by H_2O_2 in basic solution was described in equation (21.18).

4. The precipitate from step 3 may be analyzed in a number of ways. One simple method involves dissolving the precipitate in HCl(aq), followed by treatment with NH_3(aq). This causes $Fe(OH)_3$ to reprecipitate and Co^{2+} and Ni^{2+} to be converted to the ammine complexes, $[Co(NH_3)_6]^{3+}$ and $[Ni(NH_3)_6]^{2+}$. Manganese also appears in the solution as Mn^{2+}.

Note: MnO_2(s) is reduced to Mn^{2+}(aq) by HCl(aq). In an alkaline solution Co(II) is oxidized to Co(III) by O_2(g), and the presence of NH_3 stabilizes the Co(III) through the formation of the ammine complex ion. (The stabilization of oxidation states as a result of complex ion formation is discussed in Chapter 22.)

5. The $Fe(OH)_3$ precipitate is filtered off and dissolved in HCl. The presence of Fe^{3+} in the resulting solution is confirmed by the qualitative tests previously outlined in Table 21-5.

6. The filtrate containing $[Co(NH_3)_6]^{3+}$, $[Ni(NH_3)_6]^{2+}$, and Mn^{2+} is treated with $(NH_4)_2S$ to reprecipitate the sulfides. Because MnS is considerably more soluble than CoS and NiS, it can be dissolved in HCl(aq), leaving a residue of CoS and NiS. Tests for Mn^{2+} can be performed on the filtrate.

7. The sulfide precipitate is dissolved in aqua regia, followed by treatment with NH_3(aq). Thus, the species $[Co(NH_3)_6]^{3+}$ and $[Ni(NH_3)_6]^{2+}$ are obtained once again. A portion of the solution containing these ions is treated with dimethylglyoxime (DMG). Formation of a scarlet precipitate indicates the presence of nickel (see structure 21.36). A test for cobalt is performed on a second portion of the solution, which is treated with H_2SO_4, NaF, and NH_4SCN. The presence of cobalt is disclosed by the formation of the blue complex ion $[Co(SCN)_4]^{2-}$.

Note: Acidification with H_2SO_4 causes Co(III) to decompose back to Co(II). The presence of NaF is required to complex Fe^{3+} as $[FeF_6]^{3-}$, so that any traces of Fe^{3+} present in the solution will not react with SCN^- and interfere with the test for cobalt.

Summary

More than half of the elements fit the classification of transition element. This chapter is concerned mostly with the elements of the first transition series, from Sc to Zn. The transition elements are all metals; those of the first transition series (except copper) are more active than hydrogen. Most of the transition elements can exist in several different oxidation states in their compounds, but not all of these oxidation states are equally stable. The transition elements also form a variety of complex ions (see Chapter 22).

f electrons are rather ineffective in shielding outer-shell electrons from the attractive force of the nucleus. This accounts for the fact that atoms of the same group in the *second* and *third* transition series are nearly identical in size. These elements are more similar to each other than to the group member in the first transition series. Another type of resemblance is that which occurs among certain adjacent members of the same transition series. One such group considered in this chapter is the iron triad—Fe, Co, Ni. A particular property that these elements share is ferromagnetism.

The possibility of different oxidation states for the transition metals means that oxidation and reduction are involved in many of their reactions. Some of the most important oxidizing agents used in the laboratory are transition metal compounds, notably dichromates and permanganates. Dichromate ion participates in acid-base and hydration-dehydration equilibria with chromate ion. These equilibria favor the formation of dichromate ion at low pH and chromate ion at higher pH values. Chromate ion is more important as a pre-cipitating agent than as an oxidizing agent. Most of the oxides and hydroxides of the transition elements are basic. This is typical of fairly active metals. However, certain oxides and hydroxides also have acidic properties; that is, they are amphoteric. Two cases considered in this chapter are the oxides and hydroxides of Zn and Cr(III). Several of the first series transition metals fall within the same group of cations in the qualitative analysis scheme, the ammonium sulfide group. Many of the acid-base, oxidation-reduction, and precipitation reactions of this chapter are essential to the experimental procedures for separating and identifying these cations.

The transition metals are the most widely used of the elements for the manufacture of structural materials. Their desirable properties range from the ready availability, low cost, and structural strength of iron to the excellent electrical conductivity of copper. Several of the transition metals—V, Cr, Mn, Co, Ni, Mo, W—are added to iron in steelmaking.

Learning Objectives

As a result of studying Chapter 21, you should be able to

1. Cite ways in which the transition elements differ from representative elements.

2. Describe the lanthanoid contraction and explain how it affects the properties of the transition elements.

3. Describe sources and uses of the elements in the first transition series and name some of their important compounds.

4. State which are the most common oxidation states of the first transition series elements, which of these are stable and which are unstable.

5. Write equations for some important oxidation-reduction reactions involving permanganate ion and some, involving dichromate ion.

6. Discuss the chromate–dichromate ion equilibrium and the effect of pH on the concentrations of these ions in aqueous solution.

7. Write chemical equations to illustrate the amphoteric properties of the oxides and hydroxides of Zn and Cr(III).

8. Describe the property of ferromagnetism and the features of atomic structure that lead to it.

9. Describe the formation of metal carbonyls.

10. Describe qualitative tests for the detection of Fe^{2+} and Fe^{3+}.

11. Discuss some of the separations and tests for ions in the ammonium sulfide group of the qualitative analysis scheme.

12. Draw conclusions about the presence or absence of ions in a qualitative analysis "unknown," based on the results of certain laboratory tests.

Some New Terms

Ferromagnetism is a property that permits materials (notably Fe, Co, and Ni) to be made into permanent magnets. The magnetic moments of individual atoms are aligned into domains. In the presence of a magnetic field, these domains orient themselves to produce a permanent magnetic moment.

Iron triad is a term used for the group of elements Fe, Co, Ni, to emphasize similarities in their physical and chemical properties.

The **lanthanoid contraction** results from the ineffectiveness of *f* electrons in shielding outer-shell electrons from the nucleus. Atomic sizes decrease in a series of elements in which an *f* subshell fills (an inner transition series). Also, atomic sizes for members of the third transition series (which follows the lanthanoid series) are about the same as for corresponding members of the second transition series.

Metal carbonyls are compounds formed between certain

metal atoms and carbon monoxide molecules, for example, nickel carbonyl, $Ni(CO)_4$.

Exercises

Properties of the transition elements

1. What is the characteristic of their electron configurations that most distinguishes the transition elements from the representative elements?

2. How do the transition elements compare with representative metals (e.g., group IIA) with respect to oxidation states, formation of complexes, colors of compounds and magnetic properties?

3. Why do the atomic radii vary so much more for two representative elements that differ by one unit in atomic number than they do for two transition elements that differ by one unit?

4. For the series of elements listed in Table 21-1, explain why zinc has the highest values of the first and third ionization energies and copper has the highest value for the second.

5. The melting point of zinc is much lower than those of the transition elements preceding it in the periodic table (see Table 21-1). Why should this be so?

6. What significance is attached to the Roman numerals of the "B group" elements in the periodic table?

7. Explain briefly the meaning of the following terms: **(a)** "d block" elements; **(b)** iron triad; **(c)** lanthanoid contraction; **(d)** ferromagnetism; **(e)** metal carbonyl.

Chemistry of chromium and chromium compounds

8. Use information from equations (21.14) and (21.15) to determine $E°$ for the half-reaction

$$Cr^{3+}(aq) + 3\,e^- \longrightarrow Cr(s)$$

9. Write balanced equations to represent the reaction of chromium metal with **(a)** $HCl(aq)$; **(b)** $H_2SO_4(aq)$.

10. Suggest chemical reactions that might be used to obtain the following compounds from Na_2CrO_4: **(a)** $Na_2Cr_2O_7$; **(b)** Cr_2O_3; **(c)** $CrCl_3$; **(d)** $NaCrO_2$.

11. The principal chemistry of dichromate ion involves oxidation-reduction reactions, and that of chromate ion, precipitation reactions. Explain why this is so.

12. How long would an electric current of 5.0 A have to pass through a chrome-plating bath to produce a deposit 0.0010 mm thick on an object with a surface area of 35.5 cm²? (The density of Cr is 7.14 g/cm³.)

Chromate–dichromate equilibrium

13. What are the relative concentrations of $Cr_2O_7^{2-}$ and CrO_4^{2-} in a solution with a pH of **(a)** 5.0; **(b)** 9.3?

14. The ionization constant for the species $HCrO_4^-$ is listed as $K_a = 3.2 \times 10^{-7}$. Calculate a value of K for the hydration-dehydration equilibrium.

$$2\,HCrO_4^- \rightleftharpoons Cr_2O_7^{2-} + H_2O \qquad K = ?$$

15. Show that in a solution that is 0.10 M in Ba^{2+}, 0.10 M in Sr^{2+}, 0.10 M in Ca^{2+}, 1.0 M in $HC_2H_3O_2$, 1.0 M in $NH_4C_2H_3O_2$, and 0.001 M in $Cr_2O_7^{2-}$, $BaCrO_4$ will precipitate but not $SrCrO_4$ or $CaCrO_4$. Use data from this chapter and from Table 16-1, as necessary.

The iron triad

16. Explain why the iron triad elements resemble each other so strongly.

17. How does the property of ferromagnetism differ from ordinary paramagnetism? Why is this property so limited in its occurrence among the elements?

18. Which of the iron triad ions, M^{2+}, would you expect to be oxidized to M^{3+} by $O_2(g)$ in acidic solution?

$$4\,M^{2+} + O_2(g) + 4\,H^+ \longrightarrow 4\,M^{3+} + 2\,H_2O$$

19. Discuss the chemical reactions involved in the formation of the blue precipitate in Figure 19-14 representing the corrosion of iron.

Carbonyls

20. Use methods outlined in the chapter to predict formulas and structures of the carbonyls of **(a)** molybdenum; **(b)** osmium; **(c)** rhenium.

21. Iron and nickel carbonyls are liquids at room temperature, but that of cobalt is a solid. Why should this be so?

22. The compound $Na[V(CO)_6]$ has been reported. Discuss the probable nature of chemical bonding in this compound.

Oxidation-reduction

23. Write half-equations to represent **(a)** dichromate ion acting as an oxidizing agent in acidic solution; **(b)** the action of $Cr^{2+}(aq)$ as a reducing agent; **(c)** the oxidation of $Fe(OH)_3(s)$ to FeO_4^{2-} in basic solution; **(d)** the reduction of $[Ag(CN)_2]^-$ to Ag metal.

24. Balance the following oxidation-reduction equations.
   (a) $Fe_2S_3(s) + H_2O + O_2(g) \longrightarrow Fe(OH)_3(s) + S(s)$
   (b) $Ag(s) + CN^- + O_2(g) + H_2O \longrightarrow$
   $$[Ag(CN)_2]^- + OH^-$$
   (c) $Mn^{2+} + S_2O_8^{2-} + H_2O \longrightarrow$
   $$MnO_4^- + SO_4^{2-} + H^+$$

25. When $Cr_2O_7^{2-}$ is reduced with Zn in the presence of $HCl(aq)$ the following sequence of color changes is noted:

orange $\rightarrow$ green $\rightarrow$ blue $\rightarrow$ green. Suggest a sequence of chemical reactions corresponding to these color changes. (*Hint:* The final color change from blue to green occurs after the reduction reaction is completed.)

26. The electrode potential diagram (21.4) does not include a value of $E°$ for the reduction $MnO_4^- \rightarrow Mn^{2+}$ in acidic solution. Establish this value using other data in the diagram. Compare your result with the value listed in Table 19-2.

Qualitative analysis

27. What single reagent solution could be employed to effect the separation of the following pairs of solids?
    (a) $NaOH$ and $Fe(OH)_3$
    (b) $Ni(OH)_2$ and $Fe(OH)_3$
    (c) Cr_2O_3 and $Fe(OH)_3$
    (d) MnS and PbS

28. With respect to the qualitative analysis scheme for the ammonium sulfide group outlined in the text, write equations to represent   (a) dissolving FeS in $HCl(aq)$; (b) dissolving CoS in aqua regia;   (c) the action of excess $NaOH(aq)$ and H_2O_2 on $Cr^{3+}(aq)$;   (d) the action of $NH_3(aq)$ on a mixture of $Fe(OH)_3(s)$ and $Ni(OH)_2(s)$;   (e) a qualitative test for $Fe^{2+}(aq)$;   (f) the action of $HCl(aq)$ on $MnO_2(s)$.

29. Describe what might happen in the qualitative analysis of the ammonium sulfide group if one did the following, in error.

(a) failed to include NH_4Cl in the reagent used to treat the original solution in the flow sheet of Figure 21-5.
(b) failed to add H_2O_2 to the reagent used in step 3
(c) used $NaOH(aq)$ in step 4 rather than $NH_3(aq)$
(d) used aqua regia in step 6 instead of $HCl(aq)$
(e) neglected to use NaF in the test for cobalt in step 7

30. A particular stainless steel sample is known to contain some combination of Ni, Cr, and Mn alloyed with Fe. Outline a simplified qualitative analysis scheme that would enable you to determine which of these metals are present.

Quantitative analysis

31. Nickel can be determined by the precipitation of nickel dimethylglyoximate.
    (a) What is the formula of this compound (see structure 21.36)?
    (b) A 1.502-g sample of steel yields 0.259 g of nickel dimethylglyoximate. What is the percent Ni in the steel?

32. A 0.589-g sample of pyrolusite ore (impure MnO_2) is treated with 1.651 g of oxalic acid ($H_2C_2O_4 \cdot 2H_2O$) in an acidic medium. Following this reaction the excess oxalic acid is titrated with 0.1000 M $KMnO_4$, 30.06 ml being required. What is the percent MnO_2 in the ore?

$$H_2C_2O_4 + MnO_2 + 2\,H^+ \longrightarrow Mn^{2+} + 2\,H_2O + 2\,CO_2$$

$$5\,H_2C_2O_4 + 2\,MnO_4^- + 6\,H^+ \longrightarrow 2\,Mn^{2+} + 8\,H_2O + 10\,CO_2$$

Additional Exercises

1. Complete the oxidation-reduction reactions outlined for vanadium species in expression (21.1). (Assume acidic solution.)

2. Describe a simple chemical test to distinguish between $Fe^{2+}(aq)$ and $Fe^{3+}(aq)$.

3. Suggest a reaction or a series of reactions, using common chemicals, by which each of the following syntheses can be performed: (a) $Fe(OH)_3(s)$ from $FeS(s)$;   (b) $BaCrO_4(s)$ from $BaCO_3(s)$ and $K_2Cr_2O_7(aq)$;   (c) $CrCl_3(s)$ from $(NH_4)_2Cr_2O_7(s)$;   (d) $MnCO_3(s)$ from $MnO_2(s)$.

4. When *yellow* $BaCrO_4$ is dissolved in $HCl(aq)$, a *green* solution is obtained. Write a chemical equation to account for this color change.

5. In the text three conclusions were reached about certain ions of manganese by using the electrode potential diagrams of (21.4) and (21.5). Show that the following statements are true as well.
    (a) MnO_4^- in acidic solution oxidizes Mn^{2+} to $MnO_2(s)$; in basic solution MnO_4^- oxidizes $MnO_2(s)$ to MnO_4^{2-}.

(b) MnO_4^- will slowly liberate $O_2(g)$ from water, either in acidic or basic solution.

*6. Show that the corrosion reaction in which copper is converted to the basic carbonate (21.11) can be thought of as a combination of oxidation-reduction, acid-base, and precipitation reactions.

*7. In an atmosphere with industrial smog [mainly $SO_2(g)$] copper corrodes to a basic sulfate, $Cu_2(OH)_2SO_4$. Propose a series of chemical reactions to describe this corrosion.

*8. Suppose a solution is made 0.50 M in OH^- and 0.10 M in MnO_4^- in the presence of $MnO_2(s)$. Show that a significant concentration of manganate ion, MnO_4^{2-} exists in the solution.

*9. Use data from expressions (21.4) and (21.5), together with equations from elsewhere in the text, to estimate K_{sp} for $Mn(OH)_2$.

*10. A steel sample is to be analyzed for Cr and Mn, simultaneously. By suitable treatment the Cr is oxidized to $Cr_2O_7^{2-}$ and the Mn to MnO_4^-. A 10.000-g sample of steel is used to produce 250.0 ml of a solution containing $Cr_2O_7^{2-}$ and

MnO_4^-. A 10.00-ml portion (aliquot) of this solution is added to a $BaCl_2$ solution, and by proper adjustment of the acidity, the chromium is completely precipitated as $BaCrO_4$; 0.0549 g is obtained. A second 10.00-ml aliquot of this solu-

tion requires exactly 15.95 ml of 0.0750 M standard Fe^{2+} solution for its titration (in acid solution). Calculate the percent Mn and percent Cr in the steel sample. (*Hint:* Both MnO_4^- and $Cr_2O_7^{2-}$ are reduced by Fe^{2+}.)

Self-Test Questions

For questions 1 through 5 select the single item that best completes each statement.

1. A property generally expected of the transition elements is **(a)** low melting points; **(b)** high ionization energies; **(c)** variable oxidation states; **(d)** positive standard electrode (reduction) potentials.

2. Only one of the following ions is diamagnetic. That ion is **(a)** Cr^{2+}; **(b)** Fe^{3+}; **(c)** Cu^{2+}; **(d)** Zn^{2+}.

3. Of the following elements the one that is *not* expected to display an oxidation state of +6 in any of its compounds is **(a)** Ti; **(b)** Cr; **(c)** Mn; **(d)** Fe.

4. One might expect $Cl_2(g)$ to be produced if an HCl(aq) solution is heated strongly in the presence of **(a)** Zn; **(b)** MnO_2; **(c)** Cr^{3+}; **(d)** Fe^{2+}.

5. The disproportionation (decomposition) of manganate ion (MnO_4^{2-}) to permanganate ion (MnO_4^-) and manganese dioxide (MnO_2) is expected to occur most readily in a solution that is **(a)** acidic; **(b)** basic; **(c)** neutral; **(d)** neutral or basic.

6. A handbook lists the following first ionization energies: K, 4.341 eV; Ca, 6.113 eV; Zn, 9.394 eV; and Rb, 4.177 eV.

Explain the relationship among these four values. That is, why is I_1 greatest for Zn, lowest for Rb, and so on?

7. What are the products obtained when Mg^{2+}(aq) and Cr^{3+}(aq) are each treated with a limited amount of NaOH(aq)? With an excess of NaOH(aq)? Why are the results with excess NaOH(aq) different in these two cases?

8. When a soluble lead compound is added to a solution containing primarily *orange* dichromate ion, *yellow* lead chromate precipitates. Describe the equilibria involved.

9. Which of these reagents—H_2O, NaOH(aq), HCl(aq)— can be used to dissolve the following compounds. **(a)** $ZnSO_4$; **(b)** $Ni(OH)_2$; **(c)** $AgNO_3$; **(d)** FeS?

10. A solution is believed to contain one or more of the following ions: Cr^{3+}, Zn^{2+}, Fe^{3+}, Ni^{2+}. When the solution is treated with NaOH(aq) and H_2O_2, a precipitate is formed. The solution separated from this precipitate is colorless. The precipitate is dissolved in HCl(aq) and the resulting solution is treated with NH_3(aq). No precipitation occurs. Based solely on these observations, what conclusions can you draw about the ions in the solution? That is, are any proved to be present? to be absent? Are further tests needed?

Complex Ions and Coordination Compounds

Cobalt forms a simple ionic chloride, $CoCl_3$, in which three electrons are transferred from a Co atom to Cl atoms. But in the presence of $NH_3(aq)$ this cobalt(III) chloride can form a series of distinct compounds with such formulas as

$CoCl_3 \cdot 6\,NH_3$ $CoCl_3 \cdot 5\,NH_3$ $CoCl_3 \cdot 4\,NH_3$

    (I)   (II)   (III)

These are called coordination compounds.

Compound I yields 3 mol AgCl(s) per mole of compound when treated with $AgNO_3(aq)$, but compound II yields only 2 mol, and compound III only 1 mol. The answer to this puzzling situation lies in understanding the nature of chemical bonding in these compounds. In this chapter, in addition to applying bonding theory to coordination compounds, we will discover additional applications of fundamental concepts of equilibrium, acid-base theory, and oxidation-reduction.

22-1 Werner's Theory

Our present conception of coordination compounds is based on ideas proposed by the Swiss chemist Alfred Werner in 1893. Werner's main proposition was that certain metal atoms have *two types of valence*. One, the primary valence, is based on the number of electrons lost in forming the metal ion. A secondary or auxiliary valence is responsible for bonding coordinated groups, called **ligands,** to the central metal ion. NH_3 molecules are the ligands in the compound $CoCl_3 \cdot 6\,NH_3$. A more appropriate formula for the compound is $[Co(NH_3)_6]Cl_3$, indicating that the NH_3 molecules are linked directly to the central Co^{3+} ion. A metal ion together with its ligands is called a **complex ion.** The region surrounding the metal ion where ligands are found is known as the **coordination sphere.** The number of positions in the coordination sphere at which ligand attachment can occur is the **coordination number** of the central metal ion.

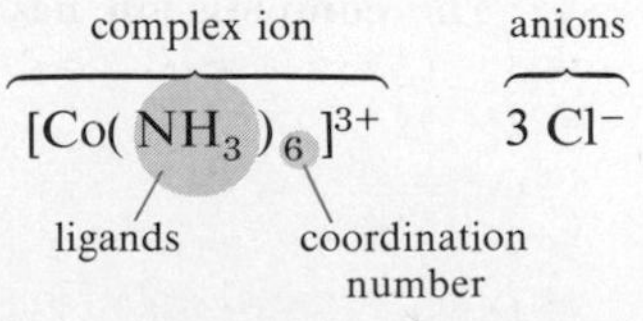

complex ion   anions

$[Co(\ NH_3\ )_6]^{3+}$ $3\ Cl^-$

ligands   coordination number

a coordination compound

FIGURE 22-1
Structures of some
complex ions.

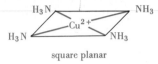

linear

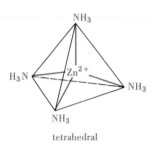

square planar

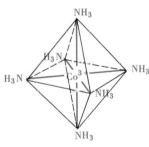

tetrahedral

octahedral

These are among the
most commonly observed
structures for complex
ions.

TABLE 22-1
Coordination numbers of some common metal ions

Cu^+	2, 4	Ca^{2+}	6	Al^{3+}	4, 6
Ag^+	2	Fe^{2+}	6	Sc^{3+}	6
Au^+	2, 4	Co^{2+}	4, 6	Cr^{3+}	6
		Ni^{2+}	4, 6	Fe^{3+}	6
		Cu^{2+}	4, 6	Co^{3+}	6
		Zn^{2+}	4, 6	Au^{3+}	4

The distribution of ligands in the coordination sphere of a metal ion produces a complex ion with a distinctive geometrical shape. The four most common structures are pictured in Figure 22-1.

22-2 The Coordination Number

The auxiliary or secondary valence proposed by Werner is now called the coordination number. There is no generally acceptable method for predicting coordination numbers, however. For one thing, some ions display more than a single coordination number. About as good a "rule" as any is that most ions display a coordination number that is twice the ionic charge. Also, the majority of transition metal ions will display a coordination number of six. These "rules" are seen to apply to most, but not all, of the ions listed in Table 22-1.

One practical use of the coordination number of a metal ion is to assist in the writing of formulas for complex ions and coordination compounds. In Example 22-1 formulas are written for the three compounds described in the introduction to this chapter. Moreover, from these formulas the difference in their behavior toward $AgNO_3(aq)$ is easily understood. In compound I all the Cl^- ions are free; in compounds II and III only two thirds and one third, respectively, of the Cl^- ions are free.

Example 22-1 Use the fact that the coordination number of Co^{3+} is six to write formulas for the coordination compounds I, II, and III on page 548.

In every case there must be *six* ligands. In compound I these are all NH_3 molecules. In compound II one Cl^- replaces an NH_3 as a ligand. This reduces the charge on the complex ion to $+2$. In compound III a second NH_3 molecule is replaced by Cl^-, resulting in a complex ion with a charge of $+1$.

$$[Co(NH_3)_6]Cl_3 \qquad [Co(NH_3)_5Cl]Cl_2 \qquad [Co(NH_3)_4Cl_2]Cl$$
$$(I) \qquad\qquad (II) \qquad\qquad (III)$$

SIMILAR EXAMPLE: Exercise 4.

Example 22-2 What are the coordination number and oxidation state of aluminum in the complex ion $[Al(H_2O)_4(OH)_2]^+$?

The complex ion has *four* H_2O molecules and *two* OH^- ions as ligands. The coordination number is *six*. Of these six ligand groups, *two* carry a charge of -1 each (the hydroxide ions) and four are *neutral* (the water molecules). The total contribution of the OH^- ions to the net charge on the complex ion is -2. Since the net charge on the complex ion is $+1$, the charge of the central alumi-

the charge on the central ion is assessed by determining the total charge of all the anions it is joined to (either by primary valence or coordination sphere)

num ion, and hence its oxidation state, must be $+3$. Diagrammatically, we can write

charge $= x$ — charge of -1 on OH^-; total negative charge $= -2$

$$[Al(H_2O)_4(OH)_2]^+$$

coordination number $= 6$ — net charge on complex ion: $x - 2 = +1$ $x = +3$

SIMILAR EXAMPLE: Exercise 4.

22-3 Ligands

A **ligand** is a species capable of donating an electron pair to a central metal ion at a particular site in a geometrical structure. Thus, a ligand is a Lewis base and the metal ion is a Lewis acid. If a ligand is capable of donating a single eletron pair, as in NH_3, it is called a **unidentate** ligand. These ligands may be monoatomic anions (but not neutral atoms) such as halide ions, polyatomic anions such as NO_2^-, simple molecules like NH_3, or more complex molecules like pyridine, C_6H_5N. A listing of some unidentate ligands is presented in Table 22-2.

TABLE 22-2
Some common unidentate ligands

Formula	Name	Formula	Name
Neutral molecules		Anions	
H_2O	aqua[a]	F^-	fluoro
NH_3	ammine	Cl^-	chloro
CO	carbonyl	Br^-	bromo
NO	nitrosyl	I^-	iodo
CH_3NH_2	methylamine	OH^-	hydroxo
C_6H_5N	pyridine	CN^-	cyano

[a] Formerly called aquo

metal ion - plus charge
gains e⁻ pair

The term "dentate" is derived from the Latin word *dens*, meaning tooth. Figuratively speaking, a unidentate ligand has one tooth, a bidentate ligand, two teeth, and so on. The ligand attaches itself to a metal ion in accordance with the number of teeth it possesses.

CHELATES. Some ligands are capable of donating more than a single electron pair, to different sites in the geometric structure of a complex ion. These are called **multidentate** ligands. The molecule, **ethylenediamine (en)**, $NH_2CH_2CH_2NH_2$, can donate *two* electron pairs, one from each N atom. It is a **bidentate** ligand.

$$H—\overset{\uparrow}{\underset{H}{\ddot{N}}}—CH_2CH_2—\overset{\uparrow}{\underset{H}{\ddot{N}}}—H \qquad (22.1)$$

A number of common multidentate ligands are listed in Table 22-3. The attachment of two en ligands to a Pt^{2+} ion is pictured in Figure 22-2. That multiple attachment of ligands does occur can be established in two ways. *This is how we conclude en has multiple attachments to Pt^{2+}:*
1. The complex ion $[Pt(en)_2]^{2+}$ has no capacity to coordinate additional ligands, that is, not NH_3, H_2O, Cl^-, and so on. Each en ligand must be attached at two points to account for the Pt coordination number of 4.
2. The en ligands show no further base properties once attached to the Pt^{2+} ion. That is, they do not accept protons from water to produce OH^- ions in solution, as would be expected of a weak base. Both $—NH_2$ groups of an en molecule must be tied up in the complex ion.

Note that in $[Pt(en)_2]^{2+}$ there exist two five-membered rings of Pt, N, and C atoms. When bonding between a

2 rings formed

Pt²⁺, remember: ligands don't change charge of ion unless they are anions themselves

FIGURE 22-2
Two representations of the chelate $[Pt(en)_2]^{2+}$.

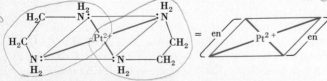

The ligands attach at adjacent corners along an edge of the square. They do *not* bridge the square by attaching to opposite corners.

a neutral atom — cannot be a ligand.

If I see this ligand in complex ion, these is its contributing C.N.

TABLE 22-3
Some common multidentate ligands (chelating agents)

Abbreviation	Name	Formula	
en	ethylenediamine *bidentate*		2
ox	oxalato	*42 carbons have completed octet*	2
o-phen	*o*-phenanthroline		2
dien	diethylenetriamine *is a tridentate*		3
trien	triethylenetetramine *is a tetradentate*		4
EDTA	ethylenediaminetetraacetato		6

The expression "chelate" is derived from the Greek word *chela*, meaning a crab's claw. The mode of attachment of a chelating agent to a metal ion bears a certain resemblance to a crab's claw.

metal ion and multidentate ligands results in ring formation (usually five- or six-membered), the process is called **chelation**. The species produced is called a **chelate**, and the multidentate ligand is a **chelating agent**.

22-4 Nomenclature

In Werner's time no attempt was made to develop a systematic nomenclature. The colors of some compounds were used as a basis for naming them. The compound $[Co(NH_3)_5Cl]Cl_2$ is purple in color and was called purpureocobaltic chloride. By analogy, any other complex ion with the formula $[M(NH_3)_5Cl]^{n+}$ was referred to as a "purpureo" complex ion (even if its color was not purple). Another approach was simply to name the compound after the investigator who first prepared it. For example, $NH_4[Cr(NH_3)_2(NCS)_4]$ was called Reinecke's salt. Current practice is to assign distinctive names for complex ions and coordination compounds based on a set of rules. Most can be named by the seven rules listed below and illustrated in Example 22-3.

1. In a coordination compound, as in a simple ionic compound, the cation is named first and then the anion.

2. In a complex ion the name(s) of the ligand group(s) is given first, followed by the name of the central metal ion.

3. In naming ligands an "o" ending is used for negative groups. Specifically, "ide" is changed to "*o*," "ate" to "*ato*," and "ite" to "*ito*." In rare instances where a ligand is a positive ion, the ending "ium" is used. If a ligand is a neutral molecule, the unmodified name of the molecule is generally used, except for the four molecules listed in Table 22-2—H_2O, NH_3, CO, NO.

4. The Greek prefixes mono, di, tri, tetra, penta, and hexa (representing 1, 2, 3, 4, 5, and 6) are used to denote the number of simple ligand groups of any given type. For certain <u>complex</u> <u>ligands</u> the terms bis, tris, tetrakis, ... are used.

5. The oxidation state of the central metal ion is denoted by a Roman numeral placed in parentheses following the name of the ion; for example, $[Cu(NH_3)_4]^{2+}$ is called tetraamminecopper(II) ion.

6. If more than one type of ligand is present, the ligands are named in the order negative, neutral, and positive, although formulas are generally written with the neutral ligands first. Also, multiatom ligands are named before single-atom ligands. For example, $[Co(NH_3)_4(NO_2)Br]^+$ is called nitrobromotetraamminecobalt(III) ion.

7. If the complex ion has a net negative charge, an "ate" ending is attached to the name of the central metal ion (and sometimes the metal is referred to by a Latin name); for example, $[CuCl_4]^{2-}$ is called tetrachlorocuprate(II) ion.

A proposed new rule 6 which is gaining acceptance is simply to name the ligands alphabetically. Thus, $[Co(NH_3)_4(NO_2)Br]^+$ would be named tetraamminebromonitrocobalt(III) ion.

Example 22-3 **(a)** What is the formula of the compound chloropentaaquachromium(III) chloride? **(b)** What is the name of the compound $K_3[Fe(CN)_6]$? **(c)** What is the name of the complex ion $[Pt(en)_2Cl_2]^{2+}$?

(a) The central metal ion is Cr^{3+}. There are one Cl^- ion and five H_2O molecules as ligands. The complex ion must carry a net charge of $+2$. Two Cl^- ions are required to neutralize this charge. The formula of the coordination compound is $[Cr(H_2O)_5Cl]Cl_2$

(b) This compound consists of K^+ cations and complex anions having the formula $[Fe(CN)_6]^{3-}$. Each cyanide ion carries a charge of -1, so the oxidation state of the iron must be $+3$. The Latin name "ferrum" is used for iron; however, an "ate" ending is also employed because the ion carries a net negative charge. The complex ion is called hexacyanoferrate(III) ion. The coordination compound is potassium hexacyanoferrate(III)

(c) The ethylenediamine ligands are neutral molecules and they are *bidentate*. The ligand Cl^- is uninegative, and two Cl^- ions are present in the complex ion. The coordination number is *six*; the net charge on the complex ion is $+2$; and the charge on the central metal ion is $+4$. The negatively charged ligands are named first and then the neutral molecules. Because there are two en ligands, the prefix "bis" is also required.

$[Pt(en)_2Cl_2]^{2+}$ = dichlorobis(ethylenediamine)platinum(IV) ion

SIMILAR EXAMPLES: Exercises 5, 6, 7.

Although most complex ions and coordination compounds are named in the manner just outlined, some common or "trivial" names are still in use. Chief among these are the ferr*o*cyanide, $[Fe(CN)_6]^{4-}$, and ferr*i*cyanide, $[Fe(CN)_6]^{3-}$, ions. These common names suggest the oxidation states of the central metal ions through the "o" and "i" designations, but they do not indicate coordination numbers. Clearly, systematic names—hexacyanoferrate(II) and hexacyanoferrate(III)—are superior.

22-5 Isomerism

Two or more species having the same composition (thus the same formula) but different structures and properties are said to be **isomers.** Isomerism arises among complex ions and coordination compounds for a variety of reasons.

IONIZATION ISOMERISM. The same numbers and types of groups are present in each of the two coordination compounds (22.2). Thus, they have the same formulas. However, the compounds are not identical. In compound (22.2a) a sulfate is coordinated directly to Cr^{3+} and is part of the octahedral complex ion; the Cl^- is the free anion of the coordination compound. In compound (22.2b) the roles of the SO_4^{2-} and Cl^- are reversed.

$$(a) \qquad\qquad (b)$$
$$[Cr(NH_3)_5SO_4]Cl \qquad\qquad [Cr(NH_3)_5Cl]SO_4 \qquad\qquad (22.2)$$

sulfatopentaamminechromium(III)   chloropentaamminechromium(III)
chloride   sulfate

COORDINATION ISOMERISM. A similar situation to that just described arises when a coordination compound is made up of both complex cations and complex anions. The ligands may be distributed differently between the complex ions, yet the compositions and empirical formulas of the coordination compounds remain the same.

$$(a) \qquad\qquad\qquad (b)$$
$$[Co(en)_3][Cr(ox)_3] \qquad\qquad [Cr(en)_3][Co(ox)_3] \qquad\qquad (22.3)$$

tris(ethylenediamine)cobalt(III)   tris(ethylenediamine)chromium(III)
tris(oxalato)chromate(III)   tris(oxalato)cobaltate(III)

LINKAGE ISOMERISM. Some ligands may attach to the central metal ion of a complex ion in different ways. For example, the nitrite ion has electron pairs available for coordination both on the N and O atoms.

$$\left[\; \overset{\displaystyle \ddot{N}}{:\!\underset{\displaystyle \cdot\cdot}{O}\cdot \qquad \underset{\displaystyle \cdot\cdot}{O}\!:} \;\right]^{-} \qquad\qquad (22.4)$$

Whether attachment of this ligand is through the N or the O atom, the formula of the complex ion is unaffected. However, the properties of the complex ion may be affected. Again, this is a case of isomerism. When attachment occurs through the N atom, the ligand is referred to as "nitro." Coordination through the O atom produces a "nitrito" complex.

$$(a) \qquad\qquad\qquad\qquad (b)$$
$$[Co(NH_3)_4(NO_2)Cl]^+ \qquad\qquad [Co(NH_3)_4(ONO)Cl]^+ \qquad\qquad (22.5)$$

nitrochlorotetraamminecobalt(III) ion   nitritochlorotetraamminecobalt(III) ion

GEOMETRIC ISOMERISM. If a single Cl^- ion is substituted for an NH_3 molecule in the complex ion $[Pt(NH_3)_4]^{2+}$, the point at which substitution occurs is immaterial. All four possibilities are alike, as shown in Figure 22-3a. Substitution of a second Cl^- produces two distinct possibilities, however (Figure 22-3b). The two Cl^- ions can

FIGURE 22-3
Geometric isomerism illustrated.

For the square planar complexes shown here, isomerism exists only when two Cl⁻ ions have replaced NH₃ molecules.

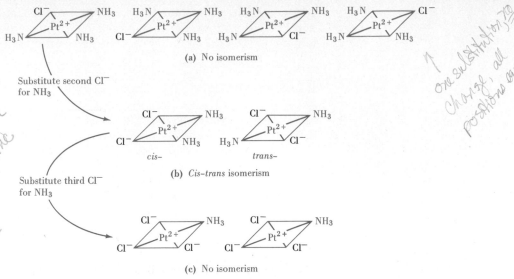

(a) No isomerism

Substitute second Cl⁻ for NH₃

cis-   *trans-*

(b) Cis–trans isomerism

Substitute third Cl⁻ for NH₃

(c) No isomerism

[handwritten margin notes:]
When you substitute ligands, various isomers occur according to geometric positions of substituted

one substitution no change, all positions equal.

either be along the same edge of the square planar structure (*cis*) or on opposite corners (*trans*). To distinguish between the two isomers, one must either draw a structure or refer to the appropriate name. The formula alone does not distinguish between them.

Note that the complex (22.6) is a neutral species, indicated by the zero charge.

$$[Pt(NH_3)_2Cl_2]^0 \qquad (22.6)$$

[handwritten: a zero complex ions are possible]

cis-dichlorodiammineplatinum(II)
or
trans-dichlorodiammineplatinum(II)

With the substitution of a third Cl⁻, again there is but a single structure; isomerism disappears (Figure 22-3c).

Geometrical isomerism is very common among octahedral complexes, but here the situation is just a bit more complicated. Take the complex ion $[Co(NH_3)_6]^{3+}$ as an example. Substitution of one Cl⁻ for an NH₃ molecule produces no isomers. Substitution of two Cl⁻ ions again leads to *cis-trans* isomerism. The *cis* isomer has Cl⁻ ions along the same edge of the octahedron. The *trans* isomer has Cl⁻ ions on opposite corners of the octahedron, that is, at opposite ends of a line drawn through the central metal ion. These two isomers are pictured in Figure 22-4. To illustrate the difference in properties of isomers, the *cis* isomer in Figure 22-4 is a blue-violet color and the *trans* is a bright green.

Another type of isomerism based on the spatial arrangement of ligand groups in a complex ion is *optical isomerism*. The particular property involved, called *optical activity*, is the ability to rotate the plane of polarization of light. This subject is discussed in Section 25-2 because of its special significance to a number of biochemical topics.

Substitution of a third Cl⁻ for one of the NH₃ molecules again leads to two possibilities. Refer to Figure 22-4a. If substitution occurs either at position 1 or 6, the result is the same—three Cl⁻ on the same face of the octahedron. This is also called a *cis* isomer. If the substitution is made either at position 4 or 5, the result is three Cl⁻ around a perimeter of the octahedron. This is referred to as a *trans* form. Substitution of a fourth Cl⁻ for an NH₃ again leads to two isomers, this time based on the positions occupied by the two remaining NH₃ molecules. *[handwritten: counts for 2]*

Example 22-4   Sketch all the possible isomers of $[Co(ox)(NH_3)_3Cl]^-$. *[handwritten circle around 6 in ox]* *[handwritten: counts for 2]*

The Co²⁺ ion exhibits a coordination number of *six*. The structure is octahedral. Recall that ox (oxalate ion) is a bidentate ligand carrying a double-negative charge (Table 22-3). Recall also that such a ligand must be attached in *cis* positions, not *trans* (Figure 22-2). Once the ox ligand has been placed, we see that there are *two* possibilities: The three NH₃ molecules can be situated (1) on the same face of the octahedron (a *cis* isomer) or (2) around a perimeter of the octahedron (a *trans* isomer).

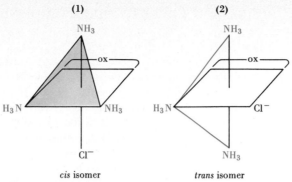

(1)	(2)
cis isomer	*trans* isomer

SIMILAR EXAMPLES: Exercises 14, 15, 17.

22-6 Bonding in Complex Ions—Valence Bond Theory

FIGURE 22-4
Cis-trans isomers of an octahedral complex.

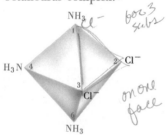

(a) *cis*-[Co(NH$_3$)$_4$Cl$_2$]$^+$

(b) *trans*-[Co(NH$_3$)$_4$Cl$_2$]$^+$

The Co^{3+} ion is at the center of the octahedron and one NH$_3$ ligand is on the far corner (corner 5), out of view.

One view of the metal ion–ligand bond is based on coordinate covalency—the central ion furnishes the orbitals and the ligands furnish electron pairs. Although this view is now considered inadequate, it does answer such questions as, "Why is there a discrete number of ligands associated with the central ion?" and "Why does a complex ion have a particular geometric structure?"

A scheme consistent with the formula and structure of the complex ion [Co(NH$_3$)$_6$]$^{3+}$ involves the hybridization of two 3*d* orbitals with a 4*s* and three 4*p* orbitals, yielding six *d*^{2}*sp*^{3} orbitals with an octahedral distribution.

Co: [Ar]   (22.7)

Co^{3+}: [Ar]   (22.8)

[Co(NH$_3$)$_6$]$^{3+}$: [Ar]   (22.9)

Three additional hybridization schemes are given in (22.10), (22.11), and (22.12) for [Zn(NH$_3$)$_4$]$^{2+}$, [Cu(NH$_3$)$_4$]$^{2+}$, and [Ag(NH$_3$)$_2$]$^+$, respectively. Geometrical structures of these four complex ions were presented in Figure 22-1.

[Zn(NH$_3$)$_4$]$^{2+}$: [Ar]   (22.10)

[Cu(NH$_3$)$_4$]$^{2+}$: [Ar]   (22.11)

[Ag(NH$_3$)$_2$]$^+$: [Kr]   (22.12)

22-7 Inner and Outer Orbital Complexes

The species $[Co(NH_3)_6]^{3+}$ is diamagnetic; $[CoF_6]^{3-}$ has a similar octahedral structure, but it is paramagnetic and appears to have *four* unpaired electrons. One explanation of this observation assumes that there are two ways in which an equivalent set of hybrid orbitals can be formed. In the first case, exemplified by $[Co(NH_3)_6]^{3+}$, the *d* orbitals come from an inner electronic shell (the third), and the resulting hybrid orbitals are d^2sp^3. The complex ion is called an inner orbital complex. Furthermore, because all electrons are paired and the complex ion is diamagnetic, it is referred to as a "low spin" complex. In $[CoF_6]^{3-}$ it is assumed that the *d* orbitals come from the outer electronic shell (the fourth), and the resulting hybrid orbitals are sp^3d^2. The complex ion is called an outer orbital complex and and is a "high spin" complex. These bonding schemes are consistent with the observed magnetic data.

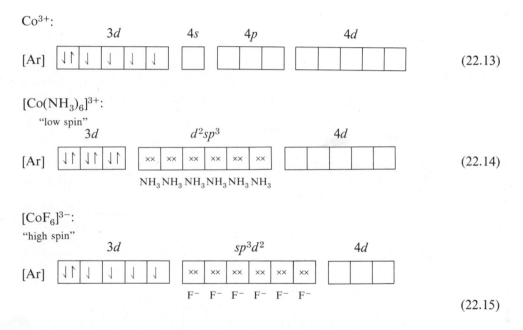

22-8 Bonding in Complex Ions—Crystal Field Theory

The valence bond theory of bonding in complex ions has some shortcomings. For instance, it does not provide insight into the origin of the characteristic colors of complex ions. Neither does it explain why $[Co(NH_3)_6]^{3+}$ is an inner orbital complex and $[CoF_6]^{3-}$ is an outer orbital complex. An alternative that is more successful in doing so is the crystal field theory. In the crystal field model, bonding in a complex ion is considered to be an electrostatic attraction between the positively charged nucleus of the central metal ion and electrons in the ligands. Repulsion occurs between the ligand electrons and the electrons of the central ion. The theory focuses on these repulsive forces, particularly as they affect *d* electrons of the central metal ion.

The *d* orbitals first presented in Figure 7-19 are not all alike in their spatial orientations, but for an isolated atom or ion they do have equal energies. One of them, d_{z^2}, is directed along the *z* axis and another, $d_{x^2-y^2}$, has lobes along the *x* and *y* axes. The remaining three have lobes extending into regions between the perpendicular *x*, *y*, and *z* axes.

Figure 22-5 pictures six anions (ligands) approaching a central metal ion along

Modifications of the simple crystal field theory that take into account such factors as the partial covalency of the metal–ligand bond are called ligand field theory. Often the single term "ligand field theory" is used to signify both the purely electrostatic crystal field theory and its modifications.

FIGURE 22-5
Approach of six anions to a metal ion to form a complex ion with octahedral structure.

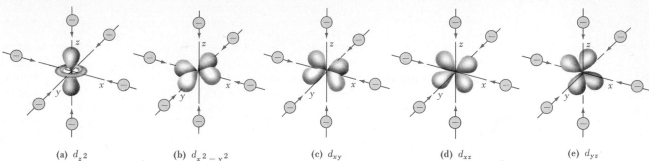

(a) d_{z^2} (b) $d_{x^2-y^2}$ (c) d_{xy} (d) d_{xz} (e) d_{yz}

The ligands (anions in this case) approach the central metal ion along the x, y, and z axes. Maximum interference occurs with the d_{z^2} and $d_{x^2-y^2}$ orbitals, and their energies are raised. Interference with the other d orbitals is not as great. A difference in energy results between the two sets of d orbitals.

the x, y, and z axes. This direction of approach leads to an octahedral complex. As a result of electrostatic repulsion between the negatively charged ligands and the d electrons of the metal ion, the d energy level is split into two groups. One group, consisting of $d_{x^2-y^2}$ and d_{z^2}, has its energy raised with respect to a hypothetical free ion in the field of the ligands. The other group, consisting of d_{xy}, d_{xz}, and d_{yz}, has its energy lowered. The difference in energy between the two groups is represented by the symbol Δ, as shown in Figure 22-6. The pattern of splitting of the d energy levels is different for complex ions with other geometric shapes. The pattern for tetrahedral complexes is described in Figure 22-7.

    Ligands differ in ability to produce a splitting of the d energy levels. Strong Lewis bases, such as CN^- and NH_3, produce a "strong" field. They repel d electrons most strongly and cause a greater separation of the two groups of d energy levels than do weak bases like H_2O and F^-. Concerning the distribution of d electrons of the metal

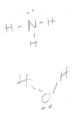

FIGURE 22-6
Splitting of d energy levels in the formation of an octahedral complex ion.

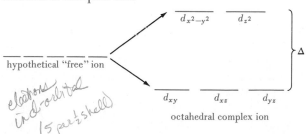

FIGURE 22-7
Crystal field splitting in a tetrahedral complex ion.

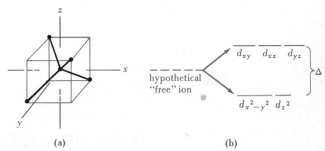

(a) The positions of attachment of ligands to a metal ion leading to the formation of a tetrahedral complex ion. Note that, unlike in the formation of an octahedral complex (Figure 22-5), the ligands do *not* approach along the x, y, and z axes.
(b) Interference with the d orbitals directed along the x, y, and z axes is not as great as with those that lie between the axes (again, see Figure 22-5). As a result, the pattern of crystal field splitting is reversed from that of an octahedral complex.

Strong field — when ligands enter along d_{z^2} & $d_{x^2-y^2}$, a strong fielded one will repel metal's ions so much they get bumped into the other orbitals (d_{yz}, d_{xz}, d_{xy}) already containing e⁻'s

Co [Ar] $3d^7 4s^2$

Co³⁺ [Ar] $3d^6$

6 electrons in outer orbital

ion, we should expect orbitals to be singly occupied, as in the complex ion $[CoF_6]^{3-}$ shown in (22.16). However, in some cases, the energy separation between the lower and higher energy states (Δ) is larger than the energy required to pair electrons. Then the energetically favored arrangement is for electrons to remain paired in the lower-energy d orbitals, as in $[Co(NH_3)_6]^{3+}$.

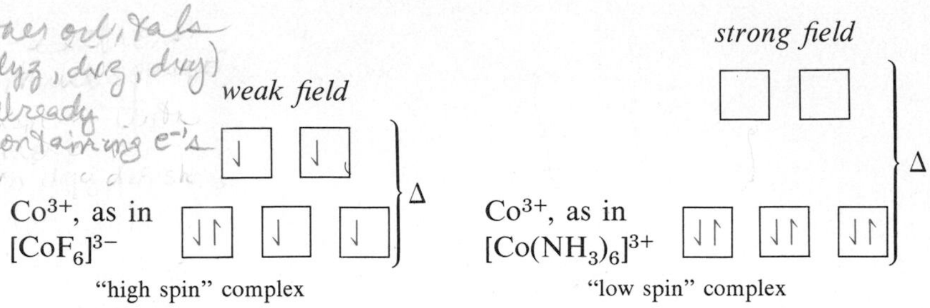

(22.16)

weak field / *strong field*

Co³⁺, as in $[CoF_6]^{3-}$ "high spin" complex

Co³⁺, as in $[Co(NH_3)_6]^{3+}$ "low spin" complex

str. base str. field large Δ

We have just seen how crystal field theory offers an explanation of the magnetic properties of the inner and outer complexes described by structures (22.14) and (22.15). Predictions of this same sort can be made for other complex ions using the following listing, called the **spectrochemical series,** to describe the degree of splitting produced by ligands.

When NO_2^- is attached through an O atom or SCN^- through the S atom, a different placement in the spectrochemical series is found than the one given here.

strong field

$$CN^- > -NO_2^- > en > py \simeq NH_3 > SCN^-$$

weak field

$$> H_2O > OH^- > F^- > Cl^- > Br^- > I^- \qquad (22.17)$$

Example 22-5 How many unpaired electrons are present in the octahedral complex $[Fe(CN)_6]^{3-}$?

The Fe atom has the electron configuration $[Ar]3d^6 4s^2$. The Fe³⁺ ion has the configuration $[Ar]3d^5$. CN^- is a strong-field ligand. Because of the large energy separation produced by this ligand in the d levels of the metal ion, all the electrons are found in the lowest energy levels. There is *one* unpaired electron.

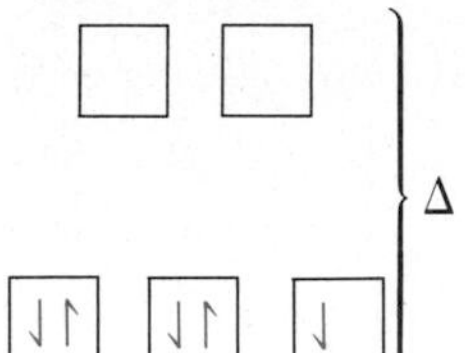

Fe [Ar] $3d^6 4s^2$

Fe³⁺ [Ar] $3d^5$

5 electrons in outer orbital

thiocyanate can link to metal atom @ more than one spot — linkage isomerism

SIMILAR EXAMPLES: Exercises 21 through 24.

22-9 Color of Complex Ions

The absorption of electromagnetic radiation by an ionic species in solution requires that electrons within the ion undergo a transition from one energy state to another. The light quanta absorbed must have energies just equal to the difference in the two energy states involved in the transition. Where this transition energy corresponds to a wavelength component in visible light, that component is absorbed and the transmitted light is colored.

Ions having a noble gas electron configuration, an outer shell of 18 electrons, or the "18 + 2" configuration do not possess electronic transitions in the energy range

The color of light transmitted by a solution, that is, the "color" of the solution, is the *complement* of the color that is absorbed. A solution containing $[Cu(H_2O)_4]^{2+}$ absorbs yellow light (about 580 nm) and transmits blue (about 450 nm); a solution with $[CuCl_4]^{2-}$ absorbs blue light and transmits yellow. The principal combinations of colors and their complements are

blue ⟷ yellow
red ⟷ blue-green
green ⟷ purple

corresponding to visible light. "White" light passes through solutions of these ions without being absorbed; these ions are colorless in solution. Examples are the alkali and alkaline earth metal ions, the halide ions, Zn^{2+} and Bi^{3+}.

Crystal field splitting of the *d* energy levels produces the energy difference, Δ, that accounts for the colors of complex ions. Promotion of an electron from a lower to a higher *d* level results from the absorption of the appropriate components of white light; the transmitted light is colored.

The colors of some complex ions of chromium are given in Table 22-4. The complex ions in question are all octahedral, so the pattern of the crystal field splitting of the *d* energy levels is that of Figure 22-6. Chromium(III) complexes all have a d^3 configuration. In each case there are three unpaired electrons in the lower energy group.

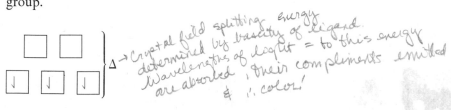

The magnitude of Δ, the crystal field splitting, is determined by the particular ligands attached to the Cr^{3+} ion. This magnitude determines what wavelength light must be absorbed to produce an electronic transition, and, in turn, what color light is transmitted by the solution. For example, in $[Cr(NH_3)_6]^{3+}$ the crystal field splitting (Δ) is greater than in $[Cr(H_2O)_6]^{3+}$. As a result, $[Cr(NH_3)_6]^{3+}$ absorbs light of a shorter wavelength (violet) than does $[Cr(H_2O)_6]^{3+}$ (yellow). The light *transmitted* by $[Cr(NH_3)_6]^{3+}$—and hence its color—is yellow; that of $[Cr(H_2O)_6]^{3+}$ is violet.

22-10 Equilibria Involving Complex Ions

If a moderately concentrated solution of $NH_3(aq)$ is added to solid silver chloride, the solid dissolves.

$$AgCl(s) + 2\,NH_3(aq) \longrightarrow [Ag(NH_3)_2]^+(aq) + Cl^-(aq) \tag{22.18}$$

Ag^+ from AgCl combines with NH_3 to form the complex ion $[Ag(NH_3)_2]^+$. The coordination compound $[Ag(NH_3)_2]Cl$ is soluble in $NH_3(aq)$. AgBr(s) is only slightly soluble in $NH_3(aq)$ and AgI(s) is essentially insoluble.

To understand these observations, we need to think of reaction (22.18) as involving two equilibria simultaneously.

$$AgCl(s) \rightleftharpoons Ag^+(aq) + Cl^-(aq) \tag{22.19}$$

$$Ag^+(aq) + 2\,NH_3(aq) \rightleftharpoons [Ag(NH_3)_2]^+(aq) \tag{22.20}$$

Because $[Ag(NH_3)_2]^+$ is a stable complex ion, the equilibrium concentration of Ag^+ in reaction (22.20) is very low. K_{sp} for AgCl is not exceeded even in the presence of

TABLE 22-4
Some complex ions of Cr^{3+} and their colors

Isomer	Color	Isomer	Color
$[Cr(H_2O)_6]Cl_3$	violet	$[Cr(NH_3)_6]Cl_3$	yellow
$[Cr(H_2O)_5Cl]Cl_2$	blue-green	$[Cr(NH_3)_5Cl]Cl_2$	purple
$[Cr(H_2O)_4Cl_2]Cl$	green	$[Cr(NH_3)_4Cl_2]Cl$	violet

moderately high Cl^- concentrations. Thus, $AgCl(s)$ is soluble in $NH_3(aq)$. Additional predictions concerning the aqueous chemistry of $[Ag(NH_3)_2]^+$ can be made through Le Châtelier's principle, as illustrated in Example 22-6.

Example 22-6   Predict the effect of adding $HNO_3(aq)$ to a saturated solution of $[Ag(NH_3)_2]Cl$ in $NH_3(aq)$.

HNO_3 neutralizes the free NH_3 in solution.

$$H^+(aq) + NO_3^-(aq) + NH_3(aq) \longrightarrow NH_4^+(aq) + NO_3^-(aq)$$

Loss of NH_3 upsets the equilibrium in reaction (22.20). To replace the neutralized NH_3, $[Ag(NH_3)_2]^+$ dissociates—the reverse of reaction (22.20). This, in turn, produces an increase in $[Ag^+]$—a disturbance that is relieved by the reverse of reaction (22.19). $AgCl(s)$ precipitates.

SIMILAR EXAMPLES: Exercises 27, 28.

FORMATION CONSTANTS OF COMPLEX IONS.   To provide a *quantitative* description of equilibria involving complex ions requires knowledge of the **formation constant, K_f.** For the reaction

$$Ag^+(aq) + 2\,NH_3(aq) \rightleftharpoons [Ag(NH_3)_2]^+(aq) \tag{22.20}$$

the equilibrium constant describing equilibrium among the complex ion and the central metal ion and ligands from which it is formed is

$$K_f = \frac{[[Ag(NH_3)_2]^+]}{[Ag^+][NH_3]^2} = 1.6 \times 10^7 \tag{22.21}$$

A more complete description of the formation of a complex ion views this to occur in a *stepwise* fashion, with a formation constant for each step. In the manner introduced in Chapter 14, these individual steps can be added to yield the net formation reaction. The overall formation constant, K_f, is simply the product of the stepwise formation constants. For example,

$Ag^+ + NH_3 \rightleftharpoons [Ag(NH_3)]^+$	$K_{f_1} = 7.7 \times 10^3$
$[Ag(NH_3)]^+ + NH_3 \rightleftharpoons [Ag(NH_3)_2]^+$	$K_{f_2} = 2.1 \times 10^3$
$Ag^+ + 2\,NH_3 \rightleftharpoons [Ag(NH_3)_2]^+$	$K_f = 1.6 \times 10^7$

For our purposes we need to use only overall formation constants. Selected data are listed in Table 22-5.

Sometimes complex ion equilibria are written in the reverse of the method used in equation (22.20) (i.e., $[Ag(NH_3)_2]^+ \rightleftharpoons Ag^+ + 2\,NH_3$). In this case the equilibrium constant is the reciprocal of the type shown in equation (22.21). Written in this way the constant is called a dissociation constant, K_D. Thus, for $[Ag(NH_3)_2]^+$,

$$K_D = \frac{[Ag^+][NH_3]^2}{[[Ag(NH_3)_2]^+]}$$

$$= \frac{1}{1.6 \times 10^7}$$

$$= 6.2 \times 10^{-8}$$

Example 22-7   A 0.10-mol sample of $AgNO_3$ is dissolved in 1.00 L of 1.00 M NH_3. If 0.010 mol $NaCl$ is added to this solution, will a precipitate form?

The total silver concentration in the solution is 0.10 mol/L, found partly as Ag^+ and partly as $[Ag(NH_3)_2]^+$. In the setup shown we assume that practically all the silver is complexed. If 0.10 mol Ag^+ is complexed, then twice this amount of NH_3 must also be complexed.

	Ag^+	$+\ 2\,NH_3$	$\rightleftharpoons$	$[Ag(NH_3)_2]^+$
Dissolve:	0.10 M	1.00 M		
At equilibrium (assume $x \ll 0.10$):	$x = [Ag^+]$	0.80 M		0.10 M

$$\frac{[[Ag(NH_3)_2]^+]}{[Ag^+][NH_3]^2} = \frac{0.10}{[Ag^+](0.80)^2} = 1.6 \times 10^7$$

$$[Ag^+] = 1.0 \times 10^{-8}$$

TABLE 22-5
Formation constants for some complex ions

Complex ion	Equilibrium reaction	K_f
$[AlF_6]^{3-}$	$Al^{3+} + 6\,F^- \rightleftharpoons [AlF_6]^{3-}$	6.7×10^{19}
$[Cd(CN)_4]^{2-}$	$Cd^{2+} + 4\,CN^- \rightleftharpoons [Cd(CN)_4]^{2-}$	7.1×10^{18}
$[Co(NH_3)_6]^{3+}$	$Co^{3+} + 6\,NH_3 \rightleftharpoons [Co(NH_3)_6]^{3+}$	4.5×10^{33}
$[Cu(CN)_3]^{2-}$	$Cu^+ + 3\,CN^- \rightleftharpoons [Cu(CN)_3]^{2-}$	2×10^{27}
$[Cu(NH_3)_4]^{2+}$	$Cu^{2+} + 4\,NH_3 \rightleftharpoons [Cu(NH_3)_4]^{2+}$	2.1×10^{14}
$[Fe(CN)_6]^{4-}$	$Fe^{2+} + 6\,CN^- \rightleftharpoons [Fe(CN)_6]^{4-}$	1×10^{37}
$[Fe(CN)_6]^{3-}$	$Fe^{3+} + 6\,CN^- \rightleftharpoons [Fe(CN)_6]^{3-}$	1×10^{42}
$[PbCl_3]^-$	$Pb^{2+} + 3\,Cl^- \rightleftharpoons [PbCl_3]^-$	2.4×10^1
$[HgCl_4]^{2-}$	$Hg^{2+} + 4\,Cl^- \rightleftharpoons [HgCl_4]^{2-}$	1.2×10^{15}
$[HgI_4]^{2-}$	$Hg^{2+} + 4\,I^- \rightleftharpoons [HgI_4]^{2-}$	1.9×10^{30}
$[Ni(CN)_4]^{2-}$	$Ni^{2+} + 4\,CN^- \rightleftharpoons [Ni(CN)_4]^{2-}$	1×10^{22}
$[Ag(NH_3)_2]^+$	$Ag^+ + 2\,NH_3 \rightleftharpoons [Ag(NH_3)_2]^+$	1.6×10^7
$[Ag(CN)_2]^-$	$Ag^+ + 2\,CN^- \rightleftharpoons [Ag(CN)_2]^-$	5.6×10^{18}
$[Ag(S_2O_3)_2]^{3-}$	$Ag^+ + 2\,S_2O_3{}^{2-} \rightleftharpoons [Ag(S_2O_3)_2]^{3-}$	1.7×10^{13}
$[Zn(NH_3)_4]^{2+}$	$Zn^{2+} + 4\,NH_3 \rightleftharpoons [Zn(NH_3)_4]^{2+}$	2.9×10^9
$[Zn(CN)_4]^{2-}$	$Zn^{2+} + 4\,CN^- \rightleftharpoons [Zn(CN)_4]^{2-}$	1×10^{18}

We must compare $[Ag^+][Cl^-]$ with $K_{sp}(AgCl) = 1.6 \times 10^{-10}$.

$[Ag^+] = 1.0 \times 10^{-8}$ $[Cl^-] = 1.0 \times 10^{-2}$

$(1.0 \times 10^{-8})(1.0 \times 10^{-2}) < 1.6 \times 10^{-10}$

AgCl will not precipitate.

SIMILAR EXAMPLES: Exercises 30, 32.

Example 22-8 What is the minimum concentration of aqueous NH_3 required to prevent $AgCl(s)$ from precipitating from 1.00 L of a solution containing 0.10 mol $AgNO_3$ and 0.010 mol $NaCl$?

The $[Cl^-]$ that must be maintained in solution is 1.0×10^{-2}. If no precipitation is to occur, $[Ag^+][Cl^-] \leqslant K_{sp}$.

$[Ag^+](1.0 \times 10^{-2}) \leqslant K_{sp} = 1.6 \times 10^{-10}$ $[Ag^+] \leqslant 1.6 \times 10^{-8}$

The maximum concentration of *free, uncomplexed* Ag^+ permitted in solution is 1.6×10^{-8} M. This means that essentially all the Ag^+ (0.10 mol) must be in the form of the complex ion, $[Ag(NH_3)_2]^+$. We need to solve the following expression for $[NH_3]$.

$$K_f = \frac{[[Ag(NH_3)_2]^+]}{[Ag^+][NH_3]^2} = \frac{1.0 \times 10^{-1}}{1.6 \times 10^{-8}[NH_3]^2} = 1.6 \times 10^7$$

$[NH_3]^2 = 0.39$ $[NH_3] = 0.62\ M$

The concentration calculated above is that of *free, uncomplexed* NH_3. Considering as well the 0.20 mol NH_3/L consumed to produce $[Ag(NH_3)_2]^+$, the total concentration of $NH_3(aq)$ required is

$[NH_3]_{tot} = 0.62 + 0.20 = 0.82\ M$

SIMILAR EXAMPLE: Exercise 31.

22-11 Some Kinetic Considerations

When $NH_3(aq)$ is added to a solution containing Cu^{2+}, there is an immediate change in color from pale blue to a very deep blue. The reaction involved is the substitution of NH_3 molecules as ligands for H_2O.

$$[Cu(H_2O)_4]^{2+} + 4\,NH_3 \longrightarrow [Cu(NH_3)_4]^{2+} + 4\,H_2O \qquad (22.22)$$
(pale blue)   (very deep blue)

This reaction occurs very rapidly—as rapidly as the two reactants can be brought together. The addition of $HCl(aq)$ to an aqueous solution of Cu^{2+} produces an immediate color change from pale blue to green, or even yellow if the $HCl(aq)$ is sufficiently concentrated.

$$[Cu(H_2O)_4]^{2+} + 4\,Cl^- \longrightarrow [CuCl_4]^{2-} + 4\,H_2O \qquad (22.23)$$
(pale blue)   (yellow)

The terms "labile" and "inert" are not related to the thermodynamic stabilities of complex ions, nor to the equilibrium constants for ligand-substitution reactions. The terms are kinetic terms, referring to the rates at which ligands may be exchanged.

Complex ions in which ligands can be interchanged rapidly are said to be **labile.** $[Cu(H_2O)_4]^{2+}$, $[Cu(NH_3)_4]^{2+}$, and $[CuCl_4]^{2-}$ are all labile.

The violet solution obtained by dissolving $[Cr(H_2O)_6]Cl_3$ in water gradually turns green in color when it is boiled. This color change results from the very slow exchange of Cl^- for H_2O ligands. A complex ion that exchanges ligands slowly is said to be nonlabile or **inert.** In general, complex ions of the first transition series, except Cr(III) and Co(III), are kinetically labile. Those of the second and third transition series are generally kinetically inert. Whether a complex ion is labile or inert affects the ease with which it can be studied experimentally. The inert ones are generally easiest to obtain and characterize, explaining perhaps why so many of the early studies of complex ions used Cr(III) and Co(III).

22-12 Applications of Coordination Chemistry

The number and variety of applications of coordination chemistry are impressive, ranging from analytical chemistry to biochemistry. The examples given here are intended simply to convey an idea of this diversity.

HYDRATES. Often when a compound is crystallized from an aqueous solution of its ions the crystals obtained are hydrated. A hydrate is a substance that has associated with each formula unit a certain number of water molecules. In some cases the water molecules are ligands bonded directly to a metal ion. The coordination compound, $[Cr(H_2O)_6]Cl_3$, may also be represented as $CrCl_3 \cdot 6\,H_2O$ and called chromium(III) chloride hexahydrate. In the hydrate, $CuSO_4 \cdot 5\,H_2O$, four H_2O molecules are associated with copper in the complex ion, $[Cu(H_2O)_4]^{2+}$, and the fifth with the SO_4^{2-} anion through hydrogen bonding. Another possibility for hydrate formation is that the water molecules may be incorporated into definite positions in the solid crystal but not associated with any particular cations or anions. This is referred to as lattice water, as in $BaCl_2 \cdot 2\,H_2O$. Finally, part of the water may be coordinated to an ion and part of it may be lattice water. Apparently, this is the case with the alums, such as $KAl(SO_4)_2 \cdot 12\,H_2O$.

AMPHOTERISM. Equations (22.24) through (22.27) represent the stepwise ionization of protons from $[Cr(H_2O)_6]^{3+}$.

$$[Cr(H_2O)_6]^{3+} + H_2O \rightleftharpoons [Cr(H_2O)_5OH]^{2+} + H_3O^+ \qquad (22.24)$$

$$[Cr(H_2O)_5OH]^{2+} + H_2O \rightleftharpoons [Cr(H_2O)_4(OH)_2]^+ + H_3O^+ \qquad (22.25)$$

notice how written when neutral

$$[Cr(H_2O)_4(OH)_2]^+ + H_2O \rightleftharpoons \boxed{Cr(OH)_3 \cdot 3\,H_2O(s)} + H_3O^+ \tag{22.26}$$

$$Cr(OH)_3 \cdot 3\,H_2O(s) + H_2O \rightleftharpoons [Cr(H_2O)_2(OH)_4]^- + H_3O^+ \tag{22.27}$$

In reaction (22.24) a proton from a *ligand* water molecule in hexaaquochromium(III) ion is transferred to a *solvent* water molecule. The H_2O ligand is converted to OH^-. The process involved is illustrated in Figure 22-8. That reaction (22.24) actually does occur is seen by the fact that aqueous solutions of chromium(III) are somewhat acidic. The reactions represented by equations (22.25) through (22.27) do not occur to any significant extent in pure water. However, they can be brought about by adding a strong base to neutralize the H_3O^+ produced in each dissociation step (Le Châtelier's principle). Thus, if any of the species, $[Cr(H_2O)_6]^{3+}(aq)$, $Cr(OH)_3(s)$, or $Cr_2O_3(s)$, is treated with a strong base, reaction occurs to produce $[Cr(H_2O)_2(OH)_4]^-$. The *chromite* ion, CrO_2^-, can be thought of as resulting from the elimination of four H_2O molecules from this complex ion.

all have +3 oxid. state

$$\begin{array}{c}
[Cr(H_2O)_6]^{3+} \xrightarrow{\;OH^-\;} \\[4pt]
Cr(OH)_3 \xrightarrow{\;OH^-,\,H_2O\;} [Cr(H_2O)_2(OH)_4]^- \xrightarrow{\;-4\,H_2O\;} CrO_2^- \\[4pt]
Cr_2O_3 \xrightarrow{\;OH^-,\,H_2O\;}
\end{array} \tag{22.28}$$

solids · *leaves 2 oxys*

If a strong acid is added to any of the species, $[Cr(H_2O)_2(OH)_4]^-$, CrO_2^-, $Cr(OH)_3$, or $Cr_2O_3(s)$, the reactions are all reversed leading to $[Cr(H_2O)_6]^{3+}$ (Le Châtelier's principle again).

$$\begin{array}{c}
[Cr(H_2O)_2(OH_4)]^- \xrightarrow{\;H^+\;} \\[4pt]
CrO_2^- \xrightarrow{\;H^+,\,H_2O\;} \\[4pt]
Cr(OH)_3 \xrightarrow{\;H^+,\,H_2O\;} [Cr(H_2O)_6]^{3+} \\[4pt]
Cr_2O_3 \xrightarrow{\;H^+,\,H_2O\;}
\end{array} \tag{22.29}$$

Similar schemes can be used to represent amphoterism in other elements. Al^{3+} and Zn^{2+} behave in an analogous fashion to that described here. For instance, with Al^{3+} the corresponding species encountered are $[Al(H_2O)_6]^{3+}$, $Al(OH)_3$, Al_2O_3, $[Al(H_2O)_2(OH)_4]^-$, and AlO_2^- (aluminate ion).

This description of amphoterism based on the ionization of a complex ion yields the same results as the method introduced in Section 20-4.

STABILIZATION OF OXIDATION STATES. The standard electrode potential for the reduction of Co(III) to Co(II) is

$$Co^{3+}(aq) + e^- \longrightarrow Co^{2+}(aq) \qquad E° = +1.82\ V \tag{22.30}$$

This suggests that $Co^{3+}(aq)$ is a strong oxidizing agent, strong enough in fact to oxidize water to $O_2(g)$.

$$4\,Co^{3+}(aq) + 2\,H_2O \longrightarrow 4\,Co^{2+}(aq) + 4\,H^+ + O_2(g) \qquad E°_{cell} = +0.59\ V \tag{22.31}$$

FIGURE 22-8
Ionization of $[Cr(H_2O)_6]^{3+}$.

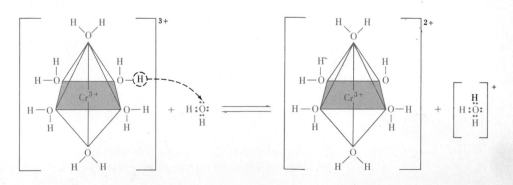

Yet, one of the complex ions featured in this chapter has been $[Co(NH_3)_6]^{3+}$. This ion is stable in water solution, even though it contains cobalt in the oxidation state, $+3$. Reaction (22.31) will not occur if the concentration of Co^{3+} is sufficiently low, and $[Co^{3+}]$ is kept very low because of the great stability of the complex ion.

$$Co^{3+}(aq) + 6 NH_3(aq) \rightleftharpoons [Co(NH_3)_6]^{3+}(aq) \qquad K_f = 4.5 \times 10^{33} \qquad (22.32)$$

The formation of stable complexes affords a means of attaining certain oxidation states that might otherwise be difficult or impossible.

THE PHOTOGRAPHIC PROCESS. A photographic film is basically an emulsion of silver bromide in gelatin. When the film is exposed to light, silver bromide granules become activated according to the intensity of the light striking them. When the exposed film is placed in a developer solution [a mild reducing agent such as hydroquinone, $C_6H_4(OH)_2$], the activated granules of silver bromide are reduced to black metallic silver. The unactivated granules in the unexposed portions of the film are practically unaffected. This developing process produces the photographic image.

The photographic process cannot be terminated at this point, however. The unactivated granules of silver bromide would eventually be reduced to black metallic silver upon exposing the film to light again. The image on the film must be "fixed." This requires that the black metallic silver that results from the developing be left on the film and the remaining silver bromide be removed. The "fixer" commonly employed is sodium thiosulfate (also known as sodium hyposulfite or "hypo"). In the fixing process $AgBr(s)$ is dissolved and the complexed silver ion is washed away.

$$AgBr(s) + 2 S_2O_3^{2-} \longrightarrow [Ag(S_2O_3)_2]^{3-} + Br^- \qquad (22.33)$$

QUALITATIVE ANALYSIS. In the separation and detection of cations in the qualitative analysis scheme, Ag^+, Pb^{2+}, and Hg_2^{2+}, are first precipitated as chlorides (recall Figure 16-3). All the other common cations form soluble chlorides. $PbCl_2(s)$ is removed from $AgCl(s)$ and $Hg_2Cl_2(s)$ by its greater solubility in hot water. $AgCl(s)$ is separated from $Hg_2Cl_2(s)$ by its solubility in $NH_3(aq)$, previously described in equation (22.18).

At another point in the qualitative analysis scheme it is desired to precipitate Cd^{2+} as the sulfide in the presence of Cu^{2+}. Normally, Cu^{2+} would precipitate along with the Cd^{2+} since K_{sp} for CuS is smaller than for CdS, 6.3×10^{-36} compared to 8.0×10^{-27}. However, by treating the solution of Cu^{2+} and Cd^{2+} with an excess of CN^- prior to saturation with H_2S, this separation can be achieved. The following reactions occur.

$$Cd^{2+} + 4 CN^- \rightleftharpoons [Cd(CN)_4]^{2-} \qquad (22.34)$$
$$K_f = 7.1 \times 10^{18}$$

$$2 Cu^{2+} + 8 CN^- \rightleftharpoons 2[Cu(CN)_3]^{2-} + C_2N_2(g) \qquad (22.35)$$
$$K_f = 2 \times 10^{27}$$

Reaction (22.35) is an oxidation-reduction reaction in which Cu^{2+} is reduced to Cu^+ and complexed with CN^- as well. The complex ion, $[Cu(CN)_3]^{2-}$, is very stable. That is, the concentration of *free* Cu^+ in equilibrium with the complex ion is very low. When a solution containing this complex ion is later saturated with H_2S, K_{sp} for Cu_2S is not exceeded. By contrast, $[Cd^{2+}]$ in equilibrium with $[Cd(CN)_4]^{2-}$ is sufficiently great that K_{sp} of CdS is exceeded under these same conditions.

ELECTROPLATING. Electrolyte solutions used in commercial electroplating are quite complex. Each component plays a role in achieving the final objective of a bright, smooth fine-grained metal deposit. A number of metals, for example, copper, silver, and gold, are generally plated from a solution of their cyano complex ions. In the electrolysis reaction shown in equations (22.36) and (22.37), the object to be plated is made the cathode and a piece of copper metal is the anode.

$$\text{anode:} \quad Cu + 3\,CN^- \longrightarrow [Cu(CN)_3]^{2-} + e^- \tag{22.36}$$

$$\text{cathode:} \quad [Cu(CN)_3]^{2-} + e^- \longrightarrow Cu + 3\,CN^- \tag{22.37}$$

The net change simply involves the transfer of copper metal from the anode to the cathode through the formation, migration, and decomposition of the complex ion $[Cu(CN)_3]^{2-}$. An additional advantage of electroplating copper from a solution of $[Cu(CN)_3]^{2-}$ is that 1 mol of copper is obtained per Faraday, in contrast to $\frac{1}{2}$ mol per Faraday that would be obtained from a solution of Cu^{2+}.

METALLURGY OF GOLD.   Gold is found in nature in the free state, but most of its ores have a very low gold content, as low as 0.001%. The basic difficulty is how to extract such little gold from so much waste rock. The process of **cyanidation** is useful to accomplish this. The crushed ore is treated with an aqueous cyanide solution in the presence of air. The free metal is simultaneously oxidized to Au^+ and coordinated by CN^- in a very stable complex ion.

[handwritten: CN oxidizes gold & forms complex]

$$4\,Au(s) + 8\,CN^- + O_2(g) + 2\,H_2O \longrightarrow 4\,[Au(CN)_2]^- + 4\,OH^- \tag{22.38}$$

The pure metal can then be displaced from solution by an active metal.

$$2\,[Au(CN)_2]^-(aq) + Zn(s) \longrightarrow 2\,Au(s) + [Zn(CN)_4]^{2-}(aq) \tag{22.39}$$

WATER TREATMENT.   Metal ions may act as catalysts in promoting undesirable chemical reacions in a manufacturing process, or they may alter in some way the properties of the material being manufactured. Thus, for many industrial purposes it is imperative to remove mineral impurities from water. Often these impurities are present only in trace amounts, for example, Cu^{2+}. Precipitation of metal ions from solution is feasible only if K_{sp} for the precipitated solid is very small.

The need to complex Cu^{2+} in the Raschig process for the production of hydrazine, N_2H_4, was mentioned in Section 4-6.

One method of water treatment involves chelation. Among the chelating agents widely employed are the salts of **ethylenediaminetetraacetic acid (EDTA),** such as the sodium salt.

[handwritten: again stability of complex leaves little free ions left in soln to ppt]

$$4\,Na^+ \begin{bmatrix} {}^-OOCCH_2 \\ \\ {}^-OOCCH_2 \end{bmatrix} NCH_2CH_2N \begin{bmatrix} CH_2COO^- \\ \\ CH_2COO^- \end{bmatrix} \tag{22.40}$$

[handwritten: has 6 places w/which to fill coordination spaces, so 1 EDTA may completely complex metal ion & render it unreactive]

FIGURE 22-9
Sequestering action of polyphosphates on metal ions.

For example, iron(II) in the presence of EDTA is converted to $[Fe(EDTA)]^{2-}$. The concentration of Fe^{2+} is held small enough that iron will not precipitate, not even as the hydroxide if a base is later added to the water. *[handwritten: →CN=6]*

For ions present in higher concentrations, such as Ca^{2+}, treatment with EDTA may be too costly. In these cases polyphosphate salts have been widely used. One such salt is sodium tripolyphosphate, $Na_5P_3O_{10}$. When present in a detergent the tripolyphosphate ions sequester or complex heavy metal ions in the manner suggested by Figure 22-9. These ions would otherwise interfere with the cleaning action of the detergent. Unfortunately, polyphosphates are believed to have some adverse environmental impacts and their use has been curtailed (see Section 26-6). Sodium carbonate and sodium silicate seem to be suitable substitutes, however.

PORPHYRINS.   The structure illustrated in Figure 22-10 is commonly found in both plant and animal matter. If the groups substituted at the bonds shown in black are all

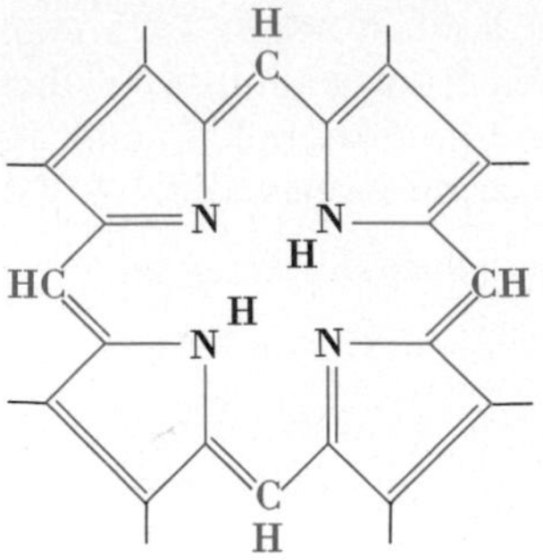

FIGURE 22-10
The porphyrin structure.

H atoms, the molecule is called porphin. By substituting other groups at these eight positions, the structures obtained are called porphyrins. Metal ions can replace the two H atoms on the central N atoms and coordinate simultaneously with all four N atoms. The porphyrin is a tetradentate ligand or chelating agent for the central metal ion. In chlorophyll the central ion is Mg^{2+}, and in hemoglobin, Fe^{2+} (see Chapter 25).

Summary

Many metal ions, particularly those of the transition elements, have the ability to form bonds with electron-donor groups (ligands). The number of ligands coordinated and the geometrical distribution of these ligands about the metal ion are distinctive features of a complex ion. Some ligands have more than one electron pair for donation. These multidentate ligands are able to attach simultaneously to two or more positions in the coordination sphere of the central metal ion. This multiple attachment produces complexes with five- or six-membered rings of atoms—chelates.

To relate the names and formulas of complex ions and of compounds containing them (coordination compounds) requires application of a set of rules. These rules deal with such matters as denoting the number and kind of ligands, the oxidation state of the central metal ion, and whether the complex ion is a cation or an anion. The positions in the coordination sphere of the central ion at which attachment of ligands may occur are not always equivalent. In geometrical isomerism different structures with different properties result depending on where this attachment occurs. Other forms of isomerism depend on such factors as the atom of the ligand group through which linkage occurs, and whether groups in a coordination compound are present within or outside the coordination sphere of the metal ion.

The structures of complex ions can be rationalized by appropriate orbital hybridization schemes using the valence bond theory. A more successful theory for explaining the magnetic properties and characteristic colors of complex ions is the crystal field theory. This theory emphasizes the splitting of the *d* energy level as a result of repulsions between electrons of the central ion and of the ligands. Different splitting patterns are obtained for different geometrical structures. Strong Lewis bases produce a "strong field" and a comparatively large energy separation between the two groups of *d* orbitals. Weaker Lewis bases produce a "weak field" and a smaller energy separation. A prediction of the magnitude of *d*-level splitting produced by a ligand can be made through the spectrochemical series.

The formation of a complex ion from a central metal ion and appropriate ligands can be viewed as an equilibrium process with an equilibrium constant known as the formation constant, K_f. In general, if the formation constant of a complex ion is large, the concentration of *free* metal ion in equilibrium with the complex ion is very small. Also important is the rate at which a complex ion exchanges ligands between its coordination sphere and the solution. Exchange is rapid in a labile complex and slow in an inert complex.

The formation of complex ions can be used to stabilize certain oxidation states, such as that of Co(III). Other applications include dissolving precipitates, such as AgCl by $NH_3(aq)$ in the qualitative analysis scheme and AgBr by $Na_2S_2O_3(aq)$ in the photographic process. With K_f values in conjunction with K_{sp} values, one may calculate whether precipitation should occur from a solution of complex ions and a precipitating reagent. Alternatively, one may calculate the solubility of a slightly soluble solute in the presence of a complexing reagent.

Learning Objectives

As a result of studying Chapter 22, you should be able to

1. Identify the central ion and the ligands in a complex ion; determine the coordination number and oxidation state of the central ion; and establish the net charge on the complex ion.

2. List the coordination numbers of some common metal ions and write the names and formulas of some common unidentate and multidentate ligands.

3. Write distinctive names based on the formulas of complex ions and coordination compounds, and distinctive formulas based on names.

4. Draw plausible structures for complex ions from information conveyed by their names and formulas.

5. Describe the types of isomerism found among complex ions and identify the possible isomers in specific cases.

6. Use the valence bond method to describe bonding and structures of complex ions.

7. Explain the basis of the crystal field theory of bonding in complex ions.

8. Use the spectrochemical series to make predictions about

d-level splitting and the number of unpaired electrons in complex ions.

9. Explain the origin of color in aqueous solutions of complex ions.

10. Write equations showing the effect of complex ion formation on other equilibrium processes, such as solubility equilibrium.

11. Use complex ion formation constants, K_f, to calculate the concentrations of free ion, ligands, and complex ions in solution.

12. Use K_f values in conjunction with K_{sp} values to make predictions about the solubilities of slightly soluble solutes in the presence of complexing reagents.

13. Explain amphoterism from the standpoint of complex ion equilibria.

14. Cite ways in which complex ion equilibria are used in the qualitative analysis scheme.

15. Describe applications of complex ion formation in the photographic process, in electroplating, in metallurgy, and in water treatment.

Some New Terms

A **chelate** results from the attachment of multidentate ligands to a metal ion. Chelates are characterized by five- and six-membered rings which include the central metal ion and atoms of the ligands.

A **chelating agent** is a *multidentate* ligand in a complex ion. That is, the chelating agent is simultaneously attached to two or more positions in the coordination sphere of the central metal ion.

A **complex ion** is a combination of a central metal ion and surrounding groups called ligands.

A **coordination compound** is a substance containing complex ions.

Coordination isomerism arises in coordination compounds consisting of both a complex cation and a complex anion. An interchange of ligands between the two complex ions leaves the composition of the compound unchanged.

Coordination number is the number of positions available for the attachment of ligands to a central metal ion in a complex ion.

The **coordination sphere** is the region around a metal ion where linkage to ligands can occur to produce a complex ion.

Crystal field theory describes bonding in complex ions in terms of electrostatic attractions between ligand electrons and the nucleus of the central metal ion. Particular attention is focused on the splitting of the d energy level of the

central metal ion that results from electron repulsions.

The **formation constant, K_f**, of a complex ion is the equilibrium constant describing equilibrium among a complex ion, the free metal ion, and ligands. The larger the value of a formation constant, the more stable the complex ion.

Geometric isomerism refers to the formation of nonequivalent structures based on the particular positions at which ligands are attached to a central metal ion in a complex ion.

Inert is the term used to describe a complex ion in which the exchange of ligands occurs only very slowly.

Ionization isomerism arises when a ligand from the coordination sphere of a metal ion is exchanged for an ion outside the coordination sphere.

Labile is the term used to describe a complex ion in which rapid exchange of ligands occurs.

Ligands are the electron-donating groups which are coordinated to a central metal ion in a complex ion.

Linkage isomerism applies to complex ions with the same compositions but with one or more ligands bonded differently. For example, the nitrite ion, NO_2^-, may bond through the N atom (nitro) or through one of the O atoms (nitrito).

The **spectrochemical series** is a ranking of ligand abilities to produce a splitting of the d energy level of a central metal ion in a complex ion.

Exercises

Definitions and terminology

1. The following terms relate to the constituents of a complex ion. What is the meaning of each? **(a)** ligand; **(b)** coordination number; **(c)** multidentate ligand; **(d)** aqua complex.

2. The following terms relate to the structure of a complex ion. What is the meaning of each? **(a)** square planar geometry; **(b)** d^2sp^3 hybrid bonding; **(c)** *cis* isomer.

3. What characteristics distinguish a chelate complex from an ordinary complex ion?

Nomenclature

4. What is the coordination number and the oxidation state of the metal ion in each of the following complex ions?
  (a) $[Ni(NH_3)_6]^{2+}$ **(b)** $[AlF_6]^{3-}$
  (c) $[Cu(CN)_4]^{3-}$ **(d)** $[Cr(NH_3)_3Br_3]$
  (e) $[Fe(ox)_3]^{3-}$

5. Name the following complex ions:
  (a) $[Ag(NH_3)_2]^+$ **(b)** $[Fe(H_2O)_6]^{3+}$
  (c) $[CuCl_4]^{2-}$ **(d)** $[Pt(en)_2]^{2+}$
  (e) $[Co(NH_3)_4(NO_2)Cl]^+$

6. Name the following coordination compounds.
  (a) $[Co(NH_3)_5Br]SO_4$ **(b)** $[Co(NH_3)_5SO_4]Br$
  (c) $[Cr(NH_3)_6][Co(CN)_6]$ **(d)** $Na_3[Co(NO_2)_6]$
  (d) $[Co(en)_3]Cl_3$

7. Write appropriate formulas for the following species.
  (a) dicyanosilver(I) ion
  (b) tetrachlorodiamminenickelate(II) ion
  (c) hexachloroplatinate(IV) ion
  (d) sodium tetrachlorocuprate(II)
  (e) potassium hexacyanoferrate(II)
  (f) bis(ethylenediamine)copper(II) ion
  (g) dihydroxotetraaquaaluminum(III) chloride
  (h) chlorobis(ethylenediamine)amminechromium(III) sulfate.

Bonding and structure of complex ions

8. What type of geometric structure would you predict for the complex ion $[Au(CN)_2]^-$? Explain.

9. The complex ion $[Ni(CN)_4]^{2-}$ has a square planar structure. The free Ni^{2+} ion has the electron configuration $[Ar]3d^8$.
  (a) Draw diagrams to represent bonding in this complex ion and its structure.
  (b) What would you expect its magnetic properties to be?

10. The complex ion $[FeCl_4]^-$ has a tetrahedral structure. Propose a bonding scheme for this complex ion using the valence bond theory. How many unpaired electrons are present in the structure?

11. In what way does the bonding scheme for the ion $[Co(NH_3)_6]^{2+}$ differ from that given in the text for $[Co(NH_3)_6]^{3+}$?

12. Draw structures to represent these complex ions.
  (a) $[Fe(en)Cl_4]^-$ **(b)** $[Fe(en)(ox)Cl_2]^-$
  (c) $[Fe(ox)_3]^{3-}$

13. Draw plausible structures of the following chelate complexes.
  (a) $[Pt(en)_2]^{2+}$ **(b)** $[Cr(dien)_2]^{3+}$
  (c) $[Fe(EDTA)]^{2-}$

14. Draw a structure to represent the complex ion *cis*-$[Cr(NH_3)_4ClOH]^+$.

15. How many isomers are possible for each of the following complex ions?
  (a) $[Co(NH_3)_5H_2O]^{3+}$ **(b)** $[Co(NH_3)_4(H_2O)_2]^{3+}$
  (c) $[Co(NH_3)_3(H_2O)_3]^{3+}$ **(d)** $[Co(NH_3)_2(H_2O_4)]^{3+}$

16. Would you expect *cis-trans* isomerism to occur in a complex ion with a **(a)** tetrahedral; **(b)** square planar; **(c)** linear structure? Explain.

17. **(a)** Draw a structure to represent the complex ion $[Co(en)_3]^{3+}$ (where *en* = ethylenediamine, a bidentate ligand).
  (b) Draw a structure for *cis*-dichlorobis(ethylenediamine)cobalt(III) ion.

18. Indicate whether isomerism is possible in each of the following cases. If so, indicate what type of isomerism is found.
  (a) $[Zn(NH_3)_4][CuCl_4]$ **(b)** $[Fe(CN)_5SCN]^{4-}$
  (c) $[Ni(NH_3)_5Cl]^+$ **(d)** $[Pt(py)Cl_3]^-$
  (e) $[Cr(NH_3)_3(OH)_3]^-$

Electrical conductivity

19. Explain why a difference in electrical conductivity should be expected for compounds I, II, and III on page 548. Which would you expect to be the best and which the poorest conductor?

20. Another member of the series of compounds referred to in Exercise 19 is $CoCl_3 \cdot 3 NH_3$. How would you expect its conductivity to compare with those of compounds I, II, and III?

Crystal field theory

21. In both $[Fe(H_2O)_6]^{2+}$ and $[Fe(CN)_6]^{4-}$ the iron is present as Fe^{2+}, yet $[Fe(H_2O)_6]^{2+}$ is paramagnetic whereas $[Fe(CN)_6]^{4-}$ is diamagnetic. Explain this difference.

22. One of the following ions is paramagnetic and one is diamagnetic; which is which? **(a)** $[MoCl_6]^{3-}$; **(b)** $[Co(en)_3]^{3+}$

23. If the ion Cr^{2+} is linked with strong-field ligands to produce an octahedral complex ion, the complex ion has *two* unpaired electrons. If Cr^{2+} is linked with weak-field ligands, the complex ion has *four* unpaired electrons. How do you explain this fact?

24. In contrast to the case of Cr^{2+} considered in Exercise 23, no matter what ligand is linked to Cr^{3+} to form an octahedral complex ion, the complex ion always has *three* unpaired electrons. Explain this fact.

25. Cyano complexes of transition metal ions (e.g., Fe^{2+} and Cu^{2+}) are often yellow in color, whereas aquo complexes are often green or blue. Why is there this difference?

26. Based on the energy level diagram for a tetrahedral complex ion, predict the number of unpaired electrons expected for the ion $[FeCl_4]^-$. Compare this with the result of Exercise 10.

Complex ion equilibria

27. Write equations to represent the following observations.
   (a) A mixture of $Mg(OH)_2(s)$ and $Zn(OH)_2(s)$ is treated with $NH_3(aq)$. The $Zn(OH)_2$ dissolves but the $Mg(OH)_2(s)$ is left behind.
   (b) When $NaOH(aq)$ is added to $CuSO_4(aq)$, a pale blue precipitate forms. If $NH_3(aq)$ is added, the pale blue precipitate redissolves, producing a solution with an intense deep-blue color. If this solution is made acidic with $HNO_3(aq)$, the deep-blue color is converted to a light blue.
   (c) A quantity of $CuCl_2(s)$ is dissolved in concentrated $HCl(aq)$ and produces a yellow solution. The solution is diluted to twice its volume with water and assumes a green color. Upon dilution to 10 times its original volume, the solution becomes pale blue in color.

28. Explain why the aqueous solubility of $PbCl_2(s)$ is unaffected by nitric acid but increases in hydrochloric acid.

29. Refer to Example 22-7. How much KI could be dissolved in 1.00 L of the $AgNO_3-NH_3$ solution described before AgI(s) would precipitate?

30. Should $PbI_2(s)$ precipitate if 0.020 g KI is added to 0.400 L of a solution that is 0.10 M in $PbCl_3^-$ and 1.5 M in *free* Cl^-? $[K_{sp}(PbI_2) = 7.1 \times 10^{-9}.]$

31. What concentration of *free* CN^- must be maintained in a solution that is 2.5 M $AgNO_3$ and 0.410 M NaCl to prevent AgCl(s) from precipitating? $[K_{sp}(AgCl) = 1.6 \times 10^{-10}.]$

32. A solution is prepared that is 0.10 M in *free* NH_3, 0.10 M NH_4Cl, and 0.15 M in $[Cu(NH_3)_4]^{2+}$. Should $Cu(OH)_2(s)$ precipitate from this solution? $(K_{sp}[Cu(OH)_2] = 1.6 \times 10^{-19}.)$

$\star$**33.** What mass of AgCl can be dissolved in 0.250 L of 0.100 M NH_3? (*Hint:* The molar solubility is equal to $[Cl^-]$, and $[Cl^-]$ is equal to the *total* silver concentration in solution.)

Applications

34. Write formulas for the following compounds.
   (a) iron(III) chloride hexahydrate
   (b) cobalt(II) hexachloroplatinate(IV) hexahydrate

35. Write simple chemical equations to show how the complex ion $[Cr(H_2O)_5OH]^{2+}$ acts as **(a)** an acid; **(b)** a base.

36. In an aqueous solution containing NH_3 and NH_4Cl, $CoCl_2$ can be oxidized to hexaamminecobalt(III) chloride using H_2O_2. Write a balanced equation to represent this reaction.

37. Why do you suppose $NH_3(aq)$ cannot be used in the "fixing" step in the photographic process?

38. A current of 1.50 A is passed for 0.333 h between a pair of copper electrodes. What mass of copper is deposited at the cathode if the electrolyte is **(a)** $[Cu(H_2O)_4]^{2+}$; **(b)** $[Cu(CN)_3]^{2-}$? Why are these quantities different?

39. What minimum mass of NaCN (in aqueous solution) must be used to leach the gold from 1000 kg of an ore containing 0.005% Au? Why must the process be carried out using more than this minimum mass?

Additional Exercises

1. What are the oxidation state and coordination number of the metal ion in each of the following?
   (a) $[Au(CN)_2]^-$ **(b)** $[Zn(NH_3)_4]I_2$
   (c) $[Co(NH_3)_2(NO_2)_4]^-$ **(d)** $[Cr(H_2O)(en)_2OH]^+$
   (e) $[Pt(en)(ONO)Cl]$

2. Write formulas for the following coordination compounds.
   (a) diaquatetraamminechromium(II) sulfate
   (b) sodium hexafluoroaluminate(III)
   (c) potassium tetracyanonickelate(II)
   (d) dichlorobis(ethylenediamine)cobalt(III) chloride

3. Name the following complex ions and coordination compounds.
   (a) $[Cr(NH_3)_4(OH)_2]Br$ **(b)** $K_3[Co(NO_2)_6]$
   (c) $[Fe(H_2O)_2(en)_2]^{2+}$ **(d)** $[Pt(en)_2Cl_2]SO_4$

4. Sketch all the possible isomers of $[Co(ox)(NH_3)_2Cl_2]^-$.

5. The common complex ion of Hg^{2+} and I^- is $[HgI_4]^{2-}$, but also encountered is $[HgI_3]^-$. What type of structure would you predict for **(a)** $[HgI_4]^{2-}$; **(b)** $[HgI_3]^-$?

$\star$**6.** The solubility of AgCN(s) in 0.200 M NH_3 is found to be 8.8×10^{-6} mol/L. What is the value of K_{sp} for AgCN(s)?

⋆7. **(a)** Draw a diagram similar to (22.16) to represent the distribution of *d* electrons in the octahedral complex ion $[Ti(H_2O)_6]^{3+}$.

(b) The ion $[Ti(H_2O)_6]^{3+}$ absorbs light of wavelength 490 nm. What is the energy separation, Δ, expressed in kJ/mol?

⋆8. With reference to the stability of $[Co(NH_3)_6]^{3+}$ in aqueous solution

(a) Verify that E_{cell}° for reaction (22.31) is +0.59 V.

(b) Calculate $[Co^{3+}]$ in a solution that has a total concentration of cobalt of 1.0 *M* and $[NH_3] = 0.10$ *M*.

(c) Show that for the value of $[Co^{3+}]$ calculated in part (b), reaction (22.31) will not occur. (*Hint:* What is $[H_3O^+]$ in 0.10 *M* NH_3? Assume a low, but reasona-

ble, concentration of Co^{2+}, say 1×10^{-4} *M*, and a partial pressure of $O_2(g)$ of 0.2 atm.)

⋆9. Use data provided in the text to demonstrate that if a solution is 0.10 *M* in total copper, 0.10 *M* in total cadmium, 0.1 *M* in free CN^-, and saturated in H_2S, CdS will precipitate but not Cu_2S. [Assume a pH of 7. $K_{sp}(CdS) = 8.0 \times 10^{-27}$; $K_{sp}(Cu_2S) = 1.2 \times 10^{-49}$.]

⋆10. A copper electrode is immersed in a solution that is 1.00 *M* in NH_3 and 1.00 *M* in $[Cu(NH_3)_4]^{2+}$. If a standard hydrogen electrode is the anode, the potential difference between it and the copper electrode is found to be -0.08 V. What is the value obtained for the formation constant of the complex ion by this method?

Self-Test Questions

For questions 1 through 6 select the single item that best completes each statement.

1. The oxidation state of nickel in the complex ion $[Ni(CN)_4I]^{3-}$ is   **(a)** -3;   **(b)** -2;   **(c)** $+2$;   **(d)** $+5$.

2. The coordination number of gold in the complex ion $[Au(dien)Cl_3]^{2-}$ is   **(a)** 3;   **(b)** 4;   **(c)** 5;   **(d)** 6.

3. Of the following complexes, one exhibits isomerism. That one is

(a) $[Ag(NH_3)_2]^+$ **(b)** $[Co(NH_3)_5NO_2]^{2+}$
(c) $[Pt(en)Cl_2]$ **(d)** $[Co(NH_3)_5Cl]^{2+}$

4. Of the following complex ions, one is a Brønsted–Lowry acid. That one is

(a) $[Cu(NH_3)_4]^{2+}$ **(b)** $[FeCl_4]^-$
(c) $[Al(H_2O)_6]^{3+}$ **(d)** $[Zn(OH)_4]^{2-}$

5. The most soluble of the following compounds in $NH_3(aq)$ is   **(a)** $BaSO_4$;   **(b)** $Cu(OH)_2$;   **(c)** SiO_2;   **(d)** $MgCO_3$.

6. The number of unpaired electrons expected for the complex ion $[Cr(NH_3)_6]^{2+}$ is   **(a)** 2;   **(b)** 3;   **(c)** 4;   **(d)** 5.

7. Sketch plausible geometric structures of the following complexes.

(a) $[Co(NH_3)_5Cl]^{2+}$ **(b)** $[Cr(en)_2(ox)]^+$
(c) *cis*-dinitrodiammineplatinum(II)
(d) *trans*-trichlorotripyridinecobalt(III)

8. Explain the following observations in terms of complex ion formation.

(a) $Al(OH)_3(s)$ is soluble in $NaOH(aq)$ but insoluble in $NH_3(aq)$.

(b) $ZnCO_3(s)$ is soluble in $NH_3(aq)$ but $ZnS(s)$ is not.

(c) $CoCl_3$ is unstable in water solution, being reduced to $CoCl_2$ and liberating $O_2(g)$. On the other hand, $[Co(NH_3)_6]Cl_3$ can be easily maintained in aqueous solution.

9. Describe how the crystal field theory makes possible an explanation of the fact that so many transition metal compounds are colored.

10. Determine which of the following compounds is appreciably soluble in aqueous NH_3:   **(a)** CuS, $K_{sp} = 6.3 \times 10^{-36}$;   **(b)** $CuCO_3$, $K_{sp} = 1.4 \times 10^{-10}$. Also use the fact that $K_f[Cu(NH_3)_4]^{2+} = 2.1 \times 10^{14}$.

23 Nuclear Chemistry

The ordinary chemistry of the elements is based on phenomena in which electrons play a major role. Both chemical properties and those physical properties related to intermolecular forces are derived from the electronic structures of atoms, ions, and molecules. The primary functions of atomic nuclei in the phenomena we have considered to this point are in establishing the masses of atoms and molecules and in furnishing a center of positive charge to keep electrons in position.

In this chapter we consider a variety of phenomena stemming *directly* from the nuclei of atoms. We will refer to these phenomena, collectively, as nuclear chemistry. Nuclear chemical phenomena are of primary importance in dealing with the actinoid (actinide) elements. However, these phenomena are also encountered with certain nuclides of the lighter elements, in particular, with artificial nuclides. Yet another part of nuclear chemistry is a study of the effects of ionizing radiation on matter—one of the central issues in the current "nuclear debate."

23-1 The Phenomenon of Radioactivity

The term "radioactivity" was proposed by Marie Curie to describe the most readily observable phenomenon that accompanies transformations of certain atomic nuclei—the emission of ionizing radiation. Ionizing radiation, as the name implies, interacts with matter to produce ions. This means that the radiation is sufficiently energetic to break chemical bonds. Some ionizing radiation is particulate (consists of particles) and some is nonparticulate. We will return to a discussion of the effects of ionizing radiation on matter in Section 23-10. For the present, let us comment briefly on the types of ionizing radiation associated with radioactivity. This will take the form of a review and extension of ideas first encountered in Chapter 2.

ALPHA (α) RAYS. This radiation consists of a stream of alpha (α) particles. These particles are identical to the nuclei of helium-4 atoms. Alpha rays produce large numbers of ions as they penetrate matter, even though their penetrating power is low. (They can generally be stopped by a sheet of paper.) Because of their positive charge, α particles are deflected by magnetic and electric fields (see Figure 23-1). The production of α particles by a radioactive nucleus can be represented through a nuclear equation. A **nuclear equation** is written in a fashion similar to a chemical equation. However, the condition of balance in a nuclear equation requires the sum of atomic numbers and the sum of mass numbers on the left side of the equation to

FIGURE 23-1
Three types of radiation from radioactive materials.

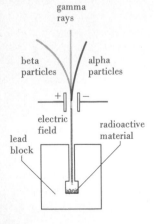

The radioactive material is enclosed in a lead block with a narrow opening. All the radiation except for that passing through the opening is absorbed by the lead. When this radiation is passed through an electric field, it splits into three beams. One beam is undeflected—these are gamma (γ) rays. A second beam is attracted toward the negatively charged plate; these are the positively charged alpha (α) particles. The third beam is deflected toward the positive plate. This is the beam of beta (β) particles—electrons. Because of their greater momentum (product of mass and velocity), α particles are deflected to a smaller extent than β particles. A similar situation exists if a magnetic field is substituted for an electric field.

equal the corresponding sums on the right. In equation (23.1), mass numbers are written as superscript numerals; atomic numbers are subscript numerals; and the α particle is represented as ^4_2He. Mass numbers total to 238, and atomic numbers to 92. The loss of an α particle by a nucleus causes a decrease of two in its atomic number and four in its mass number.

$$^{238}_{92}\text{U} \longrightarrow \,^{234}_{90}\text{Th} + \,^4_2\text{He} \tag{23.1}$$

BETA (β^-) RAYS. These rays are also comprised of particles, and β^- particles are identical to electrons. Beta rays have a greater penetrating power but a lower ionizing power than α rays. (They can pass through aluminum foil up to about 2 to 3 mm thick.) Beta particles are also deflected by electric and magnetic fields but in the *opposite* direction from α particles (see Figure 23-1). Moreover, β^- particles suffer greater deflections in these fields than do α particles, because the β^- particles have so much smaller a mass than do α particles. The nucleus of $^{234}_{90}\text{Th}$, which was a product of reaction (23.1), is itself unstable. $^{234}_{90}\text{Th}$ undergoes radioactive decay by β^- particle emission. When a nucleus loses a β^- particle, represented as $_{-1}\text{e}$, the atomic number of the nucleus *increases* by one unit and its mass number is unchanged. Thus, a β^- particle is treated as if it had an atomic number of -1 and a mass number of 0.

$$^{234}_{90}\text{Th} \longrightarrow \,^{234}_{91}\text{Pa} + \,^{0}_{-1}\text{e} \tag{23.2}$$

FIGURE 23-2
Production of gamma rays.

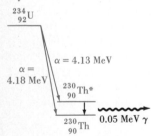

A transition of a $^{230}_{90}\text{Th}$ nucleus between the two energy states shown results in the emission of 0.05 MeV of energy in the form of γ rays.

GAMMA (γ) RAYS. Some radioactive decay processes yielding α or β particles leave a nucleus in an energetic state. The nucleus then loses energy in the form of electromagnetic radiation—a gamma (γ) ray. Gamma rays are a highly penetrating form of radiation. They are undeflected by electric and magnetic fields. In the radioactive decay of $^{234}_{92}\text{U}$, 77% of the nuclei emit α particles having an energy of 4.18 MeV. The remaining 23% of the $^{234}_{92}\text{U}$ nuclei produce α particles with energies of 4.13 MeV. In the latter case the $^{230}_{90}\text{Th}$ nuclei are left with an excess energy of 0.05 MeV. This energy is released as γ rays. If we denote the unstable, energetic Th nucleus as $^{230}_{90}\text{Th}^*$, we can write

The energy unit electron volt, eV, was introduced in Section 8-7. The unit MeV is a million electron volts. This unit and its relationship to other energy units are discussed further in Section 23-6.

$$^{234}_{92}\text{U} \longrightarrow \,^{230}_{90}\text{Th}^* + \,^4_2\text{He} \tag{23.3}$$

$$^{230}_{90}\text{Th}^* \longrightarrow \,^{230}_{90}\text{Th} + \gamma \tag{23.4}$$

This γ emission process is represented diagrammatically in Figure 23-2.

POSITRONS (β^+). The emission of β^- rays, as we shall discover in Section 23-7, is characteristic of nuclei in which the ratio of number of neutrons to number of protons is too large for stability. If this ratio is too small for stability, radioactive decay may occur by positron emission. A positron is a *positively* charged particle that is otherwise identical to a β^- particle or electron. It is designated as β^+ or as $_{+1}^{0}\text{e}$. Positron emission is commonly encountered with artifically radioactive nuclei of the lighter elements (see Section 23-3). For example,

$$^{30}_{15}P \longrightarrow {}^{30}_{14}Si + {}^{0}_{+1}e \qquad (23.5)$$

ELECTRON CAPTURE (E.C.). A second process that often occurs along with positron emission and achieves the same effect is electron capture (E.C.). In this process an electron from an inner electron shell (usually the K or L shell) is absorbed by the nucleus. (Inside the nucleus the electron is used to convert a proton into a neutron.) The dropping of an electron from a higher quantum level into that vacated by the captured electron results in the emission of x radiation. For example,

$$^{202}_{81}Tl \xrightarrow{\text{E.C.}} {}^{202}_{80}Hg \quad \text{(followed by x radiation)} \qquad (23.6)$$

Example 23-1 Write nuclear equations to represent **(a)** α particle emission by ^{222}Rn; **(b)** radioactive decay of bismuth-215 to polonium; **(c)** decay of a radioactive nucleus to produce ^{58}Ni and a positron.

Note that if a name or chemical symbol is given it is not necessary also to state the atomic number in order to identify an element.

(a) Two of the species involved in this process can be written directly from the information given. The remaining species is identified by using the basic principle for balancing nuclear equations.

$$^{222}_{86}Rn \longrightarrow (?) + {}^{4}_{2}He \qquad and \qquad {}^{222}_{86}Rn \longrightarrow {}^{218}_{84}Po + {}^{4}_{2}He$$

(b) Bismuth has the atomic number 83, and polonium, 84. The type of emission that leads to an increase of one unit in atomic number is the β^- ray.

$$^{215}_{83}Bi \longrightarrow {}^{215}_{84}Po + {}^{0}_{-1}e$$

(c) We are given the products of the radioactive decay process. The only radioactive nucleus that can produce them is $^{58}_{29}Cu$.

$$^{58}_{29}Cu \longrightarrow {}^{58}_{28}Ni + {}^{0}_{+1}e$$

SIMILAR EXAMPLES: Exercises 5, 6.

23-2 Naturally Occurring Radioactive Nuclides

$^{209}_{83}Bi$ is the nuclide of highest atomic and mass number that is stable. All known nuclides beyond it in atomic and mass numbers are radioactive. Naturally occurring $^{238}_{92}U$ is radioactive and disintegrates by the loss of α particles.

$$^{238}_{92}U \longrightarrow {}^{234}_{90}Th + {}^{4}_{2}He \qquad (23.7)$$

$^{234}_{90}Th$ is also radioactive; its decay is by β^- emission.

$$^{234}_{90}Th \longrightarrow {}^{234}_{91}Pa + {}^{0}_{-1}e \qquad (23.8)$$

$^{234}_{91}Pa$ also decays by β^- emission.

$$^{234}_{91}Pa \longrightarrow {}^{234}_{92}U + {}^{0}_{-1}e \qquad (23.9)$$

$^{234}_{92}U$ is radioactive also.

RADIOACTIVE DECAY SERIES. The chain of radioactive decay begun with $^{238}_{92}U$ continues through a number of steps of α and β emission until it eventually terminates with a stable isotope of lead—$^{206}_{82}Pb$. The entire scheme is outlined in Figure 23-3. All naturally occurring radioactive nuclides of high atomic number belong to one of three decay series—the **uranium series** just described, the **thorium series,** or the **actinium series.**

By the "age of a rock" we mean the time elapsed since molten magma froze to become a rock.

One application of these radioactive decay schemes is determining the ages of rocks and thereby the age of the earth. This method assumes that the initial radioactive nuclide, the final stable nuclide, and all the products of a decay series remain in

FIGURE 23-3
The natural radioactive decay series for $^{238}_{92}$U.

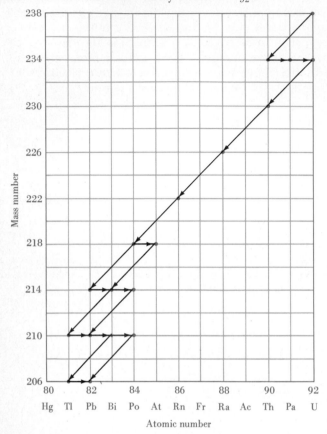

The long arrows pointing down and to the left correspond to α particle emissions. The short horizontal arrows represent β^- emissions.

the rock sample (see Section 23-5). The appearance of certain radioactive substances in the environment can also be explained through radioactive decay series.

^{222}Rn produced by the decay of ^{238}U is now known to be the cause of a fatal lung disease afflicting miners in central Europe for centuries. This disease was lung cancer and the uranium responsible for it was present in the gold, silver, and platinum ores that were being mined. ^{210}Po and ^{210}Pb have been detected in cigarette smoke. These radioactive nuclides are derived from ^{238}U, found in trace amounts in the phosphate fertilizers used in tobacco fields. Recently, it has been established that these α-emitting nuclides are implicated in the link between cigarette smoking and cancer and heart disease.

Radioactivity, which is so common among the nuclides of high atomic number, is a relatively rare phenomenon among the *naturally occurring* lighter nuclides. ^{40}K is a radioactive nuclide, as are ^{50}V and ^{138}La. ^{40}K decays by β^- emission and by electron capture.

$$^{40}_{19}K \longrightarrow {}^{40}_{20}Ca + {}^{0}_{-1}e \quad and \quad {}^{40}_{19}K \xrightarrow{\text{E.C.}} {}^{40}_{18}Ar$$

At the time the earth was formed ^{40}K was much more abundant than it is now. It is believed that the high argon content of the atmosphere (0.934%, by volume, and almost all of it as ^{40}Ar) has been derived from the radioactive decay of ^{40}K. Aside from ^{40}K, the most important radioactive nuclides of the lighter elements are those that are *artificially* produced.

23-3 Nuclear Reactions and Artificial Radioactivity

We described in Chapter 2 how Rutherford discovered the existence of the proton outside the nucleus of an atom. He found that bombardment of nitrogen nuclei by α particles produced protons, the other product being O atoms.

$$^{14}_{7}N + {}^{4}_{2}He \longrightarrow {}^{17}_{8}O + {}^{1}_{1}H \tag{23.10}$$

Although the principles involved in writing equation (23.10) are similar to those used to represent radioactive decay, there is this difference: Two "reactants" are written on the left. Instead of a nucleus disintegrating spontaneously, it must be struck by another small particle to induce a **nuclear reaction.** In the more condensed representation given by (23.11), the target and product nuclei are represented on the left and right of a parenthetical expression. Within the parentheses the bombarding particle is written first, followed by the ejected particle.

$$^{14}\text{N}(\alpha,\text{p})^{17}\text{O} \tag{23.11}$$

$^{17}_{8}\text{O}$ is a naturally occurring *nonradioactive* nuclide of oxygen (0.037% natural abundance). The situation with $^{30}_{15}\text{P}$, which can also be produced by a nuclear reaction, is somewhat different.

In 1934, when bombarding aluminum with α particles, Irene Curie and her husband Frédéric Joliot observed the emission of two types of particles—neutrons and positrons. The Joliots observed that when bombardment by α particles was terminated, the emission of neutrons also stopped; that of positrons, however, continued. Their conclusion was that the nuclear bombardment produces $^{30}_{15}\text{P}$, which undergoes radioactive decay by the emission of positrons.

$$^{27}_{13}\text{Al} + ^{4}_{2}\text{He} \longrightarrow ^{30}_{15}\text{P} + ^{1}_{0}\text{n} \qquad \text{or} \qquad ^{27}\text{Al}(\alpha,\text{n})^{30}\text{P} \tag{23.12}$$

$$^{30}_{15}\text{P} \longrightarrow ^{30}_{14}\text{Si} + ^{0}_{+1}\text{e} \tag{23.13}$$

$^{30}_{15}\text{P}$ was the first artificially produced radioactive nuclide. Since the time of its discovery over 1000 more have been obtained. The number of known radioactive nuclides now exceeds considerably the number of nonradioactive nuclides (about 280). A few of the many applications of artificial radioactivity are considered in Section 23-11.

Example 23-2 Write **(a)** a condensed representation of the nuclear reaction, $^{14}_{7}\text{N} + ^{1}_{0}\text{n} \rightarrow ^{14}_{6}\text{C} + ^{1}_{1}\text{H}$; **(b)** an expanded representation of the nuclear reaction, $^{59}\text{Co}(\text{n},?)^{56}\text{Mn}$.
(a) The bombarding particle is simply a neutron, n, and the emitted particle is a proton, p. The condensed equation is $^{14}\text{N}(\text{n},\text{p})^{14}\text{C}$.
(b) The missing particle in parentheses must have $A = 4$ and $Z = 2$; it is an α particle.

$$^{59}_{27}\text{Co} + ^{1}_{0}\text{n} \longrightarrow ^{56}_{25}\text{Mn} + ^{4}_{2}\text{He}$$

SIMILAR EXAMPLES: Exercises 10, 11.

23-4 Transuranium Elements

Until 1940 the only known elements were those that occur naturally. In 1940, the first synthetic element was produced by bombarding $^{238}_{92}\text{U}$ atoms with neutrons. First the unstable nucleus $^{239}_{92}\text{U}$ is formed. This nucleus then undergoes β^{-} decay, yielding the element neptunium, with $Z = 93$.

$$^{238}_{92}\text{U} + ^{1}_{0}\text{n} \longrightarrow ^{239}_{92}\text{U} \tag{23.14}$$

$$^{239}_{92}\text{U} \longrightarrow ^{239}_{93}\text{Np} + ^{0}_{-1}\text{e} \tag{23.15}$$

Bombardment by neutrons is a particularly effective way to produce nuclear reactions because these heavy, uncharged particles are not repelled as they approach a nucleus.

Since 1940 all the elements from $Z = 93$ to 106 have been produced artificially. For example, an isotope of the element, $Z = 105$, was produced in 1970 by bombarding atoms of $^{249}_{98}\text{Cf}$ with $^{15}_{7}\text{N}$ nuclei.

$$^{249}_{98}\text{Cf} + ^{15}_{7}\text{N} \longrightarrow ^{260}_{105}\text{Ha} + 4\,^{1}_{0}\text{n} \tag{23.16}$$

Elements 104 and 105 are the first in a series of transition elements following the actinoid series; they might be called transactinoid elements. From our knowledge of the periodic table, we should predict these elements to resemble hafnium and tantalum, respectively. Moreover, there is every reason to believe that more elements will be synthesized.

FIGURE 23-4
A charged-particle accelerator—the cyclotron.

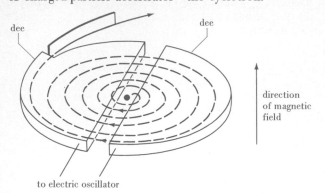

to electric oscillator

The accelerator consists of two hollow, flat, semicircular boxes, called dees, which are kept electrically charged. The entire assembly is maintained within a magnetic field. The particles to be accelerated, in the form of positive ions, are produced at the center of the opening between the dees. They are then attracted into the negatively charged dee and forced into a circular path by the magnetic field. When the particles leave the dee and enter the gap, the electric charges on the dees are reversed, so that the particles are attracted into the opposite dee. The particles are accelerated as they pass the gap, and travel a wider circular path in the new dee. This process is repeated many times over until the particles are brought to the required energy to induce the desired nuclear reaction. They are then brought out of the accelerator and made to strike the target.

A simple analogy to the nuclear accelerator is found in the action of a playground swing. If the rider is given a push each time the swing reaches the end of its arc, the swing accelerates and the arc grows wider with each push.

CHARGED-PARTICLE ACCELERATORS. To bring about nuclear reactions such as (23.16) requires that energetic particles be used as projectiles for bombarding atomic nuclei. Such energetic particles can be obtained in an accelerator. A type of accelerator known as a cyclotron is pictured and described in Figure 23-4.

A charged-particle accelerator, as the name implies, can only produce beams of charged particles as projectiles (e.g., $_1^1H^+$). In many cases neutrons are more effective as projectiles for nuclear bombardment. The neutrons required can themselves be produced through a nuclear reaction using a charged-particle beam. In the following reaction $_1^2H$ represents a beam of deuterons (actually $_1^2H^+$) from an accelerator.

$$_4^9Be + _1^2H \longrightarrow _5^{10}B + _0^1n$$

Another important source of neutrons for nuclear reactions is a nuclear reactor (see Section 23-8).

23-5 Rate of Radioactive Decay

In time, we can expect every atomic nucleus of a radioactive nuclide to disintegrate, but it is impossible to predict when a given nucleus will do so. Radioactivity is a random process. However, this all-important observation has been made: *The rate of disintegration of a radioactive material, that is, the decay rate, is directly proportional to the number of atoms present in a sample.* Consider a case where, *on the average,* 100 atoms undergo disintegration each second in a 1-million-atom sample, we would expect the average decay rate to be 200 atoms/s in a 2×10^6 atom sample, 50 atoms/s in a 5×10^5 atom sample, and so on. In mathematical terms,

rate of decay $\propto N$
and   (23.17)
rate of decay $= \lambda \cdot N$

The decay rate is expressed in atoms per unit time, such as atoms/s. N is the number of atoms in the sample being observed. λ is the **decay constant;** its unit is (time)$^{-1}$.

Equation (23.17) represents a first-order process. By simply substituting N for [A] and λ for k in the several equations presented in Chapter 13, we establish the following useful equations.

$$\log N_t - \log N_0 = \log \frac{N_t}{N_0} = \frac{-\lambda t}{2.303} \qquad (23.18)$$

$$t_{1/2} = \frac{0.693}{\lambda} \qquad (23.19)$$

Here N_t represents the number of atoms remaining at time t and N_0, the number at some initial time ($t = 0$).

TABLE 23-1
Some representative half-lives

Nuclide	Half-life[a]	Nuclide	Half-life[a]
$_1^3H$	12.26 y	$_{38}^{90}Sr$	28.1 y
$_6^{14}C$	5730 y	$_{53}^{131}I$	8.070 d
$_8^{13}O$	8.7×10^{-3} s	$_{55}^{137}Cs$	30.23 y
$_{12}^{28}Mg$	21 h	$_{84}^{214}Po$	1.64×10^{-4} s
$_{15}^{32}P$	14.3 d	$_{86}^{222}Rn$	3.823 d
$_{16}^{35}S$	88 d	$_{88}^{226}Ra$	1600 y
$_{19}^{40}K$	1.28×10^9 y	$_{90}^{234}Th$	24.1 d
$_{35}^{80}Br$	17.6 min	$_{92}^{238}U$	4.51×10^9 y

[a] s, second; min, minute; h, hour; d, day; y, year.

Instead of characterizing radioactive decay through the decay constant λ, it is customary to do so through the **half-life, $t_{1/2}$**. The shorter the half-life, the larger the value of λ and the faster the decay process. Half-lives of radioactive nuclides range over periods of time from extremely short to very long, as suggested by the representative data in Table 23-1.

Example 23-3 $^{238}_{92}U$ has a half-life of 4.5×10^9 y. It decays by the process shown in equation (23.1). How many α particles are produced per second in a sample containing 1×10^{20} atoms of $^{238}_{92}U$?

From $t_{1/2}$ we obtain λ, using equation (23.19).

$$\lambda = \frac{0.693}{4.5 \times 10^9 \text{ y}} = 1.5 \times 10^{-10} \text{ y}^{-1}$$

We use equation (23.17) to solve for the rate of decay.

$$\text{rate of decay} = (1.5 \times 10^{-10} \text{ y}^{-1}) \times 10^{20} \text{ atoms} = 1.5 \times 10^{10} \text{ atoms/y} \qquad (23.20)$$

One α particle is produced for every atom that disintegrates. The rate of α-particle production is expressed by equation (23.20). On a *per second* basis, this becomes

$$\text{rate of decay} = 1.5 \times 10^{10} \text{ } \alpha \text{ particles/y} \times \frac{1 \text{ y}}{365 \text{ d}} \times \frac{1 \text{ d}}{24 \text{ h}} \times \frac{1 \text{ h}}{60 \text{ min}} \times \frac{1 \text{ min}}{60 \text{ s}}$$

$$= 480 \text{ } \alpha \text{ particles/s}$$

SIMILAR EXAMPLES: Exercises 13, 14, 15.

Radioactive decay is subject to statistical fluctuations, and the result obtained here is simply the average decay rate. It should not be taken to mean, literally, that for each and every second exactly 480 α particles are produced.

$^{14}_{6}C$ is formed at a constant rate in the upper atmosphere by the bombardment of $^{14}_{7}N$ with neutrons.

$^{14}_{7}N + ^{1}_{0}n \longrightarrow ^{14}_{6}C + ^{1}_{1}H$

The neutrons are produced by cosmic rays. $^{14}_{6}C$ disintegrates by β^- emission.

The rate of decay just prior to the ^{14}C equilibrium being destroyed is 15 dis min^{-1}/g C and at the time of the measurement, 10 dis min^{-1}/g C. The corresponding numbers of atoms are equal to these decay rates divided by λ.

RADIOCARBON DATING. Carbon-containing compounds in living organisms maintain an equilibrium with ^{14}C in the atmosphere. The ^{14}C nuclide is radioactive and has a half-life of 5730 y. The activity associated with carbon in this equilibrium is about 15 disintegrations per minute (dis/min) per gram of carbon. When an organism ceases to live (e.g., a tree is felled), this equilibrium is destroyed and the disintegration rate falls off. From the measured disintegration rate at some later time, an estimate of the age (i.e., elapsed time since death of the organism) can be made.

Example 23-4 A wooden object is found in an Indian burial mound and subjected to radiocarbon dating. The decay rate associated with its ^{14}C content is 10 dis min^{-1}/g C. What is the age of the object (i.e., time elapsed since the tree was cut down)?

In this example all three equations, (23.17), (23.18), and (23.19), are required. Equation (23.19) is again used to determine the decay constant.

$$\lambda = \frac{0.693}{5730 \text{ y}} = 1.21 \times 10^{-4} \text{ y}^{-1}$$

Next we use equation (23.17) to represent the actual number of atoms: N_0 at $t = 0$ (the time when the ^{14}C equilibrium was destroyed) and N_t at time t (the present time).

$$N_0 = \frac{\text{decay rate (at } t = 0)}{\lambda} = \frac{15}{\lambda}$$

$$N_t = \frac{\text{decay rate (at time } t)}{\lambda} = \frac{10}{\lambda}$$

Finally, we substitute into equation (23.18).

$$\log \frac{N_t}{N_0} = \log \frac{10/\lambda}{15/\lambda} = \log \frac{10}{15} = \frac{-(1.21 \times 10^{-4} \text{ y}^{-1})t}{2.303}$$

$$-0.18 = -(5.3 \times 10^{-5} \text{ y}^{-1})t$$

$$t = \frac{0.18}{5.3 \times 10^{-5} \text{ y}^{-1}} = 3.4 \times 10^3 \text{ y}$$

SIMILAR EXAMPLES: Exercises 21, 22.

AGE OF THE EARTH. The natural radioactive decay scheme of Figure 23-3 suggests the eventual fate awaiting all the $^{238}_{92}U$ found in nature—conversion to lead. Naturally occurring uranium minerals always have associated with them some nonradioactive lead formed by radioactive decay. From the weight ratio of $^{206}_{82}Pb$ to $^{238}_{92}U$ in such a mineral it is possible to make an estimate of the age of the earth. One assumption of this method is that any lead present in the rock initially consisted of the several isotopes of lead in their naturally occurring abundances. Another is that the mineral has undergone no chemical changes that would have allowed any of the radioactive nuclides in the decay series to escape.

An exact treatment of this subject would require some discussion of the relationship between the rates of decay of a radionuclide called a "parent" and the product nuclide called a "daughter." A discussion of these relationships is beyond the scope of this text, but an indication of the method is still possible.

The half-life of $^{238}_{92}U$ is 4.5×10^9 y. According to the natural decay scheme of Figure 23-3, the basic changes that occur as atoms of $^{238}_{92}U$ and its daughters pass through the entire sequence of steps is

$$^{238}_{92}U \longrightarrow {}^{206}_{82}Pb + 8\,^4_2He + 6\,^0_{-1}e \tag{23.21}$$

Discounting the mass associated with the β^- particles, we can see that for every 238 g of uranium that undergoes complete decay, 206 g of lead and 32 g of helium are produced.

Suppose that in a rock containing no lead initially, 1.000 g $^{238}_{92}U$ had disintegrated through one half-life period, 4.5×10^9 y. At the end of that time there would be present in the sample

0.500 g $^{238}_{92}U$ undisintegrated

and

$$0.500 \times \frac{206}{238} = 0.433 \text{ g } ^{206}_{82}Pb$$

with the ratio

$$\frac{^{206}_{82}Pb}{^{238}_{92}U} = \frac{0.433}{0.500} = 0.866 \tag{23.22}$$

A $^{206}_{82}Pb/^{238}_{92}U$ ratio smaller than that shown in (23.22) would suggest the solid mineral had not been in existence for as long as one half-life period of $^{238}_{92}U$. A higher ratio would suggest a greater age for the rock. The best estimates of the age of the earth are in fact about 4.5×10^9 y. These estimates are based on the $^{206}_{82}Pb$ to $^{238}_{92}U$ ratio and on ratios for other pairs of nuclides from natural radioactive decay series.

Example 23-5 The thorium radioactive decay series produces one atom of ^{208}Pb as the final disintegration product of an atom of ^{232}Th. The half-life of ^{232}Th is 1.39×10^{10} y. A certain rock is found to have a $^{208}Pb/^{232}Th$ mass ratio of $0.14:1.00$. Use these data to estimate the age of the rock.

The decay constant, λ, is obtained in the usual fashion.

$$\lambda = \frac{0.693}{1.39 \times 10^{10} \text{ y}} = 4.99 \times 10^{-11} \text{ y}^{-1}$$

Let us base our calculation on an amount of mineral containing 1.00 g ^{232}Th at the present time, t.

$$N_t = 1.00 \text{ g } ^{232}Th \times \frac{1 \text{ mol } ^{232}Th}{232 \text{ g } ^{232}Th} = 4.31 \times 10^{-3} \text{ mol } ^{232}Th$$

To determine the initial number of moles of ^{232}Th, N_0, we might reason as follows. The total mass of ^{232}Th present in the sample of mineral when it was formed must have been the 1.00 g present currently plus the mass of ^{232}Th required to produce 0.14 g ^{208}Pb.

$$\text{no. g } ^{232}\text{Th} = 0.14 \text{ g } ^{208}\text{Pb} \times \frac{232 \text{ g } ^{232}\text{Th}}{208 \text{ g } ^{208}\text{Pb}} = 0.16 \text{ g } ^{232}\text{Th}$$

The total mass of ^{232}Th present initially was $1.00 + 0.16 = 1.16$ g ^{232}Th. The initial number of moles of ^{232}Th was

$$N_0 = 1.16 \text{ g } ^{232}\text{Th} \times \frac{1 \text{ mol } ^{232}\text{Th}}{232 \text{ g } ^{232}\text{Th}} = 5.00 \times 10^{-3} \text{ mol } ^{232}\text{Th}$$

We now have the necessary data to substitute into equation (23.18) and solve for t.

The number of atoms in a sample of an element is directly proportional to the mass of the sample. We could simply have substituted 1.00 for N_t and 1.16 for N_0.

$$\log \frac{N_t}{N_0} = \log \frac{4.31 \times 10^{-3}}{5.00 \times 10^{-3}} = \frac{-4.99 \times 10^{-11}t}{2.303}$$

$$-0.0645 = -2.17 \times 10^{-11}t$$

$$t = \frac{-0.0645}{-2.17 \times 10^{-11} \text{ y}^{-1}} = 2.97 \times 10^9 \text{ y}$$

SIMILAR EXAMPLES: Exercises 19, 20.

23-6 Energetics of Nuclear Reactions

A complete assessment of a nuclear reaction requires use of the mass–energy equivalence given by Albert Einstein.

$$E = mc^2 \tag{23.23}$$

In chemical reactions energy changes are so small that the equivalent mass changes are undetectable (though real nevertheless). We say that mass is conserved in a chemical reaction. In nuclear reactions, energies involved are orders of magnitude greater. Perceptible changes in mass do occur.

If the exact masses of nuclides are known, it is possible to calculate the energy of a nuclear reaction using equation (23.23). The term m corresponds to the net change in mass, in kg, and c, the velocity of light, is expressed in m/s. The resulting energy is in joules. Another common unit for expressing nuclear energy is the MeV (million electron volt).

$$1 \text{ MeV} = 1.602 \times 10^{-13} \text{ J} \tag{23.24}$$

Example 23-6 What is the energy associated with the α decay of ^{238}U **(a)** in MeV; **(b)** in kJ/mol?

$$^{238}_{92}\text{U} \longrightarrow \ ^{234}_{90}\text{Th} + \ ^{4}_{2}\text{He}.$$

The nuclidic masses, in amu, are

$$^{238}_{92}\text{U} = 238.0508 \qquad ^{234}_{90}\text{Th} = 234.0437 \qquad ^{4}_{2}\text{He} = 4.0026$$

(a) The net change in mass that accompanies the decay of a single nucleus of ^{238}U is $234.0437 + 4.0026 - 238.0508 = -0.0045$ amu. This loss of mass appears as kinetic energy carried away by the α particle.

The conversion factor between amu and g can be established most readily from the fact that 1 amu is exactly $\frac{1}{12}$ the mass of a carbon-12 atom: 1 amu $= \frac{1}{12} \times (12.00/6.02 \times 10^{23}) = 1.66 \times 10^{-24}$ g.

$$E = 0.0045 \text{ amu} \times \frac{1.66 \times 10^{-24} \text{ g}}{\text{amu}} \times \frac{1 \text{ kg}}{1000 \text{ g}} \times (3.00 \times 10^8)^2 \frac{\text{m}^2}{\text{s}^2}$$

$$= 6.7 \times 10^{-13} \text{ J}$$

$$E = 6.7 \times 10^{-13} \text{ J} \times \frac{1 \text{ MeV}}{1.602 \times 10^{-13} \text{ J}} = 4.2 \text{ MeV}$$

In the combustion of fuels, heats of reaction are of the order of a few thousand kJ/mol. The disintegration of 1 mol ^{238}U produces about 10^5 times more energy than an ordinary combustion reaction.

The energy equivalent to exactly 1 amu is 931.2 MeV. This is a useful quantity in calculating mass–energy relationships.

FIGURE 23-5
Nuclear binding energy in $^4_2$He.

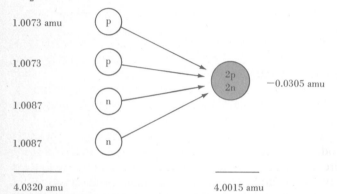

The mass of a helium nucleus ($^4_2$He) is 0.0305 amu (atomic mass unit) less than the combined masses of two protons and two neutrons. The energy equivalent to this loss of mass (called the mass defect) is the nuclear energy that binds the nuclear particles together.

(b) The calculation in part (a) is for a single disintegration. The energy in kJ/mol is based on the disintegration of 1 mol of atoms.

$$E = \frac{6.7 \times 10^{-13}\,\text{J}}{\text{atom}} \times \frac{6.02 \times 10^{23}\,\text{atoms}}{1\,\text{mol}} \times \frac{1\,\text{kJ}}{1000\,\text{J}}$$

$$= 4.0 \times 10^8\,\text{kJ/mol}$$

SIMILAR EXAMPLES: Exercises 26, 27.

NUCLEAR BINDING ENERGY. Figure 23-5 depicts a process in which the nucleus of a $^4_2$He atom is produced from two protons and two neutrons. In the formation of this nucleus there is a **mass defect** of 0.0305 amu. That is, the experimentally determined mass of a $^4_2$He nucleus is 0.0305 amu *less* than the combined masses of two protons and two neutrons. This "lost" mass is liberated as energy. By a calculation similar to Example 23-6, it can be shown that 0.0305 amu of mass is equivalent to an energy of 28.47 MeV. Since this is the energy released in forming an $^4_2$He nucleus, it is referred to as the **binding energy** of the nucleus. (Viewed in another way, an $^4_2$He nucleus would have to absorb 28.47 MeV to cause its protons and neutrons to become separated.) If we consider the binding energy to be apportioned among the two protons and two neutrons in $^4_2$He, we obtain a binding energy per nucleon of 7.1 MeV. Similar calculations can be made for other nuclei, leading to the graph shown in Figure 23-6.

Figure 23-6 indicates that the maximum binding energy per nucleon is calculated for a nucleus with a mass number of approximately 60. This leads to two interesting conclusions: (a) If small nuclei are combined into a heavier one (up to about $A = 60$), the binding energy per nucleon increases and a certain quantity of mass must be converted to energy. The

FIGURE 23-6
Average binding energy per nucleon as a function of atomic number.

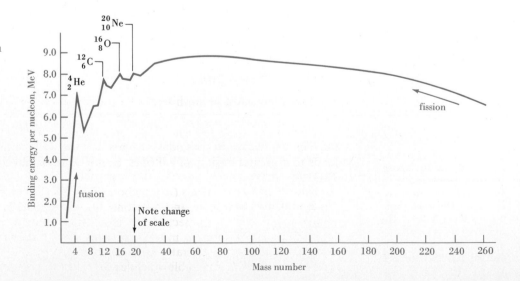

TABLE 23-2
Magic numbers for nuclear stability

Number of protons	Number of neutrons
2	2
8	8
20	20
28	28
50	50
82	82
114	126
	184
	196

TABLE 23-3
Stable isotopes of a few elements

Element	Z	Number of stable isotopes
H	1	2
O	8	3
F	9	1
Ne	10	3
Cl	17	2
Ca	20	6
Cu	29	2
Sn	50	10
I	53	1
Hg	80	7

The Z odd–N odd combination is found only in the lighter elements: $^{2}_{1}H$, $^{6}_{3}Li$, $^{10}_{5}B$, and $^{14}_{7}N$.

nuclear reaction is highly exothermic. This is a fusion process, and serves as the basis of the hydrogen bomb. (b) For nuclei having mass numbers above 60, the addition of extra nucleons to the nucleus would require the expenditure of energy (since the binding energy per nucleon decreases). On the other hand, the *disintegration* of heavier nuclei into lighter ones is accompanied by the release of energy. This is a nuclear fission process and serves as the basis of the atomic bomb and conventional nuclear power reactors. Nuclear fission and fusion are considered in greater detail in Sections 23-8 and 23-9. But first let us see what insights Figure 23-6 provides into the question of nuclear stability.

23-7 Nuclear Stability

A number of basic questions have probably occurred to you as we have been describing nuclear decay processes: Why do some radioactive nuclei decay by α emission, some by β^- emission, and so on? Why do the lighter elements have so few naturally occurring radioactive nuclides, whereas those of the heavier elements all seem to be radioactive?

Our first clue to answers for such questions comes from Figure 23-6, where several nuclides have been specifically noted. These nuclides have higher binding energies per nucleon than those of their neighbors. Their nuclei are especially stable. This observation is consistent with a theory of nuclear structure known as the **shell theory.** In the formation of a nucleus, protons and neutrons are believed to occupy a series of nuclear shells. This process is analogous to building up of the electronic structure of an atom by the successive addition of electrons to electronic shells. Just as the Aufbau process produces, periodically, electron configurations of exceptional stability, so do certain nuclei acquire a special stability as nuclear shells are closed. This condition of special stability of an atomic nucleus comes for certain numbers of protons and/or neutrons known as **magic numbers.** These magic numbers are listed in Table 23-2.

Analogous to the situation with electrons, nucleons possess the property of spin, and a pairing of spins also occurs in the filling of nuclear shells. A nucleus that contains an odd number of nucleons will have a resultant nuclear spin. There are fewer stable nuclides with odd numbers of nucleons than even numbers. This situation is illustrated by the distribution of numbers of protons (Z) and neutrons (N) among the known stable nuclides.

Z even; N even: 163 nuclides
Z even; N odd: 55
Z odd; N even: 50
Z odd; N odd: 4

Another manifestation of this pairing phenomenon is that elements of *odd* atomic number generally have only one or two stable isotopes, whereas those of even atomic number have several. When all the protons are paired (even atomic number), the nucleus is able to accommodate a greater range in the number of neutrons. This leads to a greater variety of isotopes. Some representative data are shown in Table 23-3.

For the lighter elements (up to about $Z = 20$) the common stable nuclides have equal numbers of protons and neutrons, for example, $^{4}_{2}He$, $^{12}_{6}C$, $^{16}_{8}O$, $^{28}_{14}Si$, $^{40}_{20}Ca$. For higher atomic numbers, because of increasing repulsive forces between protons, larger numbers of neutrons must be present to stabilize a nucleus and the n/p ratio increases. For bismuth the ratio is about 1.5 : 1. Figure 23-7 indicates an approximate range of n/p ratios for stable nuclides as a function of atomic number.

FIGURE 23-7

Neutron-to-proton ratio and the stability of nuclides.

(a) The belt of naturally occurring stable nuclides, ranging from $_1^1H$ to $_{83}^{209}Bi$.
(b) Naturally occurring and man-made radioactive nuclides of the heavier elements.
(c) Possible heavy nuclides of high stability (long radioactive half-lives).

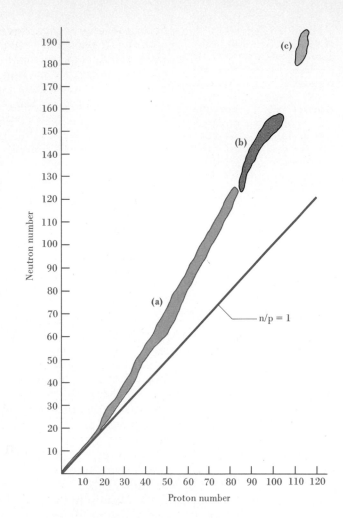

In mid-1976 a scientific report was made of the energies of x rays emitted by the proton bombardment of a particular monazite mineral (a mineral containing Ce, La, Th, and U). The atomic numbers associated with these energies appear to include 126, 116, 124, and 127 (recall Moseley's work described in Section 2-8). Thus, the first evidence suggesting the existence of "superheavy" elements has now been produced. Investigations are currently underway to verify these findings.

Example 23-7   Which of the following nuclides would you expect to be stable and which, radioactive? **(a)** ^{76}As; **(b)** ^{120}Sn; **(c)** ^{214}Po.

(a) ^{76}As has $Z = 33$ and $N = 43$. This is an odd–odd combination that is found only in four of the lighter elements. ^{76}As is radioactive. (Note also that this nuclide is outside the belt of stability in Figure 23-7.)

(b) Sn has an atomic number of 50—a magic number. The neutron number is 70 in the nuclide ^{120}Sn. This is an even–even combination and we should expect the nucleus to be stable. Moreover, Figure 23-7 shows that this nuclide is within the belt of stability. ^{120}Sn is a stable nuclide.

(c) ^{214}Po has an atomic number of 84. All known nuclides with $Z > 83$ are radioactive. ^{214}Po is radioactive.

SIMILAR EXAMPLES: Exercises 28, 30.

Using the ideas presented here (i.e., nuclear shell theory, magic numbers, etc.), nuclear scientists have predicted the possible existence of nuclides of high atomic number. These nuclides would have very long half-lives. Currently, a search is on to find such nuclides, either naturally or by creating them in an accelerator. (In order for a nuclide still to be present following the creation of the earth, its half-life would have to be greater than about 10^8 y.) Figure 23-7 suggests the general range of proton and neutron numbers for these nuclides.

Equations (23.25) and (23.26) are oversimplifications of the actual case. In order that certain properties be conserved it is necessary to postulate the presence of other extremely tiny particles (about 0.0004 times the mass of an electron) in β decay processes. These are the neutrino in equation (23.25) and the antineutrino in (23.26). The existence of these particles has been confirmed.

MECHANISM OF RADIOACTIVE DECAY.

The emission of an α particle by a nucleus is not difficult to visualize. A bundle of four nucleons (two protons, two neutrons) is ejected from a nucleus and the nucleus becomes energetically more stable. Alpha particle emission is pretty much limited to nuclides with $Z > 82$, though there are a few α emitters among the lanthanoid nuclides. Gamma ray emission can be visualized simply in terms of some rearrangement of nucleons with the release of energy as electromagnetic radiation. Emission of β^- and β^+ particles and electron capture are harder to picture. We can think in the following terms, however.

β^- emission: $\qquad {}^1_0n \longrightarrow {}^1_1p + {}^{\;\;0}_{-1}e \qquad\qquad (23.25)$

β^+ emission: $\qquad {}^1_1p \longrightarrow {}^1_0n + {}^{\;\;0}_{+1}e \qquad\qquad (23.26)$

electron capture: $\quad {}^1_1p + {}^{\;\;0}_{-1}e \longrightarrow {}^1_0n \qquad\qquad (23.27)$

In general, if a nuclide lies above the belt of stability in Figure 23-7 the n/p ratio is too high. A neutron is converted to a proton and a β^- particle is emitted (23.25). If a nuclide lies below the belt of stability, the n/p ratio is too low. Either a proton is converted to a neutron, followed by β^+ emission (23.26) or electron capture occurs (23.27). The situation for a series of isotopes of fluorine is presented in Table 23-4.

Example 23-8 By what mode will the nuclide ^{74}Br decay?

That this nuclide is radioactive can be seen in two ways. It lies below the belt of stability in Figure 23-7, and it has an odd–odd combination of protons and neutrons. The nuclide has too few neutrons to be stable. A proton must be converted to a neutron, either by positron (β^+) emission (23.26) or by electron capture (23.27).

SIMILAR EXAMPLE: Exercise 29.

TABLE 23-4
Radioactive properties of isotopes of fluorine

Isotope	Mode of decay	Half-life
^{17}F	β^+	66 s
^{18}F	β^+, E.C.	109.7 min
^{19}F	stable	—
^{20}F	β^-	11.4 s
^{21}F	β^-	4.4 s
^{22}F	β^-	4.0 s

23-8 Nuclear Fission

In 1934, the Italian physicist Enrico Fermi proposed that transuranium elements might be produced by the bombardment of uranium with neutrons. He reasoned that the successive loss of β^- particles would cause the atomic number to increase, perhaps as high as to 96. When such experiments were carried out, it was found that in fact the product did emit β^- particles. But in 1938, two chemists, Hahn and Strassman, found by chemical analysis that the products of the neutron bombardment of uranium did not correspond to elements with $Z > 92$. Neither were they the neighboring elements of uranium—Ra, Ac, Th, and Pa. Instead, the products consisted of radioisotopes of much lighter elements, such as strontium and barium. Neutron bombardment of uranium nuclei causes certain of them to undergo **fission** into smaller fragments. A fission process is depicted in Figure 23-8.

The energy equivalent of the mass destroyed in a fission process is somewhat variable because a variety of fission fragments is possible. However, the average energy for each fission event is approximately 3.20×10^{-11} J (200 MeV).

$${}^{235}_{92}U + n \longrightarrow {}^{236}_{92}U \longrightarrow \text{fission fragments} + \text{neutrons} + 3.20 \times 10^{-11} \text{ J} \quad (23.28)$$

FIGURE 23-8
Nuclear fission of ${}^{235}_{92}U$ with thermal neutrons.

A ${}^{235}_{92}U$ nucleus is struck by a neutron possessing ordinary thermal energy. First the unstable nucleus ${}^{236}_{92}U$ is produced; this then breaks up into a light and a heavy fragment and several neutrons. A variety of nuclear fragments is possible, but the most probable mass number for the light fragment is 97, and for the heavy one 137.

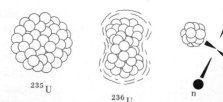

FIGURE 23-9
Light water nuclear
reactor.

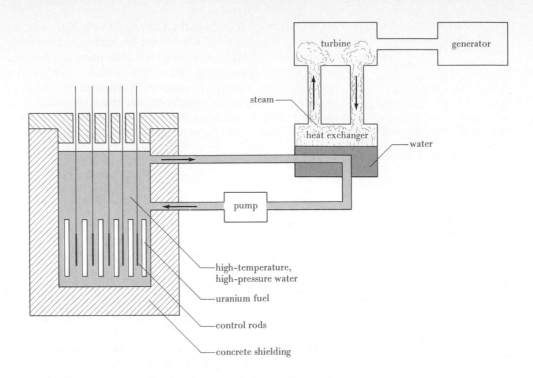

The energy in equation (23.28) seems small, but this energy is for the fission of a *single* $^{235}_{92}U$ nucleus. What if 1 g $^{235}_{92}U$ were to undergo fission?

$$\text{no. kJ} = 1.00 \text{ g } ^{235}U \times \frac{1 \text{ mol } ^{235}U}{235 \text{ g } ^{235}U} \times \frac{6.02 \times 10^{23} \text{ atoms } ^{235}U}{1 \text{ mol } ^{235}U} \times \frac{3.20 \times 10^{-11} \text{ J}}{1 \text{ atom } ^{235}U}$$

$$= 8.20 \times 10^{10} \text{ J} = 8.20 \times 10^7 \text{ kJ}$$

This is an enormous amount of energy! By contrast, the complete combustion of 1.00 g of carbon releases only 32.8 kJ.

NUCLEAR REACTORS.   In the fission of $^{235}_{92}U$, on average, 2.5 neutrons are released per fission event. These neutrons, on average, produce two or more fission events. The neutrons produced by the second round of fission produce another four or five events, and so on. The result is a **chain reaction.** If the reaction is uncontrolled, the total released energy causes an explosion; this is the basis of the atomic bomb. In a nuclear reactor this energy is released in a controlled fashion.

One common design for a nuclear reactor, called the light water reactor (LWR), is pictured in Figure 23-9. In the core of the reactor, rods of uranium-rich fuel are suspended in liquid water maintained under a pressure of from 70 to 150 atm. The water serves a dual purpose. First, it slows down the neutrons given off in the fission process so that they possess only normal thermal energy. These so-called thermal neutrons are more able to induce fission than highly energetic ones. In this capacity the water is said to act as a **moderator.** The second function of the water is as a heat-transfer medium. The energy of the fission reaction maintains the water at a high temperature (about 300°C). The high-temperature water is brought in contact with colder water in a heat exchanger. The colder water is converted to steam, which drives a turbine, which in turn drives an electric generator. A final component of the nuclear reactor is a set of **control rods,** usually cadmium metal, whose function is to absorb neutrons. When the rods are lowered into the reactor, the fission process is slowed down. When the rods are raised, the density of neutrons and the rate of fission increase.

Spontaneous fission resulting in an uncontrolled explosion occurs only if the quantity of ^{235}U exceeds the **critical mass.** With subcritical masses, neutrons escape from the ^{235}U at too great a rate to sustain a chain reaction.

A nuclear reactor based on the fission of $^{235}_{92}U$ is referred to as a nuclear burner. In the nuclear burner the fissionable nuclide is consumed (perhaps at the rate of 1 to 3 kg/d), and highly radioactive waste products accumulate. The separation, concentration, and disposal of these radioactive wastes is a difficult problem, requiring a considerable amount of chemical technology.

A completely satisfactory method for the disposal of radioactive wastes is yet to be found.

Another important aspect of nuclear burning is the high rate of consumption of a relatively rare fissionable material. $^{235}_{92}U$ accounts for only approximately 0.71% of naturally occurring uranium. To extract pure $^{235}_{92}U$ from uranium ores requires that high-grade ores be employed, usually U_3O_8. Estimated world reserves of U_3O_8, together with the anticipated growth of nuclear power, suggest that the available supply of $^{235}_{92}U$ may be depleted by the end of the century. Conventional nuclear reactors are of limited potential in the long-term production of energy.

BREEDER REACTORS. All that is required to initiate the fission of $^{235}_{92}U$ are neutrons of ordinary thermal energies. Nuclei of $^{238}_{92}U$, the abundant nuclide of uranium (99.28%), undergo the following reactions when struck by energetic neutrons.

$$^{238}_{92}U + ^{1}_{0}n \longrightarrow ^{239}_{92}U$$

$$^{239}_{92}U \longrightarrow ^{239}_{93}Np + ^{0}_{-1}e$$

$$^{239}_{93}Np \longrightarrow ^{239}_{94}Pu + ^{0}_{-1}e$$

A fissionable nuclide such as $^{235}_{92}U$ is called a fissile nuclide; $^{239}_{94}Pu$ is also fissile. A nuclide such as $^{238}_{92}U$, which can be converted into a fissile nuclide, is said to be fertile. In a breeder nuclear reactor a small quantity of fissile nuclide provides the neutrons that convert a large quantity of fertile nuclide into a fissile one. (The fissile nuclide then participates in a self-sustaining chain reaction.)

An obvious advantage of the breeder reactor, if developed, is that the amount of uranium "fuel" available would immediately jump by a factor of about 100. This is the ratio of naturally occurring $^{238}_{.02}U$ to $^{235}_{.92}U$. But the advantage is even greater than this. Breeder reactors may be able to use as nuclear fuels materials that have even very low uranium contents. For example, shale deposits exist in the western Appalachian Mountains that contain about 0.006% U by weight. These deposits extend for several hundred square miles, and all of this material is potential nuclear fuel.

There are, however, important disadvantages to the use of breeder reactors. This is especially true of the type that is currently being developed most vigorously—the liquid-metal-cooled fast breeder reactor (LMFBR). Systems must be perfected for handling a liquid metal, such as sodium, which becomes highly radioactive in the reactor. Also, the rate of heat and neutron production are both greater in the LMFBR than in the LWR, resulting in a more rapid deterioration of materials. These factors will complicate greatly the design and operation of the reactor. Perhaps the greatest unsolved problems are those of handling radioactive wastes and reprocessing plutonium fuel. Plutonium is one of the most toxic substances known. It can cause lung cancer when inhaled even in microgram (10^{-6} g) amounts. Federal health standards limit exposures to this substance to a total body burden of only 0.6 μg. Furthermore, because of its long half-life (24,000 y), any accident involving plutonium could leave an affected area almost permanently contaminated.

23-9 Nuclear Fusion

The fusion of atomic nuclei is the process whereby energy is produced on the sun. An uncontrolled fusion reaction is the basis of the hydrogen bomb. If a fusion reaction can be controlled, this will provide an almost unlimited source of energy. The nuclear reaction that holds the most immediate promise is the deuterium-tritium reaction:

$$\ce{^2_1H + ^3_1H -> ^4_2He + ^1_0n}$$ (23.29)

The difficulties in developing a fusion reaction are probably without parallel in the history of technology. In fact, the feasibility of a controlled fusion reaction has yet to be demonstrated. The basic problems are these:

In order to permit their fusion, the nuclei of deuterium and tritium must come into close proximity. Because atomic nuclei repel one another, this close approach requires the nuclei to have very high thermal energies. At the temperatures necessary to initiate a fusion reaction, gases are completely ionized into a mixture of atomic nuclei and electrons known as a **plasma.** Still higher plasma temperatures—over 40,000,000 K—are required to initiate a *self-sustaining* reaction (one that releases more energy than is required to get it started). Obviously, there is no container that can withstand these fantastically high temperatures. A method must be devised to confine the plasma out of contact with other materials and at a sufficiently high density and for a sufficient period of time to permit the fusion reaction to occur.

In the hydrogen bomb these high temperatures are attained by exploding an atomic bomb, which triggers the fusion reaction.

The two methods receiving greatest attention currently are confinement in a magnetic field and the heating of a frozen deuterium-tritium pellet with a laser beam. The critical conditions for a sustained fusion reaction have not yet been attained by either method. In addition to these difficulties another series of technical problems that must be solved involves the handling of liquid lithium, which is the anticipated heat transfer medium and tritium ($\ce{^3_1H}$) source.

$$\underset{\text{(fast)}}{\ce{^7_3Li}} + \ce{^1_0n} \longrightarrow \ce{^4_2He} + \underset{\text{(slow)}}{\ce{^3_1H}} + \ce{^1_0n}$$ (23.30)

Finally, for the magnetic containment method the magnetic field must be produced by superconducting magnets maintained at temperatures near absolute zero. Thus, the fusion reactor must have regions in which the temperature ranges from near 0 K to tens of millions of degrees kelvin, separated by distances of perhaps only 2 m.

The advantages of fusion over fission should be very great. Since deuterium comprises about one in every 6500 H atoms, the oceans of the world can supply an almost limitless amount of nuclear fuel. It is estimated that there is sufficient lithium on the earth to provide a source of tritium for about 1 million years.

23-10 Effect of Radiation on Matter

We now turn our attention briefly to the fate of the radiation or emanations of radioactive nuclei. Although there are substantial differences in the way in which α, β, and γ rays interact with atoms and molecules as they pass through samples of matter, they share an important feature: They tend to dislodge electrons from atoms and molecules to produce ions. The ionizing power of radiation may be described in terms of the number of ion pairs it forms. An ion pair consists of an ionized electron and the resulting positive ion. Alpha particles have the greatest ionizing power, followed by β particles and then γ rays. The ionized electrons produced directly by the collisions of particles of radiation with atoms are called primary electrons. These electrons may themselves possess sufficient energies to cause secondary ionizations. Thus, even though some radiation, such as γ rays, may produce few primary electrons, the total ionization associated with its passage through matter may be considerable.

A practical example of this effect is found in luminescent watch dials. The dial numerals are painted with a mixture of a fluorescent material such as zinc sulfide, and traces of a radioactive material such as radium (an α emitter). Excitation of the zinc sulfide by α particles is accompanied by light emission; the dial numerals are visible in the dark.

Not all interactions between radiation and matter cause ion formation. In some cases electrons may simply be raised to higher atomic or molecular energy levels. The return of these electrons to their normal states is then accompanied by radiation—x rays, ultraviolet light, or visible light, depending on the energies involved. Finally, some of the energy of the particles of radiation may ultimately appear as heat.

FIGURE 23-10
Some interactions of radiation with matter.

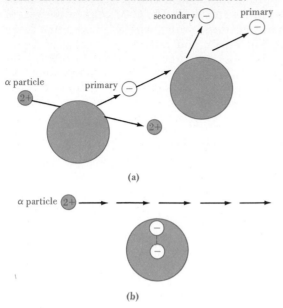

(a) The production of primary and secondary electrons by collisions.
(b) The excitation of an atom by the passage of an α particle. An electron is raised to a higher energy level within the atom. The excited atom reverts to its normal state by emitting radiation.

Some of the possibilities described here are pictured in Figure 23-10.

RADIATION DETECTORS. The interactions of radiation with matter can serve as bases for the detection of radiation and the measurement of its intensity. One of the simplest methods is that used by Becquerel in his discovery of radioactivity—the exposure of a photographic plate. The effect of α, β, and γ rays on a photographic emulsion is similar to that of x rays.

A device that has played an important role in the development of knowledge about radioactivity is the **cloud chamber** pictured in Figure 23-11. The principle involved is quite simple. If a vapor is brought to a saturated condition, say by expansion and cooling, and if there are no nuclei (such as dust particles) on which condensation can occur, the vapor may become supersaturated. If ionizing radiation passes through such a vapor, the ions produced become nuclei for the production of liquid. Droplets are produced in the form of a cloudlike track. The type of ionizing radiation and some of its characteristics can be inferred from such tracks. For instance, α particles produce short, straight thick tracks, whereas β^- particles produce long, thin meandering tracks. If the cloud chamber is placed in a magnetic field, the tracks are curved—one direction for α particles and the opposite for β^- particles.

A modification of the cloud chamber that is particularly useful in detecting high-energy radiation such as γ rays is the **bubble chamber.** In this device a substance, usually hydrogen, is kept just at its boiling point. As ion pairs are produced by the transit of an ionizing ray, bubbles of vapor form around the ions. Again tracks result that can be photographed and analyzed.

FIGURE 23-11
Wilson cloud chamber.

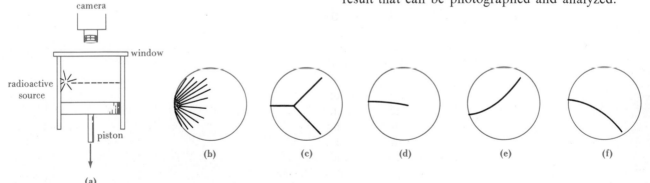

(a) Saturated vapor (say, ethanol) is cooled by sudden expansion. As a result it becomes supersaturated. Ion pairs produced in the supersaturated vapor serve as nuclei for condensation tracks.
(b) α particle tracks.
(c) An α particle in collision with an atomic nucleus. Two tracks emanate from the point where the original track disappears (one is the α particle track, and the other is that of the atom that is struck).
(d) α particle deflected in a magnetic field.
(e) β^- particle deflected in a magnetic field.
(f) Positron (β^+) deflected in a magnetic field.

FIGURE 23-12
Geiger–Müller counter.

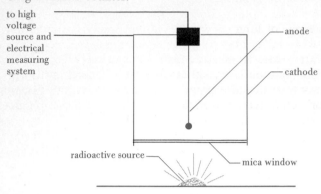

to high voltage source and electrical measuring system

anode

cathode

radioactive source

mica window

Radiation enters the G–M tube through the mica window. Ions produced by the radiation cause an electrical breakdown of the gas in the tube (usually argon). A pulse of electric current passes through the electric circuit and is counted.

TABLE 23-5
Units of radiation dosage[a]

Unit	Definition
Curie	An amount of radioactive material decaying at the same rate as 1 g of radium (3.7×10^{10} dis/s).
Rad	A dosage of radiation able to deposit 1×10^{-2} J of energy per kilogram of matter.
Rem	A unit related to the rad, but taking into account the varying effects of different types of radiation of the same energy on biological matter. This relationship is through a "quality factor," which may be taken as equal to one for x rays, γ rays, and β particles. For protons and slow neutrons the factor has a value of 5; and for α particles, 10. Thus, an exposure to 1 rad of x rays is about equal to 1 rem, but 1 rad of α particles is about equal to 10 rem.

[a] Sources of α radiation are relatively harmless when external to the body and extremely hazardous when taken internally, as in the lungs or stomach. Other forms of radiation (x rays, γ rays), because they are highly penetrating, are hazardous even when external to the body.

The most common device for detecting and measuring ionizing radiation is the **Geiger–Müller counter** pictured in Figure 23-12. The G–M counter consists of a cylindrical cathode with a wire anode running along its axis. The anode and cathode are sealed in a gas-filled glass tube. The tube is operated in such a way that ions produced by radiation passing through the tube trigger pulses of electric current. It is these pulses that are counted.

EFFECT OF IONIZING RADIATION ON LIVING MATTER. All life exists against a background of naturally occurring ionizing radiation—cosmic rays, ultraviolet light, and emanations from radioactive elements such as uranium in rocks. The level of this radiation varies from point to point on earth, being greater, for instance, at higher elevations. Only in recent times have human beings been able to create situations where living organisms might be exposed to radiation at levels significantly higher than natural background.

The interactions of radiation with living matter are the same as with other forms of matter—ionization, excitation, and dissociation of molecules. There is no question of the effect of large dosages of ionizing radiation on living organisms—the organisms are killed. But even slight exposures to ionizing radiation can cause changes in cell chromosomes. Thus it is commonly held that even at low dosage rates ionizing radiation can result in birth defects, leukemia, bone cancer, and other forms of cancers. The nagging question that has eluded any definitive answers is how great an increase in the incidence of birth defects and cancers is caused by certain levels of radiation.

RADIATION DOSAGE. Several different units are used to describe radiation dosage, that is, the amount of radiation to which matter is exposed. A summary is provided in Table 23-5.

It is thought that exposure in a single short time interval to 1000 rem of radiation would kill 100% of the population exposed. Exposure to 450 rem would probably result in death within 30 days in about 50% of the population. A single dosage of 1 rem delivered to 1 million people would probably produce about 100 cases of cancer within 20 to 30 years. The total body radiation received by most of the world's population from normal background sources is about 0.13 rem [130 millirem (mrem)] per year. The exposure in a chest x ray examination is about 20 mrem; and the estimated maximum exposure to the general population associated with the production of nuclear power is about 5 mrem/y.

Some of the foregoing statements about radiation

exposures and their anticipated effects are based on (1) medical histories of the survivors of the Hiroshima and Nagasaki atomic blasts, (2) the incidence of leukemia and other cancers in children whose mothers received diagnostic radiation during pregnancy, and (3) the occurrence of lung cancers among uranium miners in the United States. What does all of this tell us about a "safe" level of radiation exposure? One approach has been to extrapolate from these high dosage rates to the lower dosages affecting the general population. This has led the United States National Council on Radiation Protection and Measurements to recommend that the dosage rate for the general population be limited to 0.17 rem (170 mrem) per year from all sources above background level.

23-11 Applications of Radioisotopes

Both the potential of nuclear reactions to provide new sources of energy and the destructive capacity of these same nuclear reactions have been cited. Less heralded but also important are a variety of practical applications of radioisotopes. We close this chapter with a brief survey of a few of these applications.

CANCER THERAPY. We have noted how ionizing radiation in low dosages can induce cancers, but this same radiation, particularly γ rays, can also be used in the treatment of cancer. The basis of such treatment is that, although the radiation tends to destroy all cells, cancerous cells are more easily destroyed than normal ones. Thus, a carefully directed beam of γ rays or high-energy x rays of the appropriate dosage may be used to arrest the growth of cancerous cells. Also coming into use for some forms of cancer is radiation therapy using beams of neutrons.

RADIOACTIVE TRACERS. Radioactive isotopes are different from nonradioactive ones only in the instability of their nuclei, not in physical or chemical properties. Thus, in any physical or chemical process radioactive and nonradioactive isotopes are expected to behave in the same way. This fact serves as the basic principle in the use of radioactive tracers or "tagged" atoms. For example, if a small quantity of artificially radioactive ^{32}P (as a phosphate) is added to a nutrient solution that is fed to plants, the uptake of the phosphorus can be followed by charting the regions of the plant that become radioactive. Similarly, the fate of iodine in the human body can be determined by having an individual drink a solution of dissolved iodides containing a small quantity of radioactive iodine as a tracer. Abnormalities in the thyroid gland can be detected in this way.

Industrial applications of tracers are also numerous. The fate of a catalyst in a chemical plant can be followed by incorporating a radioactive tracer in the catalyst, for example, ^{192}Ir in a Pt-Ir catalyst. By monitoring the activity of the ^{192}Ir one can determine the rate at which the catalyst is being carried away and to which parts of the plant.

STRUCTURES AND MECHANISMS. Often detailed knowledge of the mechanism of a chemical reaction or the structure of a species can be inferred from experiments using radioisotopes as tracers. Consider the following experimental proof of the statement made in Section 20-5 that the two S atoms in the thiosulfate ion, $S_2O_3^{2-}$, are not equivalent.

$S_2O_3^{2-}$ is prepared from radioactive sulfur (^{35}S) and sulfite ion containing the nonradioactive isotope ^{32}S.

$$^{35}S + {}^{32}SO_3^{2-} \longrightarrow {}^{35}S{}^{32}SO_3^{2-} \qquad (23.31)$$

When the thiosulfate ion is decomposed by acidification, all the radioactivity appears in the precipitated sulfur and none in the $SO_2(g)$. The ^{35}S atoms must be bonded in a different way than the ^{32}S atoms (see Figure 20-11).

$$^{35}S^{32}SO_3^{2-} + 2\,H^+ \longrightarrow H_2O + {}^{32}SO_2(g) + {}^{35}S(s) \qquad (23.32)$$

In reaction (23.33) nonradioactive KIO_4 is added to a solution containing iodide ion labeled with the radioisotope, ^{128}I. All the radioactivity appears in the I_2 and none in the IO_3^-. This proves that all the IO_3^- is produced by reduction of IO_4^- and none by oxidation of I^-.

$$IO_4^- + 2\,{}^{128}I^- + H_2O \longrightarrow {}^{128}I_2 + IO_3^- + 2\,OH^- \qquad (23.33)$$

If a chromium(III) salt containing radioactive ^{51}Cr is added to a chrome-plating bath, no radioactivity shows up in the chrome plate. This proves that the reduction of Cr(VI) to Cr(0) at the cathode does not proceed through the oxidation state Cr(III). (Chrome plating was discussed in Section 21-3.)

ANALYTICAL CHEMISTRY. We have learned how a substance can be analyzed by precipitation from solution. The usual procedure involves filtering, washing, drying, and weighing a pure precipitate. An alternative is to react the substance to be analyzed with a reagent containing a radioisotope. By measuring the activity of the precipitate and comparing it with that of the original solution, it is possible to calculate the amount of precipitate without having to purify, dry, and weigh it. (For example, Ag^+ in solution can be precipitated as radioactive AgCl by treatment with a solution containing radioactive Cl^-.)

Another method of importance in analytical chemistry is **neutron activation analysis.** In this procedure the sample to be analyzed, normally nonradioactive, is bombarded with neutrons; the element of interest is converted to a radioisotope. The radioactivity of this radioisotope is measured. This measurement is combined with a knowledge of such factors as the rate of neutron bombardment, the half-life of the radioisotope, and the efficiency of the radiation detector to calculate the quantity of the element in the sample. The method is especially attractive because (1) trace quantities of elements can be determined (sometimes in parts per billion or less); (2) a sample can be tested without destroying it; and (3) the sample can be in any state of matter, including biological materials. Among its many uses, neutron activation analysis has been used to study archeological artifacts and to determine the authenticity of old paintings. (Old masters formulated their own paints. Differences between formulations are easily detected through the trace elements they contain.)

Summary

Radioactivity refers to the ejection of particles (α, β^-, β^+), the capture of electrons from an inner shell, or the emission of electromagnetic radiation (γ) by unstable nuclei. Often these processes are part of a radioactive decay series. That is, they occur through several steps until a stable (nonradioactive) nucleus is finally produced. With the exception of γ ray emission, radioactive decay processes lead to the transformation (transmutation) of one element into another.

All nuclides having atomic number greater than 83 are radioactive. Although a few occur naturally, most radioactive nuclides of lower atomic number are produced artificially, by bombarding appropriate target nuclei with energetic particles. The high-energy particles induce a nuclear reaction that yields the desired nuclide. Equations can be written for nuclear reactions, whether they occur through particle bombardment or by spontaneous radioactive decay. In either case the sum of the atomic numbers and the sum of the mass numbers must be constant between the two sides of the equation.

The rate of radioactive decay is directly proportional to the number of atoms in a sample. Each radioactive nuclide has a characteristic decay constant and half-life, and calculations of decay rates can be made using equations similar to those for first-order chemical kinetics. Measurements of decay rates of radioactive nuclides have a number of practical applications, ranging from determining the ages of rocks to the dating of wooden objects (carbon-14 dating).

Large quantities of energy are associated with nuclear reactions, sufficiently large that mass changes occur. The relationship between mass and energy, $E = mc^2$, can be used to calculate these energy changes. A similar calculation can be made of the quantity of energy released in the formation of a nucleus from protons and neutrons. When these nuclear binding energies per nucleon are plotted as a function of mass number a distinctive graph is obtained (Figure 23-6). From this graph one can establish that fission of heavy nuclei or fusion of lighter nuclei yield large quantities of energy. The fission process is the basis of the atomic bomb and nuclear reactors. Fusion is the energy-producing process of the stars, the hydrogen bomb, and the, as yet unperfected, thermonuclear reactor.

The stability of a nucleus depends on several factors, among them being whether the numbers of protons and neutrons are odd or even, and whether either of these is a "magic number." The most important factor is the neutron-to-proton ratio in the nucleus and whether this ratio lies within the belt of stable nuclides (Figure 23-7). Nuclides outside the belt of stability are radioactive. Their placement with respect to this belt can generally serve to indicate whether radioactive decay will occur by α, β^-, or β^+ emission or by electron capture.

One of the principal effects of the interaction of radiation with matter is the production of ions. This phenomenon can be used to detect radiation, and it is also the basis of radiation damage to living matter. Several methods have been developed to measure radiation dosages and to predict the biological effects of these dosages, but much uncertainty remains. Despite the hazards associated with radioactivity, many beneficial applications exist as well. Radioactive nuclides are used in cancer therapy, in basic studies of chemical structures and mechanisms, in analytical chemistry, and in chemical industry.

Learning Objectives

As a result of studying Chapter 23, you should be able to

1. Name the different types of radioactive decay processes and describe the characteristics of their radiation.

2. Write nuclear equations for radioactive decay processes.

3. Describe the three natural radioactive decay series, using the uranium series as an example; and give some examples of where the natural radioactive decay series are encountered in the environment.

4. Write equations for nuclear reactions produced artificially.

5. Name some of the transuranium elements and describe how they are made.

6. Describe the principles involved in the operation of a charged-particle accelerator.

7. Calculate the rate of radioactive decay, the half-life, or the number of atoms in a sample of a radioactive nuclide if two of the three quantities are known.

8. Determine the ages of rocks from a measured mass ratio of a stable nuclide to a radioactive one (such as $^{206}Pb/^{238}U$),

and the ages of carbon-containing materials from the decay rate of carbon-14.

9. Calculate the energies associated with nuclear reactions.

10. Calculate the average nuclear binding energy per nucleon for a nuclide.

11. Describe the factors that determine nuclear stability, determine whether a particular nuclide is likely to be stable or radioactive, and predict the type of decay process expected for a radioactive nuclide.

12. Describe the processes of nuclear fission and nuclear fusion, including the problems with using them as sources of energy.

13. Explain the effects of ionizing radiation on matter and describe several radiation-detection devices based on these effects.

14. Discuss methods used to express radiation dosages, some of the biological hazards of ionizing radiation, and sources of radiation to which the general population is exposed.

15. Discuss some of the practical, beneficial uses of radioisotopes.

Some New Terms

An **alpha (α) particle** is a combination of two protons and two neutrons identical to the nucleus of an ordinary helium atom, that is, $^4_2\text{He}^{2+}$.

A **beta (β⁻) particle** is an electron emitted as a result of the conversion of a neutron to a proton in a radioactive nucleus.

A **breeder reactor** is a nuclear reactor that creates more nuclear fuel than it consumes, for example by converting ^{238}U to ^{239}Pu.

A **charged-particle accelerator** is a device that imparts high energies to charged particles to be used in nuclear reactions.

A **cloud chamber** is a device used to detect ionizing radiation. It is based on the formation of a condensation trail of droplets along the path traveled by radiation through a supersaturated vapor.

A **curie** is a quantity of radioactive material decaying at the same rate as 1 g of radium (3.7×10^{10} dis/s).

Electron capture (E.C.) is a form of radioactive decay in which an electron from an inner electronic shell is absorbed by a nucleus. In the nucleus the electron is used to convert a proton to a neutron.

A **gamma (γ) ray** is a form of electromagnetic radiation emitted by certain radioactive nuclei.

A **Geiger–Müller counter** is a device used to detect ionizing radiation. Each ionizing event that occurs in the counter produces an electric discharge that can be recorded.

The **half-life** of a radioactive nuclide is the time required for one half of the atoms present in a sample to undergo radioactive decay.

Magic numbers is a term used to describe numbers of protons and neutrons that confer a special stability to an atomic nucleus.

Nuclear binding energy is the energy released to the surroundings when nucleons (protons and neutrons) are fused into an atomic nucleus. This energy replaces an equivalent quantity of matter in the formation of the nucleus.

Nuclear fission is a radioactive decay process in which a heavy nucleus breaks up into two lighter nuclei and several neutrons.

In **nuclear fusion** small atomic nuclei are fused into larger ones, with some mass being converted to energy in the process.

A **nuclear reactor** is a device in which nuclear fission is carried out as a controlled chain reaction. That is, neutrons produced in one fission event trigger the fission of another nucleus, and so on.

A **positron (β⁺)** is a *positive* electron emitted as a result of the conversion of a proton to a neutron in a radioactive nucleus.

A **rad** is a quantity of radiation able to deposit 1×10^{-2} J of energy per kilogram of matter.

A **radioactive decay series** refers to a succession of individual steps whereby an initial radioactive nuclide (e.g., ^{238}U) is ultimately converted to a stable nuclide (e.g., ^{206}Pb).

Radiocarbon dating is a method of determining the age of a carbon-containing material based on the rate of decay of radioactive carbon-14.

A **rem** is a unit of radiation related to the rad, but taking into account the varying effects of different types of radiation of the same energy on biological matter.

A **transuranium element** is one with an atomic number $Z > 92$.

Exercises

Definitions and terminology

1. Describe briefly the meaning of each of the following concepts or terms introduced in this chapter: **(a)** neutron-to-proton ratio; **(b)** nucleon; **(c)** mass–energy relationship; **(d)** background radiation; **(e)** radioactive decay series; **(f)** nuclear accelerator.

2. Explain the difference in meaning between the following pairs of terms: **(a)** naturally occurring and artificial radioisotope; **(b)** electron and positron; **(c)** primary and secondary ionization; **(d)** transuranium and transactinoid element; **(e)** nuclear fission and nuclear fusion.

3. What is the meaning of each of these symbols in describing nuclear phenomena? **(a)** α; **(b)** γ; **(c)** $t_{1/2}$; **(d)** λ; **(e)** β^+.

4. Supply a name or symbol for each of the following nuclear particles: **(a)** ^4_2He; **(b)** beta particle; **(c)** neutron; **(d)** ^1_1H; **(e)** $^0_{+1}\text{e}$; **(f)** tritium.

Radioactive processes

5. What is the nucleus obtained in each process?
(a) $^{234}_{94}\text{Pu}$ decays by α emission.
(b) $^{248}_{97}\text{Bk}$ decays by β^- emission.
(c) ^{196}Pb goes through two successive electron capture processes.
(d) $^{214}_{82}\text{Pb}$ decays through two successive β^- emissions.
(e) $^{226}_{88}\text{Ra}$ decays through three successive α emissions.
(f) $^{69}_{33}\text{As}$ decays by β^+ emission.

6. Supply the missing information in each of the following nuclear equations representing a radioactive decay process.
(a) $^{35}_{16}\text{S} \longrightarrow {}^?_{17}\text{Cl} + {}^0_{-1}\text{e}$ **(b)** $^{14}_8\text{O} \longrightarrow {}^{14}_7\text{N} + ?$
(c) $^{235}_?\text{U} \longrightarrow {}^?_?\text{Th} + ?$ **(d)** $^{214}_?? \longrightarrow {}^?_?\text{Po} + {}^0_{-1}\text{e}$

7. Both β^- (electron) and β^+ (positron) emission are observed for artificially produced radioisotopes of low atomic numbers, but only β^- (electron) emission is observed with naturally occurring radioisotopes of high atomic number. What is the reason for this observation?

Radioactive decay series

8. The natural decay series starting with the radionuclide $^{232}_{90}$Th follows the sequence represented below. Construct a graph of this series, similar to Figure 23-3.

$$^{232}_{90}\text{Th}-\alpha-\beta-\beta-\alpha-\alpha-\alpha \overset{\beta}{\underset{\beta-\alpha}{\diagup}} \overset{\beta}{\underset{\beta-\alpha}{\diagup}} {}^{208}_{82}\text{Pb}$$

9. The uranium series described in Figure 23-3 is also known as the "$4n + 2$" series because the atomic mass of each nuclide in the series can be expressed by the equation $A = 4n + 2$, where n is an integer. Show that this equation is indeed applicable to the uranium series.

Nuclear reactions

10. Write out the nuclear equations represented by the following symbolic notation: **(a)** $^7\text{Li}(p,\gamma)^8\text{Be}$; **(b)** $^{33}\text{S}(n,p)^{33}\text{P}$; **(c)** $^{239}\text{Pu}(\alpha,n)^{242}\text{Cm}$; **(d)** $^{238}\text{U}(\alpha,3n)^{239}\text{Pu}$.

11. Complete the following nuclear equations.
 (a) $^{23}_{11}\text{Na} + ^2_1\text{H} \longrightarrow ? + ^1_1\text{H}$
 (b) $^{59}_{27}\text{Co} + ? \longrightarrow ^{56}_{25}\text{Mn} + ^4_2\text{He}$
 (c) $^{238}_{92}\text{U} + ^2_1\text{H} \longrightarrow ? + ^0_{-1}\text{e}$
 (d) $^{246}_{96}\text{Cm} + ^{13}_6\text{C} \longrightarrow ^{254}_{102}\text{No} + ?$
 (e) $^{238}_{92}\text{U} + ^{14}_7\text{N} \longrightarrow ^{246}_?\text{Es} + ? ^1_0\text{n}$

12. Write an equation for each of the nuclear reactions represented by Figure 23-3.

Rate of radioactive decay

13. A sample containing radioactive $^{35}_{16}\text{S}$ is found to disintegrate at a rate of 1000 atoms/min. The half-life of $^{35}_{16}\text{S}$ is 87.9 d. How long will it take for the activity of this sample to decrease to the point of producing **(a)** 125, **(b)** 100, and **(c)** 50 dis/min?

14. The disintegration rate for a sample containing $^{60}_{27}\text{Co}$ as the only radioactive nuclide is found to be 240 atoms/min. The half-life of $^{60}_{27}\text{Co}$ is 5.2 y. Estimate the number of atoms of $^{60}_{27}\text{Co}$ in the sample.

15. How long must the radioactive sample of Exercise 14 be maintained before the disintegration rate falls to 100 dis/min?

16. The radioisotope $^{32}_{15}\text{P}$ is used extensively in biochemical studies. Its half-life is 14.2 d. Suppose that a sample containing this isotope has an activity 1000 times the detectable limit. For how long a time could an experiment be run with this sample before the radioactivity could no longer be detected? (*Hint:* Use $N_0 = 1000$ and $N_t = 1$.)

17. A sample containing $^{234}_{88}\text{Ra}$, which decays by α particle emission, is observed to disintegrate at the following rate, expressed as disintegrations per minute or counts per minute (cpm). What is the half-life of this nuclide? $t = 0$, 1000 cpm; $t = 1$ h, 992 cpm; $t = 10$ h, 924 cpm; $t = 100$ h, 452 cpm; $t = 250$ h, 138 cpm.

18. The unit **curie** is defined as a disintegration rate of 3.7×10^{10} dis/s. What mass of ^{226}Ra, with a half-life of 1602 y, is required to produce 1.00 millicurie of radiation?

19. If a meteorite is approximately 4.5×10^9 y old, what should be the mass ratio $^{208}\text{Pb}/^{232}\text{Th}$ in the meteorite? The half-life of ^{232}Th is 1.39×10^{10} y.

20. One method of dating rocks is based on their $^{87}\text{Sr}/^{87}\text{Rb}$ ratio. The ^{87}Rb is a β^- emitter with a half-life of 5×10^{11} y. A certain rock is found to have a mass ratio $^{87}\text{Sr}/^{87}\text{Rb}$ of $0.004:1.00$. What is the age of the rock?

Radiocarbon dating

21. A wooden art object is claimed to have been found in an Egyptian pyramid and is offered for sale to an art museum. Nondestructive radiocarbon dating of the object reveals a disintegration rate of 12 dis min^{-1}/g C. Do you think the object is authentic?

22. The lowest level of ^{14}C activity that seems possible for experimental detection is 0.03 dis min^{-1}/g C. What is the maximum age of an object that can be determined by the carbon-14 method?

★**23.** The carbon-14 dating method is based on the assumption that the rate of production of ^{14}C by cosmic ray bombardment has remained constant for thousands of years and that the ratio of ^{14}C to ^{12}C has also remained constant. Can you think of any effects of human activities that could invalidate this assumption in the future?

Energetics of nuclear reactions

24. Use data from Table 2-1, together with equation (23.23), to establish **(a)** equation (23.24) and **(b)** that the energy equivalent to 1.000 amu is 931.2 MeV.

25. The measured mass of the nuclide $^{20}_{10}\text{Ne}$ is 19.99244 amu. Determine the binding energy per nucleon (in MeV) in this atom.

26. Calculate the energy (in MeV) released in the nuclear reaction

$$^{10}_5\text{B} + ^4_2\text{He} \longrightarrow ^{13}_6\text{C} + ^1_1\text{H}$$

The nuclidic masses are $^{10}_5\text{B} = 10.01294$; $^4_2\text{He} = 4.00260$; $^{13}_6\text{C} = 13.00335$; $^1_1\text{H} = 1.00783$.

27. You are given the following exact atomic masses; $^6_3\text{Li} = 6.01513$; $^4_2\text{He} = 4.00260$; $^3_1\text{H} = 3.01604$; $^1_0\text{n} = 1.008665$. How much energy is released in the nuclear reaction $^6_3\text{Li} + ^1_0\text{n} \rightarrow ^4_2\text{He} + ^3_1\text{H}$, expressed in MeV?

Nuclear stability

28. Which member of the following pairs of nuclides would you expect to be most abundant in natural sources? Explain your reasoning. **(a)** $^{20}_{10}Ne$ or $^{22}_{10}Ne$; **(b)** $^{17}_{8}O$ or $^{18}_{8}O$; **(c)** $^{6}_{3}Li$ or $^{7}_{3}Li$.

29. One member each of the following pairs of radioisotopes decays by β^- emission and the other by positron (β^+) emission. Which is which? Explain your reasoning. **(a)** $^{29}_{15}P$ and $^{33}_{15}P$; **(b)** $^{120}_{53}I$ and $^{134}_{53}I$.

30. Sometimes the most abundant isotope of an element can be established by rounding off the atomic weight to the nearest whole number, for example, ^{39}K, ^{85}Rb, and ^{88}Sr. But at other times the isotope corresponding to the "rounded-off" atomic weight does not even occur naturally, for example, ^{36}Cl and ^{64}Cu. Explain the basis of this observation.

31. Some nuclides are said to be "doubly magic." What do you suppose this term means? Postulate some nuclides that might be doubly magic and locate them in Figure 23-7.

Fission and fusion

32. Describe briefly what is meant by the following types of nuclear reactors: **(a)** burner; **(b)** breeder; **(c)** thermonuclear.

33. Based on Figure 23-6, can you explain why more energy is released in a fusion than in a fission process?

34. Use data from the text to determine how many metric tons (1 metric ton = 1000 kg) of bituminous coal (85% C) would have to be burned to release as much energy as is produced by the fission of 1.00 kg $^{235}_{92}U$.

Effect of radiation on matter

35. Explain why the rem is more satisfactory than the rad as a unit for measuring radiation dosage.

36. The Geiger–Müller counter is a much more efficient device for detecting and measuring γ rays than is a cloud chamber. Why do you think this is so?

37. Discuss briefly the basic difficulties in establishing the physiological effects of low-level radiation.

38. ^{90}Sr is both a product of radioactive fallout and a radioactive waste in a nuclear reactor. This radioisotope is a β^- emitter with a half-life of 27.7 y. Suggest reasons why ^{90}Sr is such a potentially hazardous substance.

Application of radioisotopes

39. Describe how you might go about finding a leak in the $H_2(g)$ supply line in an ammonia synthesis plant by using radioactive materials.

40. Explain why neutron activation analysis is so useful in determining trace elements in a sample, in contrast to ordinary methods of quantitative analysis such as precipitation or titration.

41. A small quantity of NaCl containing radioactive $^{24}_{11}Na$ is added to an aqueous solution of $NaNO_3$. The solution is cooled and $NaNO_3$ is crystallized from the solution. Would you expect the $NaNO_3$ to be radioactive? Explain.

42. The following reactions are carried out using HCl(aq) containing some tritium ($^{3}_{1}H$) as a tracer. Would you expect any of the tritium radioactivity to appear in the $NH_3(g)$? In the H_2O? Explain.

$$NH_3(aq) + HCl(aq) \longrightarrow NH_4Cl(aq)$$

$$NH_4Cl(aq) + NaOH(aq) \longrightarrow$$
$$NaCl(aq) + H_2O + NH_3(g)$$

Additional Exercises

1. Two of the following isotopes do not occur naturally. Which are they? **(a)** ^2H; **(b)** ^{32}S; **(c)** ^{80}Br; **(d)** ^{132}Cs; **(e)** ^{184}W.

2. Each of the following isotopes is radioactive. Which would you expect to decay by β^- emission and which by positron (β^+) emission? **(a)** $^{28}_{15}P$; **(b)** $^{45}_{19}K$; **(c)** $^{72}_{30}Zn$.

3. Write an equation to represent each of the following nuclear processes.
  (a) The reaction of two deuterium nuclei (deuterons) to produce a nucleus of ^3He.
  (b) The production of $^{243}_{97}Bk$ by the α particle bombardment of $^{241}_{95}Am$.
  (c) The bombardment of ^{121}Sb by α particles to produce $^{124}_{53}I$, followed by its radioactive decay by positron (β^+) emission.

4. The half-life of tritium is 12.26 y. What would be the rate of decay of tritium atoms, per second, in 1.00 L of hydrogen gas at STP containing 0.15% tritium atoms?

5. Explain the similarities and differences between a cloud chamber and a bubble chamber.

6. Use data from Table 2-1, together with the measured mass of the nuclide $^{16}_{8}O$, 15.99491 amu, to determine the binding energy per nucleon (in MeV) in this atom.

7. If you follow the same description as in Exercise 9, the thorium series may be called the "$4n$" series and the actinium series the "$4n + 3$" series. A "$4n + 1$" series has also been established with $^{237}_{93}Np$ as the parent nuclide. To which radioactive series does each of the following nuclides belong? **(a)** $^{214}_{83}Bi$; **(b)** $^{216}_{84}Po$; **(c)** $^{215}_{85}At$; **(d)** $^{235}_{92}U$.

*8. Use the definition of Exercise 18 to determine how many millicuries of radiation are produced by a sample containing 5.10 mg ^{229}Th, which has a half-life of 7340 y?

*9. Calculate the minimum kinetic energy (in MeV) that α particles must possess to produce the nuclear reaction

$$^4_2\text{He} + \,^{14}_7\text{N} \longrightarrow \,^{17}_8\text{O} + \,^1_1\text{H}$$

The nuclidic masses are $^4_2\text{He} = 4.00260$; $^{14}_7\text{N} = 14.00307$; $^1_1\text{H} = 1.00783$; $^{17}_8\text{O} = 16.99913$. (*Hint:* What is the increase in mass in this process?)

*10. The packing fraction of a nuclide is related to the fraction of the total mass of a nuclide that is converted to nuclear binding energy. It is defined as the fraction $(M - A)/A$, where M is the actual nuclidic mass and A is the mass number. Use data from a handbook (such as *The Handbook of Chemistry and Physics,* published by the CRC Press) to determine the packing fractions of some representative nuclides. Plot a graph of packing fraction versus mass number and compare it to Figure 23-6. Explain the relationship between the two.

*11. ^{40}K undergoes radioactive decay both by electron capture to ^{40}Ar and β^- emission to ^{40}Ca. The fraction of the decay that occurs by electron capture is 0.110. The half-life of ^{40}K is 1.27×10^9 y. Assuming that a rock in which ^{40}K has undergone decay retains all of the ^{40}Ar produced, what would be the ^{40}Ar/^{40}K mass ratio in a rock that is 1.5×10^9 y old?

*12. Reference is made in the text to using a certain shale deposit in the Appalachian Mountains containing 0.006% U as a potential fuel in a breeder reactor. Assuming a density of 2.5 g/cm³, how much energy could be released from 1.00×10^3 cm³ of this material? Assume a fission energy of 3.20×10^{-11} J per fission event (i.e., per U atom).

Self-Test Questions

For questions 1 through 6 select the single item that best completes each statement.

1. Of the following types of radiation, the only one to be deflected in a magnetic field is (a) x ray; (b) γ ray; (c) β ray; (d) neutrons.

2. A process that produces a one-unit increase in atomic number is (a) electron capture; (b) β^- emission; (c) α emission; (d) γ ray emission.

3. Of the following nuclides, the one most likely to be radioactive is (a) ^{31}P; (b) ^{66}Zn; (c) ^{37}Cl; (d) ^{108}Ag.

4. One of the following elements has eight naturally occurring *stable* isotopes. We should expect that one to be (a) Ra; (b) Au; (c) Cd; (d) Br.

5. Of the following nuclides, the one most likely to decay by positron (β^+) emission is (a) ^{59}Cu; (b) ^{63}Cu; (c) ^{67}Cu; (d) ^{68}Cu.

6. Among the following nuclides, the highest nuclear binding energy per nucleon is found for (a) $^3_1$H; (b) $^{16}_8$O; (c) $^{56}_{26}$Fe; (d) $^{235}_{92}$U.

7. Write nuclear equations to represent (a) the decay of ^{230}Th by α particle emission; (b) the decay of ^{54}Co by positron emission; (c) the nuclear reaction ^{232}Th $(\alpha,4n)^{232}$U.

8. Iodine-129 is a product of nuclear fission, whether from an atomic bomb or a nuclear power plant. It is a β^- emitter with a 1.7×10^7 y half-life. How many disintegrations per second would occur in a sample containing 1.00 mg ^{129}I?

9. ^{223}Ra has a half-life of 11.4 d. How long would it take for the radioactivity associated with a sample of ^{223}Ra to decrease to 1% of its current value?

10. Explain why
   (a) Radioactive nuclides with intermediate half-lives are generally more hazardous than those with extremely short or extremely long half-lives.
   (b) Some radioactive substances are hazardous from a distance, whereas others must be taken internally to constitute a hazard.
   (c) Argon is the most abundant of the noble gases in the atmosphere.
   (d) Francium is such a rare element (less than about 30 g present in the earth's crust at any one time), and it cannot be extracted from minerals containing the alkali metals.
   (e) Such extremely high temperatures will be required to develop a self-sustaining thermonuclear (fusion) process as an energy source.

24 Organic Chemistry

More than a million compounds exist containing carbon atoms in combination with hydrogen, oxygen, nitrogen, or certain other elements. Organic chemistry is the chemistry of these compounds. The element carbon is singled out for special study because of the ability of carbon atoms to form strong covalent bonds with one another. Carbon atoms may join together into straight chains, branched chains, and rings. The nearly infinite number of possible bonding arrangements of carbon atoms into these chains and rings accounts for the vast number and variety of carbon compounds.

Originally, organic chemistry dealt only with compounds derived from living matter. Living matter was thought to possess a "vital force" necessary for the synthesis of these compounds. In 1828, the German chemist Friedrich Wöhler heated ammonium cyanate, derived from inorganic substances, and obtained the organic compound urea.

$$KOCN + NH_4Cl \longrightarrow KCl + NH_4OCN$$

$$\underset{\text{ammonium cyanate}}{NH_4OCN} \xrightarrow{\text{heat}} \underset{\text{urea}}{H_2NCONH_2}$$

The urea formed in this way proved to be identical to urea isolated from urine.

24-1 The Nature of Organic Compounds and Structures

The simplest organic compounds are those of carbon and hydrogen—hydrocarbons. The simplest of the hydrocarbons is methane, CH_4, the principal constituent of natural gas. Shown below for methane are three ways of representing an organic molecule. A Lewis structure shows the distribution of all valence electrons in a molecule. A structural formula focuses on the electrons involved in bond formation, using a dash to represent a single bond (double and triple dashes for double and triple bonds). A condensed formula conveys pretty much the same information as a structural formula but in a single line.

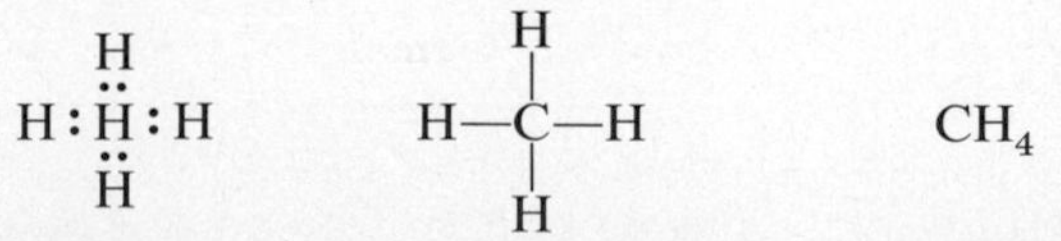

Lewis structure	structural formula	condensed formula
H:H:H with H above and H below	H—C—H with H above and H below	CH_4

FIGURE 24-1
Structural representation of the methane molecule.

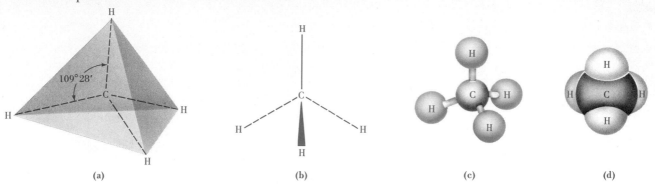

(a)   (b)   (c)   (d)

(a) Tetrahedral structure showing bond angle.
(b) Convention used to suggest a three-dimensional structure through a structural formula. The solid line represents a bond in the plane of the page. The dashed lines project *away* from the viewer and the heavy wedge projects *toward* the viewer.
(c) Ball-and-stick model.
(d) Space-filling model.

None of the foregoing structures describes the geometrical shape of the CH_4 molecule. However, both from VSEPR theory (Section 9-8) and valence bond theory (Section 10-2) we expect the distribution of the four electrons pairs around the central carbon atom to be tetrahedral. In a CH_4 molecule the four H atoms are equivalent: They are equidistant from the C atom and attached to it by covalent bonds of equal strength. The angle between any two C—H bonds is 109°28′. Molecular models are often used to represent organic molecules. Two widely used forms are illustrated in Figure 24-1. By increasing the number of carbon atoms in a molecule, other members of a hydrocarbon series can be represented, as in Figure 24-2.

SKELETAL ISOMERISM. From Figure 24-2 we see that there are *two* ways of assembling a hydrocarbon molecule with four carbon and ten hydrogen atoms. There are *two* different compounds with the formula C_4H_{10}. One is called butane and the other, isobutane. Compounds having the same molecular formula but different structural formulas are called **isomers.** Numerous possibilities for isomerism are found among organic compounds. In the case considered here the isomers differ in their carbon chains—one is a straight chain and the other, a branched chain. This type of isomerism is called **chain** or **skeletal isomerism.**

The names given to the first four members of the hydrocarbon series in Figure 24-2 are common names. As the length of the carbon chain increases, a root name is used that reflects the number of carbon atoms in the chain. To this root is attached a characteristic "ane" ending. Certain branched chain compounds are often referred to by the prefix "iso." The hydrocarbon series initiated with methane continues beyond the four-carbon compound with these characteristic names: pentane (C_5H_{12}), hexane (C_6H_{14}), heptane (C_7H_{16}), octane (C_8H_{18}), nonane (C_9H_{20}), and decane ($C_{10}H_{22}$). All of the longer chain hydrocarbons display skeletal isomerism, and the more carbon atoms present the greater the number of possible isomers. There are 18 isomers of octane, 35 of nonane, 75 of decane, and so on.

Our first encounter with isomerism was with coordination compounds in Chapter 22.

Example 24-1   Write structural formulas for all the possible isomers of hexane, C_6H_{14}.
   The basic question is: In how many different ways can six C atoms be bonded together? The key to this question is the word *different*. For example, the following formulas are not different. (Think of all the C atoms as if they were "hinged." Each structure can be rearranged into a six-carbon chain.)

FIGURE 24-2
Representations of some additional hydrocarbons.

Ethane:

H   H
| |
H—C—C—H
| |
H   H

$CH_3—CH_3$

Propane:

H   H   H
| | |
H—C—C—C—H
| | |
H   H   H

$CH_3—CH_2—CH_3$

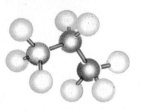

Butane:

H   H   H   H
| | | |
H—C—C—C—C—H
| | | |
H   H   H   H

$CH_3—(CH_2)_2—CH_3$

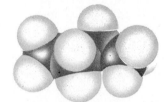

Isobutane:

H   H   H
| | |
H—C—C—C—H
| | |
H | H
    H—C—H
       |
       H

$HC(CH_3)_3$

(a)   (b)   (c)   (d)

(a) Structural formulas. **(b)** Condensed formulas. **(c)** Ball-and-stick models.
(d) Space-filling models.

C—C—C—C—C—C   C—C C—C   C—C—C   and so on

(with branched skeletons:
C—C
| |
C—C C—C   and   C—C—C
                   |
                   C—C—C)

We start with the *one* straight chain molecule, and for simplicity show only the carbon skeleton. (The complete structure requires adding the appropriate number of H atoms to produce four bonds at each C atom.)

C—C—C—C—C—C
          (1)

Next we look for the possibilities involving a five-carbon chain with one C atom as a side chain. There are only *two* possibilities.

Can you see that this structure is identical to (2)?

        C
        |
C—C—C—C—C

           C   C
           | |
C—C—C—C—C   C—C—C—C—C
        (2)   (3)

Now let us consider four-carbon chains with two one-carbon side chains. There is *one* possibility for side chains attached to *different* C atoms of the main chain (structure 4), and *one* possibility for side chains attached to the *same* C atom (structure 5).

Can you see that these structures are identical to (5)?

$$
\begin{array}{c}
\text{C} \\
| \\
\text{C—C—C—C} \\
| \\
\text{C}
\end{array}
\qquad
\begin{array}{c}
\text{C} \\
| \\
\text{C} \\
| \\
\text{C—C—C} \\
| \\
\text{C}
\end{array}
$$

$$
\begin{array}{cc}
\text{C} \quad \text{C} & \qquad \qquad \text{C} \\
| \quad | & \qquad \qquad | \\
\text{C—C—C—C} & \text{C—C—C—C} \\
& \qquad \quad | \\
& \qquad \quad \text{C} \\
\quad (4) & \qquad \quad (5)
\end{array}
$$

Any additional possibilities for a carbon skeleton prove to be identical to others already considered. The number of isomers of hexane is 5.

SIMILAR EXAMPLES: Exercises 9, 10.

NOMENCLATURE. Early in the history of organic chemistry, chemists assigned names of their own choosing to new compounds. Often these names were related to the origin or certain properties of the compounds, and some of these names are still in common use. For example, citric acid is found in citrus fruit; uric acid is present in urine; formic acid is found in ants (from the Latin word for ant, *formica*); and morphine induces sleep (from *Morpheus,* the ancient Greek god of sleep). As thousands upon thousands of new compounds were synthesized it became apparent that a system of nomenclature based on common names would be unworkable. Following several interim systems, one recommended by the International Union of Pure and Applied Chemistry (IUPAC or IUC) was adopted. A few of the more important rules for the nomenclature of hydrocarbons of the type, C_nH_{2n+2}, are these.

1. The generic (family) name of a saturated hydrocarbon is *alkane*.
2. Select the *longest* continuous carbon chain in the molecule and use the parent hydrocarbon name of this chain as the base name.
3. Every branch of the main chain is considered to be a substituent derived from another hydrocarbon. For these substituents the ending of the base name is changed from "anc" to "yl."
4. Number the carbon atoms of the continuous base chain so that the substituents appear *at the lowest numbers* possible.
5. Each substituent receives a name and number. For identical substituents use di, tri, tetra, and so on, and *repeat the numbers*.
6. Numbers are separated from other numbers by commas and from letters by dashes.
7. Arrange the substituents alphabetically by name, regardless of the numbers they carry or their complexity.
8. Whenever alternative base chains of equal length are possible, always name a compound so as to have the maximum number of side chains.

In applying rule 3, hydrocarbon substituents or alkyl groups should be named as follows.

$$CH_3—\qquad CH_3CH_2—\qquad CH_3CH_2CH_2—\qquad CH_3\underset{|}{C}HCH_3$$

methyl   ethyl   propyl   isopropyl
           (also called *n*-propyl
           or normal propyl)

$$
CH_3CH_2CH_2CH_2—\qquad
\overset{\displaystyle CH_3}{\underset{|}{C}}\!\!\!\!\!\begin{array}{c} \\ CH_3CHCH_2— \end{array}
\qquad CH_3CHCH_2CH_3
\qquad \overset{\displaystyle CH_3}{\underset{|}{C}}\!\!\!\!\!\begin{array}{c} \\ CH_3CCH_3 \end{array}
$$

butyl   isobutyl   *s*-butyl   *t*-butyl
(or *n*-butyl)   (*sec*-butyl or (*tert*-butyl or
            secondary butyl) tertiary butyl)

Example 24-2   Give appropriate IUC names for the following compounds.

(a)
$$CH_3-\underset{2}{\overset{CH_3}{\underset{|}{\overset{|}{C}}}}-\underset{3}{CH_2}-\underset{4}{\overset{CH_3}{\overset{|}{CH}}}-\underset{5}{CH_3}$$
$$\underset{1}{CH_3} \qquad\qquad CH_3$$

(b)
$$\underset{1}{CH_3}-\underset{2}{CH_2}-\underset{3}{\overset{CH_3}{\overset{|}{CH}}}-\underset{4}{CH_2}$$
$$\underset{5}{|}{CH_2}$$
$$\underset{6}{|}{CH_3}$$

If the carbon atoms in structure (a) were numbered from right to left, the name obtained would be 2,4,4-trimethylpentane. But this is *not* an acceptable name. It does not use the smallest numbers possible.

(a) The side chain substituents to be named are shown in color. Each is a methyl group, —CH_3. Two methyl groups are on the second carbon atom and one methyl, on the fourth. The main carbon chain has five atoms. The correct name is

2,2,4-trimethylpentane

(b) The chain length is six, not four. The methyl group is on the third carbon atom. The correct name is

3-methylhexane

SIMILAR EXAMPLES: Exercises 14, 16, 17.

Example 24-3   Write structural formulas and condensed formulas for the following compounds: **(a)** 4-*t*-butyl-2-methylheptane; **(b)** 2,6-dimethyl-3-ethylheptane.

(a) The substituent group, *t*-butyl, is attached to the fourth C atom in a seven-carbon chain. A methyl group is attached to the second C atom.

In arranging the substituent groups alphabetically as required in nomenclature rule 7, the symbols *n*, *s*, and *t* are not considered. Thus, butyl (even though *t*-butyl) precedes methyl in naming structure (a).

$$\overset{\qquad\qquad\qquad CH_3}{\underset{}{\qquad\qquad\qquad |}}$$
$$\overset{CH_3 \quad H_3C-\overset{|}{C}-CH_3}{H_3C-\overset{|}{CH}-CH_2-\overset{|}{CH}-CH_2-CH_2-CH_3}$$

or

$(CH_3)_2CHCH_2CH[C(CH_3)_3]CH_2CH_2CH_3$
(condensed formula)

(b) Methyl groups are substituted at the second and sixth C atoms and an ethyl group at the third.

It appears that structure (b) could also have been named 5-isopropyl-2-methylheptane. This is *not* done because only two side chain substituents would be involved instead of three (see rule 8).

$$\overset{\qquad\qquad\qquad\qquad CH_3}{\qquad\qquad\qquad\qquad |}$$
$$\overset{CH_3 \qquad\qquad CH_2 \quad CH_3}{H_3C-\overset{|}{CH}-CH_2-CH_2-\overset{|}{CH}-\overset{|}{CH}-CH_3}$$

or

$(CH_3)_2CHCH_2CH_2CH(C_2H_5)CH(CH_3)_2$
(condensed formula)

SIMILAR EXAMPLES: Exercises 15, 17, 18.

POSITIONAL ISOMERISM.   A variety of atoms or groups of atoms can be substituents on carbon chains, for example, Br. The three monobromopentanes are isomeric but they possess the same carbon skeleton. Because they differ in the position of the bromine atom on the carbon chain, they are called **positional isomers.**

$CH_3CH_2CH_2CH_2CH_2Br$ $\qquad$ $CH_3CH_2CH_2\overset{}{\underset{|}{CH}}CH_3$ $\qquad$ $CH_3CH_2\overset{}{\underset{|}{CH}}CH_2CH_3$
$$\qquad\qquad\qquad\qquad\qquad\qquad\qquad\qquad Br \qquad\qquad\qquad\qquad\qquad Br$$

1-bromopentane $\qquad\qquad\qquad$ 2-bromopentane $\qquad\qquad\qquad$ 3-bromopentane

Example 24-4   Represent all the possible isomers of C_4H_9Cl.

Perhaps the simplest approach is to consider the number of positional isomers for each of the skeletal isomers of butane. There are *two* possibilities based on *n*-butane (structures 1 and 2) and *two* based on isobutane (structures 3 and 4).

$$CH_3CH_2CH_2CH_2Cl \qquad CH_3CH_2\overset{\overset{\displaystyle Cl}{|}}{C}HCH_3 \qquad CH_3\overset{\overset{\displaystyle CH_3}{|}}{C}HCH_2Cl \qquad CH_3\overset{\overset{\displaystyle CH_3}{|}}{\underset{\underset{\displaystyle Cl}{|}}{C}}CH_3$$

$$(1) \qquad\qquad\qquad (2) \qquad\qquad\qquad (3) \qquad\qquad\qquad (4)$$

SIMILAR EXAMPLE: Exercise 10.

FUNCTIONAL GROUPS.   The form in which elements other than carbon and hydrogen are typically found in organic compounds is as distinctive groupings of one or several atoms. In some cases these groupings are substituted for H atoms in hydrocarbon chains or rings, and in others, for C atoms themselves. These groupings of atoms are called **functional groups,** and the remainder of the molecule is referred to by the symbol **R**. Table 24-1 lists some of the functional groups most frequently encountered.

Example 24-5   Use information from Table 24-1 and elsewhere in this section to derive an acceptable name for the compound $(CH_3)_2CHOCH_2CH_2CH_3$.

First we convert the condensed formula into a structural formula.

TABLE 24-1
Some classes of organic compounds

Type of compound	General structural formula	Example	Name
alkanes	$R-H$	$CH_3CH_2CH_2CH_2CH_3$	pentane
alkenes	$\underset{R}{\overset{R}{\diagdown}}C{=}C\underset{\diagdown R}{\overset{\diagup R}{}}$	$CH_3CH_2CH_2CH{=}CH_2$	1-pentene
alkynes	$R-C{\equiv}C-R$	$CH_3CH_2C{\equiv}CCH_3$	2-pentyne
alcohols	$R-OH$	$CH_3CH_2CH_2CH_2CH_2OH$	1-pentanol
alkyl halides	$R-X$	$CH_3CH_2\underset{\underset{\displaystyle Br}{\textstyle\mid}}{C}HCH_2CH_3$	3-bromopentane
ethers	$R-O-R$	$CH_3CH_2OCH_2CH_2CH_3$	ethyl propyl ether
aldehydes	$R-\overset{\overset{\displaystyle O}{\|}}{C}-H$	$CH_3CH_2CH_2CH_2\overset{\overset{\displaystyle O}{\|}}{C}H$	pentanal
ketones	$R-\overset{\overset{\displaystyle O}{\|}}{C}-R$	$CH_3CH_2\overset{\overset{\displaystyle O}{\|}}{C}CH_2CH_3$	3-pentanone
acids	$R-\overset{\overset{\displaystyle O}{\|}}{C}-OH$	$CH_3CH_2CH_2CH_2\overset{\overset{\displaystyle O}{\|}}{C}OH$	pentanoic acid
esters	$R-\overset{\overset{\displaystyle O}{\|}}{C}-O-R$	$CH_3CH_2CH_2CH_2\overset{\overset{\displaystyle O}{\|}}{C}OCH_3$	methyl pentanoate
amines	$R-NH_2$	$CH_3CH_2CH_2CH_2CH_2NH_2$	pentylamine

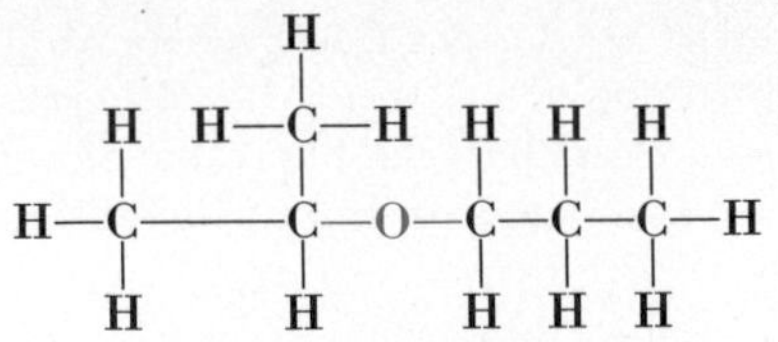

$$
\begin{array}{c}
\text{H} \\
| \\
\text{H}\quad\text{H—C—H}\quad\text{H}\quad\text{H}\quad\text{H} \\
|\qquad\quad|\qquad\quad|\quad\,|\quad\,| \\
\text{H—C————C—O—C—C—C—H} \\
|\qquad\quad|\qquad\quad|\quad\,|\quad\,| \\
\text{H}\qquad\quad\text{H}\qquad\,\text{H}\quad\text{H}\quad\text{H}
\end{array}
$$

The presence of the group C—O—C signifies that this is an ether. The group shown on the left is isopropyl; the one on the right, *n*-propyl or simply propyl. The compound can be called isopropyl propyl ether.

SIMILAR EXAMPLES: Exercises 11, 29.

24-2 Alkanes

In this section we explore further some properties of the alkanes. The essential characteristic of alkane hydrocarbon molecules is that only single covalent bonds are present. In these compounds the bonds are said to be saturated; the alkanes are known as **saturated hydrocarbons.**

The alkanes range in complexity from methane, CH_4 (accounting for over 90% of natural gas) to molecules containing 50 carbon atoms or more (found in petroleum). Each alkane differs from the preceding one by a —CH_2— or methylene group. This constant unit difference forms the basis of a **homologous series.** Members of such a series are usually closely related in chemical and physical properties. For example, a gradual increase in boiling point is noted with an increase in molecular weight. This trend of boiling points is understandable in terms of London-type forces introduced in Chapter 11. These forces increase in strength with increasing molecular weight. Also understandable from ideas presented in Chapter 11 are the data of Table 24-2, which show that the more branching on a carbon chain the lower the boiling point of the isomer (recall Figure 11-16).

CONFORMATIONS. An important type of motion in alkane molecules is suggested by ball-and-stick models. This is a motion in which one group rotates with respect to the other. Two of the many possible orientations of the two —CH_3 groups in ethane are shown in Figure 24-3.

In one of these configurations, when the molecule is viewed head-on along the C—C bond, one set of C—H bonds is directly behind the other. This structure is referred to as the **eclipsed conformation.** In this conformation the distance between H atoms on adjacent C atoms is at a minimum, leading to a condition of maximum repulsion between H atoms. This conformation is slightly less stable than the stag-

Methane is produced by the action of bacteria on decaying organisms. When encountered in bogs and swamps it is known as "marsh gas." Methane gas, referred to as "fire damp," also occurs in pockets in coal mines. The explosive nature of methane–air mixtures is one of the many hazards of coal mining.

TABLE 24-2
Boiling points of some isomeric alkanes

Family	Isomer	Boiling point, °C	Family	Isomer	Boiling point, °C
butane	*n*-butane	−0.5	hexane	*n*-hexane	68.7
	isobutane	−11.7		3-methylpentane	63.3
				isohexane	60.3
pentane	*n*-pentane	36.1		2,3-dimethylbutane	58.0
	isopentane	27.9		2,2-dimethylbutane	49.7
	2,2-dimethylpropane (neopentane)	9.5		(neohexane)	

FIGURE 24-3
Rotation about the C—C bond in ethane.

(a) Rotation about the C—C bond.
(b) Conformations of C_2H_6, in a representation known as a Newman projection. The view is along the C—C bond, end on.

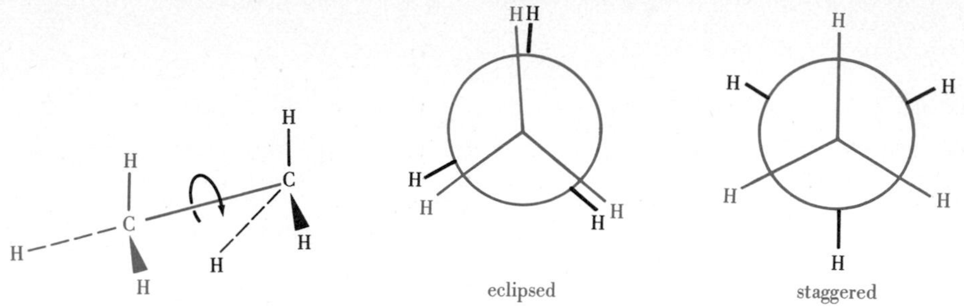

eclipsed

staggered

(a)

(b)

gered conformation. In the **staggered conformation** the hydrogen atoms are located a maximum distance apart. Although we expect the staggered conformation to be the more stable, at room temperature the thermal energy of an ethane molecule is sufficient that free rotation of the methyl groups about the C—C bond takes place. At lower temperatures, however, ethane does occur mostly in the staggered conformation. A similar situation is also encountered with higher alkanes.

RING STRUCTURES.   Alkanes in chain structures have the formula C_nH_{2n+2} and are called **aliphatic.** Alkanes can also exist in ring or cyclic structures called **alicyclic.** These rings can be thought of as resulting from the joining together of two ends of an aliphatic chain by the elimination of a hydrogen atom from each end. Simple alicyclic compounds have the formula C_nH_{2n}.

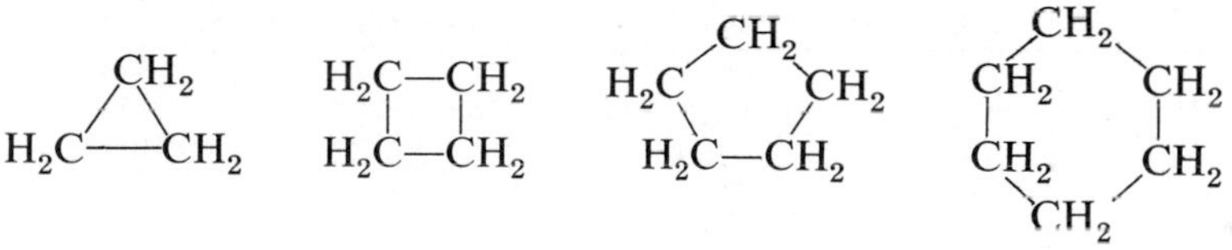

cyclopropane
C_3H_6

cyclobutane
C_4H_8

cyclopentane
C_5H_{10}

cyclohexane
C_6H_{12}

The naming of alicyclic compounds follows the rules established in the preceding section. Thus, the name of

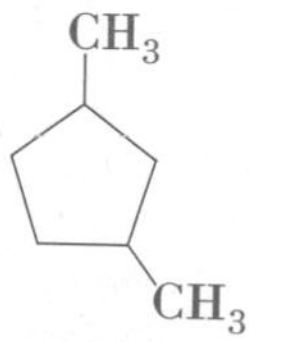

is 1,3-dimethylcyclopentane.

By convention, when a ring structure is drawn, neither the ring C atoms nor the H atoms bonded to them are written out.

The bond angles in cyclopropane are 60° compared to the normal 109.5°; the bonds are quite strained. As a result there are numerous reactions in which the ring will break open to yield a chain molecule—propane or a propane derivative. Cyclopropane is more reactive than most alkanes.

If the four carbon atoms in cyclobutane were to exist in the same plane, the C—C bond angles would be 90°. In reality the molecule buckles slightly to relieve the bond strain. If the cyclopentane molecule existed as a planar structure, the bond angles would be 108°, which is quite close to the normal 109.5°. However, in this planar structure all the H atoms would be eclipsed. A more stable arrangement is for one of the carbon atoms to be buckled out of the plane of the other four and for the H atoms to adopt a more staggered conformation.

Using ball-and-stick models it can be seen that there are two possible conformations of cyclohexane. These are the **"boat"** and the **"chair"** conformations, pictured in Figure 24-4. The models demonstrate clearly that H atoms on the first and fourth C atoms in the boat form come close enough together to repel one another. In the chair form all the H atoms are staggered. The chair form is the more stable conformation of cyclohexane. Also evident from Figure 24-4 is the fact that the twelve H atoms in cyclohexane are not quite equivalent. Six of them extend outward from the ring and are called **equatorial** H atoms. Of the other six, three are directed above and three below the ring; these are the six **axial** H atoms. When another group (say

FIGURE 24-4
Conformations of cyclohexane.

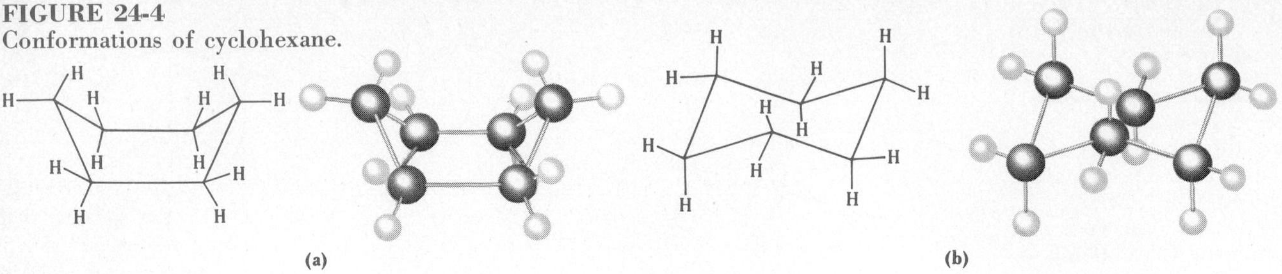

(a)

(b)

(a) Boat form. (b) Chair form; the equatorial H atoms are shown in color; the other H atoms are axial.

—CH_3) is substituted for an H atom on the ring, the preferred position is an equatorial one. This causes a minimum interference with other groups on the ring.

PREPARATION OF ALKANES. The primary source of alkanes is petroleum, but several laboratory methods have also been developed for their preparation. Some of these utilize organic substances of types that are described later in this chapter. Unsaturated hydrocarbons, whether containing double or triple bonds, may be converted to saturated hydrocarbons by the addition of hydrogen to the multiple bond system in the presence of a metal catalyst (equations 24.1 and 24.2). In the Würtz reaction halogenated hydrocarbons may be reacted with alkali metals to produce alkanes of double the carbon content (24.3). Alkali metal salts of carboxylic acids may be fused with alkali hydroxides. Here a hydrocarbon is obtained containing one carbon atom less than the original acid salt (24.4).

Organic reactions are generally represented in a manner different from that employed previously. Frequently, only the organic reactant is shown on the left. Inorganic reagents and reaction conditions (and sometimes even organic reagents) are written above the arrow. Only the important products are shown on the right. The numerous by-products of organic reactions are usually omitted. Equations may or may not be written in balanced form.

$$CH_2{=}CH_2 + H_2 \xrightarrow[\text{heat/pressure}]{\text{Pt or Pd}} CH_3{-}CH_3 \qquad (24.1)$$
$$\text{ethane}$$

$$HC{\equiv}CH + 2\,H_2 \xrightarrow[\text{heat/pressure}]{\text{Pt or Pd}} CH_3{-}CH_3 \qquad (24.2)$$
$$\text{ethane}$$

$$2\,CH_3CH_2Br + 2\,Na \longrightarrow 2\,NaBr + CH_3{-}CH_2{-}CH_2{-}CH_3 \qquad (24.3)$$
ethyl bromide
or bromoethane
butane

$$\underset{\text{sodium acetate}}{CH_3\overset{\overset{\displaystyle O}{\|}}{C}ONa} + NaOH \xrightarrow{\text{heat}} Na_2CO_3 + \underset{\text{methane}}{CH_4} \qquad (24.4)$$

Ordinary paraffin wax, used in candles, waxed paper, and in home canning, is a mixture of long-chain paraffin hydrocarbons.

SUBSTITUTION REACTIONS. Saturated hydrocarbons have little affinity for most chemical reagents. Because of this they have become known as paraffin hydrocarbons (Latin, *parum,* little; *affinis,* reactivity). Paraffin hydrocarbons are insoluble in water and do not react with aqueous solutions of acids, bases, or oxidizing agents. Halogens react only slowly with alkanes at room temperature; but at higher temperatures, particularly in the presence of light, halogenation occurs. In this reaction a halogen atom *substitutes* for a hydrogen atom. The mechanism of this substitution reaction appears to be by a chain reaction, written as follows for the chlorination of methane.

$$Cl:Cl \xrightarrow[\text{light}]{\text{heat or}} 2\,Cl\cdot$$

For emphasis only the electrons involved in bond breakage or formation are shown in the reaction mechanism.

propagation: $H_3C:H + Cl\cdot \longrightarrow H_3C\cdot + H:Cl$

$H_3C\cdot + Cl:Cl \longrightarrow H_3C:Cl + Cl\cdot$

termination: $Cl\cdot + Cl\cdot \longrightarrow Cl:Cl$

$H_3C\cdot + Cl\cdot \longrightarrow H_3C:Cl$

$H_3C\cdot + H_3C\cdot \longrightarrow H_3C:CH_3$

The reaction is initiated when some chlorine molecules absorb sufficient energy to dissociate into Cl atoms (represented above as $Cl\cdot$). Chlorine atoms collide with methane molecules to produce methyl radicals ($H_3C\cdot$), which combine with chlorine molecules to form the product, CH_3Cl. When any or all of the last three reactions proceed to the extent of consuming the radicals present, the reaction will stop. This free-radical chain reaction usually yields a mixture of products. The net equation for the formation of chloromethane is

Polyhalogenation can also occur to form CH_2Cl_2, dichloromethane or methylene dichloride (solvent); $CHCl_3$, trichloromethane or chloroform (solvent, anesthetic); and CCl_4, tetrachloromethane or carbon tetrachloride (solvent, fire extinguisher, dry cleaner).

$$CH_4 + Cl_2 \xrightarrow[\text{light}]{\text{heat or}} CH_3Cl + HCl \tag{24.5}$$

Oxidation is the reaction of hydrocarbons underlying their important use as fuels. For example,

$$C_7H_{16}(l) + 11\,O_2(g) \longrightarrow 7\,CO_2(g) + 8\,H_2O(l)$$

$$\Delta\bar{H}° = -4812\,\text{kJ/mol} \tag{24.6}$$

24-3 Alkenes and Alkynes

Note that simple alkenes and alicyclic alkanes have the same general formula: C_nH_{2n}.

Unsaturated hydrocarbons differ from the paraffins or saturated hydrocarbons in that they contain multiple bonds between carbon atoms. The simple alkenes or olefins contain one double bond and have the general formula C_nH_{2n} in their straight chain or branched chain forms. Simple alkynes or acetylenes have one triple bond between carbon atoms and can be represented by the general formula C_nH_{2n-2}.

The basic rules for naming alkanes apply, with slight modifications, to alkenes and alkynes. The base chain is taken to be the longest chain containing the multiple bond. The carbon atoms of the chain are numbered in such a way as to give the multiple bond the lowest possible number. The ending "ene" is used for alkenes and "yne" for alkynes. These rules are applied in the several examples that follow. (Common names are established by considering alkenes to be derivatives of ethylene and alkynes, of acetylene.)

Only the lowest-numbered carbon atom at which a multiple bond is found is referred to in naming the bond. It is understood that the multiple bond is between this and the next-highest-numbered carbon atom. For example, 4-methyl-2-pentyne has a triple bond between the second and third carbon atoms of a five-carbon chain.

$CH_2{=}CH_2$ $CH_3CH_2CH{=}CH_2$

ethene
(ethylene)

1-butene
(ethylethylene)

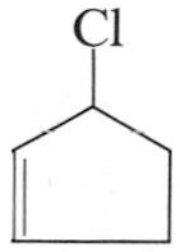

3-chlorocyclopentene

$HC{\equiv}CH$ $CH_3CH_2C{\equiv}CH$ $CH_3CHC{\equiv}CCH_3$

CH_3

ethyne
(acetylene)

1-butyne
(ethylacetylene)

4-methyl-2-pentyne
(isopropylmethylacetylene)

Example 24-6 What is the systematic name of the following structure?

$$
\begin{array}{c}
CH_3 \\
| \\
CH_2 \quad CH_3 \\
| \quad\quad | \\
CH_3\!-\!CH\!-\!C\!-\!C\!\equiv\!C\!-\!CH_3 \\
| \\
CH_3
\end{array}
$$

The longest chain is numbered to place the triple bond at the lowest possible number—2. Having made this decision, we now establish that there are three methyl groups at the positions 4, 4, and 5. The name of the compound is

4,4,5-trimethyl-2-heptyne

SIMILAR EXAMPLES: Exercises 14, 17, 18.

The alkenes are similar to the alkanes in physical properties. At room temperature, those containing two to four carbon atoms are gases; those with 5 to 18 are liquids; and those with more than 18 are solids. In general, alkynes have higher boiling points than their alkane and alkene counterparts.

GEOMETRICAL ISOMERISM. The compounds 2-butene, $CH_3CH\!=\!CHCH_3$, and 1-butene, $CH_2\!=\!CHCH_2CH_3$, are isomers. The difference between these two isomers is in the position of the double bond. This is another example of a positional isomer. Further consideration of the structure of 2-butene suggests another kind of isomerism.

$$
\begin{array}{ccc}
\underset{CH_3}{\overset{H}{\diagdown}}C\!=\!C\underset{CH_3}{\overset{H}{\diagup}} & \quad & \underset{H}{\overset{CH_3}{\diagdown}}C\!=\!C\underset{CH_3}{\overset{H}{\diagup}} \\
(a) & & (b)
\end{array}
$$

In a double bond there occurs an overlap of $2p$ orbitals to form a π bond, *in addition to* the formation of a σ bond. Rotation about the double bond at room temperature is severely restricted. As a result the foregoing molecules (a) and (b) are distinct and different (see Figure 24-5). To distinguish them by name, (a) is designated *cis*-2-butene (*cis*, Latin, on the same side) and (b) is called *trans*-2-butene (*trans*, Latin, across). This type of isomerism is called **geometrical isomerism.**

Geometrical isomerism is only one type of a general kind of isomerism called **stereoisomerism** (from the Greek word *stereos*, meaning solid or three-dimensional in nature). In stereoisomerism the number and types of atoms and bonds are the same, but certain atoms are oriented differently in space. Another type of stereoisomerism, called optical isomerism, is described in Chapter 25. Stereoisomerism has profound implications in the specificity of the chemical reactions giving uniqueness to living organisms.

Can you see why the name 4,4-dimethyl-5-ethyl-2-hexyne is incorrect?

FIGURE 24-5
Geometrical isomerism in 2-butene.

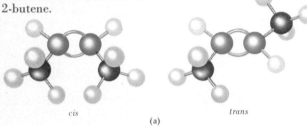

cis   *trans*
(a)

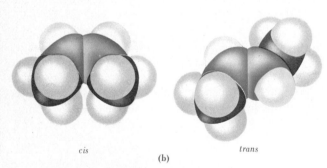

cis   *trans*
(b)

(a) Ball-and-stick models. **(b)** Space-filling models.

The terms *cis* and *trans* have the same meaning here as in the geometrical isomerism of complex ions (Section 22-5).

PREPARATION. Two general reactions for the preparation of olefins use alcohols and alkyl halides as starting materials (substrates). These are **elimination reactions,** processes in which atoms are removed from adjacent positions on a carbon chain. A small molecule is produced and an additional bond is formed between the C atoms. H_2O is eliminated in equation (24.7) and HBr in (24.8).

$$CH_3-\underset{\underset{HO}{|}}{\overset{\overset{H}{|}}{C}}-\underset{\underset{H}{|}}{\overset{\overset{H}{|}}{C}}-H \xrightarrow[\text{heat}]{H_2SO_4} CH_3CH{=}CH_2 + H_2O \tag{24.7}$$

$$CH_3\underset{\underset{H}{|}}{\overset{\overset{H}{|}}{C}}-\underset{\underset{Br}{|}}{\overset{\overset{H}{|}}{C}}-H \xrightarrow{\text{alcoholic KOH}} CH_3CH{=}CH_2 + KBr + HOH \tag{24.8}$$

Acetylene, the simplest alkyne, has been one of the most important organic raw materials used in industry. It is prepared from coal, water, and limestone.

Currently, the commercial importance of acetylene is declining because of the high energy costs required for its manufacture.

$$CaCO_3 \xrightarrow{\text{heat}} CaO + CO_2 \tag{24.9}$$

$$CaO + 3\,C \xrightarrow[\substack{\text{furnace,}\\ 2000°C}]{\text{electric}} CaC_2 + CO \tag{24.10}$$

calcium
carbide

Calcium carbide is a common name. The systematic name is calcium acetylide.

$$CaC_2 + 2\,H_2O \longrightarrow HC{\equiv}CH + Ca(OH)_2 \tag{24.11}$$

Most other alkynes are prepared from acetylene itself by taking advantage of the acidity of the C—H bond. In the presence of a very strong base, such as sodium amide, the acetylene donates protons to amide ions and forms a sodium salt, sodium acetylide (equation 24.12). This acetylide can then react with an alkyl halide (24.13).

$$H{-}C{\equiv}C{-}H + Na^+ NH_2^- \longrightarrow NH_3 + H{-}C{\equiv}C^-\,Na^+ \tag{24.12}$$

$$H{-}C{\equiv}C^-\,Na^+ + CH_3Br \longrightarrow HC{\equiv}C{-}CH_3 + NaBr \tag{24.13}$$

By continuing this reaction, the triple bond can be positioned as desired in the chain, as in the synthesis of 2-pentyne.

$$H{-}C{\equiv}C{-}CH_3 \xrightarrow{NaNH_2} Na^+\,{}^-C{\equiv}C{-}CH_3 + NH_3 \tag{24.14}$$

$$Na^+\,{}^-C{\equiv}C{-}CH_3 + CH_3CH_2Br \longrightarrow CH_3CH_2C{\equiv}CCH_3 + NaBr \tag{24.15}$$

ADDITION REACTIONS. The most significant distinction between alkanes and alkenes is that alkanes react by *substitution* and alkenes by *addition*.

$$\text{alkane:} \quad CH_3{-}CH_3 + Br_2 \longrightarrow CH_3{-}CH_2{-}Br + HBr \tag{24.16}$$

$$\text{alkene:} \quad CH_2{=}CH_2 + Br_2 \longrightarrow \underset{\underset{Br}{|}}{CH_2}{-}\underset{\underset{Br}{|}}{CH_2} \tag{24.17}$$

The splitting of a Br_2 molecule produces two identical Br atoms; $CH_2{=}CH_2$ yields two —CH_2 groups. Br_2 and $CH_2{=}CH_2$ are symmetrical reagents. The two parts that result when HBr or $CH_2CH{=}CH_2$ are split are not identical. These are unsymmetrical reagents.

When *unsymmetrical* HBr is added to *unsymmetrical* propene, a problem arises. Which of the following products will form?

$$CH_3CH{=}CH_2 + H{-}Br \longrightarrow \underset{\underset{Br}{|}\;\underset{H}{|}}{CH_3CH{-}CH_2} \quad \text{or} \quad \underset{\underset{H}{|}\;\underset{Br}{|}}{CH_3CH{-}CH_2}$$

2-Bromopropane is the sole product obtained. This fact can be explained in terms of the electron-donating abilities of H atoms and alkyl groups. However, rather than go into further detail on this matter, we simply present an empirical rule proposed by Markovnikov in 1871.

In the addition of an unsymmetrical reagent (HX, HOH, HCN, HOSO$_3$H) *to an unsymmetrical olefin, such as* CH$_3$CH=CH$_2$, *the more positive fragment of the reagent* (*usually hydrogen*) *adds to the carbon atom with the greater number of attached hydrogen atoms.*

Note how Markovnikov's rule applies to several of the addition reactions that follow.

The addition of H$_2$O to a double bond (as in equation 24.18) is the reverse of the reaction in which a double bond is formed by the elimination of H$_2$O (as in equation 24.7). An equilibrium is involved that favors the addition reaction in dilute acid and the elimination reaction in concentrated H$_2$SO$_4$(aq).

$$CH_3-\underset{\underset{CH_3}{|}}{C}=CH_2 + H_2O \xrightarrow{10\% \ H_2SO_4} CH_3-\underset{\underset{OH}{|}}{\overset{\overset{CH_3}{|}}{C}}-CH_3 \qquad (24.18)$$

t-butyl alcohol

$$CH_3-C\equiv CH + HCl \longrightarrow CH_3-\underset{\underset{Cl}{|}}{C}=\overset{\overset{H}{|}}{C}-H \xrightarrow{HCl} CH_3-\underset{\underset{Cl}{|}}{\overset{\overset{Cl}{|}}{C}}-\underset{\underset{H}{|}}{\overset{\overset{H}{|}}{C}}-H \qquad (24.19)$$

methylacetylene (propyne)   2-chloropropene   2,2-dichloropropane

$$HC\equiv CH + HCN \longrightarrow H-\underset{\underset{H}{|}}{\overset{\overset{CN}{|}}{C}}=\overset{}{C}-H \qquad (24.20)$$

cyanoethylene (acrylonitrile)

$$CH_2=CHCH_3 \xrightarrow{MnO_4^-, \ H_2O} CH_2-\underset{\underset{OH}{|}}{\overset{}{C}H}-CH_3 \\ \underset{OH}{|} \qquad (24.21)$$

propylene (propene)   1,2-propanediol (propylene glycol)

Purple-colored permanganate is decolorized by olefins (reaction 24.21). This is in contrast to the nonreactivity of alkanes and provides a qualitative test for distinguishing between alkanes and alkenes (Baeyer test). The addition reactions of HCl and HCN to alkynes are used commercially to synthesize intermediates for polymer production.

POLYMERIZATION. Ethylene and other olefins can enter into reactions in which the net effect is the opening up of the C=C double bond and the formation of giant molecules. This type of reaction is called a polymerization reaction. The key to the reaction is a free-radical initiator. In the first step of the process outlined in equations (24.22) through (24.25), an organic peroxide dissociates into two radicals (equation 24.22). The radicals add to the C=C double bonds of ethylene molecules, forming radical intermediates (24.23). These radical intermediates successively attack more ethylene molecules, forming new intermediates of longer and longer length (24.24). The reaction proceeds by a chain mechanism. The chains terminate as a result of reactions such as (24.25).

This is not our first encounter with polymerization. We have already considered the formation of silicones (Section 20-9) by a process known as *condensation polymerization*. The mechanism described here is known as *free-radical addition polymerization*. Additional examples of both types are discussed in Chapter 26.

$$R-O:O-R \longrightarrow 2 \ R-O \cdot \qquad (24.22)$$

organic peroxide

initiation: $\quad CH_2=CH_2 + RO \cdot \longrightarrow R-O-CH_2-CH_2 \cdot \qquad (24.23)$

propagation:

$$ROCH_2CH_2 \cdot + CH_2{=}CH_2 \longrightarrow ROCH_2CH_2CH_2CH_2 \cdot \tag{24.24}$$

$$RO(CH_2)_3CH_2 \cdot + CH_2{=}CH_2 \longrightarrow RO(CH_2)_5CH_2 \cdot \longrightarrow \longrightarrow$$

termination: $RO(CH_2)_xCH_2 \cdot + RO \cdot \longrightarrow RO(CH_2)_xCH_2OR \tag{24.25}$

or $2\ RO(CH_2)_xCH_2 \cdot \longrightarrow RO(CH_2)_xCH_2CH_2(CH_2)_xOR$

24-4 Aromatic Hydrocarbons

Aromatic hydrocarbon molecules have structures based on the molecule, benzene, C_6H_6. In Section 10-5 we discussed chemical bonding in benzene in some detail and arrived at various representations of the benzene molecule, including

Kekulé structures   molecular orbital
representation

Of these two possibilities we will choose the molecular orbital representation. Other aromatic hydrocarbons can be viewed as derivatives of benzene.

toluene   *o*-xylene   naphthalene   anthracene

Toluene and *o*-xylene are substituted benzenes and naphthalene and anthracene feature fused benzene rings. Whenever two rings are fused together, there is a loss of two carbon and four hydrogen atoms. Thus, naphthalene has the formula $C_{10}H_8$ and anthracene, $C_{14}H_{10}$.

When one of the six equivalent H atoms of a benzene molecule is removed, the species that results is called a **phenyl** group. Two phenyl groups may bond together as in biphenyl, or phenyl groups may be substituents on aliphatic hydrocarbon chains.

phenyl group   biphenyl   triphenylmethane

Other groups may be substituted for H atoms, raising problems in nomenclature; these are handled by a numbering system for the C atoms in the ring.

If the name of an aromatic compound is to be based on a common name other than benzene, for example, toluene, the characteristic substituent group ($-CH_3$) is assigned position "1."

3-bromotoluene
(*m*-bromotoluene)

2-bromochlorobenzene
(*o*-bromochlorobenzene)

1,4-dichlorobenzene
(*p*-dichlorobenzene)

2-chlorotoluene
(*o*-chlorotoluene)

DDT was at one time a widely used pesticide. Its use in the United States has been discontinued, however, because of environmental hazards.

Cl—⟨benzene ring⟩—$\overset{2}{C}H$—⟨benzene ring⟩—Cl

$\underset{1}{C}Cl_3$

2,2-di-(*p*-chlorophenyl)-1,1,1-trichloroethane (DDT)

The terms "ortho," "meta," and "para" (*o*-, *m*-, *p*-) may be used when there are two substituents on the benzene ring. **Ortho** refers to substituents on adjacent carbon atoms, **meta** to substituents with one carbon atom between them, and **para** to substituents opposite to one another on the ring.

Benzene and its homologs are similar to other hydrocarbons in that they are insoluble in water but soluble in organic solvents. The boiling points of the aromatic hydrocarbons (arenes) are slightly higher than those of the alkanes of similar carbon content. For example, *n*-hexane, C_6H_{14}, boils at 69°C, whereas benzene boils at 80°C. The planar structure and highly delocalized electron density in aromatic hydrocarbons increases the attractive forces acting between molecules, resulting in higher boiling points. The symmetrical structure of benzene permits closer packing in the crystalline state, resulting in a higher melting point than for *n*-hexane. Benzene melts at 5.5°C and *n*-hexane melts at −95°C.

Aromatic hydrocarbons are highly flammable and should always be handled with care. Prolonged inhalation of benzene vapor results in a decreased production of both red and white blood corpuscles, and this can prove fatal. Also, benzene is a carcinogen. Benzene should be used only under well-ventilated conditions.

One of the principal hazards in handling certain aromatic hydrocarbons is that they are carcinogenic (cancer producing). 3,4-Benzpyrene is one of the most active of these.

3,4-benzpyrene

Fused ring systems, such as 3,4-benzpyrene, are most commonly encountered when organic materials are heated to high temperatures in limited contact with air, a process known as *pyrolysis* (thermal decomposition). 3,4-Benzpyrene has been isolated in the tar formed by burning cigarettes and as a decomposition product of grease in the charcoal grilling of meat.

AROMATIC SUBSTITUTION REACTIONS. Alkenes and alkynes have localized regions of high electron density (multiple bonds). The electron density associated with the unsaturation in an aromatic ring is *delocalized* into a π electron cloud (recall the doughnut-shaped regions in Figure 10-20). Simple reactions of the aromatic ring involve not the addition of a reagent but the substitution of other atoms or groups for H atoms. Among the reactions of this type are halogenation, nitration, sulfonation, and alkylation. Examples of these are presented in Figure 24-6.

In the halogenation reaction mechanism below, the key step is believed to be the formation of a bond between a Lewis acid called an electrophile (an electron-seeking species) and the π electron cloud of the benzene (equation 24.26). The electrophile, Br^+, results from the reaction of Br_2 and $FeBr_3$. Following attack of the electrophile on the π electron cloud, a redistribution of charge and the elimination of H^+ occur. $FeBr_3$ is regenerated in the process. The net reaction is shown in equation (24.27).

$$Br_2 + FeBr_3 \longrightarrow Br^+ + FeBr_4^-$$

⟨benzene⟩ + Br^+ ⟶ ⟨benzene with H and Br, +⟩ ⟶ H^+ + ⟨benzene with Br⟩ (24.26)

$$H^+ + FeBr_4^- \longrightarrow HBr + FeBr_3$$

net reaction: ⟨benzene⟩ $\xrightarrow[FeBr_3]{Br_2}$ ⟨benzene with Br⟩ + HBr (24.27)

The substitution of a single group, X, can occur at any of the six positions of the benzene ring; they are all equivalent. For the introduction of a second substituent, however, the question arises as to which of the remaining five sites the new group

FIGURE 24-6
Some substitution
reactions of benzene
illustrated.

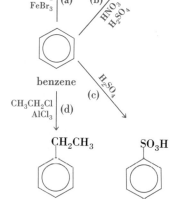

(a) Halogenation. (b)
Nitration. (c)
Sulfonation. (d)
Alkylation.

will occupy. If all of these sites were equally preferred, the distribution of products would be this purely statistical one.

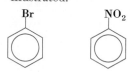

40% ortho (2/5)	40% meta (2/5)	20% para (1/5)

The following scheme describes the products resulting from nitration followed by halogenation (equation 24.28) and halogenation followed by nitration (equation 24:29). It shows that *the substitution is not random*. The —NO₂ group directs Cl to a meta position and the —Cl group is an ortho-para director.

$$\text{(24.28)}$$

$$\text{(24.29)}$$

Whether a group is an ortho-para or a meta director depends on how the presence of one substituent alters the electron distribution in the benzene ring. This makes attack by a second group more likely at one type of position than another. Examination of a large number of reactions leads to the following order:

ortho-para directors: —NH₂, —OR, —OH, —OCOR, —R, —X
(from strongest to weakest)

meta directors: —NO₂, CN, —SO₃H, —CHO, —COR, —COOH, —COOR
(from strongest to weakest)

Example 24-7 Predict the products of the mononitration of

(a) CHO

(b) OH NO₂

(a) Since —CHO is a meta director, we should expect the nitration of benzaldehyde (a) to produce 3-nitrobenzaldehyde almost exclusively.

CHO

NO₂

(b) The —OH group is an ortho-para director and we should expect the products

2,6-dinitrophenol   2,4-dinitrophenol

Considering the —NO$_2$ group as a meta director leads to the same conclusion. Exercise 31 at the end of the chapter illustrates a case where a conclusion is not quite as simple.

SIMILAR EXAMPLES: Exercises 30, 31.

24-5 Alcohols, Phenols, and Ethers

The presence of —OH, the hydroxyl group, characterizes the alcohols and phenols. Depending on the nature of the carbon atom to which the hydroxyl group is attached, three classes of alcohols are obtained.

1-butanol   2-butanol   2-methyl-2-propanol
(*n*-butyl alcohol)   (*s*-butyl alcohol)   (*t*-butyl alcohol)
(a *primary* alcohol)   (a *secondary* alcohol)   (a *tertiary* alcohol)

There may be more than one —OH group present in a molecule, resulting in a polyhydric alcohol.

1,2-ethanediol   1,2,3-propanetriol
(ethylene glycol)   (glycerol)

In phenols the hydroxyl group is attached to an aromatic ring.

phenol   2,4,6-trinitrophenol
(carbolic acid)   (picric acid)

Hexachlorophene, a phenol, was once used as an antiseptic in toothpastes, soaps, and deodorants. However, in view of laboratory observations that small concentrations of hexachlorophene produce brain damage in rats, it is currently restricted to certain medical applications.

As a group of compounds, the aliphatic alcohols are liquids whose properties are strongly influenced by hydrogen bonding. As the chain length increases, the influence of the polar hydroxyl group on the properties of the molecule diminishes. The molecule becomes less like water and more like a hydrocarbon. As a consequence, low-molecular-weight alcohols tend to be water soluble; high-molecular-weight alcohols are not. The boiling points and solubilities of the phenols vary widely, depending on the nature of the other substituents on the benzene ring.

PREPARATION AND USES. Alcohols can be obtained by the hydration of alkenes or the hydrolysis of alkyl halides.

$$\text{hydration:} \quad CH_3CH{=}CH_2 + H_2O \xrightarrow{H_2SO_4} CH_3\underset{\underset{OH}{|}}{C}HCH_3 \qquad (24.30)$$

propene   2-propanol
(propylene)   (isopropyl alcohol)

$$\text{hydrolysis:} \quad CH_3CH_2CH_2Br + OH^- \longrightarrow CH_3CH_2CH_2OH + Br^- \qquad (24.31)$$

n-propyl bromide   *n*-propyl alcohol

Methanol is known as wood alcohol because it can be produced by the destructive distillation of wood. This substance is highly toxic and can lead to blindness or death if ingested. Most methanol is manufactured synthetically from carbon monoxide and hydrogen.

$$CO(g) + 2\,H_2(g) \xrightarrow[\substack{200\ atm \\ ZnO,\ Cr_2O_3}]{350°C} CH_3OH(g) \qquad (24.32)$$

Ethanol is the common "alcohol" to the layperson. It is obtainable by the fermentation of blackstrap molasses, the residue from the purification of sugar cane, or from other materials containing natural sugars. The principal synthetic method is the hydration of ethylene with sulfuric acid.

Ethylene glycol is water soluble and has a higher boiling point (197°C) than water. Because of these properties it makes an excellent permanent nonvolatile antifreeze for use in automobile radiators. It is also used in the manufacture of solvents, paint removers, and plasticizers (softeners). Propylene glycol is used in suntan lotions and, in conjunction with fluorocarbons, to produce nonaqueous foams in aerosol products.

Glycerol (glycerin) is obtained commercially as a by-product in the manufacture of soap. It is a sweet, syrupy liquid that is miscible with water in all proportions. Because it has the ability to take up moisture from the air, it can be used to keep skin moist and soft, accounting for its use in lotions and cosmetics. It is also used to maintain the moisture content of tobacco and candy.

REACTIONS OF THE OH GROUP. The reactivity of the —OH group results from (1) the unshared electron pairs on the O atom, making the molecule basic in the Lewis sense, or (2) the polarity of the O—H bond, causing the molecule to act as a proton donor, to be acidic. Reactions (24.33) and (24.34) illustrate the first point and (24.35), the second.

Nitroglycerine is an incorrect name. The compound is an ester and should be so named. A nitro compound has a —NO$_2$ group directly on a carbon atom, as in nitrobenzene. Nitroglycerine was first prepared in 1846 by Sobreno, an Italian, but it remained for Alfred Nobel, a Swedish chemist (1861), to mix it with diatomaceous earth and produce a material less sensitive to shock. This material is called dynamite.

$$CH_3\underset{\underset{CH_3}{|}}{C}H{-}\overset{..}{\underset{..}{O}}H + CH_3\overset{O}{\overset{\|}{C}}{-}Cl \longrightarrow CH_3\overset{O}{\overset{\|}{C}}OCH(CH_3)_2 + HCl \qquad (24.33)$$

isopropyl alcohol   acetyl chloride   isopropyl acetate
(an acid halide)   (an ester)

$$\begin{matrix} CH_2{-}OH \\ | \\ CH{-}OH \\ | \\ CH_2{-}OH \end{matrix} + 3\,HONO_2 \longrightarrow 3\,H_2O + \begin{matrix} CH_2ONO_2 \\ | \\ CHONO_2 \\ | \\ CH_2ONO_2 \end{matrix} \qquad (24.34)$$

glycerol   glyceryl trinitrate
(nitroglycerine)

$$CH_3{-}O{-}H \xrightleftharpoons{NaOH} CH_3O^- Na^+ + H_2O \qquad (24.35)$$

$K_a = 1 \times 10^{-16}$

Ethers. Ethers are compounds with the general formula R—O—R. Structurally, they can be pure aliphatic, pure aromatic, or mixed.

$CH_3{-}O{-}CH_3$

dimethyl ether

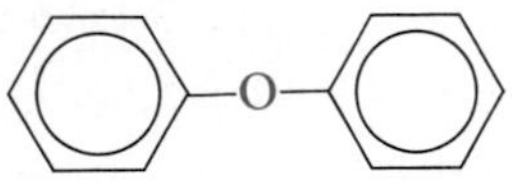

diphenyl ether

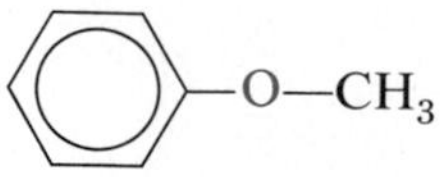

methyl phenyl ether (anisole)

Ethers can be prepared by the elimination of water from between two alcohol molecules using a strong dehydrating agent, such as concentrated H_2SO_4.

$$CH_3CH_2OH + HOCH_2CH_3 \xrightarrow[\text{conc.}]{H_2SO_4} CH_3CH_2OCH_2CH_3 + H_2O \qquad (24.36)$$

diethyl ether

Chemically, the most notable property of ethers is their comparative lack of reactivity. The ether linkage is stable to most oxidizing and reducing agents and to action by dilute acids and alkalies.

Diethyl ether has been used extensively as a general anesthetic. It is easy to administer and produces excellent relaxation of the muscles. Also, the pulse rate, rate of respiration, and blood pressure are affected only slightly. However, it is somewhat irritating to the respiratory passages and produces a nauseous after effect. More recently methyl propyl ether (neothyl) has been used, and it is reported to be less irritating. Methyl ether, a gas at room temperatures, is used as a propellant for aerosol sprays. Higher ethers have found extensive use as solvents for varnishes and lacquers.

24-6 Aldehydes and Ketones

Aldehydes and ketones contain the carbonyl group.

$$\rangle C{=}O$$

If both groups attached to a carbonyl group are carbon residues, the resulting compound is called a ketone. If one of the groups is a carbon residue and the other a hydrogen atom, the compound is called an aldehyde.

$$H{-}\overset{O}{\overset{\|}{C}}{-}H$$

methanal (formaldehyde)

$$\overset{\gamma}{CH_3}\underset{\underset{Cl}{|}}{\overset{\beta}{C}H}{-}\overset{\alpha}{CH_2}{-}\overset{O}{\overset{\|}{C}}{-}H$$

3-chlorobutanal (β-chlorobutyraldehyde)

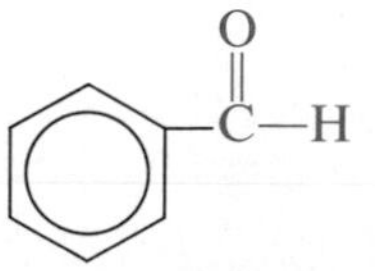

benzaldehyde

$$CH_3{-}\overset{O}{\overset{\|}{C}}{-}CH_3$$

propanone (acetone)

$$CH_3CH_2{-}\overset{O}{\overset{\|}{C}}{-}CH_2CH_3$$

3-pentanone (diethyl ketone)

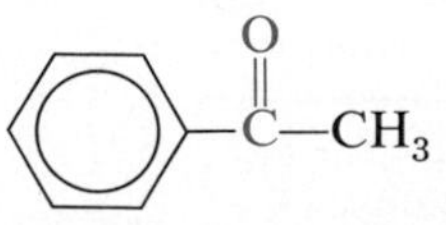

methyl phenyl ketone (acetophenone)

PREPARATION AND USES.   Partial oxidation of a primary alcohol produces an aldehyde (further oxidation yields a carboxylic acid). Oxidation of a secondary alcohol produces a ketone.

$$CH_3CH_2OH \xrightarrow[H^+]{Cr_2O_7{}^{2-}} CH_3CHO \xrightarrow[H^+]{Cr_2O_7{}^{2-}} CH_3CO_2H \tag{24.37}$$

ethanol
(a primary alcohol)   acetaldehyde
(an aldehyde)   acetic acid
(an acid)

A balanced equation to represent the oxidation of 2-propanol (a secondary alcohol) to propanone (a ketone) by dichromate ion in acidic solution is obtained in a familiar way (recall Section 19-2).

oxidation:

$$C_3H_8O \longrightarrow$$
$$C_3H_6O + 2\,H^+ + 2\,e^-$$

reduction:

$$Cr_2O_7{}^{2-} + 14\,H^+$$
$$+ 6\,e^- \longrightarrow$$
$$2\,Cr^{3+} + 7\,H_2O$$

net:

$$3\,C_3H_8O + Cr_2O_7{}^{2-}$$
$$+ 8\,H^+ \longrightarrow 3\,C_3H_6O$$
$$+ 2\,Cr^{3+} + 7\,H_2O$$

$$CH_3CHOHCH_3 \xrightarrow[H^+]{Cr_2O_7{}^{2-}} CH_3\overset{\displaystyle O}{\overset{\|}{C}}CH_3 \tag{24.38}$$

2-propanol
(a secondary alcohol)   propanone
(a ketone)

Formaldehyde, a colorless gas, dissolves readily in water. A 40% solution in water, called formalin, is used as an embalming fluid and a tissue preservative. Formaldehyde is also used in the manufacture of synthetic resins. A polymer of formaldehyde, called paraformaldehyde, is used as an antiseptic and an insecticide. Acetaldehyde is an important raw material for the production of acetic acid, acetic anhydride, and the ester, ethyl acetate.

Acetone is the most important of the ketones. It is a volatile liquid (boiling point, 56°C) and highly flammable. Acetone is a good solvent for a variety of organic compounds; and because of this it is widely used in solvents for varnishes, lacquers, and plastics. Unlike many common organic solvents, acetone is miscible with water in all proportions. This property, combined with its volatility, makes acetone a useful drying agent for laboratory glassware. Residual water is removed through several rinses with acetone, and the remaining liquid acetone film quickly evaporates. One method of producing acetone involves the *dehydrogenation* of isopropyl alcohol in the presence of a copper catalyst.

$$CH_3-\overset{\displaystyle OH}{\overset{|}{C}H}-CH_3 \xrightarrow[300°C]{Cu} CH_3-\overset{\displaystyle O}{\overset{\|}{C}}-CH_3 + H_2 \tag{24.39}$$

Aldehydes and ketones occur widely in nature. Typical natural sources are

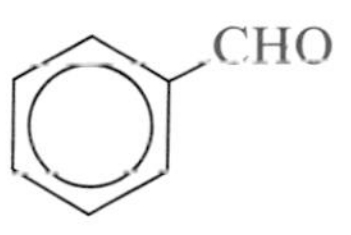

benzaldehyde
(almonds) 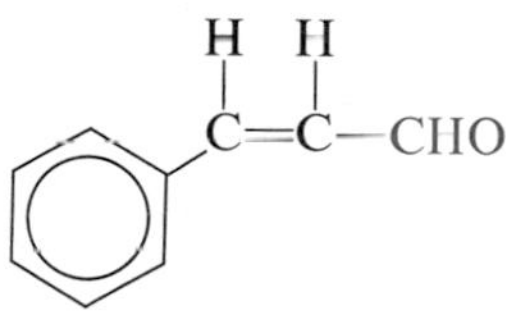   cinnamaldehyde
(cinnamon) 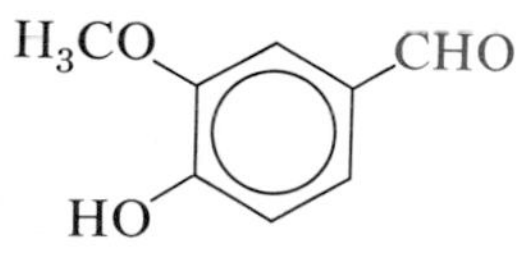   vanillin
(vanilla)

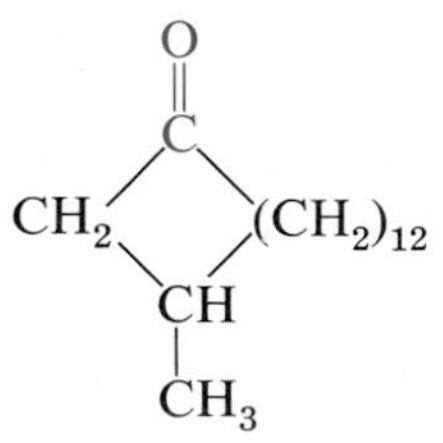

muscone
(obtained from musk deer;
used in perfumes) 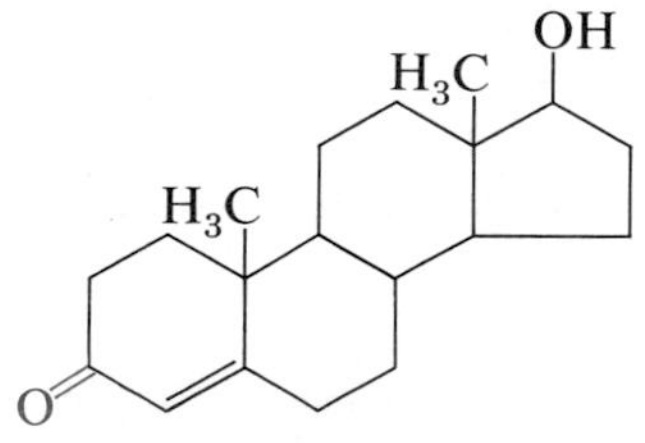   testosterone
(male sex hormone) 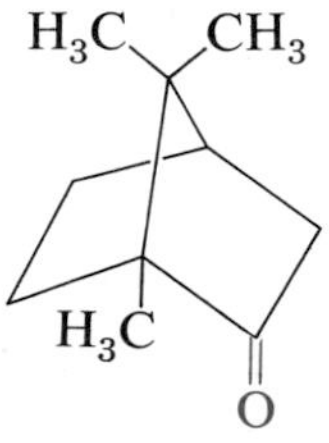   camphor
(obtained from
camphor tree)

24-7 Carboxylic Acids and Their Derivatives

Compounds that contain the carboxyl group (*carb*onyl and hydr*oxyl*)

$$\underset{O}{\overset{\displaystyle O}{\underset{\|}{-C}}}-OH$$

are called carboxylic acids; they have the general formula $R-CO_2H$. Many compounds are known where R is an aliphatic residue. These are called fatty acids since compounds of this type are readily available from naturally occurring fats and oils. The carboxyl group can also be found attached to the benzene ring. If two carboxyl groups are found on the same molecule the acid is called a dicarboxylic acid.

o-Hydroxybenzoic acid or salicylic acid occurs in nature in the willow tree (genus *Salix*). The acetyl derivative of this acid is aspirin, an analgesic (pain killer) and antipyretic (fever reducer).

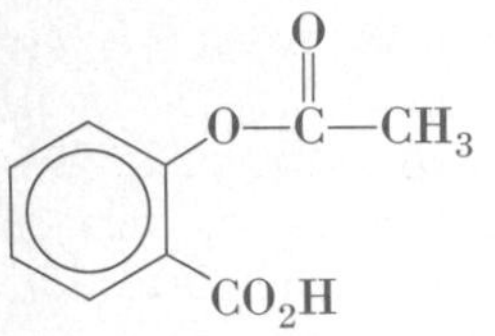

aspirin
acetylsalicylic acid

$CH_3\overset{O}{\overset{\|}{C}}-OH$	COOH	$HO-\overset{O}{\overset{\|}{C}}-\overset{O}{\overset{\|}{C}}-OH$	COOH COOH
acetic acid (an aliphatic acid)	benzoic acid (an aromatic acid)	oxalic acid (an aliphatic dicarboxylic acid)	phthalic acid (an aromatic dicarboxylic acid)

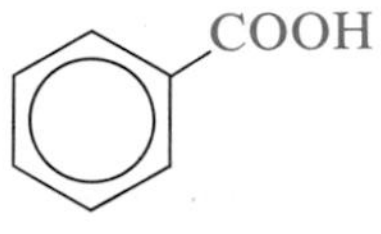

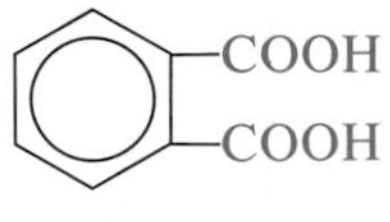

The carboxylic acids have widely used common names as well as systematic names. Some examples are given in Table 24-3.

Substituted aliphatic acids can be named either by their IUPAC names or by using Greek letters in conjunction with common names. Aromatic acids are named as derivatives of benzoic acid.

3-chlorobutanoic acid
β-chlorobutyric acid

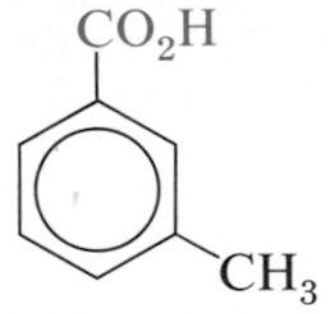

3-methylbenzoic acid
m-methylbenzoic acid
(also *m*-toluic acid)

Because many derivatives of the carboxylic acids involve simple replacement of the hydroxyl groups, special names have been developed for the remaining portion

of the molecule, $R-\overset{O}{\overset{\|}{C}}-$. The group —COR is given the general name **acyl**. Some

TABLE 24-3
Some common carboxylic acids

Structural formula	Common name	IUC name	K_a
HCO_2H	formic acid	methanoic	1.78×10^{-4}
CH_3CO_2H	acetic acid	ethanoic	1.74×10^{-5}
$CH_3CH_2CO_2H$	propionic acid	propanoic	1.35×10^{-5}
$CH_3(CH_2)_2CO_2H$	butyric acid	butanoic	1.48×10^{-5}
$CH_3(CH_2)_{16}CO_2H$	stearic acid	octadecanoic	
$CH_3(CH_2)_7CH{=}CH(CH_2)_7CO_2H$	oleic acid	9-octadecenoic	
$C_6H_5CO_2H$	benzoic	benzoic	6.4×10^{-5}
$O_2NC_6H_4CO_2H$	*p*-nitrobenzoic	4-nitrobenzoic	3.8×10^{-4}
HO_2CCO_2H	oxalic	ethanedioic	$\begin{cases} (K_{a_1})3.5 \times 10^{-2} \\ (K_{a_2})6.1 \times 10^{-5} \end{cases}$

specific examples of its use are

| formyl | acetyl | β-chloropropionyl | benzoyl |

Among the important acid derivatives are

| acyl halide (an acid halide) (X = halogen) | an ester | an acid anhydride | an amide |

Unlike the pungent odors of the carboxylic acids from which they are derived, esters have very pleasant aromas. The characteristic fragrances of many flowers and fruits can be traced to the esters they contain. Esters are used in perfumes and in the manufacture of flavoring agents for the confectionery and soft drink industries. Most esters are colorless liquids, insoluble in water. Their melting points and boiling points are generally lower than those of alcohols and acids of comparable carbon content. This is because of the absence of hydrogen bonding in esters.

PREPARATION. Two methods for the preparation of the carboxylic acids are illustrated through equations (24.40) and (24.41).

oxidation of an alcohol:

$$CH_3CH_2OH \xrightarrow[H^+]{K_2Cr_2O_7} CH_3CO_2H \tag{24.40}$$

oxidation of an aldehyde:

$$CH_3CH_2CHO \xrightarrow[OH^-]{KMnO_4} CH_3CH_2CO_2K \xrightarrow{H^+} CH_3CH_2CO_2H + K^+ \tag{24.41}$$

REACTIONS OF THE CARBOXYL GROUP. The carboxyl group displays the chemistry of both the carbonyl and the hydroxyl group. Donation of a proton to a base leads to salt formation. The sodium and potassium salts of long-chain fatty acids are known as **soaps,** for example, sodium stearate (see Chapter 25).

Heating the ammonium salt of a carboxylic acid causes the elimination of water and the formation of an **amide.** Further loss of water occurs if the amide is heated with a strong dehydrating agent such as P_2O_5. The final product, which contains the group $-C\equiv N$, is called a **nitrile.** These reactions can be reversed by the step-wise treatment with water.

$$CH_3-\overset{\overset{\displaystyle O}{\|}}{C}-O^- NH_4^+ \xrightarrow{heat} CH_3-\overset{\overset{\displaystyle O}{\|}}{C}-NH_2 + H_2O \tag{24.42}$$

ammonium acetate   acetamide

$$heat \Big\downarrow P_2O_5$$

$$CH_3C\equiv N \tag{24.43}$$

acetonitrile

The product of the reaction of an acid and an alcohol is called an ester. It, too, can be viewed as forming by the elimination of H_2O through the respective hydroxyl groups of an acid and an alcohol. The mechanism of this reaction is such that the —OH of the resulting water comes from the *acid* and the —H from the *alcohol*.

$$\underset{\substack{\text{methylpropionic acid}\\\text{(isobutyric acid)}}}{CH_3CH-\overset{\overset{\displaystyle O}{\|}}{C}-OH} \; \underset{\substack{\text{1-butanol}\\(\textit{n}\text{-butyl alcohol})}}{+ \; CH_3(CH_2)_3OH} \longrightarrow H_2O + \underset{\substack{\text{butyl methylpropionate}\\(\textit{n}\text{-butyl isobutyrate})}}{CH_3CH-\overset{\overset{\displaystyle O}{\|}}{C}-O(CH_2)_3CH_3} \qquad (24.44)$$

24-8 Amines

Amines are organic derivatives of ammonia in which one or more organic residues (R) are substituted for H atoms. Their classification is based on the number of hydrogen atoms remaining bonded to the nitrogen atom—two for primary amines, one for secondary, and none for tertiary. This classification scheme is illustrated in Figure 24-7.

Amines of low molecular weight are gases that are readily soluble in water, yielding basic solutions. The volatile members have odors similar to ammonia but more "fishlike." Amines form hydrogen bonds, but these bonds are weaker than are those in water because nitrogen is less electronegative than oxygen. As in ammonia (see Figure 10-6), the nitrogen atoms in the amines use sp^3 hybrid orbitals, resulting in pyramidal structures. The lone pair electrons occupy one of the sp^3 orbitals. Amines, like ammonia, owe their basicity to these lone pair electrons. Some properties of amines are listed in Table 24-4.

Dimethylamine is used as an accelerator in the removal of hair from hides in the processing of leather. Butyl- and amylamines are used as antioxidants, corrosion inhibitors, and in the manufacture of oil-soluble soaps. Dimethyl- and trimethylamines are used in the manufacture of ion exchange resins. Additional uses are found in the manufacture of disinfectants, insecticides, herbicides, drugs, dyes, fungicides, soaps, cosmetics, and photographic developers.

Several methods may be used to synthesize amines, but we limit our concern to

In aromatic amines, because of unsaturation in the benzene ring, electrons are drawn into the ring and this reduces the electron density on the nitrogen atom. As a result, aromatic amines are weaker bases than ammonia. In aliphatic amines the effect is the opposite, and these amines are somewhat stronger bases than ammonia.

FIGURE 24-7
A classification scheme for amines.

Amines are derivatives of ammonia. Quaternary salts are analogous to ammonium salts.

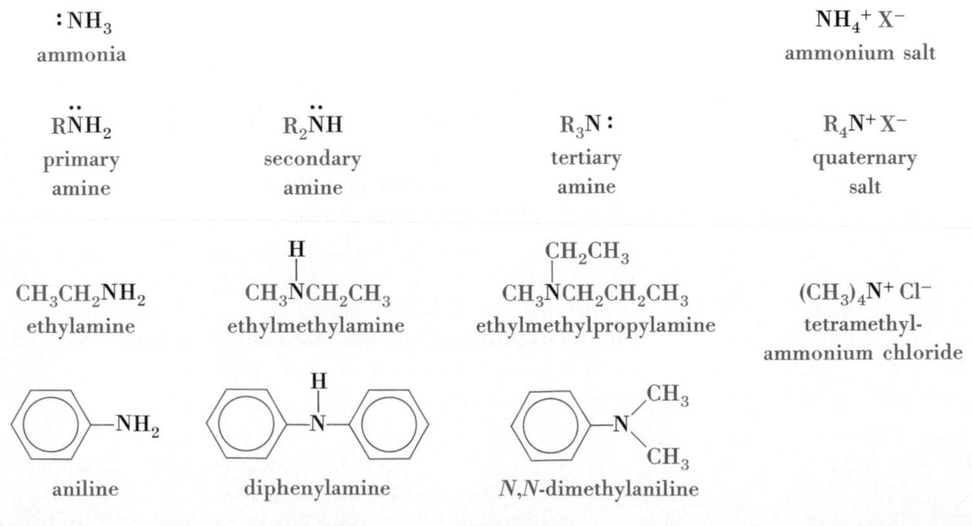

TABLE 24-4
Some properties of selected amines

Name	Formula	Boiling point, °C	K_b
ammonia	NH_3	−33.4	1.8×10^{-5}
methylamine	CH_3NH_2	−6.5	44×10^{-5}
ethylamine	$CH_3CH_2NH_2$	16.6	47×10^{-5}
butylamine	$CH_3CH_2CH_2CH_2NH_2$	77.8	40×10^{-5}
aniline	$C_6H_5NH_2$	184	4.2×10^{-10}
N-methylaniline	$C_6H_5NHCH_3$	196	7.1×10^{-10}

The designation "*N*" in *N*-methylaniline signifies that the methyl group is attached to the N atom and not the benzene ring.

an especially important one—the reduction of nitro compounds.

$$\langle\!\!\!\bigcirc\!\!\!\rangle\!-NO_2 \xrightarrow[HCl]{Fe} \langle\!\!\!\bigcirc\!\!\!\rangle\!-NH_3{}^+ Cl^- \xrightarrow{NaOH} \langle\!\!\!\bigcirc\!\!\!\rangle\!-NH_2 \qquad (24.45)$$

24-9 Heterocyclic Compounds

In the ring structures considered to this point, all the ring atoms have been carbon; these structures are said to be carbocyclic. Many compounds are encountered, both naturally and synthetically, in which one or more of the atoms in a ring structure is not carbon. These ring structures are said to be heterocyclic. The heterocyclic systems most commonly encountered contain N, O, and S atoms, and the rings are of various sizes.

Pyridine is a nitrogen analog of benzene (see Figure 24-8), but unlike benzene it is water soluble and it has basic properties (the unshared pair of electrons on the N atom is not part of the π electron cloud of the ring system). Pyridine was once obtained exclusively from coal tar, but it is now used so extensively that several synthetic methods have been developed for its production. It is a liquid with a disagreeable odor used in the production of pharmaceuticals such as sulfa drugs and antihistamines, as a denaturant for ethyl alcohol, as a solvent for organic chemicals, and in the preparation of waterproofing agents for textiles. Other examples of heterocyclic compounds will be encountered in Chapter 25.

FIGURE 24-8
Pyridine.

(a)

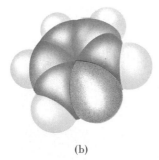

(b)

(a) Structural formula.
(b) Space-filling model.

24-10 Synthesis of Organic Compounds

Originally, all organic compounds were isolated from natural sources. However, as the chemical behavior of these compounds came to be better understood, chemists began to devise *synthetic* methods of producing them from simple starting materials. Now, organic synthesis is one of the most important aspects of organic chemistry. Equipped with a knowledge of a wide variety of reaction types, together with an understanding of the mechanisms of organic reactions, the organic chemist can devise schemes for assembling simple molecular species into more complex structures. A simple example follows:

Example 24-8 Suggest a synthesis for ethyl acetate, $CH_3CO_2CH_2CH_3$, using only inorganic substances as starting materials.

The compound in question is an ester. In equation (24.44) we noted that acids react with alcohols to produce esters. Our problem is to devise syntheses for *ethyl alcohol* and *acetic acid*. Recall that

$$CaCO_3 \xrightarrow{\text{heat}} CaO + CO_2(g) \tag{24.9}$$

$$CaO + 3\ C \xrightarrow[2000°]{\text{electric furnace}} CaC_2 + CO(g) \tag{24.10}$$

$$CaC_2 + 2\ H_2O \longrightarrow HC{\equiv}CH + Ca(OH)_2 \tag{24.11}$$

Ethylene is produced by adding H_2 to C_2H_2,

$$HC{\equiv}CH + H_2 \xrightarrow[\text{heat/pressure}]{\text{Pt or Pd}} H_2C{=}CH_2$$

Ethanol is obtained by the addition of H_2O to C_2H_4.

$$H_2C{=}CH_2 + H_2O \xrightarrow{H_2SO_4} CH_3CH_2OH$$

Some of the ethanol is oxidized to acetic acid.

$$CH_3CH_2OH \xrightarrow[H^+]{K_2Cr_2O_7} CH_3CO_2H$$

Finally, ethanol and acetic acid are combined to form ethyl acetate.

$$CH_3COOH + HOCH_2CH_3 \longrightarrow H_2O + CH_3{-}\overset{\displaystyle O}{\overset{\displaystyle \|}{C}}{-}O{-}CH_2CH_3$$

SIMILAR EXAMPLES: Exercises 39, 40.

Summary

Organic chemistry deals with the compounds of carbon. Simplest among these are carbon-hydrogen compounds—hydrocarbons. In hydrocarbons the C atoms are bonded to one another in straight or branched chains or in rings; H atoms are bonded to the C atoms. Some hydrocarbon molecules contain only single bonds (alkanes); others have some double bonds (alkenes); and still others, triple bonds (alkynes). Yet another class of hydrocarbons—aromatic hydrocarbons—is based on the benzene molecule, C_6H_6. For a given class of hydrocarbons, physical properties (e.g., melting and boiling points) generally follow a regular pattern with increasing molecular weight.

Greater variety among organic compounds results with the incorporation of certain atoms or groupings of atoms (functional groups) into hydrocarbon structures. These substituent groups are introduced through chemical reactions. With alkanes and aromatic hydrocarbons these reactions are based on *substitution:* A functional group replaces an H atom in the hydrocarbon. With alkenes and alkynes chemical reaction occurs by *addition:* Functional group atoms are joined to the C atoms at points of unsaturation (double or triple bonds).

Isomerism is frequently encountered among organic compounds. One form of isomerism is based on molecules with the same total number of C atoms but with different branching of the carbon chain. Another stems from the different positions on a hydrocarbon chain or ring at which functional groups may be attached. Still other isomers (for example, *cis-trans*) arise from different orientations of substituent groups in space.

Among the reactions that can be used for preparing alkanes are the addition of H_2 to an alkene or alkyne and the reaction of an alkyl halide with sodium. Alkenes can be prepared by *elimination* reactions. For example, the elimination of H_2O through the removal of an H atom and an —OH group from adjacent C atoms leaves a double bond between the C atoms. The principal alkyne, acetylene, is produced by the reaction of calcium carbide with water. Other alkynes can be prepared from acetylene based on the acidity of the C—H bond in $HC{\equiv}CH$.

Compounds of the general formula ROH are alcohols (phenols if R is C_6H_5). Alcohols can be prepared by the *hydration* of an alkene (addition of HOH) or the hydrolysis of an alkyl halide. Ethers (R′OR) result from the elimination of HOH from between two alcohol molecules. Aldehydes, RCHO, and ketones, RCOR′, feature the carbonyl group, $\diagdown$C$=$O. They can be prepared by the controlled oxidation of alcohols. Carboxylic acids are weak acids having the general formula, RCOOH, and featuring the carboxyl group,

$$-C\diagup^{O}_{\diagdown OH}$$

. In addition to their typical acid-base reac-

tions, carboxylic acids react with alcohols to form esters. Other reactions include the formation of amides and nitriles. Carboxylic acids can be prepared by the oxidation of an alcohol or an aldehyde. Amines are organic derivatives of ammonia, and, like ammonia, they have basic properties. They can be prepared by the reduction of nitro compounds.

The substitution of other atoms (such as N, O, or S) for C atoms in ring structures yields heterocyclic compounds. These are widely encountered among molecules of the living state (see Chapter 25).

Learning Objectives

As a result of studying Chapter 24, you should be able to

1. Give examples of alkane, alkene, alkyne, and aromatic hydrocarbons.

2. Draw structural formulas and condensed formulas for hydrocarbon molecules for which systematic (IUPAC) names are given.

3. Name hydrocarbon molecules for which structural or condensed formulas are given.

4. Determine all the possible skeletal isomers of a hydrocarbon of given formula (e.g., all the isomers of C_5H_{12}).

5. Describe the conformations of molecules, specifically the eclipsed and staggered forms of ethane and the boat and chair forms of cyclohexane.

6. Discuss the physical properties of aliphatic and aromatic hydrocarbons in relation to their bonding, structures, and molecular weights.

7. Identify examples in which *cis-trans* isomerism occurs.

8. Write equations for several reactions used in the preparation of alkanes, alkenes, and alkynes.

9. Explain why alkanes and aromatic hydrocarbons react by substitution and alkenes and alkynes by addition.

10. Write equations for the reactions of alkanes with halogens and with oxygen, emphasizing the radical chain nature of the alkane–halogen reaction.

11. Use Markovnikov's rule to predict the products of an addition reaction at a multiple bond.

12. Describe the free-radical polymerization of an alkene, as in the preparation of polyethylene from ethylene.

13. Name the common functional groups and give examples of compounds containing them.

14. Write structural formulas, identify possible isomers, and name organic compounds containing functional groups.

15. Describe methods of preparing alcohols, ethers, aldehydes, ketones, acids, and amines; and write equations for some typical reactions of these functional groups.

16. Propose schemes for synthesizing some common organic compounds.

Some New Terms

The **acyl** group is $-\overset{O}{\overset{\|}{C}}-$**R.** If R = H, this is the **formyl** group; R = CH_3, **acetyl;** and R = C_6H_5, **benzoyl.**

In **addition reactions** functional group atoms are joined to the C atoms at points of unsaturation in alkene and alkyne hydrocarbon molecules. **Markovnikov's rule** helps to establish the way in which addition of a reagent occurs.

Alcohols contain the functional group —OH and have the general formula ROH.

Aldehydes have the general formula $R-\overset{O}{\overset{\|}{C}}-H.$

Alicyclic hydrocarbon molecules have their carbon atom skeletons arranged in rings and resemble aliphatic (rather than aromatic) hydrocarbons.

Aliphatic hydrocarbon molecules have their carbon atom skeletons arranged in straight or branched chains.

Alkane hydrocarbon molecules have only single covalent bonds between C atoms. In their chain structures alkanes have the general formula C_nH_{2n+2}.

Alkene hydrocarbons have one or more carbon-to-carbon double bonds in their molecules. The simple alkenes have the general formula C_nH_{2n}.

Alkyne hydrocarbons have one or more carbon-to-carbon triple bonds in their molecules. The simple alkynes have the general formula C_nH_{2n-2}.

An **amide** has the general formula $R{-}\overset{\displaystyle O}{\overset{\|}{C}}{-}NH_2$.

An **amine** is an organic base having the formula RNH_2 (primary), R_2NH (secondary), or R_3N (tertiary), depending on the number of H atoms of an NH_3 molecule that are replaced by R groups.

Aromatic hydrocarbon molecules have carbon atoms arranged in hexagonal rings, based on the structure of benzene, C_6H_6.

Chain or **skeletal isomers** have the same number of C and H atoms in their hydrocarbon chains but a different pattern of branching in the chain.

A **condensed formula** is a simplified representation of a structural formula.

Conformations refer to the different spatial arrangements possible as a result of rotation about single bonds in a hydrocarbon molecule. Examples are the eclipsed and staggered conformations of hydrocarbon chains and the "boat" and "chair" forms of cyclohexane.

An **elimination reaction** is one in which atoms are removed from adjacent positions on a hydrocarbon chain, producing a small molecule (e.g., H_2O) and an additional bond between C atoms.

An **ester** is the product of the elimination of H_2O from between an acid and an alcohol molecule. Esters have the general formula $R{-}\overset{\displaystyle O}{\overset{\|}{C}}{-}O{-}R'$.

Ethers have the general formula ROR'.

A **functional group** is an atom or grouping of atoms attached to a hydrocarbon residue, R. The functional group often confers specific properties to an organic molecule.

Heterocyclic compounds are based on hydrocarbon ring structures in which one or more of the C atoms is replaced by atoms such as N, O, or S.

A **homologous series** is a group of compounds that differ in composition by some constant unit ($-CH_2$ in the case of alkanes).

IUPAC (or **IUC**) refers to a system of relating the names and structural formulas of organic compounds.

Ketones have the general formula $R{-}\overset{\displaystyle O}{\overset{\|}{C}}{-}R'$.

A **meta** (*m-*) **isomer** has two substituents on a benzene ring separated by one C atom.

A **nitrile** has the general formula $R{-}C{\equiv}N$.

An **ortho** (*o-*) **isomer** has two substituents attached to adjacent C atoms in a benzene ring.

A **para** (*p-*) **isomer** has two substituents located opposite to one another on a benzene ring.

Phenols have the functional group $-OH$ as part of an aromatic hydrocarbon structure.

The **phenyl group** is a benzene ring from which one H atom has been removed: $-C_6H_5$.

A **polycyclic aromatic hydrocarbon** is obtained whenever two or more benzene rings are fused together (with an appropriate loss of C and H atoms).

Polymerization is the process of producing a giant molecule (polymer) from simpler molecular units (monomers).

Positional isomers differ in the position on a hydrocarbon chain or ring where a functional group(s) is attached.

Saturated hydrocarbon molecules contain only single bonds between carbon atoms.

In **stereoisomerism** the number and types of atoms and bonds in molecules are the same, but certain atoms are oriented differently in space. ***Cis-trans*** **isomerism** is a form of stereoisomerism.

A **structural formula** for a compound indicates which atoms in a molecule are bonded together, and whether by single, double, or triple bonds.

Substitution reactions are typical of those involving alkane and aromatic hydrocarbons. In such a reaction a functional group replaces an H atom on a chain or ring.

Unsaturated hydrocarbon molecules contain one or more carbon-to-carbon multiple bonds.

Exercises

Definitions and terminology

1. What are the essential features that characterize (a) an organic compound; (b) an alkane; (c) an aromatic hydrocarbon?

2. What is the difference in meaning of the following terms? (a) aliphatic and alicyclic; (b) aliphatic and aromatic; (c) paraffin and olefin; (d) alkane and alkyl; (e) normal (*n-*) and iso-; (f) primary, secondary, tertiary; (g) axial and equatorial.

3. Give a definition and a well-chosen example of each of the following: (a) condensed formula; (b) homologous series; (c) olefin; (d) free radical; (e) ortho-para director.

Organic structures

4. Write Lewis structures of the following simple organic molecules.
  (a) $CH_3CHClCH_3$
  (b) $HOCH_2CH_2OH$
  (c) CH_3CHO

5. Write structural formulas corresponding to these condensed formulas.

(a) $CH_3CH_2CH_2CHBrCH_3$
(b) $(CH_3)_2CHCH_2CH_2CH(CH_3)CH_2CH_3$
(c) $(CH_3)_3CCH_2CH(CH_3)CH_2CH_2CH_3$
(d) $CH_3CH_2CH(CH_3)C(C_2H_5)=CH_2$

6. With appropriate sketches represent chemical bonding in terms of the overlap of pure or hybridized atomic orbitals in the following molecules.

(a) C_2H_6 (b) $H_2C=CHCl$

(c) $CH_3C\equiv CH$ (d) $CH_3\overset{O}{\overset{\|}{C}}CH_3$

(e) $CH_3CH_2NH_2$

Isomers

7. Which of the following pairs of molecules are isomers and which are not? Explain.

(a) $CH_3CH_2CH_2CH_3$ and $CH_3CH=CHCH_3$
(b) $CH_3(CH_2)_5CH(CH_3)_2$ and
    $CH_3(CH_2)_4CH(CH_3)CH_2CH_3$
(c) $CH_3CHClCH_2CH_3$ and $CH_3CH_2CHClCH_3$

(d) 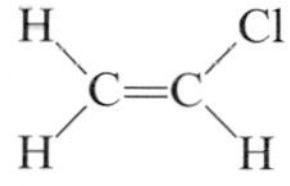 $\begin{smallmatrix} H \\ \\ H \end{smallmatrix} C=C \begin{smallmatrix} Cl \\ \\ H \end{smallmatrix}$ and $\begin{smallmatrix} H \\ \\ Cl \end{smallmatrix} C=C \begin{smallmatrix} H \\ \\ H \end{smallmatrix}$

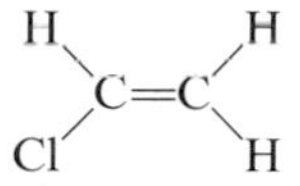

(e) 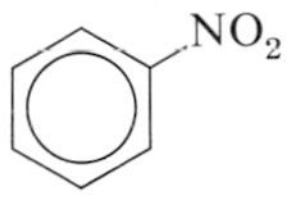 [benzene ring with NO$_2$]   and   [benzene ring with NO$_2$]

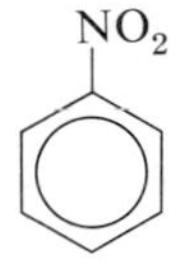

(f) 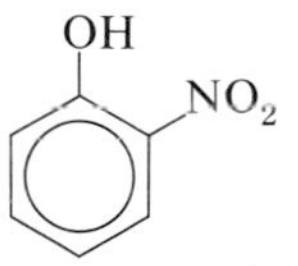 [benzene ring with OH and NO$_2$]   and   [benzene ring with OH and NO$_2$]

8. Indicate the difference in these three types of isomers: skeletal, positional, and geometrical. Which term best describes each of the following pairs of isomers?

(a) $CH_3CH_2CH_2Cl$ and $CH_3CHClCH_3$
(b) $CH_3CH(CH_3)CH_2CH_3$ and $CH_3(CH_2)_3CH_3$
(c) $CHCl=CHCl$ and $CH_2=CCl_2$

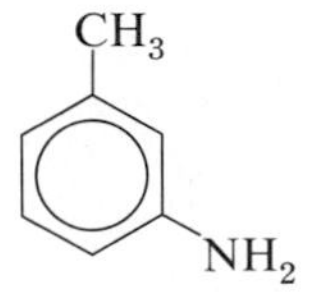

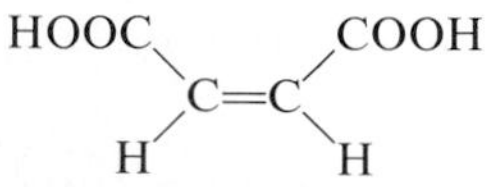

(d) 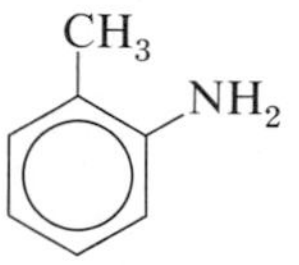 [benzene ring with CH$_3$ and NH$_2$]   and   [benzene ring with CH$_3$ and NH$_2$]

(e) $\begin{smallmatrix} HOOC \\ \\ H \end{smallmatrix} C=C \begin{smallmatrix} COOH \\ \\ H \end{smallmatrix}$ and $\begin{smallmatrix} HOOC \\ \\ H \end{smallmatrix} C=C \begin{smallmatrix} H \\ \\ COOH \end{smallmatrix}$

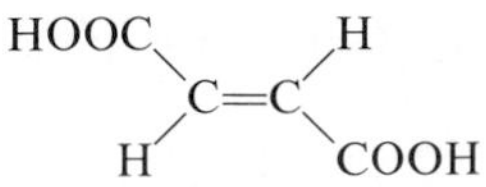

9. Draw structural formulas for all the isomers of (a) pentane; (b) heptane.

10. By drawing suitable structural formulas, establish that

there are 17 isomers of $C_6H_{13}Cl$. (*Hint:* Refer to Example 24-1.)

Functional Groups

11. Identify the functional group in each compound (i.e., whether an alcohol, amine, etc.).

(a) $CH_3CHBrCH_2CH_3$ (b) CH_3CH_2COOH
(c) $C_6H_5CH_2CHO$ (d) $(CH_3)_2CHCH_2OCH_3$
(e) $CH_3COCH_2CH_3$ (f) $CH_3CH(NH_2)CH_2CH_3$

(g) 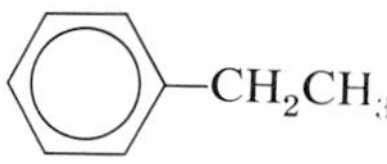 [benzene ring]$-CH_2CH_3$ (h) $CH_3CO_2CH_3$

12. The functional groups in each of the following pairs have certain features in common, but what is the essential difference between them?

(a) carbonyl and carboxyl
(b) amine and amide
(c) acid and acid anhydride
(d) aldehyde and ketone

13. Give one example of each of the following types of compounds: (a) aliphatic nitro compound;   (b) aromatic amine;   (c) chlorophenol;   (d) aliphatic diol;   (e) unsaturated aliphatic alcohol;   (f) alicyclic ketone;   (g) halogenated alkane;   (h) aromatic dicarboxylic acid.

Nomenclature and formulas

14. Give an acceptable name for each of the following structures.

(a) $CH_3CH_2\overset{CH_3}{\underset{CH_3}{\overset{|}{\underset{|}{C}}}}CH_3$ (b) $CH_3\overset{CH_3}{\overset{|}{C}}=CH_2$

(c) 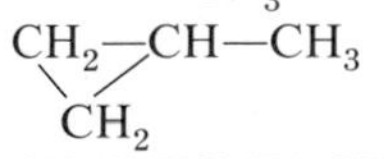 $\underset{CH_2}{\overset{|}{CH_2}}-CH-CH_3$ (d) $CH_3C\equiv CCH(CH_3)_2$

(e) $CH_3CH(C_2H_5)CH(CH_3)CH_2CH_3$
(f) $CH_3CH(CH_3)CH(CH_3)C(C_3H_7)=CH_2$

15. Draw a structure to correspond to each of the following names.

(a) isopentane   (b) cyclohexene
(c) 3-hexyne   (d) 2-butanol
(e) isopropyl methyl ether
(f) propionaldehyde
(g) *t*-butyl chloride   (h) diethylmethylamine
(i) isobutyric acid   (j) isobutyl propionate

16. Does each of the following names convey sufficient information to suggest a specific structure? Explain.

(a) pentene   (b) butanone
(c) butyl alcohol   (d) methylaniline
(e) methylcyclopentane

17. Is each of the following names correct? If not, indicate why not and give a correct name.

(a) 3-pentene   (b) pentadiene
(c) 1-propanone   (d) bromopropane
(e) 2,6-dichlorobenzene   (f) 2-methyl-3-pentyne
(g) 2-methyl-4-*n*-butyloctane
(h) 4,4-dimethyl-5-ethyl-1-hexyne
(i) 1,3-dimethylcyclohexane
(j) 3,4-dimethyl-2-pentene

18. Supply formulas for the following chemical substances.
   (a) isopropyl alcohol (rubbing alcohol)
   (b) tetraethyllead (an antiknock component of gasoline)
   (c) 2-methyl-1,3-butadiene (isoprene—monomer in the manufacture of synthetic rubber)
   (d) 2,2,4-trimethylpentane (isooctane—a constituent of gasoline)
   (e) 2,4,6-trinitrotoluene (TNT—an explosive)
   (f) methyl salicylate (oil of wintergreen) (*Hint:* Recall that salicylic acid is *o*-hydroxybenzoic acid.)
   (g) 1-phenyl-2-aminopropane (benzedrine—an amphetamine, ingredient in "pep pills")
   (h) 2-methylheptadecane (a sex pheromone of tiger moths—a chemical used for communication among members of the species) (*Hint:* "Heptadeca" means 17.)
   (i) 3,7,11-trimethyl-2,6,10-dodecatriene-1-ol (farnesol—odor of lily-of-the-valley) (*Hint:* "Dodeca" means 12.)

Experimental determination of formulas

19. Combustion of 184 mg of a hydrocarbon gave 577 mg CO_2 and 236 mg H_2O. Calculate the empirical formula of the compound. What is the molecular formula if the molecular weight is subsequently found to be 56?

Alkanes

20. Draw structural formulas for all the isomers listed in Table 24-2 and show that, indeed, the more compact structures yield lower boiling points.

21. What is the most stable conformation of the molecule *t*-butylcyclohexane? (*Hint:* Is the ring in the boat or chair form? Is the *t*-butyl group in an axial or equatorial position?)

22. Write the structure of each alkane.
   (a) Molecular weight = 44; forms two different monochlorination products.
   (b) Molecular weight = 58; forms two different monobromination products.

23. Name the principal products obtained in the reaction of
   (a) $CH_3CH_2CH=CH_2$ with H_2 in the presence of a catalyst
   (b) *n*-propyl bromide with sodium
   (c) sodium butyrate with sodium hydroxide
   (d) propane with chlorine gas in the presence of ultraviolet light

24. In the chlorination of CH_4 some CH_3CH_2Cl is obtained as a product. Explain why this should be so.

Alkenes

25. Why is it not necessary to refer to ethene and propene as 1-ethene and 1-propene? Can the same be said for butene?

26. Alkenes (olefins) and cyclic alkanes (alicyclics) each have the generic formula C_nH_{2n}. In what important ways do these types of compounds differ structurally?

27. Draw the structures of the products of each of the following reactions.
   (a) propylene + hydrogen (Pt/heat)
   (b) 2-butanol + heat (in the presence of sulfuric acid)
   (c) sodium acetylide with *t*-butyl bromide

28. Use Markovnikov's rule to predict the product of the reaction of
   (a) HCl with $CH_3CCl=CH_2$
   (b) HCN with $CH_3C≡CH$
   (c) HCl with $CH_3CH=C(CH_3)_2$

   (d) 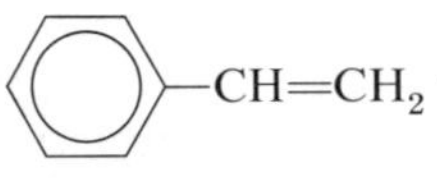 —$CH=CH_2$ with HBr

Aromatic compounds

29. Supply a correct name or formula for each of the following.

(a)

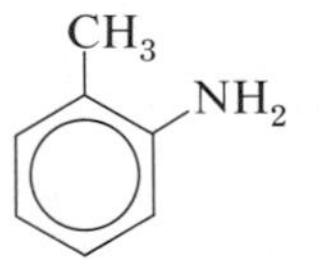

(b)

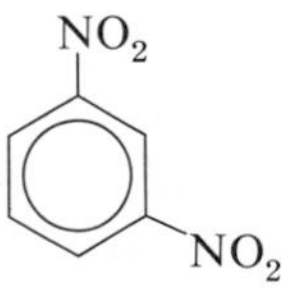

(c)

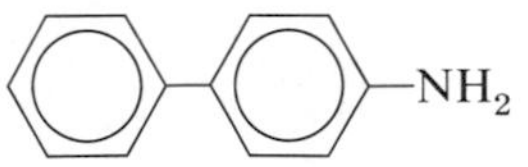

(d) *p*-nitrophenol

(e) 1,3-5-trimethylbenzene   (f) phenylacetylene
(g) 2-hydroxy-4-isopropyltoluene (thymol—flavor constituent of the herb thyme)

30. Predict the products of the monobromination of (a) *m*-dinitrobenzene; (b) aniline; (c) *p*-bromoanisole.

31. Write the isomers to be expected from the mononitration of *m*-methoxybenzaldehyde.

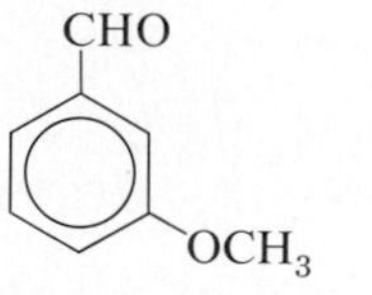

In actual fact no 3-methoxy-5-nitrobenzaldehyde is obtained. What does this fact imply about the strength of meta and ortho-para directors?

32. What principal product would you expect to obtain

when toluene is allowed to react with (a) $HNO_3 + H_2SO_4$; (b) $Cl_2 + FeCl_3$; (c) Cl_2 without $FeCl_3$ but in the presence of ultraviolet light?

*33. The symbol

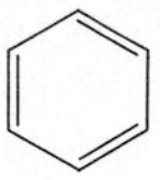

which is often used to represent the benzene molecule, is also the structural formula of cyclohexatriene. Are benzene and cyclohexatriene the same substance? Explain.

*34. In the representation for benzene

the inscribed circle represents electrons in a π bonding system (recall Figure 10-20). How many electrons are represented by the circles in this representation of naphthalene?

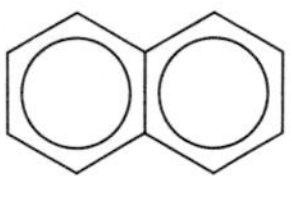

Organic reactions

35. Describe what is meant by each of the following reaction types and illustrate with an example from the text. (a) Aliphatic substitution reaction; (b) aromatic substitution reaction; (c) addition reaction; (d) elimination reaction.

36. Draw a structure to represent the principal product of each of the following reactions.
   (a) 1-pentanol + excess dichromate (acid catalyst)
   (b) butyric acid + ethanol (acid catalyst)
   (c) o-nitrophenol + sodium hydroxide
   (d) $CH_3CH_2C(CH_3)=CH_2 + H_2O$ (in the presence of H_2SO_4)

37. Use the half-reaction method to balance the following oxidation-reduction equations.

(a) $-NO_2 + Fe + H^+ \longrightarrow$

$\langle\!\!\langle\bigcirc\rangle\!\!\rangle-NH_3^+ + Fe^{3+} + H_2O$

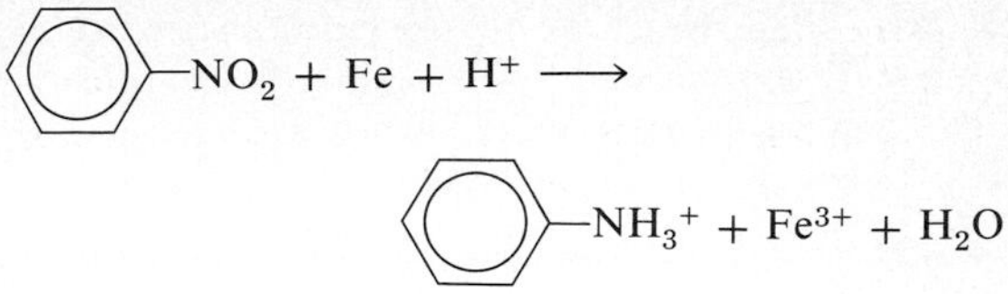

(b) $CH_3CH=CH_2 + MnO_4^- + H_2O \longrightarrow$

$\underset{OH\ \ OH}{CH_3CH-CH_2} + MnO_2 + OH^-$

(c) $\overset{OH}{\langle}-OH + Pb(C_2H_3O_2)_4 \longrightarrow$

$\overset{CHO}{\langle}CHO + Pb(C_2H_3O_2)_2 + HC_2H_3O_2$

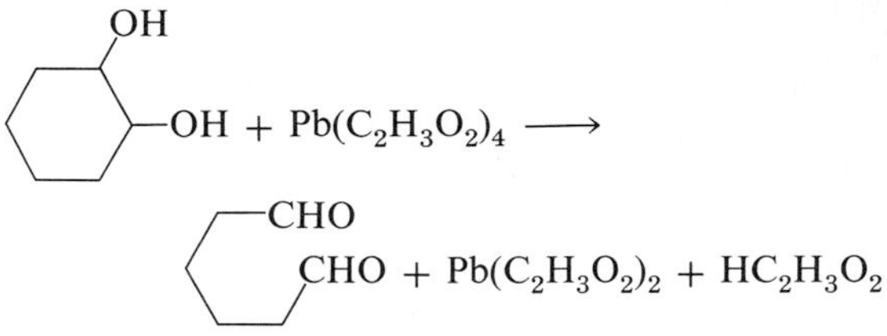

(d) $H_3C-\langle\!\!\langle\bigcirc\rangle\!\!\rangle-CH_3 + H^+ + Cr_2O_7^{2-} \longrightarrow$

$HO_2C-\langle\!\!\langle\bigcirc\rangle\!\!\rangle-CO_2H + Cr^{3+} + H_2O$

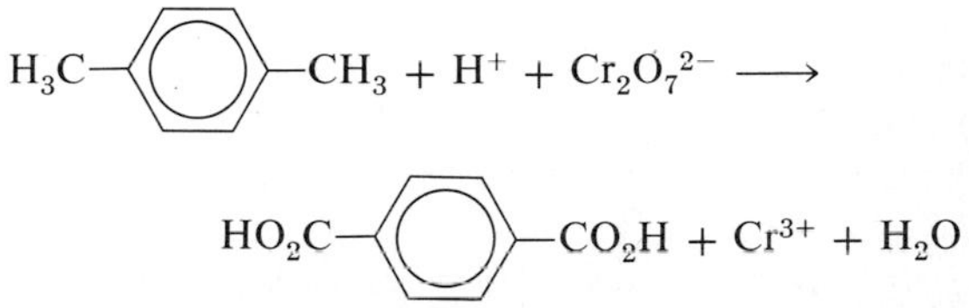

38. A 10.6-g sample of benzaldehyde was reacted with 5.9 g $KMnO_4$ in an excess of $KOH(aq)$. After filtration of the $MnO_2(s)$ and acidification of the solution, 6.1 g of benzoic acid was isolated. What was the percent yield of this reaction?

Organic synthesis

39. Starting with benzene and any aliphatic or inorganic reagents required, how would you synthesize (a) m-bromonitrobenzene; (b) p-aminotoluene?

40. Starting with acetylene as the only source of carbon, together with any inorganic reagents desired, devise syntheses for (a) acetaldehyde; (b) 1,1,2,2-tetrabromoethane; (c) acetonitrile; (d) isopropyl acetate.

Additional Exercises

1. Write structural formulas for the following compounds.
   (a) ethylisobutylmethylamine
   (b) 3-chloro-2,3-dimethylbutanoic acid
   (c) 1-phenyl-2-butanone
   (d) t-butyl isobutyrate
   (e) 1-chloro-2,4-octadiene
   (f) trans-1,4-dibromobutadiene

2. Give an acceptable name for each of the following.
   (a) $(CH_3)_2CHOCH_2CH_2CH_3$
   (b) $(CH_3)_2CHCOCH_3$
   (c) $CH_2=CHCH=CH_2$
   (d) $(CH_3)_3CCH_2CH(CH_3)CH_2CH_2CH_3$

(e) $\overset{Br}{\underset{Br}{\langle\!\!\langle\bigcirc\rangle\!\!\rangle}}-CH_3$

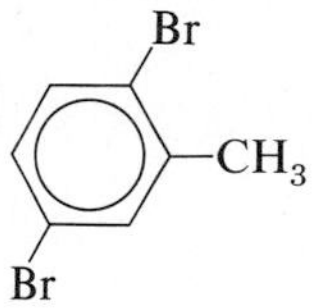

(f) $\langle\!\!\langle\bigcirc\rangle\!\!\rangle-CH_2CHBrCO_2CH(CH_3)_2$

3. Draw and name all the isomers for (a) C_6H_{14}; (b) C_4H_8; (c) C_4H_6. (Hint: Do not forget rings, double bonds, and combinations of these.)

4. Write the structure of each alkane. **(a)** Molecular weight = 72; forms four monochlorination products. **(b)** Molecular weight = 72; forms a single monochlorination product.

5. Why is *cis-trans* isomerism encountered with olefins but not with paraffins?

6. Methanol is a weaker acid than water, whereas phenol is stronger. Methylamine is a stronger base than ammonia, whereas aniline is weaker. Explain these observations.

7. Outline a series of reactions that could be used to synthesize ethanol from coal, water, and other inorganic materials.

8. Draw a structure to represent the principal product of each of the following reactions.
   (a) dimethylamine + hydrochloric acid
   (b) isopropanol + sodium
   (c) *t*-butyl bromide + NaOH(aq)

*9. Combustion of a 0.1908-g sample of a compound gave 0.2895 g CO_2 and 0.1192 g H_2O. Combustion of a second sample, weighing 0.1825 g, yielded 40.2 ml of N_2(g), collected over 50% KOH(aq) (vapor pressure = 9 mmHg) at 25°C and 735 mmHg barometric pressure. When 1.082 g of compound was dissolved in 26.00 g benzene (m.p. 5.50°C, K_f = 5.12), the solution had a freezing point of 3.66°C. What is the molecular formula of the compound?

*10. The three isomeric tribromobenzenes, I, II, and III,

when nitrated, form three, two, and one mononitrotribromobenzenes, respectively. Assign correct structures to I, II, and III.

*11. Write the name and structure of each aromatic hydrocarbon.
   (a) Formula: C_8H_{10}; forms three ring monochlorination products.
   (b) Formula: C_9H_{12}; forms one ring mononitration product.
   (c) Formula: C_9H_{12}; forms four ring mononitration products.

*12. In the molecule 2-methylbutane, the organic chemist distinguishes the different types of hydrogen and carbon atoms as being primary (1°), secondary (2°), and tertiary (3°). For the monochlorination of hydrocarbons the following ratio of reactivities has been found; 3°/2°/1° = 4.3:3:1. How many different monochloro derivatives of 2-methylbutane are possible and what percent of each would you expect to find?

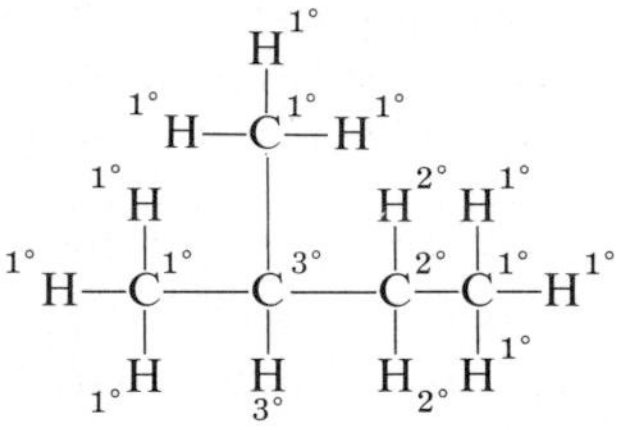

Self-Test Questions

For questions 1 through 8 select the single item that best completes each statement.

1. The compound isoheptane has the formula
   (a) C_7H_{14} **(b)** $(CH_3)_2CH(CH_2)_3CH_3$
   (c) $CH_3(CH_2)_5CH_3$ **(d)** $C_6H_5CH_3$

2. *Three* isomers exist of the hydrocarbon **(a)** C_3H_8; **(b)** C_4H_{10}; **(c)** C_6H_6; **(d)** C_5H_{12}.

3. The hydrocarbon cyclobutane has the same carbon-to-hydrogen ratio as **(a)** C_4H_{10}; **(b)** $CH_3CH=CHCH_3$; **(c)** $CH_3C\equiv CCH_3$; **(d)** C_6H_6.

4. *Cis-trans* isomerism is expected in the compound **(a)** $ClCH=CHCl$; **(b)** $CH_2=CCl_2$; **(c)** $ClCH_2CH_2Cl$; **(d)** $Cl_2C=CCl_2$.

5. The compound 2-chloro-3-methyl-1-butanol has the formula
   (a) $CH_2ClC(CH_3)_2CH_2OH$
   (b) $CH_3CHOHCH(CH_3)CH_2Cl$
   (c) $CH_3CH(CH_3)CHClCH_2OH$
   (d) $CH_3CHClCH(CH_3)CH_2OH$

6. The compound

is named   **(a)** *o*-aminotoluene;   **(b)** *p*-methylaniline;   **(c)** *m*-methylbenzene;   **(d)** 3-methylaniline.

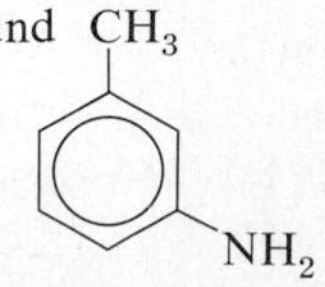

7. The most acidic of the following substances is

(a) [benzene ring with NH_2]   **(b)** [benzene ring with COOH]

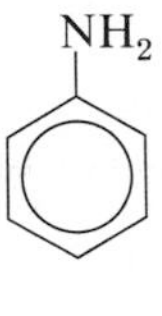

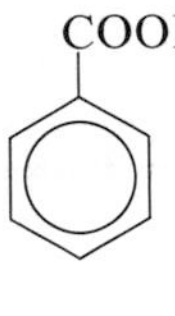

(c) [benzene ring with OH]   **(d)** CH_3CHO

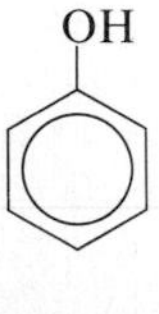

8. To prepare methyl ethyl ketone one should oxidize **(a)** 2-propanol;   **(b)** 1-butanol;   **(c)** 2-butanol;   **(d)** *t*-butyl alcohol.

9. Draw structural formulas for the following compounds: **(a)** dichlorodifluoromethane (Freon 12—a refrigerant); **(b)** *p*-bromophenol;   **(c)** 3-hydroxy-2-butanone.

10. Draw structural formulas for all the possible isomers of $C_5H_{11}Br$ and name them.

25

Chemistry of the Living State

You shall gain your bread by the sweat of your brow until you
return to the ground; for from it you were taken.
Dust you are, to dust you shall return.

Genesis 3:19

Even though not stated in scientific terms, this commentary certainly suggests man's relationship to nature and its laws. But there exists a fascinating interlude between dust and dust—that which we call life. A living organism maintains a low entropy within itself and thus resists, for a time, the universal tendency to approach equilibrium where entropy and disorder reach a maximum. To maintain the high degree of order characteristic of the living state requires the information of heredity, the energy of biochemical reactivity, and raw materials to build cells. These are some of the topics considered in this chapter.

25-1 Structure and Composition of the Cell

TABLE 25-1
Chemical elements found
in the human body

Element	Percent, by mass
O	65
C	18
H	10
N	3
Ca	2
P	1.2
K	0.20
S	0.20
Cl	0.20
Na	0.11
Mg	0.04

trace elements: Mn, Fe,
Co, Cu, Zn, B, Al, V,
Mo, I, Si

From one-celled plants and animals to the form of greatest complexity, *Homo sapiens,* there is a bewildering variety of life forms. Yet there are many characteristics that these forms share. Of the known elements, about 50 occur in living matter in measurable concentrations. Of these, about 22 have functions that are definitely known. The 11 most abundant elements in living organisms and their percentages in the human body are listed in Table 25-1. Four elements—oxygen, carbon, hydrogen, and nitrogen—together account for 96% of body mass.

In addition to water, which is the most abundant compound in all living organisms, the important constituents of the cell are compounds of three types: lipids, carbohydrates, and proteins.

The cell is the fundamental unit of all life. Cells combine to form tissues; tissues may be grouped into organs; and organs combine into organisms. Two views of a typical animal cell are presented in Figures 25-1 and 25-2. Figure 25-1 describes the types and complexity of molecules found in cells and Figure 25-2 pictures the substructure of a cell. You may find it helpful to refer back to these figures from time to time as you proceed through this chapter.

25-2 Principal Constituents of the Cell

LIPIDS. A precise definition of lipids is not possible. They are simply those constituents of plant and animal tissue that are soluble in solvents of low polarity, such as chloroform, carbon tetrachloride, diethyl ether, or benzene. Many compounds fit this description. The following categories, though arbitrary, are widely accepted:

627

FIGURE 25-1
Cellular organization.

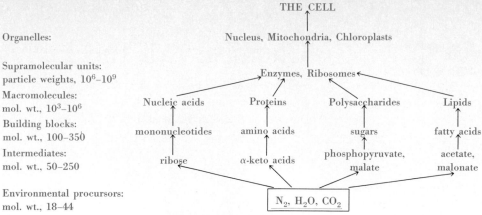

THE CELL

Organelles: Nucleus, Mitochondria, Chloroplasts

Supramolecular units:
particle weights, 10^6–10^9 Enzymes, Ribosomes

Macromolecules:
mol. wt., 10^3–10^6 Nucleic acids   Proteins   Polysaccharides   Lipids

Building blocks:
mol. wt., 100–350   mononucleotides   amino acids   sugars   fatty acids

Intermediates:
mol. wt., 50–250   ribose   α-keto acids   phosphopyruvate,   acetate,
                                                              malate   malonate

Environmental procursors:
mol. wt., 18–44 N_2, H_2O, CO_2

FIGURE 25-2
A typical animal cell.

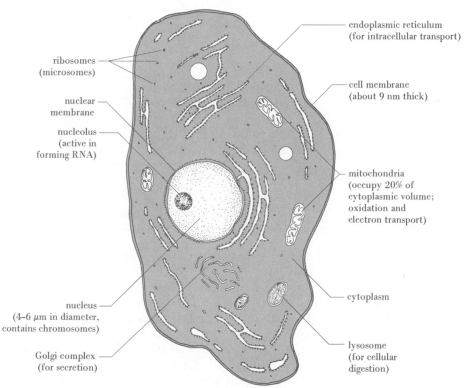

ribosomes
(microsomes)

nuclear
membrane

nucleolus
(active in
forming RNA)

endoplasmic reticulum
(for intracellular transport)

cell membrane
(about 9 nm thick)

mitochondria
(occupy 20% of
cytoplasmic volume;
oxidation and
electron transport)

cytoplasm

nucleus
(4–6 μm in diameter,
contains chromosomes)

Golgi complex
(for secretion)

lysosome
(for cellular
digestion)

TABLE 25-2
Some common fatty acids

Common name	IUPAC name	Formula
Saturated acids		
lauric acid	dodecanoic acid	$C_{11}H_{23}CO_2H$
myristic acid	tetradecanoic acid	$C_{13}H_{27}CO_2H$
palmitic acid	hexadecanoic acid	$C_{15}H_{31}CO_2H$
stearic acid	octadecanoic acid	$C_{17}H_{35}CO_2H$
Unsaturated acids		
oleic acid	9-octadecenoic acid	$C_{17}H_{33}CO_2H$
linoleic acid	9,12-octadecadienoic acid	$C_{17}H_{31}CO_2H$
linolenic acid	9,12,15-octadecatrienoic acid	$C_{17}H_{29}CO_2H$
eleostearic acid	9,11,13-octadecatrienoic acid	$C_{17}H_{29}CO_2H$

Triacylglycerol is the systematic name for an ester in which all three —OH groups of glycerol are esterified with fatty acids

$$\overset{O}{\underset{\|}{}}$$

(acyl groups, $R-\overset{O}{\underset{\|}{C}}-$). Triglyceride is a common name that has been widely used to describe these esters. The common name, triglyceride, is used in the discussion in this text, but the systematic name, triacylglycerol, is likely gradually to replace it.

1. The Triglycerides. This is the most common group of lipids in plants and animals. These lipids are esters of glycerol (1,2,3-propanetriol) with long-chain monocarboxylic acids (fatty acids).

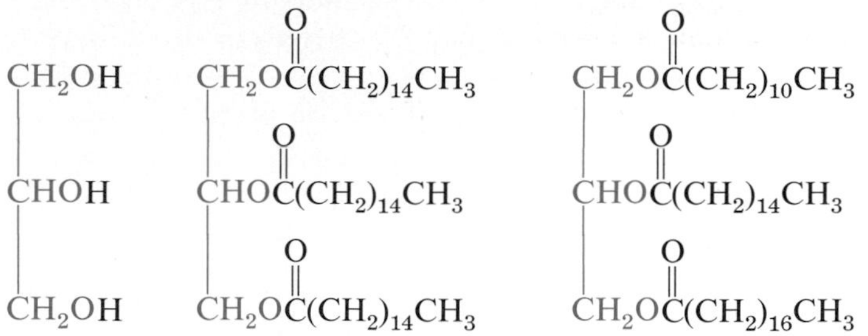

$$CH_2OH$$
$$|$$
$$CHOH$$
$$|$$
$$CH_2OH$$
glycerol

$$CH_2O\overset{O}{\overset{\|}{C}}(CH_2)_{14}CH_3$$
$$|$$
$$CHO\overset{O}{\overset{\|}{C}}(CH_2)_{14}CH_3$$
$$|$$
$$CH_2O\overset{O}{\overset{\|}{C}}(CH_2)_{14}CH_3$$
glyceryl tripalmitate
tripalmitin
(a simple glyceride; a fat)

$$CH_2O\overset{O}{\overset{\|}{C}}(CH_2)_{10}CH_3$$
$$|$$
$$CHO\overset{O}{\overset{\|}{C}}(CH_2)_{14}CH_3$$
$$|$$
$$CH_2O\overset{O}{\overset{\|}{C}}(CH_2)_{16}CH_3$$
glyceryl lauropalmitostearate
(a mixed glyceride; a fat)

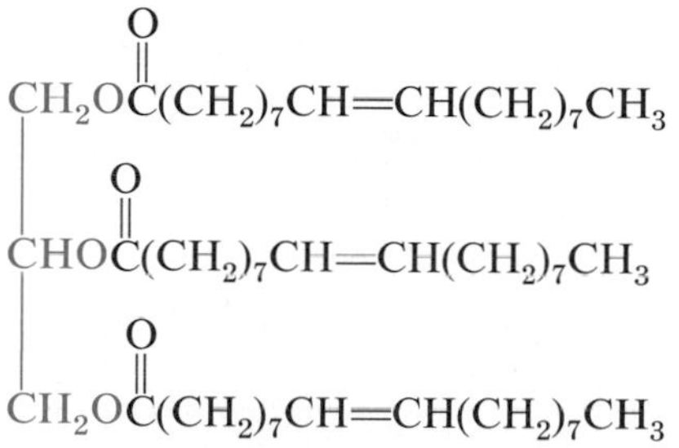

$$CH_2O\overset{O}{\overset{\|}{C}}(CH_2)_7CH{=}CH(CH_2)_7CH_3$$
$$|$$
$$CHO\overset{O}{\overset{\|}{C}}(CH_2)_7CH{=}CH(CH_2)_7CH_3$$
$$|$$
$$CH_2O\overset{O}{\overset{\|}{C}}(CH_2)_7CH{=}CH(CH_2)_7CH_3$$
glyceryl trioleate
triolein
(a simple glyceride; an oil)

In a simple glyceride all acid groups are the same; in a mixed glyceride they are not. Some long chain or fatty acids commonly encountered in glycerides are listed in Table 25-2.

Example 25-1 Write a structural formula for glyceryl butyropalmitooleate.
   This is a mixed glyceride in which the acid groups are

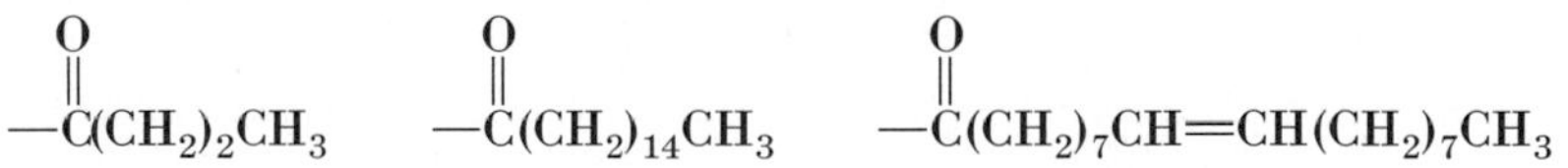

$$-\overset{O}{\overset{\|}{C}}(CH_2)_2CH_3 \qquad -\overset{O}{\overset{\|}{C}}(CH_2)_{14}CH_3 \qquad -\overset{O}{\overset{\|}{C}}(CH_2)_7CH{=}CH(CH_2)_7CH_3$$
   butyric   palmitic   oleic

The complete structure is thus

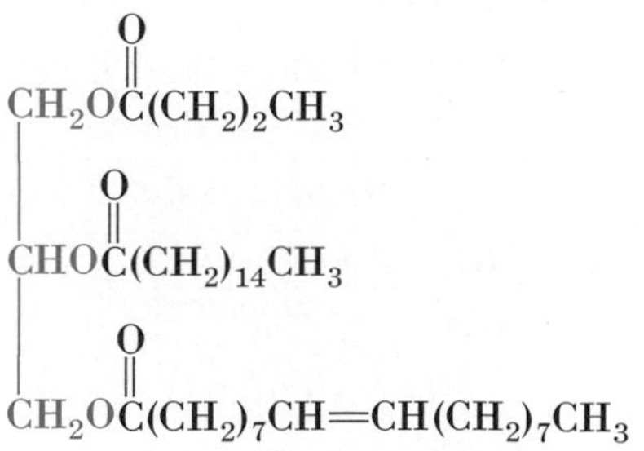

$$CH_2O\overset{O}{\overset{\|}{C}}(CH_2)_2CH_3$$
$$|$$
$$CHO\overset{O}{\overset{\|}{C}}(CH_2)_{14}CH_3$$
$$|$$
$$CH_2O\overset{O}{\overset{\|}{C}}(CH_2)_7CH{=}CH(CH_2)_7CH_3$$

SIMILAR EXAMPLES: Exercises 9, 10.

The **fats** are glyceryl esters in which saturated acid components predominate; they are solids at room temperature. **Oils** have a predominance of unsaturated fatty acids and are liquids at room temperature. The composition of fats and oils is variable and depends not only on the particular plant or animal species involved but also on dietary and climatic factors.

When pure, fats and oils are colorless, odorless, and tasteless. The characteristic colors, odors, and flavors commonly associated with them are imparted by other organic substances that are present in the impure materials. The yellow color of butter is caused by the presence of β-carotene (a yellow pigment also found in carrots and marigolds). The taste of butter is attributed to these two compounds,

$$\underset{\text{3-hydroxy-2-butanone}}{CH_3-\overset{\overset{\displaystyle O}{\|}}{C}-\underset{\underset{\displaystyle OH}{|}}{C}HCH_3} \qquad \underset{\text{diacetyl}}{CH_3-\overset{\overset{\displaystyle O}{\|}}{C}-\overset{\overset{\displaystyle O}{\|}}{C}-CH_3}$$

both produced in the aging of cream.

Glycerides can be hydrolyzed in alkaline solution to produce glycerol and the alkali metal salts of the fatty acids. These salts are commonly known as **soaps,** and the hydrolysis process is called **saponification.**

glyceride + alkali $\longrightarrow$ glycerol + soap

An early process for making soap required first that wood ashes be soaked in a ceramic pot to produce KOH. These "pot ashes" gave the element "potassium" its name.

$$\underset{\substack{\text{tristearin}\\(MW = 890)}}{\begin{matrix}CH_2OC(CH_2)_{16}CH_3\\|\\CHOC(CH_2)_{16}CH_3\\|\\CH_2OC(CH_2)_{16}CH_3\end{matrix}} + 3\,KOH \longrightarrow \underset{\text{glycerol}}{\begin{matrix}CH_2OH\\|\\CHOH\\|\\CH_2OH\end{matrix}} + \underset{\substack{\text{potassium stearate}\\(\text{a soap})}}{3\,CH_3(CH_2)_{16}COK} \qquad (25.1)$$

Saponification reactions can be used in the laboratory to yield information about the structures of glycerides. This is done through the **saponification value**—the number of milligrams of KOH required to saponify 1 g of the glyceride.

Example 25-2   What is the saponification value of tristearin?

The balanced equation for the saponification reaction is given in (25.1). We need simply to calculate the number of moles of tristearin in 1 g, and then, successively, the number of moles, grams, and milligrams of KOH required.

$$\text{no. mg KOH} = 1.00 \text{ g tristearin} \times \frac{1 \text{ mol tristearin}}{890 \text{ g tristearin}} \times \frac{3 \text{ mol KOH}}{1 \text{ mol tristearin}}$$

$$\times \frac{56.1 \text{ g KOH}}{1 \text{ mol KOH}} \times \frac{1000 \text{ mg KOH}}{1.00 \text{ g KOH}}$$

$$= 189 \text{ mg KOH}$$

SIMILAR EXAMPLES: Exercises 11, 12.

Another useful quantity in characterizing a glyceride is the **iodine number**—the number of grams of I_2 reacting with 100 g of a glyceride as a result of the addition of iodine to any double bonds that are present.

Example 25-3   What is the iodine number of triolein?

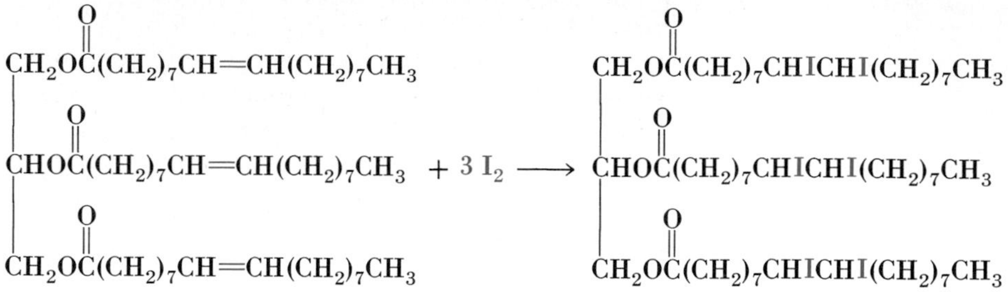

$$
\begin{array}{l}
\text{CH}_2\text{OC(CH}_2)_7\text{CH}{=}\text{CH(CH}_2)_7\text{CH}_3 \\[2pt]
\text{CHOC(CH}_2)_7\text{CH}{=}\text{CH(CH}_2)_7\text{CH}_3 \quad + 3\ \text{I}_2 \longrightarrow \\[2pt]
\text{CH}_2\text{OC(CH}_2)_7\text{CH}{=}\text{CH(CH}_2)_7\text{CH}_3
\end{array}
\qquad
\begin{array}{l}
\text{CH}_2\text{OC(CH}_2)_7\text{CHICHI(CH}_2)_7\text{CH}_3 \\[2pt]
\text{CHOC(CH}_2)_7\text{CHICHI(CH}_2)_7\text{CH}_3 \\[2pt]
\text{CH}_2\text{OC(CH}_2)_7\text{CHICHI(CH}_2)_7\text{CH}_3
\end{array}
$$

triolein
(MW = 884)

$$
\text{no. g I}_2 = 100\ \text{g triolein} \times \frac{1\ \text{mol triolein}}{884\ \text{g triolein}} \times \frac{3\ \text{mol I}_2}{1\ \text{mol triolein}} \times \frac{254\ \text{g I}_2}{1\ \text{mol I}_2}
$$

$$
= 86\ \text{g I}_2
$$

SIMILAR EXAMPLE: Exercise 12.

The compositions, saponification values, and iodine numbers of some common fats and oils are listed in Table 25-3.

Unsaturation in a fat or oil may be removed by the catalytic addition of hydrogen. Thus, oils or low-melting fats can be changed to higher melting fats. These fats, when mixed with skim milk, fortified with vitamin A, and artificially colored, are known as margarines. Edible fats and oils both hydrolyze and cleave at the double bonds by oxidation on exposure to heat, air, and light. The low-molecular-weight fatty acids produced give off offensive odors, the condition known as rancidity. Antioxidants, such as 3-*t*-butyl-4-hydroxyanisole (BHA), retard this oxidative rancidity. They are commonly added to oils in the high-temperature cooking of potato chips and other foods.

2. Phosphatides. The phosphatides (phospholipids) occur in all vegetable and animal cells and are especially prevalent in nerve tissue. They are derived from glycerol, fatty acids, phosphoric acid, and a nitrogen-containing compound. (In the following structures, R and R' are long-chain alkyl groups.)

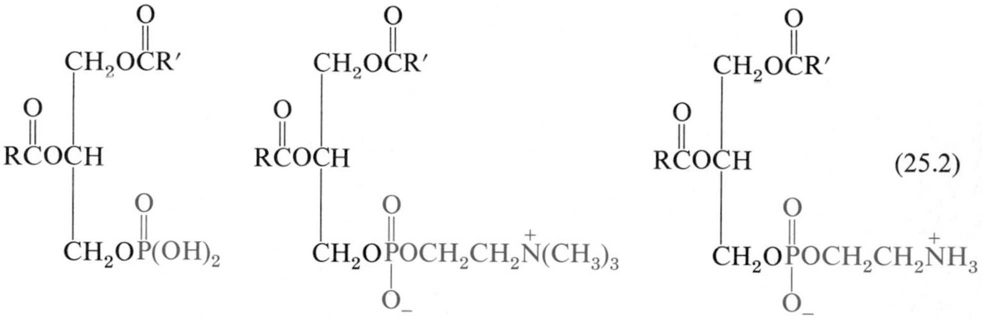

a phosphatidic acid   phosphatidylcholine (a lecithin)   phosphatidylethanolamine (a cephalin)   (25.2)

Choline, which is a quaternary ammonium compound, has the structure $[\text{HOCH}_2\text{CH}_2\text{N}^+(\text{CH}_3)_3]\text{OH}^-$. Its basicity in aqueous solution is comparable to that of KOH. When the alcohol group of choline is esterified with a phosphatidic acid, the product is a phosphatidylcholine or a **lecithin.** The lecithins are found in brain and nerve tissue and in egg yolk. In contrast to simple fats and oils, lecithins form stable colloidal suspensions in water. They are obtained from soybeans and are used as emulsifiers in the dairy and confectionery industries.

Medical evidence seems to suggest a relationship between a high intake of saturated fats and the incidence of coronary heart disease. For this reason many diets call for the substitution of unsaturated for saturated fatty acids in foods. In general, mammal fats are saturated, whereas those derived from vegetables, seafood, and poultry are unsaturated.

Removal of the fatty acid residue on the central carbon atom of a lecithin produces a lysolecithin. If this compound comes in contact with red blood cells, disintegration of the cells (hemolysis) occurs. The venom of poisonous snakes contains an enzyme that converts lecithins to lysolecithins. This accounts for the sometimes fatal effects of snakebites. Some spiders and insects also produce toxic results by the same mechanism.

TABLE 25-3
Some common fats and oils

Lipid	Component acids, % by mass						Saponification value	Iodine number
	Saturated			Unsaturated				
	Myristic	Palmitic	Stearic	Oleic	Linoleic	Linolenic		
Fats								
butter	7–10	24–26	10–13	28–31	1–3	0.2–0.5	210–230	26–28
lard	1–2	28–30	12–18	40–50	7–13	0–1	195–203	46–70
Edible oils								
corn	1–2	8–12	2–5	19–49	34–62	—	187–196	109–133
safflower	—	6–7	2–3	12–14	75–80	0.5–1.5	188–194	140–156

Ethanolamine has the formula $HOCH_2CH_2NH_2$ and serine, $HOCH_2CHNH_2CO_2H$.

FIGURE 25-3
Optical isomerism in glyceraldehyde.

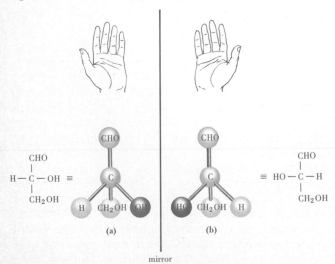

(a)   (b)

mirror

The structure in (a) is not superimposable on (b), just as a right and a left hand are not superimposable. (A right-handed glove cannot be worn on a left hand.)

Lecithins have both a highly polar and a nonpolar component. [The polar portion of the molecule is shown in color (25.2).] Lecithins are associated with membranes enclosing cell nuclei and mitochondria. Their physiological role is apparently to associate water-insoluble lipids and water-soluble components of an organism, such as in the transport of lipids in the bloodstream or in the movement of fats from one tissue to another.

Phosphatidylethanolamines or phosphatidylserines are known as **cephalins.** They are found in brain tissue, and they are also intimately involved in the blood-clotting process. Phospholipids are involved in the transport of ions across cell membranes, in certain secretory processes, and in the electron transport processes of respiration.

3. Waxes. When fatty acids form esters with long-chain *mono*hydric alcohols, the products are rather high-melting solids (35 to 100°C) called waxes. Beeswax is largely ceryl myristate, $C_{13}H_{27}CO_2C_{26}H_{53}$. It melts between 62 and 65°C and is used in shoe polish, candles, and wax coatings. Carnauba wax, a plant wax from a Brazilian palm tree, is largely myricyl cerotate, $C_{25}H_{51}CO_2C_{31}H_{63}$. It melts between 80 and 87°C and is used in polishes and to coat mimeograph stencils. Spermaceti wax (also called whale oil although not really an oil) consists mainly of cetyl palmitate, $C_{15}H_{31}CO_2C_{16}H_{33}$, and cetyl alcohol, $C_{16}H_{33}OH$. This material, obtained from the head cavity of a whale, has been used as a softening agent in ointments and in cosmetics.

CARBOHYDRATES. The literal meaning of the term "carbohydrate" is hydrate of carbon: $C_x(H_2O)_y$. Thus, sucrose or cane sugar, with the formula $C_{12}H_{22}O_{11}$ might be represented as $C_{12}(H_2O)_{11}$. A more useful definition, however, is that carbohydrates are polyhydroxy aldehydes, polyhydroxy ketones, their derivatives, and substances that yield them upon hydrolysis.

FIGURE 25-4
Optical activity.

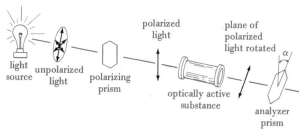

Light from an ordinary source consists of electromagnetic waves vibrating in all planes; it is unpolarized. Some substances (e.g., polaroid) possess the ability to screen out all light waves except those vibrating in a particular plane. Other substances— optically active materials—have the ability to rotate the plane of polarized light. In the illustration the light is rotated to the right by the angle α.

The term used generally to describe an object not superimposable on its mirror image is *chiral*. If two objects are superimposable, they are said to be *achiral*.

Carbohydrates that are ketones are called ketoses; those that are aldehydes are called aldoses. If the compound contains five carbon atoms it is a pentose, six carbon atoms, a hexose, and so on.

The simplest carbohydrates are the monosaccharides. Oligosaccharides contain from two to ten monosaccharide units bonded together. Names can be assigned to reflect the actual number of such units present, such as *di*saccharide and *tri*saccharide. Mono- and oligosaccharides are also called **sugars.** Polysaccharides contain more than ten monosaccharide units. The general term for all carbohydrates is glycoses. In summary,

Glycoses
- Monosaccharides
  - aldoses (aldotriose, aldotetrose, . . .)
  - ketoses (ketotriose, ketotetrose, . . .)
- Oligosaccharides (from 2 to 10 monosaccharide units)
  - disaccharides (e.g., sucrose)
  - trisaccharides (e.g., raffinose)
  - and so on.
- Polysaccharides (more than 10 monosaccharide units)
  - (e.g., starch and cellulose)

The simplest glycose is 2,3-dihydroxypropanal (glyceraldehyde), an aldotriose. If a ball-and-stick model of this molecule is made, an interesting form of stereoisomerism becomes evident. Figure 25-3 illustrates that there are *two* distinctly different structures for glyceraldehyde. These structures are not superimposable but bear the same relationship as an object and its image in a mirror. They are called **enantiomers** or mirror images. They are related to each other like a right and a left hand.

Optical Activity.   The type of isomerism pictured in Figure 25-3 is called optical isomerism. Optical isomers have the same number and kinds of atoms and groupings of atoms, but they differ in their configurations or arrangements in space. Specifically, their structures are not superimposable. One enantiomer rotates the plane of polarized light to the right and is said to be *dextro*rotatory (designated + or *d*); the other rotates the plane of polarized light to the left and is levorotatory (designated − or *l*) (see Figure 25-4). The test for mirror images and nonsuperimposability of structures is not so easily done as in Figure 25-3 when more complex molecules are involved. In these cases it is helpful to note the following: Almost all molecules exhibiting optical isomerism possess at least one carbon atom with four different groups attached to it. Such a carbon atom is said to be **chiral** or **asymmetric.**

Which arrangement of groups at an asymmetric car-

A more generalized system for absolute configurations uses R (Latin, *rectus*, right) in place of D and S (Latin, *sinister*, left) for L. However, D and L symbols are still widely used for carbohydrates.

bon atom, that is, which **absolute configuration** leads to dextrorotatory and which to levorotatory properties? In 1930, Rosanoff made a bold guess with a 50:50 chance of being correct. He proposed that the structure in Figure 25-3a was the (+) rotating one. To this species he assigned a small capital letter D, and to the structure in Figure 25-3b, the letter L. Nineteen years later, an x ray structure determination of a closely related compound proved that the guess was correct; that is, D-glyceraldehyde is indeed dextrorotatory.

Based on D-glyceraldehyde, the configurational designation D is used to indicate that the groups H, CHO, and OH, *in that order,* are situated in a clockwise fashion about the asymmetric carbon atom, when the CH_2OH group is directed away from the viewer (see Figure 25-3). In the L configuration the order is counterclockwise. The asymmetric carbon atom is shown in color in structures (25.3).

$$
\begin{array}{ccc}
\text{CHO} & & \text{CHO} \\
| & & | \\
\text{HCOH} & & \text{HOCH} \\
| & & | \\
\text{CH}_2\text{OH} & & \text{CH}_2\text{OH}
\end{array}
\qquad (25.3)
$$

D-(+)-glyceraldehyde   L-(−)-glyceraldehyde

It is difficult to represent a three-dimensional structure in a plane (two-dimensional) drawing. A useful convention that has been adopted to simplify this representation is called the **Fischer projection.** The structural formula is arranged on the page so that the backbone of the molecule extends from top to bottom and attached groups to the sides. The end groups of the backbone extend *behind* the plane of the page, *away* from the viewer. The attached groups to the left and right extend in *front* of the page, *toward* the viewer. The D-L convention just introduced is applied to the next-to-last (penultimate) carbon atom (shown in color) in the Fischer projections of the four-carbon aldoses.

$$
\begin{array}{ccccc}
\text{CHO} & & & & \text{CHO} \\
| & & & & | \\
\text{H—C—OH} & & & & \text{HO—C—H} \\
| & & | & & | \\
\text{H—C—OH} & & | & & \text{HO—C—H} \\
| & & | & & | \\
\text{CH}_2\text{OH} & & | & & \text{CH}_2\text{OH} \\
\end{array}
$$

D-erythrose   mirror   L-erythrose

A mixture of equal parts of the two enantiomers is designated DL-erythrose. Any such *dl* mixture, called a **racemic** modification, does not rotate the plane of polarized light either to the left or to the right.

But there are two more possibilities for 2,3,4-trihydroxybutanal—D-threose and L-threose. They too are enantiomers.

$$
\begin{array}{ccccc}
\text{CHO} & & & & \text{CHO} \\
| & & & & | \\
\text{HO—C—H} & & | & & \text{H—C—OH} \\
| & & | & & | \\
\text{H—C—OH} & & | & & \text{HO—C—H} \\
| & & | & & | \\
\text{CH}_2\text{OH} & & | & & \text{CH}_2\text{OH} \\
\end{array}
$$

D-threose   mirror   L-threose

If we compare the configurations of D-erythrose and D-threose, we note that these two molecules are *not* mirror images. Molecules that are optical isomers but *not* mirror images of one another are called **diastereomers.**

$$\begin{array}{ccc}
\text{CHO} & & \text{CHO} \\
\text{H—C—OH} & & \text{HO—C—H} \\
\text{H—C—OH} & & \text{H—C—OH} \\
\text{CH}_2\text{OH} & & \text{CH}_2\text{OH} \\
\text{D-erythrose} & \text{mirror} & \text{D-threose}
\end{array}$$

not mirror images

A final note of importance is that the designations D and L refer to the configuration of the groups bonded to an asymmetric carbon atom. The terms (+) and (−) represent the actual direction of rotation of polarized light. Thus, a molecule can have a D configuration and be levorotatory(−). The complete designations of the four aldo-tetroses are D-(−)-erythrose, L-(+)-erythrose, D-(−)-threose, L-(+)-threose.

Enantiomers (shortened form, antiomers) have the same physical and chemical properties. They differ only in the *direction,* not the extent to which they rotate plane-polarized light. Diastereomers (shortened form, diamers) do differ both in physical and chemical properties. Also they differ in the extent to which they rotate plane-polarized light.

Usually, when molecules with centers of asymmetry are synthesized, the product is a racemic modification. This is because the creation of these centers is a random process like flipping a coin (an equal probability for "heads" or "tails"). If optically pure isomers are desired, the racemic modification must be separated into the component enantiomers by a process called **resolution.**

Monosaccharides. Of the 16 possible aldohexoses only three occur widely in nature: D-glucose, D-galactose, and D-mannose.

$$\begin{array}{ccc}
\text{H—C=O} & \text{H—C=O} & \text{H—C=O} \\
\text{H—C—OH} & \text{H—C—OH} & \text{HO—C—H} \\
\text{HO—C—H} & \text{HO—C—H} & \text{HO—C—H} \\
\text{H—C—OH} & \text{HO—C—H} & \text{H—C—OH} \\
\text{H—C—OH} & \text{H—C—OH} & \text{H—C—OH} \\
\text{CH}_2\text{OH} & \text{CH}_2\text{OH} & \text{CH}_2\text{OH} \\
\text{D-glucose} & \text{D-galactose} & \text{D-mannose}
\end{array}$$

To some extent these sugar molecules do exist in the straight chain forms shown, but, in the main, all three occur as cyclic **hemiacetals.** We need to consider briefly the subject of hemiacetal formation.

An aldehyde and an alcohol can react in this manner.

$$R\text{—}\underset{\displaystyle\parallel}{\overset{\displaystyle O}{C}}\text{—H} + R'OH \rightleftharpoons R\text{—}\underset{\displaystyle OR'}{\overset{\displaystyle OH}{C}}\text{—H} \tag{25.4}$$

hemiacetal

Straight chain hemiacetals are unstable and can undergo further reaction with a second alcohol molecule to form an **acetal.**

$$R\text{—}\underset{\displaystyle OR'}{\overset{\displaystyle OH}{C}}\text{—H} + R'OH \xrightarrow{\text{H}^+} R\text{—}\underset{\displaystyle OR'}{\overset{\displaystyle OR'}{C}}\text{—H} + H_2O \tag{25.5}$$

acetal

Optical activity was discovered by Biot in 1815. Starting in 1848, Louis Pasteur conducted a series of investigations that established the basis of modern stereochemistry. Pasteur observed that optically inactive sodium ammonium tartarate actually consists of a mixture of two different kinds of crystals that are mirror images of one another. By a tedious process, using a magnifying glass and tweezers, Pasteur separated these crystals into two piles by hand. Each set of crystals, when dissolved in water, was found to be optically active. Solutions derived from one type of crystal rotated plane polarized light to the left; solutions of the other type produced the same degree of rotation to the right. Because this phenomenon was observed in solution, Pasteur postulated that the optical activity of the crystals was derived from an intrinsic property of the *molecules* in the crystals themselves.

FIGURE 25-5
Models of the glucose molecule.

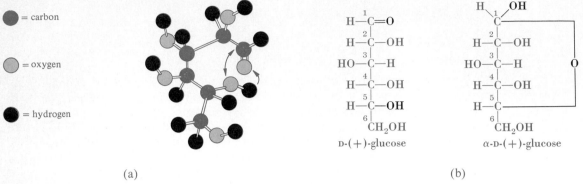

● = carbon

● = oxygen

● = hydrogen

(a)

(b)

(a) Ball-and-stick model indicating the atoms involved in ring closure.
(b) Ring closure represented through Fischer projection formulas.

FIGURE 25-6
α and β forms of D-glucose.

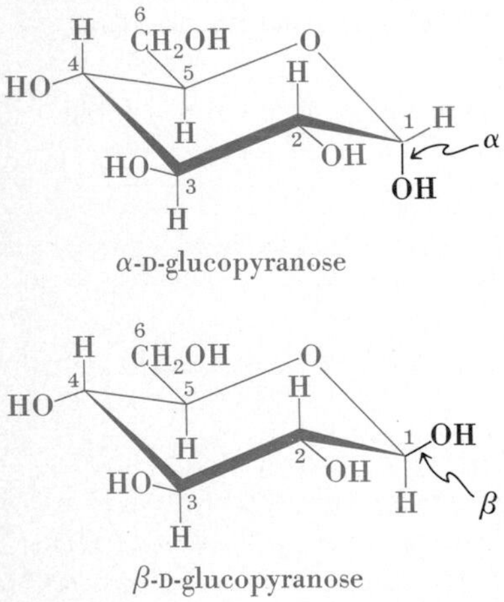

α-D-glucopyranose

β-D-glucopyranose

A still more precise name for α-D-(+)-glucose is α-D-(+)-glucopyranose. The term *pyranose* signifies a six-membered, oxygen-containing heterocycle of the pyran type.

The reactions just described involved two different molecules. With long-chain polyhydroxy aldehydes like glucose, the opportunity exists for *intra*molecular hemiacetal formation. The —OH group of the fifth carbon atom (C-5) adds to the carbonyl of the C-1 atom and produces a ring structure composed of five C atoms and one O atom, as illustrated in Figure 25-5. The configuration of this six-membered ring is of the "chair" type. Moreover, unlike the straight chain molecule, the ring hemiacetal is stable.

Closure of the ring in hemiacetal formation creates a new center of asymmetry (at the C-1 atom), with two possible orientations to consider. In the α form the OH at C-1 is axial (directed down); in the β form it is equatorial (extends out from the ring). The α and β forms of D-glucose are pictured in Figure 25-6.

The naming of monosaccharides is complicated by the fact that ring formation occurs. However, each term in a name conveys precise information about the molecule. Thus, D-(+)-glucose refers to the straight chain form of glucose in the D configuration; this form is dextrorotatory (+). The name α-D-(+)-glucose denotes the ring form derived from D-glucose in which the α configuration is found at the C-1 atom.

With some sugars a sufficient amount of the straight chain form is in equilibrium with the cyclic form so that the sugar engages in an oxidation-reduction reaction with Cu^{2+}(aq). The Cu^{2+}(aq) is reduced to red Cu_2O, and the aldehyde portion of the sugar is oxidized (to an acid). These sugars are known as **reducing sugars.** This test for a reducing sugar is conducted with alkaline copper ion complexed as the tartrate (Fehling's solution) or the citrate (Benedict's solution).

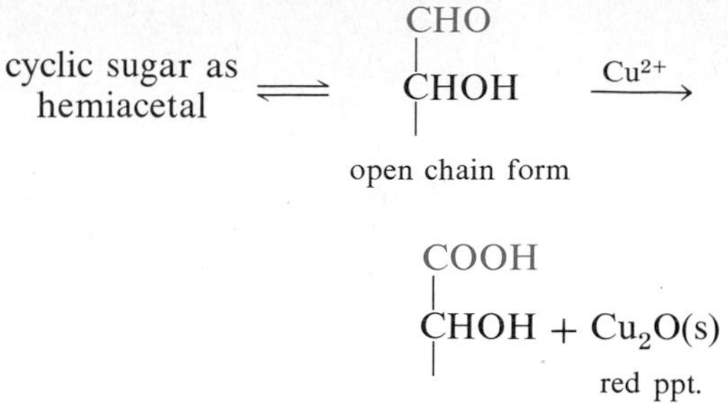

$$\text{cyclic sugar as} \atop \text{hemiacetal} \rightleftharpoons \quad \underset{\text{open chain form}}{\overset{\text{CHO}}{\underset{|}{\overset{|}{\text{CHOH}}}}} \quad \xrightarrow{\text{Cu}^{2+}}$$

$$\underset{\text{red ppt.}}{\overset{\text{COOH}}{\underset{|}{\overset{|}{\text{CHOH}}}}} + \text{Cu}_2\text{O(s)} \qquad (25.6)$$

Disaccharides. In equations (25.4) and (25.5) we saw that one molecule of an alcohol converts an aldehyde to a hemiacetal and a second molecule of alcohol converts the hemiacetal to an acetal. Acetals of the glycoses are called glycosides. D-Glucose, for example, can form a glycoside (more specifically for glucose, glucoside) with the —OH group supplied by a second monosaccharide unit. In this case, because two monosaccharide units are joined together, the product is a disaccharide. Hydrolysis of a disaccharide by acid or enzymatic catalysis yields two molecules of monosaccharides.

In considering disaccharides, attention is focused on these questions.

1. What are the component monosaccharide units?
2. Is the configuration of the linkage between the monosaccharide units α or β?
3. What is the ring size in each monosaccharide unit?

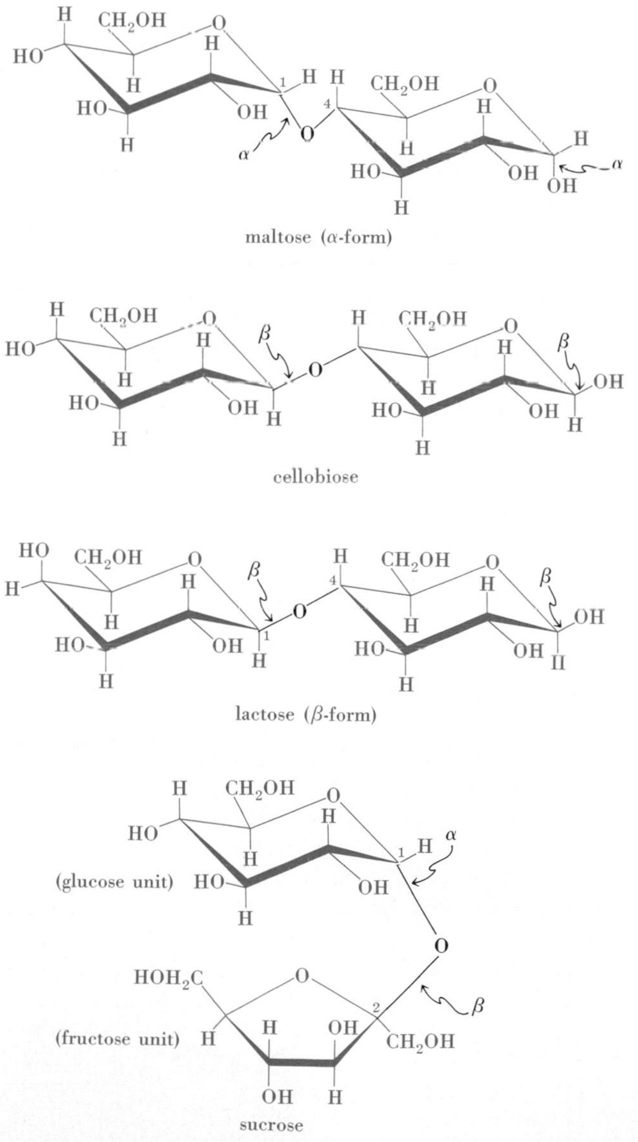

The important naturally occurring disaccharides—maltose, cellobiose, lactose, and sucrose—are presented in Figure 25-7. In maltose the monosaccharide unit on the left, originally in the hemiacetal form, is converted to the acetal by reacting with the hydroxyl group on the C-4 of a second unit. Linkage of the two units is in the α manner. The glucose unit on the right is present as a hemiacetal in its α form. This hemiacetal exists in equilibrium with the open chain form and can be oxidized by Cu^{2+}(aq). Maltose is a reducing sugar. [The full-acetal glucose unit on the left, however, is stable to Cu^{2+}(aq).] Maltose is produced by the action of malt enzyme on starch. It undergoes fermentation, in the presence of yeast, to glucose, and then to ethanol and CO_2(g).

Cellobiose is obtained by careful hydrolysis of cellulose. It is a glucose-glucose disaccharide with β linkages; it is also a reducing sugar. Lactose is the reducing sugar present in milk (4 to 6% in cow's milk and 5 to 8% in human milk). It is a galactose-glucose disaccharide having β linkages. Sucrose is ordinary table sugar (cane or beet sugar). It is a glucose-fructose disaccha-

FIGURE 25-7
Some common disaccharides.

maltose (α-form)

cellobiose

lactose (β-form)

(glucose unit)

(fructose unit)

sucrose

ride linked $1\alpha, 2\beta$. Since the glucose unit is tied up as an acetal and the fructose as a ketal, sucrose is a *nonreducing* sugar.

Polysaccharides.   Polysaccharides are composed of monosaccharide units joined into long chains by oxygen linkages. **Starch,** with a molecular weight between 20,000 and 1,000,000, is the reserve carbohydrate of many plants and is the bulk constituent of cereals, rice, corn, and potatoes. Its structural features are brought out in Figure 25-8. **Glycogen** is the reserve carbohydrate of animals, occurring in liver and muscle tissue. It has a higher molecular weight than starch, and the polysaccharide chains are more branched. **Cellulose** is the main structural material of plants. It is the chief component of wood pulp, cotton, and straw. Complete hydrolysis gives glucose. Partial hydrolysis yields cellobiose, indicating that cellulose is a β-linked polymer (see Figure 25-8). Cellulose has a molecular weight between 300,000 and 500,000, corresponding to 1800 to 3000 glucose units.

Photosynthesis.   Green plants conduct the process of photosynthesis in the presence of light. They use carbon dioxide as the only source of carbon, together with water, inorganic salts, and a catalytic agent called chlorophyll.

$$n\,CO_2 + n\,H_2O \xrightarrow[\text{chlorophyll}]{\text{sunlight}} (CH_2O)_n + n\,O_2 \tag{25.7}$$

Photosynthesis has been estimated to account for the annual conversion of 200 billion tons of carbon, as carbon dioxide, to carbohydrates. Concurrently, about 400 billion tons of $O_2(g)$ are released. Between 10 and 20% of this photosynthetic production occurs on land and the remainder in the oceans.

The photosynthesis reaction just given is greatly oversimplified. The currently accepted mechanism, proposed by Melvin Calvin (Nobel prize, 1963), involves as many as 100 sequential steps for the conversion of 6 mol of carbon dioxide to 1 mol of glucose. The elucidation of this mechanism was aided greatly by the use of the radioactive tracer carbon-14. For simplicity, the overall photosynthetic process is divided into two phases: (1) the conversion of solar energy to chemical energy, the light reaction; and (2) the synthesis, promoted by enzymes, of carbohydrate intermediates. This latter reaction can occur in the absence of light and is called the dark

Most animals, including human beings, do not possess the necessary enzymes to hydrolyze β linkages; as a result they cannot digest cellulose. Certain bacteria in ruminants (cows, sheep, horses) and termites can hydrolyze cellulose, allowing them to use it as a food. Termites, as we know, subsist on a diet of wood.

FIGURE 25-8
Two common polysaccharides.

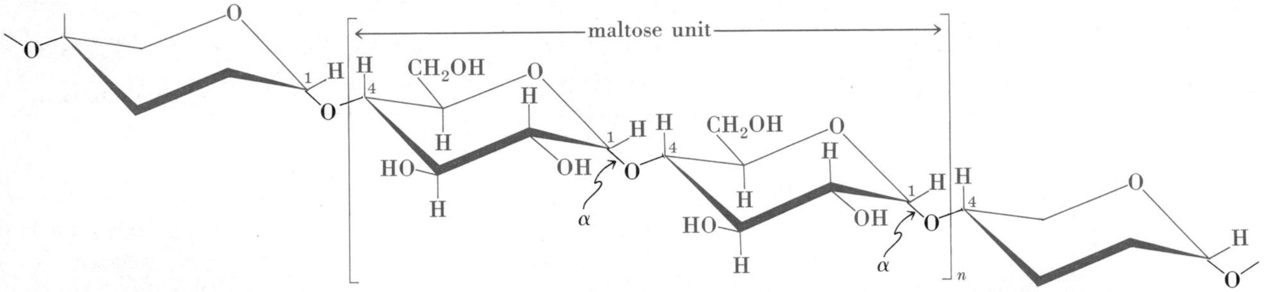

starch

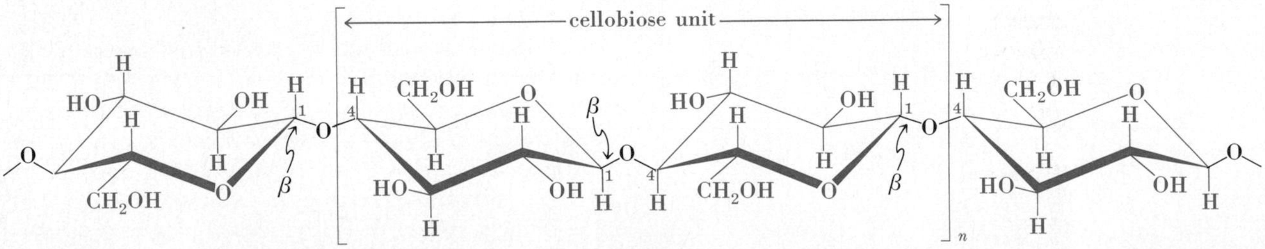

cellulose

FIGURE 25-9
Structure of
chlorophyll a.

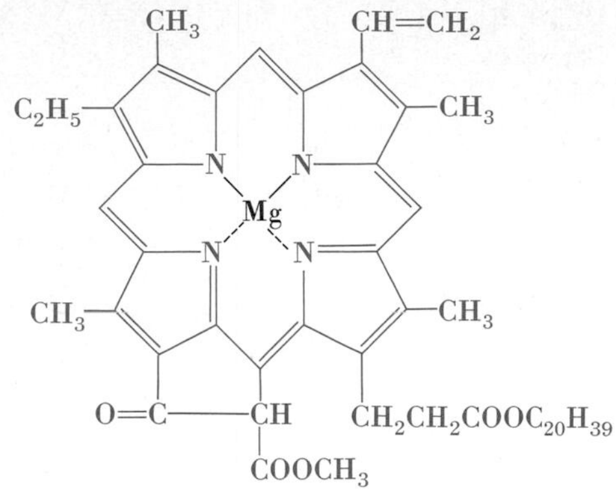

chlorophyll a

reaction. A key role in the light reaction is served by the chlorophyll. This substance, whose structure is shown in Figure 25-9, is a chelate complex. The four nitrogen atoms form a square about the central magnesium ion and the entire structure is planar. Structures of this type are called **porphyrins** and are commonly found in both plants and animals.

PROTEINS. Proteins are probably the most complex organic materials found in nature. They are the basis of protoplasm and are found in all living organisms. Proteins, as muscle, skin, hair, and other tissue, make up the bulk of the body's nonskeletal structure. As enzymes they catalyze biochemical reactions. As hormones they regulate metabolic processes; and as antibodies they counteract the effect of invading species and substances.

When a protein is hydrolyzed by dilute acids, alkalis, or hydrolytic enzymes, a mixture of α-amino acids results. (The term α means that the NH_2 group is on the carbon atom adjacent to the carboxyl group.) Proteins are high-molecular-weight long-chain polymers composed of these α-amino acids. Of the known α-amino acids, about 20 have been identified as building blocks of most plant and animal proteins. These are listed in Table 25-4.

Other than glycine ($H_2NCH_2CO_2H$), naturally occurring amino acids are optically active, mostly with an L configuration.

The name "protein" is derived from the Greek word *proteios*, meaning "of first importance" (similar to the derivation of "proton").

The reference structure for establishing the configurations of amino acids is again glyceraldehyde, with the —NH_2 group substituting for —OH and —CO_2H, for —CHO. The molecule shown here has an L configuration because the order of the groups, H, CO_2H, and NH_2, is counterclockwise.

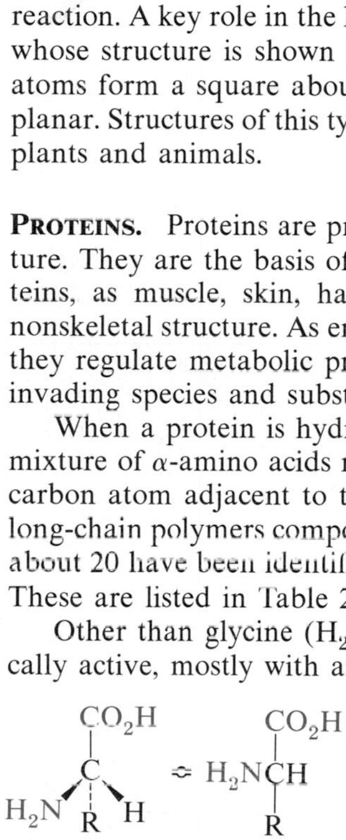

Certain amino acids are required for proper health and growth in human beings, yet the body is unable to synthesize them. These must be ingested through foods, and are called **essential** amino acids. Eight are known to be essential; the case of three others is less certain (see Table 25-4).

The amino acids are colorless, crystalline, high-melting solids that are moderately soluble in water. In an acidic solution the amino acid exists as a cation, with a proton attaching itself to the unshared pair of electrons on the nitrogen atom in the group —NH_2. In a basic solution an anion is formed. At the neutral point a proton is transferred from —CO_2H to —NH_2. The product is a dipolar ion or a "zwitterion."

$$R-CH-CO_2H \underset{H^+}{\overset{OH^-}{\rightleftharpoons}} R-CH-CO_2^- \underset{H^+}{\overset{OH^-}{\rightleftharpoons}} R-CH-CO_2^- \qquad (25.8)$$

$\quad\quad\ |$ $\qquad\qquad\qquad\quad\ |$ $\qquad\qquad\qquad\quad\ |$

$\quad NH_3^+$ $\qquad\qquad\qquad NH_3^+$ $\qquad\qquad\qquad NH_2$

acidic soln.   isoelectric point   basic soln.

TABLE 25-4
Some common amino acids

Name	Symbol	Formula	pI
		Neutral amino acids	
glycine	Gly	$HCH(NH_2)CO_2H$	5.97
alanine	Ala	$CH_3CH(NH_2)CO_2H$	6.00
valine[a]	Val	$(CH_3)_2CHCH(NH_2)CO_2H$	5.96
leucine[a]	Leu	$(CH_3)_2CHCH_2CH(NH_2)CO_2H$	6.02
isoleucine[a]	Ileu or Ile	$CH_3CH_2CH(CH_3)CH(NH_2)CO_2H$	5.98
serine	Ser	$HOCH_2CH(NH_2)CO_2H$	5.68
threonine[a]	Thr	$CH_3CHOHCH(NH_2)CO_2H$	5.6
phenylalanine[a]	Phe	$C_6H_5CH_2CH(NH_2)CO_2H$	5.48
methionine[a]	Met	$CH_3SCH_2CH_2CH(NH_2)CO_2H$	5.74
cysteine	Cys	$HSCH_2CH(NH_2)CO_2H$	5.05
cystine	(Cys)$_2$	$[SCH_2CH(NH_2)CO_2H]_2$	4.8
tyrosine	Tyr	$4\text{-}HOC_6H_4CH_2CH(NH_2)CO_2H$	5.66
tryptophan[a]	Trp	(indole ring)$-CH_2CH(NH_2)CO_2H$	5.89
proline	Pro	(pyrrolidine ring)$-CO_2H$	6.30
hydroxyproline	Hyp	(hydroxy pyrrolidine ring)$-CO_2H$	
		Acidic amino acids	
aspartic acid	Asp	$HO_2CCH_2CH(NH_2)CO_2H$	2.77
glutamic acid	Glu	$HO_2CCH_2CH_2CH(NH_2)CO_2H$	3.22
		Basic amino acids	
lysine[a]	Lys	$H_2N(CH_2)_4CH(NH_2)CO_2H$	9.74
arginine	Arg	$H_2NCNHNH(CH_2)_3CH(NH_2)CO_2H$	10.76
histidine	His	(imidazole ring)$-CH_2CH(NH_2)CO_2H$	

[a] Essential amino acids. In addition to these, arginine and glycine are required by the chick, arginine by the rat, and histidine by human infants.

Amino acids are amphoteric. The pH at which the dipolar structure predominates is called the **isoelectric point** or pI. At this pH the molecule does not migrate in an electric field. At a pH above the pI the molecule migrates to the anode (positive electrode), below the pI to the cathode (negative electrode) in an electrophoresis apparatus (recall Figure 12-17). Basic amino acids have a pI above 7, acidic ones below 7, and neutral ones near 7. In most amino acids the basicity of the amino group is about equal to the acidity of the carboxyl group. The largest group of amino acids are essentially pH neutral.

FIGURE 25-10
Amino acid sequence in beef insulin—primary structure of a protein.

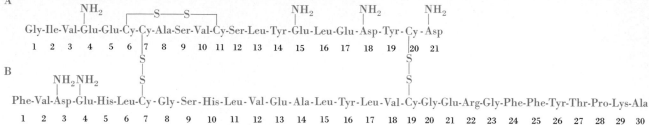

There are two poly-peptide chains joined by disulfide (—S—S—) linkages. One chain has 21 amino acids, and the other 30. In chain A the Gly at the left end is N-terminal and the Asp is C-terminal. In chain B Phe is N-terminal, and Ala is C-terminal.

Peptides. Amino acid molecules can be joined together by the elimination of water molecules between them. Two amino acids thus joined form a dipeptide.

$$R\!-\!\underset{\substack{|\\ NH_2}}{CH}\!-\!\underset{\substack{\|\\ O}}{C}\!-\!OH + HN\!-\!\underset{\substack{|\\ H}}{CH}\!-\!R' \longrightarrow H_2O + R\!-\!\underset{\substack{|\\ NH_2}}{CH}\!-\!\underset{\substack{\|\\ O}}{C}\!-\!NH\!-\!\underset{}{CH}\!-\!R'$$

a dipeptide

(25.9)

A tripeptide has three amino acid residues and two peptide linkages. A large number of amino acid units may join to form a *poly*peptide.

A simple convention is used to write the structures and to name polypeptides. The amino acid unit present at one end of the polypeptide chain has a free —NH₂ group; this is the "N-terminal" end. The other end of the chain has a free —CO₂H group; it is the "C-terminal" end. The structure is written with the N-terminal end to the left and C-terminal to the right. The base name of the polypeptide is that of the C-terminal amino acid. All the other amino acid units in the chain are named as substituents of this acid; as such their names are changed from the "ine" to the "yl" ending. Abbreviations are also commonly used in polypeptide names, as illustrated in Example 25-4.

Example 25-4 What is the name of the polypeptide whose structure is shown?

$$\underset{\substack{|\\ NH_2\\ (a)}}{H_2C}\!-\!\underset{\substack{\|\\ O}}{C}\!-\!NH\!-\!\underset{\substack{|\\ CH_3\\ (b)}}{CH}\!-\!\underset{\substack{\|\\ O}}{C}\!-\!NH\!-\!\underset{\substack{|\\ CH_2OH\\ (c)}}{CH}\!-\!\underset{\substack{\|\\ O}}{C}\!-\!OH$$

The three amino acids in this tripeptide are identified through Table 25-4. (a) = glycine; (b) = alanine; (c) = serine. The C-terminal amino acid is serine. The name is

glycylalanylserine (Gly-Ala-Ser)

SIMILAR EXAMPLES: Exercises 25, 26.

The sequence of amino acids in a polypeptide can have profound rami-fications. Hemoglobin contains four polypeptide chains, each with 146 amino acid units. The substitution of valine for glutamic acid at one site in two of these chains gives rise to the some-times fatal blood disease known as sickle cell anemia. Apparently, the altered hemoglobin has a reduced ability to trans-port oxygen through the blood.

Suppose that a tripeptide is known to consist of the three amino acids: A, B, and C. What is the correct structure: ABC?, ACB?, . . . Can you see that there are six possibilities? For longer chains, of course, the number of possibilities is enormous. Determining the sequence of amino acids in a polypeptide chain is one of the most significant problems in all of biochemistry. The Nobel prize in 1958 was awarded to Frederick Sanger for elucidation of the structure of beef insulin (see Figure 25-10). The method employed is outlined in Figure 25-11.

FIGURE 25-11
Experimental determination of amino acid sequence.

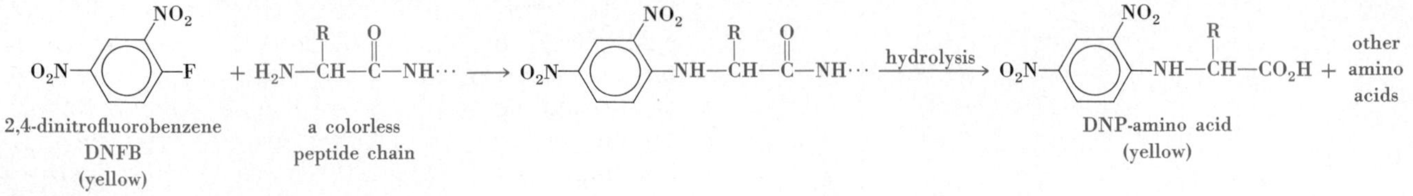

2,4-dinitrofluorobenzene
DNFB
(yellow)

a colorless
peptide chain

DNP-amino acid
(yellow)

In the reaction outline the N-terminal amino acid ends up with the yellow "marker" (a dinitrophenyl group, DNP) attached to it. By gentle hydrolysis and repeated use of the marker, a polypeptide chain can be broken down and the sequence of the individual units determined.

Example 25-5 A polypeptide, on complete hydrolysis, yielded the amino acids A, B, C, D, and E. Partial hydrolysis and sequence proof gave single amino acids together with the following larger fragments: AD, DC, DCB, BE, and CB. What must be the sequence of amino acids in the polypeptide?

By arranging the fragments in the following manner,

AD
 DC
 DCB
   BE
 CB

we see that only the sequence A-D-C-B-E is consistent with the fragments observed.

SIMILAR EXAMPLES: Exercise 27.

Examples of denaturation of proteins are numerous. The frying or boiling of an egg involves the denaturation (coagulation) of the egg albumin, a protein. The beauty shop "permanent wave" takes advantage of a denaturation process that is reversible. The proteins found in hair (e.g., keratin) contain disulfide linkages (—S—S—). Treatment of hair with a reducing agent causes cleavage of these linkages—a denaturation process. Following this step the hair is set into the desired shape. Next, the hair is treated with a mild oxidizing agent. The disulfide linkages are reestablished and the hair remains in the style in which it was set (see also Figures 25-13 and 25-14).

The distinction between large polypeptides and proteins is arbitrary. It is generally accepted that if the molecular weight is over 10,000 (roughly 50 to 75 amino acid units) the substance is a protein. Proteins, like amino acids, are amphoteric. They possess characteristic isoelectric points, and their acidity or basicity depends on their amino acid composition. When proteins are heated, treated with salts, or exposed to ultraviolet light, profound and complex changes occur. This **denaturation** usually brings about lowering of solubility and loss of biological activity (see also page 648).

Structure of proteins. The **primary structure** of a protein, as we have already seen, refers to the exact sequence of amino acids in the polypeptide chains that make up the protein. The most complex protein that has been analyzed is gamma globulin. It consists of four polypeptide chains, contains 1320 amino acid units (19,996 atoms), and has a molecular weight of 100,000.

What are the shapes of the long polymeric chains themselves? Are they simply limp and entangled like a plate of spaghetti or is there some order within chains and among chains? The structure or shape of an individual protein chain is referred to as **secondary structure.** The first work on this subject was published in 1951 by Pauling and Corey, describing x ray diffraction studies on polylysine, a synthetic polypeptide. They postulated that the orientation of this polypeptide and thus of protein chains is *helical.* A spiral, helical, or springlike shape can be either left- or right-handed; but because proteins are composed of L-amino acids their helical structure is right-handed (see Figure 25-12).

As we have learned previously, highly ordered materials produce distinctive x ray diffraction patterns, and from these patterns much can be learned about their structure. Conversely, materials in which there is little order—amorphous materials—do not have distinctive x ray diffraction patterns. From x ray studies it is clear that not all proteins have a helical configuration. Gamma globulin is one that does not.

FIGURE 25-12
An alpha helix—secondary structure of a protein.

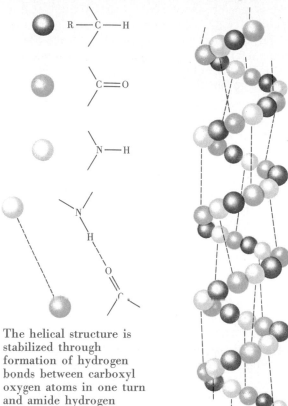

The helical structure is stabilized through formation of hydrogen bonds between carboxyl oxygen atoms in one turn and amide hydrogen atoms in the next turn above. The bulky R groups are directed outward from the atoms in the spiral.

Hemoglobin and myoglobin are helical in some portions of their chains and randomly oriented in others. Finally, other types of orientations are possible. For example, β-keratin and silk fibroin are arranged in pleated sheets. In these proteins the side chains extend above and below the pleated sheets and hydrogen bonding is between *different* molecules (interpeptide bonding) lying next to each other and about 0.47 nm apart in the same sheet. These sheets are stacked on top of one another about 1.0 nm apart, rather like a pile of sheets of corrugated roofing (see Figure 25-13).

Does a helical protein molecule possess any additional structural features? That is, are the helices elongated, twisted, knotted, or what? The final statement regarding the shape of a protein molecule lies in a description of its tertiary structure. Because the internal hydrogen bonding that occurs between atoms in successive turns of a protein helix is weak, these hydrogen bonds ought to be easily broken. In particular, we should expect them to be replaced by hydrogen bonds to water molecules when the protein is placed in water. That is, the α helix should open up and become a randomized structure when placed in water (recall the analogy of the limp spaghetti). But experimental evidence indicates that this does not happen. We are led to the conclusion that other forces must be involved in compressing the long α-helical chains into definite geometric shapes. Each protein has its own three-dimensional shape or **tertiary structure.** Three types of linkages involved in tertiary structures are described in Figure 25-14.

By x ray diffraction studies Kendrew and Perutz (Nobel prize, 1963) were able to elucidate the primary, secondary, and tertiary structures of myoglobin. The primary structure is that of a peptide of 153 units in a single chain. Secondary structure involves 70% coiling

FIGURE 25-13
Pleated sheet model of β-keratin.

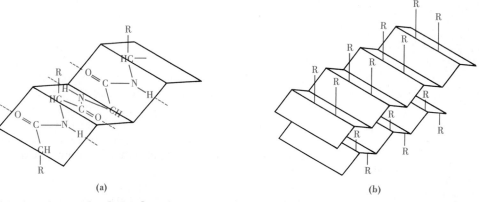

(a)   (b)

(a) A polypeptide chain showing the direction of interpolypeptide hydrogen bonds (other polypeptide chains lie to the left and to the right of the chain shown). Bulky R groups extend above and below the pleated sheet.
(b) The stacking of pleated sheets.

FIGURE 25-14
Linkages contributing to the tertiary structure of proteins.

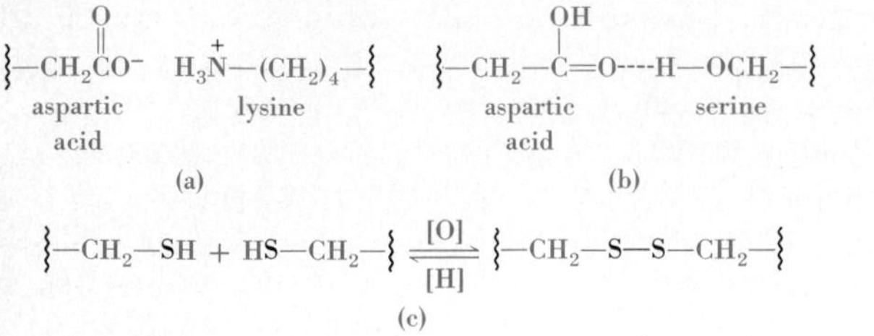

$$\}{-}CH_2CO^-\quad H_3\overset{+}{N}{-}(CH_2)_4{-}\}$$
aspartic   lysine
acid

(a)

$$\}{-}CH_2{-}\overset{OH}{\underset{}{C}}{=}O{---}H{-}OCH_2{-}\}$$
aspartic   serine
acid

(b)

$$\}{-}CH_2{-}SH + HS{-}CH_2{-}\} \xrightleftharpoons[\text{[H]}]{\text{[O]}} \}{-}CH_2{-}S{-}S{-}CH_2{-}\}$$

(c)

(a) Salt linkages. Acid-base interactions between different coils. In the example shown, the carboxyl group of an aspartic acid unit on one coil donates a proton to the free amine group of a lysine unit on another.
(b) Hydrogen bonding. Interactions between side chains of certain amino acids, for example, aspartic acid and serine.
(c) Disulfide linkages. Oxidation of the highly reactive thioalcohol (—SH) group of cysteine to a disulfide (—S—S—) can occur (as in beef insulin).

The folding of polypeptide chains into a tertiary structure is influenced by an additional factor. Hydrophobic hydrocarbon portions of the chains (R groups) tend to be drawn into close proximity in the interior of the structure, leaving ionic groups at the exterior.

FIGURE 25-15
Representation of the tertiary structure of myoglobin.

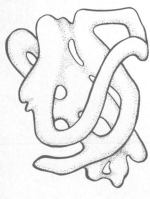

of the chain into an α helix. The tertiary structure is depicted in Figure 25-15.

The hemoglobin molecule consists of four separate polypeptide chains or subunits. The arrangement of these four subunits constitutes a still higher order of structure referred to as the **quaternary structure** (see Figure 25-16). Because it is a single polypeptide, myoglobin has no quaternary structure.

25-3 Biochemical Reactivity

Although the outward appearances of organisms vary remarkably, the similarities that exist in the chemical reactions that occur within organisms are equally striking. The totality of these reactions is referred to as **metabolism.** The part of this process in which molecules are broken down or degraded is called catabolism, and that part in which molecules are synthesized is called anabolism. Reactions for which the standard free energy change is positive are endergonic, and those for which it is negative are exergonic. The chemical substances involved in metabolism are called metabolites.

The raw materials for metabolic transformations are organic molecules, but the reactions involved are very dissimilar to those considered in the preceding chapter. In fragile biological systems, heat and pressure cannot be used to force chemical reactions to go, nor can strongly acidic or basic catalysts. The substances that do control the processes of metabolism are the catalysts known as enzymes. Herein lies the basis of biological uniqueness. An organism develops into what it is because of the particular set of enzymes it possesses.

Metabolism is a complex subject and anything more than a brief overview is much beyond the scope

FIGURE 25-16
Quaternary protein structure—the structures of heme and hemoglobin.

$$CH_2{=}CH{-}\qquad\overset{CH_3\qquad\qquad CH_3}{}\qquad{-}CH_2CH_2COOH$$

(a)

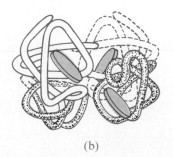

(b)

Hemoglobin is a conjugated protein. It consists of nonprotein groups (called prosthetic groups) bonded to a protein portion. **(a)** The structure of heme. **(b)** Four heme units and four polypeptide coils are bonded together in a molecule of hemoglobin.

of this text. The discussion that follows centers on the summary of metabolic processes outlined in Figure 25-17.

CARBOHYDRATE METABOLISM. Foods containing starch are the principal sources of carbohydrates for man and many animals. Digestion of starch begins in the mouth through the action of salivary enzymes, the amylases. Starch is converted to maltose and polysaccharides known as dextrins. This process continues in the acidic medium of the stomach, and the maltose and polysaccharides pass on to the small intestine. Here, amylase from the pancreas completes the conversion of polysaccharides to maltose, and the enzyme maltase converts maltose to glucose. Glucose is absorbed

FIGURE 25-17
Metabolism outline.

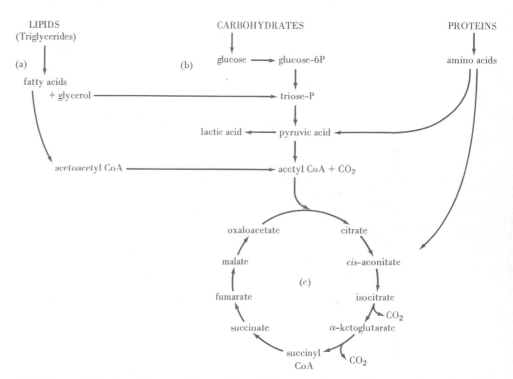

(a) Fatty acid section. Fatty acids are degraded two carbon atoms at a time. Acetyl units are fed into the citric acid cycle (c) as acetyl CoA.
(b) Glycolysis section (Embden–Meyerhof pathway). These reactions are anaerobic (no oxygen required). Carbohydrates are degraded to the six-carbon sugar glucose, and then to the three-carbon triose-P (glyceraldehyde-3-phosphate). Next the three-carbon acid, pyruvic acid, is formed from triose-P. Pyruvic acid loses a molecule of CO_2, yielding the two-carbon acetyl unit, which combines with coenzyme A (CoA) to form acetyl CoA.
(c) Citric acid cycle (Krebs cycle). A two-carbon acetyl unit from acetyl CoA joins with the four-carbon oxaloacetate unit to produce the six-carbon tricarboxylic acid, citric acid (designated here as citrate). A two-step conversion to isocitrate occurs, followed by the loss of a molecule of CO_2 and the formation of the five-carbon α-ketoglutarate. Another CO_2 molecule is lost in the formation of succinyl CoA. The remainder of the cycle involves a succession of four-carbon acids leading to oxaloacetate. The oxaloacetate regenerated at the end of the cycle now joins with another acetyl unit and the cycle is repeated. The net change occurring in the cycle, then, is that a two-carbon acetyl unit enters the cycle and two molecules of CO_2 leave.

Emphasis in this outline is on the degradation of large molecules to small ones (catabolism). Some of the species shown in the citric acid cycle are also involved in the biosynthesis of larger molecules (anabolism). For example, the amino acid asparagine is synthesized from oxaloacetate.

As the name suggests, in glucose 6-phosphate a phosphate group is substituted for the —OH group on the C-6 atom, which is a side chain of the pyranose ring (see Figure 25-6).

$$H_2C-O-\overset{\overset{\displaystyle O^-}{|}}{\underset{\underset{\displaystyle O}{||}}{P}}-OH$$

The structure of glyceraldehyde 3-phosphate is

$$
\begin{array}{c}
\overset{\overset{\displaystyle O}{||}}{C}-H \\
H-\overset{|}{\underset{|}{C}}-OH \\
H_2-\overset{|}{C}-O-\overset{\overset{\displaystyle O^-}{|}}{\underset{\underset{\displaystyle O}{||}}{P}}-OH
\end{array}
$$

through the wall of the small intestine into the bloodstream, from which it is distributed to other organs.

Glucose is ultimately oxidized to carbon dioxide and water with the liberation of energy; the principal intermediate in this process is glucose 6-phosphate (glucose-6P). Its formation is controlled by the pancreatic hormone, insulin. Once formed glucose-6P may be converted to glycogen (a polysaccharide stored in the liver), back to glucose, or it may be metabolized. The major route for this metabolism involves the anaerobic (absence of air) Embden–Meyerhof pathway, followed by an aerobic cycle (Krebs cycle). These interrelationships are suggested diagrammatically in Figure 25-17.

LIPID METABOLISM. Fats stored in the body represent a rich source of energy. The digestion of fats and oils occurs primarily in the small intestine, through the action of a combination of lipase enzymes. The products of this enzyme hydrolysis are glycerol, mixtures of mono- and diglycerides, and fatty acids. These are absorbed into the bloodstream through the wall of the intestine. Glycerol is converted to glyceraldehyde-3-phosphate (triose phosphate) and joins into the glucose metabolism route previously described. Fatty acids are oxidized to carbon dioxide and water, with the release of energy, in a series of reactions known as β oxidation. In this process oxidation occurs at the β carbon atom of a fatty acid, followed by cleavage. This means that two-carbon pieces (acetic acid) are split off. The process requires the presence of coenzyme A (CoA). For example, with palmitic acid ($C_{15}H_{31}CO_2H$) the process must be repeated seven times, with the formation of eight molecules of acetyl coenzyme A, which enter the Krebs cycle (Figure 25-17).

PROTEIN METABOLISM. In the stomach, hydrochloric acid and the enzyme pepsin hydrolyze about 10% of the amide linkages in proteins and produce polypeptides in the molecular weight range of 500 to several thousand. In the small intestine peptidases such as trypsin and chymotrypsin (from the pancreas) cleave the polypeptides into very small fragments. These are then acted upon by aminopeptidase and carboxypeptidase. The resulting free amino acids pass through the wall of the intestine, into the bloodstream, and on to various organs. Each amino acid has its own characteristic metabolic reactions, but in general each is converted to an intermediate that enters the Krebs cycle. Proteins may also be synthesized from amino acids in body cells under directions supplied by nucleic acids (see Section 25-4). However, the eight essential amino acids mentioned previously cannot be synthesized in the body and must be obtained from digested proteins.

ENERGY RELATIONSHIPS IN METABOLISM. Reactions in which molecules are synthesized in an organism, anabolic reactions, must acquire energy from reactions in which molecules are degraded, catabolic reactions. The fundamental agents responsible for these energy exchanges are adenosine triphosphate (ATP) and adenosine diphosphate (ADP). The energy released in exergonic reactions is stored by the conversion of ADP to ATP.

$$\text{ADP} + \text{inorganic phosphate (P}_\text{i}) + 30\text{–}50 \text{ kJ} \underset{\substack{\text{energy} \\ \text{released}}}{\overset{\substack{\text{energy} \\ \text{absorbed}}}{\rightleftharpoons}} \text{ATP} + H_2O \qquad (25.10)$$

Energy is released from foods by oxidation processes. The energy released in the oxidation is picked up by ADP, which is converted to ATP. Enzymes catalyze each conversion every step along the way. ADP, ATP, and two important intermediates,

Perhaps the complexity of the metabolic process can be better appreciated in terms of the data given here. If the metabolism of glucose occurred in a single step, with one ADP converted to ATP in that step, only $(33.5/2870) \times 100 = 1\%$ of the available energy would be conserved. Because metabolism occurs in many steps there is an opportunity for a much greater quantity of energy to be stored.

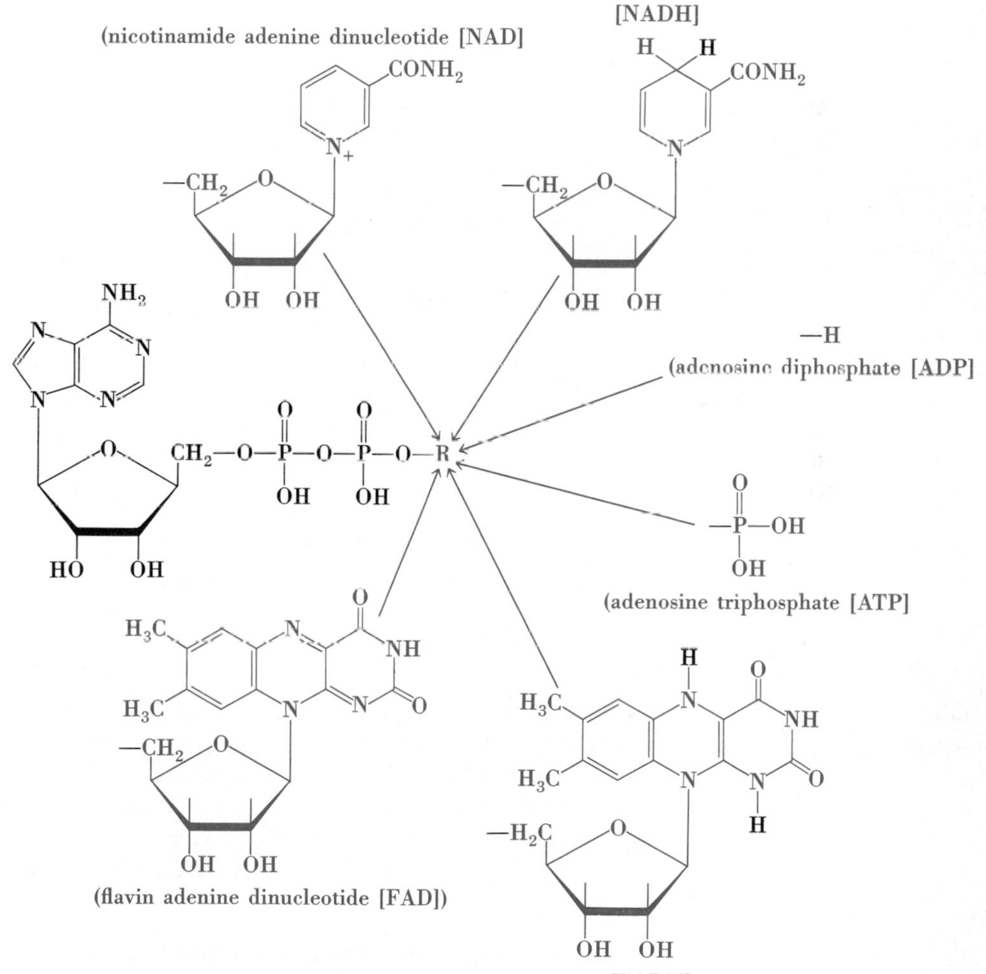

nicotinamide adenine dinucleotide (NAD) and flavin adenine dinucleotide (FAD), are pictured in Figure 25-18.

The exact way in which ADP and ATP enter into metabolic processes was not detailed in Figure 25-17. However, it is known that the conversion of 1 mol of glucose to CO_2 and H_2O is accompanied by the conversion of 38 mol ADP to ATP.

$$C_6H_{12}O_6 + 6\,O_2 + 38\,ADP + 38\,P_i \longrightarrow 6\,CO_2 + 6\,H_2O + 38\,ATP \qquad (25.11)$$

Assuming that 33.5 kJ of energy is absorbed for each mole of ADP converted to ATP, the energy stored in ATP as a result of the metabolism of 1 mol glucose is $38 \times 33.5 = 1270$ kJ. The total energy released when 1 mol glucose is converted to CO_2 and H_2O is 2870 kJ. Thus, the efficiency of energy storage in the high-energy bonds of ATP is $(1270/2870) \times 100 = 44\%$. Compared to the efficiency of heat engines in converting heat to work (recall Section 15-8), the metabolic process is seen to be an efficient one. Nearly half the energy of glucose can be stored by the body for later use.

FIGURE 25-18
Some important chemical intermediates in metabolism.

Various R groups can be joined to the structure shown in black. The reduced form of NAD, called NADH, contains one additional H atom, shown in black. The reduced form of FAD, called FADH, contains two additional H atoms, also shown in black.

ENZYMES. An enzyme is a biological catalyst that contains protein. Some enzymes are made up only of protein; some need cofactors or coenzymes to function (e.g., FAD and NAD). Enzymes are specific for each biological transformation and catalyze a reaction without requiring a change in temperature or pH. Originally, enzymes were assigned common or trivial names, such as pepsin and catalase. Present practice, however, is to name them after the processes they catalyze, usually employing an "ase" ending.

In 1913, Michaelis and Menten proposed a model of enzyme reactivity based on the formation of an enzyme-substrate complex. According to this model, an enzyme can exert its catalytic activity only after combining with the reacting substance, the **substrate** to form a complex. There appears to be a definite site on the enzyme where the substrate combines; for some enzymes there is evidence that more than one active site exists. Reaction of the substrate (S) with the enzyme (E) to form a complex (ES) permits the reaction to proceed via a path of lower activation energy than the noncatalyzed path. When the complex decomposes, products (P) are formed and the enzyme is regenerated (see Figure 25-19). This general reaction scheme and a specific example are considered below:

$$E + S \rightleftharpoons ES \longrightarrow E + P$$

$$\text{sucrase} + \text{sucrose} \rightleftharpoons \text{sucrase–sucrose complex} \xrightarrow{\text{H}_2\text{O}}$$
$$\text{sucrase} + \text{glucose} + \text{fructose}$$

> This mechanism was first introduced in Section 13-9, where its implications for the rates of enzyme-catalyzed reactions were explored.

Generally, a 10°C temperature rise produces an approximate doubling of a reaction rate, but with enzymes a certain temperature is reached beyond which a decrease in rate sets in. The optimum temperature for enzyme activity is about 37°C (98°F) for the enzymes present in warm-blooded animals. The decrease in rate beyond this temperature results from the fact that enzymes are proteins and proteins are denatured by heat. This denaturation disrupts the secondary and tertiary structure and distorts the active site on the enzyme.

Protein behavior is extremely sensitive to changes in pH. At high and low pH values complete denaturation of enzymes occurs, but even milder changes in pH cause drastic changes in the rate of enzyme action. Most enzymes in the body have their maximum activity between pH 6 and 8, wth gastric enzymes being notable exceptions. In addition to the effects of temperature and pH, specific inhibition can occur when a molecule other than the substrate competes for an active enzyme site. Also heavy metal ions (Hg^{2+}, Pb^{2+}, and Ag^+) may combine in a nonreversible way with active site groups, such as $-OH$, $-SH$, $-CO_2^-$, and $-NH_3^+$, and deactivate the enzyme. Oxalic and citric acids inhibit blood clotting by competing for the cal-

FIGURE 25-19
The lock-and-key model of enzyme action.

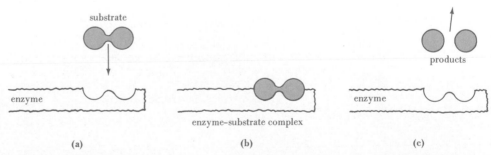

(a) The substrate attaches itself to an active site on the enzyme molecule. (b) Reaction occurs. (c) Product species detach themselves from the site, freeing the enzyme molecule to attach another substrate molecule. The substrate and enzyme must have complementary structures to produce a complex, hence the term lock-and-key.

cium ions necessary for the activation of the enzyme thromboplastin. It has also been suggested that antibiotics function by inhibiting enzyme-coenzyme reactions in microorganisms.

HORMONES. A hormone is a secretion of a ductless or endocrine gland, such as the thyroid, the pituitary, the pancreas (in part), the adrenals, and parts of the testes and ovaries. These glands secrete their products directly into the blood stream through which these products reach all parts of the body. The hormones appear to aid in the control of biological reactions, but their exact role is not clearly understood. Hormones are sometimes referred to as "chemical messengers."

A well-known protein hormone is insulin. Its function is to lower the blood sugar level by increasing the rate of conversion of glucose into muscle and liver glycogen. There is considerable evidence that insulin acts by controlling the phosphorylation of glucose. In the absence of a sufficient amount of insulin, the condition of diabetes mellitus results. Among its clinical symptoms are high levels of sugar in the blood and urine and the formation of excess ketones, giving a distinctive odor to the breath.

VITAMINS. Vitamins are substances necessary to maintain normal health, growth, and nutrition; however, they are not used in building cells or as an energy source. Their apparent function is as catalysts for biological processes. The vitamins are sometimes classified as fat soluble (vitamins A, D, E, and K) or water soluble (vitamins B and C). Vitamin A, although not found in plants itself, can be formed from β-carotene, a yellow pigment found in plants. A deficiency of vitamin A causes night blindness and xerophthalmia, a disease of the eyes in which the tear glands cease to function. Vitamin D is associated with the proper deposition of calcium phosphate, which in turn is related to normal teeth and bone development and the prevention of rickets.

Vitamin E is sometimes called the fertility factor and is involved in the proper functioning of the reproductive system. It is found in vegetable oils, such as corn germ oil, cottonseed oil, peanut oil, and wheat germ oil. There is more than one form of vitamin E and only the structure for α-tocopherol, the most active, is shown below.

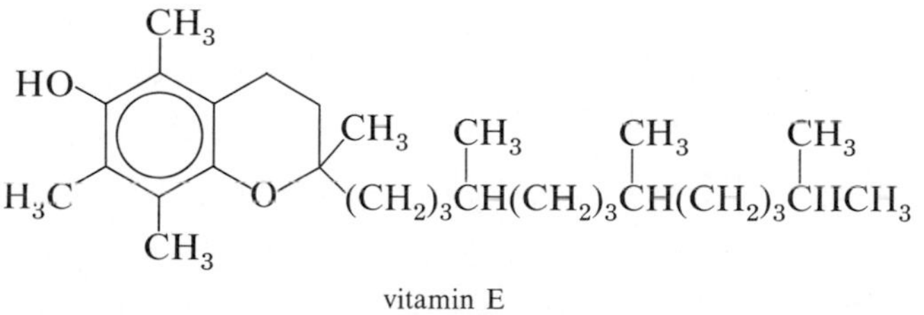

vitamin E
α-tocopherol

Vitamin K is the antihemorrhagic factor involved in bloodclotting. There are two K vitamins. Vitamin K_1 is obtained from alfalfa and vitamin K_2 by bacterial action in the intestine.

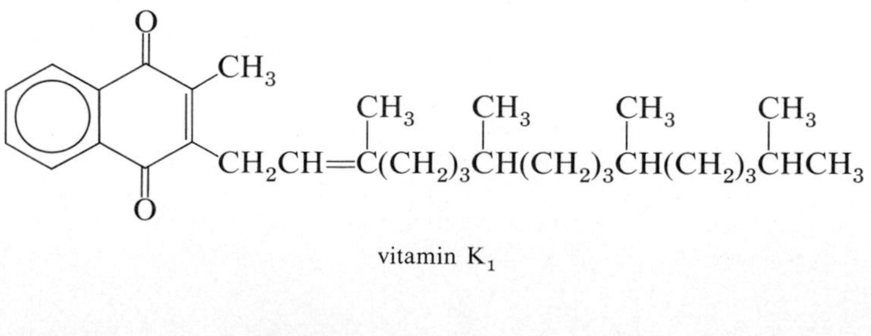

vitamin K_1

It has recently been claimed that large doses of vitamin C are effective in treating and preventing common colds, but there is conflicting evidence also. The question remains a controversial one.

Vitamin C, ascorbic acid, is the vitamin that prevents scurvy. It is found in citrus fruits, green peppers, parsley, and tomatoes. The vitamin B complex has been shown to consist of many substances, most of which seem to be involved in energy transformations related to metabolism. A deficiency of vitamin B_1, thiamine, leads to the disease called beriberi. Lack of vitamin B_2, riboflavin, causes inflammation of the lips, dermatitis, a dryness and burning of the eyes, and sensitivity to light. Both of these B vitamins are distributed widely in nature, in lean meat, nuts, and leafy vegetables.

FIGURE 25-20
Hydrolysis products of nucleic acids.

Tracing the hydrolysis reactions in the reverse direction, the combination of a pentose sugar and a purine or pyrimidine base yields a nucleoside. A nucleoside, in combination with phosphoric acid, yields a nucleotide. A nucleic acid is a polymer of nucleotides.

If the sugar is 2-deoxyribose and the bases A, G, T, and C, the nucleic acid is DNA. If the sugar is ribose and the bases A, G, U, and C, the nucleic acid is RNA. (The term 2-deoxy means without an oxygen atom on the second carbon atom.)

vitamin C, ascorbic acid

vitamin B_2, riboflavin

Vitamin B_5, niacin, is called the antipellagra factor. Pellagra is a disease characterized by dermatitis, a pigmentation and thickening of the skin, and soreness of the tongue and mouth. Niacin occurs widely distributed in nature. One excellent source is cereal grain, but unfortunately most of this is lost in the milling process. The absence of vitamin B_6 leads to dermatitis, anemia, and epileptic seizures.

vitamin B_5, niacin

vitamin B_6, pyridoxine

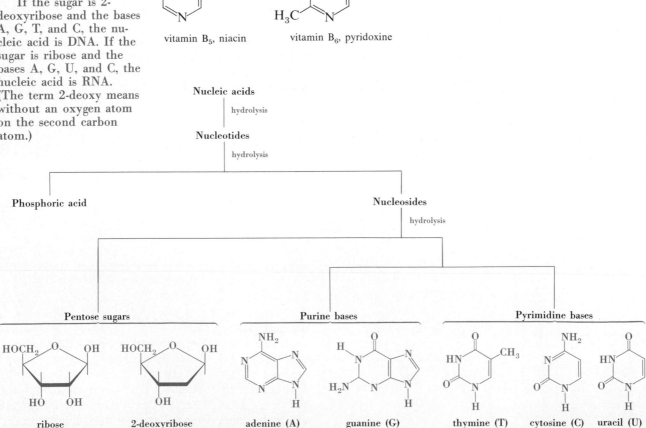

Nucleic acids

hydrolysis

Nucleotides

hydrolysis

Phosphoric acid   Nucleosides

hydrolysis

Pentose sugars   Purine bases   Pyrimidine bases

ribose   2-deoxyribose   adenine (A)   guanine (G)   thymine (T)   cytosine (C)   uracil (U)

25-4 The Nucleic Acids

As we noted earlier, lipids, carbohydrates, and proteins, taken together with water, constitute about 99% of most living organisms. The remaining 1% includes some compounds of vital importance to the existence, development, and reproduction of all forms of life. Among these are the nucleic acids. Nucleic acids carry the information that directs the metabolic activity of cells.

The nucleus of the cell contains **chromosomes** that cause replication from one generation to the next. The individual portions of the chromosomes that carry specific traits are known as genes. It has now been established that DNA or **deoxyribonucleic acid** is the actual substance constituting the genes. Figure 25-20 traces the steps that may be followed in degrading nucleic acids into their simpler constituents—heterocyclic amines known as purines and pyrimidines, a five-carbon sugar (ribose or 2-deoxyribose), and phosphoric acid. Figure 25-21 represents a portion of a nucleic acid chain.

The usual form for DNA is a **double helix.** The postulation of this structure of DNA by Crick and Watson in 1953 was one of the great scientific breakthroughs of modern times. Their work was critically dependent on two other contemporary scientific achievements. One was the precise x ray diffraction studies on DNA by Wilkins and Franklin. The other was the discovery by Chargaff (1950–1953) of a set of regularities regarding the occurrence of the purine and pyrimidine bases in nucleic acids. For DNA derived from a particular organism, these regularities, known as the **base-pairing rules,** require the following.

1. The amount of adenine is equal to the amount of thymine (A = T).
2. The amount of guanine is equal to the amount of cytosine (G = C).
3. The total amount of purine bases is equal to the total amount of pyrimidine bases (G + A = C + T).

Since large measures of intuition and inspiration were involved in Crick and Watson's work, it is not easy to show how their model was established from the facts just described. Instead, let us simply demonstrate how their model, shown in Figure 25-22, is consistent with the base-pairing rules.

In order to maintain the structure of a double helix, it is necessary that a force exist between the two single strands. As in the α-helical structure of a protein, the postulated force is based on hydrogen bonds, involving hydrogen, nitrogen, and oxygen atoms on the purine and pyrimidine bases. The necessary conditions for

FIGURE 25-21
A portion of a nucleic acid chain.

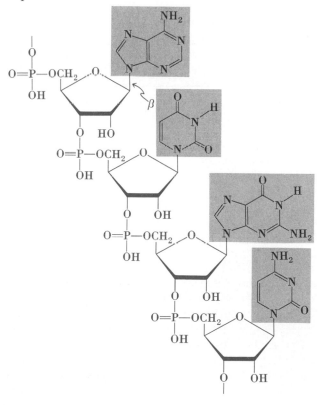

FIGURE 25-22
DNA model.

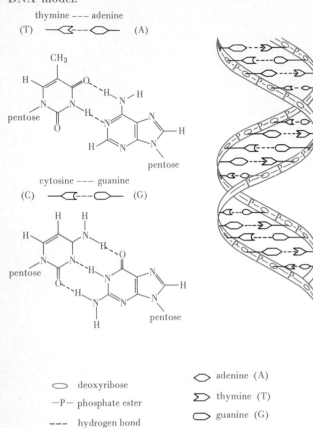

deoxyribose

—P— phosphate ester

--- hydrogen bond

adenine (A)

thymine (T)

guanine (G)

cytosine (C)

FIGURE 25-23
Replication of a DNA molecule visualized.

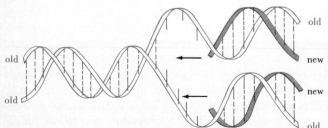

old

old

old

new

new

old

As the unzipping process occurs, hydrogen bonds between the old DNA strands are broken. The new strands grow in the direction of the arrows by attaching nucleotides which then form hydrogen bonds to the old DNA strands.

hydrogen bonding will exist only if an A on one strand appears opposite a T on the other, or if a G is bonded to a C. No other combinations will work. For example, C cannot be paired with T. Both are relatively small molecules (single ring) and would not approach each other closely enough between the strands. The combination of G and A cannot occur because the molecules are too large (double rings). With the combinations C + A and G + T, the conditions for hydrogen bonding are not right. Thus, it follows rather directly that the total amount of A must equal the total amount of T, and so on. The Chargoff rules are explained.

Proposal of the DNA structure was the important first step in the development of the theory of DNA, but at least three other questions must be explained by this theory: (1) How does a DNA molecule reproduce itself during cell division? (2) How does DNA direct the synthesis of proteins in the cell? (3) How is the information required to obtain the exact sequence of amino acids in a protein coded into the DNA structure? We cannot go into detail on these questions, but let us elaborate a bit.

The critical step in the replication of a DNA molecule requires the molecule to unwind into single strands. As the unwinding occurs, nucleotides present in the cell nucleus, through the action of enzymes, become attached to the exposed portions of the two single strands, converting each to a new doube helix of DNA. As suggested by Figure 25-23, when the original DNA (parent) molecule is completely unwound, two new molecules (daughters) appear in its place!

There are two pieces of evidence, each very convincing, that the process outlined here does indeed occur. First, electron micrographs of the DNA molecule have now been obtained, including some that capture DNA in the act of replication. Another elegant experiment involves growing bacteria in a medium containing ^{15}N atoms, so that all the N atoms of the bases of the DNA molecules are ^{15}N. The bacteria are then transferred to a nutrient with nucleotides containing normal ^{14}N. Here the bacteria are allowed to divide and reproduce themselves. The DNA of the offspring cells are then analyzed. Those of the first generation, for example, consist of DNA molecules with one strand having ^{15}N and the other ^{14}N atoms. This is exactly the result to be expected if replication occurs by the unzipping process described in Figure 25-23.

The puzzle of protein synthesis is that this occurs in the cytoplasm of the cell, not in the nucleus. Yet the molecule that directs the synthesis, DNA, is found only in the nucleus. Here is where the different forms of RNA play a fundamental role. The several steps involved in this process are represented, collectively, in Figure 25-24.

First, a molecule of **messenger RNA, mRNA,** is syn-

FIGURE 25-24
A representation of protein synthesis.

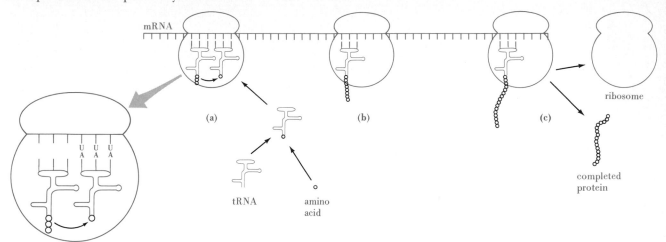

(a) Through the action of an enzyme, a tRNA molecule brings a single amino acid to a site on a ribosome. The anticodon of the tRNA (AAA for the example shown) must be complementary to the codon of the mRNA (UUU). (The amino acid carried by this tRNA is phenylalanine.) The amino acid is added to the chain in the manner shown; the chain moves from the tRNA on the left to the one on the right.
(b) As the ribosomes move along the mRNA strand, more and more amino acid units are added through the proper matching of tRNA molecules with the code on the mRNA.
(c) When the ribosome reaches the end of the mRNA strand, it and the completed protein are released. The ribosome is free to repeat the process.

Since there are only four different bases possible in an mRNA molecule—U, A, G, and C—but 20 different amino acids, it is clear that the code cannot correspond to individual base molecules. There are $4^2 = 16$ combinations of base molecules taken two at a time (i.e., UU, UA, UG, UC, etc.). But these could only account for 16 amino acids. When the base molecules are taken three at a time, there are $4^3 = 64$ possible combinations.

thesized on a portion of a DNA strand in the nucleus. This probably involves an unzipping of the DNA molecule similar to the process of DNA replication described above. The mRNA migrates out of the nucleus into the cytoplasm. mRNA has a high affinity for ribosomes and gathers them up along the chain. The combination of the mRNA and its ribosomes is called a **polysome. Transfer RNA, tRNA,** refers to a variety of rather short RNA chains, each of which is capable of attaching only a specific amino acid. The function of tRNA is to bring a specific amino acid to a site on the ribosome where the amino acid can form a polypeptide bond and become part of a growing polypeptide chain. The ribosome moves along the mRNA chain, attaching different tRNA molecules and incorporating their amino acids into the polypeptide chain. When the ribosomes reaches the end of the mRNA chain it falls off and releases the protein molecule that has been synthesized. The entire process is like the stringing of beads.

The code (set of directions) that determines the exact sequence of amino acids in the synthesis of a protein is incorporated in the chromosomal DNA. It is found in the particular pattern of base molecules on the double helix. A group of *three* base molecules in a DNA strand, called a **triplet,** causes a complementary set of base molecules to appear in the mRNA formed on it. This triplet or **codon** on the mRNA must be matched by a complementary triplet, called an **anticodon,** in a tRNA molecule. The particular tRNA with this anticodon carries a specific amino acid to the site of protein synthesis.

Through an ingenious set of experiments, which led to a Nobel prize in 1968 for Nirenberg, Holley, and Khorana, the genetic code has now been cracked. The significant features of the genetic code are the following.

TABLE 25-5
The genetic code

		Second base			
		U	C	A	G
First base	U	UUU ⎫ Phe UUC ⎭ UUA ⎫ Leu UUG ⎭	UCU ⎫ UCC ⎪ Ser UCA ⎪ UCG ⎭	UAU ⎫ Tyr UAC ⎭ UAA Nonsense UAG Nonsense	UGU ⎫ Cys UGC ⎭ UGA Nonsense UGG Trp
	C	CUU ⎫ CUC ⎪ Leu CUA ⎪ CUG ⎭	CCU ⎫ CCC ⎪ Pro CCA ⎪ CCG ⎭	CAU ⎫ His CAC ⎭ CAA ⎫ Gln CAG ⎭	CGU ⎫ CGC ⎪ Arg CGA ⎪ CGG ⎭
	A	AUU ⎫ AUC ⎬ Ile AUA ⎭ AUG Met	ACU ⎫ ACC ⎪ Thr ACA ⎪ ACG ⎭	AAU ⎫ Asn AAC ⎭ AAA ⎫ Lys AAG ⎭	AGU ⎫ Ser AGC ⎭ AGA ⎫ Arg AGG ⎭
	G	GUU ⎫ GUC ⎪ Val GUA ⎪ GUG ⎭	GCU ⎫ GCC ⎪ Ala GCA ⎪ GCG ⎭	GAU ⎫ Asp GAC ⎭ GAA ⎫ Glu GAG ⎭	GGU ⎫ GGC ⎪ Gly GGA ⎪ GGG ⎭

Initiation of a polypeptide chain appears to require the presence of the codon AUG, and termination of the chain, either UAA or UAG or UGA.

- There is more than one triplet code for most amino acids.
- The first two letters of the codon are most significant. There is considerable variation in the third.
- There are three codons that do not correspond to any amino acids. Although they are referred to as nonsense codons, these seem to play a role in stopping protein synthesis (rather like the word "STOP" used to separate phrases in a telegram).
- The various codons direct the *same* protein synthesis, whether in bacteria, plants, lower animals, or humans.

The genetic code is presented in Table 25-5.

Summary

Four categories of substances found in living organisms are considered in this chapter—lipids, carbohydrates, proteins, and nucleic acids.

One familiar group of lipids are the triglycerides. These are esters of glycerol with long-chain monocarboxylic (fatty) acids. If saturated fatty acids predominate, the triglyceride is a fat. If unsaturation (in the form of double bonds) occurs in some of the fatty acid components, the triglyceride is an oil. The catalytic addition of hydrogen to double bonds in a triglyceride converts an oil to a fat. Hydrolysis of a triglyceride with a strong base (saponification) yields glycerol and metal salts of the fatty acids—soaps. In phosphatides (phospholipids) phosphoric acid and a nitrogen-containing compound are substituted for one of the fatty acid components of a triglyceride.

The simplest carbohydrate molecules are five- and six-carbon-chain polyhydroxy aldehydes and ketones. However, these molecules convert readily into cyclic structures, that is, five- and six-membered rings. The subject is further complicated by the existence of asymmetric carbon atoms in these molecules, rendering the molecules optically active, that is, able to rotate the plane of polarized light. Moreover, optical isomers are encountered—molecules of identical composition which differ in their spatial orientations and in their optical activity. Monosaccharides or simple sugars can be readily joined into polysaccharides containing from a few to a few thousand monosaccharide units.

The basic building blocks of proteins are some 20 different α-amino acids. Two amino acid molecules may join by eliminating a H_2O molecule between them. The result is a peptide bond, —CONH—. Long polypeptide chains are formed as this process is repeated. Additional structural features include the twisting of a polypeptide chain into a helical coil and bonding between coils.

Lipids, carbohydrates, and proteins are complex molecules. Their metabolism by the body involves breaking them down into their simplest units. Carbohydrates are broken down into monosaccharides; pro-

teins, into amino acids; and lipids, into glycerol and two-carbon-chain acids. Ultimately, these products are decomposed into still simpler molecules, such as CO_2, H_2O, NH_3, and urea. Energy released in these processes is stored in the substance adenosine triphosphate (ATP).

DNA molecules found in cell nuclei are able to synthesize RNA molecules, which are the framework on which protein synthesis occurs. Proteins acquire the correct sequence of amino acids because each RNA molecule carries a code imparted to it in its synthesis by DNA. DNA molecules also have the ability to replicate themselves during cell division, which ensures that the correct protein-building code is passed on from one generation of cells to the next.

Learning Objectives

As a result of studying Chapter 25, you should be able to

1. List the four principal types of substances found in cells—lipids, carbohydrates, proteins, and nucleic acids—and describe the basic chemical composition of each.

2. Write structural formulas and names of triglycerides and indicate whether their constituent fatty acids are saturated or unsaturated.

3. Relate the structures, saponification values, and iodine numbers of triglycerides.

4. Explain how phosphatides and waxes, which are also lipids, differ from triglycerides.

5. Classify carbohydrates as monosaccharides, oligosaccharides, and polysaccharides; and give examples of each.

6. Describe the phenomenon of optical activity and the structural features of a molecule that produce it.

7. Outline the conversion of a straight chain monosaccharide into its cyclic form, and the joining of monosaccharide units into polysaccharides.

8. Show how the form adopted by an amino acid varies with solution pH.

9. Describe the formation of peptide bonds between amino acid molecules, and assign systematic names and structures to polypeptides.

10. Explain how the sequence of amino acids in a polypeptide chain is determined from information acquired in the hydrolysis of the polypeptide.

11. Explain the meaning of secondary, tertiary, and, where applicable, quaternary structure of a protein.

12. Describe metabolism and how the energy released in metabolism is stored.

13. Explain enzyme action in terms of the "lock-and-key" model.

14. Name the principal constituents of the nucleic acids and indicate how these constituents are linked together into chains and, in the case of DNA, a double helix.

15. Explain how DNA replicates itself during cell division; and outline the process of protein synthesis, indicating the role of DNA, mRNA, and tRNA.

Some New Terms

Absolute configuration refers to the spatial arrangement of the groups attached to an asymmetric carbon atom. The two possibilities are D and L.

An **α-amino acid** has a amino group (—NH_2) attached to the C atom adjacent to a carboxyl group (—COOH).

A **carbohydrate** is a polyhydroxy aldehyde, a polyhydroxy ketone, a derivative of these, or a substance that yields them upon hydrolysis.

Denaturation refers to the loss of biological activity of a

protein brought about by changes in its secondary and tertiary structure.

Deoxyribonucleic acid (DNA) is the substance that makes up the genes of the chromosomes in the nuclei of cells.

Dextrorotatory means the ability to rotate the plane of polarized light to the right, designated *d* or **+**.

Diastereomers are optically active isomers of a compound, but their structures are *not* mirror images (as are enantiomers).

Enantiomers are molecules whose structures are not super-imposable. The structures are mirror images, and the molecules are optically active.

An **enzyme** is a substance containing protein that catalyzes biological reactions.

Fats are triglycerides in which saturated fatty acid components predominate.

The **genetic code** describes the sequences of bases in DNA molecules which determine complementary sequences in mRNA and, ultimately, sequences of amino acids in proteins.

The **iodine number** of a triglyceride is the number of grams of I_2 reacting with 100 g of the triglyceride. Since reaction involves addition of iodine to double bonds, the iodine number indicates the degree of unsaturation in the fatty acid components.

The **isoelectric point** or **pI** of an amino acid or a protein is the pH at which the dipolar structure or "zwitterion" predominates.

Levorotatory means the ability to rotate the plane of polarized light to the left, designated *l* or −.

Lipids include a variety of substances sharing the property of solubility in solvents of low polarity [$CHCl_3$, CCl_4, C_6H_6, $(C_2H_5)_2O$].

Metabolism refers to the totality of chemical reactions occurring within an organism, reactions in which large molecules are broken down (catabolism) or small ones are synthesized into larger ones (anabolism).

A **monosaccharide** is a single, simple molecule having the structural features of a carbohydrate. It is called a **sugar.**

Nucleic acids are cell components comprised of purine and pyrimidine bases, pentose sugars, and phosphoric acid.

Oils are triglycerides in which unsaturated fatty acid components predominate.

Oligosaccharides are carbohydrates consisting of from two to ten simple monosaccharide units. They are also called **sugars.**

Optical activity refers to the ability of a substance to rotate the plane of polarized light.

A **peptide bond** is formed by the elimination of a water molecule from between two amino acids. The H atom comes from the —NH_2 group of one amino acid, and the —OH group from the —COOH group of the other acid.

A **polypeptide** is formed by the joining together of a large number of amino acid units through peptide bonds.

A **polysaccharide** is a carbohydrate consisting of more than ten monosaccharide units (e.g., starch and cellulose).

Primary structure refers to the sequence of amino acids in the polypeptide chains that make up a protein.

A **protein** is a large polypeptide, that is, having a molecular weight of 10,000 or more.

A **racemic modification** is a mixture of D and L isomers that does not rotate the plane of polarized light.

A **reducing sugar** is one that is able to reduce Cu^{2+}(aq) to red Cu_2O. The sugar must have available an aldehyde group (which is oxidized to an acid).

Ribonucleic acid (RNA), through its **messenger RNA (mRNA)** and **transfer RNA (tRNA)** forms, is involved in the synthesis of proteins in cells.

Saponification is the hydrolysis of a triglyceride by a strong base. The products of saponification are glycerol and a soap. The number of mg KOH required to saponify 1 g of the triglyceride is known as the **saponification value.**

Secondary structure of a protein describes the structure or shape of a polypeptide chain, for example, coiling into a helix.

A **soap** is a metal salt of a fatty acid (e.g., sodium stearate).

A **substrate** is a substance that undergoes a biological reaction through the action of an enzyme.

Tertiary structure of a protein describes the types of linkages between polypeptide chains that give a protein its three-dimensional structure.

Triglycerides are esters of glycerol (1,2,3-propanetriol) with long-chain monocarboxylic (fatty) acids.

Exercises

Structure and composition of the cell

Exercises 1 to 5 refer to a typical *E. coli* bacterium. This is a cylindrical cell about 2 μm long and 1 μm in diameter, weighing about 2×10^{-12} g and containing about 80% water, by volume. (See Figure 25-2.)

1. The intracellular pH is 6.4 and $[K^+] = 1.5 \times 10^{-4} M$. Determine the number of **(a)** H^+ ions and **(b)** K^+ ions in a typical cell.

2. The *E. coli* cell contains about 1.5×10^4 ribosomes. Assuming a ribosome to be a sphere with a diameter of 18 nm, what percentage of the cell volume do the ribosomes occupy?

3. Calculate the number of lipid molecules present, assuming their average molecular weight to be 700 and the lipid content to be 2%.

4. The cell is about 15% protein, by mass, with 90% of this protein in the cytoplasm. Assuming an average molecular weight of 3×10^4, how many protein molecules are present in the cytoplasm?

5. A single chromosomal DNA molecule contains about 4.5 million mononucleotide units. If this molecule were extended so that the mononucleotide units were 450 pm apart, what would be the length of the molecule? How does this compare with the length of the cell itself? What does this result suggest about the shape of the DNA molecule?

Lipids

6. Describe briefly what is meant by each of the following terms, using specific examples where appropriate: **(a)** lipid; **(b)** triglyceride; **(c)** simple glyceride; **(d)** mixed glyceride; **(e)** fatty acid; **(f)** soap.

7. Explain the essential distinction between the following pairs of materials: **(a)** a fat and a lipid; **(b)** a fat and an oil; **(c)** a fat and a wax; **(d)** butter and margarine.

8. Explain why phospholipids are more water soluble than simple or mixed glycerides.

9. Write structural formulas for the following.
   (a) glyceryl lauromyristolinoleate
   (b) trilaurin
   (c) potassium palmitate
   (d) cetyl linoleate [cetyl alcohol = $CH_3(CH_2)_{14}CH_2OH$]

10. Name the following compounds.

(a)
$$H_2CO-\overset{\overset{O}{\|}}{C}-C_{17}H_{35}$$
$$HCO-\overset{\overset{O}{\|}}{C}-C_{17}H_{33}$$
$$H_2CO-\overset{\overset{O}{\|}}{C}-C_{11}H_{23}$$

(b)
$$H_2CO-\overset{\overset{O}{\|}}{C}-C_{17}H_{31}$$
$$HCO-\overset{\overset{O}{\|}}{C}-C_{17}H_{31}$$
$$H_2CO-\overset{\overset{O}{\|}}{C}-C_{17}H_{31}$$

(c) $C_{13}H_{27}CO_2^- Na^+$

11. What is the saponification value of glyceryl lauropalmitostearate?

12. What simple triglyceride has a saponification value of 193 and an iodine number of 174 (see Table 25-2).

13. Oleic acid is a moderately unsaturated fatty acid. Linoleic acid belongs to a group called **polyunsaturated.** What structural feature characterizes polyunsaturated fatty acids? Is stearic acid polyunsaturated? Is eleostearic acid? Why do you suppose safflower oil is so highly recommended in dietary programs?

14. In light of present medical knowledge, which is a more desirable lipid for human consumption, one with a high saponification value or a high iodine number? Which is the "best" of those listed in Table 25-3 from this standpoint?

Carbohydrates

15. Describe what is meant by each of the following terms, using specific examples where appropriate: **(a)** monosaccharide; **(b)** disaccharide; **(c)** oligosaccharide; **(d)** polysaccharide; **(e)** sugar; **(f)** glycose; **(g)** aldose; **(h)** ketose; **(i)** pentose; **(j)** hexose.

16. The following terms are all related to stereoisomers and their optical activity. Explain the meaning of each:

(a) dextrorotatory; **(b)** levorotatory; **(c)** racemic mixture; **(d)** diastereomers; **(e)** (+); **(f)** (−); **(g)** D configuration; **(h)** *l*; **(i)** *d*.

17. Write the structure for the straight chain form of L-glucose. Is this isomer dextrorotatory or levorotatory?

18. The structure of L-(+)-arabinose is given below. From this structure derive the structure of **(a)** D-(−)-arabinose; **(b)** a diastereomer of L-(+)-arabinose.

$$
\begin{array}{c}
H-C=O \\
H-C-OH \\
HO-C-H \\
HO-C-H \\
CH_2OH
\end{array}
$$

L-(+)-arabinose

19. The pure α and β forms of D-glucose rotate the plane of polarized light to the right by 112° and 18.7°, respectively (denoted as +112° and +18.7°). Are these two forms of glucose enantiomers or diastereomers? (Consider also the structures shown in Figure 25-6.)

20. When a mixture of the pure α and β forms of glucose is allowed to reach equilibrium in solution, the rotation changes to +52.7° (a phenomenon known as mutarotation). What are the percentages of the α and β forms in the equilibrium mixture? [*Hint:* Refer to Exercise 19. If the mixture were 50:50, the rotation would be 0.50(+112) + 0.50(+18.7) = +65.4°.]

21. Why is fruit sugar (D-fructose) a reducing sugar while sucrose is not? (*Hint:* See Figure 25-7.)

Amino acids, polypeptides, and proteins

22. Describe what is meant by each of the following terms, using specific examples where appropriate: **(a)** α-amino acid; **(b)** zwitterion; **(c)** isoelectric point; **(d)** peptide bond; **(e)** polypeptide; **(f)** protein; **(g)** N-terminal amino acid; **(h)** α helix; **(i)** denaturation.

23. Write the formulas of the species expected if the amino acid phenylalanine is maintained in **(a)** 1.0 M HCl; **(b)** 1.0 M NaOH; **(c)** a buffer solution with pH 5.5.

24. A mixture of the amino acids lysine, proline, and glutamic acid is subjected to electrophoresis at a pH of 6.3. In what direction will each amino acid migrate?

25. Write the structures of: **(a)** glycylmethionine; **(b)** isoleucylleucylserine; **(c)** the different tripeptides that can be obtained from a combination of alanine, serine, and lysine; **(d)** the tetrapeptides containing two serine and two alanine amino acid units.

26. For the polypeptide Gly-Ala-Ser-Thr; **(a)** write the structural formula; **(b)** name the polypeptide. (*Hint:* Which is the N-terminal and which is the C-terminal amino acid?)

27. Upon complete hydrolysis a polypeptide yields the following amino acids: Gly, Leu, Ala, Val, Ser, Thr. Partial hydrolysis yields the following fragments: Ser-Gly-Val, Thr-Val, Ala-Ser, Leu-Thr-Val, Gly-Val-Thr. An experiment using a marker establishes that Ala is the N-terminal amino acid.

(a) Establish the amino acid sequence in this polypeptide.

(b) What is the name of the polypeptide?

28. Describe what is meant by the primary, secondary, and tertiary structure of a protein. What is the quaternary structure? Do all proteins have a quaternary structure? Explain.

29. A 1.00-ml solution containing 1.0 mg of an enzyme was deactivated by the addition of 0.346 μmol $AgNO_3$, (1 μmol $= 1 \times 10^{-6}$ mol.) What is the *minimum* molecular weight of the enzyme? Why does this calculation yield only a minimum value? (*Hint:* How many active sites are present in each enzyme molecule?)

30. Sickle cell anemia is sometimes referred to as a "molecular" disease. Comment on the appropriateness of this term. (*Hint:* See marginal note on page 641.)

Biochemial reactivity

31. Describe briefly the meaning of each of the following terms as they apply to metabolism: (a) metabolite; (b) anabolism; (c) catabolism; (d) endergonic; (e) ADP; (f) ATP.

32. The metabolism of a particular metabolite has a theoretical free energy change of -837 kJ/mol. The metabolism of 1 mol of this material in a living organism results in the conversion of 15 mol ADP to ATP. What is the percent efficiency of this metabolism?

33. Calculate the equilibrium constant for the hydrolysis of glucose 6-phosphate to glucose and phosphoric acid if $\Delta \bar{G}^\circ = -13.8$ kJ/mol.

34. Explain why the action of an enzyme is so dependent on pH.

Nucleic acids

35. What are the two major types of nucleic acids? List their principal components.

36. With reference to Figure 25-21, identify the purine bases, the pyrimidine bases, the pentose sugars, and the phosphate groups. Is this a chain of DNA or RNA? Explain.

37. DNA has been called the "thread of life." Comment on the appropriateness of this expression.

38. What are the principal functions of each of the following in protein synthesis? (a) DNA; (b) mRNA; (c) tRNA.

39. A ribosome is sometimes said to *read* an mRNA strand. Suggest a meaning for this expression?

40. What polypeptide would be synthesized by the following coding on a mRNA strand? ACCCAUCCCUU-GGCGAGUGGUAUGUAA

41. Propose a plausible polypeptide sequence on a DNA strand that would code for the synthesis of the polypeptide Ser-Gly-Val-Ala. Why is there more than one possible sequence for the DNA strand?

*42. If one strand of a DNA molecule has the sequence of bases AGC, what must be the sequence on the opposite strand? Draw a structure of this portion of the double helix, showing all hydrogen bonds.

Additional Exercises

1. Write structural formulas for the following.

(a) hydrogenation product of triolein;

(b) saponification products of trilaurin;

(c) iodination product of glyceryl lauromyristolinoleate.

2. Castor oil is a mixture of triglycerides having about 90% of its fatty acid content as the unsaturated hydroxy aliphatic acid, ricinoleic acid

$$CH_3(CH_2)_5CHOHCH_2CH{=}CH(CH_2)_7COOH.$$

Estimate the saponification value and iodine number of castor oil. (*Hint:* What triglyceride should you assume?)

3. There are eight aldopentoses. Draw their structures and indicate which are enantiomers.

4. The term **epimer** is used to describe diastereomers that differ in the configuration about a *single* carbon atom. Which pairs of the three naturally occurring aldohexoses shown on page 635 are epimers (i.e., are D-galactose and D-mannose epimers, etc.)?

5. The protein molecule hemoglobin contains four atoms of iron (recall Figure 25-16). The mass percent of iron in hemoglobin is 0.34%. What is the molecular weight of hemoglobin?

*6. Draw complete structures for each of the chemical intermediates shown in Figure 25-18, that is, ADP, ATP, NAD, NADH, FAD, and FADH.

*7. What is a nucleoside? Draw structures of the following nucleosides using information from Figure 25-20: (a) cytidine; (b) uridine; (c) guanosine; (d) deoxycytidine; (e) deoxyadenosine.

*8. In the experiment described on page 652, the first generation offspring of DNA molecules each contained one strand with ^{15}N atoms and one with ^{14}N. If the experiment were carried through a second, third, and fourth generation, what fractions of the DNA molecules would still have strands with ^{15}N atoms?

*9. Bradykinin is a nonapeptide that is obtained by the partial hydrolysis of blood serum protein. It causes a lowering of the blood pressure and an increase in capillary permeability. Complete hydrolysis of bradykinin yields three proline (Pro), two arginine (Arg), two phenylalanine (Phe), one glycine (Gly), and one serine (Ser) amino acid units. The N-terminal and C-terminal units are both arginine (Arg). In a hypothetical experiment partial hydrolysis and sequence proof reveals the following fragments: Gly-Phe-Ser-Pro; Pro-Phe-Arg; Ser-Pro-Phe; Pro-Pro-Gly; Pro-Gly-Phe; Arg-Pro-Pro; Phe-Arg. Deduce the sequence of amino acid units in bradykinin.

*10. A pentapeptide was isolated from a cell extract and purified. A portion of the compound was treated with 2,4-dinitrofluorobenzene (DNFB) and the resulting material hydrolyzed. Analysis of the hydrolysis products revealed 1 mol of DNP-methionine, 2 mol of methionine, and 1 mol each of serine and glycine. A second portion of the original compound was partially hydrolyzed and separated into four products. Separately, the four products were hydrolyzed further, giving the following four sets of compounds: **(a)** 1 mol of DNP-methionine, 1 mol of methionine, and 1 mol of glycine; **(b)** 1 mol of DNP-methionine and 1 mol of methionine; **(c)** 1 mol of DNP-serine and 1 mol of methionine; **(d)** 1 mol of DNP-methionine, 1 mol of methionine, and 1 mol of serine. What is the structural formula of the pentapeptide?

Self-Test Questions

For 1 through 6 select the single item that best completes each statement.

1. The substance *glyceryl trilinoleate* (linoleic acid: $C_{17}H_{31}COOH$) is best described as a **(a)** fat; **(b)** oil; **(c)** wax; **(d)** fatty acid.

2. One can most easily distinguish between *glyceryl tristearate* (stearic acid: $C_{17}H_{35}COOH$) and *glyceryl trioleate* (oleic acid: $C_{17}H_{33}COOH$) by measuring their **(a)** molecular weights; **(b)** saponification values; **(c)** iodine numbers; **(d)** hydrolysis constants.

3. The mixture of sugars referred to as DL (or *dl*)-erythrose rotates the plane of polarized light **(a)** to the left; **(b)** to the right; **(c)** first to the left and then to the right; **(d)** neither to the left nor to the right.

4. The coagulation of egg whites by boiling is an example of **(a)** saponification; **(b)** inversion of a sugar; **(c)** hydrolysis of a protein; **(d)** denaturation of a protein.

5. A molecule in which the energy of metabolism is stored is **(a)** ATP; **(b)** glucose; **(c)** CO_2; **(d)** glycerol.

6. Of the following, the one that is not a constituent of a nucleic acid chain is **(a)** purine base; **(b)** phosphate group; **(c)** glycerol; **(d)** pentose sugar.

7. Calculate the maximum mass of a sodium soap that could be prepared from 125 g of the triglyceride *glyceryl tripalmitate* (palmitic acid: $C_{15}H_{31}COOH$).

8. Upon complete hydrolysis of a pentapeptide the following amino acids are obtained: valine (Val), phenylalanine (Phe), glycine (Gly), cysteine (Cys), and tyrosine (Tyr). Partial hydrolysis yields the following fragments: Val-Phe, Cys-Gly, Cys-Val-Phe, Tyr-Phe. Glycine is found to be the N-terminal acid. What is the sequence of the amino acids in the polypeptide?

9. Explain why enzyme action is so dependent on factors such as temperature, pH, and the presence of metal ions. Why are enzymes so specific in the reactions they catalyze? (That is, why don't they catalyze a variety of reactions, as does platinum metal, for example?)

10. Explain what is meant by the primary, secondary, and tertiary structure of a protein.

26 Natural Resources, the Environment, and Synthetic Materials

Very few of the common materials of modern life occur naturally. Instead, they are obtained by transforming a small number of naturally occurring raw materials into desired forms. Most of these transformations are chemical. Until recent times the natural resources on earth have been treated as if they were available in unlimited quantities. But now, even though estimates of potential reserves vary widely, we know that limits do exist. In this chapter we describe the basic raw materials for creating the modern world, explore the prospect of continued use of these materials, and consider substitute materials.

Among the several categories of chemical substances we must consider are metals, inorganic chemicals, organic chemicals, and polymers (plastics and fibers). In discussing the production of these materials we will look both at traditional methods and at some new methods likely to supplant them. In the past the basic question of chemical technology has been: How can the chemical properties of materials be altered most cheaply? A question of equal importance for the future must be: How can natural resources be altered with a minimum of damage to the environment? A few of the many environmental problems resulting from human activities are also described in this chapter.

26-1 Occurrence of the Elements

Of the 106 known elements, about 90 are obtained from natural sources. The remainder can only be produced in the laboratory by nuclear reactions. Moreover, most of the elements that occur naturally are not found *free,* that is, in the pure elemental state. About 20 of the elements can be found free in nature, but the remainder occur exclusively in chemical combination.

Table 26-1 lists some of the more common elements and their abundances in the earth's crust. The earth's crust includes the atmosphere and terrestial waters along with the solid crust (about 30 miles deep). The abundances of the elements in the earth's crust are considerably different from those of the entire earth. They are even more different from abundances in the solar system or the entire universe. For example, the core of the earth, with an approximate radius of 3400 km, is believed to consist largely of iron and nickel. Also, although not a major constituent on earth, hydrogen is estimated to account for 90% of all the atoms and three fourths of the mass of the universe, with helium accounting for most of the rest.

The data in Table 26-1 suggest which elements are readily available for human use, but these figures can be misleading. Aluminum, which is the most abundant of the metals, cannot be produced as cheaply as iron. Economic feasibility requires not only that an element be present in the earth's crust but that it be in a concentrated

TABLE 26-1
Abundance of the elements in the earth's crust[a]

Element	Abundance, mass %	Some commonly occurring forms
oxygen	49.3	water; silica; silicates; metallic oxides; molecular oxygen (in earth's atmosphere)
silicon	25.8	silica (sand, quartz, agate, flint); silicates (feldspar, clay, mica)
aluminum	7.6	silicates (clay, feldspar, mica); oxide (bauxite)
iron	4.7	oxide (hematite, magnetite)
calcium	3.4	carbonate (limestone, marble, chalk); sulfate (gypsum); fluoride (fluorite); silicates (feldspar, zeolites)
sodium	2.7	chloride (rock salt, ocean waters); silicates (feldspar, zeolite)
potassium	2.4	chloride; silicates (feldspar, mica)
magnesium	1.9	carbonate; chloride; sulfate (Epsom salts)
hydrogen	0.7	oxide (water); organic matter
titanium	0.4	oxide
chlorine	0.2	common salt (NaCl); sylvite (KCl); carnallite (KCl $\cdot$ MgCl$_2$ $\cdot$ 6H$_2$O)
phosphorus	0.1	phosphate rock [Ca$_3$(PO$_4$)$_2$]; organic matter
all others[b]	0.8	

[a] The earth's crust is taken to consist of the solid crust, terrestrial waters, and the atmosphere.
[b] This figure includes C, N, and S—all essential to life; less abundant, although common metals, such as Cu, Sn, Pb, Zn, and Cr; and some nonmetals with a variety of important uses, for example, F, Br, I and B.

form from which extraction is possible by simple physical or chemical means. These concentrated deposits are called **ores.** Aluminum is more widely distributed than iron, but iron is found more abundantly in concentration in its ores than is aluminum. Of course, as methods are found to process low-grade aluminum ores (clay minerals), and if abundant energy sources can be developed, aluminum may some day be used extensively as a structural metal.

Some elements are found in only minor quantities in the earth's crust, yet they are widely known and used. Such is the case of copper (abundance, approximately 0.005%). On the other hand, some elements that are relatively abundant are simply not accessible because they occur widely dispersed in common minerals, having no characteristic ores of their own. This is the case with rubidium which, although thought to be the sixteenth most abundant element, is very expensive either in pure or compound form.

A more realistic estimate of the availability of a resource than its simple terrestial abundance is the ratio of *potential* resources to *known* reserves. The larger this ratio, the greater the prospect of the continued availability of the element. On a worldwide basis, these ratios range from as little as 10 for Cu, to 23 for Zn, 30 for Hg, 34 for P, 42 for Zn and W, 112 for U, 600 for F, 2000 for Ti, and 3000 for Al.

26-2 The Atmosphere as a Natural Resource

One of the most accessible sources of certain chemical elements is the atmosphere that envelops the earth. Although the number of different elements found in abundance in the atmosphere is limited, there are more substances present in ordinary air than is generally realized. Among the free elements in the atmosphere are nitrogen, oxygen, and the noble gases.

FIGURE 26-1
The separation of atmospheric gases.

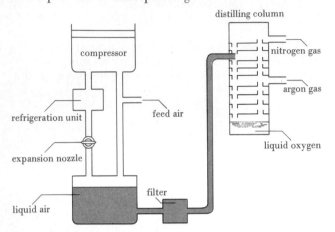

Clean air is fed into a compressor and then cooled by refrigeration. The cold air expands through a nozzle, and as a result is cooled still further—sufficiently to cause it to liquefy. The liquid air is filtered to remove $CO_2(s)$ and hydrocarbons and then distilled. Liquid air enters the top of the column where nitrogen, the most volatile component, passes off as a gas. In the middle of the column gaseous argon is removed, and liquid oxygen, the least volatile component, collects at the bottom. The normal boiling points of nitrogen, argon, and oxygen are -195.8, -185.7, and $-183.0°C$, respectively.

> The units used in this chapter are mainly those of the American chemical industry, where SI (metric) units have not been adopted to the extent that they have elsewhere in the world. One (short) ton = 2000 lb; one metric ton = 1000 kg = 2200 lb.

> Why are nitrogen compounds not present to a greater extent in the earth's solid crust? (Recall the solubility rules of Chapter 16.)

Although they form numerous chemical compounds, including compounds with one another, nitrogen and oxygen occur largely uncombined in the earth's atmosphere. This lack of reactivity can be attributed to the great strength of the bond between nitrogen atoms in N_2; 946.4 kJ of energy is required to rupture 1 mol of these bonds.

$$N{\equiv}N(g) \longrightarrow 2\,N(g) \qquad \Delta \bar{H}° = +946.4 \text{ kJ/mol}$$

Because the $N{\equiv}N$ bond is so stable, many nitrogen containing molecules have positive enthalpies and free energies of formation. The thermodynamics of their formation is unfavorable, and in many cases so too is the kinetics. For example, in the reaction of nitrogen and oxygen to form nitric oxide,

$$\tfrac{1}{2}N_2(g) + \tfrac{1}{2}O_2(g) \longrightarrow NO(g)$$
$$\Delta \bar{G}° = +88.69 \text{ kJ/mol}$$

The activation energy for this reaction is also high, in excess of 200 kJ.

The separation of oxygen, nitrogen, and the noble gases can be achieved by fractional distillation. This process, described in Figure 26-1, is a purely physical one—no chemical reactions are involved.

OXYGEN.   Current annual production of oxygen gas in the United States exceeds 17,000,000 tons. About 70% of this is consumed in the manufacture of steel. Mixtures of oxygen gas and acetylene or hydrogen are used to generate high temperatures for metal welding. Liquid oxygen ("lox") is used as an oxidizer in rocket systems. Anticipated future uses of oxygen include direct injection into streams and sewage systems as a means of oxidizing wastes and as a reactant in coal gasification processes.

NITROGEN.   Current consumption of nitrogen exceeds 15,000,000 tons annually in the United States. One important use of elemental nitrogen is to provide inert (blanketing) atmospheres in the metals, electronics, and chemical process industries. Also, liquid nitrogen is used as a freezing agent in the food processing industry. Other important uses include the manufacture of nitrogen-containing compounds. With the exception of large deposits of sodium nitrate found in Chile and Peru (Chile saltpeter), nitrogen compounds do not occur in significant quantities in the earth's crust. Chemically combined nitrogen is called "fixed" nitrogen, and any process that converts nitrogen into its compounds is referred to as the "fixation" of nitrogen.

Nitrogen is one of the elements essential to living matter. Since most plants and all animals can only use fixed nitrogen, natural nitrogen fixation processes are of extreme importance. Normally, nitrogen consumed

FIGURE 26-2
The nitrogen cycle in nature.

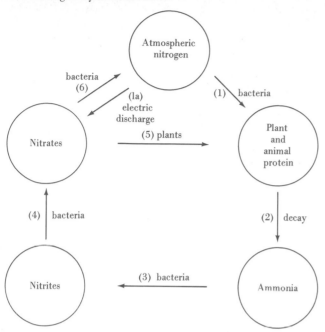

(1) Certain bacteria that reside as parasites in the root nodules of leguminous plants (beans, peas, clover, and alfalfa) can fix atmospheric nitrogen directly for conversion into plant proteins. Animal requirements are met by feeding on these and other plants. (2) The decay of plant and animal protein leads to the formation of ammonia. (3) and (4) Successive bacterial actions convert ammonia into nitrates. (5) The natural forms of fixed nitrogen that most plants require are the nitrates. This consumption again leads to plant and animal protein. (6) Certain denitrifying bacteria are capable of decomposing nitrates into elemental nitrogen, which returns to the atmosphere. (1a) As a result of electric discharges in rainstorms (lightning), a series of chemical reactions results in the direct production of nitrates as nitric acid, HNO_3.

Helium occurs naturally as a result of radioactive decay. Recall that all the natural radioactive decay schemes produce helium. For example, the complete decay of one $^{238}_{92}U$ atom results in one atom of $^{206}_{82}Pb$ and eight atoms of $^{4}_{2}He$.

by plants and animals is returned to the environment. As a result a natural "nitrogen cycle" exists, with nitrogen passing from one form to another, as pictured in Figure 26-2.

The delicate balance of the nitrogen cycle can be easily upset by human activities. When land is cultivated extensively, fixed nitrogen is removed from the soil at a greater rate than it can be replenished naturally. Nitrogen compounds must be added to the soil as fertilizers. Also, the need for nitrogen compounds for explosives, plastics and fibers, and industrial chemicals requires the artificial fixation of nitrogen. The principal industrial method of nitrogen fixation is the Haber process, considered in Section 14-8.

It has been estimated that the introduction of fixed nitrogen into the nitrogen cycle through human activities now equals or exceeds that from natural sources. Furthermore, some estimates place the quantity of nitrogen being fixed annually at 92 million metric tons. The rate of denitrification (return of nitrogen to the atmosphere) is estimated at 83 million metric tons. The difference, 9 million metric tons annually, is accumulating as nitrogen compounds in soil, groundwater, surface water, and the oceans.*

HELIUM. Although the other noble gases are extracted from the atmosphere, helium is currently being obtained from helium-bearing natural gas. A large quantity has been stored in underground reservoirs in the United States, but that storage program has now been discontinued. Since only a few natural gas sources contain sufficient quantities of helium (0.3% or more) to make its extraction feasible, a shortage of helium could result as these natural gas supplies are depleted. As shortages of materials occur in the future only two options will exist—find substitute materials or develop methods for the recovery and recycling of such materials. In the case of helium only the latter option can be considered; there can be no substitute.

Helium's properties are unique. Its most versatile uses are based on its inertness, its low molecular weight, its ability to remain as a liquid to temperatures approaching the absolute zero (liquid helium boils at 4 K), and certain unusual properties possessed by liquid helium and no other liquid.

Possible future requirements for large quantities of helium are based on its use as a refrigerant. Metals become superconducting at liquid helium temperatures. The underground transmission of electric power through metal cables suspended in liquid helium could result in less loss of power than conventional transmis-

*Delwiche, C. C., The nitrogen cycle. In *The Biosphere*, W. H. Freeman, San Francisco, 1970.

sion systems. The large electromagnets that would be required in a nuclear fusion reactor also consume large quantities of liquid helium, again to render metals superconducting (recall Section 23-9).

26-3 Atmospheric Pollution

Increasingly, human activities are leading to pollution of the atmosphere. The threat of air pollution is not to the atmosphere as a source of raw materials but directly to living organisms—plants, animals, and human beings. The atmosphere naturally contains a number of components which, in high concentrations, are known to be harmful to living organisms and other materials. These substances occur in the atmosphere as a consequence of natural events. It is only when their concentrations in local regions exceed their "normal" values that we can appropriately call them **pollutants.**

It has been hypothesized that a photochemical reaction, involving terpenes and other natural products released by trees, produces an aerosol (a colloidal dispersion of submicroscopic particles in air) that is responsible for the characteristic blue haze from which the famous Blue Ridge Mountains of Virginia derive their name.

For example, it has been estimated that human activities produce only 15% of the total global emission of hydrocarbons. The remaining sources are natural ones: Trees and other plants emit terpene hydrocarbons, and some hydrocarbons, chiefly methane, are produced by the bacterial decay of organic matter. The importance of the 15% contributed by human beings is that (1) it tends to be concentrated in selected regions on earth, principally in urban areas, and (2) it includes species associated specifically with gasoline and its combustion.

Table 26-2 lists some air pollutants resulting from human activities. Figure 26-3 depicts the environmental balance of sulfur, including both natural and artificial sources. Similar environmental cycles can be constructed for other elements. (The nitrogen cycle shown in Figure 26-2 deals only with natural sources.)

CARBON DIOXIDE. $CO_2(g)$ is not normally considered an air pollutant because it is

TABLE 26-2
Some air pollutants resulting from human activities

Pollutant	Principal sources
carbon monoxide	incomplete combustion of fossil fuels. About 80% is believed to originate from the internal combustion engine.
carbon dioxide	combustion of fossil fuels.
hydrocarbons	emission of unburned or partially burned gasoline and evaporation of industrial solvents. The internal combustion engine accounts for about 60% of the total.
lead	for example, lead compounds exhausted by automobiles using gasoline containing tetraethyl lead and other antiknock additives. There are indications that the lead concentration from human activities may exceed natural levels by 1000-fold or more.
nitrogen oxides	combustion processes in which $N_2(g)$ and $O_2(g)$ from the air combine to form $NO(g)$. Further oxidation of $NO(g)$ to $NO_2(g)$ occurs in the presence of $O_2(g)$ or $O_3(g)$.
particles	various agricultural, mechanical, and industrial processes produce dust, ash, and sulfuric acid mist. Ammonium sulfate particles result from the reaction of $NH_3(g)$ with this sulfuric acid mist. Artificially produced particles probably account for only about 10% of the total atmospheric burden.
sulfur oxides	some originates as H_2S in industrial operations. Most is produced as $SO_2(g)$, with further oxidation to $SO_3(g)$ and sulfates occurring in the atmosphere. About 50% comes from coal-burning power plants.

FIGURE 26-3
The sulfur cycle in the environment.

The figures are in millions of tons per year and were calculated on a worldwide basis. Sulfur circulates between land, sea, and air with a net transfer of about 95 million tons of sulfur to the oceans each year. [Source: E. Robinson and R. C. Robbins, Where does it all go?, *Stanford Res. Inst. J.*, No. 23, 8 (1968).]

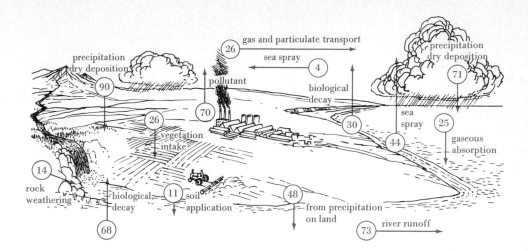

nontoxic. However, its potential effects on the environment are disturbing. The problem lies in the fact that $CO_2(g)$ absorbs infrared radiation from the earth and reradiates it back to the earth. This "greenhouse effect" could have the effect of raising the temperature of the earth (see Figure 26-4). With an increase in average temperature of the earth could come changes in climate, agricultural productivity, and, possibly even, melting of the polar ice caps.

The concentration of $CO_2(g)$ in the atmosphere rose about 6% during the first half of the century (from 296 to 315 ppm). From 1958 to 1979 the concentration increased another 20 ppm, to a current figure in excess of 330 ppm. These increases are attributed to the combustion of fossil fuels and slash-and-burn agricultural methods. Other factors, such as increased low cloud coverage and particulate concentration in the atmosphere might offset some of the effects of a buildup of $CO_2(g)$ concentration. Yet best estimates now are that the earth's average temperature might increase by as much as 6°C over the next 200 years.

It is for some of the reasons just cited that some scientists are greatly concerned about the "synfuel" (synthetic fuel) approach to the energy crisis. Such an approach implies large increases in the combustion of fuels. However, scientific opinion on the significance of increased $CO_2(g)$ production varies, primarily because the oceans serve as vast reservoirs for dissolved $CO_2(g)$, and in ways that are not completely understood.

Additional Exercise 12 in Chapter 1 presents data with which to calculate the increase in mean sea level if the polar ice caps were to melt completely. What would happen to coastal cities around the world in this event?

FIGURE 26-4
The "Greenhouse Effect."

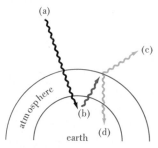

(a) Sunlight received by earth. **(b)** Infrared radiation (heat) from surface of earth absorbed by $CO_2(g)$ in atmosphere. **(c)** Infrared radiation emitted to space by $CO_2(g)$. **(d)** Infrared radiation emitted back to earth by $CO_2(g)$.

SMOG.   The term "smog" is used rather generally to describe any atmospheric condition that reduces visibility and causes a variety of minor and major irritations to most people. Although revealed most dramatically in southern California, air pollution is a worldwide phenomenon.

Local smog conditions, of course, depend on the rate at which pollutants are produced. Of equal or greater importance is the rate of their dispersal. Dispersal by winds carries pollutants in a horizontal direction along the earth's surface. In the absence of winds, mixing must occur vertically. Air at the surface of the earth is heated, expands, and rises. Cooler air aloft descends, itself becomes heated, and rises once again. Thus, a constant circulation occurs through a large air mass, and pollutants produced at ground level are dispersed.

The dispersal of pollutants is seriously hampered by atmospheric phenomena known as **temperature inversions.** Air temperature normally decreases at a regular rate with increased altitude above the surface of the earth; the normal temperature–altitude profile is shown in Figure 26-5a. There are times, however, when a different situation may prevail. For example, a mass of cool air from the ocean may move

FIGURE 26-5
Air temperature as a function of altitude.

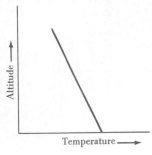

(a) Normal condition

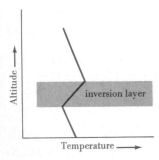

inversion layer

(b) Temperature inversion

(a) Under normal conditions air temperature decreases with altitude.
(b) Under conditions of temperature inversion a layer of warmer air is suspended aloft. At times the bottom of the inversion layer may be at ground level.

onshore and displace a layer of warmer air, holding it aloft. Another possibility is that, in a developing high-pressure atmospheric condition, layers of air may descend or subside. This air is warmed by compression and rests atop a cooler layer of air at the earth's surface. In either case the temperature–altitude profile assumes the shape shown in Figure 26-5b. Temperature inversions are particularly common in latitudes of about 30 to 35° in the summer months when a belt of high-pressure air encircles the earth.

Under conditions of temperature inversion the volume of air into which pollutants can be mixed is reduced. The inversion layer acts as a lid on the smoggy air below. Because the Los Angeles basin is surrounded by mountains on three sides, there is little horizontal dispersion of pollutants and the phenomenon of temperature inversion is of special importance.

Industrial smog is experienced in most urban areas of the United States, especially along the Atlantic seaboard. It consists primarily of particles (ash and smoke), sulfur dioxide, and sulfuric acid mist. Among the sources of the $SO_2(g)$ are industrial operations such as occur in oil refineries, smelters, coke plants, and sulfuric acid plants. The main contributors, however, are power plants burning bituminous coal or high-sulfur-content fuel oils. The $SO_2(g)$ can undergo air oxidation to $SO_3(g)$, especially when catalyzed on the surface of airborne particles (smoke) or through reaction with $NO_2(g)$. Then $SO_3(g)$ reacts with atmospheric water vapor to produce H_2SO_4 mist. A further reaction with $NH_3(g)$ may occur to produce particles of $(NH_4)_2SO_4$. The exact physiological effects of $SO_2(g)$ and H_2SO_4 mist at low concentrations are not well understood, but these substances are respiratory irritants and levels above 0.10 ppm are generally considered to be unhealthy.

Photochemical smog, the type experienced in southern California, involves a complex series of chemical reactions based on products of the automobile internal combustion engine. A simplified scheme is presented in Table 26-3. The sequence of steps is thought to begin with an NO_2 molecule absorbing near-ultraviolet radiation and dissociating to NO and O (26.1). This is followed by a reaction between O and O_2, producing ozone, O_3 (26.2). The O_3 in turn reacts with NO to reform NO_2 and O_2 (26.3).

If these were the only reactions occurring, the O_3 level would not be nearly so high as it becomes under smog conditions. The remaining reactions in Table 26-3 account for the primary production of O_3 (26.8). These reactions involve unburned hydrocarbons (26.4 and following equations).

The effects of photochemical smog are not much better understood than for industrial smog, but some of the effects are quite evident. The characteristic eye irritation associated with this type of smog is probably due to formaldehyde and acrolein ($CH_2{=}CHCHO$) produced in reaction (26.6) and peroxyacyl nitrates (PAN)

TABLE 26-3
Simplified reaction scheme for the production of photochemical smog

nitrogen–oxygen reactions:	$NO_2 + h\nu \longrightarrow NO + O$	(26.1)
	$O + O_2 \longrightarrow O_3$	(26.2)
	$O_3 + NO \longrightarrow NO_2 + O_2$	(26.3)
hydrocarbon reactions:	$RH + O \longrightarrow RO\cdot$	(26.4)
	$RO\cdot + O_2 \longrightarrow RO_3\cdot$	(26.5)
	$RO_3\cdot + RH \longrightarrow$ aldehydes + ketones	(26.6)
	(RCHO) (RCOR)	
	$RO_3\cdot + NO \longrightarrow RO_2\cdot + NO_2$	(26.7)
	$RO_3\cdot + O_2 \longrightarrow RO_2\cdot + O_3$	(26.8)
	$RO_3\cdot + NO_2 \longrightarrow$ peroxyacyl nitrates	(26.9)
	(PAN)	

A general measure of the severity of photochemical smog is gained through the **oxidant level.** The oxidants in a smog are those substances capable of oxidizing iodide ion to free iodine (the standard test employed). The principal oxidants are $O_3(g)$, $NO_2(g)$, and organic peroxides.

Other approaches to reducing photochemical smog, of course, include reducing nonessential driving, especially in cities; improving automobile engines to achieve better gasoline mileage; and carpooling and the use of public transportation systems.

produced in reaction (26.9). The incidence of all respiratory diseases (e.g., bronchitis and emphysema) is much higher than normal under smog conditions. Also, heavy crop losses have been experienced in regions of high smog. The deterioration of various substances is also well substantiated. Rubber goods, for example, are especially susceptible to degradation by ozone.

CONTROL OF AIR POLLUTION.   Measures to control air pollution must be viewed both in short-range and long-range terms. For example, one of the surest controls for photochemical smog would be to phase out of use, as quickly as possible, the internal combustion engine. Such a long-range solution may well be adopted, but for the present approaches must be based on improving the performance of these engines and on removing smog components from automotive exhaust.

Catalytic systems have been developed to reduce the emission of hydrocarbons and carbon monoxide by promoting their complete oxidation in exhaust fumes. For these catalysts to function they must be used in conjunction with lead-free gasoline, since lead tends to "poison" most catalysts. This, in turn, has required changes in petroleum-refining processes.

To remove oxides of nitrogen (NO, NO_2) from automotive exhaust requires their reduction, preferably to $N_2(g)$. However, the optimum catalyst composition to promote the reduction of oxides of nitrogen is not the same as for the oxidation of hydrocarbons and $CO(g)$. Dual catalyst systems are currently being developed to deal both with hydrocarbons and $CO(g)$ and with oxides of nitrogen.

Shortly after the introduction of catalytic systems for the oxidation of hydrocarbons and $CO(g)$, it was realized that such catalysts might also promote the oxidation of $SO_2(g)$. This would lead to $SO_3(g)$ and H_2SO_4 mist. Such reactions do in fact occur, but fortunately they are not as significant as first feared. As expected, the reduction in atmospheric hydrocarbons led to a reduction of smog levels in central Los Angeles. However, because oxides of nitrogen have not been controlled, the ozone-forming reactions still occur, although more slowly. These reactions take place as the coastal air mass moves eastward, and they have caused increases in oxidant levels in cities to the east of Los Angeles. These observations point out graphically that it is difficult to anticipate all the environmental impacts of a technological change.

The control of industrial smog hinges on the removal of sulfur from fuels and the control of $SO_2(g)$ emissions where they do occur. Also necessary is a reduction of the emission of oxides of nitrogen and particulate matter. Several methods exist for the removal of particles from industrial exhaust gases. In one method the gas is passed through a fine water spray; the particles settle with the water. Appropriately enough, this process is called "scrubbing." Another method involves passing the particle-laden gas into a device that induces a twisting flow, much like a cyclone. Particles are thrown out of the gas and settle to the bottom while clean gas emerges from this cyclone separator. Sometimes the cyclonic action and scrubbing are combined into a single separator. One of the most effective methods for removing particles from a gas is **electrostatic precipitation.** The gas is passed between electrically charged plates where gaseous ions are produced. These ions are adsorbed on solid particles and the particles, now electrically charged, are attracted to one of the electrodes where they are discharged and settle out from the gaseous stream.

Dozens of different processes have been proposed for the removal of SO_2 from stack gases. None has yet proved superior to all others, and none, in fact, has proved totally effective in removing SO_2 under all circumstances. One of the simplest processes is the injection of dry limestone into a boiler in which a fuel is burned. The limestone decomposes to quicklime. This reacts with $SO_2(g)$ in the stack gases to produce $CaSO_3$ and $CaSO_4$.

$$CaCO_3(s) \longrightarrow CaO(s) + CO_2(g) \tag{26.10}$$

$$CaO(s) + SO_2(g) \longrightarrow CaSO_3(s) \tag{26.11}$$

$$CaSO_3(s) + \tfrac{1}{2}O_2(g) \longrightarrow CaSO_4(s) \tag{26.12}$$

In some processes the $SO_2(g)$ (a Lewis acid) is allowed to react with a basic slurry of limestone or lime in a wet scrubbing action. For example, the following reactions (simplified) occur when quicklime (CaO) is used in the slurry:

$$CaO + H_2O \longrightarrow Ca(OH)_2 \tag{26.13}$$

$$Ca(OH)_2 + SO_2 \longrightarrow CaSO_3 + H_2O \tag{26.14}$$

$$CaSO_3 + SO_2 + H_2O \longrightarrow Ca(HSO_3)_2 \tag{26.15}$$

$$Ca(HSO_3)_2 + Ca(OH)_2 \longrightarrow 2\,CaSO_3 + 2\,H_2O \tag{26.16}$$

Some wet scrubbing processes are effective in removing more than 90% of the SO_2 in stack gases.

Removal of $SO_2(g)$ from exhaust gases adds a considerable expense to the burning of fossil fuels. It may be that processes to remove sulfur from fuels before burning will prove to be more feasible.

NO and NO_2 are produced in any combustion process conducted with air. These oxides result from the direct combination of $N_2(g)$ and $O_2(g)$ at high temperatures. Processes have been developed to remove the oxides by scrubbing, but they are not too effective. Alternative measures involve a careful control of the temperature and oxygen availability during the combustion to minimize the formation of oxides of nitrogen.

26-4 The Oceans as a Natural Resource

More than half the elements are found in measurable concentrations in seawater. A few of the more abundant ones are listed in Table 26-4. Since the volume of the oceans is so great (approximately 330 million cubic miles), seawater contains enormous quantities even of those elements present only in trace amounts. It is estimated that there is 10 billion tons of gold in seawater! However, despite repeated attempts no significant amount of gold has yet been extracted from seawater. We should look upon the oceans as an unlimited source of only a few substances. The oceans will always be able to meet the need for NaCl and the elements and compounds derived from it. Also, the oceans one day are likely to become a significant source of fresh water and H_2 and O_2 derived by electrolysis or other means. Two elements currently being extracted from seawater in large quantities are magnesium (Section 26-7) and bromine (Section 20-3). There is no reason to believe that the reserves of either of these elements will ever be depleted. In fact, the availability of magnesium in seawater is so great that it could conceivably supplant iron and aluminum as the most important structural metal. The only limitation to the production of magnesium is in the high consumption of electric power required. Some additional elements that have been mentioned as possible candidates for extraction from seawater are sulfur, potassium, iodine, fluorine, strontium, and boron.

MINING THE OCEAN FLOOR.   An additional possibility lies in mining materials from the ocean floor or beneath it. The principal natural resource currently being sought from beneath the ocean floor is petroleum. Offshore oil and gas drilling is becoming an increasingly important activity.

In recent years a new resource has been discovered on the ocean floor—**manga-**

A method in which $SO_2(g)$ from power-plant flue gases is recovered involves the following reactions:

$$SO_2(g) + MgO(s) \underset{750^\circ C}{\overset{150^\circ C}{\rightleftharpoons}} MgSO_3(s)$$

The $SO_2(g)$ released when $MgSO_3(s)$ is heated can be used in the manufacture of H_2SO_4.

TABLE 26-4
Principal constituents in seawater

Constituent	Concentration present, g/ton
Cl^-	18,980
Na^+	10,561
SO_4^{2-}	2,649
Mg^{2+}	1,272
Ca^{2+}	400
K^+	380
HCO_3^-	140
Br^-	65
H_3BO_3	26
Sr^{2+}	8
F^-	1

Some estimates place the quantity of offshore oil at 25% of the world's recoverable reserves.

nese nodules. These are rocklike objects composed of layers of material consisting of oxides of manganese and iron, with admixtures of small amounts of other metals such as cobalt, copper, and nickel. The nodules are roughly spherical in shape with diameters ranging from a few millimeters to about 15 cm. They are believed to grow at a rate of a few millimeters per million years. It has been proposed that marine microorganisms play a role in their formation. Possibly the deposition of manganese from solution proceeds through enzyme systems in bacteria. Estimates of the total quantity of these nodules are very great, perhaps billions of tons. However, several challenges exist to developing manganese nodules as a significant raw material. Methods must be perfected to explore the seabed, dredge for the nodules, and transport them through several thousand meters of seawater. Also, new metallurgical processes must be devised for extracting the desired metals. Nontechnical, but problems nevertheless, are the political and legal issues involved in mining in international waters. (Who owns the nodules?) The largest deposits currently known are in an area southeast of the Hawaiian Islands.

DESALINATION.   The desalination of seawater is quite simple in theory. The basic change that must be accomplished in any desalination scheme is

seawater (salt conc. c_1) $\longrightarrow$ pure water + seawater (salt conc. c_2)

where $c_2 > c_1$. Since the original seawater becomes more concentrated, the desalination process must be *nonspontaneous;* the free energy change is positive. For example, for a typical seawater containing 3.5% NaCl, by mass, $\Delta \overline{G} = 49.5$ J/mol H_2O transferred from seawater to pure water. If seawater and pure water are in contact with the same vapor at the same temperature, molecules pass from the pure water (which has the higher vapor pressure) to the seawater. However, if the seawater is maintained at a high temperature (say its boiling point), its water vapor pressure will be greater than that of pure water at room temperature. Water molecules will pass through the vapor from seawater to pure water. Seawater can be desalinized through this process of **distillation.**

Other important desalination schemes include reverse osmosis (recall Figure 12-12) and freezing (when dilute salt solutions are frozen, the solid phase obtained is pure ice). Unconventional schemes have also been proposed for obtaining fresh water, such as towing icebergs from polar regions to port cities.

26-5   The Water Environment

Aside from a relatively small quantity of water vapor in the atmosphere, there are four important types of water on the earth's surface: the "fresh" water of rivers and lakes, groundwater, continental ice sheets, and the salt waters of the seas and oceans (and a few inland bodies). Salt water accounts for about 98% of the water on the earth's surface and the continental ice sheets for most of the remainder. Freshwater bodies contain only a tiny fraction of the earth's water, but they are of extreme importance.

Many areas on the earth's surface receive so little rainfall that a chronic shortage of fresh water exists. Furthermore, in highly advanced nations such as the United States, where the per capita consumption of fresh water is high (about 1900 gal per day per person for household, industrial, and agricultural use), deficiencies in the availability of fresh water already exist and will become more critical in the future. The most abundant supply of water is in the oceans of the world, but to use this water requires developing effective desalination schemes, as discussed in the preceding section.

HARDNESS IN WATER.   Because of its solvent action on rocks, soil, and certain atmospheric gases, fresh water is never chemically pure. It may contain anywhere from a few to perhaps 1000 ppm of dissolved substances. Water containing dissolved minerals in any appreciable quantities is said to be **hard.**

Some rainwater has been reported to have pH values as low as 4 or even less, probably as a result of dissolving such atmospheric pollutants as SO_2 and SO_3.

One process whereby water becomes hard begins with dissolving $CO_2(g)$, which makes rainwater slightly acidic. When rainwater seeps through limestone beds, it dissolves $CaCO_3(s)$, converting it to soluble $Ca(HCO_3)_2$. Water that owes its hardness primarily to HCO_3^- and associated cations is said to be **temporary hard water.** The chemistry of temporary hard water is effectively summarized by equations (26.17) through (26.19).

$$CO_2 + H_2O \rightleftharpoons H_2CO_3 \tag{26.17}$$

$$H_2CO_3 + H_2O \rightleftharpoons H_3O^+ + HCO_3^- \tag{26.18}$$

$$HCO_3^- + H_2O \rightleftharpoons H_3O^+ + CO_3^{2-} \tag{26.19}$$

The addition of an acid to a carbonate-bicarbonate system produces this sequence of events:

$$CO_3^{2-} \xrightarrow{H_3O^+} HCO_3^- \xrightarrow{H_3O^+} H_2O + CO_2(g)$$

Addition of a base results in the complete conversion to carbonate.

$$CO_2(g) + H_2O \longrightarrow H_2CO_3 \xrightarrow{OH^-} HCO_3^- \xrightarrow{OH^-} CO_3^{2-}$$

Heating or evaporating water from an aqueous solution of HCO_3^- ion leads to a series of changes summarized in equation (26.20).

These equations show that bicarbonate ion, HCO_3^-, has both acidic and basic properties. For every HCO_3^- ion that acts as an acid (producing H_3O^+ and CO_3^{2-}), another accepts a proton and is converted to H_2O and CO_2.

$$
\begin{aligned}
HCO_3^- + H_2O &\longrightarrow H_3O^+ + CO_3^{2-} \\
HCO_3^- + H_3O^+ &\longrightarrow H_2O + H_2CO_3 \\
H_2CO_3 &\longrightarrow H_2O + CO_2(g) \\
\hline
2\,HCO_3^- &\longrightarrow CO_3^{2-} + H_2O + CO_2(g)
\end{aligned}
\tag{26.20}
$$

Reaction (26.20) represents the principal deleterious effect of temporary hard water. When the water is heated, the carbonate ion that is regenerated reacts with multivalent cations in the water to form a mixed precipitate of $MgCO_3$, $CaCO_3$, and $FeCO_3$ called **boiler scale.** The formation of boiler scale is a very serious problem in many industrial applications. For example, in steam power plants the formation of scale can cause a boiler to overheat, presenting an explosion hazard.

The carbonate-bicarbonate equilibria that characterize temporary hard water are also responsible for other natural phenomena. The dissolving of limestone by CO_2-charged rainwater can produce underground caves. Evaporation of water from $Ca(HCO_3)_2(aq)$ leads to the familiar limestone formations of these caves (as a result of reaction 26.20).

WATER SOFTENING.   Hard water is acceptable for some applications, but for others it is not. The term "water softening" refers to the removal of natural mineral impurities, and there are several ways in which this can be accomplished. Simple boiling will soften temporary hard water, but with the formation of unwanted boiler scale. Water containing significant concentrations of anions other than HCO_3^-, for example SO_4^{2-}, together with associated cations, is called **permanent hard water.** This is because such water cannot be softened simply by heating.

Another suitable method for softening temporary hard water is to treat it with a base and filter off the precipitated metal carbonate.

$$HCO_3^- + OH^- \longrightarrow H_2O + CO_3^{2-} \tag{26.21}$$

$$CO_3^{2-} + M^{2+} \longrightarrow MCO_3(s) \tag{26.22}$$

The source of OH^- may be slaked lime, $Ca(OH)_2$, or washing soda, Na_2CO_3. (Can you write an equation to show how Na_2CO_3 produces OH^- by hydrolysis?) Perma-

FIGURE 26-6
Ion exchange process.

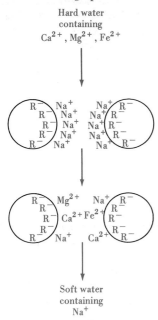

Hard water
containing
Ca^{2+}, Mg^{2+}, Fe^{2+}

Soft water
containing
Na^+

The resin pictured here is a cation exchange resin—multivalent cations (Ca^{2+}, Mg^{2+}, Fe^{2+}) are exchanged for univalent cations (Na^+). This resin can be represented as RNa. Other ion exchange resins, designated as ROH, exchange OH^- for other anions; these are anion exchange resins.

Water softening by this process is not without some disadvantages. Replacement of multivalent cations by Na^+ in drinking water would be detrimental to anyone on a low-sodium diet.

nent hard water can also be softened with Na_2CO_3. Cations such as Ca^{2+} and Mg^{2+} precipitate as carbonates and Na_2SO_4 remains in solution.

Example 26-1   A typical hard water contains 180 ppm HCO_3^-. How many kg CaO would be required to soften 1.00×10^6 gal of this water?

We start by writing the equations for the water softening reactions,

$$CaO + H_2O \longrightarrow Ca(OH)_2 \qquad Ca(OH)_2 \longrightarrow Ca^{2+} + 2\,OH^-$$

followed by equations (26.21) and (26.22). For every mole of CaO consumed, 2 mol HCO_3^- is removed from solution. The HCO_3^- content of the hard water is taken to be 180 g HCO_3^- per 1×10^6 g H_2O.

$$no. \text{ kg CaO} = 1.00 \times 10^6 \text{ gal} \times \frac{3.78 \text{ L}}{1 \text{ gal}} \times \frac{1000 \text{ cm}^3}{1 \text{ L}} \times \frac{1.00 \text{ g } H_2O}{1 \text{ cm}^3}$$

$$\times \frac{180 \text{ g } HCO_3^-}{10^6 \text{ g } H_2O} \times \frac{1 \text{ mol } HCO_3^-}{61.0 \text{ g } HCO_3^-} \times \frac{1 \text{ mol } OH^-}{1 \text{ mol } HCO_3^-}$$

$$\times \frac{1 \text{ mol } Ca(OH)_2}{2 \text{ mol } OH^-}$$

$$\times \frac{1 \text{ mol CaO}}{1 \text{ mol } Ca(OH)_2} \times \frac{56.1 \text{ g CaO}}{1 \text{ mol CaO}} \times \frac{1 \text{ kg CaO}}{1000 \text{ g CaO}}$$

$$= 313 \text{ kg CaO}$$

SIMILAR EXAMPLES: Exercises 13, 14, 15.

One of the most satisfactory methods of softening water is through **ion exchange.** The ion exchange medium may be a natural sodium aluminosilicate, called a **zeolite,** or a synthetic resinous material. Ion exchange materials consist of macromolecular (polymer) particles capable of ionizing to produce fixed ions, which remain attached to the particle surfaces, and free or mobile counterions. It is the counterions that may exchange positions with other ions in a solution passed through a bed of these particles.

In the resin pictured in Figure 26-6, the fixed ions, R, are negatively charged and the counterions are positive. At the start the counterions in the resin bed are Na^+. When hard water is passed through the bed Ca^{2+}, Mg^{2+}, and Fe^{2+}, because of their higher charge, displace Na^+ as counterions. To regenerate the resin concentrated NaCl(aq) is passed through. In high concentration Na^+ ions displace the multivalent ions from the resin particles, restoring the resin to its original condition. The ion exchange material has an indefinite lifetime. The only material consumed in softening water by this method is the sodium chloride used in the regenerating process.

Ion exchange processes can be represented through simple chemical equations. If a zeolite is involved, the symbol Z is used, if a synthetic resin, R.

$$Na_2Z + M^{2+} \longrightarrow MZ + 2\,Na^+ \tag{26.23}$$

$$Na_2R + M^{2+} \longrightarrow MR + 2\,Na^+ \tag{26.24}$$

SOAPS AND DETERGENTS.   One of the effects of hard water that is most frequently encountered in the average household is its effect on soaps. As we have noted before (Section 25-2), ordinary soaps are the sodium salts of fatty acids, represented as $RCOO^-Na^+$, where the R group is an aliphatic hydrocarbon chain ranging from about 12 to 18 C atoms in length. Figure 26-7 suggests how soap molecules can solubilize or emulsify oils in water.

Although alkali metal soaps are water soluble, the soaps of multivalent cations are not, as indicated in the familiar reaction (26.25)

FIGURE 26-7
Representation of the cleaning action of soap.

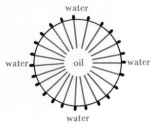

The cleaning action results from the structure of the soap molecules— they have both a long nonpolar portion (shown in color) and a polar carboxyl group. The interface between an oil droplet and the aqueous medium in which it is suspended is lined with soap molecules having their nonpolar ends in the oil and polar ends in the water. The oil droplet is solubilized.

$$2\,Na^+ + 2\,RCOO^- + Ca^{2+} \longrightarrow Ca(RCOO)_2(s) + 2\,Na^+ \qquad (26.25)$$

soap   "bathtub ring"

Thus, soaps are effective (but expensive) water softeners; but unfortunately the object being cleansed becomes coated with the precipitated heavy metal soaps. One solution to this problem is achieved through the use of synthetic detergents. Synthetic detergents are salts of organic sulfonic acids, $ROSO_3^-Na^+$. The R group may be an aliphatic hydrocarbon in the kerosene range, as in some heavy-duty cleaners, or it may have aromatic character as in most laundry products. The great advantage of a detergent over a soap is that its calcium and other heavy metal salts are soluble. Detergents do not form precipitates in hard water.

26-6 Water Pollution

The existence of air pollution and its effects are revealed rather dramatically. Water pollution is not always recognized as clearly, but it is every bit as pervasive and serious a problem and is urgently in need of attention. The sources of pollution in fresh water, and in the oceans as well, are so varied that only an indication of the nature of the problem is possible here. Human activities that lead to the pollution of water are many as is also the nature of the pollutants themselves—dissolved inorganic compounds, dissolved organic compounds, biological materials (bacteria and viruses), and suspended solids.

Because all the impurities present in water can rarely be identified exactly, it is customary to perform tests that are indicative of the presence of large classes of substances. The **biochemical oxygen demand (BOD)** is a measure of the quantity of dissolved oxygen consumed in the biological processes associated with the degradation of organic matter in water. This determination must be carried out according to standardized procedures. For example, a sample of water is injected with microorganisms and the mass of dissolved oxygen consumed as organic substances are degraded is determined after a 5-day period. The higher the BOD, the more polluted a water sample is considered to be. For one thing, the more oxygen consumed by microorganisms in degrading organic matter, the less dissolved oxygen is available to support fish life.

The 5-day BOD test normally accounts for only about 70 to 80% of the true biochemical oxygen demand of a sample. In addition, the sample may contain dissolved organic substances that cannot be degraded by microorganisms. Another test, the **chemical oxygen demand (COD),** measures the quantity of a strong oxidizing agent (dichromate ion in acidic solution) required to oxidize all the oxidizable substances in a sample to carbon dioxide and water. The results can be expressed in terms of an equivalent quantity of dissolved oxygen. As might be expected, the COD test gives higher values than the BOD test. A typical, unpolluted natural water sample has a BOD of less than 1 mg O_2/L. By contrast, water from the Mississippi River at about its midpoint has a yearly average BOD of about 6 mg O_2/L and a COD of 25 mg O_2/L.

The magnitude of wastewater production in the United States is suggested by Table 26-5, but the true magnitude is probably much greater. Among the sources not included in the table are agricultural runoff, acid mine drainage, and livestock wastes. For example, it has been estimated that the waste produced by livestock in the Midwest, unserved by sewers, is alone equivalent to that of a human population of 350 million!

WATER PURIFICATION.   In some localities the quality of groundwater is sufficiently high that it can be used for drinking and other commercial and industrial purposes

TABLE 26-5
Estimates of industrial and domestic wastes, before treatment

Activity	Wastewater, billion gal	Standard BOD, million lb	Settleable and suspended solids, million lb
primary metals	4,300	480	4,700
chemicals	3,700	9,700	1,900
paper	1,900	5,900	3,000
petroleum and coal	1,300	500	460
food processing	690	4,300	6,600
transportation equipment	240	120	n.a.[b]
rubber and plastics	160	40	50
machinery	150	60	50
textiles	140	890	n.a.[b]
electrical machinery	91	70	20
all other manufacturing	450	390	930
all manufacturing	13,100	22,000	18,000
domestic waste[a]	5,300	7,300	8,800

[a] Based on 120 million people served by sewers.
[b] n.a., not available.
SOURCE: *The Cost of Clean Water, Vol. I, Summary Report*, U.S. Department of the Interior, Federal Water Pollution Control Administration, Government Printing Office, Washington, D.C., January 1968.

with no further treatment, but this is usually not the case. Following are some of the processes employed to purify municipal water supplies:

1. *Aeration.* Air is mixed with water, either by spraying or through a waterfall. This removes dissolved gases such as CO_2 and H_2S, oxidizes Fe^{2+} to Fe^{3+}, and destroys some bacteria.
2. *Removal of hardness.* Usually this treatment, when desired, is left to the user. Sometimes this is provided as a municipal activity. When this is done on a large scale, either $Ca(OH)_2$ or Na_2CO_3 is used as the water softener.
3. *Treatment with alum.* Some water supplies have suspended colloidal particles, such as clay minerals. These can be coagulated by treatment with an electrolyte, usually alum, $NH_4Al(SO_4)_2 \cdot 12 H_2O$ or $FeCl_3$ (recall Section 12-8).
4. *Filtration.* Fine sand is generally used as the filtering medium. Suspended particles are removed as the water passes through the filter bed.
5. *Chlorination.* Treatment with chlorine destroys microorganisms and is usually the most important step in rendering water fit to drink.

An interesting consequence of the chlorination of water has recently come to light. When a municipal water source is river water, the water may contain traces of industrial wastes, hydrocarbons for instance. Chlorination of such water not only kills microorganisms but chlorinates these pollutants. A total of 66 chemical pollutants has been detected in drinking water in the lower Mississippi Valley (New Orleans). These include known toxic materials, such as $CHCl_3$, CCl_4, and polychlorinated biphenyls (PCB). Water purification with ozone, widely used in Europe, is now attracting attention in the United States. Oxygen is also being tested as an agent in water purification. Still another method involves adsorption of impurities on activated carbon.

EUTROPHICATION. Eutrophication is the term used to describe the process of enrichment of freshwater bodies with nutrients. It is a natural process that occurs over

Polychlorinated biphenyls (PCB), used at one time in inks, plastics, and paper coatings and now used only in electrical transformers and capacitors, are among the most persistent synthetic chemicals released to the environment. They can withstand very high temperatures and are not readily degraded by natural agents. Controls on the production and use of these materials in the United States are now quite stringent, but the occurrence of PCB in drinking water is believed to result from an accidental chemical reaction—the chlorination of biphenyl during water purification.

Two decades ago the most popular detergents were the alkyl benzene sulfonates (ABS).

$$CH_2(CH_2)_7\overset{\displaystyle CH_3}{\underset{\displaystyle CH_3}{C}}—CH_3$$

$$O=S=O$$
$$O^-\quad Na^+$$

Because of the branched hydrocarbon chain [featuring the *tert*-butyl group $—C(CH_3)_3$] these molecules are very resistant to biological degradation in the environment.

Biodegradable detergents are based on linear alkyl sulfonates.

$$H_3C(CH_2)_7CH(CH_2)_2\overset{\displaystyle O}{C}—OH$$

$$O=S=O$$
$$O^-\quad Na^+$$

The presence of the carboxyl group (—COOH) and the absence of chain branching account for the biodegradability of these molecules.

A useful method of concentration applicable to the few ores that are magnetic is to pass the mixture of crushed ore and rock through a magnetic field. The ore and rock are collected in separate piles. The process is called magnetic beneficiation.

geologic time periods but that can be greatly accelerated by human activities. A body of water that receives large quantities of nutrients, such as nitrates and phosphates, experiences excessive growth of algae. This is followed by a depletion in the oxygen content, fish kills, growth of undesirable anerobic bacteria, and a variety of other effects. Natural sources of these nutrients include animal wastes, decomposition of dead organic matter, and natural nitrogen fixation. Human sources include industrial wastes, municipal sewage plant effluents, and fertilizer runoff.

Some years ago a problem was recognized in the use of synthetic detergents. No natural processes existed by which these detergents could be degraded. They persisted in the environment for long periods of time, causing unsightly foam in freshwater bodies and presenting difficult problems of water purification. Once the severity of the problem was recognized, detergent manufacturers developed **biodegradable** detergents; these are now in common use.

But now detergents are suspected of contributing to the problem of eutrophication. At the present time most detergents contain a constituent called a builder. Although detergents are not precipitated by multivalent metal ions, neither do they work as effectively in the presence of these ions. One of the functions of a detergent builder is to complex or sequester metal ions in solution; this keeps the free metal ion concentrations at a very low level. The sequestering action of sodium triphosphate, $Na_5P_3O_{10}$, was discussed in Section 22-12.

Currently, many municipalities have adopted ordinances limiting the phosphate content of detergents. This development appears to be improving environmental conditions, but it must be admitted that the case against phosphates is not conclusive. It may be that in some cases, even in the presence of high concentrations of phosphates, the eutrophication process is limited by some other nutrient, perhaps even by the availability of CO_2. (Photosynthetic activity by algae is dependent on the quantity of CO_2 produced by microorganisms that metabolize organic compounds.)

26-7 Metals

The production of common metals in the United States ranges from that of silver and gold, measured in the millions of ounces, to iron and steel, for which annual production exceeds 100 million tons. In this section we look first at some basic metallurgical processes and then at the production of five specific metals—iron, copper, aluminum, magnesium, and titanium.

BASIC METALLURGICAL PROCESSES.   There is no single method for extracting all metals from their ores, but certain basic operations are generally required—concentration of the ore, roasting, reduction, and refining. To illustrate these operations, let us consider the preparation of pure metallic zinc. Zinc is found principally as the oxide, the carbonate, and the sulfide.

Concentration.   In mining operations the desired mineral from which a metal is to be extracted often constitutes only a few per cent of the material mined. Thus it is necessary to separate the desired ore from waste rock before proceeding with other metallurgical operations. One important method of ore concentration, **flotation,** is described in Figure 26-8.

Roasting and Reduction.   The purpose of roasting, where necessary, is to convert an ore to the oxide which may then be reduced. Although a variety of reducing agents may serve in the reduction process, carbon, as coke or powdered coal, is generally employed. Both of the following reactions are conducted at elevated temperatures.

not just "air", nitrogen causes problems

roasting: $2\,ZnS(s) + 3\,O_2(g) \xrightarrow{\text{heat}} 2\,ZnO(s) + 2\,SO_2(g)$ (26.26)

reduction: $ZnO(s) + C(s) \xrightarrow{\text{heat}} Zn(g) + CO(g)$ (26.27)

reducing agent

FIGURE 26-8

Concentration of an ore by flotation.

Powdered ore is suspended in water in a large vat, together with suitable additives, and the mixture is agitated with air. Particles of ore become attached to air bubbles, rise to the top of the vat, and are collected in the overflow froth. Particles of the undesired waste rock (gangue) fall to the bottom.

The success of this method depends on the use of proper additives—a material that will produce a stable foam (frother) and a substance (collector) that coats the particles of ore but does not "wet" the particles to be rejected. Pine oil is widely used as a frother and sodium ethyl xanthate as a collector.

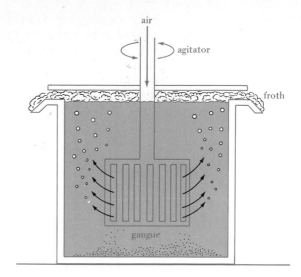

Refining   The metal that results from the roasting and reduction steps is generally not pure enough for its intended uses. The refining process removes undesirable impurities from the metal. Zinc is most commonly purified by distillation. An alternative process, in which reduction and refining are accomplished in a single operation, involves electrolysis. Zinc oxide obtained in the roasting step (26.26) is dissolved in sulfuric acid; this is an acid-base reaction.

nonspontaneous

$$ZnO(s) + 2\,H^+(aq) + SO_4^{2-}(aq) \longrightarrow Zn^{2+}(aq) + SO_4^{2-}(aq) + H_2O \qquad (26.28)$$

The aqueous zinc sulfate solution is purified and then electrolyzed. The electrode reactions are

cathode:	$Zn^{2+}(aq) + 2\,e^- \longrightarrow Zn(s)$
anode:	$H_2O \longrightarrow \frac{1}{2}O_2(g) + 2\,H^+(aq) + 2\,e^-$
unchanged:	$SO_4^{2-}(aq) \longrightarrow SO_4^{2-}(aq)$
net:	$Zn^{2+} + SO_4^{2-} + H_2O \longrightarrow Zn(s) + 2\,H^+ + SO_4^{2-} + \frac{1}{2}O_2(g)$

$$(26.29)$$

Note that in the net electrolysis reaction Zn^{2+} is reduced to pure metallic zinc and sulfuric acid is regenerated. The acid is reused in step (26.28).

Example 26-2   Write chemical equations to represent the following metallurgical processes: **(a)** roasting of galena, PbS; **(b)** reduction of $Cu_2O(s)$, using charcoal as a reducing agent; **(c)** deposition of pure silver from an aqueous solution of Ag^+.
(a) We expect this process to be essentially the same as reaction (26.26).

$$2\,PbS(s) + 3\,O_2(g) \xrightarrow{\text{heat}} 2\,PbO(s) + 2\,SO_2(g)$$

(b) The products are Cu(l) and CO(g).

$$Cu_2O(s) + C(s) \xrightarrow{\text{heat}} 2\,Cu(l) + CO(g)$$

(c) This process involves a reduction half-reaction. The accompanying oxidation half-reaction is not specified. Neither is it specified whether this is an electrolysis process or whether silver is displaced by a more active metal.

$$Ag^+(aq) + e^- \longrightarrow Ag(s)$$

SIMILAR EXAMPLES: Exercises 20, 21.

FIGURE 26-9
Typical blast furnace.

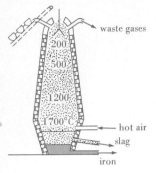

The firebrick-lined furnace is about 60 m high and 10 m in diameter at the base. Iron ore, coke, and limestone are added at the top of the furnace. Hot air is introduced near the bottom. Maximum temperatures are attained near the bottom of the furnace where molten iron and slag collect. The furnace operates continuously, producing up to 2×10^6 kg of iron per day. It needs to be shut down only every two or three years for relining and general maintenance.

TABLE 26-6
Some typical iron ores

Type of ore	Formula
oxide	
magnetite	Fe_3O_4
hematite	Fe_2O_3
ilmenite	$FeTiO_3$
limonite	$HFeO_2$
carbonate	
siderite	$FeCO_3$
sulfide	
pyrite	FeS_2
pyrrhotite	FeS

IRON AND STEEL.   Early History.   The art of ironmaking was developed to an advanced state even in ancient times. Aristotle, in 384 B.C., provided a description of the manufacture of a type of steel, called **wootz steel**, first produced in India. This is the same steel that became famous in ancient times as Damascus steel, renowned for its suppleness, its ability to hold a cutting edge, and its use in making swords. A number of technological advances have been made since ancient times. These include the introduction of blast furnaces in about A.D. 1300, the Bessemer converter in 1856, the open hearth furnace in the 1860s, and most recently, the basic oxygen furnace. However, a true understanding of the iron and steel making process has developed only within the past few decades, based on concepts of thermodynamics, kinetics, and equilibrium.

Pig Iron.   The reduction of iron ore, which is accomplished in a blast furnace, involves an impressive array of reactions. In the following simplified scheme, approximate temperatures are given so that these reactions may be keyed to regions of the blast furnace pictured in Figure 26-9.

formation of reducing agents, principally $CO(g)$ and $H_2(g)$:
$$2\,C + O_2 \longrightarrow 2\,CO \quad (1700°C)$$
$$C + CO_2 \longrightarrow 2\,CO \quad (>1000°C) \qquad (26.30)$$
$$C + H_2O \longrightarrow CO + H_2 \quad (>600°C)$$

reduction of iron oxide:
$$3\,CO + Fe_2O_3 \longrightarrow 2\,Fe + 3\,CO_2 \quad (900°C)$$
$$3\,H_2 + Fe_2O_3 \longrightarrow 2\,Fe + 3\,H_2O \quad (900°C) \qquad (26.31)$$

slag formation to remove most of the impurities:
$$CaCO_3 \longrightarrow CaO + CO_2 \quad (800-900°C)$$
$$CaO + SiO_2 \longrightarrow CaSiO_3(l) \quad (1200°C) \qquad (26.32)$$
$$3\,CaO + P_2O_5 \longrightarrow Ca_3(PO_4)_2(l) \quad (1200°C)$$

impurity formation in the iron:
$$MnO + C \longrightarrow Mn + CO \quad (1400°C)$$
$$SiO_2 + 2\,C \longrightarrow Si + 2\,CO \quad (1400°C) \qquad (26.33)$$
$$P_2O_5 + 5\,C \longrightarrow 2\,P + 5\,CO \quad (1400°C)$$

The blast furnace charge consists of iron ore, coke, a slag-forming flux, and perhaps some scrap iron. The exact proportions used depend on the composition of the iron ore and its impurities. The compositions of some typical ores are listed in Table 26-6. The purpose of the flux is to maintain the proper ratio of acidic oxides (SiO_2, Al_2O_3, and P_2O_5) to basic oxides (CaO, MgO, and MnO) to obtain an easily liquefied silicate, aluminate, or phosphate slag. Since in most ores the acidic oxides predominate, the flux generally employed is limestone, $CaCO_3$, or dolomite, $CaCO_3 \cdot MgCO_3$.

The iron obtained from a blast furnace is called **pig iron.** It contains about 95% Fe, 3 to 4% C, and varying quantities of other impurities. **Cast iron** can be obtained by pouring pig iron directly into molds of the desired shape. Cast iron is very hard and brittle and can be used only in applications where it will not be subjected to mechanical or thermal shock. For example, it is used in engine blocks, brake drums, and transmission housings in automobiles.

Steel.   The ancient method developed in India for producing wootz steel was revived in England in the 1700s, but until the 1850s the production of steel had not changed significantly from ancient times. Iron, as cast iron and wrought iron, was still the principal metal of commerce. (Wrought iron is a purified iron low in carbon content, and easily worked, that is, malleable.) In 1856, Henry Bessemer, in England,

TABLE 26-7
Some common types of iron and steel

Composition, mass %	Trade name	Density, g/cm³	Melting point, °C
98.5 Fe	wrought iron	7.7	1510
99 Fe; 1 C	steel	7.83	1430
97 Fe; 3 C	white cast iron	7.60	1150
94 Fe; 3.5 C; 2.5 Si	gray cast iron	7.0	1230
95.1 Fe; 3 Ni; 1.5 Cr; 0.4 C	nickel-chrome steel		
94.5 Fe; 5 W; 0.5 C	tungsten steel		
84.3 Fe; 14.5 Si; 0.85 C; 0.35 Mn	duriron	7.0	1265
74 Fe; 18 Cr; 8 Ni; 0.18 C	stainless N		

The invention of the Bessemer converter has been considered an event equal in importance to the inventions of printing, the magnetic compass, and the steam engine.

FIGURE 26-10
A Bessemer converter.

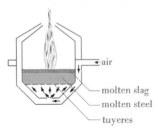

molten slag
molten steel
tuyeres
air

introduced a new, rapid, inexpensive method for converting iron to steel. The iron and steel industry was transformed completely by this and subsequent inventions. Table 26-7 lists a few of the many different types of iron and steel in use today.

The basic changes that must be accomplished in any steelmaking process are these: (1) reduction of the carbon content from an original 4 to 5% in pig iron to 0 to 1.5% in steel; (2) removal through slag formation of Si, Mn, and P (each present in pig iron to the extent of 1% or so), together with other minor impurities; and (3) addition of alloying elements (such as Cr, Ni, Mn, V, Mo, W) to give the steel its desired end properties.

The **Bessemer converter,** shown in Figure 26-10, is a steel vessel with a refractory lining. The refractory lining may be either siliceous in its composition (acid Bessemer) or it may be made of dolomite (basic Bessemer), depending on the type of impurities to be removed from the iron. The vessel can be rotated on a pair of trunions, one of which is hollow to allow for the passage of air. A blast of air is injected into the molten iron through a series of holes (tuyeres) in the bottom lining. A vigorous exothermic process occurs, involving a variety of reactions. Some of these are presented in simplified form in Table 26-8.

In the 1860s the **open hearth furnace** was developed, principally through the efforts of William Siemens in England. This furnace, pictured in Figure 26-11, works on a regenerative principle. The high temperatures required to maintain iron in a molten condition are achieved by burning a gaseous fuel (such as natural gas) over the metal. The hot gaseous products are exhausted through a network of firebricks,

TABLE 26-8
Reactions occurring in steelmaking processes

$$2\,C + O_2 \longrightarrow 2\,CO$$
$$2\,CO + Si \longrightarrow SiO_2 + 2\,C$$
$$CO + Mn \longrightarrow MnO + C$$
$$C + FeO \longrightarrow Fe + CO$$
$$Fe_3C + FeO \longrightarrow 4\,Fe + CO$$
$$2\,Fe + O_2 \longrightarrow 2\,FeO$$
$$2\,FeO + Si \longrightarrow 2\,Fe + SiO_2$$
$$FeO + Mn \longrightarrow Fe + MnO$$
$$\left.\begin{array}{l} FeO + SiO_2 \longrightarrow FeO \cdot SiO_2 \\ MnO + SiO_2 \longrightarrow MnO \cdot SiO_2 \end{array}\right\} slag$$
$$4\,P + 5\,O_2 \longrightarrow 2\,P_2O_5$$
$$3\,CaO + P_2O_5 \longrightarrow Ca_3(PO_4)_2 \quad (slag)$$

FIGURE 26-11
An open hearth furnace.

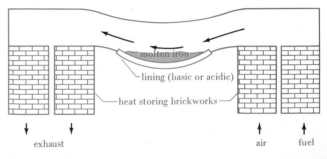

molten iron
lining (basic or acidic)
heat storing brickworks
exhaust
air
fuel

In 1962, in the United States, the quantity of steel made by the oxygen steelmaking method was only about 4% of that produced in open hearth furnaces. Today the quantity exceeds that of the open hearth method. The main factors that have prevented an even faster conversion to the new method are the need to pay off the capital investment in existing open hearth furnaces and to find the necessary capital to build the new furnaces. The conversion of discoveries and inventions into practical terms often involves considerations that extend well beyond scientific and engineering principles.

Almost 200 kg of waste is produced for every kg Cu.

FIGURE 26-12
A basic oxygen furnace.

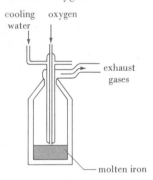

The product of reaction (26.37) is called blister copper because of the presence of frozen bubbles of $SO_2(g)$. Its principal uses are structural and for plumbing, where high purity is not required. However, high purities are essential in electrical applications.

called checkers. Periodically, the direction of gas and air flow are reversed, so that the entering gases pass through the hot checkers and become preheated before combustion. The firebrick lining of the open hearth furnace can be either of the acid or basic type, although the basic type is the only one commonly used. The usual furnace charge consists of scrap iron, molten pig iron, iron oxide, and limestone. The reactions are similar to those listed in Table 26-8. Typically, a batch of steel can be produced in about 12 hours.

The most important method in use today is the **oxygen steelmaking process.** This process is a logical extension of earlier practice. It uses pure oxygen gas rather than air to support the oxidation reactions required in refining iron. The process is carried out in a vessel much like a Bessemer converter. Oxygen gas, at about 10 atm, and a stream of powdered limestone are fed through a water-cooled lance and discharged above the molten metal (see Figure 26-12). A typical reaction time is 22 minutes.

COPPER. Copper was once rather widely available in the free state, but now native copper is being mined in the United States only in Michigan. The extraction of copper from its ores (generally as sulfides) is considerably more complex than the general scheme illustrated for zinc at the beginning of this section. This complexity arises primarily because of the presence of iron sulfides in copper ores. The usual procedures would result in iron being produced along with copper. To avoid this, iron must be removed before the final reduction to copper metal takes place. Five steps are required altogether: (1) concentration, (2) roasting, (3) smelting, (4) converting, and (5) refining.

The quality of available copper ores in the United States has been falling steadily. Currently, most copper-bearing ores being mined have only about 0.5% Cu. Concentration of these ores before further processing is essential. This is usually done by the flotation process (recall Figure 26-8), yielding an ore concentrate with about 20 to 40% Cu.

The function served by roasting the concentrated ore (when necessary) is to convert excess iron sulfide to iron oxide, with the copper remaining as the sulfide. This result can be achieved if the temperature is kept below 800°C.

$$2\,FeS(s) + 3\,O_2(g) \longrightarrow 2\,FeO(s) + 2\,SO_2(g) \qquad (26.34)$$

The roasted ore is then transferred to a smelting furnace heated to about 1400°C by the flames from burning natural gas, powdered coal, or fuel oil. In this furnace the charge melts and separates into two layers. The bottom layer is copper matte, consisting chiefly of the molten sulfides of copper and iron. The top layer is a silicate slag formed by the reaction of oxides of iron, calcium, and aluminum with SiO_2 (which either is present in the ore or is added).

$$FeO(s) + SiO_2(s) \longrightarrow FeSiO_3(l) \qquad (26.35)$$

After smelting, the copper matte is transferred to another furnace (the converter), where air is blown through the molten mass. First the remaining iron sulfide is converted to oxide (equation 26.34), followed by slag formation (equation 26.35). The slag is poured off and air is again blown through the furnace. Now the following reactions occur, producing a product that is about 98 to 99% Cu.

$$2\,Cu_2S + 3\,O_2(g) \longrightarrow 2\,Cu_2O + 2\,SO_2(g) \qquad (26.36)$$

$$2\,Cu_2O + Cu_2S \longrightarrow 6\,Cu(l) + SO_2(g) \qquad (26.37)$$

The principal method of refining copper is electrolytic. This method was discussed in Section 19-9; some uses of copper and its compounds were described in Section 21-2.

ALUMINUM. Aluminum is a newcomer to the metals scene; it remained unknown until the nineteenth century. Its existence was postulated by Humphry Davy, who attempted to isolate it by electrolyzing alumina (Al_2O_3) but failed. Friedrich Wöhler isolated the pure metal in 1827, by heating aluminum chloride with potassium metal. Wöhler's method was improved in the 1850s when sodium was substituted for potassium as a reducing agent. This development lowered the price of aluminum from about $90 to $5 per pound; but it still remained a semiprecious metal, used primarily in jewelry and artwork.

The Hall–Heroult Process. A major breakthrough in aluminum production occurred in 1886, as a result of a process developed simultaneously by Charles Martin Hall in the United States and Paul Heroult in France. Hall was a student at Oberlin College at the time of his invention, and Heroult was of the same age as Hall.

A number of interesting chemical principles are involved in the manufacture of aluminum. First, the principal ore, bauxite, consists of 40 to 60% Al_2O_3 with Fe_2O_3, SiO_2, and TiO_2 as the main impurities. The key to producing high-purity aluminum is to start with a pure raw material. Impurities are removed from bauxite ore by taking advantage of the amphoterism of Al_2O_3. Both Al_2O_3 and $Al(OH)_3$ have acidic properties; they are soluble in a strongly basic solution. The impurities are not. When a solution containing AlO_2^- is diluted with water or acidified, $Al(OH)_3$ precipitates.

$$Al_2O_3(s) + 2\ OH^-(aq) \longrightarrow$$
$$2\ AlO_2^-(aq) + H_2O \qquad (26.38)$$

$$AlO_2^-(aq) + H_3O^+ \longrightarrow$$
$$Al(OH)_3(s) \qquad (26.39)$$

Al_2O_3, obtained by heating $Al(OH)_3$, has an extremely high melting point—too high to make its direct electrolysis practicable. Instead, a small percentage of Al_2O_3 is dissolved in molten cryolite, Na_3AlF_6, which has a much lower melting point. The electrolysis cell pictured in Figure 26-13 is operated at about 950°C. Aluminum metal of 99.6 to 99.8% purity is obtained. The electrode reactions are not known with certainty but the net result is

oxidation: $3\ [C(s) + 2\ O^{2-} \longrightarrow$
$CO_2(g) + 4\ e^-]$

reduction: $4\ [Al^{3+} + 3\ e^- \longrightarrow$
$Al(l)]$

net: $3\ C(s) + 4\ Al^{3+} + 6\ O^{2-} \longrightarrow$
$4\ Al(l) + 3\ CO_2(g) \qquad (26.40)$

Uses. Pure aluminum is a malleable, ductile, silvery colored metal of low density. Its density is only

---

Napoleon III's most elegant set of flatware (outranking gold) was made of aluminum. So was the crown worn by Christian X of Denmark. The metal cap atop the Washington Monument is also made of aluminum. It was put in place in 1884, just two years before the Hall–Heroult discovery.

These insoluble impurities, called red mud, present a disposal problem. More than 10 million tons of red mud is produced annually in the United States.

FIGURE 26-13
Electrolysis cell for aluminum production.

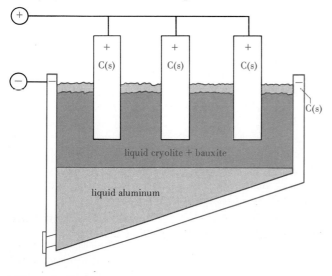

liquid cryolite + bauxite

liquid aluminum

The cathode is a carbon lining in a steel tank. The anodes are also made of carbon. Liquid aluminum is more dense than the electrolyte medium and collects at the bottom of the tank. A crust of frozen electrolyte forms at the top of the cell.

Aluminum in home electrical wiring is now regarded as a fire hazard.

about one third that of steel. The metal is not very strong; but its strength increases considerably when it is alloyed with copper, magnesium, manganese, or silicon. Aluminum-magnesium alloys, because of their very low densities, are used extensively in the aircraft industry. They also find application in the construction industry.

Another important use of aluminum is as an electrical conductor. For a given diameter wire, aluminum has only about 60% of the conductivity of copper, but because of its low density it is a better conductor than copper on a mass basis. For this reason aluminum has found considerable use in recent years in electrical transmission lines. Among the more familiar uses of aluminum is the fabrication of pots, pans, and other kitchen utensils. Both the metal and its ion (Al^{3+}) are nonpoisonous.

Aluminum is a very active metal that would dissolve slowly in water were it not for the fact that the surface of the metal is generally coated with a tight impervious coating of the insoluble oxide, Al_2O_3. This oxide, as we have learned previously, is amphoteric. Thus, aluminum is readily soluble in both acidic and basic solutions, because first the oxide coating dissolves and then the underlying metal. Its maximum resistance to corrosion is exhibited between pH 4.5 and 8.5. Much of the aluminum used commercially is treated in such a way as to build up its oxide coating artificially. In one method, called **anodizing,** an aluminum object is made the anode and a graphite rod the cathode in an electrolyte bath of $H_2SO_4(aq)$. The anode half-reaction is

$$2\,Al(s) + 3\,H_2O \longrightarrow Al_2O_3(s) + 6\,H^+ + 6\,e^- \tag{26.41}$$

Aluminum oxide coatings of varying porosities and thicknesses can be obtained. Also, the oxide can be made to absorb coloring matter or other additives.

Problems and Prospects. Despite the success of the Hall–Heroult process, problems remain. First, there is the problem of availability of raw materials. Cryolite is an essential constituent of the electrolyte bath in the Hall–Heroult process. When this process was first introduced cryolite was obtained from natural sources, but this material is quite rare and aluminum manufacturers have had to develop a *synthetic* cryolite. Several methods are available for this, including the action of hydrofluoric acid on sodium aluminate.

$$12\,HF + Al_2O_3 \cdot 3\,H_2O + 6\,NaOH \longrightarrow 2\,Na_3AlF_6 + 12\,H_2O$$

Bauxite ores are very limited in the United States, constituting only about 1% of the world's reserves. Domestic production in the United States is mainly in Arkansas. Nearly 50% of the need for bauxite in the United States is supplied by Surinam, Australia, and Jamaica.

With one or two exceptions, all the aluminum produced in the world is derived from bauxite ores. Continued production of aluminum by present methods, then, is dependent on a constant supply of these ores. Although aluminum is the most abundant metal in the earth's crust, not much of it is found as bauxite; most of it is in the form of clay minerals. For the past fifty years or more, research has been in progress to develop methods of extracting alumina from clay, but these are not economically feasible as long as bauxite ores are available.

The production of aluminum consumes a great deal of electrical energy—15,000 kilowatt-hours per ton (kWh/ton) compared to 2700 kWh/ton for steel. The implications of this fact are numerous. First, aluminum production centers have tended to develop primarily in regions where abundant electrical power is available at a low cost, and this has meant hydroelectric power chiefly. Second, if aluminum manufacturing is to continue at its present rate of growth, alternative sources of energy or alternative methods of production or both may become necessary. Worldwide aluminum production was about 6000 tons in 1900, 250,000 tons in 1930, and is about 10,000,000 tons currently.

MAGNESIUM. Of all the structural metals magnesium has the lowest density (1.74 g/cm³). Its alloys are widely used in the aircraft and other industries where light-

FIGURE 26-14
The Dow process.

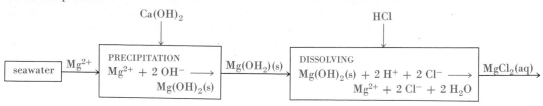

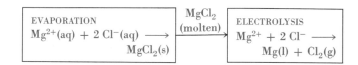

FIGURE 26-15
The electrolysis of molten $MgCl_2$.

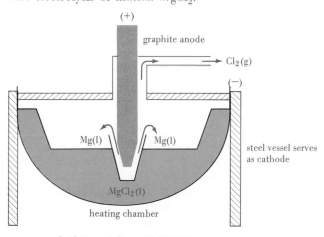

Oxidation: $2 Cl^- \rightarrow Cl_2(g) + 2e^-$
Reduction: $Mg^{2+} + 2e^- \rightarrow Mg(l)$

weight materials are required. The alloy Dowmetal, consisting of 89% Mg, 9% Al, and 2% Zn, has a tensile strength approaching that of steel.

Iron and aluminum occur more abundantly in the earth's crust than magnesium, but their production will become more difficult as high-grade ores are depleted. Magnesium, on the other hand, occurs to the extent of 0.14% in seawater and will remain at this concentration indefinitely. All the magnesium that has yet been produced could have been obtained from as little as 4 km³ of seawater.

In the **Dow process** magnesium is precipitated from seawater as the hydroxide; Mg^{2+} is the only common cation in seawater that forms an insoluble hydroxide. The source of the hydroxide ion is calcium hydroxide, which is derived by heating limestone or oyster shells and treating the resulting calcium oxide with water. (Recall equations 20.4 and 20.5.) The precipitated magnesium hydroxide is washed, filtered, and dissolved in hydrochloric acid. The concentrated $MgCl_2(aq)$ that results is evaporated nearly to dryness. The $MgCl_2$ is then melted and electrolyzed, yielding pure magnesium metal and chlorine gas. These reactions are summarized in Figure 26-14, and the electrolysis process is pictured in Figure 26-15.

TITANIUM. Although titanium is one of the more abundant elements, extensive production of the metal is a recent development, spurred at first by the requirements of the military and then by the aircraft industry. Titanium is a good alternative to aluminum and steel in the construction of aircraft because aluminum loses its strength at high temperatures and steel is too dense.

The principal ore from which the metal is produced is rutile, a rather pure form of TiO_2. The first step in its production involves converting the oxide to a gaseous chloride in the presence of carbon. Although several side reactions occur, the principal reaction is

$$TiO_2(s) + 2\ Cl_2(g) + 2\ C(s) \longrightarrow TiCl_4(g) + 2\ CO(g) \tag{26.42}$$

The purified $TiCl_4$ is next reduced to titanium metal using a good reducing agent. The **Kroll process** uses magnesium.

$$TiCl_4(g) + 2\ Mg(l) \xrightarrow[\text{He}]{850°C} Ti(s) + 2\ MgCl_2(l) \tag{26.43}$$

The reaction is carried out in a steel vessel. The liquefied $MgCl_2$ is removed and electrolyzed to produce Mg, which is recycled in reaction (26.43). The titanium is obtained as a sintered mass called titanium sponge. This material must be subjected to further treatment and alloying with other metals before it can be used. One of the requirements in the commercial development of titanium was that of devising new techniques for fabricating the metal.

In 1947, the United States production of titanium metal was only 2 tons. Today it is measured in the thousands of tons. Approximately 5 tons of titanium alloys are used in each jet engine of a Boeing 747 airplane. With the greatly expanded uses contemplated for titanium, a question arises concerning the future availability of rutile ore. It has been estimated that between 10 and 70 million tons of TiO_2 is recoverable from the world's known reserves of rutile, and this may prove adequate for some time. However, considerable interest has been expressed in obtaining TiO_2 from ilmenite ore, $FeTiO_3$, which is considerably more abundant. Methods are available for doing this, but the TiO_2 cannot be obtained as cheaply nor in as pure a condition as from rutile.

The story of titanium illustrates how a material practically unknown and unused in one decade may become a major production item in the next.

26-8 Some Raw Materials for the Inorganic Chemical Industry

Although the total number of raw materials used by the inorganic chemical industry is quite large and varied, the bulk of the production of this industry is based on just a few very common materials, principally salt, limestone, sand, sulfur, and phosphate rock.

SALT. Ordinary salt or sodium chloride is one of the most important mineral substances. Present annual consumption in the United States is about 50 million tons. Salt is used in the dairy industry, in the treatment of hides, the preservation of meat and fish, the control of ice on streets and highways, and the regeneration of ion exchange resins. In the chemical industry, salt is a source of sodium, chlorine, sodium hydroxide, hydrochloric acid, sodium carbonate, sodium sulfate, and other sodium and chlorine compounds.

Most sodium chloride is obtained from underground sources. Some (about 30% in the United States) is mined as the solid in shaft mines similar to coal mines. Some is obtained as an aqueous solution—brine—which is pumped from deep wells. Still more is obtained by dissolving underground salt in water and pumping this artificial brine. In areas located near the ocean that have warm climates, some salt is produced by the evaporation of seawater.

Sodium metal can be prepared only by the electrolysis of a molten sodium compound, such as NaCl. **Chlorine gas** can be produced either by the electrolysis of molten NaCl or an aqueous solution, mostly by the latter method (see Section 19-9).

Brines, both natural and artificial, are a source not only of sodium chloride but also of potassium compounds (such as KCl), bromine, iodine, and calcium and magnesium chlorides.

Chlorine and its compounds have myriad uses in chemical and related industries. Current consumption of chlorine exceeds 12 million tons annually in the United States. For a long time chlorine has been used in treating water to make it safe for drinking; it is also used in the treatment of swimming pools, sewage effluents, and industrial wastes. All these applications are based on the oxidizing power of chlorine and the hypochlorite ion, which forms when chlorine is dissolved in basic solutions. Hypochlorite solutions are used as bleaching agents in the household and in the production of paper, rayon, and other cellulosic materials. Chlorine is used in the production of carbon tetrachloride (an important solvent and cleaning fluid), ethylene glycol (an automobile antifreeze), and a host of other organic compounds.

Sodium hydroxide, which is produced by the electrolysis of aqueous sodium chloride, is used in petroleum refining and in the manufacture of soaps, textiles, organic chemicals, inorganic chemicals, and plastics. The annual U.S. consumption of this substance is about 12 million tons.

Sodium sulfate is a widely used inorganic chemical which can be derived from sodium chloride. Current annual production in the United States is about 1 million tons. Nearly half is obtained from natural sources, but much of it is manufactured by a process devised by J. R. Glauber (1604–1670).

$$H_2SO_4 + 2\,NaCl \xrightarrow{\text{heat}} Na_2SO_4 + 2\,HCl(g) \tag{26.44}$$

The method works well because HCl is volatile whereas H_2SO_4 is not. Sodium sulfate is also obtained as a by-product in other processes, such as the manufacture of rayon. The paper industry consumes about 70% of the United States annual production. A basic step in paper manufacturing is to convert wood to wood pulp by dissolving lignin, the noncarbohydrate portion of wood. In the kraft process this is accomplished by digesting wood in an alkaline solution of Na_2S. Na_2S is produced by reducing Na_2SO_4 with carbon.

$$Na_2SO_4 + 4\,C \xrightarrow{\text{heat}} Na_2S + 4\,CO \tag{26.45}$$

About 100 lb of Na_2SO_4 is required for every ton of paper produced.

LIMESTONE. Limestone is a naturally occurring form of calcium carbonate containing some clay and other impurities. It is the most widely used type of rock. Limestone's primary use is as a building stone (about two thirds of United States production). Other uses include the manufacture of cement (about 15%), as a flux in metallurgical processes (about 5%), as a source of quicklime and slaked lime (5%) and as an ingredient of glass.

For commercial purposes the term "limestone" includes both $CaCO_3$ and the mixed carbonate known as dolomite, $CaCO_3 \cdot MgCO_3$.

Glass. Calcium silicate is the familiar slag produced in blast furnaces. It is a liquid at high temperatures, but at normal temperatures it is a brittle water-insoluble solid. Sodium silicate, Na_2SiO_3, the product of the reaction of sodium carbonate and silica, is highly soluble in water. It is generally encountered in concentrated aqueous solution under the name "water glass."

If a mixture of sodium and calcium carbonate is fused with sand at approximately 1500°C, the product is a mixture of sodium and calcium silicates. This mixture is a liquid which, on cooling from high temperatures, becomes so viscous that, for practical purposes, it ceases to flow. (Actually, it does flow very slowly with time.) The product has the appearance of a solid except that, even in sheets of considerable thickness, it is transparent to visible light. What has just been described is ordinary glass or soda-lime glass. Variations in the proportions of the three basic ingredients, together with the admixture of other substances, can be used to alter the properties of glass. For example, borosilicate glass (Pyrex) contains about 13% of the oxide of boron, B_2O_3.

It is customary to refer to a substance in a glasslike state as a "glass," whether it is of silicate origin or not. This terminology is based on the uniqueness of the glass structure. Because there is no long-range order in the arrangements of their atoms, glasses are noncrystalline. X ray diffraction patterns of glasses resemble those of liquids; and glasses do not have definite melting points.

Portland Cement.   Portland cement is a complex mixture of calcium silicates and aluminates. It is formed by heating limestone with other materials rich in silica (SiO_2) and alumina (Al_2O_3) at about 1500°C. Minor components may also be present. The product material is pulverized before use. The reactions of cement with water and its subsequent hardening are indeed complicated. They seem to involve both hydrolysis and hydration of the cement components. Certain of the reactions occur almost immediately upon mixing cement with water. Others progress over a period of months or years. Because only water is required to set portland cement, it is referred to as a hydraulic cement. It will set even when completely submerged in water, as in the construction of bridge piers. Pure cement does not have a great deal of strength, but when mixed with sand and gravel it sets into a hard mass especially useful in a wide variety of building applications. This mixture is common concrete.

Lime.   The term "lime" is applicable to two different calcium compounds—CaO, called quicklime, and $Ca(OH)_2$, called slaked lime. The chemical reactions by which they are produced are indeed simple.

$$CaCO_3(s) \longrightarrow CaO(s) + CO_2(g) \tag{26.46}$$

$$CaO(s) + H_2O(l) \longrightarrow Ca(OH)_2(s) \tag{26.47}$$

Reaction (26.46) is reversible. At room temperature the reverse reaction occurs almost exclusively: Calcium oxide exposed to atmospheric carbon dioxide is slowly converted back to the carbonate. As the temperature is increased, the forward reaction becomes more favorable, especially if carbon dioxide is continuously removed. The commercial production of CaO is accomplished by heating limestone at about 900°C in a furnace designed to exhaust effectively the products of the limestone decomposition.

Slaked lime is the cheapest commercial alkaline substance. It is used in all applications where high water solubility is not essential. Slaked lime is used in the manufacture of other alkalis and bleaching powder, in the purification of sugar, in tanning hides, and in water softening. A mixture of slaked lime, sand, and water is the familiar mortar used in brick laying. The initial setting of mortar results from the absorption of excess water by the bricks and its loss by evaporation. The final hardening of mortar, however, involves the conversion of calcium hydroxide back to calcium carbonate by reaction with atmospheric carbon dioxide.

$$Ca(OH)_2(s) + CO_2(g) \longrightarrow CaCO_3(s) + H_2O(g) \tag{26.48}$$

SODIUM CARBONATE (SODA ASH).   Sodium carbonate is of considerable commercial value, especially in the manufacture of glass. Of the several million tons produced annually in the United States, somewhat more than half comes from natural sources such as Searles Lake and Owens Lake, both dry lakes in California, and from immense deposits in western Wyoming. The remainder is produced by a process developed by the Belgian chemist Solvay in 1863.

In the **Solvay process,** a cold concentrated solution of $NaCl$ is treated with $NH_3(g)$ and $CO_2(g)$. $NaHCO_3$ (sodium bicarbonate) precipitates.

$$Na^+ + Cl^- + NH_3 + CO_2 + H_2O \longrightarrow NaHCO_3(s) + NH_4^+ + Cl^- \tag{26.49}$$

Chemical reactions incidental to the main reaction are summarized in Figure 26-16. Sodium carbonate is produced from the bicarbonate by heating.

The manufacture of glass accounts for slightly over one half of the consumption of soda ash in the United States. Other uses include the manufacture of chemicals, about 20%; pulp and paper, 6%; soaps and detergents, 5%; and water treatment, 3%.

FIGURE 26-16
The Solvay process for the manufacture of $NaHCO_3$.

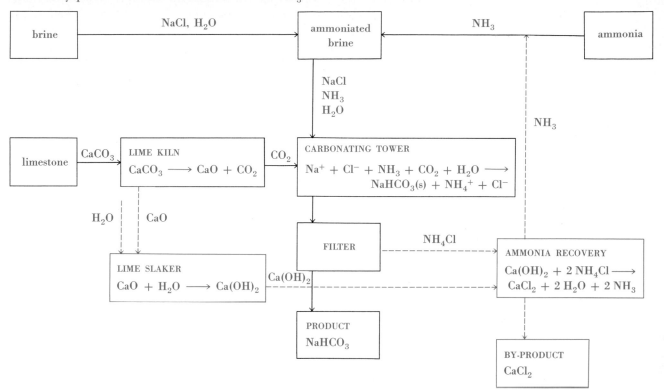

The main reaction sequence is traced by solid arrows. Recycling reactions are shown by broken arrows.

$$2\,NaHCO_3(s) \xrightarrow{\text{heat}} Na_2CO_3(s) + H_2O(g) + CO_2(g) \qquad (26.50)$$

The success of the Solvay process in the early decades of this century stemmed from the efficient use made of the raw materials through recycling. Only one by-product results, calcium chloride. The demand for calcium chloride is quite limited and only a small percentage of the several million tons produced annually is consumed. The rest has commonly been dumped, generally in streams, but this is no longer permitted. Ways must be found to recover and reuse the $CaCl_2$ or the Solvay process will have to be phased out. Here is an example of a process where the factor of economic feasibility is no longer sufficient. Environmental concerns are overriding. Fortunately, the Wyoming deposits from which Na_2CO_3 can be extracted are estimated to be sufficient for at least 3000 years.

$CaCl_2 \cdot 2\,H_2O$ is currently being studied as a possible solar energy storage medium.

SULFUR. Extensive deposits of sulfur are found in the United States in Texas and Louisiana, some of them at offshore sites. The ingenious Frasch process for mining sulfur is pictured in Figure 26-17. Superheated water is forced down the outermost of three concentric pipes into an underground bed of sulfur containing rock. The sulfur is melted and a liquid pool forms. Compressed air is passed through the innermost pipe, forcing liquid sulfur up the remaining pipe. At the surface the liquid sulfur is collected in large bins and allowed to cool. The solid sulfur obtained is approximately 99.5% pure.

Until a few years ago elemental sulfur was in short supply, but now supplies exceed demands. With the adoption of control standards for sulfur dioxide emis-

FIGURE 26-17
The Frasch process for mining sulfur.

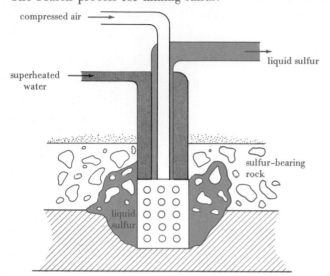

compressed air

liquid sulfur

superheated water

sulfur-bearing rock

liquid sulfur

sions, it has become necessary to remove sulfur compounds from certain natural gas supplies before burning. Currently the rate of production of by-product sulfur from natural gas is about equal to that obtained by the Frasch process.

Sulfuric Acid. Sulfuric acid has a long history; it has been produced for at least 500 years and possibly as much as 1000 years. The sulfuric acid "industry" can be dated from about the middle eighteenth century from a process in which sulfur and potassium nitrate were burned together and the gaseous products dissolved in water in lead-lined chambers, yielding $H_2SO_4(aq)$. Later it was discovered that oxides of nitrogen produced by heating KNO_3 acted as intermediates in the process—they could be recovered and re-used. We now consider intermediates of this type to be catalysts, and the manufacture of sulfuric acid by the "lead chamber" process is one of the earliest examples of *homogeneous catalysis*. A simplified description of the essential reactions, emphasizing the role of nitrogen oxides, is

$$S(s) + O_2(g) \longrightarrow SO_2(g) \tag{26.51}$$

$$NO(g) + \tfrac{1}{2}O_2(g) \longrightarrow NO_2(g) \tag{26.52}$$

$$NO_2(g) + SO_2(g) \longrightarrow SO_3(g) + NO(g) \tag{26.53}$$

$$SO_3(g) + H_2O(l) \longrightarrow H_2SO_4(aq) \tag{26.54}$$

net: $$S(s) + \tfrac{3}{2}O_2(g) + H_2O(l) \longrightarrow H_2SO_4(aq) \tag{26.55}$$

The direct conversion of $SO_2(g)$ to $SO_3(g)$ is a very slow reaction,

(slow) $$SO_2(g) + \tfrac{1}{2}O_2(g) \longrightarrow SO_3(g) \tag{26.56}$$

The pair of reactions, (26.52) and (26.53), provide an alternative mechanism whereby $SO_2(g)$ is converted to $SO_3(g)$ more rapidly. It is here that homogeneous catalysis occurs.

A method of catalyzing the direct conversion of $SO_2(g)$ to $SO_3(g)$ was patented in England in 1831. It involved heating a mixture of $SO_2(g)$ and $O_2(g)$ in the presence of platinum metal. Here *heterogeneous catalysis* is involved. Adsorption of $SO_2(g)$ and $O_2(g)$ on the platinum metal (V_2O_5 may also be used) is followed by reaction at active sites and desorption of the SO_3 produced (recall Figure 13-15). The net reaction is (26.56). The advantages of this "contact" process over the lead chamber process were not fully appreciated until later in the nineteenth century, when the need arose in the synthetic organic chemical industry for a material variously known as *oleum* or *fuming sulfuric acid*. Oleum is a solution of excess SO_3 in pure sulfuric acid (in a sense it represents "greater than 100% sulfuric acid"). Oleum cannot be produced by the lead chamber process. In

By-product sulfur (as from the treatment of natural gas) contains impurities that poison the catalysts used in the contact process for sulfuric acid. It can only be used in the older lead chamber process.

the contact process pure $SO_3(g)$ is passed into 98 or 99% sulfuric acid to form oleum. This can be diluted with water to the exact strength of H_2SO_4 desired. For example, if we use the formula $H_2S_2O_7$ (pyrosulfuric acid) for a particular oleum with 45% SO_3, by mass, the reactions are

$$SO_3(g) + H_2SO_4(l) \longrightarrow H_2S_2O_7(l) \text{ (oleum)} \tag{26.57}$$

$$H_2S_2O_7(l) + H_2O(l) \longrightarrow 2\ H_2SO_4(l) \tag{26.58}$$

$$H_2SO_4(l) + aq \longrightarrow H_2SO_4(aq) \tag{26.59}$$

The use of sulfuric acid in the United States is so widespread that the vitality of the economy as a whole can be gauged by the quantity of sulfuric acid consumed. Sulfuric acid production has been termed a "barometer" of the American economy. Sulfuric acid ranks first among all manufactured chemicals, with an annual production typically exceeding 40 million tons. Principal uses include the manufacture of fertilizers (60%) and other chemicals (12%), the refining of petroleum (6%) and metals (5%), and the production of paints and pigments (3%) and rayon and cellulose film (3%).

PHOSPHATE ROCK. The principal source of phosphorus compounds is phosphate rock—a complex material containing the mineral fluorapatite, $[3\ Ca_3(PO_4)_2 \cdot CaF_2]$. Calcium orthophosphate, $Ca_3(PO_4)_2$, can be extracted from this mineral and used in the preparation of elemental phosphorus, by the method outlined in Section 20-8. If phosphorus (usually as a liquid) is burned in air, the product is P_4O_{10}. P_4O_{10} is the acid anhydride of orthophosphoric acid.

$$P_4O_{10} + 6\ H_2O \longrightarrow 4\ H_3PO_4(aq) \tag{26.60}$$

An impure form of phosphoric acid can be prepared by the direct action of sulfuric acid on phosphate rock.

$$[3\ Ca_3(PO_4)_2 \cdot CaF_2] + 10\ H_2SO_4 + 20\ H_2O \longrightarrow$$
fluorapatite

$$6\ H_3PO_4 + 10\ [CaSO_4 \cdot 2\ H_2O] + 2\ HF \tag{26.61}$$
gypsum

Another important use is in the production of polyphosphates for detergents.

The principal use of phosphorus compounds is in fertilizers. A mixture of calcium sulfate and the more soluble calcium dihydrogen phosphate, called **normal superphosphate,** has a phosphorus content equivalent to 15 to 20% P_2O_5. It is produced by treating phosphate rock with sulfuric acid in a reaction similar to (26.61) but employing different proportions of the rock and acid.

$$[3\ Ca_3(PO_4)_2 \cdot CaF_2] + 7\ H_2SO_4 + 3\ H_2O \longrightarrow$$
$$3\ [Ca(H_2PO_4)_2 \cdot H_2O] + 7\ CaSO_4 + 2\ HF \tag{26.62}$$
normal superphosphate

If phosphate rock is treated with phosphoric acid (derived from reaction (26.61)) instead of sulfuric acid, the product is known as **triple superphosphate.** This process eliminates $CaSO_4$, and the product has a much higher phosphorus content than normal superphosphate, about 46 to 48% expressed as P_2O_5.

$$[3\ Ca_3(PO_4)_2 \cdot CaF_2] + 14\ H_3PO_4 + 10\ H_2O \longrightarrow$$
$$10\ [Ca(H_2PO_4)_2 \cdot H_2O] + 2\ HF \tag{26.63}$$
triple superphosphate

Serious environmental problems are encountered in the use of phosphate rock. Hydrogen fluoride is a product of superphosphate production. In some instances it is recovered, but in the past it was mostly released into waterways. Now, because of strict environmental controls, it has become necessary to neutralize the HF, usually with lime. Large settling ponds are required for this reaction. Because two thirds of the phosphate rock is waste, enormous deposits of waste rock are accumulated in fertilizer manufacture. The handling of this waste adds to the cost and complexity of the total operation.

26-9 Raw Materials for the Organic Chemical Industry

The two primary sources of organic compounds are coal and petroleum, mostly the latter. A smaller but still significant source is biomass. In the middle decades of this century chemical industry turned from coal to petroleum as a source of chemical raw materials. However, because of the dramatic increase in petroleum prices over the past decade, industry is once again looking to coal as an important chemical resource. Needless to say, it is unwise to overexpend either coal or petroleum as fuels because of their unique role in supplying so many other essential commodities.

COAL. Coal is an organic, rocklike material with a high ratio of carbon to hydrogen and other elements. (One proposed formula for a "molecule" of bituminous coal is $C_{153}H_{115}N_3O_{13}S_2$.) To synthesize hydrocarbons or other desired organic compounds from coal requires decreasing the C/H ratio.

In the method of **pyrolysis,** coal (usually bituminous coal) is heated to a high temperature (350 to 1000°C) in the absence of air. Volatile products are formed and an impure carbon residue called **coke** remains. Condensation of the volatile products of this destructive distillation yields black viscous **coal tar.**

$$\text{coal} \xrightarrow[\substack{\text{(absence} \\ \text{of air)}}]{\text{heat}} \text{coke} + \text{coal tar} + \text{coal gas}$$

One ton of bituminous coal yields about 1500 lb of coke, 8 gal of coal tar, and 10,000 ft^3 of coal gas. Coal gas is a mixture of H_2, CH_4, CO, C_2H_6, NH_3, CO_2, H_2S, and other components. At one time coal gas was used as a fuel. Coal tar can be distilled to yield the fractions listed in Table 26-9. From these fractions, in turn, other organic chemicals can be produced.

Pyrolysis can be thought of as a carbon-removal process. Coke is removed and the remaining products are correspondingly enriched in hydrogen and other elements. Coal gasification or liquefaction schemes involve the addition of hydrogen (and usually also oxygen). In general these schemes are based on chemical reactions that have been known for 75 years of more, but they have been updated by new

TABLE 26-9
Coal tar fractions

Boiling range	Name	Tar, mass %	Primary constituents
below 200°C	light oil	5	benzene, toluene, xylenes
200–250	middle oil (carbolic oil)	17	naphthalene, phenol, pyridine
250–300	heavy oil (creosote oil)	7	naphthalenes and methylnaphthalenes, cresols, quinoline
300–350	green oil	9	anthracene, carbazole
residue	—	62	pitch or tar

TABLE 26-10
Principal petroleum fractions

Boiling range, °C	Composition	Fractions	Uses
0–30	C_1–C_4	gas	gaseous fuel
30–60	C_5–C_7	petroleum ether	solvents
60–100	C_6–C_8	ligroin	solvents
70–150	C_6–C_9	gasoline	motor fuel
175–300	C_{10}–C_{16}	kerosene	jet fuel, diesel oil
over 300	C_{16}–C_{18}	gas-oil	diesel fuel, cracking stock
—	C_{18}–C_{20}	wax-oil	lubricating oil, mineral oil, cracking stock
—	C_{21}–C_{40}	paraffin wax	candles, wax paper
—	above C_{40} plus C	residuum	roofing tar, road materials, waterproofing

technology, particularly new catalyst systems. One approach, for example, is to burn a coal–water slurry to obtain a mixture of CO(g) and H_2(g). This gaseous mixture is converted to methanol, and with the proper catalysts, the methanol is converted to acetic acid. Heat evolved in burning the coal is used to meet heat requirements in other parts of the process. Sulfur is removed from the coal and converted to H_2SO_4(aq). The process is nonpolluting, energy efficient, and produces only the desired end products (together with some CO_2) from coal and water as starting materials.

PETROLEUM.   The principal constituents of crude oil are aliphatic hydrocarbons. Certain low-molecular-weight hydrocarbons are found dissolved in crude oil or are produced in the manufacture of gasoline. These compounds are removed and compressed into liquid form in cylinders. Propane and butane sold in this form are known as **liquefied petroleum gas (LPG).**

Crude oil is indeed a complex mixture. It has been estimated that in petroleum boiling up to 200°C there are at least 500 compounds, some aliphatic, some alicyclic, and some aromatic. Petroleum is refined by distillation into various fractions. A typical fractionation yields the products listed in Table 26-10.

Fuel Production.   Not all of the gasoline components listed in Table 26-10 are equally desirable as fuels. Some of them burn more smoothly than others. (Explosive burning results in engine "knocking.") The octane hydrocarbon, **2,2,4-trimethylpentane,** has excellent engine performance; it is given an octane rating of 100. **n-Heptane** has poor engine performance; its octane rating is set at 0. These two hydrocarbons serve as a basis for establishing the quality of automotive fuels, which are mixtures of a large number of hydrocarbons. In general, branched chain hydrocarbons have higher octane numbers than their straight chain counterparts.

A distillation column for the fractionation of liquid air was shown in Figure 26-1. The distillation columns used in petroleum refining are similar but more complex.

$$CH_3-\underset{\underset{CH_3}{|}}{\overset{\overset{CH_3}{|}}{C}}-CH_2-\underset{\underset{H}{|}}{\overset{\overset{CH_3}{|}}{C}}-CH_3 \qquad CH_3-CH_2-CH_2-CH_2-CH_2-CH_2-CH_3$$

2,2,4-trimethylpentane
(isooctane)
octane rating: 100

n-heptane
octane rating: 0

Gasoline obtained by the fractional distillation of petroleum has an octane number of 50 to 55 and is not acceptable for use in automobiles. Extensive modifications

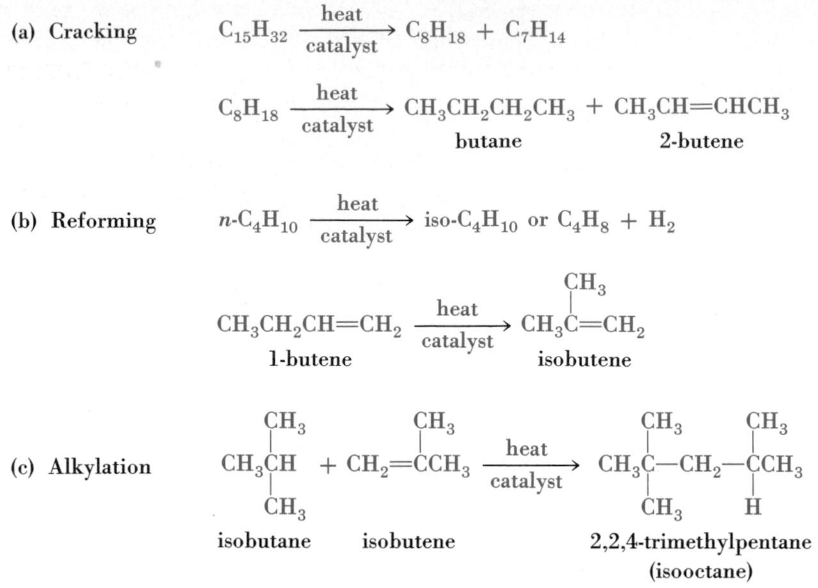

FIGURE 26-18
Some reactions associated with the production of gasoline.

(a) Cracking

$$C_{15}H_{32} \xrightarrow[\text{catalyst}]{\text{heat}} C_8H_{18} + C_7H_{14}$$

$$C_8H_{18} \xrightarrow[\text{catalyst}]{\text{heat}} \underset{\text{butane}}{CH_3CH_2CH_2CH_3} + \underset{\text{2-butene}}{CH_3CH{=}CHCH_3}$$

(b) Reforming

$$n\text{-}C_4H_{10} \xrightarrow[\text{catalyst}]{\text{heat}} \text{iso-}C_4H_{10} \text{ or } C_4H_8 + H_2$$

$$\underset{\text{1-butene}}{CH_3CH_2CH{=}CH_2} \xrightarrow[\text{catalyst}]{\text{heat}} \underset{\text{isobutene}}{CH_3\overset{\displaystyle CH_3}{\underset{}{C}}{=}CH_2}$$

(c) Alkylation

$$\underset{\text{isobutane}}{CH_3\overset{\displaystyle CH_3}{\underset{\displaystyle CH_3}{CH}}} + \underset{\text{isobutene}}{CH_2{=}\overset{\displaystyle CH_3}{C}CH_3} \xrightarrow[\text{catalyst}]{\text{heat}} \underset{\substack{\text{2,2,4-trimethylpentane}\\\text{(isooctane)}}}{CH_3\overset{\displaystyle CH_3}{\underset{\displaystyle CH_3}{C}}{-}CH_2{-}\overset{\displaystyle CH_3}{\underset{\displaystyle H}{C}}CH_3}$$

of its composition are required. The principal methods employed are of three types—thermal and catalytic cracking, reforming, and alkylation. The chemical changes involved are represented by the equations in Figure 26-18.

In **thermal cracking** large hydrocarbon molecules are broken down into molecules in the gasoline range. The presence of special catalysts promotes the production of branched chain hydrocarbons. The process known as **reforming** or isomerization converts straight chain to branched chain hydrocarbons. Also, alicyclic hydrocarbons are converted to aromatic hydrocarbons, which possess higher octane numbers. In thermal and catalytic cracking, some of the products are low molecular weight, unsaturated hydrocarbons or olefins. In the **alkylation** process these unsaturated compounds are polymerized to higher molecular weight olefins. These can be used directly as fuel components or hydrogenated to produce saturated hydrocarbons.

The octane rating of gasoline has been further improved by the addition of certain "antiknock" compounds which prevent premature combustion. Most widely used have been tetraethyllead, $(C_2H_5)_4Pb$, and tetramethyllead, $(CH_3)_4Pb$. The addition of 6 ml of tetraethyllead to 1 gal of 2,2,4-trimethylpentane raises its octane rating to 120.3. Because of the toxicity of this additive, however, only about one half as much is permitted.

In order to combine with the lead that would otherwise be deposited in the automobile engine, ethylene dibromide, $BrCH_2CH_2Br$, or ethylene dichloride, $ClCH_2CH_2Cl$, is also added to gasoline. The lead halides thus formed are exhausted to the atmosphere, creating an environmental hazard. In recent years lead-free gasolines have been appearing on the market for use in automobiles with catalytic converters. Their production requires modifications in the gasoline refining process to increase octane ratings in another way, such as increasing the proportion of aromatic hydrocarbons. These modifications account for the fact that lead-free gasoline is more expensive to produce than a gasoline with a lead additive.

Other additives to prevent fuel-line freeze-up, carburetor icing, spark-plug fouling, engine corrosion, and engine deposits make gasoline as complex a mixture as the crude oil from which it comes.

Petrochemicals. Although the principal products of the petroleum industry are fuels, chemicals produced from petroleum—petrochemicals—are essential to modern society. Current annual production of benzene in the United States exceeds 11 billion lb. Over 90% of this is produced from petroleum. The process involves cyclization and dehydrogenation of *n*-hexane to the aromatic hydrocarbon. Of the petroleum-produced benzene, about 40% is used to manufacture ethylbenzene for the production of styrene plastics, 18% to manufacture phenol, 6% to synthesize dodecylbenzene (for detergents), and 2% to make aniline. The production of aromatic compounds by dehydrogenation of alkanes yields large amounts of hydrogen gas. An important use of this hydrogen is in the Haber synthesis of ammonia.

BIOMASS.   As an energy source "biomass" is any material produced by photosynthesis, that is, plants or their principal components (cellulose, starch, sugars). Some biomass (e.g., wood) may be used directly as a fuel. Some may be converted to other gaseous, liquid, or solid materials for use as fuels or chemical raw materials.

Perhaps the best known and most widely used biomass conversion method involves the fermentation of sugars to produce ethanol. A fermentation process involves the decomposition of organic matter in the absence of air through the action of a microorganism.

$$\text{hexose sugar} \xrightarrow[\text{in yeast}]{\text{microorganisms}} 2\ C_2H_5OH + 2\ CO_2(g)$$

Disaccharides such as sucrose and polysaccharides (e.g., starch) can be hydrolyzed into monosaccharides by enzymes and then fermented to ethanol. The principal raw material for the industrial production of ethanol by fermentation is corn. Ethanol from this source is currently finding some use in the fuel "gasohol," a mixture of 10% ethanol and 90% gasoline.

The principal biomass material that may find increasing use as a chemical raw material is cellulose. This polysaccharide constitutes the skeletal material of plants. It is the most abundant organic substance in nature. Cotton, which is 98% cellulose, is the chief source of cellulose used as fiber. Cellulose accounts for about 10% of the dry mass of leaves and 50% of the content of wood. The technical uses of cellulose are dependent on its fibrous nature and on the strength and flexibility of materials produced from it.

Plants are no longer being converted in significant quantities to fossil fuels (coal, petroleum, natural gas) by geologic processes. In principle some of the same compounds now being produced from petroleum could be made directly from cellulose. Methanol (wood alcohol) is formed in the destructive distillation (pyrolysis) of wood. Cellulose can be hydrolyzed to glucose and then converted to ethanol by fermentation. Also, fermentation processes might be used to produce a series of oxygenated compounds—alcohols and ketones. These could then be converted to hydrocarbons. Thus, the entire spectrum of organic chemicals could be produced from the simple molecules CO_2 and H_2O. The required energy would be mostly solar. Combustion of the organic chemicals or products made from them would simply return CO_2 and H_2O to the environment.

Unsaturated hydrocarbons can be formed by the elimination of H_2O from an alcohol molecule (recall equation 24.7).

26-10   Polymers

A polymer is a compound of high molecular weight formed by the combination of a large number of small molecules called **monomers.** The monomers may be all of one type or of different types. The process of polymerization was introduced in Section 24-3 and a number of naturally occurring polymers—starch, cellulose, and proteins—were discussed in Chapter 25. Silicone polymers were described in Section 20-9.

It has been estimated that about one half of all chemists in the United States are working in the fields of polymer science and technology.

The name "rubber" is attributed to the chemist Joseph Priestley (1770), who discovered that it could be used to rub out pencil marks. Its chief use for several decades was in pencil erasers.

If the —CH$_2$—CH$_2$— groups in this structure are trans rather than cis, the substance is called gutta-percha, a hard brittle plastic. It has found use in transoceanic cables.

Prior to 1930 only a few synthetic polymers were known (Bakelite, for example). Nylon was first produced in 1935 by Wallace Carothers. In the 1950s a number of significant discoveries were made concerning the nature of polymerization reactions. As a result the polymer industry currently produces some 30 billion pounds of polymer materials annually.

RUBBER.   The exudate of the tree *Hevea brasiliensis,* called **latex,** is collected as a milky white colloidal fluid and coagulated with salt and acetic acid. The precipitated rubber is washed and smoked to retard the growth of mold. This crude polymer is refined and processed according to its intended use.

The structure of rubber was proposed by Pickles in 1910 to be that of a polymer of isoprene (2-methyl-1,3-butadiene). Each pair of —CH$_2$—CH$_2$— units lies on the same side of the double bond between them; rubber is a *cis* polymer.

rubber

Early rubber products were sticky in hot weather and stiff in cold weather. In 1839, Charles Goodyear, a Connecticut inventor, accidentally discovered a way to make rubber stronger, more elastic, and more resistant to heat and cold. (He spilled a sulfur-rubber mixture on a hot stove while conducting an experiment.) What Goodyear discovered was the process called vulcanization (after Vulcan, the Roman god of fire). The purpose of vulcanization is to form cross-links between long polymer chains. In the example given below, the link consists of a disulfide bridge.

Modern technology employs other additives to rubber as well as sulfur. Accelerators (zinc oxide and stearic acid) make cross-linking occur faster and more uniformly. Antioxidants (secondary aromatic amines) retard degradation by heat, sunlight, and atmospheric ozone. Reinforcing agents provide abrasion resistance and increased tensile strength.

SYNTHETIC POLYMERS.   Synthetic polymers can be classified in several ways: according to their structures, their physical properties, the processes by which they are prepared, or their uses. Most synthetic polymers in use today are prepared by one of two methods. In **condensation polymerization,** depicted in Figure 26-19, small molecules, such as water, are eliminated as monomer units attach themselves to the polymer chain. In **free-radical addition polymerization** (see Figure 26-20) monomer units add directly to a chain, with the point of attack being at a site containing an unpaired electron.

Let us consider the mechanism of free-radical addition polymerization a bit further. A small amount of benzoyl peroxide is present as an initiator. A molecule of

FIGURE 26-19
Condensation
polymerization—
formation of Dacron.

The reaction is carried
out at elevated
temperatures, reduced
pressures, and in the
presence of sodium
methoxide, CH_3ONa.

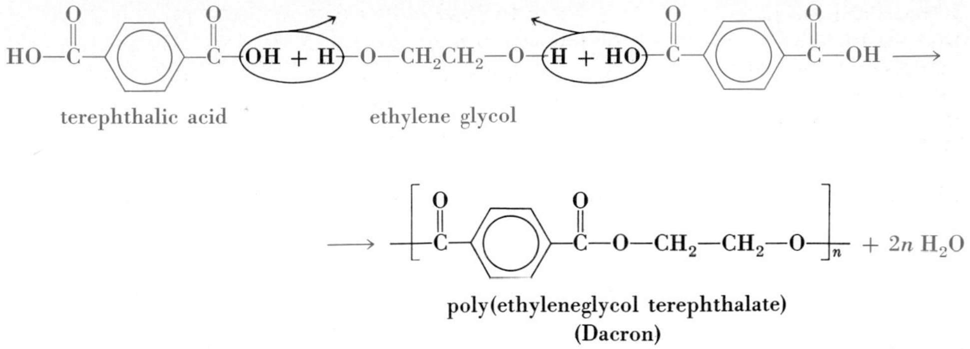

terephthalic acid   ethylene glycol

poly(ethyleneglycol terephthalate)
(Dacron)

benzoyl peroxide decomposes at 70°C to form two benzoyloxy radicals, which in turn lose CO_2 molecules to become phenyl radicals. A phenyl radical adds to a molecule of styrene to produce the more stable diphenylethane radical, which in turn adds another styrene molecule, and so on. Termination of the chain occurs when two free radicals combine. Addition polymerization generally produces longer chains and higher molecular weights than does condensation polymerization. [Because the sizes of polymer molecules are variable, the molecular weight of a polymer can only be thought of in terms of an average value (see Exercise 40).]

FIGURE 26-20
Free-radical addition
polymerization.

Free-radical formation:

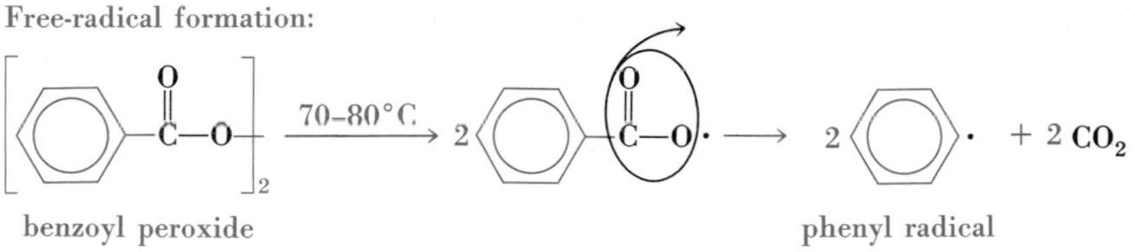

benzoyl peroxide   phenyl radical

Chain propagation:

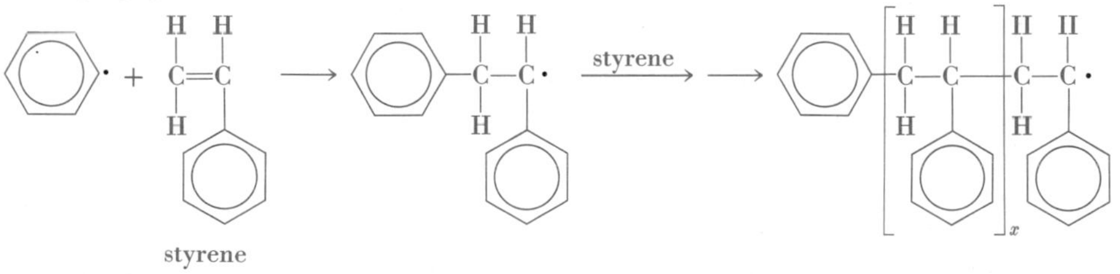

styrene

Chain termination:

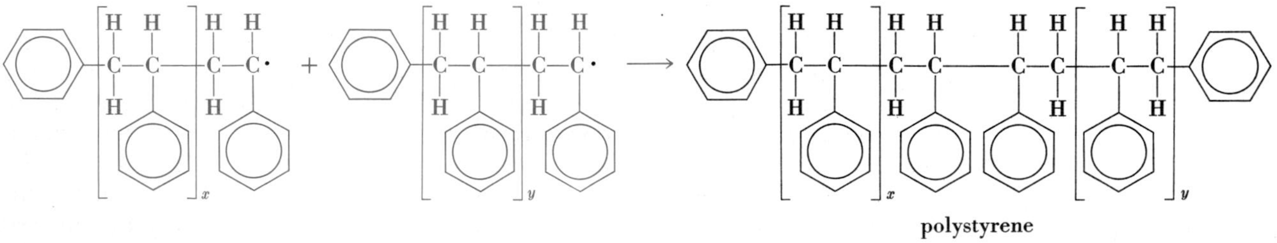

polystyrene

The method of chain termination illustrated here—the joining of two free-radical chains—is just one of several possibilities. Termination will also occur as a result of the reaction of a chain with a radical initiator (phenyl radical).

TABLE 26-11
Some synthetic carbon-chain polymers

Name	Monomer(s)	Polymer

Elastomers

neoprene
   [polychloroprene]

$$H_2C=CH-\overset{\overset{\displaystyle Cl}{|}}{C}=CH_2$$

(chloroprene)

$$\left(CH_2-CH=\overset{\overset{\displaystyle Cl}{|}}{C}-CH_2\right)_x$$

GRS, SBR, Buna S
   [poly(butadiene-co-styrene)]

$$H_2C=CHCH=CH_2 \; + \;$$ (styrene)

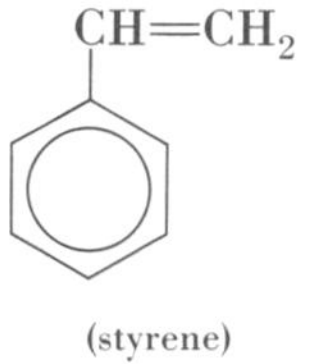

(1,3-butadiene)

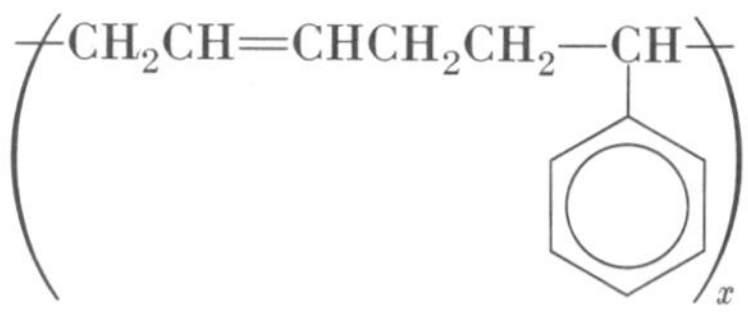

Fibers

Dacron, Terylene, Fortrel

$$HOCH_2CH_2OH + HO_2C-\bigcirc-CO_2H$$

(ethylene glycol)   (terephthalic acid)

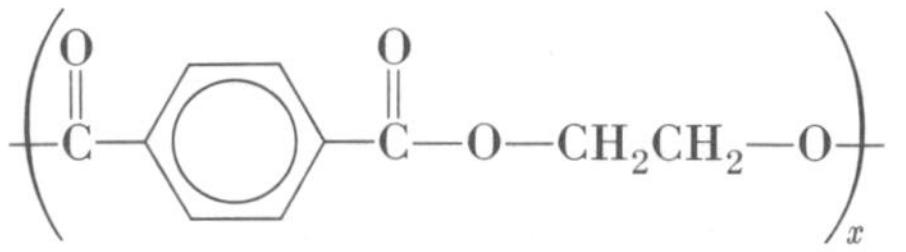

nylon 66

$$HO_2C(CH_2)_4CO_2H + H_2N(CH_2)_6NH_2$$
(adipic acid)   (1,6-hexanediamine)

$$\left(\overset{\overset{\displaystyle O}{\|}}{C}-(CH_2)_4-\overset{\overset{\displaystyle O}{\|}}{C}-NH-(CH_2)_6-NH\right)_x$$

Plastics

polyethylene

$$H_2C=CH_2$$
   (ethylene)

$$(CH_2-CH_2)_x$$

PVC, "vinyl"
   [poly(vinyl chloride)]

$$H_2C=CHCl$$
(vinyl chloride)

$$\left(CH_2-\overset{\overset{\displaystyle Cl}{|}}{CH}\right)_x$$

Teflon
   [poly(tetrafluoroethylene)]

$$F_2C=CF_2$$
(tetrafluoroethylene)

$$\left(\overset{\overset{\displaystyle F}{|}}{\underset{\underset{\displaystyle F}{|}}{C}}-\overset{\overset{\displaystyle F}{|}}{\underset{\underset{\displaystyle F}{|}}{C}}\right)_x$$

Lucite, Plexiglas
   [poly(methyl methacrylate)]

$$H_2C=\overset{\overset{\displaystyle CH_3}{|}}{C}-CO_2CH_3$$
(methyl methacrylate)

$$\left(CH_2-\overset{\overset{\displaystyle CH_3}{|}}{\underset{\underset{\displaystyle CO_2CH_3}{|}}{C}}\right)_x$$

Uses
wire and cable insulators industrial hoses and belts, shoe soles and heels, gloves
tires, hoses, flooring, shoe soles and heels
fabrics, boat sails, tire cord
fabrics, hosiery, rope
household products, bottles, tubing
bottles, records, floor tile, food wrap
insulation, gaskets, non-stick surfaces (oven-wear, frying pans)
signs, display cabinets, combs, dentures, auto tail- and signal-light lenses, reflectors

Example 26-3 The polymer Acrilan (used in fabrics, blankets, and carpets) is derived from the monomer, acrylonitrile, $CH_2{=}CHCN$, in a free-radical addition polymerization. What is the structure of this polymer?

The net result of the polymerization reaction (as in Figure 26-20) is that the double bond opens up and monomer units add to the growing chain. (The R group is the chain initiator.)

$$R{-}CH_2{-}\underset{\underset{CN}{|}}{CH}{-}CH_2{-}\underset{\underset{CN}{|}}{CH}{-}CH_2{-}\underset{\underset{CN}{|}}{CH}{-} \; \dots$$

SIMILAR EXAMPLES: Exercises 37, 39.

Polymers may be classified as elastomers, fibers, or plastics. Plastics account for about two thirds of the polymer production in the United States. The remaining one third is divided about equally between elastomers and fibers. Several examples of synthetic carbon chain polymers are given in Table 26-11.

Elastomers. The chief characteristic of elastomers is their ability to be elongated under stress and to regain their former shape when the stress is relieved. In short, they are elastic. Rubber, whether natural or synthetic, is the best known elastomer. Silicones are also elastomers. In silicones the nature of the alkyl groups has been varied so that silicone products are now available that can be vulcanized, do not swell in oils, are not attacked by ozone, and retain great flexibility at low temperatures.

Fibers. Fibers are polymers oriented to provide optimum properties, such as high tensile strength, along one axis. These are threadlike polymers that can be woven into fabrics. Cotton, wool, and silk are natural fibers. Some synthetic fibers, such as nylon, Orlon, and Dacron, have been given improved properties: increased tensile strength; lightness of weight; low moisture absorption; resistance to moths, mildew, rot, and fungus; wrinkle resistance and heat set.

Plastics. Plastics have properties that are intermediate between elastomers and fibers. Normally these polymeric materials become soft and fluid when heated, but they can be imparted with a variety of properties at room temperature. Polystyrene is stiff and brittle. Polypropylene, on the other hand, is extremely tough, impact resistant, tear resistant, and flexible in thin sheets.

26-11 Some Key Chemicals

We close Chapter 26 by reconsidering, through Table 26-12, some of the key chemicals produced in the United States. For each chemical, the table lists its relative ranking in quantity produced and its principal method of manufacture. Some of the more important uses of each chemical are listed, as is a reference to a page in the text where the production and/or use of this chemical is described. Some additional key chemicals are listed as footnotes to the table, arranged according to their relative ranking. Although some relative rankings change from year to year, the group of chemicals among the "top 50" tends to remain quite constant. (The rankings and amounts produced are those given in *Chemical and Engineering News,* May 5, 1980, p. 35.)

TABLE 26-12
The top 50 chemicals produced in the United States (1979)

Rank	Millions of tons	Chemical	Formula	Method of manufacture	Principal uses	Page reference
1	42.0	sulfuric acid	H_2SO_4	catalytic oxidation of SO_2 to SO_3, followed by reaction with H_2O	manufacture of fertilizers, chemicals; oil refining	686
2	19.4	lime	CaO	thermal decomposition of $CaCO_3$	steelmaking; manufacture of chemicals; water treatment	684
3	18.1	ammonia	NH_3	Haber process from N_2 and H_2; coal tar by-product	manufacture of fertilizers, nitric acid, ammonium compounds, cleaning agents; refrigerant	362
4	17.7	oxygen	O_2	distillation of liquid air	steelmaking; welding; medical uses	662
5	15.0	nitrogen	N_2	distillation of liquid air	blanketing atmospheres; low temperatures	662
6	14.6	ethylene	$CH_2{=}CH_2$	catalytic cracking of petroleum	manufacture of poly-ethylene, ethylene oxide and glycol, vinyl chloride, styrene	694
7	12.4	sodium hydroxide	NaOH	electrolysis of NaCl(aq)	manufacture of chemicals, pulp and paper, aluminum, textiles, soaps and detergents	683
8	12.1	chlorine	Cl_2	electrolysis of NaCl(aq)	manufacture of chemicals, pulp and paper, plastics (PVC), solvents; water treatment	504
9	10.1	phosphoric acid	H_3PO_4	reaction of phosphate rock with H_2SO_4; P_4O_{10}, with water	manufacture of fertilizers and detergents	520
10	8.6	nitric acid	HNO_3	Ostwald process: oxidation of NH_3, followed by reaction with water	manufacture of explosives, fertilizers, plastics, dyes, and lacquers	517
11	8.3	sodium carbonate	Na_2CO_3	Solvay process from NaCl, NH_3, $CaCO_3$; natural ore deposits	manufacture of glass, chemicals, pulp and paper, detergents	684

Summary

Very few of the materials of modern society are found free in nature. They must be made from a relatively small number of natural resources. Some processes for obtaining needed materials simply involve physical changes. The separation of N_2, O_2, and Ar by liquefaction and distillation of air is one example. The extraction of pure water from seawater, whether by distillation or by reverse osmosis, is another. Among the metals, a few occur free in nature, but most are present only in combined form. Reduction of metal compounds to the free metal and refining of the impure metal are metallurgical processes common to most metals. In some cases concentration and roasting of an ore are also required. One metal, magnesium, is currently derived from seawater; and a few others (e.g., titanium) are produced by rather unconventional means.

The majority of commercially important inorganic chemicals are derived from such sources as air, water, limestone, salt, sulfur, and phosphate rock. Organic materials can be produced either from petroleum, coal, or biomass, with petroleum still being the major source. However, worldwide shortages of petroleum may require shifts to the other sources in the future. In the manufacture of motor fuels, the composition of petroleum is greatly altered, with extensive use being

Rank	Millions of tons	Chemical	Formula	Method of manufacture	Principal uses	Page reference
12	7.8	ammonium nitrate	NH_4NO_3	reaction of NH_3 and HNO_3	manufacture of fertilizers, explosives	517
13	7.2	propylene	$CH_3CH{=}CH_2$	catalytic cracking of petroleum	manufacture of plastics, fibers, solvents	607
14	6.9	urea	$CO(NH_2)_2$	heating CO_2 and NH_3 under pressure	manufacture of fertilizers, plastics, pharmaceuticals	596
15	6.4	benzene	(benzene ring)	catalytic cracking and reforming of petroleum	manufacture of polystyrene, nylon, other polymers	691
16	5.9	toluene	(benzene ring)$-CH_3$	catalytic cracking and reforming of petroleum	manufacture of benzene and other organic chemicals	609
17	5.9	ethylene dichloride	CH_2ClCH_2Cl	chlorination of ethylene	manufacture of vinyl chloride	607
18	4.3	ethylbenzene	(benzene ring)$-C_2H_5$	catalytic cracking and reforming of petroleum	manufacture of styrene	691
19	3.8	vinyl chloride	$CH_2{=}CHCl$	elimination of HCl from CH_2ClCH_2Cl	manufacture of vinyl plastics	694
20	3.7	styrene	(benzene ring)$-CH{=}CH_2$	dehydrogenation of ethylbenzene	manufacture of styrene plastics, synthetic rubber	693

21 methanol	27 ethylene oxide	33 acetic acid	39 sodium sulfate	45 adipic acid
22 terephthalic acid	28 ethylene glycol	34 carbon black	40 propylene oxide	46 sodium silicate
23 carbon dioxide	29 p-xylene	35 phenol	41 calcium chloride	47 sodium tripolyphosphate
24 xylene	30 cumene	36 acetone	42 acrylonitrile	48 acetic anhydride
25 formaldehyde	31 ammonium sulfate	37 aluminum sulfate	43 vinyl acetate	49 titanium dioxide
26 hydrochloric acid	32 butadiene (1,3-)	38 cyclohexane	44 isopropyl alcohol	50 ethanol

made of the basic reactions of organic chemistry (Chapter 24). Increasingly, chemicals derived from petroleum—petrochemicals—are being used in the manufacture of polymers.

The chain of activities that starts with the production of raw materials and culminates in the use of end products can result in serious negative impacts on the environment. Probably most familiar among these environmental problems is air pollution, caused by the combustion of gasoline in automobiles and fossil fuels in electric power plants and by emissions from other industries. The treatment of water to remove the mineral constituents that cause hardness of the water is required for many of its intended uses. Other constituents now appearing in water, however, are pollutants resulting from human activities. Control and removal of these pollutants is a difficult matter. Fortunately, the body of scientific knowledge that can be applied to the detection and control of environmental pollution continues to grow. So too has there been an increase in the ability of scientists to anticipate future environmental problems (such as may result from an increased level of CO_2 in the atmosphere). More and more, a major application of scientific principles will be toward assessing environmental effects and improving the quality of the environment.

Learning Objectives

As a result of studying Chapter 26, you should be able to:

1. Name a number of the key chemicals of commerce, describe how they are manufactured, and list some of their uses.

2. Describe how air is liquefied and how O_2, N_2, and Ar are obtained from liquid air.

3. Outline the natural processes by which atmospheric nitrogen is fixed and then returned to the atmosphere—the nitrogen cycle.

4. Outline the steps required to produce pure Zn, Cu, Fe, Al, Mg, and Ti from their naturally occurring forms, and write equations for the key reactions involved.

5. Describe the processes that are used to increase the yield of gasoline from petroleum and improve its performance in internal combustion engines.

6. Illustrate the principal methods of polymer formation—free-radical addition and condensation—through structural formulas of monomers and polymers.

7. Distinguish between natural and synthetic polymers and among elastomers, fibers, and plastics.

8. Explain how temporary and permanent hardness arise in water and describe several methods of softening water.

9. Describe the cleaning action of soaps and detergents.

10. Explain how an ion exchange process works.

11. Discuss the BOD and COD tests in relation to water pollution.

12. Outline the steps involved in purifying water for drinking purposes.

13. List some common air pollutants and describe their sources and methods used to control them.

14. Discuss the factors involved in the formation of photochemical smog and the eutrophication of a lake.

Some New Terms

Alkylation is a petroleum refining process in which small hydrocarbon molecules are joined into larger ones.

Anodizing refers to the electrodeposition of Al_2O_3 on aluminum metal to impart additional protection against corrosion.

Biochemical oxygen demand (BOD) is a measure of the amount of dissolved oxygen consumed in the biological degradation of organic matter in a water sample. The greater the BOD, the more polluted is the water with organic matter.

Biomass is any material produced by photosynthesis, that is, plants or their principal components.

Chemical oxygen demand (COD) measures the amount of a strong oxidizing agent required to oxidize to CO_2 and H_2O, all the oxidizable matter in a water sample. The COD test measures the presence of certain organic substances that are not degraded by microorganisms (and hence do not show up in the BOD test).

Coke, coal tar, and **coal gas** are products of the destructive distillation (pyrolysis) of coal.

Concentration is a process by which a metal ore is freed of some of the waste rock (gangue) with which it is mixed.

Condensation is a form of polymerization in which small molecules (such as water) are eliminated from between monomer units.

Cracking refers to the breaking down of large hydrocarbon molecules into smaller ones. This may be achieved by heating and/or using catalysts.

Desalination is any process (distillation, reverse osmosis) for the separation of salts from seawater or brackish water, producing fresh water.

Detergents are salts of organic sulfonic acids, for example $ROSO_3^-Na^+$, where R is an aliphatic or aromatic hydrocarbon residue.

Eutrophication is a deterioration of a freshwater body produced by nutrients such as nitrates and phosphates. Excessive growth of algae, depletion of oxygen in the water, and fish kills all result from eutrophication.

Free-radical addition polymerization is a type of reaction in which monomer units are added to the ends of free-radical chains to produce a polymer.

Hard water contains dissolved minerals in significant concentrations. If the hardness is due principally to HCO_3^- and associated cations, the water is said to have **temporary hardness.** If hardness is due to anions other than HCO_3^-, for example, SO_4^{2-}, it is referred to as **permanent hardness.**

Industrial smog is a form of air pollution in which the principal pollutants are $SO_2(g)$, $SO_3(g)$, H_2SO_4 mist, smoke, and other emissions characteristic of industrial operations.

Ion exchange is a process in which ions in solution are exchanged for corresponding ions held to the surface of an ion exchange material. For example, Ca^{2+} and Mg^{2+} may be exchanged for Na^+; or SO_4^{2-}, for OH^-.

Monomers are certain molecules that, under the appropriate conditions, can be joined together to produce giant molecules (polymers).

The **nitrogen cycle** is a series of natural processes by which atmospheric nitrogen is fixed, enters into the food chain of animals, and is eventually returned to the atmosphere by bacteria.

Photochemical smog is an air pollution phenomenon resulting from reactions involving sunlight, oxides of nitrogen, ozone, and hydrocarbons. It occurs in an air mass trapped by a temperature inversion.

Pig iron is an impure form of iron (about 95% Fe) produced in a blast furnace.

A **polymer** is a giant molecule produced by joining together large numbers of smaller molecules called monomers.

Reduction is a metallurgical process in which a metal is obtained from one of its compounds by heating with a reducing agent [as in the reduction of Fe_2O_3 to Fe by coke (carbon)].

Refining is the final metallurgical step of removing impurities to produce a pure metal.

Reforming is a petroleum refining process in which straight

chain hydrocarbons are isomerized into branched chain ones, or in which alicyclic hydrocarbons are converted to aromatic ones.

Roasting is a metallurgical process in which gases are expelled by heating an ore (usually converting a sulfide to an oxide).

Steel is a term used to describe iron alloys containing from 0 to 1.5% C together with other key elements, such as Cr, Mn, Ni, W, Mo, and V.

Superphosphate is a mixture of $Ca(H_2PO_4)_2$ and $CaSO_4$ produced by the action of H_2SO_4 on phosphate rock.

Triple superphosphate is the product of the action of H_3PO_4 on phosphate rock; it is essentially pure $Ca(H_2PO_4)_2 \cdot H_2O$.

Exercises

Occurrence of the elements

1. Whch of the following elements would you expect to occur free in nature, which only in combined form, and which, hardly at all? Give reasons based on chemical principles. N, Br, Ag, Rn, Ga, Fr.

2. The relative abundance of an element in the atmosphere or seawater is generally a sufficient basis on which to determine the economic feasibility of its extraction. The relative abundance in the solid crust is not. What is the reason for this difference?

The atmosphere

3. The current annual production of pure oxygen gas is 17,700,000 tons. What volume of air at STP (21% O_2, by volume) must be liquefied and distilled to produce this quantity of oxygen?

4. A 55-L cylinder contains Ar at 149 atm and 25°C. What minimum volume of air at STP must have been liquefied and distilled to produce this Ar? Air contains 0.93% Ar, by volume.

5. A typical natural gas from which He can be extracted contains 0.3% He, by volume. The abundance of He in air is 5 ppm, by volume. How much more abundant is He in the natural gas than in air?

Atmospheric pollution

6. One of the "ingredients" of photochemical smog not specifically described in the text is sunlight. What is the role of sunlight?

7. Occasionally during a smog episode in the Los Angeles basin the tops of the tallest buildings are visible above the smog. Under what conditions do you suppose this might happen?

8. In 1968, over 75 billion gal of gasoline were consumed in the United States as a motor fuel.
   (a) Assuming an average of 2 ml of tetraethyl lead [$Pb(C_2H_5)_4$] per gallon of gasoline, what mass of lead, in tons, was exhausted to the atmosphere in that year? [The density of $Pb(C_2H_5)_4$ is 1.65 g/cm³]
   (b) Assuming an emission of oxides of nitrogen of 5 g per vehicle-mile and an average mileage of 15 mi/gal of gasoline, what mass of nitrogen oxides, in tons, was released to the atmosphere?

9. The combustion of 5 billion metric tons of fossil fuels (roughly the current worldwide annual rate) raises the atmospheric content of $CO_2(g)$ by about 0.7 ppm. Starting with the current level of 330 ppm, and assuming a fossil fuel reserve of 10,000 billion metric tons, what is the conceivable limit to which the CO_2 content of air might be raised by burning all this fuel?

10. In analyzing a natural gas, 25.0-L of the gas, measured at 25°C and 740 mmHg, is bubbled through Pb^{2+}(aq); a precipitate weighing 0.535 g is obtained. What is the percent H_2S, by volume, in the natural gas?

Seawater

11. Two statements are made in the text: The quantity of Mg^{2+} in seawater is 1,272 g/ton, and all the magnesium yet produced could have been obtained from as little as 4 km³ of seawater. Estimate the quantity of magnesium produced since the metal was first discovered. (The density of seawater is 1.03 g/cm³.)

The water environment

12. Explain the following observations that are made upon adding dry ice [$CO_2(s)$] to a saturated solution of $Ca(OH)_2$: A milky white precipitate first forms which then redissolves, leading to a clear colorless solution.

13. Suppose that all the cations associated with HCO_3^- in Example 26-1 are Ca^{2+}.
  (a) What is the total mass of $CaCO_3$ that would be precipitated in softening 1.00×10^6 gal of the water?
  (b) Show that in the $CaCO_3$, half of the Ca^{2+} is derived from the CaO used in the water softening and half from the water itself.

14. A particular water sample has a hardness of 130 ppm SO_4^{2-} (as $CaSO_4$).
  (a) Show how this water can be softened with Na_2CO_3.
  (b) What mass of Na_2CO_3, in grams, is required to soften 250 L of this water?

15. A particular hard water contains 96 ppm SO_4^{2-} and 183 ppm HCO_3^-, with Ca^{2+} as the only cation. How many ppm of Ca^{2+} does the water contain?

16. What mass of "bathtub ring" will form if a 5.0-L water sample having 85 ppm Ca^{2+} is treated with the soap, potassium stearate?

17. Describe (using equations similar to 26.24) how a sample of hard water containing Fe^{3+}, Mg^{2+}, Ca^{2+}, HCO_3^-, and SO_4^{2-} can be deionized by passing it first through a cation exchange resin and then an anion exchange resin. (*Hint:* What counterions are required on the resins?)

Water pollution

18. In 1968 a typical mercury cell for the production of $Cl_2(g)$ by electrolysis of NaCl(aq) suffered a loss of about 0.3 lb Hg/ton $Cl_2(g)$ produced. The production of $Cl_2(g)$ in the United States was approximately 8.5 million tons, and about 20% of this was produced in mercury cells. Estimate the quantity of mercury entering the environment from this source in 1968.

19. Dissolved phosphorus occurs naturally in seawater to the extent of 0.6 lb $CaHPO_4$ per million gal. Show that if all the phosphates discharged by the consumption of 5 billion lb of detergents in the United States in 1969 could have been uniformly distributed throughout the oceans, no pollution problem would have existed. (Assume that 40% by mass of the detergent exists as $Na_5P_3O_{10}$. Assume also a volume of seawater of 330 million cubic miles.)

Metallurgy

20. Complete and balance the following equations.
  (a) $PbS(s) + O_2(g) \xrightarrow{heat}$

  (b) $CdO(s) + C(s) \xrightarrow{heat}$

  (c) $FeCO_3(s) \xrightarrow{heat}$

  (d) $CuSO_4(aq) \xrightarrow[\text{Pt electrodes}]{\text{electrolysis}}$

  (e) $ZnO(s) + HCl(aq) \longrightarrow$

  (f) $TiCl_4(g) + Na(l) \xrightarrow{heat}$
  (g) $K^+(aq) + OH^-(aq) + Al_2O_3(s) \longrightarrow$

21. Write balanced chemical equations to represent
  (a) the roasting of $CuFeS_2$
  (b) the electrolysis of molten KCl
  (c) the action of $H_2SO_4(aq)$ on $Al_2O_3(s)$
  (d) the reduction of $Fe_3O_4(s)$ by $H_2(g)$
  (e) the action of $NH_3(aq)$ on $MgCl_2(aq)$
  (f) the removal of silicon from steel in a Bessemer converter

22. A particular ore contains 1.90% Cu, as Cu_2S. What volume of $SO_2(g)$, measured at 20°C and 740 mmHg, is produced when 3.50×10^3 kg of the ore is roasted?

23. What is the minimum mass of limestone, containing 97% $CaCO_3$, required in a blast furnace charged with 2.0×10^3 kg of a hematite iron ore containing 10.4% SiO_2?

24. Explain why the oxygen steelmaking process is superior to both the Bessemer converter and the open hearth furnace.

25. Describe a series of *simple* chemical reactions with which you could determine whether a particular metal sample is Aluminum 2S (99.2% Al) or magnalium (70% Al + 30% Mg). You are permitted to dissolve the metal sample in this testing.

26. In the purification of bauxite ore, $Al(OH)_3$ is precipitated by diluting a solution containing AlO_2^-. $Al(OH)_3$ can also be obtained if an aqueous solution of AlO_2^- is treated with $CO_2(g)$, but not if treated with HCl(aq). Explain this difference.

27. Often when an inert atmosphere is needed in an industrial process $N_2(g)$ is used. In the Kroll process for titanium (equation 26.43) He is used instead. Why is this necessary?

Inorganic chemical processes

28. A simple test to detect the presence of $CO_2(g)$ involves passing the gas into limewater [a saturated aqueous solution of $Ca(OH)_2$]. If $CO_2(g)$ is present, the limewater turns cloudy. What is this precipitate? Write a chemical equation for its formation. Would the test work as well using $Ba(OH)_2(aq)$? KOH(aq)?

29. An analysis of a particular Solvay process plant in full operation shows that for every 1.00 ton of NaCl employed, 1.03 tons $NaHCO_3$ is obtained. Only 1.5 lb NH_3 is consumed in the overall process.
  (a) What is the percent efficiency of this process for converting NaCl to $NaHCO_3$?
  (b) Why is so little NH_3 required?

30. The Frasch process for mining sulfur uses superheated water.
  (a) What is superheated water?
  (b) At what temperature must this water be maintained?
  (c) Make a *rough* estimate of the pressure under which the water must be maintained. (*Hint:* Use relevant information from previous chapters.)

31. What is the percent P, by mass, in each of the following

materials encountered in the commercial production of phosphorus compounds: **(a)** orthophosphoric acid; **(b)** calcium orthophosphate; **(c)** fluorapatite; **(d)** normal superphosphate; **(e)** triple superphosphate.

32. The term "triple superphosphate" is used to signify three times the phosphorus content of normal superphosphate. Use results from Exercise 31 to show that this is approximately true.

Organic raw materials

33. Limestone, $CaCO_3$, contains carbon, but can it be thought of as an organic raw material of importance? (*Hint:* You may find it helpful to review a few of the equations in Chapter 24.)

34. The basic components of automobile tires are rubber, a reinforcing cord, and, to add strength, a finely divided material like carbon black (carbon obtained by the incomplete combustion of a fuel, rather like soot). Describe how it would be possible to produce a tire relying almost entirely (but not exclusively) on forest products, rather than on synthetic materials.

Polymer chemistry

35. Explain why Dacron is called a polyester. What is the percent O, by mass, in Dacron?

36. Could a polymer be formed by reacting either terephthalic acid or dimethyl terephthalate with ethyl alcohol in place of ethylene glycol? Explain.

37. Nylon 66 is produced by reacting 1,6-hexanediamine with adipic acid (see Table 26-11). A different nylon polymer is obtained if sebacyl chloride

$$\underset{\text{Cl}\overset{O}{\overset{\|}{C}}(CH_2)_8\overset{O}{\overset{\|}{C}}Cl}{}$$

is substituted for the adipic acid. What is the basic repeating unit of this nylon structure?

38. Nylon belongs to a class of fibers called polyamides. What is the distinctive feature of a polyamide structure? Name some natural fibers that are also polyamides.

39. Two *different* monomers may be involved in a free-radical addition polymerization. The resulting polymer is called a **copolymer.** (See, for example, GRS in Table 26-11.) Draw structures of the following copolymers, using monomer formulas given in the chapter: **(a)** the copolymer (called Saran) of vinyl chloride and vinylidine chloride, $CH_2=CCl_2$; **(b)** poly(styrene-coacrylonitrile).

40. In referring to the molecular weight of a polymer, we can speak only of the "average molecular weight." Explain why the molecular weight of a polymer is not a unique quantity, as it is for a substance such as water.

Additional Exercises

1. Step 1a in the nitrogen cycle pictured in Figure 26-2 involves the production of nitric acid during a lightning storm. Suggest a series of reactions that could lead to the formation of HNO_3 under these conditions.

2. Write an equation(s) to suggest the role of each of the following constituents in a blast furnace: **(a)** coke; **(b)** carbon monoxide; **(c)** limestone; **(d)** iron ore.

3. Oersted (1825) produced $AlCl_3$ by passing a stream of $Cl_2(g)$ over a mixture of $Al_2O_3(s)$ and C(s). Write a balanced equation to represent this reaction. (Oersted obtained impure aluminum metal by heating the $AlCl_3$ with potassium amalgam.)

4. Write a series of chemical equations to show how triple superphosphate could be produced without the use of sulfuric acid anywhere in the process.

5. What mass of limestone ($CaCO_3$), in tons, would be consumed per year, *as a minimum*, to remove 90% of the $SO_2(g)$ from the stack gases of a power plant burning 2.2 million tons annually of coal with 3.5% S, by mass? Assume that the wet scrubbing process described on page 668 is used and that all the sulfur is recovered as $CaSO_3$. Why is the actual quantity of limestone required probably greater than that calculated here?

*6. A minimum energy figure is established in the text for the desalination of seawater: 49.5 J/mol H_2O. If electrical energy is available at 6 cents/kWh, what would be the minimum cost to desalinize 1000 m³ of seawater? Why is the real cost much higher than this?

*7. The composition of a phosphate mineral can be referred to in several different ways: %P, %P_2O_5, and %BPL [bone phosphate of lime, $Ca_3(PO_4)_2$].
   (a) Show that %P = 0.436 × (%P_2O_5) and %BPL = 2.185 × (%P_2O_5).
   (b) What is the significance of a percentage BPL greater than 100%?

*8. Explain why a polymer formed by free-radical addition generally has a higher molecular weight than a corresponding one formed by condensation polymerization.

*9. Refer to the hard water described in Exercise 15.
   (a) What mass of CaO is required to remove the HCO_3^- from a 150-L sample of this water?
   (b) How many ppm of SO_4^{2-} and Ca^{2+} remain after the treatment described in part (a)?
   (c) What mass of Na_2CO_3 is required to remove the remaining Ca^{2+} described in part (b)?

*10. Refer to Figure 26-6 and picture the counterions to be

H^+ instead of Na^+. A 100.0-ml sample of hard water is passed through the column, and as a result requires 15.17 ml of 0.0265 M NaOH for its titration. What is the hardness of the water, expressed as ppm of Ca^{2+}?

*11. An aluminum production cell of the type illustrated in Figure 26-13 operates at a current of 1.00×10^5 A and a voltage of 4.5 V. The cell is 38% efficient in converting electric energy to chemical change (the rest of the electric energy is dissipated as heat energy in the cell).

(a) What mass of Al, in lb, can be produced by this cell in 8 h?
(b) If the electric energy required to power this cell is produced by burning coal (85% C; heat of combustion of C = 32.8 kJ/g) in a power plant with 35% efficiency, what mass of coal must be burned to produce the mass of aluminum determined in part (a)?

Self-Test Questions

For questions 1 through 6 select the single item that best completes each statement.

1. The percent, by mass, of hydrogen is greatest in (a) the earth's crust; (b) the earth's core; (c) the universe as a whole; (d) the compound H_2O.

2. Production of the metal by reduction with carbon is not feasible with (a) Al_2O_3; (b) ZnO; (c) Fe_2O_3; (d) Cu_2O.

3. To soften temporary hard water, add (a) Na_2SO_4; (b) Na_2CO_3; (c) $CaCO_3$; (d) NaCl.

4. The *cheapest* raw material from which an alkaline (basic) medium can be prepared is (a) rock salt; (b) limestone; (c) phosphate rock; (d) bauxite.

5. Of the following substances the one that is *unimportant* in the production of fertilizers is (a) phosphate rock; (b) H_2SO_4; (c) NH_3; (d) Na_2CO_3.

6. The production of benzene from cyclohexane in the refining of petroleum is an example of (a) distillation; (b) cracking; (c) reforming; (d) alkylation.

7. Write a chemical equation(s) to represent each of the following reactions.
(a) the roasting of FeS;
(b) the action of HCl(aq) on $Mg(OH)_2(s)$;
(c) the effect of boiling on temporary hard water;
(d) the recovery of NH_3 in the Solvay process for $NaHCO_3$;
(e) the formation of H_3PO_4 from P

8. The abundance of F^- in seawater is 1 g F^- per ton of seawater. Suppose that a commercially feasible method could be found for extracting fluorine from seawater.
(a) What mass of $F_2(g)$, in kg, could be obtained from 1 km^3 of seawater ($d = 1.03$ g/cm^3)?
(b) Do you think the process would resemble that for extracting bromine from seawater? Explain.

9. What essential feature distinguishes between (a) monomer and polymer; (b) free-radical addition and condensation polymerization; (c) chain initiation and termination; (d) elastomer and fiber?

10. Explain why the air pollution control measures required for automobiles are not the same as those required for fossil-fuel electric power plants.

APPENDIX A

Mathematical Operations

A-1 Exponential Arithmetic

The numerical quantities encountered in this text range from very small to very large in value. For example, the mass of an individual hydrogen atom is 0.000000000000000000000000167 g; the number of molecules in 18.016 g H_2O is 602,250,000,000,000,000,000,000. In the form expressed, both of these numbers would be very cumbersome to handle in computations. To express such numbers with greater ease, the methods of exponential arithmetic may be employed.

positive powers	*negative powers*
$10^0 = 1$	$10^0 = 1$
$10^1 = 10$	$10^{-1} = \dfrac{1}{10} = 0.1$
$10^2 = 10 \times 10 = 100$	$10^{-2} = \dfrac{1}{10 \times 10} = 0.01$
$10^3 = 10 \times 10 \times 10 = 1000$	$10^{-3} = \dfrac{1}{10 \times 10 \times 10} = \dfrac{1}{10^3} = 0.001$

A number is said to be written in exponential form when it is expressed as a coefficient multiplied by a power of ten: $a \cdot 10^y$. For example, $120 = 1.2 \times 100 = 1.2 \times 10^2$.

Illustrative Examples

$24{,}100 = 2.41 \times 10{,}000 = 2.41 \times 10^4$
$0.0038 = 3.8 \times 0.001 = 3.8 \times 10^{-3}$
$6.1 \times 10^6 = 6.1 \times 1{,}000{,}000 = 6{,}100{,}000$
$4.7 \times 10^{-3} = 4.7 \times 0.001 = 0.0047$

ADDITION AND SUBTRACTION. To add or subtract numbers written in exponential form, the numbers must first be expressed to the same power of 10. Thus,

$$0.0560 + 0.0038 - 0.0152 = 5.60 \times 10^{-2} + 0.38 \times 10^{-2} - 1.52 \times 10^{-2}$$
$$= (5.60 + 0.38 - 1.52) \times 10^{-2} = 4.46 \times 10^{-2}$$

MULTIPLICATION. Consider the numbers $a \cdot 10^y$ and $b \cdot 10^z$. Their product is $a \cdot b \cdot 10^{(y+z)}$. Coefficients are multiplied and exponents are added.

703

$$0.0220 \times 0.0040 \times 750 = (2.20 \times 10^{-2})(4.0 \times 10^{-3})(7.5 \times 10^2)$$
$$= (2.20 \times 4.0 \times 7.5) \times 10^{(-2-3+2)} = 66 \times 10^{-3}$$
$$= 6.6 \times 10^1 \times 10^{-3} = 6.6 \times 10^{-2}$$

DIVISION.   Consider the two numbers, $a \cdot 10^y$ and $b \cdot 10^z$. Their quotient is $a \cdot 10^y / b \cdot 10^z = (a/b) \times 10^{(y-z)}$. The coefficients are divided and the exponent of the denominator is subtracted from the exponent in the numerator.

$$\frac{20.0 \times 636 \times 0.150}{0.0400 \times 1.80} = \frac{(2.00 \times 10^1)(6.36 \times 10^2)(1.50 \times 10^{-1})}{4.00 \times 10^{-2} \times 1.80}$$

$$= \frac{2.00 \times 6.36 \times 1.50 \times 10^{(1+2-1)}}{4.00 \times 1.80 \times 10^{-2}} = \frac{19.1 \times 10^2}{7.20 \times 10^{-2}}$$

$$= 2.65 \times 10^{(2-(-2))} = 2.65 \times 10^4$$

RAISING A NUMBER TO A POWER.   To "square" the number $a \cdot 10^y$ means to determine the value $(a \cdot 10^y)^2$ or the product $(a \cdot 10^y)(a \cdot 10^y)$. According to the rule for multiplication, this product is $(a \times a) \times 10^{(y+y)} = a^2 \cdot 10^{2y}$. When an exponential number is raised to a power, the coefficient is raised to that power and the exponent is multiplied by the power.

$$(0.0034)^3 = (3.4 \times 10^{-3})^3 = (3.4)^3 \times 10^{3 \times (-3)} = 39 \times 10^{-9} = 3.9 \times 10^{-8}$$

EXTRACTING THE ROOT OF AN EXPONENTIAL NUMBER.   To extract the root of a number is the same as raising the number to a fractional power. This means that the square root of a number is the number to the one-half power; the cube root is the number to the one-third power; and so on. Thus,

$$\sqrt{a \cdot 10^y} = (a \cdot 10^y)^{1/2} = a^{1/2} \cdot 10^{y/2}$$
$$\sqrt{156} = \sqrt{1.56 \times 10^2} = (1.56)^{1/2} \times 10^{2/2} = 1.25 \times 10^1 = 12.5$$

In the following example, where the cube root is extracted, the exponent (-5) is not divisible by 3; the number is rewritten so that the new exponent will be divisible by 3.

$$(3.52 \times 10^{-5})^{1/3} = (35.2 \times 10^{-6})^{1/3} = (35.2)^{1/3} \times 10^{-6/3} = 3.28 \times 10^{-2}$$

A-2 Logarithms

The base upon which a logarithm is formulated is denoted by a subscript number, such as $\log_{10} N = x$. If no subscript is written, assume that "log" implies that the base is 10.

The common logarithm (log) of a number (N) is that exponent (x) to which the base 10 must be raised to yield the number N.

$\log N = x$ means that $N = 10^x$

For simple powers of ten:

$$\log 1 = \log 10^0 = 0$$
$$\log 10 = \log 10^1 = 1 \qquad \log 0.10 = \log 10^{-1} = -1$$
$$\log 100 = \log 10^2 = 2 \qquad \log 0.01 = \log 10^{-2} = -2$$

MULTIPLICATION AND DIVISION.   Since the logarithms of numbers are exponents of ten, they can be handled in the same manner as other exponents. This leads to

$$\log (M \times N) = \log M + \log N$$

$$\log \frac{M}{N} = \log M - \log N$$

Illustrative Examples.   Most numbers encountered in measurements and calculations are not simple powers of 10. To allow for the determination of the logarithms of such numbers, tables of logarithms have been developed (see Table A-1). In the example below log 7.34 must be obtained from such a table.

$$\log 734 = \log (7.34 \times 10^2) = \log 7.34 + \log 10^2$$
$$= 0.866 + 2 = 2.866$$

The logarithms of numbers smaller than 1 can also be obtained easily.

$$\log 0.00130 = \log (1.30 \times 10^{-3}) = \log 1.30 + \log 10^{-3}$$
$$= 0.114 - 3 = -2.886$$

A number has a logarithm of 4.350. What is the number? The number we are seeking is said to be the **antilogarithm** of 4.350. Let us express this unknown number as $N = a \cdot 10^x$.

Think of an antilog as "the number whose logarithm is." From Table A-1 we see that the number whose logarithm is 0.3502 is 2.24.

$$\log N = 4.350 = 0.350 + 4$$
$$a = \text{antilog } 0.350 = 2.24$$
$$x = 4$$
$$N = 2.24 \times 10^4$$

Another commonly encountered example in this text is of the form: $\log N = -4.350$; what is N? The key operation here involves stating -4.350 as a difference of two numbers.

$$\log N = -4.350 = 0.650 - 5$$
$$N = (\text{antilog } 0.650) \times 10^{-5}$$
$$N = 4.47 \times 10^{-5}$$

NATURAL LOGARITHMS.   The definition of a logarithm is not limited to the base 10. For example, corresponding to $2^3 = 8$ we can write, $\log_2 8 = 3$ (read as "the logarithm of 8 to the base 2 is equal to 3"); $\log_2 10 = 3.322$; and so on. Several equations encountered in this text are derived using the methods of calculus and involve logarithmic functions. These derivations require that the logarithm be a "natural" one. This is the case only if for the base we use the quantity $e = 2.71828\ldots$ A logarithm to the base "e" is usually denoted by the symbol "ln"; that is, $\log_e x = \ln x$. The relationship between $\ln x$ and $\log_{10} x$ is a simple one. It involves the factor $\log_e 10 = 2.303$.

$$\log_e x = (\log_e 10) \times (\log_{10} x)$$

$$\ln x = 2.303 \log_{10} x = 2.303 \log x$$

In equations in this text where the factor "2.303" appears, you can assume that this was introduced to relate a natural logarithm to a logarithm to the base 10 (a common logarithm). Either use the equation as written, based on common logarithms, or omit the factor "2.303" and use natural logarithms. Most electronic calculators can handle both of these functions. One application where common logarithms *must* be used, however, is in calculations involving pH, since pH is defined in terms of a logarithm to the base 10 (see Section 17-4).

A-3 Algebraic Operations

An algebraic equation is solved when one of the quantities, the unknown, is expressed in terms of all the other quantities in the equation. This effect is achieved when the unknown is present, alone, on one side of the equation, and the rest of the terms are on the other side. To solve an equation a rearrangement of terms may be necessary. The basic principle governing these rearrangements is quite simple. *Whatever is done to one side of the equation must be done as well to the other.*

TABLE A-1
Four place logarithms

N	0	1	2	3	4	5	6	7	8	9
10	0000	0043	0086	0128	0170	0212	0253	0294	0334	0374
11	0414	0453	0492	0531	0569	0607	0645	0682	0719	0755
12	0792	0828	0864	0899	0934	0969	1004	1038	1072	1106
13	1139	1173	1206	1239	1271	1303	1335	1367	1399	1430
14	1461	1492	1523	1553	1584	1614	1644	1673	1703	1732
15	1761	1790	1818	1847	1875	1903	1931	1959	1987	2014
16	2041	2068	2095	2122	2148	2175	2201	2227	2253	2279
17	2304	2330	2355	2380	2405	2430	2455	2480	2504	2529
18	2553	2577	2601	2625	2648	2672	2695	2718	2742	2765
19	2788	2810	2833	2856	2878	2900	2923	2945	2967	2989
20	3010	3032	3054	3075	3096	3118	3139	3160	3181	3201
21	3222	3243	3263	3284	3304	3324	3345	3365	3385	3404
22	3424	3444	3464	3483	3502	3522	3541	3560	3579	3598
23	3617	3636	3655	3674	3692	3711	3729	3747	3766	3784
24	3802	3820	3838	3856	3874	3892	3909	3927	3945	3962
25	3979	3997	4014	4031	4048	4065	4082	4099	4116	4133
26	4150	4166	4183	4200	4216	4232	4249	4265	4281	4298
27	4314	4330	4346	4362	4378	4393	4409	4425	4440	4456
28	4472	4487	4502	4518	4533	4548	4564	4579	4594	4609
29	4624	4639	4654	4669	4683	4698	4713	4728	4742	4757
30	4771	4786	4800	4814	4829	4843	4857	4871	4886	4900
31	4914	4928	4942	4955	4969	4983	4997	5011	5024	5038
32	5051	5065	5079	5092	5105	5119	5132	5145	5159	5172
33	5185	5198	5211	5224	5237	5250	5263	5276	5289	5302
34	5315	5328	5340	5353	5366	5378	5391	5403	5416	5428
35	5441	5453	5465	5478	5490	5502	5514	5527	5539	5551
36	5563	5575	5587	5599	5611	5623	5635	5647	5658	5670
37	5682	5694	5705	5717	5729	5740	5752	5763	5775	5786
38	5798	5809	5821	5832	5843	5855	5866	5877	5888	5899
39	5911	5922	5933	5944	5955	5966	5977	5988	5999	6010
40	6021	6031	6042	6053	6064	6075	6085	6096	6107	6117
41	6128	6138	6149	6160	6170	6180	6191	6201	6212	6222
42	6232	6243	6253	6263	6274	6284	6294	6304	6314	6325
43	6335	6345	6355	6365	6375	6385	6395	6405	6415	6425
44	6435	6444	6454	6464	6474	6484	6493	6503	6513	6522
45	6532	6542	6551	6561	6571	6580	6590	6599	6609	6618
46	6628	6637	6646	6656	6665	6675	6684	6693	6702	6712
47	6721	6730	6739	6749	6758	6767	6776	6785	6794	6803
48	6812	6821	6830	6839	6848	6857	6866	6875	6884	6893
49	6902	6911	6920	6928	6937	6946	6955	6964	6972	6981
50	6990	6998	7007	7016	7024	7033	7042	7050	7059	7067
51	7076	7084	7093	7101	7110	7118	7126	7135	7143	7152
52	7160	7168	7177	7185	7193	7202	7210	7218	7226	7235
53	7243	7251	7259	7267	7275	7284	7292	7300	7308	7316
54	7324	7332	7340	7348	7356	7364	7372	7380	7388	7396
N	0	1	2	3	4	5	6	7	8	9

N	0	1	2	3	4	5	6	7	8	9
55	7404	7412	7419	7427	7435	7443	7451	7459	7466	7474
56	7482	7490	7497	7505	7513	7520	7528	7536	7543	7551
57	7559	7566	7574	7582	7589	7597	7604	7612	7619	7627
58	7634	7642	7649	7657	7664	7672	7679	7686	7694	7701
59	7709	7716	7723	7731	7738	7745	7752	7760	7767	7774
60	7782	7789	7796	7803	7810	7818	7825	7832	7839	7846
61	7853	7860	7868	7875	7882	7889	7896	7903	7910	7917
62	7924	7931	7938	7945	7952	7959	7966	7973	7980	7987
63	7993	8000	8007	8014	8021	8028	8035	8041	8048	8055
64	8062	8069	8075	8082	8089	8096	8102	8109	8116	8122
65	8129	8136	8142	8149	8156	8162	8169	8176	8182	8189
66	8195	8202	8209	8215	8222	8228	8235	8241	8248	8254
67	8261	8267	8274	8280	8287	8293	8299	8306	8312	8319
68	8325	8331	8338	8344	8351	8357	8363	8370	8376	8382
69	8388	8395	8401	8407	8414	8420	8426	8432	8439	8445
70	8451	8457	8463	8470	8476	8482	8488	8494	8500	8506
71	8513	8519	8525	8531	8537	8543	8549	8555	8561	8567
72	8573	8579	8585	8591	8597	8603	8609	8615	8621	8627
73	8633	8639	8645	8651	8657	8663	8669	8675	8681	8686
74	8692	8698	8704	8710	8716	8722	8727	8733	8739	8745
75	8751	8756	8762	8768	8774	8779	8785	8791	8797	8802
76	8808	8814	8820	8825	8831	8837	8842	8848	8854	8859
77	8865	8871	8876	8882	8887	8893	8899	8904	8910	8915
78	8921	8927	8932	8938	8943	8949	8954	8960	8965	8971
79	8976	8982	8987	8993	8998	9004	9009	9015	9020	9025
80	9031	9036	9042	9047	9053	9058	9063	9069	9074	9079
81	9085	9090	9096	9101	9106	9112	9117	9122	9128	9133
82	9138	9143	9149	9154	9159	9165	9170	9175	9180	9186
83	9191	9196	9201	9206	9212	9217	9222	9227	9232	9238
84	9243	9248	9253	9258	9263	9269	9274	9279	9284	9289
85	9294	9299	9304	9309	9315	9320	9325	9330	9335	9340
86	9345	9350	9355	9360	9365	9370	9375	9380	9385	9390
87	9395	9400	9405	9410	9415	9420	9425	9430	9435	9440
88	9445	9450	9455	9460	9465	9469	9474	9479	9484	9489
89	9494	9499	9504	9509	9513	9518	9523	9528	9533	9538
90	9542	9547	9552	9557	9562	9566	9571	9576	9581	9586
91	9590	9595	9600	9605	9609	9614	9619	9624	9628	9633
92	9638	9643	9647	9652	9657	9661	9666	9671	9675	9680
93	9685	9689	9694	9699	9703	9708	9713	9717	9722	9727
94	9731	9736	9741	9745	9750	9754	9759	9763	9768	9773
95	9777	9782	9786	9791	9795	9800	9805	9809	9814	9818
96	9823	9827	9832	9836	9841	9845	9850	9854	9859	9863
97	9868	9872	9877	9881	9886	9890	9894	9899	9903	9908
98	9912	9917	9921	9926	9930	9934	9939	9943	9948	9952
99	9956	9961	9965	9969	9974	9978	9983	9987	9991	9996
N	0	1	2	3	4	5	6	7	8	9

$$(x^2 \times y) + 6 = z$$

Solve for x.

$$(x^2 \times y) + \cancel{6} - \cancel{6} = z - 6$$

(1) Subtract 6 from each side.

$$(x^2 \times y) = z - 6$$

$$\frac{x^2 \times \cancel{y}}{\cancel{y}} = \frac{z - 6}{y}$$

(2) Divide each side by y.

$$x^2 = \frac{z - 6}{y}$$

$$\sqrt{x^2} = \sqrt{\frac{z - 6}{y}}$$

(3) Extract the square root of each side. (Find the quantity $\sqrt{N}$ that, when multiplied by itself, yields the number N.)

$$x = \sqrt{\frac{z - 6}{y}}$$

(4) Simplify. The square root of x^2 is simply x.

QUADRATIC EQUATIONS. A quadratic equation has the form $ax^2 + bx + c = 0$, where a, b, and c are constants (a cannot be equal to 0). A number of calculations encountered in the text require that a quadratic equation be solved. The solution of such an equation is

$$x = \frac{-b \pm \sqrt{b^2 - 4ac}}{2a}$$

In Example 14-10 the following quadratic equation is obtained.

$$x^2 + 4.28 \times 10^{-4}\, x - 1.03 \times 10^{-5} = 0$$

Its solution is

$$x = \frac{-4.28 \times 10^{-4} \pm \sqrt{(4.28 \times 10^{-4})^2 + 4 \times 1.03 \times 10^{-5}}}{2}$$

$$= \frac{-4.28 \times 10^{-4} \pm \sqrt{(1.83 \times 10^{-7}) + (4.12 \times 10^{-5})}}{2}$$

$$= \frac{-4.28 \times 10^{-4} \pm \sqrt{4.14 \times 10^{-5}}}{2} = \frac{-4.28 \times 10^{-4} \pm 6.43 \times 10^{-3}}{2}$$

$$= \frac{-4.28 \times 10^{-4} + 6.43 \times 10^{-3}}{2} = \frac{6.00 \times 10^{-3}}{2} = 3.00 \times 10^{-3}$$

Note that only the (+) value of (±) sign was used in solving for x. If the (−) value had been used, a negative value of x would have resulted. However, for the given situation a negative value of x is meaningless.

HIGHER DEGREE EQUATIONS. The highest power of x in a quadratic equation is 2; a quadratic is a second-degree equation. Some equations have the unknown appearing to a higher degree. Exact solutions of algebraic equations of higher degree than second are also possible, but a simple method that usually works quite well is the *method of successive approximations*. The following equation is obtained in the solution of Exercise 14-22; it is a cubic equation (third degree in x).

$$\frac{(0.657 + 2x)^2 \times 1.90}{(0.390 - 2x)^2(0.156 - x)} = 280$$

It can be reduced to the simpler cubic equation

$$x^3 - 0.537x^2 + 0.103x - 0.00520 = 0$$

Suppose we try the solution $x = 0.10$ and substitute into the above equation.

$$(1 \times 10^{-3}) - (5.37 \times 10^{-3})$$
$$+ (1.03 \times 10^{-2}) - (5.20 \times 10^{-3})$$
$$= (11.3 \times 10^{-3}) - (10.6 \times 10^{-3}) > 0$$

The sum of terms on the left side > 0.
   Now try $x = 1 \times 10^{-2}$.

$$(1 \times 10^{-6}) - (5.37 \times 10^{-5})$$
$$+ (1.03 \times 10^{-3}) - (5.20 \times 10^{-3}) < 0$$

The sum of terms on the left side < 0.
   The value of x we are seeking has a value $0.01 < x < 0.10$. By successive approximations the value $x = 0.076$ can be established, and these approximations can be made easily if an electronic calculator or computer is used.

A-4 Graphs

Suppose the following sets of numbers are obtained for two quantities x and y by laboratory measurement.

$$x = 0, 1, 2, 3, 4, \ldots$$
$$y = 2, 4, 6, 8, 10, \ldots$$

The relationship between these sets of numbers is not difficult to establish.

$$y = 2x + 2$$

   Ideally, the results of experimental measurements are best expressed through a mathematical equation. Sometimes, however, an exact equation cannot be written or its form is not clear from the experimental data. The graphing of data is very useful in such cases. In Figure A-1 the points listed above are located on a coordinate grid, in which x values are placed along the horizontal axis (abscissa) and y values along the vertical axis (ordinate). For each point the x and y values are indicated in parentheses.
   The data points are seen to define a straight line. A mathematical equation for a straight line always has the form

$$y = mx + b \qquad\qquad (A.1)$$

Values of m, the **slope** of the line, and b, the **intercept,** can be obtained from the straight line graph.
   When $x = 0$, $y = b$. The intercept is the point where the straight line intersects the y-axis. The slope can be obtained from two points on the graph.

$$y_2 = mx_2 + b \qquad \text{and} \qquad y_1 = mx_1 + b$$
$$y_2 - y_1 = m(x_2 - x_1)$$

$$m = \frac{y_2 - y_1}{x_2 - x_1}$$

From the straight line in Figure A-1 can you establish that $m = b = 2$?

FIGURE A-1
A straight line graph: $y = mx + b$.

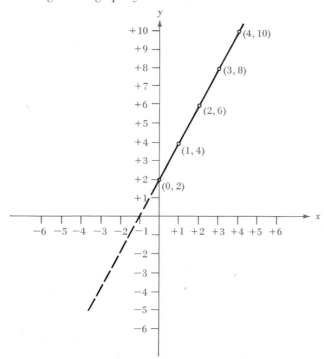

APPENDIX B

Some Basic Physical Concepts

B-1 Velocity and Acceleration

Time elapses as an object is displaced from one point to another. The **velocity** of the object is defined as the distance traveled per unit of time. An automobile that travels a distance of 60.0 km in exactly one hour has a velocity of 16.7 m/s.

Table B-1 contains experimental data on the velocity of a free-falling body. For this type of motion velocity is not constant; it increases with time—the falling body "speeds up" continuously. The rate of change of velocity with time is called **acceleration.** It has the units of distance per unit time per unit time. The constant acceleration in Table B-1, called the *acceleration due to gravity,* is 9.8 m/s².

TABLE B-1
Velocity and acceleration of a free-falling body

Time elapsed, s	Total distance, m	Velocity, m/s	Acceleration, m/s²
0	0		
1	4.9	4.9	
2	19.6	14.7	9.8
3	44.1	24.5	9.8
4	78.4	34.3	9.8

B-2 Force and Work

Newton's first law of motion states that an object at rest remains at rest, and that an object in motion remains in uniform motion, unless acted upon by an external force. The tendency for an object to remain at rest or in uniform motion is called **inertia;** a **force** is that which is required to overcome inertia. Since the application of a force to an object either gives it motion or changes its motion, the actual effect of a force is to change the velocity of an object. Change in velocity is an acceleration, so force is that which provides an object with acceleration.

Newton's second law of motion describes the force, F, required to produce an acceleration, a, in an object of mass, m.

$$F = ma \tag{B.1}$$

710

The basic unit of force in the SI system is the **newton (N).**

$$1 N = 1 \text{ kg} \times 1 \; m/s^2 \tag{B.2}$$

The acceleration listed in Table B-1 is often denoted by the symbol g. The force of gravity on an object (its weight) is the product of the mass of the object and the acceleration of gravity.

$$F = mg \tag{B.3}$$

Work is performed when a force acts through a distance. The **joule (J)** is the amount of work associated with 1 newton (N) acting through a distance of 1 m.

$$1 J = 1 N \times 1 m \tag{B.4}$$

B-3 Energy

Energy is defined as the capacity to do work, but there are further ways of categorizing energy beyond this simple statement. For example, an object in motion has the immediate capacity to do work, and its energy is called **kinetic** energy. An object at rest may also have the capacity to do work by changing its position. The energy it possesses, which can be transformed to actual work, is called **potential** energy. As a ball rolls down a hill, some of its potential energy is converted to kinetic energy.

The **kinetic** energy of an object is given by one-half the product of its mass and the square of its velocity.

$$KE = \tfrac{1}{2}mv^2 \tag{B.5}$$

Mathematical expressions for potential energy are also possible, but their exact forms depend on the manner in which this energy is "stored."

B-4 Magnetism

Attractive and repulsive forces associated with a magnet are centered at regions called **poles.** A magnet has a north and a south pole. If two magnets are aligned such that the north pole of one is directed toward the south pole of the second, an attractive force develops. If the alignment brings like poles into proximity, either both north or both south, a repulsive force develops. *Unlike poles attract; like poles repel.*

A magnetic field exists in that region surrounding a magnet in which the influence of the magnet can be felt. Internal changes produced within an iron object by a magnetic field, not produced in a field-free region, are responsible for the attractive force that the object experiences.

FIGURE B-1
Forces between electrically charged objects.

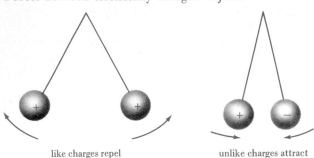

like charges repel   unlike charges attract

FIGURE B-2
Production of electric
charges by induction in a
gold leaf electroscope.

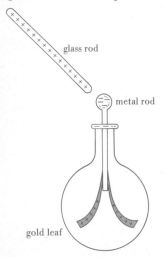

glass rod

metal rod

gold leaf

The glass rod acquires a positive electric charge by
being rubbed with a silk cloth. As the rod is brought
near the electroscope a separation of charge occurs in
the electroscope. The leaves become positively charged
and repel one another. Negative charge is attracted to
the spherical terminal at the end of the metal rod. If
the glass rod is removed, the charges on the
electroscope redistribute themselves and the leaves
collapse. If before the glass rod is removed the
spherical ball is touched by an electric conductor,
negative charge is removed from the ball, the
electroscope retains a net positive charge, and the
leaves remain outstretched.

B-5 Static Electricity

Another property with which certain objects may be
endowed is electrical charge. Analogous to the case
with magnetism, *unlike charges attract and like charges
repel* (see Figure B-1). In Coulomb's law, stated below,
a *positive* force is *repulsive;* and a *negative* force is *attractive.*

$$F = \frac{Q_1 Q_2}{\epsilon r^2} \qquad \text{(B.6)}$$

where Q_1 is the magnitude of the charge on object 1.
Q_2 is the magnitude of the charge on object 2.
r is the distance between the objects.
ϵ is a proportionality constant called the **die-
lectric constant** of the medium. The numerical value of this constant reflects the
effect that the medium separating the two
charged objects has on the force existing
between them. For vacuum, $\epsilon = 1$, and for
other media ϵ is greater than 1 (e.g., for
water $\epsilon = 78.5$).

An **electric field** exists in that region surrounding
an electrically charged object in which the influence of
the electrical charge is felt. If an uncharged object is
brought into the field of a charged object, the uncharged object may undergo internal changes that it
would not experience in a field-free region. These
changes may lead to the production of electric charges
in the formerly uncharged object, a phenomenon
called **induction** (illustrated in Figure B-2).

B-6 Current Electricity

Current electricity consists of a flow of electrically
charged particles. In electric currents in metallic conductors, the charged particles are electrons; in molten
salts or in aqueous solutions, the particles are both negatively and positively charged ions.

The unit of electric charge is called a **coulomb (C).**
The unit of electric current known as the **ampere (A)** is
defined as a flow of 1 C/s through an electrical conductor. Two variables determine the magnitude of the
electric current, I, flowing through a conductor. These
are the potential difference or voltage drop, E, along
the conductor and the electrical resistance of the conductor, R. The units of voltage and resistance are the
volt (V) and **ohm,** respectively. The relationship of electric current, voltage, and resistance is given through
Ohm's law.

$$I = \frac{E}{R} \qquad \text{(B.7)}$$

B-7 Electromagnetism

The relationship between electricity and magnetism is an intimate one. Interactions of electric and magnetic fields result in (1) magnetic fields associated with the flow of electric current (as in electromagnets), (2) forces experienced by current-carrying conductors when placed in a magnetic field (as in electric motors), and (3) electric current being induced when an electric conductor is moved through a magnetic field (as in electric generators). Numerous observations described in this text can be understood in terms of electromagnetic phenomena.

APPENDIX C

SI Units

The system of units that will in time be used universally for expressing all measured quantities is Le Système International d'Unités (The International System of Units), adopted in 1960 by the Conference Générale des Poids et Mesures (General Conference of Weights and Measures). A summary of some of the provisions of the SI convention is provided here.

C-1 SI Base Units

A single unit has been established for each of the basic quantities involved in measurement. These are

Physical quantity	Unit	Symbol
length	meter	m
mass	kilogram	kg
time	second	s
electric current	ampere	A
temperature	kelvin	K
luminous intensity	candela	cd
amount of substance	mole	mol
plane angle	radian	rad
solid angle	steradian	sr

C-2 SI Prefixes

Distinctive prefixes are attached to the base unit to express quantities that are **multiples** (greater than) or **submultiples** (less than) of the base unit. The multiples and submultiples are obtained by multiplying the base unit by powers of ten.

Multiple	Prefix	Symbol	Submultiple	Prefix	Symbol
10^{12}	tera	T	10^{-1}	deci	d
10^{9}	giga	G	10^{-2}	centi	c
10^{6}	mega	M	10^{-3}	milli	m
10^{3}	kilo	k	10^{-6}	micro	μ
10^{2}	hecto	h	10^{-9}	nano	n
10^{1}	deka	da	10^{-12}	pico	p
			10^{-15}	femto	f
			10^{-18}	atto	a

C-3 Derived SI Units

A number of quantities must be derived from measured values of the SI base quantities [e.g., volume has the unit (length)^3]. Two sets of derived units are given, those whose names follow directly from the base units and those that are given special names.

Two other SI conventions are illustrated through this table: (a) Units are written in the singular form—meter or m (*not* meters or ms); (b) negative exponents are preferred to the shilling bar or solidus (/)—$m\,s^{-1}$ and $m\,s^{-2}$ (*not* m/s and m/s/s).

Physical quantity	Unit	Symbol
area	square meter	m^2
volume	cubic meter	m^3
velocity	meter per second	$m\,s^{-1}$
acceleration	meter per second squared	$m\,s^{-2}$
density	kilogram per cubic meter	$kg\,m^{-3}$
molar mass	kilogram per mole	$kg\,mol^{-1}$
molar volume	cubic meter per mole	$m^3\,mol^{-1}$
molar concentration	mole per cubic meter	$mol\,m^{-3}$

Physical quantity	Unit	Symbol	In terms of SI units
frequency	hertz	Hz	s^{-1}
force	newton	N	$kg\,m\,s^{-2}$
pressure	pascal	Pa	$N\,m^{-2}$
energy	joule	J	$kg\,m^2\,s^{-2}$
power	watt	W	$J\,s^{-1}$
electric charge	coulomb	C	$A\,s$
electric potential difference	volt	V	$J\,A^{-1}\,s^{-1}$
electric resistance	ohm	Ω	$V\,A^{-1}$

C-4 Units to Be Discouraged or Abandoned

There are several commonly used units whose use is to be discouraged and ultimately abandoned. Their gradual disappearance is to be expected, though each is used in this text. A few such units are listed.

Another SI convention is implied here. No commas are used in expressing large numbers but spaces are left between groupings of three digits (that is, 101 325 instead of 101,325). (Decimal points are written either as periods or commas.)

Physical quantity	Unit	Symbol	Definition in SI units
length	angstrom	Å	1×10^{-10} m
force	dyne	dyn	1×10^{-5} N
energy	erg	erg	1×10^{-7} J
energy	calorie	cal	4.184 J
pressure	bar	bar	1×10^{5} Pa
pressure	atmosphere	atm	101 325 Pa
pressure	millimeter of mercury	mmHg	$(13.5951)(980.665) \times 10^{-2}$ Pa
pressure	torr	torr	133.322 Pa

C-5 Fundamental Constants

The fundamental constants introduced in this text, such as the speed of light, acceleration due to gravity, gas constant, Planck's constant, and the Faraday constant, continue to be used, but their units should be expressed as SI units. (See the inside back cover for a listing.)

APPENDIX D

Thermodynamic Properties of Substances

	$\Delta \bar{H}_f^\circ$, kJ/mol	$\Delta \bar{G}_f^\circ$, kJ/mol	$\bar{S}^\circ$, J mol^{-1} K^{-1}
$Al_2O_3(s)$	−1669.79	−1576.41	51.00
$BaCO_3(s)$	−1218.8	−1138.9	112.1
$B_2H_6(g)$	31.4	82.8	232.88
$B_2O_3(s)$	−1263.6	−1184.1	54.02
$Br(g)$	111.75	82.38	174.93
$Br_2(g)$	30.71	3.14	245.35
$Br_2(l)$	0	0	152.30
$BrCl(g)$	14.7	− 0.88	239.9
$C(g)$	718.39	672.95	157.99
$C(diamond)$	1.88	2.85	2.43
$C(graphite)$	0	0	5.69
$CCl_4(g)$	− 106.7	− 64.0	309.4
$CO(g)$	− 110.54	− 137.28	197.90
$CO_2(g)$	− 393.51	− 394.38	213.64
$CH_4(g)$	− 74.85	− 50.79	186.19
$CH_2Cl_2(g)$	− 82.0	− 58.6	234.2
$C_2H_2(g)$	226.73	209.20	200.83
$C_2H_4(g)$	52.30	68.12	219.45
$C_2H_6(g)$	− 84.68	− 32.89	229.49
$C_3H_8(g)$	− 103.85	− 23.47	269.9
$C_6H_6(g)$	82.93	129.66	269.20
$C_6H_6(l)$	49.04	124.52	172.80
$C_2H_5OH(l)$	− 277.65	− 174.77	160.67
$CaCO_3(s)$	−1207.1	−1128.76	92.88
$CaO(s)$	− 635.5	− 604.17	39.75
$CaSO_4(s)$	−1432.7	−1320.3	106.7
$Cl(g)$	121.38	105.39	165.10
$Cl_2(g)$	0	0	222.97
$CuO(s)$	− 155.2	− 127.2	43.5
$Cu_2O(s)$	− 166.69	− 146.36	100.8
$Fe_2O_3(s)$	− 822.16	− 741.0	90.0
$Fe_3O_4(s)$	−1117.13	−1014.20	146.4

	$\Delta \bar{H}_f^\circ$, kJ/mol	$\Delta \bar{G}_f^\circ$, kJ/mol	$\bar{S}^\circ$, J mol^{-1} K^{-1}
H(g)	217.94	203.26	114.60
H_2(g)	0	0	130.58
HBr(g)	− 36.23	− 53.22	198.49
HCl(g)	− 92.30	− 95.27	186.69
HF(g)	− 268.61	− 270.70	173.51
HI(g)	25.94	1.30	206.31
H_2O(g)	− 241.84	− 228.61	188.74
H_2O(l)	− 285.85	−237.19	69.96
H_2S(g)	− 20.17	− 33.01	205.64
HCHO(g)	− 115.9	− 110.0	218.7
He(g)	0	0	126.06
Hg(g)	60.84	31.76	174.89
Hg(l)	0	0	77.4
I(g)	106.61	70.17	180.67
I_2(g)	62.26	19.37	260.58
I_2(s)	0	0	116.7
KCl(s)	− 435.89	− 408.32	82.68
$MgCl_2$(s)	− 641.83	− 592.33	89.5
MgO(s)	− 601.83	− 569.57	26.8
MnO_2(s)	− 519.7	− 464.8	53.1
N(g)	472.71	455.55	153.22
N_2(g)	0	0	191.50
NH_3(g)	− 46.19	− 16.65	192.51
NH_4Cl(s)	− 315.38	− 203.89	94.6
NO(g)	90.37	86.69	210.62
N_2O(g)	81.55	103.60	219.99
NO_2(g)	33.85	51.84	240.45
N_2O_4(g)	9.67	98.28	304.30
NOCl(g)	52.59	66.36	264
NaCl(s)	− 410.99	− 384.01	72.38
O(g)	247.53	230.12	160.96
O_2(g)	0	0	205.02
O_3(g)	142.3	163.43	237.7
PCl_3(g)	− 306.4	− 286.3	311.6
PCl_5(g)	− 398.9	− 324.6	353
S(rhombic)	0	0	31.88
S(monoclinic)	0.30	0.096	32.55
SO_2(g)	− 296.90	− 300.37	248.53
SO_3(g)	− 395.18	− 370.37	256.23
SO_2Cl_2(l)	− 389	− 314	207
UO_2(s)	−1130	−1075	77.8
ZnO(s)	− 347.98	− 318.19	43.9

APPENDIX E

Electron Configurations of the Elements

Atomic number	Element	Electron configuration
1	H	$1s^1$
2	**He**	$1s^2$
3	Li	[He] $2s^1$
4	Be	____ $2s^2$
5	B	____ $2s^22p^1$
6	C	____ $2s^22p^2$
7	N	____ $2s^22p^3$
8	O	____ $2s^22p^4$
9	F	____ $2s^22p^5$
10	**Ne**	$1s^22s^22p^6$
11	Na	[Ne] $3s^1$
12	Mg	____ $3s^2$
13	Al	____ $3s^23p^1$
14	Si	____ $3s^23p^2$
15	P	____ $3s^23p^3$
16	S	____ $3s^23p^4$
17	Cl	____ $3s^23p^5$
18	**Ar**	$1s^22s^22p^63s^23p^6$
19	K	[Ar] $4s^1$
20	Ca	____ $4s^2$
21	Sc	____ $3d^14s^2$
22	Ti	____ $3d^24s^2$
23	V	____ $3d^34s^2$
24	Cr	____ $3d^54s^1$
25	**Mn**	____ $3d^54s^2$
26	Fe	____ $3d^64s^2$
27	Co	____ $3d^74s^2$
28	Ni	____ $3d^84s^2$
29	Cu	____ $3d^{10}4s^1$
30	Zn	____ $3d^{10}4s^2$
31	Ga	____ $3d^{10}4s^24p^1$

Atomic number	Element	Electron configuration
32	Ge	_____ $3d^{10}4s^24p^2$
33	As	_____ $3d^{10}4s^24p^3$
34	Se	_____ $3d^{10}4s^24p^4$
35	Br	_____ $3d^{10}4s^24p^5$
36	**Kr**	$1s^22s^22p^63s^23p^63d^{10}4s^24p^6$
37	Rb	[Kr] $5s^1$
38	Sr	_____ $5s^2$
39	Y	_____ $4d^15s^2$
40	Zr	_____ $4d^25s^2$
41	Nb	_____ $4d^45s^1$
42	Mo	_____ $4d^55s^1$
43	Tc	_____ $4d^55s^2$
44	Ru	_____ $4d^75s^1$
45	Rh	_____ $4d^85s^1$
46	Pd	_____ $4d^{10}$
47	Ag	_____ $4d^{10}5s^1$
48	Cd	_____ $4d^{10}5s^2$
49	In	_____ $4d^{10}5s^25p^1$
50	Sn	_____ $4d^{10}5s^25p^2$
51	Sb	_____ $4d^{10}5s^25p^3$
52	Te	_____ $4d^{10}5s^25p^4$
53	I	_____ $4d^{10}5s^25p^5$
54	**Xe**	$1s^22s^22p^63s^23p^63d^{10}4s^24p^64d^{10}5s^25p^6$
55	Cs	[Xe] $6s^1$
56	Ba	_____ $6s^2$
57	La	_____ $5d^16s^2$
58	Ce	_____ $4f^26s^2$
59	Pr	_____ $4f^36s^2$
60	Nd	_____ $4f^46s^2$
61	Pm	_____ $4f^56s^2$
62	Sm	_____ $4f^66s^2$
63	Eu	_____ $4f^76s^2$
64	Gd	_____ $4f^75d^16s^2$
65	Tb	_____ $4f^96s^2$
66	Dy	_____ $4f^{10}6s^2$
67	Ho	_____ $4f^{11}6s^2$
68	Er	_____ $4f^{12}6s^2$

Atomic number	Element	Electron configuration
69	Tm	_____ $4f^{13}6s^2$
70	Yb	_____ $4f^{14}6s^2$
71	Lu	_____ $4f^{14}5d^16s^2$
72	Hf	_____ $4f^{14}5d^26s^2$
73	Ta	_____ $4f^{14}5d^36s^2$
74	W	_____ $4f^{14}5d^46s^2$
75	Re	_____ $4f^{14}5d^56s^2$
76	Os	_____ $4f^{14}5d^66s^2$
77	Ir	_____ $4f^{14}5d^76s^2$
78	Pt	_____ $4f^{14}5d^96s^1$
79	Au	_____ $4f^{14}5d^{10}6s^1$
80	Hg	_____ $4f^{14}5d^{10}6s^2$
81	Tl	_____ $4f^{14}5d^{10}6s^26p^1$
82	Pb	_____ $4f^{14}5d^{10}6s^26p^2$
83	Bi	_____ $4f^{14}5d^{10}6s^26p^3$
84	Po	_____ $4f^{14}5d^{10}6s^26p^4$
85	At	_____ $4f^{14}5d^{10}6s^26p^5$
86	**Rn**	$1s^22s^22p^63s^23p^63d^{10}4s^24p^64d^{10}4f^{14}5s^25p^65d^{10}6s^26p^6$
87	Fr	[Rn] $7s^1$
88	Ra	_____ $7s^2$
89	Ac	_____ $6d^17s^2$
90	Th	_____ $6d^27s^2$
91	Pa	_____ $5f^26d^17s^2$
92	U	_____ $5f^36d^17s^2$
93	Np	_____ $5f^46d^17s^2$
94	Pu	_____ $5f^67s^2$
95	Am	_____ $5f^77s^2$
96	Cm	_____ $5f^76d^17s^2$
97	Bk	_____ $5f^86d^17s^2$
98	Cf	_____ $5f^{10}7s^2$
99	Es	_____ $5f^{11}7s^2$
100	Fm	_____ $5f^{12}7s^2$
101	Md	_____ $5f^{13}7s^2$
102	No	_____ $5f^{14}7s^2$
103	Lr	_____ $5f^{14}6d^17s^2$
104	(Ku)	_____ $5f^{14}6d^27s^2$
105	(Ha)	_____ $5f^{14}6d^37s^2$

Answers to Selected Exercises

Chapter 1

Exercises: **1.** (a) physical; (b) chemical; (c) physical; (d) chemical. **2.** (a) substance; (b) heterogeneous mixture; (c) homogeneous mixture; (d) heterogeneous mixture; (e) heterogeneous mixture; (f) homogeneous mixture; (g) substance. **3.** (a) chemical; (b) physical; (c) physical. **5.** (a) extensive; (b) intensive; (c) extensive; (d) intensive. **9.** (a) 7.5×10^3; (b) 3.17×10^5; (c) 8.152×10^6; (d) 1×10^5; (e) 6.2×10^{-3}; (f) 5.00×10^{-2}; (g) 3.8×10^{-7}; (h) 1.00×10^{-1}; (i) 3×10^0. **10.** (a) 412; (b) 0.665; (c) 0.092; (d) 628,000; (e) 4.0; (f) 29,800,000,000; (g) 0.00000193; (h) 8.30; (i) 0.00001235. **11.** (a) 1.86×10^5 mi/s; (b) 5×10^{15} to 6×10^{15} t; (c) 1.73×10^{17} W; (d) 1×10^{-6} m $= 1\ \mu$m; (e) 1×10^{-5} m $= 10\ \mu$m. **12.** (a) 1.8×10^6; (b) 6.9×10^6; (c) 3.6×10^{-4}; (d) 2.2; (e) 5.6×10^5; (f) 8.4×10^4. **13.** (a) 8.4×10^{-2}; (b) 2.0×10^3; (c) 2.3×10^{-5}; (d) 5.0×10^2. **14.** (a) exact; (b) measured; (c) measured; (d) exact; (e) measured. **15.** (a) 3; (b) 2; (c) 5; (d) 5; (e) indeterminate; (f) 1; (g) 5; (h) 1. **16.** (a) 1418; (b) 303.5; (c) 0.01404; (d) 156.3; (e) 1.800×10^5; (f) 17.60; (g) 1.500×10^3. **17.** (a) 9.01×10^4; (b) 3.9×10^9; (c) 2.0×10^{-3}; (d) 71.6; (e) 2.16×10^4. **18.** (a) 2.76×10^3 g; (b) 8.160 m; (c) 3.68×10^{-4} kg; (d) 0.725 L; (e) 167 mm; (f) 323 cm^3; (g) 2.67×10^3 mg; (h) 6.7×10^2 m. **19.** (a) 186.0 in.; (b) 24.0 lb; (c) 432 in.; (d) 5.4×10^3 s; (e) 467.0 yd; (f) 1.21×10^4 ft. **20.** (a) 3.0×10^1 cm; (b) 4.9 m; (c) 4.0×10^2 g; (d) 1.2×10^2 lb; (e) 73.8 ft; (f) 0.12 oz. **21.** 5.03 m. **22.** $8.86 for 3 lb. **23.** 10.2 s. **24.** 7.92 in. **25.** (a) 6.7×10^2 mg; (b) 8.9 mg/kg. **26.** 1×10^6 m^2. **27.** 0.29 m^3. **28.** (a) 43,560 ft^2; (b) 0.405 hm^2. **29.** 1.57×10^3 km/h. **30.** (a) 104°F; (b) 25°C; (c) 666.7°C; (d) -285°F. **31.** 47.8°C; -8.3°C. **32.** no. **33.** -459.67°F. **34.** 1.26 g/cm^3. **35.** (a) 2.78×10^2 g; (b) 0.901 L;

(c) 9.22 lb. **36.** 0.7920 g/cm^3. **37.** 104 cm^3. **38.** (a) < (b) < (c). **39.** 0.04 g. **40.** 12% A, 22% B, 46% C, 14% D, 6% F. **41.** 1.5×10^2 g. **42.** 343 g ethanol. **43.** 25.5 g. **44.** 2.58 L.

Self-Test Questions: **1.** (c). **2.** (a). **3.** (d). **4.** (b). **5.** (c). **6.** (b). **8.** 1.11×10^4. **9.** 436 lb.

Chapter 2

Exercises: **3.** (a) yes; (b) 27.3% C, 72.7% O. **4.** first experiment, 39.4% sodium; second experiment, 60.6% chlorine. **5.** both experiments, 11.2% hydrogen. **6.** (a) AB, A_2B, AB_2; (b) 33% A, 50% A, 20% A. **7.** hydrogen peroxide, HO_2; water HO. **8.** at. wt. of magnesium = 24.3 (average of three values). **9.** (a) HgO; (b) 84; (c) 192. **13.** N_2O, NO_2. **14.** HgO and Hg_2O. **20.** (a) 3.4×10^4; (b) 4.0×10^7. **21.** (a) 0; (b) -1.60×10^{-7} C; (c) 3.20×10^{-7} C. **23.** e/m (n) $= e/m$ $(^{40}_{18}\text{Ar}) < e/m$ $(^{37}_{17}\text{Cl}^-) <$ e/m $(^4_2\text{He}^{2+}) < e/m$ (proton) $< e/m$ (electron). **29.** (a) $^{22}_{10}\text{Ne} < ^{58}_{27}\text{Co} < ^{59}_{29}\text{Cu} < ^{120}_{48}\text{Cd} < ^{112}_{50}\text{Sn} < ^{122}_{52}\text{Te}$; (b) $^{22}_{10}\text{Ne} < ^{59}_{29}\text{Cu} < ^{58}_{27}\text{Co} < ^{112}_{50}\text{Sn} < ^{122}_{52}\text{Te} < ^{120}_{48}\text{Cd}$; (c) $^{22}_{10}\text{Ne} < ^{58}_{27}\text{Co} < ^{59}_{29}\text{Cu} < ^{112}_{50}\text{Sn} < ^{120}_{48}\text{Cd} < ^{122}_{52}\text{Te}$. **31.** (a) 40.6%; (b) 42.2%. **32.** (a) 5.81×10^{-11} g; (b) 6.14×10^{-11} g; (c) 5.89×10^{-11} g. **33.** (a) 0.584668; (b) 1.583200; (c) 6.99262. **34.** none. **35.** 69.7. **36.** 238.0. **37.** (a) ^7_3Li; (b) approximately 9 parts ^7_3Li to 1 part ^6_3Li; (c) 7.5% ^6_3Li and 92.5% ^7_3Li. **38.** 0.36% $^{15}_7\text{N}$. **39.** 39.964. **40.** (a) six; (b) 36, 37, 38, 38, 39, 40; (c) most abundant: $^1\text{H}^{35}\text{Cl}$; second: $^1\text{H}^{37}\text{Cl}$.

Self-Test Questions: **1.** (d). **2.** (b). **3.** (a). **4.** (c). **5.** (b). **6.** (d). **8.** 12.0127. **9.** -2.609×10^3 C/g. **10.** 108.9.

Chapter 3

Exercises: **2. (a)** 1.50×10^{24}; **(b)** 9.0×10^{21};
(c) 2.4×10^{14}.   **3. (a)** 2.7×10^{24};   **(b)** 1.3×10^{25};
(c) 3.0×10^{19}; **(d)** 1.69×10^{24}.   **4. (a)** 82.4 mol C;
(b) 247 mol H; **(c)** 41.2 mol O.
5. (a) 3.10×10^{24} Mg atoms; **(b)** 1.20×10^{2} g O_2;
(c) 3.02×10^{3} kg Fe; **(d)** 23.7 cm³ Na; **(e)** 8.72×10^{23}.
6. (a) $\frac{1}{3}$; **(b)** 10.8% S, by mass.   **7.** 3.15×10^{3} g.
8. 2.05×10^{23} Ag atoms.   **9.** 4.35×10^{24} Cl⁻ ions.
10. 4.37×10^{12} Pb atoms.   **11. (a)** 3.35×10^{-5} mol S_8;
(b) 1.61×10^{20} S atoms.
12. 1.26×10^{15} $CHCl_3$ molecules.
13. (b) and (c) are correct.   **14. (a)** 149.2;
(b) 2.02×10^{24} C atoms.   **15. (a)** 21; **(b)** 2:1;
(c) 0.060 g H/g C; **(d)** oxygen.
16. 72.2% C, 7.1% H, 4.7% N, 16.0% O.   **17.** Li_2S.
18. guanidine.   **19.** 4.7 lb $(NH_4)_2SO_4$.   **20.** Cl_2O_7.
21. B_2H_6.   **22.** $C_8H_{10}N_4O_2$.
23. cobalt(II) oxide, CoO; cobalt(III) oxide, Co_2O_3.
24. $C_6H_4Cl_2$.   **25. (a)** potassium iodide;
(b) calcium chloride; **(c)** magnesium nitrate;
(d) potassium chromate; **(e)** cesium sulfate;
(f) chromium(III) oxide; **(g)** iron(II) sulfate;
(h) zinc sulfide;
(i) calcium hydrogen carbonate (bicarbonate);
(j) potassium cyanide;
(k) potassium hydrogen phosphate;
(l) ammonium iodide; **(m)** copper(II) hydroxide.
26. stannous fluoride; $Pb(C_2H_3O_2)_4$; $Co_2(SO_4)_3$;
potassium periodate; aurous chloride.
27. iodine (mono)chloride; iodine trichloride;
chlorine trifluoride; bromine pentafluoride.
28. (a) CaO; **(b)** SrF_2; **(c)** $Al_2(SO_4)_3$; **(d)** $(NH_4)_2CrO_4$;
(e) $Mg(OH)_2$; **(f)** K_2CO_3; **(g)** $Zn(C_2H_3O_2)_2$;
(h) $Hg(NO_3)_2$; **(i)** Fe_2O_3; **(j)** $CrCl_2$; **(k)** Li_2S;
(l) $Ca(H_2PO_4)_2$; **(m)** $Mg(ClO4)_2$; **(n)** $KHSO_4$.
29. (a) ClO_2; **(b)** SiF_4; **(c)** C_3S_2; **(d)** B_2Br_4.
30. 43.8% H_2O.   **31.** (c).   **32.** $MgSO_4 \cdot 7 H_2O$.
33. 1.27 g.   **34.** $CuSiF_6 \cdot 6 H_2O$.
35. (a) 93.71% C, 6.31% H; **(b)** C_5H_4; **(c)** $C_{10}H_8$.
36. $C_4H_{10}O$.   **37.** CH_4N.   **38.** C_2H_5.
39. 0.2214 g AgI.
40. 77.9% Cu, 11.3% Sn, 4.6% Pb, 6.2% Zn.
41. $ZnSO_4 \cdot 7 H_2O$.
42. element is P; compounds are PCl_3 and PCl_5.
43. at. wt. of M = 24.4.   **44.** at. wt. of M = 27.
45. at. wt. of M = 56.

Self-Test Questions: **1.** (c).   **2.** (b).   **3.** (d).
4. (a).   **5.** (c).   **6.** 494 cm³.   **7. (a)** calcium iodide;
(b) $Fe_2(SO_4)_3$; **(c)** SO_3; **(d)** BrF_5;
(e) ammonium cyanide; **(f)** calcium chlorite;
(g) $LiHCO_3$.   **8. (a)** 57.5% Cu; **(b)** 720 g CuO.
9. N_2O_5.   **10.** $Na_2SO_3 \cdot 7 H_2O$.

Chapter 4

Exercises: **1. (a)** no reaction;
(b) $RaCl_2(aq) + Na_2SO_4(aq) \rightarrow 2 NaCl(aq) + RaSO_4(s)$;
(c) the black residue is Fe_3O_4:
$3 Fe_2O_3(s) + H_2(g) \rightarrow 2 Fe_3O_4(s) + H_2O(g)$.
2. (a) $2 Mg(s) + O_2(g) \rightarrow 2 MgO(s)$;
(b) $S(s) + O_2(g) \rightarrow SO_2(g)$;
(c) $CH_4(g) + 2 O_2(g) \rightarrow CO_2(g) + 2 H_2O(l)$;
(d) $Ag_2SO_4(aq) + BaI_2(aq) \rightarrow BaSO_4(s) + 2 AgI(s)$.
3. (a) $C(s) + H_2O(g) \rightarrow CO(g) + H_2(g)$;
(b) $2 Al(s) + 3 Cu^{2+}(aq) \rightarrow 2 Al^{3+}(aq) + 3 Cu(s)$;
(c) $ZnS(s) + 2 H^+(aq) \rightarrow Zn^{2+}(aq) + H_2S(g)$;
(d) $2 Cl_2(g) + 2 H_2O(g) \rightarrow 4 HCl(g) + O_2(g)$.
4. (a) $1, 2 \rightarrow 1, 2$; **(b)** $1, 1 \rightarrow 1, 1$; **(c)** $1, 3 \rightarrow 1, 3$;
(d) $6 \rightarrow 8, 1$; **(e)** $3, 2 \rightarrow 3, 1, 3$; **(f)** $3, 1 \rightarrow 2, 1$;
(g) $6, 16 \rightarrow 1, 12, 1$; **(h)** $1, 6 \rightarrow 3, 2$; **(i)** $1, 8 \rightarrow 1, 2, 2, 4$;
(j) $1, 12 \rightarrow 4, 2, 6$.   **5. (a)** $1, 2 > 1, 2$; **(b)** $1, 1 \rightarrow 1, 2$;
(c) $2, 6 \rightarrow 2, 3$; **(d)** $1, 2 \rightarrow 1, 1, 1$; **(e)** $1, 4, 2 \rightarrow 1, 2, 1$.
6. (a) $C_5H_{12} + 8 O_2 \rightarrow 5 CO_2 + 6 H_2O$;
(b) $2 C_6H_6 + 15 O_2 \rightarrow 12 CO_2 + 6 H_2O$;
(c) $C_6H_{12}O_6 + 6 O_2 \rightarrow 6 CO_2 + 6 H_2O$;
(d) $2 C_3H_6O_2 + 5 O_2 \rightarrow 4 CO_2 + 6 H_2O$;
(e) $2 C_3H_7OH + 9 O_2 \rightarrow 6 CO_2 + 8 H_2O$.
7. 6.6 mol Cl_2.   **8.** 6.75 mol Cl_2, 1.12 mol P_4.
9. (a) 5.94 mol H_2; **(b)** 107 g H_2O; **(c)** 35.0 g CaH_2.
10. (a) 0.122 mol O_2; **(b)** 7.34×10^{22} O_2 molecules;
(c) 6.07 g KCl.   **11.** 2.62 g CO_2.
12. 86.3% Fe_2O_3.   **13.** 94.4% Ag_2O.
14. 1.49 g H_2.   **15.** 90.4 ml.
16. (a) 0.312 M C_2H_5OH; **(b)** 7.81 M CH_3OH;
(c) 0.547 M $C_3H_8O_3$; **(d)** 4.10 M $CO(NH_2)_2$;
(e) 1.79×10^{-5} M $(C_2H_5)_2O$.   **17. (a)** 41.2 mol KCl;
(b) 5.31 g Na_2SO_4; **(c)** 15.2 cm³ CH_3OH;
(d) 6.75 gal C_2H_5OH.   **18.** 1.35 M $C_{12}H_{22}O_{11}$.
19. 604 ml.   **20.** 556 ml.   **21.** 0.756 M $MgSO_4$.
22. 29.53 ml.   **23. (a)** 4.94 g $Ca(OH)_2$;
(b) 53.4 kg $Ca(OH)_2$.   **24. (a)** 6.24 M NH_3;
(b) 11.0% NH_3.   **25. (a)** 404 ml; **(b)** 0.2408 M HCl.
26. 58.3% Fe.   **27.** 72 mg Mg.
28. (a) 0.33 mol Na_2CS_3, 0.17 mol Na_2CO_3, 0.50 mol H_2O;
(b) 1.70×10^{2} g Na_2CS_3.   **29.** 4.76 g NH_3.
30. (a) $2 H_2 + O_2 \rightarrow 2 H_2O$;
(b) 0.105 mol H_2, 0.0 mol O_2, 2.276 mol H_2O.
31. 43 g Cl_2.   **32.** 161 g HCl.   **33.** 15.3 mol CO_2.
34. a mixture.   **35.** 60.0 mol Cl_2.
36. 66.5 g $BaCO_3$.   **37.** 1.34 g $AgNO_3$.
38. 2.03×10^{2} kg Fe.   **39.** 62% yield.
40. 56 g $C_2H_4O_2$.   **41.** 0.92 g Cu.
42. 6.1×10^{2} L.   **43.** 6.95 kg NH_3.
44. (a) $2 CH_2CHCH_3 + 2 NH_3 + 3 O_2 \rightarrow 2 CH_2CHCN +$
$6 H_2O$; **(b)** 58%; **(c)** 1100 lb NH_3.

Self-Test Questions: **1.** (c).   **2.** (d).   **3.** (a).
4. (a).   **5.** (b).   **6.** (c).
7. (a) $Hg(NO_3)_2 \rightarrow Hg(l) + 2 NO_2(g) + O_2(g)$;

(b) $Na_2CO_3(aq) + 2 HCl(aq) \rightarrow 2 NaCl(aq) + H_2O +$ $CO_2(g)$; **(c)** $2 C_7H_6O_2 + 15 O_2 \rightarrow 14 CO_2 + 6 H_2O$. **8.** 25.8 ml. **9.** 0.719 g Na.

Chapter 5

Exercises: **1. (a)** 0.971 atm; **(b)** 3.02 atm; **(c)** 4.8 atm; **(d)** 0.979 atm; **(e)** 1.65 atm. **2. (a)** 10 cm; **(b)** 20 cm; **(c)** 76 cm; **(d)** 76 cm. **3. (a)** 116 cm; **(b)** 3.03 m; **(c)** 1.28 m; **(d)** 1.54 g/cm³. **4.** 1.24 atm. **5.** 746.9 mmHg. **6. (a)** 17.5 L; **(b)** 8.24 L. **7.** 514 mmHg. **8. (a)** 137 cm³; **(b)** 113 cm³. **9.** 105°C. **10. (a)** $V_f = 3V_i$; **(b)** $V_f = V_i/4$; **(c)** $V_f = V_i$. **11.** 275 K. **12.** 3.65 g. **13.** 32.7 L. **14.** 3.98 atm. **15.** 3.36 L. **16.** 75 g N_2. **17.** 5.7×10^2 K. **18.** 0.349 atm. **19.** 30.0 g/mol. **20.** 42.2 g/mol. **21.** C_2ClF_5. **22.** 1.73 g CO_2/L. **23.** P_4. **24.** oxygen. **25.** 7.6×10^6 L $SO_2(g)$. **26.** 309 L $CO_2(g)$. **27.** 13.8% $KClO_3$. **28. (a)** 1.2×10^4 L $H_2(g)$; **(b)** 123 L $NH_3(g)$; **(c)** 2.17×10^4 L NH_3 (at STP). **29. (a)** 0.0597 mol $SO_2(g)$; **(b)** 6.82 L **30.** 3.92 L. **31.** 2.52×10^3 g Ne. **32.** 754 mmHg. **33.** 12 g He. **34. (a)** 217 mmHg; **(b)** 652 mmHg. **35. (a)** apparent mol. wt. = 28.6; **(b)** less; **(c)** about 130:1. **36.** 3.80 L. **37.** 0.108 g O_2. **38.** 151 cm³. **39.** 741 mmHg. **42. (a)** 1090 K; **(b)** 493 m/s. **43.** 324 m/s. **44.** $\bar{u} = 6.5 \times 10^3$ m/s; $u_{rms} = 7.1 \times 10^3$ m/s. **45. (a)** 4.0:1.0; **(b)** 1.4:1.0; **(c)** 1.004:1.000. **46.** mol. wt. = 59. **47. (a)** nonideal; **(b)** ideal; **(c)** nonideal. **48. (a)** 29.1 atm; **(b)** 25.7 atm.

Self-Test Questions: **1.** (b). **2.** (a). **3.** (c). **4.** (a). **5.** (b). **6.** (d). **7.** 6.1 L. **9.** 77.3 L $H_2(g)$. **10.** 0.126 atm.

Chapter 6

Exercises: **1.** 2.8×10^3 cal. **2. (a)** -1.2×10^5 J; **(b)** 3.6×10^2 g. **3.** 0.0929 cal g⁻¹ °C⁻¹. **5.** 2.3×10^2 J mol⁻¹ °C⁻¹. **6.** 26.2°C. **7. (a)** 0; **(b)** +25 cal; **(c)** −47 J; **(d)** +395 J; **(e)** −125 J. **8. (a)** yes; **(b)** yes; **(c)** temperature remains constant. **10.** (c). **11. (a)** −6.24 kJ/g CaO; **(b)** −2.36 × 10⁵ kJ; **(c)** Yes. **12. (a)** 776 L; **(b)** 2.76 × 10⁴ L. **13.** $\Delta H = q_p = -6.24 \times 10^3$ kJ. **14. (a)** $\Delta \bar{H} = -56$ kJ/mol KOH. **15.** 81°C. **16. (a)** $\Delta \bar{E} = -2008$ kJ/mol C_3H_7OH; **(b)** $\Delta \bar{H} = -2012$ kJ/mol C_3H_7OH. **17. (a)** $\Delta H > \Delta E$; **(b)** $\Delta H < \Delta E$; **(c)** $\Delta H < \Delta E$; **(d)** $\Delta H > \Delta E$. **18.** 8.4×10^2 J/°C. **19.** 1.91°C. **20. (a)** $\Delta \bar{H} = -5.65 \times 10^3$ kJ/mol $C_{12}H_{22}O_{11}$; **(b)** $C_{12}H_{22}O_{11}(s) + 12 O_2(g) \rightarrow 12 CO_2(g) + 11 H_2O(l)$; **(c)** $\Delta \bar{E} = \Delta \bar{H} = -5.65 \times 10^3$ kJ/mol $C_{12}H_{22}O_{11}$; **(d)** 19 Cal. **21.** 32.7 metric tons.

22. about 33°C above room temperature. **26.** $\Delta \bar{H} = -56.52$ kJ/mol. **27.** $\Delta \bar{H} = -236.6$ kJ/mol. **28.** $\Delta \bar{H} = -747.6$ kJ/mol. **29.** $\Delta \bar{H} = -128.1$ kJ/mol. **31.** $\Delta \bar{H}° = +202.5$ kJ/mol. **32.** $\Delta \bar{H}° = -562.58$ kJ/mol. **33.** $\Delta \bar{H}°_f[CCl_4(g)] = -108$ kJ/mol. **34.** $\Delta \bar{H}° = +15.32$ kJ/mol. **35.** $\Delta \bar{H}° = -1322.8$ kJ/mol. **36.** $\Delta \bar{H}° = -1026$ kJ/mol. **37. (a)** 1.78×10^6 kJ; **(b)** 4.94×10^4 L $CH_4(g)$. **38. (b)** temperature increases; **(c)** $\Delta t = +0.4$°C. **39.** 87.0% CH_4, by volume. **40.** $\Delta \bar{H}° = -4.85 \times 10^3$ kJ/mol C_7H_{16}.

Self-Test Questions: **1.** (c). **2.** (b). **3.** (a). **4.** (d). **5.** (c). **6.** (d). **8.** 101°C. **9. (a)** $C_6H_5OH(s) + 7 O_2(g) \rightarrow 6 CO_2(g) + 3 H_2O(l)$; **(b)** $\Delta \bar{E} = -3.063 \times 10^3$ kJ/mol C_6H_5OH; **(c)** $\Delta \bar{H} = -3.065 \times 10^3$ kJ/mol C_6H_5OH. **10.** $\Delta \bar{H}°_f[COCl_2(g)] = -219$ kJ/mol.

Chapter 7

Exercises: **1. (a)** 300 nm; **(b)** 1.56×10^{-4} cm; **(c)** 3.6×10^7 nm; **(d)** 2.18×10^4 Å; **(e)** 6.2×10^{-7} m; **(f)** 4.70×10^{-7} m. **2. (a)** 3.00×10^{-6} m; **(b)** 3.5×10^{-5} m; **(c)** 1.5×10^{-1} m. **3. (a)** 1.7×10^{13} s⁻¹; **(b)** 2.38×10^{13} s⁻¹; **(c)** 6.25×10^{15} s⁻¹; **(d)** 9.84×10^{14} s⁻¹. **4.** 3.261226 cm. **5. (a)** and **(d)**. **6.** 8.3 min; **7.** 656.2 nm; 486.1 nm; 434.0 nm; 410.1 nm. **8.** no. **9.** 364.56 nm. **10. (a)** 7.62×10^{-19} J; **(b)** 4.59×10^5 J/mol photons. **11.** 286 nm (ultraviolet). **12.** from 2.58×10^{-19} to 5.10×10^{-19} J. **13. (a)** 8.6×10^{-19} J; **(b)** ultraviolet, yes; infrared, no. **14. (a)** 1.09×10^{15} s⁻¹; **(b)** 275 nm; **(c)** no. **16. (a)** 1.3 nm; **(b)** -8.716×10^{-20} J. **17. (a)** 0.42 nm; **(b)** 1.937×10^{-18} J. **18. (a)** 7.40×10^{13} s⁻¹; **(b)** 4.05×10^{-4} cm; **(c)** infrared. **19. (a)**. **20. (a)** -8.716×10^{-18} J; **(b)** -2.179×10^{-18} J. **25.** $\Delta x \geqslant 2.10 \times 10^{-13}$ m. **26.** 7.27×10^5 m/s. **29.** (d). **31.** (b); (e); (f). **32. (a)** $2p$; **(b)** $4s$; **(c)** $5d$. **33.** $n = 3$, $l = 0$; $n = 4$, $l = 1$; $n = 5$, $l = 2$. **34. (a)** one; **(b)** none; **(c)** three; **(d)** five. **35. (a)** and **(b)**. **36.** $3p$, $3d$, $4p$, $5s$, $6s$, $6p$, $5f$. **37.** (e) < (b) < (a) < (c) = (d). **38. (a)** B; **(b)** V; **(c)** Si. **39. (b)** P ($Z = 15$); **(c)** $[Kr]4d^25s^2$; **(d)** Te ($Z = 52$), $[Kr]4d^{10}5s^25p^4$; **(e)** I ($Z = 53$), $[Kr]4d^{10}5s^25p^5$; **(f)** $[Xe]4f^{14}5d^{10}6s^26p^3$. **40.** (c) is correct. **41. (a)** 2; **(b)** 0; **(c)** 2; **(d)** 2; **(e)** 14. **43. (a)** $1s^32s^32p^93s^33p^93d^{15}4s^34p^95s^1$; **(b)** $1s^21p^62s^22p^62d^{10}3s^23p^63d^{10}3f^{14}4s^24p^65s^2$.

Self-Test Questions: **1.** (b). **2.** (a). **3.** (d). **4.** (b). **5.** (d). **6.** (c).

7. 1.60×10^2 kJ/mol of photons.
8. greater energy: 589.0 nm;
energy difference $= 4 \times 10^{-22}$ J/photon.
9. $n = 5$.   **10. (a)** Se, $1s^2 2s^2 2p^6 3s^2 3p^6 3d^{10} 4s^2 4p^4$;
(b) I, $1s^2 2s^2 2p^6 3s^2 3p^6 3d^{10} 4s^2 4p^6 4d^{10} 5s^2 5p^5$.

Chapter 8

Exercises: **1.** 2.6 g/cm^3.   **2.** 15 g/cm^3.
5. $Z = 118$; $Z = 119$; at. wt. about 310.   **6. (a)** In;
(b) O, Se, Te, and Po are similar to S;
elements outside group VIA are dissimilar;
(c) Cs (and Ba); **(d)** I; **(e)** Xe or Rn.   **10. (a)** 0;
(b) IVA; **(c)** IA; **(d)** IB; **(e)** VIB.
11. (a) $[Ar]3d^{10}4s^2 4p^1$; **(b)** $[Kr]4d^1 5s^2$;
(c) $[Kr]4d^{10}5s^2 5p^2$; **(d)** $[Kr]4d^{10}5s^1$.   **13. (a)** 3;
(b) 6;   **(c)** 5; **(d)** 3; **(e)** 24.   **14. (a)** 1; **(b)** 5; **(c)** 10;
(d) 6; **(e)** 14; **(f)** 8   **15. (a)** $[Ar]3d^{10}4s^2 4p^6$;
(b) $[Ar]3d^{10}4s^2 4p^6$; **(c)** $1s^2 2s^2 2p^6$; **(d)** $[Kr]4d^{10}5s^2 5p^6$;
(e) $[Ar]3d^{10}$; **(f)** $[Kr]4d^{10}$; **(g)** $[Xe]4f^{14}5d^{10}6s^2$.
16. Pb, $[Xe]4f^{14}5d^{10}6s^2 6p^2$; $Z = 114$, $[Rn]5f^{14}6d^{10}7s^2 7p^2$.
18. (a) B; **(b)** Te.
20. $Y^{3+} < Sr^{2+} < Rb^+ < Kr < Br^- < Se^{2-}$.
21. $Li^+ < Se < Y < Br^-$.   **25.** 76.3 J.
26. 408.9 eV.   **27.** -10 kJ.   **28.** $I_1 = 13.60$ eV/atom.
29. $Cs < Sr < As < S < F$.
30. Se, 2.5; Te, 2.4; Ge, 2.0; Sn, 1.9.   **31.** 1.2 to 2.0.
32. (a) Ba; **(b)** S; **(c)** Bi.
33. $Rb > Ca > Sc > Fe > Te > Br > O > F$.   **35.** Fe^{2+}.
36. diamagnetic: K^+, Zn^{2+}, Sn^{2+}; paramagnetic: Cr^{3+}, Co^{3+}.
37. V^{3+}, $[Ar]3d^2$; Cu^{2+}, $[Ar]3d^9$; Cr^{3+}, $[Ar]3d^3$.
41. (a) 5.7 g/cm^3; **(b)** Ga_2O_3, 74% Ga.
42. (a) $Sr(NO_3)_2$; **(b)** $BaBr_2$; **(c)** Ag_2O; **(d)** Cu_2Te;
(e) $Al_2(SO_4)_3$; **(f)** Ga_2O_3; **(g)** Li_3N.   **43.** CO.
44. (a) $Mg(s) + 2 HCl(aq) \rightarrow MgCl_2(aq) + H_2(g)$;
(b) $2 Cs(s) + 2 H_2O(l) \rightarrow 2 CsOH(aq) + H_2(g)$;
(c) $Be(s) + H_2O(l) \rightarrow$ no reaction;
(d) $2 Na(s) + 2 HI(aq) \rightarrow 2 NaI(aq) + H_2(g)$;
(e) $Si(s) + HCl(aq) \rightarrow$?
(f) $Ba(s) + 2 H_2O(l) \rightarrow Ba(OH)_2(aq) + H_2(g)$.
45. (a) Cs, Rb, and K should exhibit photoelectric effect
with visible light; less active metals would not.
(b) Rn; **(c)** about 5 eV; **(d)** about 25°C;
(e) about 6 g/cm^3; **(f)** E.N. about 2.0.

Self-Test Questions: **1.** (b).   **2.** (c).   **3.** (d).
4. (b).   **5.** (c).   **6.** (b).   **7. (a)** 34; **(b)** 45; **(c)** 18;
(d) 2; **(e)** 4; **(f)** 6.   **8.** Br.   **9. (a)** C; **(b)** Rb; **(c)** At.

Chapter 9

Exercises: **8. (a)** RbCl, ionic, f. wt. $= 120.9$;
(b) H_2Se, covalent, f. wt. $= 80.98$;
(c) BCl_3, covalent, f. wt. $= 117.2$;
(d) Cs_2S, ionic, f. wt. $= 297.9$;

(e) SrO, ionic, f. wt. $= 103.6$;
(f) OF_2, covalent, f. wt. $= 54.00$.
13. formal charges on the atoms are **(a)** 0 for each I;
(b) $+1$ for S, -1 for one O atom, 0 for the other O atom;
(c) 0 for all atoms; **(d)** 0 for all atoms;
(e) 0 for H and one O atom, -1 for the other O atom;
(f) $+1$ for N, 0 for one O atom, -1 for the other O atom;
(g) 0 for all atoms.   **14. (a)** H_2NOH; **(b)** SCS;
(c) ONCl; **(d)** NNO.   **23.** diamagnetic.
24. diamagnetic, (a), (d), (e); paramagnetic, (b), (c), (f).
27. sulfur-to-nitrogen triple bond.   **29. (a)** linear;
(b) tetrahedral; **(c)** trigonal bipyramidal; **(d)** V-shaped;
(e) T-shaped; **(f)** octahedral.   **30. (a)** linear; **(b)** linear;
(c) V-shaped; **(d)** trigonal planar.
31. AXE_3, linear (e.g., HF).   **32.** tetrahedral.
34. linear.   **36. (a)** 137 pm; **(b)** 149 pm; **(c)** 177 pm;
(d) 141 pm; **(e)** 205 pm.   **39. (a)** -100 kJ/mol;
(b) -129 kJ/mol.   **40. (a)** endothermic;
(b) exothermic; **(c)** exothermic.
41. $\Delta H_f^\circ[NH_3(g)] = -42$ kJ/mol.   **42.** 631 kJ/mol.
43. 11% ionic.
44. $AsH_3 < AsI_3 < AsBr_3 < AsCl_3 < AsF_3$.
45. resultant dipole moments in SO_2, NO, HBr, NH_3,
H_2S, CH_2Cl_2.
47. ionic resonance energies: HF, 270 kJ/mol;
HBr, 50 kJ/mol.
48. $C-H < Br-H < F-H < Na-Cl < K-F$.
49. (a) 1.68; **(b)** 0.72.   **50.** 209 kJ/mol.   **52. (a)** $+3$;
(b) -2; **(c)** $+4$; **(d)** -2; **(e)** $+5$; **(f)** $+7$; **(g)** $+2$;
(h) $+6$; **(i)** $+7$.
53. $HS^- < S_2O_3^{2-} < S_4O_6^{2-} < SO_3^{2-} < HSO_4^- < S_2O_8^{2-}$.
54. (a) iron(III) bromide; **(b)** chromium(III) iodide;
(c) calcium hypochlorite; **(d)** sodium bromate;
(e) potassium periodate; **(f)** sodium persulfate.
55. (a) HIO_4; **(b)** SnO_2; **(c)** Na_2SeO_4; **(d)** $Mg(ClO_4)_2$;
(e) $Au(CN)_3$; **(f)** KI; **(g)** BaTe.

Self-Test Questions: **1.** (b).   **2.** (d).   **3.** (c).
4. (a).   **5.** (a).   **6. (a)** CH_2Cl.
7. plausible Lewis structure is (c).   **10. (a)** V-shaped;
(b) trigonal planar; **(c)** tetrahedral.   **11.** -759 kJ/mol.

Chapter 10

Exercises: **2.** (c).   **6. (a)** sp^3d^2; **(b)** sp; **(c)** sp^3;
(d) sp^2; **(e)** sp^3d.   **7.** tetrahedral.   **11.** sp^3d.
13. linear, (a), (b); planar (d).   **14.** 875 kJ/mol.
15. stable diamagnetic: Li_2, C_2, N_2, F_2;
stable paramagnetic, B_2, O_2.
16. both N_2^- and N_2^{2-}.   **17.** O_2^+ has a stronger bond.
18. (a) diamagnetic;
(b) paramagnetic, one unpaired electron;
(c) paramagnetic, one unpaired electron.
24. isoelectronic: NO^+, CO, CN^-; isoelectronic: CN^+, BN.
28. (b), (c).
32. 6.02×10^{22} energy levels and conduction electrons.

Self-Test Questions: **1.** (d).   **2.** (b).   **3.** (b).
4. (d).   **6. (a)** 6; **(b)** 2.

Chapter 11

Exercises: **5.** 30.9 kJ/mol C_6H_6.
6. 60.4 L $CH_4(g)$.   **9. (a)** 300 mmHg; **(b)** 35°C.
10. (a) 93.5°C; **(b)** 507 mmHg.   **11.** 226 mmHg.
12. 0.904 mmHg.   **13.** 15.3 L.   **14. (a)** 184°C;
(b) 257 mmHg.   **15.** 280°C.
16. (a) 4.45×10^4 J/mol; **(b)** 81°C.   **17.** 59°C.
18. CO_2, HCl, NH_3, SO_2, and H_2O.   **20.** 308 kJ.
21. 43.6 kJ.   **22. (a)** 31.8 mmHg;   **(b)** 80.6 mmHg;
(c) 85.6 mmHg.   **23. (b)** 97°C.   **24.** 26.7 mmHg.
25. 18.3°C.   **31.** $N_2 < F_2 < Ar < O_3 < Cl_2$.
33. CH_3OH.   **35. (a)** $C_{10}H_{22}$; **(b)** $H_3C—O—CH_3$;
(c) CH_3CH_2OH.   **47.** $CsI < MgBr_2 < CaO$.
49. (a) BaF_2; **(b)** $MgCl_2$.   **51.** -805 kJ/mol KF.
52. -704 kJ/mol NaI.   **56. (a)** 362 pm;
(b) 4.74×10^{-23} cm³; **(c)** 4; **(d)** 4.22×10^{-22} g;
(e) 8.90 g/cm³.   **57.** 1.75 g/cm³.   **59. (a)** six;
(b) eight.   **62. (a)** 1.68×10^{-22} cm³; **(b)** 3.89×10^{-22} g;
(c) 2.32 g/cm³.

Self-Test Questions: **1.** (d).   **2.** (b).   **3.** (c).
4. (a).   **5.** (c).   **8.** (c).

Chapter 12

Exercises: **5.** 59.0% KI.   **6. (a)** 15.1%; **(b)** 11.9%;
(c) 12.2%.   **7.** 30.8 g $HC_2H_3O_2$.
10. 3.73 M CH_3OH.   **11.** 0.639 g.   **12.** 73.0 ml.
13. at 15°C: 2.134 M C_2H_5OH; at 25°C: 2.128 M C_2H_5OH.
14. 0.292 m $C_6H_4Cl_2$.   **15.** 17.7 g I_2.
16. 21.4 m HF; 16.5 M HF.   **17.** 682 g H_2O.
18. For C_7H_{16}, C_8H_{18}, and C_9H_{20}, mole fractions are
0.215, 0.277, and 0.507, respectively; corresponding mole
percents are 21.5%, 27.7%, and 50.7%.
19. (a) 0.0223; **(b)** 0.04166.   **20.** 4.3 g C_2H_5OH.
21. 450 ml $C_3H_8O_3$.   **22.** 80°C.
23. 16 g $KClO_4$ can be crystallized.   **24. (a)** yes;
(b) 18 g NH_4Cl; **(c)** no.
25. Sample will dissolve more H_2S.   **26.** 30 g.
29. $P_{C_6H_6} = 51.4$ mmHg; $P_{C_7H_8} = 13.0$ mmHg;
$P_{total} = 64.4$ mmHg.
30. $\chi_{C_6H_6} = 0.798$; $\chi_{C_7H_8} = 0.202$.   **31.** $\chi_{C_6H_6} = 0.328$.
32. 23.5 mmHg.   **33.** 1.2×10^2 g/mol.
34. $C_6H_4N_2O_4$.
35. (a) $K_f = 20$°C kg solv. (mol solute)$^{-1}$.
36. 3 parts $C_2H_6O_2$ to 10 parts H_2O, by volume.
38. 100.02°C.   **41.** 0.31 mol $C_6H_{12}O_6$/L.
42. mol. wt. $= 6.48 \times 10^4$.   **43.** mol. wt. $= 2.82 \times 10^5$.
44. 0.10 M K$^+$; 0.20 M Mg^{2+}; 0.50 M Cl$^-$.
45. 0.465 M Cl$^-$.   **48. (a)** -0.19°C; **(b)** -0.38°C;
(c) -0.57°C; **(d)** -0.38°C; **(e)** -0.19°C; **(f)** -0.38°C;
(g) -0.20°C.   **51. (b)** $AlCl_3$.   **52. (a)** 16 cm²;

(b) 8.35×10^{-3} cm².

Self-Test Questions: **1.** (b).   **2.** (c).   **3.** (a).
4. (d).   **5.** (c).   **6.** (b).   **7. (a)** 2.22% $C_{10}H_8$;
(b) 0.178 m $C_{10}H_8$; **(c)** 4.64°C.   **8.** 5.0% NaCl.
10. 85 ml.

Chapter 13

Exercises: **3.** rate $= k[CH_3CHO]^2$.
4. (a) mol L^{-1} s^{-1}; **(b)** s^{-1}; **(c)** L mol^{-1} s^{-1}.
5. (a) 0.531 M; **(b)** 2.2×10^{-4} mol L^{-1} s^{-1};
(c) 0.524 M.   **6. (a)** 29.5 ml;
(b) 1.1×10^{-4} mol O_2 L^{-1} s^{-1}; **(c)** 4.4 cm³ O_2.
7. 4.5×10^{-4} mol L^{-1} s^{-1}.   **8. (a)** 3000 mmHg;
(b) 1400 mmHg.   **9. (a)** II; **(b)** I; **(c)** III.
10. 70 s for first-order reaction.
11. (a) [A] $= 0$; **(b)** [A] $\simeq 0.33$; **(c)** [A] $\simeq 0.48$.
12. (a) 1.0×10^{-2}; **(b)** 4.6×10^{-3}; **(c)** 3.7×10^{-3}.
13. 9.2×10^{-6} mol L^{-1} s^{-1}.
14. (a) first order in $HgCl_2$; second order in $C_2O_4^{2-}$;
third order overall;
(b) $k = 7.6 \times 10^{-3}$ L² mol^{-2} min^{-1};
(c) initial rate $= 7.4 \times 10^{-6}$ mol L^{-1} min^{-1}.
15. (a) first order in H_2; second order in NO; third order
overall; **(b)** rate $= k \times P_{H_2} \times (P_{NO})^2$.
16. (a) first order in I$^-$ and in OCl$^-$, -1 order in OH$^-$;
(b) first order overall; **(c)** $k = 60$ s^{-1}.   **17. (a)** 0.017 M;
(d) 4.3×10^{-3} mol L^{-1} min^{-1}; **(f)** 6.5×10^{-2} min^{-1};
(g) 11 min; **(h)** 22 min.   **18. (a)** 67 min;
(b) 1.0×10^{-2} min^{-1}.   **19. (a)** 4400 s;
(b) 7.8 L $O_2(g)$ at STP.   **20.** 1.6×10^3 s.
21. (b) 3.5×10^{-2} min^{-1}; **(c)** 4.9×10^{-3} mol L^{-1} s^{-1}.
22. 19.9 min.   **23.** 23.3 min.
24. (b) 160 min; 193 min.   **25. (b)** 4.42×10^{-4} s^{-1};
(c) 408 mmHg; **(d)** 931 mmHg; **(e)** 534 mmHg.
27. (a) 63 kJ/mol.   **29. (b)** 52 kJ/mol;
(c) 1.9×10^{-2} L mol^{-1} s^{-1};
(d) 8.0×10^{-5} mol L^{-1} s^{-1}.   **30.** $E_a = 161$ kJ/mol.
31. 405 K.   **32. (a)** 53 kJ/mol.   **33.** 287 K.
36. zero order.

Self-Test Questions: **1.** (a).   **2.** (c).   **3.** (b).
4. (c).   **5.** (d).   **6.** (d).   **7.** (b).   **8.** (c).
9. (d).   **10.** (b).   **11. (a)** 10 min; **(b)** 0.160 g.

Chapter 14

Exercises: **4. (a)** 4.31×10^{-2}; **(b)** 3.1×10^3;
(c) 1.5×10^{-2}; **(d)** 7.5×10^2.   **5.** $K_c = 9.1 \times 10^{-16}$.
6. $K_c = 1.1 \times 10^6$.   **7.** $K_c = 0.51$.   **8.** $K_c = 1.14$.
9. $K_c = 3.8 \times 10^{-2}$.   **11.** $K_c = 1.49 \times 10^2$.
12. (b).   **13. (b)** $K_c = 6.59 \times 10^{-1}$.
14. (a) 0.100 mol O_2; **(b)** 0.400 mol O_2.   **15.** 25 L.
16. no; reaction occurs in reverse direction.
17. (b) mixture is not at equilibrium; **(c)** 6.0 L.

18. 1.2×10^{-3} mol Cl_2.

19. 1.50 mol each of CO, H_2O, CO_2, and H_2.

20. 0.95 mol $SbCl_5$; 2.05 mol $SbCl_3$; 0.05 mol Cl_2.

21. (a) mixture is not at equilibrium; (b) reaction proceeds in the forward direction.

22. 0.240 mol SO_2; 0.081 mol O_2; 0.807 mol SO_3.

23. $[Ag^+] = [Fe^{2+}] = 0.44\ M$; $[Fe^{3+}] = 0.56\ M$.

24. $[Cr^{3+}] = 0.03\ M$; $[Cr^{2+}] = 0.27\ M$; $[Fe^{2+}] = 0.11\ M$.

25. 23.2; 6.5; 2.7×10^{-2}; 6.8×10^{-5}.

26. (a) 1.2×10^{-3}; (b) 5.55×10^5; (c) 0.429.

27. $K_p = 2.1 \times 10^{-3}$.   28. 0.673 atm.

29. (a) 0.96 atm; (b) less; (c) $P_{CO_2} = 1.33$ atm; $P_{H_2O} = 0.17$ atm.   30. increase.

31. 49.6%.   32. (a) $K_c = 1.1 \times 10^{-2}$; (b) $K_p = 1.3$.

33. 19%.   35. (a) 80.0%; (b) 176 atm.

36. $K_p = 5.8 \times 10^{22}$.   37. $\Delta H° = -11.1$ kJ/mol.

38. $K_p = 12.5$.   39. (b) 380 K.

40. (a) displace to left; (b) no effect; (c) displace to right.   42. (a) decrease; (b) decrease; (c) increase (d) none.   43. (a) true; (b) false; (c) false; (d) true.   45. (b), (d).

Self-Test Questions: 1. (d).   2. (c).   3. (a).
4. (c).   5. (d).   6. (b).   7. (b).   8. (a).
11. $K_c = 2.0 \times 10^{-19}$.   12. (a) reverse direction; (b) 0.30 atm.

Chapter 15

Exercises: 1. (a) decrease; (b) increase; (c) increase.   2. (a) increase; (b) decrease; (c) indeterminate; (d) decrease; (e) indeterminate.
3. (a) 1 mol H_2(g, 1 atm, 50°C); (b) 50.0 g Fe(s, 1 atm, 20°C); (c) 1 mol Br_2(l, 1 atm, 58°C); (d) 0.10 mol O_2(g, 0.10 atm, 25°C).   7. (a) case 3; (b) case 2; (c) case 1; (d) case 4.
8. $\Delta H = 0$; $\Delta S > 0$; $\Delta G < 0$.
9. $\Delta H = 0$; $\Delta S > 0$; $\Delta G < 0$.   10. 399 K.
11. (a) -33.30 kJ/mol; (b) -242.09 kJ/mol; (c) 99.76 kJ/mol; (d) -60.83 kJ/mol.
12. -285 J mol^{-1} K^{-1}.   13. 162.53 kJ/mol.
14. -5.20×10^3 kJ/mol.   15. (a) 100°C; (b) $\Delta G = 0$.   17. solid iodine.   18. CO_2(g).
19. $\Delta \bar{H}°_{vap} = 44.01$ kJ/mol; $\Delta \bar{S}°_{vap} = 118.78$ J mol^{-1} K^{-1}.
20. (b).   21. (a) 30.71 kJ/mol; (b) 349 K.
22. 691 K.   23. -2.14×10^3 J/mol.
24. (b) 134 mmHg.   25. (a) $K_p = 3.4$; (b) $\Delta \bar{G}° = -10.2$ kJ/mol; (c) forward direction.
29. 126.6 kJ/mol.   30. $K_p = 1.6 \times 10^{12}$.   31. (a) no; (b) raising the temperature.   32. -341 kJ/mol.
35. (a) $\Delta \bar{H}° = -41.13$ kJ/mol; $\Delta \bar{S}° = -42.42$ J mol^{-1} K^{-1}; $\Delta \bar{G}° = -28.49$ kJ/mol; (b) $K_p = 0.546$.   37. (a) high; (b) 977 K.
38. $\Delta \bar{H}°_f = -238.24$ kJ/mol; $\Delta \bar{G}°_f = -168.77$ kJ/mol; $\bar{S}° = 272.8$ J mol^{-1} K^{-1}

39. (a) 489 K.   40. 28.8 atm.

Self-Test Questions: 1. (d).   2. (b).   3. (a).
4. (c).   5. (b).   6. (d).   9. (a) 326 K; (b) 2.5 kJ/mol.   10. (a) exothermic; (b) -186.81 kJ/mol; (c) $K = 5.0 \times 10^{32}$; (d) all temperatures.

Chapter 16

Exercises: 5. 3.98×10^{-14}.   6. 9.6×10^{-9}.
7. 1.7×10^{-5}.   8. (a) $1.2 \times 10^{-3}\ M$; (b) $7.1 \times 10^{-5}\ M$; (c) $2.4 \times 10^{-4}\ M$.   9. (b).   10. 7.2 ppm F$^-$.
11. 0.68 or 68%.   12. $7.4 \times 10^{-3}\ M$.
13. 4.1×10^{-9}.   14. 5.30×10^{-5}.
15. (a) $1.7 \times 10^{-4}\ M$; (b) $1.3 \times 10^{-4}\ M$; (c) $2.6 \times 10^{-9}\ M$; (d) $3.3 \times 10^{-3}\ M$; (e) $1.6 \times 10^{-20}\ M$.   19. (a) $1.7 \times 10^{-4}\ M$; (b) $2.8 \times 10^{-5}\ M$; (c) $0.058\ M$; (d) $1.6 \times 10^{-9}\ M$.
20. 1.7 ppm F$^-$.   21. 1.9×10^{-3} mol MgF_2/L.
22. $[OH^-] = 5.4 \times 10^{-13}\ M$.   23. yes.   24. no.
25. (a) yes; (b) yes; (c) no.   26. yes.   27. (a) yes; (b) $[Ag^+] = 4.0 \times 10^{-6}\ M$.
28. (a) $[OH^-] = 4.5 \times 10^{-3}\ M$; (b) $[OH^-] = 1.4 \times 10^{-4}\ M$.   29. 98%.
30. $[Pb^{2+}] = 1.6 \times 10^{-2}\ M$.   38. (a) no; (b) yes.
39. (a) AgBr; (b) no.   40. (b)

Self-Test Questions: 1. (d).   2. (a).   3. (c).
4. (b).   5. (c).   6. (a).   7. 5.8×10^{-4} mg Pb^{2+}/ml.
8. $AgI < AgCl < Ag_2CO_3 < Ag_2SO_4 < AgNO_3$.
9. (a) $PbCrO_4$; (b) $[Pb^{2+}] = 1.6 \times 10^{-6}\ M$; (c) yes.

Chapter 17

Exercises: 1. (a) acidic; (b) basic; (c) basic; (d) acidic; (e) acidic.   2. (a) H_2O; (b) HCl; (c) HOCl; (d) HCN.   3. (a) IO_4^-; (b) $C_3H_5O_2^-$; (c) $C_6H_5COO^-$; (d) $C_6H_5NH_2$.   4. (b), (c), (d), (f), (g).   6. (a) base; (b) either; (c) either; (d) acid.
10. (a) $[H_3O^+] = 1.0 \times 10^{-3}\ M$, $[OH^-] = 1.0 \times 10^{-11}\ M$; (b) $[OH^-] = 4.0 \times 10^{-2}\ M$, $[H_3O^+] = 2.5 \times 10^{-13}\ M$; (c) $[H_3O^+] = 4.0 \times 10^{-3}\ M$, $[OH^-] = 2.5 \times 10^{-12}\ M$; (d) $[OH^-] = 2.6 \times 10^{-3}\ M$, $[H_3O^+] = 3.8 \times 10^{-12}\ M$.
11. $[H_3O^+] = 3.88 \times 10^{-4}\ M$.   12. (a) 4; (b) 3.7; (c) 2.15; (d) 10.40.   13. (a) 3; (b) 1.30; (c) 3.10; (d) 11.18.   14. pH = 12.46.   15. pH = 1.50.
16. 2.44 ml.   17. $[Mg^{2+}] = 0.02\ M$.
20. 1.3×10^{-5}.   21. $[CHO_2^-] = 8.4 \times 10^{-3}\ M$.
22. 1.0×10^{-8}.   23. 2.1 g/L.   24. 3.20.
25. 5.0×10^{-2}.   26. 5.4 mg.   27. 2.70.
28. 11.38.   29. 1.9%.   30. $0.172\ M$.
32. -0.095°C.
33. pH = 3.92; $[CO_3^{2-}] = 5.6 \times 10^{-11}\ M$.
34. (a) $[H_3O^+] = [HS^-] = 9.1 \times 10^{-5}\ M$, $[S^{2-}] = 1.0 \times 10^{-14}\ M$;

(b) $[H_3O^+] = [HS^-] = 2.3 \times 10^{-5}\,M$,
$[S^{2-}] = 1.0 \times 10^{-14}\,M$;
(c) $[H_3O^+] = [HS^-] = 1.0 \times 10^{-6}\,M$,
$[S^{2-}] = 1.0 \times 10^{-14}\,M$.
36. $[H_3O^+] = [H_2PO_4^-] = 0.022\,M$;
$[HPO_4^{2-}] = 6.2 \times 10^{-8}\,M$; $[PO_4^{3-}] = 1.4 \times 10^{-18}\,M$.
37. (b) 9.62.   **39. (a)** neutral; **(b)** acidic; **(c)** neutral;
(d) neutral; **(e)** basic.   **40.** 8.0.
41. $1.7\,M\,KNO_2$.   **42.** pH $= 2.96$.
45. (a) forward; **(b)** reverse; **(c)** reverse; **(d)** forward;
(e) reverse; **(f)** forward.
46. (c) $<$ **(f)** $<$ **(d)** $<$ **(e)** $<$ **(b)** $<$ **(a)**.
47. propylamine.   **50. (a)** $K_a \simeq 1 \times 10^{-2}$.

Self-Test Questions: **1.** (c).   **2.** (a).   **3.** (b).
4. (d).   **5.** (d).   **6.** (a).   **7.** (c).   **8.** (b).
9. (e) $>$ (c) $>$ (j) $>$ (b) $>$ (i) $>$ (g) $>$ (d) $>$ (a) $>$ (f) $>$ (h).
10. 3.0 g.

Chapter 18

Exercises: **2. (a)** $0.100\,M$; **(b)** $0.100\,M$; **(c)** $0.200\,M$;
(d) $5.22 \times 10^{-5}\,M$; **(e)** $2.18 \times 10^{-5}\,M$.
3. 1.39×10^{-3}.   **4.** 5.34.   **5. (a)** 27.8 ml; **(b)** 3.5 ml;
(c) 250.0 ml.   **6.** (c), (e).   **7. (c)** pH $= 7.69$.
8. (a) 3.56; **(b)** 7.65; **(c)** 9.30.   **9.** 1.3 g.
10. $1.00\,L\,0.100\,M\,HCHO_2 + 0.575\,L\,0.100\,M\,NaCHO_2$.
11. $59.7\,ml\,1.00\,M\,HCl + 940.3\,ml\,0.100\,M\,NaCHO_2$.
14. (a) 4.91; **(b)** 1.02.   **15. (a)** 9.13; **(b)** 9.46;
(c) 16 drops.   **16. (a)** 4.77; **(b)** 4.95; **(c)** $5.1\,g\,Ba(OH)_2$;
(d) 11.6.   **19. (a)** colorless; **(b)** red; **(c)** blue;
(d) yellow; **(e)** blue; **(f)** green.
20. (a) 39% in anion form; **(b)** acid form (red).
21. $[OH^-] = 0.0900\,M$.   **22.** 10.7 ml.   **23.** 2.96.
29. (a) 2.28; **(b)** 5.71.   **30. (a)** no; **(b)** 18.16%.
31. (a) 24.50 ml; **(b)** 19.0 ml; **(c)** 16.61 ml.
32. (a) yes; **(b)** no; **(c)** yes.   **33.** 9.5.
34. (b) 1.6 g.   **35.** yes.   **37.** $0.18\,mol\,Mg(OH)_2/L$.
38. no.   **39.** pH < 2.0.   **41.** $0.23\,g\,FeS/L$.
42. $1.3 \times 10^{-3}\,mol\,CoS/L$.   **44. (a)** yes; **(b)** no;
(c) no; **(d)** yes.   **46. (a)** $0.17\,M\,HC_2H_3O_2$;
(b) $2.24 \times 10^{-2}\,M\,Ba(OH)_2$; **(c)** $0.18\,M\,C_6H_5NH_2$;
(d) $0.018\,M\,NH_4Cl$.   **47. (a)** no; **(b)** no; **(c)** yes;
(d) no; **(e)** yes.   **49. (a)** 100.5; **(b)** 29.2; **(c)** 74.1.
50. 50.1 g.   **51. (a)** $0.24\,N\,KOH$; **(b)** $0.001\,N\,HI$;
(c) $4.0 \times 10^{-3}\,N\,Ca(OH)_2$;
(d) $0.15\,N$ or $0.30\,N$ or $0.45\,N\,H_3PO_4$;
(e) $0.01\,N\,C_6H_5COOH$.   **52.** $0.46\,N$.   **53.** 16.3 ml.
54. (a) $0.151\,N$; **(b)** $0.302\,N$; **(c)** $0.453\,N$.
55. 32.1% H_2SO_4.

Self-Test Questions: **1.** (c).   **2.** (b).   **3.** (b).
4. (d).   **5.** (a).   **6.** (c).   **7. (a)** basic; **(b)** acidic;
(c) acidic; **(d)** independent of pH;

(e) independent of pH; **(f)** acidic.   **8. (a)** 32.0 g;
(b) 3.91.   **9. (a)** 3.10; **(b)** 4.20; **(c)** 8.00; **(d)** 11.08.
10. (b) $[C_2H_3O_2^-] = 1.7\,M$.

Chapter 19

Exercises: **3. (a)** $3, 8, 2 \rightarrow 3, 2, 4$;
(b) $4, 10, 1 \rightarrow 4, 1, 3$; **(c)** $5, 2, 6 \rightarrow 2, 8, 5$;
(d) $3, 1, 8 \rightarrow 3, 2, 7$; **(e)** $3, 10, 10, 4 \rightarrow 6, 10, 9$;
(f) $1, 1 \rightarrow 1, 1, 2$.   **4. (a)** $3, 6 \rightarrow 5, 1, 3$;
(b) $4, 1, 4 \rightarrow 4, 1, 4$; **(c)** $1, 12, 14 \rightarrow 2, 20, 3$;
(d) $2, 27, 10 \rightarrow 2, 6, 32$; **(e)** $1, 3, 3 \rightarrow 3, 1$.
5. (a) $2, 2 \rightarrow 4, 1$; **(b)** $3, 2 \rightarrow 3, 1, 3$;
(c) $3, 8, 2 \rightarrow 3, 4, 2$; **(d)** $4, 10, 2 \rightarrow 4, 5, 1$;
(e) $2, 5, 6 \rightarrow 2, 5, 3$.
8. $Na^+ < Zn^{2+} < I_2 < IO_3^- < PbO_2 < F_2$.
9. oxidizing agent only: (d), (g); reducing agent only: (a),
(b); either: (c), (e), (f).
10. $E° = +0.99\,V$.   **11. (a)** $0.337 < E° < 0.799\,V$;
(b) $-0.440 < E° < 0.000\,V$.   **13.** $E° = -0.256\,V$.
14. (a) no; **(b)** yes; **(c)** no; **(d)** no; **(e)** yes.
15. (a) yes; **(b)** yes; **(c)** no; **(d)** no; **(e)** yes.
20. (a) $Cu(s)|Cu^{2+}(aq)\|Fe^{2+}(aq), Fe^{3+}(aq)|Pt(s)$;
(b) $Pt|Sn^{2+}(aq), Sn^{4+}(aq)\|Cr^{3+}(aq), Cr_2O_7^{2-}(aq)|Pt$;
(c) $Pt|O_2(g)|H_2O, H^+\|Cl^-(aq)|Cl_2(g)|Pt$.
21. (a) $Sn(s) + Zn^{2+} \rightarrow Sn^{2+} + Zn(s)$;
(b) $H_2O_2(aq) + 2\,Ag^+ \rightarrow 2\,Ag(s) + 2\,H^+ + O_2(g)$;
(c) $2\,Cu^{2+} + 2\,H_2O \rightarrow 2\,Cu(s) + O_2(g) + 4\,H^+$.
22. (a) $Fe(s)|Fe^{2+}(aq)\|Cl^-(aq)|Cl_2(g)|Pt(s)$;
(b) e.g., $Mg(s)|Mg^{2+}(aq)\|Zn^{2+}(aq)|Zn(s)$;
(c) $Pt|Cu^+(aq), Cu^{2+}(aq)\|Cu^+(aq)|Cu(s)$.
27. (a) $[Cu^{2+}] = 7 \times 10^{-38}\,M$;   **(b)** yes.
28. (a) $E_{cell} = +0.83\,V$.   **29. (a)** lower; **(b)** 0.69 V.
30. $E_{cell} = 0.031\,V$.   **31.** $[Cl^-] = 2.5 \times 10^{-3}\,M$.
32. (a) $-520\,kJ/mol$; **(b)** $-48\,kJ/mol$;
(c) $+5.79\,kJ/mol$.   **33.** $K = 1.3 \times 10^{91}$.
34. $[Fe^{2+}] = 0.42\,M$.
41. (a) anode: $Cl_2(g)$, cathode: $Cu(s)$; **(b)** $Cl_2(g), H_2(g)$;
(c) $O_2(g)$ and H^+, $H_2(g)$; **(d)** $Cl_2(g), Ba(l)$;
(e) I_2, OH^- and $H_2(g)$;
(f) $O_2(g)$ and H^+, $H_2(g)$ and OH^-.   **43. (a)** 0.852 g Cu;
(b) 2.90 g Ag; **(c)** 0.749 g Fe; **(d)** 0.241 g Al.
44. 73.1 min.   **45. (a)** $166\,cm^3\,O_2(g)$;
(b) $172\,cm^3\,O_2(g)$.   **46. (a)** $1.86 \times 10^3\,C$; **(b)** 2.21 A.
47. (a) 114; **(b)** 15.8; **(c)** 5.5; **(d)** 27.8; **(e)** 12.6.
48. 58.4% Fe.   **49. (a)** $5.594 \times 10^{-3}\,M$;
(b) $3.356 \times 10^{-2}\,N$.   **50.** 41.4 ml.

Self-Test Questions: **1.** (d).   **2.** (c).   **3.** (b).
4. (b).   **5.** (c).   **6.** (a).
7. $3\,PbO(s) + 2\,MnO_4^- + H_2O \rightarrow 3\,PbO_2(s) +$
$2\,MnO_2(s) + 2\,OH^-$.
8. $Zn(s)|Zn^{2+}(aq)\|H^+(aq), NO_3^-(aq)|NO(g)|Pt(s)$.
9. pH $= 2.82$.   **10.** $[Cu^+] = 1.08\,M$; $[Sn^{4+}] = 1.04\,M$;
$[Cu^{2+}] = 0.92\,M$; $[Sn^{2+}] = 0.96\,M$.

Chapter 20

Exercises: **4.** (a) no; (b) Mg_3N_2.   **5.** 175 g.
6. pH = 11.55.
8. oxidizing agent: $SO_4{}^{2-}$; reducing agent: Al.
10. (a) -423 kJ/mol Fe; (b) -594 kJ/mol Mn;
(c) $+45$ kJ/mol Mg.
13. $ClO^- + H_2O + 2\,e^- \rightarrow Cl^- + 2\,OH^-$.   **14.** yes.
15. (a) $E° = +1.49$ V; (b) no: $E°_{cell} = -0.14$ V.
17. (a) high P; (b) low T.
18. $IO_3{}^-/I^-$ couple, $E° = +1.09$ V;
HIO/I_2 couple, $E° = +1.45$ V;
$IO_3{}^-/HIO$ couple, $E° = 1.14$ V.
22. basic.
23. (a) $O_3 + 2\,I^- + 2\,H^+ \rightarrow O_2 + I_2 + H_2O$;
(b) $3\,O_3 + S + H_2O \rightarrow 2\,H^+ + SO_4{}^{2-} + 3\,O_2$;
(c) $O_3 + 2\,[Fe(CN)_6]^{4-} + H_2O \rightarrow 2\,[Fe(CN)_6]^{3-} + O_2 +$
$2\,OH^-$.   **25.** 4×10^{37} O_3 molecules.
26. oxidation state of S:
$+4, +6, +8, +1, +3, +7$, respectively.
31. 0.426 g.   **32.** 6.49% Cu.   **34.** (a) 9.18; (b) 3.89.
36. (a) $NH_3OH^+ + 2\,H^+ + 2\,e^- \rightarrow NH_4{}^+ + H_2O$;
(b) $E° = +1.35$ V; (c) yes.   **37.** -622.33 kJ/mol.
38. $[NO_3{}^-] = 0.1716$ M.   **39.** 863 kg rock.
42. (a) $Ca(PO_3)_2$; (b) $K_4P_2O_7$;
(c) sodium metaantimonite; (d) sodium metabismuthate;
(e) Na_3BiO_4.

Self-Test Questions: **1.** (b).   **2.** (a).   **3.** (b).
4. (d).   **5.** (c).   **7.** (a) Mg_3N_2;
(b) calcium hydroxide; (c) rubidium superoxide;
(d) chlorite ion; (e) $HBrO_3$; (f) $S_2O_3{}^{2-}$; (g) Si_3H_8;
(h) calcium metasilicate; (i) magnesium pyrophosphate;
(j) $Ca(H_2PO_4)_2$.
8. (a) $MgCl_2 \xrightarrow{\text{electrolysis}} Mg(l) + Cl_2(g)$;
(b) $BaCO_3(s) \rightarrow BaO(s) + CO_2(g)$;
(c) $CaO(s) + H_2O \rightarrow Ca(OH)_2(s)$;
(d) $2\,Al(s) + 3\,Cu^{2+}(aq) \rightarrow 2\,Al^{3+}(aq) + 3\,Cu(s)$;
(e) $2\,IO_3{}^- + 5\,HSO_3{}^- \rightarrow I_2 + 5\,SO_4{}^{2-} + 3\,H^+ + H_2O$;
(f) $H_2O_2 + Cl_2(g) \rightarrow 2\,H^+ + 2\,Cl^- + O_2(g)$;
(g) $3\,ZnS(s) + 8\,H^+ + 2\,NO_3{}^- \rightarrow 3\,Zn^{2+} + 3\,S(s) +$
$2\,NO(g) + 4\,H_2O$;
(h) $4\,NH_3(g) + 5\,O_2(g) \rightarrow 4\,NO(g) + 6\,H_2O$.
10. $E° = +0.38$ V.

Chapter 21

Exercises: **8.** $E° = -0.74$ V.
9. (a) $Cr(s) + 2\,H^+ \rightarrow Cr^{2+} + H_2(g)$;
(b) $2\,Cr(s) + 12\,H^+ + 3\,SO_4{}^{2-} \rightarrow 2\,Cr^{3+} + 6\,H_2O +$
$3\,SO_2(g)$.
12. 56 s.   **13.** (a) $[Cr_2O_7{}^{2-}]/[CrO_4{}^{2-}]^2 = 3.2 \times 10^4$;
(b) $[Cr_2O_7{}^{2-}]/[CrO_4{}^{2-}]^2 = 8.0 \times 10^{-5}$.   **14.** $K = 33$.
18. Fe^{2+}.   **20.** (a) $Mo(CO)_6$; (b) $Os(CO)_5$;
(c) $Re_2(CO)_{10}$.
23. (a) $Cr_2O_7{}^{2-} + 14\,H^+ + 6\,e^- \rightarrow 2\,Cr^{3+} + 7\,H_2O$;

(b) $Cr^{2+} \rightarrow Cr^{3+} + e^-$;
(c) $Fe(OH)_3(s) + 5\,OH^- \rightarrow FeO_4{}^{2-} + 4\,H_2O + 3\,e^-$;
(d) $[Ag(CN)_2]^- + e^- \rightarrow Ag(s) + 2\,CN^-$.
24. (a) 2, 6, 3 $\rightarrow$ 4, 6; (b) 4, 8, 1, 2 $\rightarrow$ 4, 4;
(c) 2, 5, 8 $\rightarrow$ 2, 10, 16.   **26.** $E° = +1.51$ V.
27. (a) H_2O; (b) NH_3(aq); (c) NaOH(aq);
(d) HCl(aq).   **28.** (a) $FeS + 2\,H^+ \rightarrow Fe^{2+} + H_2S$;
(b) $CoS + 4\,H^+ + 2\,NO_3{}^- + 4\,Cl^- \rightarrow [CoCl_4]^{2-} + S +$
$2\,H_2O + 2\,NO_2$;
(c) $2\,Cr^{3+} + 3\,H_2O_2 + 10\,OH^- \rightarrow 2\,CrO_4{}^{2-} + 8\,H_2O$;
(d) $Ni(OH)_2 + 6\,NH_3 \rightarrow [Ni(NH_3)_6]^{2+} + 2\,OH^-$;
(e) $Fe^{2+} + K_3[Fe(CN)_6] \rightarrow$ Turnbull's blue;
(f) $MnO_2 + 4\,H^+ + 2\,Cl^- \rightarrow Mn^{2+} + 2\,H_2O + Cl_2$.
31. (a) $NiC_8H_{14}N_4O_4$; (b) 3.50% Ni.   **32.** 82.3% MnO_2.

Self-Test Questions: **1.** (c).   **2.** (d).   **3.** (a).
4. (b).   **5.** (a).
7. with limited NaOH(aq): $Mg(OH)_2(s)$ and $Cr(OH)_3(s)$;
with excess NaOH(aq): $Mg(OH)_2$ and $CrO_2{}^-$.
9. (a) all three; (b) HCl(aq); (c) H_2O; (d) HCl(aq).
10. ion present: Ni^{2+}; ions absent: Cr^{3+}, Fe^{3+};
further tests needed: Zn^{2+}.

Chapter 22

Exercises: **4.** (a) coordination number = 6,
oxidation state = $+2$;
(b) 6, $+3$; (c) 4, $+1$; (d) 6, $+3$; (e) 6, $+3$.
5. (a) diamminesilver(I); (b) hexaaquairon(III);
(c) tetrachlorocuprate(II);
(d) bis(ethylenediamine)platinum(II);
(e) nitrochlorotetraamminecobalt(III).
6. (a) bromopentaamminecobalt(III) sulfate;
(b) sulfatopentaamminecobalt(III) bromide;
(c) hexaamminechromium(III) hexacyanocobaltate(III);
(d) sodium hexanitrocobaltate(III);
(e) tris(ethylenediamine)cobalt(III) chloride.
7. (a) $[Ag(CN)_2]^-$; (b) $[Ni(NH_3)_2Cl_4]^{2-}$; (c) $[PtCl_6]^{2-}$;
(d) $Na_2[CuCl_4]$; (e) $K_4[Fe(CN)_6]$; (f) $[Cu(en)_2]^{2+}$;
(g) $[Al(H_2O)_4(OH)_2]Cl$; (h) $[Cr(en)_2(NH_3)Cl]SO_4$.
8. linear.   **15.** (a) one; (b) two; (c) two; (d) two.
16. (a) no; (b) yes; (c) no.
18. (a) coordination isomerism; (b) linkage isomerism;
(c) no isomerism; (d) no isomerism;
(e) *cis-trans* isomerism.   **19.** best: I; poorest: III.
20. poorest of all—a nonconductor.
22. (a) paramagnetic; (b) diamagnetic.   **26** five.
29. 1.4×10^{-6} g KI.   **30.** no.
31. $[CN^-] = 3.4 \times 10^{-5}$ M.   **32.** no.
33. 0.18 g AgCl.   **34.** (a) $FeCl_3 \cdot 6\,H_2O$;
(b) $Co[PtCl_6] \cdot 6\,H_2O$.
36. $2\,Co^{2+} + H_2O_2 + 12\,NH_3 \rightarrow 2\,[Co(NH_3)_6]^{3+} + 2\,OH^-$.
38. (a) 0.592 g; (b) 1.18 g.   **39.** 25 g NaCN.

Self-Test Questions: **1.** (c).   **2.** (d).   **3.** (b).
4. (c).   **5.** (b).   **6.** (a).

10. insoluble: CuS; highly soluble: $CuCO_3$.

Chapter 23

Exercises: **4. (a)** α particle; **(b)** β^- or $_{-1}^{0}e$; **(c)** $_0^1n$;
(d) proton; **(e)** positron; **(f)** $_1^3H$.   **5. (a)** $_{92}^{230}U$; **(b)** $_{98}^{248}Cf$;
(c) $_{80}^{196}Hg$; **(d)** $_{84}^{214}Po$; **(e)** $_{82}^{214}Pb$; **(f)** $_{32}^{69}Ge$.   **6. (a)** 35;
(b) $_1^0e$; **(c)** 92, $_{90}^{231}Th$, $_2^4He$; **(d)** Bi, $_{84}^{214}Po$.   **11. (a)** $_{11}^{24}Na$;
(b) $_0^1n$; **(c)** $_{94}^{240}Pu$; **(d)** 5 $_0^1n$; **(e)** $_{99}^{246}Es + 6\ _0^1n$.
13. (a) 264 d; **(b)** 292 d; **(c)** 380 d.   **14.** 9.6×10^8.
15. 6.6 y.   **16.** 142 d.   **17.** 3.63 d.   **18.** 1.0×10^{-3} g.
19. 0.226 g $^{208}Pb/1.00$ g ^{232}Th.   **20.** 3×10^9 y.
21. no.   **22.** 5×10^4 y.   **25.** 8.10 MeV.
26. 4.06 MeV.   **27.** 4.80 MeV.   **28. (a)** $_{10}^{20}Ne$; **(b)** $_8^{18}O$;
(c) $_3^7Li$.   **29.** β^-: $_{15}^{33}P$ and $_{53}^{134}I$; positron: $_{15}^{29}P$ and $_{53}^{120}I$.
34. 2.94×10^3.

Self-Test Questions: **1.** (c).   **2.** (b).   **3.** (d).
4. (c).   **5.** (a).   **6.** (c).
7. (a) $_{90}^{230}Th \rightarrow\ _{88}^{226}Ra +\ _2^4He$; **(b)** $_{27}^{54}Co \rightarrow\ _{26}^{54}Fe +\ _{+1}^{0}e$;
(c) $_{90}^{232}Th +\ _2^4He \rightarrow\ _{92}^{232}U + 4\ _0^1n$.   **8.** 6.1×10^3 dis/s.
9. 75.8 d.

Chapter 24

Exercises: **7. (a)** no; **(b)** yes; **(c)** no; **(d)** no; **(e)** no;
(f) yes.   **8. (a)** positional; **(b)** skeletal; **(c)** positional;
(d) positional; **(e)** geometrical.
11. (a) halide (bromide); **(b)** carboxylic acid;
(c) aldehyde; **(d)** ether; **(e)** ketone; **(f)** amine;
(g) phenyl; **(h)** ester.   **14. (a)** 2,2-dimethylbutane;
(b) 2-methyl-1-propene or, simply, methylpropene;
(c) methylcyclopropane; **(d)** 4-methyl-2-pentyne;
(e) 2-ethyl-3-methylpentane;
(f) 3,4-dimethyl-2-*n*-propyl-1-pentene.   **16. (a)** no;
(b) yes; **(c)** no; **(d)** no; **(e)** yes.
17. (i) and (j) are correct.
18. (a) $CH_3CH(OH)CH_3$
(b) $Pb(C_2H_5)_4$ **(c)** $CH_2{=}C(CH_3)CH{=}CH_2$

(d) $(CH_3)_3CCH_2CH(CH_3)_2$ **(e)**

(f)

(g)

(h) $(CH_3)_2CH(C_{15}H_{31})$
(i) $(CH_3)_2C{=}CH(CH_2)_2C(CH_3){=}CH(CH_2)_2C(CH_3){=}$
$CHCH_2OH$
19. C_4H_8 **22. (a)** $CH_3CH_2CH_3$
(b) $CH_3CH_2CH_2CH_3$ or $CH(CH_3)_3$
23. (a) *n*-butane; **(b)** *n*-hexane and NaBr;
(c) *n*-propane and Na_2CO_3;
(d) 1-chloropropane and 2-chloropropane.
28. (a) $CH_3CCl_2CH_3$ **(b)** $CH_3C(CN){=}CH_2$

(c) $CH_3CH_2CI(CH_3)_2$ **(d)**

29. (a) 2-methylaniline; **(b)** *m*-dinitrobenzene;
(c) 4-aminobiphenyl;

(d)

(e)

(f)

(g)

30. (a) 1-bromo-3,5-dinitrobenzene;
(b) 2-bromoaniline, 4-bromoaniline;
(c) 2,4-dibromoanisole.   **34.** 10.
37. (a) 1, 2, 7 → 1, 2, 2; **(b)** 3, 2, 4 → 3, 2, 2;
(c) 1, 1 → 1, 1, 2; **(d)** 1, 16, 2 → 1, 4, 10.   **38.** 90%.

Self-Test Questions: **1.** (b).   **2.** (d).   **3.** (b).
4. (a).   **5.** (c).   **6.** (d).   **7.** (b).   **8.** (c).

9. (a)

(b)

(c)

Chapter 25

Exercises: **1. (a)** 300 H^+ ions; **(b)** 1×10^5 K^+ ions.
2. 3%.   **3.** 3×10^7.   **4.** 5×10^6.
5. length of extended DNA molecule: $2 \times 10^3\ \mu$m.
10. (a) glyceryl laurooleostearate; **(b)** glyceryl trilinoleate;
(c) sodium myristate.   **11.** 216.   **12.** trilinolein.
19. diastereomers.   **20.** 36% α, 64% β.
23. (a) $C_6H_5CH_2\underset{\underset{NH_3^+\ Cl^-}{|}}{C}HCO_2H$
(b) $C_6H_5CH_2\underset{\underset{NH_2}{|}}{C}HCO_2^-\ Na^+$
(c) $C_6H_5CH_2\underset{\underset{NH_3^+}{|}}{C}HCO_2^-$
24. lysine to cathode; glutamic acid to anode;
proline does not migrate.
26. (b) glycylalanylserylthreonine.
27. (a) Ala-Ser-Gly-Val-Thr-Leu;
(b) alanylserylglycylvalylthreonylleucine.
29. minimum mol wt = 2.9×10^3.
32. about 60% efficiency.   **33.** $K = 2.6 \times 10^2$.
40. Thr-His-Pro-Leu-Ala-Ser-Gly-Met.
41. possible DNA base sequence: AGA-CCA-CAA-CGA.

Self-Test Questions: **1.** (b).   **2.** (c).   **3.** (d).

4. (d).   **5.** (a).   **6.** (c).   **7.** 129 g soap.
8. Gly-Cys-Val-Phe-Tyr.

Chapter 26

Exercises: **1.** free: N, Ag, Rn; combined: Br, Ag, Ga;
very rare: Fr.
3. 5.3×10^{13} L.   **4.** 7.9×10^5 L.   **5.** 600 times.
8. (a) 1.7×10^5 t; **(b)** 6.2×10^6 t.   **9.** about 1700 ppm.
10. 0.225% H_2S.   **11.** 6.4×10^6 t.   **13.** 1.12×10^6 g.
14. (a) $Ca^{2+} + 2 Na^+ + CO_3^{2-} \rightarrow CaCO_3(s) + 2 Na^+$;
(b) 35.8 g.   **15.** 100 ppm.   **16.** 6.4 g.
18. about 5×10^5 lb Hg.   **22.** 1.29×10^4 L.
23. 3.6×10^2 kg.   **29. (a)** 72% eff.   **31. (a)** 31.6% P;

(b) 20.0% P; **(c)** 18.4% P; **(d)** 10.9% P; **(e)** 24.6% P.
35. 33.3% O.   **37.** $+CO(CH_2)_8CO(NH)(CH_2)_6(NH)+_x$.

Self-Test Questions: **1.** (c).   **2.** (a).   **3.** (b).
4. (b).   **5.** (d).   **6.** (c).
7. (a) $2 FeS(s) + 3 O_2(g) \rightarrow 2 FeO(s) + 2 SO_2(g)$;
(b) $Mg(OH)_2(s) + 2 H^+ \rightarrow Mg^{2+}(aq) + 2 H_2O$;
(c) $2 HCO_3^-(aq) \rightarrow H_2O + CO_2(g) + CO_3^{2-}(aq)$ *and*
$Ca^{2+}(aq) + CO_3^{2-}(aq) \rightarrow CaCO_3(s)$;
(d) $2 NH_4Cl + Ca(OH)_2 \rightarrow CaCl_2 + 2 NH_3(g) + 2 H_2O$;
(e) $P_4(s) + 5 O_2(g) \rightarrow P_4O_{10}(s)$ *and*
$P_4O_{10}(s) + 6 H_2O \rightarrow 4 H_3PO_4(l)$.
8. (a) 1.14×10^6 kg.

Index

Boldface page numbers indicate end-of-chapter definitions. Abbreviations in italics after page numbers indicate exercise (*ex.*), figure (*f.*), marginal note or footnote (*n.*), and table (*t.*).

Selected Physical Constants

acceleration due to gravity:	g	9.8067 m s^{-2}		electronic charge:	e	1.6021×10^{-19} C
speed of light (in vacuum):	c	2.997925×10^{8} m s^{-1}		electronic rest mass:	m	9.1091×10^{-28} g
gas constant:	R	0.082056 L atm mol^{-1} K^{-1}		Planck's constant:	h	6.6256×10^{-34} J s
		8.3143 J mol^{-1} K^{-1}		Faraday constant:	$\mathcal{F}$	9.6487×10^{4} C/mol e^{-}
				Avogadro's number:	N	6.02252×10^{23} mol^{-1}

Multiples and Submultiples in the SI System

(See also Appendix C)

Multiple	Prefix	Symbol	Submultiple	Prefix	Symbol
10^{12}	tera	T	10^{-1}	deci	d
10^{9}	giga	G	10^{-2}	centi	c
10^{6}	mega	M	10^{-3}	milli	m
10^{3}	kilo	k	10^{-6}	micro	μ
10^{2}	hecto	h	10^{-9}	nano	n
10^{1}	deka	da	10^{-12}	pico	p
			10^{-15}	femto	f
			10^{-18}	atto	a

Some Common Conversion Factors

Length
1 meter (m) = 39.37 inches (in.)
1 in. = 2.54 centimeters (cm)

Volume
1 liter (L) = 1000 ml = 1000 cm^{3}
1 L = 1.057 quart (qt)

Mass
1 kilogram (kg) = 2.205 pounds (lb)
1 lb = 453.5 grams (g)

Energy
1 joule (J) = 1 N m = 1 kg m^{2} s^{-2}
1 calorie (cal) = 4.1840 J
1 electron volt (eV) = 1.6021×10^{-19} J
1 eV/atom = 96.49 kJ/mol
1 kilowatt hour (kW hr) = 3600.0 kJ

Force
1 newton (N) = 1 kg m s^{-2}

mass–energy equivalence: 1 atomic mass unit (amu) = 1.66×10^{-24} g = 931 MeV

Some Common Geometric Formulas

perimeter of a rectangle = $2l + 2w$; area of a rectangle = $l \times w$;
volume of a parallelepiped = $l \times w \times h$; area of a triangle = $\frac{1}{2}$(base $\times$ height);
circumference of a circle = $2\pi r$; area of a circle = πr^{2}; area of a sphere = $4\pi r^{2}$;
volume of a sphere = $\frac{4}{3}\pi r^{3}$; volume of a cylinder or prism = (area of base) $\times$ (height)